The SAGE
Handbook *of*

Biogeography

The SAGE
Handbook *of*

Biogeography

Edited by
Andrew C. Millington,
Mark A. Blumler
and Udo Schickhoff

Los Angeles | London | New Delhi
Singapore | Washington DC

SAGE Publications Ltd
1 Oliver's Yard
55 City Road
London EC1Y 1SP

SAGE Publications Inc.
2455 Teller Road
Thousand Oaks, California 91320

SAGE Publications India Pvt Ltd
B 1/I 1 Mohan Cooperative Industrial Area
Mathura Road, Post Bag 7
New Delhi 110 044

SAGE Publications Asia-Pacific Pte Ltd
33 Pekin Street #02-01
Far East Square
Singapore 048763

Library of Congress Control Number: 2011920449

British Library Cataloguing in Publication data

A catalogue record for this book is available from the British Library

ISBN 978-1-4129-1951-7

Typeset by Cenveo Publisher Services.
Printed in Great Britain by MPG Books Group, Bodmin, Cornwall
Printed on paper from sustainable resources

MIX
Paper from responsible sources
FSC
www.fsc.org
FSC® C018575

Contents

Acknowledgements

The editors would like to thank the following people for advising at various stages in the preparation of the handbook, and in preparing Chapter 1: Roy Alexander, Carl Beierkuhnlein, Niall Burnside, Dave Cairns, Margaret Davis, Anke Jentsch, Jon Fjeldsa, John Flenley, Guan Kalyun, Peter Hugill, Peter Kershaw, Tim Killeen, Clarissa Kimber, Woo-Seok Kong, Charles Lafon, Mariella Leo, Glen MacDonald, Mike Meadows, Sue Page, Vladimir Onipchenko, Richard Pott, Karsten Rahbek, Michel Richter, Lily Rodriguez, Qingshan Ron, Doug Sherman, R.B. Singh, Helen Watson, and Ken Young. This is also the first opportunity to thank the hundreds of respondents to the IGU Survey of Biogeography Teaching reported on in Chapter 1. Alas they are too numerous to mention individually, but we are extremely grateful that they took time to complete the questionnaires they were sent. Finally, we also want thank a number of anonymous referees who commented on various chapters. They provided a valuable service for the editors and authors.

The research in Chapter 3 was supported by a grant from the National Commission on Science for Sustainable Forestry (NCSSF) and made possible by a fellowship from the Udall Center for Studies in Public Policy at the University of Arizona. George Malanson and Scott Franklin are thanked for their input to this research. The authors of Chapter 3 thank Keith Bennett for providing comments and helpful suggestions. Shonil A. Bhagwat's position at the School of Geography and the Environment is funded by NERC-QUEST and the Leverhulme Trust. Wilhelm Barthlott contributed to early versions of Chapter 9, and V. St. Louis, V. Radeloff, D. Waller, J. Zedler and J. Zhu form provided valuable comments and suggestions for Chapter 22. On behalf of all the authors, we thank the many organizations and universities that have funded research that they précised in their individual chapters.

In the College of Geosciences at Texas A&M University a number of Aggies provided valuable contributions to the production of this book, and we take this opportunity to thank them for their help. Indumathi Srinath converted many figures into formats suitable for publication, as well as drawing some individual diagrams from scratch; Jennifer Rumford prepared photographs for publication; and two B.S. Environmental Programs students—Krista Burns and Erin Rush—indexed a number of chapters.

The editorial team at Sage Publications in London have exhibited great patience and been extremely helpful as we have worked on this task. In particular we thank Robert Rojek for commissioning the book, and our editors—Sarah-Jayne Boyd and, in the early stages, David Mainwearing. In terms of the final stages of production we are very grateful to Nicola Marshall at Sage Publications, and Rahul Raghavan and Deepa Joshith at Cenveo Publisher Services. They have kept us on track despite the obstacles and delays along the way.

Finally, we make the make the two oft seen, but nonetheless very important, acknowledgements. First, we thank all those people who have helped the editors and authors who have been inadvertently omitted above. Secondly, the editors would like to thank their families for being forgiving as we tilted the work-life balance slightly too far toward the former while preparing this book.

List of Contributors

Thomas P. Albright is an Assistant Professor in Geography and is affiliated with the program in Ecology, Evolution, and Conservation Biology at the University of Nevada, Reno. His background includes training in geography (B.S. University of Wisconsin-Madison, MA University of California at Santa Barbara) and biology (PhD and postdoctoral scholarship at UW-Madison) and works at the US Geological Survey's Earth Resources and Observation Science Center. At UNR, he directs the Laboratory for Conservation Biogeography, where research is focused on applying geospatial, statistical, and computational tools to understand biotic responses to climate and environmental change.

Yves Bergeron is Professor at the Université du Québec à Montréal and at the Université du Québec en Abitibi-Témiscamingue. A specialist in forest ecology, he has dedicated his career to studying the dynamics of the Canadian boreal forest and has shown how Canada's boreal forest is controlled by fire and insect epidemics. His work has helped develop new forest management approaches that are inspired by natural dynamics.

Carl Beierkuhnlein holds the Chair of Biogeography at the University of Bayreuth where he directs the Global Change Ecology graduate program within the Elite Network of Bavaria. The research interests of the chair of Biogeography spread over a broad range of biogeographic research areas including the biotic and functional diversity of ecosystems; the spatial organisation of plant species and communities; bioindication; island biogeography focussing on endemic species as well as climate change effects on isolated systems; nature conservation that addresses questions concerning applied ecology. Within the context of climate change, research is conducted for the German Federal Agency for Nature Conservation on the impacts of climate changes on species and biotopes, and within the framework of the Bavarian Climate Program 2020 the concentration is on two projects examining the impacts of climate change on ecosystems. This research is integrated in FORKAST (Climatic Impacts on Ecosystems and Climatic Adaptation Strategies) funded by the Bavarian State Ministry of Sciences, Research and the Arts. Continuing the climate change theme, research is also conducted on vector-borne infectious diseases in climate change for the Bavarian State Ministry of the Environment and Public Health. In collaboration with the chair of disturbance ecology at Bayreuth, experimental research is being conducted on the impacts of extreme weather events on the vegetation.

Shonil A. Bhagwat is a Senior Research Fellow at the School of Geography and the Environment at the University of Oxford. His research interests focus on three areas of investigation: conservation beyond protected-area boundaries; the cultural and spiritual values in conservation; and climate change and conservation.

Mark A. Blumler is Associate Professor in Geography at SUNY-Binghamton in upstate New York. He has wide ranging research and teaching interests in biogeography (life-form and distribution; mediterranean-type ecosystems; invasions; geogenetics; conservation-biogeographical aspects; climate change and ecosystems); plant ecology (succession theory; seed ecology; species-diversity; plant strategies); environmental history (agricultural origins; the Columbian exchange; desertification; history of environmental thought; California; the Middle East) and social theory (diffusion vs. independent invention and cultural change; environmental vs. social factors in cultural change; Western influences on critical social theory; Carl Barks) which befits somebody trained in the wake of Carl Sauer.

Daniel G. Brown is a Professor in the School of Natural Resources and Environment and faculty associate in the Center for the Study of Complex Systems, Program in the Environment, and the Population Studies Center at the University of Michigan. He has been on the editorial board of *Landscape Ecology*, currently sits on the editorial board of *Journal of Land Use Science*, and has served as guest editor for issues of the *International Journal of Geographical Information Science*, *Computers Environment and Urban Systems*, *Ecological Applications*, *Plant Ecology*, and *GeoCarto International*. A member of the NASA's Land Cover and Land Use Change Science Team since 1997, he also chairs the Land Use and Land Cover Change Steering Group for the US Climate Change Science Program. His research focuses on linking observable landscape patterns and dynamics, obtained through remote sensing, ecological mapping, and digital terrain analysis, with ecological and social processes.

Niall G. Burnside is a Senior Lecturer in Biogeography at the University of Brighton, and is currently Secretary of the Biogeography Research Group of the Royal Geographical Society (UK). Niall has research interests in grassland and woodland ecology, the use of GIS for environmental management and predictive modeling. His expertise covers the development and use of Geographical Information Systems for broad-scale biogeographical and ecological evaluation, and the combination of field-based surveys techniques, remotely sensed data and habitat classification. His PhD and much of his subsequent research has centered on grassland systems and the restoration and management of these under agricultural intensification and climate change.

Jeffrey Cardille is an Assistant Professor at the Université de Montréal. He has a PhD and MS in Environmental Studies (University of Wisconsin-Madison), an MS in Operations Research (Georgia Institute of Technology), and a BS in Applied Mathematics and Computer Science (Carnegie Mellon). He is interested in a wide array of terrestrial and aquatic systems from a landscape ecology perspective, frequently employing landscape metrics, remote sensing, and spatial statistics to solve real-world research questions.

Anthony Cole is a Research Scientist at the New Zealand Centre for Ecological Economics, Palmerston North., New Zealand; and an adjunct Associate Professor in the School of People, Environment and Planning at Massey University.

Mary E. Edwards is Professor of Physical Geography at the University of Southampton. Her interests are centred on global environmental change: understanding climate-driven changes in landscape, vegetation, and ecosystem processes over a range of timescales through palaeoenvironmental records and data-model comparisons. Her main geographic area of interest is the boreal-arctic region, and recent work relates to boreal forest dynamics, including fire disturbance and hydrological regimes, and thermokarst dynamics in relation to biogeochemical processes. Recent collaborative work involves the estimation of past and present plant diversity using novel molecular techniques. A second area of research is concerned with vegetation ecology, biodiversity conservation and land management in both Europe and Madagascar.

Neal J. Enright is Professor of Plant Ecology in the School of Environmental Science at Murdoch University in Perth, Western Australia. He is particularly interested in fire ecology and management, plant species co-existence in high diversity ecosystems, and the demography of woody plants. He has published more than 100 papers and book chapters in the ecological sciences and has co-authored/edited several monographs including *Ecology of the Southern Conifers* (Melbourne University Press/Smithsonian Press 1995).

John Flenley is Professor Emeritus at Massey University, New Zealand, having moved there from Hull University in the late 1970s. He gained his PhD from Australian National University and has had a long and distinguished career using palynological techniques to unravel questions of archaeology, palaeobotany and palaeoenvironmental science. His field work has mainly been in the Pacific Basin, and his dossier of research includes pioneering research on Easter Island.

Giles M. Foody is Professor of Geographical Information Science at the University of Nottingham, and has a PhD from the University of Sheffield. He applies remote sensing and spatial science to biogeographical and land use issues. He has published over 150 refereed journal articles, 8 books and >175 other publications (book chapters, conference papers etc). He currently serves as Editor-in-chief of International Journal of Remote Sensing (Letters editor 2001-8; EiC 2007–) Editor-in-chief Remote Sensing Letters (Founding editor-in-chief, 2010–) 'Co-ordinating' editor Landscape Ecology (2005–) Associate editor Ecological Informatics & guest editor, vol 2(2), (2005–) Editorial boards (6): Geocarto International (1996–); Remote Sensing of Environment (2005–); International Journal of Applied Remote Sensing (2006–); International Journal of Applied Earth Observation and Geoinformation (2007–); Remote Sensing (2009–); Scottish Geographical Journal (2010–).

Claire Freeman is Associate Professor in Geography and Director of the Planning Programme at the University of Otago, Dunedin New Zealand. Claire came to Otago via lecturing posts at Massey University, Leeds Metropolitan University and the University of the North West in South Africa and has been a planner for the Urban Wildlife Trust in Birmingham, England. Her research interests are in planning for nature, children's environments and community planning. She has undertaken substantial research looking at planning for nature in the city and the need to rethink ecological values when addressing the urban context. Together with colleagues she has recently been involved with a major project to develop a detailed GIS habitat data base for Dunedin city to enhance ecological protection and planning for the city.

Duane A. Griffin is Associate Professor of Geography at Bucknell University in Pennsylvannia. He is actively involved in a number of research projects including fine-scale habitat structure and plant diversity, mapping upland vernal pools of the ridge-and-valley province which is linked a collaborative research on landscape spatial structure and amphibian diversity and development, and a global wildlands project. He also professes an interest in 'The Hollow Earth'.

Geoffrey Griffiths is a Senior Lecturer in Physical Geography at the University of Reading. His research interests cover landscape and biodiversity, GIS and habitat restoration, biodiversity of tropical forests, GIS/remote sensing techniques for predictive vegetation mapping in Brazil and the Mediterranean, understanding the relationship between landscape character and biodiversity, and methods for evaluating ecological quality.

Michael J. Hill is a Professor in Earth System Science and Policy at the University of North Dakota. His research interests include remote sensing, biogeochemical processes and land-use change in savanna and grassland ecosystems, and analysis of coupled human-environment systems using spatial multi-criteria analysis.

Anke V. Jentsch is Professor of Disturbance Ecology and Vegetation Dynamics at the Helmholtz Centre for Environmental Research in the Department of Conservation Biology at the University of Leipzig. Her research includes the response of vegetation to extreme weather events in central Europe (research conducted with the University of Bayreuth); the roles of disturbance in the creation, maintenance and loss of biodiversity; changing land use and biodiversity in cultural landscapes; natural hazards, fire ecology and extreme events; ecosystem dynamics in dry acidic grasslands; and *Corynephorus canescens* as decision support for a habitat connectivity programme in German dry acidic grasslands.

Daniel Kneeshaw is member of the Centre for Forest Studies and a Professor at the Université du Québec à Montréal. He has been active in studying the response of boreal forests to natural disturbances;

primarily disturbances that partially affect the forest canopy such as gaps, windthrow and insect outbreaks. He uses both field techniques and landscape modeling to explore forest responses, to confront public perceptions on different scenarios and to develop forest management and conservation strategies. He is currently editor of *Canadian Journal of Forest Research*.

Anselm Kratochwil is Professor of Ecology at the University of Osnabrück, Germany in the Faculty of Biology and Chemistry. He studied zoology, geobotany, limnology and pedology at the University of Freiburg where he got his Diploma and PhD in Biology. Since 1992 he has been the chair of the Ecology Department and member of the directory board of the Institute of Environmental Systems Research of the University of Osnabrück. He has held lectures, seminars, practical courses and ecological excursions in the field of general ecology, animal ecology, plant ecology, vegetation science and biocenology. His research interest currently focuses on flower-visitor communities of different types of ecosystems, the structure and dynamics of wild-bee communities in different landscapes and the restoration of alluvial landscapes and inland sand-dune communities. Together with Angelika Schwabe he published the textbook *Ökologie der Lebensgemeinschaften* (Ecology of Communities).

John A. Kupfer is a Professor in the Department of Geography at the University of South Carolina. His research examines the manner by which spatial patterns of ecological phenomena interact with and constrain ecological processes. Much of his work stresses vegetation dynamics in human-modified systems, and his projects regularly incorporate quantitative methods and computer techniques, including ecosystem modeling, GIS, remote sensing and geostatistics. Much of his work has focused on the effects of forest fragmentation, and his paper (co-authored by George Malanson and Scott Franklin) 'Not seeing the ocean for the islands: The influence of matrix-based processes on forest fragmentation effects' was awarded the 2007 Henry Cowles Award for Excellence in Publication by the Biogeography Specialty Group of the Association of American Geographers. He received his BA from Valparaiso University and his MA and PhD from the University of Iowa.

Timo Kuuluvainen is an Associate Professor in the Department of Forest Sciences at the University of Helsinki. His research has focused on the fundamental interactions between structure, dynamics and biodiversity of the boreal forest, using both empirical and modeling approaches. He has published more than 100 peer-reviewed papers and text-book chapters. His current research is centered on using ecological understanding to develop methods of restoration and ecosystem management of the boreal forest.

Pedro J. Leitão is a Postdoctoral Researcher at the Geomatics Laboratory of the Humboldt-Universität zu Berlin. Pedro's main research interest is on spatial data analysis in ecology and biogeography. He is particularly interested in the use of spatial and remotely sensed data for ecological research and biodiversity and ecosystem monitoring. He is experienced with most remote sensing data products (optical, thermal, hyperspectral, laser scanning, etc.), and he has a strong focus both on methods development for information extraction and data analysis and on their application for addressing specific research questions. Amongst other things, he has done extensive research on steppe and farmland birds, from regional to global scales, studying their patterns of distribution and habitat selection, and respective effects of climatic and resource availability changes.

George P. Malanson is the Coleman-Miller Professor in the Department of Geography at the University of Iowa. His research is primarily in biogeography, addressing how spatial patterns and processes affect vegetation dynamics in response to human-induced changes such as climatic change and altered disturbance regimes. His research integrates fieldwork, computer simulations, and statistical analyses. Much of the fieldwork is based in the northern Rocky Mountains. Recently, he has directed more attention toward aspects of land-use and land-cover change. He is Editor-for-Biogeography of *Physical Geography*, Associate Editor of *Arctic, Antarctic, and Alpine Research* and is on the editorial boards of *Annals of the Association of American Geographers* and *Geography Compass*.

Rob Marchant has been a Senior Lecturer in the Environment Department, University of York, UK since 2005. Previously he held posts at the University of Leicester; Trinity College, Dublin; and the University of Amsterdam. His PhD was on the Late Quaternary montane vegetation dynamics in Bwindi-Impenetrable Forest, southwest Uganda. His research continues to focus on vegetation dynamics and ecosystem change. In particular, he uses ecological modelling, archaeological and biogeographical data to determine the role of past events in shaping the present day composition of vegetation, especially in the tropics.

Carlos F. Mena is an Ecuadorian doctoral student in the Department of Geography and a pre-doctoral trainee in the NSF IGERT Program at the Carolina Population Center, University of North Carolina at Chapel Hill. He is also the recipient of a NASA Fellowship (2004-2007) from the Land Cover/Land Use Change Program. His research involves the study of land-use and land-cover change in the Northern Ecuadorian Amazon, and biocomplexity in coupled human-natural systems. Specifically, his research deals with the relationships among socioeconomic factors, and demographic variables, and forest and agricultural transitions in frontier environments. He has an undergraduate degree in Geographic Engineering from Escuela Politecnica del Ejercito, Quito, Ecuador and a graduate degree in Environmental Sciences from Florida International University, Florida, USA.

Annette Menzel has been Professor of Ecological Cimatology at the Techncial University of Munich since 2007. She studied forest science at Ludwig Maximilians University Munich and worked in the Bavarian state forest administration. Her research interests focus on plant phenology and climate change.

Joseph P. Messina is a Professor of Geography at Michigan State University. He holds a joint appointment with the Department of Geography and the Center for Global Change and Earth Observations. He is also a faculty member of the Environmental Science and Policy Program, and the Center for Water Sciences. He completed his PhD in Geography at the University of North Carolina at Chapel Hill in 2001 focusing on land-use and land-cover change in the Ecuadorian Amazon. His current research includes continuing work in Ecuador and Northeast Thailand, new work in northern China, and basic GIScience research on spatial error models, simulation models, and software interface development.

Andrew C. Millington is Dean of the School of the Environment at Flinders University, Adelaide. His current research focuses on land-use change and incorporates both its drivers, and its consequences for various types of ecosystems: he relies on remote sensing as a major research tool. He has worked in a diverse array of biomes covering the tropics and sub-tropics of South America, Africa, western Asia and Australia. He has worked for the World Bank, UN and EU; and has held academic positions at Texas A&M, Leicester, Reading and Sierra Leone Universities, and obtained an MA from the University of Colorado and a DPhil from the University of Sussex. He edited *The Geographical Journal* and is currently on the editorial boards of *Journal of Land Use Science,* and *Arid Land Restoration and Rehabilitation.* He was formerly chair of the IGU Commission on Biogeography and Biodiversity.

Jayalaxshmi Mistry is Senior Lecturer of Geography at Royal Holloway, University of London. Her main research interests are in natural resource and land management. Her current work is looking at ways of linking local livelihoods with biodiversity conservation in the Rupununi savannas of Guyana and in the use of participatory video for empowering local communities to take action about their natural resource management issues. She is also concerned with fire management in tropical savannas, particularly decision-making for fire use and indigenous and traditional fire management in the cerrado of Brazil. She is the author of two books on tropical savannas; a student textbook '*World Savannas*' and an edited volume '*Savannas and dry forests: linking people with nature*'.

Ingo Möller studied geography, botany, soil science and geology. His PhD thesis covers the plant sociology and vegetation ecology of north-western Svalbard. He held a couple of positions in German universities and research institutions, and is presently as Senior Scientist at the Federal Institute of Geosciences and Natural Resources, Hannover. Particular research interests are in arctic and tropical vegetation and ecology as well as carbon budgets and cycles on local to regional scales. Field work has led him several times to Svalbard and the Canadian Arctic.

Jens Mutke is Curator (Akademischer Oberrat) at the Nees Institute for Biodiversity of Plants at the University of Bonn. He is a member of the German National Scientific Committee of DIVERSITAS within the Earth System Science Partnership. He holds a Diploma in Biology and a PhD in Botany. His main research interests are macroecology, conservation biogeography, and tropical ecology. Current research projects focus on continental to global patterns of land plant diversity and their implications for nature conservation. He did field studies on the ecology of tropical montane forests and on the vegetation of rock outcrops (inselbergs) in South America.

Joanne M. Nightingale currently works at NASA's Goddard Space Flight Center, and is responsible for coordinating EOS global land product validation activities and currently chairs the CEOS Working Group on Calibration and Validation, sub-group for Land Product Validation. Joanne obtained her Ph.D. in geography from the University of Queensland in Australia and has completed two post-doctoral research positions at universities within the United States. Her research interests center on remote sensing and ecosystem/biogeochemical modeling with emphasis on incorporating remotely sensed information into regional ecosystem models that assess seasonal inter- and intra-annual trends in carbon storage and terrestrial productivity in addition to fostering validation efforts for improving the accuracy of satellite-derived land products.

Patrick E. Osborne is a Senior Lecturer in the School of Civil Engineering and the Environment at the University of Southampton. His recent research focuses on applied ecological issues, especially in relation to spatial ecology and the conservation of agricultural birds. In the last decade he has worked on a multi-scale study of habitat selection by steppe birds using remote sensing and GIS, and is currently starting work on modelling the distributions of steppe birds at the global scale. He is also training Russian researchers to survey bustards using Distance Sampling and to build a spatial model of the bustard's distribution in Saratov Oblast, Russia.

Josep Peñuelas is a Research Professor of the National Research Council of Spain (CSIC), and Director of the Global Ecology Unit CREAF-CEAB-CSIC located at CREAF (Center for Ecological Research and Forestry Applications)-Universitat Autònoma de Barcelona. He works on plant ecophysiology and atmosphere-biosphere interactions; and has recently investigated global change, climate change, atmospheric pollution, biogenic VOC emissions, remote sensing, and functioning and structure of Mediterranean terrestrial plants and ecosystems. Alongside his academic publications he tries to popularize science issues in journals and Spanish newspapers. He has been a visiting professor at over 20 universities in North and Central America, Europe, Asia and Australia. He has received national (Conde de Barcelona 1990, Medi Ambient Institut d'Estudis Catalans-Caixa Sabadell 2008, I Premi Recerca de Catalunya 2010) and international (NASA 1993, Science Ministry of Japan 1998) awards. He has been president of Institució Catalana d'Història Natural and adviser of the Consell Asessor per al Desenvolupament Sostenible of the Generalitat of Catalonia.

Stuart R. Phinn is a Professor of Geography in the School of Geography, Planning and Environmental Management at the University of Queensland. He is currently the Director of Australia's Terrestrial Ecosystem Research Network (www.tern.org.au). He also directs the Centre for Spatial Environmental Research (www.gpem.uq.edu.au/cser), which focuses on collaborative research and training in all areas of geographic information science. His own work develops remote sensing solutions for environmental monitoring and management applications by combining field and image data sets.

Brett R. Riddle is a Professor in the School of Life Sciences at the University of Nevada, Las Vegas, and has a PhD from the University of New Mexico. His research focuses primarily on the history of biodiversity in western North America, with ongoing projects including: historical assembly of the warm desert biotas; phylogeography of Great Basin montane island biotas; and molecular systematics and biogeography of diverse North American rodent groups. He is co-founder and past President of the International Biogeography Society, an editor of the Journal of Biogeography, and an associate editor of Systematic Biology.

Udo Schickhoff has been Professor of Biogeography in the Institute of Geography at the University of Hamburg since 2004, prior to which he held professorial positions in geography at the University of Bonn, landscape ecology at the University of Rostock, and was foundation professor in the Institute of Botany at the University of Griefswald. He is currently working on two research projects funded by the Volkswagen-Stiftung Research Foundation, in one he coordinates a project on the impact of the transformation process on human-environmental interactions in southern Kyrgyzstan for the universities of Hamburg, Berlin, Bonn and several institutions in Kyrgyzstan. He also works on a project on changing pastoral ecosystems in western Mongolia (Khovd Aimag) using a landscape-ecological approach for assessing grazing land degradation and grazing capacity. He has previously worked on research projects in Germany and the Karakoram. He is currently chair of the IGU Commission on Biogeography and Biodiversity.

Angelika Schwabe is Professor of Botany with main emphasis on vegetation ecology and geobotany at Darmstadt University of Technology, Germany. She studied biology and geography and carried out her PhD in Geobotany at the University of Freiburg, Germany. After her habilitation in Freiburg she was awarded a Heisenberg Grant by the German Research Foundation (DFG). Her research focuses on vegetation complexes, scale-dependent processes and plant-animal interactions. She has experience both in phytosociological classification approaches as well as in ecological field experiments. She studied vegetation complexes of extensive grassland, running waters and alpine dry grassland. Presently she is especially involved in projects dealing with strategies for the protection of endangered sandy grassland, e.g., by extensive livestock grazing, and in restoring endangered habitats. She is a member of the Advisory Council of the International Association for Vegetation Science.

Tim H. Sparks is a Research Fellow in the Institute of Advanced Studies at the Technical University of Munich. Until 1991 he worked in agricultural research switching to ecological research at the Monks Wood Research Station until 2009. In 1998 he founded the UK Phenology Network and has represented the UK in collaborative research programmes in phenology. Tim is a visiting professor at the Poznań University of Life Sciences, and at the University of Liverpool and was a contributing author to the last IPCC report. He is an editor of *Climate Research*, and on the editorial boards of five other journals. Immediately prior to starting his fellowship he was working on biodiversity indicators at the University of Cambridge.

Chris Stoate is Head of Research at the Game and Wildlife Conservation Trust's 'Allerton Project', a research and demonstration farm in Leicestershire, England. He has contributed to numerous UK government and EU-funded research projects and is the author of more than eighty research papers, and a book, '*Exploring a Productive Landscape*'. He has researched the ecology and management of farming systems in West Africa and southern Europe, as well as in England and is a farmer in his own right.

David M. Taylor is currently University Professor of Geography at Trinity College, University of Dublin, a position he has held since January 2001. Prior to his appointment at Trinity College, David was a member of faculty at the National University of Singapore (1997-2001) and at the University of Hull, UK (1989-1997). David's DPhil is on the dynamics of tropical vegetation in central Africa during the Late Quaternary (entitled The vegetation history of southwestern Uganda over the last 40,000 years). His main research interest continues to be the dynamics of ecosystems at low latitudes over the last few thousands of years.

Dietbert Thannheiser studied geography, biology and geology. His PhD thesis is on the vegetation of northern Norway. Afterwards he focused on coastal vegetation and arctic environments. From 1970 to 1982 he was Associate Professor at the University of Münster, and in 1983 he was appointed as Professor at the Institute of Geography of the University of Hamburg, He has gained extensive polar field experience in the Canadian Arctic, Svalbard, Iceland, and Greenland.

Piotr Tryjanowski is Director of the Zoological Institute at the University of Poznan, and one of the most active avian ecologists and researchers in Poland. He was previously a Trustee of Birdlife Poland and is an advisor to EcoFund. His research interests cover a wide spectrum of behavioural ecological and evolution issues. He leads research on different ecological groups and taxa, though birds of agricultural environments are undoubtedly closest to his heart. In particular he works on population ecology; synchronization phenomenon; environmental effects, and the influence climate on behavioural ecology.

Monica G. Turner is the Eugene P. Odum Professor of Ecology in the Department of Zoology, University of Wisconsin-Madison. She obtained her PhD in ecology in 1985 from the University of Georgia, spent seven years as a research scientist at Oak Ridge National Laboratory, and joined the faculty of UW-Madison in 1994. She has published over 190 scientific papers and has authored or edited six books, and she is co-editor in chief of the journal, *Ecosystems*. She was elected to the National Academy of Sciences in 2004. Her research emphasizes the causes and consequences of spatial heterogeneity in ecological systems, focusing largely on ecosystem and landscape ecology in forested systems. Research areas include: interactions of disturbance regimes (fire and insects), vegetation dynamics, and nutrient cycling in the Greater Yellowstone Ecosystem; effects of historic and contemporary land-use patterns on southern Appalachian landscapes; and land-water interactions in north temperate landscapes.

Yordan Uzunov is Associate Professor of Hydrobiology and Head of the Department of Bio-Indication and Environmental Risk Assessment of the Central Laboratory of General Ecology at the Bulgarian Academy of Sciences. He is an aquatic ecologist devoted to study and implement knowledge about the structure and functions of riverine and lacustrine ecosystems, in particular bottom invertebrates, in environmental monitoring, stream/river typology and classification of ecological status of water bodies. He has introduced modern approaches within hydrobiological studies of Bulgarian rivers, lakes and swamps/wetlands, and developed the methodology and methods for the national bio-monitoring of rivers. He was formerly Deputy Minister of the Environment (1992-1997); since 2004 has been a member of the UN Secretary General's Advisory Board on Water and Sanitation.

Robert Voeks is a Professor of Geography at California State University—Fullerton and Chair of the Environmental Studies Graduate Program. His research focuses on tropical ethnobotany and cultural ecology—especially medicinal plants and healing systems—in Brazil and Borneo. He is currently vice-president of the California Geographical Society, and serves on the editorial board for the *International Journal of Environmental Science and Technology*.

Ioannis Vogiatzakis is a Research Fellow in the Centre for Agri-Environmental Studies at the Univeristy of Reading and has research interests in plant ecology and biogeography of Mediterranean islands and mountains, GIS and remote sensing for vegetation and habitat modeling, landscape character assessment and habitat restoration, the effects of landscape structure and habitat quality on biodiversity, and the integration of species and habitat data for nature conservation evaluation.

Stephen Waite is a Principal Lecturer in Ecology at the University of Brighton. He has a BSc Hon. in Biology and a DPhil in plant population ecology from the University of Sussex and a BA in mathematics from the Open University. He has written an advanced text on the statistical analysis of environmental data and has published extensively on various aspects of applied ecology. He has particular research interests in the relationships between plant reproductive strategies, germination ecology and species distribution and population viability at a local and landscape scale. Current research interests include the biogeography of fragmented woodland and chalk-heath plant communities.

Stephen J. Walsh is a Professor of Geography, member of the Ecology Curriculum, and Research Fellow of the Carolina Population Center at the University of North Carolina – Chapel Hill, USA. He is on the editorial boards of *Plant Ecology, GeoCarto International*, and the *Annals of the Association of American Geographers*, he has co-edited special issues in the *Journal of Vegetation Science, Geomorphology*, and *Photogrammetric Engineering and Remote Sensing*. Since 2001, he has co-edited a series of books for Kluwer Academic Publishers—*GIS and Remote Sensing Applications in Biogeography and Ecology* (2001, Millington, Walsh, Osborne); *Linking People, Place, and Policy: A GIScience Approach* (2002, Walsh and Crews-Meyer); and *People and the Environment: Approaches for Linking Household and Community Surveys to Remote Sensing and GIS* (2003, Fox, Rindfuss, Walsh, Mishra); and in 2003 for Elsevier, *Mountain Geomorphology – Integrated Earth Systems* (Butler, Walsh, Malanson). Specific research foci are on pattern and process at the alpine treeline, biocomplexity in coupled human-natural systems, scale dependence and information scaling, and land-use and land-cover dynamics using spatial simulation models, GIS, and remote sensing.

Katherine J. Willis works at Oxford University where she holds the Tasso Leventis Chair of Biodiversity and is Director of the Biodiversity Institute in the Zoology Department. She established the Oxford Long-term Ecology Laboratory in 2002 and was made a Professor of Long-term Ecology in 2008. She moved to her current position in Zoology in October 2010. She is an Adjunct Professor (Professor II) in the Department of Biology, University of Bergen, Norway. She has recently been elected to the position of Director-at-Large of the International Biogeography Society; was awarded the Lyell Fund for 2008 by the Geological Society of London, elected as a Fellow of the Royal Geological Society in 2009, and made a Foreign Member of the Norwegian Academy of Sciences and Letters in 2010. Her research interests focus on the relationship between long-term (>50 years) ecosystem dynamics and environmental change.

Kenneth R. Young has traveled in and researched the tropics of Central and South America since the 1970s. Currently, he is focusing his research on the dynamism of tropical landscapes, the alterations caused by climate change, and the challenges in carrying out biodiversity conservation. He is currently Professor and Head of the Department of Geography and the Environment of the University of Texas at Austin.

Foreword

Imagine the scene. Four biogeographers – an American, two Germans and the fourth British – on a field excursion, bumping through the wilds of the Caucasus in a minibus accompanied by Armenian geographers and biologists. Between stops, with dust billowing from under the bus, talk turned to what our undergraduates were being taught in upstate New York, the English Midlands and two German universities – one in the former East, the other in the former West. As the afternoon heat became increasingly stifling, talk shifted effortlessly to contemporary biogeographical research and (post)graduate training in those three countries, and the differences in what we had presented at the pre-field course symposium in Yerevan State University a few days earlier compared to the talks of our Armenian colleagues who had been trained in the Soviet School of Geography. The differences between our biogeographies appeared as distant as the political doctrines in the Soviet Republic of Armenia and East Germany were from those in West Germany, the UK and the USA a decade earlier. It was June 2000.

Such conversations are commonplace when academics gather – the small talk between stops on field trips, or tales of academia told and retold in bars at conferences the world over. Such conversations, we tell our graduate students, are "…the real business of conferences". Conversations they remain, to be recalled, embellished and rehearsed again the next time old friends meet. But not always.

In this case the conversation took on a life of its own when the British biogeographer was approached not long afterwards by Robert Rojek who was developing a series of physical geography titles for Sage Publication's Handbook Series.

It's spring in New York City. Sitting in a bar in the Avenue of the Americas during the annual conference of the Association of American Geographers are the four biogeographers who had met in Armenia, two other German biogeographers, and Robert Rojek and David Mainwearing from Sage. The heated topic of conversation was the *Handbook of Biogeography*. Why the heat? Simple, we were a group of academics discussing the chapters that each of us felt *had* to be included in the *Handbook of Biogeography*. Our experiences were so different and our feelings so strong, that the list of chapters was well over fifty. Scrawled on Andrew Millington's notes from this discussion is "70?" and "Can we publish a two-volume handbook?" It was April 2001.

Left to the academics the debate would probably never have been resolved. Business sense prevailed. Discussions continued for months. "Around 40 chapters" was the strong advice from Robert Rojek. An editorial team was established. Glenn MacDonald was consulted over matters palaeoecological. We set about finalizing topics and potential authors. It was 2004 when authors were first approached and up to two years before we had our final list. As we write this introduction years later we understand the enormity of the task that that field trip conversation a decade ago in Armenia led prompted. Should it have remained one of the conversations that would be recalled, embellished and rehearsed again the next time we met? We think not. We hope you agree as you continue reading.

Andrew Millington, Mark Blumler and Udo Schickhoff
Adelaide, Binghamton, College Station and Hamburg.

Situating Contemporary Biogeography

Andrew C. Millington, Mark A. Blumler, and Udo Schickhoff

1.1 A SHARED ENDEAVOR

Biogeography is shared between three globally recognized and broad disciplines: biology, geography, and geology. Though all biogeographers share a common, natural history or ancestry dating back to the eighteenth and nineteenth centuries, the twentieth century has witnessed divergence as its practitioners have evolved to populate different niches within each of their three disciplinary homes. While we do not go as far as Mike Meadows, who opens his 1997 review of the sub-discipline with "Biogeography is as diverse as the organisms that form its subject matter" (583), we agree with the biogeographical analogy alluded to in the title of his article—there has been adaptive radiation of biogeography.

Biogeography's history has been partly written (e.g., Cox and Moore, 2005; Ebach and Tangey, 2007; Lomolino et al., 2004), but a truly comprehensive analysis of how it has evolved from its common origin to something distinctive in each of the three disciplines from a global perspective still eludes us. This book, with its aim of reflecting on contemporary biogeographical trends within geography, is not the place for such an analysis. Nonetheless in the following chapter (Blumler et al., this volume), we review how biogeographical thought has evolved (with a geographical emphasis) and in this chapter we offer an analysis of contemporary 'difference' in biogeography. By this, we mean the differences between styles of biogeography within geography in different parts of the academy. Having identified these, we suggest how they have evolved and what the

divergence they have fostered may mean for the future of biogeography.

The aim of this chapter is primarily to define what contemporary biogeography is within its 'geographical niche'. We do not attempt to do this for the 'geological' and 'biological' niches because we are introducing a book on a sub-discipline of geography, as interesting as comparing these other niches might be. In writing it this way, we acknowledge this does provide a highly biased perspective. First, we describe the main differences in the teaching of biogeography in geography departments around the world. We then look at the main contemporary biogeographical research themes in geography before providing a preliminary analysis as to how these patterns have emerged. Next, we consider what the apparent fragmentation within biogeography might mean for the sub-discipline within geography. In this, we pay particular attention to the challenges and opportunities it might pose for the future of geographical biogeography. We conclude by outlining the structure of this book and considering the rationale behind it.

1.2 BIOGEOGRAPHY IS THRIVING—BUT NOT EVERYWHERE

Biogeography is recognised, but is it thriving? Were it not recognised as a valid subdivision of our subject, it is unlikely that SAGE would have included it in its handbook survey of geography's sub-disciplines. However, this has not always been the case in Anglo-American academia (see, for example, Gregory, 2000). The status of biogeography in English-speaking geography departments dramatically increased from the mid-1960s to the early 1980s (e.g., Tivy, 1982; Watts' comment on its status in the United Kingdom (1978); for North America, see Rogers, 1983). Before then it was generally weakly developed and non-existent in most departments. However, the English-speaking world is but one geographical arena in which biogeography is taught and researched. The common ancestry of biogeography lies in the geographical borders of modern-day Germany. Biogeography is currently a strong sub-discipline in German geography departments and has been so throughout the twentieth century. The influence of German-speaking natural historians, geologists, biologists, and geographers is fundamental to the story of biogeography (see, for example, the following chapter; Cox and Moore, 2005; Ebach and Tangey, 2007; Lomolino et al., 2010). The 'Continental School' of biogeography, which evolved mainly in German-speaking

Europe, remains strong in Central and Eastern Europe and in the countries of the former Soviet Union. It is a diversion at this point to speculate that if SAGE had launched a geography handbook series in 1965—when it was founded—it is unlikely it would have included biogeography. By contrast, it is very likely that a German academic publisher would have included something that we would recognize today as a biogeography handbook.

A global survey of undergraduate and graduate biogeography teaching (in geography departments) carried out for the International Geographical Union between 1998 and 2002 and a mid-1990s survey of biogeography in British geography departments (Table 1.1) revealed the following:

- Biogeography was well established and actively pursued in the majority of the geography departments in Canada, Germany, the United Kingdom, and the United States.
- It was also well developed in Australian and New Zealand geography departments. Many of these departments had added biogeography to their

Table 1.1 Summary of IGU survey of biogeography teaching, 1998–2002

Area	Proportion of people teaching biogeography who declare themselves primarily biogeographers (%)	Countries with > three geography departments teaching biogeography	Notes
Europe	66	Germany, United Kingdom[1]	Little[2] biogeography in geography departments in Scandinavia, Netherlands, Belgium, France, Italy, Portugal, Spain
Asia	48	India, Russia, China	China, Japan, and Russia are underrepresented in the returns. Little biogeography in geography departments in western Asia
Africa	36	South Africa	Almost no biogeography taught in geography departments in francophone and lusophone African universities, or in North Africa
North America	55	Canada, United States	
South and Central America	36	Argentina, Chile	Brazil and Mexico are underrepresented in the returns
Pacific	54	Australia, New Zealand	

1 The United Kingdom was not included in the IGU survey as the BSG (see Table 1.2) had surveyed the status of biogeography in U.K. universities a few years previously.

2 The qualitative descriptor is used in relation to the number of geography departments in each country.

curricula around the same time as the expansion in Great Britain and North America in the 1960s and 1970s. Biogeography was a core element of most curricula at the time of the survey; often more than one person was active in teaching biogeography.

- The situation in continental Europe (outside Germany) was split between two camps at least. In eastern and central Europe (e.g., Austria, Romania, Poland, and German-speaking Swiss universities), biogeography was taught in a number of geography departments. They generally tended to have one biogeographer, whereas some departments in Germany had two or more: for example, Bayreuth, Erlangen, Münster, and Bonn, as did many departments in Canada, the United Kingdom, and the United States. By contrast, very little biogeography was taught in geography departments in France and in French-speaking departments in Belgium and Switzerland and many of the countries of the Mediterranean littoral. For example, no biogeography was taught in Greek geography departments, and there was little taught in Italy, Portugal or Spain.

- In English-speaking universities in the Caribbean, Africa, Asia, and the Pacific, biogeography was considered to be a legitimate and an even important part of curricula, but its inclusion appeared to depend on the efforts of one person in small departments who was not necessarily a biogeographer. This gave it a rather precarious foothold in the education of geographers and, therefore, shaky foundations for research and application to pressing 'development' issues. This was particularly so for countries with only one university, but the general observation applies to countries with many universities (e.g., India, Nigeria, and South Africa). We consider that the place of biogeography in curricula in these universities is due to their establishment by, and close links with, other Commonwealth universities throughout the twentieth century. In fact, some African universities were established as satellite campuses of British universities: for example, Fourah Bay College (part of the University of Sierra Leone) was part of Durham. Additionally, many people teaching in these universities obtained their higher degrees in Australia, Canada, New Zealand, the United Kingdom, and the United States, and then from often a very narrow range of universities in those countries (e.g., Cambridge, Durham, Liverpool, Oxford, and Swansea in Great Britain). Further evidence of the links with Anglo-American biogeography in the developed world is the use of undergraduate textbooks published in the United Kingdom and United States, most notably Cox and Moore's *Biogeography: An Ecological and Evolutionary*

Approach[1], though in India at least two biogeography textbooks have been published in English by Indian geographers (Bhattacharyya, 2003; Methani and Sinha, 2010) for Indian geography undergraduates, and Meadows (1985) has published a textbook in South Africa.

- The lack of biogeography teaching that was reported in French geography was reflected also in an apparent dearth of biogeography in geography departments throughout the francophone zone, though there were exceptions, for example, Universitié Nationale de Bénin. A similar trend was observed in lusophone geography departments in Africa and Asia.

- The situation in Central and South America was the most complex of the developing world regions. Biogeography is well embedded in a number of geography departments in Argentina[2] and Chile, where geography is a strong discipline. This is related to its European (mainly Germanic) roots, and it probably applies to southern Brazil as it is a phenomenon of the southern cone countries of South America (Gade, 2006a, 2006b). However, in the rest of Brazil,[3] where geography in general is also strong, the curriculum is dominated by human geography themes and biogeography is poorly represented. In other Latin American countries where geography is taught at university, biogeography was either not represented (e.g., Colombia, Guatemala) or taught in only one of a number of geography departments (e.g., Costa Rica, Ecuador). There were no geography departments in Bolivia and Paraguay.

- Departments in China, Mongolia, Japan, and the former Soviet Union were not adequately surveyed at the turn of the millennium. Subsequently, we have interviewed geographers and biologists about the status of biogeography in these countries in writing this chapter, and consulted gazetteers of geographers and, in the case of China consulted Leng et al. (2009). Biogeography is weakly developed in Japanese geography departments, in a somewhat dated directory of Japanese geographers only six geographers in 225 departments listed their specialist area as biogeography: four described themselves as vegetation geographers, one a biogeographer and another a palaeobiologist (The Association of Japanese Geographers, 1991). The dearth of Japanese biogeographers is stark when compared to work on vegetation science emanating from Japanese biology departments. This body of work is strongly influenced by German geobotanists, and it is epitomized by Akira Miyawaki (1924–present), who established a school of phytosociology, syntaxonomy, and plant geography at Yokohama National University and who edited the ten-volume *The Vegetation of Japan* (1980–89). Around 1950, phytogeography and zoogeography courses were launched in geography departments in universities in China, and simultaneously, teaching materials were published. These courses were merged later into biogeography courses that were taught as part of contemporary geography curricula but were also offered in biology, ecology, natural resource management, and environmental science departments. Geographers have responded by publishing biogeography textbooks (e.g., Northeast Normal University, 1989; Yin, 2004; Chen, 2007). In 2007, the Higher Education Press translated the seventh edition of Cox and Moore's textbook. Biogeography in Russia is so well developed that some universities have biogeography departments: for example, Moscow State University. There are also several biogeography textbooks in Russian, but complete information on biogeography teaching in the former Soviet Union still eludes us and is a great lacuna in this handbook.

What we teach is not necessarily what we research, and our teaching may also be contingent on what courses other biogeographers, in biology and geology departments, in our universities are already offering. Notwithstanding this, and the fact that any global survey is likely to suffer from geographical differences in response rates, the >300 responses obtained paint a picture of uneven training in biogeography to undergraduate and graduate students globally. While it was a core element in many departments where the Anglo-American and Continental (German) School traditions hold sway, it was poorly represented elsewhere. A further concern is the low proportion of people actually teaching biogeography who declared their primary (research/scholarship) interest to be biogeography (Table 1.1). This ranged from 66% in Europe to 36% in Africa and Latin America. The high proportion in Europe can be attributed to the long history of the sub-discipline in most European countries and the larger size of many departments, compared to departments in Africa and Latin America. However, even in Europe, these proportions were lower than anticipated. There were (and still are) departments where significant numbers of graduates are being trained and undergraduate programs are buoyant (e.g., Oxford; University College London in the United Kingdom; British Columbia; Tennessee; UCLA; Wisconsin in North America; Bayreuth, Bonn, Erlangen, Münster, and Trier in Germany; Basel in Switzerland). The global unevenness in teaching evident in the survey is mainly an outcome of four things: (a) the strength of physical geography within geography generally within a country—where it is strong, biogeography is more likely to be strong; (b) the perceived importance,

relevance, and utility of biogeography to national research and teaching goals; (c) the links between biogeographers in biology, geography, and geology within a country; and (d) the roles that influential biogeographers play or have played in the past in promoting the subject.

The situation with respect to research is more complex. Many biogeographers, perhaps the majority, conduct research in the countries where they teach. But biogeographers' enduring interests in other parts of the world, which we consider date back to the investigations of Darwin, Humboldt, Linneaus, and Wallace, and the expansion of Victorian natural history and science, mean that a significant amount of research has been and still is prosecuted in countries where biogeography's foothold in geography is precarious. For example, this book's editors have conducted research in countries where geography is not a university-level subject (e.g., Bolivia) or where biogeography is poorly represented in geography departments (e.g., Israel, Pakistan, Kyrgyzstan, and Yemen). Biogeographers who reside outside such countries—like the editors—have been able to obtain funds for research and (our own) doctoral training in these countries relatively easily. Yet it appears that in doing so developed world biogeographers have generally neglected to use their influence to ensure that biogeography attains a permanent place in geography curricula in the countries where geography is taught. We are wealthy researchers but poor missionaries: biogeographers appear to neglect the reproduction of our subject outside their own countries.

Progressing beyond the 1990's survey has been possible because of detailed reviews carried out by Young et al. (2004) and Veblen (1989) for North America, and Joyce (2009) for the United Kingdom. As far as we could ascertain no such reviews exist for other countries. In Germany, biogeography has gained a strong foothold within physical geography during the twentieth century. This can be attributed to influential geographers, such as Carl Troll and Josef Schmithüsen, who developed the concept of landscape ecology out of their biogeographical perspectives. In spite of the decline of geography (including biogeography) at some German universities due to fiscal constraints, biogeography has recently become a thriving sub-discipline and is expanding at universities such as Trier and Bayreuth. The significant upturn in German biogeography is related to the growing importance of research on implications of climate change and land degradation, but it is also due to the change from a more descriptively aligned discipline into a modern environmental science with strong reference to organisms and their spatial patterns. The increasing use of biogeographical experiments, augmenting the acceptance of biogeography within biology, has also contributed to the 'biogeography boom'. Interestingly, a few years after the establishment of the Working Group Biogeography within the Association of Geographers at German Universities (VGDH) in 1998, a Specialist Group on Macroecology was founded within the Ecological Society of Germany, Austria, and Switzerland (GfÖ). Scrutinizing the program of the macroecology meetings, it is thinly disguised biogeography, although the macroecology proponents who are biologists (mostly zoologists) consider it to be a 'young discipline'. Apparently, Middle European biologists doing biogeography hesitate to call it biogeography and instead prefer to use the term macroecology. Thus, there is recent biogeographical research that is not labeled biogeography, a situation that might hold for other regions as well.

Within the United States, the recent growth in biogeography has elevated it to more-or-less equal status with the more traditional branches of physical geography—geomorphology and climatology; the membership of the Association of American Geographers (AAG) biogeography specialty group (441) now exceeds that of the climate and geomorphology groups (411 and 393 respectively, 2010 data). In the United Kingdom, biogeography plays a secondary role to geomorphology within physical geography. In Germany, the number of physical geographers who declare biogeography as their topical specialization (36, 2010 data) ranks fifth after geomorphology (134), climate (79), soils (75), and hydrology/glaciology (55). However, the German biogeography community is increasing and catching up with the other sub-disciplines, reflected inter alia in the establishment of the Working Group Biogeography within the Association of Geographers at German Universities (VGDH). Internationally, the formation of the interdisciplinary International Biogeography Society in 2003 and the Biogeography and Biodiversity Commission of the International

Geographical Union (IGU) is raising the profile of biogeography within its three disciplines, though perhaps least so in geography (Table 1.2).

1.3 WHAT IS BIOGEOGRAPHY?

Notwithstanding the exceptions noted above, biogeography is generally well represented in geography departments around the world. It is also practised in biology and geology, but it was beyond our purview to comment on their recent development and global status in this book. Beyond these three disciplines, biogeography is taught and researched in environmental science and environmental studies departments, and in several branches of applied biology such as forestry, range management, horticulture, and pest management. Given these different disciplinary foci, the answer to the fundamental question 'What is biogeography?' will differ according to one's disciplinary home or department. The answer to the same question, asked specifically of geographers, could also vary according to where one has been trained, what teaching and research experiences one has had, where one currently is employed, and the length of one's career. If you were trained in and then spent your career in German geography departments, your biogeographical preoccupations would be different from those of a biogeographer who had trained and worked simultaneously in the United States. Evidence for this observation will be readily apparent as you read this book; compare, for example, the biogeographical emphases evident in the chapters on classification, biodiversity gradients, disturbance, and polar and mountain environments in this book (all written by German biogeographers) and consider how they might have been written by an American or a Canadian biogeographer. Biogeography evolves. So,

Table 1.2 National affinity groups for biogeographers organised by geographers, and other key organisations

National organisation	Affinity group	Year established
Royal Geographical Society with the Institute of British Geographers	Biogeography Research Group	1974
Association of American Geographers	Biogeography Speciality Group	1981
International Geographical Union	Biogeography and Biodiversity Commission	1996
Association of Geographers at German Universities	Working Group Biogeography	1998
International Biogeography Society		2003

somebody with a recently acquired Ph.D. entering a British geography department as a biogeographer will have a different set of research questions and skills from those of a biogeographer who has been in the same department for 30 years or more. A European- or North American-trained biogeographer who takes up a university position in Australia, New Zealand, or South Africa will find different, as well as similar, biogeographical concerns to those of colleagues they left behind. Geography is a mosaic of geographical and historical contingencies (Livingstone, 1994; Martin, 2005); these maxims can be applied to many, if not all, of geography's sub-disciplines, not just biogeography. In biogeography, historical contingency arises among late-nineteenth century natural historians at which time different subject material gained prominence in different countries due to the establishment of schools of geographical thought; and the presence of biogeographers in other disciplines (especially biology and geology) and the levels of interaction (bio)geographers[4] had with them.

Maybe the fundamental question we should then ask in *this* book is not 'What is biogeography?' but 'What is contemporary biogeography to a geographer?' As already noted, practicing biogeography within geography differs between countries, and there may even be differences in biogeography between geography departments in the same country. But are these differences significant, that is, are there biogeographies (both within geography *and* between disciplines)? If differences exist, are they so great that collaboration is impossible, or are they small enough to suggest the possibility of synergistic opportunities? To use a more biological analogy, is crossbreeding impossible and can we benefit from hybrid vigor?

Take two definitions of biogeography:

The study of the geographical distribution of plants, animals and other organisms. (Spellerberg and Sawyer, 1999: 1)

The study of the past and present geographic distributions of plants and animals and other organisms. (MacDonald, 2003: 1)

The first—a typical definition—stresses 'distribution' of all living things. The authors are biologists. The second definition is little different from the first—it also focuses on 'distribution' of all living things but explicitly mentions 'past' and 'present'. It is from a key text written for geographers, by a geographer with a strong historical biogeography interest. What is apparent from these two contemporary definitions (and many other in textbooks over the last three decades) is

that they are essentially the same. At one level, we do all agree on what biogeography is.

Is that, however, the reality? We argue there is only partial universal agreement. In geography, there are activities under the name of 'biogeography' that hardly fit either definition. In fact, there are different biogeographies within geography, and there are some biogeographies that (bio)geographers (compared to biological and geological biogeographers) contribute very little to at all. One way we have attempted to define 'biogeography' in geography is to undertake a contextual analysis of key words in section headings and chapters of a series of textbooks that have introduced biogeography to undergraduates since the 1970s from the Anglo-American and 'Continental' schools (Table 1.3). The analysis of the most frequently occurring words (Table 1.4) is revealing. If we omit 'biogeography', a number of tendencies in Anglo-American biogeography are revealed. First, we see the close relationship with ecology. The most frequently occurring word is 'ecosystem'—perhaps the must fundamental concept in ecology (Tansley, 1935)—and words that signal the influences of abiotic factors on (mainly) plant distributions such as 'environment(al)', 'soil', and 'change' (e.g., when used in the context of environmental change). The secondmost frequently occurring word is 'distribution', which, when considered alongside 'realm', 'communities', and 'community', illustrates what many biogeographers (in all three disciplines) would see as their main academic pursuit and which ties in with the definitions above. Further analysis of the use of this word group shows that it is more strongly aligned to spatial patterns of distribution, rather than temporal changes in distributions: this is perhaps not surprising as these books have been written by geographers. The final theme (represented by 'vegetation' and 'plant') to emerge from this analysis is the preoccupation of (bio)geographers with the plant kingdom at the expense of the animal kingdom. Two lacunae are historical (Quaternary) biogeography and the biogeography-society interface, despite both being well represented in contemporary research amongst (bio)geographers. The analysis integrates books published over 35 years, with 10 published before 1983. This may partly explain the dominance of the word 'ecosystem' and the underrepresentation of the human theme, but it does little to explain the lack of historical biogeography.

The German-language textbooks were in some ways more varied in the words they used (we recorded approximately 220 key words, against approximately 70 in English-language books). There are similarities and differences to the English key words: (a) the preoccupation with vegetation is more strongly developed, 'vegetation' occurs in

Table 1.3 Introductory biogeography texts written by geographers or used in geography in the Anglo-American and 'Continental' traditions that were used to extract key words for the context analysis

Title	Authors and publication dates	Country of first publication
Anglo-American School		
Systematic and Regional Biogeography	Morain (1970)	United States
A Geography of Plants and Animals	De Laubenfels (1970)	United States
Introduction to Biogeography	Seddon (1971)	United Kingdom
Principles of Biogeography	Watts (1971)	United Kingdom
Biogeography	Robinson (1972)	United Kingdom
Basic Biogeography	Pears (1977)	United Kingdom
Biogeography: Natural and Cultural	Simmons (1979)	United Kingdom
Plant Geography	Kellman (1980)	United States
Biogeography: Structure, Process, Pattern and Change Within the Biosphere	Jones (1980)	United Kingdom
Geography of the Biosphere	Furley and Newey (1983)	United Kingdom
Introduction to World Vegetation	Collinson (1988)	United Kingdom
Biogeography: A Study of Plants in the Ecosphere	Tivy (1993) (1st ed., 1971; 2nd ed., 1982)	United Kingdom
Fundamentals of Biogeography	Huggett (1998)	United Kingdom
Biogeography	Bhattacharyya (2003)	India
Biogeography: Space, Time and Life	MacDonald (2003)	United States
Environmental Biogeography	Ganderton and Coker (2005)	United States
Continental School		
Einführung in die Biogeographie von Mitteleuropa	Freitag (1962)	(West) Germany
Allgemeine Vegetationsgeographie	Schmithüsen (1968)	(West) Germany
Vegetationsgeographie auf Ökologisch—Soziologischer Grundlage	Schmidt (1969)	(East) Germany
Biogeographie	Aario and Illies (1970)	(West) Germany
Arealsysteme und Biogeographie	Müller (1981)	(West) Germany
Vegetationsgeographie	Klink and Mayer (1983)	(West) Germany
Biogeographie und Landschaftsökologie	Hoffmann (1985)	(West) Germany
Biogeographie, Artbildung, Evolution.	Sedlag and Weinert (1987)	(East) Germany
Vegetationsgeographie	Reichelt and Wilmanns (1989)	(West) Germany
Ökologie der Lebensräume	Tischler (1990)	Germany
Tiergeographie	Sedlag (1995)	Germany
Allgemeine Pflanzengeographie.	Richter (1997)	Germany
Vegetation und Klimazonen	Walter and Breckle (1999) (1st edn. 1970)	Germany
Handbuch der Ökozonen	Schultz (2000)	Germany
Vegetationszonen der Erde	Richter (2001)	Germany
Biogeographie	Beierkuhnlein (2007)	Germany

Table 1.4a The ten most frequently occurring key words in chapter and section titles in biogeography textbooks published in the Anglo-American biogeography tradition. Equivalent words that also occur in the Continental School list (Table 1.4b) in italics. See Table 1.3 for titles of books analysed.

Key words (in descending order of occurrence)	Number of occurrences in total	Number of books that used key word (n = 16)
Ecosystem	40	9
Distribution	26	8
Biogeography	23	7
Plant	19	6
Environment	16	8
Soil	14	5
Vegetation	14	5
Realm	11	4
Communities/ Community	10	5
Change	10	4

Table 1.4b The eleven most frequently occurring key words in chapter and section titles in biogeography textbooks published in the 'Continental School' biogeography tradition. Equivalent words that also occur in Anglo-American tradition list (Table 1.4a) in italics. See Table 1.3 for titles of books analysed

Key words (in descending order of occurrence)	Number of occurrences in total	Number of books that used key word (n = 11
Vegetation	26	8
Zonobiom and Ökozone/n	16	5
Vegetation*	14	8
Verbreitung (distribution)	9	5
Pflanzen	9	5
Biosphäre	9	2
Tropen	8	2
Differenzierung	7	2
Ökologisch	6	4
Erde	6	4
*formen	6	3

*indicates word used as a root or suffix, e.g., Vegetationsgürtel.

both lists and there are also nine occurrences of 'Pflanzen' on the German list. The distribution-spatial pattern theme is also strongly developed again with 'Verbreitung', 'Zonobiom', and 'Ökozone/n'. The strongest theme in English books, ecology, is weakly represented in German books (e.g., only six occurrences of ökologisch). This might be partly explained by the fact that the list consists predominantly of books written by geographers, a choice that was purposeful. The word frequency would undoubtedly change if books written by biologists are included, as it would for the English-language books. Again, the historical biogeography and biogeography-society themes are poorly represented. But themes we might have expected to see well represented in German list such as 'Landschaftsökologie' and 'Pflanzensoziologie' were not.

In Table 1.4a, four of the top ten key words—ecosystem, biogeography, distribution, and communities/community—also appear in the seventh edition of Cox and Moore's *Biogeography: An Ecological and Evolutionary Approach* (2005). We have already noted that this was the most frequently used English-language textbook in the survey of biogeography teaching. But we also include it here because it is interesting to speculate if any other sub-discipline of geography has given up its 'own' influence in training its undergraduates in the way that biogeography appears to have done. Peter Moore—a frequent contributor to *Progress in Physical Geography*—is a biologist, as was Barry Cox.

We have used other evidence to try and unravel what biogeography is to contemporary geography. These include Cowell and Parker's (2004) analysis of biogeography publishing in the *Annals of the Association of American Geographers*; Veblen's (1989) and Young et al.'s (2004) review of biogeography in North America; review articles in *Progress in Physical Geography*; and inputs from a range of colleagues. Four biogeographies clearly survive alongside each other in geography departments globally and we suggest a fifth might be emerging:

1   A biogeography that is in essence ecological, which focuses mainly on the distribution of contemporary and recent vegetation (and in the latter context hybridizes with the palaeoecological theme in historical biogeography). Major contemporary research themes include the role of climate (and to a lesser extent other environmental factors) in determining spatial distribution patterns; vegetation dynamics, plant disturbance ecology and succession; and land-use and land-cover change. It is often called 'ecological biogeography', but is not ecology.[5] At one time it would also have been recognized in almost

all biology departments, but developments in molecular biology and the status afforded biomedical research have reduced the importance of organismic and population biologists, who mainly contribute to this area. In a number of biology departments in the United Kingdom, for example, departing ecologists have not been replaced. Those remaining have developed close alliances with (bio)geographers; sometimes they have even moved to geography departments. Yet, in many areas of applied biology (e.g., wildlife and fisheries, range management and forestry departments in North America) 'ecological biogeography' remains credible. Fellow biogeographers in biology and applied biology departments remain a major opportunity for peer-to-peer research collaboration. Indeed the similarities in questions asked and methods used by 'biological' and 'geographical' ecological biogeographers are so similar that there is a convincing argument that such collaborations are not even cross-disciplinary.

2   The second theme is 'historical biogeography', which, in geography, focuses mainly on the changing distribution of vegetation during the Quaternary. Also practised in biology and geology, it derives from the geological tradition of paleobiology (see the following chapter, Blumler et al.), which can be divided into paleontology (concerned with deep time and, primarily, evolutionary questions) and Quaternary studies (concerned more with environmental reconstruction). In geology departments, the focus has mainly been on plant evolution over a longer geological time scale than the Quaternary, which most (bio)geographers focus on. In biology, the main focus until the advent of cladistic biogeography and phylogeny was, like geography, mainly on the Quaternary. But developments in cladistics and phylogeny (see Riddle, this volume) have led many biologists to extend their time scales of interest, and geographers have been encouraged to engage with these debates (e.g., Young, 2003). It can be argued that whereas (bio)geographers are mainly interested in the past distributions of individual plant species or plant communities, geologists have been interested in a more holistic view of changing distribution patterns *and* evolutionary trends in individual taxa, and biologists mainly in the evolutionary trends. Historical biogeography is part of an environmental reconstruction tradition, and it is intellectually broader than changing vegetation distributions. For example, it has made major contributions to archaeological investigations and climate change science.

3   A third biogeography has a very strong spatial component and relies heavily on a group of geospatial technologies: remote sensing, geographic information sciences, and spatial modeling (see Section IV, this volume). We call this 'spatial biogeography'. Though it is clear that these techniques are, in a biogeographical context, used to help answer important questions within an ecological and, to a lesser extent, historical biogeography we do not accept the argument that these technologies simply provide part of the toolkit for biogeography. Developments in remote sensing technologies, algorithms for information extraction (from remote sensing data), spatial modeling, and geographic information science as well as increased computing power—have enabled biogeographers to (a) acquire information they could not have done using other methods, (b) store, retrieve, and analyse data in ways that could not be done previously, and (c) analyze these data by harnessing developments in geostatistics and spatial modeling alongside immense computing power. These developments have enabled new types of biogeographical study: for example, global biome mapping and monitoring, and biocomplexity modeling. In our opinion, this is now a clearly recognizable type of biogeography.

4   The fourth type of biogeography is that which links biogeography with societal issues. This has always overlapped with the aspects of historical biogeography that contribute to the archaeological agenda, but nowadays most scholarship derives from a series of contemporary 'jumping off' points (as well as some in recent history, say the last half millennium). These mainly include major global environmental issues, such as global climate change, biodiversity, deforestation, and desertification, which have led biogeographers to become active in topics like conservation and ethnobiology and to engage closely with environmental historians, ecological economists, and political ecologists in social science. There are important sub-areas within this field. For example, animal and crop domestication literature, when written by geographers, is inherently biogeographical. Conservation biogeography is the latest, high-profile sub-field within the biogeography-society area. Not only are (bio)geographers playing an important role in this sub-field (see, for example, the compilations edited by Zimmerer and Young (1998), and Adams (2008)), but they are engaging with human geographers and a plethora of natural and social scientists, and placing biogeography firmly in the psyche of the wide range of people reading conservation literature. This field is potentially very broad and we concluded the book with a section linking biogeography and society.

5   We speculate about the emergence of a fifth type of 'biogeography', one that has strong currency

within geography and is being led by social scientists. That is research into relationships between 'the human'—at scales ranging from the individual to society—and 'nature'—from individual biological entities to a wider nature. In general, nature-society research is not being led by (bio) geographers but by human geographers (e.g., Philo and Wilbert, 2003; Whatmore, 2002, 2007; Wolch 2007; Wolch and Emel, 1998). Unlike much of the scholarship in the themes identified above, some conservation ecology excepted, there is greater focus on animals than plants: "Flora … remains an even more ghost-like presence in contemporary theoretical approaches" (Jones and Cloke 2002: 4). Perhaps our speculation only arises because this biogeography is, in general, not being led by (bio)geographers, though there are exceptions (e.g., Head and Aitchison, 2009).

The contemporary global synthesis of biogeography can be compared to the situation that Watts (1978) saw emerging in the 'new' biogeography in Britain of the 1970s. He identified five trends:

(a) investigations of soil-vegetation-environment complexes;
(b) relationships between major vegetation types and particular animal species;
(c) analyses of distributions of individual species and of the influencing processes;
(d) Quaternary community or ecosystem change; and
(e) mankind-ecosystem-community relationships.

The most striking differences in our list and Watts's trends are the dearth of work on vegetation-animal relationships that he predicted (some types of landscape ecology excepted) and the spectacular growth in the use of remote sensing, GIS, and spatial modeling in biogeography.

Spellerberg and Sawyer's (1999) five major 'schools' of biogeography or areas of specialist study (Table 1.5) provide the opportunity for a more recent comparison. They also provide a comprehensive view of biogeography, which, when compared to the five biogeographies identified above, clearly indicates the niches that (bio) geographers occupy within the whole of biogeography, that is, ecological, historical, analytical, and applied. The majority of (bio)geographers could hardly be considered active in Spellerberg and Sawyer's dispersalist (some U.S. West Coast scientists excepted), vicariance, regional, and taxonomic biogeographies.

Some aspects of biogeography are poorly represented in contemporary geography. The most obvious relate to the 'all living organisms' part of

Table 1.5 Spellerberg and Sawyer's (1999) Schools of Biogeography and the overlap with the 'biogeographies' of contemporary geography (in bold in first column)

Spellerberg and Sawyer's Schools of Biogeography	Level of development in contemporary geography (parentheses refer to chapters in this book related to this school)
Ecological Biogeography	Strongly developed (all Chapters in Sections II and III)
Historical Biogeography	Very weakly developed (if at all) over evolutionary and many geological timescales; strongly developed over Quaternary timescales (Willis et al., Marchant and Taylor)
Dispersal Biogeography	Inherent in historical studies of vegetation and in vegetation reconstruction; received prominence in the Berkeley School (Blumler)
Vicariance Biogeography Panbiogeography Cladistic Biogeography	Weakly developed in contemporary geography
Analytical Biogeography	Well developed in geography: e.g., remote sensing, landscape ecological, and spatial modeling (all chapters in Section IV)
Regional Biogeography	Weakly developed generally; best developed in the 'Continental School'
Taxonomic Biogeography	Very weakly developed, if at all
Applied Biogeography	Very strongly developed in geography: e.g., conservation, ethnobotany, invasive species, fire ecology (all chapters in Section V)

the definitions of biogeography. For many decades, most (bio)geographers have lacked significant engagement with animals—zoogeography. In what was, arguably, the first biogeography textbook in English written by somebody closely connected with geography, Marion Newbigin (1936) gave plants and animals almost equal weighting in '*Plant and Animal Geography*', considering their distributions and the links between their distributions. The main area where we engage with animals is in some types of landscape ecology, though even this aspect of landscape ecology is not well developed in geography. We do not espouse necessarily revisiting animal distributions in the ways that Newbigin outlined, albeit that this is possible. Beyond the burgeoning nature-society area that (bio)geographers could contribute too, we consider very fruitful areas in zoogeography for (bio)geographers to make an impact lie in the consideration of pathogens and their vectors—an area that is generally under researched in biogeography—and the biogeography of marine organisms. The world's oceans and seas have been a major lacuna for geography, and marine biodiversity is a vastly under-researched area.

1.4 DIVERGENCE

A century ago, biogeographers were generally familiar with each other's literatures, whatever their disciplinary home. But as disciplines emerged in the early twentieth century biologists, geographers, and geologists developed particular emphases within a broad biogeography. Consequently the broad communion of biogeographers became less connected. Divergence has increased in the last half century, not least in part because of the seemingly exponential proliferation of scientific publications.

In Anglo-American geography, there was a significant expansion in biogeography from the mid-1960s through the 1970s. In the United Kingdom, this period was the peak time in biogeography textbook publishing (Table 1.3); the Biogeography Research Group of the IBG (now RGS-IBG) was established with Len Curtis, John Taylor, and Bob Pullan as its first officers; and, importantly, David Watts and John Flenley (at that time both at Hull University) become the first editors of the *Journal of Biogeography* in 1974. There were parallel expansions in North America, Australia, and New Zealand. For example, the Biogeography Speciality Group of the AAG was established in 1981 with Jerry MacDonald, Lee Henry Slorp, Tom Veblen, Hartmut Walter, and Dean Wilder as its first officers.

Despite the growth in biogeography in the English-speaking academy in the 1960s and 1970s, by the early 1980s Rogers (1983) noted that the growth of biogeography teaching, graduate training, and the number of biogeographical faculty in North American departments had fallen behind that in the United Kingdom. Nonetheless, he identified three very influential departments—the California system campuses at Berkeley and Los Angeles (UCLA), and Wisconsin. McGill University provided the greatest concentration of biogeography within Canadian geography. As we argue, biogeography in North America has overtaken that in the United Kingdom in volumetric terms, mainly due to a late 1980s and early 1990s slump in the United Kingdom. Ten years prior to our writing this, Agnew and Spencer (1999) identified biogeography as a core sub-discipline in British physical geography. Yet others writing at the same time disagreed, arguing that other disciplines had taken it over (Gregory et al., 2002). Joyce (2009) and Whittaker and Sax (2003) suggest that this was due to ecology (in particular) being better organized than biogeography, but this is possibly only part of the reason for the decline in British biogeography during the 1990s.

What stimulated the expansion of biogeography in Anglo-American consciousness in the late 1960s and 1970s? A combination of interviews with biogeographers active in the 1970s conducted for this book and other data suggest that there may have been four main factors. First, there was the demand for increased access to higher education starting in the 1970s (and which continues to the present day). Second, the number of higher education establishments began to expand in the 1960s in both the United Kingdom and United States, and higher education participation rates increased. The same happened in Germany, with biogeography positions being created during the expansion of existing, and the foundation of new, universities during the 1970s. Third, was rising student demand for courses related to the 'environment', the profile of which had been heightened by the environmentalism of the 1960s. This was a strong trend in both North America and Europe: for example, in Germany during the aftermath of the Limits to Growth report, there were heated public debates about the fate of tropical rainforests and nuclear power. The fourth and final factor is the role of agency—both individual and institutional.

In the United Kingdom, the expansion of higher education establishments, student numbers, and the strong position of geography in secondary school curricula created a demand for geography, with biogeography being seen as integral to physical geography. For many North American departments in the United States, expanding into biogeography was relatively straightforward for many departments that were dominated by climatologists and geomorphologists. In essence, it was a demand-led phenomenon. It had parallels in Britain, where the publication of 10 biogeography textbooks by U.K.-based authors in the 1970s and the early 1980s is testament to the demand for course materials in geography departments in universities and polytechnics at that time. In the United States, the environmental movement of the 1960s and 1970s increased academic and popular interest in environmental issues. Biogeography clearly falls under that umbrella, and it may have led some geography departments to offer courses that were relevant to these national debates, but the biggest beneficiaries of the environmental wave were biology departments.

The role of individual and institutional agency is also important in the development of biogeography in the late twentieth century. In North America, many biogeographers, including many who work at the biogeography-society nexus (some of whom may not consider themselves biogeographers), have influences that can be traced back to the Berkeley School of American Geography (see Blumler et al., this volume). Developments in biogeography were part of the wider Berkeley School, which originated in Carl Sauer's (1899–1975) interests in human-environment interaction (e.g., Sauer, 1941; 1952; 1956). Key biogeographical themes researched by the Berkeley School included fire, invasive species, ethnobotany, conservation, historical ecology, agricultural origins, and crop genetics. Many themes were influenced by developments outside the Berkeley School—most notably, historical ecology, which was well advanced in Europe—and some, such as agricultural origins and crop genetics, have waned in importance within geography. The broader cultural geography that the Berkeley School espoused, and which influenced Berkeley School 'biogeography', has been usurped by a newer cultural geography that had its origins in sociology and which first gained prominence in British geography in the 1980s (e.g., Valentine, 2001). New cultural geography has studied the natural world in different ways to most (bio)geographers, let alone biogeographers in biology and geography, and has only recently engaged with the natural world in a way that (bio)geographers can confidently and comfortably contribute to (e.g., Phillips et al., 2008): this is the speculative fifth type of biogeography we alluded to above.

Members of the Berkeley School advised a generation of biogeographers who later dispersed throughout U.S. geography. Key among these were Jonathan Sauer (see Blumler et al., this volume) and Thomas Vale (Wisconsin), Thomas

Veblen (Colorado), and Robert Frenkel (Oregon). Biogeographers such as these went on to train many of the current mid-career and older biogeographers in North American geography departments. As a consequence, the Berkeley School has promoted an enduring heritage amongst the teaching and research of some biogeographers to the present day, even though its currency within human geography has waned.

A parallel line of agency in North American biogeography dates back to the time when, as biogeography in geography departments expanded further, many turned their attention to graduate students trained in biology in their search for a second biogeographer. This influence has a shorter history than the Berkeley School but nonetheless has had important influences on the composition of biogeographers within contemporary North American biogeography. This trend was initiated by the geography department at UCLA when they appointed Walt Westman—a student of Robert Whitaker's at Cornell—who went on to train a number of biogeographers in a hybrid biology-geography tradition. Like those influenced by Berkeley School professors, many of Westman's students have gone on to train significant numbers of practising biogeographers in geography departments in North America.

At the start of the expansion of biogeography in the 1960s and 1970s historical biogeography was a well-developed area in the United Kingdom, . The key influence here was the Sub-Department of Quaternary Research (SDQR) at Cambridge, and its two directors up to that time—Sir Harry Godwin (1901–85) and Richard West. This interdisciplinary center was established after the Second World War, though a group of Cambridge-based scientists were working on archaeological themes in the English fenlands before then. These included Godwin, a plant ecologist who introduced palynology, which had developed in Scandinavia in the early twentieth century, to Britain and was influenced by the work of the Danish palynologist Johannes Iversen (1904–72) (West, 1988). His counterpart in North America was the botanist Paul Sears (1891–1990). Godwin went on to become the SDQR's first director in 1948. He was influenced by phytosociology, having close contact with both A.G. Tansley (also at Cambridge) and European plant ecologists through international phytogeography excursions. Godwin trained a generation of interdisciplinary Quaternary scientists who worked in the United Kingdom, Australia, and New Zealand, one of whom—Richard West—succeeded him as the SDQR director and in turn trained a cohort of PhD students, including influential historical biogeographers in geography, biology, and geology. SDQR also created an environment of hybridization with, for example, geographers using facilities in SDQR that were not available to most geographers elsewhere in the 1970s. Many of them were seeking jobs when universities began their expansion in the 1970s (Turner and Gibbard, 1996). As U.K. geography departments expanded in the 1970s, they also hired from people with biological training, especially in the area of historical biogeography (e.g., John Flenley, who co-wrote the following chapter). This trend was parallel with that in the United States but hiring from the biological sciences was not as strong in the United Kingdom as it was on the other side of the Atlantic.

Other agents influencing British biogeography in the second half of the twentieth century were either less dominant or less clear. The Berkeley School had little influence—though Ian Simmons (1979) acknowledges the influence of a sabbatical at Berkeley; David Harris (Institute for Archaeology, University of London) was trained by Carl Sauer; and David Watts at McGill researched in the 'Sauerian mould'—and very few geographers adopted European plant sociology. Two 'domestic' influences do seem to have had some traction. First, there was the local and landscape studies school promoted by W.G. Hoskins, see for example Hoskins (1955), which influenced biogeographers such as Robert Eyre (Eyre, 1963) at Leeds into an ecological biogeography that acknowledged human occupation of the landscape. Second, there was the cadre of scientists, including biogeographers, who had worked on natural resource issues (e.g., Stobbs, 1963) and in overseas universities (for example, Monica Cole; Royal Geographical Society, n.d.)

Divergence in biogeography occurred in two forms during the twentieth century. First, biogeography took on increasingly different stances in biology, geography, and geology. Second, differences emerged among the major groupings of (bio)geographers in geography—in continental Europe, Britain, and North America.

1.5 THE IVORY ARCHIPELAGO

Joyce (2009: 1) states that divergence may have led to isolation and lack of recognition of a "shared, cognate science built upon a considerable lineage of conceptual achievements." That may be too strong a statement, nonetheless: the proliferation of topic areas, concerns, and theoretical structures in the broad arena of biogeography means that nobody today is truly 'on top of the field'. This is, of course, an issue across the entire spectrum of academia. As the evolutionary biologist David Wilson puts it:

It is easy for scientists and intellectuals to smile at the ignorance of . . . the general public, but the fact is that they're not much better. The Ivory Tower would be more aptly named the Ivory Archipelago. It consists of hundreds of isolated subjects, each divided into smaller subjects in an almost infinite progression. . . . Each perspective has its own history and special assumptions. One person's heresy is another's commonplace. (2007: 2)

We appreciate the biogeographical analogy—courses in biogeography often include discussion of how archipelagos differ from both single islands and continents in terms of evolutionary pathways and in terms of biodiversity. We also agree with the implication that academic disciplines are not quite islands unto themselves: typically there are a few, transdisciplinary connections but these may be quite limited and only between cognate fields. Alternatively, it might be appropriate to speak of an 'Ivory Tower of Babel' (pun intended), because of the pronounced differences in language, terminology, and jargon that have evolved in different fields, sometimes making them mutually incomprehensible.

Contributing factors are the rise of electronic publishing, the 'gold standard' of research, and the ratings of departments (and individuals) according to bibliometrics. This has had the effect of constraining the literature cited to a narrower range concentrated in 'select' journals (Agnew, 2009; Evans, 2008). In biogeography that has elevated journals, such as the *Journal of Biogeography* and its spin-off journals, major ecology journals published by national societies (e.g., the Ecological Society of America and the British Ecological Society), and top-line journals like *Nature, Science, Proceedings of the National Academy of Sciences*, and *Philosophical Transactions of the Royal Society B* at the expense of smaller, highly specialized journals in biogeography, such as those published by natural history museums and regional societies (e.g., *Castanea*, Southern Appalachian Botanical Society; *The Great Basin Naturalist*, Brigham Young University; *American Midland Naturalist*, Notre Dame University). Agnew (2009: 3) commented, "Who reads journals any more, looking to experience the joy of serendipity, particularly when those journals have low impact factors?" We add to this our broad inability to process materials in other languages as evinced by the bibliographies of many papers. English-speaking biogeographers are simultaneously at an advantage and disadvantage because all high-impact factor journals are in English (which attract non-native English speakers to submit their best material to them) and affords those who read and speak the language fluently the dubious luxury of ignoring biogeographical literature not written in English. The strength of biogeography in German-speaking geography allows a vibrant peer-reviewed literature to be published; unfortunately, much of it fails to stimulate the minds of English-speaking biogeographers. Given the nature of the human brain, which is great at mimicry but not terribly good at "thinking outside of the box" (cf. Pinker, 2003), the long-range impact is likely to be the decreased ability to achieve paradigmatic breakthroughs. Thinking outside of the box requires reading outside of the (disciplinary) box, paying attention to specialist journals with low citation indices, and reading outside of the English language.

Land degradation is a cogent example of the type of problem that can arise because the ivory archipelago exists. It is widely perceived as a serious global environmental problem, especially as manifested in topics like deforestation and desertification and their impacts on ecosystem services and people's livelihoods. Biogeographical evidence of vegetation change (e.g., forest loss, decline in forest quality, and forest recovery—the so-called forest transition) is highly relevant to 'deforestation' discourses. Vegetation dynamics (and therefore biogeographical contributions) have not played such an important role in desertification as they have in deforestation, probably due to lower amounts of vegetation in ecosystems thought susceptible to 'desertification' and much weaker links to another major global problem—biodiversity—which is lower in arid and semi-arid ecosystems, and was not on the global environmental issues menu when desertification rose to prominence in the 1970s. In fact, at the 1992 Rio Earth Summit, political trade-offs were made between 'desertification' and 'biodiversity' to enable the Convention to Combat Desertification and the Convention on Biological Diversity to proceed simultaneously.

While some scientists agree with Jared Diamond (2006) that land degradation has become so serious that it threatens our civilization with collapse, many nonequilibrium ecologists and biogeographers, geomorphologists, and human geographers argue that the degradation thesis has, for the most part, been disproven (e.g., Blaikie and Muldavin, 2004; Blumler, 2002, 2006). Many scientists—Diamond among them—appear to be unaware of the existence of this critique; others, we assume, presume that it must not be empirically based. Several geographers who would not categorize themselves as biogeographers have carried out excellent, empirical 'biogeographical' research that has chipped away at the degradation edifice (e.g., Bassett and Bi Zueli, 2000; Brower and Dennis, 1998; Davis, 2007; Fairhead and Leach,

1996; Turner, 1998). Interestingly, most of these studies focus on arid and semi-arid ecosystems: ecosystems in which the geomorphologist David Thomas (1993; Thomas and Middleton, 1994) demonstrates convincingly that there is little empirical evidence to support the existence of widespread desertification.

While quite a few mainstream environmental scientists—some biogeographers among them—appear to accept the viewpoint that land degradation has little empirical support, other critics, who at times are vitriolic in their antiscience stance, nonetheless have gathered some solid scientific data that counter the orthodox views on land degradation. The irony here is that some of the 'successful biogeographical research' in this area has been carried out by human geographers—typically political ecologists or cultural ecologists—whose influences stem from critical social theory. Most academics would presume that critical social theory and science cannot merge, but as F. Scott Fitzgerald (1945) pointed out, "the test of a first-rate intelligence is the ability to hold two opposed ideas in the mind at the same time, and still retain the ability to function." A final irony is that, although this situation cries out for hypothesis testing and debate, there is very little debate, let alone hypothesis testing. Have we reached an impasse where each camp is convinced that its understanding is correct, so much so that they rarely engage in face-to-face debate but rather proclaim their positions in 'their' journals? This is an outcome of the current failure to read outside of one's sub-discipline, so that one person's heresy can indeed become another's commonplace, to paraphrase Wilson (2007: quoted above).

1.6 AIM AND STRUCTURE OF THE HANDBOOK

Given the divergent trends in biogeography, it appears we have embarked on a Sisyphean task in editing this book. That would be true if we were cataloging all aspects of biogeography, but the aim of the handbook is to reflect (as comprehensively as possible) the current state of biogeography as a sub-discipline of geography. The differences in biogeography within geography meant that obtaining agreement on the 'main' areas, and then identifying 'acknowledged' leaders was extremely difficult (see the foreword). Furthermore, as we wanted to take a global approach to biogeography, some topics do not appear as full chapters and some 'acknowledged' leaders (in say one country) do not appear as chapter authors. We have attempted to celebrate our differences in both topic and author choice. A further aim was to link to other disciplines, in particular biology, where we felt there were important developments that biogeographers in geography should be more actively engaged with. Chapters on classification (Angelika Schwabe and Anselm Kratochwil), phylogeny (Brett Riddle), biodiversity gradients (macroecology) (Jens Mutke), and bioindication (Yordan Uzonov) are chapters written by biologists in the hope of stimulating geographers.

The book has five sections, each with an introduction. Section I reviews key *theories* and *concepts* in biogeography. Biogeography has been criticized as being light on theory, or having only one theory—the Equilibrium Theory of Island Biogeography (ETIB). We take issue with such criticisms and focus on theory and concept at the start of the handbook. If broadly defined, biogeography is rich in theories, many of which are still (rightly or wrongly) still debated: for example, evolution, vegetation succession, and ETIB. We trace the evolution of biogeographical theories in the first chapter in this section (Mark Blumler, Anthony Cole, John Flenley and Udo Schickhoff), and examine theories of biodiversity (Duane Griffin) and theory in landscape ecology (John Kupfer). Related to these theories are a series of key concepts: classification (Angelicka Schwabe and Anselm Kratochwil); phylogeny (Brett Riddle); and refugia (Kathy Willis, Shonil Bhagwat and Mary Edwards).

Section II considers current knowledge about *gradients* and the importance of *disturbances*. Most biogeographers the world over see spatial and temporal patterns as the essence of biogeography. Individual chapters explain spatial patterns (gradients) (Jens Mutke); abiotic and biotic controls on spatial patterns (Udo Schickhoff); the influence of disturbances in general (Anke Jentsch and Carl Beierkuhnlein); and the impacts of fire (Neal Enright) and climate change (Tim Sparks, Annette Menzel, Josep Peñuelas and Piotr Tryjanowski) on spatial and temporal changes in distributions.

Section III was the hardest to reach agreement on. It reconsiders what we know about selected *biomes* and *environments*. Consensus was hard to reach because all biomes are important to biogeographers. However, there are simply too many to include in this handbook, and also we did not want to ignore important environments that do not easily fit biome models. We selected four broad zonal regions that, more or less, map onto biomes and three environments. There is a bigger point here. We consider biomes to be a form of regional geography, which in itself gets poor treatment in contemporary geography. In the United States,

regional geography is taught far more extensively than it is the United Kingdom, for instance, but regional geography or area studies is not a mainstream research theme on either side of the Atlantic. Authors have written personal essays on these, rather than following a model. The polar and subpolar regions (Ingo Möller and Dietbert Thannheiser) were chosen because of debates over the use of these environments and the evidence that climate change is impacting these biomes more so than many others. Boreal forests (Daniel Kneeshaw, Yves Bergeron and Timo Kuuluvainen) were selected because of their significance in terms of the global carbon budget. Savannas (Jayalaxshmi Mistry) remain something of an ecological conundrum because of the coexistence of grasses and trees; and tropical forests (Ken Young) have witnessed significant landcover change since the 1950s. We balanced the biome essays with three 'environments'. The choice of mountains (Udo Schickhoff) reflects the significant amounts of research conducted globally on mountain biogeography and ecology and was driven by concerns about biodiversity conservation and climate change. The chapter on agricultural environments (Chris Stoate) also reflects biodiversity concerns. The choice of urban environments (Claire Freeman) was stimulated by concerns about increasing urbanization and also reflects research on urban ecology.

Section IV examines how *mapping* and *modeling* is used in biogeography. This area has seen tremendous growth since the 1980s with the increased availability of different types of remotely sensed data (as a primary data source), developments in GIS, spatial modeling, and increased computing power. Two chapters consider remote sensing for mapping biogeographical distributions (Giles Foody and Andrew Millington) and as inputs for modeling (Joanne Nightingale, Stuart Phinn and Michael Hill). Three chapters consider different approaches to modeling that are used extensively in biogeography: predictive modeling (Niall Burnside and Stephen Waite), simulation modeling (George Malanson), and biocomplexity (Steve Walsh, George Malanson, Joe Messinia, Dan Brown and Carlos Mena). The importance of landscape ecology in this area is represented by a consideration of spatial patterns by Tom Albright, Monica Turner and Jeffrey Cardille.

Section V explores the links between *biogeography and society*. We have been highly selective. The major area of contemporary concern—biodiversity conservation—is explored in detail by Rob Marchant and David Taylor, Geoff Griffiths and Ioannis Vogiatzakis, and Patrick Osborne and Pedro Leitão. We explore vegetation as resource in Robert Voeks's chapter on ethnobotany, and vegetation as a hazard in Mark Blumler's chapter on invasives. Importantly, we provide one illustration of how ecological biogeography can be utilized, through the vehicle of bioindication (Yordan Uzonov).

1.7 BIOGEOGRAPHICAL FUTURES

The role of this handbook should be not only to celebrate the diversity of biogeography and biogeographers, and applaud the very real contributions made within each tradition and subject area; but also to point out divergences, identify debates, and encourage engagement with critical, unresolved issues. The sub-discipline is in rude health in some countries but is doing little more than 'hanging in there' elsewhere. Since the 1960s biogeography has expanded and contracted in response to trends in both higher education and geography. It seems to us that biogeography could often be accused of being guilty of not promoting itself and not organizing itself within geography. Compared to geomorphology, for example, it has been a 'bit player'.

Gregory et al. (2002) point out the paradox that, while biogeographers work in geography, evolutionary biology, and the environmental sciences, most biogeographical research is conducted in biology departments and ecological research centers. The implication is clear—the frontiers of biogeographical research currently lie outside geography. Joyce (2009), in an upbeat commentary in *Area*, suggests that British biogeographers are increasing their engagements with economics, geomorphology, social sciences, archaeology, and marine biology. The first comment is unnecessarily pessimistic, but if we continue to reproduce the narrow definition of biogeography referred to earlier in this chapter, it is clear how Gregory's paradox comes about and that the death knell is being rung for biogeography in the discipline of geography. Joyce's assessment suggests that we should expand our interdisciplinary linkages, and while interdisciplinary research is an important element of any discipline nowadays, and essential in tackling important global issues, overemphasis on interdisciplinarity can devalue individual disciplines. Without a core biogeography, and (bio) geographers who self identify with it, there will be no interdisciplinary future for biogeography in geography departments.

NOTES

1 Cox and Moore's textbook, in at least one of its editions, was the most widely used undergraduate textbook in the English-speaking universities surveyed, and it has been translated into Chinese and German.

2 In fact, biogeography is so well developed in Argentina that it is represented in a series of 12 university-level textbooks, each representing a sub-discipline (Petanga de del Río, 1992).

3 The survey response rate from Brazilian geography departments was poor, and the status of biogeography in Brazil at that time comes from various individual sources.

4 At this point we introduce the term (bio)geographer to indicate those biogeographers who have trained in and/or practice their biogeography within geography departments to distinguish them from biogeographers in other disciplines.

5 Even though it is not ecology, some geographers are confused by what we do or how we label ourselves. Demeritt (2009), in an editorial on environmental geography, notes one of the contributions of physical geography/ers to environmental geography is ecology, but not biogeography.

REFERENCES

Aario, L. and Illies, J. (1970) *Biogeographie*. Braunschweig: Westermann Verlag,

Adams, W.M. (ed.) (2008) *Conservation*. London: Earthscan.

Agnew, J. (2009) 'President's column: the impact factor', *AAG Newsletter*, 44:3.

Agnew, C., and Spencer, T. (1999) 'Editorial: Where have all the physical geographers gone?' *Transactions Institute of British Geographers NS*, 24:5–9.

Association of Japanese Geographers (1991) *Departments of Geography in Japanese Universities*. Tokyo: The Association of Japanese Geographers.

Bassett, T.J., and Bi Zueli, K. (2000) 'Environmental discourses and the Ivorian savanna', *Annals of the Association of American Geographers*, 90:67–95.

Beierkuhnlein, C. (2007) *Biogeographie*. Stuttgart: Eugen Ulmer Verlag,

Bhattacharyya, N.N. (2003) *Biogeography*. New Delhi: Rajesh Publications.

Blaikie, P.M., and Muldavin, J.S.S. (2004) 'Upstream, downstream, China, India: The politics of environment in the Himalayan region', *Annals of the Association of American Geographers*, 94:520–48.

Blumler, M.A. (2002) 'Changing paradigms, wild cereal ecology, and agricultural origins'. In R.T.J., Cappers and S. Bottema (eds.), *The Dawn of Farming in the Near East*, pp. 95–111. Studies in Early Near Eastern Production, Subsistence and Environment 6, 1999. Berlin: ex-Oriente.

Blumler, M.A. (2006) 'Regional contrasts in perceptions of pastoralism: The example of environmental history', *Research in Contemporary and Applied Geography: A Discussion Series*, 30(1):1–27.

Brower, B., and Dennis, A. (1998) 'Grazing the forest, shaping the landscape? Continuing the debate about forest dynamics in Sagarmatha National Park'. In K.S. Zimmerer and K.R. Young (eds.), *Nature's Geography: New Lessons for Conservation in Developing Countries*, 184–208. Madison: University of Wisconsin Press.

Chen, K. (2007) *Biogeography*. Xinging: Qinghai Normal University Press.

Collinson, A.C. (1988) *Introduction to World Vegetation*. 2nd ed. London: George Allen and Unwin.

Cox, C.B., and Moore, P.R. (2005) *Biogeography: An ecological and evolutionary approach*. 7th edn. Oxford: Blackwell.

Cowell. C.M. and Parker, A.J. (2004) 'Biogeography in the Annals', *Annals Association of American Geographers*, 94(2): 256–268.

Davis, D. (2007) *Resurrecting the Granary of Rome: Environmental History and French Colonial Expansion in North Africa*. Athens OH: Ohio University Press.

de Laubenfels, D.J. (1970) *A Geography of Plants and Animals*. Dubuque, IA: W.C. Brown.

Demeritt, D.A. (2009) 'From externality to inputs and inference: Framing environmental research in geography', *Transactions Institute of British Geographers*, NS 34:3–11.

Diamond, J. (2006) *Collapse: How Societies Choose to Fail or Succeed*. New York: Viking Penguin.

Ebach, M.C., and Tangey, R.S. (2007) *Biogeography in a Changing World*. Boca Raton, FL: CRC Press.

Evans, J.A. (2008) 'Electronic publication and the narrowing of science and scholarship', *Science* 321: 395–99.

Eyre, R.S. (1963) *Vegetation and Soils - a world picture*. London: Edward Arnold,

Fairhead, J., and Leach. M. (1996) *Misreading the African Landscape: Society and Ecology in a Forest-savanna Mosaic*. Cambridge: Cambridge University Press.

Fitzgerald, F. S. (1945) *The Crack-up*. New York: J. Laughlin.

Furley, P.A., and Newey, W.W. (1983) *Geography of the Biosphere*. London: Butterworths.

Gade, D.W. (2006a) 'Converging ethnobiology and ethnobibliography: Cultivated plants, Heinz Brücher, and Nazi ideology', *Journal of Ethnobiology*, 26(1), 82-106.

Gade, D.W. (2006b) 'Paraguay 1975: Thinking back on the fieldwork movement', *Journal of Latin American Geography*, 5(1), 31–49.

Ganderton, P. and Coker, P. (2005) *Environmental Biogeography*. Upper Saddle River NJ: Prentice Hall.

Gregory, K.J. (2000) *The Changing Nature of Physical Geography*. London: Edward Arnold.

Gregory, K.J., Gurnell, A.M., and Petts, G.E. (2002) 'Restructuring physical geography', *Transactions Institute of British Geographers NS*, 27: 136–54.

Freitag, H. (1962) *Einführung in die Biogeographie von Mitteleuropa*. Stuttgart: Gustav Fischer Verlag.Head, L. and Aitchison, J. (2009) 'Emerging human-plant geographies', *Progress in Human Geography*, 33(2), 236-45

Jones, O. and Cloke, P. (2002) *Tree cultures. The place of trees and trees in their place*. Oxford: Berg.

Hoffman, M. (1985) *Biogeographie unde Landschaftsökolgie*. Paderborn: Schöningh Verlag.

Hoskins, W.G. (1955) *Making of the English Landscape*. Harmondsworth: Penguin Books.

Huggett, R.J. (1998) *Fundamentals of Biogeography*. London: Routledge.

Jones, R.L. (1980) *Biogeography: Structure, Process, Pattern and Change Within the Biosphere*. Amersham: Hulton Educational.

Joyce, C. (2009) 'New challenges in biogeography' , *Area*, 41(3): 354-57.

Kellman, M.C. (1980) *Plant Geography*. 2nd ed. Methuen: London.

Klink, H-J. and Mayer, E. (1983) *Vegetationsgeographie*. Braunschweig: Westermann.

Leng, S., Li, X., Li, Y., Xu, H., Kang, M., Jiang, Y., Yin, X., Tao, Y., and Xin, W., (2009) Recent Progress in Biogeography in China. *Acta Geographica Sinica,* 2009-09.

Livingstone, D.N. (1994) *The Geographical Tradition*. Oxford: Blackwell.

Lomolino, M.V., Riddle, B.H., Whittaker, R.J. and Brown, J.H. (2010). *Biogeography*. 4th ed. Sunderland, MA: Sinauer Associates.

MacDonald, G.M. (2003) *Biogeography: Introduction to Space, Time and Life*. New York: John Wiley.

Martin, G.J. (2005) *All Possible Worlds*. 4th ed. New York: Oxford University Press.

Methani, S, and Sinha, A. (2010) *Biogeograpghy*. New Delhi: Commonwealth Publishers.

Meadows, M.W. (1985) *Biogeography and Ecosystems of South Africa*. Cape Town: Juta and Company.

Meadows, M.W. (1997) 'Biogeography—adaptive radiation of a discipline', *Progress in Physical Geography*, 21:583–92.

Morain, S.A. (1970) *Systematic and Regional Biogeography*. New York: Van Nostrand Reinhold.

Müller, P. (1981) *Arealsysteme und Biogeographie*. Stuttgart: Eugen Ulmer Verlag.

Newbigin, M.I. (1936) *Animal and Plant Geography*. London: Methuen and Co. Ltd.

Northeast Normal University (1989) *Biogeography*. Changchun: Press of Northeast Normal University.

Pears, N.V. (1977) *Basic Biogeography*. London: Longman.

Petanga de del Río, A.M. (1992) *Biogeografia*. San Isidro, Argentina: Editorial Ceyne.

Phillips, M., Page, S., Saratsi, E., Tansey, K., and Moore, K. (2008) 'Diversity, scale and green landscapes in the gentrification process: Traversing ecological and social science perspectives', *Journal of Applied Geography*, 28: 54–76.

Philo, C., and Wilbert, C. (2000) *Animal Spaces, Beastly Places*. London: Routledge.

Pinker, S. (2003) *The Blank Slate: The Modern Denial of Human Nature*. New York: Penguin.

Reichert, G. and Wilmanns, O. (1973) *Vegetationsgeographie*. Braunschweig: Georg Westermann Verlag.

Richter, M. (1997) *Allgemeine Pflanzengeographie*. Stuttgart: B.G.Teubner.

Richter, M. (2001) *Vegetationszonen der Erde*. Gotha: Klett-Perthes.

Robinson, H. (1972) *Biogeography*. London: MacDonald and Evans.

Rogers, G.F. (1983) 'Growth of biogeography in Canadian and U.S. geography departments', *Professional Geographer*, 35: 219–26.

Royal Geographical Society. (n.d.) Monica Cole. http://www.rgs.org/OurWork/Grants/Grant+benefactors/Monica+Cole.htm. Accessed 29 May 2009.

Sauer, C.O. (1941) 'The Personality of Mexico', *The Geographical Review*, 21(3), 353-64.

Sauer, C.O. (1952) *Agricultural Origins and Dispersals*. New York: American Geographical Society.

Sauer, C.O. (1956) 'The agency of man on earth'. In W. Thomas (ed.), *Man's Role in Changing the Face of the Earth*. Chicago: University of Chicago Press.

Schmidt, G. (1969) *Vegetationsgeographie auf Ökologisch-Sociologischer Grundlage*. Leipzig: BSB B.G. Teubner Verlagsgesellschaft.

Schmidthüsen, J. (1968) *Allgemeine Vegetationsgeographie*. Berlin: Walter de Gruyter & Co.

Schultz, J. (2000) *Handbuch der Ökozonen*. Stuttgart: Eugen Ulmer Verlag.

Seddon, B. (1971) *Introduction to Biogeography*. London: Duckworth.

Sedlag, U. (1995) *Tiergeographie*. Berlin: Urania Verlag.

Sedlag, U. and Weinert, E. (1987) *Biogeographie, Artbildung, Evolution*. Jena: Gustav Fischer Verlag.

Simmons, I.G. (1979) *Biogeography: Natural and Cultural*. London: Edward Arnold.

Spellerberg, I.F., and Sawyer, J.W.D. (1999) *An Introduction to Applied Biogeography*. Cambridge: Cambridge University Press.

Stobbs, A.R. (1963) *Soils and Geography of the Boliland Region, Sierra Leone*. Freetown, Sierra Leone: Government Printer.

Tansley, A.G. (1935) 'The use and abuse of vegetational terms and concepts', *Ecology*, 16(3): 284-307.

Thomas, D.S.G. 1993. 'Sandstorm in a teacup? Understanding desertification', *Geographical Journal*, 159:318–31.

Thomas, D.S.G., and N.J. Middleton. (1994) *Desertification: Exploding the Myth*. Chichester: John Wiley.

Tischler, W. (1990) *Ökologie der Lebensraume*. Stuttgart: Gustav Fischer Verlag.

Tivy, J. (1993) *Biogeography: A Study of Plants in the Ecosphere*. 3rd edn. Harlow: Longman Science and Technical.

Turner, C., and Gibbard, P. (1996) 'Richard West—an appreciation', *Quaternary Sciences Reviews*, 15: 375–89.

Turner, M.D. (1998) 'The interaction of grazing history with rainfall and its influence on annual rangeland dynamics in the Sahel'. In K.S. Zimmerer and K.R. Young (eds.), *Nature's Geography: New Lessons for Conservation in Developing Countries*, pp. 237–61. Madison: University of Wisconsin Press.

Valentine, G. (2001) 'Whatever happened to the social? Reflection on the "cultural turn" in British human geography', *Norsk geografisk tidsskrift*, 55(3): 166–72.

Veblen, T. (1989) 'Biogeography'. In C.J. Willmott and G.L. Gaile (eds.), *Geography in America*, pp. 28-46. Columbus: Merrill Publishing.

Walter, H. and Breckle, S-W. (1999) *Vegetation und Klimazonen*. Stuttgart: Eugen Ulmer Verlag.

Watts, D.R. (1971) *Principles of Biogeography*. London: McGraw-Hill.

Watts, D.R. (1978) 'The new biogeography and its niche in physical geography', *Geography*, 63: 324–37.

West, R.G. (1988) 'Harry Godwin, 1901–1985', *Biographical Memoirs of Fellows of the Royal Society*, 34: 261–92.

Whatmore, S. (2002) *Hybrid Geographies*. London: Sage Publications

Whatmore, S. (2007) 'Hybrid Geographies: rethinking the human in human geographies'. In L. Kalof and A. Fitzgerald (eds) *The Animals Reader: The essential classics and contemporary writings*. Oxford: Berg.

Whitaker, R.J. and Sax, D.F. (2003) 'A 21st Century Pangea? The emergence of a new international forum for Biogeographers', *Journal of Biogeography*, 30(3): 315-17

Wolch, J. (2007) 'Green urban worlds', *Annals of the Association of American Geographers*, 97, 373-384.

Wolch, J. and Emel, J. (1998) *Animal geographies: place, politics and identity at the nature-culture borderlands*. London:Verso.

Wilson, D.S. (2007) *Evolution for Everyone: How Darwin's Theory Can Change the Way We Think About Our Lives*. New York: Random House.

Yin, X. (2004) *Biogeography*. Beijing: Higher Education Press.

Young, K.R. (2003) 'Genes and biogeographers: Incorporating a genetic perspective into biogeographical research', *Physical Geography*, 24:447–66.

Young, K.R., M.A. Blumler, L. Daniels, T.T. Veblen, and S.S. Ziegler. (2004) 'Biogeography in North America'. In G.L. Gaile and C.J. Willmott (eds.), *Geography in America at the Dawn of the 21st Century*, pp. 17–31. Oxford: Oxford University Press.

Zimmerer, K.S., and K.R. Young (eds.) (1998) *Nature's Geography: New Lessons for Conservation in Developing Countries*. Madison: University of Wisconsin Press.

Revisiting Theories and Concepts

We start with a series of chapters organized around the themes of biogeographical theory, concepts, and thought. Theory is, of course, an important and necessary part of science. Ideally, it is tested and modified in the face of contrary evidence. This is easier said than done in biogeography, given the complexity of interactions and the possible role of historical contingency. As a consequence, there are widely diverging opinions concerning several major theories in this field: a theme that resonates throughout this section.

By its very nature, theory is prone to overgeneralization. This is a necessary part of the scientific process, which involves repeated attempts to generalize from limited data, hopefully coming closer to the truth with each iteration. Scientific theory, then, is always a work in progress. But this also has the unfortunate consequence that uncritical application of scientific theory to policy and planning in the real world can be problematic (for examples, see Chapter 1).

We could not include all biogeographical theory, choosing instead to provide first an overview (Blumler et al., Chapter 2). In addition to outlining the history of biogeographical thought, and describing how the various traditions within biogeography came about, this chapter concentrates on several key theories. Next Duane Griffin (Chapter 3) reviews theories purporting to explain patterns of species diversity—a puzzle of long-standing interest to biogeographers and one that takes on much greater meaning today when so much conservation concern is focused on species declining toward extinction. Biodiversity theory

is itself extremely diverse, with little clear consensus though the proponents of differing viewpoints for the most part tolerate each other—but there are exceptions (Kaiser, 2000).

We follow with chapters treating subfields of biogeography. John Kupfer's Chapter 4 describes landscape ecology—a major but relatively new area, dating from Forman and Godron (1981), and Naveh and Lieberman (1984), though with considerable development since then, most notably the level of quantification that has paralleled developments in remote sensing and GIScience. Chapter 5 treats classification, a field that also features a highly developed terminology, much of it elaborated in Europe. Appropriately, then, our authors (Angelika Schwabe and Anselm Kratochwil) are German. This chapter demonstrates the expanding influence of phytosociology, not just within Europe but globally—though it sometimes is modified when adopted overseas. Phylogeography, treated in Brett Riddle's Chapter 6, is an even more recently derived branch of study, dating from Avise et al. (1987). Finally, we end this section with Kathy Willis et al.'s chapter by reviewing recent evidence bearing on traditional beliefs regarding refugia. Though less theoretical, it is included here both because of the importance of the refugial concept, and because much of the evidence is paleobotanical, an area otherwise underrepresented in this volume.

Chapter 2, on the history of biogeographical thought, proved difficult to write: different traditions have reached contrasting conclusions regarding key developments, concepts, or even about what matters in biogeography. Problematic areas include, but are not limited to, the nature of communities; succession and vegetation dynamics; the role of disturbance; classification; vicariance versus dispersal; island biogeography theory; contingency versus rules; and the proper use of theory and modeling in scientific study. Within this context, Mark Blumler et al. concentrated on three key theories. The first—evolution—is undeniably biogeography's greatest contribution to science. Consideration of its development raises questions about contemporary biogeographical research. Evolution came out of the natural history tradition, and Darwin made heavy use of the Comparative Method; neither is much utilized today. Theories of vegetation dynamics and succession are currently in a confused state, with highly divergent views and relatively little tolerance of alternatives. In fact, while it probably deserved a separate chapter, we decided that there might not be anyone who could cover the subject in a manner that was fair to all. Vegetation dynamics relate also to equilibrium versus non-equilibrium models in ecology and biogeography, yet another aspect of theory that remains unresolved, with divergent perspectives. Finally, the chapter examines the Equilibrium Theory of Island Biogeography (MacArthur and Wilson, 1967), perhaps the most influential yet also the most problematized biogeographical theory of recent times.

One source of tension in writing this chapter concerned the nature of science. The lead author has published on the history of thought and consequently has interacted with the history of science community. Scientists tend to see their disciplines in progressionist terms, much as Newton famously did, as a series of great minds "standing upon the shoulders" of their predecessors. In contrast, science historians view science as being constrained and channeled by social context and sometimes doubt about whether any real progress is achieved. Both perspectives are present in the chapter, which reflects the relative positions of the four authors. For what it is worth, our own opinion is that both perspectives have merit.

Biodiversity is arguably the central concern of contemporary biogeography and is a major concern of the environmental movement. Maybe then a consequence is that Griffin's Chapter 3 deserves greater comment than the others in this introduction (but see also Mutke, Chapter 9). Truly, today we have a 'tangled bank' of theories about what makes some places species-rich and others not. Griffin sees hopeful signs of increased understanding, while the review that he most frequently cites (Palmer, 1994) suggests that progress has been limited, though this article is now 15 years old. The information explosion and recycling of old theories in new guise has created a morass that no one person can hope to survey completely, though Duane Griffin has made a yeomanly effort—only to be asked to remove many of the references to save space! We note that the terms of debate have changed dramatically over the years, as have the prominent theories. Thirty years ago, for instance, much of the discussion revolved around equilibrium versus nonequilibrium theories of diversity (Connell, 1978). Both types of theory still have currency, yet Griffin's treatment does not make this distinction, suggesting that it no longer is considered terribly relevant. But there also does not seem to be any particular reason for this change, other than that scholarly interest happened to turn in new directions.

In attempting to resolve the very confusing mass of data about species diversity, one issue may be whether all species truly are equivalent, even within phyletic groups. One may recall J.B.S. Haldane's famous (possibly apocryphal) quip, when asked if he believed in God, that "the Creator has an inordinate fondness for beetles." Beetle diversity must exert an inordinate influence on insect diversity. Another concern is proximate versus ultimate causation. Environmental heterogeneity—which is later referred to as geodiversity (Mutke, Chapter 9)—is generally accepted as a promoter of high diversity, but in the classical examples of tropical rain forests and coral reefs it is life itself that creates the heterogeneity. (Admittedly, limestone on land also typically weathers to create heterogeneous microenvironments. Plant species diversity is typically greater on limestone, than on other substrates, at least in part for this reason.)

Another issue is the possible role of historical contingency. One can arrange evolutionary biologists along a continuum, from the late Stephen Jay Gould, who believed that there are few if any rules in nature, to E.O. Wilson, who tends to discount historical contingency. Gould (1989) famously claimed that if there were a million Earth-like planets, then only on one would life evolve beyond the unicellular level. Diversity would be much lower, surely, on such planets than on ours. Since Gould's death, the biogeographical community has moved further toward a Wilsonian view than it already had done (Gould never represented the consensus), but few would argue that historical contingency plays no role at all. Griffin does represent the consensus view when he states, "All systems (above the quantum level, at least) are governed by deterministic processes and thus exactly predictable in principle," while also accepting that in practice this may never be fully achievable. But if, as some physicists are arguing, the universe is in the nature of a hologram (Bekenstein, 2003), then determinism (cause and effect) may not be as characteristic as most scientists believe.

Historical contingency seems particularly apropos given that biodiversity has increased throughout time (Sepkoski and Miller, 1998). If biodiversity is as much an evolutionary as an ecological feature, diversity itself may be sufficiently an accidental by-product in that it is not explicable on theoretical grounds. During the Cambrian Explosion, life was far more diverse in the sea than on land; in all likelihood if scientists had been alive then they surely would have argued for the importance of water to life and suggested that the landmasses are too dry to support high levels of biodiversity. Similar arguments are made today about deserts, but overall the land has become far more diverse (in species, not phyla) than the sea even though the latter covers approximately two-thirds of the Earth's surface and is deeper than the continents are high. Can we be certain sure that deserts, tundra, and other relatively recent (geologically speaking) extreme environments will not eventually become as diverse as tropical rain forest? Given also that massive colonization of land occurred only after nearly four billion years of evolution, how can we state with any degree of certainty that the process must now have reached equilibrium, after only a few hundred million additional years? Furthermore, since the sun's decreasing warmth is according to one estimate likely to eliminate multicellular life within about 500 million years (Brownlee and Ward, 2007), life's evolutionary trajectory on this planet may have mostly run its course. Life may go extinct as the planet becomes uninhabitable before it has had the time to reach maximum equilibrium diversity levels.

Kupfer's (Chapter 4) thorough treatment of concepts, methods, and metrics of landscape ecology illustrates that, like phylogeography (Chapter 6), the field is both recent and rapidly expanding. Consequently, in both cases, it is difficult as yet to evaluate the claims of their practitioners. Landscape ecology seems particularly relevant to the conservation questions that arise out of habitat patchiness, especially where anthropogenic, such as forest fragmentation. Given the proliferation of metrics, it is pertinent to ask which are the most useful for different biogeographical questions? This leads to a much broader question: how do we measure the utility of our methods? This question applies to many areas of biogeography covered in this book, and it is not meant to be a sideswipe at landscape ecology.

Chapter 5 illustrates in great detail a point made in Chapter 2: the Continental School has emphasized classification. Although classification generally was abandoned in the Anglo-American world through much of the twentieth century under the influence of Gleasonian ideas, it is making a strong recovery. For the most part, Anglo-Americans rely on approaches already developed in Europe. Schwabe and Kratochwil's thorough treatment of terminology is likely to be very useful for those who have not been exposed to this literature. As they point out, the phytosociological approach with its emphasis on relevés for vegetation sampling and characterization is spreading and seems ready to become the predominant method globally. We believe its recent spread in the United States, albeit in modified form, reflects a desire among conservation organizations to identify communities for purposes of preservation. There also is a connection to the rise of biome and global change modeling, which biogeographers (and others) are actively engaged in across the world. Certainly the return to classification generally reflects both motivations, as Schwabe and Kratochwil clearly explain.

Riddle's Chapter 6 on phylogeography details the rapid rise in the use of genetic data, combined with the assumption of a constant molecular clock, to work out phylogenies that are superior to those derived solely from morphological data, and to provide new estimates of divergence times. The fossil record must tend to underestimate time since origination of taxa, because the record is so incomplete that fossils are unlikely to be found that correspond to the first generation of the taxon in question. On the other hand, it is not out of the question that the molecular record may sometimes give an overestimate, since mutation rates are not constant but tend to increase during times of environmental stress, for example, during mass extinctions. Regardless, it is undeniably true that the molecular data provide a fresh and illuminating perspective on evolutionary history.

Willis et al.'s Chapter 7 demonstrates that traditional beliefs on refugia in the Amazon and Europe need modification and that the reality is more complex than traditionally thought. Many European species found refuge during the most recent glacial periods in local microenvironments north of the Mediterranean, the traditionally favored refuge area. We note that this new information leaves us still wondering as to the cause of the extinctions of trees such as the hemlock (*Tsuga*) and tulip tree (*Liriodendron*), which left Europe with a depauperate arboreal flora compared to the eastern United States and east Asia. Because these extinctions occurred in the Pliocene or early Pleistocene, they may not be readily accessible through palynology, which is generally more informative about the recent past.

REFERENCES

Avise, J.C., Arnold, J., Ball, R.M., Bermingham, E., Lamb, T., Neigel, J.E., Reeb, C.A., and Saunders, N.C. (1987) 'Intraspecific phylogeography: The mitochondrial DNA bridge between population genetics and systematics', *Annual Review of Ecology and Systematics*, 18: 489–522.

Bekenstein, J. (2003) 'Information in the holographic universe', *Scientific American*, 289: 59–65.

Brownlee, D., and Ward, P. (2007) *The Life and Death of Planet Earth*. New York: Henry Holt.

Connell J. H. (1978) 'Diversity in tropical rain forests and coral reefs', Error! Hyperlink reference not valid. 199: 1302–10.

Forman, R.T.T., and Godron, M. (1981) 'Patches and structural components for a landscape ecology', *BioScience*, 31: 733–40.

Gould, S.J. (1989) *Wonderful Life: The Burgess Shale and Nature of History*. New York: W.W. Norton.

Kaiser, J. (2000) 'Rift over biodiversity divides ecologists', *Science*, 289:1282–83.

MacArthur, R.M., and Wilson, E.O. (1967) *The Theory of Island Biogeography*. Princeton, NJ: Princeton University Press.

Naveh, Z., and Lieberman, A.S. (1984) *Landscape Ecology, Theory and Application*. New York: Springer-Verlag.

Palmer, M.W. (1994) 'Variation in species richness—Towards a unification of hypotheses', *Folia Geobotanica and Phytotaxonomica*, 29: 511–30.

Sepkoski, J.J., and Miller, A.I. (1998) 'Analyzing diversification through time', *Trends in Ecology and Evolution*, 13: 158–59.

2

History of Biogeographical Thought

Mark A. Blumler, Anthony Cole,
John Flenley, and Udo Schickhoff

2.1 INTRODUCTION

Some 250 years of research and scholarly writing on what modern biogeography is principally about—the causes of the spatial distribution of living things—preceded the 1974 watermark of the first publication of the initial volume of the *Journal of Biogeography*. Early theoretical contributions came out of the natural history tradition, as it developed within disciplines such as biology, geography, and geology; they were not always published under the name of biogeography. While there are excellent overviews of the history of ecological, environmental, and evolutionary thought that discuss aspects of biogeography (e.g., Botkin, 1990; Glacken, 1967; McIntosh, 1985; Worster, 1977), no comprehensive treatment exists. The major textbooks (e.g., Cox and Moore, 2005; Lomolino et al., 2006) are excellent but do not cover all biogeography traditions, and different traditions perceive its history somewhat differently. So, in reviewing this history, our challenge is to correctly interpret the past. There is a danger of overstating some contributions, omitting others, and failing to see the importance of events that were far removed from the issues of our time.

2.2 BIOGEOGRAPHY IN THE ANCIENT WORLD

The origin of modern science is usually traced back to the Greeks, but it would do well to remember that it developed within a Christian context. Biology, in particular, incorporated and held on to Christian concepts for a long time (Blumler, 1996). The traditional definition of species, with its emphasis on sharp reproductive barriers, is Creationist: if God created each species separately, there would be little possibility of interbreeding. Actually, hybridization leading to speciation, or the fusion of formerly separate species through allopolyploidy, is pervasive among plants and other organisms (Anderson, 1949; Arnold, 1992). Yet even Ernst Mayr, arguably the most influential evolutionary biologist of the twentieth century, continued to defend a (modified) biological species concept until his death (Mayr, 1996). The nature of species remains a matter of discussion. The ecological niche also is in part a Christian concept, related to the 'balance of nature'. If God created each species separately, he must also have devised each for its appropriate place and function in a harmonious Nature—that is, its niche. Linnaeus (1751), for instance, employed this concept, though he termed it 'station' (Worster, 1977). In the twentieth century, the niche concept was re-invigorated by the proliferation of equilibrium models in ecology. Yet across the sciences, the past century has seen a shift from viewing nature as static, and in equilibrium, to seeing it as dynamic, and even chaotic. In biogeography, the theory of evolution and the application of plate tectonics to biotic distributions both represent this shift, as does the shift from equilibrium to nonequilibrium ecology, among many Anglo-American practitioners.

From the Greeks came varied perspectives, including conceptions of nature as mutable or dynamic. Theoprastus' approach to biogeography was inductive (Raup, 1942), while Aristotle's was deductive. One Greek influence was the division of the Earth into torrid, temperate, and frigid zones (still reflected in the *tierra caliente, templada,* and *fria* zones along the Andes). This tripartite division was carried over in the early classification of the Earth into what we now term biomes, for example, tropical rain-, temperate deciduous-, and boreal forest biomes (de Laubenfels, 1975). Via de Candolle (1855), it became the basis of Köppen's (1918; Köppen and Geiger, 1930) widely taught climatic classification system, intended to match climate with vegetation. But since we now know that at times in the past 'tropical' rain forest has extended as far as the Arctic Circle (Cox and Moore, 2005), it may be time to consider other groupings.

2.3 DISCOVERIES THAT CHALLENGED LONG-HELD VIEWS

As Europeans sent out explorers and colonizers during the centuries after 1492, discoveries in the New World, especially, raised doubts about the literal interpretation of the Bible. Were the American Indians descendants of Noah, and if not, were they even human? If two of every animal survived the Flood on Noah's Ark, how did armadillos migrate from Mt. Ararat to North America, and why didn't any stay in Asia? If God created each species for its own special niche, how could alien species invade and replace natives? Such considerations, along with the fossil record and other geological evidence for change and a long duration of the Earth's history, increasingly caused educated Europeans to regard the Bible as metaphorical rather than literal.

Modern biogeography is usually said to have originated with Humboldt, who focused on the aggregates of vegetation in nature, in relation to environment. Previously, individuals such as Carolus Linnaeus (1707–78) had organized nature taxonomically and had begun the process of determining species distributions. Linnaeus promoted a binomial nomenclature we still use today. In his work at the University of Uppsala from 1741 onward he was sent plant and animal specimens from around the globe. The tenth edition of Linnaeus's (1758) *Systema Naturae* classified some 4,400 species of animals and 7,700 species of plants. Linnaeus's intellectual descendants, such as Wildenow, Engler (1879-82), Sclater (1858), and Wallace (1876), were able to use the accumulating information on the distribution of named species to begin to delineate the world's floral and faunal provinces or realms, though only gradually was it realized that these distributions must have changed over time.

Contemporary to Linnaeus was Georges-Louis Leclerc, Comte de Buffon (1707-1788), whose thinking about plant and animal distributions had a profound influence on the next two generations of naturalists, including Charles Darwin. Buffon's thinking and therefore his contribution to biogeography differed in a number of important ways from that of Linnaeus. First, he concluded that the age of the Earth was much greater than 6,000 years as commonly believed at the time. Using the cooling rate of iron as a proxy for time, he calculated its age had to be closer to 75,000 years. In his multivolume *Histoire naturelle, générale et particuliére* (1761), Buffon drew on mounting field evidence to suggest that despite similarities in local environment and climate, different regions of the Earth had distinctive mammals and birds— a relationship later known as Buffon's Law. This seemed to imply some innate ability in living things to adjust to local environments through a process he called 'improvement' or 'degeneration'. However, reasoned Buffon, if species were immutable as generally thought, then their inability to 'improve' would have surely prevented them from migrating across inhospitable places following the Flood. This conclusion seemed to be further confirmed by the fact that some animals retained redundant or vestigial parts—suggestive of gradual development rather than spontaneous Creation. Based on this reasoning, Buffon's theory of dispersal involved climate-mediated migration from a possible center of origin in the previously warmer climes of northwestern Europe. Somehow, he theorized, populations of the New and Old World became separated on continental-scale islands during this migration phase and then changed.

> Hence, it is not impossible, that, without inverting the order of Nature, all the animals of the New World were originally the same with those of the Old, from whom they derived their existence; but that, being afterwards separated by immense seas, or impassable lands, they would in the progress of time, suffered all the effects of a climate that had become new to them, and must have had its qualities changed by the very causes which produced the separation. (Buffon, 1761)

Johann Reinhold Forster (1729–98), a scientist on Cook's second voyage, confirmed Buffon's law and extended it to include plants as well as mammals and birds across all regions of the world. He also drew attention to the fact that the diversity of plant species decreased from what he called the 'luxuriance of vegetation' in the tropical isles toward the 'rigorous frost of the Antarctic regions'. Forster (1778) developed one of the first frameworks for classifying plant assemblages around the world into regions of biotic similarity.

Alexander von Humboldt (1769–1859) is generally credited with being the first to approach the study of vegetation quantitatively and to stress the importance of environment in determining the distribution of communities. After Humboldt, there was increasing emphasis not only on defining floral and faunal provinces, but also on defining and explaining what we now call biomes, though admittedly the two were conflated for some time (de Laubenfels, 1975). Humboldt made contributions in areas we now recognize as climatology, geomorphology, and geology (Humboldt and Bonpland, 1805), and his role in the development of geography is well recognized (e.g., Martin, 2005).

Humboldt also recognized that the latitudinal gradient in biodiversity (see Mutke, this volume) described earlier by Forster could be identified in local elevational zones as a result of what we today call temperature lapse rates. This was especially evident in his surveys of the vegetation on Mount Chimborazo in modern-day Ecuador. These discoveries led him to further studies of climate-mediated change in plant communities, for which he employed mapping techniques based on isothermal lines. Humboldt's investigations inspired a series of European phytogeographers (including de Candolle, 1820; Drude, 1890; Grisebach, 1872; Schimper, 1898; Schouw, 1822; Warming, 1895), who developed a more sophisticated understanding of the world's biomes and their relationships to the environment. De Candolle (1820) already was discussing endemism and disjunctions in the early nineteenth century. By the late nineteenth century, Wagner and van Sydow's (1889) map showed all the modern biomes, except Mediterranean scrub. Schimper added the latter and argued that it represented a striking illustration of evolutionary convergence— a notion still widely promulgated today, though the reality is more complex than generally recognized (Blumler, 1991, 2005).

2.4 UNIFORMITARIANISM AND DEEP-TIME

Uniformitarianism and the notion of geological or 'deep-time' were proposed by a Scottish geologist, James Hutton (1726–97)—dubbed by some as the 'father of geology'. Field discoveries led him to challenge the prevailing 'Neptunist' theories of the time, which proposed that rocks had precipitated out of a single flood. It seemed to Hutton that the rocks formed in the sedimentary deposits he had witnessed, originated from a series of successive floods (Dean, 1992). Given that modern depositional, erosional, and uplift processes were gradual and occurred over long periods of time, it seemed only reasonable to conclude that the Earth was far more than a few thousand years old (Briggs, 2004). Widespread acceptance of uniformitarianism and deep-time, decades later, was largely due to the advocacy of the Scottish geologist Charles Lyell (1797–1875) and his colleague John Playfair (1798–1819). Ironically, while uniformitarianism was of major importance in understanding the Earth's history and in the development of the theory of evolution, it would much later serve as a barrier to acceptance of catastrophist theories of the K/T boundary and other mass extinctions.

While Lyell championed uniformitarianism, it was a contemporary French geologist, Adolphe-Théodore Brongniart (1801–76), who recognized the opportunity to use the geological record as a basis for recording the history, classification, and distribution of fossil plants. In the first volume of *Histoire des végétaux fossiles* in 1837, Brongniart demonstrated the value of studying botanical changes (especially extinctions) with respect to time. He showed that successive geological time periods could be characterized by the dominance of different plant life forms. Brongniart had reasoned that the close interrelationship between the Earth and the living things it supported implied that if the Earth's biophysical processes were mutable, then it should be possible to find evidence of the transmutation of species.

This idea of species transmutation (evolution) was not new at the time. Darwin (1809–82) began work on his theory of natural selection. Hutton (1794) had also advocated 'uniformitarianism for living creatures,' and the idea of evolution by acquired characteristics had been advocated by the French naturalist Jean-Baptiste Lamarck (1744–1829). William Paley had put forward the idea of divine design in nature. Geoffroy Saint-Hilaire had attempted to expand the evolutionary thinking of Lamarck, while Robert Chambers's writings inspired Wallace.

Darwin's initial inspiration was Humboldt, whose name he mentions more often in his journal of the *Beagle* than any other scientist (Darwin, 1845). Darwin might not have gone on the *Beagle* expedition had he not read Humboldt. As the ship's scientist, he modeled himself after Humboldt, particularly in the breadth of his observations (Worster, 1977). For instance, his explanation of the origin of coral atolls from volcanic islands remains an important contribution to

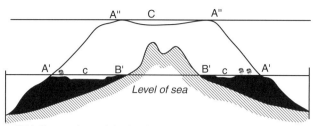

A'A', Outer edges of the barrier-reef at the level of the sea, with islets on it.
B'B', The shores of the included island.
cc, The lagoon-channel.
A"A", Outer edges of the reef, now converted into an atoll.
C', The lagoon of the new atoll.
N.B.–According to the true scale, the depths of the lagoon-channel are much exaggerated.

Section of coral-reef

Figure 2.1   Darwin's model of coral atoll formation (after Darwin, 1845)

geomorphology (Figure 2.1). His other major inspiration was Lyell.

As is well known, exploration and biogeographical observations were crucial to the development of Darwin's theory. The spectacular success of the alien plant cardoon (*Cynara cardunculus*) in the Argentinian pampas seemed counter to the perfect Creation. Patterns on the Galapagos seemed particularly contradictory to what one might have expected the Creator to have designed, in three key respects (Sauer, 1969). First, Darwin noted the occupation of diverse Galapagos niches by species closely related to South American mainland taxa occurring in very different environments; second, he noticed the lack of similarity of the Galapagos fauna to that of the climatically and geologically similar but geographically distant Cape Verde Islands; and finally, as he was about to sail he was told that the tortoises varied from one island to another. While he could not confirm this about the tortoises, he was able to use his specimens to verify that a similar differentiation occurred in other taxa, such as finches. He came to understand that all three patterns were best explained as due to chance overseas dispersal from the nearest mainland, followed by speciation and what today we call 'adaptive radiation'.

Also well known, and therefore not repeated here, is the story of how Darwin delayed publication of his ideas about natural selection, until Wallace's independent development of the theory forced him to do so. As he had anticipated, Darwin's (1859) opus faced criticism from many quarters, including the Church. However, in the long period taken to prepare his manuscript he had

been meticulous in paying attention to the internal consistency of his arguments, the faithful documenting of observations and inferences, and consideration of likely objections. He drew upon numerous subjects to advance his argument, including biogeography (Firenze, 1997). For instance, he pointed out that the southern continents have highly differentiated faunas, while the northern continents do not. Thus, he drew on the taxonomic evidence accumulated by earlier biogeographers. The pattern was best explained as due to the isolated positions of the southern continents, in contrast to the connectedness of those in the north. As Darwin put it,

> Barriers of any kind, or obstacles to free migration, are related in a close and important manner to the differences between the productions of various regions. . . . We see the same fact in the great difference between the inhabitants of Australia, Africa, and South America under the same latitude; for these countries are almost as much isolated from each other as is possible. (1859: 345)

He also pointed out that the barrier of Panama explained the great difference in the marine faunas of the Pacific and Atlantic Oceans. In contrast, when a continent extends over a long latitudinal range, one encounters what we would now term ecotypic or clinal variation, because of common origin somewhere along the latitudinal gradient:

> The naturalist in travelling . . . from north to south never fails to be struck by the manner in which

successive groups of beings, specifically distinct, yet clearly related, replace each other. He hears from closely allied, yet distinct kinds of birds, notes nearly similar, and sees their nests similarly constructed, but not quite alike, with eggs coloured in nearly the same manner. (1859: 346)

Darwin noted, too, that while large mammalian herbivores are present on every continent, they are absent from most islands, such as the Galapagos, where they are replaced by giant tortoises and iguanas (Worster, 1977). Again, this seemed to be the result of dispersal rather than of Creation. Even clearer was the greater dispersal capacity of some plants, explaining their greater ability to colonize isolated islands, compared with mammals—and he pointed out that bats were an unsurprising exception. He discussed what we now call the Great American Interchange between North and South America, giving a modern explanation for the greater success of northern animals in the south as opposed to vice versa:

I suspect that this preponderant migration from north to south is due to the greater extent of land in the north, and to the northern forms having existed in their own homes in greater numbers, and having consequently been advanced through natural selection and competition to a higher stage of perfection or dominating power, than the southern forms. (1859: 370)

In discussing island biogeography, Darwin noted both the tendency of islands to have fewer species than equivalent continental areas, and also what we now term 'disharmonic' floras and faunas. His explanation again was that isolation prevents dispersal. In this context, he even wrote about insular woodiness, more recently discussed by Carlquist (1965) with respect to the Asteraceae.

Islands often possess trees or bushes belonging to orders which elsewhere include only herbaceous species . . . trees would be little likely to reach distant oceanic islands; and an herbaceous plant, though it would have no chance of successfully competing in stature with a fully developed tree, when established on an island and having to compete with herbaceous plants alone might readily gain an advantage by growing taller and taller and overtopping the other plants. (Darwin, 1859: 381)

Finally, Darwin noted the impact of alien species on isolated islands such as St. Helena:

He who admits the doctrine of the creation of each separate species, will have to admit, that a sufficient number of the best adapted plants and animals have not been created on oceanic islands; for man has unintentionally stocked them from various sources far more fully and perfectly than has nature. (1859: 379)

He was unable to complement these dispersalist arguments with vicariant explanations for distribution (see below), because during his day no one believed that the continents moved. In arguing in support of his theory, Darwin marshaled additional evidence from several other fields besides biogeography. But as Firenze (1997) has pointed out, the biogeographical evidence is the easiest for the layperson to grasp, and it is the most effective in countering Creationist arguments, yet, unfortunately, it is seldom if ever employed today in debates over evolution.

Alfred Russell Wallace (1823–1913) also began his career with an expedition to South America, during which he studied the Rio Negro's zoology, botany, physical geography, and the languages and customs of its indigenous peoples. In short, he, too, was a geographer, as well as a naturalist. He then traveled to Indonesia, where he developed his version of the theory of evolution. (As was also true for Darwin, the theory fell into place when he connected Malthus's [1798] ideas about population growth to his biogeographical observations. It was the tendency of populations to produce more offspring than would ultimately be able to survive that gave natural selection a chance to operate.) As illustrated by the famous 'Wallace's Line,' he also contributed to the determination and mapping of faunal provinces. Subsequently, he discussed the existence of latitudinal diversity gradients.

For Wallace and Darwin it was the influence of extensive travel and time spent on 'strange' continents and oceanic islands, and seeing things that didn't fit conventional thinking, that led them to question conventional wisdom. For Darwin, it was in the year following his five-week stay on the Galapagos Islands that he began to reflect on the immutability of species.

While Wallace's work in the Malay Archipelago had largely convinced him of the verity of natural selection, it was at the urging of others, including Darwin, that he undertook a general review of the geographic distribution of animals. Initially, this work was hindered due to the immaturity of systems of classification. However, with developments in this field, by 1874 he was able to resume. He succeeded in greatly extending the regionalized system of geographic distribution developed earlier by Sclater (1858), a contribution that still forms the basis of eco-zones in use today. Wallace's (1876) two-volume epic was both a scholarly and synthetic achievement. Wallace based his global distributional scheme on a discussion of all of the current and historic factors then known to influence animal distribution patterns within each region. For example, he theorized about the extentionalist ideas of land bridges, the effects of periods of glaciation, elevational gradients on mountain ranges, and the influence of regional vegetation on potential ranges of animal dispersal. In addition to this, he provided tables of all of the families and genera of higher animals with details of their distributional locations. When completed, it served as the definitive textbook on biogeography for decades to come.

In *Island Life* (1881), Wallace greatly assisted the process of reviewing, classifying, and consolidating the thinking of the time about islands. Until it was published, thinking about the significance of oceanic islands in elucidating likely causes of biological distributions and patterns of species richness had not generally been connected with thinking about continents. Wallace drew this distinction and noted the central role that islands played in the study of biological patterns. *Island Life* achieved more than a refocusing of what was then well known about continental, environmental factors of distribution. It extended the range of natural selection and thus showed how new species emerge in small isolated populations.

Evolutionary biologists and natural historians have continued to note differences between oceanic islands and continents, and to speculate on the causes of the differences (e.g., Carlquist, 1965). Adaptive radiation, so frequently encountered on oceanic archipelagos, continues to be of great interest. The vagaries of dispersal to isolated islands continue to be investigated and discussed (e.g., Ridley, 1930; van der Pijl, 1982). This type of research receded into the background during the heyday of the Equilibrium Theory of Island Biogeography (see below), but it never died out completely.

The end of the nineteenth century brought to a close a number of remarkable and defining theoretical achievements in the history of biogeography. Geologists and paleontologists had embraced uniformitarianism, calibrated estimates of the age of the Earth against proxy geological and fossil data, and advanced theories that posited a mutable Earth. Darwin and Wallace had built upon geological discoveries by amassing observational and distributional biological data to confirm the central role of natural selection in the mutability of species. Not only were global biological distributions better documented and classified, but the historical causal factors responsible for these distributions were better understood. This in turn had thrown clearer light on what Darwin (1859) had referred to as the *Origin of Species*. Meanwhile, European phytogeographers, such as Schimper and Warming, were refining the classification of biomes in relation to climate, in terms of the ecophysiology of plant growth; and Warming (1895) had outlined a model of succession to climax that would become the dominant theory (albeit in several versions) in plant ecology for much of the twentieth century.

By the start of the twentieth century, all biogeographers were more or less in communication and were aware of developments within the various branches of the field, even if not always in full agreement. For example, the French regarded Lamarck, not Darwin, as the true father of evolutionary theory (Glick, 1974). Subsequently, several separate schools developed and in some respects their underlying paradigms diverged. Today, biogeographers frequently are poorly informed about developments within schools other than their own.

2.5 HISTORICAL BIOGEOGRAPHY (PALEOBIOLOGY)

De Candolle (1820) was perhaps the first person to distinguish historical and ecological traditions, suggesting that ecological biogeography involved physical causes acting at the present time, whereas historical biogeography depended upon past causes. For example, millions of years may be involved in the action of evolution and tectonics. One problem with a discrete classification framework of this kind lies in the difficulty of the categorization of causes like the Pleistocene glaciations of a more intermediate time horizon. Thus, it is useful to divide the paleobiology (historical biogeography) tradition into paleontology, concerned primarily with deep-time, and evolutionary questions; and Quaternary studies, concerned primarily with environmental reconstruction over the past few million years. Relatively few paleobiologists concern themselves with both, so they can be thought of as two separate traditions (see Millington et al., this volume).

As analytical methods have begun to emerge in historical biogeography over the last century it has become evident that distributional patterns are not the result of a single historical or ecological cause (Crisci et al., 2003). Yet despite the conceptual and theoretical problems associated with tightly defining historical and ecological biogeography, this distinction has persisted throughout the twentieth century.

Palynology, which developed early in the twentieth century, was particularly appropriate for studies of Pleistocene and Recent environments. A single pollen diagram contains information about many taxa through time (see Figure 29.2 in Marchant and Taylor, this volume, for an excellent example). It may be possible to derive conclusions about presence, relative or absolute abundance, biodiversity, macro- and microclimate, seasonality, migration, extinction, and human presence and impact from such diagrams. This has made it possible to test empirically many of the hypotheses put forward to explain peculiar biogeographical patterns. Most important, palynology had been applied to the former distribution of plants and was shown to be useful in the reconstruction of past environments. Recent decades have seen a proliferation of additional techniques; some (e.g., radiocarbon dating and charcoal analysis) have improved the utility of pollen cores, while other techniques, such as isotopic ratios, have provided independent lines of evidence.

Palynology and these related techniques have led to the publication of regional reviews of the reconstructed Quaternary history of vegetation in almost all the world's regions, after the pioneering works in Northern Europe established the technique's efficacy (e.g., Godwin, 1940 and references therein), including 'difficult' areas like the Arctic (Huntley and Webb, 1988) and the Tropics (Flenley, 1979). There have also been illuminating studies, based on palynology and related techniques, of many specialized aspects of biogeography (e.g., community instability—Davis, 1981; historical taxon mapping—Huntley and Birks, 1983; human-induced deforestation—Williams, 2006; agricultural origins—Yasuda, 2002; environmental history—Oldfield, 2005; and responses to climatic change—Bush and Flenley, 2007). Palynology has also made very important contributions to the data needed for the testing of mathematical models used in the predictions of global warming (e.g., Goudie and Cuff, 2002). Pre-Quaternary palynology is increasingly important in studies of the evolution of plants and in historical biogeography (e.g., Morley, 2000; Flenley, 2007).

As the Earth became more thoroughly explored by biologists, its floristic and faunal realms became better defined (Darlington, 1957; Good, 1974). However, these are viewed in the context of plate tectonics rather than seen as existing immutably since the dawn of Creation. It is becoming more accepted to approach the subject of geographic distribution from an ecological perspective (Cox and Moore, 2005; Lomolino et al., 2006, although some efforts have been made to return the field to a geographic orientation (Craw et al., 1999).

2.5.1 Plate tectonic theory

The theory of continental drift (plate tectonics) had an enormous impact on historical explanations of species distribution patterns. Although there were individuals before him who had proposed that continents moved, Alfred Wegener (1880–1930) published in 1915 a hypothesis supported with considerable evidence. His theory was largely rejected by geologists, supposedly because he did not provide a believable mechanism, but it is more likely it was rejected because his evidence was almost entirely geographical (i.e., his fossil evidence was biogeographical), and because he was an outsider—a meteorologist. This is in contrast to Darwin, who also was attacked for lacking a mechanism for natural selection before the rediscovery of Mendelian genetics, and yet gets full credit today. Wegener still is shunted to the background. In his repeated attempts to persuade the geological hegemony, Wegener suggested several possible mechanisms, including convection currents in the mantle, which would seem similar enough to what we now label plate tectonics.

Plate tectonics superseded the many alternative theories purporting to explain patterns of tectonic activity, such as the contracting and expanding Earth theories (Carey, 1976), and several others that seemed absurd to intelligent laymen (see the marvelous satire by Barks, 1956). These developments had a profound influence on thinking about historical biogeographical causes. They implied the existence of a co-evolutionary interplay between life and a changing Earth landmass, and a differing conception of land bridges than the former understanding (Croizat, 1964).

Leon Croizat (1894–1982) is the most prominent name in the study of the biogeography of the distant past. He not only linked geographic barriers and living things through co-evolution, but he also attempted to relate biological diversity to three principal factors (form, space, and time). This framework is still used for the disciplinary coordination of historical biogeography to this day. Form may be thought of as including both molecular (DNA) and morphological (phenotype) characteristics that are the principal concern of plant and animal systematics. The temporal dimension has been provided by paleontology (fossil evidence of animals and plants) and palynology (fossil pollen evidence of plant pollen), while biogeography has stressed the central role of spatial analysis. Collectively, the various disciplinary strands that make up historical biogeography currently employ at least 31 different techniques (Crisci et al., 2003), an achievement that has, in recent times, led to lively debate as to the actual epicenter of the discipline (Avise, 2000; Craw et al., 1999; Humphries and Parenti, 1986). Yet while a rather diverse methodological orientation has emerged, the scope of historical investigation has remained relatively the same—geographical arrangement and space-time processes.

While plate tectonics is generally accepted as an important factor in distribution, nonetheless there is far less consensus about the relative contributions made by dispersal and vicariance. In vicariant distributions an ancestral population is divided into subpopulations through the emergence of a barrier they cannot cross (a disjuncture). In dispersal, a preexisting barrier limits the range of an ancestral population. Ever since the early theorizing of Linnaeus, biogeographers had supported distribution by dispersal, an idea that was initially linked with the assumption of Earth immutability (see Riddle, this volume).

However, between 1940 and 1960 dispersal theories were challenged by two botanists, Stanley Cain (1944) and Croizat (1958), who promoted vicariance as an alternative explanation. The subsequent acceptance of plate tectonics made vicariance more plausible. Later, Croizat (1964) tried to resolve indecision about the relative importance of dispersal and vicariance by proposing the existence of alternating cycles of each. This theoretical model was later supported by Craw et al. (1999). Croizat's (1964) focus of attention on species diversity implied that the ecological community-scale was just as important to this type of theoretical analysis, even though geographical distribution research had been strongly dominated by an ecosystem orientation.

While it is useful to distinguish the paleobiology tradition from other branches of biogeography, its practitioners seem on the whole to be apprised of each other's work, and they are aware of developments within other traditions. In contrast, the separation of the 'Continental' and 'American' biology traditions stem from the early twentieth century, when beliefs about succession began to diverge.

2.6 ECOLOGICAL BIOGEOGRAPHY

2.6.1 Succession theory and nonequilibrium ecology

The concept of succession is usually traced to Henry Cowles's (1899) pioneering study of Lake Michigan sand dunes, but Cowles and Frederick E. Clements (1874–1945) borrowed the basic model from Warming, with the bio-utopian aspect of Clements's thought deriving more from Drude (McIntosh, 1985; Worster, 1977) and Herbert Spencer (Raup, 1942; Blumler, 1996). Clements's (1916) theory of plant succession became known as the mono-climax theory, indicating a single

equilibrium state. He believed the order and endpoint for succession in a region were determined by climate, a proposition that was challenged on theoretical (Gleason, 1926; Simberloff, 1980) and empirical (Connell and Slatyer, 1977; Walker, 1970) grounds, and it was never accepted in Europe. European thinking about succession was similar except that multiple climaxes were accepted within climate regions, corresponding especially to soil variability. Consequently, Clements and his followers displayed little interest in vegetation mapping below the biome level, while in contrast, on the continent, the Zurich-Montpellier school of phytosociology (Braun-Blanquet, 1928, 1951, 1964) and its competitors developed extremely detailed systems of classification and mapping, which continue to this day.

In contrast, within American biology in the latter half of the century even biomes increasingly were deemed unworthy of discussion, because they were descriptive rather than theoretical. Only with the rise in concerns over global change has the study of biomes undergone a resuscitation. Note that while Braun-Blanquet and other European ecologists sharply distinguished their views on succession from those of Clements, their version was equally equilibrium-based and as such, would subsequently be classified by American nonequilibrium ecologists as 'Clementsian'.

Clements's model combined the Enlightenment ideal of progress with the Romantic idealization of the 'forest primeval', thus giving it a powerful and persuasive emotive content (Blumler, 1996). Gleason's (1926) critique of the reality of communities was in part a Darwinian response to Clements's 'Lamarckian' (really, Spencerian) notion of the community as a superorganism. Gleason's argument, and the ideas of others such as Raup (1964), played an important role in overturning traditional notions of succession in the United States. The following is perhaps the earliest clear statement of the nonequilibrium perspective:

Ecological and conservation thought at the turn of the century was nearly all in what might be called closed systems of one kind or another. In all of them some kind of balance or near balance was to be achieved. The geologists had their peneplain; the ecologists visualized a self-perpetuating climax; the soil scientists proposed a thoroughly mature soil profile, which eventually would lose all trace of its geological origin and become a sort of balanced organism in itself. It seems to me that social Darwinism, and the entirely competitive models that were constructed for society by the economists of the nineteenth century, were all based on a slow development towards some kind of equilibrium. I believe that there is evidence in all of these fields that the systems are open, not closed, and that probably there is no consistent trend towards balance. Rather, we should think in terms of massive uncertainty, flexibility and adjustability. (Raup, 1964: 19)

But the existence of this perspective was not the only reason for the different trajectories that succession theory took in the United States and Europe. Important also was the difference in landscapes: human-dominated in Europe, and 'wilderness' in the United States. In Europe, succession theory was not really amenable to hypothesis testing, since any deviation from the expected climax could be explained in terms of prior human impact. In North America, too, there was little intentional testing of succession theory, but the management of wilderness areas for climax created natural experiments. It turned out that the predictions of the Clementsian model of succession were not always right (Walker, 1970). Repeated failures of management predicated upon Clementsian theory led ultimately to a re-evaluation of the theory (Botkin, 1990), increasing appreciation of the importance of natural disturbance and of the highly contingent nature of vegetation response to disturbance (Sprugel, 1991), and the recognition of additional successional pathways besides Clementsian 'facilitation' (Connell and Slatyer, 1977).

Connell and Slatyer (1977) reviewed the succession literature and concluded that 'inhibition'— where early successional species exclude those that are later in the sequence, unless and until disturbance creates openings in the early successional cover—probably is as frequent (if not more so) as facilitation or tolerance. Inhibition arguably is not really succession, since it requires additional disturbance after the disturbance that initiates succession (Blumler, 1993). 'Chronic patchiness' (Botkin and Keller, 1994), where the first species to occupy a site after disturbance is able to outcompete and exclude all subsequent arrivals, also is not succession in the normal sense. There are even cases of reverse succession, for example, when annuals outcompete and replace herbaceous perennials and woody plants (Blumler, 1993) and when herbs replace woody plants after nitrogen fertilization (e.g., Wood et al., 2006). Consequently, nonequilibrium ecologists now prefer 'vegetation dynamics' to 'succession' (Glenn-Lewin et al., 1992), and they generally accept the Gleasonian view that communities are of limited validity (Gleason, 1926; Whittaker, 1975; Davis, 1981). In the English-speaking world, 'climax' is now usually placed in quotes, because it is considered imaginary (Botkin and Keller, 1994).

2.6.2 *Ecosystem ecology*

In contrast, ecosystem ecology (Odum, 1969) incorporated Clementsian assumptions. It became increasingly popular over approximately the same time period as did nonequilibrium ecology. Ecosystem ecology caught on especially within the environmental movement and among those scientists concerned with biogeochemical cycling and simulation modeling. Succession theory was not really relevant to those studying biogeochemical cycling, but it did pertain to simulation modeling. Robert MacArthur's contributions (see below), such as the Equilibrium Theory of Island Biogeography (MacArthur and Wilson, 1967), also were equilibrium based. Equilibrium and nonequilibrium theories continue to co-exist— somewhat uncomfortably—for instance, in theorizing about species diversity (Connell, 1978). The equilibrium models of the 1900s (see the quote by Raup, above), which dominated the sciences and social sciences then, are now generally abandoned; but this is less true within ecology and biogeography. The co-existence of equilibrium and nonequilibrium perspectives does not resemble the Kuhnian model of an outmoded paradigm being overturned. It bears somewhat more similarity to Gould's (1981) example of repeated influence on scientists from the larger society. The environmental movement, and society as a whole, tends still to believe in a balance of nature, and so researchers are attracted to equilibrium models. However, this does not seem to be a complete explanation for their continuing popularity among ecologists and biogeographers.

2.6.3 *The continental school*

On the European Continent (but not the British Isles, which was influenced more by American developments), the twentieth century saw a continued emphasis on describing and analyzing vegetation-environment relationships, complemented by the development of systems of vegetation classification. In their influential textbooks, Warming (1895), Schimper (1898), Schröter (1904), and others advocated a causal approach in vegetation studies and contributed to establishing vegetation science as an important branch of ecology around the turn of the century. The advancement of physiological plant ecology (e.g., Larcher, 1994) facilitated consolidating findings on environmental relationships of vegetation types. Heinrich Walter was the most influential exponent of ecophysiologically based vegetation studies that led to his global vegetation surveys (e.g., Walter, 1973; Walter and Breckle,

1982–91, 1985). Contemporary efforts were directed into the development of standardized vegetation-analysis methods, which allowed the systematic classification of plant communities and quantitative evaluations of community data. The evolution of continental-European phytosociology—the Zurich-Montpellier School—is intrinsically tied to the name of Josias Braun-Blanquet (1884–1980). His textbooks (1928, 1951, 1964) have influenced vegetation science for decades, and his approach found acceptance in many regions of the globe (but not in the United States and Great Britain). Wrongly identified for some time as the mere description of plant communities by Anglo-American vegetation ecologists, the Braun-Blanquet approach was designed as an ecological approach and goes far beyond pure community typology (Mueller-Dombois and Ellenberg, 1974; Dierschke, 1994; see also van der Maarel, 2005). For instance, the ecological tradition of continental-European vegetation science that included experimental approaches as well, has allowed for the accumulation of particularly rich knowledge on the environment of plant communities and the ecological behavior of participating species, reflected, for example, in the system of indicator values for central European vascular plant species regarding moisture, soil nitrogen status, soil reaction, soil chloride concentration, light regime, temperature, and continentality (Ellenberg et al., 1992; see Uzunov, this volume). Ellenberg also developed a widely noted global classification system of plant formations (Ellenberg and Mueller-Dombois, 1974) and summarized the available knowledge of the ecology of central European vegetation in classical textbooks (Ellenberg, 1988, 1996).

Ecological biogeography on the European continent during the twentieth century was on the one hand heavily influenced by vegetation ecology, further developed mainly by (geo)botanists focusing on plant species and communities and their distribution and environmental relations. On the other, biogeography received special impetus from physical geography (especially plant or vegetation geography), whose exponents focused on regions and landscapes and their biocoenological setting (e.g., Schmithüsen, 1968). Based on their spatial perspective on relationships among landscape elements, physical geographers shaped the emerging discipline of landscape ecology. Carl Troll (1899–1975) introduced the term 'Landschaftsökologie' (landscape ecology), motivated by the novel perspective on landscapes offered by aerial photographs (Troll, 1939). He defined landscape ecology as the study of the complex spatial and temporal interactions between biocoenoses and their environmental conditions in a given landscape section (Troll, 1966).

Troll's concept was broadened by Schmithüsen (1976) and Neef (1967), who explicitly integrated humans and their activities into the continental-European concept of landscape ecology. Based on the theoretical fundament of a holistic landscape science, elaborated by the above 'founding fathers', efforts toward integrated, interdisciplinary research concepts pertaining to environmental issues and land-management problems (e.g., the UNESCO Man and Biosphere program) have been reinforced since the 1970s. At the same time, the central European concept of landscape ecology was adapted and modified by emerging schools of landscape ecology in other parts of Europe and elsewhere (e.g., Naveh and Lieberman, 1984; Forman and Godron, 1986; see Kupfer, this volume). Compared to vegetation ecology, plant geography, and landscape ecology, the impact of animal ecology and zoogeography on European biogeography was meager. The insufficient integration of animal ecology and vegetation ecology is still considered a methodological deficit, even though efforts toward unifying theories and concepts are being strengthened (e.g., Kratochwil and Schwabe, 2001).

Influences from continental Europe on North American and British biogeography have been idiosyncratic. Raunkiaer (1934) elaborated Warming's concept of 'life forms', an approach that received a mixed reaction both in Europe and in the United States (Raup, 1942). Subsequently, Raunkiaer's life-form classification influenced those interested in describing 'functional types' as a means of enabling predictive biome modeling (e.g., Shugart, 1997). Küchler (1947, 1967) brought the German vegetation mapping tradition to the United States and found a vacant niche because of the near absence of mapping there. His influence persists in maps of 'potential natural vegetation' and in the equilibrium succession-to-climax assumptions of some biome models. As Prentice et al. (1992: 132) put it, "the climatic control of plant distribution has not been regarded as a central research field in ecology or biogeography for fifty years or more." Their statement was true for the United States, though not for Europe. When American interest was rekindled, German approaches were available to serve as a template.

While we have emphasized German research, other European nations were on parallel tracks and were generally aware of each other's work, including the Russians. The latter influenced the Chinese (e.g., Grubov, 1969), while the imperialist powers influenced their former colonies. Consequently, the Continental school exerts great influence throughout the non-English-speaking world. For instance, UN vegetation maps reflect the Continental tradition. Only in one or two countries, notably Britain and Israel, was there any merging of the Continental and American approaches.

2.6.4 The Berkeley School

Meanwhile, a branch of biogeography had arisen within the Berkeley School of American Geography, originating in Carl Sauer's (1889–1975) interest in human-environment interaction and, especially, agricultural origins and dispersals (Sauer, 1941, 1952, 1956). Sauer was influenced by his advisors at the University of Chicago and their strong interests in the historical geography of settlement; his primary intellectual allegiance was to nineteenth-century German geographers such as Ratzel, whose own major influences were Darwin and Haeckel (who coined the term 'ecology'). Abiding biogeographical themes have included, besides agricultural origins and dispersals (Blumler, 1992; Blumler and Byrne, 1991; Sauer, 1993), crop genetics and evolution (Blumler, 2003; Zimmerer, 1996), fire (Minnich, 1983; Vale, 2002), vegetation dynamics (Veblen, 1992), historical ecology (Minnich, 2008), invading species (Mensing and Byrne, 1997), ethnobotany (Voeks, 1997), landscapes (Gade, 1999), conservation (Zimmerer and Young, 1998), environmental thought (Blumler, 2002; Lewis, 1992; Zimmerer, 1994), and, in general, human-environment interaction. The broad interest of Berkeley School members in manifold human-environment interactions is associated with a generally broader concern for conservation than the intense focus on biodiversity characteristic of the American biology school.

Nonequilibrium theory was rapidly accepted within the Berkeley School, perhaps more so than in American biology, reflecting geography's emphasis on contingency and complexity. Both Gleason (1922) and Raup (1942) had published in the *Annals*, and Raup specifically aligned himself with Sauer and with Hartshorne's idiographic orientation. Sauer's son, Jonathan, trained in biology with his father's close friend Edgar Anderson, whose interest in weeds, dumps, gardens, and so on was congruent with the Berkeley School human-environment orientation. But J. Sauer's understanding of vegetation dynamics differed sharply from Anderson's Clementsian views. Tim Brothers, one of J. Sauer's students, described his perspective this way:

> He distrusted grand theories, particularly those arrived at by reasoning from general principles, and he took delight in pointing out examples to contradict them. He did not believe in

deterministic models of plant succession, island diversity, plant migration or crop domestication. . . . Above all, Sauer was interested in the myriad interactions between people and plants. His discussions of crops and weeds were not just about the plants concerned but about the cultural contexts that produced them. . . . This emphasis continues to distinguish biogeography as done by geographers from the versions practiced by many other kinds of scientists. (Brothers et al., 2009: 170)

Widespread acceptance of nonequilibrium ecology is illustrated by the many publications of Tom Veblen (e.g., 1992) and his students, on vegetation dynamics and natural disturbance; equally numerous palynological papers (American palynologists especially in geography tend to align with Margaret Davis's [1981] Gleasonian views on communities, whereas in Europe, for instance, she is less influential); as well as occasional contributions from others (e.g., Brown, 1993).

2.7 THE AMERICAN BIOLOGY SCHOOL

2.7.1 Robert MacArthur and the theory of Island biogeography

At about the same time that plate tectonics was gaining acceptance, and just as American plant ecology was beginning to shift toward a nonequilibrium perspective, Robert MacArthur brought equilibrium models from economics into ecology and initiated a trend, which continues, of emphasizing the theoretical over the descriptive. Although the bulk of his work was ecological, not biogeographical, he did contribute some work toward the latter (e.g., MacArthur, 1972).

Much of the concern was with niche assembly of communities, using the equilibrium assumption of one niche, one species—an assumption, it is now known, that does not always hold (Shmida and Ellner, 1984). MacArthur and E.O. Wilson (1963, 1967) established an enormous influence on biogeography when they drew attention back to the role of island area in explaining species distribution. Islands had come to the attention of biogeographers in early theorizing around centers of origin, the mapping of species distributions, and in Wallace's *Island Life*. MacArthur and Wilson were puzzled by the same observations that had drawn the attention of early biogeographers as to why islands have fewer species than do sample areas of the same size on continents. This led them to propose their Equilibrium Theory of Island Biogeography (ETIB).

Islands are useful as biological laboratories because they are of limited size and have clear boundaries. Insularity is found not only in a sea. For example, there are islands of woodland in a sea of pasture, islands of pasture in a sea of arable land, ponds in a sea of land, caves in a sea of rock, cool mountains in a warm lowland, and high-UVB mountains in a low-UVB lowland (Flenley, 2007). Initially, the same principles were thought to apply, and therefore the ETIB was broadly applied in conservation and the design of nature reserves. Some pointed out, however, that oceanic islands are embedded in a much more hostile matrix than are habitat islands on land, with significant consequences, for example, for invading species.

A plot of the number of species on an island against the area of the island on a log-log scale is described by the equation:

$$S = CA^z \qquad (1)$$

where S = number of species, A = island area, C = a constant depending on taxa, region, etc., and Z is a constant (about 0.3 in this case).

The conclusion is that large islands have more species (or genera in the case quoted above). MacArthur and Wilson explained this as an equilibrium reached through time, between the immigration of new species and the extinction of species already there. When few species are present, the immigration rate is high, as all arrivals are new species. When many species are present, most new arrivals represent species already

there; therefore, the immigration rate drops. This balance between immigration and extinction is illustrated in Figure 2.2. When few species are present, only a few are available to become extinct; therefore, extinction is low. When many species are present, many are available to become extinct, and their average population size is low; therefore, extinction is high.

This is modified in two ways. A far (isolated) island receives less immigration than a near one. A large island has fewer extinctions than a small one (because larger populations are less likely to hit zero) (Figure 2.3). Therefore, large near islands usually have a high diversity, while small far islands usually have low diversity (see Figure 2.4). All of this assumes equilibrium conditions. A similar though not identical result could be produced by a nonequilibrium model in which far islands just take longer to fill up, through immigration and evolution (adaptive radiation). In fact, MacArthur and Wilson suggested that in evolutionary time the number of species on isolated islands would creep upward, a point that most have overlooked, and that Sauer (1969) suggested violated the fundamental reasoning behind the model.

The best way to detect an equilibrium is the presence of turnover (i.e., change of the species list through time, without change in the total number of species). The island is then said to be saturated. MacArthur and Wilson expected rapid turnover, which empirical studies for the most part have not verified.

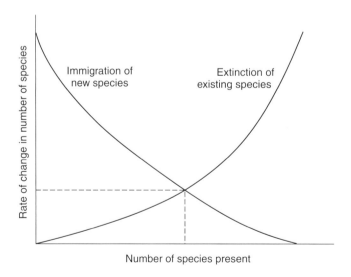

Figure 2.2   The equilibrium model of the biota of an island (after MacArthur and Wilson, 1963)

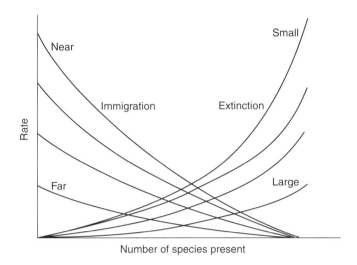

Figure 2.3 The relationships between rates of species immigration and extinction with island size (small and large) and isolation (near and far) (after MacArthur and Wilson, 1963)

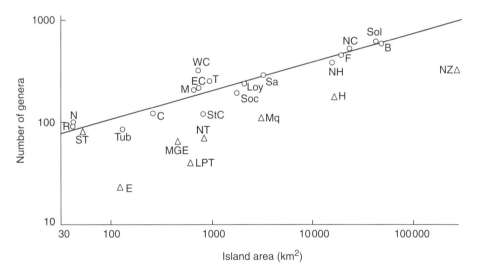

Figure 2.4 The relationships between island area and diversity of conifers and flowering plant genera in the Pacific Islands. The more isolated islands are indicated by triangles. The data from the other islands lie very close to a straight line (the regression coefficient), suggesting that generic diversity in these islands is almost entirely controlled by island area—the correlation coefficient is 0.94, indicating a very high degree of correlation. The islands are B – Bismark Archipelago; C – Cook Islands; E – Easter Island; ED – East Carolines; F – Fiji Islands; H – Hawaiian Islands; Loy – Loyalty Islands; LPT – Line, Phoenix, and Tokelau groups; M – Marianas; MGE – Marshall and Gilbert Islands, Tuvalu; Mq – Marquesas; N – Norfolk Island; NC – New Caledonia; NG – New Guinea; NH – Vanuatu; NT – Northern Tuamotu Islands; NZ – New Zealand; Ph – Philippines; R – Rapa Island; Sa – Samoa group; Soc – Society Islands; Sol – Solomon Islands; ST – Southern Tuamotu Islands; StC – Santa Cruz Islands; T – Tonga group; Tub – Tubai group; WC – West Carolines. Date from Van Balgooy, 1971 (after Cox & Moore, 1993)

There are exceptions to the overall generalization. Most of these are islands isolated in their location. The most extreme case is Easter Island, with a known native higher plant flora of 48 species (Flenley and Bahn, 2002). The depauperate nature of the flora, and especially the presence of only one tree and two shrub species, was originally explained as a result of its exceptional isolation. However, the study of fossil pollen (Flenley et al., 1991) and fossil wood and charcoal (Orliac, 2000) have shown that the island was once well forested, with numerous additional woody species, including an extinct palm. The decline of the forest and the flora occurred almost entirely within the archaeologically known time of human occupancy starting at perhaps AD 800 or earlier (Cole and Flenley, 2008), and many bird species were extirpated or made extinct also during that period. Since the people were constructing giant stone statues with weights up to 80 tons, and were moving them distances of several kilometers, it seems almost certain that a large amount of timber was used in the moving process. The combination of this with agricultural clearance for cropping easily explains the deforestation. The continuing ability of the island to support vegetation is evidenced by the approximately 160 introduced species that now grow on the island. Easter Island illustrates the need to interpret data with caution.

The ETIB assumed an ongoing or dynamic interplay between local extinction and immigration rates in maintaining optimal levels of species diversity for a given island area—a kind of index for ecological potential or carrying capacity—and relative distance from species source. The theory was effective in (a) drawing greater attention to biodiversity in general (MacArthur, 1972); (b) promoting the conservation movement by promoting theory development around (what are today much criticized) principles for reserve design; (c) theoretically underpinning the development of metapopulation biology (Levin, 1970); and (d) attempting to provide a methodological basis (generally not used today) for the estimation of extinction rates.

The theory has not been without its critics (Gilbert, 1980; McIntosh, 1980; Sauer, 1969; Slobodkin, 1996; Whittaker, 1998 Williamson, 1981). McIntosh (1985: 280) reported that the theory "was variously described as a revolution, as nomothetic, or as fairyland." In the earliest and perhaps the most withering attack, Sauer (1969) questioned almost every aspect of the ETIB, though he also praised sections of the book that dealt with ancillary matters, such as speculations about the role of stepping-stone islands. Interestingly, these sections are today almost forgotten in the intense focus over what MacArthur

and Wilson themselves acknowledged was a simplified model. Over the years, biogeography textbooks have drastically altered their evaluation of the ETIB, from entirely positive to decidedly mixed. Metapopulation theory (Hanski and Simberloff, 1997) has perhaps been the most important subsequent development. Otherwise, with the exception of Hubbell's (2001) 'unified neutral theory of biodiversity and biogeography', ETIB's theoretical development has somewhat foundered.

MacArthur's models were generally borrowed from economics, so it is unsurprising that MacArthur and Wilson's (1963) original model is identical to the supply and demand curves used by economists to determine partial equilibrium market prices based on the so-called multifactor theory of value. The same model, although applied in a different domain, suffers from similar limitations.

First, there is the problem of appropriate system time horizon and the implications this has for defining equilibrium conditions. It is evident that both economic markets and island ecosystems collapse (Flenley, 1985; Whittaker et al., 1989) and then move toward new equilibrium points or what theoretical ecologists refer to as neighboring basins of attraction (Gunderson and Holling, 2002). The ability to move between one equilibrium point and another challenges the more traditional view that ecological equilibrium points are fixed and stable (i.e., movement away from equilibrium is quickly followed by a return to the same fixed point) (see Sprugel, 1991 for a similar argument regarding 'natural' vegetation).

An additional problem for the ETIB—one that does not apply in economics—is the role of evolution. The history of life shows generally increasing biodiversity, despite occasional mass extinctions. Thus, it may be that equilibrium has not yet been achieved, even in the oldest persisting environments, and that islands, which are inherently rather short-lived, are likely to see a continuing increase in diversity due to adaptive radiation, until they begin to sink beneath the waves. Certainly, adaptive radiation is a major feature of island systems, suggesting that (a) not all niches are immediately filled and (b) evolutionary processes can fill them, though this takes time. Recently, Whittaker et al. (2008) expanded the ETIB to take account of these variables.

As will be evident from this book, the ETIB still exerts major influence on the research agenda about island biogeography and reserve design. Wilson, who unlike MacArthur is still alive, continues to be praised for development of the theory (e.g., Quammen, 1996). The phenomenon of 'relaxation' (Diamond, 1982)—the often prolonged time it takes to lose species down to the equilibrium level after habitat destruction—was recognized soon after the promulgation of the ETIB, but some conservationists continue to assume instantaneous equilibration with reduced areas after habitat destruction. In retrospect, it seems odd that so much was, and still is, made of an equilibrium model, during a period when nonequilibrium theory supposedly was replacing equilibrium theory, in the United States and other English-speaking countries. Wilson's (1959) earlier proposal of a 'taxon cycle' in ants colonizing islands remains fertile (e.g., Ricklefs, 2005), and this may ultimately prove more useful than the more ambitious ETIB, while relaxation is proving to be a concept with serious implications that reserve managers and others in conservation need to recognize and understand (e.g., Rosenzweig, 2003)—in contrast, perhaps, to the ETIB.

The American Biology School has become highly theoretical since MacArthur. Its proponents tend to assert that the field 'matured' when it became so (McIntosh, 1985). This appears to reflect the preference for deduction over induction in the sciences. But not all have found this approach useful. For instance, McIntosh (1980) reviewed developments in theoretical ecology and concluded that the search for explanation has proven 'frustrating'. Nonetheless, according to McIntosh (1985: 273): "Theoreticians . . . are a resilient lot, and the difficulty of identifying constants, laws, or rules has not inhibited them." Even Hutchinson, who was the inspiration for the rise to prominence of mathematical theory, had the following, cautionary remarks:

> Many ecologists of the present generation have great ability to handle the mathematical basis of the subject. Modern biological education may let us down if it does not insist, and it still shows too few signs of insistence, that a wide and quite deep understanding of organisms, past and present, is as basic a requirement as anything else in ecological education. (1975, quoted in McIntosh, 1985)

In this regard, it is well to remember that the greatest contribution of biogeography to science, the theory of evolution, resulted from natural history studies.

2.8 RECENT DEVELOPMENTS

Resilience theory was a natural extension of the nonequilibrium era in ecology (Holling, 1973, 1978, 1986) and eventually it drew attention to the existence of a feedback between community complexity and loss of resilience that would form the basis of a far-from-equilibrium base theory (Gunderson and Holling, 2002). It is as yet unclear how widely this theory will be accepted. Despite the repeated claims that nonequilibrium theory has won out within American biology, equilibrium concepts repeatedly reassert themselves. In part this is because equilibrium mathematical theory and modeling are more tractable and give less contingent, more confident (if you accept the assumptions) predictions. Also, the environmental movement remains deeply tied to equilibrium notions because of its bio-utopian beliefs, such as in the balance of nature (Blumler, 1996). For instance, Worster (1977) was suspicious of attempts to overthrow Clementsian succession, even impugning the motives (on the basis of precious little evidence) of Gleason and Raup, though later (Worster, 1990) he offered some grudging acceptance of nonequilibrium theory. Most biogeographers are environmentalist, so beliefs that take hold within the movement can exert great influence on scientists. The information explosion also appears to be playing a role, in that the impossibility of thorough literature review carries with it the possibility for theories considered outmoded in some circles to be treated as received wisdom in others.

Prior to 1900, all biogeographers were reading a common literature. But for much of the twentieth century there was little communication between the Continental and American biology schools. The last American to review Continental approaches to vegetation study was Whittaker (1962, 1973), and relatively little attention was paid to those works in the United States. Occasionally, continental biogeographers came to the United States and attempted to promote continental ideas (e.g., Mueller-Dombois and Ellenberg, 1974) but with little impact. In contrast, Continental biogeographers have not been terribly positive about vegetation studies in the English-speaking world. For instance, Westhoff (1970, quoted in McIntosh, 1985) divided vegetation studies as "the practical, the logical, the usual Anglo-American, and the negative" and stressed the implication that "the usual Anglo-American terminology regarding the science of vegetation research is neither practical nor logical." Nonequilibrium ecology has had little influence on the Continental School. A few attempts have been made to consider nonequilibrium theory, but on the whole these do not seem to accept its fundamental premises (e.g., Barbero et al., 1990).

The rise of concern over species extinctions (biodiversity) and global change, in parallel in Europe and in the English-speaking world, has produced a merger in recent years. A recent example of cross-tradition communication is in biome modeling (e.g., Cramer et al., 2001), which began with Box's (1981) correlational model based on functional types. Box studied in Germany with Lieth and Walter, though it is unclear how much of his work they influenced. In any case, biome modeling in the English-speaking world was given its more influential impetus by Woodward (1987), who translated the eco-physiological approach of the Germans into terms that Anglo-Americans could appreciate. Woodward also drew upon European notions of succession, though currently some modelers are attempting to incorporate nonequilibrium-based vegetation dynamics.

2.9 CONCLUSIONS

The overall decline of knowledge of the scientific literature is unfortunately fostering a decreasing awareness of alternative perspectives. All schools of biogeography have something to offer, which is not to say that their underlying beliefs are necessarily correct in all particulars. They cannot be, since the views of one school sometimes contradict those of another. We agree with Robert May (1981, quoted in McIntosh, 1985), who opposes "naïvely simple formulations of The Way To Do Science" promulgated by 'doctrinaire vigilantes'.

We find the detailed vegetation mapping, classification, and analysis of the Continental school fruitful. We also would argue that Heinrich Walter's work is a 'must-read' for any student of vegetation. While the criticisms of McIntosh (1980, 1985) and others of the theoretical bent of the American biology school undoubtedly have merit, the school evidences an undeniable and exciting creativity, repeatedly coming up with new approaches that may prove groundbreaking (e.g., Brown, 1995). Finally, the Berkeley School might be criticized for being overly concerned with the historical or the practical, but on the other hand, its intersection with the social sciences and with the abiotic physical environment has added perspective that at times has enabled its practitioners to produce illuminating, if unfortunately overlooked, critiques of existing scientific theory. American geography's respect for local peoples and its inherent orientation toward 'adaptive management', even before Holling (1978) coined the term, have had the result that it has repeatedly produced sensible and sensitive prescriptions regarding land use in poor countries (e.g., Zimmerer and Young, 1998).

REFERENCES

Anderson, E. (1949) *Introgressive Hybridization*. New York: John Wiley.

Arnold, M.L. (1992) 'Natural hybridization as an evolutionary process', *Annual Review of Ecology and Systematics*, 23:237–61.

Avise, J.C. (2000) *Phylogeography: The History and Formation of Species*. Cambridge, MA: Harvard University Press.

Barbero, M., Bonin, G., Loisel, R., and Quézel, P. (1990) 'Changes and disturbances of forest ecosystems caused by human activities in the western part of the mediterranean basin', *Vegetatio*, 87:151–73.

Barks, C. (1956) 'Land beneath the ground', *Walt Disney's Uncle Scrooge* #13.

Blumler, M.A. (1991) 'Winter-deciduous versus evergreen habit in mediterranean regions: a model', in R.B. Standiford (tech. coord.), *Proceedings of the Symposium on Oak Woodlands and Hardwood Rangeland Management*, October 31–November 2, 1990, Davis, CA, pp. 194–97. USDA Forest Service, Gen. Tech. Rep. PSW-126, Berkeley.

Blumler, M.A. (1992) 'Independent inventionism and recent genetic evidence on plant domestication', *Economic Botany*, 46:98–111.

Blumler, M.A. (1993) 'Successional pattern and landscape sensitivity in the Mediterranean and Near East', in D.S.G. Thomas and R.J. Allison (eds.), *Landscape Sensitivity*, pp. 287–305. Chichester, UK: John Wiley.

Blumler, M.A. (1996) 'Ecology, evolutionary theory, and agricultural origins', in D.R. Harris (ed.), *The Origins and Spread of Agriculture and Pastoralism in Eurasia*, pp. 25–50. London: UCL Press.

Blumler, M.A. (2002) 'Environmental management and conservation', in A.R. Orme (ed.), *Physical Geography of North America*, pp. 516–35. Oxford: Oxford University Press.

Blumler, M.A. (2003) 'Introgression as a spatial phenomenon', *Physical Geography*, 24:414–32.

Blumler, M.A. (2005) 'Three conflated definitions of Mediterranean climates', *Middle States Geographer*, 38:52–60.

Blumler, M.A., and Byrne, R. (1991) 'The ecological genetics of domestication and the origins of agriculture', *Current Anthropology*, 32:23–54.

Botkin, D. (1990) *Discordant Harmonies: A New Ecology for the Twenty-first Century*. New York: Oxford University Press.

Botkin, D. and Keller, E. (1994) *Environmental Science: Earth as a Living Planet*. New York: John Wiley.

Box, E.O. (1981) *Macroclimate and Plant Forms: An Introduction to Predictive Modeling in Phytogeography*. The Hague: Dr. W. Junk.

Braun-Blanquet, J. (1928, 1951, 1964) *Pflanzensoziologie. Grundzüge der Vegetationskunde* (1st ed., Berlin: Biologische Studienbücher 7; 2nd ed., Vienna: Springer; 3rd ed., New York: Springer).

Briggs, J.H.Y. (2004) *Ages in Chaos: James Hutton and the Discovery of Deep Time*. New York: Forge.

Brongniart, A. (1837) *Histoire des végétaux fossils*. Paris: G. Dufour et E. d'Ocagne.

Brothers, T.S., B. Friedrich, Gade, D.W. and Kimber, C.T. (2009) 'Jonathan D. Sauer (1918–2008): Perspectives on his life and work in Latin America and beyond', *Journal of Latin American Geography*, 8:165–80.

Brown, D.A. (1993) 'Early nineteenth-century grasslands of the midcontinent plains', *Annals of the Association of American Geographers*, 83:589–612.

Brown, J.H. (1995) *Macroecology*. Chicago: University of Chicago Press.

Buffon, G.L.L. (1761) *Histoire naturelle, generale et particuliere*. Paris: Academie Francaise.

Bush, M.B., and J.R. Flenley, eds. (2007) *Tropical Rainforest Responses to Climatic Change*. Chichester, UK: Praxis.

Cain, S.A. (1944). *Foundations of Plant Geography*. New York: Harper and Brothers.

Carey, S.W. (1976) *The Expanding Earth*. New York: Elsevier Scientific.

Carlquist, S. (1965) *Island Life*. New York: Natural History Press.

Clements, F.E. (1916) 'Plant succession: An analysis of the development of vegetation', *Carnegie Institute of Washington Publications*, 242:1–512.

Cole, A., and Flenley, J. (2008) 'Modelling human population change on Easter Island far-from equilibrium', *Quaternary International*, 184:150–65.

Connell, J.H. (1978) 'Diversity in tropical rain forests and coral reefs', *Science*, 199:1302–10.

Connell, J.H., and R.O. Slatyer. (1977) 'Mechanisms of succession in natural communities and their role in community stability and organization', *American Naturalist*, 111: 1119–44.

Cowles, H.C. (1899) 'The ecological relations of the vegetation of the sand dunes of Lake Michigan', *Botanical Gazette*, 27:95–117, 167–202, 281–308, 361–91.

Cox, B.C., and P.D. Moore. (1993) *Biogeography, an Ecological and Evolutionary Approach*. 5th ed. Oxford: Blackwell.

Cox, B.C., and P.D. Moore. (2005) *Biogeography, an Ecological and Evolutionary Approach*. 7th ed. Oxford: Blackwell.

Cramer, W. et al. (2001) 'Global response of terrestrial ecosystem structure and function to CO_2 and climate change: Results from six dynamic global vegetation models', *Global Change Biology*, 7:357–73.

Craw, R.C., Grehan, J.R., and M.J. Heads. (1999) *Panbiogeography: Tracking the History of Life*. Oxford: Oxford University Press.

Crisci, J.V., Katinas, L., and P. Posadas. (2003) *Historical Biogeography*. Cambridge, MA: Harvard University Press.

Croizat, L. (1958) *Panbiogeography*. Vols I, IIa-IIb. Caracas: Published by the Author.

Croizat, L. (1964) *Space, Time, Form: The Biological Synthesis*. Caracas: Published by the Author.

Darlington, P.J. (1957) *Zoogeography: The Geographical Distribution of Animals*. New York: John Wiley.

Darwin, C.R. (1845) *Journal of researches into the natural history and geology of the countries visited during the voyage of the H. M. S. Beagle round the world under the command of Capt. Fitz Roy, R. N.* London: John Murray.

Darwin, C.R. (1859) *The origin of species by means of natural selection; or, the preservation of favored races in the struggle for life*. London: John Murray.

Davis, M.B. (1981) 'Quaternary history and the stability of forest communities', in D.C. West, H.H. Shugart, and D.B. Botkin (eds.), *Forest Succession: Concepts and Applications*, pp. 132–54. New York: Springer-Verlag.

Dean, D.R. (1992) *James Hutton and the History of Geology*. New York: Cornell University Press.

de Candolle, A. (1820) *Élémentaire de Géographie Botanique*. Chicago: University of Chicago Press.

de Candolle, A. (1855) *Geographie Botanique Raisonné*. Paris: V. Masson.

de Laubenfels, D.J. (1975) *Mapping the World's Vegetation: Regionalization of Formations and Flora*. Syracuse, NY: Syracuse University Press.

Diamond, J.M. (1982) 'Biogeographic kinetics: Estimation of relaxation times for avifaunas of Southwest Pacific islands', *Proceedings of the National Academy of Sciences of the USA*, 69:3199–203.

Dierschke, H. (1994) *Pflanzensoziologie. Grundlagen und Methoden*. Stuttgart: Ulmer Verlag.

Drude, O. (1890) *Handbuch der Pflanzengeographie*. Stuttgart: Verlag von J. Engelhorn.

Ellenberg, H. (1988) *Vegetation Ecology of Central Europe*. 4th ed. Cambridge: Cambridge University Press.

Ellenberg, H. (1996) *Vegetation Mitteleuropas mit den Alpen in ökologischer, dynamischer und historischer Sicht*. 5th ed. Stuttgart: Ulmer Verlag.

Ellenberg, H., and Mueller-Dombois, D. (1974) 'Tentative physiognomic-ecological classification of plant formations of the Earth', in D. Mueller-Dombois and H. Ellenberg (eds.), *Aims and Methods of Vegetation Ecology*, pp. 466–88. New York: John Wiley.

Ellenberg, H., H.E. Weber, R. Dull, V. Wirth, W. Werner, and D. Paulissen. (1992) 'Zeigerwerte von Pflanzen in Mitteleuropa', *Scripta Geobotanica* 18. Gottingen: Goltze.

Engler, A. (1879–82) *Versuch einer Entwicklungsgeschichte der Pflanzenwelt*. Leipzig: Wilhelm Engelmann.

Firenze, R.F. (1997) The identification, assessment, and amelioration of perceived and actual barriers to teachers incorporation of evolutionary theory as a central theme in life science classes through a two-week institute and follow-up studies. PhD dissertation, Binghamton University.

Flenley, J.R. (1979) *The Equatorial Rain Forest, a Geological History*. London: Butterworths.

Flenley, J.R. (1985) 'A re-survey of the flora of the island of Krakatoa (Krakatau)', *National Geographic Society Research Reports*, 20:205–30.

Flenley, J.R. (2007) 'Ultraviolet insolation and the tropical rainforest: Altitudinal variations, Quaternary and recent change, extinctions, and biodiversity', in M.B. Bush and J.R. Flenley (eds.), *Tropical Rainforest Responses to Climatic Change*, pp. 219–35. Chichester, UK: Praxis.

Flenley, J.R., and Bahn, P. (2002) *The Enigmas of Easter Island; Island on the Edge*. Oxford: Oxford University Press.

Flenley, J.R., King, A.S.M., Teller, J.T., Prentice, M.E., Jackson, J., and Chew, C. (1991) 'The Late Quaternary vegetational and climatic history of Easter Island', *Journal of Quaternary Science*, 6:85–115.

Forman, R.T.T., and Godron, M. (1986) *Landscape Ecology*. New York: John Wiley.

Forster, J.R. (1778) *Observations made during a voyage round the world, on Geography, Natural History and Ethnic Philosophy*. London: G. Robinson.

Gade, D.W. (1999) *Nature and Culture in the Andes.* Madison WI: University of Wisconsin Press.

Gilbert, L.E. (1980) Coevolution of animals and plants: a 1979 postscript. in Coevolution of Animals and Plants. L. E. Gilbert and P. H. Raven (eds.). pp. 247–263. Austin TX: University of Texas Press.

Glacken, C.J. (1967) *Traces on the Rhodian Shore: Nature and Culture in Western Thought from Ancient Times to the End of the Eighteenth Century.* Berkeley: University of California Press.

Gleason, H.A. (1922) 'The vegetational history of the Middle West', *Annals of the Association of American Geographers,* 12: 39–85.

Gleason, H.A. (1926) 'The individualistic concept of the plant association', *Bulletin of the Torrey Botanical Club,* 53: 1–20.

Glenn-Lewin, D.C., R.K. Peet, and T.T. Veblen. (1992) *Plant Succession: Theory and Prediction.* New York: Chapman and Hall.

Glick, T.F. (1974) *The Comparative Reception of Darwinism.* Austin TX: University of Texas Press.

Godwin, H. (1940) 'Pollen analysis and forest history in England and Wales', *New Phytologist,* 39: 370–400.

Good, R. (1974) *The Geography of the Flowering Plants.* 4th ed. London: Longmans.

Goudie, A.S., and Cuff, D.J. (eds). (2002) *Encyclopedia of Global Change: Environmental Change and Human Society.* Oxford: Oxford University Press.

Gould, S.J. (1981) *The Mismeasure of Man.* New York: W. W. Norton.

Grisebach, A.R.H. (1872) *Die vegetation der erde.* 2 vols. Leipzig: Wilhelm Engelmann.

Grubov, V.I. (1969) Flora and vegetation. *The Physical Geography of China,* Vol. 2, pp. 265–364. Institute of Geography, USSR Academy of Sciences. New York: Fred Praeger.

Gunderson, L.H., and C.S. Holling. (2002) *Panarchy: Understanding Transformations in Systems of Humans and Nature.* Washington, DC: Island Press.

Hanski, I.A., and Simberloff, D.S. (1997) 'The metapopulation approach, its history, conceptual domain and application to conservation', in I.A. Hanski and M.E. Gilpin (eds.), *Metapopulation Biology: Ecology, Genetics and Evolution,* pp. 2–26. San Diego CA: Academic Press.

Holling, C.S. (1973) 'Resilience and stability of ecological systems', *Annual Review of Ecology and Systematics,* 4: 1–24.

Holling, C.S. (1978) *Adaptive Environmental Assessment and Management.* Chichester, UK: John Wiley.

Holling, C.S. (1986) 'The resilience of terrestrial ecosystems: Local surprise and global change', in W.E. Clark and R.E. Munn (eds.), *Sustainable Development of the Biosphere.* Cambridge: Cambridge University Press.

Hubbell, S.P. (2001) *The Unified Neutral Theory of Biodiversity and Biogeography.* Princeton, NJ: Princeton University Press.

Humboldt, A. von, and A. Bonpland. (1805) *Essai sur la Geographie des Plantes.* Librairie Lebrault Schoell, Paris: Levrault, Schoell & Co.

Humphries, C.J., and L.R. Parenti. (1986) *Cladistic Biogeography.* Oxford: Clarendon.

Huntley, B., and H.J.B. Birks. (1983) *An Atlas of Past and Present Pollen Maps for Europe: 0–13000 Years Ago.* Cambridge: Cambridge University Press.

Huntley, B., and T. Webb III. (1988) *Vegetation History.* Dordrecht: Kluwer.

Hutton, J. (1794) *Investigation of the principles of knowledge and of the progress of reason, from sense to science.* Edinburgh: A. Strahan and T. Cadell.

Köppen, W.P. (1918) 'Klassifikation der Klimate nach Temperatur Niederschlag, und Jahreslauf', *Petermann's Mitteilungen,* 64: 193–203.

Köppen, W.P., and R. Geiger. (1930) *Handbuch der Klimatologie.* Berlin: Gebrüder Borntraeger.

Kratochwil, A., and Schwabe, A. (2001) *Ökologie der Lebensgemeinschaften.* Stuttgart: Ulmer Verlag.

Küchler, A.W. (1947) 'A geographical system of vegetation', *Geographical Review,* 37: 233–40.

Küchler, A.W. (1967) *Vegetation Mapping.* New York: Ronald Press.

Larcher, W. (1994) *Ökophysiologie der Pflanzen.* 5th ed. Stuttgart: Ulmer Verlag

Levin, R. (1970) 'Extinction', In M. Gerstenhaber (ed.), *Some Mathematical Problems in Biology,* pp. 75–107. Providence, RI: American Mathematical Society.

Lewis, M.W. (1992) *Green Delusions: An Environmentalist Critique of Radical Environmentalism.* Durham, NC: Duke University Press.

Linnaeus, C. (1751) 'Specimen academicum de Oeconomia Naturae', *Amoenitates Acadamicae,* II: 1–58.

Linnaeus, C. (1758) *Systema naturae.* 10th ed. Stockholm: Laurentii Salvii Holmiae.

Lomolino, M.V., B.H. Riddle, and J.H. Brown. (2006) *Biogeography.* 3rd ed. Sunderland, MA: Sinauer Associates.

MacArthur, R.H. (1972) *Geographical Ecology: Patterns in the Distribution of Species.* New York: Harper & Row.

MacArthur, R.H., and E.O. Wilson. (1963) 'An equilibrium theory of insular biogeography', *Evolution,* 17: 373–87.

MacArthur, R.H., and E.O. Wilson. (1967) *The Theory of Island Biogeography.* Princeton, NJ: Princeton University Press.

Malthus, T. (1798) *An essay on the principle of population: or a view of its past and present effects on human happiness, with an inquiry into our prospects respecting the future removal or mitigation of the evils which it occasions.* London: Royal Economic Society.

Martin, G.J. (2005) *All Possible Worlds.* New York: Oxford University Press.

Mayr, E. (1996) 'What is a species, and what is not?', *Philosophy of Science,* 63: 262–77.

McIntosh, R.P. (1980) 'The background and some current problems of theoretical ecology', *Synthese,* 43: 195–255.

McIntosh, R.P. (1985) *The Background of Ecology: Concept and Theory.* Cambridge: Cambridge University Press.

Mensing, S.A., and R. Byrne. (1997) 'Pre-mission invasion of *Erodium cicutarium* in California', *Journal of Biogeography,* 25:757–82.

Minnich, R.A. (1983) 'Fire mosaics in southern California and northern Baja California', *Science*, 219: 1287–94.

Minnich, R.A. (2008) *California's Fading Wildflowers: Lost Legacy and Biological Invasions*. Berkeley: University of California Press.

Morley, R.J. (2000) *Origin and Evolution of Tropical Rain Forests*. Chichester, UK: John Wiley.

Mueller-Dombois, D., and H. Ellenberg. (1974) *Aims and Methods of Vegetation Ecology*. New York: John Wiley.

Naveh, Z., and Lieberman, A.S. (1984) *Landscape Ecology: Theory and Application*. New York: Springer.

Neef, E. (1967) *Die theoretischen Grundlagen der Landschaftslehre*. Leipzig: Gotha.

Odum, E. (1969) 'The strategy of ecosystem development', *Science*, 164: 262–70.

Oldfield, F. (2005) *Environmental Change: Key Issues and Alternative Approaches*. Cambridge: Cambridge University Press.

Orliac, C. (2000) 'The woody vegetation of Easter Island between the early 14th and the mid 17th centuries AD', in C.M.Stevenson and W.S. Ayres (eds.), *Easter Island Archaeology and Research on Early Rapanui Culture*. Los Osos, CA: Easter Island Foundation.

Pijl, L. van der. (1982) *Principles of Dispersal in Higher Plants*. Berlin: Springer Verlag.

Prentice, I.C., Cramer, W., Harrison, S.P., Leemans, R., and R.A. Monserud. (1992) 'Global biome model: Predicting global vegetation patterns from plant physiology and dominance, soil properties and climate', *Journal of Biogeography*, 19: 117–34.

Quammen, D. (1996) *The song of the dodo: island biogeography in an age of extinctions*. New York: Scribners.

Raunkiaer, C. (1934) *The Life Forms of Plants and Statistical Plant Geography*. Oxford: Clarenden Press.

Raup, H.M. (1942) 'Trends in the development of geographic botany', *Annals of the Association of American Geographers*, 32: 319–54.

Raup, H.M. (1964) 'Some problems in ecological theory and their relation to conservation', *Journal of Ecology*, 52(Suppl.): 19–28.

Ricklefs, R.E. (2005) 'Taxon cycles: insights from invasive species,' In D.F. Sax, J.J. Stachowicz, and S.D. Gaines (eds.), *Species Invasions: Insights into Ecology, Evolution, and Biogeography*, pp. 165–99. Sunderland, MA: Sinauer Associates.

Ridley, H.N. (1930) *The dispersal of Plants Throughout the World*. Ashford, UK: L. Reeve & Co. Ltd.

Rosenzweig, M.L. (2003) *Win Win Ecology: How the Earth's Species Can Survive in the Midst of Human Enterprise*. New York: Oxford University Press.

Sauer, C.O. (1941) 'The personality of Mexico', *Geographical Review* 31: 353–64.

Sauer, C.O. (1952) *Agricultural Origins and Dispersals*. New York: American Geographical Society.

Sauer, C.O. (1956) 'The agency of man on earth', In W. Thomas (ed.), *Man's Role in Changing the Face of the Earth*, pp. 49–69. Chicago: University of Chicago Press.

Sauer J.D. (1969) 'Oceanic islands and biogeographical theory: A review', *Geographical Review*, 59: 582–93.

Sauer, J.D. (1993) *Historical Geography of Crop Plants: A Select Roster*. Boca Raton, FL: CRC Press.

Schimper, A.F.W. (1898) *Pflanzengeographie auf physiologischer grundlage*. Jena: Gustav Fischer.

Schmithüsen, J. (1968) *Allgemeine Vegetationsgeographie*. 3rd ed. Berlin: De Gruyter.

Schmithüsen, J. (1976) *Allgemeine Geosynergetik. Grundlagen der Landschaftskunde*. Berlin: De Gruyter.

Schouw, J.F. (1822) *Grundtræk til en almindelig Plantegeografie*. Copenhagen: Gyldendalske Boghandels Forlag.

Schröter, C. (1904) *Das Pflanzenleben der Alpen. Eine Schilderung der Hochgebirgsflora*. Zürich: Verlag von Albert Raustein.

Sclater, P.L. (1858) 'On the general geographic distribution of the members of the class Aves', *Journal of the Linnaean Society of London, Zoology*, 2: 130–45.

Shmida, A., and S. Ellner. (1984) 'Coexistence of species with similar niches', *Vegetatio*, 58: 29–55.

Shugart, H.H. (1997) 'Plant and ecosystem functional types', in T.M. Smith, H.H. Shugart, and F.I. Woodward (eds.), *Plant Functional Types: Their Relevance to Ecosystem Properties and Global Change*, pp. 20–43. Cambridge: Cambridge University Press.

Simberloff, D. (1980) 'A succession of paradigms in ecology: Essentialism to materialism and probabilism', *Synthese*, 43: 3–39.

Slobodkin, L. B. 1996. Review of D. Quammen, *The song of the dodo: island biogeography in an age of extinctions*, and I. Thornton, *Krakatau: the destruction and reassembly of an island ecosystem. Nature* 381: 205–206.

Sprugel, D.G. (1991) 'Disturbance, equilibrium, and environmental variability—what is natural vegetation in a changing environment?', *Biological Conservation*, 58: 1–18.

Troll, C. (1939) 'Luftbildplan und ökologische Bodenforschung', *Zeitschrift der Gesellschaft für Erdkunde zu Berlin*, 7/8: 241–98.

Troll, C. (1966) 'Landschaftsökologie als geographisch-synoptische Naturbetrachtung', In C. Troll, Ökologische Landschaftsforschung und vergleichende Hochgebirgsforschung. *Erdkundliches Wissen*, 11: 1–13.

Vale, T.R. ed (2002) *Fire, Native Peoples, and the Natural Landscape*. Washington, DC: Island Press.

van Balgooy, M.M. (1971) 'Plant-geography of the Pacific as based on a census of phanerogram genera', *Blumea*, Suppl. 6: 1–222.

van der Maarel, E. (2005) 'Vegetation ecology—an overview', In E. van der Maarel (ed.), *Vegetation Ecology*, pp. 1–51. Oxford: Blackwell.

Veblen, T. (1992) 'Regeneration dynamics', in D.C., Glenn-Lewin, R. K. Peet, and T. T. Veblen (eds.), *Plant Succession: Theory and Prediction*, pp. 152–87. New York: Chapman and Hall.

Voeks, R.A. (1997) *Sacred Leaves of Candomble: African Magic, Medicine, and Religion in Brazil*. Austin TX: University of Texas Press.

Wagner, H., and E. von Sydow. (1889) *Sydow-Wagners Methodischer Schul Atlas*. Gotha: Justus Perthes.

Walker, D. (1970) 'Direction and rate in some British post-glacial hydroseres', In D. Walker and R. G. West (eds), *Studies in the Vegetational History of the British Isles, Essays in Honour of Harry Godwin*. Cambridge: Cambridge University Press.

Wallace, A.R. (1876) *The geographical distribution of animals with a study of the relations of living and extinct faunas as elucidating the past changes of the earth's surface*. 2 vols. New York: Hafne.

Wallace, A.R. (1881) *Island Life, or the phenomena and causes of insular faunas and floras including a revision and attempted solution of the problem of geological climates*. London: Harper and Brothers.

Walter, H. (1973) *Vegetation of the Earth in Relation to Climate and Ecophysiological Conditions*. London: Springer Verlag.

Walter, H., and S.W. Breckle (1982–91) *Ökologie der Erde*. 4 vols., Stuttgart-New York: Gustav Fischer.

Walter, H., and S.W. Breckle. (1985) *Ecological Systems of the Geobiosphere. Vol. 1. Ecological Principles in Global Perspective*. Berlin: Springer Verlag.

Warming, E. (1895) *Plantesamfund: Grundtrak af den Okologiska Plantegeografi*. Copenhagen: Philipsen.

Wegener, A. (1915) *Die Entstehung der Kontinente und Ozeane*. Braunschweig: Vieweg & Sohr.

Whittaker, R.H. (1962) 'Classification of natural communities', *Botanical Review*, 28: 1–239.

Whittaker, R.H., ed. (1973) *Ordination and Classification of Communities*. The Hague: Dr. W. Junk.

Whittaker, R.H. (1975) *Communities and Ecosystems*. 2nd ed. New York: Macmillan.

Whittaker, R.J. (1998) *Island Biogeography: Ecology, Evolution, and Conservation*. Oxford: Oxford University Press.

Whittaker, R.J., Bush, M.B., and Richards, K. (1989) 'Plant recolonization and vegetation succession on the Krakatau Islands, Indonesia', *Ecological Monographs*, 59: 59–123.

Whittaker, R.J., Triantis, K.A., and Ladle, R.J. (2008) 'A general dynamic theory of oceanic island biogeography', *Journal of Biogeography*, 35: 977–94.

Williams, M. (2006) *Deforesting the Earth*. Chicago: University of Chicago Press.

Williamson, M.H. (1981) *Island Populations*. Oxford: Oxford University Press.

Wilson, E.O. (1959) 'Adaptive shift and dispersal in a tropical ant fauna', *Evolution*, 13: 122–44.

Wood, Y.A., T. Meixner, P.J. Shouse, and E.B. Allen. (2006) 'Altered ecohydrologic response drives native shrub loss under conditions of elevated nitrogen deposition', *Journal of Environmental Quality*, 35: 76–92.

Woodward, F.I. (1987) *Climate and Plant Distribution*. Cambridge: Cambridge University Press.

Worster, D. (1977) *Nature's Economy: The Roots of Ecology*. San Francisco: Sierra Club.

Worster, D. (1990) 'The ecology of order and chaos', *Environmental History Review*, 14: 1–18.

Yasuda, Y., ed. (2002) *The Origins of Pottery and Agriculture*. New Delhi: Lustre Press.

Zimmerer, K.S. (1994) 'Human geography and the "new ecology": the prospect and promise of integration', *Annals of the Association of American Geographers*, 84: 108–25.

Zimmerer, K.S. (1996) *Changing Fortunes: Biodiversity and Peasant Livelihood in the Peruvian Andes*. Berkeley: University of California Press.

Zimmerer, K.S., and K.R. Young, eds. (1998) *Nature's Geography: New Lessons for Conservation in Developing Countries*. Madison WI: University of Wisconsin Press.

3

Diversity Theories

Duane A. Griffin

It is interesting to contemplate a tangled bank, clothed with many plants of many kinds, with birds singing on the bushes, with various insects flitting about, and with worms crawling through the damp earth. Charles Darwin (1859: 489)

biodiversity gradients (see Mutke, this volume), is the most pervasive and ubiquitous of biogeographical diversity patterns. I end with a brief discussion about recent attempts at theory reduction and unification and promising avenues for future research.

3.1 INTRODUCTION

Biogeography's roots lie in the insight that the diversity of organisms varies dramatically and nonrandomly among the world's tangled banks. Documenting that variability and developing robust understandings of the processes that produce and maintain it remains central to the field. Both goals have proven elusive, however. We do not yet have an order-of-magnitude estimate of global biodiversity, much less a unified and comprehensive theoretical understanding of how it works. Faced with ongoing, primarily anthropogenic, diversity losses, the task of filling these gaps in ways that will support conservation policy and planning is one of the most critical challenges facing biogeography and ecology today (Brooks and McLennan, 2002; Clark et al., 2001). Thus far, the effort has proven challenging in ways that are unique in the history of science (Palmer, 1994).

I begin this chapter by outlining some of the challenges that biodiversity presents, the history of diversity theory, scale issues, and theory development regarding community diversity. I then turn to the species-area relationship, which, along with

3.2 THE DIVERSITY PUZZLE

A thought experiment helps illustrate some of the difficulties that diversity poses. Imagine Darwin's tangled bank: a small patch of habitat in a larger landscape. How much diversity does it harbor? The question begs others. What exactly do we mean by diversity and how do we measure it? Why is there not more of it, or less? How does it compare to other tangled banks elsewhere, and how does it change through time. Why?

Most of the current literature on biodiversity is based on taxonomic counts, the simplest and most intuitive measure of diversity (Gaston and Spicer, 2004), but this is just one dimension of the phenomenon. The United Nations (1993: 147) Convention on Biological Diversity—arguably the most influential single policy document on the topic—defines it more broadly as "the variability among living organisms from all sources including, inter alia, terrestrial, marine and other aquatic ecosystems and the ecological complexes of which they are part; this includes diversity within species, between species and of ecosystems."

A full characterization of biodiversity by these criteria must account for all of the genetic, morphological, physiological, and functional variability occurring within the bank and link it to the nearby habitats, ecosystems, and regional ecological complexes. This is a tall order, and no natural habitat has ever been investigated at this level of depth and detail. Taxonomic counts, on the other hand, are readily gained, and species richness (the number of species in an area) in particular has become the standard accounting unit for diversity inventories and theory development.

There are many advantages to this approach. Species richness is relatively easy to measure in the field or to extract from museum collections, maps, and other sources, and species richness tends to correlate with variation in genetics, functional traits, and other facets of biodiversity (Gaston and Spicer, 2004). Moreover, because it represents a simple count, species richness can be applied at any spatial scale and, sampled appropriately, can be converted to density for comparisons between samples or sites.

Species counts, however, are far from ideal. What constitutes a species varies both between and within groups, so different taxonomies represent varying degrees of ecological, biogeographical, or evolutionary distinctiveness. Species richness also gives equal weight to all species, be they giant sequoias or dandelions. If intraspecific variability within a population, habitat, or region is more important than simple taxonomic richness, species counts will lead both theory and management astray. Furthermore, reliance on the species level may also obscure ecological, phylogenetic, functional, and ecological relationships that may be important to a comprehensive understanding of biodiversity (see Lomolino and Heaney, 2004; Roy et al., 2004). Weighted metrics based on species abundance distributions (Magurran, 2003) and direct measures of genetic, morphological, or functional diversity can overcome these limitations, but these require more intensive and costly sampling and measurement protocols. Molecular genetics approaches, in particular, offer more direct access to the fundamental roots of biodiversity and thus hold the promise of overcoming the limitations of simple taxonomy.

For now, however, the grip of Linnaean tradition remains strong. With its limitations in mind, and for simplicity's sake limiting the discussion to plants, let us return to the thought experiment using species richness as our measure of biodiversity: how many species does the hypothetical bank harbor?

Begin with a randomly selected square meter of tangled bank and count the species within it. The resulting value reflects something of the ecology of the locale: the plants occupying the plot have been able to germinate and grow within the range of environmental conditions they have experienced. But they represent a subset of species that *might* occupy it. The actual diversity and composition depend not only on environmental conditions and species characteristics, but also on the contingencies of past and current disturbance events, propagule dispersal, germination, and growth, and interactions with neighbors, herbivores, and parasites. More broadly, the composition also depends on the evolutionary and biogeographic history of the entire biotic province of which the small community is a part. A snapshot sample of a single square meter will thus impose a very large imprint of stochastic noise on the deterministic signals that might point to a satisfying explanation for the diversity of the plot.

A second one-meter-square plot will likely include some of the species that occur in the first plot, but others as well. So will the third, the fourth, and so on. Add enough plots and you will eventually obtain an inventory of the entire bank. Gather enough ecological data, and patterns of relationships between species numbers and environment will emerge from the noise together with hints of the processes responsible for the diversity you identify. Plotting the number of species encountered against the area sampled yields a new pattern: a curve that rises steeply at first, and then levels off as you begin to exhaust the local flora. This 'collection curve' is one manifestation of the species-area relationship (SPAR), the most robust pattern in biogeography. Slicing the data differently and plotting numbers of species against numbers of individuals representing each reveals another of the great patterns: the species-abundance distribution (SAD).

Repeating the experiment in other locales will yield different species numbers and identities relationships between diversity and environmental characteristics and different SPARs and SADs. All of these patterns demand explanation. How much is signal, and how much is noise? What generalizations can you make about them?

Based on the substantial literature on diversity patterns it is reasonable to hypothesize that on the tangled banks of the world (a) there will be more species of small organisms than large ones; (b) there will be more heterotrophs than autotrophs; (c) large, environmentally heterogeneous, warm, and humid environments with diverse regional species pools will generally harbor more species that accumulate more steadily with area than smaller, more uniform, cooler and drier environments with smaller regional species pools; and (d) there will be a relatively small number of common species, and many more rarer ones. Other more complex relationships exist between diversity and biomass, disturbance, nutrients, productivity, and predation, and many of these too will vary from place to place (see Gaston and Spicer, 2004; Huston, 1994; Lomolino et al., 2006; Rosenzweig, 1995).

What processes produce these patterns? At its most fundamental level, diversity represents the dynamic balance between species gained and lost. At broad scales and over long time periods, this means the balance between speciation or colonization and extinction or extirpation; for finer scales and shorter time frames, it is the balance between birth, death, and dispersal. Interactions and variability within and between species, between species and the abiotic environment and its geography, disturbance regimes, and regional species pools determine gain and loss rates, all of which are subject to geographic and historic contingencies (Rosenzweig, 1995).

There is enormous scope for complexity within this simple list, and efforts to wrangle that complexity into predictive models based on a few key processes have produced a surprisingly large research literature. In a cursory review, Palmer (1994) identified over 120 hypotheses for various aspects of taxonomic diversity, noting that even after taking synonyms and special cases into account, the list of potential hypotheses is substantial. More than a decade later, the list of hypotheses continues to grow, and the search for a unified theory of diversity goes on.

This is not to say we have learned nothing. Given enough information about regional biogeography and local ecology, empirical models may produce reasonably accurate predictions about the diversity a particular site may harbor. But if environmental changes or local idiosyncrasies create conditions beyond the model's parameters, such predictions will fail. A robust predictive account of biodiversity requires a firmer and more mechanistic basis, ideally rooted in first principles. This is the grail of biodiversity theory, the history of which is worth considering briefly.

3.3 THE DEVELOPMENT OF DIVERSITY THEORY

As described in Chapter 2, explorations from the fifteenth century vastly expanded Europeans' awareness of the biological riches that lay beyond their horizons, which in turn began the process of unraveling biblical literalism as an explanation for biodiversity. A sufficient alternative had to wait until 1858, when Wallace's and Darwin's papers on evolution by natural selection appeared (Browne, 1983).

From then until the 1960s, questions of diversity remained primarily questions of evolution and extinction at broad scales in geologic time and thus largely the domain of evolutionary biology, paleontology, and 'classical' biogeography rooted in systematics, including some of the first quantitative work on latitudinal gradients beginning in the 1950s. Two notable exceptions were interest in the species-area relationship early in the twentieth century (Arrhenius, 1921; Gleason, 1925; Williams, 1943) and Preston's (1948) investigations into species-abundance distributions, but otherwise ecologists paid little heed to diversity patterns until G. Evelyn Hutchinson's President's Address to the American Society of Naturalists in 1958 (Hutchinson, 1959).

Hutchinson had observed two species of water boatman in a cistern near the shrine to Sicily's patron saint—Santa Rosalia—near Palermo and wondered "why there should be two and not 20 or 200 species of the genus in the pond" (146)? His speculations on the question touched off a veritable explosion of theory building that by the 1970s had produced a substantial body of literature on the role of niches, competition and limiting similarity, community assembly, stability, saturation and invasibility, disturbance and succession, environmental correlates of diversity, and more (Ricklefs, 2004). Ecologists also expended a great deal of energy on issues related to measurement and statistical models for characterizing abundance distributions and species-area curves and summary measures for species richness and evenness (Magurran, 2003; Tjørve, 2003), though Huston (1994) notes that much of the latter effort contributed little to our understanding of diversity.

The ecological turn in diversity theory focused primarily on identifying deterministic controls on local diversity from which broader-scale diversities emerge. From this equilibrium perspective, historical and geographical contingencies represented secondary factors that warranted little attention in their own right (Ricklefs, 2004). An important exception was MacArthur and Wilson's seminal Equilibrium Theory of Island Biogeography (ETIB), first proposed in 1963 (see Chapter 2) and developed more fully in their 1967 monograph. While it had little use for historical contingency, spatial structure was a key component of the theory.

At its core, the ETIB seeks to explain diversity on islands as a dynamic equilibrium between immigration and extinction rates, with geographic distance being a primary control on the former and island size on the latter (see Lomolino et al., 2006; Whittaker, 1998). Mathematically rigorous and graphically lucid, the ETIB raised interest in dispersal and other spatial processes among ecologists and demonstrated the potential and power of quantitative and predictive theory to biogeographers (Whittaker, 1998).

The allure was strong but not universal. The ETIB rests on a number of unrealistic assumptions and methodological flaws that Sauer (1969) exposed in an early and prescient critique that attracted little attention at the time. A decade later, Gilbert (1980) conducted a careful analysis of the ETIB and studies testing it and found little to recommend it as a predictive tool. Those and other critiques did little to slow what became a juggernaut.

The ETIB was deeply flawed (Brown and Lomolino, 2000; Walter, 2004), but it was, as Whittaker (2001: 1442) describes, "wrong in interesting ways." Deployed prematurely in conservation, it led to decades-long debates over reserve design (chronicled in Quammen, 1996). However, it also spurred the development of what might be broadly referred to as the 'spatial biosciences', that is, landscape ecology, metapopulation and metacommunity ecology, and macroecology, and much of biogeography in its contemporary form (Lomolino et al., 2006). It continues to inspire research and produce insights into biogeographical processes (e.g., Losos and Ricklefs 2009).

While ecologists and ecological biogeographers were addressing questions of how diversity is maintained in contemporary biotas, historical biogeography was undergoing its own revolution in trying to understand how diversity is produced. Croizat's panbiogeography and Hennig's cladistic approach to phylogenetics brought analytical rigor to the field, and the development of plate tectonic theory provided a long-needed geophysical framework for understanding the biogeographic history of Earth's diversity. Together, these developments laid the groundwork for the development (and much bitter fighting around) of vicariance biogeography and its rival, cladistics (see Riddle, this volume).

The failure of local determinism to provide an adequate account of diversity patterns prompted ecologists to turn their attention to regional and historical processes (e.g., Ricklefs and Schluter, 1993) and spatial ecology. At the same time, the failure of narrowly dogmatic approaches to historical biogeography and advances in molecular genetics gave rise to a more integrative phylogenetic biogeography.

These developments, together with changing understandings and foci within ecology and ecological biogeography and technical and methodological advances (e.g., molecular genetics and geographic information systems), are now bringing the concerns, interests, and goals of historical and ecological biogeography into the same realm (Brooks and McLennan, 2002; Lomolino and Heaney, 2004; Lomolino et al., 2006). They are also helping to drive efforts to clarify basic concepts to integrate and unify hypotheses, as discussed below (e.g., Lomolino and Heaney, 2004; Lomolino et al., 2006; Ricklefs, 2004; Scheiner and Willig, 2005; Whittaker et al., 2001).

3.4 PATTERN AND SCALE

Nonrandom patterns are windows to understanding processes in nature, and both the detection and attribution of diversity patterns depend on spatial scale—the size of abstractions and representations of reality. Despite their importance, scale issues have not always received the careful treatment they deserve in biogeographical theory and analysis (Blackburn and Gaston, 2002; Crawley and Harral, 2001; Rahbek, 2005). Whittaker et al. (2001: 454) go so far as to argue that lack of detailed attention to scale has likely fatally weakened efforts toward a general understanding of diversity (cf. Wiens, 1989, but see Storch et al., 2007).

The simplest and probably most intuitive conception of spatial scale is the size of the area being analyzed, captured in R.H. Whittaker's (1977) inventory diversity designations. In this framework, *point diversity* refers to a small unit (e.g., a sampling quadrat) within a local ecological unit (e.g., a habitat or patch), the total species richness of which is *alpha diversity. Gamma diversity* is the sum of species in all of the local units within a landscape, which in turn sum to equal *epsilon diversity* over broad geographic areas. R.J. Whittaker et al. (2001) argue for replacing these designations with the more descriptive local (alpha), landscape (gamma), regional (epsilon), and interregional/continental, but the earlier scheme is well entrenched. Whittaker (1977) also recognized 'differentiation' diversity—the turnover in composition between inventory units at each scale: *pattern diversity* between points at the alpha level, *beta diversity* between alpha units within a landscape, and *delta diversity* between landscapes within a region.

Most critical to diversity pattern detection are the components of spatial scale, generally identified as sample unit, grain, extent, and focus (Scheiner et al., 2000; Turner et al., 2001). The spatial and temporal dimensions used to collect data in a study are its sample unit (e.g., quadrats, or trapping time). Grain and extent refer to resolution: extent is the area and time over which observations are made, and grain is the size and duration of the individual observation units. If sampling is spatially continuous (e.g., in remote sensing), the size of the spatial grain and sample unit are equal; if individual samples are combined, the grain will necessarily be larger than the sample unit. Together, grain and extent determine the upper and lower limits of detectable patterns. Scheiner et al. (2000) propose focus to refer to the scale at which grains may be aggregated into a spatial hierarchy and demonstrate how changes in grain, focus, and extent can affect pattern. Moving-window techniques and similar approaches can be used to change the focus of a data set for hierarchical analysis (e.g., Palmer, 2006).

3.5 COMMUNITY DIVERSITY THEORY

Diversity theory has achieved its most prolific flowering in the effort to explain patterns within individual ecological communities. At the most fundamental level, this "local process paradigm" (Ricklefs, 2004) seeks to answer the question of how species manage to coexist.

According to classical niche theory, species with similar resource requirements living in the same place should not be able to coexist. The intersection of exponential population growth and the limitations of finite resources forces individuals to compete for energy, nutrients, and space, and those that are best able to capture available resources survive differentially. Within species, this competition drives natural selection. Between species, its logical endpoint is the exclusion of all but one dominant species. This is the competitive exclusion principle (CEP; Hardin, 1960), and it places strict theoretical limits on the number of species that can live together in local communities. Experiments bear out the CEP's basic validity, yet competitive exclusion is exceedingly rare in real ecological communities where species do manage to coexist, sometimes in fantastic abundance. How?

3.5.1 Community diversity hypotheses

Palmer (1994) recognizes that exclusion is rare in nature because the CEP requires a set of strict conditions that must be met for it to occur. Drawing a parallel between the CEP and the Hardy-Weinberg principle of genetic equilibrium, he identifies seven conditions necessary for exclusion. These are: (1) sufficient time to allow exclusion; (2) a temporally constant and (3) spatially uniform environment; (4) only one resource limits growth; (5) rare species have no advantage over common ones; (6) species have the opportunity to compete; and (7) there is no immigration.

The corollary is that coexistence increases as these conditions are violated. Palmer organizes hypotheses on the basis of these violations. Only 24 of the 120 hypotheses in his study do not fit in his classification scheme because they either violate the premise of the CEP or do not clearly relate to it. Of the remaining 96, over half involve spatial and temporal variability, while 30% violate the single limiting resource condition. Twenty hypotheses violate more than one condition.

Spatial differentiation of available resources or limiting factors is one of the most biogeographically interesting CEP violations. Topographic or edaphic heterogeneity, for example, expands niche space and shifts competitive hierarchies, which fosters coexistence. Temporal variability in the form of disturbance events, climate fluctuations, and seasonal change induce similar competitive shifts and disrupt trajectories toward exclusion.

Few of the hypotheses that Palmer (1994) catalogues or others have introduced since have been falsified. Palmer notes that this abundance of theory is unique in the history of science and betokens a lack of progress from a Popperian standpoint. Palmer's classification scheme reduces most of the tangle to a small handful of basic controls on local diversity, which likely interact in complex ways and with varying importance depending on spatial scale and other factors (see Snyder and Chesson, 2004).

3.5.2 Testing the local process paradigm

Considering the local process paradigm more broadly, Ricklefs (2004) notes that it makes three predictions about community diversity: (1) there should be strong correlations between diversity and environment; (2) there should be an upper limit on diversity ('saturation') that is independent of regional diversity; and (3) diversity levels should converge for similar habitats in different biogeographic regions.

Empirical support for each of these predictions is mixed. Correlations between species richness and environmental factors are legion (e.g., Huston, 1994), but relationships vary in space and time, linking these correlations to specific ecological mechanisms has proven complex, and distinguishing local environmental controls from regional and evolutionary ones may be impossible, since the two are deeply interrelated (Herzog and Kessler, 2006; Partel, 2002; Ricklefs, 2004). Evidence of convergence and saturation is similarly ambiguous. Some systems show convergence (Kessler et al., 2001), but others bear the strong imprint of regional history (e.g., Montoya et al., 2007). Cornell and Lawton (1992) propose that local diversity should be independent of regional diversity if communities are saturated, and proportional to it if not. Empirical studies have yielded mixed results (e.g., Lawes et al., 2000), though Loreau (2000) cautions that the relationship between local and regional diversity is scale-dependent. Positive correlations between introduced and native species richness suggest that communities are unsaturated (e.g., Palmer, 2006; Sax et al., 2002), though the issue of local-regional interactions is still far from being resolved (Russell et al., 2004).

As noted above, more careful attention to scale issues may help clarify some ambiguities within the local process paradigm. Ricklefs (2004) offers other suggestions as well. However, complexity may place severe limits on our ability to explain local diversity with precision. I return to this theme at the end of the chapter.

3.6 THE SPECIES-AREA RELATIONSHIP (SPAR)

The seemingly trivial insight that more species occur in large areas than small ones belies an enormous amount of interesting complexity, and the mechanisms producing the pattern of increase may include the most important controls on diversity (Huston, 1994). SPARs have been documented across a wide range of environments, types of organisms, and spatial scales. They have played important roles in the development of diversity theory and hold obvious and critical implications for biodiversity inventory and conservation. Not surprisingly, therefore, they attracted a great deal of attention from (and generated no small amount of controversy among) ecologists and biogeographers (see Lomolino, 2001; Lomolino and Weiser, 2001; Martin and Goldenfeld, 2006; Rosenzweig, 1995, 1999; Scheiner, 2003; Tjørve, 2003; Whittaker, 2004).

Arrhenius (1921) and Gleason (1925) developed mathematical SPAR models in the 1920s, and while other models have been fitted to the relationship, the Gleason and Arrhenius models have seen widest use (see Lomolino et al., 2006; Tjørve, 2003, and references therein). The Gleason, or semi-log, model has been widely used in plant ecology and takes the form $S = d + k \log A$. The Arrhenius, or power function, model scales species, S, to area, A, as $S = cA^z$. In both models, the constants are estimated empirically and vary with spatial scale, type of organism, and biogeographic setting. Preston (1960, 1962) argued from a theoretical basis that the Arrhenius model with $z = 0.27$ should best fit SPARs for isolated

and independent populations. The model gained wide acceptance among biogeographers, in no small part due to its importance in the ETIB.

Because they summarize the scaling relationship between species and area, SPAR parameters are rich targets for investigation. Systematic variations in the power model coefficients and exponents provide clues to fundamental processes underlying diversity patterns. Rosenzweig reported that z values for island archipelagos are typically higher than for mainland areas, and those for areas within biotic provinces are lower than interprovincial SPARs. However, in a massive meta-analysis of nearly 800 SPARs, Drakare et al. (2006) found no significant differences among island and mainland SPARs. They found that z values are negatively correlated with latitude and positively correlated with animal body size, are similar for terrestrial and marine (but not freshwater) settings and between biogeographic realms, and that they vary considerably for different habitats.

Variation in the coefficient c has attracted far less attention, though Lyons and Willig (1999) hint that it may vary systematically with latitude (cf. Connor and McCoy, 1979). Gould commented on the lack of attention to c in 1979, but with few exceptions (e.g., Wright, 1983), it has largely been ignored (Lomolino, 2001). The oversight likely has two bases. The first is the use of regression analysis for estimating model parameters from log-transformed species counts and areas so that the power function $S = cA^z$ becomes log $S = \log C + z \log A$. While c represents the intercept and z the slope of the regression line in log-log space, they do not represent the slope and intercept of the SPAR itself (Lomolino et al., 2006). Rather, c controls the initial slope and z its rate of increase with area (Lomolino, 2001). Second, while z is not dependent on sample unit size, c is, which makes comparisons troublesome (Rosenzweig, 1995; Drakare et al., 2006).

Confusion regarding SPARs may arise from the often overlooked fact that the method used to construct the SPAR affects its shape (Drakare et al., 2006). Scheiner (2003) identifies six types of curves, distinguished by whether sample areas are unequal (as with islands) or equal, and in the latter case, whether species counts represent point samples in nested quadrats or mean densities, and whether the density calculations are spatially explicit or not (cf. Gray et al., 2004 and references therein). At the very least, there is a fundamental difference between nested sampling designs, in which smaller areas are contained within larger ones (e.g., quadrats), and independent samples that do not overlap (e.g., islands or ponds). The difference typically parallels Preston's (1962) distinction between samples and isolates. Isolates, in Preston's usage, are habitat units separated from others by an intervening matrix that reduces interaction between units, while samples are survey areas within a continuous large habitat unit that are free to interact with external areas. (The parallel is not perfect, however: species counts based on political units—counties or provinces or nonnested grid cells—represent biological samples, but statistical isolates.)

The SPAR slope in log-log space changes with spatial scale (Figure 3.1; Williams, 1943), reflecting the scale of underlying spatial processes responsible for differentiation diversity (Lomolino et al., 2006; Crawley and Harral, 2001; Drakare et al., 2006). At the finest scale range, the curve is likely to be convex for samples or flat for isolates (MacArthur and Wilson dubbed this the 'small island effect'; see Lomolino, 2001). The trajectory flattens as the species pool is exhausted (Crawley and Harral, 2001; Lomolino, 2001). At the broadest scales it may be sigmoidal upward if the study area crosses the boundaries of biogeographic provinces, reflecting the effects of dispersal limitation and differential evolutionary histories (Rosenzweig, 1995). The slope is only likely to be linear at intermediate scales, where it will reflect the texture and magnitude of environmental variability (Palmer, 2007), the values of which will depend on the organisms and the biogeographical system being examined.

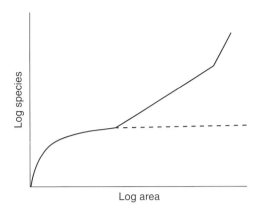

Figure 3.1   Species-area curve in log-log space based on Williams's (1964) curve for flowering plants. The domain ranges from 10 cm to the entire globe. Note that the relationship is linear only at intermediate scale lengths

differentiation diversity, as occurred after the breakup of Pangaea beginning in the Middle/Late Jurassic (Lomolino et al., 2006). Dispersal has the opposite effect, as occurred when the Isthmus of Panama arose during the Pliocene, the effects of which Flessa (1975) quantified using the SPAR.

3.7 SPECIES-AREA RELATIONSHIP HYPOTHESES

Numerous hypotheses have been proposed to account for the SPAR (see Lomolino et al., 2006; Rosenzweig, 1995; Whittaker, 1998, 2004), but six in particular stand out for their importance or promise: (1) the interprovincial hypothesis; (2) the sampling effect; (3) the environmental heterogeneity hypothesis; (4) the Equilibrium Theory of Island Biogeography; (5) the energy hypothesis; and (6) the Unified Neutral Theory.

3.7.1 The interprovincial hypothesis

Rosenzweig (1995) developed this historical hypothesis, which accounts for the steepening SPAR as area increases beyond the biogeographic province boundaries. Differentiation diversity (species turnover) between areal units in a SPAR has a strong influence on its slope at all scales, and differentiation is likely to be highest between regions with unique histories of speciation, extinction, and immigration (see, for example, Latham and Ricklefs, 1993). Regional isolation increases

3.7.2 The sampling effect, or random placement, hypothesis

Preston (1962) developed this hypothesis, which attributes the increase in species with area to the spatial and statistical distribution of individuals of each species in a community, landscape, or region. In any biota, a few species will be represented by very many individuals, and many species will be moderately common or rare. SADs generally fit a normal curve when binned by logarithmic abundance, a pattern that Preston (1962) dubbed the canonical log-normal distribution. This being the case, the SPAR emerges as a sampling artifact of species abundances, since large areas will include more individuals and sample a larger portion of the species-abundance distribution than will small areas (see Plotkin et al., 2000; Rosenzweig, 1995). Martin and Goldenfeld (2006) found that a log-normal distribution with enhanced rarity, together with conspecific clustering (Plotkin et al., 2000) produces realistic power-model SPARs. This is not to say, however, that SPARs are simply stochastic artifacts, because SADs depend on fundamental population processes that vary geographically. That being the case, the sampling effect hypothesis is incomplete in that it begs the question of what controls SADs, which has yet to be satisfactorily resolved (He, 2005).

3.7.3 The environmental variability/ habitat diversity hypothesis

That larger areas encompass a wider range of environmental conditions than smaller ones follows from Tobler's First Law of Geography (1970, 236), that is, that "everything is related to everything else, but near things are more related than distant things." Biogeographically, this 'law' is expressed as distance decay in biotic and abiotic environmental characteristics. Larger areas encompass greater geographical distances than smaller ones and are thus likely to have greater ranges of environmental variability, and thus longer resource gradients. The greater amount of niche space in large areas should support a wider range of species drawn from the regional species pool. Comparisons of homogeneous habitats of varying areas and of equal-area samples with varying degrees of habitat heterogeneity support the role of environmental variability in producing SPARs, especially in mainland settings (Rosenzweig, 1995).

3.7.4 MacArthur and Wilson's equilibrium theory of Island Biogeography (1967)

The ETIB is a neutral theory that ignores ecological details and posits steady-state species richness resulting from a dynamic equilibrium between immigration and extinction. Immigration is a function of island isolation (geographical distance) from propagule sources. Larger islands, which present larger targets for immigrants, have higher immigration rates than smaller ones, as do those near source regions. The island area controls extinction rates through its effect on population sizes: small islands can hold fewer individuals than large ones, and extinction probability is inversely related to population size. In principle, these dynamics should apply to both true islands and islandlike habitat patches on continents.

Its basic premise—that species richness represents a balance between gains and losses, and that geography plays an important role in both—is logically irrefutable. However, details matter, and as noted above, the ETIB has not held up well to empirical testing (Lomolino et al., 2006; though cf. Rosenzweig, 1995). Evidence for dynamic

equilibrium species richness is ambiguous (Sax, 2002; Walter, 1998; Whittaker, 2004), islands seem not to be saturated with species (Sax and Gaines, 2003), and islands and continental biogeographies are fundamentally different in important ways (Walter, 2004). Other critiques and weaknesses of the model include problems associated with meaningful measures of isolation and area and its failure to account for interspecific differences and ecological interactions between species, speciation, and secular environmental change at various time scales (Lomolino et al., 2006).

3.7.5 Species-energy theory

Recognizing the fundamental importance of energy for all biological activity, Wright (1983) substituted available energy for area in the MacArthur and Wilson model and proposed that $S = kE^z$, where k is a fitted constant and E is total available energy. The logic of this substitution is simple: species richness is proportional to the number of individuals living on an island, and the number of individuals is proportional to available resources. Wright tested his hypotheses for birds and angiosperms on 28 islands worldwide and found no significant relationship between species and area, though actual evapotranspiration and total productivity (proxies for available energy) explained 70% of the variation in log species richness for plants and 80% for birds. Wright's energy theory has been extended but primarily in the context of geographical diversity gradients, as discussed in Mutke, this volume.

3.7.6 Hubbell's Unified Neutral Theory (2001)

The Unified Neutral Theory (UNT) is the single most provocative, controversial, and arguably the most interesting proposal since 1967. Hubbell extends MacArthur and Wilson's ETIB by using concepts adapted from the neutral theory of evolutionary genetics (see, for example, Fox, 2006). The theory is based on five parameters and three simplifying assumptions. The parameters represent the number of trophically similar individuals in a local community (J_L) and the larger metacommunity (J_M); the individual reproduction (β) and speciation (υ) rates (i.e., the probability that an individual will reproduce or undergo speciation per unit time), which govern metacommunity dynamics; and the immigration rate (m), or the probability that an individual from the metacommunity will replace a dead individual in a local community. The assumptions are (a) that speciation is by point mutation; (b) the 'zero sum assumption' that community resources (e.g., the number of microsites) are saturated, and thus the number of individuals in the community is constant; and, most important, (c) ecological neutrality, the assumption that all individuals and species are ecologically equivalent.

Under the neutrality assumption, all species are equivalent in terms of competition, commensalism, birth, death, speciation, and extinction. Birth and death are treated as balanced stochastic events: individuals die at random and are replaced by individuals from the community, which contains all individuals within dispersal distance of the vacant microsite. More rarely (probability = m), individuals from elsewhere in the metacommunity colonize open sites, thereby linking communities into metacommunities by dispersal. The population of the local community, J_L, together with m, are the bases for the 'fundamental dispersal number', $I = 2J_Lm$, which is a more meaningful and less problematic measure of isolation than simple geographic distance, which MacArthur and Wilson (1967) use.

The core of the UNT is the 'fundamental biodiversity number', Θ, given by $\Theta = 2J_m\upsilon$, which governs species richness. Thus, Θ is an index of metapopulation (i.e., regional) diversity that is richer than S because it can be linked to birth, death, dispersal, speciation, and extinction more explicitly and in more readily meaningful and interpretable ways than the Arrhenius model's c and z parameters. Substituting population density, ρ, times area, A, for metapopulation size ($J_m = A\rho$) relates Θ directly to area: $\Theta = 2A\rho\upsilon$ (Turner and Hawkins, 2004).

The value of Θ, however, does not lie in its utility as a diversity index, but in the fact that, assuming it can be parameterized, species richness and relative species abundances can be predicted on the basis of it using neutral community models. This is accomplished by constructing a 'master equation' for species abundances and solving it analytically, numerically, or by simulation to find the stationary probability distribution at equilibrium.

Because population density and area are constant in the UNT, individuals are locked in a zero-sum game in which species populations change in a random walk 'community', or 'ecological drift', analogous to (and modeled by much of the same mathematics as) genetic drift of neutral alleles in population genetics. All random walks eventually lead to extinction, which is balanced by rare and random speciation events that result in new species.

The UNT has been controversial within community ecology because the idea that community structure arises from dispersal is directly counter to the niche assembly approach to understanding diversity that ecologists have used for the past 50 years (though see Gewin, 2006). But it was the failure of that approach that the UNT attempts to circumvent. As Hubbell (2005: 166) describes it,

> The traditional strategy has been to assume that ecological communities are inherently high-dimensional . . . and then to build rather complex models from the outset, incorporating as many of the details of the growth and interactions of each and every species and with their physical environment as possible. Neutral theory, however, adopts a fundamentally different strategy, taking virtually the opposite tack. It begins with the simplest possible hypothesis one can think of—for example, the functional equivalence of species—and then adds complexity back into the theory only as absolutely required to obtain satisfactory agreement with the data.

Remarkably, the UNT accurately describes many patterns fairly well, including species abundance distributions, geographic range and abundance distributions and relationships, and species-area relationships, particularly for tropical forest communities (Bell, 2001; Hubbell, 2001, 2005). How well depends on how closely we look, and McGill et al. (2006) argue that if we look at all closely, neutral models fail most empirical tests to date.

To dismiss the UNT because its assumptions are too simple or because it does not fit empirical data misses the point of the neutral approach. Its primary purpose is heuristic, and its value lies in its utility for constructing null hypotheses against which, for example, niche-based theory can be tested and for identifying directions from which more rigorous, complete, and predictive theory that merges neutral and niche-based models might emerge (Gewin, 2006; Holyoak and Loreau, 2006; McGill et al., 2006; Purves and Pacala, 2005). In its current form, the UNT is limited to diversity within single trophic levels, but further development, refinement, and extensions might be possible (Hubbell, 2005).

3.7.7 Other SPAR theories and future directions

Other hypotheses that have been deployed to explain the SPAR relate to disturbance, species' body and territory sizes, and the presence of internal barriers that would facilitate allopatric speciation (see Lomolino, 2001 and references therein). It would be surprising if any single hypothesis or theory could account for the full range of variation in species diversity within an area, especially over a full range of spatial scales. The convex curvature of log-log plots at the finest scales likely reflects processes that determine SADs. The small island effect is likely due, at least in part, to recurrent disturbance events, while convexity at the upper end of the curve reflects differential evolutionary and biogeographic histories between regions or provinces. The signal of gain-loss processes (i.e., speciation, immigration, and extinction) dominate at intermediate scales where the slope of the SPAR is constant (Drakare et al., 2006; Lomolino, 2001; Turner and Tjørve, 2005; Palmer, 2006).

Within the intermediate scale range, SPARs are likely to reflect combinations of factors. Kalmar and Currie (2006) extended Wright's work to test several island biogeographical hypotheses using a global data set of breeding land birds on 346 islands. They were able to explain 85–90% of global variance in species richness based on area, isolation, and annual average temperature. The slopes of their curves are controlled by temperature and annual precipitation, with no contribution from isolation. Habitat diversity contributed positively to species richness in warm regions but negatively in cold ones. Their results are, therefore, consistent with parts of both MacArthur and Wilson's and Wright's hypotheses and suggest that there is spatial direction to the habitat diversity hypothesis.

SPARs continue to yield important insights, promising applications, and interesting linkages with other biological phenomena (see May and Stumpf, 2006; Southwood et al., 2006). Finally, it is worth bearing in mind Palmer's (2006: 427) insight "that the species-area relationship is not monolithic, and possess[es] a richness that cannot be summarized by a few parameters." Further exploration into the scaling relationship between area and diversity is needed to elucidate that richness.

3.8 UNTANGLING DARWIN'S BANK?

In 2008, 4,943 papers on the topic of species diversity were added to ISI's Web of Knowledge (environmental sciences and ecology, biodiversity and conservation, zoology, plant sciences, and geography subject areas; as of April 30, 2009), which represents a rate of one paper every 1.8 hours. I have presented, at best, a partial skeletal

outline of the body of theory guiding this phenomenal output and attempting to make sense of this simplest, and in some ways most problematic, measure of biodiversity.

So where do we stand with regards to Darwin's tangled bank in terms of the prospects for a unified theory that untangles the superabundance of complexity it offers? Will it be possible to understand it in a way that would meet the needs of the UN Convention on Biodiversity and allow us to make, if not precise predictions, at least reasonable forecasts about future biodiversity that might guide conservation policy (Clark et al., 2001)? What might such a theory look like?

Insights and experience from the past 50 years of diversity research can guide us. First and foremost, it is clear that a unified theory will have to be hierarchical, both spatially and temporally, as Whittaker et al. (2001), O'Brien (2006), and others have proposed. Hierarchy theory itself (Allen and Starr, 1982) has fallen off of most maps, but it may yet afford an epistemological guide for navigating some of the conceptual morasses that have beset diversity theory in the past. Secondly, it will ideally be composed of simple, elegant, and biologically meaningful models, mechanistically rooted in first principles in the way that Allen et al. (2002) envision, and perhaps linked together in the same way that seven governing equations are interlinked to create predictive climate models. And finally, such a model will have sufficient predictive power to enable quantitative forecasts for future states of the biosphere.

This is a formidable goal, but there is no reason to believe that it is fundamentally impossible. How might we reach it? Increasing attention to the details of scale and space is a large step in the right direction, as is the growing awareness of philosophical issues relating to the production of scientific knowledge and what seems to be a growing willingness to address implicit assumptions about our methods and approaches (e.g., Gotelli and McGill, 2006). Increasing cross-fertilization between ecological and historical biogeography (Wiens and Donoghue, 2004) and between biogeography and kindred spirits in related disciplines is exciting (Lomolino and Heaney, 2004), as is the establishment and progress of the International Biogeography Society (www. biogeography.org).

There are, however, a number of impediments to creating a general understanding of diversity across scales of space and time. One of these is simply the availability of empirical data. We need more data on species, their abundances, and their distributions, and we need more compilations of the vast store of data that have been collected in the past four centuries. To the Linnaean and Wallacean shortfalls—gaps in our inventory of

Earth's biodiversity and its distribution (Whittaker et al. 2005)—we might add a 'Humboldtian shortfall' to describe the particular lack of taxonomic and environmental data from underexplored areas far from well-funded research universities and research stations, and taxonomically far from the low-hanging fruit of vascular plants, vertebrates, and their parasites. These taxa are important, but so are nematodes, tardigrades, and bacteria, which may be governed by very different processes than their larger counterparts. Metagenomic approaches (Rusch et al., 2007) may hold great promise in this regard.

It would also be useful to re-evaluate many of the dichotomies we take as givens, such as local vs. regional, ecological vs. evolutionary, and deterministic vs. historical. Regions are composed of local places and interactions between them are continuous, not discrete, and it is now clear that evolutionary processes operate on contemporary time scales (Stockwell et al., 2003). That history matters is widely recognized (e.g., Buckley and Jetz, 2007; Price, 2004), but what is more interesting are the questions of when and where it matters, how important it is relative to deterministic factors, and what sets the balance. Likewise, we should continue to critically assess received wisdoms about the meanings of such fundamental patterns and processes.

Indeed, Nekola and Brown (2007) have recently expanded on Preston's (1950) insight that SADs, SARS, and other diversity patterns have analogs in other complex dynamical systems. They suggest that these patterns may be the products of processes more general than those specific to biology, and that models capable of precise quantitative prediction may be impossible, though recent work by Dewar and Porté (2008) and Pueyo et al. (2007) show how concepts from statistical physics might provide a way toward probabilistic prediction.

In his capstone address at the inaugural meeting of the International Biogeography Society in 2003, James Brown (2004, 365) said that "it is hard to imagine a more exciting time to be a biogeographer," which perfectly captured the sense of that first meeting and the point we are at with regards to diversity theories. The challenges of finally untangling Darwin's bank are enormous, but the need is great and the grail seems a lot closer than it was in 1959. At the very least, we are gaining a better sense of where it is not.

REFERENCES

Allen, A.P., J.H. Brown, and J.F. Gillooly. (2002) 'Global biodiversity, biochemical kinetics, and the energetic-equivalence rule', *Science*, 297: 1545–48.

Allen, T.F.H., and T.B. Starr. (1982) *Hierarchy: Perspectives for Ecological Complexity.* Chicago: University of Chicago Press.

Arrhenius, O. (1921) 'Species and area', *Journal of Ecology*, 9: 95–99.

Bell, G. (2001) 'Neutral macroecology', *Science*, 293: 2413–18.

Blackburn, T.M., and K.J. Gaston. (2002) 'Scale in macroecology', *Global Ecology and Biogeography*, 11: 185–89.

Brooks, D.R., and D.A. McLennan. (2002) *The Nature of Diversity.* Chicago: University of Chicago Press.

Brown, J.H. (2004) Concluding remarks. In *Frontiers of Biogeography: New Directions in the Geography of Nature* ed. M.V. Lomolino and L.R. Heaney, 361–68. Sunderland, MA: Sinauer.

Brown, J.H., and M.V. Lomolino (2000) 'Concluding remarks: Historical prospect and the future of island biogeography theory', *Global Ecology and Biogeography*, 9: 87–92.

Browne, J. (1983) *The Secular Ark: Studies in the History of Biogeography.* New Haven CT: Yale University Press.

Buckley, L.B., and W. Jetz. (2007) 'Environmental and historical constraints on global patterns of amphibian richness', *Proceedings of the Royal Society B-Biological Sciences*, 274: 1167–73.

Clark J.S. and 15 authors (2001) 'Ecological forecasts: An emerging imperative', *Science*, 293: 657–60.

Connor, E.F., and E.D. McCoy. (1979) 'The statistics and biology of the species-area relationship', *American Naturalist*, 113: 791–833.

Cornell, H.V., and J.H. Lawton. (1992) 'Species interactions, local and regional processes, and limits to the richness of ecological communities: A theoretical perspective', *Journal of Animal Ecology*, 61: 1–12.

Crawley, M.J., and J.E. Harral. (2001) 'Scale dependence in plant biodiversity', *Science*, 291: 864–68.

Darwin, C. (1859) *On the Origin of Species.* London: John Murray. Facsimile reprint, Cambridge MA: Harvard University Press, 1964.

Dewar, R.C., and A. Porté. (2008) 'Statistical mechanics unifies different ecological patterns', *Journal of Theoretical Biology*, 251: 389–403.

Drakare, S., J.J. Lennon, and H. Hillebrand. (2006) 'The imprint of the geographical, evolutionary and ecological context on species-area relationships', *Ecology Letters*, 9: 215–27.

Flessa, K. W. (1975) 'Area, continental drift and mammalian diversity', *Paleobiology*, 1: 189–94.

Fox, C.W. (2006) *Evolutionary Genetics: Concepts and Case Studies.* Oxford: Oxford University Press.

Gaston, K.J., and J.I. Spicer. (2004) *Biodiversity: An Introduction.* Malden MA: Blackwell.

Gewin, V. (2006) 'Beyond neutrality—ecology finds its niche', *PLoS Biology*, 4: e278.

Gilbert, F.S. (1980) 'The equilibrium theory of island biogeography: Fact or fiction'? *Journal of Biogeography*, 7: 209–35.

Gleason, H.A. (1925) 'Species and area', *Ecology*, 6: 66–74.

Gotelli, N.J., and B.J. McGill. (2006) 'Null versus neutral models: What's the difference?' *Ecography*, 29: 793–800.

Gould, S.J. (1979) 'An allometric interpretation of species-area curves: The meaning of the coefficient', *American Naturalist*, 114: 335–43.

Gray, J.S., K.I. Ugland, and J. Lambshead. (2004) 'On species accumulation and species-area curves', *Global Ecology and Biogeography*, 13: 567–68.

Hardin, G. (1960) 'Competitive exclusion principle', *Science*, 131: 1292–97.

He, F. (2005) 'Deriving a neutral model of species abundance from fundamental mechanisms of population dynamics', *Functional Ecology*, 19: 187–93.

Herzog, S.K., and M. Kessler. (2006) 'Local vs. regional control on species richness: A new approach to test for competitive exclusion at the community level', *Global Ecology and Biogeography*, 15: 163–72.

Holyoak, M., and M. Loreau. (2006) 'Reconciling empirical ecology with neutral community models', *Ecology*, 87: 1370–77.

Hubbell, S.P. (2001) *The Unified Neutral Theory of Biodiversity and Biogeography*. Princeton NJ: Princeton University Press.

Hubbell, S.P. (2005) 'Neutral theory in community ecology and the hypothesis of functional equivalence', *Functional Ecology*, 19: 166–72.

Huston, M.A. (1994) *Biological Diversity: The Coexistence of Species on Changing Landscapes*. Cambridge: Cambridge University Press.

Hutchinson, G.E. (1959) 'Homage to Santa Rosalia, or why are there so many kinds of animals'? *American Naturalist*, 93: 145–59.

Kalmar, A., and D.J. Currie. (2006) 'A global model of island biogeography', *Global Ecology and Biogeography*, 15: 72–81.

Kessler, M., B.S. Parris, and E. Kessler. (2001) 'A comparison of the tropical montane pteridophyte floras of Mount Kinabalu, Borneo, and Parque Nacional Carrasco, Bolivia', *Journal of Biogeography*, 28: 611–22.

Latham, R.E., and R.E. Ricklefs. (1993) 'Global patterns of tree species richness in moist forests: Energy-diversity theory does not account for variation in species richness', *Oikos*, 67: 325–33.

Lawes, M.J., H.A.C. Eeley, and S.E. Piper. (2000) 'The relationship between local and regional diversity of indigenous forest fauna in KwaZulu-Natal Province, South Africa', *Biodiversity and Conservation*, 9: 683–705.

Lomolino, M.V. (2001) 'The species-area relationship: New challenges for an old pattern', *Progress in Physical Geography*, 25: 1–21.

Lomolino, M.V., and L.R. Heaney, eds. (2004) *Frontiers of Biogeography: New Directions in the Geography of Nature*. Sunderland, MA: Sinauer.

Lomolino, M.V., B.R. Riddle, and J.H. Brown. (2006) *Biogeography*. Sunderland, MA: Sinauer.

Lomolino, M.V., and M.D. Weiser. (2001) 'Towards a more general species-area relationship: Diversity on all islands, great and small', *Journal of Biogeography*, 28: 431–45.

Loreau, M. (2000) 'Are communities saturated? On the relationship between alpha, beta and gamma diversity', *Ecology Letters*, 3: 73–76.

Losos, J.B. and R.E. Ricklefs, eds. (2009) *The Theory of Island Biogeography Revisited*. Princeton NJ, Princeton University Press.

Lyons, K.A., and M.R. Willig. (1999) 'A hemispheric assessment of scale-dependence in latitudinal gradients of species richness', *Ecology*, 80: 2483–91.

MacArthur, R.M., and E.O. Wilson. (1967) *The Theory of Island Biogeography*. Princeton NJ: University of Princeton Press.

Magurran, A.E. (2003) *Measuring Biological Diversity*. Oxford: Blackwell.

Martin, H.G., and Goldenfeld, N. (2006) On the origin and robustness of power-law species-area relationships in ecology. *Proceedings of the National Academy of the USA*, 103: 10310–15.

May, R.M., and M.P.H. Stumpf. (2000) 'Species-area relations in tropical forests', *Science*, 290: 2084–86.

McGill, B.J., B.A. Maurer, and M.D. Weiser. (2006) 'Empirical evaluation of neutral theory', *Ecology*, 87: 1411–23.

Montoya, D., M.A. Rodriguez, M.A. Zavala, and B.A. Hawkins. (2007) 'Contemporary richness of Holarctic trees and the historical pattern of glacial retreat', *Ecography*, 30: 173–82.

Nekola, J.C., and J.H. Brown. (2007) 'The wealth of species: Ecological communities, complex systems and the legacy of Frank Preston', *Ecology Letters*, 10: 188–96.

O'Brien, E.M. (2006) 'Biological relativity to water-energy dynamics', *Journal of Biogeography*, 33: 1868–88.

Palmer, M.W. (1994) 'Variation in species richness—Towards a unification of hypotheses', *Folia Geobotanica and Phytotaxonomica*, 29: 511–30.

Palmer, M.W. (2006) 'Scale dependence of native and exotic species richness in North American Floras', *Preslia*, 78: 427–36.

Palmer, M.W. (2007) Species-area curves and the geometry of nature. In *Scaling Biodiversity*, ed. D. Storch, P.A. Marquest, and J.H. Brown. Cambridge: Cambridge University Press.

Partel, M. (2002) 'Local plant diversity patterns and evolutionary history at the regional scale', *Ecology*, 83: 2361–66.

Plotkin J.B., M. Potts, N. Leslie, N. Manokaran, J. LaFrankie, and P. Ashton. (2000) 'Species-area curves, spatial aggregation, and habitat specialization in tropical forests', *Journal of Theoretical Biology*, 207: 81–99.

Preston, F.W. (1948) 'The commonness, and rarity, of species', *Ecology*, 29: 254–83.

Preston, F.W. (1950) 'Gas laws and wealth laws', *The Scientific Monthly*, 71: 309–11.

Preston, F.W. (1960) 'Time and space and the variation of species', *Ecology*, 41: 611–27.

Preston, F.W. (1962) 'The canonical distribution of commonness and rarity', *Ecology*, 43: 185–215, 431–432.

Price, J.P. (2004) 'Floristic biogeography of the Hawaiian Islands: Influences of area, environment and paleogeography', *Journal of Biogeography*, 31: 487–500.

Pueyo, P., F. He, and T. Zillio. (2007) 'The maximum entropy formalism and the idiosyncratic theory of biodiversity', *Ecology Letters*, 10: 1017–28.

Purves, D.W., and S.W. Pacala. (2005) Ecological drift in niche-structured communities: Neutral pattern does not imply neutral process. In *Biotic Interactions in the Tropics*, ed. D. Burslem, M. Pinard and S. Hartley, 107–38. Cambridge: Cambridge University Press.

Quammen, D. (1996) *The Song of the Dodo: Island Biogeography in an Age of Extinctions*. New York: Scribner.

Rahbek, C. (2005) 'The role of spatial scale and the perception of large-scale species-richness patterns', *Ecology Letters*, 8: 224–39.

Ricklefs, R.E. (2004) 'A comprehensive framework for global patterns in biodiversity', *Ecology Letters*, 7: 1–15.

Ricklefs, R.E., and D. Schluter, eds. (1993) *Species Diversity in Ecological Communities: Historical and Geographical Perspectives*. Chicago: University of Chicago Press.

Rosenzweig, M.L. (1995) *Species Diversity in Space and Time*. Cambridge: Cambridge University Press.

Rosenzweig, M.L. (1999) 'Heeding the warning in biodiversity's basic law', *Science*, 284: 276–77.

Roy, K., D. Jablonski, and J.W. Valentine. (2004) Beyond species richness: Biogeographic patterns and biodiversity dynamics using other metrics of diversity. In *Frontiers of Biogeography: New Directions in the Geography of Nature*, ed. M. V. Lomolino and L. R. Heaney, 151–70. Sunderland, MA: Sinauer.

Rusch D.B and 38 authors (2007) 'The Sorcerer II Global Ocean Sampling expedition: Northwest Atlantic through Eastern Tropical Pacific', *PLoS Biology*, 5: 398–431.

Russell, J.C., M.N. Clout, and B.H. McArdle. (2004) 'Island biogeography and the species richness of introduced mammals on New Zealand offshore islands', *Journal of Biogeography*, 31: 653–64.

Sauer, J.D. (1969) 'Oceanic islands and biogeographical theory: A review', *Geographical Review*, 59: 582–93.

Sax, D.F. (2002) 'Equal diversity in disparate species assemblages: A comparison of native and exotic woodlands in California', *Global Ecology and Biogeography*, 11: 49–57.

Sax, D.F., and S. D. Gaines. (2003) 'Species diversity: From global decreases to local increases', *Trends in Ecology and Evolution*, 18:561–66.

Sax, D.F., S.D. Gaines, and J.H. Brown. (2002) 'Species invasions exceed extinctions on islands worldwide: A comparative study of plants and birds', *American Naturalist*, 160: 766–83.

Scheiner, S.M. (2003) 'Six types of species-area curves', *Global Ecology and Biogeography*, 12: 441–47.

Scheiner, S.M., S.B. Cox, M. Willig, G.G. Mittelbach, C. Osenberg, and M. Kaspari. (2000) 'Species richness, species-area curves and Simpson's paradox', *Evolutionary Ecology Research*, 2: 791–802.

Scheiner, S.M., and M.R. Willig. (2005) 'Developing unified theories in ecology as exemplified with diversity gradients', *American Naturalist*, 166: 458–69.

Snyder, R.E., and P. Chesson. (2004) 'How the spatial scales of dispersal, competition, and environmental heterogeneity interact to affect coexistence', *American Naturalist*, 164: 633–50.

Southwood, T.R.E., R.M. May, and G. Sugihara. (2006) 'Observations on related ecological exponents', *Proceedings of the National Academy of Sciences of the United States of America*, 103: 6931–33.

Stockwell, C.A., A.P. Hendry, and M.T. Kinnison. (2003) 'Contemporary evolution meets conservation biology', *Trends in Ecology and Evolution*, 18: 94–101.

Storch D., P.A. Marquest, and J.H. Brown, eds. (2007) *Scaling Biodiversity*. Cambridge: Cambridge University Press.

Tjørve, E. (2003) 'Shapes and functions of species-area curves: A review of possible models', *Journal of Biogeography*, 30: 827–35.

Tobler W. (1970) 'A computer movie simulating urban growth in the Detroit region', *Economic Geography*, 46: 234–40.

Turner, J.R.G., and B.A. Hawkins. (2004) The global biodiversity gradient. In *Frontiers of Biogeography: New Directions in the Geography of Nature*, ed. M.V. Lomolino and L.R. Heaney, 171–90. Sunderland, MA: Sinauer.

Turner, M.G., R.H. Gardner, and R.V. O'Neill, (2001) *Landscape Ecology in Theory and Practice: Pattern and Process*. New York: Springer.

Turner, W.R., and E. Tjørve. (2005) 'Scale-dependence in species–area relationships', *Ecography*, 28: 1–10.

United Nations. (1993) 'Convention on biological diversity (with annexes). Concluded at Rio de Janeiro on 5 June 1992', *United Nations Treaty Series* 1760: 142–382.

Walter, H.S. (1998) 'Driving forces of island biodiversity: An appraisal of two theories', *Physical Geography*, 19: 351–77.

Walter, H.S. (2004) 'The mismeasure of islands: Implications for biogeographical theory and the conservation of nature', *Journal of Biogeography*, 31: 177–97.

Whittaker, R.H. (1977) Evolution of species diversity in land communities. In *Evolutionary Biology*, ed. M.K. Hecht, W.C. Steere and B. Wallace, 250–68. New York: Plenum.

Whittaker, R.J. (1998) *Island Biogeography: Ecology, Evolution, and Conservation*. Oxford: Oxford University Press.

Whittaker, R.J. (2001) 'Review: Wrong in interesting ways', *Journal of Biogeography*, 28: 1441–42.

Whittaker, R.J. (2004) Dynamic hypotheses of richness on islands and continents. In *Frontiers of Biogeography*, ed. M.V. Lomolino and L.R. Heaney, 211–31. Sunderland, MA: Sinauer.

Whittaker, R.J and 11 authors (2005) 'Conservation biogeography: Assessment and prospect', *Diversity and Distributions*, 11: 3–23.

Whittaker, R.J., K.J. Willis, and R. Field. (2001) 'Scale and species richness: Towards a general, hierarchical theory of species diversity', *Journal of Biogeography*, 28: 453–70.

Wiens, J.A. (1989) 'Spatial scaling in ecology', *Functional Ecology*, 3: 385–97.

Wiens, J.J., and M.J. Donoghue. (2004) 'Historical biogeography, ecology and species richness', *Trends in Ecology and Evolution* ,19: 639–44.

Williams, C.B. (1943) 'Area and number of species', *Nature*, 152: 264–67.

Wright, D.H. (1983) 'Species-energy theory: An extension of species-area theory', *Oikos*, 41: 496–506.

Theory in Landscape Ecology and Its Relevance to Biogeography

John A. Kupfer

4.1 BIOGEOGRAPHY AND LANDSCAPE ECOLOGY

Traditionally defined as the study of spatial patterns of organisms and biological diversity, biogeography today explores geographic variation in all forms of life, from genetic, physiological, and morphological variation among individuals and populations to differences in the diversity and composition of communities, ecosystems, and biomes. A central focus of contemporary biogeographical research involves determining how patterns of species and communities reflect spatial and temporal variations in underlying environmental patterns and processes. Research over the last two decades in the field of landscape ecology, however, has demonstrated a more reciprocal relationship in which patterns of biotic entities are not only spatial manifestations of underlying gradients and processes but also act to affect fundamental ecological processes such as dispersal, competition, disturbance, and fluxes of energy and matter across space. Studies of deforestation and forest fragmentation (Ewers and Didham, 2006) and riparian systems (Malanson, 1993), for example, offer a multitude of examples on the interrelationships between ecological pattern and process. Thus, for the 'ecological biogeographer' interested in understanding the factors structuring current distributions of biota, an awareness of basic landscape ecological principles is crucial.

Landscape ecology can be defined as the study of spatial pattern and its relationship to ecological process at a range of scales, or as Fahrig (2005: 3) has said, "the study of how landscape structure affects the processes that determine the abundance and distribution of organisms." There are several aspects of landscape ecology that distinguish it from other sub-disciplines within ecology (Turner et al., 2001). First, it explicitly addresses the importance of spatial configuration for ecological process. In other words, rather than pattern being used solely to understand process, the interactions between pattern and process are the subject of interest. Second, landscape ecology often focuses on spatial extents that are larger in geographic scale than those in other ecological sub-disciplines, although spatial heterogeneity can be manifested at a range of scales in part because different organisms view and respond to their surroundings in different ways. In fact, the importance of scale is a central theme in landscape ecological research. Finally, landscape ecology is generally more focused on the relationship of human activity to landscape pattern, process, and change than many other areas of ecology have been. Indeed, landscape ecological principles are commonly applied in the fields of biological conservation, regional planning, and ecosystem management.

In this chapter, I provide an overview of the historical development of landscape ecology and outline its principal concepts, theories, and emerging issues. I do not attempt to present a comprehensive review of the field because several books and papers have recently done so (e.g., Farina, 2000; Turner et al., 2001; Wiens and Moss, 2005; Turner, 2005a). Rather, I outline its central themes and discuss them within the context of biogeographical theory and application. Despite the theoretical, conceptual, and methodological overlaps between biogeography and landscape ecology, questions about the nature of the relationship between the two fields persist, and many prominent textbooks in both fields make surprisingly little explicit mention of the other. As someone who considers himself both a landscape ecologist and a biogeographer, I have therefore chosen to emphasize the implicit, but sometimes unrecognized, linkages between the fields and the value of ideological cross-pollination.

4.2 HISTORICAL DEVELOPMENT OF LANDSCAPE ECOLOGY

An understanding of current and future directions in landscape ecology as well as its ties to biogeography is facilitated by reflecting on the field's roots and development (see, for example, Forman, 1995; Turner et al., 2001; Wiens, 2002). The term 'landscape ecology' was introduced in the late 1930s by the German biogeographer Carl Troll, whose thinking was influenced by the broad-scale perspective provided by technical advances in aerial photography and theoretical advances in ecosystem science. From its inception and early growth in central and eastern Europe through the 1970s, landscape ecology developed with the interweaving of ideas from human regional geography, land-use planning, resource economics, landscape architecture, and other disciplines (Schreiber, 1990). It provided a novel viewpoint for understanding the interactions of humans with their environment at spatial scales broader than those that were generally the focus of ecologists and carried with it a strong early emphasis on land evaluation, classification, and mapping as a basis for land-use recommendations, pursuits that are not surprising given the long history of intensive human land use and modification in its region of origin. To this day, the European school of landscape ecology retains a more applied, transdisciplinary approach (Drdoš, 1996; Opdam et al., 2002; Metzger, 2008).

The 1980s were marked by growth in both the geography and focus of landscape ecology. Changes in the former were driven in part by the introduction of landscape ecology to English-speaking audiences through a handful of key publications (e.g., Forman and Godron, 1981; Naveh and Lieberman, 1984) and two important meetings: the 1981 International Congress of the Netherlands Society for Landscape Ecology (Tjallingii and De Veer, 1982) and the 1983 Allerton Park workshop (Risser et al., 1984). The latter, in particular, helped to define a set of themes that established a foundation for the development of landscape ecology in North America over the following decades (Wiens, 2008). While the expansion of landscape ecology into Canada and the United States is often discussed, scientists in other parts of the world were similarly exposed to the new discipline. Translations of Troll's 'Landscape Ecology' and Neef 's 'Stages in the Development of Landscape Ecology' started the widespread introduction, evaluation, and assimilation of landscape ecology as a new discipline in China in the 1980s, for example (Fu and Lu, 2006). The globalization of landscape ecology continued with the formation of the International Association for Landscape Ecology in 1982 and the establishment of its flagship journal, *Landscape Ecology*, in 1987.

The 1980s and 1990s were also important periods because of fundamental changes in the orientation and nature of the discipline as landscape ecologists began to place greater emphasis on the development and testing of scientific theory (Farina, 2000). Some of this change can be attributed to technological advances in remote sensing, ecological modeling, geographic information systems (GIS), and spatial analysis, but many developments and novel avenues of landscape ecological research stemmed from theoretical advances in the broader field of ecology coupled with the formation and growth of the 'North American' school of landscape ecology, which was dominated by American and Canadian ecologists and biologists. In particular, the establishment of landscape ecology in North America as well as in Australia in the 1980s was concurrent with the widespread acceptance of nonequilibrium concepts in ecology and a recognition that ecosystems could not be understood without considering flows of energy and material across ecosystem boundaries (Wiens, 2002). This view of ecosystems as open systems required an understanding of how mosaics of ecosystems interact to affect ecosystem processes, the central focus of landscape ecology.

The growth and increasingly scientific orientation of landscape ecology was also shaped by the quantitative revolutions of the 1960s and 1970s in ecology and geography, especially efforts to develop prescriptive models for maximizing species diversity and protecting critical habitat for endangered species. The first quantitative links between habitat loss and species extinctions were developed out of concepts formalized within island biogeography theory, which stated that the number of species on oceanic islands is a function of island size (which was linked to extinction rates and habitat heterogeneity) and isolation (which was believed to influence the arrival of potential colonists) (MacArthur and Wilson, 1967). While the idea that species richness varied as a function of habitat area was well established at the time, the importance of habitat isolation and especially its quantification by distance to a potential colonizer source was more novel. Ecologists subsequently drew parallels between oceanic islands and terrestrial habitats that had been fragmented by human land uses and began to study the relationships among biodiversity, remnant forest area, and forest patch isolation (Kupfer, 1995). These developments helped pave the way for the development of metapopulation theory (e.g., Hanski, 1999) and other models of species dynamics in spatially heterogeneous environments.

The last few decades have been marked by a broader acceptance of landscape ecological concepts and principles within the scientific community. After a period of largely descriptive studies, landscape ecological research has been characterized by a greater diversity of research topics and study scales and an increased use of methodological, mathematical, and statistical approaches (Andersen, 2008), often with an explicit goal of testing the generality of concepts across systems and scales. Landscape ecology has been important in the development of integrative landscape research (Antrop, 2007) and has contributed to substantial advances in understanding the causes and ecological consequences of spatial heterogeneity and how relationships between pattern and process vary with scale (Turner, 2005b). Through their research, landscape ecologists have offered new perspectives on the function and management of both natural and human-dominated landscapes.

4.3 CENTRAL CONCEPTS IN LANDSCAPE ECOLOGY

While much of the terminology in landscape ecology has arisen as part of the discipline's evolution, some of the most commonly used terms have been borrowed and adapted from other fields. Discipline-specific nuances and even inconsistent usage within landscape ecology itself necessitate a brief explanation of several key concepts. This discussion also helps to lay the foundation for many of the key themes and principles in landscape ecological research.

4.3.1 Landscape

The term 'landscape' has a long and rich history rooted in the tradition of sixteenth-century European landscape painters, who recognized not only the physical characteristics of a tract of land but also its human, cultural, and aesthetic aspects. Different ways of defining or conceptualizing landscapes have since developed, each with its own perspective, concepts, and methods. Three groups in particular can be recognized: the human sciences (including historical geography, historical ecology, philosophy, and many others), the applied sciences (including landscape design, architecture, and planning), and the natural sciences (Antrop, 2005). The term was introduced to biogeography more than a half century ago by the Canadian ecologist Pierre Dansereau (1957) and is one of the fundamental concepts in landscape ecology.

With dozens of slightly varying definitions in the landscape ecological literature, it might seem that 'landscape' is the most-defined, rather than best-defined, term in the field. Most differences in the usage of the term, however, stem from whether landscape is treated as a spatial scale, a level within an ecological hierarchy, or a functional perspective defined by the causes and consequences of spatial patterns in the environment (Wiens, 2002). The most common usage of the term is to denote a spatial scale (the 'landscape scale'). Scale in general refers to the physical dimensions of observed entities and phenomena and thus involves measures and measurement units. Scale in this sense also refers to the temporal and spatial dimensions over which phenomena are observed. With respect to the spatial dimensions of a landscape, there is no intrinsic scale that defines the existence of a landscape or a particular scale inherent in the definition of a landscape. The landscape scale, however, is often defined with respect to human perceptions of their surroundings and thus encompasses areas on the order of tens to thousands of square kilometers. While this selection of a reference scale is open to criticism, it is understandable because the ways in which humans perceive, conceptualize, relate to, and behave toward the natural world have been shaped by the linkage between nature and human cultures that is a central idea in many long-held definitions of landscape—a debate that is well rehearsed in geography.

The second definition of landscape, often termed the 'landscape level', refers not to the spatial dimensions of an entity or study area but rather to a level within an ecological hierarchy. More generally, a level is a class to which various entities belong and that is characterized by its rank ordering within a hierarchy; the level of organization provides a relative characterization of system organization (Urban et al., 1987). A level of organization is not itself a scale, but it can be characterized by a spatial scale that is in large part dictated by how it is defined. The landscape level is a human-prescribed definition that refers to a biological organization more inclusive than an ecosystem but less inclusive than a biome. Landscapes in this context are recognized as heterogeneous collections of different ecosystem types and often identified by the dominant ecosystem or land-use cover type (e.g., agricultural or forested landscapes).

In theory, these two definitions address different things: the landscape scale refers to the spatial domain of an entity while the landscape level is scale-independent and depends on the criteria used to define the system. There is even an extensive body of literature discussing the inappropriateness of using these definitions interchangeably (e.g., papers in Peterson and Parker, 1998). In practice, the two definitions may overlap in scope and scale because of the intrinsic way in which humans visualize and compartmentalize our surroundings and our tendency to order things hierarchically (Wiens, 2002). Both definitions share a common cultural root that focuses on aspects of the human-oriented approach to landscape as stated more than 200 years ago by the German geographer Alexander von Humboldt, who regarded landscapes as "the total character of a part of the Earth's surface" (Farina, 2000). Many of the questions addressed in landscape-level or landscape-scale research are also the same ones traditionally addressed by ecologists but carried out over a larger spatial extent, thereby providing a link to studies in biogeography. Thus, while landscape as a level of ecological organization and landscape as a spatial dimension are not synonymous, their meanings converge when placed within the predefined context of human perceptions and actions.

The third definition of landscape focuses more explicitly on the effects of spatial pattern on process as determined by characteristics of the organisms or ecological systems of interest. From an ecological standpoint, this definition avoids a primary limitation inherent in the typical implementation (if not conceptualization) of the first two, that the landscape is defined on the basis of human perceptions. Instead, it recognizes that the scales at which spatial patterns and processes are expressed differ among organisms because they perceive and respond to heterogeneity differently. Streambeds may be considered landscapes for stream invertebrates, and spatial heterogeneity in soils may be characterized at scales relevant to individual plants or even microbes (Turner, 2005b). Accordingly, there is greater interest in quantifying patterns of variability in time and space, understanding how patterns change with scale, and explaining the causes and consequences of patterns (e.g., Bossenbroek et al., 2005; Urban, 2005).

While distinct, these definitions are not mutually exclusive. All three recognize that landscapes are composed of landscape elements (e.g., different ecosystems, habitat types, successional stages, land uses) that have a particular spatial configuration ('landscape structure') within a landscape mosaic, and that landscape structure affects what goes on within the elements themselves as well as the interactions among elements within the larger mosaic (Plate 4.1). They do, however, emphasize fundamentally different conceptualizations of landscape. The first two definitions, which reflect the roots of landscape ecology in the European tradition, emphasize large areas or regions and humans and their activities. As such, they are consistent with modern approaches to land management in which actions at the landscape level or landscape scale are advocated because they encompass more variety than do actions focused on individual habitats, land-cover types, or administrative units. The third definition stresses the causes and consequences of spatial patterns at variable spatial scales defined by the organism or process of interest, reflecting the more bio-physical orientation of landscape ecology that was pioneered in North America. In contrast to the first two definitions, this definition is applicable across scales and adaptable to different systems.

4.3.2 Spatial heterogeneity

Central to the definitions of landscape just discussed is the importance of spatial heterogeneity. Spatial heterogeneity has two distinct components:

1. composition—the kinds and amounts of different elements that make up the landscape (e.g., deciduous forest, evergreen forest, row crops, wetlands, low-density residential developments); and
2. configuration—the spatial arrangement of those landscape elements across the landscape. (Gustafson, 1998)

A central tenet of landscape ecology is that these components, both their types and their arrangement, are important in that they influence landscape processes such as the flows of organisms, materials, energy, or disturbances across the landscape and thus shape the structure and function of the landscape mosaic through time.

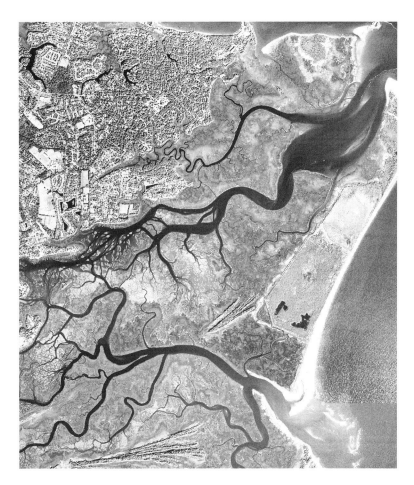

Plate 4.1   A coastal landscape near Charleston, South Carolina, in the United States, composed of a mosaic of maritime forest types, tidal freshwater and brackish water marshes and flats, estuarine habitats, ponds, beach and dune systems, and human-modified habitats varying in intensity of usage. Vegetation patterns reflect the interplay of natural environmental gradients shaped by climatic and geomorphic processes (e.g., variations in salinity, tidal effects, and soil characteristics) as well as natural (e.g., hurricanes) and anthropogenic (e.g., agriculture, residential development) disturbances

Spatial heterogeneity exists in part due to gradients in resource distribution, with climate, geology, and topography serving as the broad-scale constraints that determine the distribution of resources, especially in terms of soils, water, and solar radiation (Swanson et al., 1988). Against this background, natural disturbances continually alter the mosaic by creating new patches in sizes, shapes, and places that do not necessarily reproduce the previous mosaic pattern (e.g., Turner and Romme, 1994). The effects of human activities are further superimposed upon and interact with natural patterns and processes, resulting in a complex and dynamic landscape mosaic whose structure reflects the influences of natural and anthropogenic agents. The degree of stability of this mosaic in time and space, which itself is a function of scale, varies among landscape elements, with change being a constant feature (Baker, 1989).

Landscape structure, and thus spatial heterogeneity, is generally quantified from maps, remotely sensed images, and GIS coverages meant to capture the compositional and spatial aspects of landscapes. In many cases, landscape elements are defined as discrete entities or patches, and landscape pattern is described using landscape metrics developed to quantify patch-level attributes (e.g., size, shape, isolation) and mosaic-level characteristics (e.g., patch richness and diversity, connectivity, contagion) (Haines-Young and Chopping, 1996). Spatial patterns of some ecological phenomena have also been characterized using spatio-analytical methods and tools such as geostatistical and spatial statistical techniques, including variogram analysis, spatial autocorrelation, wavelets, and texture analysis (Bunn et al., 2005; Fortin and Dale, 2005; Wagner and Fortin, 2005). In either case, the ultimate goal of landscape analyses should be to link measures of landscape structure to specific effects on ecological processes rather than treating quantitative descriptions of spatial pattern as an end unto itself (Li and Wu, 2004). In studies of forest loss and fragmentation, for example, landscape metrics quantifying spatial patterns of deforestation and fragmentation are linked with measures of biodiversity and ecological processes at a range of scales, both as a means of validating landscape-level fragmentation effects (i.e., by linking changes in pattern to ecological responses) and as an effort to develop tools and indices that can be used as aids in forest management and monitoring (Ripple et al., 1991; Ranta et al., 1998; Millington et al., 2003; Munroe et al., 2007).

Finally, because the analysis of spatial pattern and heterogeneity is central to the field of landscape ecology and necessary for understanding biogeographic distributions in existing landscapes, it is important to recognize that representations and analyses of heterogeneity are defined in part by decisions relating to aspects of spatial scale. For example, the numbers, sizes, and shapes of patches within a landscape change with the spatial resolution of the data used to define the patches and the extent of the study area (Wickham and Riitters, 1995; Wu et al., 2002). Conclusions derived from an individual map or data set are thus contingent on the resolution of the data, the detail of the thematic classification, and the selection of the specific boundary for the study area, all of which are influenced and shaped by human perceptions of nature and other constraints.

4.3.3 Scale

The problem of relating ecological phenomena across scales—including ways of quantifying patterns of variability in space and time, understanding how patterns change with scale, and clarifying the causes and consequences of pattern—is one of the central problems in ecology (Wiens, 1989) and one that has a long tradition in biogeography (Whittaker et al., 2001). The earlier

discussion concerning the definition of the term 'landscape' provides one example of the importance of studying scale, that is, the identification of the scales at which spatial patterns and processes are expressed and significant for individual organisms and species. While the study of scale and its relevance to ecological patterns and processes is not solely the province of landscape ecologists, scale is at the heart of the discipline, and landscape ecologists have provided fundamental insights into the importance of scale in ecological research, for instance, through the application of hierarchy theory to the study of landscape mosaics (e.g., Kotliar and Wiens, 1990).

In landscape ecology, scale is often defined as the spatial or temporal dimension in which an object or process is recognizable. Operationally, 'spatial scale' is defined by the following:

1. grain: the finest level of spatial resolution within a data set (e.g., the size of a pixel in a satellite image or the area of a sample unit in field studies); and
2. extent: the defined size of a study area.

The terms 'fine-scale' and 'broad-scale' are often used to describe the spatial extent of a study or the dimensions at which certain processes act or are important (e.g., fine-scale vs. broad-scale processes). Forman (1995) also provided a link between scale and spatial heterogeneity by emphasizing the coarseness in texture or granularity of spatial elements composing an area: a fine-grained landscape is one containing many small landscape elements while a coarse-grained landscape is mainly composed of larger landscape elements.

Grain and extent are important because they define the implied relevance of a study, influence the conclusions drawn by an observer, and dictate whether results can be extrapolated to other times or locations (MacNally, 2005). It is well known that ecological phenomena tend to have characteristic spatial and temporal scales and that different patterns emerge at different scales of investigation for virtually any aspect of an ecological system. Biogeographical distributions provide many examples of such scale effects. Regionally and globally, plant species patterns reflect the roles of physical processes associated with global atmospheric circulation patterns, soil-forming regimes, and plate tectonic activity. At local scales, where such factors effectively become constants, microtopographic and microclimatic factors become more important, and local biological interactions may introduce temporal and spatial lags in system dynamics, for example, through competitive effects (Lomolino et al., 2006). The spatial distribution of two species may thus appear to be negatively related at the local scale due to competition but positively associated at the regional level because of the broad-scale influences of habitat selection.

For more than two decades, the importance of scale and the development of conceptual frameworks for understanding scale dependence have prompted ecologists to critically examine the effect of scale and whether understanding could be translated from one scale to another (e.g., Willig et al., 2003). Some of the impetus for these studies was the growing number and awareness of problems that were regional and global in nature (e.g., acid rain, climate change, habitat fragmentation). Scientists found themselves challenged to use data and understanding obtained from fine-scale studies (e.g., small field plots) to infer or project consequences that would occur at broad scales (Turner et al., 2001). A large number of field and simulation studies conducted at multiple spatial scales have since clearly demonstrated that the processes and parameters important in predicting or explaining ecological phenomena at one scale are not typically the same as those at other scales (e.g., Meentemeyer, 1984; Levin, 1992).

Landscape ecology is in essence an applied discipline, and applied ecological challenges require the interfacing of phenomena that occur on different scales or levels of space, time, and ecological organization. While many multiscale studies specifically examine the effects of changing the spatial resolution of the data or the spatial extent of the study area, there is a corresponding necessity for studies and management actions that are scaled appropriately for specific organisms or processes of interest. This type of organism-centric approach recognizes that species differ in the scales at which they use resources or perceive their environment (their ecological neighborhoods), and consequently, the definition of fundamental landscape elements such as habitat patches or dispersal corridors are species- and process-specific (D'eon et al., 2002; Girvetz and Greco, 2007). Understanding the responses of different organisms to spatial pattern at multiple scales, particularly the development of generalizable principles based on factors such as body mass or resource requirements (e.g., functional groups), remains a high priority for ecological studies.

4.4 PRINCIPLES AND DIRECTIONS IN LANDSCAPE ECOLOGY

Wiens (2005) captured the central issues and themes that define landscape ecology in three fundamental questions:

1. What creates pattern in landscapes? In other words, what are the sources of the spatial variation in the quantitative or qualitative properties of systems?
2. How does landscape pattern affect process? That is, how do gradients or discontinuities in landscape mosaics affect the flows of energy, materials, individuals, or information through space?
3. How are the answers to the first two questions affected by scale?

Within this broader framework, reviews and landscape ecology texts identify subtly different concepts, principles, and future directions for the field. There are, however, a number of recurrent themes that provide useful insights into where landscape ecology is as a discipline and where it is headed in the decades ahead.

4.4.1. Quantifying and understanding landscape patterns and spatial heterogeneity

Collectively, Wiens's first two questions address the sources and consequences of spatial heterogeneity. Answering these questions involves understanding the factors that structure the pattern of biotic phenomena across landscapes and how ecological processes are shaped by and interact with the configuration of landscape elements. The latter objective often necessitates linking observed or modeled ecological patterns and processes to quantitative measures of landscape pattern (e.g., Lindenmayer et al., 2002), but the interpretations of such results are subject to the manners by which spatial heterogeneity is conceptualized and represented. Thus, much of the past, current, and future research in landscape ecology involves quantifying and understanding the implications of spatial heterogeneity.

Causes and consequences of landscape pattern

Understanding the factors that structure the pattern of biotic phenomena across landscapes is a goal for not only landscape ecologists but also for biogeographers, spatial ecologists, and ecologists in general. Such studies emphasize how patterns of species, ecosystems, or biodiversity are related to underlying environmental patterns, typically as they are expressed at the human-perceived landscape scale. Consequently, the goal is often to define the relationships between biotic phenomena and direct gradients (e.g., soil moisture, solar radiation) or indirect gradients (e.g., topography, elevation) in environmental factors (e.g., Lookingbill and Urban, 2005). The role of disturbance in shaping the landscape mosaic is also a common theme in studies of ecological patterns, including the effects of both natural disturbances (e.g., fire) and anthropogenic activities (e.g., fragmentation and land use change). While ecosystem patterns have long been studied as a function of current conditions, research is also clarifying the importance of past disturbances (Kulakowski and Veblen, 2007) and ecological legacies—the types, durations, and extents of persistent effects of prior land use on ecological pattern and process (Foster et al., 1998).

The study of how landscape pattern affects the distribution and movement of organisms, energy, materials, or information through space is distinctly the province of landscape ecologists in that it focuses on the reciprocal interactions between pattern and process. Landscape ecologists emphasize how organisms use resources that are spatially heterogeneous and how they live, reproduce, disperse, and interact in landscape mosaics so the importance of patch size (or some similar measure of habitat availability) and patch isolation are central to many studies (Turner, 2005a). As scientists and managers continue to wrestle with issues regarding how organisms respond to spatial patterns, it is clear that there will be opportunities for synthesis among population ecologists, biogeographers, conservation biologists, and landscape ecologists (Turner et al., 2001; Boutin and Hebert, 2002).

Finally, understanding the patterns, causes, and consequences of spatial heterogeneity for ecosystem function is a research frontier in landscape ecology and ecosystem ecology (Turner, 2005b). Past research on watershed geomorphology and hydrology, for example, has shown the importance of landscape composition and configuration in controlling basin hydrology and patterns of sediment erosion and deposition (e.g., the role of riparian buffers: Peterjohn and Correll, 1984). Similarly, patterns of deforestation and retention have been shown to alter energy fluxes and microclimatic patterns across a landscape and constrain the types, extents, frequencies, and even intensities of disturbances; these changes in turn affect the dynamics of species and the resulting biodiversity (Cochrane, 2001).

Quantifying landscape patterns and spatial heterogeneity

The quantification of landscape pattern is central to clarifying relationships between ecological processes and spatial patterns, and thus the measurement, analysis, and interpretation of spatial patterns has received much attention in landscape ecology (Turner, 2005a). Driven by rapid advances in geospatial technologies and data availability, dozens of landscape indices or metrics meant to capture aspects of landscape structure have been developed and applied at scales ranging from landscapes to regions, continents, and even the globe over the last two decades. Research on the statistical properties and sensitivity of metrics to changing landscape patterns (Hargis et al., 1998) as well as methods for documenting empirical relationships between metric values and ecological patterns and processes still continues (e.g., Gillespie and Walter, 2001). The creation of neutral landscape models that produce expected landscape patterns in the absence of specific processes can yield insights into what spatial patterns to expect from random processes, generate benchmarks against which to compare real landscape patterns, and provide replicate landscapes that share statistical similarities with real landscapes and thus be used to generate statistical generalizations from a set of landscapes (Milne, 1992; With and Crist, 1995; Wang and Malanson, 2008).

Tools such as GIS have not only facilitated the quantification of landscape properties, but they have also simplified the classification, evaluation, and mapping of biotic and abiotic features at the landscape scale. One common approach to developing ecological land classifications and carrying out land evaluations involves coupling coverages of biophysical variables and biotic phenomena (e.g., species, ecosystems) with statistical models to develop homogeneous 'land units' or 'land types' (Host et al., 1996). The delineation of individual units is based on intrinsic characteristics of the system and can vary from a single selection criterion (e.g., soil, vegetation type) to a range of properties that collectively define individual units (e.g., characteristic species, topographic setting, disturbance history, soil characteristics). The resulting units provide a template for interpreting the spatial distribution and variability of biotic phenomena in the landscape, facilitate the assessment of ecosystem health or vulnerability, serve as a framework for broad-scale management, and provide the basis for monitoring networks (e.g., Abella et al., 2003). Landscape classification thus forms the basis for land evaluation, which in turn is the basis for land use planning and management.

Conceptualizing and representing spatial variability in habitat quality

From the birth of landscape ecology, landscapes have been depicted as aggregations of discrete landscape elements. This representation is inherent in the corridor-patch-matrix and landscape mosaic models popularized by Forman and

Godron (1986) and Forman (1995). This conceptualization is appealing and intuitive in that it is consistent with the way in which humans compartmentalize our surroundings, particularly in landscapes with a long history of human land uses and where such uses contrast greatly with the structure of natural ecosystems (e.g., row crops in forested landscapes). This treatment of landscapes is also easy to implement using remote sensing and GIS, allowing scientists and managers to classify landscapes into discrete landscape elements and facilitating the quantitative analysis of landscape patterns.

Despite the prevalence of a patch-based landscape mosaic approach, it has been suggested that the ways in which spatial heterogeneity of landscape components are conceptualized and put into practice need to be broadened. The definition of a patch is scale- and species-dependent, and distinct patch boundaries are not always easy to define, even at a chosen scale. There is the potential for neglecting small patches and habitat features not recognized as patches by humans, and although a patch is implicitly defined as being homogeneous with respect to certain criteria, there is typically some degree of variability present, for example, fine-scale heterogeneity that is too small to be captured at the minimum mapping unit selected to define the patches (i.e., a variation of the Modifiable Area Unit Problem; Unwin, 1996). Finally, populations in patches are influenced not only by characteristics of the patches themselves but also the nature of the surrounding habitat; this type of context-specific patch property can be difficult to integrate into the landscape mosaic model.

Two alternative models for representing spatial variability in landscapes have been proposed. The landscape continuum (or 'variegation') model (McIntyre and Hobbs, 1999) focuses on variations in conditions across a landscape. Landscapes may, for example, be represented as continuous surfaces of a habitat variable (e.g., forest cover) so that small elements of habitat that might otherwise be classified as unsuitable in a larger matrix can be better accounted for (e.g., individual trees within a grassland matrix). More recently, Fischer et al. (2004) proposed the use of a 'habitat contour model' in which contours representing changes of habitat suitability over space are created and mapped for multiple species. These maps may also be overlaid to define overall patterns of landscape heterogeneity.

In general, the corridor-patch-matrix, variegation, and habitat contour models have varying advantages and disadvantages with respect to their degree of realism, degree of complexity, ease of quantification and interpretation, and ability to deal with multiple species and multiple scales (Fischer et al., 2004). Considering landscapes from the perspectives of multiple models could help to clarify relationships between patterns of landscape structure and their effects on ecological processes.

4.4.2. The importance of connectivity

Understanding the effects of landscape pattern on process is at the heart of landscape ecology so it is not surprising that connectivity—the connectedness of landscape elements or the landscape mosaic as a whole—has been and continues to be a primary research topic and one that transcends landscape ecology to fields such as biogeography and conservation biology. While the focus of connectivity is often on how it influences the dispersal of organisms across landscapes, thereby affecting gene flow, extinction, and recolonization rates in habitat patches, the term in fact applies to the movement or propagation of a much broader range of phenomena (e.g., water and disturbances: Miller and Urban, 2000).

Interest in landscape connectivity is perhaps best evidenced by the long-standing attention paid to habitat corridors, linear strips of habitat connecting patches of similar habitat (Anderson and Jenkins, 2006). Forested riparian corridors and hedgerows in an agricultural landscape, for instance, are thought to increase the movement of some individuals among habitat patches, promote genetic exchange, and facilitate recolonization of suitable habitat patches (Tewksbury et al., 2002). Corridors may also (a) reduce mortality during interpatch movement; (b) direct the movement of a broad range of taxa across the landscape; (c) provide additional habitat area; (d) increase the foraging area for wide-ranging species; and (e) serve as refugia from large disturbances (Beier and Noss, 1998; Haddad et al., 2003). Despite the widespread acceptance of corridors as management tools because of their intuitive and logical appeal, corridor benefits are species-specific, and debate continues over the ability of corridors to actually increase connectivity for many species (e.g., Davies and Pullin, 2007). Further, there is concern about the potential for the increased spread of disturbances, diseases, and predators of species of concern; by serving as dispersal corridors, road verges and trails, for example, may facilitate the dispersal of nonnative species.

The other primary area of landscape ecological research concerning connectivity involves the inter-patch movement of organisms across areas of 'unsuitable' ecosystem types (e.g., forest patches embedded within an agricultural matrix). Populations in heterogeneous or fragmented landscapes are sometimes studied as metapopulations, subsets of habitat patches that are linked by infrequent between-patch migrations. Each population has a finite probability of extinction, and each local extinction may be balanced by recolonization from neighboring populations. Recolonization potential is based on the size and suitability of patches as well as their isolation from potential colonist patches. Both field and simulation studies have suggested that inter-patch distance influences the dispersal of species, but specific effects are highly variable among species and affected by the structure and composition of the intervening matrix (e.g., Ricketts, 2001; Bender and Fahrig, 2005).

Much of the past work on corridors and landscape connectivity has focused on structural connectivity, the degree to which habitat patches are physically contiguous or connected. The movement of organisms across a landscape, however, is determined by a landscape's functional connectivity, the degree to which it facilitates or impedes movement (Tischendorf and Fahrig, 2000). A habitat need not be structurally connected to be functionally connected because functional connectivity is influenced by an organism's vagility, the presence of corridors and 'stepping stones', matrix characteristics, and the willingness of an individual to cross habitat gaps (Bélisle, 2005). While structural connectivity is easy to determine and thus widely used, it is important only to the extent that it coincides with the underlying functional connectivity of a landscape. Consequently, there has been considerable recent work on measuring and modeling functional connectivity through innovative approaches such as movement cost surfaces (Adriaensen et al., 2003) and graph theory (Pascual-Hortal and Saura, 2006).

4.4.3 Boundaries, edges, ecotones, and transitions

Given that the configuration of landscape elements affects what goes on within the elements as well as their interactions, boundaries between elements should be expected to play a particularly important role in governing patch and mosaic dynamics (Wiens, 2002). The study of various types of boundaries and ecotones, including forest edges, land-water interfaces, and other types of transitions, has thus been a common focus of landscape ecological research.

The earliest studies of boundaries focused on the 'edge effect': changes in microclimate, forest structure, biotic composition, and ecological function along forest edges exposed to nonforested habitats such as agricultural fields, clear cuts, roads, or pastures. Edge characteristics have since been documented for a range of geographic locations, abiotic variables (e.g., temperature, light, moisture balance), forest characteristics (e.g. tree density, species diversity, abundance of nonnative species), and processes (e.g., increased predation, modified energy fluxes, altered competitive relationships), and the roles of mitigating factors such as edge age, repeat disturbance, and geographic orientation (e.g., north vs. south) have all been explored (Harper et al., 2005). The patch- and landscape-scale implications of edge habitats have been evaluated through empirical measures (e.g., core-area indices: Laurance and Yensen, 1991) and field-based studies (e.g., Fletcher, 2005).

While edges have often been characterized as sharp, static boundaries between forested and nonforested systems, many boundaries are dynamic and multidimensional, resulting in significant differences in edge characteristics and complexity. Studies of edges now not only document characteristics within edges but also examine their dynamic and functional properties, including their roles as sinks and sources (e.g., Weathers et al., 2001; Cadenasso and Pickett, 2001), their ability to mediate cross-boundary movements (Duelli et al., 1990), and their effects on biotic interactions and species dynamics (Fagan et al., 1999; Kupfer and Runkle, 2003). Comparing the structures and functions of boundaries across systems and ecological scales can foster a better understanding of the diverse dynamic ecological roles that boundaries play (Cadenasso et al., 2003).

Finally, the location and pattern of ecosystem transitions and ecotones and their changes through time continue to be studied in landscape ecological research and have become an area of overlapping interest between landscape ecologists and biogeographers. For example, a large number of studies have addressed the responses of ecotones to realized or projected climate changes (e.g., alpine treeline ecotones: Kullman, 1998), including factors beyond climate that may act to mediate ecotone responses (e.g., seed rain, herbivory: Malanson et al., 2007; Cairns et al., 2007). Further, while a shift in ecotone location or pattern in response to environmental changes is often the phenomenon of interest, it has also been shown that responses can be driven by the form and pattern of the ecotone itself, and ecotones have recently been examined as complex self-organizing systems (Malanson et al., 2006).

4.4.4 Scale

A number of important topics related to scale have already been addressed, but it bears reiterating that the emergence of landscape ecology has done much to increase awareness of the importance of scale. Some scale effects are well recognized, but there is still much to be done, for example, identifying the correct scales for studying and understanding particular patterns and processes and applying knowledge gained at one scale to others (Turner et al., 2001; Schmitz, 2005). One of the primary lessons from scale-oriented research in landscape ecology is that studying a system at an inappropriate scale can result in identifying patterns that are solely artifacts of scale. Key themes of ongoing research include the following:

1. the influence of scale on system openness and boundary definition and function (e.g., Strand et al., 2007);
2. space-time scaling and predictability of ecosystem attributes and properties, including research on scale-invariant processes (Gamarra and He, 2008) and the continued development and testing of general scaling principles (e.g., Hay et al., 2001); and
3. the integration of hierarchy theory, complex adaptive systems, and generative landscape science, which involves combining models of candidate processes that are believed to give rise to observed patterns with empirical observations (Brown et al. 2006).

One final area of scale effects currently receiving research interest involves the occurrence and importance of cross–scale interactions: these interactions occur when processes at one spatial or temporal scale interact with processes at another scale to result in nonlinear dynamics driven by threshold responses. Such interactions (a) change the pattern-process relationships across scales, for example, allowing fine-scale processes to propagate and thereby influence a broad spatial extent or a long time period; and (b) pose formidable challenges for understanding and forecasting ecosystem dynamics (Peters et al., 2007). Allen (2007) provided examples of cross-scale pattern-process relationships for interactions among forest dieback, fire, and erosion in northern New Mexico, including cases where environmental stress, operating on individual trees, can cause tree death that is amplified by insect mortality agents to propagate to patch and then landscape or even regional-scale forest dieback.

4.5. FORGING CLOSER TIES BETWEEN LANDSCAPE ECOLOGY AND BIOGEOGRAPHY

The emergence of ecological biogeography as a major sub-field within biogeography over the last few decades has paralleled the maturation of landscape ecology. It is clear that landscape ecological methods and theory have much to offer biogeographers, particularly those focusing on contemporary species patterns, and that there has been some infiltration of ideas and insights from landscape ecology into biogeographic research. One tangible manifestation of this has been the increasing percentage of articles in the flagship biogeography journals *Journal of Biogeography* and *Global Ecology and Biogeography* that mention terms such as 'landscape', 'heterogeneity', and 'scale' (Figure 4.1). Conversely, Young and Aspinall (2006) highlighted how the fusion of geographical thinking with geospatial technologies that address coupling of spatial pattern and process can contribute to a more biogeographically and ecologically focused landscape ecology. Below, I outline four areas where overlap between the two fields is particularly strong and where researchers are likely to benefit from the cross-pollination of ideas.

4.5.1 Studies of population dynamics in heterogeneous landscapes

Debate beginning in the 1970s about the applicability of island biogeography theory to terrestrial habitats led to a scientific consensus that its direct translation to nature reserve design was problematic for a number of reasons, but the basic ideas are nonetheless highly relevant. Spatial structure is now recognized as an essential element in contemporary theories of population dynamics in heterogeneous areas, and theories associated with metapopulations and the role of connectivity in fragmented landscapes have grown directly out of the emphasis on isolation that was central to island biogeography theory. More recently, researchers have documented the importance of matrix quality and heterogeneity in dictating population dynamics in forest remnants (Kupfer et al., 2006), resulting in a greater emphasis on studies that focus on the value of the landscape matrix as foraging habitat or dispersal routes for certain forest taxa (Lindenmayer and Franklin, 2002; Umetsu et al., 2008). Studies addressing issues of gene flow, area-diversity linkages, and other topics of central biogeographical importance in spatially heterogeneous areas will continue to benefit from advances made in landscape ecology. For example, research at the intersection of conservation genetics, biogeography, and landscape ecology has grown rapidly into the field of landscape genetics, which focuses on understanding how landscape characteristics structure populations and the processes

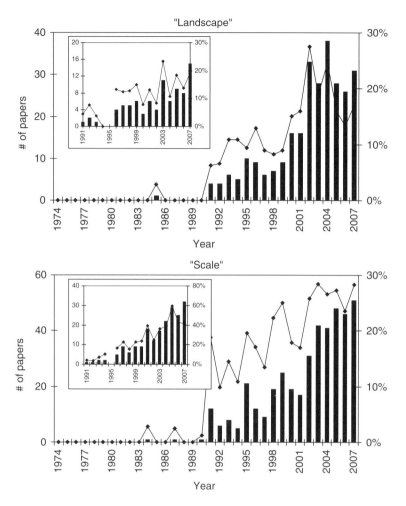

Figure 4.1   The total number (bars) and percentage (lines) of articles in *Journal of Biogeography* (1974–2007) and *Global Ecology and Biogeography* (1991–2007) (inset) with 'landscape', 'scale', or 'heterogen*' (for heterogeneous or heterogeneity) in their titles, abstracts, or key words. The search was conducted on July 17, 2008, using the ISI Web of Science database and does not include corrections to previously published papers

and patterns of gene flow and local adaptation (Manel et al., 2003; Parker and Markwith, 2007). Similarly, studies from the nascent field of landscape epidemiology have provided insights into the effects of landscape pattern and its changes on disease vectors, hosts, and pathogens (Meentemeyer et al., 2008).

4.5.2 Habitat distribution modeling

Predictive modeling of species and community distributions has become an increasingly

important tool for addressing issues in ecology, biogeography, and evolution (Guisan and Zimmerman, 2000). Most approaches have their roots in quantifying species–environment (or community–environment) relationships and extrapolating those relationships in space (e.g., through the use of GIS). Such models range from correlative or physiologically based models of species distributions (Anderson et al., 2002; Robertson et al., 2003) to simulations of species pattern and dispersal on hypothetical landscapes (Malanson and Cramer 1999; Nesslage et al., 2007) (see Burnside and Waite, and Malanson, this volume). Common topics that link the interests of biogeographers and landscape ecologists include the identification and protection of habitat for threatened and endangered species (Rayner et al., 2007), the spread of nonnative species (Schussman et al., 2006), and the effects of climate change onspecies ranges (Thuiller et al., 2008). Output from predictive models may also serve as neutral landscape models for assessing biotic patterns in actual landscapes, evaluating spatial metrics, and studying landscape connectivity, for instance, in relation to organism movement or dispersal. Approaches that incorporate the use of spatial models and information (Miller et al., 2007; Kupfer and Farris, 2007) provide a natural bridge between landscape ecology and biogeography.

4.5.3 Applied Biogeography and conservation

One common thread to the two sections above is the increasingly applied nature of many biogeographic studies. In introducing the field of conservation biogeography, Lomolino (2004) noted that scientists and managers need to develop not only a better understanding of static biotic distributions but also the biogeographic dynamics of imperiled species and those agents that threaten biological diversity. The advanced tools and techniques for identifying and analyzing patterns of biological diversity that biogeographers and landscape ecologists share include GIS, spatial statistics, remote sensing, and rapidly advancing abilities to interpolate and predict patterns of species occurrence, abundance, and richness. Principles and theories from not only biogeography but also landscape ecology, conservation biology, and other fields should help to guide this research. For example, studies quantifying and predicting patterns of landscape change (e.g., Griffith et al., 2003; Messina et al., 2006) as well as empirical studies linking landscape metrics to ecological processes or organism responses in fragmented landscapes (e.g., Westphal et al., 2003; Hernández-Stefanoni and Dupuy, 2008) represent an area of shared mutual interest between biogeographers and landscape ecologists.

4.5.4 Complexity, thresholds, and feedbacks

Studies documenting the existence and importance of feedbacks, thresholds, nonlinear responses, criticality, and local contingencies have given rise to the evolution and application of complex systems theory (or complexity theory) in a number of fields (Stallins and Parker, 2003; Phillips, 2006; Walsh and McGinnis, 2008), including ecology (Grimm et al., 2005; Proulx, 2007). Ecosystems are ideal examples of complex adaptive systems, in which higher-level system properties such as trophic structure and diversity–productivity relationships emerge from often complex interactions among components and may feed back to influence the subsequent development of those interactions (Levin, 1998). The location of alpine treeline ecotones, for example, is shaped by feedbacks and nonlinearities involving seed dispersal, tree germination, and mortality, and the dynamics of treeline landscapes are self-organizing wherein landscape-scale linear correlations between spatial pattern and the rate of advance of trees into tundra arise from localized nonlinear interactions (Zeng and Malanson, 2006). Studies that clarify the degree to which system features are the result of self-organization and address whether ecosystems are buffered to changes over their ecological and evolutionary development or subject to critical states and rapid change will benefit by blending insights from population biology, biogeography, landscape ecology, and ecosystems science (Levin, 1998). Ryan et al. (2007) present a conceptual framework, a complex adaptive landscape that unifies research in complex adaptive systems and landscape ecology on system properties, features, and dynamics of natural and human-modified systems.

4.6 CONCLUSION

In response to her own question "When is a landscape perspective necessary?" Fahrig (2005: 7) answered by saying "whenever landscape structure can be expected to have a significant effect on the response variable (abundance/distribution/pattern)." Developments in the field of landscape ecology have shown that many aspects of biotic distribution and pattern (and the underlying processes associated with those distributions) are a function of the types and amounts of ecosystems in heterogeneous areas as well as their arrangement. This realization is not new to biogeographers because the effects of habitat amount and isolation on extinction and immigration rates and thus species diversity were the central components of island biogeography theory. Given their many shared interests, goals, tools, and methodologies, ecological biogeographers and landscape ecologists will undoubtedly find many more opportunities for collaboration over the upcoming decades.

REFERENCES

Abella, S.R., Shelburne, V.B., and MacDonald, N.W. (2003) 'Multifactor classification of forest landscape ecosystems of Jocassee Gorges, southern Appalachian Mountains, South Carolina', *Canadian Journal of Forest Research*, 33: 1933–46.

Adriaensen, F., Chardon, J.P., De Blust, G., Swinnen, E., Villalba, S., Gulinch H., and Matthysen, E. (2003) 'The application of "least-cost" modelling as a functional landscape model', *Landscape and Urban Planning*, 64: 233–47.

Allen, C.D. (2007) 'Interactions across spatial scales among forest dieback, fire, and erosion in northern New Mexico landscapes', *Ecosystems*, 10: 797–808.

Andersen, B.J. (2008) 'Research in the journal landscape ecology, 1987–2005', *Landscape Ecology*, 23: 129–34.

Anderson, A.B., and Jenkins, C.N. (2006) *Applying Nature's Design: Corridors as a Strategy for Biodiversity Conservation*. New York: Columbia University Press.

Anderson, R.P., Laverde, M., and Peterson, A.T. (2002) 'Geographical distributions of spiny pocket mice in South America: Insights from predictive models', *Global Ecology and Biogeography*, 11: 131–41.

Antrop, M. (2005) 'From holistic landscape synthesis to transdisciplinary landscape management', in B. Tress, G. Tress, G. Fry, and P. Opdam (eds.), *From Landscape Research to Landscape Planning: Aspects of Integration, Education and Application*. Heidelberg, Germany: Springer. pp. 27–50.

Antrop, M. (2007) 'Reflecting upon 25 years of landscape ecology', *Landscape Ecology*, 22: 1441–43.

Baker, W.L. (1989) 'Landscape ecology and nature reserve design in the Boundary Waters Canoe Area, Minnesota', *Ecology*, 70: 23–35.

Beier, P., and Noss, R.F. (1998) 'Do habitat corridors provide connectivity?' *Conservation Biology*, 12: 1241–52.

Bélisle, M. (2005) 'Measuring landscape connectivity: The challenge of behavioral landscape ecology', *Ecology*, 86: 1988–95.

Bender, D.J., and Fahrig, L. (2005) 'Matrix structure obscures the relationship between interpatch movement and patch size and isolation', *Ecology*, 86: 1023–33.

Bossenbroek, J.M., Wagner, H.H., and Wiens, J.A. (2005) 'Taxon-dependent scaling: Beetles, birds, and vegetation at four North American grassland sites', *Landscape Ecology*, 20: 675–88.

Boutin, S., and Hebert, D. (2002) 'Landscape ecology and forest management: Developing an effective partnership', *Ecological Applications*, 12: 390–97.

Brown, D.G., Aspinall, R., and Bennett, D.A. (2006) 'Landscape models and explanation in landscape ecology—A space for generative landscape science?' *The Professional Geographer*, 58: 369–82.

Bunn, A.G., Waggoner, L.A., and Graumlich, L.J. (2005) 'Topographic mediation of growth in high elevation foxtail pine (*Pinus balfouriana* Grev. et Balf.) forests in the Sierra Nevada, USA', *Global Ecology and Biogeography*, 14: 103–14.

Cadenasso, M.L., and Pickett, S.T.A. (2001) 'Effect of edge structure on the flux of species into forest interiors', *Conservation Biology*, 15: 91–97.

Cadenasso, M.L., Pickett, S.T.A., Weathers, K.C., and Jones, C.G. (2003) 'A framework for a theory of ecological boundaries', *BioScience*, 53: 750–58.

Cairns, D.M., Lafon, C., Moen, J., and Young, A. (2007) 'Influences of animal activity on treeline position and pattern: Implications for treeline responses to climate change', *Physical Geography*, 28: 419–33.

Cochrane, M.A. (2001) 'Synergistic interactions between habitat fragmentation and fire in evergreen tropical forests', *Conservation Biology*, 15: 1515–21.

Dansereau, P. (1957) *Biogeography: An Ecological Perspective*. New York: Ronald Press.

Davies, Z.G., and Pullin, A.S. (2007) 'Are hedgerows effective corridors between fragments of woodland habitat? An evidence-based approach', *Landscape Ecology*, 22: 333–51.

D'eon, R.G., Glenn, S.M., Parfitt, I., and Fortin, M.J. (2002) 'Landscape connectivity as a function of scale and organism vagility in a real forested landscape', *Conservation Ecology*, 6(2): 10. http://www.consecol.org/vol6/iss2/art10/.

Drdoš. J. (1996) 'A reflection on landscape ecology', *Ekologia*, 15: 369–75.

Duelli, P., Studer, M., Marchand, I., and Jakob, S. (1990) 'Population movements of arthropods between natural and cultivated areas', *Biological Conservation*, 54: 193–207.

Ewers, R.M., and Didham, R.K. (2006) 'Confounding factors in the detection of species responses to habitat fragmentation', *Biological Review*, 81: 117–42.

Fagan, W.E., Cantrell, R.S., and Cosner, C. (1999) 'How habitat edges change species interactions', *American Naturalist*, 153: 165–82.

Fahrig, L. (2005) 'When is a landscape perspective important?' in J.A. Wiens and M.R. Moss (eds.), *Issues and Perspectives in Landscape Ecology*. Cambridge: Cambridge University Press. pp. 3–10.

Farina, A. (2000) *Landscape Ecology in Action*. Dordrecht: Kluwer.

Fischer, J., Lindenmayer, D.B., and Fazey, I. (2004) 'Appreciating ecological complexity: Habitat contours as a conceptual landscape model', *Conservation Biology*, 18: 1245–53.

Fletcher, R.J. (2005) 'Multiple edge effects and their implications in fragmented landscapes', *Journal of Animal Ecology*, 74: 342–52.

Forman, R.T.T. (1995) *Land Mosaics: The Ecology of Landscapes and Regions*. Cambridge: Cambridge University Press.

Forman, R.T.T., and Godron, M. (1981) 'Patches and structural components for a landscape ecology', *BioScience*, 31: 733–40.

Forman, R.T.T., and Godron, M. (1986) *Landscape Ecology*. New York: Wiley.

Fortin, M.J., and Dale, M.R.T. (2005) *Spatial Analysis: A Guide for Ecologists*. Cambridge: Cambridge University Press.

Foster, D.R., Knight, D.H., and Franklin, J.F. (1998) 'Landscape patterns and legacies resulting from large, infrequent forest disturbances', *Ecosystems*, 1: 497–510.

Fu, B.J., and Lu, Y.H. (2006) 'The progress and perspectives of landscape ecology in China', *Progress in Physical Geography*, 30: 232–44.

Gamarra, J.G.P., and He, F. (2008) 'Spatial scaling of mountain pine beetle infestations', *Journal of Animal Ecology*, 77: 796–801.

Gillespie, T.W., and Walter, H. (2001) 'Distribution of bird species richness at a regional scale in tropical dry forest of Central America', *Journal of Biogeography*, 28: 651–62.

Girvetz, E.H., and Greco, S.E. (2007) 'How to define a patch: A spatial model for hierarchically delineating organism-specific habitat patches', *Landscape Ecology*, 22: 1131–42.

Griffith, J.A., Stehman S.V., and Loveland, T.R. (2003) 'Landscape trends in Mid Atlantic and Southeastern United States ecoregions', *Environmental Management*, 32: 572–88.

Grimm, V., Revilla, E., Berger, U., Jeltsch, F., Mooij, W.M., Railsback, S.F., Thulke, H.H., Weiner, J., Wiegand, T., and DeAngelis, D.L. (2005) 'Pattern-oriented modeling of agent-based complex systems: Lessons from ecology', *Science* 310: 987–91.

Guisan, A., and Zimmermann, N.E. (2000) 'Predictive habitat distribution models in ecology', *Ecological Modelling*, 135: 147–86.

Gustafson, E.J. (1998) 'Quantifying landscape spatial pattern: What is the state of the art?' *Ecosystems*, 1: 143–56.

Haddad, N.M., Bowne, D.R., Cunningham, A., Danielson, B.J., Levey, D.J., Sargent, S., and Spira, T. (2003) 'Corridor use by diverse taxa', *Ecology*, 84: 609–15.

Haines-Young, R., and Chopping, M. (1996) 'Quantifying landscape structure: A review of landscape indices and their application to forested landscapes', *Progress in Physical Geography*, 20: 418–45.

Hanski, I. (1999) *Metapopulation Ecology*. Oxford: Oxford University Press.

Hargis, C.D., Bissonette, J.A., and David, J.L. (1998) 'The behavior of landscape metrics commonly used in the study of habitat fragmentation', *Landscape Ecology*, 13: 167–86.

Harper, K.A., Macdonald, S.E., Burton, P.J., Chen, J., Brosofske, K.D., Saunders, S.C., Euskirchen, E.S., Roberts, D., Jaiteh, M.S., and Esseen, P. (2005) 'Edge influence on forest structure and composition in fragmented landscapes', *Conservation Biology*, 19: 768–82.

Hay, G.J., Marceau, D.J., Dube, P., and Bouchard, A. (2001) 'A multiscale framework for landscape analysis: Object-specific analysis and upscaling', *Landscape Ecology*, 16: 471–90.

Hernández-Stefanoni, J.L., and Dupuy, J.M. (2008) 'Effects of landscape patterns on species density and abundance of trees in a tropical subdeciduous forest of the Yucatan Peninsula', *Forest Ecology and Management*, 255: 3797–805.

Host, G.E., Polzer, P.L., Mladenoff, D.J., White, M.A., and Crow, T.R. (1996) 'A quantitative approach to developing regional ecosystem classifications', *Ecological Applications*, 6: 608–18.

Kotliar, N.B., and Wiens, J.A. (1990) 'Multiple scales of patchiness and patch structure: A hierarchical framework for the study of heterogeneity', *Oikos*, 59: 253–60.

Kulakowski, D., and Veblen, T.T. (2007) 'Effect of prior disturbances on the extent and severity of wildfire in Colorado subalpine forests', *Ecology*, 88: 759–69.

Kullman, L. (1998) 'Tree-limits and montane forests in the Swedish Scandes: Sensitive biomonitors of climate change and variability', *Ambio*, 27: 312–21.

Kupfer, J.A. (1995) 'Landscape ecology and biogeography', *Progress in Physical Geography*, 19: 18–34.

Kupfer, J.A., and Farris, C.A. (2007) 'Incorporating spatial non-stationarity of regression coefficients into predictive vegetation models', *Landscape Ecology*, 22: 837–52.

Kupfer, J.A., Malanson, G.P., and Franklin, S.B. (2006) 'Not seeing the ocean for the islands: The influence of matrix-based processes on forest fragmentation effects', *Global Ecology and Biogeography*, 15: 8–20.

Kupfer, J.A., and Runkle, J.R. (2003) 'Edge-mediated effects on stand dynamics in forest interiors: A coupled field and simulation approach', *Oikos*, 101: 135–46.

Laurance, W.F., and Yensen, E. (1991) 'Predicting the impacts of edge effects in fragmented habitats', *Biological Conservation*, 55: 77–92.

Levin, S.A. (1992) 'The problem of pattern and scale in ecology', *Ecology*, 73: 1943–67.

Levin, S.A. (1998) 'Ecosystems and the biosphere as complex adaptive systems', *Ecosystems*, 1: 431–36.

Li, H.B., and Wu, J.G. (2004) 'Use and misuse of landscape indices', *Landscape Ecology*, 19: 389–99.

Lindenmayer, D.B., Cunningham, R.B., Donnelly, C.F., and Lesslie. R. (2002) 'On the use of landscape surrogates as ecological indicators in fragmented forests', *Forest Ecology and Management*, 159: 203–16.

Lindenmayer, D.B., and Franklin, J.F. (2002) *Conserving Forest Biodiversity*. Washington, DC: Island Press.

Lomolino, M.V. (2004) 'Conservation biogeography', in M.V. Lomolino and L.R. Heaney (eds.), *Frontiers of Biogeography*. Sunderland, MA: Sinauer. pp. 293–296.

Lomolino, M.V., Riddle, B.R., and Brown, J.H. (2006) *Biogeography*. Sunderland, MA: Sinauer.

Lookingbill, T.R., and Urban, D.L. (2005) 'Gradient analysis, the next generation: Towards more plant-relevant explanatory variables', *Canadian Journal of Forest Research*, 35: 1744–53.

MacArthur, R.H., and Wilson, E.O. (1967) *The Theory of Island Biogeography*. Princeton, NJ: Princeton University Press.

MacNally, R. (2005) 'Scale and an organism-centric focus for studying interspecific interactions in landscapes', in J.A. Wiens and M.R. Moss (eds.), *Issues and Perspectives in Landscape Ecology*. Cambridge: Cambridge University Press. pp. 52–69.

Malanson, G.P. (1993) *Riparian Landscapes*. Cambridge: Cambridge University Press.

Malanson, G.P. and 15 authors (2007) 'Alpine treeline of western North America: Linking organism-to-landscape dynamics', *Physical Geography*, 28: 378–96.

Malanson, G.P., and Cramer, B.E. (1999) 'Ants in labyrinths: Lessons for critical landscapes', *The Professional Geographer*, 51: 155–70.

Malanson, G.P., Zeng, Y., and Walsh, S.J. (2006) 'Landscape frontiers, geography frontiers: Lessons to be learned', *The Professional Geographer*, 58: 383–96.

Manel, S., Schwartz, M.K., Luikart, G., and Taberlet, P. (2003) 'Landscape genetics: Combining landscape ecology and population genetics', *Trends in Ecology & Evolution*, 18: 189–97.

McIntyre, S., and Hobbs, R. (1999) 'A framework for conceptualizing human effects on landscapes and its relevance to management and research models', *Conservation Biology*, 13: 1282–92.

Meentemeyer, R.K., Rank, N.E., Anacker, B.L., Rizzo, D.M., and Cushman, J.H. (2008) 'Influence of land-cover change on the spread of an invasive forest pathogen', *Ecological Applications*, 18: 159–71.

Meentemeyer, V. (1984) 'The geography of organic decomposition rates', *Annals of the Association of American Geographers*, 74: 551–60.

Messina, J.P., Walsh, S.J., Mena, C.F., and Delamater, P.L. (2006) 'Land tenure and deforestation patterns in the Ecuadorian Amazon: Conflicts in land conservation in frontier settings', *Applied Geography*, 26: 113–28.

Metzger, J.P. (2008) 'Landscape ecology: Perspectives based on the 2007 IALE World Congress', *Landscape Ecology*, 23: 501–4.

Miller, C., and Urban, D.L. (2000) 'Connectivity of forest fuels and surface fire regimes', *Landscape Ecology*, 15: 145–54.

Miller, J., Franklin, J., and Aspinall, R. (2007) 'Incorporating spatial dependence in predictive vegetation models', *Ecological Modelling*, 202: 225–42.

Millington, A.C., Velez-Liendo, X.M., and Bradley, A.V. (2003) 'Scale dependence in multitemporal mapping of forest fragmentation in Bolivia: Implications for explaining temporal trends in landscape ecology and applications to biodiversity conservation', *ISPRS Journal of Photogrammetry & Remote Sensing*, 57: 289–99.

Milne, B.T. (1992) 'Spatial aggregation and neutral models in fractal landscapes', *American Naturalist*, 139: 32–57.

Munroe, D.K., Nagendra, H., and Southworth, J. (2007) 'Monitoring landscape fragmentation in an inaccessible mountain area: Celaque National Park, Western Honduras', *Landscape and Urban Planning*, 83: 154–67.

Naveh, Z., and Lieberman, A.S. (1984) *Landscape Ecology, Theory and Application*. New York: Springer-Verlag.

Nesslage, G.M., Maurer, B.A., and Gage, S.H. (2007) 'Gypsy moth response to landscape structure differs from neutral model predictions: Implications for invasion monitoring', *Biological Invasions*, 9: 585–95.

Opdam, P., Foppen, R., and Vos, C. (2002) 'Bridging the gap between ecology and spatial planning in landscape ecology', *Landscape Ecology*, 16: 767–79.

Parker, K., and Markwith, S. (2007) 'Expanding biogeographic horizons with genetic approaches', *Geography Compass*, 1: 246–74.

Pascual-Hortal, L., and Saura, S. (2006) 'Comparison and development of new graph-based landscape connectivity indices: Towards the prioritization of habitat patches and corridors for conservation', *Landscape Ecology*, 21: 959–67.

Peterjohn, W.T., and Correll, D.L. (1984) 'Nutrient dynamics in an agricultural watershed: Observations on the role of a riparian forest', *Ecology*, 65: 1466–75.

Peters, D.P.C., Bestelmeyer, B.T., and Turner, M.G. (2007) 'Cross–scale interactions and changing pattern-process relationships: Consequences for system dynamics', *Ecosystems*, 10: 790–96.

Peterson, D.L., and Parker, V.T., eds. (1998) *Ecological Scale: Theory and Applications*. New York: Columbia University Press.

Phillips, J.D. (2006) 'Evolutionary geomorphology: Thresholds and nonlinearity in landform response to environmental change', *Hydrology and Earth System Sciences*, 10: 731–42.

Proulx, R. (2007) 'Ecological complexity for unifying ecological theory across scales: A field ecologist's perspective', *Ecological Complexity*, 4: 85–92.

Ranta, P., Blom, T., Niemela, J., Joensuu, E., and Siitonen, M. (1998) 'The fragmented Atlantic rain forest of Brazil: Size, shape and distribution of forest fragments', *Biodiversity and Conservation*, 7: 385–403.

Rayner, M.J., Clout, M.N., Stamp, R.K., Imber, M.J., Brunton, D.H., and Hauber, M.E. (2007) 'Predictive habitat modelling for the population census of a burrowing seabird: A study of the endangered Cook's petrel', *Biological Conservation*, 138: 235–47.

Ricketts, T.H. (2001) 'The matrix matters: Effective isolation in fragmented landscapes', *American Naturalist*, 158: 87–99.

Ripple, W.J., Bradshaw, G.A., and Spies, T.A. (1991) 'Measuring forest landscape patterns in the Cascade Range of Oregon, USA', *Biological Conservation*, 57: 73–88.

Risser, P.G., Karr, J.R., and Forman, R.T.T. (1984) *Landscape Ecology: Directions and Approaches*. Champaign, IL: Illinois Natural History Survey.

Robertson, M.P., Peter, C.I. Villet, M.H., and Ripley, B.S. (2003) 'Comparing models for predicting species' potential distributions: A case study using correlative and mechanistic predictive modelling techniques', *Ecological Modelling*, 164: 153–67.

Ryan, J.G., Ludwig, J.A., and Mcalpine, C.A. (2007) 'Complex adaptive landscapes (CAL): A conceptual framework of multi-functional, non-linear ecohydrological feedback systems', *Ecological Complexity*, 4: 113–27.

Schmitz, O.J. (2005) 'Scaling from plot experiments to landscapes: Studying grasshoppers to inform forest ecosystem management', *Oecologia*, 145: 225–34.

Schreiber, K.F. (1990) 'The history of landscape ecology in Europe', in I.S. Zonneveld and R.T.T. Forman (eds.), *Changing Landscapes: An Ecological Perspective*. New York: Springer-Verlag. pp. 21–33.

Schussman, H., Geiger, E., Mau-Crimmins, T., and Ward, J. (2006) 'Spread and current potential distribution of an alien grass, *Eragrostis lehmanniana* Nees, in southwestern USA: Comparing historical data and ecological niche models', *Diversity and Distributions*, 12: 582–92.

Stallins, J.A., and Parker, A.J. (2003) 'The influence of complex systems interactions on barrier island dune vegetation pattern and process', *Annals of the Association of American Geographers*, 93: 13–29.

Strand, E.K., Robinson, A.P., and Bunting, S.C. (2007) 'Spatial patterns on the sagebrush steppe/Western juniper ecotone', *Plant Ecology* 190: 159–73.

Swanson, F.J., Kratz, T.K., Caine, N., and Woodmansee, R.G. (1988) 'Landform effects on ecosystem patterns and processes', *BioScience*, 38: 92–98.

Tewksbury, J.J., Levey, D.J., Haddad, N.M., Sargent, S., Orrock, J.L., Weldon, A., Danielson, B.J., Brinkerhoff, J., Damschen, E.I., and Townsend, P. (2002) 'Corridors affect plants, animals, and their interactions in fragmented landscapes', *Proceedings of the National Academy of Sciences of the United States of America*, 99: 12923–26.

Thuiller, W. and 11 authors (2008) 'Predicting global change impacts on plant species distributions: Future challenges', *Perspectives in Plant Ecology, Evolution and Systematics*, 9: 137–52.

Tischendorf, L., and Fahrig, L. (2000) 'On the usage and measurement of landscape connectivity', *Oikos*, 90: 7–19.

Tjallingii, S.P., and De Veer, A.A., eds. (1982) *Perspectives in Landscape Ecology. Proceedings of the 1st International Congress in Landscape Ecology, Veldhoven, The Netherlands, April 6–11 1981*. Wageningen, Netherlands: Center for Agricultural Publishing and Documentation.

Turner, M.G. (2005a) 'Landscape ecology: What is the state of the science?' *Annual Review of Ecology, Evolution and Systematics*, 36: 319–44.

Turner, M.G. (2005b) 'Landscape ecology in North America: Past, present, and future', *Ecology*, 86: 1967–74.

Turner, M.G., Gardner, R.H., and O'Neill, R.V. (2001) *Landscape Ecology in Theory and Practice*. New York: Springer-Verlag.

Turner, M.G., and Romme, W.H. (1994) 'Landscape dynamics in crown fire ecosystems', *Landscape Ecology*, 9: 59–77.

Umetsu, F., Metzger, J.P., and Pardini, R. (2008) 'Importance of estimating matrix quality for modeling species distribution in complex tropical landscapes: A test with Atlantic forest small mammals', *Ecography*, 31: 359–70.

Unwin, D.J. (1996). 'GIS, spatial analysis and spatial statistics', *Progress in Human Geography*, 20: 540–51.

Urban, D.L. (2005) 'Modeling ecological processes across scales', *Ecology*, 86: 1996–2006.

Urban, D.L., O'Neill, R.V., and Shugart, H.H. (1987) 'Landscape ecology', *BioScience*, 37: 119–27.

Wagner, H.H., and Fortin, M.J. (2005) 'Spatial analysis of landscapes: Concepts and statistics', *Ecology*, 86: 1975–87.

Walsh, S.J., and McGinnis, D. (2008) 'Biocomplexity in coupled human-natural systems: The study of population and environment interactions', *Geoforum*, 39: 773–75.

Wang, Q., and Malanson, G.P. (2008) 'Neutral landscapes: Bases for exploration in landscape ecology', *Geography Compass*, 2: 319–39.

Weathers, K.C., Cadenasso, M.L., and Pickett, S.T.A. (2001) 'Forest edges as nutrient and pollutant concentrators: Potential synergisms between fragmentation, forest canopies, and the atmosphere', *Conservation Biology*, 15: 1506–14.

Westphal, M.I., Field, S.A., Tyre, A.J., Paton, D., and Possingham, H.P. (2003) 'Effects of landscape pattern on bird species distribution in the Mt. Lofty Ranges, South Australia', *Landscape Ecology*, 18: 413–26.

Whittaker, R.J., Willis, K., and Field, R. (2001) 'Scale and species richness: Towards a general hierarchical theory of species diversity', *Journal of Biogeography*, 28: 453–70.

Wickham, J.D., and Riitters, K.H. (1995) 'Sensitivity of landscape metrics to pixel size', *International Journal of Remote Sensing*, 16: 3585–94.

Wiens, J.A. (1989) 'Spatial scaling in ecology', *Functional Ecology* 3: 385–97.

Wiens, J.A. (2002) 'Central concepts and issues of landscape ecology', in K.J. Gutzwiller (ed.), *Applying Landscape Ecology in Biological Conservation*. New York: Springer-Verlag. pp. 3–21.

Wiens, J.A. (2005) 'Towards a unified landscape ecology', in J.A. Wiens and M.R. Moss (eds.), *Issues and Perspectives in Landscape Ecology*. Cambridge: Cambridge University Press. pp. 365–73.

Wiens, J.A. (2008) 'Allerton Park 1983: The beginnings of a paradigm for landscape ecology?' *Landscape Ecology*, 23: 125–28.

Wiens, J.A., and Moss, M.R., eds. (2005) *Issues and Perspectives in Landscape Ecology*. Cambridge: Cambridge University Press.

Willig, M.R., Kaufman, D.M., and Stevens, R.D. (2003) 'Latitudinal gradients of biodiversity: Pattern, process, scale, and synthesis', *Annual Review of Ecology, Evolution and Systematics*, 34: 273–309.

With, K.A., and Crist, T.O. (1995) 'Critical thresholds in species responses to landscape structure', *Ecology*, 76: 2446–59.

Wu, J., Shen, W., Sun, W., and Tueller, P.T. (2002) 'Empirical patterns of the effects of changing scale on landscape metrics', *Landscape Ecology*, 17: 761–82.

Young, K.R., and Aspinall, R. (2006) 'Kaleidoscoping landscapes, shifting perspectives', *The Professional Geographer*, 58: 436–47.

Zeng, Y., and Malanson, G.P. (2006) 'Endogenous fractal dynamics at alpine treeline ecotones', *Geographical Analysis*, 38: 271–87.

5

Classification of Biogeographical and Ecological Phenomena

Angelika Schwabe and Anselm Kratochwil

5.1 INTRODUCTION

Biogeography is a highly interdisciplinary discipline, and one of the fields to which it is strongly linked is ecology. Therefore, the classification approaches of both disciplines should be reconsidered together. According to a commonly accepted definition, biogeography analyzes spatial distributions of organisms in the past, at present, and in the future, taking into account different scales. According to the first definition of ecology, by the zoologist Ernst Haeckel in the year 1866, it is the science of the relations of organisms among themselves and to their surrounding environments, but a modern textbook (Begon et al., 1996) states that ecology deals with 'the distribution and abundance of different types of organism over the face of the earth, and about the physical, chemical but especially the biological features and interactions that determine these distributions and abundances; see also Begon et al. (2006).' As a consequence it is clear that we now consider there to be a broad overlap between biogeographical and ecological topics.

We introduce classification approaches for species, species groups, functional groups, communities, community groups, habitats, biomes, biogeographical realms, ecozones, and ecoregions. We regard classification as a tool for structuring data—in our case, especially data with spatial relevance. The aim of classification is to obtain groups of objects in discrete classes (often in a hierarchical structure). Within a given group, the objects are homogeneous, allow group-specific generalization, and are distinct from those in other groups. Classification can start from the entity (divisive classification) or the single elements (agglomerative classification). The most common and often most successful approach is the divisive-polythetic one (polythetic = referring to many elements, e.g., species). Ecological and biogeographical data can be compared by univariate procedures (regarding only one factor) or by multivariate procedures with different variables (ordination). The latter are not classification sensu stricto but help to understand, for example, different clusters of classification. As biogeographical and ecological elements 'from the community to the biome' often follow gradients, there is no one 'true' and no single classification, but mostly there are several possible ways to typify elements. The same is true for the scale problem: not all classification types are appropriate for all scale levels. In this chapter we concentrate on terrestrial systems and deal with both fine-scale and broad-scale views.

There is no doubt that progress in the area of taxonomic classification is essential for biogeographical and ecological research. The following question arises: how useful is classification, from the level of communities up to the level of the whole geobiosphere, for answering biogeographical questions?

5.2 TAXA-BASED CLASSIFICATION

5.2.1 Introduction

The diagnosis of the structure of area systems and the evolutionary, historical, and/or ecological reasons for taxa distribution patterns are investigated by the discipline of chorology. The basic elements for classification are taxa and their geographical ranges (areas of distribution) or groups of taxa with similar geographical ranges (area types). Factors limiting geographical ranges are not only floral and faunal history and climate, soil, and topographical boundaries but also dispersal limits and, in the case of plant species, for example, herbivore attacks (Bruelheide and Scheidel, 1999), or, in the case of animals, for example, the lack of specific food resources. Nowadays, attention is focussed especially on the dynamics of invader-area systems and area dynamics caused by global change.

Taxa-based hierarchical classification systems refer to floral and/or faunal contrasts. The contrast is calculated by summing up the taxa occurring in area A but not in area B, and vice versa. Calculations for sections of 100 km^2 between areas A and B allow a specific gradient to be determined.

The similarities of species composition in specific areas are worked out, for example, by cluster analysis and other numerical classification approaches. Molecular data give new insights into taxa differentiation and distribution (Heywood and Watson, 1995; Cox and Moore, 2005).

5.2.2 Historical aspects

Centuries ago, the 'fathers of zoo- and phytogeography', Georges-Louis Leclerc de Buffon, later Comte de Buffon (1707–88), and Gottfried Reinhold Treviranus (1776–1837), described species formations in different parts of the world and tried to classify the distribution and structural pattern (see 'classical papers', compiled by Lomolino et al., 2004; Schroeder, 1998 and references therein). Buffon was the first to recognize the regional biogeographical differentiation of species in what is now referred to as Buffon's law. What is not commonly known is that Treviranus (1803) had already elaborated the first classification of eight main flora types: Nordic (plant species found in the Old and New World north of 50°N); Virginian (North America between 50°N to 35°N); West Indian (Americas between 35°N and 35°S); Orient (Eurasia from the Mediterranean to Japan, but excluding southeast Asia); East Indian (southeast Asia); Africa (Africa); Austroasian

(Australia and the Pacific islands, excluding New Zealand); and Antarctic (South America south of 35°S and New Zealand).

5.2.3 Phytogeographic and zoogeographic regions of the world

Linking traditional taxa-based phyto- and zoogeographical maps to form one unified biogeographic map of the world is problematic, as the dispersal mechanisms and evolutionary characteristics are different in each taxonomic group. Therefore, we introduce phytogeographic and zoogeographic regions of the world (i.e., for vascular plants and mammals) in two different maps (Figure 5.1). Unified biogeographic maps are possible if mainly structural elements are in focus, as in the case of biomes (see Section 5.6).

5.2.4 Phytogeographic classification

As a result of its younger earth history, floral exchange was possible in the northern hemisphere up to the start of the Pleistocene period. The similarities are so strong that North America and northern Eurasia form the Holarctic floral kingdom. The differences between the tropical areas are considerable and there are two distinct Neotropical and Palaeotropical kingdoms. The same is true for the southern part of Africa (Capensis kingdom) as well as for the Australian and Antarctic kingdoms. There are large transition zones, especially between the Holarctic and Neotropical kingdoms. The lines of evidence for this differentiation including the Capensis kingdom refer to Rikli (1913).

The nomenclature for the hierarchically divided subsystems, from broad scale to fine scale, is: kingdoms, regions, subregions, provinces, subprovinces, districts, and subdistricts. The districts are characterized by floral elements. In Europe, for example, the Arctic, Boreal, Atlantic, Central European, Mediterranean, Pontic, and Turanian floral regions can be distinguished. Combinations with orographic characteristics (e.g., boreoalpine) and the use of 'eu-' and 'sub-' prefixes (e.g., eu-, subatlantic) are used for further differentiation. The regions are differentiated with chorological groups of species, which form similar distribution clusters. Many examples of different area types are presented in the classical chorological atlas of Meusel et al. (1965 ff).

Meusel et al. (1965 ff) identified ten floral belts globally, which correspond to latitudinal belts (i.e., arctic, boreal, temperate, submeridional,

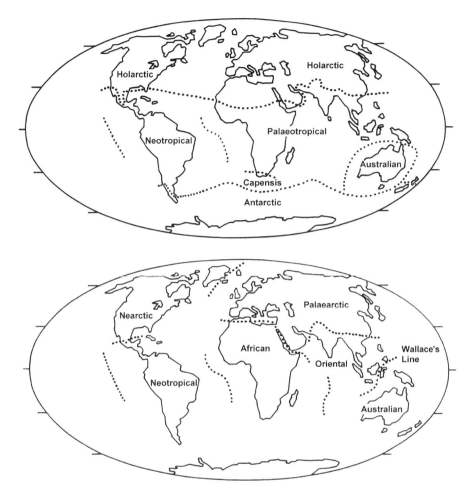

Figure 5.1   Phytogeographic (1a) and zoogeographic regions (1b) of the world; 1b refers mainly to mammals (modified from various authors)

meridional, boreosubtropical, tropical, austrosub-tropical, australic, antarctic). They are modified by gradients of continental and maritime climates. Floral kingdoms and floral belts can be combined. In Eurasia, for example, the biogeographical relations between the boreoalpine and the boreal element sensu stricto are strong, not only for the flora but for the bird fauna as well, reflecting the shared floral and faunal history. This was, for example, worked out by Mattes (1988), comparing the subalpine *Picea abies*- and *Pinus cembra*-*Larix decidua* forests of the Central Alps with boreal forests. The forests of the Alps share 72% of its bird species with the boreal zone of northeastern Fennoscandia, but on the level of subspecies the areas are more separated (only 51% shared taxa).

Other authors use 'centers of plant species endemism' for a subdivision of areas (phytocho-ria); this approach has been used in South America, Africa, Asia, and New Zealand by various authors (see references in Jäger, 1995).

5.2.5 Zoogeographic classification

In comparison with the floral kingdoms, the zoogeographical regions are more differentiated in the case of the Holarctic (Nearctic and Palaearctic are specified as distinct regions by different authors, for example, de Lattin, 1967; Morrone, 2002; Cox and Moore, 2005) and the palaeotropical region (African and Oriental), but there is no separate zoogeographical region corresponding to the Capensis floral kingdom. Wallace's Line between the Australian-Oriental region corresponds to a remarkable faunal boundary. The differentiation was mainly worked out by Philip Lutley Sclater in the year 1858 and Alfred Russel Wallace in 1876 (see Lomolino et al., 2004 and references therein). In general the differentiation of the two regions on either side of Wallace's Line is strongly taxa-based (in this case, mammal taxa).

If so-called wandering families of mammals (i.e., those with mainly worldwide distribution, e.g., sciurids and leporids) are excluded from the analysis, the interrelationships of the terrestrial mammal families show that especially high specificity characterises the African, Neotropical, and Australian regions, in which 16, 15, and 10 families, respectively, are not shared with any other region (Cox and Moore, 2005: Figure 9.3). The nomenclature of the hierarchically ordered classification from broad-scale to fine-scale is similar to phytogeography: for instance, subregions and provinces are subdifferentiated. However, the term 'district' is not used in a uniform manner.

The nomenclature of faunal elements follows on the one hand the actual area types, an approach that is widely accepted. On the other hand, and often hypothetically, historical-genetic analysis is used (de Lattin, 1967; subsequently adapted by Müller, 1974). 'Centers of taxa dispersal' (i.e., centers of taxa origin and/or differentiation) are distinguished (e.g., holomediterranean faunal element, including refuge areas). This is confusing and criticized by some authors (Cox and Moore, 2005). The assumption that present-day biodiversity hot spots of certain taxa have been the cores for radiation of wider areas is problematic (see, for example, the discussion of the 'New Zealand school of panbiogeography' summarized in Cox and Moore, 2005).

5.3 STRUCTURE-BASED CLASSIFICATION AND FUNCTIONAL-ECOLOGICAL APPROACHES

5.3.1 Plant growth forms, life-form types, and functional types

The physiognomy of plants, their specific growth form, had been recognized early as a classification feature, which makes it possible to describe structural types without identifying the specific taxa and therefore to work out global comparisons, for example, of neo- and palaeotropical rain forests with markedly different taxa. Alexander von Humboldt (1769–1859) in 1805 had already differentiated growth forms such as palms, banana form, malvaceous form, bombaceous form, mimosa form, heather, cactus form, orchids, casuarinas, conifer, arum form, lianas, aloe form, grass form, ferns, lilies, willow form, myrtle form, Melastoma form, and laurel form (see Lomolino et al., 2004).

Growth forms can hardly be regarded without considering functional aspects of adaptation— for instance, to water or salt stresses, extreme temperatures, and wind. Such adaptation strategies lead to similarities of growth forms in different plant taxa. In 1855 Alphonse de Candolle proposed, in his *Géographie botanique raisonnée*, that the main factors influencing the distribution of plant growth forms are heat and drought tolerances of plant species; he had already differentiated types according to temperature and moisture requirements (e.g., Megatherms in the case of high temperature and sufficient moisture, and Mesotherms in the case of moderate temperature and moisture; see Archibold, 1995 and references therein). In retrospect he was the first person who identified physiologically adapted plant forms, which we now call plant functional types (PFTs).

The understanding of growth forms and their ecological significance was reinforced in particular by Eugen Warming from 1895 onwards, and by Andreas F.W. Schimper (1898). Warming (1909) wrote "the greatest advance, not only in biology in its wider sense, but also in oecological phytogeography, will be the oecological interpretation of the various growth forms."

Christen Raunkiaer drew up a system of plant life-form types in 1904, which considered in particular plants' strategies to survive cold or dry periods during a year. This system was later supplemented by Ellenberg and Mueller-Dombois (1974, see references therein). Raunkiaer divided plants into phanerophytes (trees and shrubs), chamaephytes (dwarf shrubs), hemicryptophytes (buds are protected by the litter layer), cryptophytes (storage organs persist in the soil), therophytes (regenerate from seeds), and helo-/hydrophytes (storage organs persist in swamp or water); phanerophytes may be subdifferentiated into phanerophytes sensu stricto (trees) and nanophanerophytes (shrubs). Therophytes are often seed-bank species. There are remarkable relations between climatic types and life-form spectra; for example, therophytes are favoured not only in (semi)-deserts with episodic precipitation but also by mediterranoid climates with a seasonal change between the wet west-wind and dry trade-wind zone (Figure 5.2). The plant life-form types already bridge structural and functional aspects. In their functional characteristics communities are

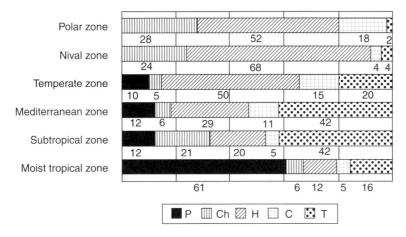

Figure 5.2   Raunkiaer's plant life-form spectra from different regions of the world (from Kratochwil and Schwabe, 2001). P, Phanerophytes; Ch, Chamaephytes; H, Hemicryptophytes; C, Cryptophytes; T, Therophytes.

comparable, which are floristically dissimilar but have the same traits and trade-offs.

In plant and vegetation ecology, the importance of functional aspects of classification has been discussed extensively since the 1970s, having mainly been inspired by Grime's (1974) proposed triangular model of competitors, stress-tolerant species and ruderal-strategists (the C-S-R-model), the latter occurring in open, disturbed habitats. Many authors have discussed this model, and there is some agreement that only three and some combined strategies: C-R, S-R, C-S (see Grime, 1979) is a simplification of the reality (e.g., Ellenberg, 1996; Wilson and Lee, 2000 and references therein). Nevertheless, the C-S-R-model had stimulating results for vegetation classification and, especially, in understanding the effects of grazing (intermediate levels of C-S-R) and spontaneous succession (increase of competitors).

To identify the type of function it is useful to differentiate between two types of functional groups:

1. guilds sensu Root (1967), which are linked to resource use (see below),
2. PFTs, which in a wider sense are groups of plants with similar structural and functional characteristics.

Meanwhile, extensive trait bases for plant species with functional characteristics were worked out in some parts of the world (e.g., LEDA trait base for the northwest European flora; Walker et al., 2005

for the Arctic vegetation map). Genome sizes may be important characteristics for PFTs, for example, in the total angiosperm sample woody growth form is characterized by a smaller genome size compared with the herbaceous growth form (Ohri, 2005). Life-form types and other different traits can be used for classification, if the data are linked, for example, with relevés (see below). Box (1996) proposed a system of 15 dominant plant structural-functional types, which explains the main global vegetation pattern of mature ecosystems and is compatible with the biome classification by Walter and Breckle (1999) (see below). Plant functional types may be linked to climate change processes (Woodward and Cramer, 1996). In general, there is a need for a unified global framework for PFTs.

5.3.2 Animal life-form types, functional groups, and guilds

Adapted to specific environmental conditions and with similar structural, physiological, developmental, and/or ethological features, different organization types (structural types) of animals are characterized as life-form types. Species of the same life-form type—in many cases with no phylogenetic or systematical relationship—have evolved convergences (convergent evolution) and, analogously, adaptations according to a corresponding similarity in environmental selection pressures (adaptive syndrome).

Animal life-form types may be specified according to the following:

1. different locomotion: for example, sessile, hemisessile, and vagile animals; the latter can be burrowing, climbing, creeping, jumping, running, or flying with many subtypes;
2. different foraging or feeding mode: for example, particle feeder: suspension feeder, filterer, tentacle feeder; substrate feeder; sap feeder: licker, piercer-and-sucker; macrophageous animals: 'swallower', comminutor, decomposer; collector, grazer, predator; and
3. different substrate: 'pedo-, geo- and phytophilous animals'; for example, limicolous animals (mud), terricolous a. (ground/soil: epi-, endo-, mesogaion), arenicolous a. (sand: epi-, endo-, mesopsammon), herbicolous a. (on plants: epiphytobios, in plant tissue: endophytobios), lignicolous a. (on dry trunks: epidendrobios, in woody tissue: endodendrobios).

Hundreds of different life-form types are present, for example, in forests (including soils), with the key importance of invertebrates. Therefore, it is not possible to give a scientifically satisfying worldwide overview in a table or figure.

In animal ecology a functional group refers to species groups with similar structural and functional characteristics; they may be grouped, for example, according to similar locomotion mode (see above). This definition corresponds to the PFTs in the wider sense (see above).

Guilds sensu Root (1967) are composed of species with similar resource utilization (e.g., phytophageous animals with the subtypes sap feeder, ectophageous leaf-eater, or miner). Guild classifications were elaborated, for example, for soil predators, for flower-visiting insects, for different groups of birds and mammals (see Kratochwil and Schwabe, 2001 and references therein). Especially the guild concept sensu Root (1967) leads to structural and functional insights and—in view of the enormous species richness—it is mostly linked to taxonomic groups, such as ungulate browsers versus grazers in the savanna biome, semiaquatic herbivore mammals, flower-visiting bees feeding on nectar and pollen, and so forth. Unfortunately, for many existing guild classifications it is unclear what concept the classification is based on (Wilson, 1999).

5.3.3 Formations

Heinrich August Grisebach was the first to differentiate the vegetation of the earth according to physiognomic types, which he called 'formations'. In 1838 he defined a phytogeographical formation as a group of plants that built up a physiognomic type, for example, a meadow or deciduous woodland. These formations are characterized either by only one dominant species or by different species of the same physiognomic type. There are numerous formation classifications of the Earth, for example, those elaborated by Schimper (1898); Rübel (1930); Whittaker (1970, 1973); and others. These formation types are integrated in the biome approach (Section 5.6).

5.4 COMMUNITY-BASED CLASSIFICATION

5.4.1 Clarifying terminology

Groups of different species or species-groups that occur regularly together in a spatially limited area are defined as coenosis or community; both terms do not imply any hierarchical rank. Most of the species interact and form the biotic part of an ecosystem. As a general term for the occurrence of different species in a certain spatial unit—either with or without interspecific relations—the word 'assemblage' is proposed. Some authors use 'community' and 'assemblage' synonymously (e.g., Cox and Moore, 2005). Other authors suppose that plant communities have no fixed boundaries (see discussion in van der Maarel, 2005).

The term 'biocoenosis' was first described by Karl August Möbius in 1877, including the specification that all organisms in a biocoenosis require one another. Since then a biocoenosis has come to be seen as a species composite in which the individuals have similar requirements for abiotic and biotic conditions at a site. Interactions occur, at least for some of the species. Trophic relations between the individuals are concentrated in the biocoenosis, but not fully restricted to it. In particular, most of the animals of a biocoenosis, as well as the cryptogams and microorganisms found in it, depend on the structure built up by the plants (Kratochwil and Schwabe, 2001). Corresponding to this definition, a phytocoenosis is characterized by a definable and repeated grouping of plant species, which reflects the abiotic and biotic site conditions (including anthropogenic factors such as grass mowing, management of forests, harvesting or anthropo-zoogenic factors as livestock grazing). A classification for the phytocoenosis-part of a biocoenosis is possible, and we will introduce some approaches. The zoocoenosis has to be divided into different zootaxocoenoses with reference to the phytocoenosis, to parts of the phytocoenosis or to vegetation complexes. Such parts may be different strata (e.g., litter layer, components of woodland and forest structure such as canopy, understorey) or minor habitats (choriotope = biochorion, e.g., tree stumps).

Ten years before Tansley (1935) introduced the term 'ecosystem', Sukatschew (1926) combined the biocoenosis and the abiotic site conditions (clima-, hydro-, pedo-, and edaphotope) in the term 'biogeocoenosis'. A biogeocoenosis is defined as the complex of homogeneous natural environmental factors covering a distinct part of the Earth's surface (atmosphere, rock, soil, hydrological conditions, vegetation, animals, and microorganisms). According to Sukatschew, the 'biogeocoenosis' represents, as a classification unit, the result of biotic and abiotic interaction emphasizing climatological, hydrological, pedological, and geological factors. In contrast, the ecosystem describes a functional and dimensionless unit consisting of biotic and abiotic characteristics and processes. Flows of energy in food webs with producers, consumers, and decomposers and cycling of carbon, mineral nutrients, and water characterize the open systems and should guarantee ecosystem functioning. In principle, ecosystems are not objects for classification.

5.4.2 Classification of plant communities

Historical aspects

Theoprastus (372–287 BC) in his *Historia Plantarum* had already described the regular occurrence of different plant species at particular sites. The first steps toward the modern investigation of plant communities were made especially by Warming in 1895; Schroeter and Kirchner in 1902 (see Westhoff and van der Maarel, 1973), Sukatschew (1926), and Braun-Blanquet (1928). In the 1920s a controversy was initiated between the approaches of Clements (1916) and Gleason (1926). Clements was convinced that communities are so strictly organized that they act like a 'superorganism' and can only survive if the specific organisms coexist. Gleason (1926) believed just the opposite with his individualistic concept. The superorganism-concept has been fully rejected (first by Tansley, 1939; see Austin, 2005 for a modern view), but there is evidence—documented by many research results—that community structures exist (see, e.g., Westhoff and van der Maarel, 1973; Ellenberg, 1996, Rodwell et al., 2002; van der Maarel, 2005). However, there are assemblages without community structure, for example, pioneer stages on disturbed sites as road margins or debris accumulations with a lot of stochastical dispersal processes in the course of recovery by

plant and animal species. Another example for such assemblages without community structure is insects, flying in the night close to a lighting system.

In continental Europe and Russia in particular, plant community-based classification systems had been elaborated as early as the first half of the twentieth century, based on relevé samples with exact floristic census and cover-abundance scales (e.g., the Braun-Blanquet method with character plant species; see, e.g., Knapp, 1984 and references therein). The latter method has now become widely used globally.

Different classification approaches for plant communities

There are differences in classification systems with respect to the importance of the analytical characteristics 'dominance' and 'character species', which had been compiled, for example, by Whittaker (1973) and Westhoff and van der Maarel (1973). The vegetation classification based on dominant species was established by Ragnar Hult, Rutger Sernander, and Einar Du Rietz between 1881 and 1921 and used in the Scandinavian and Baltic areas. It is mostly applied to communities that contain few species but have extensive layers of dominant species, and hence are differentiated according to the dominant plant species. The basic unit is the 'sociation', defined by a homogeneous species composition with dominant species in each stratum of the vegetation (e.g., *Pinus sylvestris-Vaccinium myrtillus-Cladonia alpestris*-Sociation). Aimo Cajander in Finland developed a system from the year 1909 onwards by using the field layer in a forest as the indicator for abiotic conditions. The vegetation types of the 'Russian School' of Sukatschew (1926) and other authors are also defined by the dominance of species; the term 'association' corresponds here to the Scandinavian sociation (see Westhoff and van der Maarel, 1973 and references therein). From 1980 onwards the floristic association concept has been widely adopted in the (former) Soviet Union (Korotkov et al., 1991).

The floristic association concept, with its first representatives being Schröter and Flahault (see Westhoff and van der Maarel, 1973 and references therein), was mainly established by Josias Braun-Blanquet and was later called the 'Zürich-Montpellier school'. Already Flahault and Schröter (1910) defined the association "as a plant community type of definite floristic composition, uniform habitat conditions, and uniform physiognomy." The Braun-Blanquet approach was elaborated in species-rich communities, especially in the Alps, and is based on floristic similarities between plot areas, which are sampled by relevés (all occurring plant species and macroscopic cryptogams). Relevés are made in the field in homogeneous plots, which differ in size according to the structure (richness in strata) and species richness of a community type, using the Braun-Blanquet cover-abundance scale (Braun-Blanquet, 1928, 1964). For vascular plant communities the minimal areas range from 0.5 1 m² (Lemnetea communities) to about 10,000 m² in tropical rain forests (further data are given by Braun-Blanquet, 1964; van der Maarel, 2005). The relevés are grouped according to their floristic similarity (today by polythetic-divisive approaches, such as TWINSPAN, or by multivariate ordination, see below).

A hierarchical system of character and differential species is the fundamental principle of the Braun-Blanquet approach. The basic unit is the association: the lowest-level unit in a hierarchical system that still has its own character species. A character species should be restricted to a particular, defined phytosociological unit, whereas differential species indicate special, for example, abiotic factors, such as soil moisture and nutrient supply. The highest-level phytosociological unit is the class (ending -etea), followed by the order (-etalia), the alliance (-ion), and the association (-etum). Subassociations and the lowest-level unit (variant) only have differential species (see Pignatti et al., 1994). All levels from class to association have their specific character species. The general term for types at all hierarchical levels is 'syntaxon'. As an example from the vegetation of eastern Siberia, the following syntaxa have been identified or described: Festuco-Brometea, Thymetalia gobici, Festuco-Thymion gobici, and Hemerocalletum minoris (Korotkov et al., 1991).

In handbooks of phytosociology (e.g., Ellenberg, 1963, 1996; Wilmanns 1973, 1998) the order of plant communities in the books follows the 'sociological progression': from simply structured types with one stratum (e.g., Lemnetea) to structurally diverse woodland types. Ellenberg (1963, 1966) and Wilmanns (1973, 1998) combined the sociological progression with formations in the course of their books, which are both excellent examples for a combination of the Braun-Blanquet classification approach and the functional-ecological aspects.

Ellenberg (1974, 1992) developed a system of 'indicator values for vascular plants' based on the large central European phytosociological databases and the relative abundance of species in specific communities in the field. For the parameters that follow here, moisture, light, temperature, continentality, soil reaction, salts, and nitrogen, there are relative values (mostly on 9- to 12-part scales: higher numbers indicate the higher intensity of the factor) for each plant species. In phytosociological relevés it is possible to calculate the quantitative or qualitative value of the whole relevé (according to either cover/abundance or presence/absence data of each plant species). In the case of moisture and soil reaction values in particular, the approach reflects quite well the abiotic conditions. Meanwhile, such indicator values are available for different countries in Europe: for example, Italy (Pignatti et al., 2005), the southern Aegean region (Böhling et al., 2002), Hungary (Borhidi, 1993), Switzerland (Landolt et al., 2010), Poland (Zarzycky, 1984), Great Britain and Eire (Hill et al., 2004), and the Faroë Islands (Lawesson et al., 2003). There is also some criticism, for example, concerning the nitrogen values (which are values for soil fertility according to Hill et al., 1997, 2004; or in some systems for phosphorus, Chytrý et al., 2009) and the empirical data for nutrient values are often insufficient. Further, there are pitfalls that indicator values may interact and certain processes cannot be separated (Schwabe et al., 2007). On the other side many authors have shown that the pH-gradient correlates well with the R-values (e.g., reviewed by Diekmann, 2003; see also Schaffers and Sýkora, 2000). Nevertheless, the approach helps to develop generalizations (van der Maarel, 1993) and is especially useful to interpret ordinations and to use the values as benchmarks (Hill et al., 2004). In Figure 5.3 we present an example of dry grassland from three separated valleys in the southern inner Alps in Italy. According to macroclimatological data (Schwabe and Kratochwil, 2004) gradients of moisture, temperature, and continentality in particular should be reflected by the indicator values. A Detrended Correspondence Analysis (DCA) of about 90 Braun-Blanquet relevés of 50 square-meter-plots is shown in Figure 5.3a. The net diagrams of the average indicator values of the plots and communities underline the xero-mesothermic gradient on axis one. The upper parts of axis two are characterized by more mesohygric conditions. The net diagrams (Figure 5.3b) explain decisive factors for the position of the communities (Pignatti et al., 2005; Schwabe et al., 2007). This data set was also applied to compare relevés from 1990 to 1995 with older ones of Braun-Blanquet, sampled from 1930 to 1950 (Schwabe et al., 2007), and has shown high stability of the dry grassland communities; it is useful not only for spatial but also for temporal approaches.

Criticisms of the Braun-Blanquet method refer mainly to possible observer bias in sampling and classification (Westfall et al., 1997). This can be excluded by improved objectivity (e.g., by representative sampling, no selection of plot areas with specific species) and by polythetic-divisive classification methods. Nonetheless, in the case of rare types (e.g., rock-fissure vegetation), preferential sampling is necessary and objectivity is

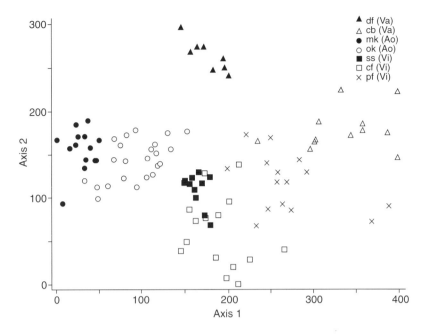

Figure 5.3a   DCA of dry grassland communities in a climatical gradient of the inner Alps. Valleys: Va Valtellina, Ao Valle d'Aosta, Vi Vinschgau, Valle Venosta. Small letters refer to distinct plant communities: mk, Melico-Kochietum prostratae; ok, Onosmo-Koelerietum vallesianae (okp acidophytic subassociation); ss, Stipo-Seselietum variae; df, Diplachno-Festucetum valesiacae; pf, Poo-Festucetum valesiacae; cb, Centaureo-Brachypodietum. Eigenvalues axis 1: 0.53, axis 2: 0.28 (from Schwabe et al., 2007)

compromised. Because of this difficulty, the method had not been widely adopted in Britain and the United States up to the 1980s. Heywood and Watson (1995: 97) stated, "Outside continental Europe, phytosociology has attracted less interest and as a result much less is known about the types of vegetation." Nevertheless, various phytosociological publications originated from Great Britain as early as 1955 (Poore, 1955a, 1955b) and later (e.g., Shimwell, 1968 and others). Since then an elaborated classification system has been developed by Rodwell (1991–2000), covering all British plant communities in five volumes according to a National Vegetation Classification (NVC) summarizing 35,000 relevés. This classification is used in, for instance, the journals of the British Ecological Society. The plot areas are chosen solely on the basis of homogeneity. The communities are grouped into 12 differentiated formations: woodland and scrub, mires, heaths, and so forth, and are named according to diagnostic species, for example, *Fraxinus excelsior–Acer campestre–Mercurialis perennis* woodland.

In the United States, meanwhile, a Vegetation Classification Panel was set up by the Ecological

Society of America and adapted by the Federal Geographic Data Committee (National Vegetation Classification, see also Grossman et al., 1998). The classification is based on a system with higher-level floristic-physiognomic units (alliances) and lower-level floristic units (associations). The definition for association used here combines floristic and physiognomic aspects: "A vegetation classification unit defined on the bases of a characteristic range of species composition, diagnostic species occurrence, habitat conditions and physiognomy." In general the Braun-Blanquet approach is more floristically defined than the U.S. system, but the two approaches overlap to a considerable extent regarding the associations and alliances, with the exception that there are no fixed endings in the U.S. system. Here is an example for the U.S. system: Formation: Temperate Grassland, Meadow and Shrubland, Alliance: *Sporobolus heterolepis-(Deschampsia caespitosa, Schizachyrium scoparium)* herbaceous alliance, Association: *Sporobolus heterolepis-Schizachyrium scoparium-(Carex scirpoidea)/(Juniperus horizontalis)* herbaceous association.

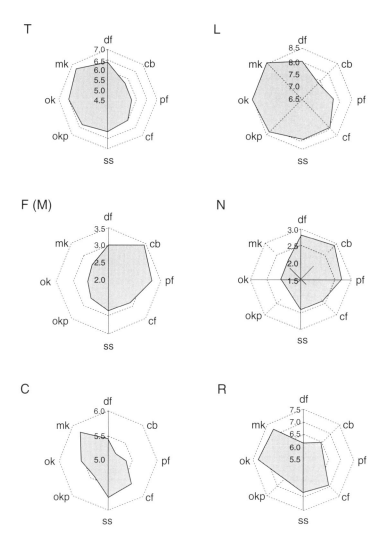

Figure 5.3b   Net diagrams of the average of Pignatti-Ellenberg indicator values of the data set of Figure 5.3a. T, Temperature; L, Light; F(M), Moisture; N, Nutrients; C, Continentality; R, Reaction. The T values of mk, ok, df reflect the top position of the summer hot valleys, additionally df (near Lago di Como with insubric climate) shows higher F (M) values. The highest L values are present in the very open communities of mk and ok. The intrinsic parts of Ao and Vi show the highest C values. Reaction values reflect quite well the acidity of the substrate. Communities with higher soil moisture show higher N values (interaction of two types of indicator values) (from Schwabe et al., 2007)

National vegetation classifications using the Braun-Blanquet approach are available for different countries. For example, in the Netherlands there is a nearly complete census of all plant communities and their distribution with an excellent phytosociological database of more than 300,000 relevés and a permanent plot system (*De Vegetatie van Nederland* 1995 ff). Large amounts of

summarized data also exist, for example, for Japan (e.g. Miyawaki, 1980–1989), Austria (e.g., Willner and Grabherr, 2007), the Czech Republic (Chytrý, 2007), Hungary (Borhidi, 1996), Germany (Oberdorfer, 1977 ff.; Dierschke et al., 1996 ff.), France (e.g. Géhu, 1973; Julve, 1993), Italy (Biondi et al., 1997), Romania (e.g., Coldea et al., 1997) Spain and Portugal (e.g., Rivas-Martínez et al., 2001), and South Africa (Mucina et al., 2000; Mucina and Rutherford, 2006).

The 'European Vegetation Survey' (Rodwell et al., 2002) summarizes the phytosociological alliances for Europe; and the Prodromus of Korotkov et al. (1991) summarizes the syntaxa for the territory of the former Soviet Union.

Numerical and multivariate methods have become available for the ordination and classification of vegetation relevés. As ordination methods, principal components and correspondence analysis are used as well as multidimensional scaling (MDS) and nonmetric MDS (NMDS). An example for a DCA for species-rich communities with a clear gradient is given in Figure 5.3a; further examples for different ordination methods are introduced by Quinn and Keough (2007). There is still discussion about the best ordination method; often DCA ordinations produce results that can be ecologically interpreted very well (see Figure 5.3a) but are from a theoretical approach not as elegant as NMDS. But even NMDS produces in some cases, for example, horseshoe effects (see Leyer and Wesche, 2007). A classification method with an agglomerative approach is the cluster analysis. TWINSPAN is a polythetic-divisive classification method. It is widely used in classification approaches, because relevés and species are classified. Nevertheless, there is a lot of criticism, especially concerning the transparency of the TWINSPAN approach (van Groenewould, 1992; Bruelheide and Chytrý, 2000). An example of a classification for a large area using TWINSPAN was worked out in Australia, where 5,000 relevés were sampled in the major plant communities. Presence/absence data were classified by TWINSPAN into 338 overstorey and 60 understorey floristic groups. The types belong to eight biogeographical regions with 45 subdivisions. Furthermore, this is an excellent basis for correlations with animal communities: in this case, species richness of nonarboreal vertebrates correlates with the plant species richness of the understorey stratum (Specht and Specht, 2001).

Classification of plant communities as a basis for vegetation mapping

Classified data with differentiation of species-groups are the bases for community-based vegetation maps. Extensive map material is available, for example, for Japan, France, Spain, Italy, Germany, Poland, whole Europe, the Arctic region, and others. The abiotic-biotic potential of a site is often assessed by using a map of potential natural vegetation; this is defined as the vegetation that would become established if succession processes were to be completed abruptly under the present abiotic and biotic conditions (Tüxen, 1956). Geobotanical mapping approaches are summarized in Pedrotti (2004). Maps of the actual vegetation and the potential natural vegetation have been elaborated, for example, by Miyawaki (1980–1989), for all provinces in Japan. This has proven to be an excellent basis for planning purposes. For the *Map of the Natural Vegetation of Europe* (2004; scale 1:2,500,000), spatial data are based on the exact classification of 700 mapping units. In Europe the data are essential for the per se protected areas 'Fauna-Flora-Habitat' of the European Union. The vegetation map for the whole Arctic region 1:7,500,000 is an excellent example of the combined mapping of vegetation, plant functional types, and an abiotic template (Raynolds et al., 2005; Walker et al., 2005). From tropical regions there are only few maps available (e.g., Hueck and Seibert, 1981 for South America; Navarro and Ferreira, 2004 for Bolivia).

By linking the results of community mapping with the biome approach (see below), plant communities may have a zonal, an extrazonal, an azonal, or an intrazonal distribution. 'Zonal' applies to the large biomes ordered according to the latitudinal belts, such as the boreal zone. If there are plant communities, for example, dominated by *Picea abies* in the nemoral zone on mesoclimatically cold sites, they occur extrazonally. When extreme edaphic conditions are indicated, for example, by salt marsh vegetation, they are always termed azonal. Intrazonal distribution of communities occurs only scattered in one zone (e.g., specific Central European dry grassland types in the temperate zone).

5.4.3 Is it possible to classify animal communities?

There is a long tradition of describing animal communities as units characterized by the similarity of typical and/or dominant species. The pioneers in the nineteenth century (see references in Kratochwil and Schwabe, 2001) focussed primarily on marine systems. For terrestrial systems, for example, Shelford (1913) proposed a comprehensive characterization of animal communities of the temperate-zone Americas.

A classification of animal communities is in principle difficult: the species and life-form diversity is much higher than in plants, there is high variation of body sizes between species, animals exhibit high mobility, there is a high diversity of habitat preferences, often there are different life-history stages with different habitats, and many species are short-lived or characterized by small population sizes.

A pragmatic approach is to focus on zootaxo-coenoses and guilds. Specific coincidences and affinities exist between plant formations (e.g., savannas, steppes, boreal coniferous forests) and the species composition of animal communities. But fluctuations in animal guilds between different years may be high even in relatively stable vegetation types, as was shown for the subcontinental sand vegetation (class Koelerio-Corynephoretea) and the wild bee pollinator community (Kratochwil et al., 2009). Almost all terrestrial animal species prefer (at least in one developmental stage) habitats dominated by vegetation. Therefore the classification of vegetation can serve as a matrix to work out coincidences and noncoincidences between different zootaxo-coenoses, guilds, and vegetation. This approach has been successfully applied, for example, to soil fauna, nematodes, annelids, spiders, mollusks, millipedes, insects (e.g., flower-visiting insects), birds, and mammals (there are many examples, see Kratochwil and Schwabe, 2001). However, classification of the whole biocoenosis is unfeasible.

5.4.4 Microhabitats and micro-communities

Especially when plant communities are rich in different structures they compose sub-communities, which belong to the same life-form type and which inhabit certain microhabitats—for instance, tree stumps with their bryophyte layer, which are called 'synusiae', 'microcoena', or 'microcommunity'. These are 'dependent communities'. There are different approaches to the ranking of synusial units (summarized, for example, by Barkman, 1973), especially for lichen and bryophyte communities. Animal synusiae are in most cases characterised by guilds. Biological crusts are often good examples of microorganism-dominated microcoena built up by Cyanobacteria, green algae and macro-cryptogams. Phanerogams and cryptogams interact: for example, cyanobacteria provide nitrogen by fixation. Especially in cold and/or very arid regions there are extensive crusts without phanerogams. Often biological crusts are classified according to floristic similarity (Belnap and Lange, 2001).

Bültmann (2005) studied lichen-dominated microcoena in southeast Greenland. These often only extend a few square centimetres but occur over vast areas. Fine-scale abiotic factors such as snow-cover duration and wind exposure are reflected in their occurrence. There are specific correlations between phytocoena and microcoena: for example, Solorinion croceae communities are mainly restricted to Salicetea herbaceae communities.

In this case, as in others, the biogeographer gains deeper insights into the ecology of phanerogam communities by understanding the types of microcoena and how they indicate environmental variables and vice versa.

5.5 COMMUNITY COMPLEX-BASED CLASSIFICATION

Repeated combinations of plant communities in landscape sections form vegetation complexes, which constitute a bridge between the community level and the landscape level. Groups of vegetation complexes link the level of the landscape (topic level) to the choric level.

A systematic registration and analysis of the vegetation complexes by mainly inductive methods began around 1970. The first concepts for inductive recording and analysis were developed by Reinhold Tüxen, following suggestions made by the vegetation geographer Josef Schmithüsen (Schwabe, 1997 and references therein).

The young scientific field of vegetation complex research (or sigmasociology, sigma = sum of communities) has been applied in different countries and regions. Examples include Japan (e.g., Miyawaki, 1978); Canada (Béguin et al., 1994; Thannheiser, 1989); Bolivia (Navarro, 2003); Poland (Matuszkiewicz, 1979; Wojterski et al., 1994); France (e.g., Géhu, 1977); Italy (e.g., Pignatti, 1980); Spain (e.g., Rivas-Martínez, 1987, 1994); Scandinavia (e.g., Dierssen and Dierssen, 1980); Germany (e.g., Tüxen, 1978; Schwabe, 1989, 1991); and Switzerland (e.g., Theurillat, 1992; Zoller et al., 1978). Important applications of sigmasociology to vegetation complexes and biogeography so far have been the differentiation of animal habitats (e.g., Béguin et al., 1977; Schwabe and Mann, 1990), the comparison of different cultural landscapes (e.g., southern Spain and northern Morocco: Deil, 1997, 2003), as a component of multi-layer models in landscape ecology (Navarro, 2003; Thannheiser, 1988), and in identifying and explaining altitudinal gradients (Theurillat, 1992; Schmidtlein, 2003) (see Schwabe, 1997 and further references therein).

It is generally impossible to determine ecosystem boundaries at the landscape level. Such an ecosystem "should be uniform regarding the biogeochemical turnover, and contain all fluxes above and below the ground area under consideration" (Schulze et al., 2005). Certain vegetation complexes occur in a regular order: for example, complexes around springs, along the margins of running water, in fens or bogs, and on *Calluna*-heathland. To study the plant communities of such complexes, methods have been elaborated that use definable spatial units. These are relatively homogeneous geotopes (= physiotopes), specified as plot areas. Physiotopes, such as rock complexes with steep slopes, designate a unit having an approximate homogeneity, which can be defined geomorphologically and topographically. Physiotopes are built up by ever-repeated combinations of (micro)-habitats, which is reflected in the occurrence of plant communities that often have indicator value. Forman and Godron (1981) call them "clusters of interacting stands which are repeated in similar form in a landscape." In anthropogenically influenced landscapes, different types of land use mark the boundaries of plot areas in addition to the physiotope factors.

The method for making vegetation complex relevés is relatively similar to the Braun-Blanquet approach (cf. Section 5.4.2). Homogeneous plot areas are determined and all vegetation units (which must carefully be studied beforehand) are sampled on a scale adapted to the Braun-Blanquet scale (but referring to communities, not to species). In open communities (e.g., extensively grazed grassland) also, single individuals or groups of shrubs or trees may be added. The relevés are classified afterwards by using a divisive-polythetic approach such as TWINSPAN, by cluster analysis, and/or ordinated by correspondence analysis (see, e.g., Schwabe and Kratochwil, 2004 for inneralpine vegetation complexes). The denomination of the complexes is based on the communities that characterize them most. For near-natural site complexes, a unit of the natural vegetation is used by preference.

After studying large landscape sections it is possible to work out landscape-typical distribution patterns (e.g., community complexes showing different levels of eutrophication). The method was applied, for example, to large transects in the driest parts of the Alps, reflecting the indicator value of community complexes influenced by humidity and temperature (Schwabe and Kratochwil, 2004).

Biogeographers in Europe used the sigmasociological approach for mapping purposes (Sigmachorology, see the review by Schwabe, 1997 and references therein). Rivas-Martínez (1987) includes, apart from the potential natural vegetation, all substitute communities to describe a sigmetum. The next spatial level is to regard whole geoseries (geosigmetum) (e.g., Rivas-Martínez, 1987).

In Switzerland an approach has been elaborated that employs areas of a certain prespecified size as a base unit, and therefore includes a deductive element to sampling. These areas are characterized by phytosociological units (e.g., associations/communities or alliances). On the level of phytosociological alliances, this procedure was applied to draw up a remarkable atlas of the whole of Switzerland (Hegg et al., 1993).

5.6 COMBINED PLANT STRUCTURAL, ABIOTIC, AND PARTLY TAXA-BASED APPROACHES: BIOMES AND BIOGEOGRAPHIC REALMS

Regarding higher levels of complexity, Clements (1916) proposed a classification approach for the whole geobiosphere, including phyto- and zoogeographical aspects—biomes. The biome approach is the key concept of the global work of Heinrich Walter (1898–1989). According to Walter and Breckle (1970 ff, 1983 ff) biomes are characterized by similar plant formations that include their animals and microorganisms. Decisive is the plant formation that occurs as the terminal stage in the macroclimate in question. The biome differentiation published by Walter and Breckle is based on the pattern of aridity, humidity, and extreme temperatures, which are essential for the ecophysiological conditions of plant growth and correspond to plant formations and soil types. The most important information for plant growth is depicted in the 'Ecological climate diagrams', the 'Walter diagrams' (Walter and Breckle 1970 ff, 1983 ff), which show very clearly the humid and arid periods in a year (Figure 5.4).

The authors differentiate 'zonobiomes' ('ecological climatic zones' with sub-zonobiomes), which are large and climatically uniform zones within the geobiosphere. Additionally there are orobiomes (OB X) with altitudinal belts in the different zonobiomes and extreme types of soil with azonal vegetation such as swamp soils (pedobiomes). All sharp borderlines are often artificial, and therefore transitional zones occur (zonecotones). The nine main zonobiomes (ZB) are presented in Figure 5.5.

The most important PFTs of the biomes are described in Table 5.1. In some biomes large herbivore mammals with high browsing or grazing impact are common (e.g., ZB IIa, VII, IX, orobiomes). In others, small mammals such as *Dipodomys* sp. (ZB III: neotropical), *Marmota* species (ZB VII, orobiomes), or invertebrates such as termites (ZB II, III and others) influence biome structures.

Physiognomic similarities are often high between ecological-equivalent types in different continents as a result of convergent evolution and adaptive syndromes. For instance, striking floristic differences and structural similarities are present in the five mediterranean areas of the world: the Mediterranean proper, California,

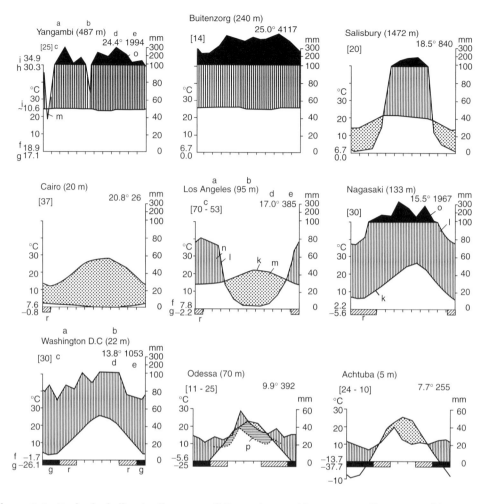

Figure 5.4   Ecological climate diagrams of the main zonobiomes according to Breckle (Walter and Breckle, 1999), slightly modified. Horizontal axis: northern hemisphere January to December, southern hemisphere July to June. Vertical axis: Temperature in °C, precipitation in mm. Letters indicate a, station; b, height above sea level; c, number of years of observation (first temperature, second precipitation); d, mean annual temperature; e, mean annual precipitation; f, mean daily temperature of the coldest month; g, absolute minimum temperature; h, mean daily maximum temperature of the warmest month; i, absolute maximum temperature; j, mean daily temperature fluctuation (h, I, j only for tropical stations); k, curve of mean monthly temperature; l, curve of mean monthly precipitation; m, arid period (dotted); n, humid period (vertical hatching); o, mean monthly precipitation >100 (scale reduced, dark areas indicate perhumid season; p, supplementary precipitation curve, reduced to 10°C = 30 mm, horizontal area above = relative dry period (only for steppe stations); q, month with a mean daily minimum below 0°C (black) = cold season; r, months with absolute minimum below 0°C (diagonally hatched), i.e., late or early frosts possible; s, number of days with mean temperature above + 10°C (duration of vegetation period); t, number of days with mean temperature above -10°C. Zonobiomes according to Figure 5.5: ZB I, Yangambi (Congo), Buitenzorg (Java); ZB II, Salisbury (Zimbabwe); ZB III, Cairo (lower Nile); ZB IV, Los Angeles (California); ZB V, Nagasaki (Japan); Washington D.C.; ZB VII, Odessa (Black Sea); ZB VIIa, Achtuba (lower Volga); ZBVII (rIII) (extreme arid desert with cold winters), Nukuss (Central Asia); ZB VIII, Archangelsk (Siberian boreal zone); ZB IX, Karskije Vorota (Island Vaigatsch, Russian tundra)

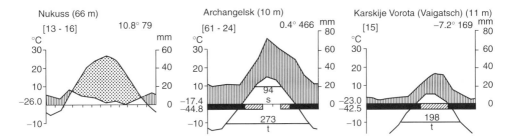

Figure 5.4   Cont'd

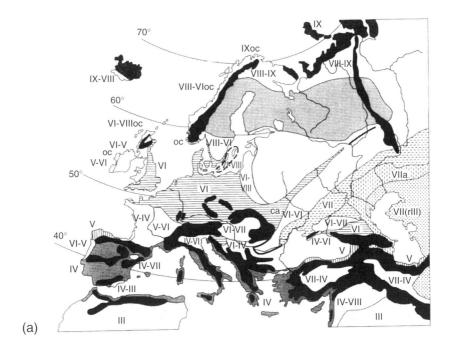

(a)

Figure 5.5   Main zonobiomes of the earth (after Breckle, from Walter and Breckle 1999; slightly modified); a, Europe; b, Asia; c, Australia, New Zealand; d, Africa; e, North and Central America; f, South America: ZB I, evergreen tropical rain forest: equatorial humid diurnal climate; ZB II, tropical semi-evergreen and wet-season green forests and (IIa) savannas, grassland, dry woodlands: humido-arid tropical summer rain region; ZB III, warm deserts and semi-deserts: subtropical arid climates; ZB IV, sclerophyllic mediterranean woodlands: arido-humid winter rain region, Mediterranean regions; ZB V, temperate rain forests, evergreen broad-leaved laurophyll forests: warm-temperate, humid climate; ZB V, deciduous nemoral forests: temperate climate; ZB VII, steppes and (VIIa) (semi)-deserts with cold winters: arid-temperate climate; ZB VIII, boreal forest (evergreen or deciduous coniferous): coldtemperature boreal climate; ZB IX, tundra and polar deserts: arctic climate. OB X Orobiomes, mountains. White spaces between zonobiomes (ZB) are zonoecotones. Further abbr.: a/h, relatively arid or humid for a specific ZB; oc/co, climate with oceanic or continental tendency; fr, frequent frost in tropical mountain regions; wr/sr, prevailing winter or summer rain; swr, two rainy seasons; ep, episodic rain; nm, dew or fog precipitation (nonmeasurable); (rIII), rain as sparse as in ZB III; (tI), temperature curve as in ZB I

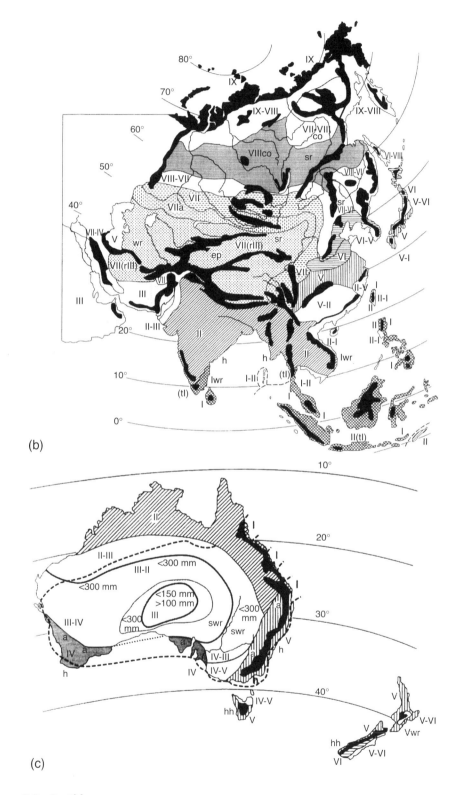

(b)

(c)

Figure 5.5   Cont'd

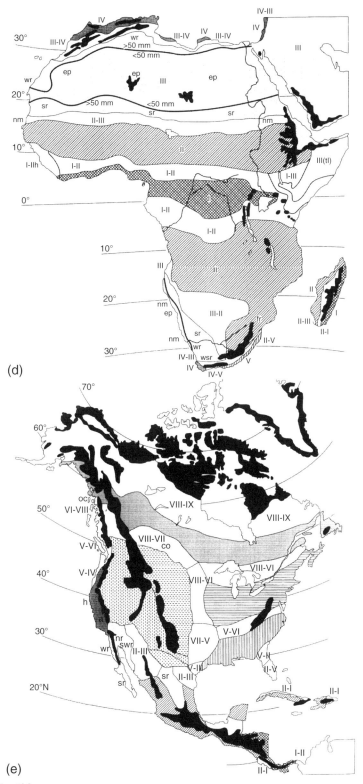

(d)

(e)

Figure 5.5   Cont'd

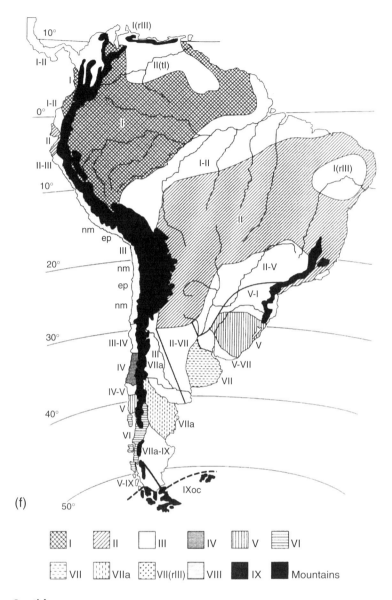

Figure 5.5   Cont'd

central Chile, the Cape Province of South Africa, and southwest Australia. This is also true for the guild of nectar-feeding birds and mammals.

The eight 'biogeographical realms' of Udvardy (1975, updated 1982) are principally based on floral kingdoms of the world: Nearctic, Western and Eastern Palaearctic, Afrotropical, Indomalaya, Australian, Neotropical, Oceania, and Antarctica are differentiated (the Capensis region is integrated into the Afrotropical realm). In a second step, 14 biomes are differentiated, including some mixed mountain and island

systems and lakes; the other types correlate to a large extent with the Walter-Breckle biome approach introduced above, with the exception that boreal forests are classified as temperate needle-leaf forests. The 14 biomes are subdivided into biogeographical provinces. This approach bridges taxic- and structure-based classifications. The classification is used by the International Union for Conservation of Nature (e.g., for the World Heritage Areas) and was prepared as part of the UNESCO 'Man and the Biosphere Programme (MAB)'; see also Lomolino et al. (2005).

Table 5.1 Dominant plant structural-functional types of the world, related to the biome types of Figure 5.5; ZB zonobiomes, OB orobiomes (after Box, 1996; slightly modified)

	Dominant plant type	Biome type(s)	Structure, life-form	Other characteristics
a	Tropical evergreen/ broad-leaved trees	Tropical rain forests ZB I	Tall woody, evergrowing	Mesomorphic
b	Tropical deciduous broad-leaved trees, arborescents	Raingreen forests, woodlands, scrub ZB II	Woody (p.p. facultative deciduous)	
c	Extra-tropical evergreen broad-leaved trees (mainly laurophyll)	Evergreen broad-leaved forests, temp. rain forests ZB V	Woody, evergreen	Mesomorphic, shade-tolerant
d	Temperate deciduous broad-leaved trees	Summergreen broad-leaved forests and woodlands ZB VI	Woody, obligate deciduous	Winter-dormant
e	Temperate/boreal needle-leaved evergreen trees	Needle-leaved evergreen forest/open woodlands ZB (VI) VIII	Woody, evergreen	Winter-dormant (cold-tolerant)
f	Boreal/cool-temperate deciduous needle-leaved trees	Deciduous boreal needle-leaved forest/ open woods ZB VIII	Woody, obligate deciduous	Winter-dormant (cold-tolerant)
g	Sclerophyll trees/arborescents	Arido-humid woodlands, scrub ZB IV	Short woody, (semi-) evergreen	Xeromorphic, light-demanding
h	Sclerophyll/coriaceous shrubs/dwarf shrubs	Shrublands, krummholz, semi-deserts ZB III, OB X	Basally determinate, (semi-) evergreen	Xeromorphic, light-demanding
i	Deciduous shrubs/ dwarf shrubs	Shrublands, krummholz, semi-deserts ZB VIIa, OB X	Basally determinate, deciduous	Rapid growth, seasonally dormant
k	Short-season broad-leaved dwarf shrubs	Tundra: dwarf shrub, graminoid, etc. ZB IX	Basally ramifying, evergreen/deciduous	Winter-dormant (cold-tolerant)
l	Diurnally active tuft- arborescents/frutescents/ forbs	Tropical alpine scrub OB X	Rosettes (diurnal), evergreen	Tolerant to diurnal frost, high UV, etc.
m	Grasses and related graminoids	Savannas and grasslands ZB II, VII	Herbaceous, opportunistic	Rapid growth, spreading
n	Stress-tolerant succulents	Semi-desert scrub ZB III	Stem/leaf/root-succulents, evergreen	Slow growth, water storage in tissue
o	Ephemeral herbs	Semi-desert scrub ZB III	Annual, perennial, ephemeral	Short life cycle/growing season
p	Stress-tolerant cryptogams	Tundra, deserts ZB III, VIIa, IX	Cryptogams	Slow growth

Biome types were chosen as reference areas for experimental-ecological research in the International Biological Program (IBP); see, for example, Archibold (1995) and examples in Goodall (1977 ff). For some biomes, therefore, ecosystem studies with a great deal of flux- and other ecosystem-based data are available; nevertheless, these data are never sufficient to enable an extrapolation to the whole biome.

5.7 CLASSIFICATIONS WITH PREDOMINANCE OF ABIOTIC FACTORS

Climate and vegetation as a basis for classification of the world's vegetation were already used in the classification system drawn up by Wladimir Köppen (1846–1940). Köppen (1931) referred to the plant groups of Alphonse de Candolle (see Section 3.1) and classified five major climate zones, using annual and monthly precipitation and temperature: A (tropical/megathermal) climates, B dry (arid and semiarid) climates, C (temperate/ mesothermal) climates, D (continental/microthermal) climates, and E polar climates. Further subgroups are differentiated (climate types and subtypes).

The life-zone system of Holdridge (1947) uses annual precipitation and 'biotemperature' (average days/year without temperatures $<0°C$ and $>30°C$ and potential evapotranspiration ratio, combining precipitation and biotemperature in an index). The life-zone system differentiates 38 life-zone types referring to 'terminal communities'. As average data are used, the regions characterized by seasonal pattern are not depicted in an appropriate way. The ecoregions of Bailey and Hogg (1986) integrate landform, soil, drainage system, and altitude. The global differentiation of four domains (polar, humid temperate, humid tropical, and dry), 30 divisions, and 98 provinces is based on the macroclimate.

Ecozones are characterized by climatic factors, morphodynamics, soil-forming processes, production, and potential for agriculture and forestry. Different plant formations, biomes, and land-use systems reflect these characteristics (Schultz, 1988, 2000). The borderlines of the ecozones refer to the climatic differentiation of Troll and Paffen (1964), focusing especially on the climate to vegetation impact. The ecozone map elaborated by Schultz (1988, 2000) has overlaps to a great extent with the Walter-Breckle biome map (Figure 5.5).

Lauer et al. (1996) worked out a global classification that focusses on the heat and water budget, especially on the lengths of wet and dry seasons. Four principal zones, marked by daylength variation of irradiation: tropics, subtropics, mid-latitudes, and polar regions, are differentiated, which is problematic, for example, when winter-cold steppe areas are characterized as being subtropical. The aim is to classify the climates of the earth by ecophysiological characteristics of the real vegetation.

5.8 OUTLOOK: THE IMPORTANCE OF CLASSIFICATION FOR BIOGEOGRAPHERS AND ECOLOGISTS

As discussed above, there are taxa-, structure-, community/community complex-, and biome/ ecoregion-based classification approaches to classification. Vegetation analysis at the landscape level can be carried out by traditional vegetation

mapping, by analysis of transects and gradients (see Mutke, this volume) or by studying vegetation complexes. It can be hypothesized that the understanding of vegetation complexes as spatial units will lead to new insights on a high complexity level. Three main levels of spatial vegetation classification lead from communities to vegetation-complexes in landscapes and to biomes.

Though the antecedents for classification go back in some cases over two centuries, there are three compelling reasons why we still need classification:

1. Classified data are essential as reference units for questions concerning subjects such as global environmental change, environmental impact assessment, in the context of nature conservation and resource management (especially of threatened habitat types, for example, the 'Flora-Fauna-Habitat-directive' of the European Union), for planning purposes, in restoration ecology, and also as a reference framework for expensive experimental research.
2. Many research results demonstrate that 'ecological rules' are dependent on the type of ecosystem investigated. An example would be biodiversity–productivity rules. Often a unimodal, hump-shaped relationship with a diversity maximum at medium productivity level is proposed (e.g. García et al., 1993 for a Mediterranean saltmarsh; Süss et al., 2007 for temperate sand ecosystems). For different grassland communities some authors described positive relationships (e.g., Pfisterer et al., 2004; Huston and DeAngelis, 1994) and others negative ones (e.g., Goldberg and Miller, 1990). It is important to classify the system in question, with the aim of producing transferable results.
3. Using classification, new relationships have been worked out (e.g., by correlation of indicator species- or indicator community groups and plot-based abiotic or biotic data, nowadays with the help of cluster analysis and multivariate data analysis). The example of the 'Ellenberg' indicator values shows that most of these values, which had been worked out by analysing phytosociological databases, are very useful (Figure 5.3).

A number of plant and vegetation ecologists are interested in 'how vegetation works' and not in distinguishing plant communities (Rodwell, 1991–2000). Nonetheless a large, worldwide system of GIS-referenced permanent plots in classified communities, which are repeated for each community type, are essential for the identification of effects of climate change, atmospheric pollution, species invasions, and other disturbance factors. Meanwhile large parts of the world use floristically based classifications, which are relatively similar. A goal for the future is to standardize national vegetation classification systems by developing a system that is acceptable worldwide. The integration of zoological and functional-ecological data will be easier if such a database were established and homogeneous.

The first period of classification can be defined between the eighteenth and early twentieth centuries. A second period can be identified in the second half of the twentieth century in which the study of ecological and biogeographical processes has dominated biogeography and ecology leading to the idea that, to some extent, classification is not essential. Now its importance is again becoming generally accepted, especially in the context of long-term studies, global change, and applied biogeography and ecology. In the future the combination of classification and functional aspects has to be developed. Functional aspects should have a macroecological scale in order to be useful for prediction and modelling.

REFERENCES

Archibold, O.W. (1995) *Ecology of World Vegetation*. London: Chapman & Hall.

Austin, M.P. (2005) 'Vegetation and environment: Discontinuities and continuities', in E. van der Maarel (ed.), *Vegetation Ecology*. Oxford: Blackwell. pp. 52–84.

Bailey, R.G., and Hogg, H.C. (1986) 'A world ecoregions map for resource reporting', *Environmental Conservation*, 13: 195–202.

Barkman, J.J. (1973) 'Synusial approaches to classification', in R.H. Whittaker (ed.), Ordination and Classification of Communities. *Handbook of Vegetation Science* 5. The Hague: Dr. W. Junk. pp. 435–91.

Begon, M., Harper, J.L., and Townsend, C.R. (1996) *Ecology: Individuals, Populations and Communities*. 3rd ed. Oxford: Blackwell (4th ed. 2006).

Béguin, C., Grandtner, M., and Gervais, C. (1994) 'Analyse symphytosociologique de la végétation littorale du Saint-Laurent près de Cap-Rouge, Québec', *Phytocoenologia*, 24: 27–51.

Béguin, C., Matthey, W., and Vaucher, C. (1977) 'Faune et sigmassociation', in R. Tüxen (ed.) *Berichte Internationale Symposien der Internationalen Vereinigung für Vegetationskunde 1976*. Vaduz: Cramer. pp. 9–19.

Belnap, J. and Lange, O.L. (eds.). (2001) 'Biological soil crusts: Structure, function, and management', in *Ecological Studies*, 150. Berlin: Springer.

Biondi, E., Bracco, F., and Nola, P. (1997) 'Lista delle unità sintassonomiche della vegetazione italiana', *Fitosociologica*, 32: 1–227.

Böhling, N., Greuter, W., and Raus, T. (2002) 'Zeigerwerte der Gefässpflanzen der Südägäis', *Braun-Blanquetia*, 32: 3–107.

Borhidi, A. (1993) *Social Behaviour Types of the Hungarian Flora, its Naturalness and Relative Ecological Indicator Values*. Pécs: Janus Pannonius Tudom. Kiadv.

Borhidi, A. (ed.). (1996) *Critical Revision of the Hungarian Plant Communities*. Veszprém: Janus Pannonius University Press.

Box, E.O. (1996) 'Plant functional types and climate at the global scale', *Journal of Vegetation Science*, 7: 309–20.

Braun Blanquet, J. (1928) 'Pflanzensoziologie. Grundzüge der Vegetationskunde', in W. Schoenichen, *Studienbücher 7*. Berlin: Springer (3rd ed. 1964).

Bruelheide, H., and Chytrý, M. (2000) 'Towards unification of national vegetation classifications: A comparison of two methods for analysis of large datasets', *Journal of Vegetation Science*, 11: 295–306.

Bruelheide, H., and Scheidel, U. (1999) 'Slug herbivory as a limiting factor for the geographical range of Arnica montana', *Journal of Ecology*, 87: 839–48.

Bültmann, H. (2005) 'Syntaxonomy of arctic terricolous lichen vegetation, including a case story from Southeast Greenland', *Phytocoenologia*, 35(4): 909–49.

Chytrý, M. (ed.). (2007) *Vegetace České republiky 1*. Prague: Academia.

Chytrý, M., Hejcman, M., Hennekens, S.M., and Schellberg, J. (2009) 'Changes in vegetation types and Ellenberg values after 65 years of fertilizer application in the Rengen Grassland Experiment, Germany', *Applied Vegetation Science*, 12: 167–176.

Clements, F.E. (1916) *Plant Succession: An Analysis of the Development of Vegetation*. Carnegie Institution of Washington. Publ. No. 242.

Coldea, G., Sanda, V., Popescu, A., and Ştefan, N. (1997) *Les associations végétales de Roumanie. Tome 1. Les associations herbacées naturelles*. Cluj-Napoca: Presses Universitaires de Cluj.

Cox, C.B., and Moore, P.D. (2005) *Biogeography. An Ecological and Evolutionary Approach*. 7th ed. Oxford: Blackwell.

De Vegetatie van Nederland. 5 vols. Uppsala, Leiden: Opulus. Vol. 1: Schaminée, J., Stortelder, A.F.H., and Westhoff, P. (1995) 295 pp., Vol. 2: Schaminée, J., Weeda, E.J., and Westhoff, V. (1995) 357 pp., Vol. 3: Schaminée, J., Stortelder, A.F.H., and Weeda, E.J. (1996) 355 pp. Vol. 4: Schaminée, J., Weeda, E.J., and Westhoff, V. (1998) 345 pp., Vol. 5: Stortelder, A.F.H., Schaminée, J.H.J., and Hommel, P.W.F.M. (1999) 375 pp.

Deil, U. (1997) 'Zur geobotanischen Kennzeichnung von Kulturlandschaften'. *Erdwiss. Forschung*, 36: 1–189. Stuttgart: Steiner.

Deil, U. (2003) 'Traditional and modern vegetation landscapes—a comparison of northern Morocco and southern Spain', *Phytocoenologia*, 33(4): 819–60.

Diekmann, M. (2003) 'Species indicator values as an important tool in applied plant ecology: A review'. *Basic and Applied Ecology*, 4: 493–506.

Dierschke, H., et al. (eds.). (1996 ff.) *Synopsis der Pflanzengesellschaften Deutschlands*. Vols. 1–10. Printed by "Floristisch-soziologischen Arbeitsgemeinschaft e.V.". Göttingen: Goltze.

Dierssen, K., and Dierssen, B. (1980) 'The distribution of communities and community complexes of oligotrophic mire sites in Western Scandinavia', *Colloques phytosociologiques*, 7: 95–119. Vaduz: Cramer.

Ellenberg, H. (1963) *Vegetation Mitteleuropas mit den Alpen*. Stuttgart: Ulmer (5th ed. 1996).

Ellenberg, H. (1974) 'Zeigerwerte der Gefäßpflanzen Mitteleuropas', *Scripta Geobotanica* (Göttingen) 9 (2nd ed. 1992).

Ellenberg, H., and Mueller-Dombois, D. (1974) *Aims and Methods of Vegetation Ecology*. New York: John Wiley.

Flahault, C., and Schröter, C. (1910) *Phytogeographische Nomenklatur. Berichte und Anträge*. 3rd International Congress of Botany. Brussels, Belgium. pp. 14–22.

Forman, R.T.T., and Godron, M. (1981) 'Patches and structural components for a landscape ecology', *Bioscience*, 31: 733–40.

García, L. V., Maranon, T., Morena, A., and Clemente, L. (1993) 'Above-ground biomass and species richness in a Mediterranean salt marsh', *Journal of Vegetation Science*, 4: 417–24.

Géhu, J.-M. (1973) 'Unités taxonomiques et vegetation potentielle naturelle du Nord de la France', *Documents Phytosociologiques*, 4: 1–22.

Géhu, J.-M. (1977) 'Le concept de sigmassociation et son application à l'ètude du paysage végétal des falaises atlantiques francaises', *Vegetatio*, 14(2): 117–25.

Gleason, H.A. (1926) 'The individualistic concept of the plant association', *Bulletin of the Torrey Botanical Club*, 53: 7–26.

Goldberg, D.E., and Miller, T.E. (1990) 'Effects on different resource additions on species diversity in an annual plant community', *Ecology*, 71: 213–25.

Goodall, D.W. (ed.). (1977 ff) *Ecosystems of the World*. Amsterdam: Elsevier.

Grime, J.P. (1974) 'Vegetation classification by reference to strategies', *Nature*, 250: 26–31.

Grime, P. (1979) *Plant Strategies and Vegetation Processes*. New York: John Wiley.

Grisebach, A. (1838) 'Über den Einfluß des Klimas auf die Begrenzung natürlicher Floren', *Linnaea*, 12: 159–200.

Grossman, D.H., Faber-Langendoen, D., Weakley, A.S., et al. (1998) *International Classification of Ecological Communities: Terrestrial Vegetation of the United States*. Vol. 1. The national vegetation classification system: development, status, and applications. Arlington, VA: The Nature Conservancy.

Hegg, O., Béguin, C., and Zoller, H. (1993) *Atlas schutzwürdiger Vegetationstypen der Schweiz*. Bundesamt Umwelt, Wald u. Landschaft (ed.). Bern.

Heywood, V.H., and Watson, R.T. (eds.). (1995). *Global Biodiversity Assessment*. Cambridge: Cambridge University Press.

Hill, M.O., and Carey, P.D. (1997) 'Prediction of yield in the Rothamsted park grass experiment by Ellenberg indicator values', *Journal of Vegetation Science*, 8: 579–86.

Hill, M.O., Preston, C.D., and Roy, D.B. (2004) *PLANTATT— Attributes of British and Irish Plants: Status, Size, Life History, Geography and Habitats*. Monks Wood: Institute of Terrestrial Ecology. Centre for Ecology and Hydrology Publication.

Holdridge, L.R. (1947) 'Determination of world plant formations from simple climatic data', *Science*, 105: 367–68.

Hueck, K., and Seibert, P. (1981) *Vegetationskarte von Südamerika*. Stuttgart: Springer.

Huston, M.A., and DeAngelis, D.L. (1994) 'Competition and coexistence: The effects of resource transport and supply rates', *American Naturalist*, 144 (6): 954–77.

Jäger, E.J. (1995) 'Plant geography II', *Progress in Botany*, 56: 397–415.

Julve, P. (1993) 'Synopsis phytosociologique de la France (Communautés de plantes vasculaires)', *Lejeunia N.S.*, 140: 1–160.

Knapp, R. (1984) 'Sampling methods and taxon analysis in vegetation science', in H. Lieth (ed.), *Handbook of Vegetation Science* 1(4): 1–370 Berlin-Heidelberg: Springer.

Köppen, W. (1931) *Grundriss der Klimakunde. Klimakarte der Erde*. 2nd ed. Berlin: Walter de Gruyter.

Korotkov, K.O., Morozova, O.V., and Belonovskaja, E.A. (1991) *The USSR Vegetation Syntaxa Prodromus*. Moscow Dr. Gregory E. Vilchek, Moscow.

Kratochwil, A., Beil, M., and Schwabe, A. (2009) 'Complex structure of a plant-bee community—random, nested, with gradients or modules?' *Apidologie,* 40(6): 634–50.

Kratochwil, A., and Schwabe, A. (2001) *Ökologie der Lebensgemeinschaften: Biozönologie*. Stuttgart: Ulmer.

Landolt, E. et al. (2010) *Flora Indicativa*. Bern: Haupt.

de Lattin, G. (1967) *Grundriss der Zoogeographie*. Jena: Gustav Fischer.

Lauer, W., Rafiqpoor, M., and Frankenberg, P. (1996) 'Die Klimate der Erde. Eine Klassifikation auf ökophysiologischer Grundlage der realen Vegetation' *Erdkunde*, 50: 275–300.

Lawesson, J.E., Fosaa, A.M., and Olsen, E. (2003) 'Calibration of Ellenberg indicator values for the Faroë Islands', *Journal of Vegetation Science*, 6: 53–62.

Leyer, I., and Wesche, K. (2007) *Multivariate Statistik in der Ökologie*. Berlin: Springer.

Lomolino, M. V., Riddle, B. R., and Brown, J. H. (2005*) Biogeography*. 3rd ed. Sunderland, MA: Sinauer.

Lomolino, M. V., Sax, D.F., and Brown, J.H. (2004) *Foundations of Biogeography. Classic Papers with Commentaries*. Chicago: University of Chicago Press.

Map of the Natural Vegetation of Europe (2004). Federal Agency for Nature Conservation (Compilation: Bohn, U). CD-ROM. Bonn, Germany.

Mattes, H. (1988) 'Untersuchungen zur Ökologie und Biogeographie der Vogelgemeinschaften des Lärchen-Arvenwaldes im Engadin', *Münstersche Geographische Arbeiten* 30: 1–138.

Matuszkiewicz, J.M. (1979) 'Landscape phytocomplexe and vegetation landscapes, real and typological landscape units of vegetation', *Documents phytosociologiques N.S.*,4: 663–672.

Meusel, H., Jäger, E., and Weinert, E. (1965 ff) 'Vergleichende Chorologie der zentraleuropäischen', *Flora*. Vols. 1–3. Jena: Gustav Fischer.

Miyawaki, A. (1978) 'Sigmassoziationen in Mittel- und Süd-Japan', in R. Tüxen (ed.), *Berichte Internationale Symposien der Internationalen Vereinigung für Vegetationskunde* 1977. Vaduz: Cramer. pp. 241–263.

Miyawaki, A. (ed.) (1980–1989*) Nippon Shokusei-Shi (Vegetation of Japan)*. 10 vols. with colour maps. Tokyo: Shibundo.

Morrone, J.J. (2002) 'Biogeographical regions under track and cladistic scrutiny', *Journal of Biogeography*, 29: 149–52.

Mucina, L., Bredenkamp, G. J., Hoare, D.B., and McDonald, D.J. (2000) 'A natural vegetation database for South Africa', *South African Journal of Science*, 96: 497–98.

Mucina, L. and Rutherford, M.C. (eds.). (2006) 'The vegetation of South Africa, Lesotho and Swaziland', *Strelitzia* 19: 1–816.

Müller, P. (1974) *Aspects of Zoogeography*. The Hague: Dr. W. Junk.

Navarro, G. (2003) 'Tipología fluvial y vegetacón riparia amazónica en el departemento de Pando (Bolivia)', *Revista Biologia y Ecologia*, 13: 3–29.

Navarro, G., and Ferreira, W. (2004) 'Zonas de vegetación potencial de Bolivia: Una base para el análisis de vacíos de conservación', *Revista Biological y Ecologia*, 15: 40 pp.

Oberdorfer, E. (ed.). (1977 ff.) *Süddeutsche Pflanzenge-sellschaften* I (1977, 311 pp.), II (1978, 355 pp.), III (1983, 455 pp.), IV (1992, 282 pp. + 580 pp). Jena: Fischer.

Ohri, D. (2005) 'Climate and growth form: The consequences for genome size in plants', *Plant Biology*, 7: 449 58.

Pedrotti, F. (2004) *Cartografia Geobotanica*. Bologna: Pitagora Editrice.

Pfisterer, A.B., Jushi, J., Schmid, B., and Fischer, M. (2004) Rapid decay of diversity-productivity relationships after invasion of experimental plant communities. *Basic and Applied Ecology*, 5: 5–14.

Pignatti, S. (1980) 'I complessi vegetazionali del Triestino', *Studia Geobotanica*, 1(1): 131–47.

Pignatti, S., Menegoni, P., and Pietrosanti, S. (2005) 'Valori di bioindicatione delle piante vascolari della Flora d'Italia. Bioindicator values of vascular plants of the Flora of Italy', *Braun-Blanquetia*, 39: 3–95.

Pignatti, S., Oberdorfer, E., Schaminée, J.H.J., and Westhoff, V. (1994) 'On the concept of vegetation class in phytosociology', *Journal of Vegetation Science*, 6: 143–52.

Poore, M.E.D. (1955a) 'The use of phytosociological methods in ecological investigations. I. The Braun-Blanquet system', *Journal of Ecology*, 43: 226–44.

Poore, M.E.D. (1955b) 'The use of phytosociological methods in ecological investigations. II. Practical issues involved in an attempt to apply the Braun-Blanquet system', *Journal of Ecology*, 43: 245–69.

Quinn, G.P., and Keough, M. J. (2007) *Experimental Design and Data Analysis for Biologists*. Cambridge: Cambridge University Press.

Raynolds, M.K., Walker, D.A., and Maier, H.A. (2005) 'Plant community level mapping of arctic Alaska based on the Circumpolar Arctic Vegetation Map', *Phytocoenologia*, 35(4): 821–48.

Rikli, M. (1913) 'Florenreiche', in *Handbuch der Naturwissenschaften* 4: 776–857. Jena: Fischer.

Rivas-Martínez, S. (1987) *Mapa de series de vegetatcion de Espana y memoria* 1 : 400.000. Ministeria Agricultura, Pesca y Alimentación (ed.) ICONA. Serie Tecnica Madrid.

Rivas-Martínez, S. (1994) 'Dynamic-zonal phytosociology as landscape science', *Phytocoenologia*, 24: 23–25.

Rivas-Martínez, S., Fernández-Gonzales, F., Loidi, J., Lousã, M., and Penas, A. (2001) 'Syntaxonomical checklist of vascular plant communities of Spain and Portugal to association level', *Itinera Geobotanica*, 14: 5–341.

Rodwell, J.S. (ed.). (1991–2000) *British Plant Communities*. Vols. 1–5. Cambridge: Cambridge University Press.

Rodwell, J.S., Schaminée, J.H.J., Mucina, L., Pignatti, S., Dring, J., and Moss, D. (2002) *The Diversity of European Vegetation. An Overview of Phytosociological Alliances and Their Relationships to EUNIS Habitats*. Wageningen: Rapport Expertise Centrum LNV 2002/054.

Root, R.B. (1967) 'The niche exploitation pattern of the blue-gray gnatcatcher', *Ecological Monographs*, 37: 317–50.

Rübel, E.F. (1930) *Pflanzengesellschaften der Erde*. Bern: Hans Huber.

Schaffers, A.P., and Sýkora, K.V. (2000) 'Reliability of Ellenberg indicator values for moisture, nitrogen and soil reaction: A comparison with field measurements', *Journal of Vegetation Science*, 11: 225–44.

Schimper, A.F.W. (1898) *Pflanzen-Geographie auf ökolo-gischer Grundlage*. Jena: Fischer.

Schmidtlein, S. (2003) 'Raster-based detection of vegetation patterns at landscape scale levels', *Phytocoenologia*, 33 (4): 603–21.

Schroeder, F.-G. (1998) *Lehrbuch der Pflanzengeographie*. Wiesbaden: Quelle and Meier.

Schultz, J. (1988) *Die Ökozonen der Erde. Die physiologischer Gliederung der Geosphäre*. Stuttgart: Ulmer.

Schultz, J. (2000) *Handbuch der Ökozonen*. Stuttgart: Ulmer.

Schulze, E.-D., Beck, E., and Müller-Hohenstein, K. (2005) *Plant Ecology*. Berlin: Springer.

Schwabe, A. (1989) 'Vegetation complexes of flowing-water habitats and their importance for the differentiation of landscape units', *Landscape Ecology*, 2: 237–53.

Schwabe, A. (1991) 'A method for the analysis of temporal changes in vegetation pattern on a landscape level', *Vegetatio*, 95: 1–19.

Schwabe, A. (1997) 'Sigmachorology as a subject of phyto-sociological research: A review', *Phytocoenologia*, 27: 463–507.

Schwabe, A., and Kratochwil, A. (2004) 'Festucetalia valesi-acae communities and xerothermic vegetation complexes in the Central Alps related to environmental factors', *Phytocoenologia*, 34: 1–118.

Schwabe, A., Kratochwil, A., and Pignatti, S. (2007) 'Plant indicator values of a high-phytodiversity country (Italy) and their evidence, exemplified for model areas with climatic gradients in the southern inner Alps', *Flora*, 202: 339–49.

Schwabe, A., and Mann, P. (1990) 'Eine Methode zur Beschreibung und Typisierung von Vogelhabitaten, gezeigt am Beispiel der Zippammer (Emberiza cia)', *Ökologie der Vögel (Ecology of birds)*, 12: 127–157. Stuttgart.

Shelford, V.E. (1913) 'Animal communities in temperate America as illustrated in the Chicago region. A study in animal ecology', *Geographical Society of Chicago, Bulletin*, 5: 1–362.

Shimwell, D.W. (1968) 'The phytosociology of calcareous grasslands in the British Isles', University of Durham, PhD thesis.

Specht, R., and Specht, A. (2001) 'Ecosystems of Australia', in S.A. Levin (ed.), *Encyclopaedia of Biodiversity* 1: 307–24 San Diego CA: Academic Press.

Sukatschew, W.N. (Sukachev, V.N.) (1926) *Bolota, ikh obrazovanie, razvitie i svojstva*. 3rd ed. Leningrad: Leningradskogo Lesnogo Instituta.

Süss, K., Storm, C., Zimmermann, K., and Schwabe, A. (2007) 'The interrelationship between productivity, plant species richness and livestock diet: A question of scale', *Applied Vegetation Science*, 10: 169–82.

Tansley, A.G. (1935). 'The use and abuse of vegetational concepts and terms', *Ecology*, 16: 284–307.

Tansley, A.G. (1939) 'British ecology during the past quarter-century: The plant community and the ecosystem', *Journal of Ecology*, 27: 513–30.

Thannheiser, D. (1988) 'Eine landschaftsökologische Studie bei Cambridge Bay, Victoria Island, N.W.T., Canada', *Mitteilungen Geographische Gesellschaft Hamburg*, 78: 1–52.

Thannheiser, D. (1989) 'Landschaftsökologische Untersuchungen im Bereich der Prä-Dorset-Station Umingmak (Banks Island, Kanada)', *Polarforschung*, 59(1/2): 61–78.

Theurillat, J.-P. (1992) 'Etude et cartographie du paysage végétal (Symphytocoenologie) dans la région d'Aletsch (Valais, Suisse)', *Beiträge geobotanische Landesaufnahme der Schweiz*, 68: 1–384.

Treviranus, G.R. (1803) *Biologie oder Philosophie der lebenden Natur für Naturforscher und Ärzte*. Vol. 2. Göttingen.

Troll, C., and Paffen, K.H. (1964) 'Karte der Jahreszeitenklimate der Erde', *Erdkunde*, 18: 5–28.

Tüxen, R. (1956) 'Die heutige potentielle natürliche Vegetation als Gegenstand der Vegetationskartierung', *Angewandte Pflanzensoziologie*, 13: 5–42.

Tüxen, R. (1978) 'Bemerkungen zu historischen, begrifflichen und methodischen Grundlagen der Synsoziologie', in R. Tüxen (ed.) *Berichte Internationale Symposien der Internationalen Vereinigung für Vegetationskunde* 1977: 3–11. Vaduz: Cramer.

Udvardy, M.D.F. (1975, updated 1982) 'A classification of the biogeographical provinces of the world', *Occasional Paper* 18. Morges, Switzerland: IUCN.

van der Maarel, E. (1993) 'Relations between sociological-ecological species groups and Ellenberg indicator values', *Phytocoenologia*, 23: 343–62.

van der Maarel, E. (2005) 'Vegetation ecology—an overview', in E. van der Maarel (ed.), *Vegetation Ecology*. Oxford: Blackwell. pp. 1–51.

van Groenewould, H. (1992) 'The robustness of correspondence, detrended correspondence, and TWINSPAN analysis', *Journal of Vegetation Science*, 3: 239–46.

Walker, D.A., Raynolds, M.K., Daniëls, F.J.A., Einarsson, E., Elvebakk, A., Gould, W.A. Katenin, A.E., Kholod, S.S.,

Markon, C.J., Melnikov, E.S. Moskalenko, N.G. Talbot, S.S., Yurtsev, B.A., and CAVM Team. (2005) 'The circumpolar Arctic vegetation map', *Journal of Vegetation Science*, 16: 267–82.

Walter, H., and Breckle, S.-W. (1970) *Vegetation und Klimazonen*. 1th ed. Stuttgart: Ulmer (7th ed. 1999).

Walter, H., and Breckle, S.-W. (1983 ff) *Ökologie der Erde*. Vols. 1–4. Stuttgart: Fischer.

Warming, E. (1909) *Ecology of Plants*. Oxford: Clarendon Press.

Westfall, R. H., Theron, G.K., and Rooyen, N. (1997) 'Objective classification and analysis of vegetation data', *Plant Ecology*, 132: 137–54.

Westhoff, V., and van der Maarel, E. (1973) 'The Braun-Blanquet approach', in R.H. Whittaker (ed.) Ordination and Classification of Communities. *Handbook Vegetation Science* 5: 617–726. The Hague: Dr. W. Junk.

Whittaker, R.H. (1970) *Communities and Ecosystems*. New York: Macmillan.

Whittaker, R.H. (1973) 'Approaches to classifying vegetation', in R.H. Whittaker (ed.), 'Ordination and classification of communities', *Handbook Vegetation Science* 5: 235–354. The Hague: Dr. W. Junk.

Willner, W., and Grabherr, G. (eds.). (2007) *Die Wälder und Gebüsche Österreichs*. 2 vols. München: Spektrum Elsevier.

Wilmanns, O. (1973) *Ökologische Pflanzensoziologie*. 1st ed. Heidelberg: Quelle and Meyer (6th ed. 1998).

Wilson, J.B. (1999) 'Guilds, functional types and ecological groups', *Oikos*, 86: 507–22.

Wilson, J.B., and Lee, W.G. (2000) 'C—S—R triangle theory: community-level predictions, tests, evaluation of criticisms, and relation to other theories', *Oikos*, 91: 77–96.

Wojterski, T., Wojterska, H., and Wojterska, M. (1994) 'Geobotanical division of Pomorze Gdanskie based on the map of potential natural vegetation and of potential landscape phytocomplexes and vegetation landscapes', *Badania Fitzjograf*, 43 B: 9–49.

Woodward, F.I., and Cramer, W. (eds.). (1996) 'Plant functional types and climatic change. Special feature', *Journal of Vegetation Science*, 7: 306–430.

Zarzycky, K. (1984) 'Indicator values of vascular plants in Poland [in Polish]', *Nauk Krakow Institut Botaniki Polskiej Akademii*. (W Szafer Institute of Botany, Polish Academy of Sciences).

Zoller, H., Béguin, C., and Hegg, O. (1978) 'Synsoziogramme und Geosigmeta des submediterranen Trockenwaldes in der Schweiz', in R. Tüxen (ed.), *Berichte Internationale Symposien der Internationalen Vereinigung für Vegetationskunde* 1977: 287–302. Vaduz: Cramer.

6

The Expanding Role of Phylogeography

Brett R. Riddle

6.1 HISTORICAL OVERVIEW

The goal of historical biogeography is to reconstruct the geography of speciation, dispersal, and extinction of lineages, and the assembly and disassembly of biotas, within the context of a dynamic Earth history. As such, historical biogeography has long been of central importance within evolutionary biology and is assuming a revitalized role within modern ecology and conservation biology (Wiens and Donoghue, 2004; Whittaker et al., 2005).

Naturalists of the nineteenth century—including Wallace, Sclater, Hooker, and Darwin—made remarkable leaps in our understanding of the distribution and diversification of life (see Blumler et al., this volume, for a more detailed discussion of nineteenth-century naturalists), but their insights suffered from the absence of a robust model of geological and climatic history on a dynamic Earth. In modern times, a revolution in historical biogeography occurred in concert with the development of the theory of plate tectonics (Dietz, 1961; Hess, 1962; Lomolino et al., 2006), which stimulated the development of *vicariance biogeography* (Nelson, 1974; Platnick and Nelson, 1978; Rosen, 1978)—an approach and associated methods that highlight both the passive transport of organisms to far reaches of the Earth on drifting continents as well as their passive separation across newly formed barriers such as mountain ranges or rivers. An important prediction of a vicariance model is that species or other taxa that share a common geographic distribution will exhibit congruent spatial and temporal patterns of isolation and divergence as continents drift apart or other sorts of barriers arise. Vicariance biogeography originated as an alternative to the dispersalist biogeography of Matthew (1915), Simpson (1940), and Darlington (1957), which emphasized construction of idiosyncratic and ad hoc dispersal scenarios (Brundin, 1966). A goal of the vicariance biogeographers was to eliminate the nontestable 'just-so stories' approach that was the hallmark of the dispersalists by grounding historical biogeography within a framework that combined Croizat's (1964) goals of searching for general patterns of distribution with Hennig's (1966) method, which required the use of monophyletic groups to reconstruct phylogenetic and, by extension, biogeographic histories (Croizat et al., 1974).

Subsequent to the tumultuous paradigm shift induced by the vicariance biogeographers, historical biogeography has witnessed the development of a somewhat bewildering number of different approaches, methods, and questions (Table 6.1). However, it continues to languish in controversies over approaches and fundamental goals (Brooks et al., 2004; Funk, 2004; Ebach and Tangney, 2006), with surprisingly little progress having been made in explaining fundamental biogeographic patterns and processes associated with a dynamic Earth history. The heated earlier controversies between vicariance and dispersal (e.g., Darlington, 1965; Brundin, 1966) have transformed into more sophisticated, albeit still heated, debates on such fundamental issues as the role of phylogenetic systematics in reconstructing taxon and biotic histories (e.g., panbiogeography

Table 6.1 A representative array of either historically or currently important approaches and methods in historical biogeography (after Crisci et al., 2003; as modified in Lomolino et al., 2006)

Approaches	Goal and selected methods	Original authors and general references
Descriptive biogeography	Comparing species lists	Sclater (1858); Hooker (1844–60)
Evolutionary biogeography	Center of origin-dispersal	Matthew (1915); Cain (1944)
Phylogenetic biogeography I	Phylogenetic systematics	Henning (1966); Brundin (1966)
Ancestral areas analysis	Areas of origin prior to dispersal	Bremer (1992, 1995)
	Weighted ancestral areas analysis	Hausdorf (1998)
Panbiogeography	Generalized tracks on a dynamic Earth	Croizat (1958)
	Track analysis	Croizat (1958)
Cladistic (vicariance) biogeography	Vicariance on a dynamic Earth	Nelson (1974)
	Reduced area cladogram	Rosen (1978)
	Component analysis (CA)	Nelson and Platnick (1981); Humphries and Parenti (1999)
	Three-area statement (TASS)	Nelson and Ladiges (1992)
	Paralogy-free subtrees	Nelson and Ladiges (1992)
Phylogenetic biogeography II	Vicariance, dispersal, geography of speciation	Wiley (1980)
	Brooks Parsimony Analysis	Wiley (1980)
	Primary and secondary BPA	Van Geller and Brooks (2000); Brooks et al. (2001)
Parsimony Analysis of Endemicity (PAE)	Natural distribution patterns of taxa	B. Rosen (1988)
	Areas of endemism	Craw (1988); Morrone (1994)
Event-based methods	Benefit/costs modeling of events	Ronquist and Nylin (1990)
	Dispersal vicariance analysis (DIVA)	Ronquist (1997)
	Parsimony-based tree fitting	Page (1994); Ronquist (2002)
Phylogeography	Geography of genealogical lineages	Avise et al. (1987); Avise (2000)
	Phylogeny of gene trees	various
	Nested clade analysis	Templeton et al. (1995); Templeton (2004)
	Coalescent-based approaches	various; Knowles (2003)
	Comparative phylogeography	Zink (1996); Arbogast and Kenagy (2001)

vs. most other methods; Craw et al., 1999); the tractability of dispersal as a biogeographic process (e.g., cladistic [i.e., vicariance] biogeography vs. most other methods; Ebach, 2001); and whether to limit the goals of historical biogeography to recovering vicariance histories, or to expand them to incorporate the geographies of lineage divergence, speciation, and the identification of ancestral areas (e.g., cladistic biogeography vs. phylogenetic biogeography, ancestral areas analysis, and even-based methods; e.g., Bremer, 1992; Brooks et al., 2004; Ronquist, 1997; Siddall, 2005).

Recently, however, investigators have increasingly begun to call for and witness the emergence of a revitalized, more integrative, and richer historical biogeography with a renewed conceptual and analytical vigor, and as a critical anchor

within modern ecology, evolutionary biology, conservation, and global change biology (e.g., Wiens and Donoghue, 2004; Riddle, 2005; Riddle et al., 2008; see also the preceding chapters in Section I). This revitalization is being fueled by:

1. an explosive increase in the numbers of molecular-based phylogenetic and population genetic datasets available for analysis (e.g., any issue of the journals *Journal of Biogeography, Molecular Ecology, Molecular Phylogenetics and Evolution, Systematic Biology*; Figure 6.1);
2. the development of approaches that allow for the estimation of the relative likelihoods of ancestral areas, dispersal events, and vicariance events (ideally, including the time frames of their occurrences; Donoghue and Moore, 2003; Ree et al., 2005; Ree and Smith, 2008);

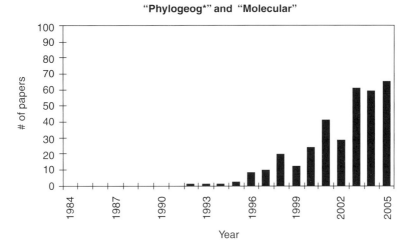

Figure 6.1   The expanding role of molecular genetics, including phylogenetics and population genetics, within historical biogeography, 1975–2008, as summarized by searching the ISI Web of Science (March 2009) using topic searches for "historical biogeograph+ and molecular", and "phylogeograph* not historical biogeograph*". Much of this growth has been fueled by the introduction of phylogeography in 1987, as discussed further in this chapter

3. the availability of geographic information systems (GIS)—based niche modeling and increasingly detailed paleoclimatic and geological models (Hugall et al., 2002; Graham et al., 2004; Kidd and Ritchie, 2006; Waltari et al., 2007), and;
4. the emergence in 1987 of *phylogeography* (Avise, 2000)—an approach that has expanded the scope of historical biogeography into more recent temporal, smaller spatial, and finer taxonomic contexts than had generally been sought within the original vicariance paradigm (Riddle and Hafner, 2004, 2007).

In the remainder of this chapter, I provide an overview of phylogeography, with attention to the expansion of its role within historical biogeography, ecology, evolutionary biology, and conservation/global change biology.

6.2 WHAT IS PHYLOGEOGRAPHY?

Arguably, the most influential development within historical biogeography in the late twentieth century (Figure 6.2a) was the fusion of molecular genetics with increasingly sophisticated methods in phylogenetic systematics and population genetics into phylogeography (Avise et al., 1987), defined as "a field of study concerned with the principles and processes governing the geographic distributions of genealogical lineages, especially those within and among closely related species" (Avise, 2000: 3). One can see immediately from this definition that phylogeography is positioned explicitly to bridge between the traditionally disparate concerns of microevolution (evolutionary changes within species) and macroevolution (evolutionary changes among species and higher taxa), and serves also to establish a temporal context for phylogeographic studies, which are almost completely constrained within a Neogene time frame, with the majority centered on the Pleistocene (Figure 6.2c)—a time frame of critical interest to evolutionary biologists, ecologists, and conservation/global change biologists. The focus on genealogical lineages, or gene trees,

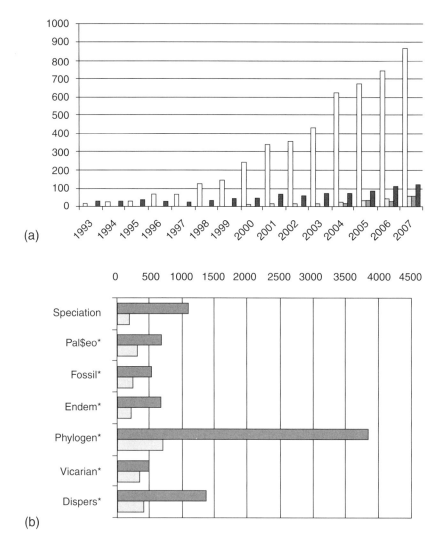

(a)

(b)

Figure 6.2   An array of comparisons between phylogeography and historical biogeography using topic searches on the ISI Web of Science, 1993–2007. (a) Total number of hits for, from left to right in each year, "phylogeograph*", "comparative phylogeograph*"; "statistical phylogeograph*", or "historical biogeograph* not phylogeograph*". (b—e), searches used either "phylogeograph* and [the word or phrase on the graph]" (dark gray bars) or "historical biogeograph* not phylogeograph*" and [the word or phrase on the graph]" (light gray bars)

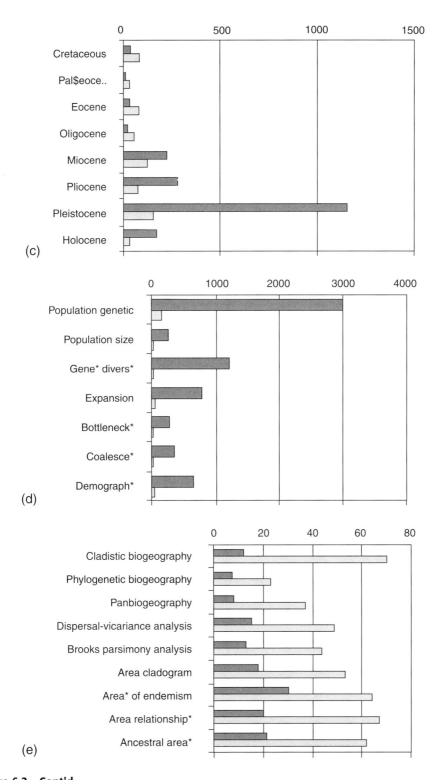

Figure 6.2   Cont'd

conceptually and empirically decouples the phylogenetic histories of populations or species from those of the genes and genomes embedded within them. To the extent that different portions of a genome (or genomes) within an organism represent independent gene trees, each will reflect a separate history of ancestor-descendant relationships with its own rates of lineage sorting—the process through which ancestral gene lineages are lost as a function of historical population size and structure, mutation, selection, or drift (Edwards and Beerli, 2000; Edwards et al., 2007). One advantage of estimating organism relationships from gene trees is that one can use different gene sequences that are evolving either slower or faster for assessment of older or more recent relationships, respectively (Avise, 2000).

Throughout the 1990s, much of phylogeography was concerned with the discovery and analysis of geographic structure within single gene lineages—mainly mitochondrial DNA (mtDNA) in animals and chloroplast DNA (cpDNA) in plants (Avise, 2000). Each of these organellar genomes offers the advantage of being relatively small (compared, for example, to the nuclear genome), with little or no decipherable recombination during replication and transmission, making the construction of a gene tree from any or all parts of the mtDNA or cpDNA a relatively straightforward process. An added advantage, for most animal mtDNA at least, is that the rates of evolutionary change—including both mutation/fixation as well as linage-sorting processes—are much higher than for comparable stretches of nuclear DNA (nDNA), thus increasing the degree of resolution of recovered mtDNA gene trees within species and within late Neogene time frames.

6.2.1 Why worry about gene-tree discordance?

Sole reliance on either mtDNA or cpDNA as a single gene-tree 'snapshot' of population and species histories is not without problems (Figure 6.3). Transmission of different genomes and different gene segments in sexually differentiated species theoretically can, and, empirically, at least occasionally does bias the depiction of organism-level historical relationships generated from a single gene tree.

For example, Lindell et al. (2005) reported a case of *cytonuclear discordance* between deep mitochondrial divergences versus considerably shallower nuclear divergence in the zebra-tailed lizard (*Callisaurus draconoides*) along the Baja California peninsula, concluding that such discordance resulted either from differential rates of ancestral lineage sorting across genomes, or from introgression of nuclear genes across a secondary contact zone without parallel mitochondrial introgression (see also examples in leopard frogs, Di Candia and Routman, 2007; and black-tailed brush lizards, Lindell et al., 2008). These sorts of discordance suggest that caution is warranted when inferring biogeographic histories from a single gene tree, although Zink and Barrowclough (2008) make a case, using birds as an example, for the continued theoretical and empirical advantages of mtDNA-based inferences in much of phylogeography over those generated from nuclear DNA.

Second, a growing number of studies have inferred instances of cytonuclear discordance resulting from historical introgression of a mitochondrial genome into a different species (e.g., mammals, Alves et al., 2006; lizards, McGuire et al., 2007; beetles, Nagata et al., 2007; flies, Linnen and Farrell, 2008; fish, Slechtova et al., 2008; toads, Vogel and Johnson, 2008). These cases of discordance suggest caution in using mtDNA alone to reconstruct phylogenetic histories across closely related species—indeed, the use of multiple independent datasets to do so is usually advocated by systematists generally and so is not a problem unique to organellar-based phylogenetics.

Third, the stochastic effects of lineage sorting has resulted in a recent trend in phylogeography toward the analysis of multiple, independent gene trees so as to not bias the estimate of a population-level or species-level history of relationships from any one gene tree alone (Edwards and Beerli, 2000; Ballard and Whitlock, 2004; Carstens and Knowles, 2007b). This concern has become particularly pronounced as phylogeography increasingly explores the influences of the late Pleistocene glacial-interglacial climatic oscillations. These influences during the most recent few hundred thousand years on population and species divergence occur because of the increasing likelihood of discordance across independent gene trees due to stochastic lineage sorting as time, since lineages diverged from a common ancestor decrease (Morando et al., 2004; Carstens and Knowles, 2007a; Knowles and Carstens, 2007).

6.2.1 What is unique about phylogeography?

One very important difference between phylogeographic and other historical biogeographic approaches lies in sampling design—rather than sampling one or a few exemplars from a phylogenetic lineage and/or area of endemism, the backbone of the phylogeographic approach is the sampling of a large number of localities, and often

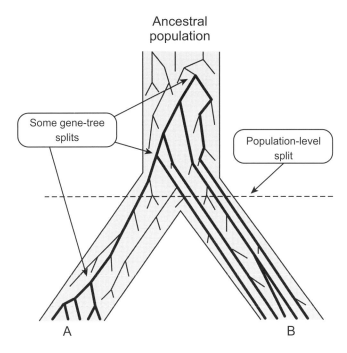

Figure 6.3 The concept of a set of gene trees (light and dark lines) embedded within a single population or species tree (bracketing shaded area). A number of the ancestral gene lineages (light lines) have gone extinct through the process of lineage sorting, while the fate of a single gene tree (bold lines) demonstrates that population B continues to contain genealogical lineages that are more closely related to those in population A, leading to phylogenetic incongruence between the gene tree and population tree. Note also that the gene-tree splits might considerably predate the population-level split, and that the gene tree in population A does not become monophyletic relative to population B until well after the population split (reprinted from Avise, 2000)

multiple individuals per locality, throughout the range of a species and frequently across closely related species. Phylogeographic data sets therefore tend to be quite large and contain detailed information about not only the distributions of genealogical lineages on a landscape or seascape, but also the distributions of genetic variation embedded within such lineages (Avise, 2000; Lomolino et al., 2006). This sort of information is particularly valuable for addressing two kinds of questions.

First, as mentioned above, phylogeographic studies are often concerned with the historical response of species and populations to shifting distributions and connectivity of habitats during Pleistocene glacial-interglacial climatic oscillations. For example, many phylogeographic studies seek to test hypotheses about the existence and locations of refugia during the latest Pleistocene glacial period and to address the related question of directions and magnitudes of range expansions

out of such refugia as species track the postglacial expansion or relocation of their habitats (see Willis et al., this volume for further discussion of refugia). A good phylogeographic sampling design allows for the employment of a variety of population genetics approaches to ascertaining histories of population decreases and expansions—as would accompany, for example, refugial persistence with post-refugial range expansion.

Second, whereas phylogeography is often correctly considered to be primarily a study of intraspecific pattern and process, many phylogeographic studies have discovered previously unknown (cryptic) phylogenetic breaks within the distribution of what had previously been considered to be a single species (a March 2009 topic search of the ISI Web of Science for "phylogeograph*" and "cryptic species" revealed 358 such studies). When coupled with the application of molecular clocks to estimate divergence times, the elucidation of cryptic species and geographically

Independent taxon histories

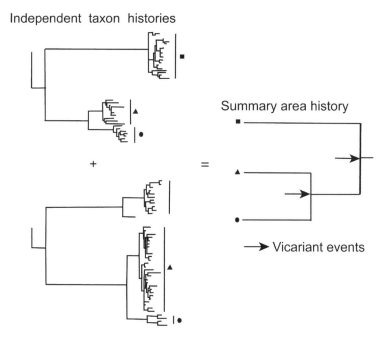

Figure 6.4   Simplified illustration of one reason that comparative phylogeography provides an important linkage between phylogeography and other approaches in historical biogeography. Here, the trees on the left are built using variable alleles (i.e., haplotypes) and a phylogeographic sampling design for each of two different species. The reciprocally monophyletic lineages marked by circle, triangle, and square symbols can be inferred to represent congruent responses from each species to a shared biogeographic history, perhaps owing to the vicariant events shown on the area cladogram on the right. In this simplified example, while the data and question rest at a phylogeographic scale, the approach and logic are those that underlie cladistic (or vicariance) biogeography (reprinted from Lomolino et al, 2006)

congruent phylogeographic breaks across co-distributed taxa within a comparative phylogeography framework (Figure 6.4; e.g., Bermingham and Moritz, 1998; Arbogast and Kenagy, 2001; Riddle and Hafner, 2004, 2007) have allowed for postulation of plausible geological and/or paleoclimatic events in Earth history as causal agents of within-species and species group gene-tree divergence and distribution patterns. For example, the geographic congruence of nine cryptic lineages of mammals and reptiles distributed south and north of a midpeninsular region on the Baja California peninsula, originally postulated to support a 'cryptic vicariant event' associated with a seaway connecting the Pacific Ocean with the Sea of Cortez across the midpeninsula (Riddle et al., 2000) has grown to now include cryptic signatures in additional vertebrates (Zink et al., 2001; Lindell et al., 2005, 2006, 2008; Recuero et al., 2006; Leaché and Mulcahy, 2007), as well as invertebrates (Crews and Hedin, 2006) and even marine fishes (Riginos, 2005). This allows for increasing

sophistication and rigor in the comparative phylogeographic and biogeographic analysis (Riddle and Hafner, 2006; Leaché et al., 2007).

Third, as mentioned previously, a particular strength of phylogeography (Avise, 2000) is the connection it creates between two different views of history—the population genetics of microevolutionary processes (e.g., historical population size and structure, migration rates, gene-tree coalescence times) and the phylogenetics of macroevolution (e.g., the geography of speciation and assembly of biotas). This conceptual and analytical duality has challenged phylogeographers to find an integrative architecture between phylogenetics and population genetics approaches to research design and data analysis (e.g., see Table 3 in Kidd and Ritchie, 2006). This duality is increasingly being integrated into a more synthetic, hierarchical context, suggesting that different approaches are likely to be optimal at different spatial and temporal scales of phylogeographic structure (e.g., Althoff and Pellmyr, 2002;

Zarza et al., 2008). The population genetic approach includes an array of methods that can begin with a rooted phylogenetic tree but is more often with an unrooted haplotype network (haplotype is a term that refers either to variable loci in a haploid genome such as animal mtDNA, or to a single copy of a variable locus from a diploid genome). Once constructed, a network can be analyzed for significant phylogeographic structure using either standard population genetics parameters (e.g., Fst, haplotype and nucleotide diversity, analysis of molecular variance) or computationally intensive coalescence-based methods. Coalescent approaches recently have been incorporated into a model-based and hypothesis-testing framework termed statistical phylogeography (Knowles, 2004; Templeton, 2004), which incorporates estimations of sources of error in the derivation of molecular evolutionary and historical population parameters that are used to infer historical and ongoing evolutionary processes. Methods built upon a maximum likelihood or Bayesian framework are becoming increasingly popular. For example, Excoffier and Heckel (2006) list 11 available programs that utilize a Markov chain Monte Carlo and Bayesian inference framework to assess a range of phylogeographic population attributes, including recent migration rates, gene flow, and colonization histories; population subdivisions; hybrid populations; and demographic histories within and divergence times between populations.

are becoming increasingly popular as either a means of testing the veracity of phylogeographic hypotheses for the locations of past lineage distributions or as a way of generating a priori hypotheses of the locations and numbers of past refugia to then be evaluated using phylogeographic assessments (Hugall et al., 2002; Graham et al., 2004; Kidd and Ritchie, 2006; Carstens and Richards, 2007; Richards et al., 2007; Waltari et al., 2007; Kozak et al., 2008). For example, Waltari et al. (2007) contrasted ecological niche modeling (ENM) predictions of past Last Glacial Maximum (LGM) (21,000 years ago) refugia in 20 North American terrestrial vertebrate species with phylogeographically derived predictions of refugial locations and found significant spatial correlation between the two approaches for 14 of the species.

While the ENM approach appears to offer a promising complement to phylogeographic reconstructions, a current weakness lies in the lack of paleoclimatic models predating the LGM, although extant species and phylogeographic lineages often have histories extending back into early Pleistocene, Pliocene, or Miocene time frames. Likewise, the lack of paleoclimate models postdating the LGM into early and mid-Holocene time frames (e.g., during the Holocene climatic warm maximum about 6,000 years ago) currently restrict the potential to use these methods for predicting the distributional responses of species to ongoing and future global warming.

6.3 RECENT VARIATIONS ON A PHYLOGEOGRAPHIC THEME

Phylogeography continues to integrate newer research tools and link to related approaches that provide opportunities to explore questions in some exciting new ways—a few examples of the more promising avenues currently generating a good deal of interest are summarized briefly here.

6.3.1 Phylogeography and GIS-based ecological niche modeling

With the increasing accessibility of GIS and the capacity to store and analyze large geographic data sets, and the large numbers of museum specimens now geo-referenced and available through digital databases, the popularity of a variety of approaches to modeling present, past, and future ecological niches, or bioclimatic envelopes, has grown rapidly. Several of these, including Bioclim, DesktopGARP, and MaxEnt,

6.3.2 Geophylogenies and phylogeographic information systems

Kidd and Ritchie (2006) recognized a need to place more 'geography' into the phylogeographic research arena by calling for a more explicit incorporation of GIS-based capacities to visualize and analyze phylogeographic data across space and through time. As such, a *geophylogenetic* GIS data model, currently being built within the program *Geophylobuilder*, offers exciting possibilities to examine the historical development of phylogeographic structure upon an evolving landscape in both 2D and 3D models. Several examples illustrating these capabilities are provided at https://www.nescent.org/wg_EvoViz/GeoPhyloBuilder.

6.3.3 Landscape genetics

Perhaps the most extreme recent expansion of phylogeographic approaches away from a traditional historical biogeographic context is represented by the emerging arena of *landscape genetics*, which seeks to develop a geographic population genetics approach to addressing questions formulated within the spatial and temporal purview of landscape ecology (Manel et al., 2003; Storfer et al., 2007; Holderegger and Wagner, 2008). More specifically, the overarching goal of this emerging approach is to utilize molecular population genetics to address the role that current landscape attributes (e.g., see Kupfer and Albright et al., this volume for relevant discussions of landscape ecology) play in structuring the spatial pattern of population architectures (e.g., spatial distributions of distinct population segments, patterns of gene flow or genetic isolation, metapopulation structure). Examples include a landscape genetic approach to addressing adaptive radiation in Darwin's finches (Petren et al., 2005) and an assessment of landscape attributes that underlie gene flow between timber rattlesnakes within hibernacula (Clark et al., 2008).

6.3.4 Ancient DNA and phylochronology

A number of sites around the world have accumulated fossils, within a time frame spanning the Last Glacial Maximum some 18,000 years ago into the Holocene, that have preserved enough intact fragments of ancient DNA molecules to allow for their amplification and sequencing (Willerslev and Cooper, 2005). Phylochronology seeks to utilize serial genetic data to assess fluctuating changes in population genetic structure within a species at a single locality across a series of time slices, with such studies to date extending back several thousands of years (Hadly et al., 2004). To the extent that such studies incorporate an assessment of spatial as well as temporal variation in genetic structure, they can be considered as an extension of phylogeography that complements the estimation of population histories across space by measuring directly the dynamics of populations (e.g., population bottlenecks, expansions, and isolation) across time. These sorts of genetic data are preserved only under conditions of low humidity, low temperature, and low acidity (Hofreiter et al., 2001; Smith et al., 2003), and rarely will extend beyond 100,000 years (Lindahl, 2000; Willerslev and Cooper, 2005). Therefore, localities with well-preserved serial stacked fossils are expected to be quite rare and are unique and valuable snapshots into a species' evolutionary history (Ramakrishnan et al., 2005; Ramakrishnan and Hadly, 2009). For example, Chan et al. (2006) used a 10,000-year serial record of genetic data to relate a population bottleneck in tuco-tucos (fossorial rodents) to a notable earlier Holocene volcanic eruption and ashfall in Andean South America.

6.3.5 Phylogeography in the age of phylogenomics

Phylogenomics, broadly construed, is the use of whole genomic sequences to address biological, genomic, and bioinformatic problems (Eisen and Fraser, 2003; DeSalle, 2005). A subset of this burgeoning field uses whole genomes and data derived from them in a comparative framework to address phylogenetic and phylogeographic questions (Delsuc et al., 2005; DeSalle, 2005; Wiens, 2008). The array of whole nuclear, mitochondrial, and chloroplast genomes available for comparative analyses is growing rapidly; for example, as of July 2008, the National Center for Biotechnology Information (NCBI) listed whole mitochondrial genomes for 1,301 animals, of which 917 were vertebrates.

While its sampling design makes it unlikely that most future phylogeographic studies will utilize a complete array of fully sequenced whole genomes for every individual sample in an analysis, studies are likely to benefit from phylogenomics data and approaches because the utilization of whole genomes aligned across a broad array of taxa provides the opportunity to search efficiently for multiple gene regions in both nuclear and organelle genomes that have an appropriate level of variability to be informative across target populations and species (i.e., enabling efficient utilization of multilocus gene-tree approaches). For example, if a researcher were interested in using a multilocus approach to assessing phylogenetic and phylogeographic structure of populations (and possible 'cryptic species') within a geographic subset of Mole salamanders in the genus *Ambystoma*, he or she might begin by aligning whole mitochondrial genomes available for six species of *Ambystoma* from the NCBI database to search for potentially informative mitochondrial genes and nuclear introns, single nucleotide polymorphisms (SNPs), or microsatellites.

6.4 THE PHYLOGEOGRAPHIC EXPANSION OF HISTORICAL BIOGEOGRAPHY

6.4.1 Temporally deep versus shallow biogeographies

The conceptually and analytically expansive nature of phylogeography positions it as a bridge between what have become two essentially separate foci that occupy disparate spatial and temporal regimes in biogeography. Most directly, the phylogeographically based biogeography of the Quaternary has focused on questions such as locations of refugia (see Willis et al., this volume) and the geographic dynamics of populations, species, and biotas (Avise, 2000; Riddle and Hafner, 2004, 2007), with process explanations often invoking the major glacial-interglacial climatic oscillations of the Pleistocene, and occasionally intersecting with interests of ecological biogeographers, including latitudinal effects on biogeographic structure (e.g., Lessa et al., 2003). Another important and contentious topic has been the relative importance of Pleistocene climatic perturbations on the generation of new species, a debate that has been of considerable interest to avian phylogeographers (Klicka and Zink, 1997, 1999; Arbogast and Slowinski, 1998; Johnson and Cicero, 2004; Weir and Schluter, 2004; Zink et al., 2004). For example, Johnson and Cicero (2004), and Weir and Schluter (2004) have attributed appreciable levels of diversity in North American birds to accelerated rates of speciation during the Pleistocene; whereas Zink et al. (2004) concluded that there was no evidence for a spike in speciation rates during the Pleistocene (see Lovette, 2005 for a more synthetic reconciliation between advocates of Pleistocene and pre-Pleistocene speciation in birds).

On the other hand, phylogeography has not featured prominently in the largely pre-Quaternary realm of a historical biogeography that frequently references geological processes such as drifting continents, emerging islands, and the appearance and disappearance of land bridges (Riddle and Hafner, 2004, 2007). Questions within this realm have included the locations, originations, composition, and relationships between areas of endemism; the locations of and expansion of lineages and biotas away from ancestral areas; the relative importance of dispersal versus vicariance in shaping the Earth's biotas; and the tempo, mode, and location of major episodes of diversification (Figure 6.2).

The leading theoretical historical biogeographers of the 1970s and 1980s, including Gareth Nelson, Norman Platnick, and Donn Rosen, created methods based on the assumption that vicariance is more likely than dispersal and that only vicariance generates a pattern that is recoverable through hypothesis-driven analysis—an approach that has more recently been called the 'maximum vicariance' paradigm (Brooks et al., 2004). Advocates of maximum vicariance have criticized phylogeography for resurrecting the 'just-so' stories of earlier dispersalist biogeography (Humphries, 2000; Ebach and Humphries, 2002). However, Riddle and Hafner (2004, 2007) contradicted this view by showing that phylogeographic studies do frequently reference vicariance (Figure 6.2b). Even so, they also showed that phylogeographers, who have commonly conducted their studies a single taxon at a time, had not learned an important lesson from the vicariance biogeographers—that deciphering vicariance from dispersal often requires the examination of biogeographic structure across multiple, co-distributed taxa within a biota. In reality, therefore, a strong conceptual disconnect does appear to still separate phylogeography from much of the rest of historical biogeography (see Table 6.1), but the increasingly popular approach of comparative phylogeography—most commonly described as "the geographical comparison of evolutionary subdivision across multiple co-distributed species or species complexes" (Arbogast and Kenagy, 2001: 819)—is beginning to fill this gap (Figure 6.2a).

6.4.2. Comparative phylogeography

As originally conceived, comparative phylogeography was similar to those goals in historical biogeography that seek to reconstruct the history of biotas (as opposed to single taxa) in several aspects. First, they both generally rely on examination of phylogenetic hypotheses from two or more co-distributed taxa to recover a general history of biotic responses to events in Earth history. In addition to vicariance as an accepted general biotic response, we have been recently reminded that dispersal, with subsequent isolation and divergence, can be a general biotic event that is recoverable from the analysis of area cladograms (e.g., 'geo-dispersal', Lieberman, 2004). Well-known examples might include the Great American Biotic Interchange between North and South America following the completion of the Panamanian Land Bridge about 3.5 million years ago (Stehli and Webb, 1985) and the development of the Beringian Land Bridge between the eastern Palearctic and western Nearctic periodically throughout the Cenozoic (Donoghue and Moore, 2003). Comparative phylogeography has also been utilized to develop a molecular perspective on the assembly and disassembly of communities in a more explicitly ecological framework (e.g., Wares, 2002; Webb et al., 2002).

Second, taxon cladograms derived from phylogeographic data can be used within any historical biogeographic approach that seeks to investigate the geography of lineage divergence and speciation through vicariance, peripheral isolates dispersal, post-speciation dispersal, or sympatric divergence (Riddle and Hafner, 2006). Therefore, comparative phylogeography provides a basis for unifying the strengths of phylogeography (sampling designs and statistically rigorous assessment of dispersal) with those of historical biogeography (rigorous approaches to sorting

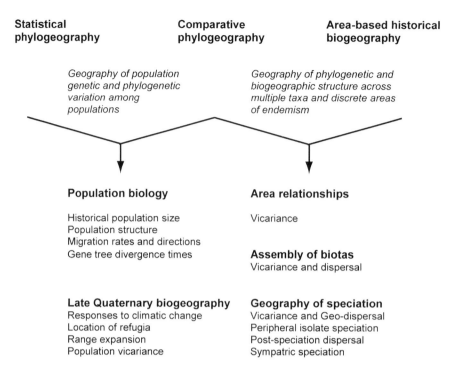

Figure 6.5 One view of the various domains of statistical phylogeography, comparative phylogeography, and area-based historical biogeography within an expanded and integrated biogeography (reprinted from Riddle and Hafner, 2007)

vicariance from various forms of dispersal, and elucidating ancestral areas) (Figure 6.5).

Importantly, comparative phylogeography is expanding the spatio-temporal realm of historical biogeography. Because phylogeography has emphasized divergence among populations within species and among closely related species, many of the questions of interest involve causal events in Earth history within time frames frequently focusing on the Pleistocene, although often extending into earlier Pliocene and Miocene times—primarily those associated with the Neogene cooling trend culminating in the large 100,000-year glacial and interglacial cycles of the last half of the Pleistocene (Ruddiman, 2001). The recovery of relationships between evolutionary lineages that have originated within these relatively recent time frames often requires the use of a rapidly evolving portion of an organism's genome—such as the mtDNA control region in many vertebrates (Taberlet, 1996), and intron regions in many nuclear genes (DeSalle, 2005)— and the geographic distribution of those lineages requires a phylogeographic sampling strategy (multiple localities and individuals). Comparative

phylogeography therefore offers the potential to expand the major goals of historical biogeography into the realm of populations, closely related species, and young biotas (Arbogast and Kenagy, 2001; Riddle and Hafner 2006, 2007).

Another advantage of integrating the strengths of comparative phylogeography and historical biogeography derives from the fact that phylogeographic data can be analyzed using either phylogenetic or population genetic approaches— comparative and statistical phylogeographic frameworks can be utilized reciprocally in a stepwise approach to create and evaluate hypotheses of taxon and biotic histories (Riddle and Hafner, 2006). Specifically, one might use a co-distributed, multi-taxon analysis (e.g., primary and secondary Brooks Parsimony Analysis [BPA]; Brooks et al., 2001) to develop phylogenetically based hypotheses of vicariance, geo-dispersal, peripheral isolates divergence, and post-divergence dispersal on an area cladogram. Biogeographers generally recognize that hypotheses of biotic histories based on area cladogram reconstructions are generated with an often large degree of uncertainty about the influence of an

Table 6.2 Outline of a possible five-step procedure for combining phylogeographic and phylogenetic biogeographic approaches to postulating a range of vicariance and dispersal events within the historical assembly of a biota. Methods are illustrative only, being those applied by Riddle and Hafner (2006)—an array of alternative methods (particularly in steps 2, 3, and 4) are also available (e.g., see Table 6.1)

Step	Procedures
1	Delineate the biota of interest, units of analysis, and distributional areas
	Data: Phylogeographic-scale; phylogroups
	Distributional Areas: Based on phylogeographic studies and physiographic features
2	Delineate initial set of areas of endemism
	Data: Phylogroup distributions across distributional areas
	Method: Parsimony Analysis of Endemicity (PAE) to identify core areas of endemism
3	Determine general divergence structure (vicariance or geo-dispersal)
	Data: Cladograms with phlyogroups as terminal units of analysis
	Method: Phylogenetic biogeography (Brooks Parsimony Analysis: primary BPA)
4	Resolve departures from general divergence (resolve reticulate area relationships)
	Data: Cladograms as in step 3, re-coded after duplicating reticulate areas
	Method: Phylogenetic biogeography (secondary BPA)
5	Test hypotheses of taxon and biotic histories generated in steps 3 and 4
	Data: Phylogroup distributions: geographic-genetic architecture within phylogroups
	Methods: Phylogeographic statistics: additional phlyogenetic analyses

array of variables difficult to evaluate (e.g., extinction events, timing of events, association with specific events in Earth history). However, each of these hypotheses is amenable to further evaluation by then switching to a population genetics—based analysis employing theory and methods from statistical phylogeography (Table 6.2). For example, a well-supported association between two areas might be interpreted as supporting biotic responses to a vicariant event, or alternatively, a geo-dispersal event from one area into the other (biotic dispersal followed by isolation and divergence). Statistical phylogeographic theory would predict a distinctly different geographic population genetic architecture under each of these hypotheses, and could therefore be used to discriminate between them (see Riddle and Hafner, 2006 for examples and further discussion).

6.5 THE PHYLOGEOGRAPHIC EXPANSION OF ECOLOGY, EVOLUTION, AND CONSERVATION AND GLOBAL CHANGE BIOLOGY

Biogeography generally, and more specifically historical biogeography, has matured theoretically, analytically, and empirically within the past

several decades to a point at which it promises to assume a fundamental role within highly integrative approaches to solving long-standing as well as newer problems in ecology and evolutionary biology (e.g., Wiens and Donoghue, 2004; Riddle, 2005; Lomolino et al., 2006). As argued above, phylogeography is playing an increasingly important role in the revitalized relevance of historical biogeography, particularly because of the power it brings to building bridges between disciplines with intersecting interests in a Neogene time frame (approximately the last 24 million years). It is within this time frame—incorporating profound changes in global climates, including the Middle Miocene Climatic Optimum about 14 million years ago, and proceeding through the Plio-Pleistocene cooling trend leading into the profound glacial and interglacial cycles of the latter half of the Pleistocene (Ruddiman, 2001; Zachos et al., 2001)—that evolutionary biologists, ecologists, and conservation biologists have become increasingly interested in the influences of geological and climate changes on the diversity and distributions of species, and the composition and function of biotas.

Phylogeography has provided a rigorous approach to addressing the relative importance of lineage divergence and extinction in the production of modern species (e.g., for a range of examples from the large literature on times of diversification in mammals, see Vrba, 1992; Riddle, 1995; Vrba et al., 1995; Steppan et al., 2004). Moreover, a trend in community ecology has been to address the long-term temporal dynamics of nonequilibrium associations between species and biotas precipitated primarily by range shifting, and episodes of range contraction and expansion responses to climate changes during a Late Pleistocene to Recent time frame (Graham et al., 1996; Riddle, 1998; Wares, 2002; Lyons, 2003; Kozak et al., 2005; Cadena, 2007). An applied extension of the consequences of range shifting on lineages and biotas addresses the issue of species invasions, and here, too, phylogeography is being utilized to reconstruct events (Vellend et al., 2007; Muirhead et al., 2008).

Many of the sorts of studies mentioned above are currently being addressed by combining phylogeography with ENM (see previous discussion) to provide complementary genetic and modeling assessments of species and biotic responses to past, and potentially future, global climate changes. The combination of phylogeographic with ENM studies is increasingly being suggested as an approach that is appropriate to address more explicit and applied conservation and global climate change problems because of the strength of the former to uncover morphologically cryptic lineages as well as genetic signatures of range expansion and contraction, and of the latter to add an ecological dimension to understanding lineage divergence and distributions (Hijmans and Graham, 2006; Rissler and Apodaca, 2007; Sattler et al., 2007). However, the influence of phylogeography on conservation biology as a means of recovering novel evolutionary lineages in fact extends back to the early 1990s when it assumed a fundamental role in the development of the genetically based definitions of Evolutionarily Significant Units (ESUs) and Management Units (MUs; Moritz, 1994). While the process of debating the actual entities and historical processes that an ESU designation should represent originated somewhat earlier (Woodruff, 1989), and continued to be debated into this decade (e.g., Crandall et al., 2000; Green, 2005; Fallon, 2007), most modern uses in conservation attempt to align ESU designations with a metric of 'neutral' genetic variation (Moritz, 2002; de Guia and Saitoh, 2007), thus making mtDNA in animals and cpDNA in plants still the most frequent units used to delineate ESUs. The use of such a genetically based ESU more generally throughout ecology and evolutionary biology in place of more traditionally defined species was advocated by Riddle and Hafner (1999), and, indeed, is essentially the premise behind the more recent Barcode of Life initiatives (deWaard et al., 2008).

6.6 CONCLUSIONS AND PROSPECTS

The future of phylogeographically based studies in biogeography, ecology, and evolution is bright owing to the increasingly integrative nature of research across these disciplines, and the contribution that phylogeography makes in building analytical and conceptual bridges among them and outward to the earth sciences. Moreover, the creation of a fully integrated environment for understanding past processes that underlie the emergence of distributional and diversification patterns in extant lineages and biotas will provide the critical framework required to be able to predict their future responses to an Earth changing rapidly under the combined stressors of habitat alteration, invasive species, and global climate change. In order to begin to fully realize the level of cross-disciplinary integration envisioned, phylogeographic approaches will continue to mature in a number of ways, as discussed throughout much of this chapter, and as reiterated and emphasized here:

1. Analyses within a comparative phylogeography framework will continue to expand to include a sufficient number of co-distributed taxa to allow for tests of historical assembly and congruence across 'whole ecosystems' (Riddle et al., 2008), allowing researchers to use modern, genetically based approaches to test classical (Diamond, 1975) and recent (Hubbell, 2001; Tilman, 2004) hypotheses in community ecology.
2. Analyses will continue to utilize a growing number of independent gene trees drawn from both organellar and nuclear genomes in order to resolve gene-tree versus species-tree incongruence, to estimate with greater confidence times since common ancestry, and to decipher species boundaries in the absence of complete lineage sorting (Funk and Omland, 2003; Carstens and Knowles, 2007b).
3. Analyses will continue to advance in being able to rigorously disentangle complexity in the phylogeographic histories of multiple co-distributed taxa (e.g., Hickerson et al., 2006; Leaché et al., 2007).
4. Analyses that incorporate both phylogeographic and ENM components will continue to grow in sophistication as bioclimatic models for various past and future time slices accumulate and are refined, and as researchers develop novel hypothesis-testing frameworks that synergistically optimize the value of both approaches.
5. Analyses will become appreciably more integrated across vast expanses of time, space, and biodiversity as teams of investigators merge and begin to address 'really big' research

questions—utilizing growing networks of cyber-infrastructure and databases.

6. All of the above will become critical components of building an architecture within an emerging 'conservation biogeography' that has the capacity to address consequences for biodiversity of current and future habitat alterations, introductions of invasive species into native biotas, and global climate change.

REFERENCES

Althoff, D.M., and Pellmyr, O. (2002) 'Examining genetic structure in a bogus yucca moth: A sequential approach to phylogeography', *Evolution*, 56: 1632–43.

Alves, P.C., Harris, D.J., Melo-Ferreira, J., Branco, M., Ferrand, N., Suchentrunk, F., and Boursot, P. (2006) 'Hares on thin ice: Introgression of mitochondrial DNA in hares and its implications for recent phylogenetic analyses', *Molecular Phylogenetics and Evolution*, 40: 640–41.

Arbogast, B.S., and Kenagy, G.J. (2001) 'Comparative phylogeography as an integrative approach to historical biogeography', *Journal of Biogeography*, 28: 819–25.

Arbogast, B.S., and Slowinski, J.B. (1998) 'Pleistocene speciation and the mitochondrial DNA clock', *Science*, 282: 1955a.

Avise, J. (2000) *Phylogeography: The History and Formation of Species*. Cambridge, MA: Harvard University Press.

Avise, J.C., Arnold, J., Ball, R.M., Bermingham, E., Lamb, T., Neigel, J.E., Reeb, C.A., and Saunders, N.C. (1987) 'Intraspecific phylogeography: The mitochondrial DNA bridge between population genetics and systematics', *Annual Review of Ecology and Systematics*, 18: 489–522.

Ballard, J.W.O., and Whitlock, M.C. (2004) 'The incomplete natural history of mitochondria', *Molecular Ecology*, 13: 729–44.

Bermingham, E., and Moritz, C. (1998) 'Comparative phylogeography: Concepts and applications', *Molecular Ecology*, 7: 367–69.

Bremer, K. (1992) 'Ancestral areas: A cladistic reinterpretation of the center of origin concept', *Systematic Biology*, 41: 436–45.

Brooks, D.R., Dowling, A.P.G., van Veller, M.G.P., and Hoberg, E.P. (2004) 'Ending a decade of deception: A valiant failure, a not-so-valiant failure and a success story', *Cladistics—the International Journal of the Willi Hennig Society*, 20: 32–46.

Brooks, D.R., van Veller, M.G.P., and McLennan, D.A. (2001) 'How to do BPA, really', *Journal of Biogeography*, 28: 345–58.

Brundin, L. (1966) 'Transantarctic relationships and their significance as evidenced by midges', *Kungliga Svenska Vetenskapsakademiens Handlinger (series 4)*, 11: 1–472.

Cadena, C.D. (2007) 'Testing the role of interspecific competition in the evolutionary origin of elevational zonation: An example with Buarremon brush-finches (Aves,

Emberizidae) in the neotropical mountains', *Evolution*, 61: 1120–36.

Carstens, B.C., and Knowles, L.L. (2007a) 'Shifting distributions and speciation: Species divergence during rapid climate change', *Molecular Ecology*, 16: 619–27.

Carstens, B.C., and Knowles, L.L. (2007b) 'Estimating species phylogeny from gene-tree probabilities despite incomplete lineage sorting: An example from Melanoplus grasshoppers', *Systematic Biology*, 56: 400–411.

Carstens, B.C., and Richards, C.L. (2007) 'Integrating coalescent and ecological niche modeling in comparative phylogeography', *Evolution*, 61: 1439–54.

Chan, Y.L., Anderson, C.N.K., and Hadly, E.A. (2006) 'Bayesian estimation of the timing and severity of a population bottleneck from ancient DNA', *PLoS Genetics*, 2: 451–60.

Clark, R.W., Brown, W.S., Stechert, R., and Zamudio, K.R. (2008) 'Integrating individual behaviour and landscape genetics: The population structure of timber rattlesnake hibernacula', *Molecular Ecology*, 17: 719–30.

Crandall, K.A., Bininda-Emonds, O.R.P., Mace, G.M., and Wayne, R.K. (2000) 'Considering evolutionary processes in conservation biology', *Trends in Ecology & Evolution*, 15: 290–95.

Craw, R.C., Grehan, J.R., and Heads, M.J. (1999) *Panbiogeography: Tracking the History of Life* New York: Oxford University Press.

Crews, S.C., and Hedin, M. (2006) 'Studies of morphological and molecular phylogenetic divergence in spiders (Araneae: Homalonychus) from the American southwest, including divergence along the Baja California peninsula', *Molecular Phylogenetics and Evolution*, 38: 470–87.

Crisci, J.V., Katinas, L., and Posadas, P. (2003) *Historical Biogeography: An Introduction*. Cambridge, MA: Harvard University Press.

Croizat, L. (1964) *Space, Time, Form: The Biological Synthesis*. Caracas: Published by the author.

Croizat, L., Nelson, G., and Rosen, D.E. (1974) 'Centers of origin and related concepts', *Systematic Zoology*, 23: 265–87.

Darlington, P.J., Jr. (1957) *Zoogeography: The Geographical Distributions of Animals*. New York: John Wiley.

Darlington, P.J., Jr. (1965) *Biogeography of the Southern End of the World*. Cambridge, MA: Harvard University Press.

de Guia, A.P.O., and Saitoh, T. (2007) 'The gap between the concept and definitions in the Evolutionarily Significant Unit: The need to integrate neutral genetic variation and adaptive variation', *Ecological Research*, 22: 604–12.

Delsuc, F., Brinkmann, H., and Philippe, H. (2005) 'Phylogenomics and the reconstruction of the tree of life', *Nature Reviews Genetics*, 6: 361–75.

DeSalle, R. (2005) 'Animal phylogenomics: Multiple interspecific genome comparisons', *Molecular Evolution: Producing the Biochemical Data, Part B*, 104–33.

deWaard, J.R., Ivanova, N.V., Hajibabaei, M., and Hebert, P.D.N. (2008) 'Assembling DNA barcodes. Analytical protocols', *Methods in Molecular Biology*, 410.

Di Candia, M.R., and Routman, E.J. (2007) 'Cytonuclear discordance across a leopard frog contact zone', *Molecular Phylogenetics and Evolution*, 45: 564–75.

Diamond, J.M. (1975) Assembly of species communities. In Cody, M.L., and Diamond, J.M. (eds.), *Ecology and Evolution of Communities*. Cambridge, MA: Belknap Press. pp. 342–444.

Dietz, R.S. (1961) 'Continental and ocean basin evolution by spreading of the sea floor', *Nature*, 190: 854–57.

Donoghue, M.J., and Moore, B.R. (2003) 'Toward an integrative historical biogeography', *Integrative and Comparative Biology*, 43: 261–70.

Ebach, M.C. 2001: Extrapolating cladistic biogeography: A brief comment on van Veller et al. (1999, 2000, 2001). *Cladistics—the International Journal of the Willi Hennig Society*, 17: 383–88.

Ebach, M.C., and Humphries, C.J. (2002) 'Cladistic biogeography and the art of discovery', *Journal of Biogeography*, 29: 427–44.

Ebach, M.C., and Tangney, R.S. (2006) *Biogeography in a Changing World*. Boca Raton, FL: CRC Press.

Edwards, S.V., and Beerli, P. (2000) 'Perspective: Gene divergence, population divergence, and the variance in coalescence time in phylogeographic studies', *Evolution*, 54: 1839–54.

Edwards, S.V., Liu, L., and Pearl, D.K. (2007) 'High-resolution species trees without concatenation', *Proceedings of the National Academy of Sciences of the United States of America*, 104: 5936–41.

Eisen, J.A., and Fraser, C.M. (2003) 'Phylogenomics: Intersection of evolution and genomics', *Science*, 300: 1706–7.

Excoffier, L., and Heckel, G. (2006) 'Computer programs for population genetics data analysis: A survival guide', *Nature Reviews Genetics*, 7: 745–58.

Fallon, S.M. (2007) 'Genetic data and the listing of species under the U.S. endangered species act', *Conservation Biology*, 21: 1186–95.

Funk, D.J., and Omland, K.E. (2003) 'Species-level paraphyly and polyphyly: Frequency, causes, and consequences, with insights from animal mitochondrial DNA', *Annual Review of Ecology, Evolution, and Systematics*, 34: 397–423.

Funk, V.A. (2004) Revolutions in historical biogeography. In Lomolino, M.V., Sax, D.F. and Brown, J.H. (eds.), *Foundations of Biogeography: Classic Papers with Commentaries*. Chicago: University of Chicago Press.

Graham, C.H., Ron, S.R., Santos, J.C., Schneider, C.J., and Moritz, C. (2004) 'Integrating phylogenetics and environmental niche models to explore speciation mechanisms in dendrobatid frogs', *Evolution*, 58: 1781–93.

Graham, R.W. and 19 authors (1996) 'Spatial response of mammals to late quaternary environmental fluctuations', *Science*, 272: 1601–6.

Green, D.M. (2005) 'Designatable units for status assessment of endangered species', *Conservation Biology*, 19: 1813–20.

Hadly, E.A., Ramakrishnan, U., Chan, Y.L., van Tuinen, M., O'Keefe, K., Spaeth, P.A., and Conroy, C.J. (2004) 'Genetic response to climatic change: Insights from ancient DNA and phylochronology', *PLoS Biology*, 2: 1600–1609.

Hennig, W. (1966) *Phylogenetic Systematics*. Urbana IL: University of Illinois Press.

Hess, H.H. (1962) History of ocean basins. In Engel, A.E.J., James, H.L. and Leonard, B.F. (eds.), *Petrological Studies: A Volume in Honor of A. F. Buddington*. New York: Geological Society of America. pp. 599–620.

Hickerson, M.J., Stahl, E.A., and Lessios, H.A. (2006) 'Test for simultaneous divergence using approximate Bayesian computation', *Evolution*, 60: 2435–53.

Hijmans, R.J., and Graham, C.H. (2006) 'The ability of climate envelope models to predict the effect of climate change on species distributions', *Global Change Biology*, 12: 2272–81.

Hofreiter, M., Serre, D., Poinar, H.N., Kuch, M., and Paabo, S. (2001) 'Ancient DNA', *Nature Reviews Genetics*, 2: 353–59.

Holderegger, R., and Wagner, H.H. (2008) 'Landscape genetics', *Bioscience*, 58: 199–207.

Hubbell, S.P. (2001) *The Unified Neutral Theory of Biodiversity and Biogeography*. Princeton: Princeton University Press.

Hugall, A., Moritz, C., Moussalli, A., and Stanisic, J. (2002) 'Reconciling paleodistribution models and comparative phylogeography in the Wet Tropics rainforest land snail Gnarosophia bellendenkerensis (Brazier 1875)', *Proceedings of the National Academy of Sciences of the United States of America*, 99: 6112–17.

Humphries, C.J. (2000) 'Form, space and time; which comes first'? *Journal of Biogeography*, 27: 11–15.

Johnson, N.K., and Cicero, C. (2004) 'New mitochondrial DNA data affirm the importance of Pleistocene speciation in North American birds', *Evolution*, 58: 1122–30.

Kidd, D.M., and Ritchie, M.G. (2006) 'Phylogeographic information systems: Putting the geography into phylogeography', *Journal of Biogeography*, 33: 1851–65.

Klicka, J., and Zink, R.M. (1997) 'The importance of recent ice ages in speciation: A failed paradigm', *Science*, 277: 1666–69.

Klicka, J., and Zink, R.M. (1999) 'Pleistocene effects on North American songbird evolution', *Proceedings of the Royal Society of London Series B*, 266: 695–700.

Knowles, L.L. (2004) 'The burgeoning field of statistical phylogeography', *Journal of Evolutionary Biology*, 17: 1–10.

Knowles, L.L., and Carstens, B.C. (2007) 'Estimating a geographically explicit model of population divergence', *Evolution*, 61: 477–93.

Kozak, K.H., Graham, C.H., and Wiens, J.J. (2008) 'Integrating GIS-based environmental data into evolutionary biology', *Trends in Ecology & Evolution*, 23: 141–48.

Kozak, K.H., Larson, A.A., Bonett, R.M., and Harmon, L.J. (2005) 'Phylogenetic analysis of ecomorphological divergence, community structure, and diversification rates in dusky salamanders (Plethodontidae: Desmognathus)', *Evolution*, 59: 2000–2016.

Leaché, A.D., Crews, S.C., and Hickerson, M.J. (2007) 'Two waves of diversification in mammals and reptiles of Baja California revealed by hierarchical Bayesian analysis', *Biology Letters*, 3: 646–50.

Leaché, A.D., and Mulcahy, D.G. (2007) 'Phylogeny, divergence times and species limits of spiny lizards (Sceloporus magister species group) in western North

American deserts and Baja California', *Molecular Ecology*, 16: 5216–33.

Lessa, E.P., Cook, J.A., and Patton, J.L. (2003) 'Genetic footprints of demographic expansion in North America, but not Amazonia, during the Late Quaternary', *Proceedings of the National Academy of Sciences of the United States of America*, 100: 10331–34.

Lieberman, B.S. (2004) Range expansion, extinction, and biogeographic congruence: A deep time perspective. In Lomolino, M.V. and Heaney, L.R. (eds.), *Frontiers of Biogeography: New Directions in the Geography of Nature*. Sunderland, MA: Sinauer. pp. 111–24.

Lindahl, T. (2000) 'Fossil DNA', *Current Biology*, 10: R616.

Lindell, J., Mendez-de la Cruz, F.R., and Murphy, R.W. (2005) 'Deep genealogical history without population differentiation: Discordance between mtDNA and allozyme divergence in the zebra-tailed lizard (Callisaurus draconoides)', *Molecular Phylogenetics and Evolution*, 36: 682–94.

Lindell, J., Mendez-De La Cruz, F.R., and Murphy, R.W. (2008) 'Deep biogeographical history and cytonuclear discordance in the black-tailed brush lizard (Urosaurus nigricaudus) of Baja California', *Biological Journal of the Linnean Society*, 94: 89–104.

Lindell, J., Ngo, A., and Murphy, R.W. (2006) 'Deep genealogies and the mid-peninsular seaway of Baja California', *Journal of Biogeography*, 33: 1327–31.

Linnen, C.R., and Farrell, B.D. (2008) 'Phylogenetic analysis of nuclear and mitochondrial genes reveals evolutionary relationships and mitochondrial introgression in the sertifer species group of the genus Neodiprion (Hymenoptera: Diprionidae)', *Molecular Phylogenetics and Evolution*, 48.

Lomolino, M.V., Riddle, B.R., and Brown, J.H. (2006) *Biogeography*. 3rd ed. Sunderland, MA: Sinauer.

Lovette, I.J. (2005) 'Glacial cycles and the tempo of avian speciation', *Trends in Ecology & Evolution*, 20: 57–59.

Lyons, S.K. (2003) 'A quantitative assessment of the range shifts of Pleistocene mammals', *Journal of Mammalogy*, 84: 385–402.

Manel, S., Schwartz, M.K., Luikart, G., and Taberlet, P. (2003) 'Landscape genetics: Combining landscape ecology and population genetics', *Trends in Ecology & Evolution*, 18: 189–97.

Matthew, W.D. (1915) 'Climate and evolution', *Annals of the New York Academy of Sciences*, 24: 171–318.

McGuire, J.A., Linkem, C.W., Koo, M.S., Hutchison, D.W., Lappin, A.K., Orange, D.I., Lemos-Espinal, J., Riddle, B.R., and Jaeger, J.R. (2007) 'Mitochondrial introgression and incomplete lineage sorting through space and time: Phylogenetics of crotaphytid lizards', *Evolution*, 61: 2879–97.

Morando, M., Avila, L.J., Baker, J., and Sites, J.W. (2004) 'Phylogeny and phylogeography of the Liolaemus darwinii complex (Squamata: Liolaemidae): Evidence for introgression and incomplete lineage sorting', *Evolution*, 58: 842–61.

Moritz, C. (1994) 'Defining "evolutionarily significant units" for conservation', *Trends in Ecology & Evolution*, 9: 373–75.

Moritz, C. (2002) 'Strategies to protect biological diversity and the evolutionary processes that sustain it', *Systematic Biology*, 51: 238–54.

Nagata, N., Kubota, K., Yahiro, K., and Sota, T. (2007) 'Mechanical barriers to introgressive hybridization revealed by mitochondrial introgression patterns in Ohomopterus ground beetle assemblages', *Molecular Ecology*, 16: 4822–36.

Nelson, G. (1974) 'Historical biogeography—alternative formalization', *Systematic Zoology*, 23: 555–58.

Petren, K., Grant, P.R., Grant, B.R., and Keller, L.F. (2005) 'Comparative landscape genetics and the adaptive radiation of Darwin's finches: The role of peripheral isolation', *Molecular Ecology*, 14: 2943–57.

Platnick, N.I., and Nelson, G. (1978) 'A method of analysis for historical biogeography', *Systematic Zoology*, 27: 1–16.

Ramakrishnan, U., and Hadly, E.A. (2009) 'Using phylochronology to reveal cryptic population histories: Review and synthesis of 29 ancient DNA studies', *Molecular Ecology*, 18: 1310–30.

Ramakrishnan, U., Hadly, E.A., and Mountain, J.L. (2005) 'Detecting past population bottlenecks using temporal genetic data', *Molecular Ecology*, 14: 2915–22.

Recuero, E., Martinez-Solano, I., Parra-Olea, G., and Garcia-Paris, M. (2006) 'Phylogeography of Pseudacris regilla (Anura: Hylidae) in western North America, with a proposal for a new taxonomic rearrangement', *Molecular Phylogenetics and Evolution*, 39: 293–304.

Ree, R.H., Moore, B.R., Webb, C.O. and Donoghue, M.J. (2005) 'A likelihood framework for inferring the evolution of geographic range on phylogenetic trees', *Evolution* 59: 2299–311.

Ree, R.H., and Smith, S.A. (2008) 'Maximum likelihood inference of geographic range evolution by dispersal, local extinction, and cladogenesis', *Systematic Biology*, 57: 4–14.

Richards, C.L., Carstens, B.C., and Knowles, L.L. (2007) 'Distribution modelling and statistical phylogeography: An integrative framework for generating and testing alternative biogeographical hypotheses', *Journal of Biogeography*, 34: 1833–45.

Riddle, B.R. (1995) 'Molecular biogeography in the pocket mice (*Perognathus* and *Chaetodipus*) and grasshopper mice (*Onychomys*): The Late Cenozoic development of a North American aridlands rodent guild', *Journal of Mammalogy*, 76: 283–301.

Riddle, B.R. (1998) 'The historical assembly of continental biotas: Late Quaternary range-shifting, areas of endemism, and biogeographic structure in the North American mammal fauna', *Ecography*, 21: 437–46.

Riddle, B.R. (2005) 'Is biogeography emerging from its identity crisis'? *Journal of Biogeography*, 32: 185–86.

Riddle, B.R., Dawson, M.N., Hadly, E.A., Hafner, D.J., Hickerson, M.J., Mantooth, S.J., and Yoder, A.D. (2008) 'The role of molecular genetics in sculpting the future of integrative biogeography', *Progress in Physical Geography*, 32: 173–202.

Riddle, B.R., and Hafner, D.J. (1999) 'Species as units of analysis in ecology and biogeography: Time to take the blinders off', *Global Ecology and Biogeography*, 8: 433–41.

Riddle, B.R., and Hafner, D.J. (2004) 'The past and future roles of phylogeography in historical biogeography. In Lomolino, M.V. and Heaney, L.R. (eds.), *Frontiers of Biogeography*, Sunderland, MA: Sinauer. pp. 93–110.

Riddle, B.R., and Hafner, D.J. (2006) 'A step-wise approach to integrating phylogeographic and phylogenetic biogeographic perspectives on the history of a core North American warm deserts biota', *Journal of Arid Environments*, 65: 435–61.

Riddle, B.R., and Hafner, D.J. (2007) Phylogeography in historical biogeography: Investigating the biogeographic histories of populations, species, and young biotas. In Ebach, M.C. and Tangney, R.S. (eds.), *Biogeography in a Changing World*. Boca Raton, FL: CRC Press. pp. 161–76.

Riddle, B.R., Hafner, D.J., Alexander, L.F., and Jaeger, J.R. (2000) 'Cryptic vicariance in the historical assembly of a Baja California Peninsular Desert biota', *Proceedings of the National Academy of Sciences of the United States of America*, 97: 14438–43.

Riginos, C. (2005) 'Cryptic vicariance in Gulf of California fishes parallels vicariant patterns found in Baja California mammals and reptiles', *Evolution*, 59: 2678–90.

Rissler, L.J., and Apodaca, J.J. (2007) 'Adding more ecology into species delimitation: Ecological niche models and phylogeography help define cryptic species in the black salamander (Aneides flavipunctatus)', *Systematic Biology*, 56: 924–42.

Ronquist, F. (1997) 'Dispersal-vicariance analysis: A new approach to the quantification of historical biogeography', *Systematic Biology*, 46: 195–203.

Rosen, D.E. (1978) 'Vicariant patterns and historical explanation in biogeography', *Systematic Zoology* 27: 159–88.

Ruddiman, W.F. (2001) *Earth's Climate: Past and Future*. New York: W.H. Freeman.

Sattler, T., Bontadina, F., Hirzel, A.H., and Arlettaz, R. (2007) 'Ecological niche modelling of two cryptic bat species calls for a reassessment of their conservation status', *Journal of Applied Ecology*, 44: 1188–99.

Siddall, M.E. (2005) 'Bracing for another decade of deception: The promise of Secondary Brooks Parsimony Analysis', *Cladistics*, 21: 90–99.

Simpson, G.G. (1940) 'Mammals and land bridges', *Journal of the Washington Academy of Sciences*, 30: 137–63.

Slechtova, V., Bohlen, J., and Perdices, A. (2008) 'Molecular phylogeny of the freshwater fish family Cobitidae (Cypriniformes: Teleostei): Delimitation of genera, mitochondrial introgression and evolution of sexual dimorphism', *Molecular Phylogenetics and Evolution*, 47: 812–31.

Smith, C.I., Chamberlain, A.T., Riley, M.S., Stringer, C., and Collins, M.J. (2003) 'The thermal history of human fossils and the likelihood of successful DNA amplification', *Journal of Human Evolution*, 45: 203–17.

Stehli, F.G., and Webb, S.D., eds. (1985) *The Great American Biotic Interchange*. New York: Plenum.

Steppan, S.J., Adkins, R.M., and Anderson, J. (2004) 'Phylogeny and divergence-date estimates of rapid radiations in muroid rodents based on multiple nuclear genes', *Systematic Biology*, 53: 533–53.

Storfer, A., Murphy, M.A., Evans, J.S., Goldberg, C.S., Robinson, S., Spear, S.F., Dezzani, R., Delmelle, E., Vierling, L., and Waits, L.P. (2007) 'Putting the "landscape" in landscape genetics', *Heredity*, 98: 128–42.

Taberlet, P. (1996) The use of mitochondrial DNA control region sequencing in conservation genetics. In Smith, T.B. and Wayne, R.K. (eds.), *Molecular Genetic Approaches in Conservation*. New York: Oxford University Press. pp. 125–42.

Templeton, A.R. (2004) 'Statistical phylogeography: Methods of evaluating and minimizing inference errors', *Molecular Ecology*, 13: 789–809.

Tilman, D. (2004) 'Niche tradeoffs, neutrality, and community structure: A stochastic theory of resource competition, invasion, and community assembly', *Proceedings of the National Academy of Sciences of the United States of America*, 101: 10854–61.

Vogel, L.S., and Johnson, S.G. (2008) 'Estimation of hybridization and introgression frequency in toads (Genus: Bufo) using DNA sequence variation at mitochondrial and nuclear loci', *Journal of Herpetology*, 42: 61–75.

Vrba, E.S. (1992) 'Mammals as a key to evolutionary theory', *Journal of Mammalogy*, 73: 1–28.

Vrba, E.S., Denton, G.H., Partridge, T.C., and Burckle, L.H. (1995) *Paleoclimate and Evolution, with Emphasis on Human Prigins*. New Haven, CT: Yale University Press.

Waltari, E., Hijmans, R.J., Peterson, A.T., Nyari, A.S., Perkins, S.L., and Guralnick, R.P. (2007) 'Locating pleistocene refugia: Comparing phylogeographic and ecological niche model predictions', *PLoS ONE* 2: e563.

Wares, J.P. (2002) 'Community genetics in the Northwestern Atlantic intertidal', *Molecular Ecology*, 11: 1131–44.

Webb, C.O., Ackerly, D.D., McPeek, M.A., and Donoghue, M.J. (2002) 'Phylogenies and community ecology', *Annual Review of Ecology and Systematics* 33: 475–505.

Weir, J.T., and Schluter, D. (2004) 'Ice sheets promote speciation in boreal birds', *Proceedings of the Royal Society of London Series B*, 271: 1881–87.

Whittaker, R.J., Araujo, M.B., Paul, J., Ladle, R.J., Watson, J.E.M., and Willis, K.J. (2005) 'Conservation biogeography: Assessment and prospect', *Diversity and Distributions*, 11: 3–23.

Wiens, J.J. (2008) 'Systematics and herpetology in the age of genomics', *Bioscience*, 58: 297–307.

Wiens, J.J., and Donoghue, M.J. (2004) 'Historical biogeography, ecology and species richness', *Trends in Ecology & Evolution*, 19: 639–44.

Willerslev, E., and Cooper, A. (2005) 'Ancient DNA', *Proceedings of the Royal Society of London Series B-Biological Sciences*, 272: 3–16.

Woodruff, D.S. (1989) The problems of conserving genes and species. In Western, D. and Pearl, M.C. (eds.), *Conservation for the Twenty-First Century.* New York: Oxford University Press. pp. 76–88.

Zachos, J., Pagani, M., Sloan, L., Thomas, E., and Billups, K. (2001) 'Trends, rhythms, and aberrations in global climate 65 Ma to present', *Science*, 292: 686–93.

Zarza, E., Reynoso, V.H., and Emerson, B.C. (2008) 'Diversification in the northern neotropics: Mitochondrial and nuclear DNA phylogeography of the iguana Ctenosaura pectinata and related species', *Molecular Ecology*, 17.

Zink, R.M., and Barrowclough, G.F. (2008) 'Mitochondrial DNA under siege in avian phylogeography', *Molecular Ecology*, 17: 2107–21.

Zink, R.M., Kessen, A.E., Line, T.V., and Blackwell-Rago, R.C. (2001) 'Comparative phylogeography of some aridland bird species', *Condor*, 103: 1–10.

Zink, R.M., Klicka, J., and Barber, B.R. (2004) 'The tempo of avian diversification during the Quaternary', *Philosophical Transactions of the Royal Society of London Series B*, 359: 215–19.

The Biogeographical Importance of Pleistocene Refugia

Katherine J. Willis, Shonil A. Bhagwat, and Mary E. Edwards

7.1 INTRODUCTION

The concept of populations of plants and/or animals surviving the extremes of the cold stages of the Pleistocene ice ages in refugia (isolated clusters in microenvironmentally favourable locations) is well established in the literature. It forms the backbone to a number of key biogeographical theories used to explain observed present-day diversity and distributions of plants and animals including patterns varying in scale from a biome through to genetic variation between populations, and with examples ranging from the tropics to the Arctic.

Over the past 10 years there has been the emergence of a number of new palaeoecological and genetic studies specifically aimed at examining the biogeographical importance of Pleistocene refugia. These have been predominantly based around sites in the Neotropics, North America, and Europe, and almost all of these studies have led to critical re-evaluation of the biogeographical theories relating to the role of refugia in determining present-day diversity and distribution. This chapter reviews the original biogeographical theories relating to Pleistocene refugia and examines them alongside these newly emerging studies. Examples will be taken from the Neotropics, North America, and Europe, although it is acknowledged that there are many other regions in the world that have also been studied in this respect, for example, South America (Premoli et al., 2000), Africa (Maley, 1991; Maley and Brenac, 1998), and Australia (Kirkpatrick and Fowler, 1998; Schneider and Moritz, 1999). From all of these studies it is apparent that the refugial paradigm is currently undergoing a paradigm shift (sensu Kuhn, 1970) and that in many regions a more complex and different model needs to be constructed to understand the biogeographical importance of Pleistocene refugia.

7.2 NEOTROPICS

Literature about low-latitude species diversity is replete with refugial theories, particularly those relating to the Neotropics (Hill and Hill, 2001). The original theory, first put forward by Haffer in 1969, was based on studies of present-day patterns of bird endemism in the Amazon basin. Haffer (1969), on the basis that there appeared to be pockets of high levels of endemism in mid-altitude sites, proposed refugial isolation, and subsequent allopatric speciation of birds during the cold stages of the Pleistocene (1.8 Ma [million years ago]) ice ages. He argued that during the aridity-limited environment of the cold stages, the forested habitats able to support such bird populations were restricted to mid-altitude locations where moisture levels would have been higher due to orographic precipitation. In contrast, the arid climate in the lowland Amazon basin would have resulted in the replacement of forests by a 'sea of savanna'. The prolonged isolation of fauna and flora during the Pleistocene cold stages in mid-altitude refugia and their separation from other populations by a 'sea of savanna' resulted in allopatric speciation and ultimately pockets of endemism.

Although Haffer's (1969) theory was originally used to explain pockets of bird endemism in the Amazon basin, it has since become a widely utilized model for low-latitude environments and has been used a number of times to explain centers of endemism in other groups within the Amazon basin, including lizards, frogs, plants, and butterflies (Figure 7.1a). However, one of the underlying problems with Haffer's theory is circularity in his line of argument. He used evidence from the current distribution of endemic bird species in the Amazon basin to identify regions of refugia, and then proposed that isolation in these refugia was responsible for their high levels of endemism. It was impossible to verify this line of reasoning without independent evidence to support the fact that during the cold stages of the Pleistocene the lowland Amazon basin turned into a sea of savanna.

Over the past 10 years, two lines of independent evidence (fossil and genetic) have been used to test the assumptions underlying this hypothesis, and results from these studies have started to reveal a rather different picture of the redistribution of vegetation in the Neotropics during the cold stages of the Pleistocene. Each will be addressed in turn.

7.2.1 Fossil evidence

A number of key pollen diagrams from sedimentary sequences in the Neotropics (for a review, see Mayle et al., 2004) have indicated that the response of the vegetation to the cold stages of the Pleistocene was far more complex than previously thought. There are now a number of good sedimentary sequences from the Amazon basin that contain fossil pollen that detail vegetation changes over the cold stages of the Pleistocene. These include a record that spans 170,000 years from Lake Pata in northwest Amazonia (Bush et al., 2002; Colinvaux et al., 1996) and a 50,000-year record from the Amazon fan sediments (Haberle and Maslin, 1999: Figure 3). The latter sequence is of particular interest because it represents a basin-wide record of vegetation change in the Amazon basin. In all the sequences from the Amazon basin there is no evidence for the establishment of a 'sea of savanna' in the lowland basin during the cold stages but rather that it remained forested throughout. The composition of the forest was different within warm stages, with many more cold Andean taxa than at present (Colinvaux et al., 2000). This in itself is interesting because it suggests that the total opposite of the Haffer model occurred: trees migrated downward from the mid-altitude sites into the lowland basin rather than the other way around.

To understand why forest taxa moved down into the basin rather than upward as predicted by Haffer it is necessary to examine evidence from the fossil palaeo-climate proxies. Reconstruction of the climate in Amazonia during the Last Glacial Maximum (LGM: between 22,000–19,000 years ago) by using fossil proxies, including evidence from stable isotopes (Huang et al., 2001), gas concentrations in fossil groundwater (Stute et al., 1995), and tropical corals (Guilderson et al., 1994), supports the suggestion that glacial temperatures in the Amazon would have been up to 5°C cooler than present. Fossil proxies also indicate that there was a significant reduction in atmospheric CO_2 concentrations to as much as 180–200 ppm. (Cowling et al., 2001). The fossil evidence does not, however, support the suggestion that the Amazon basin became much drier.

Based on palaeolimnological evidence obtained from a number of lakes in northwest Amazonia (Bush et al., 2002), it is now widely recognized that lake levels were relatively high during the LGM and that humid conditions existed. Some estimates suggest that precipitation was even higher than the present day (Baker et al., 2001; Mayle et al., 2004). Thus the most limiting climatic factors to vegetation during the cold stages in Amazonia would have been cooler temperatures and a carbon-depleted atmosphere. Since the impact of both on the vegetation would have increased with altitude (Bennett and Willis, 2000; Huang et al., 2001), it is highly likely that these were the principal mechanisms responsible for restricting many mid-altitude Andean forest taxa to the lowland Amazon basin during the LGM. As a result, the forest that formed in the Amazon basin during the LGM has no present-day analogue (Colinvaux et al., 2000).

From a biogeographical viewpoint, evidence for a nonanalogue forest in the Amazon lowland basin during the cold stages of the Quaternary is particularly interesting because it would have had exactly the opposite effect to that predicted by Haffer (1969). Rather than the cold stages of the ice ages resulting in small pockets of isolated forest, creating conditions for allopatric speciation, the greater mixing of forest communities in the lowland Amazon basin would rather have resulted in increased opportunities for genetic mixing and interbreeding.

(a)

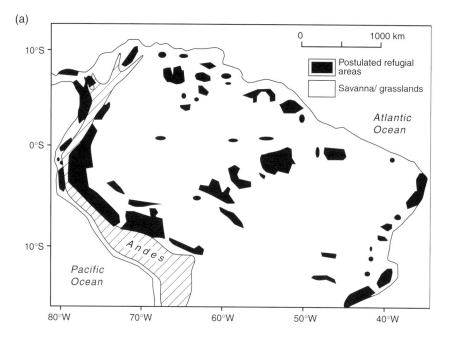

(b)

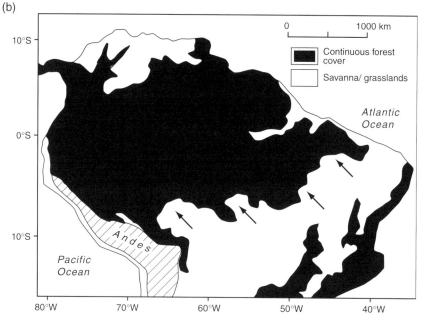

Figure 7.1   Pleistocene refugia in the Neotropics: (a) Haffer's (1969) proposed 'sea of savanna' (marked in white) in the Amazon basin during the last full glacial with scattered mid-altitude refugia (marked in black); (b) Alternative model for the Neotropics (e.g., Colinvaux et al., 2000) during the Pleistocene ice ages. Here the Amazonian region is considered to have been under continuous forest cover during the last glacial but the composition of the forest was markedly different. In addition, other studies have indicated that there was significant reduction in forests at the forest/savanna boundaries (schematically marked in as arrows) to be replaced by savanna and seasonally dry tropical forests (Mayle et al., 2004)

The question, therefore, is whether or not the cold stages of the Quaternary had a real impact on the distribution of forested vegetation in the Neotropics. Whilst the majority of studies for the Neotropics have been carried out in the rain forest, recent work in Neotropical seasonally dry forests and ecotonal boundaries of the forest—savanna and upper-cloud forest—open grassland suggest that this is not necessarily a representative picture for the whole of the Neotropics. For example, recent work by Mayle and others (Behling and Hooghiemstra, 2000; Mayle, 2004; Mourguiart and Ledru, 2003) is starting to indicate that in these 'periphery' forests there was far greater dynamism in these vegetation formations in response to the cold stages of the Quaternary than in the Amazon lowland. In particular, palaeoecological work from sites in Bolivia suggest that in this region much of the edge of the forest was replaced by savanna during the Pleistocene cold stages and that the forest/savanna boundary moved northward by as much as 30 km, resulting in far greater expanses of savanna relative to today (Figure 7.1b). The effect that this had on population distributions was somewhat different from the mid-altitude refugial localities suggested by Haffer. It is through genetic evidence that this difference is now starting to emerge (Pennington et al., 2004).

7.2.2. Genetic evidence

In support of a Pleistocene isolation model are genetic studies based on rattlesnakes (Wüster et al., 2005), parrots (Eberhard and Bermingham, 2004), and a number of plant taxa (Pennington et al., 2004). Interestingly all three studies are based on types that presently occupy dry formation vegetation including savanna woodland and seasonally dry tropical forest. For example, *Crotalus durissus*, a Neotropical rattlesnake, presently occupies primarily seasonally dry landscapes from Mexico to Argentina, but it avoids the rain forests of Central and tropical South America. Studies of the mitochondrial DNA gene sequence of this species indicate that there are distinct genetic differences between populations from north and south of the Amazon rain forest and that this split occurred at approximately 1.1 Ma. Therefore, this suggests that at some time in the early stages of the Pleistocene, Amazonian rain forests must have become fragmented or shrunk considerably to allow intermixing of these now disjunct and genetically distinctive populations. Eberhard and Bermingham (2004) reach a similar conclusion in a study of a species complex of neotropical parrots, *Amazona ochrocephala*. This complex includes eleven named subspecies that are distributed from Mexico to the Amazon basin and are usually found associated with dry or deciduous woodland, gallery forest, and savanna woodland. A number of different populations were sampled across their present-day distribution and their genetic differentiation was determined by using mitochondrial DNA. Results from this study suggest that members of this complex are very closely related with short genetic distances among the different subspecies. This indicates a relatively recent and rapid colonization, probably from South America. Given that the habitat preference of these birds is for lowland open vegetation formations (and indeed presently they avoid continuous moist forest), Eberhard and Bermingham (2004) have suggested that this colonization into Central America occurred during the cold stages of the Pleistocene when open landscapes were more prevalent.

Work by Pennington et al. (2004) focused on a number of taxonomically unrelated plant groups presently found in the seasonally dry tropical forests (SDTF). The SDTF are currently located in a number of isolated areas within the Neotropics and have a high level of endemism. Molecular analyses of these taxa indicated that while the South American populations probably diversified during the Tertiary with lineage splits occurring between 20–11 Ma, a much more recent diversification pattern is apparent in the Central American populations. Here most of the species appear to have originated during the Pleistocene with splitting ages of between 1.2–1 Ma or less. Therefore, this study suggests that Pleistocene climate fluctuations resulted in the intermixing and then isolation of these seasonally dry vegetation formations in Central America, leading to allopatric speciation. Once again, such results do not support a continuous moist lowland rain forest through all of the cold stages of the Pleistocene.

In contrast to these three examples, however, a wealth of other studies exist that do not indicate evidence for Pleistocene refugial isolation and genetic diversification (Table 7.1). These include studies of populations of Amazonian Neotropical tree frogs (Chek et al., 2001), *Heliconius* butterflies (Brower, 1994), the *Pionopsitta* parrots (Ribas et al., 2005), the Atlantic Forest spiny rat *Trinomys* (Lara and Patton, 2000), and the central American tropical tree *Poulsenia armata* (Moraceae) (Aide and Rivera, 1998). In studying the molecular phylogenies of these organisms three broad research approaches were adopted to look for evidence of (a) greater genetic diversity in postulated refugial areas in contrast to nonrefugial areas; (b) a steplike pattern of decreasing genetic diversity away from the refugial areas (representing the warm-stage migration of populations from these refugia); and (c) lineage splits that have occurred over the past 1.8 Ma

Table 7.1 Approximate dates for lineage splits of some species that were thought to have undergone Pleistocene refugial isolation in the Neotropics

Taxa	Scientific names	Distribution	Glacial refugia	Data used	Level of divergence	Time of divergence	Reference
Butterflies	Heliconius erato	Neotropics	Amazonian region; east of Andes	Mitochondrial sequences	Clades	1.5–2 Ma	(Brower, 1994)
Frogs	Genus Hyla	Mainly neotropics (80% species)	Amazonian basin	16S rRNA, 12S rRNA, and cytochrome-b gene sequence	Species, populations	More than 3.8 Ma	(Chek et al., 2001)
Atlantic Forest spiny rats	Genus Trinomys	Central and South America	Atlantic forest	Cytochrome-b gene sequences of species	Species	7–1.6 Ma	(Lara and Patton, 2000)
Trees	Poulsenia armata (Moraceae)	Central America	Riparian zones along the Caribbean coast	DNA, RAPD markers	Populations	More than 25 Ma (?)	(Aide and Rivera, 1998)
Arboreal ants	Pseudomyrmex viduus	South America	Amazon basin	Morphological characteristics (no genetic evidence)	Species	Pre-Pleistocene, possibly Tertiary	(Ward, 1999)
Small mammals	Various	Amazonia	Amazonian region	Mitochondrial cytochrome-b gene sequence	Species	Various, but all pre-Pleistocene	(Lessa et al., 2003)
Birds, butterflies, lizards, scorpions, trees	Various	Neotropics (mainly Amazonia)	Amazonian region	Distribution ranges, published molecular data	Species	Various, but all pre-Pleistocene	(Nores, 1999)
Neotropical rattlesnake	Crotalus durissus	Neotropics: Mexico to Argentina	Amazonian region	Mitochondrial DNA gene sequences	Populations	1.1 Ma	(Wüster et al., 2005)
Neotropical parrots	Amazona acrocephala species complex	Neotropics: Mexico to Amazon basin	Amazonian region	Mitochondrial DNA gene sequences	Subspecies	During the Pleistocene	(Eberhard and Bermingham, 2004)
Neotropical parrots	Genus Pionopsitta and Gyropsitta	Pinopsitta in Neotropics; Gyropsitta in the Choco´, Central America, and Amazonia	Amazonian region	Mitochondrial DNA gene sequences	Species	8.7–0.6 Ma	(Ribas et al., 2005)

(i.e., during the cold stages). None of these studies provided convincing evidence to support a model of Pleistocene refugial isolation and diversification. To date, most work has been carried out on dating of lineage splits (for a review, see Moritz et al., 2000) and for many taxa previously assumed to be of Pleistocene age, it is apparent that their lineage splits occurred much earlier, usually in the Tertiary (Table 7.1).

In summary, the question of the biogeographical importance of Pleistocene refugia in the Neotropics is still a topic of much debate. Increasingly the fossil and genetic evidence for a sea of savanna fragmenting the lowland rain forests does not appear to have occurred. If there was fragmentation, then this appears to have had little impact on the genetic diversity of the lowland rain-forest populations. However, at the margins of the rain forest—in regions currently occupied

by dry forest and savanna, there does appear to have been much more dynamism during the Pleistocene. Evidence suggests that in these environments, isolation of populations may well have resulted in genetic diversification. A similar conclusion was reached by Hooghiemstra and van der Hammen (1998), who suggested that in different parts of the Amazon basin the biogeographical importance of Pleistocene refugia might have been markedly different.

7.3 NORTH AMERICA

During the Pleistocene ice ages large ice sheets covered much of North East America (the Laurentide ice sheet) and North West America (the Cordilleran ice sheet) (Figure 7.2). At the

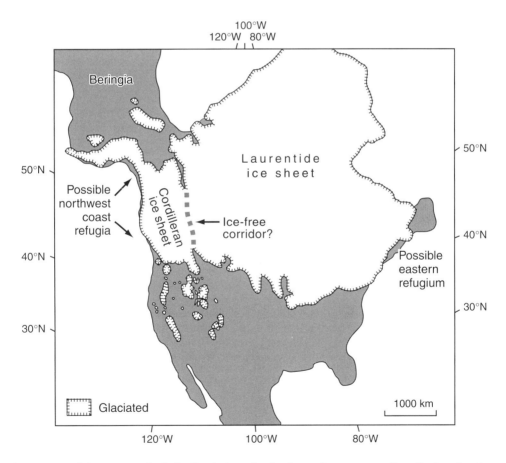

Figure 7.2   Pleistocene refugia in North America in the regions surrounding the ice sheets: (a) a Beringian refugium, including much of Alaska and the Bering sea shelf exposed by lower sea levels; (b) an eastern refugium; (c) a large and complex southern refugium incorporating all the areas south of the ice sheet; and (d) a Pacific Northwest coast refugium

margins of these ice sheets, however, were considerable unglaciated regions and up to four different Pleistocene refugial areas have been identified in (a) the large region south of the ice margin; (b) Beringia; (c) areas along the Pacific northwest coast west of the ice; and (d) an area east of the ice (Brown and Lomolino, 1998; Jackson et al., 2000; Rogers et al., 1991; Figure 7.2). These varied greatly both in extent and the likelihood of temporal continuity throughout the Pleistocene glaciation. For example, while the Laurentide and Cordilleran ice sheets (Figure 7.2) at times covered up to 60% of the continent, Beringia, stretching from northwest Canada to Eastern Siberia, actually increased in area due to glacial sea-level lowering, providing expanded habitat for arctic and alpine taxa (Figure 7.2). In contrast, the Pacific northwest refugial area would have been a region of coastal strips and islands, its geography continually changing as a function of Cordilleran ice dynamics and isostatic and eustatic sea-level changes. To the south of the ice, unglaciated North America remained a continental-sized region containing a wide range of habitats, albeit under radically different climate regimes from the present.

As in other parts of the world, evidence of the dramatic fluctuations of Quaternary environments in North America has led biogeographers to consider whether not only changes in population dynamics but also allopatric speciation resulted from the fragmentation of species ranges into refugia. It was originally proposed, for example, that the isolation between East Coast and West Coast populations of *Dendroica* warblers resulted in a new species of warbler in western North America after each event (there are currently three western species and two eastern species of warblers; reviewed in Brown and Lomolino, 1998). However, as with the Neotropics model of Haffer (1969), most North American examples of this type were untested in the fossil record and based on present-day distributions alone, albeit with better geological evidence on the changing environmental conditions during the last glacial-interglacial cycle. Key questions are whether, for fragmented species ranges, the length of isolation is sufficient for genetic diversification to occur, and whether indeed there is evidence of Quaternary speciation. Palaeoecological and genetic studies have moved our understanding of these issues forward over the past two decades.

7.3.1 Fossil evidence

Eastern North America

Evidence from pollen records for temperate flora surviving in refugia in Eastern North America remains ambiguous. Most fossil studies

(e.g., Davis, 1981; Davis and Shaw, 2001) suggest that trees were displaced far from their present ranges with populations of temperate taxa located in the south, both in the interior and along the Atlantic coastal plain (e.g., *Quercus* [oaks], Cupressaceae [cypresses and junipers], *Acer* [maples], and *Fagus* [beech]). In the early postglacial period, it is postulated that these tree species then dispersed northward again into deglaciated terrain following unique spatial and temporal trajectories (Davis, 1981). However, Bennett (1988) argued that some taxa, such as *Fagus*, which is poorly represented in the Holocene fossil pollen record, may nevertheless have been present in small numbers over larger regions than those expected from conventional interpretations of pollen values, and Jackson et al. (1997) report plant-macrofossil evidence of trees, particularly conifers, close to the Laurentide ice margins in late- and even full-glacial time. More recently, McLachlan and Clark (2004) carried out a modern representation study that showed that neither pollen nor macrofossils adequately record the full distributional range of, for example, *Fagus*. While not affecting major conclusions about the displacements of the main populations of tree species, these findings (Bennett, 1985; Jackson et al., 1997; McLachlan and Clark, 2004) indicate the probability of small populations of trees surviving in Eastern North America through periods of generally unfavorable climates.

Western North America

In Western North America during the LGM, the climate was far moister due to southerly displacement of the jet stream (COHMAP, 1988). In glacial periods, the extent of desert habitats was reduced and forests were expanded; in contrast, interglacial periods such as the Holocene have seen a reduction in forest habitats and desert expansion. Thompson (1988) provides an excellent summary of the fossil evidence for various geographic patterns of response of western taxa to such climate fluctuations. The fossil data largely come from packrat middens, which contain abundant macrofossils of local vegetation. Records across the West date from the present to the limits of radiocarbon dating (~45,000 years ago), and when mapped at a subcontinental scale, show several clear patterns. Some ranges shifted (big sagebrush; *Artemisia tridentata*); others were greatly reduced in glacial times but expanded in the Holocene (ponderosa pine; *Pinus ponderosa*), and yet others were widespread in glacial times but their ranges diminished in the Holocene (bristlecone pine; *Pinus longaeva*). Subsequent studies have confirmed the varied and individualistic responses of species to Quaternary climate change and provide useful hypotheses for testing with genetic data.

Beringia

Beringia was largely treeless during glacial stages (Hopkins et al., 1982; Edwards et al., 2000). Arctic-alpine taxa displaced from much of North America by ice sheets presumably survived during numerous glacial intervals, but until recently it has generally been assumed that conifers were eliminated from the region during full-glacial conditions and that they migrated from the south after the ice sheets retreated (Hopkins et al., 1981; Ritchie and MacDonald, 1986). A recent synthesis of 149 fossil pollen records, however, suggests that at least six boreal tree and shrub taxa (*Populus, Larix, Picea, Pinus, Betula, Alnus/Duschekia*) may have survived the LGM within the region, albeit in relatively small populations (Brubaker et al., 2005). Persistent, low levels of pollen of these taxa occur from the LGM through the Late Glacial, with late-glacial or early Holocene expansions on the Alaskan side occurring prior to the opening of the 'ice-free corridor' between the Laurentide and Cordilleran ice sheets (Szeicz and MacDonald, 2001; Figure 7.2).

Pacific Northwest Coast

During the Pleistocene ice ages, the dynamic paleogeography of the Northwest Coast may have provided opportunities for island-hopping migration between Beringia and more southerly latitudes and also for the periodic isolation of populations on islands and on coastal margins hemmed in by glaciers. There is good evidence that colonizable land existed as early as 16,300 years ago in the Queen Charlotte Islands (Haida Gwaii; Hetherington et al., 2003). Tree populations also occurred near the ice margins surprisingly early. For example, near the southwest limits of the ice on Vancouver Island wooded vegetation dominated by lodgepole pine (*Pinus contorta*) occurred at about 14,000 years ago or earlier (Brown and Hebda, 2003). Much further north, Peteet (1991) recorded macrofossils of *P. contorta* at Yakutat, Alaska, by 11,500 years ago, while the taxon was virtually absent from the local pollen record, which suggests either survival in a glacial refugium or a remarkably fast migration.

7.3.2 Genetic evidence

In looking for genetic evidence to support an allopatric speciation model linked to the Pleistocene refugial isolation, research in North America has broadly focused on two questions.

First, can a geographically defined genetic pattern (phylogeography) be recognized as being indicative of refugial isolation? Second, do significant lineage splits (of the group in question) occur during the Pleistocene?

Recent phylogeographic studies tend to support the contention based on palaeoecological data that species did indeed survive in refugial populations and that this refugial isolation has left a lasting genetic imprint on both plants and animals. In plant populations, for example, high haplotype diversity in Alaska of *Picea glauca* (white spruce) in relation to other regions of North America argues for survival of this species in a Berigium refugium during the LGM, rather than postglacial migration from the south, supporting the Brubaker et al. (2005) fossil-based reconstruction (Anderson et al., 2006). A similar study on *P. mariana* (black spruce; the modern ranges of the two species are similar) revealed several different populations south of the Laurentide ice sheet, suggesting a fragmentation into different refugial populations from west to east (Jaramillo-Correa et al., 2004).

Also, a study of the haplotype distributions of *Acer* and *Fagus* (maple and beech) indicated that populations of these taxa may have survived in refugia in Eastern North America beyond the refugial ranges commonly ascribed from pollen distribution patterns (McLachlan et al., 2005), thus supporting the contention of Bennett (1988; see above).

Many species in the terrestrial fauna so far studied also demonstrate distinctive clades according to refugial isolation. Again, the east-west separation is apparent in many patterns. Clear genetic distinction linked to refugial isolation has been demonstrated between eastern and western populations of black bears, *Ursus americanus* (Wooding and Ward, 1997); martens, *Martes americana* (Stone et al., 2002); and southern red-backed voles, *Clethrionomys gapperi* (Runck and Cook, 2005).

In Beringia, studies indicate a range of patterns of genetic variation in mammalian taxa, reflecting the link to Asia in glacial periods and the complex palaeogeography of glaciated coastal Alaska. Patterns in the tundra vole (*Microtus oeconomus*), for example, suggest several interacting processes: expansion of low-diversity eastern Beringian populations from a colonisation from Asia; shared genes across Beringia, suggesting periodic mixing between Asia and Alaska in glacial periods; and regional postglacial expansions from refugia within Beringia, showing low diversity as a result of bottlenecking (Galbreath and Cook, 2004).

Isolation as a consequence of fluctuating ice age environments therefore appears to have left a much more clearly discernible genetic imprint on the populations of both plants and animals in North America than those found in the Neotropics. Most studies, however, have focused on genetic variation within populations of the same species, which begs the question as to whether refugial isolation led to allopatric speciation. To determine this requires estimates of between-species diversity and the dating of lineage splits of sister species—and here the evidence is ambiguous (Table 7.2).

A phylogenetic study of sister species of North American warblers whose East Coast and West Coast species diversity had been linked to Pleistocene isolation (see above) demonstrated that the majority of the lineage splits occurred before the Pleistocene (Klicka and Zink, 1997). Klicka and Zink (1997, p. 1668) concluded that "the most recent glaciations were not, it seems, the force driving songbird diversification so much as they functioned as an ecological obstacle course through which some species were able to persist." A similar conclusion was also reached by Steele et al. (2005) concerning *Dicamptodon copei* and *D. aterrimus*, two species of giant salamanders in the Pacific Northwest. Their results suggested that speciation is attributable to pre-Pleistocene events and thus that glacial isolation played little part in the diversification of the genus. Therefore for most organisms studied thus far in North America, lineage splits appear to have occurred on timescales longer than that of the Pleistocene (Table 7.2). Cycles of isolation and coalescence during the Pleistocene have certainly led to increasing, decreasing, and redistribution of genetic diversity in populations, but it is far from clear that the formation of new species has occurred to any degree.

7.4 EUROPE

During the Pleistocene ice ages, large areas of northern Europe were covered by ice sheets (Figure 7.3), while glaciers and icefields developed in many mountain regions of central and southern Eurasia, including the Alps, the Pyrenees, the Carpathians, and the Caucasus (Ehlers and Gibbard, 2004).

Beyond the ice sheets and glaciers, the landscape of much of northern and western Europe was covered in permafrost, and aeolian sands were widespread. Further south, in particular the three southern peninsulas of Europe (namely the Balkan, Iberian, and Italian), however, the landscape remained relatively ice-free and probably supported a relict soil.

Palaeoclimatic simulations indicate that strong climatic gradients existed across Europe during the LGM with a north-south gradient from arctic to cold temperate climates north of the trans-European mountain barrier and a west-east transition from the maritime Atlantic climate to the continental climate of eastern Europe (Barron and Pollard, 2002; Barron et al., 2003). This resulted in strong regional differences in surface temperatures, with estimates suggesting that winter temperatures in northern Europe were up to 10 to 20°C cooler; in southern Europe and parts of central Europe it was 7 to 10°C cooler; and in more southwesterly regions, such as Spain, it was only 2 to 4°C cooler than present. Winter and summer precipitation also had a strong east-west gradient, bringing to central and south-eastern Europe more precipitation than to the three southern peninsulas of Europe.

The 'traditional' refugial model for Europe is that temperate fauna and flora survived in refugia in the three southern peninsulas of Europe and also in the Near East. This model is based on the premise that the apparently inhospitable climate and soils of central and northern Europe would have rendered them incapable of supporting woody temperate vegetation and the fauna associated with them (Huntley and Birks, 1983; Bennett et al., 1991). Much interest (fossil and genetic evidence) has therefore been focussed on the impact of glacial isolation on the populations in these three southern peninsulas. More recently, however, some studies have started to suggest that other regions may also have supported small populations of plants and animals, including regions of central Europe (Willis et al., 2000; Willis and van Andel, 2004; Stewart and Lister, 2004; Bhagwat and Willis, 2008) and also in the Alps (Birks and Willis, 2008). In the latter region it has been proposed that ice-free mountain tops (nunataks) protruding above the glaciers may have harbored populations of higher plants. This nunatak hypothesis and its counterpart, the tabula rasa hypothesis (i.e., total eradication of plant life and subsequent recolonization from refugia outside the ice shield), have been debated since the early twentieth century (for a review, see Schonswetter et al., 2005).

7.4.1 Fossil evidence

Mediterranean
Various syntheses of pollen records have led to the suggestion that during the Pleistocene ice ages much of the Mediterranean regions of southern Europe (including the three southern peninsulas) were predominantly open landscapes, composed

Table 7.2 Approximate dates for lineage splits of some species that were thought to have undergone Pleistocene refugial isolation in North America

Taxa	Scientific names	Distribution	Glacial refugia	Data used	Level of divergence	Time of divergence	Reference
Songbirds	Various	North America	Various across North America	Mitochondrial DNA sequence	Species	Up to 5.5 Ma	(Klicka and Zink, 1997)
Sea cucumber	Cucumaria miniata and C. pseudocurata	Northwestern Pacific	Cordilleran ice-sheet margin	Mitochondrial DNA sequence	Populations	Not specified, but thought to be during Pleistocene	(Arndt and Smith, 1998)
Crustacean	Sida crystallina	North America	Four separate regions across North America	Allozyme, mitochondrial DNA, nuclear DNA	Populations	3.7–2.4 Ma	(Cox and Hebert, 2001)
Pygmy salamander	Desmognathus wrighti	Southern Appalachian mountain peaks	Four separate locations in the southern Appalachians	Allozymic loci and 12S rRNA gene in the mtDNA genome	Populations	17.8–1.34Ma	(Crespi et al., 2003)
Eastern chipmunk	Tamias striatus	Central and Northern United States	Laurentide ice-sheet margin	DNA	Populations	Up to 200,000 years ago	(Rowe et al., 2004)
Southern red-backed vole	Clethrionomys gapperi	North America	Three separate regions in the Rockies and Appalachians; and receding ice sheet	Cytochrome-b gene sequences	Clades	Prior to and through Pleistocene	(Runck and Cook, 2005)
Giant salamanders	Dicamptodon copei and D. aterrimus	Pacific Northwest of North America	Separate refugia in north and south of range	Cytochrome-b gene sequences	Species	2 Ma	(Steele et al., 2005)

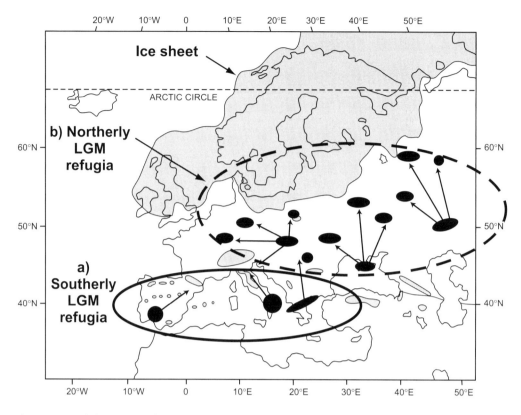

Figure 7.3   Pleistocene refugia in Europe: (a) the Iberian, Italian and Balkan peninsulas as postulated refugia during the last full glacial; (b) additional refugia, 'cryptic northern refugia', in Europe (redrawn from Birks and Willis, 2008). Also indicated are the proposed extent of the ice sheets and mountain glaciers at the LGM (Svendsen et al., 2004)

of low shrubs, grasses, and other open-ground herbaceous plants, including Chenopodiaceae and *Artemisia* (Huntley et al., 2003). However, trees were also intermittently present, probably existing in small refugial 'pockets', especially in mountain regions (Willis and Whittaker, 2002). The occurrences of such microenvironmentally favourable locations have been found in numerous localities in the Mediterranean (Figure 7.3a); for example, in western Spain, fossil evidence for tree taxa surviving the full-glacial present include *Juniperus*, *Betula*, and *Quercus* in both mountain and middle and lower altitude regions (Carrion, 2002). In France (Massif Central) and in Italy (Follieri et al., 1998; Allen and Huntley, 2000) continuous populations of *Pinus* have been recorded in the fossil record extending well back into the full glacial, while further east, in Greece and the Balkans, deciduous *Quercus* and *Pinus* were present in mountain regions (Tzedakis et al., 2002). Similarly, there is evidence for the

distribution of fir (*Abies sp.*) in a number of refugia; a recent study by Terhürne-Berson et al. (2004), for example, using fossil evidence from over 208 pollen sequences and 38 macrofossil sites, revealed Pleistocene refugial localities in Italy, Greece, the Pyrenees in Spain, southeast France, and northwest Italy.

In addition, a number of other tree taxa have been recorded but in smaller and less-persistent quantities (reviewed in Bennett et al., 1991). For example, Krebs et al. (2004) used sweet chestnut (*Castanea sativa*) pollen records from sites in Europe to identify possible glacial refugia for sweet chestnut. Their results suggest numerous small chestnut refugia were located in mountainous regions of southern Europe, or at low to mid-altitudes and at the margins of mountain chains. It would thus appear that mixed populations of trees survived in small refugial populations throughout the Mediterranean region.

Central and southeastern Europe

It has been widely assumed that the full-glacial environments of central and southeastern Europe were too harsh to be able to support any woody vegetation; until recently, few pollen diagrams have been available to confirm or dispute this interpretation. The full-glacial landscapes of central and southeastern Europe have therefore usually been classified as treeless steppes or even polar deserts (e.g., Huntley et al., 2003). This interpretation is, however, at odds with results from paleoclimatic modeling, which indicates that this part of Europe during the LGM would have had more precipitation than the three southern peninsulas of Europe and temperatures that should have been able to support taiga forest (Barron and Pollard, 1993; Alfonso et al., 2003). Recently emerging fossil evidence that indicates the survival of trees throughout central and southeastern Europe (Figure 7.3b) therefore makes an interesting contribution to this debate (Willis and van Andel, 2004; Bhagwat and Willis, 2008).

Pollen and macrofossil analysis from buried peat in the Czech Republic dated to about 28,000 years ago (Rybníčová and Rybníček, 1991), for example, indicates a coniferous forest containing *Pinus sylvestris, P.* cf. *mugo, P. cembra, Picea, Larix, Juniperus communis*, and *Betula*. There is also evidence for the scattered presence of temperate deciduous trees, including *Ulmus, Acer, Corylus, Quercus*, and *Tilia*. In addition, three lake sites that contain pollen diagrams extending back into the end of the LGM at Švarcenberk in the Czech Republic (Pokorný and Jankovská, 2000), Sarret in Hungary (Willis et al., 2000), and Steregoiu in northwest Romania (Björkman et al., 2003) indicate that open-forested vegetation with pine and birch were predominant in this environment.

The presence of trees in central and eastern Europe during the LGM is also confirmed by the macrofossil charcoal record. Evidence of over 100 ^{14}C full-glacial dated macrofossil charcoal remains of woody species (including *Pinus, Betula, Picea, Larix, Salix*) have been obtained from sites throughout central and eastern Europe (Willis et al., 2000; Willis and van Andel, 2004). Similar results have also been found by Cheddadi et al. (2006). They used an extensive data set of pollen and macrofossil remains of *Pinus sylvestris* to locate the extent of its glacial refugia during the Pleistocene. In addition they used a vegetation model to simulate the extent of the potential refugia during the last glacial period. Their pollen and macrofossil data confirmed the existence of glacial refugia around the Iberian Peninsula and Italy, as well as in the Hungarian Plain and the Danube area (cf. Willis and van Andel, 2004). The overall results suggested that during the last glacial period *P. sylvestris* survived in scattered and restricted refugial areas throughout most of Europe.

The fossil evidence therefore supports the traditional refugial model of populations of temperate trees surviving in the Mediterranean region (including the three southern peninsulas during the Pleistocene ice ages). These populations were probably isolated for long periods of time (up to 100,000 years at a time) in small microenvironmentally favorable locations. There were, however, also some tree types that were able to survive much further north in central and southeastern Europe in contrast to the refugial model. These were predominantly (but not exclusively) types that are presently found in the southern margins of the boreal forests (e.g., *Pinus, Salix, Betula, Picea*, etc.) and were able to withstand the climatic extremes and semipermanently frozen soils that would have been characteristic of these full-glacial landscapes (Willis and Niklas, 2004; Willis and van Andel, 2004). These more northerly regions where these plants grew (and some fauna lived) have been termed 'cryptic refugia' (Stewart and Lister, 2004) (Figure 7.3b) although it is debatable whether these can even been classified as 'refugia' in the strictest sense because they were probably not necessarily isolated populations but rather scattered populations across the central and southeastern European landscape (Bhagwat and Willis, 2008).

7.4.2 Genetic evidence

A large body of molecular data now exists for the genetic patterns of various European species and the impact of refugial isolation both within and between populations (Hewitt, 2000 and references therein). From these studies it has become apparent that the Pleistocene ice ages left a lasting genetic legacy—not only in differences between refugial areas but also in terms of postglacial migration routes; there are distinctive genetic differences apparent between northern European populations of a number of different fauna and floral species, which can be traced back to the specific refugial region that the species migrated from (Hewitt, 1996).

Similar to the fossil evidence, however, there is emerging evidence from the genetic data for several different types of refugia: Mediterranean refugia; northerly cryptic refugia; and alpine refugia. Evidence to support isolation in various Mediterranean refugia has been demonstrated for a number of flora and fauna: both trees and herbs and large and small mammals (Hewitt, 2000, 2004). In a study of common and pygmy shrews (*Sorex araneus* and *S. minutus*; Insectivora) and the bank vole (*Clethrionomys glareolus*; Rodentia), for example, Bilton et al. (1998) demonstrated that the genetic compositions of these three small mammals on the Mediterranean peninsulas are distinctive from those found elsewhere in Europe, and this is probably indicative of their full-glacial isolation in Mediterranean refugia. A good herbaceous plant example is provided in a study by Trewick et al. (2002), who examined chloroplast DNA sequences from 331 individual plants of rockfern (*Asplenium ceterach*), representing 143 populations from throughout its range in Europe, plus outlying sites in North Africa and the near East. Here it was demonstrated that tetraploids of *A. ceterach* were encountered throughout Europe; diploids predominated in the Pannonian-Balkan region; and hexaploids were encountered only in southern Mediterranean populations. Based on their results, Trewick et al. (2002) suggested that the Pannonian-Balkan region formed a northern Pleistocene refugium while populations persisted for a long time in the southern Mediterranean as suggested by the presence of diploids, tetraploids, and hexaploids in the populations from Greece.

Genetic evidence for the impact of more diffuse, northerly survival of species has been demonstrated for a number of tree species mentioned above in the fossil record (e.g. *Salix, Betula, Pinus*) (Lascoux et al., 2004) and also a number of animal species (reviewed in Stewart and Lister, 2003; Bhagwat and Willis, 2008). A study of the Quaternary history of European beech (*Fagus sylvatica*) populations (Magri et al., (2006), however, indicated that the genetic evidence can also provide information on more northerly refugia that were not identified in the fossil record. From the fossil record it had always been assumed that beeches were located in the southerly Mediterranean refugia. However, this study (Magri et al., 2006) suggests that the beech survived the last glacial period in multiple refugia—a central European refugia and a Mediterranean refugia. Interestingly, it is also apparent that the Mediterranean refugia did not contribute to the colonization of central and northern Europe.

Evidence for alpine refugia is almost entirely based on molecular phylogenetic records rather than on fossil records. A number of studies have demonstrated genetically distinct populations of the Alps (attributed to glacial refugia) along the southwestern, southern, eastern, and northern borders of the Alps (Schonswetter et al., 2005). Therefore, even within the strongly glaciated central parts of the Alps, high-elevation plants apparently survived on ice-free mountain tops and the postglacial recolonization of the Alps would not only have started from peripheral refugia, but

also from source areas within the ice sheet (Birks and Willis, 2008).

It is apparent from both the fossil and molecular records that there were at least three refugial regions in Europe during the Pleistocene ice ages; Mediterranean, cryptic northerly, and Alpine. This isolation led to genetic distinctiveness within and between populations and there is an imprint of this isolation on present-day populations in Europe. However, a key question to be addressed is whether this isolation was long enough for the development of new lineages. The maximum amount of time in isolation (glacial interval) would probably have been approximately 100,000 years, which is a very short interval of time for speciation to occur (Willis and Niklas, 2004). There is then the added 'problem' of intermixing in intervening interglacials, thus undoing any accumulated change. Evidence from the fossil and molecular records display very little evidence for the appearance of new species/lineages during the Pleistocene; from the fossil record the predominant mode during the Pleistocene is one of local extinction and redistribution, not speciation (Willis and Niklas, 2004). In the molecular record most of the lineage splits occur before the Pleistocene and are usually in the mid- to Late Tertiary (Table 7.3). In Europe, therefore, the Pleistocene ice ages have certainly impacted the genetic distinctiveness of the species but this had not tended to lead to speciation.

7.5 CONCLUSIONS

Three main conclusions emerge from this review of the biogeographical importance of Pleistocene refugia:

1. In the light of recent fossil and molecular evidence, the geographical distribution of Pleistocene refugia is far more complex than originally envisaged. In the Neotropics, there is little evidence to support the original Haffer (1969) model for a 'sea of savanna' replacing the lowland tropical rain forest. Rather, the majority of the rain forest remained but the composition of this forest altered to include montane forest species. However, significant contraction of the rain forest probably did occur at its boundaries and was replaced by savanna and/or seasonally dry forest. These margins may well have provided refugia for species adapted to dry, open environments. In North America there is evidence for a number of refugia, including tree refugia as far north as Beringia and evidence for trees growing right up to the edge of the Laurentide ice sheet. Finally in Europe, in addition to the 'traditional'

Table 7.3 Approximate dates for lineage splits of some species that were thought to have undergone Pleistocene refugial isolation in Europe

Taxa	Scientific names	Distribution	Glacial refugia	Data used	Level of divergence	Time of divergence	Reference
Newts	*Triturus vulgaris* and *Triturus montandoni*	Eurasia	Caucasus, Anatolia, Balkans, Italy, central Europe	Mitochondrial DNA	Clades	4.5–1.0 Ma	(Babik et al., 2005)
Shrews (common and pygmy) and bank vole	*Sorex araneus, S. minutus;* and *Clethrionomys glareolus*	Europe	Central Europe, western Asia	Mitochondrial DNA	Populations	Pre-Pleistocene (precise time not stated)	(Bilton et al., 1998)
Butterfly	*Erebia triaria*	NW Iberian peninsula	More than one location on Iberian Peninsula	Mitochondrial DNA	Populations	238,000 years ago for one of the isolated populations; later for others	(Vila et al., 2005)
Plant (annual)	*Microthlaspi perfoliatum* (Brassicaceae)	Europe	Primary in Iberia, Italy Balkans; secondary in many other areas	Isozyme and chloroplast DNA	Populations	Pre-Pleistocene (precise time not stated)	(Koch and Bernhardt, 2004)
Dwarf red deer	*Cervus elaphus jerseyensis*	Jersey (United Kingdom)		Not specified	Subspecies	Less than 6,000 years ago	(Lister, 1995; Lister, 2004)

Mediterranean refugia, there is emerging evidence for widespread cryptic refugia across central and northern Europe.

2. The isolation of fauna and flora in these refugia resulted in distinctive genetic differentiation between populations. These genetic differences are still apparent in many populations presently, both in their original refugial localities and also in the populations that migrated from them in the early postglacial. Many refugial regions, as well as being genetically distinct, have high levels of genetic diversity.

3. Despite the apparent genetic differences, there is little evidence to indicate that the length of time in isolation is sufficient for allopatric speciation to occur. Molecular evidence indicates that lineage splits for most taxa occur well before the Pleistocene (Tables 7.1, 7.2, and 7.3) and there is little fossil evidence to indicate the emergence of new species. In fact, from the fossil record, the Pleistocene ice ages appear to have been a time of redistribution of populations resulting in local extinction rather than speciation (Willis and Niklas, 2004).

Finally, with predicted increasing climatic variability, an understanding of the location of Pleistocene refugia is highly relevant to the long-term conservation of flora and fauna (Willis and Birks, 2006). These refugial regions, with their high levels of genetic diversity and distinctiveness, hold the evolutionary potential for the future.

REFERENCES

Aide, T. M., and E. Rivera. (1998) 'Geographic patterns of genetic diversity in Poulsenia armata (Moraceae): implications for the theory of Pleistocene refugia and the importance of riparian forest', *Journal of Biogeography*, 25:695–705.

Alfano, M.J., Barron, E.J., Pollard, D., Huntley, B. and Allen, J.R.M. (2003) 'Comparison of climate model results with European vegetation and permafrost during Oxygen Isotope Stage 3', *Quaternary Research*, 59: 97–107.

Allen, J.R.M. and Huntley, B. (2000) 'Weichselian palynological records from southern Europe: correlation and chronology', *Quaternary International*, 73/74: 111–125.

Anderson, L.L., F.S. Hu, D.M. Nelson, R.J. Petit, and K.N. Paige. (2006) 'Ice-age endurance: DNA evidence of a white spruce refugium in Alaska' *Proceedings National Academy of Sciences*, 103: 12447–12450

Arndt, A., and M. J. Smith. (1998) 'Genetic diversity and population structure in two species of sea cucumber: differing patterns according to mode of development', *Molecular Ecology*, 7:1053–1064.

Babik, W., W. Branicki, J. Crnobrnja-Isailovic, D. Cogalniceanu, I. Sas, K. Olgun, N. A. Poyarkov, M. Garcia-Paris, and

J. W. Arntzen. (2005) 'Phylogeography of two European newt species - discordance between mtDNA and morphology', *Molecular Ecology*, 14:2475–2491.

Barron, E. and Pollard, D. (2002) 'High-resolution climate simulations of Oxygen Isotope State 3 in Europe', *Quaternary Research*, 58: 296–309.

Barron, E., van Andel. T.H. and Pollard, D. (2003) 'Glacial environments II. Reconstructing the climate of Europe in the Last Glaciation', In: van Andel, T.H. Davies, S.W. (Eds), *Neanderthals and Modern Humans in the European Landscape during the Last Glaciation*. Cambridge: McDonald Institute for Archaeological Research. pp 57–78.

Behling, H., and H. Hooghiemstra. (2000) 'Holocene Amazon rainforest-savanna dynamics and climatic implications: high-resolution pollen record from Laguna Loma Linda in eastern Colombia', *Journal of Quaternary Science*, 15:687–695.

Bennett, K.D. (1988) 'Holocene geographic spread and population expansion of *Fagus grandifolia* in Ontario, Canada', *Journal of Ecology*,76: 547–557.

Bennett, K. D., P. C. Tzedakis, and K. J. Willis. (1991) 'Quaternary Refugia of North European Trees', *Journal of Biogeography*, 18:103–115.

Bennett, K. D., and K. J. Willis. (2000) 'Effect of global atmospheric carbon dioxide on glacial-interglacial vegetation change', *Global Ecology and Biogeography*, 9:355–361.

Bhagwat, S.A. and Willis, K.J. (2008) Species persistence in northerly glacial refugia of Europe: a matter of chance or biogeographical traits? Journal of Biogeography, **35**: 464–482

Bilton, D. T., P. M. Mirol, S. Mascheretti, K. Fredga, J. Zima, and J. B. Searle. 1998. Mediterranean Europe as an area of endemism for small mammals rather than a source for northwards postglacial colonization. Proceedings of the Royal Society of London Series B-Biological Sciences **265**:1219–1226.

Birks, H.J.B. and Willis, K.J. 2008. Alpines, trees and refugia in Europe. Plant Ecology and Diversity **1**: 146–160.

Björckman, L., Feurdean, A. and Wohlfarth, B. 2003. Late-Glacial and Holocene forest dynamics at Steregoiu in the Gutaiului Mountains, Northwest Romania. Review of Palaeobotany and Palynology, **124**: 79–111

Brower, A. V. Z. 1994. Rapid Morphological Radiation and Convergence among Races of the Butterfly Heliconius-Erato Inferred from Patterns of Mitochondrial-DNA Evolution. Proceedings of the National Academy of Sciences of the United States of America **91**:6491–6495.

Brown, J. H., and M. V. Lomolino 1998. Biogeography. Sinauer, Sunderland, MA.

Brown, K.J. and R.J. Hebda. 2003. Coastal rainforest connections disclosed through a Late Quaternary vegetation, climate, and fire history investigation from the Mountain Hemlock Zone on southern Vancouver Island, British Colombia, Canada. Review of Palaeobotany and Palynology 123: 247–269

Brubaker, L. B., P. M. Anderson, M. E. Edwards, and A. V. Lozhkin. 2005. Beringia as a glacial refugium for boreal trees and shrubs: new perspectives from mapped pollen data. Journal of Biogeography **32**:833–848.

Bush, M. B., M. C. Miller, P. E. De Oliveira, and P. A. Colinvaux. 2002. Orbital forcing signal in sediments of two Amazonian lakes. Journal of Paleolimnology **27**:341–352.

Carrion, J.S. 2002. Patterns and processes of Late Quaternary environmental changes in a montane region of southwestern Europe. Quaternary Science Reviews, **23**: 2369–2387.

Cheddadi, R., G. G. Vendramin, T. Litt, L. Francois, M. Kageyama, S. Lorentz, J. M. Laurent, J. L. de Beaulieu, L. Sadori, A. Jost, and D. Lunt. 2006. Imprints of glacial refugia in the modern genetic diversity of Pinus sylvestris. Global Ecology and Biogeography **15**:271–282.

Chek, A. A., S. C. Lougheed, J. P. Bogart, and P. T. Boag. 2001. Perception and history: Molecular phylogeny of a diverse group of neotropical frogs, the 30-chromosome Hyla (Anura: Hylidae). Molecular Phylogenetics and Evolution **18**:370–385.

COHMAP Members. (1988). Climatic changes of the last 18,000 years. observations and model simulations. Science 241, 1043–1052.

Colinvaux, P. A., P. E. De Oliveira, and M. B. Bush. 2000. Amazonian and neotropical plant communities on glacial time-scales: The failure of the aridity and refuge hypotheses. Quaternary Science Reviews **19**:141–169.

Colinvaux, P. A., P. E. DeOliveira, J. E. Moreno, M. C. Miller, and M. B. Bush. 1996. A long pollen record from lowland Amazonia: Forest and cooling in glacial times. Science **274**:85–88.

Cowling, S. A., M. A. Maslin, and M. T. Sykes. 2001. Paleovegetation simulations of lowland Amazonia and implications for neotropical allopatry and speciation. Quaternary Research **55**:140–149.

Cox, A. J., and P. D. N. Hebert. 2001. Colonization, extinction, and phylogeographic patterning in a freshwater crustacean. Molecular Ecology **10**:371–386.

Crespi, E. J., L. J. Rissler, and R. A. Browne. 2003. Testing Pleistocene refugia theory: phylogeographical analysis of Desmognathus wrighti, a high-elevation salamander in the southern Appalachians. Molecular Ecology **12**:969–984.

Davis, M. B. 1981. Quaternary history and the stability of forest communities. Pages 132–153 in D. C. West, H. H. Shugart, and D. B. Botkin, editors. Forest succession: concepts and application. Springer-Verlag, New York, New York, USA.

Davis, M.B. and R.B. Shaw. 2001. Range Shifts and Adaptive Responses to Quaternary Climate Change. Science 692:673–679.

Eberhard, J. R., and E. Bermingham. 2004. Phylogeny and biogeography of the Amazona ochrocephala (Aves: Psittacidae) complex. Auk **121**:318–332.

Edwards, M.E., Anderson, P.M., Brubaker, L.B., Ager, T., Andreev, A.A., Bigelow, N.H., Cwynar, L.C., Eisner, W.R., Harrison, S.P., Hu, F.-S., Jolly, D., Lozhkin, A.V., MacDonald, G.M., Mock, C.J., Ritchie, J.C., Sher, A.V., Spear, R.W., Williams, J. and Yu, G. (2000). Pollen-based biomes for Beringia 18,000, 6000 and 0 14C yr B.P. Journal of Biogeography 27: 521–554.

Ehlers, J. and Gibbard, P.L. (eds.) 2004. Quaternary Glaciations - Extent and Chonology. Volume 1. Europe. Elsevier, Amsterdam.

Follieri, M., Giardini, M., Magri, D., and Sadori, L. 1998. Palynostratigraphy of the last glacial period in the volcanic region in central Italy. Quaternary International, **47**: 3–20.

Guilderson, T. P., R. G. Fairbanks, and J. L. Rubenstone. 1994. Tropical Temperature-Variations since 20,000 Years Ago - Modulating Interhemispheric Climate-Change. Science **263**:663–665.

Haberle, S. G., and M. A. Maslin. 1999. Late Quaternary vegetation and climate change in the Amazon basin based on a 50,000 year pollen record from the Amazon fan, ODP site 932. Quaternary Research **51**:27–38.

Haffer, J. 1969. Speciation in Amazonian Forest Birds. Science **165**:131–137.

Hewitt, G. M. 1996. Some genetic consequences of ice ages, and their role in divergence and speciation. Biological Journal of the Linnean Society **58**:247–276.

Hewitt, G. 200 The genetic legacy of the Quaternary Ice Ages. Nature, **405**: 907–913.

Hewitt, G. M. 2004. Genetic consequences of climatic oscillations in the Quaternary. Philosophical Transactions of the Royal Society of London Series B-Biological Sciences **359**:183–195.

Hill, J. L., and R. A. Hill. 2001. Why are tropical rain forests so species rich? Classifying, reviewing and evaluating theories. Progress in Physical Geography **25**:326–354.

Hooghiemstra, H., and T. van der Hammen. 1998. Neogene and Quaternary development of the neotropical rain forest: the forest refugia hypothesis, and a literature overview. Earth-Science Reviews **44**:147–183.

Hopkins, D.M., Matthews, J.V., Jr., Schweger, C.E., and Young, S.B. (1982). (Eds.) Paleoecology of Beringia. New York, Academic Press.

Hopkins, D.M., Smith, P.A., and Matthews, J.V., Jr. (1981). Dated Wood from Alaska and the Yukon: Implications for Forest Refugia in Beringia: Quaternary Research 15, 217–249.

Huang, Y., F. A. Street-Perrott, S. E. Metcalfe, M. Brenner, M. Moreland, and K. H. Freeman. 2001. Climate change as the dominant control on glacial-interglacial variations in C-3 and C-4 plant abundance. Science **293**:1647–1651.

Huntley, B. and Birks, H.J.B. 1983. An atlas of past and present pollen maps for Europe: 0–13,000 years ago. Cambridge University Press, Cambridge.

Huntley, B., Alfano, M.J., Allen, J.R.M., Pollard, D., Tzedakis, P.C., de Beaulieu, J-L., Gruger, E., Watts, W. 2003. European vegetation change during Marine Oxygen Isotope Stage 3. Quaternary Research **59**: 195–212.

Jackson, S. T., R. S. Webb, K. H. Anderson, J. T. Overpeck, T. Webb, J. W. Williams, and B. C. S. Hansen. 2000. Vegetation and environment in Eastern North America during the Last Glacial Maximum. Quaternary Science Reviews **19**:489–508.

Jackson,S.T., J.T. Overpeck, T.Webb-111, S. E. Keattch, and K. H. Anderson. 1997. Mapped Plant-Macrofossil And Pollen Records Of Late Quaternary Vegetation Change In Eastern North America. Quaternary Science Reviews 16: l–70.

Jaramillo-Correa, J. P., J. Beaulieu, and J. Bousquet. 2004. Variation in mitochondrial DNA reveals multiple distant glacial refugia in black spruce (Picea mariana), a transcontinental North American conifer. Molecular Ecology **13**:2735–2747.

Jaramillo-Correa, J.P., J. Beaulieu, F.T. Ledig and J. Bousquet. 2006. Decoupled mitochondrial and chloroplast DNA population structure reveals Holocene collapse and population isolation in a threatened Mexican-endemic conifer. Molecular Ecology 15: 2787–2800.

Kirkpatrick, J. B., and M. Fowler. 1998. Locating likely glacial forest refugia in Tasmania using palynological and ecological information to test alternative climatic models. Biological Conservation **85**:171–182.

Klicka, J., and R. M. Zink. 1997. The importance of recent ice ages in speciation: A failed paradigm. Science **277**:1666–1669.

Koch, M., and K. G. Bernhardt. 2004. Comparative biogeography of the cytotypes of annual Microthlaspi perfoliatum (Brassicaceae) in europe using isozymes and cpDNA data: Refugia, diversity centers, and postglacial colonization. American Journal of Botany **91**:115–124.

Krebs, P., M. Conedera, M. Pradella, D. Torriani, M. Felber, and W. Tinner. 2004. Quaternary refugia of the sweet chestnut (Castanea sativa Mill.): an extended palynological approach. Vegetation History and Archaeobotany **13**:145–160.

Kuhn, T. S. 1970. The Structure of Scientific Revolutions. University of Chicago Press, Chicago.

Lara, M. C., and J. L. Patton. 2000. Evolutionary diversification of spiny rats (genus Trinomys, Rodentia: Echimyidae) in the Atlantic Forest of Brazil. Zoological Journal of the Linnean Society **130**:661–686.

Lascoux, M. Palme, A.E., Cheddadi, R. And Latta, R.G. 2004. Impact of Ice Ages on the genetic structure of trees and shrubs. Philosophical Transactions of the Royal Society B, **359**: 1471–2907

Lessa, E. P., J. A. Cook, and J. L. Patton. 2003. Genetic footprints of demographic expansion in North America, but not Amazonia, during the Late Quaternary. Proceedings of the National Academy of Sciences of the United States of America **100**:10331–10334.

Lister, A. M. 1984. Evolutionary and Ecological Origins of British Deer. Proceedings of the Royal Society of Edinburgh Section B-Biological Sciences **82**:205–229.

McLachlan, J.S. and J.S. Clark. 2004. Reconstructing historical ranges with fossil data at continental scales. Forest Ecology and Management 197:139–147.

Magri, D. and 13 authors (2006). A new scenario for the Quaternary history of European beech populations: palaeobotanical evidence and genetic consequences. New Phytologist **171**:199–221.

Maley, J., and P. Brenac. 1998. Vegetation dynamics, palaeoenvironments and climatic changes in the forests of western Cameroon during the last 28,000 years BP. Review of Palaeobotany and Palynology **99**:157–187.

Mayle, F. E. 2004. Assessment of the Neotropical dry forest refugia hypothesis in the light of palaeoecological data and vegetation model simulations. Journal of Quaternary Science **19**:713–720.

Mayle, F. E., D. J. Beerling, W. D. Gosling, and M. B. Bush. 2004. Responses of Amazonian ecosystems to climatic and atmospheric carbon dioxide changes since the last glacial maximum. Philosophical Transactions of the Royal Society of London Series B-Biological Sciences **359**:499–514.

Moritz, C., J. L. Patton, C. J. Schneider, and T. B. Smith. 2000. Diversification of rainforest faunas: An integrated molecular approach. Annual Review of Ecology and Systematics **31**:533–563.

Mourguiart, P., and M. P. Ledru. 2003. Last Glacial Maximum in an Andean cloud forest environment (Eastern Cordillera, Bolivia). Geology **31**:195–198.

Nores, M. 1999. An alternative hypothesis for the origin of Amazonian bird diversity. Journal of Biogeography **26**:475–485.

Pennington, R. T., M. Lavin, D. E. Prado, C. A. Pendry, S. K. Pell, and C. A. Butterworth. 2004. Historical climate change and speciation: neotropical seasonally dry forest plants show patterns of both Tertiary and Quaternary diversification. Philosophical Transactions of the Royal Society of London Series B-Biological Sciences **359**:515–537.

Peteet, D.M. 1991. Postglacial migration history of lodgepole pine near Yakutat, Alaska. Can. J. Bot., 69: 786–796.

Polly, P. D. 2001. On morphological clocks and paleophylogeography: towards a timescale for Sorex hybrid zones. Genetica **112**:339–357.

Premoli, A. C., T. Kitzberger, and T. T. Veblen. 2000. Isozyme variation and recent biogeographical history of the long-lived conifer Fitzroya cupressoides. Journal of Biogeography **27**:251–260.

Ribas, C. C., R. Gaban-Lima, C. Y. Miyaki, and J. Cracraft. 2005. Historical biogeography and diversification within the Neotropical parrot genus Pionopsitta (Aves: Psittacidae). Journal of Biogeography **32**:1409–1427.

Rybnícová, E. and Rybníček, K. 1991. The environment If the Pavlovian: palaeoecological results from Bulhary, South Moravia. In: Palaeoevegetational Developments in Europe, Proceedings of the Pan-European Palaeobotanical Conference, 1991, Vienna Museum of Naturla History, pp 73–79.

Ritchie, J. C. and MacDonald, G. M. (1986). The patterns of post-glacial spread of white spruce. J. Biogeogr. 13, 527–540.

Rogers, R. A., L. A. Rogers, R. S. Hoffmann, and L. D. Martin. 1991. Native-American Biological Diversity and the Biogeographic Influence of Ice-Age Refugia. Journal of Biogeography **18**:623–630.

Rowe, K. C., E. J. Heske, P. W. Brown, and K. N. Paige. 2004. Surviving the ice: Northern refugia and postglacial colonization. Proceedings of the National Academy of Sciences of the United States of America **101**:10355–10359.

Runck, A. M., and J. A. Cook. 2005. Postglacial expansion of the southern red-backed vole (Clethrionomys gapperi) in North America. Molecular Ecology **14**:1445–1456.

Schneider, C., and C. Moritz. 1999. Rainforest refugia and evolution in Australia's Wet Tropics. Proceedings of the Royal Society of London Series B-Biological Sciences **266**:191–196.

Schonswetter, P., I. Stehlik, R. Holderegger, and A. Tribsch. 2005. Molecular evidence for glacial refugia of mountain plants in the European Alps. Molecular Ecology **14**:3547–3555.

Stewart, J.R. and Lister, A.M. 2001. Cryptic northern refugia and the origins of modern biota. Trends in Ecology and Evolution, **16**: 608–613.

Szeicz, J.M. and G.M. MacDonald. 2001. Montane climate and vegetation dynamics in easternmost Beringia during the Late Quaternary. Quaternary Science Reviews 20: 247–257.

Steele, C. A., B. C. Carstens, A. Storfer, and J. Sullivan. 2005. Testing hypotheses of speciation timing in Dicamptodon copei and Dicamptodon aterrimus (Caudata: Dicamptodontidae). Molecular Phylogenetics and Evolution **36**:90–100.

Stone, K. D., R. W. Flynn, and J. A. Cook. 2002. Post-glacial colonization of northwestern North America by the forest-associated American marten (Martes americana, Mammalia: Carnivora: Mustelidae). Molecular Ecology **11**:2049–2063.

Stute, M., M. Forster, H. Frischkorn, A. Serejo, J. F. Clark, P. Schlosser, W. S. Broecker, and G. Bonani. 1995. Cooling of Tropical Brazil (5-Degrees-C) During the Last Glacial Maximum. Science **269**:379–383.

Svendsen, J.I. and 30 authors (2004) 'Late quaternary ice sheet history of northern Eurasia', Quaternary Science Reviews, **23**:1229–1271.

Terhürne-Berson, R., T. Litt, and R. Cheddadi. 2004. The spread of Abies throughout Europe since the last glacial period: combined macrofossil and pollen data. Vegetation History and Archaeobotany **13**:257–268.

Thompson, R.S. 1988. Western North America. In Vegetation History, B. Huntley and T. Webb III (Eds.). Kluwer Academic, Dordrecht, Netherlands; pp 415–458.

Trewick, S. A., M. Morgan-Richards, S. J. Russell, S. Henderson, F. J. Rumsey, I. Pinter, J. A. Barrett, M. Gibby, and J. C. Vogel. 2002. Polyploidy, phylogeography and Pleistocene refugia of the rockfern Asplenium ceterach: evidence from chloroplast DNA. Molecular Ecology **11**:2003–2012.

Tribsch, A., and P. Schonswetter. 2003. Patterns of endemism and comparative phylogeography confirm palaeoenvironmental evidence for Pleistocene refugia in the Eastern Alps. Taxon **52**:477–497.

Tzedakis, P. C., I. T. Lawson, M. R. Frogley, G. M. Hewitt, and R. C. Preece. 2002. Buffered tree population changes in a quaternary refugium: Evolutionary implications. Science **297**:2044–2047.

Vila, M., J. R. Vidal-Romani, and M. Bjorklund. 2005. The importance of time scale and multiple refugia: Incipient speciation and admixture of lineages in the butterfly Erebia triaria (Nymphalidae). Molecular Phylogenetics and Evolution **36**:249–260.

Walker, M. and Willis, K.J. in press. Environmental Framework in The Population History of the Late Glacial: Founders and Farmers of Europe and the Near East. eds. William Davies, Clive Gamble, Paul Pettit and Martin Richards (eds) Cambridge University Press, Cambridge, 2005.

Ward, P. S. 1999. Systematics, biogeography and host plant associations of the Pseudomyrmex viduus group (Hymenoptera: Formicidae), Triplaris- and Tachigali-inhabiting ants. Zoological Journal of the Linnean Society **126**:451–540.

Willis, K.J., Rudner, E., and Sumegi, P. 2000. The Full-Glacial Forests of Central and Southeastern Europe. Quaternary Research, **53**:203–213.

Willis, K.J. and Whittaker, R.J. 2002. The Refugial Debate. Science, **287**:1406–1407.

Willis, K. J., and K. J. Niklas. 2004. The role of Quaternary environmental change in plant macroevolution: the exception or the rule? Philosophical Transactions of the Royal Society of London Series B-Biological Sciences **359**:159–172.

Willis, K. J., and T. H. van Andel. 2004. Trees or no trees? The environments of recentral and eastern Europe during the Last Glaciation. Quaternary Science Reviews **23**:2369–2387.

Willis, K.J. and H.J.B. Birks. 2006. What is natural? The need for a long-term perspective in biodiversity conservation. Science **314**:1261–1265.

Wooding, S., and R. Ward. 1997. Phylogeography and Pleistocene evolution in the North American black bear. Molecular Biology and Evolution **14**:1096–1105.

Wüster, W., J. E. Ferguson, J. A. Quijada-Mascarenas, C. E. Pook, M. D. Salomao, and R. S. Thorpe. 2005. Tracing an invasion: landbridges, refugia, and the phylogeography of the Neotropical rattlesnake (Serpentes: Viperidae: Crotalus durissus). Molecular Ecology **14**:1095–1108.

Explaining Distributions, Gradients, and Disturbances

Biogeography has diverse origins, and biogeographers have different backgrounds. On the one hand, concepts and approaches stem from disciplines such as geography, botany, zoology, ecology, geology, palaeontology, taxonomy, and genetics. On the other hand, biogeography is a science of synthesis, merging approaches and findings from all of these disciplines. Seeking a unifying bracket among biogeographers, one inevitably encounters the problem of distributional patterns of biota in space and time, and how they are produced by biological and ecological processes, environmental factors, gradients, and disturbances. The chapters in the previous section have argued that the evolution of biogeographical thought generated the vital need to understand both the distribution patterns of organisms over the surface of the Earth and why different regions accommodate different species and species assemblages, even though environmental conditions might be similar. From the old Greeks to the Darwinian revolution, the geographical variation of life on Earth was a matter of prime interest. It still is today—modern biogeography still studies how organisms and communities vary along environmental gradients and analyzes the underlying processes that result in the geographical variation in nature. Humans have increasingly influenced environmental gradients and disturbance regimes and have provoked distinct species responses to altered environmental conditions. To comprehend, and particularly to predict, species responses to global environmental change—a major issue of contemporary biogeography—it is indispensable

to first consider the evolution of general patterns of biogeographical distributions before arriving at deeper understandings of principal species-environment relationships. This section of the handbook serves to relate the distribution of species in space and time to environmental conditions by elucidating the role of abiotic and biotic factors and processes, gradients, and disturbance regimes. Using a wide range of examples from different regions and environments, the scholarship presented here builds on the theoretical framework elaborated in the first section, in order to consider fundamental, but still hotly debated, issues in ecological and, to a lesser extent, historical biogeography.

In the introductory Chapter 8, Udo Schickhoff treats present-day biogeographical distributions as being affected significantly by past geographical change, the physical environment, and biotic interactions. Species ranges and their variations in space and time precede a review of the significance of these explanatory variables. Since it became obvious in the eighteenth and nineteenth centuries that species are not randomly distributed over the globe, species ranges have been used to establish phytogeographical divisions or to differentiate between biogeographical regions. Surprisingly, delineations of floral kingdoms are still debated as exemplified by the recently proposed enlargement of the South African (Cape) floral region (Born et al., 2007), after generations of plant geographers had been taught that biogeographical lines are static rather than dynamic. Such refinements of floristic contrasts between biogeographical regions are today overshadowed by dynamic responses of species ranges to environmental changes as well as by human-induced species invasions. Recent range dynamics caused by climate change, land degradation, or anthropogenic introductions contrast with classical approaches in biogeography perceiving species ranges as spatially static entities. The spatially dynamic view of modern biogeographers assumes as a matter of course that biological and geographical patterns and processes have to be examined at many temporal and spatial scales.

It is not long ago that biogeographical theories shifted from perceived rather stable topographic and climatic conditions, and static biogeographical patterns, to ideas centered around dynamic environments and disequilibria. The acceptance of plate tectonics in the 1960s (Wilson, 1963)— today considered the key factor in explaining distribution patterns of biota—gave a fresh impetus to biogeographical studies and made a major contribution to this paradigm shift. Plate tectonics theory provided fresh perspectives in the discussion of the processes of vicariance and dispersal, that is, whether spatially separated distribution patterns have to be attributed to the break-up of a once-continuous range or to long-distance dispersal. The deeper understanding of terrestrial distributions during the science era of plate tectonics has put this discussion on a firmer footing and facilitated the emergence of modern vicariance biogeography. The impact of the theory of plate tectonics on modern foundations of biogeography, built largely upon the work of the two schools of vicariance and dispersal biogeography (Giller et al., 2004), has been reinforced by the development of analytical methods elucidating the roles of dispersal and vicariance in distribution patterns of biota. In particular, molecular-based technologies have brought many achievements and novel perspectives to biogeography. Phylogeography is today a fundamental component of a core modern biogeography (Riddle, this volume; Lomolino et al., 2006), and at the same time an ever-expanding bridge between biogeography and related disciplines (Riddle et al., 2008).

The actual geographical range limits of the vast majority of species are controlled by combined effects of the physical environment and biotic interactions. On a macroscale, species ranges are often limited by climatic or topographical barriers, whereas the distribution of specific habitat conditions in relation to the dispersal ability or physiology of the species as well as interactions with other species are important drivers at regional and local scales. Using a wide range of examples in terms of species and environments and looking at various spatial scales, Schickhoff (Chapter 8)

stresses in particular the role of climate as a physical factor explaining distribution patterns. The awareness that climate exerts a strong controlling influence on the geographical distribution of plants dates back to Ionian philosophy and was first elaborated by Theophrastus (370 to 285 B.C.). Today, it is a central thesis in plant ecology and biogeography (Woodward, 1987), and the correlation of range limits with climatic parameters is increasingly used in ecological modeling. Since significant species migrations are to be expected as a response to climatic change, it is an increasingly important ecological challenge to predict warming-induced range shifts. Particularly important tools in this respect are bioclimatic models. These models, showing a general ability to fit geographical ranges, have become very popular in recent years. One of the shortcomings of these models, in particular of mechanistic models, is that biotic interactions are not explicitly considered or even assumed to be unimportant for species distributions (e.g., Jeschke and Strayer, 2008). Actually, as shown in the last section of Chapter 8, the often underestimated interspecific interactions directly affect the competitiveness of species leading to a difference between the fundamental niches (physiological behavior in the laboratory) and the realized niches (ecological behavior in the field) of species. Thus, distribution patterns are the result of combined physical and biotic controls.

In Chapter 9, Jens Mutke augments the explanations of species distributions by theories and models explaining biodiversity patterns and gradients. Highlighting the 'Linnean shortfall' and the 'Wallacean shortfall' in his introductory remarks, he shows the scientific community quite plainly how deficient our knowledge still is with regard to (a) the numbers of microorganisms, plants, and animals that actually inhabit the Earth; and (b) their global, regional, and even local distributions—a confession of failure after more than 200 years of systematic research. The first detailed map of the global distribution of plant species richness (Barthlott et al., 1996) was a milestone in research on spatial diversity patterns, and an uninvolved observer might have wondered with incredulity why such a map had not been produced much earlier. This world map and the subsequently revised versions may serve as principal cartographic starting points for biogeographers to develop theories on biodiversity centers, hotspots, and gradients, all the more in view of the fact that species richness of higher plants as primary producers in ecosystems is congruent with those of many other taxonomic groups. However, research on geographical variation in species diversity that can be traced back at least to the first European naturalists who explored the tropics, had intensified well before, in particular since the 1950s. Mutke discusses possible explanations for latitudinal and elevational gradients in diversity including ecological, evolutionary, and historical hypotheses, suggesting that interactions of causative factors rather than single mechanistic hypotheses provide the best explanation of biodiversity patterns. It is expected that causal explanations of biodiversity gradients will be further improved by new large digital data sets, geostatistical tools, and new insights and methods from historical biogeography, phylogeography, and macroecology.

In addition to past geographical changes, physical factors, and biotic interactions, disturbances influence distribution patterns of biota at various spatial scales. In Chapter 10, Anke Jentsch and Carl Beierkuhnlein develop the significance of disturbance regimes in detail for biogeographical patterns and ecosystem dynamics. Disturbances such as fires, volcanic eruptions, hurricanes, floods, mass movements, and other extreme events are present in all ecosystems, occur across a wide range of spatial and temporal scales, and affect all levels of biological organization. Being inherent to ecosystems, disturbances have to be incorporated into concepts of nature conservation as Jentsch and Beierkuhnlein rightly postulate in their introduction. Concentrating on natural processes of disturbance, they present a geography of disturbance regimes, introduce the main types of disturbances, and elaborate the ecological importance of disturbance events and their influences on biogeographical distributions. The subsequent thought-provoking discussion of aspects of fundamental significance such as rhythms and synchronization, timing of events, and succession and inertia reveals that understanding the role of temporal variability of events, processes, and patterns is crucial to understand ecosystem dynamics as a whole as well as to manage novel ecosystems in a rapidly changing world. Their outlook gives cause for serious concern. In addition to the pervasiveness of anthropogenic disturbances, the probability and amplitude of extreme weather events and related disturbances is increasing with climate change, thus enhancing the role of disturbances in many ecosystems in the twenty-first century. Critical thresholds for species, populations, and communities might be exceeded. Unknown consequences for ecosystem functioning and services require a new orientation of scientific investigations into extreme disturbance events (Jentsch et al., 2007).

In Chapter 11, Neal Enright focuses on fire as the most pervasive form of disturbance in terrestrial ecosystems. He reviews the relevance of fire as a disturbance factor in different ecosystem types and details the responses of ecosystems and species to fire. The history of fire is to some extent a history of human-environmental interactions. For hundreds of thousands of years people have used fire, mostly for conversion of forest to nonforest, and today humans are the leading ignition source of fires globally. In his introductory overview of fire regimes in major terrestrial ecosystems, Enright emphasizes ecosystem-specific relationships between timing, frequency, and intensity of fires and various adaptations of plants. Shortest return intervals (1–5 years) occur in savanna ecosystems where all perennial plant species are capable of vegetative regrowth. Recruitment strategies in general are discussed in the section on fitness in fire-prone environments. Survival of dormant seeds in canopy and soil seed banks, and the cueing of germination to physical and chemical triggers associated with fire are other vital and widespread adaptations in fire-prone ecosystems. Enright's treatment of ecosystem changes with time since fire and in relation to fire history reveals fundamental differences between ecosystems in terms of species composition, species richness, and biomass levels, thus corroborating the need for understanding the role of temporal variability of ecosystem processes stressed in Chapter 10. Reviewing the objectives and effectiveness of fire management, Enright traces the changing institutional policies from fire suppression to the managed use of fire. Such policy changes reflect deficient knowledge of fire-environment relationships. Asset protection and biological conservation objectives might be more effectively achieved on the basis of traditional practices of fire use that are employed by indigenous people. Potential effects of climate change on fire, particularly in relation to the impacts on rates and levels of biomass and fuel accumulation, are not fully understood and represent an important area of future investigation.

As stated above, significant species migrations and changing range boundaries are to be expected as responses to climatic change. The ongoing trend of increasing global average surface temperatures triggers significant changes in biological systems, being detected to an increasing degree on all continents and in most oceans (Parmesan, 2006). In the final chapter of this handbook section, Tim Sparks and his co-authors provide evidence for temperature-related changes in the populations, distribution, and phenology of plants and animals. They stress that impact assessments regarding population increases or decreases, demographics, or community composition heavily rely on good long-term data, which are increasingly becoming available. Increasing asynchrony between interacting species (e.g., between herbivorous insects and their host plants) due to different responses to climate warming is assumed to have negative fitness consequences and will greatly influence population dynamics. They underline that poleward and altitudinal upward shifts of species ranges have occurred across a wide range of taxonomic groups and geographical locations during the twentieth century and that range boundary changes may facilitate the invasion of nonnative species from adjacent areas. However, adaptation may fail to match the pace or magnitude of predicted changes in climate, leading to an increase in extinction risk. Asynchronous species responses will create no-analogue communities and changes in ecosystem structure and function, reinforced by different phenological responses to climate change. To accurately predict responses and future changes of species, communities, and habitats will remain a prime task of biogeographical research throughout the twenty-first century.

REFERENCES

Barthlott, W., Lauer, W., and A. Placke. (1996) 'Global distribution of species diversity in vascular plants: Towards a world map of phytodiversity', *Erdkunde*, 50: 317–28.

Born, J., Linder, H.P., and P. Desmet. (2007) 'The Greater Cape floristic region', *Journal of Biogeography*, 34: 147–62.

Giller, P.S., Myers, A.A., and Riddle, B.R. (2004) 'Earth history, vicariance, and dispersal', in Lomolino, M.V., Sax, D.F., and J.H. Brown (eds.), *Foundations of Biogeography*. Chicago: University of Chicago Press. pp. 267–76.

Jentsch, A., Kreyling, J., and Beierkuhnlein, C. (2007) 'A new generation of climate change experiments: Events, not trends', *Frontiers in Ecology and the Environment*, 5: 365–74.

Jeschke, J.M., and Strayer, D.L. (2008) 'Usefulness of bioclimatic models for studying climate change and invasive species', *Annals of the New York Academy of Sciences*, 1134: 1–24.

Lomolino, M.V., Riddle, B.R., and Brown, J.H. (2006) *Biogeography*. 3rd ed. Sunderland, MA: Sinauer.

Parmesan, C. (2006) 'Ecological and evolutionary responses to recent climate change', *Annual Review of Ecology, Evolution, and Systematics*, 37: 637–69.

Riddle, B.R., Dawson, M.N., Hadly, E.A., Hafner, D.J., Hickerson, M.J., Mantooth, S.J., and Yoder, A.D. (2008) 'The role of molecular genetics in sculpting the future of integrative biogeography', *Progress in Physical Geography*, 32: 173–202.

Wilson, J.T. (1963) 'Evidence from islands on the spreading of ocean floors', *Nature*, 197: 536–38.

Woodward, F.I. (1987) *Climate and Plant Distribution*. Cambridge: Cambridge University Press.

8

Biogeographical Distributions: The Role of Past Environments, Physical Factors, and Biotic Interactions

Udo Schickhoff

8.1 INTRODUCTION: DISTRIBUTION PATTERNS OF BIOTA

One of the fundamental questions of biogeography refers to distribution patterns of biota. Each definition of biogeography centers on detecting, analyzing, and interpreting distribution patterns in space and time. How geographical, environmental, and historical factors have led to the great diversity of organisms and their distributions over the surface of the globe has been one of the core issues from the origins of biogeography to the present day. Biogeography has evolved as a discipline by tackling the problems of recognizing and defining distribution patterns and identifying the mechanisms and processes that produce those patterns. The subdisciplines of ecological, historical, and analytical biogeography are equally concerned with the question of distribution patterns. The origins of modern biogeography can be traced back to the advent of the Renaissance in Europe, when interest in natural sciences was reawakened. After the Swiss doctor and zoologist Gesner had described elevation belts of the European Alps in the sixteenth century (Zoller and Steinmann, 1987/91), Buffon (1761), Forster (1778), and

Linnaeus (1781) were among the first naturalists who studied and tried to explain distribution patterns. Buffon discovered that regions in the Old World and the New World were inhabited by different species even though habitat conditions were similar, a phenomenon that became known as 'Buffon's Law'. Forster, a companion of Captain Cook's second expedition to the South Seas, related changes in the plant cover along the route to changes in the physical characteristics of the environment. He even observed and explained species-area relationships and latitudinal gradients in diversity. Linnaeus, the founder of modern systematics and the binary nomenclature of species, developed the 'center of origin/dispersal' hypothesis stating that the world's biota originated from a single landmass and spread out to colonize virgin land areas emerging from the sea. Later, von Humboldt (1805), the 'father of phytogeography', provided detailed analyses of plant distributions controlled by physical environments The influential works of de Candolle (1855), Darwin (1859), Grisebach (1872), Haeckel (1868), Wallace (1876), Schimper (1898), Wulff (1943), Troll (1966), MacArthur and Wilson (1967), Walter (1973), Ellenberg (1988), and Ellenberg and Leuschner (2010), among others, followed his pioneering stance.

During the nineteenth century it became obvious that species are not randomly distributed over the globe and that the land surface can be differentiated into biogeographical regions with distinct sets of plants and animals. Largely based on the earlier phytogeographical divisions of Engler (1879–82), Drude (1884), Diels (1908), and Rikli (1913), six floral kingdoms are commonly delineated (Kratochwil and Schwabe, this volume; cf. Good, 1974; Takhtajan, 1986; for faunal regions, see Smith, 1983; Cox and Moore, 2005). Pronounced floral contrasts, that is, accumulations of range limits, characterize the boundaries of adjacent floral kingdoms, which can be further subdivided into floral regions and floral provinces (e.g., Meusel et al., 1965 for the Eurasian part of the Holarctic floral kingdom). Such boundaries, also called biogeographical lines, are often transition zones that allow some species to pass and extend their geographical range, while for others, for example, the passage might be too difficult for physiologically less tolerant or less competitive species. In other cases various kinds of barriers such as mountains, water gaps, or deviating environmental conditions act as biogeographical filters and make adjacent regions inaccessible. The same principles apply to smaller geographical ranges within biogeographical regions.

Each species occupies a distinct geographical area, the species range that differs from the range of other species with other habitat requirements. The common oak (*Quercus robur*), for instance, is widely distributed throughout the central and eastern European plains and is restricted to mountains in southern Europe, whereas other oak species (e.g., *Q. pubescens, Q. coccifera*) occupy far smaller species ranges. The actual geographical range of plant species is usually smaller than their potential geographical range (Figure 8.1). Macroclimate and topography are the most important factors determining the approximate shape of the potential range. Reaching the limits of the potential range requires sufficient time periods. However, even when there are no time constraints, dispersal barriers prevent plants from taking up the entire potential range. Even in the accessible parts of the potential range a plant will only establish where the other habitat requirements are fulfilled. Here, edaphic and biotic factors can be effective obstacles for plant establishment (cf. Figure 8.1). Dispersal and establishment processes result in a smaller actual range, the limits of which may depend on climatic, geomorphological, biotic, edaphical, or temporary factors.

Species ranges vary in terms of size, form, geographical position, and population density. Some plants and animals have extremely large species ranges that may span the entire terrestrial

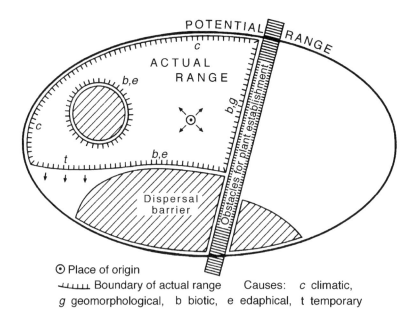

Figure 8.1   Potential range, dispersal, and range boundaries of a newly evolved plant species (after Schroeder, 1998, modified)

globe, whereas others are restricted to a location of a few square meters. However, as early biogeographers noted (e.g., Willis, 1922), the majority of species have small ranges and only very few species have very large ones. Species with smaller ranges are often limited by their narrow tolerances regarding abiotic and biotic conditions, whereas widely distributed species must be able to tolerate a wide range of physical conditions and to coexist with many other species. Thus, contrasting range sizes must be attributed to differing ecological amplitudes (niche breadths) and competitive pressure as well as to dispersal barriers and historical factors (see below).

Taxa with very large geographical ranges and a wide distribution over most continents or biogeographical regions are called cosmopolitans. Easily dispersed microorganisms often show a cosmopolitan distribution. Among higher plants, cosmopolitans are rare. Even plant species with very large range sizes, for example, reed (*Phragmites australis*) or bracken (*Pteridium aquilinum*), may show considerable distribution gaps (Figure 8.2). Cosmopolitans are widespread, but in contrast to ubiquists they are confined to specific site conditions. Species with restricted ranges, for example, taxa confined to one biogeographical region, are classified as endemics regardless of the size of their geographical range. A species can be endemic to a continent or to one mountain valley or even smaller habitats. Thus, the term 'endemic' is very relative. For terrestrial plants and animals, endemic distributions are common. The degree of endemism decreases with increasing taxonomic level: that is, genera and families are far less endemic than species. The Cape Floristic Region in South Africa is comparatively rich in endemic flowering-plant families. Five families of angiosperms (Penaeaceae, Roridulaceae, Geissolomataceae, Grubbiaceae, and Lanariaceae) are endemic to that region (Rebelo et al., 2006). On the other hand, the Asteraceae and the Poaceae are examples of flowering-plant families that are very widespread and have almost worldwide distributions. Palaeoendemics are taxa that once had larger geographical ranges, whereas species that have recently evolved and are still in the process of being distributed to wider areas are referred to as neoendemics. Closely related species that differ ecologically and replace each other in different habitats (ecotypes) or in different areas

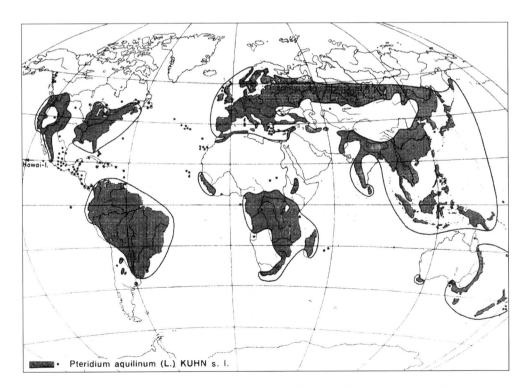

Figure 8.2   The world distribution of bracken (Pteridium aquilinum), a cosmopolitan species (after Meusel et al., 1965)

(geographical races) are termed vicariant species. Well-known are vicariant species on limestone and silicate rocks in the European Alps (e.g., *Rhododendron hirsutum* on calcareous and *Rh. ferrugineum* on acidic soils).

Generally, the range of a genus is larger than the ranges of single species belonging to this genus. A particularly large number of species ranges often concentrate in a certain area of the genus range (not necessarily in the center). Such areas are called centers of dispersal. It is often assumed that species have evolved in those centers and have spread from them into different directions. Family ranges, containing all the ranges of respective genera, are accordingly larger than genera ranges. Many family ranges show continuous zonal distributions (i.e., their ranges form broad latitudinal bands following climatic zones). The range of the maple family (Aceraceae) with some 150 species, for instance, is largely confined to the temperate zone (Figure 8.3), whereas the range of the Palmae, consisting of about 2,800 species, is pantropical (Figure 8.4). Other common zonal patterns include circumpolar, boreal, austral, and amphitropical distributions. Ranges of circumpolar taxa circle the Arctic or Antarctic regions, boreal taxa are distributed in higher northern latitudes, austral taxa are in higher southern latitudes, and amphitropical taxa occur on either side of the tropics but not in the tropics itself. It is often observed that range sizes tend to increase with increasing latitude: geographical ranges at higher latitudes are larger compared to the tropics, and species ranges further north are distributed over an increasingly wider range of latitudes. This pattern, known as 'Rapoport's Rule', also holds for altitudinal distributions where the elevational ranges of species increase with the elevation of the midpoint of their ranges (Stevens, 1989, 1992). Range limitations by interspecific interactions in the tropics, the greater severity of dispersal barriers at lower latitudes, and the selection of broadly tolerant, wide-ranging species by the Pleistocene climatic cycles at higher latitudes are discussed as underlying mechanisms (Brown, 1995).

Another major distinction of common distributional patterns is between continuous and disjunct ranges. In the case of a continuous range, the encircling of all localities of a taxon results in a single distribution area. By contrast, geographical ranges that are divided into two or more geographically separated parts are referred to as disjunct ranges. The shape of range disjunctions can be quite varied. Range parts may have the same size, or smaller range parts (exclaves) are separated from the main range (Figure 8.5). Sometimes localities are greatly dispersed and totally discontinuous. Underlying causes of disjunct ranges include geological or climatic change, evolution, and jump dispersal by natural and human agencies. It is not always clear where to separate continuous from disjunct distributions since localities commonly tend to disperse towards the

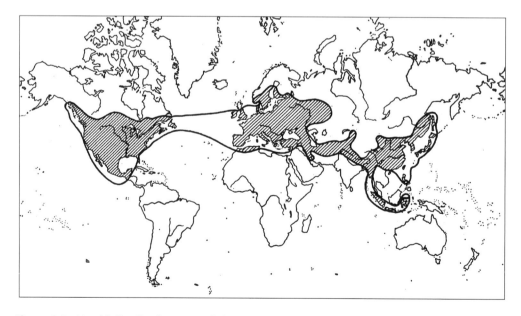

Figure 8.3   World distribution map of the maple family (Aceraceae), an example of a temperate range (after Walter and Straka, 1970)

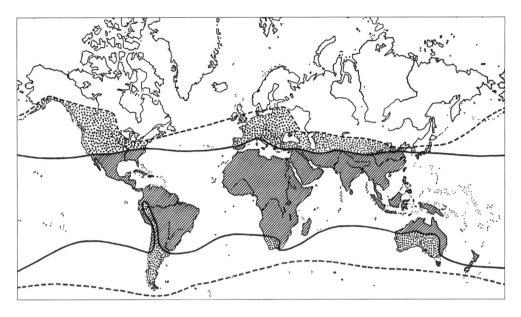

Figure 8.4 World distribution map of the palm family (Palmae), an example of a pantropical range (after Walter and Straka, 1970)

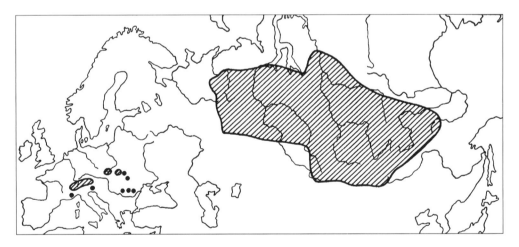

Figure 8.5 Disjunct range of Pinus cembra with main range in Siberia and exclaves in the Carpathians and the European Alps (after Meusel et al., 1965)

boundaries of species ranges. Within the majority of species ranges, the abundance of localities decreases from a main distributional area (abundance center) to the range periphery. To term a range as disjunct, the distance between the nearest range parts must be sufficiently large to prevent the exchange of diaspores.

Species ranges are not constant in space and time. Instead of being spatially static entities as early biogeographers postulated, they are continuously altering as a result of ecological responses of species and individuals to environmental changes (cf. Hengeveld et al., 2004). Range boundaries are subject to change as the spatial configuration of determining abiotic and biotic conditions changes. Dispersal processes lead to range expansions; local extinctions result in range contractions. Range expansions coupled with differential extinctions in parts of the former range are termed range shifts. Range boundaries may change slowly, for example, in response to climate warming as observed at the polar treeline in recent decades. In other cases range boundaries alter rapidly, for example, when plants and animals spread across large distances by jump dispersal. Today, humans are the most successful dispersal agents. Introducing species to all corners of the globe means crossing biogeographical barriers and enabling exotic species to colonize distant regions (see Blumler, this volume for a detailed treatment of exotic colonization). Range contractions and local extinctions have occurred in an unprecedented way as a result of human activities in recent decades (Beierkuhnlein, 2007). Land use and land cover change (cf. Foody and Millington, this volume), destruction of habitats, and species invasions and climate change (cf. Sparks et al., this volume) will continue to increasingly alter species ranges. Detecting and analyzing response processes and changing distributional patterns of biota will be the prime future research tasks of biogeography.

8.2 PAST GEOGRAPHICAL CHANGES AND PRESENT-DAY SPECIES DISTRIBUTIONS

Understanding present biogeographical patterns of species and ecosystems requires a sound knowledge of the past physical changes of the Earth's surface. Many aspects of modern distributions of species are to be attributed to continuous changes in the geography of continents and oceans and in the Earth's climate over geological time scales. Processes in the physical environment such as continental drift, expansion and contraction of seas, rising and erosion of mountains, appearance and disappearance of islands and land bridges, or advance and retreat of ice masses and other climate-driven effects, profoundly influence dispersal and extinction processes and hence distributions of species, ecosystems, and biomes.

Plate tectonics is the key factor for explaining today's patterns of distribution of life. Sea-floor spreading and continental drift have caused splitting and collision of landmasses and their movement across latitudinal bands of climate so that evolutionary development in various parts of the Earth has followed divergent routes. In particular, the plate tectonic history of the last 250 million years is crucial to an understanding of modern biogeography. Tectonic events such as the formation of the super continent of Pangaea and the subsequent rifting and splitting of Laurasia and Gondwanaland, along with associated continental movements in the Mesozoic and Cenozoic, have substantially influenced distribution and evolution of biota. Taxa of the phylogenetically old group of conifers, for instance, show hemispherical distribution patterns with the Podocarpaceae representing the Gondwana part and the Pinaceae the Laurasia part. The distribution of the currently dominating angiosperms, whose main development took place in the Tertiary period when the Pangaea landmasses had already split up into different continents (Willis and McElwain, 2002), is by contrast much more sharply differentiated.

The configuration of today's floral and faunal realms is a direct biogeographical consequence of long-term changes in the positions of continents relative to each other and to the poles. For example, the evolution of the recently enlarged Cape (Born et al., 2007), Antarctic, and Australian floral kingdoms commenced after the complete separation of Gondwanaland. Australia has been effectively isolated by surrounding oceans after the final separation from the Antarctic about 50 million years ago, resulting in a high degree of endemism reflected today in the Australian floristic and zoogeographical realms (White, 1990). Owing to the separation of Africa and South America in the late Cretaceous, rather large floristic differences exist between the tropical floras of the New and Old Worlds (Africa, Southeast Asia) so that two floral kingdoms (Neotropic and Paleotropic) can be distinguished. Contrastingly, floristic differences in the extratropical northern hemisphere are relatively low since North America and Greenland separated from Eurasia as late as the Pleistocene. These continents are considered as one floristic realm: the Holarctic (cf. Kratochwil and Schwabe, this volume).

The slow movement of continents and oceans over millions of years has been accompanied by tremendous variations of surface features having substantially affected the spread of organisms. Changing areas of landmasses and marine basins, large-scale orogenic mountain ranges, and island chains that are formed as a result of volcanic activity may lead to isolation or provide opportunities for biotic exchange. Extensive mountain ranges often constituted effective barriers to lowland species or served as migration corridors for taxa adapted to montane habitats. For example, the collision of the Indian and the Eurasian plates created the Himalayas, which effectively isolate many taxa on either side of this topographic barrier. At the same time, hygrophilous Sino-Japanese floristic elements have spread along the monsoon-influenced south slope of this mountain chain to constitute a western appendix of the Sino-Japanese floral region (Figure 8.6; Schickhoff, 1993). Moreover, plate movements affect dispersal, extinction, and distribution patterns of biota indirectly through alterations of regional and global

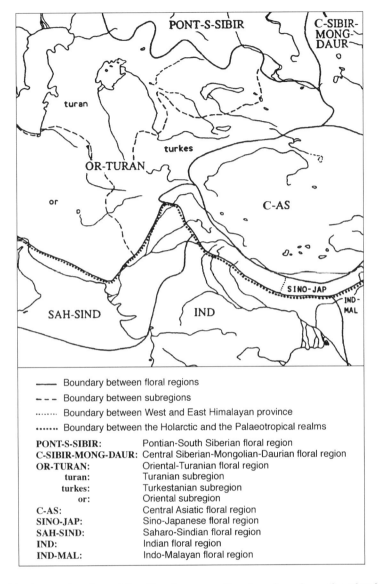

Figure 8.6 Western extension of the Sino-Japanese floral region along the Himalayan south slope (after Meusel et al., 1965; Schickhoff, 1993, modified)

climates. Landmasses drifting over the poles can potentially trigger glacial cycles and subsequent sea-level fluctuations. Even mass extinctions are related to climate changes induced by plate tectonics (e.g., the great Permian event; cf. Hallam and Wignall, 1997). Changes in the areal extent of continents and oceans, their latitudinal positions as well as orogenesis cause large alterations in atmospheric and oceanic circulations, and substantial shifts of marine currents, humidity/aridity, precipitation patterns, temperature, wind, albedo, and other climate parameters (Ruddiman, 1997).

The Indian summer monsoon, of extraordinary importance for vegetation patterns in South Asia, may serve well as an example for the significance of orogenesis: stages in the monsoon's evolution are linked to phases of Himalayan-Tibetan plateau uplift (cf. Zhisheng et al., 2001).

Evidence from paleoecological studies shows climatic changes have always exerted a strong controlling influence on the distribution of biota. Temperature reconstructions based on oxygen isotope ratios in coral and foraminiferal carbonate indicate that early Tertiary climates were much warmer than they are today (Jansen et al., 2007). Accordingly, the early Tertiary flora in central Europe are represented by palms and other species occurring today in the tropics. At the same time, the so-called Arcto-Tertiary flora, corresponding to a warm temperate climate, prevailed in what are now polar regions. One of those species found north of 80°N was the bald cypress (*Taxodium*), which is restricted nowadays to south of 35°N in North America. Many temperate and subtropical forest genera such as *Ginkgo*, *Sassafras*, *Diospyros*, *Castanea*, *Platanus*, and *Vitis* occurred at latitudes at least 20–25° north of their present range limits (Walter, 1986). As the climate cooled in the mid-Tertiary, plants responded by migrating southward. Macrofossils from the Pliocene (5–2 million years ago) already indicate a temperate climate in central Europe. However, the species richness of the flora was much higher than today; many tree genera that are now only found in eastern North America or East Asia were still present.

The climate deterioration in the late Tertiary culminated in successive advances of continental glaciers during the Pleistocene and associated environmental changes that account for the species impoverishment of the European flora. Environmental stresses for taxa included dramatic climate and site-ecological alterations, geographical shifts of their prime habitats, and changed dispersal abilities. Successive cold and warm phases triggered large-scale species migrations, with species being able or unable to track their shifting environments. Species that remained in situ had to adapt to the altered conditions or became extinct. Thus, plant communities disappeared or were disrupted and changed into novel communities with substantially changed species compositions. Broadly, vegetation zones shifted toward the equator during glacial periods and toward the poles during interglacials, and in mountain environments toward lower or higher elevations. However, vegetation types did not move as entire entities. The shifts were complicated by topoclimatic and topographic influences such as mountain ranges and ocean basins as well as by individual species responses (cf. Lomolino et al., 2006). During the Last Glacial Maximum in Eurasia (c. 20,000 years B.P.), when mean annual temperatures were probably 8–10°C lower than today, major vegetation types were shifted southward of their present locations by 10° to 20° of latitude.

The deciduous forest biome of Europe became fragmented during each glaciation and restricted to protected locations in southern Europe, and the forests were repeatedly displaced by tundra and steppe vegetation. At the beginning of each interglacial, temperate forests expanded and became established again, however, with rather different species assemblages. Not all of the more thermophilous species could persist in southern refugia since mountain ranges (i.e., the Pyrenees, Alps, Carpathians, and Balkan Mountains) and the Mediterranean Sea blocked southward migrations of species and latitudinal shifts of biomes to a large extent (for a more detailed discussion of European refugia, see Willis et al., this volume). In addition, colder and drier glacial climates affected the competitive capacity of species that was already reduced due to higher sensitivity to water stress under lower atmospheric CO_2 conditions (cf. MacDonald, 2003). As a result, many taxa such as *Sequoia*, *Taxodium*, *Nyssa*, *Liquidambar*, *Tsuga*, *Carya*, *Pterocarya*, and *Eucommia* were successively eliminated by alternating glacial and nonglacial conditions. Many of these genera and hence the Pliocene character have been preserved in contemporary North American and East Asian floras: the number of tree genera of temperate deciduous and conifer forests in eastern North America (124) is much higher compared to Europe (53) (Ellenberg, 1988). In North America, north-south-running mountain ranges and rivers facilitated climate-driven latitudinal species migrations, whereas ice sheets originating in Siberia were less extensive and climatic alterations less severe than those in Europe. In addition, present-day higher summer temperatures contribute to higher tree species richness in North American and East Asian temperate forest biomes.

As a result of the geographical range shifts during the Pleistocene, many taxa show present-day disjunct distributions. Examples include large-scale disjunctions of more than 100 tree genera that were formerly distributed all over the northern hemisphere but are now restricted to North America and East Asia. Another typical disjunct pattern of the northern hemisphere is exemplified by arctic-alpine species (e.g., *Arabis alpina*, *Bistorta vivipara*, *Dryas octopetala*, *Salix herbacea*, and *Silene acaulis*). During glacial periods, alpine species migrated down to lower elevations and adjacent plains where they mixed with arctic species moving southward. After climatic amelioration these species followed glacial recession that provided dispersal routes both to the north and to higher elevations in southern mountains (European Alps, Tian Shan, Himalayas). Resulting arctic-alpine disjunctions indicate present-day close floristic relationships between these distant regions. Pleistocene glaciations have also greatly influenced the distribution pattern of endemics in Europe: localities of endemic species are almost exclusively confined to areas south of the limit of maximum glaciation (cf. Ellenberg and Leuschner 2010).

Analyses of fossil pollen allow reconstruction of the vegetation history during the late- and postglacial periods that indicates again a strong correlation between vegetation distribution and climate (Woodward, 1987). The colonization of deglaciated areas in Europe started with steppe and tundra vegetation that was replaced by pioneer forests (willow, birch, and pine forests) during warmer phases of the Late Glacial. With further climate warming after the onset of the Holocene (c. 11,500 years B.P.), there had been a considerable expansion of closed forests into zonal habitats throughout central Europe (Lang, 1994). Pine forests (*Pinus sylvestris*) with an understory of hazel (*Corylus avellana*) were widely established during the early Holocene (Preboreal/Boreal) followed by mixed oak forests (*Quercus, Ulmus, Tilia, Acer*) during the Atlantic period (8000–4300 years B.P.). Subsequent cooler and moister conditions in the Subboreal (4300–3000 years B.P.) enabled beech (*Fagus sylvatica*) to become the most competitive and dominant tree species in central Europe. The beech had spread from various Pleistocene refugia in the Mediterranean region reaching the central highland regions of central Europe by 7000–6500 years B.P. and the coastal areas of the North Sea and the Baltic by c. 1800 years B.P. (Pott, 2000). Beech forests represent the potential natural vegetation of zonal habitats in central Europe. Their present-day distribution would be more extensive if their expansion had not been greatly influenced by human activity since Neolithic times.

In recent years, the glacial history of arctic and alpine plants and postglacial colonization routes are increasingly detected by comparative phylo-geographical approaches (Taberlet et al., 1998; Schönswetter et al., 2003). Based on molecular genetics, the analysis of genetic variation of geographically distinct populations can be used to reconstruct distribution and migration patterns in response to climatic alterations of the ice ages. In general, phylogeography is now a funda-mental component of modern biogeography primarily due to its vast potential for analyzing evolutionary and climatic responses of species and populations to geological and climatic dynamics in Earth history (Riddle et al., 2008; Avise, 2009; Riddle, this volume).

8.3 THE PHYSICAL ENVIRONMENT AND PLANT SPECIES DISTRIBUTIONS

Paleoecological studies provide an essential but only a single link in the explanation of present-day distribution patterns of biota. Correlations between spatial distributions of species and spatial distributions of environmental variables suggest that differences in the physical environment play a major role in controlling where certain species occur and where they do not exist. The multitude of environmental factors affecting individuals, species, or populations in a particular place is called a habitat or a site (Walter, 1960). Among site factors, various climatic factors exert the most prominent environmental constraints on species distributions. Macroclimate is crucial to the understanding of global vegetation patterns and geographical ranges of single species. Macroclimate affects microclimate, soils, and vegetation as an independent factor. Soils and vegetation are closely interrelated and, in turn, influence microclimatic patterns. Microclimate and soil physical and chemical properties are greatly influenced by parent rock and weathering processes, and these processes are often decisive small-scale site factors. Vegetation is again affected by the fauna, above-ground as well as below-ground, where the decomposition of dead organic substances is initiated, followed by the final mineralization by microorganisms. However, the complex interrelationships between vegetation and site factors at a particular habitat, depicted in a simplified scheme in Figure 8.7, are modified and often highly disturbed by human impact. Moreover, the competitive factor influences vegetation-environment relationships (see below).

Ecological tolerance, that is, the tolerance of a species to the multitude of site factors, determines whether it lives only under very specific habitat conditions (habitat specialists) or occurs over a wider range of environmental conditions (habitat generalists). From the perspective of an individual plant, factor groups affecting the physiological functioning of plants (e.g., growth, development, morphology, resistance) can be differentiated into primary and secondary site factors (Figure 8.8). Climate, topography, and soils as well as biotic influences represent complex (secondary) site or terrain factors exerting indirect influences on plants via primary site factors. The latter include light, heat (temperature), water (moisture), vari-ous chemical factors, and various disturbance-related or stress-inducing mechanical factors. Topographic variations, for example, may cause strongly contrasting topo- and microclimates that again control light, temperature, and humidity conditions of a specific habitat. Erosion and accu-mulation processes on mountain slopes determine soil depth and may strongly influence humus content, soil texture, and hence factors such as soil moisture and nutrient supply. The short-term variable, interrelated primary site factors have ecophysiologically direct effects on plants

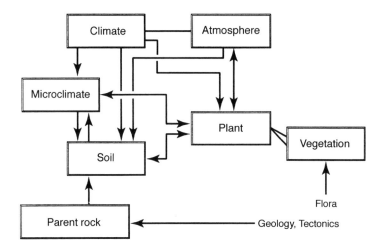

Figure 8.7 Simplified sketch of interrelationships between vegetation and site factors (after Walter, 1986; Walter and Breckle, 1999, modified)

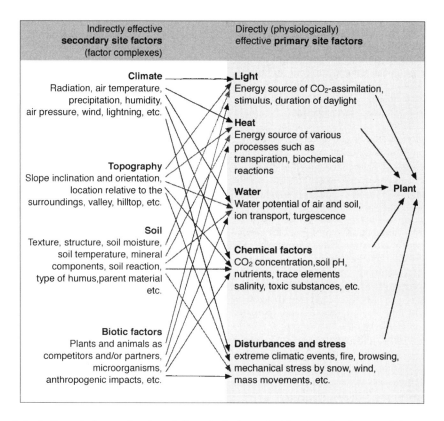

Figure 8.8 Indirectly (secondary) and directly (primary) effective site factors (after Walter, 1960; Ellenberg, 1968; Klink, 1996, modified)

(environment in an ecophysiological sense). Plants, in turn, have multiple feedbacks to these factors. The effectiveness of primary site factors results from energy and matter transformation processes in the ecosphere and varies accordingly within different habitats.

In order to characterize the environment of a plant community, indicator values assigned to the participating plant species can be used. In recent decades, a system of indicator values has been developed for central European vascular plants with respect to distinct environmental factors including light regime, temperature, continentality, moisture, soil reaction (acidity/lime content), soil nitrogen, and soil chloride concentration (Ellenberg et al., 1992). The relations to these factors are expressed in a, typically ordinal, nine-point scale. The figures are symbols for the ecological behavior of a species, that is, for its occurrence under field conditions (realized niche). The indicator values have proved a practical tool to allow quick estimates of vegetation-environment relationships. Average indicator values calculated on the basis of species lists often showed significant correlations to measured data. Such calculations cannot, of course, replace exact measurements. Ecological indicator values are increasingly used for varied spatial applications in plant geography and landscape ecology (e.g., Schmidtlein and Ewald, 2003).

8.3.1 Climate

The correlation of observed distribution patterns of vegetation formations with climatic zones already prompted early biogeographers (e.g. von Humboldt and Bonpland, 1805) to stress the fundamental importance of climate regarding geographical distributions of species. The increasing awareness of the climatic control of plant distribution resulted in the interlinked and parallel development of vegetation classification and mapping and climatic classification and mapping (Daubenmire, 1978; Keddy, 2007). From a global perspective, the Earth's plant cover is a reflection of climatic conditions, only regionally modified by topographic, geological, and edaphic influences and disturbances. Thus, terms used for vegetation zones closely follow terms used for climatic zones (cf. Walter and Breckle, 1985; Archibold, 1995; Richter, 2001; Pott, 2005; Schultz, 2005). Latitude largely controls the overall distribution of climatic conditions since it determines the amount of solar energy intercepted over a given area and the seasonal distribution of insolation. The spatio-temporal pattern of insolation and the insolation-driven global atmospheric and oceanic circulation, strongly modified by the rotating Earth and its extremely varied surface, involves latitudinal and longitudinal heat transfer and major semipermanent wind and air pressure conditions. These controls produce the spatial and temporal differentiation of temperatures and humidity on the Earth's surface.

The dominant control that climate exerts on the distribution of plant life is basically reflected in morphological and functional adaptations to the environment. Similar climatic conditions in widely separated regions result in physiognomically and structurally similar plant growth forms and vegetation formations, irrespective of the taxonomic affinity of involved species. This phenomenon of ecologically equivalent vegetation is exemplified, for instance, by sclerophyllous shrubs and trees in the widely disjunct mediterranean regions or by stem succulent species of the unrelated families of Euphorbiaceae and Cactaceae in Old and New World arid regions. Consistently, classifications of life forms, growth forms, and plant functional types (de Candolle, 1820; Raunkiaer, 1934; Dansereau, 1957; Box, 1996) as well as global classifications of plant formations, for example, the physiognomic-ecological classification of plant formations of the Earth (in Mueller-Dombois and Ellenberg, 1974), are based on climate-related plant traits or mirror climatic requirements of plants. For instance, the widely used system proposed by Raunkiaer (1934) divides plants growing under different climatic conditions into five main life-form categories (phanerophytes, chamaephytes, hemicryptophytes, cryptophytes, therophytes) according to the arrangement of their perennating tissues and their life-history strategies. From the number of species in each category, biological spectra of communities can be calculated that provide insight into climatic habitat conditions. Different plant formations can be distinguished by the relative dominance of life forms. A tropical rain forest, for instance, has over 70% phanerophytes (trees and large shrubs with persistent above-ground stems and buds), whereas therophytes (annual plants that regenerate from seeds each year) occur abundantly in desert areas.

Based on ecophysiological, physiognomic, and climate-statistical approaches and models (e.g., Holdridge, 1947; Walter, 1960; Whittaker, 1975; Box, 1981; Woodward, 1987), it has been established that temperature and water availability are decisive climatic elements for the differentiation of the plant cover into biomes. The classification of global vegetation formations according to mean annual precipitation and mean annual temperature (where temperature is also a measure of potential evapotranspiration) is a vivid illustration of the relationship between climax vegetation, its characteristic life forms, and major climatic elements

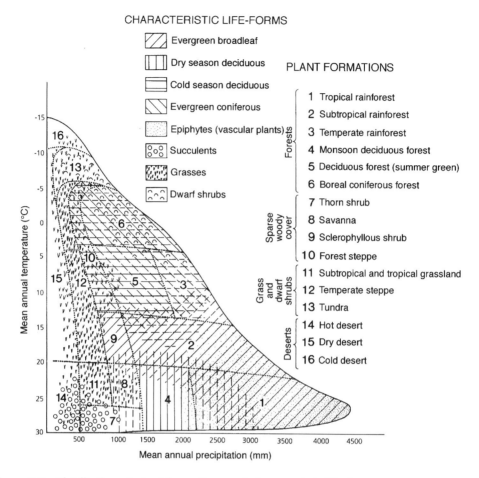

Figure 8.9 Distribution of plant formations and characteristic life forms according to mean annual temperatures and mean annual precipitation (after Whittaker, 1975; Körner, 2002, modified)

(Figure 8.9), albeit seasonal differences in temperature and precipitation are not depicted. In recent years, statistical relationships between species ranges and climatic data (climate envelopes) or between distribution of biomes and climatic threshold values are increasingly used in modeling studies to project the response to climate change or to reconstruct paleodistributions (cf. Hijmans and Graham, 2006; Prentice et al., 2007).

Favorable heat and moisture conditions in the tropics enhance net primary production and are the precondition for high species richness (Griffin, this volume). Due to the continually warm, moist climatic conditions the tropical rain forest (lower right, Figure 8.9) has the highest annual net primary productivity (Berlekamp et al., 2001) and contains the greatest amount of standing biomass of any biome (Whittaker, 1975). Moreover,

tropical forests harbor approximately 40% of the world's flora and contain the highest diversities of insects, birds, and other faunal groups. Extraordinarily high species diversity is particularly observed in areas with high geodiversity, that is, spatial heterogeneity of the environment (Mutke, this volume).

For a single plant, temperature determines the state of the protoplasm and the vital processes taking place therein. For instance, the rate of photosynthesis is dependent on temperature. Tropical plants often require temperatures greater than 20°C for efficient photosynthesis, whereas some arctic and alpine species may still conduct photosynthesis at temperatures around 0°C (Larcher, 1994). As poikilothermic organisms and being continuously exposed to the weather, plants have a limited tolerance for low temperatures, which slows down metabolic activity. Freezing cellular water causes tissue damage and other physical disruptions. The spatial and temporal distribution of low temperatures and freezing is thus an important determinant of plant distribution. Autochthonous species of the tropical rain forest biome, for example, have not developed any adaptations to periods of freezing that set a latitudinal limit to their distribution. By contrast, species in temperate regions often need chilling during winter in order to grow well the following summer. Arctic and alpine species manage to thrive in extremely cold, short-growing season habitats where they had to develop remarkable strategies to overcome environmental and biotic limitations to growth and reproduction (Crawford, 2008). Conspicuous vegetation limits such as polar and alpine treelines are mainly caused by heat deficiency, that is, insufficient air and soil temperatures during growing season and insufficient duration of growing season (cf. Holtmeier, 2009; Körner, 2007). Resistance of trees and tree seedlings at alpine treelines to freezing, frost drought, ice particle abrasion, and related phenomena is critical for their growth and survival. Range boundaries of species, communities, or biomes are often related to certain isotherms. The polar treeline lies very close to the 10°C July isotherm and serves well as an example for the approximate coincidence of vegetation limits and threshold temperatures (Kneeshaw et al., this volume). Range boundaries of single species may also show a remarkable correspondence to certain isotherms, exemplified, for instance, by the eastern range limits of *Fagus sylvatica* (-2°C January isotherm) or *Ilex aquifolium* (0°C January isotherm) in Europe (Walter and Straka, 1970). However, such correlations do not indicate direct causal relationships. As a rule, distributions are determined by multiple abiotic and biotic environmental factors and not by a single temperature line. In addition, extreme climatic events, especially their frequency and duration, may substantially influence survival and hence microdistribution of plants and animals. For instance, the Mediterranean oak species *Quercus ilex* tolerates a short period of extreme frost with temperatures as low as -20°C, but not a long period of -1°C (Stoutjesdijk and Barkman, 1992).

Broad distribution patterns of vegetation are governed even more significantly by the availability of moisture. First, water is essential to plant physiological processes. Only sufficient water content keeps the protoplasm physiologically active. Vital processes (e.g., photosynthesis, uptake and movement of nutrients, cell growth, protein synthesis, turgor maintenance, and transpiration) depend to a large extent on the degree of protoplasm hydration. Higher terrestrial plants,

able to maintain highly hydrated protoplasm independent of the moisture in the surrounding air (homoiohydric plants), are well adapted to changing water availability and have even penetrated into extreme desert regions. A wide variety of adaptations allow plants to conduct photosynthesis and to survive under dry conditions. However, only very few water stress tolerators can tolerate desiccation of tissue to the point of being nearly completely dry. With respect to their water economy, plants are classified into xerophytes showing many specialized adaptations to dry environments and mesophytes adapted to wetter and more shaded environments. The latter typically close their stomata when subjected to drought and temperature stress. Many xerophytes have small, often sclerophyllous leaves and are able to reduce their transpiration to a minimum when water is scarce. The drier the region, the further apart xerophytes grow. An extensive root system has to be developed in order to take up water from a larger soil volume. Whereas xerophytes require the presence of some water in the soil, succulents are able to survive long periods of drought without taking up any water. This subgroup of xerophytes is characterized by special water-storage organs (succulent leaves, stems, or roots). Many succulents separate the process of photosynthesis, conducted during daylight when stomata are closed to retard transpiration, from CO_2 uptake during nighttime when stomata are open (i.e., CAM photosynthetic pathway).

At a global scale, distribution patterns of mesophytes and xerophytes and respective biomes basically correspond to the precipitation conditions over land areas. Precipitation is spatially unevenly distributed. The amount of precipitation is determined by the nature of the air mass involved and the degree to which that air is uplifted. Continentality (precipitation) gradients are well reflected in vegetation patterns displaying a successive change from hygrophilous to more xerophilous floristic elements in continental interiors; for example, in Europe, the changing dominance from Atlantic species in the west to continental desert species in the east vividly illustrates decreasing nearness to moisture sources and increasing continentality (Figure 8.10). Moreover, seasonal precipitation patterns are fundamental to the understanding of distribution patterns of vegetation. For instance, central Europe and southern Europe receive about the same amount of annual precipitation. However, the more balanced seasonal precipitation regime with a summer maximum in central Europe contrasts with the pronounced dry summer climate of the mediterranean region. Accordingly, the floristic composition of deciduous forests, for example, beech forests that prevail in central Europe and

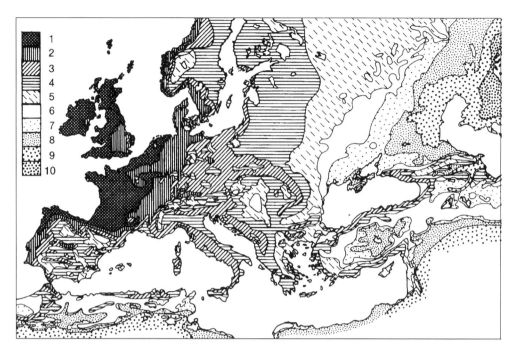

Figure 8.10   West-east increasing continentality of climate in Europe as reflected in the vascular plant species spectrum (based on the entire flora). 1 = oceanic and suboceanic species abundant; 2 = less abundant; 3 = to the eastern limit of many suboceanic species, subcontinental species already abundant; 4 = numerous subcontinental and some continental species, still a few suboceanic species; 5 = last occurrences of some suboceanic species, still many broadleaved tree and shrub species; 6 = many widely distributed continental species, steppe forest species prevalent; 7–9 = continental species increasingly abundant, increasing number of desert plants; 10 = continental desert plants prevalent (after Jäger, 1968; Ellenberg and Leuschner, 2010)

occur at higher elevations in southern Europe, differs between these regions. Mediterranean beech forests contain a considerable percentage of submediterranean and mediterranean elements adapted to the mediterranean seasonality (cf. Horvat et al., 1974).

The dominant control of climate on plant distribution is not only expressed in horizontal vegetation patterns. Mountain environments are characterized by steep climatic gradients over a very short vertical distance, causing a distinct altitudinal zonation of vegetation. The vertical thermal gradient represents a change in temperature conditions otherwise only observed over a vast latitudinal distance. Certain similarities between altitudinal belts and horizontal vegetation zones, however, may not obscure the fact that there are substantial differences between mountain climates and the climates of higher latitudes. For instance, the seasonless climate of the alpine belt of tropical mountains with almost uniform

temperatures throughout the year provides a strong contrast to the thermal and solar seasonal climate of the arctic tundra with the semiannual change from polar day to polar night. When ascending mountains of temperate midlatitudes, the colline (submontane), montane, subalpine, alpine, subnival, and nival vegetation belts can be differentiated according to structural patterns of plant formations (Figure 8.11). The timberline separates the subalpine from the alpine belt. It is not useful, however, to transfer these terms to mountains in other ecozones: for example, to arid mountains without any tree growth. Specific regional terms of altitudinal vegetation belts, as proposed, for example, by Richter (2001), are more suitable for global comparisons. At a global scale, the zonation into altitudinal vegetation belts is primarily a response to the altitudinal lapse rate of temperature. The approximately parallel course of the snow line, the upper timberline, and other altitudinal vegetation limits (Figure 8.12) and the

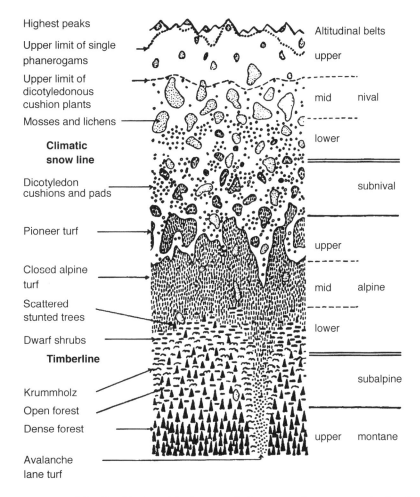

Figure 8.11   Upper altitudinal belts with characteristic vegetation patterns in the European Alps (after Ellenberg and Leuschner 2010, modified)

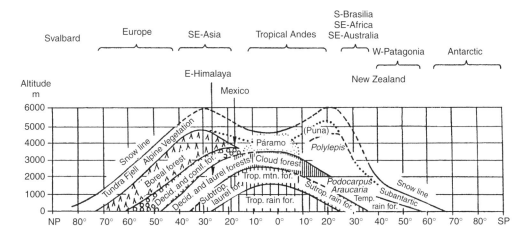

Figure 8.12   Latitudinal transect showing altitudinal vegetation belts and snow line (after Troll, 1948; Ellenberg, 1996; Holtmeier, 2009; modified)

correspondence to temperature conditions suggest that heat deficiency generally determines this altitudinal configuration. Snow line and timberline are highest in subtropical mountains and drop from there to the equator and to higher latitudes.

At regional and local scales, however, vertical moisture gradients, exposure and other effects—for example topography, soil properties, and biotic influences—may considerably influence the altitudinal zonation of vegetation. Sequence types of altitudinal vegetation belts north of the main Himalayan range in northern Pakistan (Figure 8.13) illustrate the impact of a humidity gradient at a regional scale. The altitudinal vegetation zonation corresponds to the spatial differentiation of the influx of monsoonal humidity. The high mountain chains of the West Himalaya and Karakorum act as barriers against moist monsoonal air masses from the south, resulting in a gradual decrease of humidity along a SW-NE gradient. Accordingly, closed coniferous forests of the upper montane belt (subhumid sequence) become restricted to northern aspects (semihumid sequence), and give way to open juniper forests (semiarid sequence), to *Artemisia* shrublands with isolated trees (subarid sequence), and finally to dwarf scrub and steppes (eu-arid sequence) toward the inner mountains (Schickhoff, 2000).

Exposure differences in altitudinal belts (cf. Figure 8.13) already indicate that topo-and microclimates may differ considerably from the regional climate and cause respective variation in vegetation patterns. As a local variant of the macroclimate, topoclimate describes effects due to the topography of the landscape. Microclimate, on the other hand, refers to conditions in the lower two meters of the atmosphere and the upper 0.5 to one meter of the soil, strongly influenced by the condition of the soil surface and the vegetation (Stoutjesdijk and Barkman, 1992). The climate plants experience at the ground surface or within the vegetation canopy is quite different from that measured by a climate station. The microclimate at the south side and at the north side of a dense scrub in temperate regions may show the same magnitude of difference over a few meters as the macroclimate does over several thousand kilometers.

Particularly above the timberline in high mountains, small-scale distribution patterns of plants are largely determined by the combination of topo- and microclimatic effects. Even small topographic irregularities and differences of slope angle and aspect affect radiation, temperature, evaporation, wind speed, and snow accumulation, resulting in marked contrasts in structure and floristic composition of plant communities. Moreover, vegetation cover itself modifies microclimatic

parameters to a large extent. Vegetation patterns of an alpine dwarf scrub heath (Figure 8.14) vividly illustrate these complex interrelationships: effects of topography and microclimate modify other abiotic and biotic site factors so that a small-scale mosaic of microsites with differing habitat conditions is created, regardless of the homogeneous macroclimate. The microsite pattern induces a characteristic vegetation pattern with some species indicating distinct microhabitat conditions (ecological indicator species). Striking contrasts between north-and south-facing slopes in many midlatitude and subtropical high mountains indicate that the most obvious topoclimatic effects must be attributed to exposure. Differences in solar radiation and related site conditions (e.g., thermal regime, humidity, duration of snow cover, and soil moisture) often cause a characteristic pattern of species distribution at northern and southern aspects. For example, closed coniferous and birch forests on north-facing slopes in the upper montane/subalpine belt of the Karakorum Mountains benefit from the more equable topoclimate, the long-lasting snow cover, and the favorable soil water budget. By contrast, juniper trees on south-facing slopes, becoming free of snow three to four months earlier, have to cope with a combination of unfavorable climatic effects such as high irradiation, high evaporation, and drought stress (Schickhoff, 2005). In general, the altitudinal position of timberlines in northern hemisphere subtropical, temperate, and boreal high mountains is higher on south-facing slopes. In the equatorial zone, contrasts are more pronounced between east-facing slopes, exposed to the sun for the entire morning, and west-facing slopes, which are often rainy in the afternoon.

In contrast to physical controls on the distribution of plants such as temperatures and moisture, the amount of sunlight varies to a far lesser extent at the surface of the Earth. There is no region where plant growth would not be possible due to the lack of light intensity. Even in polar regions with the annual rhythm of polar day and polar night, sparse or missing plant cover is not a response to lack of light but to unfavorable temperatures. The more homogeneous distribution of light implies that differences in light intensity affect the large-scale geographical distribution of biota only to a limited extent. By contrast, the differing amount of sunlight may heavily influence small-scale species distributions. Floristic differences between sunny and shady sites reflect deviating light conditions as well as related microclimatic effects (i.e., heat and moisture). As for the light requirements, plants are basically differentiated into heliophytes, which grow best in full sunlight, and sciophytes, which grow best in shade. The latter are marked by low

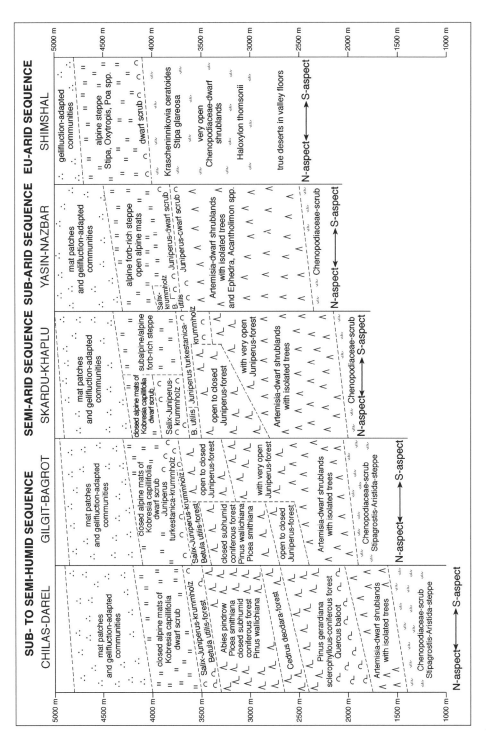

Figure 8.13 Humidity-dependent sequence types of altitudinal vegetation zonation in the East Hindukush and Karakorum. Mean annual precipitation in montane/subalpine belts of sequences: sub- to semihumid (350–700 mm), semiarid (200–350 mm), subarid (150–300 mm), eu-arid (80–200 mm) (after Schickhoff, 2000)

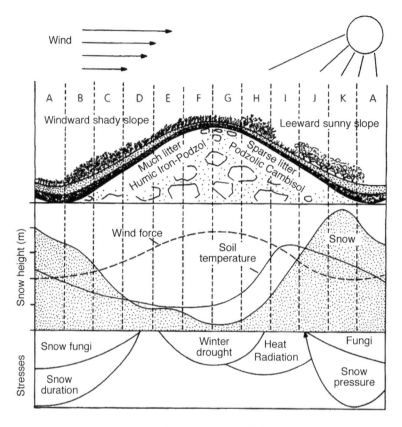

Figure 8.14   Vegetation pattern of an alpine dwarf scrub heath in the European Alps as a response to topography, microclimate, and further interrelated abiotic and biotic site factors. A: shallow hollow with grasses, mosses and Soldanella; B: Rhododendron ferrugineum scrub; C: Rhododendron ferrugineum scrub with Vaccinium myrtillus; D: Vaccinium uliginosum scrub; E: Loiseleuria procumbens heath; F: Lichen heath with bare spots due to wind erosion; G: open vegetation with rosette and cushion plants and Juncus trifidus; H: Arctostaphylos uva-ursi—Vaccinium vitis-idaea scrub; I: sun bare spot; J: Juniperus communis ssp. alpina scrub with Calluna vulgaris, Vaccinium vitis-idaea; K: Rhododendron ferrugineum scrub with Juniperus communis ssp. alpina (after Aulitzky, 1963; Larcher, 1994; modified)

light saturation levels (5–10% of full sunlight or even less), allowing them to survive on very shady sites. Sciophytes, typically found on the floor of dense forests, are almost always perennials, whereas annual plants require high light intensities to complete their life cycles within a single year. The seeds of many plant species will only germinate when exposed to sufficient levels of light intensity.

A basic principle of climate-related plant distribution patterns refers to compensatory effects of small-scale habitat conditions that enable plants to grow outside their preferred climatic zone. Hygrophilous species, for instance, may extend their range to areas with less precipitation as long as the groundwater table is high. Continental

elements growing in zonal habitats in eastern Europe are restricted to dry south-facing slopes at their western range margin since they depend on a favorable habitat under more humid conditions. Boreal species avoid higher temperatures at southern range margins and switch to higher elevations. Thus, the less favorable the regional climate, the more plants attempt to compensate for this climatic alteration by a habitat change in order to keep habitat conditions relatively constant. This ecological law of 'relative habitat constancy and changing biotope' (Walter, 1973) applies particularly to distribution patterns in mountain environments. If temperature and moisture conditions remain unmodified by topography or soils, that is, if the natural vegetation grows in correspondence with the macroclimate, it is termed zonal vegetation. This vegetation can occur as extrazonal vegetation in favorable habitats in other climatic zones, for example, when steppe vegetation is encountered in the more humid forest zone on dry southern slopes or on porous limestone soils. The term azonal vegetation is applied to vegetation for which edaphic factors (e.g., soil moisture) are more important than the climate, and which occurs in several climatic zones (e.g., vegetation of lakes and saline soils).

Mapped geographical distributions of plants, classified into types of characteristic range configurations, provide indirect indications of climatic conditions. Several patterns of distribution (location and extension of range boundaries) frequently recur and indicate a geographical alliance among taxa. Such distributions of species can be grouped together, even if range size and extent vary and boundaries often do not coincide, and they are regarded as members of the same geographical element (geoelements according to Meusel et al., 1965 and Walter and Straka, 1970). Geoelements can be used as a tool for the climate-ecological interpretation of site conditions by constructing floristic-chorological spectra of plant communities based on the evaluation of species lists. An example of a geoelement-based climate indication in the Kaghan Valley, located on the south slope of the West Himalaya, Pakistan, is provided in Figure 8.15. Floristic-chorological spectra of two plant communities are compared: the *Pinus wallichiana-Androsace rotundifolia* community is distributed on upper montane south-facing slopes of the lower valley, characterized by high humidity, high annual precipitation, and a pronounced monsoonal influence. The *Artemisia brevifolia-Juniperus excelsa* community occurs on upper montane south-facing slopes of the upper valley under increased thermal and hygric continentality in the rain shade of the outer mountain chains (Schickhoff, 1996). The contrasting range types of participating species reflect this climatic gradient and the changing site conditions. The proportion of hygrophilous monsoon elements (1a, 1b) considerably decreases in the upper Kaghan, whereas species strongly increase that are also distributed in the Karakorum (1c), in Central Asia (2), and in the Turkestanian subregion (3a). Thus, the upper Kaghan already shows intensive floristic relationships to regions beyond the main Himalayan range that are indicative of the changing climate and landscape character along the valley profile.

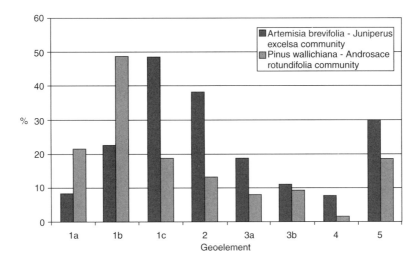

Figure 8.15   Change of the floristic-chorological spectrum of plant communities from humid lower to semiarid upper Kaghan Valley, exemplified by communities from upper montane south-facing slopes. Distribution types of the Kaghan flora: 1a: only on the south slope of the West Himalaya (East Afghanistan to Central Nepal); 1b: also in the East Himalaya (East Nepal—SE Tibet—SW China eastward); 1c: also in the Karakorum and NE Hindukush; 2: also in the Central Asian region; 3a: also in the Turkestanian subregion; 3b: also in the Mediterranean, Oriental, or Turanian subregions; 4: also in the Saharo-Sindian region; 5: all over temperate-boreal-arctic Eurasia or the northern hemisphere or cosmopolitan (after Schickhoff, 1993; modified)

8.3.2 Soils

Soils represent one of the complex secondary site factors (cf. Figure 8.8) that exert indirect influences on plants via primary site factors such as chemical factors, soil moisture, and heat. Effects of soils on plant growth and distribution patterns are closely interrelated to effects of other abiotic and biotic factors since soil formation relies on climate, organisms, topography, parent material, and time. Soil formation starts with physical and biochemical weathering of parent rock and the accumulation of organic material from dead and decaying organisms. During the course of a primary succession, complex interactions between temperature and moisture conditions, the nature of the parent material and topography, and the development of microbial, plant, and animal assemblages and their activities determine further soil development. Thus, the rate of soil formation varies, and different pedogenic regimes (laterization, podzolization, gleization, calcification, and salinization) produce various soil types with differentiating properties (Brady and Weil, 2007). Soil types can basically be differentiated into pedocals of semiarid and arid regions and pedalfers of humid regions. Global distribution patterns of soil types are complex, but major (zonal) soil types show a close correlation with climatic zones. Within the world's zonal biomes, however,

areas of intrazonal and azonal soils (e.g., saline soils or soils that have little or no profile development) exist that are characterized by distinctive vegetation. Such areas with soils that do not correspond to climatic (zonal) soil types are referred to as pedobiomes (Walter and Breckle, 1985). Pedobiomes may cover extensive areas on stony, sandy, saline, water-logged, or nutrient-poor soils, but they commonly form a mosaic of small areas. In some geographical regions, for example, dryland zonal soil types may be distributed over vast areas, but in others, especially mountains, they are characterized by a small-scale mosaic of diverse soil types.

At landscape and local scales, soil physical and chemical properties often exert a strong control on plant distribution and floristic composition of communities. Differences in soil texture, structure, mineralogy, water content, and chemistry are commonly reflected in a changing composition of vegetation types. Soil moisture and nutrient richness, which is more or less equivalent to soil alkalinity, figure most prominently among edaphic factors differentiating vegetation in climatically homogeneous areas. The ecogram depicting the distribution of broadleaved forest communities in the submontane belt of central Europe (Figure 8.16) is a classic example of the importance of soil moisture and nutrient status (acidity conditions) for the competitive capacity

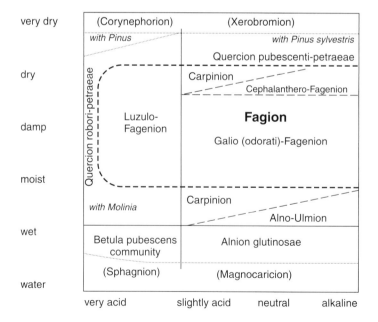

Figure 8.16   Distribution of central European submontane broadleaved forest communities according to soil moisture status and soil acidity (after Ellenberg and Leuschner 2010; modified)

of forest-forming tree species and the distributional range of respective forest communities (Ellenberg and Leuschner, 2010). The alliance of beech forests (phytosociological term: Fagion sylvaticae) is separated into true beech forests (Galio-Fagenion) occupying zonal habitats that are damp to moist and slightly acid to alkaline, woodrush-beech forests (Luzulo-Fagenion) tolerating more acid conditions, and orchid-beech forests (Cephalanthero-Fagenion) occurring on relatively dry, calcareous slopes. Mixed oak forests (Quercion pubescenti-petraeae) occupy even drier habitats and occur at the drought limit of forests, whereas birch-oak forests (Quercion robori-petraeae) are forced onto very acid, dry to wet sites. Oak-hornbeam forests (Carpinion) occur on slightly acid to neutral soils being too dry or too wet for beech forests. Along the moisture gradient on neutral to alkaline wet soils, alder-elm-ash forests (Alno-Ulmion) give way to alder swamp forests (Alnion glutinosae) that occur along the moisture limit of forests, together with birch swamp forests (*Betula pubescens* community within Piceo-Vaccinienion uliginosi) on acid to very acid soils.

The decisive role that soil moisture and soil nutrient levels play in differentiating the vegetation of a region has been elaborated in numerous vegetation ecological studies. Even slight variations in soil moisture and acidity often result in floristically distinct vegetation types. In phytosociological terms, associations can often be differentiated into subassociations or variants along soil moisture and acidity gradients. The ordination analysis of riparian willow communities on the arctic slope of Alaska (Figure 8.17) serves as an example. Spatial patterns and floristic composition of these riparian *Salix* shrublands are primarily controlled by a complex edaphic gradient representing soil pH and soil moisture (Schickhoff et al., 2002). The pioneer community Epilobio-Salicetum alaxensis indicates sites with frequent disturbances and coarser-textured, relatively dry, initial alluvial soils with basic reaction along mountain creeks (subassociation polemonietosum acutiflori; average soil pH: 8.8; average soil moisture: 25.4% by volume) and on gravel bars, floodplains, and lower terraces of rivers (subass. parnassietosum kotzebuei; pH 6.6; 23.2%). With decreasing river influence on higher terraces, decreasing active layer depth, higher soil moisture, and the transition to finer-textured, more nutrient-rich, less basic soils, the Anemono-Salicetum richardsonii (subass. lupinetosum arctici; pH 6.8; 45.1%) replaces the pioneer association. Another subassociation of the Anemono-Salicetum (salicetosum pulchrae; pH 5.8; 53.1%) occurs on upland tundra stream banks where environmental conditions and floristic composition

already lead up to the Valeriano-Salicetum pulchrae. This association is distributed along smaller tundra streams and creeks on paludified acid soils with massive ground ice and thick moss layers, resulting in decreased depth of thaw and increased soil moisture (pH 5.0; 61.7%).

According to the occurrence along the gradient of soil acidity and lime content, plant species are commonly differentiated into calcicoles (or calciphiles) and calcifuges (or calciphobes). Calcicolous species are basic reaction and lime indicators, often growing only on calcareous soils. By contrast, calcifuges are mainly found on acid soils. However, plants do not respond to the calcium content of the soil, but to soil reaction. Calcicolous species may occur on alkaline soils without free calcium, for example, on gypsum or basalt soils; therefore it is more appropriate to use the terms basophytic and acidophytic species (Pott and Hüppe, 2007). Some species grow over a wide soil pH range, for example, *Pinus sylvestris* (Scots pine) or *Juniperus communis* (common juniper) occurring on calcareous soils and very acid sandy soils. These eurytopic species contrast with stenotopic species only growing in a very narrow soil pH range (e.g., *Erica tetralix* [bog heather] on acid substrates). The ecological behavior of species, however, reflects field conditions including the competition of many other plants. Most species show a broader physiological amplitude (without competition) compared to the ecological amplitude. Another common distinction is made according to species occurrence along a gradient of available nitrogen. Nitrophiles, indicators of sites rich in available nitrogen such as *Sambucus nigra* (elder), *Urtica dioica* (common nettle), or *Galium aparine* (goosegrass) in Europe, are favored by the increasing pollution levels of airborne mineral nitrogen in recent decades. Nitrophytic species constitute the 'Lägerflora' on excessively fertilized sites such as ruderal or cattle-resting places. On the other hand, species indicating nutrient-poor soils (e.g., typical species of oligotrophic grassland communities) are on the decline.

Special physiological or morphological adaptations allow species to colonize intrazonal and azonal soils and cope with their peculiar physical structure or unusual chemical composition. Examples include adaptations to dryland soils (reduction of transpiring surfaces, thick cuticles, small and sunken stomata, succulence), to waterlogged soils or submergence (aerenchyma, anoxia tolerance), to rocky or mobile substrates (taproots, elastic roots, potential to regenerate below-ground organs), to serpentinite soils or sites rich in heavy metals (immobilization of metal ions in cell walls, absorption in cytoplasm, active excretion). Special adaptations are also required by halophytes

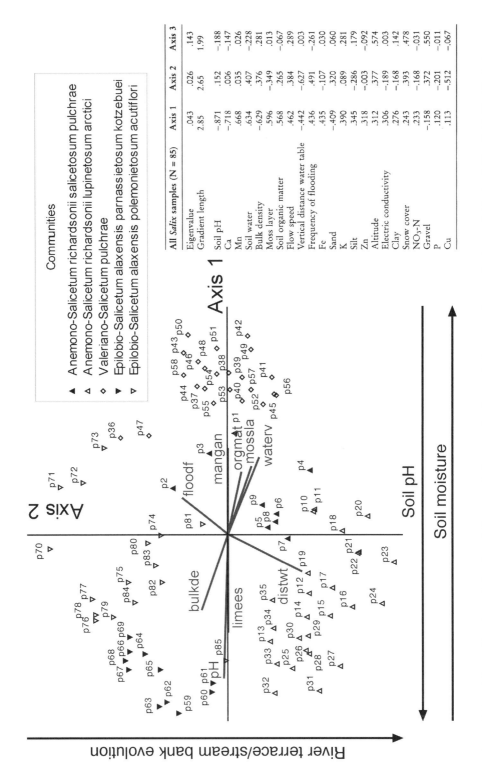

Figure 8.17   DCA ordination of riparian Salix associations and subassociations on the arctic slope of Alaska (after Schickhoff et al., 2002; modified)

(i.e., plants growing on saline and halomorphic soils, which contain high concentrations of sodium, chlorides, and sulfates). Halophytes are adapted to widely varying external and internal concentrations of salt. Mechanisms that prevent cytoplasmic salt concentrations from exceeding tolerance levels include (a) ion accumulation and increasing cellular osmotic potential in order to restrict or exclude salt uptake; (b) salt dilution in cells by the accumulation of water (succulence); and (c) salt extrusion via salt glands or by the shedding of organs that have accumulated salt. Saline and halomorphic soils typically occur at coastal sites, in estuaries and salt marshes, and in arid inland basins where salts accumulate in the topsoil due to high evaporation (Strauss and Schickhoff, 2007). Halobiomes may cover extensive areas, for example, in tropical estuaries and deltas supporting mangrove formations.

Corresponding to the steep gradient of salinity in the transition zone from saline to nonsaline habitats, it is common to find a characteristic vegetation zonation. The coastal habitats in the East German part of the Baltic Sea, one of the largest brackish water systems in the world, are a typical example. In the Prerow Bay, Arkona Basin, and Pomeranian Bay regions, a landscape that has been formed by glacial deposits drowned due to Holocene eustatic sea level rise and which is now characterized by numerous bays and lagoons with shallow water bodies, a distinctive vegetation sequence from outer dune coasts to inner lagoon coasts can be observed (Figure 8.18). The successional stages of white, gray, and brown dunes show distinct edaphic conditions and species assemblages. Approaching the coastal wetlands and grasslands at the inner lagoon coast, the change to halomorphic soils is reflected in the transition to halophytic vegetation. The potential natural reed belt at flat or gently sloping shorelines has been largely transformed to salt grasslands in recent centuries under the influence of cattle grazing and aperiodic flooding. Somewhat similar vegetation sequences are developed on the Wadden Sea islands at the North Sea coast when proceeding from the shoreline of the

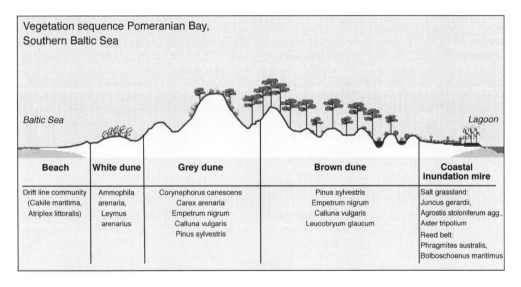

Figure 8.18 Vegetation sequence from outer dune coasts to inner lagoon coasts in the Pomeranian Bay, southern Baltic Sea. The pioneer vegetation of sandy beaches consists of salt-tolerant, nitrophilous species along drift lines of dead algae and eelgrass. Pioneering dune vegetation facilitates the formation of embryo dunes, which grow up to white dunes (also referred to as yellow dunes), characterized by initial freshwater lenses and nitrogen accumulation. Older white dunes turn into gray dunes, now stabilized by a dense plant cover that enhances humus enrichment and further soil development. The gray dune stage is marked by a cover of mosses and lichens and/or poor grasslands. Later, dwarf shrubs replace typical gray dune plants and initiate the brown dune stage, in which pine woodlands dominate on further developed, often podzolic soils. The sequence ends with halophytic grasslands or reed at the inner lagoon coast

outer coast over white, gray, and brown dune complexes to the salt marshes of the landward coasts (Doing, 1983; Schickhoff and Seiberling, 2003).

8.3.3 Disturbances and stresses

Among primary site factors (cf. Figure 8.8) a group of factors can be subsumed under the category 'disturbances and stresses', which exert influence on local and geographical distributions of species via mechanical impact. Mechanical factors include damages and injuries caused by herbivory (see below), mowing, snow load, wind, fire, ice and sand corrasion, avalanches, mudflows, and landslides. Differences in exposure to persistent or periodic mechanical stress are reflected in ˜lant distributions and floristic composition of communities. For instance, special adaptations to the abrasive and shaping effects of wind (plant stature, canopy structure) allow plants to thrive in wind-exposed habitats, such as wind edges in alpine terrain where the only growth forms are attached to the ground and thus decoupled from atmospheric conditions (cushion plants, prostrate or creeping plants) are competitive. Likewise, mechanical influences of snow (snow pressure, snow creep, sliding snow, avalanches) require adapted growth forms. For example, at the upper treeline of the West Himalaya, growth forms of birch (*Betula utilis*) become stunted, depressed, and deformed. On steeper slopes and along avalanche lanes, snow creep and sliding snow induce strong stem deformation leading to the characteristic curved birch stems. The ability of developing growth forms to adjust to mechanical stresses of snow is a major factor for the higher competitiveness of birch at the upper treeline compared to conifers (Schickhoff, 2005).

Mechanical influences on species distributions that result from sudden widespread disturbance and destruction are referred to as 'disturbances'. The effects of disturbance events such as fire (see Enright, this volume), lava flows, flash floods, droughts, landslides and other mass movements, hurricanes, other wind and ice storms, insect outbreaks, burrowing and grazing animals. and the like are potentially dramatic, including the complete destruction of habitats and their inhabitants. Disturbances cause spatial heterogeneity and patchiness; influence competition, environment, and substrate and resource availability; and are thus major shaping forces for composition, structure, and functioning in ecosystems (White and Jentsch, 2001). Since life-history attributes evolved in response to disturbances (e.g., fire resistance), disturbances have also been evolutionary forces.

Responses of biota to disturbances are extensively summarized in Jentsch and Bierkuhnlein, this volume.

8.4 BIOTIC INTERACTIONS AND PLANT SPECIES DISTRIBUTIONS

Biotic interactions such as competition, cooperation, or herbivory represent a complex secondary site factor (cf. Figure 8.8) the influence of which on geographical ranges and population densities of plant species is occasionally underestimated. It is often assumed that species distributions are more controlled by physical factors and that the presence or absence of competitors only plays a minor role. However, with the exception of range limits in arid, cold or salt deserts, where single plants grow isolated, physical factors are only indirectly effective by affecting the competitiveness of species. The most conspicuous evidence for the importance of interspecific interactions in limiting distributions arises from the fact that plant species can thrive far beyond their geographical range if protected from the competition of other plants, animal herbivores, and pathogens. For instance, the northern range limit of the evergreen Mediterranean holly oak (*Quercus ilex*) is located in the southern Rhone Valley, but cultivated trees survive in botanical gardens in Bonn, Copenhagen, or Leipzig. Beech (*Fagus sylvatica*) still grows in the botanical gardens of Kiev and Helsinki, whereas competition restricts the northeastern range limit to the Weichsel region in Poland (Walter and Breckle, 1999).

Biotic interactions are common in all ecosystems and occur between competing individuals of the same species (intraspecific) as well as between individuals of two or more species (interspecific) (Table 8.1).

Competition is the mutually inhibiting influence of individuals of a site that use the same resources (in particular, light, soil moisture, nutrients). If essential resources are in sufficiently short supply, competing plants suffer from reduced growth, fertility/reproduction, and survival. Intraspecific competition maintains competitive species populations by sorting out weaker individuals of a species so that strong individuals remain. Interspecific competition is accompanied by the suppression of less competitive species and may even result in competitive exclusion. However, one species can only outcompete another and cause its extinction in the case of a very strong competitive superiority. In late successional plant communities, mixed stands commonly develop in which species proportions correspond to their competitive capacity.

Table 8.1 Interspecific biotic interactions between two species (A and B). The species influences' on each other are positive (+), negative (–), or nonexistent (0) (modified from Körner, 2002)

A > B	B > A	Kind of interaction
–	–	Competition
+	–	Parasitism, herbivory
+	+	Symbiosis (mutualism, cooperation)
+	0	Commensalism
0	–	Amensalism
0	0	Neutralism

An experiment conducted with a varying groundwater table and three central European meadow grasses (Walter, 1960) vividly illustrated the decisive role of competition for the distribution of plant species. Meadow foxtail (*Alopecurus pratensis*) prevails in moist grasslands, oatgrass (*Arrhenatherum elatius*) dominates under medium soil moisture, and brome grass (*Bromus erectus*) grows in dry grasslands. However, if sown in pure culture on plots with varying groundwater levels from 0 to 150 cm, all three grasses showed a growth optimum at an intermediate groundwater depth. Differences in their ecological behavior became only apparent when sown in mixture (i.e., under conditions of competition). Oatgrass as the most competitive species occupied the optimum range of intermediate groundwater depth and displaced meadow foxtail to the wetter part of the gradient and brome grass to the dry one. The same principle applies to soil reaction. Species that occur in a narrow pH range under stiff competition may thrive over a wide pH range without competition. Thus, the ecological behavior under field conditions has to be distinguished from the physiological behavior, that is, the optimum and total tolerance range of the same taxon without competitors. It follows that geographical ranges of species do not necessarily indicate their physiological requirements.

Competitive interactions are commonly differentiated into resource exploitation in which species or individuals use up resources and make them unavailable to others and into direct interference competition in which physical interaction or active inhibition restricts other individuals to get access to and use resources. However, exploitation is often turning into interference competition and effects of both kinds of interaction may be separated only with difficulty (Keddy, 2007). Exploitation competition (i.e., plants reducing light, soil moisture, or nutrients available to other plants) is well documented. For instance, resource

competition affects the growth conditions in forests with multiple canopy layers. The upper tree layer dominates the capture of light, and lower tree, shrub, and field layers can only intercept whatever is transmitted downward through the canopy. The importance of light competition is exemplified by the spring-flowering geophytes in temperate deciduous forests, which use the temporal niche before the frondescence of trees when the forest floor is heated by irradiation in order to develop from their below-ground perennial buds and complete growth, flowering, and seed set within a few weeks. Tree layer individuals are also superior to field layer individuals in competing for soil moisture since tree roots have a higher suction pressure compared to herbaceous plants. During prolonged drought periods, the fine roots of trees may withdraw water from the soil to the extent that the uptake of water by herbaceous plants suddenly declines and the plants begin to wilt. Likewise, tree roots are more efficient in nutrient uptake. Tree root competition is significant, for instance, in spruce (*Picea abies*) forests. 90% of the spruce root biomass is located together with the roots of the field layer species in the upper 20 cm of the soil profile. Root-trenching experiments showed that tree roots use up most of the available nitrogen and cause the insufficient nitrogen supply of the undergrowth and that nitrogen uptake of herbaceous plants like wood sorrel (*Oxalis acetosella*) considerably increases when spruce root competition is eliminated (Walter, 1986).

Interference competition, for example, plant species interfering with competitors by chemical means (allelopathy), is probably less important. Some plants exude chemicals that are deleterious to other organisms and may serve to inhibit germination of competitors. Although the effectiveness of allelopathic interactions is still contentious among some plant ecologists, there are numerous case studies showing that allelopathy plays a significant role in plant dominance, succession, and formation of plant communities (Reigosa et al., 2006). For instance, the conspicuous inhibition of undergrowth in *Eucalyptus* plantations or in stands of *Juglans*, *Robinia*, and some conifers is attributed to allelopathic impacts on competing species.

Biotic interactions between plants and animals such as herbivory or parasitism, in which the plant species suffer and the animal species benefit, potentially cause considerable changes in plant cover and plant species distribution. In most cases, herbivores (i.e., animals that feed on living plants) act to the plant's detriment by grazing, browsing, rasping, sucking, mining, etc., regardless of being invertebrate or vertebrate. The percentage of net primary productivity (NPP), which is consumed by herbivores, varies with vegetation: 2 to 3% for desert scrub or arctic/alpine tundra, 4 to 7% for forest, 10 to 15% for temperate grasslands with minimal grazing, and 30 to 60% for African grasslands (Barbour et al., 1998). The existence of specialized herbivore niches (pollen eaters, nectar eaters, nut eaters, fruit and berry eaters, leaf eaters, grain and seed eaters, grazers, and browsers) indicates that all parts of a plant provide important food sources, including those which are essential for plant dispersal. In noninteractive herbivore-plant systems (seed predation), for example, birds eating plant seeds, the long-term survival of a plant population may be influenced by the consumption of a large portion of the annual seed production. In interactive herbivore-plant systems (grazing), herbivores affect plant growth as well as the plants affecting the herbivores through plant abundance, quality, and distribution (Huggett, 2004).

Effects of herbivory on species distributions are often restricted to structural changes in plant communities, for example, changes in vertical stand structure or shifts in species compositions. Stand structural changes are clearly detectable where a fence separates grazed from ungrazed areas. However, continued grazing may also result in range contractions of host plants and ultimately in landscape changes. Grazing animals have particularly detrimental effects on the young growth of trees and shrubs and thus indirectly favor easily regenerating grassland species. In mixed forests, selective grazing pressure leads to the increase of coniferous trees at the expense of broadleaved tree species, which are preferentially browsed. Under intense grazing pressure, replacement communities evolve in the course of regressive successions (retrogressions) that largely consist of unpalatable, toxic, and often thorny or spiny species. The cultural landscape in central Europe is inter alia the result of many centuries of grazing by cattle, horses, goats, and pigs, which has eliminated forests and prevented reforestation. Uncontrolled grazing continues to cause vegetation degradation and landscape changes in many parts of the world, in particular in mountains and drylands. For instance, long-term grazing-induced failure of tree regeneration has contributed to the depression of the upper timberline on many Himalayan south slopes by several hundred meters (Schickhoff, 2005). In response to the selection pressure caused by herbivores, plants have developed various effective antiherbivore defense mechanisms including structural adaptations such as thorns, spikes, stinging hairs, raphides, as well as the production of many distasteful and poisonous chemical deterrents.

Parasitism is another biotic interaction that potentially alters the geographical distribution of species. Parasites wholly depend on other organisms for nutrients, normally causing harm but not causing death immediately. The geographical ranges of many specific parasites often correspond to a large extent with those of their plant host. At the same time, parasites continuously affect local abundances and geographical distributions of host plants. Attacks of parasites, including pathogenic microbes and fungi, on tree species may lead to an extensive loss of populations and thus to considerable vegetation changes. Probably the largest change wrought in the structure of communities by a parasite has been the elimination of the chestnut (*Castanea dentata*) in North American forests, where it had been a dominant tree over large areas until the introduction of the fungal pathogen *Endothia parasitica* from China (Begon et al., 2006). The European field elm (*Ulmus minor*) has suffered large population reductions in recent decades from the attack of another fungal pathogen, *Ophiostoma ulmi* (for further details, see Uzonov, this volume). The effectiveness of parasites in controlling species distributions is also used to regulate pest populations. For example, only the introduction of the South American cactus moth (*Cactoblastis cactorum*) in Australia kept the fast spreading populations of prickly pear cactus (*Opuntia stricta*) under control, an invasive species introduced from America as a hedge plant.

Likewise, symbiotic interactions (cooperation, mutualism, and commensalism) can have particular relevance for the geographical distribution of plant species. For instance, relationships between plants and pollinators often tightly control the distribution of the involved plants and insects or birds. Moths of the genus *Tegeticula* feed only on *Yucca* plants in North American deserts. In turn, the *Yucca* depend on the moth to transfer their pollen (MacDonald, 2003). In this case, neither species is able to extend its range into areas that cannot support the other. Mutualistic relationships that are obligatory for one or both partners imply that these interactions must have a significant influence on distribution. The symbiosis between neotropical *Acacia* trees and aggressive ants of the genus *Pseudomyrmex* is an apposite example. The trees provide the ants with microhabitats and specialized foods rich in sugars, oils, and proteins, whereas the ants attack herbivorous predators and clear away competing vegetation. Similar associations exist between *Cecropia* trees and ant species. The coevolved specializations between these interdependent mutualists resulted in virtually identical ranges (Lomolino et al., 2006). However, many mutualistic relationships are not obligately species-specific, at least outside the tropics, that is, several different species may supply the same service as in the case of mycorrhizae and temperate zone trees. Thus, the influence of symbiotic interactions on species ranges is often less obvious.

8.5 OUTLOOK: DISTRIBUTION PATTERNS AND CONTEMPORARY BIOGEOGRAPHY

Explanations of the geographical distributions of biota are far from complete as long as anthropogenic interferences are not considered. Current changes in distribution patterns of plants and animals result primarily from human activities. Anthropogenic direct drivers such as habitat change, loss and fragmentation, introduced alien species, exploitation, rapid climate change, pollution, and pathogens cause biodiversity loss and often range declines in species and populations. It is thus one of the major research fields of contemporary biogeography and concurrently of fundamental importance for effective management and conservation to understand and predict changes in distribution patterns of species, communities, and formations caused by ongoing pervasive human impacts. A great number of recently published biogeographical studies use bioclimatic modeling approaches to generate and test hypotheses of shifting species distributions as influenced by global environmental change.

At the same time, however, there is another major recent development: the revitalization of the interest in plant disjunctions, endemism, and other distribution patterns, driven by the availability of new phylogenetic methods and the increasing application of phylogeographical approaches. Phylogeography has contributed significantly to the emergence of biogeography in recent decades and continues to play an important role through giving fresh impetus to exploring the diversity of distribution patterns and their causal forces. The seminal relevance of both emerging approaches— bioclimatic modeling and phylogeography—that can also be combined is reflected in the exponentially increasing number of publications in respective journals as well as in the programs of contemporary biogeography meetings and conferences. Explaining the geographic distribution of species continues to be a core issue for contemporary biogeography.

REFERENCES

Archibold, O.W. (1995) *Ecology of World Vegetation*. London: Chapman and Hall.

Aulitzky, H. (1963) 'Grundlagen und Anwendung des vorläufigen Wind-Schnee-Ökogrammes', *Mitteilungen der Forstlichen Bundesversuchsanstalt, Mariabrunn*, 60: 765–834.

Avise, J.C. (2009) 'Phylogeography: Retrospect and prospect', *Journal of Biogeography*, 36: 3–15.

Barbour, M.G., Burk, J.H., Pitts, W.D., Gilliam, F.S., and M.W. Schwartz. (1998) *Terrestrial Plant Ecology*. 3rd ed. Menlo Park, CA: Benjamin Cummings.

Begon, M., Townsend, C.R., and J.L. Harper. (2006) *Ecology: From Individuals to Ecosystems*. 4th ed. Oxford: Blackwell.

Beierkuhnlein, C. (2007) *Biogeographie*. Stuttgart: Eugen Ulmer.

Berlekamp, J., Stegmann, S., and H. Lieth. (2001) 'Global net primary productivity maps' (http://www.usf.uni-osnabrueck.de/~hlieth).

Born, J., Linder, H.P., and P. Desmet. (2007) 'The Greater Cape floristic region', *Journal of Biogeography*, 34: 147–62.

Box, E.O. (1981) *Macroclimate and Plant Forms: An Introduction to Predictive Modeling in Phytogeography*. The Hague: Dr. W. Junk.

Box, E.O. (1996) 'Plant functional types and climate at the global scale', *Journal of Vegetation Science*, 7: 309–20.

Brady, N.C., and R.R. Weil. (2007) *The Nature and Properties of Soils*. 14th ed. Upper Saddle River, NJ: Prentice-Hall.

Brown, J.H. (1995) *Macroecology*. Chicago: University of Chicago Press.

Buffon, G.L.L. (1761) *Histoire naturelle, generale et particuliere*. Paris: Academie Française.

Candolle, A.P. de. (1820) 'Essai elementaire de geographie botanique', in *Dictionnaire des sciences naturelles*, vol. 18. Strasbourg: Flevrault.

Candolle, A.P. de. (1855) *Géographie botanique raisonnée ou exposition des faits principaux et des lois concernantes la distribution géographique des plantes de l'epoque actuelle*. 2 vols. Paris: Masson.

Cox, C.B., and P.D. Moore (2005) *Biogeography: An Ecological and Evolutionary Approach*. 7th ed. Oxford: Blackwell.

Crawford, R.M.M. (2008) *Plants at the Margin: Ecological Limits and Climate Change*. New York: Cambridge University Press.

Dansereau, P. (1957) *Biogeography: An Ecological Perspective*. New York: Ronald Press.

Darwin, C. (1859) *On the Origin of Species by Means of Natural Selection, or the Preservation of Favoured Races in the Struggle for Life*. London: John Murray.

Daubenmire, R.F. (1978) *Plant Geography. With Special Reference to North America*. New York: Academic Press.

Diels, L. (1908) *Pflanzengeographie*. Leipzig: Göschen.

Doing, H. (1983) 'The vegetation of the Wadden sea islands in Niedersachsen and the Netherlands', in K.S. Dijkema and W.J. Wolff (eds.), *Flora and Vegetation of the Wadden Sea Islands and Coastal Areas*. Rotterdam: Balkema. pp. 165–85.

Drude, O. (1884) 'Die Florenreiche der Erde', *Petermanns Geogr. Mitt., Ergänzungsbd*. 16. Gotha-Leipzig.

Ellenberg, H. (1968) 'Wege der Geobotanik zum Verständnis der Pflanzendecke', *Naturwissenschaften*, 55: 462–70.

Ellenberg, H. (1988) *Vegetation Ecology of Central Europe.* Cambridge: Cambridge University Press.

Ellenberg, H. (1996) 'Páramos und Punas der Hochanden Südamerikas, heute größtenteils als potentielle Wälder anerkannt', *Verhandlungen der Gesellschaft für Ökologie*, 25: 17–23.

Ellenberg, H., and Leuschner, C. (2010) *Vegetation Mitteleuropas mit den Alpen in ökologischer, dynamischer und historischer Sicht.* 6th ed. Stuttgart: Ulmer Verlag.

Ellenberg, H., Weber, H.E., Düll, R., Wirth, V., Werner, W., and Paulissen, D. (1992) 'Zeigerwerte von Pflanzen in Mitteleuropa', *Scripta Geobotanica,* 18. Göttingen.

Engler, A. (1879–82) *Versuch einer Entwicklungsgeschichte der Pflanzenwelt.* Leipzig: Wilhelm Engelmann.

Forster, J.R. (1778) *Observations Made during a Voyage Round the World, on Geography, Natural History and Ethnic Philosophy.* London: G. Robinson.

Good, R. (1974) *The Geography of the Flowering Plants.* 4th ed. London: John Wiley.

Grisebach, A. (1872) *Die Vegetation der Erde nach ihrer klimatischen Anordnung.* 2 Bde. Leipzig: Wilhelm Engelmann.

Haeckel, E. (1868) *Natürliche Schöpfungsgeschichte.* Berlin: Georg Reimer.

Hallam, A., and Wignall, P. (1997) *Mass Extinctions and Their Aftermath.* Oxford: Oxford University Press.

Hengeveld, R., Giller, P.S., and Riddle, B.R. (2004) 'Species ranges', in M.V. Lomolino, D.F. Sax, and J.H. Brown (eds.), *Foundations of Biogeography. Classic Papers With Commentaries.* Chicago: University of Chicago Press. pp. 449–55.

Hijmans, R.J., and Graham, C.H. (2006) 'The ability of climate envelope models to predict the effect of climate change on species distributions', *Global Change Biology*, 12: 2272–81.

Holdridge, L.R. (1947) 'Determination of world plant formations from simple climatic data', *Science*, 105: 367–68.

Holtmeier, F.K. (2009) *Mountain Timberlines. Ecology, Patchiness, and Dynamics.* Dordrecht: Kluwer Academic.

Horvat, I., Glavac, V., and Ellenberg, H. (1974) *Vegetation Südosteuropas.* Stuttgart: Gustav Fischer.

Huggett, R.J. (2004) *Fundamentals of Biogeography.* 2nd ed. London: Routledge.

Humboldt, A. von, and Bonpland, A. (1805) *Essai sur la Geographie des Plantes.* Paris: Levrault, Schoell and Co.

Jäger, E. (1968) 'Die pflanzengeographische Ozeanitätsgliederung der Holarktis und die Ozeanitätsbindung der Pflanzenareale', *Feddes Repertorium*, 79: 157–335.

Jansen, E., et al. (2007) 'Palaeoclimate', in S. Solomon, D. Qin, M. Manning, Z. Chen, M. Marquis, K.B. Averyt, M. Tignor, and H.L. Miller (eds.), *Climate Change 2007: The Physical Science Basis.* Cambridge: Cambridge University Press. pp. 433–97.

Keddy, P.A. (2007) *Plants and Vegetation: Origin, Processes, Consequences.* New York: Cambridge University Press.

Klink, H.J. (1996) *Vegetationsgeographie.* 2nd ed. Braunschweig: Westermann.

Körner, C. (2002) 'Ökologie', in P. Sitte, E.W. Weiler, J.W. Kadereit, A. Bresinsky, and C. Körner (eds.), *Lehrbuch der Botanik für Hochschulen.* 35th ed. Heidelberg-Berlin: Spektrum Akademischer Verlag. pp. 887–1043.

Körner, C. (2007) 'Climatic treelines: Conventions, global patterns, causes', *Erdkunde*, 61: 316–24.

Lang, G. (1994) *Quartäre Vegetationsgeschichte Mitteleuropas. Methoden und Ergebnisse.* Jena: Gustav Fischer.

Larcher, W. (1994) *Ökophysiologie der Pflanzen.* 5th ed. Stuttgart: Ulmer Verlag.

Linnaeus, C. (1781) 'On the increase of the habitable earth', *Amoenitates Academicae* 2: 17–27. Transl. by F.J. Brandt.

Lomolino, M.V., Riddle, B.R., and Brown, J.H. (2006) *Biogeography.* 3rd ed. Sunderland, MA: Sinauer.

MacArthur, R.H., and Wilson, E.O. (1967) *The Theory of Island Biogeography.* Princeton, NJ: Princeton University Press.

MacDonald, G.M. (2003) *Biogeography: Introduction to Space, Time and Life.* New York: John Wiley.

Meusel, H., Jäger, E., and Weinert, E. (1965) *Vergleichende Chorologie der zentraleuropäischen Flora*, Vol. 1. Jena: Gustav Fischer.

Mueller-Dombois, D., and Ellenberg, H. (1974) *Aims and Methods of Vegetation Ecology.* New York: Blackburn Press.

Pott, R. (2000) 'Palaeoclimate and vegetation—long-term vegetation dynamics in central Europe with particular reference to beech', *Phytocoenologia*, 30: 285–333.

Pott, R. (2005) *Allgemeine Geobotanik. Biogeosysteme und Biodiversität.* Berlin-Heidelberg: Springer.

Pott, R., and Hüppe, J. (2007) *Spezielle Geobotanik. Pflanze—Klima—Boden.* Berlin-Heidelberg: Springer.

Prentice, I.C., Bondeau, A., Cramer, W., Harrison, S.P., Hickler, T., Lucht, W., Sitch, S., Smith, B., and Sykes, M.T. (2007) 'Dynamic global vegetation modeling: quantifying terrestrial ecosystem responses to large-scale environmental change', in J.G. Canadell, D.E. Pataki, and L.F. Pitelka (eds.), *Terrestrial Ecosystems in a Changing World.* Berlin-Heidelberg: Springer. pp. 175–92.

Raunkiaer, C. (1934) *The Life Forms of Plants and Statistical Plant Geography.* Oxford: Oxford University Press.

Rebelo, A.G., Boucher, C., Helme, N., Mucina, L., and Rutherford, M.C. (2006) 'Fynbos biome', in L. Mucina and M.C. Rutherford (eds.), 'The Vegetation of South Africa, Lesotho and Swaziland', *Strelitzia*, 19: 52–219.

Reigosa, M.J., Pedrol, N., and Gonzalez, L. (eds.). (2006) *Allelopathy: A Physiological Process with Ecological Implications.* Dordrecht: Springer.

Richter, M. (2001) *Vegetationszonen der Erde.* Gotha: Klett-Perthes.

Riddle, B.R., Dawson, M.N., Hadly, E.A., Hafner, D.J., Hickerson, M.J., Mantooth, S.J., and Yoder, A.D. (2008) 'The role of molecular genetics in sculpting the future of integrative biogeography', *Progress in Physical Geography*, 32: 173–202.

Rikli, M. (1913) *Die Florenreiche.* Handwörterbuch der Naturwissenschaften IV. Jena: Gustav Fischer.

Ruddiman, W.F. (ed.) (1997) *Tectonic Uplift and Climate Change.* New York: Plenum.

Schickhoff, U. (1993) 'Das Kaghan-Tal im Westhimalaya (Pakistan). Studien zur landschaftsökologischen Differenzierung und zum Landschaftswandel mit vegetationskundlichem Ansatz', *Bonner Geographische Abhandlungen,* 87. Bonn.

Schickhoff, U. (1996) 'Contributions to the synecology and syntaxonomy of West Himalayan coniferous forest communities', *Phytocoenologia*, 26: 537–81.

Schickhoff, U. (2000) 'The impact of Asian summer monsoon on forest distribution patterns, ecology, and regeneration north of the main Himalayan range (E-Hindukush, Karakorum)', *Phytocoenologia*, 30: 633–54.

Schickhoff, U. (2005) 'The upper timberline in the Himalayas, Hindu Kush and Karakorum: A review of geographical and ecological aspects', in G. Broll and B. Keplin (eds.), *Mountain Ecosystems: Studies in Treeline Ecology*. Berlin-Heidelberg: Springer. pp. 275–354.

Schickhoff, U., and Seiberling, S. (2003) 'Zwischen Land und Meer—Küstenvegetation an Nord- und Ostsee', in Leibniz-Institut für Länderkunde (ed.), *Nationalatlas Bundesrepublik Deutschland, Bd. 3, Klima, Pflanzen- und Tierwelt*. Heidelberg-Berlin: Springer. pp. 116–17.

Schickhoff, U., Walker, M.D., and Walker, D.A. (2002) 'Riparian willow communities on the Arctic slope of Alaska and their environmental relationships: A classification and ordination analysis', *Phytocoenologia*, 32: 145–204.

Schimper, A.F.W. (1898) *Pflanzengeographie auf physiologischer Grundlage*. Jena: Gustav Fischer.

Schmidtlein, S., and J. Ewald. (2003) 'Landscape patterns of indicator plants for soil acidity in the Bavarian Alps', *Journal of Biogeography*, 30: 1493–503.

Schönswetter, P., Paun, O., Tribsch, A., and Nikfeld, H. (2003) 'Out of the Alps: Colonization of northern Europe by east alpine populations of the glacier buttercup *Ranunculus glacialis* L. (Ranunculaceae)', *Molecular Ecology*, 12: 3373–81.

Schroeder, F.G. (1998): Lehrbuch der Pflanzengeographie. Wiesbaden: Quelle and Meyer.

Schultz, J. (2005) *The Ecozones of the World: The Ecological Divisions of the Geosphere*. Berlin-Heidelberg: Springer.

Smith, C.H. (1983) 'A system of world mammal faunal regions. I. Logical and statistical derivation of the regions', *Journal of Biogeography*, 10: 455–66.

Stevens, G.C. (1989) 'The latitudinal gradient in geographic range: How so many species coexist in the tropics', *American Naturalist*, 133: 240–56.

Stevens, G.C. (1992) 'The elevational gradient in altitudinal range: An extension of Rapoport's latitudinal rule to altitude', *American Naturalist*, 140: 893–911.

Stoutjesdijk, P., and Barkman, J.J. (1992) *Microclimate: Vegetation and Fauna*. Uppsala: Opulus Press.

Strauss, A., and Schickhoff, U. (2007) 'Influence of soil-ecological conditions on vegetation zonation in a western Mongolian lake shore—semi-desert ecotone', *Erdkunde*, 61: 72–92.

Taberlet, P., Fumagalli, L., Wust-Saucy, A.G., and Cosson, J.F. (1998) 'Comparative phylogeography and postglacial colonization routes in Europe', *Molecular Ecology*, 7: 453–64.

Takhtajan, A.L. (1986) *The Floristic Regions of the World*. Berkeley: University of California Press.

Troll, C. (1948) 'Der asymmetrische Aufbau der Vegetationszonen und Vegetationsstufen auf der Nord- und Südhalbkugel', *Jahresbericht des Geobotanischen Instituts Rübel in Zürich für*, 1947, 46–83.

Troll, C. (1966) *Ökologische Landschaftsforschung und vergleichende Hochgebirgsforschung*. Erdkundliches Wissen 11. Wiesbaden: Franz Steiner Verlag.

Wallace, A.R. (1876) *The Geographical Distribution of Animals*. 2 vols. London: Harper and Brothers.

Walter, H. (1960) *Standortslehre. Analytisch-ökologische Geobotanik*. 2nd ed. Stuttgart: Eugen Ulmer.

Walter, H. (1973) *Vegetation of the Earth in Relation to Climate and Ecophysiological Conditions*. London: English Universities Press.

Walter, H. (1986) *Allgemeine Geobotanik*. Stuttgart: Eugen Ulmer.

Walter, H., and Breckle, S.W. (1985) *Ecological Systems of the Geobiosphere, Vol. 1. Ecological Principles in Global Perspective*. Berlin: Springer-Verlag.

Walter, H., and Breckle, S.W. (1999) Vegetation und Klimazonen. *Grundriss der globalen Ökologie*. 7th ed. Stuttgart: Ulmer.

Walter, H., and Straka, H. (1970) *Arealkunde. Floristisch-historische Geobotanik*. 2nd ed. Stuttgart: Ulmer.

White, M.E. (1990) 'The flowering of Gondwana', in *The 400 Million Year History of Australia's Plants*. Princeton, NJ: Princeton University Press.

White, P.S., and A. Jentsch. (2001) 'The search for generality in studies of disturbance and ecosystem dynamics', *Progress in Botany*, 63: 399–449.

Whittaker, R.H. (1975) *Communities and Ecosystems*. 2nd ed. New York: Macmillan.

Willis, J.C. (1922) *Age and Area*. Cambridge: Cambridge University Press.

Willis, K.J., and McElwain, J.C. (2002) *The Evolution of Plants*. New York: Oxford University Press.

Woodward, F.I. (1987) *Climate and Plant Distribution*. Cambridge: Cambridge University Press.

Wulff, E.V. (1943) *An Introduction to Historical Plant Geography*. Waltham, MA: Chronica Botanica Co.

Zhisheng, A., Kutzbach, J.E., Prell, W.L., and Porters, S.C. (2001) 'Evolution of Asian monsoons and phased uplift of the Himalayas-Tibetan Plateau since late Miocene times', *Nature*, 411: 62–66.

Zoller, H., and Steinmann, M. (1987/91) *Conrad Gesner: Conradi Gesneri Historia plantarum*. Dietikon-Zürich: Urs Graf Verlag.

Biodiversity Gradients

Jens Mutke

9.1 INTRODUCTION

The distribution of biological diversity across the Earth is highly uneven. A single hectare of Amazonian rain forest might harbor the same number of plant species as all of Ireland (e.g., Valencia, 1994; Davis et al., 1994–1997). Tropical Africa has more than twice the number of amphibian species compared to cold and temperate Europe and Asia together (Stuart et al., 2004). These extreme differences have fascinated naturalists and scientists since the time of von Humboldt, Forster, and Wallace (von Humboldt, 1808; Wallace, 1878; Hawkins, 2001). However, they also have important consequences in the context of conservation, management, and use of natural resources (Redford et al., 2003; Brooks et al., 2006), and there is a rapidly increasing number of analyses of broad-scale spatial biodiversity patterns. These have become possible by the increasing amount of (digitally) available data and large synoptical studies such as, for example, the Global Amphibian Assessment (Stuart et al., 2004). Moreover, increasing computational power and GIS capacities, as well as further methodological improvements, for example, by molecular phylogenetic analyses (see Riddle, this volume), has led to a deeper understanding of the factors and mechanisms shaping spatial biodiversity patterns.

9.2 THE UNEVEN DISTRIBUTION OF BIODIVERSITY DATA—'KNOWLEDGE GRADIENTS'

It is often surprising to nonbiologists how little we actually know about our cohabitants on Earth.

Only 1.75 million species of microorganisms, plants, and animals are currently registered by scientists (Hawksworth and Kalin-Arroyo, 1995). But almost all analyses trying to assess the existing overall species richness on our planet have estimated between 5 and 30 million species. If we accept the 13.6 million species used as a working figure in the Global Biodiversity Assessment (Hawksworth and Kalin-Arroyo, 1995), some 90% of Earth's biological diversity still awaits to be described. The percentage of species already known differs substantially among groups and among regions. Large, conspicuous, or economically important organisms such as the vertebrates, butterflies, or plants are generally relatively well known. On the other hand, for most groups of insects or microorganisms more than 95% of the species are currently undescribed. For instance, half of the species of spiders known to science have been described in the last 30 years (Simon, 1995), whereas half of the vascular plant genera were already known around 1850 and half of the species were described around 1900 (Mutke and Barthlott, 2005). Recently, new tools like DNA barcoding led to discoveries of large numbers of so-called cryptic species that have not been identified as separate species based on classical morphological data (Hebert et al., 2004; Smith et al., 2008). The fact that much of the biological diversity has yet to be formally described and catalogued has been named the 'Linnean shortfall' (Lomolino et al., 2006). In addition, there is the 'Wallacean shortfall': for many taxa we have only inadequate knowledge of their global, regional, and even local distributions (Whittaker et al., 2005). Schatz (2002) analyzed one of the largest botanical databases, TROPICOS, with more than 1.7 million plant distribution records based on herbarium data. More than three-quarters of all species were represented by fewer than 10 records, whereas only a few species were represented by hundreds or even thousands. Kier et al. (2005) found that especially the plant diversity of flooded grasslands and savannahs is poorly documented. Frodin (2001) presented a map of 'areas that most need floras', highlighting areas like Colombia, the Congo Basin, Angola, and several areas in the Indo-Malayan region, such as Sulawesi and Papua New Guinea. Several studies documented the highly heterogeneous sampling intensity, especially for tropical floras and faunas (Nelson et al., 1990; Cumming, 2000; Küper et al., 2006).

Recently, there has been considerable effort to at least make the information available that is physically housed in natural history collections. Today, distribution data based on some 5–10% of the 2.5 billion specimens are already digitized (Graham et al., 2004). There are large programs, such as the Global Biodiversity Information Facility (www.GBIF.org), that make this data accessible. Moreover, modern molecular methods have allowed first biogeographical analyses for groups for which only a tiny part of their overall diversity has been currently been identified. For example, Pommier et al. (2005) used bacterial RNA data from in situ sequencing to analyze biogeographic patterns of marine bacterioplankton.

9.3 A BRIEF HISTORY OF BIODIVERSITY GRADIENT STUDIES

By the eighteenth and nineteenth centuries, the earliest naturalists and biogeographers had mentioned the stark differences between the overwhelmingly diverse floras and faunas of the tropical countries they were traveling through and that of their temperate homes (von Humboldt, 1808; Wulff, 1935; Hawkins, 2001). In 1808, Alexander von Humboldt wrote (from a translation by Otté and Bohn, 1850), "Thus, the nearer we approach the tropics, the greater the increase in the variety of structure, grace of form, and mixture of colors, as also in perpetual youth and vigour of organic life."

Actually, von Humboldt (1817) was the first to provide detailed statistical analyses of the regional floras of the world, including total species numbers as well as the relative importance of various plant families. In the *Berghaus Atlas* accompanying the *Kosmos* written by von Humboldt (1845), rough latitudinal gradients of the relative importance of various plant families in relation to the overall flora are shown, based on these statistics (Berghaus, 1837–1847). It is fascinating to see that von Humboldt even discussed possible explanations for the extraordinary diversity such as the variety of suitable climatic conditions or the complex landscape, which together we would term today 'geodiversity' (Barthlott et al., 1996), 'environmental heterogeneity' (Ricklefs, 1977), or 'environmental diversity' (Faith and Walker, 1996; Faith, 2003). Wallace (1878) emphasized the relative stability of the tropics as one reason for its high biodiversity. In 1935, the Russian botanist Eugenii Wladimirowich Wulff was the first to publish a world map of plant species richness with almost global coverage (Wulff, 1935). In the accompanying publication he presented a short review of studies on the latitudinal gradient. Additionally, he presented a discussion on the role of the main parameters governing diversity patterns, especially temperature, humidity, continentality of the climate, increased species richness in mountain areas due to a higher diversity of habitats and their potential role as refugia, and the 'age' of a flora. For higher plants, the work of Malyshev (1975) as well as our own work (Barthlott et al., 1996, 2005; Mutke and Barthlott, 2005; Kier et al., 2009) resulted in even more detailed maps of the global distribution of species richness and endemism, as well as additional macroecological analyses (Mutke et al., 2001; Kreft and Jetz, 2007; Kreft et al., 2008; Jetz et al., 2008).

This is in contrast to animal data. It was only recently that global species richness patterns have been mapped for vertebrates (Brooks et al., 2004; Ceballos et al., 2005; Ceballos and Ehrlich, 2006; Grenyer et al., 2006; Lamoreux et al., 2006; Buckley and Jetz, 2007). These are now complemented with several more detailed analyses of different aspects of global patterns of vertebrate diversity (Orme et al., 2005, 2006; Storch et al., 2006; Davies et al., 2007; Jetz et al., 2007; Schipper et al., 2008; Kier et al., 2009). Nonetheless, analyses of global diversity patterns for groups other than vertebrates and higher plants are still scarce (e.g., Mutke and Barthlott, 2005; Feuerer and Hawksworth, 2007; Konrat et al., 2008).

In the second half of the twentieth century the discussion about the underlying mechanisms of the latitudinal increase of species richness toward the equator intensified (see below, e.g., Fischer, 1960; Pianka, 1966; Terborgh, 1973; Rhode, 1992; Rosenzweig, 1995). Several reviews have been published (e.g., Gaston, 1996, 2000; Field, 2002; Willig et al., 2003; Hillebrand, 2004; Turner, 2004; Mutke et al., 2005; Sarr et al., 2005; Lomolino et al., 2006).

The second important biodiversity gradient that has been studied for a long time is the altitudinal or elevational gradient (Rahbek, 1995, 1997; Lomolino, 2001; Grytnes and McCain, 2007).

A very early description and first analysis of this gradient had already been provided by von Humboldt (1808, 1845). There are many studies on altitudinal zonation of vegetation belts on mountains (e.g., Whittaker, 1960; Lauer, 1976; Cleef et al., 1984; Frahm and Gradstein, 1991). For gradients of species richness in relation to altitude, Terborgh's (1977) study has been widely cited. It shows a monotonic decrease of bird species richness with elevation in the Peruvian Andes. However, especially since the mid-1990s there is an increasing amount of literature recognizing a hump-shaped diversity gradient with richness maxima at mid-elevations (Rahbek, 1995, 1997, 2005). A search in the Science Citation Index in May 2011 with the key words 'altitudinal (or elevational) and gradient and diversity' produced a list of 665 papers and the number of papers per year increase constantly. Almost 170 papers were published in the years 2009 to 2010.

9.4 DIFFERENT MEASURES OF BIOLOGICAL DIVERSITY

Though most analyses of biodiversity gradients refer to species richness or species numbers, there are several other quantitative measures as well as qualitative aspects that could be included. Particularly in the context of conservation planning, measures of the rarity and uniqueness of species or taxa might be at least equally important. The classical publication of Rabinowitz (1981) refers to seven forms of rarity: looking at range size, local abundance, and habitat breadth there are eight combinations of being either rare or not regarding each of these parameters. As one of the eight combinations would be a locally abundant species with a large range size and a broad habitat breadth, like *Chenopodium album*, only the other seven are rare in some way (Table 9.1). These include, for example, *Lodoicea maldivica*, a palm species occurring only on one island of the Seychelles, where it forms almost monodominant stands with high local abundances (Edwards et al., 2002).

The information on the range size or degree of endemism is often included in spatial analyses, as these are parameters that can be directly deduced from distributional data. There are different approaches to map indices of range size rarity (Williams et al., 1996) or endemism richness (Kier and Barthlott, 2001; Kier et al., 2009). Several studies have shown that centers of species richness and endemism are only partially congruent (Jetz and Rahbek, 2002; Orme et al., 2005; Kreft et al., 2006).

A species might as well be rare or unique with regard to its phylogeny if it belongs to a highly distinct, maybe monotypic genus or family. The well-known *Ginkgo biloba*, the only remaining species of the gymnosperm class Ginkgopsida, is a prominent example. Such species are often rated as especially important due to the unique genetic information and evolutionary heritage they contain (Williams, 1993; Mooers et al., 2005; Isaac et al., 2007). Another approach to represent as much of the overall diversity of a given group is to use a higher-taxon approach, for example, to map families or genera instead of species (Williams et al., 1994; Francis and Currie, 2003).

For conservation purposes it is also important to question the intactness of biological diversity. Different priority-setting strategies focus either on hotspots of biodiversity (Myers et al., 2000; Küper et al., 2004b; Mittermeier et al., 2005) where high human impact requires immediate action, or to preserve wilderness or 'good news' areas (Mittermeier et al., 1998, 2003), with low human impact resulting in low costs for conservation measures (Brooks et al., 2006).

Another important focus is the current or potential economic value. The same is true and might become even more important for the relevance of biodiversity to ecosystem functioning and ecosystem services for human well-being (Costanza et al., 1997; Millennium Ecosystems Assessment, 2005).

9.5 SCALE DEPENDENCE OF BIODIVERSITY PATTERNS

At least since the classical work by Whittaker (1972, 1977), it is common textbook knowledge that biodiversity is highly scale-dependent. According to Whittaker, at least three different scales of biodiversity can be defined: alpha diversity addresses within-habitat diversity, whereas gamma diversity refers to the diversity of larger areas containing several different habitats. In contrast to these two 'inventory diversities', beta diversity as a 'differentiation diversity' describes the between-habitat diversity, measured as the species turnover between different sites or along environmental gradients. The relationship between these three types of diversity can be described by the formula gamma = alpha x beta (Whittaker, 1977). Thus, high biodiversity values at a larger area might be the result of either a high diversity of different habitats and niches within that area (beta) or a high number of species within each of the occurring habitats (alpha)—or both. If we look at the tropical Amazon forest in the eastern lowlands of Ecuador, study sites of, for example, one

Table 9.1 Classification of rare species after Rabinowitz (1981), modified from Gaston (1994). All listed combinations of geographic range size, habitat specificity, and local population size are rare at least in one sense—except for the first one

Geographic range size	Habitat specificity	Local population size	Examples
Large	Wide	Large, dominant somewhere	Locally abundant over a large range in several habitats: *not rare!* (*Chenopodium album*)
Large	Wide	Small, nondominant	Constantly sparse over a large range and in several habitats (e.g., the European orchid *Epipactis helleborine*)
Large	Narrow	Large, dominant somewhere	Locally abundant over a large range in a specific habitat (Red mangrove *Rhizophora mangle*; many water lilies *Nymphaea* spp.)
Large	Narrow	Small, nondominant	Constantly sparse in a specific habitat but over a large range (the fern *Botrychium lunaria*, the carnivorous plant *Utricularia subulata*)
Small	Wide	Large, dominant somewhere	Locally abundant in several habitats but restricted geographically (some island endemics, e.g., *Teline microphylla* or *Myrica faya* from the Canary Islands)
Small	Wide	Small, nondominant	Constantly sparse and geographically restricted in several habitats (*nonexistent?*)
Small	Narrow	Large, dominant somewhere	Locally abundant in a specific habitat but restricted geographically (Venus flytrap, *Dionaea muscipula*; California pitcher plant, *Darlingtonia californica*; several island endemics, e.g., *Euphorbia canariensis* or *Lodoicea maldivica*)
Small	Narrow	Small, nondominant	Constantly sparse and geographically restricted in a specific habitat (the Sundew *Drosera intermedia*; many Cycad species; the alpine Devil's-Claw *Physoplexis comosa*)

hectare in size have the highest documented woody plant species richness (473 species) in the world (Valencia, 1994). Looking at the overall plant species richness in this one hectare plot including herbs and epiphytes, the number (>1,000 species) is comparable to the total flora of the Netherlands. In contrast, Andean forest plots above c. 1,500–2,000 meters above sea level (m.a.s.l) show woody plant species numbers below 50 species per hectare (Braun et al., 2002; Mutke, 2002). But if we compare the total number of plant species for the Ecuadorean Andes (an area of 94,000 km²) and the Amazon lowlands of the same country (84,000 km²), the picture is quite the opposite. The Andes, with almost 9,900 documented species, have a flora twice as species rich as the lowlands with c. 4,900 species (Jørgensen and León-Yánez, 1999). While alpha diversity is higher in the lowland, beta diversity at this scale is much higher in the mountains. Scale effects like this might lead to important misinterpretations of biodiversity gradients and their according determinants if the scale and extent of the analysis are

not taken into account (Rahbek and Graves, 2001; Whittaker et al., 2001; Rahbek, 2005).

The Ecuador example illustrates that the relation between species number and area size is quite different in these landscapes. The fact that 5% of the plant species richness of the whole country can be found within a single hectare (or 0.000004% of the land area) shows that the relation between species richness and area size is not a linear one. This relation has been studied at least since the time of de Candolle (1855, cited in Wulff, 1935: 61; Rosenzweig, 1995: 8) and formally described by the model $S = cA^Z$ by Arrhenius (1920, 1921), where S is the species number of the area A, c is the species number of an area of the size 1, and z describes the slope of the resulting species area curve, which can be interpreted as floristic turnover or floristic heterogeneity (Malyshev, 1991). Several reviews of the species-area relationship and respective mathematical models have been published (Connor and McCoy, 1979; Malyshev, 1991; Rosenzweig, 1995; Caley and Schluter, 1997; Plotkin et al., 2000; Crawley and Harral,

2001; Whittaker et al., 2001; Willis and Whittaker, 2002). The species-area relationship and the mechanisms associated with area that determine species richness play an important role in Island Biogeography as well (MacArthur and Wilson, 1967; Kalmar and Currie, 2006; Heaney, 2007; Whittaker et al., 2007; Kreft et al., 2008; see also Griffin, this volume).

Willis and Whittaker (2002) demonstrated that it is not only species richness patterns that are scale-dependent, but the explanatory variables governing the patterns, and even the slope of the species-area curve, are highly scale-dependent as well. In an analysis of diversity patterns of British birds, Lennon et al. (2001) showed that the diversity patterns calculated with 10-km resolution and with 90-km resolution are statistically unrelated. Willis and Whittaker (2002) summarized that, for example, at the local scale biotic interactions, disturbances, or habitat structure are most important, whereas at the landscape scale soil and altitude may be dominant parameters. At a regional scale, diversity patterns often seem to be governed by climatic parameters and area. A similar hierarchical model of the importance of different drivers at micro-, meso-, and macroscale has been proposed by Sarr et al. (2005). Analyzing diversity patterns of South American birds with multiple regression models at spatial resolution ranging from 1x1° to 10x10° grid cells, Rahbek and Graves (2001) found precipitation as the most important model parameter at fine scales. At coarser spatial scales, topography was the dominant factor. Together with the interaction term 'topography x latitude' it was the only parameter included in the models at all spatial scales.

9.6 DIFFERENT KINDS OF BIODIVERSITY GRADIENTS

Several types of environmental gradients have been investigated in the context of variation in biological diversity. Though the latitudinal and the altitudinal gradients of biodiversity are probably the most cited, other gradients exist, such as climatic gradients (Currie, 1991) or gradients of depth in aquatic systems (Smith and Brown, 2002).

On the other hand, different aspects of biodiversity in addition to pure taxon richness are studied in gradient analyses, such as spatial gradients of plant traits (Kühn et al., 2006; Moles et al., 2007), morphological diversity (Shepherd, 1998), or the number of invasive species (Sax, 2001). The rapidly increasing possibilities to analyze genetic diversity have resulted in many recent studies of

diversity gradients mainly within species (e.g., Serre and Pääbo, 2004).

In the following sections, we discuss broad-scale biodiversity gradients starting with global centers of diversity and then focus on latitudinal gradients, altitudinal gradients, and marine gradients. Though not covered in our chapter, one should mention the set of so-called biogeographic or ecogeographic rules relating to latitude like the Bergmann's rule (increasing body mass with increasing latitude), Allen's rule (endotherms of lower latitudes have longer body appendages), or Rapoport's rule (increasing range size with increasing latitude) (Lomolino et al., 2006).

9.6.1 Centers and Hotspots of Biodiversity

Though only few groups of organisms have been sufficiently documented so far, it seems that some parts of the globe are general centers of biological diversity irrespective of the organism group investigated. In the terrestrial realm, there are important congruent centers of high diversity of a variety of taxa, such as plants (Barthlott et al., 2005; Mutke and Barthlott, 2005) or several groups of vertebrates (Brooks et al., 2004; Ceballos and Ehrlich, 2006; Grenyer et al., 2006) (compare Figures 9.1 and 9.2 as well as Jetz et al., 2008; Kier et al., 2009): the tropical Andes, the western Amazon, parts of Central America, southeast Brazil, Mount Cameroon, the East African Mountains, and several areas in southeast Asia and the Indomalayan Region (especially northern Borneo and the Malayan Peninsula). (For a discussion of taxon-specific differences in these patterns, cf. the section on taxon-specific patterns in Section 9.6.2.)

Figure 9.1 shows 20 global centers of vascular plant species richness with species densities of at least 3,000 species per 10,000 km^2. The five most important centers with more than 5,000 species/10,000 km^2 are Costa Rica-Chocó, Atlantic Brazil, tropical eastern Andes, northern Borneo, and New Guinea. These five areas account for only 0.2% of the terrestrial surface area of the globe but harbor at least 6.2% of all vascular plant species as endemics (Barthlott et al., 2005). This high concentration of endemic species within a few centers or hotspots has already been shown by Myers and coworkers (Myers, 1988; Myers et al., 2000; Mittermeier et al., 2005). Combining data on plant endemism and human threats, they mapped 'biodiversity hotspots' on a global scale. These have been defined as regions that harbor at least 0.5 percent (1,500 species) of the global flora as endemics and have lost at least 70% of their

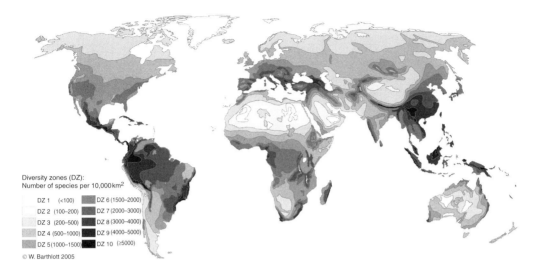

Diversity zones (DZ):
Number of species per 10,000 km²

DZ 1 (<100)
DZ 2 (100–200)
DZ 3 (200–500)
DZ 4 (500–1000)
DZ 5 (1000–1500)
DZ 6 (1500–2000)
DZ 7 (2000–3000)
DZ 8 (3000–4000)
DZ 9 (4000–5000)
DZ 10 (≥5000)

© W. Barthlott 2005

Figure 9.1 World map of vascular plant species richness (after Barthlott et al., 2005; Mutke and Barthlott, 2005)

natural vegetation cover. Most of the currently recognized 35 global biodiversity hotspots correspond with the global or regional maxima of species richness shown in Figure 9.1.

Another global assessment of centers of restricted-range species is represented by the work of Birdlife International on endemic bird areas (EBAs) (Bibby et al., 1992; Stattersfield et al., 1998). These regions are defined by the co-occurrence of at least two (up to 80) bird species with range sizes of less than 50,000 km². The size of an EBA ranges between few square kilometers to more than 100,000 km². Almost 80% of the EBAs are located in the tropics or subtropics, especially on islands or mountain ranges. The same is true for the 595 centers of imminent extinction (Ricketts et al., 2005), which are home to 794 highly threatened species of mammals, birds, selected reptiles, amphibians, and conifers that are restricted to just a single site. As the EBAs, these centers are concentrated in tropical forests, on islands, and in mountainous areas.

9.6.2 Latitudinal gradients

The gradient of increasing biological diversity with decreasing latitude has received enormous attention by biogeographers and ecologists. A search in the Science Citation Index in May 2011 with the key words 'latitudinal AND gradient AND diversity' produced a list of 785 papers with the number of papers per year increasing

constantly. More than 70 publications were added to this data in 2009, and almost 100 in 2010. As shown by most of these publications and as documented in the meta-analysis by Hillebrand (2004), the pattern of increasing species richness with decreasing latitude is a common pattern for a wide range of different groups of organisms. However, it is still one of the most discussed and yet not fully understood patterns in ecology. Huston (1994) even termed it the 'the holy grail of ecology'. Theories aiming to explain this pattern are discussed below in the next section as well as in the chapter by Griffin (this volume).

Though the latitudinal diversity gradient is a common pattern, it is not a universal one. Exceptions include some marine groups, bumble bees, and gymnosperms, which have their main diversity centers outside the tropics (Rhode, 1992; Williams, 1993; Mutke and Barthlott, 2005). In many other cases we do not find a monotonic increase in species richness with decreasing latitude but secondary maxima, for example, in Mediterranean climate regions or minima, for example, in low latitude deserts, as shown for swallowtail butterflies (Collins and Morris, 1985) and for vascular plants (Linder, 2001; Mutke et al., 2001; Mutke and Barthlott, 2005). In addition, the latitudinal gradient in many analyses is not symmetric around the northern and southern hemispheres (Gaston, 1996; Chown et al., 2004). For many groups, species richness declines more rapidly with latitude in the north, despite the fact that land area is much larger in the northern than in the southern hemisphere.

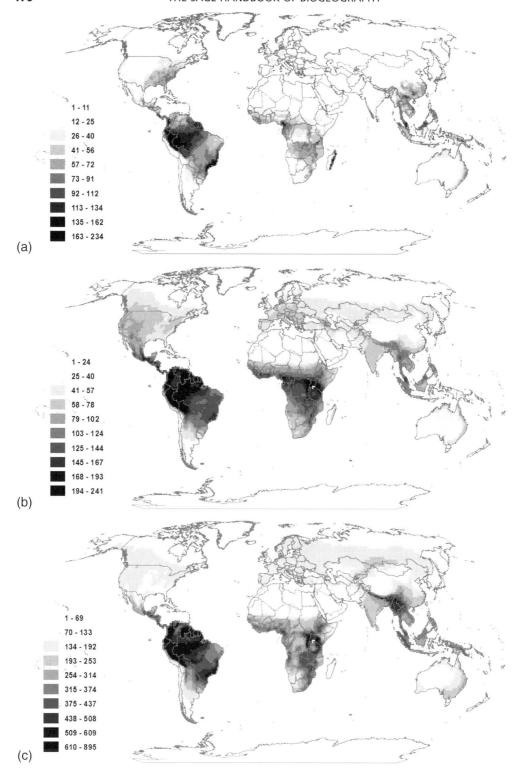

Figure 9.2   World maps of species richness of (a) amphibians (Stuart et al., 2004; Buckley and Jetz, 2007), (b) mammals (Ceballos and Ehrlich, 2006, Jetz, unpubl.), and (c) birds (Jetz et al., 2007) on an equal area grid of 110 x 110 km

Taxon-specific patterns

Species richness of vertebrates and insects is often closely correlated to plant diversity (compare Figures 9.1 and 9.2) (Barthlott et al., 1999; Nic Lughadha et al., 2005; Proches and Cowling, 2006). Plants are the most important primary producers and structuring elements of terrestrial ecosystems. Novotny et al. (2006) and Kitching (2006) showed the high dependence of insect diversity on plant diversity. Kissling et al. (2007) found that the species richness of *Ficus* trees has the strongest direct effect on richness of avian frugivores in tropical Africa. Variables related to water, energy, and habitat heterogeneity have mainly indirect influence. This pattern is similar for bird and woody plant species richness in Kenya (Kissling et al., 2008).

Whether such a direct linkage between the diversity of producers and consumers can be found at the global scale as well is still a matter of debate, though the pure statistical correlation of plant and vertebrate species richness at this scale is high (Barthlott et al., 1999; Qian and Ricklefs, 2008; Jetz et al., 2008). Despite the congruency in Figures 9.1 and 9.2, important differences exist between the patterns of plant and vertebrate diversity. One of the most striking disparities is the high species richness of plants in the South African Cape region, which has the status of being one of only seven global plant kingdoms (Linder, 2003; Kreft and Jetz, 2007). As shown, for example, by Jetz and colleagues (Buckley and Jetz, 2007; Jetz et al., 2007, see Figure 9.2), Brooks et al. (2004), Ceballos et al. (2005), and Grenyer et al. (2006), this is in sharp contrast to birds, mammals, or amphibians. Highest species richness values for mammals are found in northern South America, especially in the Amazonian lowlands and the Andes, as well as in the East African Mountains. The regional centers in southeast Asia are less species rich (Ceballos et al., 2005; Ceballos and Ehrlich, 2006), though Papua New Guinea is an important center for restricted-range and threatened species. For mammals, especially the relatively high diversity in tropical Africa compared to the Neotropics, is in contrast to the situation for plants and the other vertebrate groups. Mammal species richness of the Afrotropics with c. 1,170 species is about the same as in the Neotropics with c. 1,300 species. In contrast, Neotropical bird species richness is 70% higher (c. 3,800 spp.) than in the Afrotropics (c. 2,230 spp.), and there are three times as many known amphibian species in the Neotropics (c. 2,730 spp.) compared to the Afrotropics (c. 930 spp.) (Baillie et al., 2004). Regarding the flora of the two continents, Africa hosts only about half the number of plant species (Beentje et al., 1994) estimated to occur in South America (Gentry, 1982).

Regarding the diversity of plants within the major biomes as defined by Olson et al. (2001), highest species richness can be found in tropical and subtropical moist broadleaf forests. Sixteen out of the 20 centers of plant diversity in Figure 9.1 belong to this biome. Almost 92% of the area of the 20 centers belongs to forest biomes in general. High mean species richness can be found as well in tropical and subtropical coniferous forests and Mediterranean forests, woodlands, and scrub. The biomes with the lowest mean species richness values are tundra and taiga regions (Figure 9.3).

The diversity patterns of amphibians differ in some regions from what is shown in Figures 9.1 and 9.2 for the other taxa, though incomplete knowledge might still play a role. More than 40% of the amphibian species known today have been described between 1992 and 2003 (Köhler et al., 2005). Some of the global centers such as southeast Brazil, the western Amazon basin, southern Central America, or Mount Cameroon are congruent for amphibians with that for plants, mammals, and birds. However, the latitudinal gradient of amphibian diversity is much steeper than that for other taxa, and amphibian diversity is much lower on islands due to their poor dispersal abilities (Brooks et al., 2004).

Focusing on sub-Saharan Africa, Küper and colleagues (cited in Scholes et al., 2006) showed that the differences regarding the centers of species richness of plants and animals would result in enormous amounts of area needed if one aims to conserve all species in protected areas. However, if one focused on the 1% of the area of sub-Saharan Africa with the most important overlapping cross-taxon diversity centers, it would already be possible to represent 58% of all plant and vertebrate species of the region. Similar results have been found by Grenyer et al. (2006) for birds, mammals, and amphibians at the global scale.

However, generalizations that arise from such studies are challenged by the fact that the results are still based only on a very small subsample of the overall biodiversity. It is quite probable that we will never be able to map the total species richness of such megadiverse clades like insects.

9.6.3 Explaining broad-scale diversity patterns

Over the last decades there have been numerous studies trying to explain or at least to correlate biodiversity patterns with several environmental parameters (compare, for example, Rhode, 1992; Gaston, 1996; Willig et al., 2003). Many studies have found high correlations between species

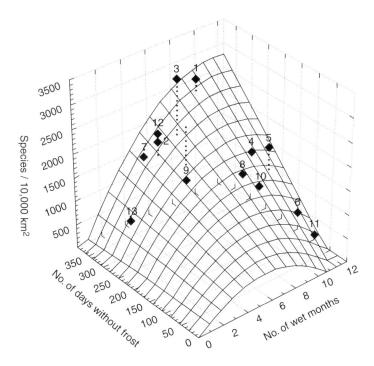

Figure 9.3   Mean vascular plant species richness per 10,000 km² of the 14 terrestrial biomes in relation to frost-free days and number of wet months (based on the data set used in Mutke and Barthlott (2005)). Biome definitions after Olson et al. (2001). 1. Tropical and subtropical moist broadleaf forests; 2. Tropical and subtropical dry broadleaf forests; 3. Tropical and subtropical coniferous forests; 4. Temperate broadleaf and mixed forests; 5. Temperate coniferous forests; 6. Boreal forests/taiga; 7. Tropical and subtropical grasslands, savannas, and shrublands; 8. Temperate grasslands, savannas, and shrublands; 9. Flooded grasslands and savannas; 10. Montane grasslands and shrublands; 11. Tundra; 12. Mediterranean forests, woodlands, and scrub; 13. Deserts and xeric shrublands

richness and current climate (Hawkins et al., 2003). Most analyses relate species richness to measures of environmental energy, water availability, productivity, environmental heterogeneity, environmental stability, or simply area (e.g. Kreft and Jetz, 2007). Mittelbach et al. (2007) distinguish three main sets of explanations for the latitudinal gradient: (a) ecological hypotheses focusing on mechanisms of species coexistence; (b) evolutionary hypotheses focusing on rates of diversification; and (c) historical hypotheses that focus on changes of the environments in Earth's history. The chapter by Griffin in this volume provides some additional background in this area.

Area and the latitudinal gradient
The hypothesis that differences in areas of the major biomes might play a major role for the species richness gradient was introduced by

Terborgh (1973) and further developed by Rosenzweig (1995) looking at the sizes of nine latitudinal zones or biomes. They showed that with respect to their classification the tropics was the largest zone and argued that extinction rates should be lower in larger biomes and speciation rates higher. Evaluating alternative biome classifications, Hawkins and Porter (2001) found no significant relation between either the size of local biomes (biomes subdivided by biogeographical regions) or biogeographical realms and their respective bird species richness. However, Turner (2004) suggested that metacommunities or the species pool within the tropics may be larger than that at other latitudes as species at higher latitudes disperse mostly in an east-west direction within their climate zones. The latitudinal extent of the species range of a widespread bird species in Eurasia might be around 20°. The relatively uniform temperature across the intertropical belt in contrast might allow isotropical dispersal and lead to species ranges such as that of the butterfly *Heliconius erato* that spans over 60° latitude (Turner, 2004). However, this explanation is in contrast to Rapoport's rule, which states that seasonal variation at high latitudes favors species with broad ecological tolerances, resulting in larger distribution ranges. In the tropics, lower ecological tolerance of species might result in higher spatial heterogeneity or species turnover, which then leads to higher species richness at a biogeographical scale (Stevens, 1989). Though patterns differ between taxonomic groups, the global analysis by Orme et al. (2006) for birds shows a weak tendency for areas with high species richness to house species with significantly smaller median range areas. However, Orme et al. (2006) did not find really clear global rules governing the patterns. Lowest median range sizes and lowest median latitudinal range extents can be found on islands and in low-latitude mountain areas in particular. Globally, these data show an increase in median geographic range area from southern to northern latitudes, which is strongly correlated with the almost linear increase of global land area from southern to northern latitude.

Fine and Ree (2006) argued that analyses of species richness against area on this scale have to integrate the change of the area of a biome through time—a time-integrated species-area effect. They found a positive correlation of a composite parameter integrating the area of a given biome over geological time with current tree species richness in that biome. On the basis of this analysis, the latitudinal gradient with highest species richness in the tropics might be caused by the greater area that tropical forests have had over geological time.

Climate and productivity

Two centuries ago, the first biogeographers like von Humboldt noticed the dependence of biological diversity on climate parameters such as temperature and rainfall (von Humboldt, 1808). These relations were increasingly studied during the last decades (Lauer and Frankenberg, 1979; Currie and Paquin, 1987; O'Brien, 1993; Mutke et al., 2001; Hawkins et al., 2003; Currie et al., 2004; Woodward et al., 2004; Mutke and Barthlott, 2005; Storch et al., 2006; Kreft and Jetz, 2007) and led to the formulation of the species-energy hypothesis (Wright, 1983), water-energy hypothesis (Hawkins et al., 2003; Hawkins and Porter, 2003), and the ambient energy hypothesis (Turner et al., 1987). The predictive value of energy- and water-related variables is itself linked to latitude (Kerr and Packer, 1997; Hawkins et al., 2003; Mutke and Barthlott, 2005; Kreft and Jetz, 2007). Whereas species richness is mainly correlated with measures of energy at high latitudes, at lower latitudes water supply or combined measures become more important where species are not limited by energy.

For the explanation of the close correlation of species richness to energy, three hypotheses are often cited (Currie et al., 2004):

1. The 'energy-richness hypothesis' (Wright, 1983) suggests that the energy supply limits the number of individuals in a given community. Thus, the higher number of individuals might allow more species to occur at abundances large enough to maintain viable populations. This might translate to a higher number of species at more productive sites. However, Currie et al. (2004) found largely inconsistent evidence for the predictions made by this hypothesis.
2. The 'physiological tolerance hypothesis' states that diversity patterns are governed by the tolerances of individual species or taxa for different sets of climatic conditions (Currie et al., 2004). In the context of the phylogenetic niche conservatism, this is highlighted as an important mechanism in shaping today's floras and faunas (Wiens and Graham, 2005; Donoghue, 2008).
3. The 'speciation rate hypothesis' states that evolutionary rates are faster at higher ambient temperatures, due to shorter generation times, higher mutation rates, and/or faster physiological processes (Rhode, 1992; Allen et al., 2002; compare next section).

Evolution and history

Important mechanisms governing the species richness at biogeographical scales like speciation act far too slowly to explain biodiversity patterns at these scales on the basis of current climate (McGlone, 1996). The latitudinal gradient of biodiversity extended back to the Mesozoic and into the Paleozoic (Crame, 2001). Using especially the increasing amount of phylogenetic data available, many recent studies tested for differences in diversification rates at different latitudes (Cardillo, 1999; Cardillo et al., 2005; Ricklefs, 2006b). Other authors related biodiversity gradients to historical changes in the environment (Fine and Ree, 2006; Mittelbach et al., 2007).

Wiens and Donoghue (2004) advocated for an integration of the ecological explanations discussed above with historical biogeography. One important aspect of this is the 'tropical conservatism hypothesis': the tropics had a greater geographical extent until some 30–40 million years ago, resulting in the fact that many groups of organisms originated in the tropics and thus had more time for speciation in tropical regions (i.e., the time-for-speciation effect, Stephens and Wiens,

2003). Actually, most families of modern plants initially diversified in times when much of the world was tropical in the early Tertiary (Ricklefs, 2005). Still today, more than half the families of flowering plants are restricted to the tropics. In contrast, only 15% are distributed primarily in temperate latitudes (Ricklefs and Renner, 1994).

Jablonski et al. (2006) relate comparable patterns for marine bivalves to differences in the rates of origination, extinction, and of changes in geographic distribution at different latitudes. They studied fossil records for this group over the last 11 million years searching for the time and place of first occurrence of a taxon. Instead of favoring one of the classical scenarios with the tropics either as a cradle or as a museum of biodiversity, they propose an 'out of the tropics' model, in which taxa preferentially originate in the tropics and expand toward the poles without losing their tropical presence. However, there is no conclusion about the reason for the higher rates of diversification in the tropics (Jablonski et al., 2006). The question of whether these rates are mainly caused by higher origination rates or whether higher extinction rates in the more seasonal and climatically more unstable higher latitudes play a role, is still not solved and might as well be taxon-specific (Martin et al., 2007; Wiens, 2007). Within the amphibians, the number of hylid frogs per region closely correlates with the time of colonization of that region without major differences in net diversification rates per unit time, as predicted by the tropical conservatism hypothesis and the time-for-speciation effect (Wiens et al., 2006). However, frogs and salamanders in general are of temperate origin. The clades within these groups that entered the tropics now contain most of the species as a result of higher net diversification rates. This seems to be due to lower extinction rates in the tropics as compared to that in higher latitudes (Wiens, 2007).

Looking at global patterns of mammals, birds, reptiles, amphibians, and vascular plants, Jansson (2003) found that climatic stability since the Last Glacial Maximum was a good predictor of endemism values. He argues that smaller climatic shifts result in a higher probability that paleoendemics survive and that diverging gene pools persist without going extinct or merging.

Geodiversity

Spatial heterogeneity of the environment at broad spatial scales is often linked to mountain areas though steep climatic gradients might occur in other regions as well. The correlation of mountains with biodiversity and species richness patterns might have at least three causes: First, looking at the heterogeneity linked to the topography,

mountains have larger surface areas than planar areas that have the same size on the map. A theoretical slope of 45° increases the real area compared to the planar one by 40%. However, using a 220 x 220 km grid and a digital elevation model (DEM) with 1 km resolution for continental Europe, Nogués-Bravo and Araújo (2006) found only 6% of the grid cells with significantly increased surface area (5–25% area deviation) and there were only minor effects when this parameter was included in models to explain species richness patterns. Comparable results have been found by Triantis et al. (2008) using high resolution DEM for several oceanic islands. Second, mountain areas harbor a higher diversity of climatic zones and habitats than lowland areas because of, for example, the strong altitudinal gradient of temperature and windward-leeward effects on precipitation. This has led to concepts and terminologies of geodiversity (Barthlott et al., 1996; Leser, 1997; Jedicke, 2001), environmental heterogeneity (Ricklefs, 1977; Tuomisto and Ruokolainen, 2005), or environmental diversity (Faith and Walker, 1996; Faith, 2003). Third, linked to geodiversity is the role of mountain environments during climatic fluctuations. Mountains might act as 'museums' or refugia for species (Fjeldså and Lovett, 1997). Especially at lower latitudes—where the highest correlation of species richness with geodiversity is found (Kerr and Packer, 1997; Rahbek and Graves, 2001; Kreft and Jetz, 2007)—mountains harbor vegetation zones starting at tropical rain forests going up to permanent snow. Instead of migrating hundreds of kilometers north or south in reaction, for example, to an increase in temperature, species might reach just a higher elevational zone with suitable climate. This does not only allow species to survive climatic changes and thus to accumulate higher biodiversity. It also leads to allopatric splitting of populations on separate mountain peaks and increased diversification rates, especially given, for example, alternating glacial and interglacial periods (Janzen, 1967; Roy, 1997; Rull, 2005).

Several authors found significant correlations between biological diversity and the spatial heterogeneity of the environment (Kerr and Packer, 1997; Mutke et al., 2001; Rahbek and Graves, 2001; Kühn et al., 2004). This relation depends on the range sizes of the species involved. Centers of the diversity of species with restricted range sizes correlate well with mountain areas whereas richness of species with larger range sizes show closer correlation to climatic parameters or primary productivity (Jetz and Rahbek, 2002; Kreft et al., 2006).

Geodiversity or the spatial heterogeneity of the abiotic environment is not always a result of a heterogeneous topography. The high edaphic heterogeneity in the lowland fynbos ecosystems in Southern Africa is an important driver of high plant species richness in these systems (Cowling and Lombard, 2002). However, the fact that Kerr et al. (2001) found land cover diversity to be a better predictor of the species richness of Canadian butterflies than topographical variation might be influenced by the extent of the study. A positive effect of a complex topography and mountains can be found especially at lower latitudes where energy is not the main limiting factor and where temperature allows a higher diversity of different vegetation types along the elevational gradient (Kerr and Packer, 1997; Kreft and Jetz, 2007).

Null models and neutral models

Mainly based on the work of Colwell and colleagues (Colwell and Hurtt, 1994; Colwell and Lees, 2000; Colwell et al., 2004), it has been proposed that the pattern of the latitudinal and altitudinal gradients is only partly governed by environment and history. Instead, the mid-domain peak often found along these gradients might be explained by geometric constraints of the random placements of species ranges within a bounded domain (Colwell et al., 2004). Several studies have demonstrated in one-dimensional (Brehm et al., 2007) and two-dimensional (Jetz and Rahbek, 2001) analyses that geometric constraints indeed might play a role for biodiversity patterns (but see Currie and Kerr, 2008). Others have found only weak or no correlations between the observed patterns and predictions of pure null models (e.g., Kerr et al., 2006), or better performance of models combining mid-domain effects with environmental drivers (Fu et al., 2006; Storch et al., 2006). One important but also controversial contribution was the development of the 'unified neutral theory of biodiversity' by Stephen Hubbell (2001). This theory assumes that all organisms are ecologically equivalent, and it considers only factors such as random dispersal, the birth and death of individuals, and the total number of organisms in the community to explain the structure of ecological communities (Hubbell, 2001; Whitfield, 2002). Though being based on only a few simple parameters, the theory makes predictions as well for broad-scale diversity patterns (Turner, 2004). Discussion about the validity of the theory and tests of its possible predictions are still vivid (e.g., Ostling, 2005; Ricklefs, 2006a; Adler et al., 2007).

9.6.4 Altitudinal Gradients

The fact that the environment, including its biological components, shows transitions and changes related to altitude had probably been rec-

ognized by human beings long before the origins of biogeography (Lomolino, 2001). Regarding quantitative analyses of altitudinal diversity gradients, a widely cited study by Terborgh (1977) showed a monotonic decrease of bird species richness with increasing elevation. During the last decades, many analyses have instead showed unimodal or hump-shaped gradients with highest diversity at mid-elevations (Rahbek, 1995, 1997, 2005; Grytnes and McCain, 2007)—the so-called mid-elevation bulge (Gentry and Dodson, 1987; Bhattarai et al., 2004; Küper et al., 2004a; Herzog et al., 2005; Krömer et al., 2005; Fu et al., 2006). However, Ibisch et al. (1996) showed that the shape of the altitudinal gradient of epiphyte species richness in the Peruvian Andes changed completely with the spatial resolution of the elevational zones in the analyses. Using intervals of 500-m elevations, they found a monotonic decrease of species richness with each level of elevation. If the elevational gradient is only classified in four classes, the mid-elevations of the 'tierra templada' possess the highest species richness, though the surface area is more than four times smaller compared to the 'tierra caliente' ('tierra caliente', 0–1,000 m.a.s.l. 'tierra templada', 1000–2,500 m.a.s.l.; 'tierra fria', 2,500-4,000 m.a.s.l. m; and 'tierra helada', > 4,000 m.a.s.l.).

Nogués-Bravo et al. (2008) show that the observed pattern of plant species richness along an altitudinal gradient in the Pyrenees changes completely with the extent of the gradient included in the analyses.

Moreover, these patterns are highly taxon- and life form-specific (Kessler, 2001a, 2002). In an analysis of species richness gradients of plant life forms, such as trees, shrubs, epiphytes, or herbs in the Peruvian Andes, we have shown that trees show a monotonic decrease of species richness with elevation. All other life forms show secondary species richness maxima at mid-elevations or at least only a moderate decrease of species richness from the lowland to these elevations (Braun et al., 2002; Mutke, 2002). In addition, all groups besides trees showed elevated species-to-family ratios in mid-elevations. These might indicate important evolutionary radiation events in these highly geodiverse areas (see, e.g., Hughes and Eastwood, 2006). Figure 9.4 shows comparable gradients for endemic species of Ecuador and the relationship between overall species richness and available area per altitudinal zone.

With regard to environmental parameters and biotic processes, several causes and drivers of altitudinal diversity gradients have been proposed. These include area (Körner, 2000; Braun et al., 2002), the generally decreasing temperature and productivity with increasing elevation, mid-elevation maxima of precipitation and air humidity

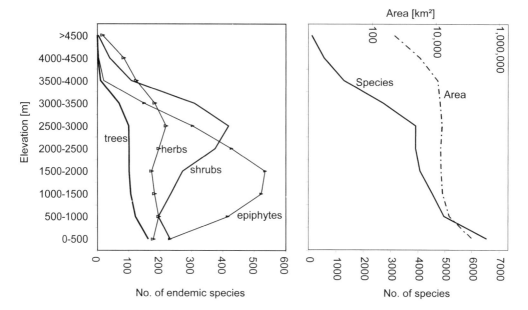

Figure 9.4   Altitudinal gradients of species richness and endemism in Ecuador. (left) Numbers of endemic species for trees, shrubs, herbs, and epiphytes based on data in Valencia et al., 2000. (right) Total species richness versus area per 500-m elevational zone (modified from Braun et al., 2002 based on data in Jørgensen and León-Yánez, 1999)

(Lauer, 1976; Kessler, 2001b; Krömer et al., 2005), high levels of geodiversity resulting in the isolation of small populations (Brown, 2001), or recent orogeny driving speciation in some areas (Gentry, 1982; Hughes and Eastwood, 2006). A combination of various factors seems likely.

Geometric constraints are also discussed as the cause for mid-elevational bulge of diversity. Null models based on a random position of species along the elevational gradient without taking into account any environmental parameter correlated well with many documented hump-shaped gradients (Colwell and Hurtt, 1994; Rahbek, 1997; Colwell and Lees, 2000; Grytnes and Vetaas, 2002; Colwell et al., 2004). Depending on the taxonomic group studied and, for example, the range size and ecological specialization of the taxa the importance of environmental and geometric factors might vary (Fu et al., 2006; Brehm et al., 2007).

9.7 BIODIVERSITY GRADIENTS IN MARINE ENVIRONMENTS

Though 70% of the Earth's surface is covered by oceans, our knowledge of the diversity of life in marine habitats is still very limited. Until the beginning of this millennium, less than 5 km² of the floor of the deep sea had been studied with respect to its fauna (Türkay, 2001). The biodiversity of marine ecosystems still exhibits great surprises. Using molecular analyses, Sogin et al. (2006) found up to 20,000 different types of microbes in samples of 1 liter of oceanic water—despite the fact that until today only about 5,000 different bacteria have been named by scientists for the entire planet (Hawksworth and Kalin-Arroyo, 1995). Since the year 2000, the 'Census of Marine Life', an international initiative with researchers from 80 countries, aims to document and understand biodiversity in all ocean realms (Martínez Arbizu and Brix, 2008, compare www.coml.org).

Despite these large gaps in our knowledge about marine biodiversity, several analyses of global marine diversity patterns had been published recently. An analysis of global tuna and billfish diversity patterns showed peaks at intermediate latitudes with lower species richness toward the equator and the higher latitudes. Species richness maxima lie off the U.S. and Australian east coasts, south of the Hawaiian Islands chain, east of Sri Lanka, and in the southeastern Pacific (Worm et al., 2005). This pattern of highest diversity at intermediate latitudes is the same for foraminiferan zooplankton (Rutherford et al., 1999), as well as for marine mammals (Schipper et al., 2008). Centers of diversity of benthic marine macroalgae are located mostly in temperate regions, like eastern Asia, especially around Japan or southern Australia, each of which contains 350–450 genera of algae. Lowest diversity occurs in the polar regions where fewer than 100 genera have been recorded. Tropical areas show only moderate numbers of macroalgae genera with the Indomalayan region having a higher diversity than, for example, the Atlantic or the eastern Pacific at the same latitudes (Kerswell, 2006).

An analysis of the diversity patterns of more than 3,000 reef fish, corals, snails, and lobsters, however, showed the highest diversity for all of these groups in the 'coral triangle' of Southeast Asia centered around the equator. Species richness declines with increasing latitude, as well as toward the Pacific and the Indian Oceans. Compared to the southeast Asian center, the much less species rich regional center of the Atlantic Ocean is located in the Caribbean (Roberts et al., 2002). Estimates of the overall species richness of coral ecosystems range from 1 to 3.5 million species—twice the number of all species documented by scientists for all groups of organisms on Earth so far (Adey, 1998; Adey, 2005). The diversity hotspot of large benthic foraminifera, which today is as well concentrated in the Indo-Australian Archipelago, shifted across almost half the globe during the last 50 million years (Renema et al., 2008). While the highest generic richness was located in the West Tethyan and Arabian region in the Eocene, the hotspot shifted to the Arabian and Southeast Asia in the Miocene and finally to the Indo-Australian Archipelago.

A unique habitat with a highly specialized fauna are deep sea vents, which are mainly situated along the boundaries of the tectonic plates. More than 80% of the species occur nowhere else and 75% of the species are only known from one locality (WWF and IUCN, 2001).

9.8 GRADIENTS OF BIOCULTURAL DIVERSITY

Cultural diversity is closely linked to the diversity of the environment—both the abiotic (geodiversity) and the biotic (biodiversity). Especially good data are available for the diversity and distribution of languages (Grimes, 2000; UNESCO et al., 2003). Based on an earlier version of our dataset shown in Figure 9.1, Stepp et al. (2004) found a close correlation between plant species richness and language diversity at the global scale.

Seven out of nine megadiverse countries in regard to languages are also on the list of biological megadiversity countries (UNESCO et al., 2003). Papua New Guinea—one of the 20 global centers of high plant species richness—is home to more than 850 of the 7,000 different human languages (UNESCO et al., 2003). However, the relation between biological and cultural diversity might be scale-dependent (Manne, 2003; Pautasso, 2007). Manne (2003) found a significant correlation of species richness of passerine birds and language diversity in Latin America at a coarse scale but a much weaker relation at finer scales. One reason for this finding and one of the main conclusions of her study is that languages are less able to coexist sympatrically compared to species. This results in generally smaller range sizes and thus potentially in a higher vulnerability.

Generally, the highest density of languages can be found in tropical areas with low human population density and especially in tropical mountain areas (Maffi, 2005; Stepp et al., 2005). It is argued that higher population densities result in more frequent contacts between human populations and thus in some kind of linguistic homogenization (Maffi, 2005; Stepp et al., 2005). High geodiversity in mountain areas with barriers to migration might have a similar effect as low population density and is also an important driver of allopatric speciation (Stepp et al., 2005). Additionally, one might argue that aseasonal tropical environments with a high productivity provide humans with all the necessities without the need for migrations or trade—and thus again this results in fewer contacts and less linguistic homogenization.

9.9 CONCLUSIONS

Reviewing the extensive literature on biodiversity gradients, it seems that the theories and models to explain these gradients are almost as plentiful as the species that are the basis of these patterns. It might be too optimistic to expect patterns that are built on the ecology and distribution of thousands of individual species to be explained by a few parameters. However, one could argue as well that trends that are found for so many different groups in different places and ecosystems should have rather simple explanations (Rhode, 1992). The increasing amount of large digital data sets on species distributions, phylogenies, environmental data, and so on, together with increasing computational power and new geostatistical tools, has rapidly increased our understanding of the distribution of life on Earth during the last decade. Integrating these new data sets as well as new insights and methods from historical biogeography, phylogeographic studies and (macro)ecology will lead us from pure statistical correlations to causal explanations of the biodiversity gradients on Earth (Wiens and Donoghue, 2004).

REFERENCES

Adey, W.H. (1998) 'Coral reefs: Algael structured and mediated ecosystems in shallow, turbulent, alkaline waters', *Journal of Phycology*, 34: 393–406.

Adey, W.H. (2005) 'Marine plant diversity', in Krupnick, G.A. and Kress, J.W. (eds.), *Plant Conservation: A Natural History Approach*. Chicago: University of Chicago Press. pp. 25–34.

Adler, P.B., HilleRisLambers, J., and Levine, J.M. (2007) 'A niche for neutrality', *Ecology Letters*, 10: 95–104.

Allen, A.P., Brown, J.H., and Gillooly, J. (2002) 'Global biodiversity, biochemical kinetics and the energetic-equivalence rule', *Science*, 297: 1545–48.

Arrhenius, O. (1920) 'Distribution of the species over the area', *Meddeland. Vetenskapsakad. Nobelinst*, 4: 1–6.

Arrhenius, O. (1921) 'Species and area', *Journal of Ecology*, 9: 95–99.

Baillie, J.E.M., Hilton-Taylor, C., and Stuart, S. (eds.). (2004) *A Global Species Assessment*. Gland, Cambridge: IUCN.

Barthlott, W., Lauer, W., and Placke, A. (1996) 'Global distribution of species diversity in vascular plants: Towards a world map of phytodiversity', *Erdkunde*, 50: 317–28.

Barthlott, W., Kier, G., and Mutke, J. (1999) 'Globale Artenvielfalt und ihre ungleiche Verteilung', *Courier Forschungsinstitut Senckenberg*, 215: 7–22.

Barthlott, W., Mutke, J., Rafiqpoor, M.D., Kier, G., and Kreft, H. (2005) 'Global centres of vascular plant diversity', *Nova Acta Leopoldina*, 92: 61–83.

Beentje, H.J., Adams, B., Davis, S.D., and Hamilton, A.C. (1994) 'Regional overview: Africa', in Davis, S.D., Heywood, V.H. and Hamilton, A.C. (eds.), *Centres of Plant Diversity. A Guide and Strategy for Their Conservation. Vol. 1: Europe, Africa and the Middle East*. Cambridge: IUCN Publications Unit. pp. 101–48.

Berghaus, H.K.W. (1837–1847) *Berghaus physikalischer Atlas*. Gotha: Justus Perthes Geographische Anstalt.

Bhattarai, K.R., Vetaas, O.R., and Grytnes, J.-A. (2004) 'Fern species richness along a central Himalayan elevational gradient, Nepal', *Journal of Biogeography*, 31: 389–400.

Bibby, C.J., Collar, N.J., Crosby, M.J., Heath, M.F., Imboden, C., Johnson, T.H., Long, A.J., Stattersfield, A.J., and Thirgood, S.L. (1992) *Putting Biodiversity on the Map: Priority Areas for a Global Conservation*. Cambridge: ICBP.

Braun, G., Mutke, J., Reder, A., and Barthlott, W. (2002) 'Biotope patterns, phytodiversity and forestline in the Andes, based on GIS and remote sensing data', in Körner, C. and Spehn, E.M. (eds.), *Mountain Biodiversity: A Global Assessment*. London: Parthenon. pp. 75–89.

Brehm, G., Colwell, R.K., and Kluge, J. (2007) 'The role of environment and mid-domain effect on moth species richness along a tropical elevational gradient', *Global Ecology and Biogeography*, 16: 205–19.

Brooks, T.M. and 14 authors (2004) 'Coverage provided by the Global Protected-Area System: Is it enough?' *BioScience*, 54: 1081–91.

Brooks, T.M., Mittermeier, R.A., da Fonseca, G.A.B., Gerlach, J., Hoffman, M., Lamoreux, J.F., Mittermeier, C.G., Pilgrim, J.D. and Rodrigues, A.S.L. (2006) 'Global biodiversity conservation priorities', *Science*, 313: 58–61.

Brown, J.H. (2001) 'Mammals on mountainsides: Elevational patterns of diversity', *Global Ecology and Biogeography*, 10: 101–9.

Buckley, L.B., and Jetz, W. (2007) 'Environmental and historical constraints on global patterns of amphibian richness', *Proceedings of the Royal Society of London Series B*, 274: 1167–73.

Caley, M.J., and Schluter, D. (1997) 'The relationship between local and regional diversity', *Ecology*, 78: 70–80.

Cardillo, M. (1999) 'Latitude and rates of diversification in birds and butterflies', *Proceedings of the Royal Society of London B*, 266: 1221–25.

Cardillo, M., Orme, C.D.L., and Owens, I.P.F. (2005) 'Testing for latitudinal bias in diversification rates: An example using new world birds: Latitudinal gradients', *Ecology*, 86: 2278–87.

Ceballos, G., and Ehrlich, P.R. (2006) 'Global mammal distributions, biodiversity hotspots, and conservation', *Proceedings of the National Academy of Sciences of the United States of America*, 103: 19374–79.

Ceballos, G., Ehrlich, P.R., Soberón, J., Salazar, I., and Fay, J.P. (2005) 'Global mammal conservation: What must we manage?' *Science*, 309: 603–7.

Chown, S.L., Sinclair, B.J., Leinaas, H.P., and Gaston, K.J. (2004) 'Hemispheric asymmetries in biodiversity—A serious matter for ecology', *PLoS Biology*, 2: 1701–7.

Cleef, A.M., Rangel Ch., O., van der Hammen, T., and Jaramillo, M.R. (1984) 'La vegetación de las selvas del transecto Buritaca', in Van der Hammen, T. and Ruiz, P.M. (eds.), *La Sierra Nevada de Santa Marta (Colombia) transecto Buritaca—La Cumbre*. Berlin: Cramer, J. pp. 267–406.

Collins, N.M., and Morris, M.G. (1985) *Threatened Swallowtail Butterflies of the World. IUCN Red Data Book*. Cambridge: IUCN.

Colwell, R.K., and Hurtt, G.C. (1994) 'Nonbiological gradients in species richness and a spurious Rapoport effect', *The American Naturalist*, 144: 570–95.

Colwell, R.K., and Lees, D.C. (2000) 'The mid-domain effect: Geometric constraints on the geography of species richness', *Trends in Ecology and Evolution*, 15: 70–76.

Colwell, R.K., Rahbek, C., and Gotelli, N.J. (2004) 'The mid-domain effect and species richness patterns: What have we learned so far?' *American Naturalist*, 163: E1–E23.

Connor, E.F., and McCoy, E.D. (1979) 'The statistics and biology of the species-area relationship', *American Naturalist*, 113: 791–833.

Costanza, R., and 12 authors. (1997) 'The value of the world's ecosystem services and natural capital', *Science*, 387: 253–60.

Cowling, R.M., and Lombard, A.T. (2002) 'Heterogeneity, speciation/extinction history and climate: Explaining regional plant diversity patterns in the Cape Floristic Region', *Diversity and Distributions*, 8: 163–69.

Crame, J.A. (2001) 'Taxonomic diversity gradients through geological time', *Diversity and Distributions*, 7: 175–89.

Crawley, M.J., and Harral, J.E. (2001) 'Scale dependence in plant biodiversity', *Science*, 291: 864–68.

Cumming, G.S. (2000) 'Using habitat models to map diversity: Pan-African species richness of ticks (Acari : Ixodida)', *Journal of Biogeography*, 27: 425–40.

Currie, D.J. (1991) 'Energy and large-scale patterns of animal- and plant-species richness', *American Naturalist*, 137: 27–49.

Currie, D.J., and Kerr, J.T. (2008) 'Tests of the mid-domain hypothesis: A review of the evidence', *Ecological Monographs*, 78: 3–18.

Currie, D.J., and Paquin, V. (1987) 'Large-scale biogeographical patterns of species richness of trees', *Nature*, 329: 326–27.

Currie, D.J., and 10 authors. (2004) 'Predictions and tests of climate-based hypotheses of broad-scale variations in taxonomic richness', *Ecology Letters*, 7: 1121–34.

Davies, R.G., Orme, C.D.L., Webster, A.J., Jones, K.E., Blackburn, T.M., and Gaston, K.J. (2007) 'Environmental predictors of global parrot (Aves: Psittaciformes) species richness and phylogenetic diversity', *Global Ecology and Biogeography*, 16: 220–33.

Davis, S.D., Heywood, V.H., Herrera-MacBryde, O., Villa-Lobos, J.L. and Hamilton, A.C. (eds.). (1994–1997) *Centres of Plant Diversity. A Guide and Strategy for Their Conservation*. Cambridge: IUCN Publications Unit.

Donoghue, M.J. (2008) 'A phylogenetic perspective on the distribution of plant diversity', *Proceedings of the National Academy of Sciences*, 105: 11549–55.

Edwards, P.J., Kollmann, J., and Fleischmann, K. (2002) 'Life history evolution in Lodoicea maldivica (Arecaceae)', *Nordic Journal of Botany*, 22: 227–37.

Faith, D.P. (2003) 'Environmental diversity (ED) as surrogate information for species-level biodiversity', *Ecography*, 26: 374–79.

Faith, D.P., and Walker, P.A. (1996) 'Environmental diversity: On the best-possible use of surrogate data for assessing the relative biodiversity of sets of areas', *Biodiversity and Conservation*, 5: 399–415.

Feuerer, T., and Hawksworth, D.L. (2007) 'Biodiversity of lichens, including a world-wide analysis of checklist data based on Takhtajan's Floristic regions', *Biodiversity and Conservation*, 16: 85–98.

Field, R. (2002) 'Latitudinal diversity gradients', *Encyclopedia of Life Sciences*, 1–8.

Fine, P.V.A., and Ree, R.H. (2006) 'Evidence for a time-integrated species-area effect on the latitudinal gradient in tree diversity', *American Naturalist*, 168: 796–804.

Fischer, A.G. (1960) 'Latitudinal variations in organic diversity', *Evolution*, 14: 64–81.

Fjeldså, J., and Lovett, J.C. (1997) 'Geographical patterns of old and young species in African forest biota: The significance of specific montane areas as evolutionary centres', *Biodiversity and Conservation*, 6: 325–46.

Frahm, J.P., and Gradstein, R.S. (1991) 'An altitudinal zonation of tropical rain forests using bryophytes', *Journal of Biogeography*, 18: 669–78.

Francis, A.P., and Currie, D.J. (2003) 'A globally consistent richness-climate relationship for angiosperms', *The American Naturalist*, 161: 1–37.

Frodin, D.G. (2001) 'Floras in retrospect and for the future', *Plant Talk*, 25: 36–39.

Fu, C., Hua, X., Li, J., Chang, Z., Pu, Z., and Chen, J. (2006) 'Elevational patterns of frog species richness and endemic richness in the Hengduan Mountains, China: Geometric constraints, area and climate effects', *Ecography*, 29: 919–27.

Gaston, K.J. (1994) *Rarity*. London: Chapman and Hall.

Gaston, K.J. (1996) 'Biodiversity—Latitudinal gradients', *Progress in Physical Geography*, 20: 466–76.

Gaston, K.J. (2000) 'Global patterns in biodiversity', *Nature*, 404: 220–27.

Gentry, A.H. (1982) 'Neotropical floristic diversity: Phytogeographical connections between Central and South America, Pleistocene climatic fluctuations, or an accident of the Andean orogeny?' *Annals of the Missouri Botanical Garden*, 69: 557–93.

Gentry, A.H., and Dodson, C.H. (1987) 'Diversity and biogeography of neotropical vascular epiphytes', *Annals of the Missouri Botanical Garden*, 74: 205–33.

Graham, C.H., Ferrier, S., Huettman, F., Moritz, C., and Peterson, A.T. (2004) 'New developments in museum-based informatics and applications in biodiversity analysis', *Trends in Ecology and Evolution*, 19: 497–503.

Grenyer, R., and 15 authors. (2006) 'Global distribution and conservation of rare and threatened vertebrates', *Nature*, 444: 93–96.

Grimes, B.F. (ed.). (2000) *Ethnologue, Volume 1: Languages of the World*. 14th ed. Dallas, TX: Summer Institute of Linguistics International.

Grytnes, J.-A., and McCain, C.M. (2007) 'Elevational trends in biodiversity', in Levin, S.A. (ed.), *Encyclopedia of Biodiversity*. New York: Elsevier. pp. 1–8.

Grytnes, J.A., and Vetaas, O.E. (2002) 'Species richness and altitude: A comparison between null models and interpolated plant species richness along the Himalayan altitudinal gradient, Nepal', *American Naturalist*, 159: 294–304.

Hawkins, B.A. (2001) 'Ecology's oldest pattern?' *Endeavour*, 25: 133–34.

Hawkins, B.A., and Porter, E.E. (2001) 'Area and latitudinal diversity gradient for terrestrial birds', *Ecology Letters*, 4: 595–601.

Hawkins, B.A., and 11 authors. (2003) 'Energy, water, and broad-scale geographic patterns of species richness', *Ecology*, 84: 3105–17.

Hawkins, B.A., and Porter, E.E. (2003) 'Water-energy balance and the geographic pattern of species richness of western Palearctic butterflies', *Ecological Entomology*, 28: 678–86.

Hawksworth, D.L., and Kalin-Arroyo, M.T. (1995) 'Magnitude and distribution of biodiversity', in Heywood, V.H. and

Watson, R.T. (eds.), *Global Biodiversity Assessment*. Cambridge: Cambridge University Press. pp. 107–91.

Heaney, L.R. (2007) 'Is a new paradigm emerging for oceanic island biogeography?' *Journal of Biogeography*, 34: 753–57.

Hebert, P.D.N., Penton, E.H., Burns, J.M., Janzen, D.H., and Hallwachs, W. (2004) 'Ten species in one: DNA barcoding reveals cryptic species in the neotropical skipper butterfly Astraptes fulgerator', *Proceedings of the National Academy of Sciences of the United States of America*, 101: 14812–17.

Herzog, S.K., Kessler, M., and Bach, K. (2005) 'The elevational gradient in Andean bird species richness at the local scale: A foothill peak and a high-elevation plateau', *Ecography*, 28: 209–22.

Hillebrand, H. (2004) 'On the generality of the latitudinal diversity gradient', *The American Naturalist*, 163: 192–211.

Hubbell, S.P. (2001) *The Unified Neutral Theory of Biodiversity and Biogeography*. Princeton, NJ: Princeton University Press.

Hughes, C., and Eastwood, R. (2006) 'Island radiation on a continental scale: Exceptional rates of plant diversification after uplift of the Andes', *Proceedings of the National Academy of Sciences of the United States of America*, 103: 10334–39.

Huston, M.A. (1994) *Biological Diversity*. Cambridge: Cambridge University Press.

Ibisch, P.L., Boegner, A., Nieder, J., and Barthlott, W. (1996) 'How diverse are neotropical epiphytes? An analysis based on the "Catalogue of flowering plants and Gymnosperms of Peru" ', *Ecotropica*, 2: 13–28.

Isaac, N.J.B., Turvey, S.T., Collen, B., Waterman, C., and Baillie, J.E.M. (2007) 'Mammals on the EDGE: Conservation priorities based on threat and phylogeny', *PLoS ONE*, 2: e296.

Jablonski, D., Roy, K., and Valentine, J.W. (2006) 'Out of the tropics: Evolutionary dynamics of the latitudinal diversity gradient', *Science*, 314: 102–6.

Jansson, R. (2003) 'Global patterns in endemism explained by past climatic change', *Proceedings of the Royal Society of London Series B*, 270: 583–90.

Janzen, D.H. (1967) 'Why mountain passes are higher in the tropics', *American Naturalist*, 101: 233–49.

Jedicke, E. (2001) 'Biodiversität, Geodiversität, Ökodiversität. Kriterien zur Analyse der Landschaftsstruktur—ein konzeptioneller Diskussionsbeitrag', *Naturschutz und Landschaftsplanung*, 33: 59–68.

Jetz, W., Kreft, H., Ceballos, G., and Mutke, J. (2008) 'Global associations between terrestrial producer and vertebrate consumer diversity', *Proceedings of the Royal Society of London Series B*, 276: 269–78.

Jetz, W., and Rahbek, C. (2001) 'Geometric constraints explain much of the species richness pattern in African birds', *Proceedings of the National Academy of Sciences of the United States of America*, 98: 5661–66.

Jetz, W., and Rahbek, C. (2002) 'Geographic range size and determinants of avian species richness', *Science*, 297: 1548–51.

Jetz, W., Wilcove, D.S., and Dobson, A.P. (2007) 'Projected impacts of climate and land-use change on the global diversity of birds', *PLoS Biology*, 5: 1211–19.

Jørgensen, P.M., and León-Yánez, S. (eds.). (1999) *Catalogue of the Vascular Plants of Ecuador*. St. Louis: Missouri Botanical Garden Press.

Kalmar, A., and Currie, D.J. (2006) 'A global model of island biogeography', *Global Ecology and Biogeography*, 15: 72–81.

Kerr, J.T., and Packer, L. (1997) 'Habitat heterogeneity as a determinant of mammal species richness in high-energy regions', *Nature*, 385: 253–54.

Kerr, J.T., Southwood, T.R.E., and Cihlar, J. (2001) 'Remotely sensed habitat diversity predicts butterfly species richness and community similarity in Canada', *Proceedings of the National Academy of Sciences of the United States of America*, 98: 11365–70.

Kerr, J.T., Perring, M., and Currie, D.J. (2006) 'The missing Madagascan mid-domain effect', *Ecology Letters*, 9: 149–59.

Kerswell, A.P. (2006) 'Global biodiversity patterns of benthic marine algae', *Ecology*, 87: 2479–88.

Kessler, M. (2001a) 'Patterns of diversity and range size of selected plant groups along an elevational transect in the Bolivian Andes', *Biodiversity and Conservation*, 10: 1897–921.

Kessler, M. (2001b) 'Pteridophyte species richness in Andean forests in Bolivia', *Biodiversity and Conservation*, 10: 1473–95.

Kessler, M. (2002) 'The elevational gradient of Andean plant endemism: Varying influences of taxon-specific traits and topography at different taxonomic levels', *Journal of Biogeography*, 29: 1159–65.

Kier, G., and Barthlott, W. (2001) 'Measuring and mapping endemism and species richness: A new methodological approach and its application on the flora of Africa', *Biodiversity and Conservation*, 10: 1513–29.

Kier, G., Mutke, J., Dinerstein, E., Ricketts, T.H., Küper, W., Kreft, H., and Barthlott, W. (2005) 'Global patterns of plant diversity and floristic knowledge', *Journal of Biogeography*, 32: 1107–16.

Kier, G., and Kraft, H., Lee, T.M., Jetz, W., Ibisch, P.L., Nowicki, C., Mutke, J. and Barthlott, W. (2009) 'A global assessment of endemism and species richness across island and mainland regions', *Proceedings of the National Academy of Sciences of the United States of America*, 106(23): 9322–9327.

Kissling, D.W., Field, R., and Böhning-Gaese, K. (2008) 'Spatial patterns of woody plant and bird diversity: Functional relationships or environmental effects?' *Global Ecology and Biogeography*, 17: 327–39.

Kissling, W.D., Rahbek, C., and Böhning-Gaese, K. (2007) 'Food plant diversity as broad-scale determinant of avian frugivore richness', *Proceedings of the Royal Society of London Series B*, 274: 799–808.

Kitching, R.L. (2006) 'Crafting the pieces of the diversity jigsaw puzzle', *Science*, 313: 1055–57.

Köhler, J., Vieites, D.R., Bonett, R.M., García, F.H., Glaw, F., Steinke, D., and Vences, M. (2005) 'New amphibians and global conservation: A boost in species discoveries in a highly endangered vertebrate group', *BioScience*, 55: 693–96.

Konrat, M.V., Renner, M., Söderström, L., Hagborg, A., and Mutke, J. (2008) 'Early land plants today: Liverwort species diversity and the relationship with higher taxonomy and higher plants', *Fieldiana Botany*, 47: 91–104.

Körner, C. (2000) 'Why are there global gradients in species richness? Mountains might hold the answer', *Trends in Ecology and Evolution*, 15: 513–14.

Kreft, H., and Jetz, W. (2007) 'Global patterns and determinants of vascular plant diversity', *Proceedings of the National Academy of Sciences of the of United States of America*, 104: 5925–30.

Kreft, H., Jetz, W., Mutke, J., Kier, G., and Barthlott, W. (2008) 'Global diversity of island floras from a macroecological perspective', *Ecology Letters*, 11: 116–27.

Kreft, H., Sommer, J.H., and Barthlott, W. (2006) 'The significance of geographic range size for spatial diversity patterns in Neotropical palms', *Ecography*, 29: 21–30.

Krömer, T., Kessler, M., Gradstein, S.R., and Acebey, A. (2005) 'Diversity patterns of vascular epiphytes along an elevational gradient in the Andes', *Journal of Biogeography*, 32: 1799–809.

Kühn, I., Bierman, S.M., Durka, W., and Klotz, S. (2006) 'Relating geographical variation in pollination types to environmental and spatial factors using novel statistical methods', *New Phytologist*, 172: 127–39.

Kühn, I., Brandl, R., and Klotz, S. (2004) 'The flora of German cities is naturally species rich', *Evolutionary Ecology Research*, 6: 749–64.

Küper, W., Kreft, H., Nieder, J., Köster, N. and Barthlott, W. (2004a) 'Large-scale diversity patterns of vascular epiphytes in Neotropical montane rain forests', *Journal of Biogeography*, 31: 1477–87.

Küper, W., and 9 authors. (2004b) 'Africa's hotspots of biodiversity redefined', *Annals of the Missouri Botanical Garden*, 91: 525–36.

Küper, W., Sommer, J.H., Lovett, J.C., and Barthlott, W. (2006) 'Deficiency in African plant distribution data— missing pieces of the puzzle', *Botanical Journal of the Linnean Society*, 150: 355–68.

Lamoreux, J.F., Morrison, J.C., Ricketts, T.H., Olson, D., Dinerstein, E., McKnight, M., and Shugart, H.H. (2006) 'Global tests of biodiversity concordance and the importance of endemism', *Nature*, 440: 212–14.

Lauer, W. (1976) 'Klimatische Grundzüge der Höhenstufung tropischer Gebirge', in Uhlig, H. and Ehlers, E. (eds.), Wiesbaden: Franz Steiner. pp. 76–90.

Lauer, W., and Frankenberg, P. (1979) 'Zur Klima- und Vegetationsgeschichte der westlichen Sahara', *Abhandlungen der Mathematisch-Naturwissenschaftlichen Klasse*, 1: 61 pp.

Lennon, J.J., Koleff, P., Greenwood, J.J.D., and Gaston, K.J. (2001) 'The geographical structure of British bird distributions: diversity, spatial turnover and scale', *Journal of Animal Ecology*, 70: 966–79.

Leser, H. (1997) 'Von der Biodiversität zur Landschaftsdiversität. Das Ende des disziplinären Ansatzes der Diversitätsproblematik', in Erdmann, K.H. (ed.), *Internationaler Naturschutz*. Heidelberg. pp.145–75.

Linder, H.P. (2001) 'Plant diversity and endemism in sub-Saharan tropical Africa', *Journal of Biogeography*, 28: 169–82.

Linder, H.P. (2003) 'The radiation of the Cape flora, southern Africa', *Biological Reviews*, 78: 597–638.

Lomolino, M.V. (2001) 'Elevation gradients of species-density: Historical and prospective views', *Global Ecology and Biogeography*, 10: 3–13.

Lomolino, M.V., Riddle, B.R., and Brown, J.H. (2006) *Biogeography*. 3rd edn. Sunderland, MA: Sinauer.

MacArthur, R.H., and Wilson, E.O. (1967) *The Theory of Island Biogeography*. Princeton, NJ: Princeton University Press.

Maffi, L. (2005) 'Linguistic, cultural, and biological diversity', *Annual Review of Anthropology*, 29: 599–617.

Malyshev, L.I. (1975) 'The quantitative analysis of flora: Spatial diversity, level of specific richness, and representativity of sampling areas (in Russian)', *Botanicheskiy Zhurnal*, 60: 1537–50.

Malyshev, L.I. (1991) 'Some quantitative approaches to problems of comparative floristics', in Nimis, P.L. and Crovello, T.J. (eds.), *Quantitative Approaches to Phytogeography*. Dordrecht: Kluwer Academic. pp. 15–33.

Manne, L.L. (2003) 'Nothing has yet lasted forever: Current and threatened levels of biological and cultural diversity', *Evolutionary Ecology Research*, 5: 517–27.

Martin, P.R., Bonier, F., and Tewksbury, J.J. (2007) 'Revisiting Jablonski (1993): Cladogenesis and range expansion explain latitudinal variation in taxonomic richness', *Journal of Evolutionary Biology*, 20: 930–36.

Martínez Arbizu, P., and Brix, S. (2008) 'Bringing light into deep-sea biodiversity', *Zootaxa* 1866: 5–6.

McGlone, M.S. (1996) 'When history matters: Scale, time, climate and tree diversity', *Global Ecology and Biogeography Letters*, 5: 309–14.

Millennium Ecosystems Assessment. (2005) *Ecosystems and Human Well-being: Biodiversity Synthesis*. Washington, DC: World Resources Institute.

Mittelbach, G.G., and 21 authors. (2007) 'Evolution and the latitudinal diversity gradient: Speciation, extinction and biogeography', *Ecology Letters*, 10: 315–31.

Mittermeier, R.A., Mittermeier, C.G., Brooks, T.M., Pilgrim, J.D., Konstant, W.R., da Fonseca, G.A.B. and Kormos, C. (2003) 'Wilderness and biodiversity conservation', *Proceedings of the National Academy of Sciences of the United States of America*, 100: 10309–13.

Mittermeier, R.A., Myers, N., da Fonseca, G.A.B., Olivieri, S. and Thomson, J.B. (1998) 'Biodiversity hotspots and major tropical wilderness areas: Approaches to setting conservation priorities', *Conservation Biology*, 12: 516–20.

Mittermeier, R.A., and 7 authors. (2005) *Hotspots Revisited: Earth's Biologically Richest and Most Endangered Terrestrial Ecoregions*. Sierra Madre: University of Virginia.

Moles, A.T., Ackerly, D.D., Tweddle, J.C., Dickie, J.B., Smith, R., Leishmann, M.R., Mayfield, M.M., Pitman, A., Wood, J.T. and Westoby, M. (2007) 'Global patterns in seed size', *Global Ecology and Biogeography*, 16: 109–16.

Mooers, A.O., Heard, S.B., and Chrostowski, E. (2005) 'Evolutionary heritage as a metric for conservation', in Purvis, A., Brooks, T.L., and Gittleman, J.L. (eds.),

Phylogeny and Conservation. Oxford: Oxford University Press. pp. 120–38.

Mutke, J. (2002) *Räumliche Muster Biologischer Vielfalt—die Gefäßpflanzenflora Amerikas im globalen Kontext*. PhD thesis, Rheinische Friedrich-Wilhelms-Universität, Bonn.

Mutke, J., and Barthlott, W. (2005) 'Patterns of vascular plant diversity at continental to global scales', *Biologiske Skrifter*, 55: 521–37.

Mutke, J., Kier, G., Braun, G., Schultz, C., and Barthlott, W. (2001) 'Patterns of African vascular plant diversity—a GIS based analysis', *Systematics and Geography of Plants*, 71: 1125–36.

Mutke, J., Kier, G., Krupnick, G.A., and Barthlott, W. (2005) 'Terrestrial plant diversity', in Krupnick, G.A. and Kress, W.J. (eds.), *Plant Conservation: A Natural History Approach*. Chicago: University of Chicago Press. pp. 15–25.

Myers, N. (1988) 'Threatened biotas: "Hot spots" in tropical forests', *Environmentalist*, 8: 187–208.

Myers, N., Mittermeier, R.A., Mittermeier, C.G., da Fonseca, G.A.B., and Kent, J. (2000) 'Biodiversity hotspots for conservation priorities', *Nature*, 403: 853–58.

Nelson, B.W., Ferreira, C.A.C., da Silva, M.F. and Kawasaki, M.L. (1990) 'Endemism centres, refugia and botanical collection density in Brazilian Amazonia', *Nature*, 345: 714–16.

Nic Lughadha, E., and 19 authors. (2005) 'Measuring the fate of plant diversity: Towards a foundation for future monitoring and opportunities for urgent action', *Philosophical Transactions of the Royal Society of London Series B*, 360: 359–72.

Nogués-Bravo, D., Araujo, M.B., Romdal, T., and Rahbek, C. (2008) 'Scale effects and human impact on the elevational species richness gradients', *Nature*, 453: 216–19.

Nogués-Bravo, D., and Araújo, M.B. (2006) 'Species richness, area and climate correlates', *Global Ecology and Biogeography*, 15: 452–60.

Novotny, V., Drozd, P., Miller, S.E., Kulfan, M., Janda, M., Basset, Y., and Weiblen, G.D. (2006) 'Why are there so many species of herbivorous insects in tropical rainforests?' *Science*, 313: 1115–18.

O'Brien, E.M. (1993) 'Climatic gradients in woody plant species richness: Towards an explanation based on an analysis of southern Africa's woody flora', *Journal of Biogeography*, 20: 181–98.

Olson, D.M., and 17 authors. (2001) 'Terrestrial ecoregions of the world: A new map of life on Earth', *BioScience*, 51: 933–38.

Orme, C.D.L., and 14 authors. (2005) 'Global hotspots of species richness are not congruent with endemism or threat', *Nature*, 436: 1016–19.

Orme, C.D.L., and 11 authors. (2006) 'Global patterns of geographic range size in birds', *PLoS Biology*, 4: 1276–83.

Ostling, A. (2005) 'Neutral theory tested by birds', *Nature*, 436: 635–36.

Pautasso, M. (2007) 'Scale dependence of the correlation between human population presence and vertebrate and plant species richness', *Ecology Letters*, 10: 16–24.

Pianka, E.R. (1966) 'Latitudinal gradients in species diversity: A review of concepts', *American Naturalist*, 100: 33–46.

Plotkin, J.B., and 11 authors. (2000) 'Predicting species diversity in tropical forests', *Proceedings of the National Academy of Sciences of the United States of America*, 97: 10850–54.

Pommier, T., Pinhassi, J., and Hagström, A. (2005) 'Biogeographic analysis of ribosomal RNA clusters from marine bacterioplankton', *Aquatic Microbial Ecology*, 41: 79–89.

Proches, S., and Cowling, R.M. (2006) 'Insect diversity in Cape fynbos and neighbouring South African vegetation', *Global Ecology and Biogeography*, 15: 445–51.

Qian, H., and Ricklefs, R.E. (2008) 'Global concordance in diversity patterns of vascular plants and terrestrial vertebrates', *Ecology Letters*, 11: 547–53.

Rabinowitz, D. (1981) 'Seven forms of rarity', in Synge, H. (ed.), *The Biological Aspects of Rare Plant Conservation*. Chichester U.K.: John Wiley. pp. 205–17.

Rahbek, C. (1995) 'The elevational gradient of species richness—a uniform pattern', *Ecography*, 18: 200–205.

Rahbek, C. (1997) 'The relationship among area, elevation, and regional species richness in Neotropical birds', *American Naturalist*, 149: 875–902.

Rahbek, C. (2005) 'The role of spatial scale and the perception of large-scale species-richness patterns', *Ecology Letters* 8: 224–39.

Rahbek, C., and Graves, G.R. (2001) 'Multiscale assessment of patterns of avian species richness', *Proceedings of the National Academy of Sciences of the United States of America*, 98: 4534–39.

Redford, K.H., and 13 authors. (2003) 'Mapping the conservation landscape', *Conservation Biology*, 17: 116–31.

Renema, W., and 14 authors. (2008) 'Hopping hotspots: Global shifts in marine biodiversity', *Science*, 321: 654–57.

Rhode, K. (1992) 'Latitudinal gradients in species diversity: The search for the primary cause', *Oikos*, 65: 514–27.

Ricketts, T.H., and 30 authors. (2005) 'Pinpointing and preventing imminent extinctions', *Proceedings of the National Academy of Sciences of the United States of America*, 102: 18497–501.

Ricklefs, R.E. (1977) 'Environmental heterogeneity and plant species-diversity—hypothesis', *American Naturalist*, 111: 376–81.

Ricklefs, R.E. (2005) 'Historical and ecological dimensions of global patterns in plant diversity', *Biologiske Skrifter*, 55: 583–603.

Ricklefs, R.E. (2006a) 'The unified neutral theory of biodiversity: Do the numbers add up?' *Ecology*, 87: 1424–31.

Ricklefs, R.E. (2006b) 'Global variation in the diversification rate of passerine birds', *Ecology*, 87: 2468–78.

Ricklefs, R.E., and Renner, S.S. (1994) 'Species richness within families of flowering plants', *Evolution*, 48: 1619–36.

Roberts, C.M., and 11 authors. (2002) 'Marine biodiversity hotspots and conservation priorities for tropical reefs', *Science*, 295: 1280–84.

Rosenzweig, M.L. (1995) *Species Diversity in Space and Time*. Cambridge: Cambridge University Press.

Roy, M.S. (1997) 'Recent diversification in African greenbuls (Pycnonotidae: Andropadus) supports a montane speciation model', *Proceedings of the Royal Society of London Series B*, 264: 1337–44.

Rull, V. (2005) 'Biotic diversification in the Guayana Highlands: A proposal', *Journal of Biogeography*, 32: 921–27.

Rutherford, S., D'Hondt, S., and Prell, W. (1999) 'Environmental controls on the geographic distribution of zooplankton diversity', *Nature*, 400: 749–53.

Sarr, D.A., Hibbs, D.E., and Huston, M.A. (2005) 'A hierarchical perspective of plant diversity', *The Quarterly Review of Biology*, 80: 188–212.

Sax, D.F. (2001) 'Latitudinal gradients and geographic ranges of exotic species: Implications for biogeography', *Journal of Biogeography*, 28: 139–50.

Schatz, G.E. (2002) 'Taxonomy and herbaria in service of plant conservation: Lessons from Madagascar's endemic families', *Annals of the Missouri Botanical Garden*, 89: 145–52.

Schipper, J., and 130 authors. (2008) 'The status of the world's land and marine mammals: Diversity, threat, and knowledge', *Science*, 322: 225–30.

Scholes, R.J., Küper, W., and Biggs, R. (2006) 'Biodiversity', in UNEP (ed.), *Africa Environment Outlook II. Our Environment, Our Wealth*. Earthprint. pp. 226–61.

Serre, D., and Pääbo, S. (2004) 'Evidence for gradients of human genetic diversity within and among continents', *Genome Research*, 14: 1679–85.

Shepherd, U. (1998) 'A comparison of species diversity and morphological diversity across the North American latitudinal gradient', *Journal of Biogeography*, 25: 19–29.

Simon, H.-R. (1995) 'Arteninventar des Tierreiches. Wieviele Tierarten kennen wir?' *Naturwissenschaftlicher Verein Darmstadt-Berichte N. F.*, 17: 103–21.

Smith, K.F., and Brown, J.H. (2002) 'Patterns of diversity, depth range and body size among pelagic fishes along a gradient of depth', *Global Ecology and Biogeography*, 11: 313–22.

Smith, M.A., Rodriguez, J.J., Whitfield, J.B., Deans, A.R., Janzen, D.H., Hallwachs, W., and Hebert, P.D.N. (2008) 'Extreme diversity of tropical parasitoid wasps exposed by iterative integration of natural history, DNA barcoding, morphology, and collections', *Proceedings of the National Academy of Sciences of the United States of America*, 105: 12359–64.

Sogin, M.L., Morrison, H.G., Huber, J.A., Welch, D.M., Huse, S.M., Neal, P.R., Arrieta, J.M., and Herndl, G.J. (2006) 'Microbial diversity in the deep sea and the under-explored "rare biosphere"', *Proceedings of the National Academy of Sciences of the United States of America*, 103: 12115–20.

Stattersfield, A.J., Crosby, M.J., Long, A.J., and Wege, D.C. (1998) *Endemic Bird Areas of the World. Priorities for Biodiversity Conservation*. Cambridge: Bird Life International.

Stephens, P.R., and Wiens, J.J. (2003) 'Explaining species richness from continents to communities: The time-for-speciation effect in emydid turtles', *American Naturalist*, 161: 112–28.

Stepp, J.R., Castaneda, H., and Cervone, S. (2005) 'Mountains and biocultural diversity', *Mountain Research and Development*, 25: 223–27.

Stepp, J.R., Cervone, S., Castaneda, H., Lasseter, A., Stocks, G., and Gichon, Y. (2004) 'Development of a GIS for global biocultural diversity', *Policy Matters*, 13: 267–70.

Stevens, G.C. (1989) 'The latitudinal gradient in geographical range: How so many species coexist in the Tropics', *American Naturalist*, 133: 240–56.

Storch, D., and 12 authors. (2006) 'Energy, range dynamics and global species richness patterns: Reconciling mid-domain effects and environmental determinants of avian diversity', *Ecology Letters*, 9: 1308–20.

Stuart, S.N., Chanson, J.S., Cox, N.A., Young, B.E., Rodrigues, A.S.L., Fischman, D.L., and Waller, R.W. (2004) 'Status and trends of amphibian declines and extinctions worldwide', *Science*, 306: 1783–86.

Terborgh, J. (1973) 'On the notion of favorableness in plant ecology', *American Naturalist*, 107: 481–501.

Terborgh, J. (1977) 'Bird species diversity on an Andean elevational gradient.' *Ecology*, 58: 1007–19.

Triantis, K.A., Nogués-Bravo, D., Hortal, J., Borges, P.A.V., Adsersen, H., Fernandez-Palacios, J.M., Araujo, M.B., and Whittaker, R.J. (2008) 'Measurements of area and the (island) species-area relationship: New directions for an old pattern', *Oikos*, 117: 1555–59.

Tuomisto, H., and Ruokolainen, K. (2005) 'Environmental heterogeneity and the diversity of pteridophytes and Melastomataceae in western Amazonia', *Biologiske Skrifter*, 55: 37–56.

Türkay, M. (2001) 'Die Tiefsee, der größte Lebensraum', in Türkay, M. (ed.), *Leben ist Vielfalt*. E. Schweizerbart. pp. 9–30.

Turner, J.R.G. (2004) 'Explaining the global biodiversity gradient: Energy, area, history and natural selection', *Basic and Applied Ecology*, 5: 435–48.

Turner, J.R.G., Gatehouse, C.M., and Corey, C.A. (1987) 'Does solar energy control organic diversity? Butterflies, moths and the British climate.' *Oikos*, 48: 195–205.

UNESCO, Terralingua, and WWF (eds.). (2003) *Sharing a World of Difference. The Earth's Linguistic, Cultural, and Biological Diversity*. Paris: UNESCO.

Valencia, R. (1994) *Composition and Structure of Three Ecuadorian Forests at Different Elevations*. PhD thesis, University of Aarhus, Aarhus/Dänemark.

Valencia, R., Pitman, N., León-Yánez, S., and Jørgensen, P.M. (2000) *Libro rojo de las plantas endémicas del Ecuador*. Quito: Herbario QCA, Pontificia Universidad Católica del Ecuador.

von Humboldt, A. (1808) *Ansichten der Natur*. Tübingen: Cotta.

von Humboldt, A. (1817) *De Distributione geographica plantarum secundum coeli temperiem et altitudinem montium, prolegomena*. Paris.

von Humboldt, A. (1845) *Kosmos. Entwurf einer physischen Weltbeschreibung, Erster Band*. Stuttgart und Tübingen: Cotta.

Wallace, A.R. (1878) *Tropical Nature and Other Essays*. New York: Macmillan.

Whitfield, J. (2002) 'Neutrality versus the niche', *Nature*, 417: 480–81.

Whittaker, R.H. (1960) 'Vegetation of the Siskiyou Mountains, Oregon and California', *Ecological Monographs*, 30: 279–338.

Whittaker, R.H. (1972) 'Evolution and measurement of species diversity', *Taxon*, 21: 213–51.

Whittaker, R.H. (1977) 'Evolution of species diversity in land communities', *Evolutionary Biology*, 10: 1–67.

Whittaker, R.J., Araújo, M.B., Paul, J., Ladle, R.J., Watson, J.E.M., and Willis, K.J. (2005) 'Conservation biogeography: Assessment and prospect', *Diversity and Distributions*, 11: 3–23.

Whittaker, R.J., Ladle, R.J., Araujo, M.B., Maria Fernandez-Palacios, J., Domingo Delgado, J., and Ramon Arevalo, J. (2007) 'The island immaturity—speciation pulse model of island evolution: An alternative to the "diversity begets diversity" model', *Ecography*, 30: 321–27.

Whittaker, R.J., Willis, K.J., and Field, R. (2001) 'Scale and species richness: Towards a general, hierarchical theory of species diversity', *Journal of Biogeography*, 28: 453–70.

Wiens, J.J. (2007) 'Global patterns of diversification and species richness in amphibians', *American Naturalist*, 170: S86–S106.

Wiens, J.J., and Donoghue, M.J. (2004) 'Historical biogeography, ecology and species richness', *Trends in Ecology and Evolution*, 19: 639–44.

Wiens, J.J., and Graham, C.H. (2005) 'Niche conservatism: Integrating evolution, ecology, and conservation biology', *Annual Review of Ecology, Evolution and Systematics*, 36: 519–39.

Wiens, J.J., Graham, C.H., Moen, D.S., Smith, S.A., and Reeder, T.W. (2006) 'Evolutionary and ecological causes of the latitudinal diversity gradient in hylid frogs: Treefrog trees unearth the roots of high tropical diversity', *Amercian Naturalist*, 168: 579–96.

Williams, P.H. (1993) 'Measuring more of biodiversity for choosing conservation areas, using taxonomic relatedness', in Moon, T.-Y. (ed.), *International Symposium on Biodiversity and Conservation*. Seoul: Korean Entomological Institute. pp. 194–227.

Williams, P.H., Gibbons, D., Margules, C., Rebelo, A., Humphries, C., and Pressey, R.L. (1996) 'A comparison of richness hotspots, rarity hotspots and complementary areas for conserving diversity of British birds', *Conservation Biology*, 10: 155–74.

Williams, P.H., Humphries, C.J., and Gaston, K.J. (1994) 'Centers of seed-plant diversity—the family way', *Proceedings of the Royal Society of London Series B*, 256: 67–70.

Willig, M.R., Kaufman, D.M., and Stevens, R.D. (2003) 'Latitudinal gradients of biodiversity: Patterns, process, scale and synthesis', *Annual Review of Ecology, Evolution and Systematics*, 34: 273–309.

Willis, K.J., and Whittaker, R.J. (2002) 'Species diversity—scale matters', *Science*, 295: 1245–46.

Woodward, F.I., Lomas, M.R., and Kelly, C.K. (2004) 'Global climate and the distribution of plant biomes', *Philosophical Transactions of the Royal Society of London Series B*, 359: 1465–76.

Worm, B., Sandow, M., Oschlies, A., Lotze, H.K., and Myers, R.A. (2005) 'Global patterns of predator diversity in the open oceans', *Science*, 309: 1365–69.

Wright, D.H. (1983) 'Species-energy theory: An extension of species area-theory', *Oikos*, 41: 496–506.

Wulff, E.W. (1935) 'Versuch einer Einteilung der Vegetation der Erde in pflanzengeographische Gebiete auf Grund der Artenzahl', *Repertorium Specierum Novarum Regni Vegetabilis*, 12: 57–83.

WWF and IUCN. (2001) *The Status of Natural Resources on the High Seas*. Gland, Switzerland: WWF/IUCN.

Explaining Biogeographical Distributions and Gradients: Floral and Faunal Responses to Natural Disturbances

Anke V. Jentsch and Carl Beierkuhnlein

10.1 INTRODUCTION

In this chapter we discuss the importance of disturbance regimes for biogeographical patterns and ecosystem dynamics on a global basis. We stress the roles of disturbances at various spatial and temporal scales ranging from long-term evolutionary processes to short-term successional trajectories. We introduce (a) the general causes of biogeographical patterns; (b) the geography of disturbance regimes; and (c) the ecological importance of disturbance for biogeographical patterns, such as species diversity. We then discuss aspects of fundamental significance, such as (d) rhythms and synchronization; (e) the timing of events; and (f) successional pathways and inertia, which we feel are underrepresented in disturbance ecology but which may lead to fresh perspectives in the future.

At the global scale, biodiversity patterns vary along latitudinal, altitudinal, and environmental gradients (see Mutke, this volume). These are related to climate (see Schickhoff, Chapter 8, this volume), habitat diversity, and human influences. In addition to gradients, an important global-scale phenomenon is that specific types of natural disturbances dominate different biomes (e.g., see Figure 10.2). Furthermore, within each climate zone and ecozone, different altitudinal belts are

prone to characteristic types and intensities of mechanical processes, for example, landslides and avalanches, which act as triggers of disturbance. Alterations to ecosystems due to disturbances such as floods, fire, and traditional land uses may cause greater species loss than gradual shifts in the environment due to climate change (Tilman, 1996).

Disturbances, then, are ubiquitous, inherent, and unavoidable, and they affect all levels of biological organization. Ecosystems are influenced by various kinds of disturbances, the range of which is enormous; for example, fires (see Enright, this volume), windstorms, landslides, flooding, logging, grazing, burrowing animals, and outbreaks of pathogens. Natural and anthropogenic disturbances cause ecosystems to undergo changes that can range from sudden to gradual and whose effects can range from dramatic to subtle. Therefore, the presence of disturbances in all ecosystems, their occurrence across a wide range of spatial and temporal scales, and their continuity across all levels of ecological organization is the essence of their importance (Pickett and White, 1985; White and Jentsch, 2001). In the following paragraphs, we develop further the argument for the importance of understanding disturbance regimes to gain a better understanding of floral and faunal distributions and biodiversity gradients.

Since the global crisis of biodiversity came to the forefront of scientific and public debate during the 1990s, there have been very many conservation actions. However, these often draw on static views of ecosystems and biomes. Dynamic perspectives—ones that concentrate on the protection of processes rather than on particular target animal or plant communities—are comparatively rare. The vast majority of conservation efforts focus on designating spatial reserves, that is, the many types of protected areas. Despite our comment about the lack of a dynamic perspective in protected area planning, we acknowledge that nature protection is accepted as a major societal goal and in our minds this raises two questions:

1. What kinds of nature should be protected?
2. Does nature protection also include the partial destruction of communities as a consequence of disturbance-induced ecosystem dynamics?

The concept of 'naturalness' is implicit in these two questions. However, it is rarely defined well. The perception about what is 'natural' differs significantly between people because of different cultural values. Nonetheless, it is hard to find a time in human history—recent human history at least—when humans were not altering most ecosystems or regions substantially. The length of human disturbance history in any one region also bears on the consideration of value. For example, what is considered worthy of being preserved in many European landscapes would hardly fulfill the criteria in landscapes that have experienced less human alteration. The concept of naturalness has been incorporated into the term 'wilderness'. Although wilderness areas have been mapped globally, the concept is particularly prevalent in North America. A closer look at wilderness areas shows that they are highly dynamic, and often prone to natural disturbances such as wildfires and floods. Neither the concept of naturalness nor that of wilderness excludes ecological disturbance by natural agents (cf. Sprugel, 1991).

At the present time anthropogenic disturbance regimes and their impacts on ecological systems appear to be the strongest drivers of biodiversity patterns. Indeed, ongoing land-use change is considered to be the major driver of biodiversity loss (Sala et al., 2000). However, in this chapter we concentrate on natural processes while acknowledging that human activities are superimposed on almost all natural disturbance regimes. Ancient cultural landscapes, such as those in eastern China, northern India, and the Mediterranean Basin; or strongly invaded ecosystems, such as New Zealand and California (see Blumler, this volume), are examples of once natural disturbance regimes that have been very strongly modified by human activities. Apart from its direct effects, climate change will have many indirect effects on ecosystems, one of which will be to force change in disturbance regimes. We anticipate that the indirect effects of altered disturbance regimes will have stronger impacts on regional biodiversity, on local predictability of environmental conditions, and on biogeographical patterns than on climate change itself (e.g., McGradySteed et al., 1997).

10.2 CAUSES OF BIOGEOGRAPHICAL PATTERNS

In general, biogeographical patterns and distributions can be explained by recent and historical processes (Figure 10.1). These in turn can be attributed to biotic and abiotic mechanisms that define the environments of species and

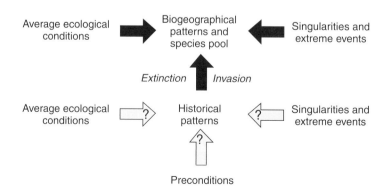

Figure 10.1 Explanation of biogeographical patterns. Singularities and extreme events also play a role in ecological history

communities (Beierkuhnlein and Jentsch 2004). In addition, spatial patterns of disturbances at various scales can be important in determining processes such as resource availability, seasonal migration, and the maintenance of minimum regional or local population sizes.

Biogeographers have developed various theories about the spatial drivers of species abundance patterns including the importance of resource constraints, the role of area size, and the rule of distance decay. Yet temporal drivers of species abundance patterns such as evolutionary time span, land-use history, or contingency of disturbance regime are rarely considered. Descriptions and analyses of biogeographical patterns have traditionally been based on the collection of environmental data that characterize typical, average, or major conditions within known or mapped distributions (for examples, see Schickhoff, Chapter 8, this volume). Shifts in key parameters across thresholds that are limited in time (e.g., late frost events) have been considered important for setting the boundaries of some distributions. Nonetheless, many ecological explanations of species distribution use climatic envelopes that tend to ignore these short-term (climate) events. We argue that the main reasons short-term events are ignored in mapping and explaining distributions is due to the short duration of rare, occasional events and because past events can rarely be detected, recorded, or reproduced. Notwithstanding the fact that their ecological and biogeographical effects may persist even though the event occurred many years previously. As long as such events are limited to small areas, they will just affect populations with limited distributional ranges. Some of these types of events occur more 'regularly' and/or are spatially widespread. But even in these cases they may be masked by 'average' environmental conditions across a region.

10.3 THE GEOGRAPHY OF DISTURBANCE REGIMES

Disturbances occur across various spatial and temporal scales. The sum of all types of disturbances, including their spatial and temporal organization, is described by the term 'disturbance regime' (White and Jentsch, 2001). Biogeographical overviews concentrate on either the influence of long-lasting ecological and environmental conditions or on long-term climatic shifts and environment fluctuations during, for instance, the Pleistocene. These approaches are quite successful in explaining general biogeographical patterns (see examples in Schickhoff and Willis

et al., this volume). However, they ignore the natural variability over time that determines the maxima and minima of ecological parameters. In some circumstances these extreme conditions may be more influential than average conditions in explaining biogeographical patterns. Irregularities and deviations from general trends—and here we include extreme maxima and minima—are usually considered to be noise in the data sets used to determine regional patterns.

Aside from the large-scale patterns alluded to above, and slow and gradual geological processes such as continental drift, nonlinear and abrupt events have always been part of the suite of geological disturbance mechanisms that influence biogeographical patterns. Good examples of nonlinear, abrupt geological events are earthquakes that can either be direct disturbance mechanisms or indirect (i.e., by triggering mass movements). However, the spatial distribution of earthquakes is anything but random, being most frequent around transform plate boundaries (cf. maps produced by national geological surveys, e.g., the USGS Earthquake Hazards program, http://earthquake.usgs.gov/eqcenter/recenteqsww/). This nonrandom spatial pattern of earthquakes indicates that seismic activity affects global biogeographical patterns in a nonrandom manner. Similar arguments can be made for many other geological phenomena, for example, volcanism and tsunami.

Disturbances in general are not distributed randomly, and different disturbance regimes characterize particular ecosystems and biomes. For example, high mountain ecosystems are characterized by geomorphological and cryospheric processes such as cryoturbation, solifluction, soil erosion, glacial and fluvial erosion, avalanches, mudflows, and other mass movements, which occur in different magnitude-frequency relationships ranging from high magnitude-low frequency (e.g., mudflows and avalanches) to low magnitude-high frequency (e.g., glacial erosion). These processes create heterogeneous patches of varying sizes, shapes, and distributions and their biotic legacies vary as well (Pimentel and Kounang, 1998). Ecological attributes like habitat complexity, susceptibility, stability, persistence, and resilience differ substantially between ecosystems and climate zones. The impacts of past disturbance processes, and the time intervals between similar magnitude events (the return or recurrence interval of a disturbance event) also differ. The effectiveness of a specific disturbance event is modified by such attributes.

As we have argued already in this chapter, neither the Earth's climate system nor its vegetation cover can be understood by adopting a static view, such as reliance on global environmental overviews such as Köppen's classification system, which concentrates on trends of average and monthly temperatures and moisture, though precipitation seasonality is also important. Even the recent update of Köppen's map based on an updated set of climate statistics (Peel et al., 2007), and the maps recently produced for the Thornthwaite system (Feddema, 2008) still adopt static views.

Using climate classifications to map and model ecological and biogeographical phenomena is a particularly pertinent example because climate is linked to important geomorphological and hydrological processes such as mass movements, slope erosion, hurricanes, floods, and other short-term events that have proven strong and direct ecological effects (e.g., Donnelly and Woodruff, 2007). The effects of such events may be enhanced or buffered by relief and vegetation. It is very likely that the distribution of natural vegetation types, which is generally assumed to be a reflection of long-term climatic means, is also influenced by these short-term phenomena. For example, the availability of large amounts of snow at high altitudes, in combination with steep slopes, increases the probability of avalanches being an important disturbance mechanism. High mountain forest ecosystems located in cold, humid regions have high avalanche frequencies and the generally accepted horizontal vegetation zonation is

modified to include vertical structure (Plate 10.1). Whilst the horizontal zonation is a response to altitudinal variations in temperature and precipitation averages, the vertical element in the vegetation patterns are determined by single events such as avalanches, landslides, mudflows, and rockfalls.

10.3.1 The main types of disturbances

As already mentioned, there are many types of ecological disturbance triggers that have their origins in the natural environment. They range from (a) sudden to slow onset (compare landslides to solifluction); (b) high magnitude, low frequency to low magnitude, high frequency (compare earthquakes to soil erosion): (c) those that occur almost anywhere on the Earth's surface globally to those that are spatially restricted (compare flooding to avalanches); and (d) those with large to severely restricted impacts (compare a regional flood to a single flood on a small river). Comprehensive treatments of the geomorphological, hydrological, and climatological triggers of disturbance can be found elsewhere (White and Jentsch, 2001; Walker and del Moral, 2003; Johnson and Myanishi, 2007). Therefore in the remainder of this section we selectively consider important types of geomorphologically,

Plate 10.1   Debris flows structuring alpine vegetation patterns vertically at 2,500 m. above sea level, Taschachtal, Austria)

hydrologically, and climatologically induced disturbance mechanisms that affect ecosystem dynamics.

Volcanism: Throughout Earth history, volcanism has led to regionally restricted, large ecological disturbances—mainly because of the effects of lava flows, lahars, and local ashfalls. Particularly large eruptions, however, have caused widespread ecological disruption through the effects of disturbances of areally extensive ashfalls (e.g., the Lava Creek and Huckleberry Falls ashbeds from the Yellowstone Supervolcano, both of which covered most of the western United States). Volcanism is concentrated along plate margins and weaker parts of the Earth's crust (cf. distributional maps at the Smithsonian Institute Global Volcanism Project, http://www.volcano.si.edu/world/).

Volcanic eruptions exert direct effects, such as destroying ecosystems through the actions of lava flows, gaseous emissions, and ashfalls, as well as creating new substrate for succession at the same time. Nutrient-rich ashes can be transported in the air over large distances. Ash deposition has both negative effects—burying vegetation and corrosive effects—and positive effects such as fertilization pulses. In high mountain areas where the upper slopes of volcanoes are covered by snow and ice, mudflows known as lahars are frequently observed after volcanic eruptions. In most eruptions, these kinds of disturbances are extreme events and most species do not survive.

However there are exceptions, for example, vegetation succession is initiated on lahar deposits. As many volcanoes with snow and ice caps are distributed in a linear pattern along high-fold mountain belts (e.g., the Andes) lahars do not have widespread global distribution, but they are characteristic of mountains along continental margins.

Mass movements: Mass movements are natural disturbances that often reoccur frequently at similar locations on the same slopes. They include rockfalls, debris flows and mudflows, landslides, soil creep, and avalanches. They can be classified in terms of velocity (ranging from high-velocity rockfalls to low-velocity soil creep) and the regolith-moisture ratio (ranging from relatively 'dry' debris flows to relatively 'wet' mudflows). They are triggered by slope instabilities caused by the interaction of slope angle and regolith or hydrological properties and are most frequent in mountainous areas where they may shape treeline ecotones (Plate 10.2) and restrict forest development beyond the normal zonal features that are anticipated if 'stable' climatic models are applied.

Drought: Drought affects a large part of the continental subtropics, and are thus spatially very important on a global scale (Mehl et al., 2000; IPCC, 2007). In recent decades, evidence of increased extreme climate and weather events such as hurricanes, drought, heat waves, heavy

Plate 10.2   Primary succession on 150-year-old lava flow with the pioneer species *Rumex lunaria* (La Palma, Canary Islands)

rainfall and snowfall, and increased freeze-thaw dynamics has accumulated, and these extremes are anticipated to increase in frequency. For instance, heat waves and droughts are expected to increase in Africa (New et al., 2006). However, increasing drought will not be limited to arid and semiarid climates. Humid tropical ecosystems, such as the vast Amazon forest, which were not considered drought-prone but rather stable in terms of contemporary climates, also suffer from temporarily limited water shortage (Hutyra et al., 2005). An elongation of heat periods is also reported for temperate climates. In Central Europe such phases have doubled in length during the last 120 years (Della-Marta et al., n.d.)

Heavy rainfall and flooding: The El Niño Southern Oscillation (ENSO) is known to be accompanied by abnormalities and deviations from ordinary precipitation regimes almost worldwide. Both extremely wet and dry conditions are found (Figure 10.2). Related to these extremes during ENSO events are flooding and drought. Even if there have been some very remarkable flood events in large valleys of continental Europe, it is currently under debate whether their frequency and magnitude is yet out of the range of the historical record and probability (Milly et al., 2002).

Wind: Mechanical stress that is related to strong winds is a common kind of disturbance and its impact on ecosystems has been well summarized by Everham and Brokaw (1996) generally, and Lugo (2008) for tropical storms. Storms cause various types of disturbances, not just those related to wind stress. For example, hail, heavy precipitation, and icing are all associated with storms, and flooding is also a common consequence. The pattern of severe tropical cyclones is directly linked to the heating of ocean surfaces, to the cooling by strong ocean currents, and to prevailing directions of atmospheric circulation. But the severest effects, even in the tropics and subtropics, are generally restricted to the tropics as they rapidly lose energy over land surfaces (Figure 10.3). Nonetheless, associated tropical storms and depressions after landfalling hurricanes along the Gulf of Mexico coastline can extend well into North America; those storms that track up the Atlantic can affect New England and the maritime provinces and can even impact northwest Europe as severe extra-tropical storms, such as the 'Great Storm of 1987' in the British Isles. The highest hurricane/tropical cyclone frequencies are found in the northwestern subtropical Pacific Ocean and in the western subtropical North Atlantic Ocean. They cause severe damage on tropical and subtropical islands in the Pacific and Caribbean respectively, and on these islands their ecological importance is greatest. However, the Atlantic and Gulf of Mexico coastlines of North America and those of east Asia are also frequently exposed to these storms.

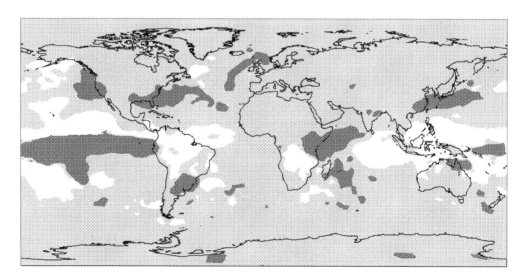

Figure 10.2   Divergence from long-term mean monthly precipitation (1980–2004) during a strong El Niño between December 1997 and February 1998 (in mm/month) based on weather observation and remote sensing. White areas were extraordinary dry; gray marks show normal conditions ± 25% of precipitation; dark shaded areas performed with extremely high rainfall during that time (DWD / GPCC)

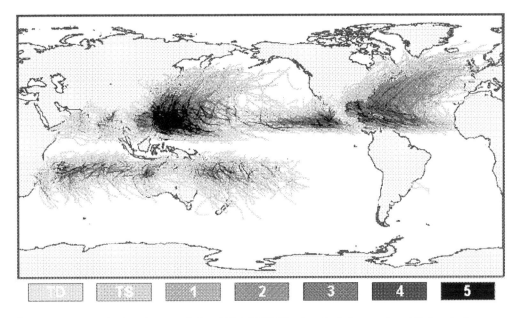

Figure 10.3   Tropical cyclone activity, 1985–2005 (TD: tropical depression; TS: tropical storm; Sturm, 1–5: hurricane categories in the Saffir-Simpson Hurricane Scale) (www. globalwarmingart.com)

From an ecological perspective, landfalling cyclones are very important in the tropics and subtropics (Lugo, 2008). Forests are severely damaged by cyclones, and this in turn stimulates succession again and again in regions with high cyclone activity. The ability of trees to resprout in such forests can be an important ecological trait in such regions. In these forests the return periods for hurricanes and cyclones are the decisive scaling factor in terrestrial ecosystem regeneration. Some authors (e.g., Webster et al., 2005) argue that the maximum wind speed and the intensity and the duration of hurricanes have increased recently, though the trends vary from one ocean region to another. Some places, for example, the Canary Islands, which were not previously impacted by hurricanes in the historical past, are now vulnerable. Ecosystems in coastal regions in the subtropics and tropics will have to adapt to changes in cyclone regimes.

Fire: As Enright (this volume) shows, fire regimes are fundamental disturbances shaping ecosystems and patterns of diversity. Large fires depend on local preconditions such as the availability of dry biomass (fuel) and weather conditions such as favourable phases of drought (Meyn et al., 2007 and references therein). Fires occur in most types of grasslands, savannas, and forests worldwide (Figure 10.4). General exceptions are found in the humid tropics and in some areas with

moist temperate climates such as Central Europe, though even in these areas there are exceptions. The duration of fires is physically limited by fuel availability and as a consequence fires are virtually absent in very arid areas (scarce vegetation) and in very humid areas (where droughts are rare and dry, cured fuel is also scarce). More detailed information on fire regimes and fire and ecosystem functioning is provided by Enright (this volume).

Anthropogenic disturbances: Though by no means the main thrust of this chapter, human actions in cultural landscapes act as agents of disturbance. Some actions that occur regularly, such as grass and hay mowing, allow more species to adapt and establish, than under regimes with irregular impacts. Disturbances that occur regularly also terminate some successional trajectories and allow some less-competitive species to coexist with strong competitors (Figure 10.5). This in turn contributes to the maintenance of high levels of community diversity, as predicted by the Intermediate Disturbance Hypothesis (Connell, 1978; Huston, 1994; Hubbel et al., 1999).

If a land-use regime is modified, either towards on the one hand more intensive land use or large-scale standardization of rhythms or on the other toward abandonment and extensification, negative consequences for biodiversity are likely occurrences. In addition, the longer a 'traditional'

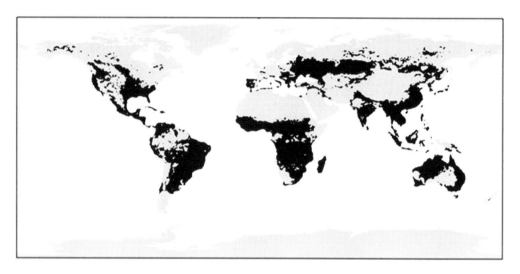

Figure 10.4   The occurrence of fires worldwide is determined by climatic conditions and human impact (fire data from MODIS ranging from 2000 until 2007)

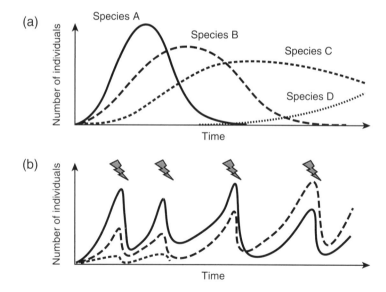

Figure 10.5   The contribution of disturbances to maintain the coexistence of species' populations with differing competitive power. Trajectories and the replacement of less competitive species are interrupted regularly. If the disturbance regime is modified or if disturbances cease to occur, the consequences for the composition of regional communities are evident

land use has been maintained, the stronger the biodiversity reaction that is likely to occur.

Heterogenous disturbance regimes: The global importance of disturbance in a biogeographical context is highlighted by the fact that spatial patterns of species diversity and evolution are closely correlated with the regional diversity of disturbance types. For example, the greater number of altitudinal belts in tropical high mountains compared to subpolar mountains means there are more types of disturbances in the former than the latter. This in turn corresponds to the general phenomenon of increased phytodiversity toward the Equator. Nevertheless, the explanation of hotspots of (phyto)diversity in tropical mountains does not yet integrate the short-term temporal dynamics of disturbances.

10.3.2 Complexity of explaining biogeographical patterns by disturbance

Environmental constraints such as minimum temperatures or the lack of substrate explain regional biogeographical patterns and altitudinal zones quite well, as has been shown for the climatologically (nontopographically) limited tree line (Körner and Paulsen, 2004). However, as we have already pointed out, disturbances are omnipresent in all ecosystems. Therefore, new dynamic approaches are needed to explain vegetation patterns. Wesche et al. (2000) have proven that in the east African high mountains the local tree line ecotone is disturbance-dependent and is fire-induced and fire-maintained rather than related to thermal limitations. Tree lines in the tropical Andes have similar controls (Joergensen et al., 1995). Isolated trees that are found above the actual limit of closed forest stands are explained in tropical high mountains as relicts of forests that formerly reached higher. Such trees might have survived in a local favourable habitat, where they are less vulnerable to fire, grazing, and woodcutting (Miehe and Miehe, 1994). Lowland limits of continuous forests are also frequently related to human land use. In the American Sierra Nevada Mountains, Richter (2001) found altitudinal forest zones separated by large gaps that he explained in terms of historical limitations. Potentially suitable tree species for these ecological niches may not have reached these places.

We now understand that temporally limited events are important in explaining biogeographical patterns at all scales ranging from local to global. Throughout geological time, catastrophic events or phases have been accompanied by abrupt shifts in the species composition of the fossil record. The alternation of periods of stability and periods of change are reflected in the traditional biostratigraphical classification of sediments. Global extinction events have been limited to relatively short time periods (on a geological time scale). However, their repercussions are still felt today. This is because evolution is stimulated by free niches after extinction events. In fact, biodiversity can reach higher levels after these phases of extinction and instability. But the time scale has to be considered. Evolution is a slow process compared to extinction and it generally takes about 10 million years to regain pre-extinction levels of diversity (Kirchner and Weil, 2000).

The echoes of large-scale disturbances also explain recent ecosystem and biome patterns. Of course, this is not only true for events in the geological past, but all short-term disturbances are followed by trajectories of succession and, for some, evolution. Fires in the boreal forest may last only for a few days. However, their effects can be seen for centuries through cohorts of tree populations that can be maintained for more or less the complete life cycle of long-lived tree species (see Kneeshaw et al., this volume). The disturbance legacies of fires are intriguing in the current era of fire suppression. Novel ecosystems may evolve if fire suppression is successful over many years and, as fuel accumulation increases, the risks of stronger fires than previously experienced are possible, and the successional pathways these may lead to are, as yet, unknown.

10.4 THE ECOLOGICAL IMPORTANCE OF DISTURBANCE

At the local scale and when looking at single organisms, disturbance events can be destructive. At larger scales and when populations, communities, and the species pool are considered, disturbances create spatial heterogeneity and temporal niche diversity (Jentsch 2004). Interactions between organisms respond to disturbances as well because species will react specifically. The interrelation between disturbances and biodiversity is addressed by the Intermediate Disturbance Hypothesis (Connell, 1978; Huston, 1994; Hubbel et al., 1999). This theory assumes that the highest species diversity is connected to intermediate levels of disturbance intensity, frequency, or spatial extent. Its relevance was proven for tropical rain forests and coral reefs. However, the IDH refers to regional species pools and to regional-scale features of disturbances and cannot be put into absolute values. Disturbances control both stability at larger scales and deflections of ecological functions and patterns at smaller scales (Turner et al., 1998; Walker et al., 1999). To assess the significance of particular disturbances for biodiversity, two scales have to be considered: the 'patch scale' related to a single disturbance event, and the 'multi-patch scale' that integrates a pattern of disturbed and undisturbed patches (Jentsch et al., 2002).

The species and community responses depend on (a) external factors, such as the magnitude of the event, spatial extent, return interval, resource availability, and environmental constraints after the event; and (b) internal factors, such as biodiversity of the regional species pool, preadaptation to the type of disturbance, and any internal functional strategies. This explains why the disturbance history of a patch and organic legacies after disturbance events determine the mode of regeneration (White and Jentsch, 2001: Jentsch and Beierkuhnlein, 2003).

Organisms develop adaptation to mechanical stresses, extraordinary temperatures, and shortages of excess water and—even if they do not or the adaptations are unsuccessful—strategies for organismic regeneration after the death of individuals and populations. For example, long-lasting soil seed-banks, high dispersal capacity, or tolerance to a wide range of ecological conditions are traits that may help organisms to cope with irregular and rare events. The degree of adaptation to a certain kind of disturbance strongly depends on the frequency of the event and on survival rates of individuals.

In many cases, the connection between a disturbance event and its ecological effect is obvious: fires in forests leave tree scars and release events in the tree ring record. However, the extinction events discussed above are more controversial. Witness the debate as to whether the impact of large extraterrestrial bodies (e.g., the Chicxuclub impact in Yucatan) or the large-scale eruptions of trap basalt (e.g., in India, Siberia, and Argentina) and their consequences for atmospheric chemistry of the stratosphere are important for explaining major extinction events in Earth history.

Cause and effect are hard to link in such circumstances, and it is usual that the longer ago the event occurred the lower the temporal resolution of dating will be. Many important measurements and data on disturbances are available only for some decades, whereas the ecological effects of disturbance events may last a much longer time (Johnson and Myanashi, 2007). Proxies, for instance, those derived through dendrochronology or palynology, may cover longer time scales. However, again they are subject to the problem of circular reasoning. For example, the strongest recorded volcanic eruptions (Tambora, 1815; Krakatau, 1883; Novarupta, 1912; Pinatubo, 1991) are reflected very well globally in tree-ring records. Other singularities in tree rings indicating short-term reduced tree growth are also observed (e.g., for 1453, 1601, and 1641), but the links to specific events cannot be traced (Figure 10.6).

The ecological effects of ENSO events are tremendous and far-reaching. Considering the

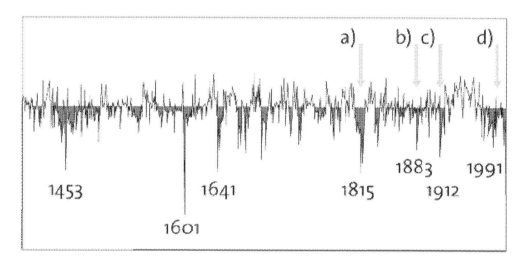

Figure 10.6 Climatic signals of globally important volcanic eruptions. The short-term cooling effect is reflected in reduced tree growth. (a) Tambora, 1815; (b) Krakatau, 1883; (c) Novarupta, 1912; (d) Pinatubo, 1991 (Beierkuhnlein, 2007)

irregularity of the phenomenon, it is important to point out that the recurrence interval is shorter than the life cycle of most long-lived key species in the affected terrestrial habitats. Organisms have developed adaptations to this irregularity. In some areas where ENSO conditions are characterized by high precipitation (e.g., Peruvian lowlands) ENSO can stimulate seed germination and vegetation regeneration after years of drought (Figure 10.7). The effects of drought during ENSO events seem to be less persistent, though when perhumid ecosystems are exposed to ENSO-related drought, effects can be considerable. The Amazon and the Indomalayan Archipelago can suffer severe droughts during ENSO years. In marine ecosystems fluctuations in biodiversity in the Pacific Ocean can be directly linked to ENSO (Worm et al., 2005). In pelagic communities, the extraordinarily low availability of nutrients and the related effects on the food web are more important than temperature itself. The latter does contribute, however, to coral reef bleaching that can also be stimulated by ENSO in tropical oceans. As ENSO events can be traced a long way back in time, it demonstrates the stability of the moisture regime in equatorial lowlands. Occasional drought signals can even serve as impulses for synchronized flowering and increased reproductive success.

A further impetus to study disturbance ecology comes from the growing awareness of extreme climatic events in connection with global climate change (IPCC, 2007). Ecological disturbances cover many types of events that are discrete in relation to the life cycle and the turnover of species or ecosystems they affect. The resulting effects may be attributed to behaviour, biomass, or species composition among a wide range of biological parameters. Effects also may be related to elemental and energy stores and fluxes and/or to information (e.g., genes, interactions). It makes sense to differentiate between the absolute impact of a certain event and its relative impact on the organisms or communities. The same event may have strong consequences or almost none, depending on the system that is affected (Table 10.1). If disturbances such as extreme climatic conditions reoccur during the life spans of organisms (e.g., trees) then these events will be determining factors in the exclusion of certain species that would otherwise tolerate the average climatic conditions indicated by the remaining species. This illustrates the importance of recording not only the magnitude of single events but also their frequency and the time intervals between disturbances. In addition, the 'selectivity' of disturbances has to be considered. By selectivity we mean that the fraction of affected individuals or species may differ, ranging from all objects to certain categories or even individuals.

Important facets of disturbances are their duration and magnitude. Both relate to the organisms and ecosystems affected. A similar magnitude event may cause only a slight response in one case but complete destruction in another. For example, the loss of a cohort of propagules may

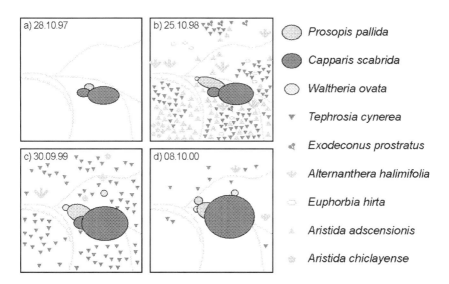

Figure 10.7 Short-term flaring-up of vegetation at the Peruvian coast after a strong ENSO event. Obviously, plant strategies are well adapted to such irregular events connected with favourable conditions (after Richter and Ise, 2005)

Table 10.1 Absolute and relative parameters to characterize and analyze disturbances

Parameter	Absolute impact	Relative impact
Type	Physical process	Adaptation strategies present
Magnitude	Value of energetic or material influence	Intensity of the effect on organisms and ecosystems
Time	Particular date or time span	Point in time during seasonal or individual development
Duration	Length of time of disturbance event	Affected phase of life cycles or development
Frequency	Occurrence per time	Related to the life span of organisms or to the turnover of ecosystems
Spatial extension	Area or altitudinal extent	Percentage of habitat
Interaction	Combined mechanisms	Feedback on biological processes and ecological complexity

be disastrous for one short-lived species whereas another species may experience this type of impact most years and be adapted to cope with frequent reproductive failure.

Generally, it is helpful to describe ecosystem dynamics as a function of two ratios: (a) the relationship of disturbance extent to landscape extent; and (b) the relationship of return interval to recovery interval (Turner et al., 1993). The effects of disturbances depend on the temporal and spatial scales of the system that is affected. Rare, large-scale events such as volcanic eruptions may be followed by a complete breakdown of a system: for example, primary succession will occur. Stability is supported when the disturbance extent is small compared to the extent of the complete habitat, and when the disturbance interval is long relative to the recovery interval of the dominant vegetation. Ecosystems that are rich in species are assumed to exhibit higher functional resilience in the face of small- to intermediate-scale disturbance.

How then can biogeographers define 'temporally limited' or 'discrete'? Some catastrophic events in Earth's history may have lasted for many years (some contemporary droughts still do) or even centuries. At the level of a particular organism such an event could hardly be described as discrete or finite; but considering their effects on the evolution of taxa they can be considered discrete. This demonstrates the fact that disturbances must always be linked to a specific biological object, process, or question.

Disturbances occur in all kinds of ecosystems and biomes at various scales. They control ecological rhythms and create temporal niches. They create spatial heterogeneity and allow species with differing competitive abilities to coexist. Thus, biodiversity is strongly dependent on the maintenance of disturbance regimes. The search

for generality in disturbance processes across ecosystems and biomes is an important scientific task (White and Jentsch, 2001). This is particularly so in a rapidly changing world, where there is an increasing need to identify general laws for the ecological role of temporal variability in order to adapt management strategies in nature reserves and to buffer the undesired effects of land-use change on ecological services. The concepts of naturalness and wilderness introduced earlier are terms with historic relevance but they hardly are applicable to ongoing or anticipated developments in the biosphere. Naturalness or wilderness never should have acquired the connotation of disturbances being absent. In fact, natural systems are highly dynamic. Conservative preservation of specific communities cannot be expected to be a reasonable approach when climate is changing (see Marchant and Taylor, this volume). Understanding the role of temporal variability should help to manage novel ecosystems in a dynamic but nonetheless sustainable way.

10.5 RHYTHMS AND SYNCHRONISATION

The temporal aspect of disturbances can be characterized in terms of stochastic variability in return period, episodic occurrence in discrete events or periodicity in the event regime. Periodic 'rhythms' occur at various time scales, ranging from diurnal through annual seasonality to longer cycles. Increasingly other temporal processes overlay these rhythms. The longer the period, the less uniform the cyclic replication seems to be. The intrinsic life cycles and the seasonal patterns of activity of organisms follow these periodic rhythms.

Diurnal rhythms are such trivial quotidian phenomena that changes in parameters such as day length or the degree of variability during a year easily can be overlooked. The diversity within certain groups of animals such as moths, however, can only be understood when the temporal niches during daytime are considered. In tropical climates the diurnal rhythm is the all-dominant pulse of nature. Temporal variability in resource availability is one factor in ecological niche creation and thus supports biodiversity. Species that are similar in their ecology may coexist simply by temporal avoidance (Clark et al., 2003).

A second rhythm, which is of great importance outside of the tropics, is seasonality. For example, the exuberant displays of geophytes in central European deciduous forests exhibit the use of temporally limited light resources in early spring because air temperatures close to the ground surface are already warm. This is in contrast to tree species that avoid the risk of rare but damaging late frosts by later leafing. For example, beech trees do not produce leaves until early summer, by which time the life cycle of many of the geophytes is almost complete.

Cyclic phenomena may also be attributed to biotic processes. Some insect populations are characterized by periodic pest outbreaks that can lead to massive economic damage (e.g., Treter, 2001). Such peaks in pest populations are regularly observed over large areas of boreal forests that are dominated by only a few tree species. Balsam fir (*Abies balsamea*) in northeastern North America is periodically affected by Eastern Spruce Budworm (*Choristoneura fumiferana*). The periodicity of population outbreaks correlated with economic damage is about 40 to 60 years (see Kneeshaw et al., this volume).

Interrelationships between population cycles have also been observed. Most prominent, though still debated, are predator-prey relationships such as that between the American snowshoe hare (*Lepus americanus*) and lynx (*Lynx canadensis*). Many tree species exhibit cycles of high seed production with periods of several years (masting) such as many oaks. This temporal synchronization supports the strengthening of the gene pool in oak species.

Obviously, the global climate is anything but constant. In fact it is characterized by episodic and abrupt changes. In part, these changes are related to rhythmic fluctuations that have interacted with the biosphere throughout Earth history and constantly stimulated the evolution of biodiversity. Most prominent is the alternation between periods of glaciations and warm climates that characterized the Pleistocene. Fluctuations in the Earth's inclination and rotation have modified the solar exposure of continents and thus led to changes in albedo. The so-called Milankovich Cycles are accompanied by the formation of glacial ice sheets at higher latitudes. In subtropical regions, atmospheric circulation patterns also change at this time. Monsoon rainfall patterns shift and are followed by shifts in vegetation types such as savanna ecosystems. Even in equatorial rain forests climatic fluctuations have been recorded. The areal extent of rain forests has expanded and contracted throughout the Pleistocene (see Willis et al, this volume). Nevertheless, the long-term persistence of tropical ecosystems has been related to the high number of species that are found there. Sanders (1968) explains this through niche separation—the time-stability hypothesis (see Griffin, this volume).

The influence of geographical position on the biogeographical effectiveness of a certain disturbance event is reflected in the intensity of the effects of Pleistocene glaciations in Central Europe, which can be catgorized as periodic, recurrent events. As a consequence of the Gulf Stream, the warm temperate climate of northwest Europe is located at higher latitudes compared to northeast America and east Asia. In addition, mountain ranges and the Mediterranean Sea have operated as efficient barriers and filters to migration and dispersal. Southward movements during glacials were restricted and re-migration after warming was filtered by these mountain barriers. This may have led to regional extinctions (see Willis et al., this volume), and with each climatic fluctuation the species pool was probably further depleted. For example, whereas Central Europe harbours three oak species, the number of oaks is about 10 times greater in North America and east Asia where modern climatic conditions are comparable. This pattern is also found in other long-lived species.

Disturbance is an important evolutionary force in the development of life-history attributes and functional traits (Figure 10.8) as it selectively promotes the ability to cope with a highly dynamic environment. Interestingly then, species that are adapted to phases of stability can coexist with species that are promoted by short phases of unstable conditions. This coexistence is a function of temporal variability, which creates niches for species in addition to ecological and spatial niches. In turn, high biodiversity contributes to functional resilience of ecosystems despite disturbance effects. Various authors address stability issues by considering the role of disturbances in ecosystem dynamics (see Grimm and Wissel, 1997). The general influence of disturbances in the development of regional biodiversity illustrates that disturbance can be seen as a major driver of evolution.

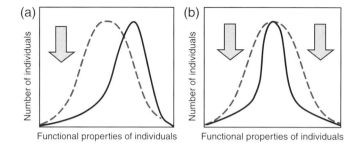
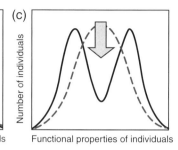

Figure 10.8 **The occurrence of deviations from traditional environmental conditions may direct shifts in the evolution of populations and species and even the disruption of populations connected with the development of new species**

Large-scale disturbance events become drivers of global biodiversity when geological time scales are considered. The crucial issue that connects the historical (regional scale) and fossil (global scale) records is that single events are hidden or ameliorated if they are not of lasting consequence. Only extremely large events with global consequences for the biosphere and for extinction versus speciation are well reflected in the fossil record. Thus it takes millions of years for global species diversity to reach pre-extinction event values (Figure 10.9). As disturbances contribute not only to dispersal but also in limiting the establishment and development of populations, they play an important role as a filter to regional biodiversity. Species with limited distributions are more likely to become extinct in the face of newly introduced species (an indirect influence of disturbances) or environmental shifts. In this respect, it appears problematic that most species or regional floras and faunas tend toward rarity. Local and regional extinctions are likely to occur when novel disturbances happen.

10.6 TIMING OF EVENTS

It is not only the duration and periodicity of a disturbance event that are important. When the disturbance occurs, time (i.e., in relation to other biological phenomena such as the timing of biotic processes and synchronization with external rhythms) has to be considered. If the timings of biotic and abiotic processes do not match in terms of 'normal' timing, organisms can be damaged even if the level of external (abiotic) stresses they are exposed to can be tolerated at other times of the year or in later phases of ontogenetic development. For example, the distribution of European

Beech (*Fagus sylvatica*) is limited strongly by late frost events in early summer. Freshly opened young leaves are easily damaged by frost damage and late frosts can cause large losses in leaf phytomass. In the central European Alps the continental-type climate and cold air masses that drain into valley bottoms enhance the risk of late frost events. On the periphery of the Alps this phenomenon is much less common and as a consequence, *F. sylvatica* is present in the Alpine periphery but absent in the central Alps.

Shifts in seasonality may disrupt interspecific interactions as species react differently to changing temperature and day-length signals. The common European butterfly *Operophtera brumata* is found in many vegetation types and is associated with flowering deciduous trees. Higher spring temperatures, however, stimulate the larvae to develop earlier: at times when flowers often have not developed. If such temporal disconnections last for several years, populations can plummet and local extinctions can occur. The early season development of many perennial plant species, especially trees, is triggered more by the photoperiod than air temperature. Genetically fixed rhythms may be maintained even when the risk factors (disturbances) that they have evolved to cope with have diminished. Species such as these have limited options for adaptation to climate change. Respiration losses related to increased temperatures may even cause damage to these species when still defoliated. Larigauderie and Körner (1995), for example, found that many alpine plants had a low potential for acclimatization. A detailed discussion of species responses to climate change is provided by Sparks et al. (this volume).

The occasional transport of diaspores and/or living organisms occur as more-or-less singular events. A combination of continuous

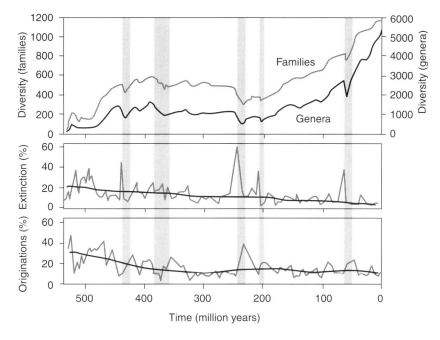

Figure 10.9   Geological documentation of the development of plant families and genera. Large extinction events were followed by phases of increased originations. Note the time lag of several million years (after Kirchner and Weil, 2000)

short-distance expansion and extraordinary long-distance dispersal events may explain current distribution patterns better than the specific dispersal properties of individual species. Reid's (1899) paradox addresses the fact that the dispersal of many species cannot be explained well by their regular capacity for dispersal and the ecological parameters that can be measured. Events that are limited to short time spans and may not re-occur are probably responsible for a large portion of the spatial patterns of global biodiversity. Processes that contribute to these events are strong winds and storms, extraordinary ocean currents and, increasingly, human action (see Blumler's discussion of invasives, this volume). For example, until recently cattle egrets (*Bubulcus ibis*) were only common in Africa and Asia. During the second half of the nineteenth century some were observed in South America, but these initial invaders did not establish themselves permanently. However, more dispersal events occurred during the twentieth century and since 1937 cattle egrets have spread across the continent. They reached North America as early as 1941 and the cattle egret is the most abundant heron in the New World at the present time (Telfair, 1994) (Figure 10.10).

10.7 SUCCESSION AND INERTIA

Postdisturbance succession is a classical field of ecological and biogeographical theory in which principal assembly rules as well as time scales of years and decades are considered (White and Jentsch, 2004). The species pool is considered constant, evolutionary processes are excluded, and invasion and extinction are ignored. The species composition develops from initial communities— characterized by short-lived and less-competitive r-strategists—to the hypothetical climax vegetation with its long-lived, competitive K-strategists. Changes in soil fertility and nutrient availability cannot be ignored but classical concepts (e.g., Clements, 1916) posit the overall environment as more or less static during succession. However, site-dependent specific developments do take place, the dynamic trajectories and the turnover of communities will differ, and long-lasting phases of transition may even support the impression of stability.

Volcanic eruptions are often followed by primary succession (e.g., Walker and del Moral, 2003), because they create completely new substrates (e.g., Bush and Whittaker, 1994; Plate 10.3). Large, intensive fires may generate

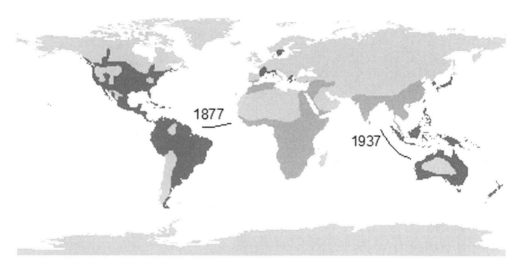

Figure 10.10   Disturbances by strong winds can drift individuals over large distances. Single episodic dispersal events of the cattle egret (*Bubulcus ibis*), for example, in 1877, are responsible for the establishment of this species in the Americas

sterile landscapes as well. In secondary succession the disturbance event is less intense. Remnants of previous ecosystems persist in situ (e.g., seeds, tubers, corms, and underground plant material from which plants can regenerate). Degradation is a regressive succession that occurs when self-accelerating mechanisms of resource depletion and loss are initiated.

Connell and Slatyer (1977) structured the theories of succession by considering the following mechanisms:

1. Facilitation: species of precedent stages enable the establishment of the succeeding populations through their functional contribution (e.g., soil development).
2. Tolerance: species are differentiated according to resource use. Those of later stages are able to better tolerate stresses related to resource depletion.
3. Inhibition: once established, strong competitors may hinder the establishment of other species. Until the completion of their life cycles, less-competitive species that can be better adapted to resource availability during later stages will be suppressed.

Early communities are dominated and controlled by external environmental factors. The species that compose these communities are generally ruderals, pioneers, colonizers, and weak competitors. They exhibit low rates of photosynthesis and are inefficient when light availability is low.

The species typically are mobile, efficient dispersers and develop a persistent seed bank.

Mature communities perform more developed networks of functional interaction. Late successional species have lower rates of photosynthesis but are shade tolerant and work more efficiently at low light availability. These strong competitors are less mobile and disperse over short distances. They can reach old age and persist over long time periods. However, besides individual life traits it is important that populations may regenerate and demographic structures develop that contribute to maintain dominance at the level of long-term ecosystem development as long as disturbances are absent. In the Mediterranean the long history of nonsustainable land use and the resulting loss of soil through land degradation processes has resulted in the former climax vegetation (e.g., evergreen sclerophyllous oak forest) being replaced by a secondary vegetation, known in different countries as Garrigue, Maccia, Maquis, or Phrygana.

In landscapes, areas of disturbance and succession are organized in spatial patterns. Rarely are large surfaces affected, for example, large forest fires in less-structured boreal forests. But even there, remnants of previous ecosystems are maintained. In fact, mosaics of freshly disturbed and mature undisturbed communities are characteristic for most ecosystems and landscapes. The coexistence of various stages of succession is a major component of landscape and ecosystem diversity. Cyclic developments occur after disturbance but are neighboured by communities

Plate 10.3   Bristlecone-Pine (*Pinus aristata*) at Bryce Canyon (USA). Some individuals of this species were recorded to be almost 5,000 years old. They have experienced a wide range of environmental conditions without lethal consequences. However, the populations of this species seem to be not very vital under recent climate conditions

of different ages. Remmert (1991) integrates this into the concept of mosaic cycles. The mosaic structure of successional phases is most evident in forest ecosystems. There, single disturbance events such as wind throw, fire, or pest outbursts may cause long-term effects connected with the arrangement of various stages of development thereafter. After severe disturbance and in areas with low-diversity species pools, there is a high probability that only a few species may contribute to the succeeding vegetation. In fact, it can even be dominated by a single species. In such mono-dominant and even aged stands,

subsequent development can be controlled by intrinsic processes. Cohort die-off can occur when the natural life cycle of trees is accomplished. In this case disturbance is not the reason for the collapse of the community. However, deterministic concepts share the shortcoming that individual developments and chance are excluded. The establishment of a certain functional group will influence the following sequence and trajectory. The concept of 'patch-dynamics' (Pickett and Thompson, 1978) assumes neighbouring phases of succession and age but is basically more individualistic.

Some plant species have very long life spans ranging from several hundred to several thousand years (Plate 10.4). Examples can be found among trees, shrubs, dwarf shrubs, and clonal grasses. Long-lived, slow-growing individuals of a species may not be able to adapt to changing environmental conditions. However, they can persist against the competitive pressure of other species over many decades. They represent ecological inertia in the face of a changing environment. The risk of not being able to cope with changing environmental conditions by adaptation or migration is that of becoming extinct. Evolutionary inertia may provide temporal refuges in repeatedly changing environments. The potential to survive suboptimal conditions is a survival option until conditions become favourable again. However, strong disturbances such as avalanches may remove long-lived organisms. Disturbance can thereby contribute to a reduction of inertia in ecosystems.

10.8 CONCLUSIONS AND FOOD FOR THOUGHT

1. *Dominance of human disturbance.* In many parts of the world, anthropogenic disturbances outperform the effects of natural disturbances. Land use and land-use change are the major drivers of biodiversity loss. Mowing and ploughing regimes and nutrient pulses determine the environment in agricultural landscapes. Agricultural land use is concentrated in areas with highly productive soils and favorable climate at least during one season of the year. In these environments natural disturbance regimes were often replaced many centuries ago.

2. *Human disturbance regimes do not replace natural disturbance regimes adequately.* In the tropics, soils are generally poor in nutrients due to intensive weathering and leaching. When forests are lost, agriculture cannot simulate the cation-pumping capacity of trees in permanent ecosystems. Forest loss in previously forested ecosystems has long-lasting negative effects that cannot be compensated for by human management. Nutrients are depleted and on some soils farming often has to be abandoned after a number of years.

3. *Frequency and magnitude of extreme disturbances events should increase in the near future.* In the last few years the global dimensions of the human impact on climate has been recognized. Global warming is expected to continue through the twenty-first century, and with an increased amount of energy in the atmosphere it is not only average temperatures that are expected to rise. There are likely to be increases in the probability and amplitude of extreme weather events (Figure 10.11) (Easterling et al., 2000; Mehl et al., 2000; Webster et al., 2005). These will interact

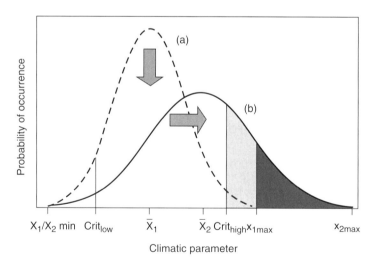

Figure 10.11 Future climate changes are expected to be accompanied not only by a shift in mean values (e.g., temperature) but also with increasing temporal variability. Hitherto, rare extreme events are likely to become more frequent (light grey). Additionally, novel extremes are expected (dark grey). Low critical temperatures that may control the occurrence of pest species will decline but not be missed completely

with many different types of disturbance regimes (e.g., Breshears et al., 2005), which in turn will enhance the role of disturbances in many ecosystems during the twenty-first century (IPCC, 2007; Holmgren et al., 2006; Jentsch et al., 2007).

4. *Threshold dynamics and nonlinear response.* Critical thresholds for organismic populations and community dynamics may be exceeded regionally due to increasing variability in climate and disturbance regimes (Fernandez and Vrba, 2005; Scheffer and Carpenter, 2003). The consequences for ecosystem stability, ecotone shifts, and ecological services to mankind are unclear (e.g., Allen and Breshears, 1998; De Boeck et al., 2007). On the one hand it is very likely that naturally established long-lived species will be stressed and their performance constricted. On the other hand, short-lived species will be able to utilize the niches that are opened. As a consequence, the stability of ecosystems could be shortened. Local and regional extinctions will happen rapidly, but the migration and dispersal of adapted species will need longer time periods.

REFERENCES

Allen, C.D., and Breshears, D.D. (1998) 'Drought-induced shift of a forest-woodland ecotone: Rapid landscape response to climate variation', *Proceedings of the National Academy of Sciences of the United States of America*, 95: 14839–42.

Beierkuhnlein, C. (2007) *Biogeographie*. Stuttgart: Verlag Eugen Ulmer.

Beierkuhnlein C., and Jentsch, A. (2004) 'Ecological importance of species diversity. A review on the ecological implications of species diversity in plant communities', in Henry, R. (ed.), *Diversity and Evolution of Plants*. Wallingford: CAB International. pp. 249–85.

Breshears, D.D. and 12 authors. (2005) 'Regional vegetation die-off in response to global-change type drought', *Proceedings of the National Academy of Science of the United States of America*.

Bush, M.B., and Whittaker, R.J. (1994) 'Krakatau: Colonization patterns and hierarchies', *Journal of Biogeography*, 18: 341–56.

Clark, J.S., Mohan, J., Dietze, M., and Ibanez, I. (2003). 'Coexistence: How to identify trophic trade-offs', *Ecology*, 84: 17–31.

Clements, F.E. (1916). *Plant Succession: An Analysis of the Development of Vegetation*. Washington, DC: Carnegie Institution of Washington.

Connell, J.H. (1978). 'Diversity in tropical rain forests and coral reefs', *Science*, 199: 1302–10.

Connell, J.H., and Slatyer, R.O. (1977). 'Mechanisms of succession in natural communities and their role in community stability and organization', *American Naturalist*, 111: 1119–44.

De Boeck, H.J., Lemmens, C., Gielen, B., Bossuyt, H., Malchair, S., Carnol, M., Merckx, R., Ceulemans, R., and Nijs, I. (2007). 'Combined effects of climate warming and plant diversity loss on above- and below-ground grassland productivity', *Environmental and Experimental Botany*, 60: 95–104.

Donnelly, J.P., and Woodruff, J.D. (2007) 'Intense hurricane activity over the past 5,000 years controlled by El Niño and the West African monsoon', *Nature*, 447: 465–68.

Easterling, D.R., Meehl, G.A., Parmesan, C., Changnon, S.A., Karl, T.R., and Mearns, L.O. (2000) 'Climate extremes: Observations, modeling, and impacts', *Science*, 289: 2068–74.

Everham III, E.M., and Brokaw, N.V.L. (1996) 'Forest damage and recovery from catastrophic wind', *Botanical Review*, 62: 113–85.

Feddema, J. (2008) 'A revised Thornthwaite-type global climate classification', *Physical Geography*, 26, 442–66.

Fernandez, M.H., and Vrba, E.S. (2005) 'Rapoport effect and biomic specialization in African mammals: revisiting the climatic variability hypothesis', *Journal of Biogeography*, 32: 903–18.

Grimm, V., and Wissel, C. (1997) 'Babel, or the ecological stability discussion: An inventory and analysis of terminology and a guide for avoiding confusion', *Oecologia*, 109: 323–34.

Holmgren, M. and 18 authors. (2006) 'Extreme climatic events shape arid and semiarid ecosystems', *Frontiers in Ecology and the Environment*, 4: 87–95.

http://www.globalwarmingart.com/wiki/Image:Tropical_ Storm_Map_png.

Hubbel, S.P., Foster, R.B., O'Brien, S.T., Harms, K.E., Condit, R., Wechsler, B., Wright, S.J., and Loo de Lao, S. (1999) 'Light-gap disturbances, recruitment limitation, and tree diversity in a neotropical forest', *Science*, 283: 554–57.

Huston, M. (1994) *Biological Diversity*. Cambridge: Cambridge University Press.

Hutyra, L.R., Munger, J.W., Nobre, C.A., Saleska, S.R., Vieira, S.A., and Wofsy, S.C. (2005) 'Climatic variability and vegetation vulnerability in Amazonia', *Geophysical Research Letters* 32, doi:10.1029/2005GL024981.

IPCC. (2007) *Climate Change 2007. The Physical Science Basi*s. Geneva: Intergovernmental Panel on Climate Change.

Jentsch, A. (2004) 'Disturbance driven vegetation dynamics. Concepts from biogeography to community ecology, and experimental evidence from dry acidic grasslands in Central Europe', *Dissertationes Botanicae*, 384: 218 pp.

Jentsch A., and Beierkuhnlein, C. (2003) 'Global climate change and local disturbance regimes as interacting drivers for shifting altitudinal vegetation patterns in high mountains', *Erdkunde*, 57: 218–33.

Jentsch, A., Beierkuhnlein, C., and White, P.S. (2002) 'Scale, the dynamic stability of forest ecosystems, and the persistence of biodiversity', *Silva Fennica*, 36/1: 393–400.

Jentsch A., J. Kreyling, and C. Beierkuhnlein. (2007) 'A new generation of climate change experiments: Events, not trends', *Frontiers in Ecology and the Environment*, 5: 365–74.

Joergensen, P.M., Ulloa Ulloa, C., Madsen, J.E., and Valencia, R. (1995) 'A floristic analysis of the high Andes of Ecuador', in Churchill, S.P. et al. (eds.), *Biodiversity and Conservation of Neotropical Montane Forests*. New York: New York Botanical Garden. pp. 221–37.

Johnson E., and Myanishi K. (2007) *Plant Disturbance Ecology—The Process and The Response*. New York: Academic Press.

Kirchner, J.W., and Weil, A. (2000) 'Delayed biological recovery from extinctions throughout the fossil record', *Nature*, 404: 177–80.

Körner, C., and Paulsen, J. (2004) 'A world-wide study of high altitude treeline temperatures', *Journal of Biogeography*, 31: 713–32.

Larigauderie, A., and Körner, C. (1995) 'Acclimation of leaf dark respiration to temperature in alpine and lowland plant species', *Annales Botanicae*, 76: 245–52.

Lugo, A.E. (2008) 'Visible and invisible effects of hurricanes on tropical forest ecosystems: An international review', *Austral Ecology*, 34(3): 368–98.

McGradySteed, J., Harris, P.M., and Morin, P.J. (1997) 'Biodiversity regulates ecosystem predictability', *Nature*, 390: 162–65.

Mehl, G.A. and 16 authors. (2000) 'An introduction to trends in extreme weather and climate events: Observations, socioeconomic impacts, terrestrial ecological impacts, and model projections', *Bulletin of the American Meteorological Society*, 81:413–16.

Meyn, A., Buhk, C., White, P.S., and Jentsch, A. (2007) 'Environmental drivers of large, infrequent wildfires: The emerging conceptual model', *Progress in Physical Geography*, 31: 287–312.

Miehe, G., and Miehe, S. (1994) 'Zur oberen Waldgrenze in tropischen Gebirgen', *Phytocoenologia*, 24: 53–110.

Milly, P.C.D., R.T. Wetherald, K.A. Dunne, and T.L. Delworth. (2002) 'Increasing risk of great floods in a changing climate', *Nature*, 415: 514–17.

New, M. and 18 authors. (2006) 'Evidence of trends in daily climate extremes over southern and west Africa', *Journal of Geophysical Research* 111 D14102, doi:10.1029/2005JD006289.

Peel, M.C., Finlayson, B.L., and McMahon, T.A. (2007) 'Updated world map of the Köppen–Geiger climate classification', *Hydrology and Earth Systems Science*, 11: 1633–44.

Pickett, S.T.A., and Thompson, J.N. (1978) 'Patch dynamics and the design of nature reserves', *Biological Conservation*, 13: 27–37.

Pickett, S.T.A., and White, P.S. (eds). (1985) *The ecology of Natural Disturbance and Patch Dynamics*. London: Academic Press.

Pimentel, D., and Kounang, N. (1998) 'Ecology of soil erosion in ecosystems', *Ecosystems*, 1: 416–26.

Reid, C. (1899) *The Origin of the British Flora*. London: Dulau

Remmert, H. (1991) *The Mosaic-Cycle Concept of Ecosystems*. *Ecological Studies*, 85. Berlin: Springer-Verlag.

Richter, M. (2001) *Vegetationszonen der Erde*. Gotha, Stuttgart: Klett-Perthes.

Richter, M., and Ise, M. (2005) 'Monitoring plant development after el Niño 1997/98 in North-Western Peru', *Erdkunde*, 59: 136–55.

Sala, O.E. and 18 authors. (2000) 'Global biodiversity scenarios for the year 2100', *Science*, 287, 1770–74.

Sanders, H.L. (1968) 'Marine benthic diversity: A comparative study', *American Naturalist*, 102: 243–82.

Scheffer, M., and Carpenter, S.R. (2003) 'Catastrophic regime shifts in ecosystems: Linking theory to observation', *Trends in Ecology and Evolution*, 18: 648–56.

Sprugel, D.G. (1991) 'Disturbance, equilibrium, and environmental variability—what is natural vegetation in a changing environment?' *Biological Conservation*, 58: 1–18.

Telfair, R.C. (1994) 'Cattle Egret (Bubulcus ibis)', in *The Birds of North America*, No. 113. Philadelphia, PA: Academy of Natural Sciences, and Washington DC: The American Ornithologists' Union.

Tilman D. (1996) 'The benefits of natural disasters', *Science*, 273: 1518.

Treter, U. (2001) 'Natürliche Regenerationsprozesse und Bestandsentwicklung der Tannenwälder (Abies balsamea) in Neufundland, Kanada', *Bremer Beiträge zur Geographie und Raumplanung*, 7: 31–45.

Turner, G.M., Baker, W.L., Peterson, C.J., and Peet, R.K. (1998) 'Factors influencing succession: Lessons from large, infrequent natural disturbances', *Ecosystems*, 1: 511–23.

Turner, G.M., Romme, R.H., Gardner, R.H., O'Neill, R.V., and Kratz, T.K. (1993) 'A revised concept of landscape equilibrium: Disturbance and stability on scaled landscapes', *Landscape Ecology*, 8: 213–27.

Walker, B., Kinzig, A., and Langridge, J. (1999) 'Plant attribute diversity, resilience, and ecosystem function: The nature and significance of dominant and minor species', *Ecosystems*, 2: 95–113.

Walker, L.R., and del Moral, R. (2003) *Primary Succession and Ecosystem Rehabilitation*. Cambridge: Cambridge University Press.

Webster, P.J., Holland, G.H., Curry, A.J., and Chang, H.R. (2005) 'Changes in tropical cyclone number, duration, and intensity in a warming environment', *Science*, 15: 1844–46.

Wesche, K., Miehe, G., and Kaeppeli, M. (2000) 'The significance of fire for afroalpine ericaceous vegetation', *Mountain Research and Development*, 20: 340–47.

White, P.S., and Jentsch, A. (2001) 'The search for generality in studies of disturbance and ecosystem dynamics', *Progress in Botany*, 63: 399–449.

White, P.S., and Jentsch, A. (2004) 'Disturbance, succession and community assembly in terrestrial plant communities', in V.M. Temperton, R. Hobbs, M. Fattorini, and S. Halle (Ed.), *Assembly Rules in Restoration Ecology—Bridging the Gap Between Theory and Practise*. Washington, DC: Island Press Books. pp. 342–66.

Worm, B., Sandow, M., Oschlies, A., Lotze, H.K., and Myers, R.A. (2005) 'Global patterns of predator diversity in the open oceans', *Science*, 309: 1365–69.

11

Fire and Ecosystem Function

Neal J. Enright

11.1 INTRODUCTION

Fire is a major source of landscape scale disturbance in many of the world's terrestrial ecosystems; it triggers processes of ecosystem recovery that vary markedly depending on the extent to which species have evolved with fire. On a geological timescale, fire has become increasingly important as a disturbance factor with the evolutionary radiation of the angiosperms, especially grasses, since the Late Cretaceous, and with increasing seasonality of climate since the Late Tertiary (Bowler, 1982; Bond and Van Wilgen, 1996; Bond et al., 2005). The use of fire by people dates back at least several hundreds of thousands of years in Africa, 50,000 years in Australia, and from a few thousands to tens of thousands of years in other parts of the world (Pyne, 1991, 1995). Whether human-induced increases in fire frequency have resulted in evolutionary responses at the species level is unclear, although perhaps possible for areas with very long histories of human habitation. People have almost certainly increased the extent of both tropical savannas and temperate grasslands at the expense of forest during the prehistoric period through their use of fire. However, while evidence for the conversion of forest to nonforest is strong for recently peopled regions (e.g., New Zealand; McGlone, 1989), it is often difficult to disentangle the roles of climate change and people where the history of human habitation stretches back to the last glaciation or beyond (Enright and Thomas, 2008).

Over recent centuries, European expansionism across the globe, the industrial revolution, human population growth, and the land-use consequences of growth, have further increased the impact of fire on ecosystems. While fire continues to reduce areas of forest in many parts of the world, policies of fire exclusion during the twentieth century in some industrialized countries began to have the opposite effect, with woody vegetation replacing more open vegetation types (Pyne, 1991). The history of fire, and its intimate association with people, reveals it as an ecological and evolutionary phenomenon of great importance in relation to the fundamental properties of terrestrial ecosystems (e.g., their distributions, composition, structure, and function).

The focus of this chapter is on the ecology of fire. It reviews the relevance of fire as a disturbance factor in different ecosystem types, and the responses of ecosystems and species to it. Current areas of scientific investigation are identified, as are landscape-level management issues linked to fire. The relationship between fire behavior and ecosystem properties is addressed only insofar as this relates on the one hand to ecosystem function and, on the other, to current issues in fire management. It is not possible to review all areas of fire and ecosystem function in this treatment, and readers interested in more detailed accounts are referred to books by Gill et al. (1981), Pyne (1991, 1995), Whelan (1995), Bond and Van Wilgen (1996), and Bradstock et al. (2002).

11.2 FIRE REGIMES OF MAJOR TERRESTRIAL ECOSYSTEMS

Plants are not adapted to fire per se but rather to a particular fire regime (Gill, 1981). It is the variation in how often fires occur (frequency), when in the year they occur (season), how fiercely they burn (intensity/severity), and how large they are (size) that determines the mix of plant and animal species that will occupy a particular site.

In tropical rainforests fires are rare, perhaps occurring once in 1,000 years or more, so the recurrence interval exceeds the longevity of even the longest-lived components of such communities. Disruption to the forest canopy due to natural (e.g., drought, hurricane) and human (e.g., logging) impacts may increase the susceptibility of rainforest to fire by reducing understory humidity and increasing the load of dead fuels (Cochrane and Schulze, 1999; Van Niewstadt and Sheil, 2005). Fire can be catastrophic in wet tropical forests, killing most trees, and since soil seed banks are dominated by short-lived secondary species (Hopkins and Graham, 1983; Enright, 1985), recovery requires the dispersal of propagules into the affected area from adjacent, undisturbed forest and a long succession without further landscape-scale disturbance. In dry tropical forests, and near forest-savanna boundaries, the majority of tree species are able to resprout (Russell-Smith and Stanton, 2002), reflecting the more frequent occurrence of fires and the volatile dynamics of the forest-savanna ecotone. The conversion of tropical rain forest to savanna due to fire spread from adjacent grasslands and deliberate burning by people represents a different process, since conditions for ignition and spread are more conducive to fire propagation. Rainforest expansion into savanna is also evident when fire frequency is reduced (Russell-Smith and Stanton, 2002).

In ecosystems where fire is likely to recur at least within the expected lifetime of the longest-lived species, for example, the boreal forests of North America (Gauthier et al., 1996), and closed *Eucalyptus* (wet sclerophyll) forests of southern Australia (Gill and Catling, 2002), the dominant trees typically show adaptations enhancing the probability of recruitment after fire (Table 11.1). Serotiny, the storage of seeds in closed cones or fruits in the tree crown for more than one year so that there is an accumulation of seeds in a canopy-stored seed bank (Lamont et al., 1991), represents one such adaptation common in species of spruce (e.g., *Picea mariana*) and pine (e.g., *P. banksiana*) in northern North America, and of eucalypts in southern Australia (e.g., *E. regnans*, *E. diversicolor*). The heat of fire causes the release of seeds into a recruitment environment characterized by increased light and nutrient availability and by reduced competition for space and water. Since the trees themselves are killed by high-severity fires, crown fires in these ecosystems are referred to as stand-replacing fires and occur about every 100 to 300 years (Gill and Catling, 2002; Schoennagel et al., 2003). Low-intensity fires restricted to the understories of such stands may occur more frequently (10- to 50-year intervals), affecting ground and shrub layer vegetation but

Table 11.1 Fire regimes, plant responses, and adaptations for major terrestrial ecosystems

Ecosystem type	Surface fire interval (y)	Crown fire interval (y)	Plant response	Main adaptations
Rain forest		>1,000	Nonadapted	None
Boreal forest		100–300	Fire-sensitive	Serotiny
Coniferous forest	5–20	80–300	Fire-sensitive	Serotiny, thick bark
Wet eucalyptus forest	10–20	80–300	Fire-sensitive	Serotiny, hard-seededness, thick bark
Dry eucalyptus forest	4–10	30–80	Fire-sensitive and fire-tolerant	Serotiny, hard seeds, thick bark, vegetative regrowth (above and below ground), smoke-stimulated seed germination
Fynbos, Kwongan, Chaparral, Maquis, Garrigue		5–40	Fire-tolerant and fire-sensitive	Serotiny, hard seeds, vegetative regrowth (mostly from below ground), smoke-stimulated seed germination, fire ephemerals
Savanna, Cerrado	1–5	1–10	Fire-tolerant	Vegetative regrowth (below ground)
Grasslands	1–5		Fire-tolerant	Vegetative regrowth (below ground)
Semideserts	10–50		Nonadapted	Soil-stored seeds. May be some savanna/ grassland species that resprout

leaving most trees alive. The absence of fire from some Australian wet sclerophyll forests for more than 300 years may allow a succession to rain forest dominated by nonadapted trees such as the southern beech (*Nothofagus* spp.). The spatial extent of these rainforest is highly dependent upon fire frequency, with rain forest expanding from wet, valley refugia if fire frequency declines, and contracting if it increases, making them particularly susceptible to human impacts on fire regime.

In systems where fire is frequent, such as woodlands and shrublands in Mediterranean-type climates, many plant species display the ability to regrow vegetatively after fire, and some of these species show fire-stimulated flowering and fruiting. Vegetative regrowth is not unique to the vegetation of fire-prone ecosystems and so should not necessarily be regarded as an adaptation to fire. However, the range of forms of vegetative regrowth in such environments does suggest an evolutionary radiation of this mechanism in relation to fire; dormant buds may be held beneath the bark above ground (epicormic) or below ground on buried stems (lignotubers) and roots, while tightly packed leaf bases protect apical meristems in some species (Plate 11.1). In all cases, meristems are protected from the extreme heat of fire, producing new growth when the plant canopy is removed by fire. Vegetative regrowth species maintain a largely intact root system, so they can regrow quickly after fire. Mature plants can resume reproduction within one to three years (this time is referred to as a secondary juvenile period).

Fire-killed species with canopy- or soil-stored seed banks are also present in these ecosystems, but they will only persist if their life histories (particularly, time to reproductive maturity and overall longevity) are compatible with the fire regime. The fire interval for such ecosystems ranges from about 10 to 40 years (Keith et al., 2002; Hobbs, 2002).

Fire is most frequent in savanna ecosystems, with return intervals of one to five years, being most frequent where higher rainfall promotes faster rates of grass recovery and fuel accumulation (Williams et al., 2002; Felderhof and Gillieson, 2006). Perennial plant species here are all capable of vegetative regrowth. Fire-killed species are largely restricted to annual herbs and grasses, which may be prominent only in the first year after fire. Perennial grasses dominate and quickly return above-ground biomass to levels capable of supporting fire. A fire-tolerant tree layer may be present and can vary markedly in density.

Fire may occur in any ecosystem type when conditions conducive to fire spread coincide with the presence of an ignition source. In arid ecosystems, a lack of continuous fuels usually precludes fire. However, following high rainfall events, arid ecosystems may quickly accumulate sufficient biomass of perennial plants plus ephemeral grasses to carry fires of high intensity once grasses have cured under subsequent dry conditions. For example, in 1983 a fire burned 20,000 square kilometers of spinifex (*Triodia* sp.) grassland in central Australia, with another fire covering 5,000 square kilometers recurring two years later within the same area (Allan and Southgate, 2002). In contrast, fires in wetlands generally occur under drought conditions, either as surface or peat fires. Turetsky et al. (2004) estimated that 1,850 square kilometers of peatland burned annually in western Canada, and Yin (1993) reported a correlation between fire size, occurrence, and the weather conditions caused by the El Niño Southern Oscillation (ENSO) in Okefenokee Swamp in Georgia in the United States. Surface fires may have only short-term impacts, with species composition and wetland structure returning to prefire patterns within 5 to 10 years (Norton and de Lange, 2003), while peat fires, and increased frequency of fire, may consume large amounts of dead organic matter and seriously disrupt the ecology of such systems for decades to centuries (Kuhry, 1994).

11.3 FITNESS IN A FIRE-PRONE ENVIRONMENT

Mutch (1970) proposed that fire-prone vegetation may have evolved to increase flammability and so increase the likelihood that fire would recur, perpetuating the fire-dependent community. High levels of cellulose, secondary compounds (including terpenes and waxes), fine foliage, stringy or ribbony bark, and fine branches (providing a high surface area to volume ratio) are common in many plant species of such ecosystems (e.g., *Eucalyptus* and *Pinus* species) and these promote fire if an ignition source is available. According to Bond and Keeley (2005), boreal forests, eucalypt woodlands, shrublands, grasslands, and savannas all qualify as flammable ecosystems, with fire potentially promoted by the physical and chemical attributes of some species. Nevertheless, Bond and Midgley (1995) argued that the Mutch hypothesis was flawed since it is based on a premise of species- or community-level selection, whereas selection operates at the level of the individual. They proposed, and tested by simulation, a modified hypothesis ('kill thy neighbor') where a flammable mutant 'torch' individual could increase in abundance so long as it spread the fire to neighboring individuals, resulting in their death.

Plate 11.1 Responses of resprouter species to fire: top left—epicormic regrowth on the trunk and major branches of *Allocasuarina fraseriana* (Casuarinaceae) in woodlands near Perth, Western Australia; top right—regrowth from buried stem tissue (a lignotuber) in *Banksia candolleana* (Proteaceae) in the northern sandplains, Western Australia (note also the fire-opened serotinous fruits); bottom left—regrowth of foliage from densely packed apical buds in *Dasypogon hookeri* (Dasypogonaceae) and *Xanthorrhoea preissii* (Xanthorrhoeaceae) in eucalypt forest, Western Australia; and bottom right—fire-stimulated flowering in *X. preissii*

Death of the mutant 'torch' individuals alone would not lead to successful development of a fire-prone system since fire could not spread under such circumstances. This view was supported by Kerr et al. (1999), who found that the evolution of flammability in one or more species within a fire-prone community could drive selection of traits in those, and other (non-flammable), species that increased their capacity to cope with fire. These models show that flammability can act as a 'niche constructing' trait that modifies the local environment to the benefit of

flammable plants (Schwilk, 2003; Bond and Keeley, 2005).

Recruitment of new individuals (genets) in fire-prone ecosystems occurs mostly in the first year after fire (postfire recruitment), with limited establishment of individuals from seed in the subsequent years between fires (interfire recruitment). Particularly in the case of fire-killed (nonsprouter) species, this behavior has a clear fitness benefit: individuals recruited in the first year after fire have the longest available time for growth and reproduction before the next fire occurs, and so, on average, they will have produced (and stored in a canopy or soil seed bank) more seeds than would any individual recruited later in the interfire period. Consequently, they will contribute more individuals to future generations. Environmental conditions for recruitment are also best in the first year after fire due to reduced competition for light and moisture and the increased provision of nutrients through the combustion of live and dead plant materials consumed by the fire.

The evolution of recruitment strategies linked to fire are primarily associated with adaptations for survival of dormant seeds in canopy and soil seed banks, and for the cueing of germination to physical and chemical triggers associated with fire. Hard-seededness in soil seed-storage species and serotiny are the two most prominent attributes of seed banks in fire-prone environments. Interestingly, while serotiny is prominent in some geographical regions (e.g., Australia, South Africa) it is absent or rare in others (e.g., Chile, California, and the Mediterranean Basin), and it seems likely that it has arisen only where ancestral floristic elements had appropriate precursor traits, such as woody seed pods, before fire became a strong selective force from about the mid-Tertiary onward. In strongly serotinous species, more than 10 years of annual seed crops may be stored in the plant canopy, creating a large store of seeds available for dispersal after fire and reducing the potentially deleterious effects of year-to-year fluctuations in seed production (Plate 11.2; Figure 11.1). Enright et al. (1998a, 1998b) showed

that the fitness benefits of serotiny in Australian shrubland species accrued quickly, so that once species were moderately serotinous there was a declining return in further investment in serotiny. They concluded also that the release of some seeds in the interfire period (i.e., incomplete serotiny) represented a bet-hedging strategy that safeguarded against species loss where fire interval might occasionally exceed the species longevity. In the genus *Pinus*, the time interval between stand-replacing fires can lead to variations in serotiny: Gauthier et al. (1996) found that serotiny in southern Canadian jack pine (*P. banksiana*) stands was lowest when lethal fire intervals were long, a result supported also by Schoennagel et al. (2003) for lodgepole pine (*P. contorta var. latifolia*) stands in Yellowstone National Park. No such within-species variation in serotiny has been reported for any of the many Australian serotinous species, other than in relation to broad climatic gradients (Cowling and Lamont, 1985). A fundamental difference between the pine and shrubland systems is that low-intensity understory fires, which do not lead to seed release from serotinous cones, can occur in the former but not the latter (where all fires affect the canopy). Benkman and Siepielski (2004) found evidence that high rates of loss of canopy-stored seeds to predators, such as the pine squirrel, can also select against strong serotiny.

Storage of seeds in a persistent soil seed bank is common to many ecosystems in biomes ranging from tropical forests to deserts (Thompson, 1978).

Plate 11.2   Canopy seed storage (serotiny) in *Banksia hookeriana*, northern sandplain shrublands, southwestern Australia: left—a 15-year-old individual immediately after fire showing the lifetime accumulation of woody fruits (two unburned inflorescences are visible in the right and bottom of the image); right—close-up of burned cones showing the woody follicles from which seeds have been released

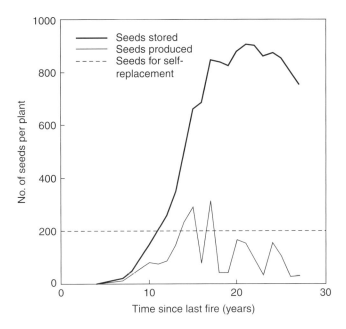

Figure 11.1 Average cumulative canopy seed store, number of viable seeds produced per year, and estimated number of viable seeds required for self-replacement for a 15-year-old *Banksia hookeriana* in the northern sandplain shrublands of southwest Australia (from Lamont and Enright, 2000)

Seeds are generally dispersed each year following ripening (although in some species, ripening may continue after dispersal), and seeds accumulate in the surface layers of the soil where they remain dormant until appropriate triggers for germination are received. In fire-prone environments, hard seed coats are characteristic of many soil-storage species (e.g., Fabaceae). Hard-seededness enhances long-term survival of seeds in the soil by protecting them against attack by decomposers. Dormancy is broken by the heat of fire, which cracks the seed coat, allowing germination in the following wet season (Auld and O'Connell, 1991). Enhancement of germination response appears greatest for seeds receiving at least a few minutes of heat in the range 80–100°C; temperatures less than 60°C are unlikely to crack the seed coat, and those greater than 120°C are likely to cause seed death (Figure 11.2). Because dry soil is a good insulator against heat, only the top one to two centimeters of the soil profile is likely to experience high temperatures during fire (that might be lethal to seeds), while depths below five centimeters may not receive sufficient heat to break dormancy (Ramsay and Oxley, 1996). In addition, seed size is critical to patterns of postfire recruitment since small seeds may not be able to emerge if buried too deep in the soil

(Brown et al., 2003). The interaction of soil heating, seed size, and burial depth is thus important in determining the spatial distribution and abundance of recruits. Many hard-seeded species are also characterized by the presence of ant-attracting attachments (eliasomes) so that seeds are collected by ants (myrmechocory) and taken to their nests. The seed may be stored, or discarded, in a location where probability of recruitment is enhanced (although evidence of effectiveness is equivocal).

The role of chemical triggers for germination in fire-prone ecosystems was first identified in California where Keeley and Pizzorno (1986) found that water-soluble leachate from charred wood promoted germination in some Chaparral species. Subsequently, Brown (1993) and other workers in South Africa reported that smoke triggered germination in a range of fynbos species. Smoke has now been shown to break dormancy in many species from fire-prone ecosystems in most parts of the world, including some that had previously proven difficult to germinate under experimental conditions (Keeley and Fotheringham, 1998; Bell, 1999; Perez-Fernandez and Rodriguez-Echeverria, 2003). The active compound in smoke that is responsible for dormancy breaking has recently been identified as

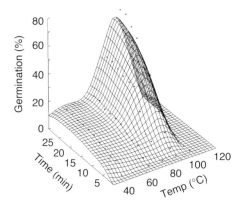

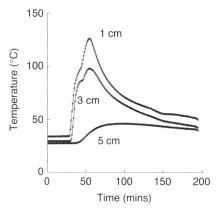

Figure 11.2 (a) Laboratory germination response (smoothed) of the hard-seeded species *Acacia longifolia* (Fabaceae) to experimental heat treatments (5 minutes at 40°, 60°, 80°, 100°, and 120°C); and (b) soil temperatures measured at 1, 3, and 5 cm depth during an experimental fire in sandplain shrublands, Little Desert National Park, Victoria, SE Australia

Seeds of some species remain difficult to germinate experimentally, yet seedlings may be abundant in the field after fire, suggesting other triggers for germination. Burial time, cold temperatures, darkness, passage through the guts of animals, or these in combination with heat or smoke, may increase germination. Tieu et al. (2001) found that smoke-triggered germination in several Australian Ericaceae (formerly Epacridaceae) and *Stylidium* species was enhanced by a period of burial, while Gilmour et al. (2000) found that heat, smoke, and storage in the dark all contributed positively to germination in the rare Tasmanian shrub *Epacris tasmanica*. Research into germination responses of species in fire-prone ecosystems is likely to remain a fruitful area for investigation, with important benefits for our understanding of the functioning of natural communities and for application in horticulture and restoration.

Some resprouting species have no seed dormancy but are characterized by fire-stimulated flowering and seed production, as in the Australian grasstrees (*Xanthorrhoea* spp.) and waratah (*Telopea speciosissima*). In *Xanthorrhoea*, thousands of seeds are produced on a large flowering spike about six to nine months after a fire (Plate 11.1) and are available to germinate in the second winter after fire. In *T. speciosissima* the first seed crop is produced in the second postfire year, with smaller seed crops for several years thereafter, so that recruitment is most likely two to three years after a fire (Denham and Auld, 2002). Delayed recruitment in relation to fire means that seedlings of these species must compete with older seedlings of nonsprouters recruited in the first year after fire, and with rapidly recovering vegetative regrowth species, including their own parents. Chances of successful recruitment are therefore greatly reduced. However, the survival of established individuals reduces the need for recruitment after every fire.

Individuals of resprouter species show a high probability of surviving fire, regrowing from buds protected by bark or tightly packed leaf bases above ground, or by soil below ground (Plate 11.1). Most resprouters also produce seeds that are stored either in a canopy or soil seed bank, but levels of seed production and storage are generally lower, and the time to reproductive maturity is much longer than for nonsprouter species (Enright and Lamont, 1989; Groom and Lamont, 1996; Lamont and Wiens, 2003). Low seed supply and strong competition from large numbers of nonsprouter seedlings mean that recruitment in resprouters may only occur occasionally after fire, probably in combination with favorable postfire weather conditions (Enright and Lamont, 1992; Keeley, 1992; Bond and Midgley, 2003). On the other hand, resprouters

the butenolide, 3-methyl-2*H*-furo[2,3-*c*]pyran-2-one (Flematti et al., 2004). Its effectiveness at very low concentrations (<1 p.p.b.) raises the possibility of commercial preparation for use in rehabilitation and restoration programs where large areas of topsoil might be treated to promote seed germination. Interestingly, smoke has also been shown to increase germination rates in some common plants from environments where fire is unlikely to have acted as a selective force (e.g., celery; Thomas and Van Staden, 1995), and the origins of the effect may relate to chemicals released during soil decomposition, where sudden changes in rates of decomposition indicate a change in environment (e.g., canopy opening) to one suitable for recruitment.

can maintain populations during periods that are unfavorable for recruitment (a storage effect) and might lead to a sharp decline in nonsprouter populations, such as those associated with poor conditions for recruitment after fire or with short fire intervals. In the former, seedlings of nonsprouters may perish, while in the latter, fire may recur before nonsprouters have reached reproductive maturity or before they have accumulated a sufficient store of seeds to ensure parent replacement (Enright et al., 1998a, 1998b).

The coexistence of resprouter and nonsprouter perennial species in the same communities has prompted questions concerning the costs and benefits of each strategy and the circumstances that might favor one over the other. In a study of the genus *Erica* in South Africa, Ojeda (1998) argued that resprouter species were better able to persist under harsh conditions for recruitment (see also Bellingham, 2000; Bond and Midgley, 2003), with nonsprouters showing an evolutionary radiation of species in regions with more reliable winter rainfall. Resprouting becomes more common as fire frequency increases, indicating an advantage for vegetative regrowth species at short fire intervals. Nonsprouters are favored at intermediate fire intervals, while evidence is equivocal concerning which strategy is favored as the interval between fires further increases. Enright et al. (1998b) suggest that resprouter populations may fail in the long-term absence of fire due to senescence of vegetative buds on extant individuals (leading ultimately to plant death) in combination with limited seedling recruitment due to low seed production and unfavorable conditions for establishment. On the other hand, nonsprouters may persist at low density in long-unburned stands through occasional recruitment of individuals in gaps caused by adult plant senescence (Enright et al., 1998a). Pausas and Lloret (2007) found no clear pattern for resprouter species from the western Mediterranean in relation to fire interval, and Keeley (1992) reported that some resprouting woody species of Californian Chaparral showed continuous basal sprouting and increased recruitment from seed in old stands (56 to 120 years since the last fire), while obligate seeders showed little recruitment other than immediately after fire.

11.4 ECOSYSTEM CHANGES WITH TIME SINCE FIRE

Change in species composition with time since fire in fire-prone ecosystems generally follows the initial floristic composition model, with most species present at the time of fire persisting either through vegetative regrowth or recruitment from seed soon after fire (Purdie and Slatyer, 1976; Purdy et al., 2002). Species richness is often at a maximum in the first one to three years after fire due to the presence of short-lived herbs and grasses. Some of these may be opportunistic species that germinate in response to changes in light and moisture, while others are fire ephemerals, requiring specific triggers (heat, smoke) for germination (e.g., *Austrostipa compressa* in southwest Australia, Smith et al., 1999; *Phacelia brachyloba* in Californian chaparral, Moreno and Oechel, 1991). Short-lived invasives, either accumulated in the soil seed bank, or dispersed after fire from adjacent land uses, may be prominent in the first few years after fire but do not generally persist (Litton and Santelices, 2002; Ghermandi et al., 2004).

In ecosystems where resprouting is dominant, above-ground biomass may approach prefire levels within 5 to 10 years (Bell et al., 1984; Clemente et al., 1996), primarily due to the rapid growth of resprouters. Biomass continues to increase as a range of nonsprouting species with differing longevities grow to maturity. Successional changes in composition and biomass after fire have been linked to habitat preferences for some animal species, although Monamy and Fox (2000) argue that some late seral specialists respond more to cover, reflecting site differences, than to changes in species composition or time since fire per se.

Where fire-killed species dominate the community (e.g., boreal and wet sclerophyll forests), the return to mature biomass levels may take from decades to centuries (Gill and Catling, 2002). Nevertheless, recruitment of major tree species can still be restricted to the first few years after fire when suitable recruitment microsites are temporarily available, for example, white spruce (*Picea glauca*) in boreal forests (Purdy et al., 2002) and mountain ash (*Eucalyptus regnans*) in southeast Australian wet sclerophyll forests (Ashton, 1981). In the continued absence of fire, species richness may decline slowly due to the senescence of nonsprouters with intermediate longevities, including many Fabaceae, for example, *Cistus* spp. in Mediterranean woodlands and shrublands (Clemente et al., 1996), *Acacia* species in Australian forests and shrublands, and perennial understory herbs in temperate pine-hardwood forests (Harrod et al., 2000). In the long-term absence of fire, change in the structure, composition, and function of fire-prone ecosystems may vary considerably between regions even within the same ecosystem type. In South Africa, forest would likely replace grassland and fynbos where rainfall is more than 650 mm yr^{-1} (Bond et al., 2003), and in Californian Chaparral long-lived resprouter tree species (including *Quercus*,

Rhamnus, and *Prunus* spp.) may replace fire-sensitive nonsprouters (Keeley, 1992). However, in drier kwongan (Australian shrublands) and fynbos there is no strong evidence for succession of shrubland to another state. The consequences of long-term fire exclusion from such ecosystems are worthy of further investigation.

The effect of fire on soil properties includes changes in soil nutrient levels through the ashing of live and dead plant tissue, in composition and abundance of soil microorganism communities, and in soil structure (Humphreys and Craig, 1981). In Australia, research has focused on the 'ash-bed effect' of substantially increased soil fertility and tree seedling growth rates associated with high severity fires in *Eucalyptus* forests. However, while available levels of phosphorus and basic cations (especially calcium, magnesium, and potassium) increase after fire, the effect is short-lived (a few months to years) and, in the case of vegetation on moderate to fertile sites, is little different quantitatively from the amounts of nutrients cycling through the decomposer system on an annual basis. On low-fertility sites the quantity of nutrients that is released represents a larger proportion of the total nutrient store and may more markedly influence fire-related system dynamics. Increased seedling growth rate was initially linked to increased phosphorus availability, but recent work suggests that this is an indirect relationship: fire heats and dries the surface layers of the soil, affecting the dominant ectomycorrhizae associated with tree seedlings. Ectomycorrhizae from dried (heated) soils show increased phosphorus uptake compared to ectomycorrhizae of seedlings in undried (e.g., prefire) forest soils (Launonen et al., 2004).

Large amounts of nitrogen may be lost from ecosystems during fires due to volatilization (at temperatures greater than $200°C$) and ash convection. Wan et al. (2001) reviewed data from 87 studies, reporting an average 58% reduction in total fuel nitrogen after fire and a positive correlation between fuel nitrogen loss and fire intensity. At the same time, available soil ammonium (NH_4^+) and nitrate (NO_3^-) pools increased twofold or more during the first year after fire, declining thereafter to prefire levels. In terms of ecosystem dynamics and management, large losses of nitrogen at the stand level through fuel combustion might affect long-term community structure and composition (DeBano et al., 1998), while short-term increases in available nitrogen (and phosphorus) after fire can favor invasives over indigenous species in low-nutrient systems (Heddle and Specht, 1975; Musil and Midgley, 1990).

Changes to soil physical conditions may include aggregation of fine particles and the development of hydrophobicity due to condensation of vaporized organic hydrophobic compounds on soil particles under high fire intensities (Humphreys and Craig, 1981). Low-intensity fires, such as fuel reduction (management) burns, are less likely to lead to the chemical, microbiological, and physical changes noted here, so variations in fire intensity (including within stands) may affect patterns of postfire vegetation change through their impact on soil properties.

11.5 ECOSYSTEM CHANGES IN RELATION TO FIRE HISTORY

While the mean fire frequencies experienced by different ecosystem types and their consequences have been considered already, variations in the return interval between successive fires within a particular vegetation type may be even more critical in terms of their impacts on community composition, structure, and function. If fires recur too frequently, then some fire-killed perennial species may be eliminated as insufficient time is available for them to reach maturity and/or to accumulate a sufficient store of seeds to ensure recruitment in the next generation. If fires do not recur frequently enough, some species may disappear through senescence and the decline of viable propagules in the seed bank.

Cary and Morrison (1995) compared sclerophyll forests and shrublands, burned a different number of times over recent decades, and with differing minimum fire intervals, in Brisbane Water National Park near Sydney, Australia. They found that areas with shorter fire intervals were associated with a reduction in the number of woody fire-sensitive species, and that where fires were very frequent (intervals of one to five years) there was a greater abundance of herbaceous fire-tolerant species. Lloret and Vila (2003) described a similar result for some Mediterranean plant communities, with more frequent fire events leading to a decrease in the diversity of species dependent on regeneration from seed. In a southeast Norwegian boreal forest, Niklasson and Drakenberg (2001) documented change from a mean fire frequency of 20 years during the period 1400–1770, to no large fires since that time, based on the cross-dating of fire scars in Scots pine (*Pinus sylvestris*) trees and stumps, resulting in a succession from pine to spruce. Studies of fire frequency in forests of the American and Canadian Rockies also report decreases in fire frequency associated both with changes from warmer/drier to cooler/more moist weather conditions over the past few centuries (Johnson and Larsen, 1991), and with fire suppression since about 1900 (Wright and Agee, 2004). More frequent, lower-intensity fires in the pre-European settlement period have been associated with open forests dominated by trees of larger sizes, while infrequent fire has been associated with increased fuel loads and tree density (Brown and Baxter, 2003).

The uses of satellite imagery and remote sensing techniques provide new opportunities for investigation of the consequences of changes in a fire regime (see Foody and Millington, this volume). The year and season of fire (since the mid-1970s) can be ascertained from Landsat and other satellite sources for areas that lack historical records or suitable surrogate measures (e.g., fire-scar histories). Diaz-Delgado et al. (2002) used normalized difference vegetation index (NDVI) patterns to interpret the effects of frequent fire on regeneration in Mediterranean pine and oak communities, noting decreased rates of regrowth in resprouting oaks for areas affected by two fires in short succession. Turner et al. (1994) used Landsat Thematic Mapper (TM) imagery to analyze fire size and intensity patterns for the Yellowstone National Park fires of 1988. They found that few high-intensity burn sites were far from potential sources of propagules for plant reestablishment. Follow-up studies showed that crown fires were characteristic of nonequilibrium systems, with strong postfire recruitment by lodgepole pine (*P. contorta*) and lower species richness than in lower-intensity burn areas (Turner et al., 1997). However, fire could not explain past recruitment (and recruitment failure) patterns in trembling aspen (*Populus tremuloides*) stands, successful establishment probably depending on a complex interaction between climate, fire, and predator-prey relationships between the main browser—elk—and their main predator—wolves (Romme et al., 1995).

Increasing power and sophistication of computer simulation models has also facilitated exploration of the behavior and effects of fire. Models can explore fire-regime scenarios in ways that are not possible experimentally, examining (among other things) the effects of fire frequency, size, patchiness, and season on species and community properties (e.g., Groeneveld et al., 2002). Models can estimate extinction risk for species in relation to altered fire regimes (Liu et al., 2005) and can test the potential ecological effects of past fire regimes (e.g., pre-European settlement in Australia and the Americas) based on reconstructions from historical and ecological (fire-scar) evidence (Piñol et al., 2007; Groeneveld et al., 2008) and of future fire regimes under climate change (e.g., Cary, 2002). This area of investigation is likely to be one of the most productive in the near future, but caution is required in the interpretation (and implementation) of results since models are simplifications of nature and are only as good as the data on which they are based. As our understanding of fire ecology increases, so too will our confidence in the outputs of such models.

11.6 FIRE BEHAVIOR, INTENSITY, AND SEASON

Fire behavior refers primarily to the rate of spread and intensity of fires, which is determined by weather conditions, fuel load, vegetation, and topography. Plant communities vary in their structure, productivity, and fuel accumulation and distribution characteristics. In forests and woodlands, fuel load generally includes surface litter (fallen leaves, twigs, and bark), near-surface live and dead biomass (e.g., grasses and herbs), and that fraction of coarse woody debris likely to be consumed by fire (Cheney et al., 1992). As levels of above-ground biomass recover with time since the last fire, leaves, twigs, and bark are shed according to normal seasonal patterns of growth and senescence. Litter builds up at the soil surface until eventually the annual amount of litter falling to the ground and the amount of litter that is decomposed reach a quasi-equilibrium, ranging from less than 10 tonnes per hectare (t ha^{-1}) in dry forests to more than 25 t ha^{-1} in some wet forests. Fuel load represents only 1 to 5% of total above-ground biomass in most forests, but it can be 90–100% in shrublands and grasslands where all of the above-ground vegetation can be regarded as a part of the fuel load (Walker, 1981).

Fires spread fastest uphill with winds during dry and windy conditions. Each increase of 10 degrees in slope approximately doubles the rate of spread and halves the threshold fuel load, giving the same intensity of fire according to Luke and McArthur (1978). To quantify the weather effect on rate of spread with the wind, forest and grassland fire danger indices have been developed in a number of countries (Luke and McArthur, 1978; Cheney and Sullivan, 1997). The Australian Forest Fire Danger Index (FFDI) is based on drought condition, air temperature, relative humidity, and wind velocity in the open at a 10-meter height (Noble et al., 1980). There are five categories in the danger rating system: low (0–5), moderate (5–12), high (12–25), very high (25–50), and extreme (50+). Under the worst possible fire weather conditions, the FFDI may exceed a score of 100, with fire intensity reaching 100,000 kilowatts per meter (kW m^{-1}) or more. The maximum intensity of forest fires that can be controlled by firefighters is about 4000 kW m^{-1} (Luke and McArthur, 1978; McCarthy and Tolhurst, 1998). Fire intensity refers to the heat yield of fire expressed in kilowatts per meter (kW m^{-1}) of fire perimeter and is calculated as the product of fuel load, rate of spread of the fire, and heat yield of the fuel (kj kg^{-1}).

Since the probability and intensity of fire will vary seasonally with weather conditions, much consideration has been given to the ecological effects and management implications of fires burned in different seasons of the year. A large number of studies have been conducted into early versus late dry-season burns in the *Eucalyptus* savanna ecosystem of northern Australia, contrasting 'traditional' early season burning by Aborigines with late season burning by pastoralists and public estate land managers (Williams et al., 1998). Early season fires were found to be of lower intensity (2,100 kW m^{-1}) than late season fires (7,700 kW m^{-1}). A positive feedback effect is reported, with high-intensity late season fires leading to increased grass production, more frequent high-intensity fires, and less tree-layer recruitment. A similar pattern of behavior is reported for Mediterranean basin woodlands in relation to the tussock grass *Ampelodesmos mauritanica* (Vila et al., 2001). In South African savannas, winter lightning-strike fires maintain open tree savannas, while dry season anthropogenic fires lead to increased woody-layer density of the dominant shrub species *Colophospermum mopane* (Kennedy and Potgeiter, 2003).

In strongly seasonal regions with a Mediterranean-type climate, management fires in spring (i.e., the end of the growing season) are generally of lower intensity than those at the end of the dry season in autumn (Enright and Lamont, 1989). Higher fire temperatures lead to the greater release of seeds from serotinous species and more dormancy breaking of seeds stored in the soil, so that seedling recruitment is positively related to fire intensity (Moreno and Oechel, 1991; Hodgkinson, 1991; Williams et al., 2004). High rates of attrition of seeds on, and in, the soil over summer may also occur following spring fires due to increased diurnal temperature range and granivore activity prior to the arrival of winter rains. However, Enright and Lamont (1989) and Lamont et al. (1993) found that high seedling densities following autumn fires were associated with high seedling mortality rates over the first one to two years, and levels of recruitment between seasons were eventually little different. Nevertheless, some species may require high-intensity fires for successful recruitment: in Californian Chaparral, the resprouter shrub *Adenostoma fasciculatum* showed decreased seedling recruitment as fire intensity increased, while the nonsprouter *Ceanothus greggii* was not affected (Moreno and Oechel, 1991). In the fire-prone pine rocklands of Florida, the endemic understory resprouter

Chamaecrista keyensis (Fabaceae) showed significantly higher fruit production after high-intensity fires, so that recurrent low-intensity management fires might adversely affect the species (Liu et al., 2005). Differences in fire intensity and associated soil heating have been closely linked to the probability of germination of hard-seeded species (Auld and O'Connell, 1991; Williams et al., 2004) and fire season may affect both the composition and abundance of new recruits. Morrison (2002) found that high-intensity fires in sclerophyll woodlands and shrublands near Sydney favored fire-tolerant monocotyledons and (hard-seeded) Fabaceae, moderate-intensity fires favored Proteaceae (including many serotinous, fire-killed shrubs), while low-intensity fires saw an increase in Ericaceae.

Where rainfall is only weakly seasonal, the fire season may be less important than site differences and year-to-year variations in rainfall in driving recruitment patterns after fire (e.g., see Whelan and York, 1998; Bradstock and Bedward, 1992 in relation to recruitment of Proteaceae species in the Sydney region). Similarly, even in strongly seasonal environments, such as the garrigue (shrublands) of France, fires burned at the same frequency and in the same season may cause little change from one fire cycle to the next (Trabaud, 1991).

11.7 FIRE MANAGEMENT

11.7.1 Asset protection

Fire-prone vegetation is most typical of climate regions that are characterized by rainfall and temperature regimes conducive both to biomass accumulation (providing fuel) and seasonal conditions conducive to ignition and spread of fire. These climatic circumstances are also favorable for crop and pasture production, timber growth, and human habitation, so that such areas often support large human populations and their associated infrastructure. Natural areas that surround, or are embedded within, this human landscape are likely to carry fires regularly, some of which will inevitably affect people and their property. The frequency and intensity of fires in natural vegetation are caused by weather conditions and fuel loads. After stand-level disturbance by fire, regrowth will gradually return living biomass and dead fine fuel loads to levels that can again sustain a fire. In many Australian forests, vegetative recovery of trees and shrubs from protected buds beneath the bark or soil means that recovery is rapid and fuel loads can return to near the mature forest equilibrium level within 7 to 15 years and are capable of carrying fire within 5 to 10 years (McCarthy and Tolhurst, 2001).

The managed use of fire to reduce fuel loads, and so lessen the risk of wildfires affecting human assets—by lowering the probability and rate of fire spread so that fires do not propagate, or are able to be extinguished quickly by firefighters—is commonplace in regions where fire and people coexist. While fuel reduction burning may lower the risk of wildfires and their impacts, such wild-fires nevertheless occur from time to time, and there is considerable debate about the objectives and effectiveness of fire-management policies (Bradstock et al., 2002; Cary et al., 2003). Under extreme weather conditions, even recently fuel-reduced areas may not slow a wildfire suffi-cient to allow its successful attack by firefighters. For example, Moritz (2003) found that fires frequently burned through young chaparral vege-tation in Los Padres National Forest, California, with extreme fire weather overriding vegetation age-related fuel characteristics. Similarly, some vegetation types cannot be fuel-reduced since the species are fire-sensitive. Where species have economic—timber or water catchment—or con-servation values, fuel reduction burns may also be inappropriate.

Modern approaches to the managed use of fire must balance the need to protect human life and property, the ecological requirements of plant and animal species and the ecosystems that they make up, and the range of other objectives that the use of fire might seek to achieve (e.g., water quality and supply, honey production, production of new growth for livestock grazing, suppression of invasive species). Changing technologies and occupational health and safety requirements are already causing a shift in the nature of firefight-ing, with less direct attack on the ground by firefighting crews. The use of fire retardants by firefighters may have as yet unknown impacts on patterns of postfire vegetation recovery, with little measured effect in some cases but germination of some species suppressed in others (Larson et al., 1999).

11.7.2 Conservation

The use of fire for conservation involves determi-nation of frequency, season, severity, and extent of fire in relation to its effects on plant and animal species and communities. It also includes the option of fire exclusion where desirable species or community properties will be adversely affected by fire. Ecological burning regimes must take into account the critical life-history attributes of spe-cies, including conditions favoring recruitment, growth and survival to maturity, time to maturity,

and longevity. Noble and Slatyer (1980) proposed a 'vital attributes' scheme to describe these critical stages, particularly in relation to fire-prone envi-ronments. This qualitative scheme is now used by a number of natural estate management organiza-tions in Australia and elsewhere to determine fire return envelopes that are compatible with manage-ment objectives for species and communities. However, such schema are only as good as the data on which they are based, and even basic life-history information is not known for all species in more than a few community types. Further, quali-tative or semiquantitative assessments cannot easily predict how abundance hierarchies might shift for fire intervals that fall within different parts of supposedly suitable fire frequency enve-lopes. Such responses are determined by actual levels of seed supply among species at the time of fire, and conditions for recruitment after fire. Here again, our knowledge is even more limited.

Ecological fire management may sometimes conflict with management for asset protection. For example, long periods without fire are advocated for the conservation of the malleefowl (*Leipoa ocellata*) in the semi-arid mallee (*Eucalyptus* spp.) woodlands of southeast Australia. This species requires large amounts of litter to build nest mounds, and given the slow litter accumulation rates in the mallee woodlands, fire-free intervals of up to 60 years or more are estimated as optimal for this purpose (Benshemesh, 1990). On the other hand, the favored food resources for the species may be most available in vegetation no more than 20 to 30 years old, while fires more frequent than every 10 years would likely lead to local extinction. The management difficulty is how to contain fire within the system, thereby avoiding damage to assets on adjacent private land while conserving the malleefowl.

Fragmentation and reduction in the size of natural vegetation remnants through land aliena-tion for a variety of human land uses provide other problems in fire-prone ecosystems. The likelihood of natural fires is reduced due to the low probabil-ity that a natural ignition will occur within the fragment. Prior to fragmentation, ignitions some distance away may have facilitated fire spread throughout such areas. The use of managed fire within remnants may be complicated by their small size and by the presence of private assets on all sides. Local community involvement and support for management are imperative if fire is to be allowed to play a role in the conservation management of such remnants. Alternatively, surrounding land uses may introduce fire to, or require managed fires within, fragments at such a high frequency that this too can lead to unwanted species compositional changes (Gill and Williams, 1996).

11.7.3 Use of fire by indigenous peoples

Changing institutional policies concerning fire management over the past century, from policies of forest-fire suppression in North America, Australia, and elsewhere in the first half of the twentieth century to those of the managed use of fire for fuel reduction thereafter, have been based on increases in our understanding of ecosystem dynamics and of our technical capacity to manage fire. Such changes represent an implicit acknowledgment that our understanding of the relationship between fire and environment was, and is still, imperfect. Various positions have been argued since the earliest days of European settlement of the Americas and Australia concerning the relationship between native peoples and their environment, including their use of fire (Pyne, 1995; Horton, 2000). Scientific assessments of the evidence for the use of fire by indigenous peoples to manage biological resources have mostly viewed them as anecdotal, or in any case irrelevant to modern fire-management objectives. Greater credence is now being given to the potential value of traditional practices in achieving asset protection and biological conservation objectives of land management as a result of new fire history reconstructions that support anthropological and historical accounts of traditional fire regimes, and the concordance between developing scientific understanding and traditional burning practices in areas such as the tropical savanna woodlands of northern Australia (Bowman, 1998; Yibarbuk et al., 2001; Bowman et al., 2004). Studies there have found that traditional fire regimes produced a fine-scale mosaic of burned and unburned areas at lower frequencies and intensities than under European fire regimes. In areas managed under the post-European settlement regime of annual to biennial, large fires are more strongly dominated by annual grasses and show a greater build-up of grassy fuels capable of carrying frequent fires. Additionally, tree-layer vegetation was found to be greater in areas subject to traditional fire management (Bowman and Prior, 2004). Based on such evidence, traditional knowledge is now incorporated into fire-management planning for natural areas in northern Australia and in some parts of North America (Kimmerer and Lake, 2001).

Increasing evidence from fire-history reconstructions for areas where traditional knowledge has been mostly lost (such as in southern Australia and many parts of North America) generally indicates frequent fires of small sizes and assumed low intensity relative to present fire regimes (Ward et al., 2001; Wright and Agee, 2004). Ward et al. (2001) report a mean fire interval of four years for dry forests and shrublands in southwest Australia in the pre-European period based on dark stem bands in grasstrees (*Xanthorrhoea* spp.) that are reputedly caused by fire. They argued that this regime represented active fire management of the landscape by Aborigines, calling upon support also from historical sources during the early stages of European settlement that describe regular use of fire by local peoples and open forest understories as a consequence (Abbott, 2003). While a similar record is reported for shrublands by Enright et al. (2005; and see Table 11.2), Enright et al. question the validity of the fire-history reconstruction since the frequent-fire, open-vegetation scenarios are not consistent with our present understanding of the ecological dynamics of shrubland systems. Many fire-killed woody plants could not survive such frequent fires since their juvenile stages are too long (Enright et al., 2005). Small fire sizes cannot resolve the problem either, since the grasstree record shows all parts of the landscape had high fire frequencies, so that there can be few patches that remain unburned long enough, or are large enough, to provide viable refugia for these fire-sensitive species. Many of these fire-sensitive woody plants now dominate shrublands and forest understories at the regional scale. It is clear that our understanding of prehistoric fire regimes in many parts of the world, and their ecological effects, is incomplete. Any notion that such reconstructed regimes may be appropriate to meet modern fire-management objectives must be tempered by the need for further verification of past fire regimes and understanding of the ecological dynamics of

Table 11.2 Mean fire intervals for consecutive 20-year periods from 1872 to 1991 at Yardanogo Nature Reserve, Western Australia, based on grasstree (*Xanthorrhoea acanthostachya*) fire-history reconstructions. Error term is SD, *n* is number of grasstrees, *i* is total number of fire intervals per period. The years with fires is the number of years per period in which a fire is recorded (from Enright et al., 2005)

Years	Fire interval	n	i	Years with fires
1872–1891	4.0 ± 1.6	3–5	14	10
1892–1911	4.4 ± 2.1	5–11	39	9
1912–1931	4.6 ± 2.4	13–14	56	8
1932–1951	8.3 ± 4.3	14	32	7
1952–1971	8.0 ± 4.4	14–15	29	7
1972–1991	10.0 ± 5.1	15	35	6

ecosystems in relation to such. Large-scale and long-term experimental studies are likely to be essential in resolving this issue.

11.7.4 Restoration ecology

Restoration ecology has become central to the rehabilitation activities of mining companies throughout the world, but large-scale disturbances have generally been excluded from the vision of how restoration might proceed. Restoration programs may have excluded fire both because the reconstructed systems may be too young to carry a fire, and because fire is seen as a threat to operational activities, for example, it may damage site infrastructure, and smoke may affect the safety of operations. However, eventually fire must be returned to restored fire-prone ecosystems if they are to converge on a composition, structure, and function similar to their natural counterparts. The experimental reintroduction of fire has begun recently for restored jarrah forests (*Eucalyptus marginata*) in bauxite-mined areas in southwest Australia, seeking to identify the best age at which to burn sites so as to minimize the impacts of invasive species and maximize the recruitment and survival of desired native species (Grant, 2000). The potential for 'broad-acre' use of the active compound in smoke responsible for dormancy breaking in many species from fire-prone ecosystems has already been mentioned.

11.7.5 Climate change

Projected global warming over the next 100 years due to increases in atmospheric levels of CO_2 and other greenhouse gases has been the focus of many studies concerning the possible impacts of climate change on the distribution of plants, animals, and communities. General circulation simulation models (GCMs) based on an assumed twofold increase in atmospheric CO_2 project that mean temperatures will increase at least 2–4°C over this time, along with shifts in rainfall, and intensity and frequency of global climate phenomena such as the ENSO, which is associated with drought in the eastern Pacific (Timmerman et al., 1999; IPCC, 2007). A number of studies have sought to identify the likely impacts of climate change on fire regimes, and in turn, the effects of changed fire regimes on vegetation (Flannigan et al., 2001; Cary, 2002, Williams et al., 2009). Results for boreal forest areas in Canada and Russia suggest longer fire seasons, more extreme fire weather days, and a larger area affected by extreme fire weather (Stocks et al., 1998), but

patterns are not necessarily uniform, with fire frequency predicted to increase in western Canada more than in eastern Canada. In Australia, GCM scenarios for fire regime in mountain forests and alpine areas of southeast Australia suggest more fires in autumn (i.e., a longer fire season) and a shorter mean fire interval at the landscape scale, due mostly to increased spread rather than more ignitions. If such scenarios are correct, then the mean age of vegetation at any site will be less than it is now, with potentially serious consequences for species with long juvenile stages and those that are already near their fire-frequency extinction thresholds (Cary, 2002). However, there are many uncertainties concerning the potential effects of climate change on fire, particularly in relation to the impacts of temperature and rainfall changes on rates and levels of biomass and fuel accumulation, and of increased CO_2 on plant water-use efficiency and productivity (Williams et al., 2009). As models of climate change continue to improve, more specific research questions concerning the distributions and conservation management of species and communities will be possible and will represent an important area of investigation.

11.7.6 Water supply

The effects of vegetation clearance and plantation timber growth on plant water use and catchment-scale water yields have been reported for many forest ecosystems, with results generally suggesting short-term increases in water yield followed by medium-term decreases of various magnitudes and durations (Cornish and Vertessy, 2001). Tree layer water use and predicted catchment level water yield in fire-killed mountain ash (*Eucalyptus regnans*) forests of southeast Australia show a decrease in the catchment water yield of 50% or more during the sapling and pole stages of forest development following stand-level fires (Kucera, 1987). Projected decreases in yield are less substantial (generally less than 20%) during the regrowth phase in fire-affected stands for resprouting *Eucalyptus* species typical of dry sclerophyll forests. While the land area occupied by fire-killed wet sclerophyll eucalypts in Australia is small relative to that occupied by dry sclerophyll species, such areas represent some of the main water-storage reservoirs servicing the major population centers of southern Australia. They are currently managed to exclude fire, but this increases the long-term risk of wildfire due to fuel load build-up. Water resource and fire managers find themselves in a conundrum; how can long-term water supply be safeguarded against the likely water losses that would occur if reservoir catchment areas are affected by large fires?

Water supply in areas of low rainfall and increasing human population will continue to prove a controversial topic for decision makers. Until now, the managed use of fire has largely been restricted to objectives associated with fuel-load reduction for asset protection and conservation management. Manipulation of the fire regime to increase water yield for human use is likely to have ecological consequences for natural ecosystems that conflict with biological conservation objectives.

11.8 CONCLUSIONS

Fire-prone ecosystems show a number of major differences from most other ecosystems both in species demographic behavior and community properties. Recruitment of individuals is largely restricted to the first year, or few years, after fire, with little recruitment subsequently during the interfire period. This reflects the massive fitness advantage associated with early establishment after fire, which provides an initial environment with increased resources and reduced competition, and the maximum time to grow and store seeds (or buds) before another fire occurs. This is particularly important for fire-killed perennial species. Species richness also peaks in the first years after fire, reflecting the persistence of most prefire species within burned areas through vegetative regrowth and recruitment from seed stores, plus the recruitment of fire-ephemeral species from the soil seed bank. Studies of fire frequency and intensity effects on communities nevertheless suggest a role for intermediate disturbance dynamics, with intermediate frequencies and intensities of fire less likely to lead to species losses than more extreme fire regimes.

The behavior and ecology of fire in plant communities continue to present major challenges for research: recent advances in our understanding of germination cues (e.g., smoke), debates and controversies over how fire can be managed in relation to biological conservation and asset protection objectives, the relevance of traditional fire regimes to modern fire management, the effects of fire on water yield, the impacts of climate change on fire regimes, and many other issues, reinforce the notion that 'the more we know, the more we need to know'.

ACKNOWLEDGMENTS

Many people have contributed to the development of my understanding of fire ecology over several decades of university-based research and teaching. Foremost among them are Byron Lamont and Malcolm Gill. An array of other colleagues and students, particularly in Australia and South Africa, have played at least some part in shaping the ideas that I present here. However, any errors of fact or interpretation remain solely my own.

REFERENCES

Abbott, I. (2003) 'Aboriginal fire regimes in south-western Australia: Evidence from historical documents', in N. Burrows and I. Abbott (eds.), *Fire in South-western Australian Ecosystems: Impacts and Management*. Leiden: Backhuys. pp. 119–46.

Allan, G.E., and Southgate, R.I. (2002) 'Fire regimes in the spinifex landscapes of Australia', in R.A. Bradstock, J.E. Williams, and M.A. Gill (eds.), *Flammable Australia: The Fire Regimes and Biodiversity of a Continent*. Cambridge: Cambridge University Press. pp. 145–176.

Ashton, D.H. (1981) 'Fire in tall open forests', in A.M. Gill, R.H. Groves, and I.R. Noble (eds.), *Fire and the Australian Biota*. Canberra: Australian Academy of Science. pp. 339–68.

Auld, T.D., and O'Connell, M.A. (1991) 'Predicting patterns of post-fire germination in 35 eastern Australian Fabaceae', *Australian Journal of Ecology*, 16: 53–70.

Bell, D.T. (1999) 'Turner review No.1: The process of germination in Australian species', *Australian Journal of Botany*, 47: 475–517.

Bell, D.T., Hopkins, A.J.M., and Pate, J.S. (1984) 'Fire in the Kwongan', in J.S. Pate and J.S. Beard (eds.), *Kwongan: Plant Life of the Sandplain*. Nedlands WA: UWA Press. pp. 178–204.

Bellingham, P.J. (2000) 'Resprouting as a life history strategy in woody plant communities', *Oikos*, 89: 409–16.

Benkman, C.W., and Siepielski, A.M. (2004) 'A keystone selective agent? Pine squirrels and the frequency of serotiny in lodgepole pine', *Ecology*, 85: 2082–7.

Benshemesh, J. (1990) 'Management of malleefowl with regard to fire,' in J.C. Noble, P.J. Joss, and G.K. Jones (eds.), *The Mallee Lands: A Conservation Perspective*. Melbourne: CSIRO Publishing. pp. 206–11.

Bond, W.J., and Keeley, J.E. (2005) 'Fire as a global "herbivore": The ecology and evolution of flammable ecosystems', *Trends in Ecology and Evolution*, 20: 387–94.

Bond, W.J., and Midgley, J.J. (1995) 'Kill thy neighbour: An individualistic argument for the evolution of flammability', *Oikos*, 73: 79–85.

Bond, W.J., and Midgley, J.J. (2003) 'The evolutionary ecology of sprouting in woody plants', *International Journal of Plant Sciences*, 164: S103–S114.

Bond, W.J., Midgley, J.J., and Woodward, F.I. (2003) 'What controls South African vegetation—climate or fire?' *South African Journal of Botany*, 69: 79–91.

Bond, W.J., and Van Wilgen, B.W. (1996) *Fire and Plants*. London: Chapman and Hall.

Bond, W.J., Woodward, F.I., and Midgley, J.J. (2005) 'The global distribution of ecosystems in a world without fire', *New Phytologist*, 165: 525–37.

Bowler, J.M. (1982) 'Aridity in the late Tertiary and Quaternary of Australia', in W.R. Parker and P.J.M. Greenslade (eds.), *Evolution of the flora and fauna of arid Australia* Glen Osmond, SA: Peacock Publications. pp. 35–45.

Bowman, D.M.J.S. (1998) 'Tansley Review No. 101. The impact of Aboriginal landscape burning on the Australian biota', *New Phytologist*, 140: 385–410.

Bowman, D.M.J.S., and Prior, L.D. (2004) 'Impact of Aboriginal landscape burning on woody vegetation in *Eucalyptus tetrodonta* savanna in Arnhem Land, northern Australia', *Journal of Biogeography*, 31: 807–817.

Bowman, D.M.J.S., Walsh, A., and Prior, L.D. (2004) 'Landscape analysis of Aboriginal fire management in Central Arnhem Land, north Australia', *Journal of Biogeography*, 31: 207–223.

Bradstock, R.A., and Bedward, M. (1992) 'Simulation of the effect of season of fire on post-fire seedling emergence of two *Banksia* species based on long-term rainfall records', *Australian Journal of Botany*, 40: 75–88.

Bradstock, R.A., Williams, J.E., and Gill, A.M. (eds.). (2002) *Flammable Australia: The Fire Regimes and Biodiversity of a Continent*. Cambridge: Cambridge University Press.

Brown, J., Enright, N.J., and Miller, B.P. (2003) 'Seed production and germination in two rare and three common, co-occurring *Acacia* species from SE Australia', *Australian Journal of Ecology*, 28: 271–80.

Brown, N.A.C. (1993) 'Promotion of germination of fynbos seeds by plant-derived smoke', *New Phytologist*, 123: 575–83.

Brown, P.M., and Baxter, W.T. (2003) 'Fire history in coast redwood forests of the Mendocino Coast, California', *Northwest Science*, 77: 147–58.

Cary, G.J. (2002) 'Importance of a changing climate for fire regimes in Australia', in R.A. Bradstock, J.E. Williams, and M.A. Gill (eds.), *Flammable Australia: The Fire Regimes and Biodiversity of a Continent*. Cambridge: Cambridge University Press. pp. 26–49.

Cary, G.J., Lindenmayer, D., and Dovers, S. (eds.). (2003) *Australia Burning: Fire Ecology, Policy and Management Issues*. Melbourne: CSIRO Publishing.

Cary, G.J., and Morrison, D.A. (1995) 'Effects of fire frequency on plant species composition of sandstone communities in the Sydney region—combinations of fire intervals', *Australian Journal of Ecology*, 20: 418–26.

Cheney, N.P., Gould, J.S., and Knight, I. (1992) *A Prescribed Burning Guide for Young Regrowth Forests of Silvertop Ash*. Sydney: Forests Commission of NSW.

Cheney, P., and Sullivan, A. (1997) *Grassfires, Fuel Weather and Fire Behavior*. Melbourne: CSIRO Publishing.

Clemente, A.S., Rego, F.C., and Correia, O.A. (1996) 'Demographic patterns and productivity of post-fire regeneration in Portuguese Mediterranean maquis', *International Journal of Wildland Fire*, 6: 5–12.

Cochrane, M.A., and Schulze, M.D. (1999) 'Fire as a Recurrent Event in Tropical Forests of the Eastern Amazon: Effects on Forest Structure, Biomass, and Species Composition', *Biotropica*, 31: 2–16.

Cornish, P.M., and Vertessy, R.A. (2001) 'Forest age-induced changes in evapotranspiration and water yield in a eucalypt forest', *Journal of Hydrology*, 242: 43–63.

Cowling, R.M., and Lamont, B.B. (1985) 'Variation in serotiny of three Western Australian *Banksia* species along a climatic gradient', *Australian Journal of Ecology*, 10: 345–50.

DeBano, L.F., Neary, D., and Ffolliott, P.F. (1998) *Fire's Effects on Ecosystems*. New York: John Wiley.

Denham, A.J., and Auld, T.D. (2002) 'Flowering, seed dispersal, seed predation and seedling recruitment in two pyrogenic flowering resprouters', *Australian Journal of Botany*, 50: 545–57.

Diaz-Delgado, R., Lloret, F., Pons, X., and Terradas, J. (2002) 'Satellite evidence of decreasing resilience in Mediterranean plant communities after recurrent wildfires', *Ecology*, 83: 2293–303.

Enright, N.J. (1985) 'Existence of a soil seed bank under rainforest in New Guinea', *Australian Journal of Ecology*, 10: 67–71.

Enright, N.J., and Lamont, B.B. (1989) 'Seed banks, fire season, safe sites and seedling recruitment in five co-occurring *Banksia* species', *Journal of Ecology*, 77: 1111–22.

Enright, N.J., and Lamont, B.B. (1992) 'Recruitment variability in the resprouting shrub *Banksia attenuata* and non-sprouting congeners in the northern sandplain heaths of south-western Australia', *Acta Oecologica*, 13: 727–41.

Enright, N.J., Lamont, B.B., and Miller, B.P. (2005) 'Anomalies in grasstree fire history reconstructions for SW Australian vegetation', *Austral Ecology*, 30: 668–73.

Enright, N.J., Marsula, R., Lamont, B.B. and Wissel, C. (1998a) 'The ecological significance of canopy seed storage in fire-prone environments: A model for non-sprouting shrubs', *Journal of Ecology*, 86: 946–59.

Enright, N.J., Marsula, R., Lamont, B.B. and Wissel, C. (1998b) 'The ecological significance of canopy seed storage in fire-prone environments: A model for resprouting shrubs', *Journal of Ecology*, 86: 960–73

Enright, N.J., and Thomas, I. (2008) 'Pre-European fire regimes in Australian ecosystems', *Geography Compass*, 2: 1–33.

Felderhof, L., and Gillieson, D. (2006) 'Comparison of fire patterns and fire frequency in two tropical savanna bioregions', *Australian Journal of Ecology*, 31: 736–46.

Flannigan, M., Campbell, I., Wotton, M., Carcaillet, C., Richard, P. and and Bergeron, Y. (2001) 'Future fire in Canada's boreal forest: paleoecology results and general circulation model – regional climate model simulations', *Canadian Journal of Forest Research*, 31: 854–864.

Flematti, G.R., Ghisalberti, E.L., Dixon, K., and Trengrove, R.D. (2004) 'A compound from smoke that promotes seed germination', *Science*, 305: 977.

Gauthier, S., Bergeron, Y., and Simon, J.P. (1996) 'Effects of fire regime on the serotiny level of jack pine', *Journal of Ecology*, 84: 539–48.

Ghermandi, L., Guthmann, N., and Bran, D. (2004) 'Early post-fire succession in northwestern Patagonia grasslands', *Journal of Vegetation Science*, 15: 67–76.

Gill, A.M. (1981) 'Adaptive responses of Australian vascular plant species to fires', in Gill, A.M., Groves, R.H., and Noble, I.R. (eds.), *Fire and the Australian Biota*. Canberra: Australian Academy of Science. pp. 273–310.

Gill, A.M., and Catling, P. (2002) 'Fire regimes and biodiversity of forested landscapes of southern Australia', in R.A. Bradstock, J.E. Williams, and M.A. Gill (eds.), *Flammable Australia: The Fire Regimes and Biodiversity of a Continent*. Cambridge: Cambridge University Press. pp. 351–69.

Gill, A.M., Groves, R.H., and Noble, I.R. (eds.). (1981) *Fire and the Australian Biota*. Canberra: Australian Academy of Science.

Gill, A.M., and Williams, J.E. (1996) 'Fire regimes and biodiversity: The effects of fragmentation of southeastern Australian eucalypt forests by urbanization, agriculture and pine plantations', *Forest Ecology and Management*, 85: 261–78.

Gilmour, C.A., Crowden, R.K., and Koutoulis, A. (2000) 'Heat shock, smoke and darkness: Partner cues in promoting germination in *Epacris tasmanica* (Epacridaceae)', *Australian Journal of Botany*, 48: 603–8.

Groeneveld, J., Enright, N.J., and Lamont, B.B. (2008) 'Simulating the effects of different spatio-temporal fire regimes on plant metapopulation persistence in a Mediterranean-type region', *Journal of Applied Ecology*, 45: 1477–85.

Groeneveld, J., Enright, N.J., Lamont, B.B., and Wissel, C. (2002) 'A spatial model of coexistence among three *Banksia* species along a habitat gradient in fire-prone shrublands', *Journal of Ecology*, 90: 762–74.

Groom, P.K., and Lamont, B.B. (1996) 'Reproductive ecology of non-sprouting and resprouting species of *Hakea* (Proteaceae) in SW Australia', in S.D. Hopper, J. Chappill, M. Harvey, and A. George (eds.), *Gondwanan Heritage: Past, Present and Future of the Western Australian Biota*. Chipping Norton: Surrey Beatty and Sons. pp. 239–48.

Harrod, J.C., Harmon, M.E., and White, P.S. (2000) 'Post-fire succession and 20th century reduction in fire frequency on xeric southern Appalachian sites', *Journal of Vegetation Science*, 11: 465–72.

Heddle, E.M., and Specht, R.L. (1975) 'Dark Island Heath (Ninety-Mile Plain, South Australia). VIII. The effect of fertilizers on composition and growth', 1950–1972. *Australian Journal of Botany*, 23: 151–64.

Hobbs, R. (2002) 'Fire regimes and their effects in Australian temperate woodlands', in Bradstock, R.A., Williams, J.E., and Gill, A.M. (eds.), *Flammable Australia: The Fire Regimes and Biodiversity of a Continent*. Cambridge: Cambridge University Press. pp. 305–26.

Hodgkinson, K.C. (1991) 'Shrub recruitment response to intensity and season of fire in a semi-arid woodland', *Journal of Applied Ecology*, 28: 60–70.

Hopkins, M.S., and Graham, A.W. (1983) 'The species composition of soil seed banks beneath lowland tropical rainforests in north Queensland, Australia', *Biotropica*, 15: 90–99.

Horton, D. (2000) *The Pure State of Nature*. Sydney: Allen and Unwin.

Humphreys, F.R., and Craig, F.G. (1981) 'Effects of fire on soil chemical, structural and hydrological properties', in Gill, A.M., Groves, R.H., and Noble, I.R. (eds.), *Fire and the Australian Biota*. Canberra: Australian Academy of Science. pp. 177–200.

Intergovernmental Panel on Climate Change (IPCC). (2007) *Climate Change 2007*—The Physical Science Basis. Contribution of Working Group I to the Fourth Assessment Report of the IPCC. Cambridge: Cambridge University Press.

Johnson, E.A., and Larsen, C.P. (1991) 'Climatically induced change in fire frequency in the southern Canadian Rockies', *Ecology*, 72: 194–201.

Keeley, J.E. (1992) 'Recruitment of seedlings and vegetative sprouts in unburned Chaparral', *Ecology*, 73: 1194–208.

Keeley, J.E., and Fotheringham, C.J. (1998) 'Smoke-induced seed germination in Californian chaparral', *Ecology*, 79: 2320–6.

Keeley, S.C., and Pizzorno M. (1986) 'Charred wood stimulated germination of two fire-following herbs of the California chaparral and the role of hemicellulose', *American Journal of Botany*, 73: 1289–97.

Keith, D.A., McCaw, W.L., and Whelan, R.J. (2002) 'Fire regimes in Australian heathlands and their effects on plants', in R.A. Bradstock, J.E. Williams, and M.A. Gill (eds.), *Flammable Australia: The Fire Regimes and Biodiversity of a Continent*. Cambridge: Cambridge University Press. pp. 199–237.

Kennedy, A.D., and Potgieter, A. (2003) 'Fire season affects size and architecture of *Colophospermum mopane* in southern African savannas', *Plant Ecology*, 167: 179–92.

Kerr, B., Schwilk, D.W., Bergman, A., and Feldman, M.W. (1999) 'Rekindling an old flame: A haploid model for the evolution and impact of flammability in resprouting plants', *Evolutionary and Ecological Research*, 1: 807–33.

Kimmerer, R.W., and Lake, F.K. (2001) 'The role of indigenous burning in land management', *Journal of Forestry*, 99: 36–41.

Kuczera, G. (1987) 'Prediction of water yield reductions following a bushfire in ash-mixed species eucalypt forest', *Journal of Hydrology*, 94: 215–236.

Kuhry, P. (1994) 'The Role of Fire in the Development of Sphagnum-Dominated Peatlands in Western Boreal Canada', *Journal of Ecology*, 82: 899–910.

Lamont, B.B., and Enright, N.J. (2000) 'Adaptive advantages of aerial seed banks', *Plant Species Biology*, 15: 157–66.

Lamont, B.B., and Wiens, D. (2003) 'Is seed set and speciation always low among species that resprout after fire, and why?' *Evolutionary Ecology*, 17: 277–92.

Lamont, B.B., Le Maitre, D.C., Cowling, R.M. and Enright, N.J. (1991) 'Canopy seed storage in woody plants', *Botanical Review*, 57: 277 – 317.

Lamont, B.B., Witkowski, E.T.F., and Enright, N.J. (1993) 'Post-fire litter microsites: Safe for seeds, unsafe for seedlings', *Ecology*, 74: 501–12.

Larson, D.L., Newton, W.E., Anderson, P.J., and Stein, S.J. (1999) 'Effects of fire retardant chemical and fire suppressant foam on shrub steppe vegetation in northern Nevada', *International Journal of Wildland Fire*, 9: 115–127.

Launonen, T.M., Ashton, D.H., Kelliher, K.J., and Keane, P.J. (2004) 'The growth and P acquisition of *Eucalyptus regnans* F. Muell. Seedlings in air-dried and undried forest soil in relation to seedling age and ectomycorrhizal infection', *Plant and Soil*, 267: 179–89.

Litton, C.M., and Santelices, R. (2002) 'Early post-fire succession in a *Nothofagus glauca* forest in the Coastal Cordillera of south-central Chile', *International Journal of Wildland Fire*, 11: 115–25.

Liu, H., Menges, E.S., Snyder, J.R., Koptur, S., and Ross, M.S. (2005) 'Effects of fire intensity on vital rates of an endemic herb of the Florida Keys, USA', *Natural Areas Journal*, 25: 71–76.

Lloret, F., and Vila, M. (2003) 'Diversity patterns of plant functional types in relation to fire regime and previous land use in Mediterranean woodlands', *Journal of Vegetation Science*, 14: 387–98.

Luke, R.H., and McArthur, A.G. (1978) *Bushfires in Australia*. Canberra: Australian Government Publishing Service.

McCarthy, G.J., and Tolhurst, K.G. (2001) *Effectiveness of Broadscale Fuel Reduction Burning in Assisting with Wildfire Control in Parks and Forests in Victoria*. Melbourne: Department of Natural Resources and Environment. Research Report 51.

McGlone, M.S. (1989) 'The Polynesian settlement of New Zealand in relation to environmental and biotic changes', *New Zealand Journal of Ecology*, 12: 115–29.

Monamy, V., and Fox, B.J. (2000) 'Small mammal succession is determined by vegetation density rather than time elapsed since disturbance', *Australian Journal of Ecology*, 25: 580–87.

Moreno, J.M., and Oechel, W.C. (1991) 'Fire intensity effects on germination of shrubs and herbs in southern Californian chaparral', *Ecology*, 72: 1993–2004.

Moritz, M.A. (2003) 'Spatiotemporal analysis of controls on shrubland fire regimes: Age dependency and fire hazard', *Ecology*, 84: 351–61.

Morrison, D.A. (2002) 'Effects of fire intensity on plant species composition of sandstone communities in the Sydney region', *Australian Journal of Ecology*, 27: 433–41.

Musil, C.F., and Midgley, C.F. (1990) 'The relative impact of invasive Australian acacias, fire and season on the soil chemical status of a sandplain lowland fynbos community', *South African Journal of Botany*, 56: 419–27.

Mutch, R.W. (1970) 'Wildland fires and ecosystems—a hypothesis', *Ecology*, 51: 1046–51.

Niklasson, M., and Drakenberg, B. (2001) 'A 600-year tree-ring fire history from Norra Kvills National Park, southern Sweden', *Biological Conservation*, 101: 63–71.

Noble, I.R., and Slatyer R.O. (1980) 'The use of vital attributes to predict successional changes in plant communities subject to recurrent disturbances', *Vegetatio*, 43: 5–21.

Noble, I.R., Bary, G.A.V., and Gill, A.M. (1980) 'McArthur's fire-danger meters expressed as equations', *Australian Journal of Ecology*, 5: 201–3.

Norton, D.A., and De Lange, P.J. (2003) 'Fire and Vegetation in a Temperate Peat Bog: Implications for the Management of Threatened Species', *Conservation Biology*, 17: 138–148.

Ojeda, F. (1998) 'Biogeography of seeder and resprouter *Erica* species in the Cape Floristic Region—Where are the resprouters?' *Biological Journal of the Linnaean Society*, 63: 331–47.

Pausas, J., and Lloret, F. (2007) 'Spatial and temporal patterns of plant functional types under simulated fire regimes', *International Journal of Wildland Fire*, 16: 484–92.

Perez-Fernandez, M.A., and Rodriguez-Echeverria, S. (2003) 'Effect of smoke, charred wood and nitrogenous compounds on seed germination of ten species from woodland in central-western Spain', *Journal of Chemical Ecology*, 29: 237–51.

Piñol, J., Castellnou, M., and Beven, K. J. (2007) 'Conditioning uncertainty in ecological models: Assessing the impact of fire management strategies', *Ecological Modelling*, 207: 34–44.

Purdie, R.W., and Slatyer, R.O. (1976) 'Vegetative succession after fire in sclerophyll woodland communities in south-eastern Australia', *Australian Journal of Ecology*, 1: 223–36.

Purdy, B.G., Macdonald, S.E., and Dale, M.R. (2002) 'The regeneration niche of white spruce following fire in the mixedwood boreal forest', *Silva Fennica*, 36: 289–306.

Pyne, S.J. (1991) *Burning Bush: A Fire History of Australia*. New York: Henry Holt.

Pyne, S.J. (1995) *World Fire: The Culture of Fire on Earth*. Seattle: University of Washington Press.

Ramsay, P.M., and Oxley, E.R.B. (1996) 'Fire temperatures and post-fire plant community dynamics in Ecuadorian grass paramo', *Vegetatio*, 124: 129–44.

Romme, W.H., Turner, M.G., Wallace, L.L., and Walker, J.S. (1995) 'Aspen, elk and fire in northern Yellowstone National Park', *Ecology*, 76: 2097–106.

Russell-Smith, J., and Stanton, P. (2002) 'Fire regimes and fire management of rainforest communities across northern Australia', in R.A. Bradstock, J.E. Williams, and M.A. Gill (eds.), *Flammable Australia: The Fire Regimes and Biodiversity of a Continent*. Cambridge: Cambridge University Press. pp. 329–50.

Schoennagel, T., Turner, M.G., and Romme, W.H. (2003) 'The influence of fire interval and serotiny on post-fire lodgepole pine density in Yellowstone National Park', *Ecology*, 84: 2967–78.

Schwilk, D.W. (2003) 'Flammability is a niche construction trait: Canopy architecture affects fire intensity', *American Naturalist*, 162: 725–33.

Smith, M.A., Bell, D.T., and Loneragan, W.A. (1999) 'Comparative seed germination ecology of *Austrostipa compressa* and *Ehrharta calycina* (Pocaeae) in a Western Australian *Banksia* woodland', *Australian Journal of Ecology*, 24: 35–42.

Stocks, B.J. and 10 authors (1998) 'Climate change and forest fire potential in Russian and Canadian boreal forests', *Climatic Change*, 38: 1–13.

Thomas, T.H., and Van Staden, J. (1995) 'Dormancy break of celery (Apium graveolens L.) seeds by plant derived smoke extract', *Plant Growth Regulation*, 17: 195–98.

Thompson, K. (1978) 'The occurrence of buried viable seeds in relation to environmental gradients', *Journal of Biogeography*, 5: 425–30.

Tieu, A., Dixon, K.W., Meney, K.A., and Sivasithamparam, K. (2001) 'Interaction of soil burial and smoke on germination patterns in seeds of selected Australian native plants', *Seed Science Research*, 11: 69–77.

Timmerman, A., Oberhuber, J., Bacher, A., Esch, M., Latif, M., and Roeckner, E. (1999) 'Increased El Nino frequency in a climate model forced by future greenhouse warming', *Nature*, 398: 694–697.

Trabaud, L. (1991) 'Fire regimes and phytomass growth in a *Quercus coccifera* garrigue', *Journal of Vegetation Science*, 2: 307–14.

Turetsky, M.R., Amiro, B.D., Bosch, E., and Bhatti J.S. (2004) 'Historical burn area in western Canadian peatlands and its relationship to fire weather indices', *Global Biogeochemical Cycles*, 18: doi:10.1029/2004GB002222.

Turner, M.G., Hargrove, W.W., Gardner, R.H., and Romme W.H. (1994) 'Effects of fire on landscape heterogeneity in Yellowstone National Park', *Journal of Vegetation Science*, 5: 731–42.

Turner, M.G., Romme, W.H., Gardner, R.H., and Hargrove, W.W. (1997) 'Effects of fire size and pattern on early succession in Yellowstone National Park', *Ecological Monographs*, 67: 411–33.

Van Nieuwstadt, M.G.L., and Sheil, D. (2005) 'Drought, fire and tree survival in a Borneo rain forest, east Kalimantan, Indonesia', *Journal of Ecology*, 93: 191–201.

Vila, M., Lloret, F., Ogheri, E., and Terradas, J. (2001) 'Positive fire-grass feedback in Mediterranean Basin woodlands', *Forest Ecology and Management*, 147: 3–14.

Wan, S., Hui, D., and Luo, Y. (2001) 'Fire effects on nitrogen pools and dynamics in terrestrial ecosystems: A meta-analysis', *Ecological Applications*, 11: 1349–65.

Ward, D.J., Lamont, B.B., and Burrows, C.L. (2001) 'Grasstrees reveal contrasting fire regimes in eucalypt forest before and after European settlement of southwestern Australia', *Forest Ecology and Management*, 150: 323–29.

Whelan, R.J. (1995) *The Ecology of Fire*. Cambridge: Cambridge University Press.

Whelan, R.J., and York, J. (1998) 'Post-fire germination of *Hakea sericea* and *Petrophile sessilis* after spring burning', *Australian Journal of Botany*, 46: 367–76.

Williams, R.J., Griffiths, A.D. and Allan, G.E. (2002) 'Fire regimes and biodiversity in the savannas of northern Australia', in R.A. Bradstock, J.E. Williams, and M.A. Gill (eds.), *Flammable Australia: The Fire Regimes and Biodiversity of a Continent*. Cambridge: Cambridge University Press. pp. 281–304.

Williams, P.R., Congdon, R.A., Grice, A.C., and Clarke P.J. (2004) 'Soil temperature and depth of legume germination during early and late dry season fires in a tropical eucalypt savanna of north-east Australia', *Australian Journal of Ecology*, 29: 258–63.

Williams, R.J., Gill, A.M., and Moore P.H.R. (1998) 'Seasonal changes in fire behaviour in a tropical savanna in northern Australia', *International Journal of Wildland Fire*, 8: 227–39.

Williams, R.J. and 13 authors. (2009) 'The impact of climate change on fire regimes and biodiversity in Australia—a preliminary assessment', report to Department of Climate Change and Department of Environment Heritage and The Arts: Canberra.

Wright, C.S., and Agee, J.K. (2004) 'Fire and vegetation history in the eastern cascade Mountains, Washington', *Ecological Applications*, 14:443–59.

Yibarbuk, D., Whitehead, P.J., Jackson, D., Godjuwa, C., Fisher, A., Cooker, P., Chonquenor, D. and Bowman, D.M.J.S. (2001) 'Fire ecology and Aboriginal land management in central Arnhem Land, northern Australia: a tradition of ecosystem management', *Journal of Biogeography*, 28: 325–343.

Yin, Z.Y. (1993) 'Fire Regime of the Okefenokee Swamp and Its Relation to Hydrological and Climatic Conditions', *International Journal of Wildland Fire*, 3: 229–240.

12

Species Response to Contemporary Climate Change

Tim H. Sparks, Annette Menzel,
Josep Peñuelas, and Piotr Tryjanowski

12.1 INTRODUCTION

There is a rapidly growing literature detailing the response of species to the temperature increases currently being experienced globally. Since 1860, when global records began, the mean air temperature of the Earth has risen; during the twentieth century the increase has been $0.7 \pm 0.1°C$, with increases reported as greater in the northern than in the southern hemisphere (IPCC, 2007). It is the responsibility of the Intergovernmental Panel on Climate Change (IPCC) to assess, among many other tasks, the level of evidence for climate impacts on the natural world. In their Fourth Assessment Report (AR4), the IPCC had a lot more literature to trawl through looking for evidence than it did for earlier reports (see www.ipcc.ch). This partly reflects the increases in scientific literature generally but also the increased awareness among the scientific community that climate change is potentially damaging to wildlife. As a consequence, many long-term data sets collected for other purposes, such as population monitoring, have been investigated to see if they contain any evidence of climate responses. The IPCC concluded that many biological and physical systems had already been affected by climate change (IPCC, 2007; Rosenzweig et al., 2008).

This stresses the importance of long-term data sets. For many years monitoring has been the poor relation of scientific research, often operating on a shoestring budget and surviving because of the determination of a few individual scientists. Now these data sets are proving to be of immense value in evaluating change and related impacts.

More resources are being put into monitoring, for example, phenological networks in Canada, the Netherlands, the United Kingdom, and the United States, and new technology is being used to both create electronic databases and to promote results via the Internet.

In this chapter we provide evidence for temperature-related changes in the populations, distribution, and phenology of plants and animals. Because of a limit on space we aim to provide an overview rather than a comprehensive picture, but we trust that we convey the message of the growing evidence of change despite the relatively modest temperature increases so far experienced. The sensitivity of the natural world documented here suggests that substantial changes can be anticipated as a consequence of climate warming in the coming century.

12.2 POPULATION RESPONSE

A general warming of the climate may profoundly affect wildlife populations in terms of their numbers and distribution and the characteristics of the individuals. Understanding and coping with this problem, therefore, is a leading new challenge for management and conservation.

However, determining the effects of weather and climate on populations is sometimes very difficult, because climatological and meteorological factors do not necessarily affect populations directly. For example, changes may manifest themselves via changes to prey or predators and through habitat change. In addition, other disturbing factors, such as land-cover and land-use change, may mask relationships.

12.2.1 Population increases and decreases

Most of the data available on population effects refer to animals, and examples are dominated by birds. However, changes to population size in plants (e.g., Sturm et al., 2001) and invertebrates (e.g., Warren et al., 2001; Beaugrand et al., 2002) are present in the literature. Predictions of the consequences of the expected changes in climate on bird population dynamics were summarized by Sæther et al. (2004). Basically, two different hypotheses have been proposed to explain in which period of the year a change in climate will have the strongest effect on fluctuations in population size. The 'tub hypothesis' suggests that fluctuations in population size are closely related to climate variation during the nonbreeding season because, in combination with density dependence (where population is regulated to some carrying capacity), weather conditions determine the number of birds surviving during this critical period of the year and this hypothesis relates to northern temperate altricial birds (species hatched without feathers and reliant on their parents for food and warmth). The 'tap hypothesis' suggests that annual variation in population size is related to the weather during the breeding season because this will influence the inflow of new recruits into the population the following year, which is characteristic of many nidifugous species (hatched already covered in down) and species living under arid conditions.

12.2.2 Demographics

Changes in population size are the net effect of changes in demographic variables; this is the balance between birth and death. Therefore, the impact of climate change should be relatively easy to predict. However, very rare weather events or climate factors directly affect the birth ratio and death of individuals. Sæther et al. (2004) clearly show that the magnitude of changes in mean demographic variables depends upon the strength of density dependence. Thus, if density dependence is weak the population will be affected more than if the regulation is strong. Moreover, in many cases climate change will result in changes in the nature of density dependence. A straightforward example is where climate change alters the area of suitable habitat (e.g., by sea-level rise altering the

area of habitat or climate change affecting the area of a given vegetation type). To clearly answer how climate changes influence populations through their demographic parameters, good long-term data are needed.

An example linking changes in population size, with demography of local population and occupation of new breeding areas, was provided by Tryjanowski et al. (2005) during research on the white stork in southern Poland. During the last century the white stork increased its elevational range; however, the drivers are likely to be complex. The species has the biological potential to colonize mountain areas but also appears to follow habitat and climate changes. Hence, it is documented that climatological factors (changes in temperatures and run-off) created new habitat and offered new foraging opportunities for storks. The impacts of climate here are clearly indirect.

12.2.3 Biometrical measurements

The probability of occupation of new areas may be connected with morphological traits, for example, body size. Temperature may influence organism body size, and this is especially so in the homeotherms (warm-blooded animals), whose basal metabolic rate is known to be related to body mass (Schmidt-Nielsen, 1984). The paleontological record provides us with striking examples of very rapid changes in the body size of mammals, which often preceded their extinctions (e.g., Guthrie, 2003). Therefore, we may assume that if body size of a homeothermic animal changes over time in a consistent way, this is a very important evolutionary event. The majority of studies focus on birds and mammals, and for animals we have good theoretical predictions and adequate statistical data. Theoretically, climate change could influence morphology of avian and mammalian species in several ways. Bergmann's rule states that taxa living in warmer climates are generally smaller than those living in colder regions (e.g., Blackburn et al., 1999; Ashton, 2002). If Bergmann's rule is valid, animals may be expected to become smaller with rising average temperatures, and this prediction is supported by studies on birds and mammals (e.g., Yom-Tov, 2003; Yom-Tov et al., 2003).

However, a shift in ambient temperature may also influence food availability. If food availability increases, animals should be expected to increase their body size (similar to the so-called secular trend in human populations: Ulijaszek et al., 1998); if food supply declines, animals might decrease in size. If food supply is affected, population density may change, and then density-dependent mechanisms may influence body size and other morphometric characteristics.

A most interesting study has been conducted in 25 mammalian species in Denmark. The body length of small species increased, that of large species decreased, and that of medium-sized species remained unaltered over the past 175 years (Schmidt and Jensen, 2003). The authors suggested that the reason for differential changes in mammals of different sizes could be due to increased habitat fragmentation. Medium-sized animals cope with it better than either large- or small-sized ones (this is called the island rule; van Valen, 1973). Thus, it is beneficial for small mammals to increase body size and large ones to decrease it. It is important to note that, in this study, all mammals, irrespective of their species-specific size, were exposed to the same climate fluctuations and trends.

12.2.4 Community composition

Changes in population size and range, even for individuals, should affect community structure. However, data on links between climate change and community patterns are very scare (Harrington et al., 1999). Good examples are based on the prediction that the reaction of species to weather events and climate is often asymmetrical (Peñuelas and Filella, 2001). Species invade faster than resident species recede. For example, plant species diversity has increased in certain regions (e.g., northwestern Europe) due to a northward movement of southern thermophilic species, whereas the effect on cold-tolerant species is still limited (Tamis et al., 2005). Moreover, changes in whole ecosystems may be the result of rapid climate change. Examples for altered or stable synchrony in ecosystems via multispecies interactions are still uncommon. Visser and Holleman (2001) determined the responses of winter moth-egg hatching and oak bud burst to temperature, describing a system with strong selection on synchronization. They reported that there has been poor synchrony in recent springs, which was due to an increase in spring temperatures without a decrease in the incidence of freezing spells in winter. This is a clear warning that such changes in temperature pattern may affect ecosystem interactions more strongly than changes in mean temperature. In the next stage these changes may influence local bird populations foraging on moths, for example, tits and flycatchers; however, both species can react in different ways (Both and Visser, 2001; Visser et al., 2002; Both et al., 2006).

Because climate change affects species and biomes not only directly, but also indirectly by influencing prey, pests, predators, and habitats, it is difficult to separate direct and indirect effects. Moreover, the situation is complicated by the fact that many species, especially large animals, react to environmental factors with time delays (Sæther et al., 2005). However, we should say that biome shifts, and later on extinction, can be quick, and all simulations predict that this will remain the same (or be even stronger) during the next 100 years. Changes would therefore be easy to detect within a single human generation, which should encourage protection of environmental resources and habitats.

12.3 DISTRIBUTION

Climatic regimes determine species distributions through species-specific physiological thresholds of temperature and water availability (Woodward, 1987). We currently are experiencing the strongest warming trend of the last millennium with average temperatures rising by about 0.7°C in the twentieth century (IPCC, 2007). Future temperature rises are likely to exceed this warming with a predicted rise of between 0.1°C and 0.4°C per decade across Europe (IPCC, 2001)—a rate that is unparalleled in recent history (Huntley, 1991). Shifts in plant and animal species and biome distribution toward the poles or at higher altitudes in response to warming have been described for past climate changes (Gates, 1993). It is thus reasonable to expect they have also occurred in response to current climate warming and some evidence has already been found.

However, factors affecting species distribution interact in complex ways, and range shifts are often episodic rather than gradual. Moreover, rates of range shifts vary greatly within and among species with different dispersal abilities. For example, elevational shifts of alpine plant species lag behind the isothermal shift, whereas butterflies appear to track decadal warming quickly (Parmesan et al., 1999). Migratory species are among the best documented but often exhibit large interannual fluctuations in their breeding sites, making it difficult to discern long-term range shifts. By contrast, it is easier to discern range changes in more sedentary species or in plants that present slower processes of population extinction and colonization. Of course, in some cases (e.g., reef-building corals), range shifts in response to changing temperatures may not occur if latitudinal distributions are also limited by other factors such as light (Hoegh-Guldberg, 1999). In spite of this diversity and complexity in distribution responses, there is now considerable evidence that poleward and altitudinal upward shifts of species ranges have also occurred across a wide range of taxonomic groups and geographical locations during the twentieth century in response to current climate warming (IPCC, 2007; Walther et al., 2002; Parmesan and Yohe, 2003).

12.3.1 Latitudinal shifts

There are species that have colonized previously 'cool' regions, including sea anemones in Monterey Bay (Sagarin et al., 1999) and lichens and butterflies in Europe (Parmesan et al., 1999; van Hark et al., 2002), whereas some arctic species have contracted their range areas (Beaugrand et al., 2002; Hersteinsson and MacDonald, 1992). There are multiple examples of latitudinal shifts such as these both in animal and plants correlated with current warming (Table 12.1). In their quantitative assessment of those available up to 2003, Parmesan and Yohe (2003) found that 80% of the 434 studied species with changing distribution during the later twentieth century had shifted in accordance with climate change predictions. They report that over the past 40 years, maximum range shifts varied from 200 kilometers (butterflies in Parmesan et al., 1999) to 1,000 kilometers (marine copepods in Sagarin et al., 1999). Since their review and other parallel reviews, such as those of Walther et al. (2002) and Root et al. (2003), new studies have appeared with further evidence (Table 12.1). As a very recent example, it has been reported that the distributions of both exploited and nonexploited North Sea fishes have responded markedly to recent increases in sea temperature, with nearly two-thirds of species shifting in mean latitude or depth or both over 25 years. For species

with northerly or southerly range margins in the North Sea, half have shown boundary shifts with warming, and all but one shifted northward (Perry et al., 2005). It has also been found that fish species with shifting distributions have faster life cycles and smaller body sizes than nonshifting species, showing once more the expected differential response among species.

Latitudinal shifts have also been reported for many plant species. Many arctic and tundra communities have been replaced by trees and dwarf shrubs while there was a latitudinal advance of arctic shrub cover (Sturm et al., 2001). In the Netherlands (Tamis et al., 2005), the United Kingdom, and central Norway, thermophilic (warmth-loving) plant species have become significantly more frequent compared with 30 years ago (in the Netherlands, by around 60%). In contrast, there has been a small decline in the presence of traditionally cold-tolerant species.

12.3.2 Altitudinal shifts

There is also much evidence for altitudinal shifts of vegetation (Table 12.1). Higher temperatures and longer growing seasons have created suitable conditions for certain plant species to move to higher altitudes. In the Alps, over the past 60 years, spruce and pine species have migrated upward into the subalpine region, and subalpine shrubs now grow on the summits. Although there are also a lot of examples of a relatively stable treeline position in the last half century, significant elevational vegetation rises have been reported in Alaska, Scandinavia (Kullman, 2003), the Alps (Grabherr et al., 1994), the Mediterranean region (Peñuelas and Boada, 2003; Peñuelas et al., 2007),

Table 12.1 Some summaries of distributional change attributed to warming

Taxa		Location	Examples of references
Invertebrates	Butterflies	Europe, North America, U.K., Czech Republic, Spain	Parmesan et al., 1999; Walther et al., 2001
	Dragonflies, damselflies	Germany, U.K.	Walther et al., 2001; Hickling et al., 2006
	Spittlebugs	California	Karban and Strauss, 2004
Plants	Shrubs	Alaska	Sturm et al., 2001
	Forbs	Europe, Montana	Lesica and McCune, 2004; Braithwaite et al., 2006
	Trees	Europe, New Zealand	Meshinev et al., 2000; Peñuelas and Boada, 2003; Peñuelas et al., 2007
Vertebrates	Birds	Costa Rica, U.K., Poland	Tryjanowski et al., 2005
	Foxes	Canada	Hersteinsson and MacDonald, 1992

New Zealand, and Bulgaria (Meshinev et al., 2000).

Endemic mountain plant species are threatened by the upward migrations of more competitive subalpine shrubs and tree species that can replace them (Grabherr et al., 1994). The net effect on the community structure, for example, on the species richness, diverges from region to region and even within single regions. While richness has increased in some places, it has declined in others. In the Alps, for example, the net effect has been an increase in species richness in 21 out of 30 summits compared with 50 to 100 years ago (Walther et al., 2001). Similar trends have occurred in the Pyrenees, Scandinavia, Bulgaria, the Ural, Chile, and Australia (e.g., Kullman, 2003; IPCC, 2001; Meshinev et al., 2000).

These reports of vegetation increases are heavily biased toward the expansion of a species at the leading edge of its distribution. Recruitment is generally more sensitive to climate than mortality, so an increase in reproduction at the expanding range-edge of a tree species' distribution occurs more rapidly than an increase in the mortality of established trees at the retreating edge. Consequently, it may take decades longer to detect range changes at the retreating edge than at the expanding edge if the same survey-based methods are used to analyze both. Instead, changes in growth and reproduction are among the primary responses of trees to environmental variation. In the Montseny Mountains, in Catalonia, it has been found that growth at the lower *Fagus* altitudinal limit is characterized by a rapid recent decline starting in approximately 1960. By 2003, the growth of mature trees had fallen by 49% when compared with predecline levels (Jump et al., 2006). Given that populations in other areas of southern Europe show very similar climate-growth responses to these, it is very likely that this decline in beech growth is a more widespread phenomenon in the lower altitudinal range in southern regions.

12.3.3 Climate-linked invasions, extinctions, and new communities

Because of these changes in species range, nonnative species from adjacent areas may become new elements of the biota. Examples reviewed by Walther et al. (2002) include warm-water species that have recently appeared in the Mediterranean and the North Sea and thermophilous plants that spread from gardens into surrounding countryside. Climate-linked invasions might also involve the immigration of unwanted neighbors such as epidemic diseases. There is much evidence that a steady rise in annual temperatures has been associated with expanding mosquito-borne diseases in the highlands of Asia, East Africa, and Latin America (Epstein et al., 1998).

But not all species can migrate as fast as needed to cope with the predicted rate of future climate change and thereby to track the climate to which they are adapted. They will suffer strong pressure to adapt to their new conditions. Although levels of climate-related variation in natural populations may be high, they are not inexhaustible in the face of intense directional selection that will result from rapid climate change (Jump and Peñuelas, 2005). Habitat fragmentation will amplify the effect of intense selection by reducing the supply of new genetic variation by gene flow from neighboring populations (Jump and Peñuelas, 2006). It is likely therefore that in many cases plant adaptation will fail to match the pace or magnitude of predicted changes in climate. The decoupling of climate and local adaptation is liable to have negative consequences for plant fitness and survival throughout a species' range. Consequences will include unpredictable changes in the presence and abundance of species within communities and a potential reduction in their ability to respond to environmental perturbations such as pest and disease outbreaks and extreme weather events. A rangewide increase in population extinction risk is likely to result (Figure 12.1) (Jump and Peñuelas, 2005). Consequences at longer time and spatial frames could be the extinction of many species (Thomas et al., 2004).

Realized migration and extinction rates will differ greatly between different species. For example, individual longevity (Peñuelas, 2005), high intra population genetic diversity (Jump and Peñuelas, 2006), and the potential for high rates of pollen flow might make tree species especially resistant to extinction. Consequently, novel plant communities may be formed in response to climate change (Walther et al., 2002). Such communities are evident in the fossil record as species assemblages with no modern analogue. These communities are believed to result from differential migration rates during past climatic changes (Huntley, 1991). Some of the examples of range shifts mentioned earlier involve community-level changes (Walther et al., 2002). There are also changes in the composition and structure of the new communities. For example, since changes in distribution are often asymmetrical, with species invading faster from lower elevations (Rusterholz and Erhardt, 1998) or latitudes (Hersteinsson and MacDonald, 1992) than resident species are receding upslope or poleward there is a (presumably transient) increase in species richness of the community in question.

Given the rapid rate at which climate change is predicted to occur, we should consider climate as

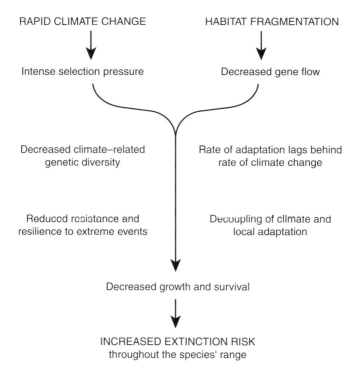

Figure 12.1   The interaction of rapid climate change and habitat fragmentation within populations leading to a rangewide increase in extinction risk in species unable to migrate at sufficient velocity. This can occur in the absence of habitat fragmentation if climate change occurs at a rate faster than the maximum rate of gene flow between populations (from Jump and Peñuelas, 2005)

a key force in driving future community composition. Directing research effort at understanding these potential effects of climate change will allow us to predict, with better accuracy, the changes that are likely to take place and possibly to prepare for some of their more extreme effects.

12.4 PHENOLOGY

Recent reviews of evidence for biological impacts of climate change have been dominated by phenological examples (e.g., Parmesan and Yohe, 2003; Rosenzweig et al., 2008). A definition may be needed for those not familiar with the term; phenology is the study of the timing of annually recurring natural events. In seasonal climates, both plants and animals have distinct seasonal cycles and it is relatively simple to record the beginning (and sometimes the end) of these events, known as phenophases. As an example, a

deciduous tree will have phenophases such as leaf budburst, leaf and flower development in spring, fruiting in summer, and leaf coloration and leaf fall in autumn. These phases will typically be simple to record and do not require any special equipment. Anybody with a reasonable ability to identify species and the time to make regular observations can contribute to phenological recording. Many phases are very temperature-dependent; thus, directional changes in temperature, for example, during a period of climate warming, will be matched by a directional change in the particular phase. Consequently many species have clearly shown changes in phenology as a consequence of rising temperatures in recent decades.

Most phenological data were not collected with the notion that they would make useful climate change indicators, but rather they may have been collected for general natural history interest, to evaluate the effect of environment on phenology, the impacts of climate on agriculture, or as a by-product of, for example, population monitoring.

However, the sheer volume and duration of phenological data have made them useful for studying the effects of a changing climate on biological systems.

Most data derive from the northern hemisphere and from the late nineteenth century onward. However, there are examples of much older data. These include cherry flowering records in Kyoto, Japan, dating back to the eighth century; vine harvest records in France, Switzerland, and Rhineland dating back to the late fourteenth century; and a tree leafing record from Geneva dating back to 1808 (http://www.meteoschweiz.admin. ch/web/de/klima/klima_schweiz/phaenologie/ Phaenobeobachtungen_seit_1808.html). Data have been collected in different ways, by different organizations and in different countries and yet appear remarkably robust (Schwartz, 2003). Observations will typically involve native or naturalized species, but in the case of the International Phenological Gardens cloned trees have been planted in locations across Europe for direct comparison of environmental effects (Menzel and Fabian, 1999).

That plant phases are related to temperatures can clearly be seen in Figure 12.2. Here, a very significant correlation ($r = -0.90$, $p < 0.001$) exists between the mean U.K. flowering date of Horse Chestnut *Aesculus hippocastanum* and the mean March-April Central England temperature. A regression suggests a 1°C increase in temperature would be associated with a 7.1 ± 0.4 day advance in flowering. But there is more that can be gained from this figure. The data derive from a former and a current phenological network and demonstrate that the two schemes are compatible, and also that recent recordings have tended to be early. Indeed a formal test shows that the mean of the current scheme (1998–2005) is significantly earlier than the mean of the former scheme (1891– 1947) by 10.7 days ($t_{62} = 3.75$, $p < 0.001$). While the correlation displayed in Figure 12.2 is very high, there is some hint of a nonlinear relationship between the two; a formal test confirms this with a quadratic regression: a significant improvement over a linear one ($F_{1,61} = 91.35$, $p < 0.001$, R^2 increased from 80.9 to 85.5%).

Such temperature responses are not unusual; in fact, they are the norm. Figure 12.3 displays the temperature response in a variety of species, events, and locations. The date on which ash (*Fraxinus excelsior*) trees became bare in autumn in Tenbury Wells, United Kingdom, in the years 1915–1931 was related to August-October mean temperatures. The relationship of 5.4 ± 1.3 days later for each 1°C warmer was highly significant ($p = 0.001$). The first capture date of the peach-potato aphid (*Myzus persicae*) in Scotland from 1971 to 2004 was strongly correlated with the January-March mean temperature ($p < 0.001$), such that each 1°C increase in temperature was

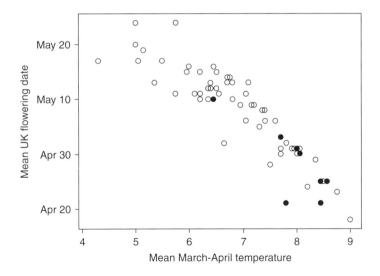

Figure 12.2   The relationship between the flowering date of horse chestnut *Aesculus hippocastanum* and the mean March-April temperature in the United Kingdom. Open symbols refer to data from the Royal Meteorological Society network (1891–1947) and solid symbols from the current U.K. Phenology Network (1998–2005)

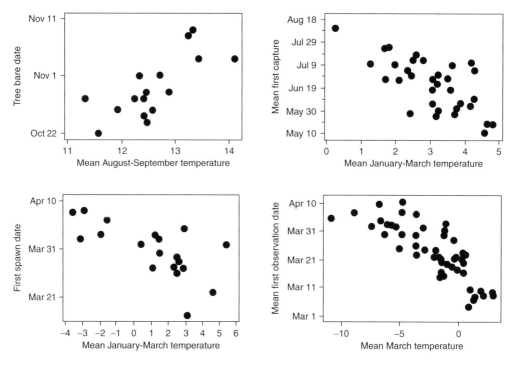

Figure 12.3   Temperature responses for (top left) ash *Fraxinus excelsior*, bare date in Tenbury Wells, United Kingdom; (top right) peach potato aphid *Myzus persicae*, first capture date in Scotland; (bottom left) common frog *Rana temporaria*, spawning date in Poland; and (bottom right) skylark *Alauda arvensis*, first appearance in Estonia

associated with a 14.6 ± 2.7 day advance in first capture. The first spawning dates of common frogs (*Rana temporaria*) in the Wielkopolska Province in Poland in 1978–2002 were strongly correlated with the January-March mean temperature (p = 0.001); each 1°C increase associated with an advance of 1.5 ± 0.4 days in first evidence of breeding. The final element in Figure 12.3 shows a strong correlation between the spring migration timing of skylark (*Alauda arvensis*) in Estonia in 1948–1996 and the March temperature (p<0.001) with a 1°C increase associated with a 2.5 ± 0.3 day advance in migration. From this example, it can be clearly seen that temperature affects the phenology of plants, invertebrates, and both cold-blooded and warm-blooded vertebrates. The examples reveal changes in spring and autumn events, in events associated with migration and breeding and in a range of locations. Furthermore it is quite clear that there is a huge range of response to temperature; in this example, there is nearly a tenfold difference between frog breeding and aphid appearance. There is little evidence that these response rates have changed over time (e.g., during the last century).

Since there seems to be clear evidence for a well-understood temperature response, then we expect that phenology is changing as a consequence of the current spate of global temperature increase. Studies have reported phenological change across the northern hemisphere from Canada, the United States (Abu-Asab et al., 2001), Europe (Menzel and Fabian, 1999; Fitter and Fitter, 2002; Lehikoinen et al., 2004), and Japan. Abu-Asab et al. (2001) summarized flowering phenology of 100 plant species in Washington, D.C., and of these, 89 showed some advance. Fitter and Fitter (2002) analyzed phenological change of 385 plant species in Oxfordshire and concluded that their average first flowering date has advanced by 4.5 days during the past decade compared with the previous four decades. Menzel and Fabian (1999) summarized spring and autumn tree phenology using evidence from a network of cloned plants grown across Europe. They found that spring events, such as leaf unfolding, have advanced by six days, whereas autumn events, such as leaf coloring, have been delayed by 4.8 days, and that the mean annual growing season has therefore lengthened by 10.8 days since the early 1960s. These shifts could be attributed to changes in air temperature. In general, an earlier spring might be associated with higher geographical and spatial variability.

The absence of reports from other parts of the globe would appear to be more associated with the absence of data, rather than the absence of effects. Terrestrial examples of change have recently been extended to marine situations. Furthermore meta-analyses of changes have clearly shown how widespread phenological advance in spring has been (Walther et al., 2002; Parmesan and Yohe, 2003; Root et al., 2003; Menzel et al., 2006). In a meta-analysis of more than 100,000 phenological series in Europe, Menzel et al. (2006) also showed that countries with greater temperature increase had experienced greater phenological advance.

Fitter and Fitter (2002) reported differential phenological changes. They suggested greater change in annual species than in perennial species, and in insect-pollinated species rather than in wind-pollinated species. In bird migration greater change has been shown in short-distance migrants than in long-distance migrants (Tryjanowski et al., 2002). The general impression is that responses in plants and invertebrates may be greater than in vertebrates (e.g., Peñuelas et al., 2002). What might the consequences of these differential changes be? Fitter and Fitter (2002) suggested there may be less opportunity for hybridization between some plant species as their flowering dates move apart and greater opportunities for those whose flowering dates were becoming closer. Changed phenology may result in changed competition; early species may be able to gain a competitive advantage if they reproduce early. In some bird species productivity is greater in early nesting individuals (e.g., white stork, Tryjanowski et al., 2004). In some birds and invertebrates additional broods may be possible if the breeding season is extended. However, the greatest problems may result through disruption of synchrony within food webs. There are, so far, few reported examples of these with the relatively modest warming so far experienced, but they may be a feature of our natural world over the coming century as predicted temperature increases continue. One example of possible disruption concerns the phenological plasticity of oak trees and moth caterpillars relative to that of great tits (*Parus major*). Oak leafing can progress rapidly in warm springs and the development time of caterpillars from egg to pupae can almost halve, but the incubation period of eggs is fixed. The bird has to anticipate the peak caterpillar crop for its chicks several weeks ahead. A suddenly warm spring may disrupt this planning, and reduced chick weights and greater mortality were the consequences of phenological asynchrony. More recently, Both et al. (2006) report population declines in flycatchers as a consequence of phenological mismatch.

Some species will undoubtedly benefit from phenological change. It is likely that some species are distributionally limited by low temperatures and thus are unable to successfully complete a life cycle. A relaxation of phenological bottlenecks may encourage some of the distributional expansions, both latitudinally and altitudinally, reported earlier in this chapter. We saw earlier that some phenological response may be nonlinear. Differential change and nonlinear change will both contribute to making predictions of future phenology less simple, but it is important that phenological research shifts its balance from identifying climate impacts to predictive modeling.

12.5 FINAL REMARKS

The detection of changes in species, whether of population size, distribution, or phenology relies on the availability of long-term data. Such data were, until very recently, scientifically unfashionable but now their potential and value are recognized. Sources of long-term data are being actively sought and investigated to determine if climate signals can be seen within these records. While much research funding is still short-term in nature, a number of new networks have recently emphasized the commitment to long-term research. However, we are also still reliant on a few dedicated individuals running long-term research projects, often on a shoestring budget.

A fairly recent phenomenon has been the use of long-term data for purposes for which they were not originally intended. Examples would include the abstraction of phenological date from schemes established to monitor populations (e.g., of migrant birds, Sokolov et al., 1998) or demography (e.g., nest record schemes, Crick and Sparks, 1999). Distributional data can also be extracted from population-monitoring schemes. Such exploitation of data sets has to be commended and gives added value to the monitoring schemes.

A posteriori studies can only examine the species and the aspects of species that have been monitored. Thus, we may be restricted in studies of synchrony within food webs because we may not have data on all aspects of the web taken from the same locality. New schemes plan to overcome these restrictions, but it will be some years before they have sufficient useful data for investigation.

As a consequence of these restrictions most of the documented results come from developed countries in the northern hemisphere. This is not to suggest that changes are not happening globally, but rather that there are restrictions on the availability of data sources.

There are a number of priority areas for research now. We still need to identify and exploit existing data resources, and we possibly need to think laterally about the 'data' we exploit. Photographic images of glacier retreat can be impressive and thought-provoking. Undoubtedly, greater use of photographic evidence (as used by Peñuelas and Boada, 2003) will provide more evidence of habitat change and museum, herbarium, and other historic records may tell us more about recent distributional shifts.

To date, research has focused on the evidence of climate impacts. It is important that this continues to identify where change is already happening and is used to help alert conservationists, policy makers, and the general public to climate-related change. However, there is a growing need to exploit this information to predict likely future changes in population status and distribution and to identify species, communities, and habitats that will both benefit and suffer as a consequence of climate warming.

REFERENCES

Abu-Asab, M.S., Peterson, P.M., Shetler, S.G., and Orli, S.S. (2001) 'Earlier plant flowering in spring as a response to global warming in the Washington, DC, area', *Biodiversity and Conservation*, 10: 597–612.

Ashton, K.G. (2002) 'Patterns of within-species body size variation of birds: Strong evidence for Bergmann's rule', *Global Ecology and Biogeography*, 11: 505–23.

Beaugrand, G., Reid, P.C., Ibañez, F., Lindley, J.A., and Edwards, M. (2002) 'Reorganization of North Atlantic marine copepod biodiversity and climate', *Science*, 296: 1692–94.

Blackburn, T.M, Gaston, K.J., and Loder, N. (1999) 'Geographic gradients in body size: a clarification of Bergmann's rule', *Diversity and Distributions*, 5: 165–74.

Both, C., Bouwhuis, S., Lessells, C.M., and Visser, M.E. (2006) 'Climate change and population declines in a long-distance migratory bird', *Nature*, 441: 81–83.

Both, C., and Visser, M.E. (2001) 'Adjustment to climate change is constrained by arrival date in a long distance migrant bird', *Nature*, 411: 296–98.

Braithwaite, M.E., Ellis, R.W., and Preston, C.D. (2006) *Change in the British Flora 1987–2004*. London: Botanical Society of the British Isles.

Crick, H.Q.P., and Sparks, T.H. (1999) 'Climate change related to egg-laying trends', *Nature*, 399: 423–24.

Epstein, P.R., et al. (1998) 'Biological and physical signs of climate change: Focus on mosquito-borne diseases', *Bulletin of the American Meteorological Society*, 79(3): 409–17.

Fitter, A.H., and Fitter, R.S.R. (2002) 'Rapid changes in the flowering time in British plants', *Science*, 296: 1689–91.

Gates, D.M. (1993) *Climate Change and the Biological Consequences*. Sunderland, MA: Sinauer.

Grabherr, G., Gottfried, M., and Pauli, H. (1994) 'Climate effects on mountain plants', *Nature*, 369: 448.

Guthrie, R.D. (2003) 'Rapid body size decline in Alaskan Pleistocene horses before extinction', *Nature*, 426: 169–71.

Harrington, R., Woiwod, I., and Sparks, T. (1999) 'Climate change and trophic interactions', *Trends in Ecology and Evolution*, 14: 146–50.

Hickling, R., Roy, D.B., Hill, J.K. Fox, R., and Thomas, C.D. (2006) 'The distributions of a wide range of taxonomic groups are expanding polewards', *Global Change Biology*, 12: 450–55.

Huntley, B. (1991) 'How plants respond to climate change—migration rates, individualism and the consequences for plant communities', *Annals of Botany*, 67: 15–22.

Hersteinsson, P., and MacDonald, D.W. (1992) 'Interspecific competition and the geographical distribution of Red and Arctic foxes *Vulpes vulpes* and *Alopex lagopus*', *Oikos*, 64: 505–15.

Hoegh-Guldberg, O. (1999) 'Climate change, coral bleaching and the future of the world's coral reefs', *Marine and Freshwater Research*, 50: 839–66.

IPCC. (2001) *Climate Change 2001: Impacts, Adaptation and Vulnerability. Contribution of Working Group II to the Third Assessment Report of the Intergovernmental Panel on Climate Change*. Cambridge: Cambridge University Press.

IPCC. (2007) *Climate Change 2007: Synthesis Report. Contribution of Working Groups I, II and III to the Fourth Assessment Report of the Intergovernmental Panel on Climate Change*. Geneva: IPCC.

Jump, A.S., Hunt, J., and Peñuelas, J. (2006) 'Rapid climate change-related growth decline at the southern range edge of *Fagus sylvatica*', *Global Change Biology*, 12: 2163–74.

Jump, A.S., and Peñuelas, J. (2005) 'Running to stand still: Adaptation and the response of plants to rapid climate change', *Ecology Letters*, 8: 1010–20.

Jump, A.S., and Peñuelas, J. (2006) 'Genetic effects of chronic habitat fragmentation in a wind-pollinated tree', *Proceedings of the National Academy of United States of America*, 103: 8096–100.

Karban, R., and Strauss, S.Y. (2004) 'Physiological tolerance, climate change, and a northward range shift in the spittlebug, *Philaenus spumarius*', *Ecological Entomology*, 29: 251–54.

Kullman, L. (2003) 'Recent reversal of Neoglacial climate cooling trend in the Swedish Scandes as evidenced by mountain birch tree-limit rise', *Global and Planetary Change*, 36: 77–88.

Lehikoinen, E., Sparks, T.H., and Zalakevicius, M. (2004) 'Arrival and departure dates', *Advances in Ecological Research*, 35: 1–31.

Lesica, P., and McCune, B. (2004) 'Decline of arctic-alpine plants at the southern margin of their range following a decade of climatic warming', *Journal of Vegetation Science*, 14: 679–90.

Menzel, A., and Fabian, P. (1999) 'Growing season extended in Europe', *Nature*, 397: 659.

Menzel, A. and 30 authors (2006) 'European phenological response to climate change matches the warming pattern', *Global Change Biology*, 12: 1969–76.

Meshinev, T., Apostolova, I., and Koleva, E. (2000) 'Influence of warming on timberline rising: A case study on *Pinus peuce* Griseb. In Bulgaria', *Phytocoenologia*, 30(3–4): 431–38.

Parmesan, C. and 13 authors (1999) 'Poleward shifts in geographical ranges of butterfly species associated with regional warming', *Nature*, 399: 579–83.

Parmesan, C., and Yohe, G. (2003) 'A globally coherent fingerprint of climate change impacts across natural systems', *Nature*, 421: 37–42.

Peñuelas, J. (2005) 'A big issue for trees', *Nature*, 437: 965–66.

Peñuelas, J., and Boada, M. (2003) 'A global change-induced biome shift in the Montseny Mountains (NE Spain)', *Global Change Biology*, 9: 131–40.

Peñuelas, J., and Filella, I. (2001) 'Phenology: Responses to a warming world', *Science*, 294: 793–95.

Peñuelas, J., Filella, I., and Comas, P. (2002) 'Changed plant and animal life cycles from 1952–2000 in the Mediterranean region', *Global Change Biology*, 8: 531–44.

Peñuelas, J., Ogaya, R., Boada, M., and Jump, A. (2007) 'Migration, invasion and decline: Changes in recruitment and forest structure in a warming-linked shift of European beech forest in Catalonia', *Ecography*, 30: 830–38.

Perry, A.L., Low, P.J., Ellis, J.R., and Reynolds, J.D. (2005) 'Climate change and distribution shifts in marine fishes', *Science*, 308: 1912–15.

Root, T.L., Price, J.T., Hall, K.R., Schneider, S.H., Rosenweig, C. and Pounds, J.A. (2003) 'Fingerprints of global warming on wild animals and plants', *Nature*, 421: 57–60.

Rusterholz, H.P., and Erhardt, A. (1998) 'Effects of elevated CO_2 on flowering phenology and nectar production of nectar plants important for butterflies of calcareous grasslands', *Oecologia*, 113: 341–49.

Rosenzweig, C. and 13 authors (2008) 'Attributing physical and biological impacts to anthropogenic climate change', *Nature*, 453: 353–58.

Sæther, B.-E. and 15 authors (2005) 'Generation time and temporal scaling of bird population dynamics', *Nature*, 436: 99–102.

Sæther, B.-E., Sutherland, W.J., and Engen, S. (2004) 'Climate influences on avian population dynamics', *Advances in Ecological Research*, 35: 185–209.

Sagarin, R., Barry, J.P., Gilman, S.E., and Baxter, C.H. (1999) 'Climate-related change in an intertidal community over short and long time scales', *Ecological Monographs*, 69: 465–90.

Schmidt, N.M., and Jensen, P.M. (2003) 'Changes in mammalian body length over 175 years—adaptations to a fragmented landscape?' *Conservation Ecology*, 7(2), article 6. [online].

Schmidt-Nielsen, K. (1984) *Scaling: Why Is Animal Size So Important*. Cambridge: Cambridge University Press.

Schwartz, M.D. (ed.). (2003) *Phenology: An Integrative Environmental Science*. Dordrecht: Kluwer.

Sokolov, L.V., Markovets, M. Yu., Shapoval, A.P., and Morozov, Yu.G. (1998) 'Long-term trends in the timing of spring migration of passerines on the Courish Spit of the Baltic Sea', *Avian Ecology and Behaviour*, 1: 1–21.

Sturm, M., Racine, C., and Tape, K. (2001) 'Climate change: Increasing shrub abundance in the arctic', *Nature*, 411: 546–47.

Tamis, W.L.M., van't Zelfde, M., van der Meijden, R., Groen, C.L.G. and Udo de Haas, H.A. (2005) 'Ecological interpretation of changes in the Dutch flora in the 20th century', *Biological Conservation*, 125: 211–24.

Thomas, C.D. and 18 authors (2004) 'Extinction risk from climate change', *Nature*, 427: 145–48.

Tryjanowski, P., Kuźniak, S., and Sparks, T. (2002) 'Earlier arrival of some farmland migrants in western Poland', *Ibis*, 144: 62–68.

Tryjanowski, P., Sparks, T., and Profus, P. (2005) 'Uphill shifts in the distribution of the white stork *Ciconia ciconia* in southern Poland: The importance of nest quality', *Diversity and Distributions*, 11: 219–23.

Tryjanowski, P., Sparks, T.H., Ptaszyk, J., and Kosicki, J. (2004) 'Do white storks *Ciconia ciconia* always profit from an early return to their breeding grounds?', *Bird Study*, 51: 222–27.

Ulijaszek, S.J., Johnston, F.E., and Preece, M.A. (eds.). (1998) *Human Growth and Development*. Cambridge: Cambridge University Press.

van Hark, C.M., Aptroot, A., and van Dobben, H.F. (2002) 'Long-term monitoring in the Netherlands suggests that lichens respond to global warming', *Lichenologist*, 34: 141–54.

van Valen, L.M. (1973) 'Pattern and the balance of nature', *Evolutionary Theory*, 1: 31–49.

Visser, M.E., and Holleman, L.J.M. (2001) 'Warmer springs disrupt the synchrony of Oak and Winter Moth phenology', *Proceedings of the Royal Society of London Series B*, 268: 289–94.

Visser, M.E., Silverin, B., Lambrechts, M.M., and Tinbergen, J.M. (2002) 'No evidence for tree phenology as a cue for the timing of reproduction in tits *Parus* spp', *Avian Science*, 2: 77–86.

Walther, G.R., Burga, C.A., and Edwards P.J. (eds) (2001) *"Fingerprints" of Climate Change—Adapted Behaviour and Shifting Species Ranges*. New York: Kluwer.

Walther, G.R., Post, E., Convey, P., Menzel, A., Parmesan, C., Beebee, T.J.C., Fromentin, J-M., Hoegh-Guldberg, O. and

Bairlein, F. (2002) 'Ecological responses to recent climate change', *Nature*, 416: 389–95.

Warren, M.S., et al. (2001) 'Rapid responses of British butterflies to opposing forces of climate and habitat change', *Nature*, 414: 65–69.

Woodward, F.I. (1987) *Climate and Plant Distribution*. Cambridge: Cambridge University Press.

Yom-Tov, Y. (2003) 'Body sizes of carnivores commensal with humans have increased over the past 50 years', *Functional Ecology*, 17: 323–27.

Yom-Tov, Y., Yom-Tov, S., and Baagøe, H. (2003) 'Increase of skull size in the red fox (*Vulpes vulpes*) and Eurasian badger (*Meles meles*) in Denmark during the twentieth century: An effect of improved diet?' *Evolutionary Ecology Research*, 5: 1037–48.

SECTION III

Reconsidering Biomes and Ecosystems

Early humans had intimate relationships with the landscapes they inhabited. On their exodus from Africa to Asia and Europe, they must have become increasingly aware of the different flora and fauna as new landscapes unfolded along their path. Whether they thought much about these changes in anything but terms of food and shelter is lost in the mists of antiquity, but they probably did not think of landscape differences in anything but local terms or, if we stretch a geographical point, regional. It was not until the Greek civilization that we first see the ecumene—their 'known' world that comprised most of modern-day Eurasia and Africa—divided into the climatic zones of torrid, temperate, and frigid. Did the Greeks think about floral and faunal changes related to these climate zones? Maybe. They certainly used arguments that we would label as environmentally deterministic in order to relate human habitation to the geographical distribution of climate in the world they knew. For example, Theophrastus (370 to 285 B.C.) postulated that climate exerted a strong control on the geographical distribution of vegetation. Such arguments may have extended quite widely to flora and fauna.

A strong thread in biogeography then has been in ordering space in terms of floral and faunal distributions at various geographical scales, but in the context of this chapter we are most concerned with the global. Biomes, as we consider them nowadays, were introduced and discussed in early works (see Blumler et al., this volume). Though biomes had fallen out of favor in some biogeographical circles by the 1970s and 1980s, interest in them has recently surged because of the need to

understand biotic responses to anthropogenic climate change at scales ranging from global (e.g., Walker and Steffen, 1996) to that of individual biomes (e.g., Hari and Kulmama, 2008; Lal et al., 2000)

Our dilemma as editors in reconsidering biomes and ecosystems was what to include and what to omit. We settled on the main drivers of contemporary global change—climate and land-use transformation—as the imperatives for selection. Some of the greatest impacts of climate change on biogeographical distributions are being experienced in the polar regions and atop the world's highest mountain ranges: regions considered by Ingo Möller and Dietbert Thannheiser (Chapter 13), and Udo Schickhoff (Chapter 17), respectively. High mountains also have a legacy of centuries of significant human occupation, a point illustrated in Chapter 17. Arguably, however, the greatest contemporary anthropogenic transformations are occurring in the boreal and tropical forests and savannas: the latter having the longest history of human occupation and modification of the three. They are dealt with by Daniel Kneeshaw, Yves Bergeron, and Timo Kuuluvainen (Chapter 14), Kenneth Young (Chapter 16), and Jayalaxshmi Mistry (Chapter 15) respectively. The climate-anthropogenic change dichotomy used in this rationalization is a simplistic editorial tool. All biomes are subject to climate change, and all have human impacts. Rather than a dichotomy, a gradient exists between biomes influenced more strongly by contemporary climate change than human drivers through to those where human influences outweigh those of climate. However, land-use transformations over the last ten millennia, that is, since early sedentary agriculture and town building, have created globally important anthropogenic landscapes that cannot be considered natural but that nonetheless have significance in terms of plant and animal distributions. Chris Stoate considers the biogeography of agricultural landscapes in Chapter 18, while Claire Freeman focuses our attention on urban landscapes in Chapter 19. The expansion of rangeland and cultivation in the last three centuries (Ellis et al., 2010) and postindustrial revolution urban growth underline the recent historical importance of these landscapes. This is reinforced by the fact that over half the world's population now lives in cities whose growth is accelerating and placing new pressures on agricultural areas.

Each author or group of authors has written about their particular region from their own personal viewpoint, rather than following a rigid set of editorial guidelines. Ingo Möller and Dietbert Thannheiser argue that, when considering polar biogeography, ocean life and marine processes are as important as land-based processes. While the Arctic polar regions cover approximately 5% of the Earth (26.4 million km²), the land component is only about 7.6 million km², making complex land-ocean interactions important and differentiating them from other biomes in this section. This aspect of polar regions is reinforced further when we turn our attention to the Antarctic, which is composed of roughly 52 million km², but it has a land area of about 14 million km² of which only 4.5% is ice-free: the ecologically important habitats are therefore restricted to the coastal fringes. These landscapes are characterized by extreme conditions, most notably a prolonged period of low temperatures and large annual variations in insolation that are the dominant ecological factors. It is in these areas where global warming has been the greatest in the last few years—particularly western Alaska—leading to permafrost melting, changing glacier regimes, and breakup of ice sheets. Importantly, we are developing an understanding of the biogeographical responses to climate change; for example, shrub expansion across the circumpolar tundra (Tape et al., 2006) and range extension of birdlife such as macaroni penguins in Antarctica (Gorman et al., 2010).

Daniel Kneeshaw and his coauthors focus on the role of disturbances in boreal forest dynamics in Chapter 14. The spatial variations in climate across the boreal zone influence strongly which disturbances characterize subregions. In North America fire is usually considered the most important disturbance vector, a fact most likely due to the pervasive nature of stand-replacing fires and of fire research among forest ecologists on that continent. Yet fire regimes differ even within

the North American boreal zone—in drier, western Canada short fire cycles are more frequent than in wetter, eastern Canada. Generally throughout the circumboreal zone, fire frequencies decrease toward alpine or taiga environments, a trend that is probably related to lower fuel availability. Some ecologists have gone as far as suggesting that fire's role in some boreal ecosystems is overemphasized. Where fire return intervals are long, or fire has not occurred since the last glaciation, disturbances such as insect outbreaks, fungi, windthrow, and senescence gain greater prominence.

Boreal forests not only cover a large area (over 20% of the world's forests), but are also globally important biogeographically (e.g., over half of the world's conifers are found in Russia) and ecologically (e.g., the biome contains five times more carbon than temperate forests and almost twice as much as tropical forests). Yet until recently they have been somewhat neglected in terms of global importance because of our focus on tropical deforestation. This has changed since the breakup of the USSR, because of the postsocialist timber industry and mineral extraction in Siberia, though these drivers of forest loss exist in other boreal forests (and did so before 1991).

In Chapter 15, Jayalaxshmi Mistry initially considers the ongoing debate surrounding the key characteristic of savannas—tree-grass coexistence. Sankaran et al. (2004) grouped the competing explanatory models into competition-based and demographic bottlenecks. The former emphasizes the role of competitive interactions and niche separation with respect to limiting resources and the second emphasizes demographic mechanisms and abiotic and biotic factors in promoting tree-grass persistence. Empirical evidence supports both types of models, but it is mainly small-scale, short-term, and site-specific. Consequently global consensus still eludes us (House et al., 2003), yet both models have helped increase our understanding of savanna dynamics. But as Sankaran et al. (2004) argue, the ecological reality of tree-grass competition will be found between "the competitive dominance of trees with respect to grasses changing with life-history stage, over time and across environmental gradients."

Savannas are spatially and temporally dynamic heterogeneous landscapes determined by factors linked in a hierarchy. There has been a paradigm shift from equilibrium dynamics to a disequilibrium view of savanna functioning around hierarchical patch dynamics that conceptualizes ecological systems as nested hierarchies of patch mosaics. Because climate change affects savannas alongside extensification and intensification of savanna land-use systems, understanding savanna dynamics in relation to the relative proportions of woody and herbaceous vegetation and applying hierarchical patch dynamics have strong potential implications for resource conservation and management.

Kenneth Young (Chapter 16) takes a different approach in his synthesis of humid tropical forests to the treatments of boreal forests and savannas in the previous two chapters. He argues that while biophysical factors associated with changes in ecosystem structure along spatial environmental gradients are important, much of the uniqueness of this biome originates with historical and evolutionary events affecting biodiversity over time scales of millennia and greater. Biogeography is a key discipline utilized for making decisions about the protection and conservation of biodiversity: this is critical in the tropics where though species diversity and endemism are very high, financial resources for conservation efforts are limited. He argues that important roles for biogeography in this arena are to document biogeographical patterns to assist conservation decision making, and understanding of biogeographical processes to predict conditions in unstudied locales.

In Chapter 17, Udo Schickhoff extends the biogeographical strictures that normally emerge when biomes are considered by highlighting the changing ecology of mountain ecosystems alongside the implications of climate change and land-use transformations for mountain people as well as the millions who depend on ecosystem services provided by mountain ecosystems. In recognizing this he follows renewed scientific interest in mountain environments, further supported by international efforts to establish mountains as a research priority (e.g., Chapter 13 of Agenda 21, Rio Earth Summit, 1992; International Year of the Mountains, 2002; Bishkek Global Mountain Summit, 2002), and that the acute sensitivity of mountain regions in providing opportunities to detect, model, and analyze global change processes and their effects on biophysical and socioeconomic systems.

Landscapes strongly altered by sedentary agriculture have existed in parts of Eurasia and South America for over four millennia, and many others have been converted in the last three hundred years. Yet, as Chris Stoate argues in Chapter 18, the concept of agricultural regions being part of biogeography is a reasonably new concept. That they can be managed for positive outcomes in terms of nature is an even more recent conceptual advance, one promulgated by an increasing band of ecologists (e.g., Daily et al., 2000). He shows that many species are associated with farmland because they have evolved with agriculture as farming systems have evolved. Species diversity is often high and many endemic species thrive in traditionally managed agroecosystems throughout the world. He notes that in some regions the argument has moved on from agriculture being cast as the villain behind biodiversity loss, to being the savior of biodiversity in traditional low-external input systems in the face of modern high-external input farming. Nowhere is this argument stronger than in Europe, where the European Union has acknowledged the validity of the argument by developing agri-environmental policies and actions covering the ecological services delivered by traditional farming systems such as northwest Atlantic moorlands, and the dehesa and montado—*Quercus*-arable-fallow mosaics (Stoate et al., 2003)—on the Iberian Peninsula. He also shows how relatively high-input farming systems can be managed to integrate production and nature outcomes, for example, perennial grassy vegetation along field boundaries support high wintering densities of predatory Carabid and Staphylinid beetles, which control crop pests such as aphids, reduce competitive annuals, and simultaneously support many other invertebrates, mammals, and birds.

Claire Freeman introduces the enormous levels of biodiversity present in the urban environment in Chapter 19, and she examines the paradox that the built environment can both alienate and embrace nature. She also looks at the ways that the biodiversity potential of urban areas can be realized and how existing biodiversity can be supported in the often-difficult battle against encroaching hard surfaces and buildings. The built environment offers a complex, challenging, changing, highly variable, and accessible environment for biogeographical investigation. The paradox is that to date few biogeographers have focused their attention on these exciting locations.

REFERENCES

Daily, G., Ceballos, G., Pacheco, J., Suzán, G., and Sánchez-Azofeifa, A. (2003) 'Countryside biogeography of Neotropical mammals: Conservation opportunities in agricultural landscapes of Costa Rica', *Conservation Biology*, 17(6): 1814–26.

Ellis, E.C., Goldewijk, K.K., Siebert, S., Lightman, D., and Ramankutty, N. (2010) 'Anthropogenic transformation of biomes, 1700–2000', *Global Ecology and Biogeography*, 19(5): 589–606.

Gorman, K.B., Erdmann, E.S., Pickering, B.C., Horne, P.J., Blum, J.R, Lucas, H.M., Patterson-Fraser, D.L., and Fraser, W.R. (2010) 'A new high-latitude record for the macaroni penguin (Eudyptes chrysolophus) at Avian Island, Antarctica', *Polar Biology*, 33(8): 1155–58.

Hari, P., and Kulmama, L. (2008) *Boreal Forest and Climate Change*. Dordrecht: Springer.

House, J.I., Archer, S., Breshears, D.D., Scholes, R.J., and NCEAS. (2003) 'Tree-grass interactions participants', *Journal of Biogeography*, 30: 1763–77.

Lal, R., Kimble, J.M., and Stewart, B.A. (2000) *Global Climate Change and Cold Region Ecosystems*. Boca Raton, FL: Lewis Publishers.

Sankaran, M., Ratnam, J., and Hanan, N.P. (2004) 'Tree-grass coexistence in savannas revisited—insights from an examination of assumptions and mechanisms invoked in existing models', *Ecology Letters*, 7: 480–90.

Stoate, C., Araújo, M., and Borralho, R. (2003) 'Conservation of European farmland birds: Abundance and species diversity', *Ornis Hungarica*, 12–13: 33–40.

Tape, K., Sturm, M., and Racine, C. (2006) 'The evidence for shrub expansion in Northern Alaska and the Pan-Arctic', *Global Change Biology*, 12(4): 686–702.

Walker, B., and Steffen, W. (1996) *Global Change and Terrestrial Ecosystems*. New York: Cambridge University Press.

Ecosystem Dynamics of Subpolar and Polar Regions

Ingo Möller and Dietbert Thannheiser

13.1 INTRODUCTION

The polar regions are landscapes that are characterized by their extreme natural conditions. In the high northern and southern latitudes around the poles it is basically the prolonged period of low temperatures and the annual variations in insolation that have to be emphasized among all other determinant ecological factors. Due to a short vegetation growing period of less than three months and reduced summer warmth (even in the warmest months the average temperatures do not rise above 10°C) trees cannot exist in the polar landscapes. Thus, the absence of trees is typic, and often used as a criterion for the delimitation of polar regions.

On a large scale the northern limit of trees and forests corresponds broadly with the 10°C isotherm in July, and for several decades it has been used as a delimitation criterion of the northern polar region, the Arctic. But when considering this criterion on a regional scale, the northern limit of tree growth and thus the southern limit of the Arctic is not congruent with the 10°C isotherm: In regions with a maritime climatic regime it occupies a more southerly position and under continental climatic regimes it is further north (Figure 13.1). Moreover, the boundary between the Arctic and the adjacent boreal forest zone is not sharp but forms a of 10 to 50 kilometers wide continuous transition, which is still dominated by some open, scattered stands of birches in oceanic regions and larches and pines in continental regions (see Chapter 15 for more details on the boreal forest biome). This transition is called

the forest-tundra ecotone or, more popularly, the subpolar region (see also Lloyd and Fastie, 2003).

The delimitation of the southern polar region, the Antarctic, is more difficult than the Arctic because of the lack of land area, which could exhibit the limit of tree growth. It is in fact the broad marine antarctic convergence zone that is commonly used to separate the southern polar region from its adjacent areas. Around the entire Antarctic continent at about 50° South an amalgamation of cold Antarctic surface waters and warm Atlantic-Pacific waters is found. This convergence leads to a marked rise in water temperatures and a change in water salinity. The northern extent of sea ice and the occurrence of icebergs as well as the extent of Antarctic marine organisms follow this delimitation (Figure 13.2).

13.2 GENERAL FEATURES AND POTENTIALS OF THE POLAR REGIONS

13.2.1 The arctic polar region

The arctic polar regions north of the tree limit cover some 26.4 million km², or about 5% of the Earth's surface. However, within this huge area the extent of the landmass is only about 7.6 million km². These arctic land areas can physiognomically be divided into three categories: The glaciers free of higher life forms, the arctic polar desert, and the arctic tundra.

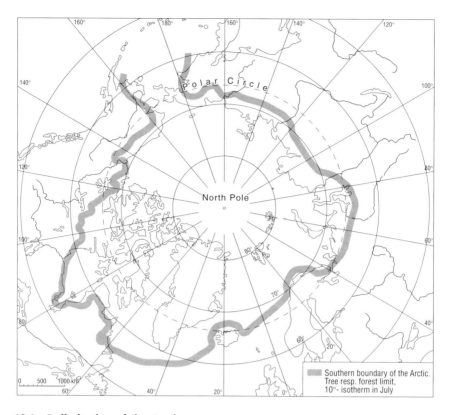

Figure 13.1   Delimitation of the Arctic

In an idealized transect from south to north the arctic tundra covers the regions north of the tree limit. The term 'tundra' means 'treeless roll-ing country' and derives from the Finnish word 'tunturi' for 'treeless plain'. Nevertheless, it is characterized by an astonishingly high abun-dance of lichens, mosses, and vascular plants. The Arctic polar desert stretches north of the arctic tundra. It is covered by a sparse vegetation (up to about 10% coverage) characterized by some cushion plants.

The ground surfaces of both regions are charac-terized by the presence of permafrost, which is commonly called the 'frozen ground'. The thick-ness of the frozen ground varies largely between 10 cm and 1,500 m. The most decisive geoeco-logical consequence of permafrost is sealing of the subsoil which reduces drainage significantly. But even in the coldest regions, during the polar summer with its permanent daylight conditions, the soil temperatures rise usually above the freez-ing point, which thaws the upper soil strata. Taken together, these strata are called the 'active layer'. Their thickness varies between 20 and 180 cen-timeters, though they fluctuate in depth from year to year. The existence of the active layer and the

annual fluctuations are driven by complex interac-tions between geoecological factors such as climatic conditions, soil structure, slope aspect and angle, vegetation cover, water content, and thickness and duration of snow cover. The active layer is very important for plant life because it represents a very important stock of water and nutrients available for plant roots. In order to exploit this stock during polar summer, root growth follows the thawing of the frozen ground at depth.

The soils of polar regions are defined by their development under low temperatures, relatively arid conditions, and weak activities of the biotic soil components (i.e. roots, micro-organisms, soil fauna). Due to the slowing down of chemical and biological processes under low temperatures, polar soils are commonly not as well differenti-ated as the soils of temperate regions. However, lithosols or skeletal soil can be observed wherever a snow-free period with temperatures above the freezing point exists for only a few days each year. In moist areas, where water percolation is reduced or blocked by the frozen ground or a compacted sediment layer, paludous soils in the form of tundra mires may develop. Organic material

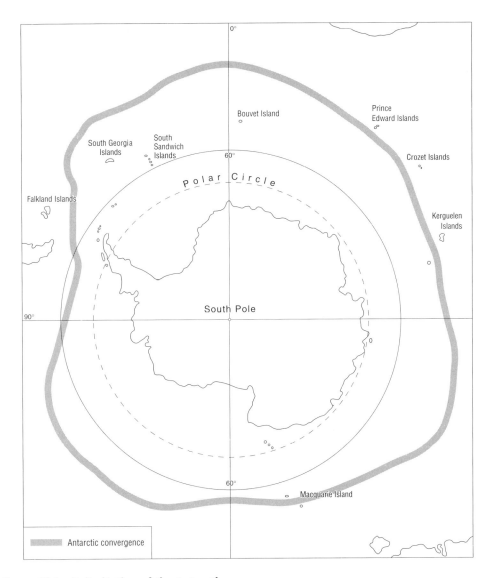

Figure 13.2 Delimitation of the Antarctic

accumulates in these conditions because of an imbalance between production and consumption. The overall carbon production by photosynthesis is higher than the respiration of carbon by micro-organisms, soil fauna, and plant roots. Thus, a certain amount of the produced phytomass is deposited as organic matter.

Arctic vegetation

With the extension of the ice sheets during the glacial periods the arctic flora migrated to the lowlands of the south where it mixed with the alpine flora of the mountainous regions, which itself migrated north. This was the origin of the 'arctic-alpine floral element', leading to a striking floristic similarity of the Arctic and the alpine belts of the northern mountain regions of the northern hemisphere (Ozenda, 1993). As an element of the Holarctic floral kingdom, three-quarters of the arctic vascular plants are present in Europe as well as in North America. The distribution of many arctic plant species is circumpolar, and closely related species replace each other throughout the region; for instance, the birches *Betula nana* in Greenland, Europe, and Western Siberia; *B. exelis* in Central Siberia; *B. midden-dorfii* in eastern Siberia; and *B. glandulosa* in Canada. Under similar geoecological conditions the flora of Arctic Siberia and Canada is nowadays richer and more diverse than the flora of the European Arctic. The number of plant species ranges from about 60 at the northern tip of the Taimyr Peninsula to 165 in Spitsbergen, and more than 450 in the arctic regions of East Siberia and Canada (Porsild and Coy, 1979; Walter and Breckle, 1994). The high number of cryptogams (lichens and mosses)—more than 500 species—is remarkable.

The structure and the composition of the arctic vegetation change according to the gradual climatic shift from south to north as well as from maritime to continental climates. While the vegetation covers more than 80% of the soil surfaces in subpolar regions close to the tree limit, it diminishes slowly to the north. Therefore, the vegetation cover in the arctic tundra regions varies slightly from almost closed, low-growing or

procumbent stands to more open patches, but throughout the arctic polar desert, vegetation is reduced to very sparse and widely open stands. This difference, and the succession of arctic tundra and arctic polar desert vegetation from south to north, can also be found in corresponding altitudinal belts in mountainous areas.

According to the surveys of Polunin (1951, 1969), Aleksandrova (1980), Edlund (1984), Daniels (1994), and CAVM (2003), three different vegetation zones can be differentiated for the arctic region, with the length of the vegetation growing period and other climatic influences being the dominating factors (Figure 13.3).

1. The Northern Arctic Vegetation Zone contains the most northern areas where the vegetation cover varies normally between 5 and 25%; however, it comprises large stretches without any vegetation at all. It embraces the arctic polar desert.

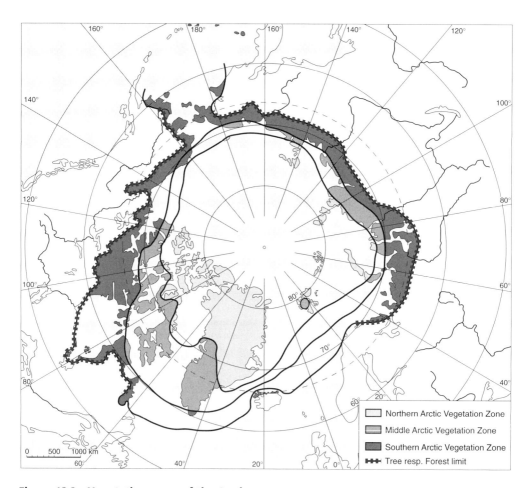

Figure 13.3   Vegetation zones of the Arctic

The low-growing vegetation stands are made up of cushion plants, grasses, and a few dwarf shrubs. Crytograms, which do not need a particular soil formation, are much more common and widespread than in other arctic vegetation zones (Thannheiser and Wüthrich, 2006).

2. In the Middle Arctic Vegetation Zone, the vegetation cover benefits from better ecological conditions and ranges usually between 25 and 50% cover. The plants can grow up to a height of 20 cm: they frequently include cryptogams, grasses, and some herbaceous plants. Dwarf shrubs gain a particular dominance with those in the genus *Dryas* being characteristic of this zone. In maritime-influenced regions their cover can increase up to 75%, but this is almost exclusively due to an expansion of mosses.

3. The Southern Arctic Vegetation Zone contains vast areas of the Arctic and borders the adjacent boreal zone. As already mentioned above, it is not a homogeneous vegetation zone, but a continuous transition zone toward the tree and forest limit. The vegetation cover ranges from 50 to 100%. It is dominated by birch and willow shrubs (*Betula* and *Salix*), which reach up to a height of 40 to 60 centimeters. Dwarf shrubs of *Cassiope*, *Empetrum*, and *Vaccinium* contribute to the vegetation cover as well as sedges (*Carex*). In contrast, it is these sedges, together with mosses, which dominate the moist areas of the lowlands. Under the mosses, some peat builders of the *Sphagnum* family can be found, and apart from the sedges, other important grass genera are *Dupontia* and *Arctophila*.

The terrestrial fauna of the Arctic

The cold, long-lasting darkness and low productivity of arctic terrestrial ecosystems are basically responsible for the relatively low faunal diversity of these landscapes. Cold-blooded (poikilothermic) terrestrial vertebrates are largely absent from the Arctic. The most successful denizens are therefore warm-blooded (homeothermic) birds and mammals. During their seasonal migrations they even reach into the northernmost parts of the Arctic.

The isolating plumes of the birds as well as a well-developed subcutaneous adipose tissues and dense pilosity of the mammals enable the preservation of a relatively high body temperature—homeothermy—and an acceleration of the reproduction cycle, which is necessary for the existence of higher life forms under the prevailing low temperature conditions. Most arctic terrestrial birds only visit the Arctic during the breeding period in summer. About 120 species of geese, ducks, divers, swans, birds of prey, wading birds, ptarmigan, and a few songbirds have been registered (Thannheiser and Wüthrich, 2006). They benefit from the vegetation (especially geese), a relatively dense population of insects (songbirds and wading birds), and, along the coasts and around the ponds, mussels, worms, and crabs. There exist interesting inter-specific relations, for example, the cyclic fluctuations of the populations of avian predators and their favorite prey, the lemming (Seldal et al., 1994).

Regarding the terrestrial mammals, only 50 of the about 4,000 species found in the world reach the Arctic, and only a dozen can be found there throughout the whole year. While small mammals such as lemmings, shrews, and voles are able to rest during the harsh arctic winter in well-protected shelters within the soils, large mammals like reindeer and caribou undertake long migrations. The terrestrial mammals represent the most important beneficiaries of the phytomass produced each year (reindeer, caribou, musk ox, moose, hares, lemmings, and voles). For example, each lemming consumes annually up to 40–50 kilograms of plant material (compare this to the mean weight—78 grams—of lemmings; Alaska Department of Fish and Game, 1994). Other characteristic mammals are well-known hunters and carnivores of the Arctic (wolves, foxes, stoats, weasels, wolverines, and shrews). The land areas of the southern Arctic, particularly the continental interiors, show a relative richness in flying insects (chafers, ants, bumblebees, and butterflies) and a massive occurrence of two-winged flies, such as biting midges and black flies.

13.2.2 Ice-free lands of the Antarctic

The polar regions of the southern hemisphere cover roughly 52 million km^2, equal to 10% of the Earth's surface. The land area of the Antarctic is about 14 million km^2, but only 4.5% is free of ice and all ecologically important habitats can be found within the coastal fringes. Some 50 species of mosses and lichens have been registered between stones or other small protected habitats like fissures or clefts of rocks. The grasses *Deschampsia antarctica* and *Colobanthus quitensis* are the only two vascular plant species found in the Antarctic (see also Frenot et al., 2005).

The terrestrial animal life of the continental Antarctic is also extremely reduced; currently there are no higher animal life forms such as amphibians, reptiles, birds, or mammals, which could be assigned ecologically to the terrestrial areas. The diversity of the invertebrates is, in comparison to the polar region of the northern hemisphere, extremely reduced and restricted to niches provided by the soil (Tilbrook, 1970; Block, 1984).

13.2.3 Polar oceans and their fauna

The polar oceans are—in striking contrast to the terrestrial polar ecosystems—highly productive areas with an enormous economic potential. However, these productive areas are restricted to the 'upwelling areas' in the polar oceans. Vast areas, both in the north and in the south, are much less productive. The high productivity of the upwelling areas is based on nutrient-rich deep waters coming to the surface. Their productivity is even higher than those of open waters of sub-tropical and tropical oceans. Nearly all primary production occurs under the permanent light conditions of the summer period. It is reduced to almost zero during the dark, winter period. The large, rich populations of marine mammals, consumable fishes, and seabirds in the productive areas have contributed to a serious economic exploitation of the polar regions.

The Arctic Ocean covers an area of 14.5 million km^2 between 70°N and the North Pole. The marine fauna is characterized by huge populations of warm-blooded vertebrates like whales, seals, and seabirds. These animals are located in a high position on the food pyramid. They are found in such high numbers because all lower elements of the food pyramid (fishes, crabs, plankton, and autotrophic primary producers like diatoms and peridinians) exist in huge numbers as well. The arctic waters are known for their whale richness, but most of the whale species are only temporarily to be found in the Arctic (i.e., in arctic summer when the Arctic Ocean contains an overabundance of nutrients). With the upcoming winter season and the shortening of daylength most of the toothed cetacenas as well as the blue, humpback, and gray whales and the roqual migrate to temperate or tropical regions, where they bear offspring. Seabirds are particular beneficiaries of the huge amount of secondary plankton and small fishes. They represent an important terminal element of the arctic food pyramid, and during their summer breeding period they inhabit the coastal fringes of the Arctic Ocean in populous colonies.

The Southern Ocean comprises all open waters around the Antarctic continent. At about 28 million km^2, it is nearly twice as large as the Arctic Ocean. It contains highly endemic populations. For instance, 90% of the individuals and 60% of the species of Antarctic fish belong to one order only: The *Notothenoidei* (which include Antarctic cod, weevers, and rapacious fishes). Like whales, some of the seabirds of the Southern Ocean belong to the world's most prominent migrating animals (e.g., albatrosses). They only appear on Antarctic land areas during their breeding periods, and spend the rest of their life by sailing over or swimming in the open waters

(Senn, 1994). Only 38 bird species breed south of the antarctic convergence: 10 species on the Antarctic continent, 15 on the Antarctic Peninsula, and the rest within the subpolar regions. Penguins and petrels are the dominant bird families of the Southern Ocean.

13.3. POLAR AND SUBPOLAR ECOSYSTEMS

13.3.1 Biodiversity

In comparison with the temperate regions, the polar regions are characterized by low biodiversity (see Mutke, this volume for a detailed discussion of global biodiversity gradients). But in both polar terrestrial and marine habitats 'deserts' and 'oases' exist. Due to their higher productivity, the oceans are able to support a higher number of organisms than the less-productive terrestrial ecosystems of polar regions. Only a small number of species contribute to the enormous richness in individuals (i.e., numbers of individuals). For example, there are thousands of seals and seabirds found along coastal fringes and around upwelling areas. Thus, the polar and subpolar regions are characterized by their low biodiversity and high numbers of individuals. Nonetheless, the Arctic and the Antarctic show a profound difference concerning their basic biotic settings. Due to their better geographic accessibility to other continental areas and an overall more moderate climate, the terrestrial areas of the Arctic are inhabited by various vascular plants and mammals, while the terrestrial areas of the Antarctic have a few terrestrial species only (no mammals and only two vascular plant species).

The existence of intra-zonal 'oases' in terms of biodiversity can be based on various, but locally different, ecological factors such as climate, soil, and snow conditions. The inner fiord areas of Spitsbergen show, for example, relatively well-developed vegetation stands with some comparatively fastidious plant species like the dwarf birch (*Betula nana*) and the crowberry (*Empetrum nigrum* ssp. *hermaphroditum*). The higher biodiversity of these areas depends particularly on the degree of continentality. Even if their winter temperatures are much lower than in maritime areas, the summers in the continental areas offer some better fundamental development possibilities for plant life: At the beginning of the vegetation period they are free of snow relatively early in the year, dry up faster, and achieve higher temperature sums.

The vertebrates and most of the vascular plants found in the Arctic have invaded in the last 10,000 to 20,000 years. Some, though, have developed there. Thus, the ecological conditions around the ice-covered areas are something new for the recent biological evolution, and all polar ecosystems in their present manifestations are young ecosystems (Stonehouse, 1971). Arctic species as elements of these ecosystems are of high importance for modern research on evolution. Phenomena such as hybridization, vicariance, spreading, or adaptation of life cycles to environmental conditions can be well studied here.

The polar regions are not suitable for bigger invertebrates and the cold-blooded terrestrial vertebrates. Warm-blooded vertebrates are in a much better position because of the mechanisms developed to keep their body temperatures constantly higher than the temperature of their environment. They dispose inner physiological settings, in which all life processes can run under optimal temperature conditions. For this reason, the warm-blooded birds and mammals of polar regions are the most important beneficiaries of the vegetation, while in temperate and tropical regions the insects represent the most important herbivores.

Cycles of mass propagation and the fast collapse of arctic animal populations are highly linked to the low biodiversity. Nearly all smaller arctic rodents show a mass propagation in a four-year period. Their predators (e.g., arctic foxes, weasels, snowy owls, skuas) consequently develop in the same way, though slightly delayed in time. In years of mass propagation of rodents, certain polar tundra birds such as shore birds are no longer attractive to their normal, natural predators. Hence, they achieve higher breeding success in these years. Mass propagation is usually a local phenomenon. For example, the production cycles in Barrow (Alaska) are not synchronous with those of Ellesmere Island (in the eastern Canadian Arctic). The reasons for the enormous population fluctuations have been debated for many years (Seldal et al., 1994). The seed production of several plants, like the overall growth behavior of the vegetation (Hope et al., 2003), does not follow an annual rhythm because the plants have to assure their survival by the accumulation of nutrients over several years. Consequently, the nutrient or food supply for certain species can be cyclic. Moreover, after a mass propagation the nutrient levels decline and hence the population also declines. Thus, certain population fluctuations depend more on the availability of nutrients than on the number of their predators. Nevertheless, the most important preconditions for these cycles are the low biodiversity combined with a high reproduction potential, especially of rodents (see also Aars and Ims, 2002).

13.3.2 Energy and water budgets and their effects

The solar radiant energy plays a much more decisive role in polar ecosystems than for instance for ecosystems of the temperate zone. This becomes particularly obvious in spring, when with temperatures below the freezing point, the first signs of plant and animal activity appear at the ground surface. At this time, thin bars of ice or snow on rocks and soil surfaces function as 'greenhouse jackets' for the underlying vegetation (Kappen, 1993). The shortwave radiation penetrates ice and snow to a certain degree, while the longwave, thermal radiation emitted from the soil does not, so that the encapsulated cavities at ground level heat up. This is, in effect, a small-scale 'greenhouse effect'. The temperatures under these protecting jackets can be up to 10°C higher than at the surrounding soil surfaces. It is not only the vegetation that benefits from these natural greenhouses. Shortly before the covering layers melt, there is a remarkable accumulation of soil fauna; particularly arthropods which search actively for these small oases with their favorable temperature and humidity conditions and protection against hostile cold winds (Wüthrich, 1989).

The thermal radiation emitted from the Earth's surface is partly absorbed by water vapor, carbon dioxide, and ozone molecules in the atmosphere, which consequently warms up. Hence, the atmosphere reduces the amount of outgoing radiation lost from the atmosphere and turns itself into a source of thermal radiation, the so-called counter-radiation of the atmosphere. In terms of the radiation balance, thermal terrestrial radiation is negative while the atmospheric counter-radiation is positive. This counter-radiation is of particular importance when areas of low pressure with relatively warm cloud covers advance from temperate to polar regions. This influx of additional warmth from temperate latitudes compensates to a certain degree the overall annual loss of radiation energy from polar regions. In this context, the ocean currents and the large Siberian rivers also play important roles as carriers of heat to the polar regions.

The key parameters for thermal conduction of heat into the ground are the density, the specific heat capacity, and the heat conductivity of the soils. These parameters depend themselves on the composition and structure of the substrate, and its water and ice contents. Additional local factors that may be of particular influence are the melting of the frozen ground (and, therefore, the thickness of the active layer), the thickness and length of winter snow cover, the thickness of organic soil layers, and the characteristics of percolating water

(Harris and Corte, 1992; see also Osterkamp and Jorgenson, 2006).

The thickness of the active layer is usually between 20 and 180 cm, but the distribution of soil organisms like springtails (*Collembola*), mites (*Acari*), enchytrea, and larvae of two-winged flies (*Diptera*) is concentrated to the upper six centimeters of the soil profile only. Additionally, maxima can be observed within the organic layers (usually the upper two centimeters), where daily temperature changes are still present. These temperature fluctuations cannot be measured at greater depths. Under a dense moss cover, for example, they do not exist at depths of more than 10 cm.

Compared to the soil fauna, plant roots are able to exploit much greater depths during the summer. Their development at depth follows a compromise between the plant's demand for water and nutrients (drying up of the upper soil layers) and their heat requirements (lowering frozen ground temperatures). In the northern arctic zone, and within dry skeletal soils with a pronounced and thick active layer, active living plant roots can be found up to a depth of 70 cm. This is similar to the processes of intensive rock weathering and soil genesis, both of which also alter seasonally.

From the climatic point of view, large parts of the Arctic and Antarctic are arid. Precipitation is low and varies normally between 100 mm in continental areas to 400 mm at the coasts. The evaporation is, due to the low temperatures, very low and almost only occurs during the snow-free vegetation growing period during summer. For example, 85% of the evaporation in the northern arctic zone occurs at this time. During the vegetation growing period, evaporation is 20 to 40% higher than precipitation. Evaporation and precipitation both increase within altitudes up to 600 m (Ohmura, 1982).

High summer runoff is typical of arctic rivers, though discharge differs slightly according to their catchment structure. In northern Alaska, for example, the summer runoff is about 66–86% of the precipitation (Lilly et al., 1998). Depending on the region, different amounts of the annual precipitation fall as snow: in northern Alaska, it is 33–47%; in Resolute (Arctic Canada), 75%; and in Spitsbergen, 40–50%. The snow cover builds up a temporary water reserve that strongly influences catchment drainage.

In contrast to what might be expected from the discussion of climate above, polar habitats are dominated by humid conditions from an ecological point of view. Most of the habitats obtain supplementary water by melting snow or water percolating above the frozen ground. In combination with the low evaporation, continuous moist conditions are common for polar ecosystems.

However, for most parts of the polar regions the low summer precipitation has to be understood as a favorable factor. During late spring, the thin snow cover melts rapidly, and an important consequence of this is that the length of the vegetation growing period is extended. Thick snow cover insulates well against very low temperatures in winter, but it drastically reduces the length of the vegetation growing period. This leads to a drop in biodiversity by reducing the development conditions for many species. Furthermore, the productivity of habitats is reduced. A shortening of the vegetation growing period is therefore a major stress factor, one that is even much more powerful and influential than low winter temperatures.

Climatic aridity is a limiting factor for only a few polar areas. Examples are the areas free of ice in central Antarctica, which receive an annual precipitation of less than 50 mm. They are actually 'polar deserts'. They exist near the antarctic coasts as well, for example, the ice-free dry valleys of the McMurdo Region (Campbell et al., 1998). With an annual precipitation of 13 mm and an average temperature of −25°C, these areas are characterized by their low level of geomorphologic activity with a stable dry frozen ground (reduced melting and freezing cycles) and relatively old sediment deposits. The existence of the dry valleys as well as related ecological structures and processes are not sufficiently explained yet.

Comparable 'polar deserts' exist in the Arctic in northern Greenland and the northwestern Canadian Arctic archipelago. Precipitation of less than 90 mm reduces plant life to very sparse and open stands without any woody species (Thannheiser, 1991). In addition to the lack of water, reduced summer warming is a limiting ecological factor (see also Barber et al., 2000). Snow cover at higher altitudes is an important water stock in these dry areas since it releases significant amounts of meltwater. Last but not least, some ecologically dry habitats exist in generally more humid parts of the Arctic as well. Their ecological dryness is caused by skeletal soils with low water capacity or their topographic position. The latter effect can particularly be observed on hilltops, where no tributary influx of water compensates the fundamental dryness of the habitats (Thannheiser and Möller, 1994).

13.3.3 Plant productivity and production

The overall productivity of arctic plant communities (even if all components such as algae, lichens, mosses, grasses, or dwarf shrubs are included) does not depend exclusively on their photosynthetic potential. Environmental factors such as available light, water, and nutrients plus the length of the productive period are at least equally influential (see also Schuur et al., 2007). Under clear skies, the permanent light during the polar summer enables a nearly continuous positive net photosynthesis (i.e., there are no respiration losses during dark nights) and fast, effective storage of the organic matter produced. Research in maritime-influenced ecosystems of the Northern Arctic Zone of Spitsbergen reveals that the carbon balance is negative for most habitats in cloudy, wet rainy summers. In such high arctic locations, the respiratory losses to the atmosphere start at light conditions with a photon flux of 300 Einstein (Wüthrich et al., 1999). Thus, the lack of light because of cloudiness can be the crucial factor for productivity. Conversely, the light compensation point, where production equals respiration, is relatively high at 300 Einstein. In subpolar peatlands, where a high water table significantly reduces soil respiration due to oxygen deficiency and coldness, most of the important plant communities are able to store carbon at 100 Einstein (Wüthrich and Schaub, 1997).

The above-ground phytomass is often used as an indicator of the production of plant communities. It is at about 140 g m^{-2} for the northernmost scattered stands of arctic vegetation. Towards the middle arctic tundra, the average rises to a little over 1 kg m^{-2}, while it ranges from 11 to 17 kg m^{-2} in the southern arctic tundra. However, large parts of the phytomass are to be found below-ground. Therefore, in order to get a real impression of vegetation production it is always necessary to record both above-ground and below-ground phytomass (Möller, 2000). The ratio of above-ground to below-ground phytomass is on average about 1:1.3 for the northernmost arctic vegetation, 1:3.2 in the middle arctic tundra, while it is approximately 1:10 in the peatlands of the southern arctic.

Research in the coastal northern arctic tundra of Spitsbergen provides further details about phytomass distribution (Thannheiser et al., 1998). The highest total phytomass (above-ground plus below-ground = 3.18 kg m^{-2}) is found in stands fertilized by birds, the so-called skua hummocks. The moss tundra stands, which cover most of the area, have a phytomass of about 1.5 kg m^{-2}. The phytomass of snow bed communities decreases according to the length of their limiting snow-free periods and range from 0.82 kg m^{-2} for stands with the longest snow-free periods to 0.15 kg m^{-2} for those with the shortest snow-free periods. The organic carbon of the soils varies between 0.5 and 7.7 kg m^{-2}, with the highest amounts for the skua hummocks and the lowest

for extremely wind-exposed, scattered habitats. When vegetation and soil data are combined, the vegetation has between 5.2 to 23.6% of the total organic carbon of these habitats.

Taking a large-scale perspective, the proportion of the total organic carbon that is phytomass decreases with increasing latitude, and in the sparse stands of the northernmost arctic tundra only a small amount of the overall carbon as well as the nutrient pool is bounded in the phytomass (see above). At a small scale, the influence of the water budget on the organic production is obvious: moist habitats enable a luxuriant plant growth and more organic material is produced than in dry habitats (see also Sitch et al., 2007).

The plant communities of moist habitats usually possess a high below-ground phytomass, which is important for plant growth and for building up soil organic matter. Sedges and mosses, which cover up to 100% of the moist habitat stand surfaces, are major contributors to production. Their below-ground organic reserves enable an effective plant growth at the beginning of the vegetation growing period. The plants can even initiate new leaves when their roots are still in the frozen ground and additional nutrient uptake is still impossible (Chapin et al., 1986).

For the development of soil organic matter it is crucial how much of the organic matter produced annually is respired as carbon dioxide by micro-organisms and soil fauna. If the annual production is equal to respiration, organic matter remains constant. But if the production is higher than the respiration, organic carbon will be stored as humus or as incorporated organic matter. As already mentioned, the respiratory activity of micro-organisms is reduced in moist habitats. Therefore, moist and cold areas are usually a natural carbon sink (Nadelhoffer et al., 1992). However, the accumulation of soil organic matter is limited and peat formations remain relatively shallow in polar regions compared to other parts of the world.

Nevertheless, polar soils can achieve similar decomposition rates of organic matter as soils of the temperate zone. Fertilized skua hummocks or soils underneath bird cliffs (Figure 13.4) have very high rates of organic matter accumulation and are used as fertilizers for agricultural purposes. This shows that an essentially high potential exists for an effective productivity. The short vegetation growing period, which also limits micro-organisms, reduces the annual turnover of carbon, and it limits the productivity even more than the oft-mentioned reduced availability of nutrients.

Most of the plant species of polar regions are perennial. Therefore, only a small part of their phytomass is built up in any one year, and

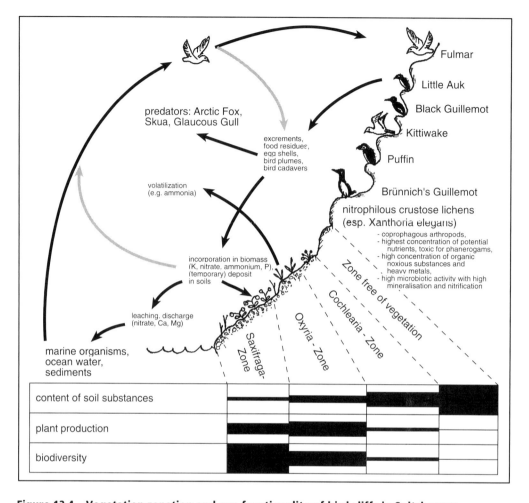

Figure 13.4   Vegetation zonation and eco-functionality of bird cliffs in Spitsbergen

individual plant growth progresses slowly. In contrast to their slow growth, life expectancy is high. Many dwarf shrubs like *Dryas* and *Arctostaphylos* live more than 100 years. A twig of *Empetrum nigrum ssp. hermaphroditum* from Spitsbergen revealed an age of more than 80 years (Elvebakk and Spjellkavik, 1995), while a trunk of *Juniperus* of the southern arctic tundra near the tree limit showed an age of 544 years. The widespread reindeer lichen (*Cladonia rangiferina*) grows annually by one to five millimeters. Lichen heaths grazed by reindeer need at least 10 years to recover (Pöhlmann, 1975; Wegener et al., 1992; Johansen et al., 1996). Dwarf shrubs like *Betula glandulosa* are affected by grazing as well. Their stands, cropped by caribou, show both reduced vegetation cover and a reduced leaf mass for several years. In northwestern Arctic Canada, where the herding of introduced caribou occurs in an

extensive manner, the herds can easily over-utilize vegetation. The relationship between nutrient requirements and the slow recovery of damaged vegetation stands is one important reason for the long seasonal migrations of hoofed animals between their low-productivity summer habitats and their high-productivity winter habitats, (i.e., usually between north and south; Manseau et al., 1996).

13.4 ECODIVERSITY

In the following discussion, ecodiversity is used in its original sense as diversity of ecosystems (Naveh, 1994), which is also known as landscape diversity (Leser and Nagel, 1998). This must be mentioned because the two, formerly clearly distinct terms of biodiversity and ecodiversity were unfortunately merged at the UNCED-Conference in Rio de Janeiro in 1992: the term biodiversity was enlarged from its proper sense of biological and genetic diversity, through the inclusion of ecosystem diversity.

13.4.1 The influence of scale on ecodiversity

The ecodiversity of the arctic tundra depends broadly on the perspective taken. Ecodiversity changes markedly from one scale to another. At a microscale of a few leaves to a few meters, the arctic tundra is well structured. Difference in the relative relief of terrain of just a few centimeters results in enormous differences in winter shelter, temperature, humidity, soil moisture, snow cover, and the length of the vegetation growing period. The structure of the biocenoses follows these small differences and varies from one patch to another. It is not only their structure that varies, species composition also alters according to the spatial patterns of the basic influences mentioned above. Thus, arctic tundra has a high microscale ecodiversity. This changes as we take a meso- or macroscale perspective. At these scales, the arctic tundra is comparatively homogeneous with few apparent variations in structure and composition, and both its mesoscale and macroscale ecodiversity are low (see also Stow et al., 2004).

13.4.2 Meso- and macroscale differentiation of ecosystem complexes

As mentioned above, the arctic tundra is relatively homogeneous at the meso- and macroscales. Changes in plant composition are only recognized through the occurrence of precise ecological perimeters. Regularly recurring vegetation units like communities and associations are clearly noticed while conducting floristic surveys of arctic vegetation. These units can themselves be grouped according to their similarities in species composition and be taken as a basis for a higher analytical plant sociology, giving precise information of the differentiation of ecosystems and ecosystem complexes. Vegetation maps based on air photos or satellite images indicate the distribution patterns of different basic vegetation units by visualizing the connectedness or spatial dispersal of areas dependent on the ecological conditions of the particular sites. But these vegetation maps can only depict an approximate reality due to their scale-related limitations in representation.

Thus, higher plant sociological analysis can rapidly obtain more precise and reliable concise knowledge of vegetation units of large survey areas than mapping from satellite imagery or aerial photography.

The organization of vegetation units cannot be solved solely by the floristic-sociological systematic approach of Braun-Blanquet (1964). In order to generalize information on arctic vegetation over larger areas, the application of a more detailed hierarchical arrangement of syntactic units of different categories is recommended. But when working with the higher plant sociological approach, it is necessary to have accurate knowledge of all the communities occurring in a study area. In the same manner as individual species (taxa) are constituents of plant communities (syntaxa), the total number of all precisely defined communities occurring in a homogenous area can be combined in a Sigmetum (Sigma syntaxa) (Tüxen, 1973; Schwabe, 1997). Braun-Blanquet (1964) stated that definite groupings of plant communities recur under similar environmental conditions under which they compete with each other. He named these steadily recurring often mosaic-like or belt-like communities 'community complexes' (= ecosystems). The basic requirement for considering a community complex as a Sigmetum is the presence of one or several characteristic communities in the complex (Tüxen, 1977). Schwabe and Kratochwil (this volume) provides a detailed account of these approaches.

In the same way that plant communities can be combined in an abstract hierarchical system (e.g., alliances, orders, classes) on the basis of their species combinations, it is also possible to combine the Sigmeta in a pertinent system (Tüxen, 1977). Sigmeta in contact within a homogeneous landscape can be combined in higher complexes of a higher rank, the so-called Geosigmeta or 'ecosystem complexes' (Géhu and Rivas-Martinez, 1981). The application of Geosigmeta provides the accentuation of the macroscale perspective, because they repeatedly cover very large areas in almost the same combination.

In order to get some precise, large-scale data on the diversity of the arctic tundra, and in order to prove the effectiveness of the higher plant sociological approach, three study areas in Spitsbergen have been evaluated (Thannheiser and Möller, 2005). The relevé sites at Eidembukta, Liefdefiord, and Sassendalen varied from 40,000 m^2 to 1,000 ha. Within a homogeneous landscape, relevés of ecological units were divided into plots of 2,000 to 5,000 m^2 and all plant communities were recorded according to Braun-Blanquet's (1964) seven-partite scale. Statistical analysis of the field data provided a raw data table, which was aggregated to create a constancy table reflecting precisely plant community occurrences, and the plant communities were divided into six constancy classes (Table 13.1). All Sigmeta sampled in Spitsbergen could be grouped into only one Geosigmetum, the *Salicetum polaris Geosigmetum* (Thannheiser and Möller, 2005). Due to its isolated position, only 165 higher natural plant species are known in Spitsbergen. Consequently, the number of plant communities is also low. Spitsbergen receives much snowfall. Furthermore, the archipelago (and especially in its coastal fringe) is affected by the maritime influence of the Arctic Ocean. The resulting humid ecological conditions are evident in the vegetation. Snow bed communities that are limited by the thickness and length of the winter snow cover are the most widespread vegetation units (Möller, 2000).

According to Table 13.1, each study area is considered a sub-Geosigmetum: *Salicetum polaris*, the typical snow bed community, dominates in all study areas but is accompanied by different co-dominating communities in each area:

1. In Eidembukta, a windy, foggy western coastal site located in the Northern Arctic Vegetation Zone (Möller et al., 1998), the co-dominate of *Salicetum* is another snow bed community, the lichen-rich *Cetrarietum delisei*. Only 16 plant communities occur here.
2. In contrast, most plant communities were found in the Liefdefiord sub-Geosigmetum

(Thannheiser, 1994). Liefdefiord is part of the favorable area of the Middle Arctic Vegetation Zone, where dwarf shrub communities are co-dominant. Its maritime influence is much less than that at Eidembukta. The occurrence of snow bed communities here reflects long-lasting snow cover due to its northerly exposure.

3. The sub-Geosigmetum of Sassendalen represents the warm, continental part of the Middle Arctic Vegetation Zone (Thannheiser et al., 2001), where dwarf shrub communities dominate slightly over the snow bed communities. The sub-Geosigmetum is therefore characterized by *Dryadetum minoris*.

The data in the constancy table (Table 13.1) demonstrate that even tundra vegetation can be differentiated over a large scale with respect to vegetation units that occur regionally or locally. The higher plant sociological approach stresses the quantitative evaluation of all units in larger areas, which otherwise is only possible through the application of resource-intensive conventional surveys and mapping. The importance of this type of vegetation analysis will undoubtedly increase in the future, because an exact distribution of vegetation units can be established rapidly for large geographical areas. In this way, changes in the patterns of diversity due to ecological disturbance and climatic change can easily be quantified and recognized over both the short-term and the long-term.

Table 13.1 Constancy table of the *Salicetum polaris* Geosigmetum

A: *Cetrarietum delisei salicetosum polaris* Subgeosigmetum (Eidembukta)
B: *Salicetum polaris drepanocladodetosum uncinati* Subgeosigmetum (Liefdefiord)
C: *Dryadetum minoris salicetosum polaris* Subgeosigmetum (Sassendalen)

Location (Spitsbergen)		A Eidembukta	B Liefdefiord	C Sassendalen
Year		1996	1992	2000
Area (km²)		0.65	10.0	1.03
Number of Sigma relevés		20	75	5
Salicetum polaris typicum Sigmetum				
Salicetum polaris typicum	[sb]	IV	IV	IV
Cetrarietum delisei-salicetosum polaris Subsigmetum				
Cetrarietum delisei salicetosum polaris	[sb]	V	.	.
Cetrarietum delisei typicum	[sb]	V	.	.
Cetrarietum delisei saxifragetosum oppositifoliae	[sb]	II	.	.
Cerastio regelii-Poetum alpinae	[sb]	III	IV	I
Phippsietum algidae-concinnae	[sb]	+	III	.
Deschampsietum alpinae	[wl]	I	III	.
Arctophiletum fulvae	[wm]	+	I	.

Continued

Table 13.1 Cont'd

Location (Spitsbergen)		A *Eidembukta*	B *Liefdefiord*	C *Sassendalen*
Salicetum polaris-drepanocladetosum uncinati Subsigmetum				
Salicetum polaris drepanocladosum uncinati	[sb]	IV	V	
Tomenthypnetum involuti	[wl]	III	IV	IV
Calliergono-Bryetum cryophili	[wl]	II	IV	I
Orthothecium chryseon-community	[sb]	.	IV	IV
Scorpidium revolvens-community	[sb]	.	.	IV
Bryo-Dupontietum pelligerae	[wm]	.	I	II
Puccinellietum phryganodis	[cc]	.	II	I
Caricetum ursinae	[cc]	.	I	I
Caricetum subspathaceae	[cc]	.	+	I
Eriophorum scheuchzeri-community	[wm]	.	+	I
Drepanoclado-Ranunculetum hyperborei	[wm]	.	I	.
Caricetum stantis	[wm]	.	+	.
Calliergono-Caricetum saxatilis	[wm]	.	+	.
Carex saxatilis-community	[wm]	.	+	.
Eripophorum angustifolium ssp. triste-community	[wm]	.	.	I
Dryadetum minoris-salicetosum polaris Sigmetum				
Dryadetum minoris salicetosum polaris	[ds]	II	IV	V
Dryadetum minoris typicum	[ds]	+	III	IV
Dryadeto-Cassiopetum tetragonae	[ds]	.	II	IV
Empetrum nigrum ssp. hermaphroditum-community	[ds]	.	II	.
Carici rupestris-Dryadetum octopedalae	[ds]	.	II	.
Caricetum nardinae	[ds]	.	I	.
Saxifraga oppositifolia-community Sigmetum				
Saxifraga oppositifolia-community	[rs]	III	III	V
Racomitrium canescens-community	[rs]	IV	.	III
Sphaerophoro-Racomitrietum lanuginosi	[rs]	.	V	.
Papaveretum dahliani	[rs]	.	III	II
Festuca rubra ssp. arctica-community	[rs]	.	II	.
vegetation of bird stands (skua hummocks)	[sh]	+	+	+
Oxyrio-Trisetum spicati	[rs]	+	+	.
Potentilletum pulchellae	[rs]	.	+	I
Puccinellietum angustatae	[rs]	.	+	I
Luzuletum arcuatae	[rs]	.	+	.

degree of presence: + = 0—5%, I = 6—20%, II = 21—40%, III = 41—60%, IV = 61—80%, V = 81—100%

vegetation types: sb = snow bed communities; wm = water and (tundra) mire communities; cc = coastal vegetation; wl = wetland communities; rs = communities of exposed ridges, slopes, and plains; ds = dwarf shrub communities; sh = vegetation of skua hummocks.

13.4.3 Predicted changes of polar and subpolar ecosystems due to climate change

The results of recent research indicate that Arctic—as well as Antarctic—climates will dramatically change within the next 100 years. Climate models predict a rise in annual average temperatures of 3–5°C (against the 1981–2000 average for the terrestrial areas (ACIA, 2005; IPCC, 2007). For several years, global warming has already led to increased evaporation, which itself accounts for higher precipitation. The annual precipitation in arctic regions is predicted to rise by 20% by the end of the twenty-first century (ACIA, 2005).

This climate change will generally promote further expansion of the vegetation, because higher temperatures favor higher and denser vegetation cover in polar regions. In particular, the northward expansion of bushes like the willow *Salix alaxensis* from subpolar regions is anticipated. The occurrence of woody plants in the contemporary Northern Arctic Vegetation Zone is also predicted (compare Epstein et al., 2004; Hinzman et al., 2005; Tape et al., 2006; Thannheiser and Wüthrich, 2006). Consequently, many of today's arctic plant species will experience competition from invasive species. As interspecific competition increases, it is likely that many arctic plant species will be replaced by others, resulting first in a change of the composition of plant communities, and later the entire vegetation setup at regional scales will be affected.

Falling permafrost tables have already been proven (Lachenbruch and Marshall, 1986). This increases the dessication of the soils along the beaches and rivers, and accelerates the wind erosion of fine mineral and organic matter. The resulting decrease in soil water capacity will provoke water stresses in vegetation.

Complex problems exist concerning the predicted increase of precipitation, especially in the arctic coastal fringes. In large depressions and areas with limited run-off, increased precipitation will lead to increased soil moisture levels combined with acidification and, therefore, a decrease in biotic respiration (i.e., the decomposition of organic matter). The subsequent local formation of peat or paludous soils will affect directly the existing structuring of arctic ecosystems and ecosystem complexes by equalizing their differences. Reduced biodiversity and ecodiversity are inevitable.

Under cloudy and rainy conditions, some polar ecosystems already show retarded and/or limited vital functions such as germination, flowering, and fruiting (Möller et al., 1998). Many ecosystems receive such small amounts of incoming radiation during the productive summer period that they already have negative carbon balances in this period (Wüthrich et al., 1999). Thus, entire habitats continuously lose organic matter and, over time, their basic fertility will decline.

Humanity has already started the biggest experiment in biological evolution—unfortunately, without taking into account either the major effects or the various and highly interlinked side effects. Even if all the impacts at local and regional scales cannot be precisely predicted, it is obvious that a new era of ecosystem evolution has been initiated in polar regions.

REFERENCES

Aars, J., and Ims, R.A. (2002) 'Intrinsic and climatic determinants of population demography: The winter dynamics of tundra voles', *Ecology*, 83: 3449–56.

ACIA (Arctic Climate Impact Assessment) (2005) 'Scientific report: Impacts of a warming Arctic'. Cambridge: Cambridge University Press.

Alaska Department of Fish and Game (1994) 'Lemmings', www.adfg.alaska.gov/static/education/wns/lemmings.pdf. Accessed 1 May 2011.

Aleksandrova, V.D. (1980) *The Arctic and Antarctic. Their Division into Geobotanical Areas.* Cambridge: Cambridge University Press.

Barber, V., Juda, G.P., and Inne, B. (2000) 'Reduced growth of Alaskan white spruce in the twentieth century from temperature-induced drought stress', *Nature*, 405: 668–73.

Block, W. (1984) 'Terrestrial microbiology, invertebrates, and ecosystems', in R.M. Laws (ed.), *Antarctic Ecology 1.* London: Academic Press. pp. 163–236.

Braun-Blanquet, J. (1964) *Pflanzensoziologie. Grundzüge der Vegetationskunde.* 3rd ed. Vienna: Springer.

Campbell, I.B., Claridge, G.G.C., Campbell, D.I., and Balks, M.R. (1998) 'Permafrost properties in the McMurdo Sound-Dry Valley region of Antarctica', in A.G. Lewkowicz and M. Allard (eds.), *7th International Conference on Permafrost.* Yellowknife: Collection Nordicana, Université Laval. pp. 121–26.

CAVM (Circumpolar Arctic Vegetation Map). Scale 1:7 500 000; Conservation of Arctic Flora and Fauna (CAFF) map, No 1 (2003). U.S. Fish and Wildlife Service, Anchorage, Alaska.

Chapin III, F.S., Shaver, G.R., and Kedrowski, R.A. (1986) 'Environmental controls over carbon, nitrogen and phosphorus fractions in Eriophorum vaginatum in Alaskan tussock tundra', *Journal of Ecology*, 74: 167–95.

Daniels, F.J.A. (1994) 'Vegetation classification in Greenland', *Journal of Vegetation Science*, 5: 781–90.

Edlund, S.A. (1984) 'High Arctic plants: New limit emerge', *Geos*, 1: 10–13.

Elvebakk, A., and Spjellkavik, S. (1995) 'The ecology and distribution of Empetrum nigrum ssp. hermaphroditum on

Svalbard and Jan Mayen', *Nordic Journal of Botany*, 15: 541–52.

Epstein, H.E., Calef, M.P., Walker, M.D., Chapin III, F.S., and Starfield, A.M. (2004) 'Detecting changes in arctic tundra plant communities in response to warming over decadal time scales', *Global Change Biology*, 10: 1325–34.

Frenot, Y., Chown, S.L., Whinam, J., Selkirk, P.M., Convey, P., Skotnicki, M., and Bergstrom, D.M. (2005) 'Biological invasions in the Antarctic: Extent, impacts and implications', *Biological Review*, 80: 45–72.

Géhu, J.-M., and Rivas-Martinez, S. (1981) 'Notions fondamentales de phytosociologie', in R. Tüxen (ed.), *Ber. Int. Sympos. Int. Ver. Vegetationskunde*. Lehre: Cramer Verlag. pp. 5–30.

Harris, S.A., and Corte, A.E. (1992) 'Interactions and relations between mountain permafrost, glaciers, snow and water', *Permafrost and Periglacial Processes*, 3: 103–10.

Hinzman, L.D., and 34 authors. (2005) 'Evidence and implications of recent climate change in northern Alaska and other Arctic regions', *Climate Change*, 72: 251–98.

Hope, A., Boynton, W., Stow, D., and Douglas, D. (2003) 'Interannual growth dynamics of vegetation in the Kuparuk river watershed based on the normalized difference vegetation index', *International Journal of Remote Sensing*, 24: 3413–25.

IPCC. (2007) 'Climate change 2007: The physical science basis. Summary for policymakers. Geneva 2007', http://www.ipcc.ch. Accessed April 1, 2009.

Johansen, B., Tømmervik, H., and Karlsen, S.R. (1996) 'Reinbeitene sett fra satellitt', *Ottar*, 211: 17–24.

Kappen, L. (1993) 'Plant activity under snow and ice, with particular reference to lichens', *Arctic*, 46: 297–302.

Lachenbruch, A.H., and Marshall, B.V. (1986) 'Changing climate: Geothermal evidence from permafrost in the Alaskan Arctic', *Science*, 234: 689–96.

Leser, H., and Nagel, P. (1998) 'Landscape diversity—a holistic approach', in W. Barthloff and M. Wininger (eds.), *Biodiversity—A Challenge for Development Research and Policy*. Berlin-Heidelberg: Springer. pp. 129–43.

Lilly, E.K., Kane, D.L., Hinzman, L.D., and Gieck, R.E. (1998) 'Annual water balance for three nested watersheds on the north slope of Alaska', in A.G. Lewkowicz and M. Allard (eds.), *7th International Conference on Permafrost*. Yellowknife: Collection Nordicana, Université Laval. pp. 669–74.

Lloyd, A.H., and Fastie, C.L. (2003) 'Recent changes in treeline forest distribution and structure in interior Alaska', *Ecoscience*, 10: 176–85.

Manseau, M., Huot, J., and Crête, M. (1996) 'Effects of summer grazing by caribou on composition and productivity of vegetation: Community and landscape level', *Journal of Ecology*, 84: 503–13.

Möller, I. (2000) *Pflanzensoziologische und vegetationsökologische Studien in Nordwestspitzbergen*. Hamburg: Mitt. Geogr. Gesellschaft in Hamburg 90.

Möller, I., Thannheiser, D., and Wüthrich, C. (1998) 'Eine pflanzensoziologische und vegetationsökologische Fallstudie in Westspitzbergen', *Geoökodynamik*, 19: 1–18.

Nadelhoffer, K.J., Giblin, A.E., Shaver, G.R., and Linkins, A.E. (1992) 'Microbial processes and plant nutrient availability in Arctic soils', in F.S. Chapin III, R.L. Jefferies, J.F. Reynolds, G.R. Shaver, and J. Svoboda (eds.), *Arctic Ecosystems in a Changing Climate: An Ecophysiological Perspective*. San Diego: Academic Press. pp. 281–300.

Naveh, Z. (1994) 'From biodiversity to ecodiversity: A holistic approach to landscape conservation and restoration', Proceedings VI. International Congress of Ecology. Manchester, 1994.

Ohmura, A. (1982) 'Regional water balance on the Arctic Tundra in summer', *Water Resources Research*, 18: 301–5.

Osterkamp, T.E., and Jorgenson, J.C. (2006) 'Warming of permafrost in the Arctic National Wildlife Refuge, Alaska', *Permafrost Periglacial Processes*, 17: 65–69.

Ozenda, P. (1993) 'Etage alpin et Toundra de montagne: parenté ou convergence?' *Fragm. Flor. Geobot., Suppl* 2: 457–71.

Pöhlmann, H. (1975) 'Ökologische Untersuchungen an Rentieren in Spitzbergen', *Verhandl. Gesellschaft für Ökologie*, 4: 89–92.

Polunin, N. (1951) 'The real Arctic: Suggestions for its delimitation, subdivision and characterization', *Journal of Ecology*, 39: 308–15.

Polunin, N. (1969) *Introduction to Plant Geography*. London: Longmans.

Porsild, A.E., and Cody, W.J. (1979) *Vascular Plants of Continental Northwest Territories, Canada*. Ottawa: National Museum of Canada.

Schuur, E., Crummer, K., Vogel, J., and Mack, M. (2007) 'Plant species composition and productivity following permafrost thaw and thermokarst in Alaskan tundra', *Ecosystems*, 10: 280–92.

Schwabe, A. (1997) 'Sigmachorology as a subject of phytosociological research: A review', *Phytocoenologia*, 27: 463–507.

Seldal, T., Andersen, K.-J., and Högstedt, G. (1994) 'Grazing-induced proteinase inhibitors: A possible cause from lemming population cycles', *Oikos*, 70: 3–11.

Senn, D.G. (1994) Segelnd auf hoher See – Zur Biologie der Albatrosse. Botmingen: R+R Verlag. 81 pp.

Sitch, S., McGuire, A.D., Kimball, J., Gedney, N., Gamon, J., Engstrom, R., Wolf, A., Zhuang, Q., Clein, J., and McDonald, K.C. (2007) 'Assessing the carbon balance of circumpolar arctic tundra using remote sensing and process modelling', *Ecological Applications*, 17: 213–34.

Stonehouse, B. (1971) *Animals of the Arctic*. London: Henry Holt.

Stow, D.A., and 23 authors. (2004) 'Remote sensing of vegetation and land-cover change in arctic tundra ecosystems', *Remote Sensing Environment*, 89: 281–308.

Tape, K., Sturm, M., and Racine, C. (2006) 'The evidence for shrub expansion in northern Alaska and the Pan-Arctic', *Global Change Biology*, 12: 686–702.

Thannheiser, D. (1991) 'Die landschaftsökologischen Verhältnisse auf dem westlichen kanadischen Arktis-Archipel', in G. Davis and A. Wieger (eds.), *Kanada: Gesellschaft, Landeskunde, Literatur*. Würzburg: Königshausen and Neumann. pp. 109–28.

Thannheiser, D. (1994) 'Vegetationsgeographisch-synsoziologische Untersuchungen am Liefdefjord (NW-Spitzbergen)', *Zeitschrift für Geomorphologie Suppl.*, 97: 205–14.

Thannheiser, D., Haacks, M., and Wüthrich, C. (2001) 'Vegetationsökologische Untersuchungen im Sassendalen (Spitzbergen)', *Geoöko*, 22: 117–39.

Thannheiser, D., and Möller, I. (1994) 'Frostbodenformen im inneren Woodfjord, NW-Spitzbergen', *Zeitschrift für Geomorphologie Suppl.*, 97: 195–204.

Thannheiser, D., and Möller, I. (2005) 'Case studies of synsociological vegetation units in Spitsbergen and the Canadian Arctic Archipelago', *Hoppea, Denkschrift Regensburger Botanische Gesellschaft*, 66: 435–45.

Thannheiser, D., Möller, I., and Wüthrich, C. (1998) 'Eine Fallstudie über die Vegetationsverhältnisse, den Kohlenstoffkreislauf und mögliche Auswirkungen klimatischer Veränderungen in Westspitzbergen', *Verhandlungen Gesellschaft für Ökologie*, 28: 475–84.

Thannheiser D., and Wüthrich C. (2006) 'Die Vegetation der terrestrischen Polarregionen und ihre Anpassungen', in Lozán et al. (eds.), *Warnsignale aus den Polarregionen. Wissenschaftliche Auswertungen*. Hamburg: Broschiert. pp. 102–6.

Tilbrook, P.B. (1970) 'The terrestrial environment and invertebrate fauna of the maritime Antarctic', in M.W. Holdgate (ed.), *Antarctic Ecology 2*. London: Academic Press. pp. 886–96.

Tüxen, R. (1973) 'Vorschlag zur Aufnahme von Gesellschaftskomplexen in potentiell natürlichen Vegetationseinheiten', *Acta Bot. Acad. Sci. Hungary*, 19: 379–84.

Tüxen, R. (1977) 'Bemerkungen zur historischen, begrifflichen und methodischen Grundlagen der Synsoziologie', in R. Tüxen (ed.), *Assoziationskomplexe (Sigmeten)*. Lehre: Cramer. S. 3–17.

Walter, H., and Breckle, S.-W. (1994) *Spezielle Ökologie der gemäßigten und arktischen Zonen Euro-Nordasiens. Ökologie der Erde 3*. Stuttgart: Ulmer Verlag.

Wegener, C., Hansen, M., and Jacobsen, L.B. (1992) *Vegetasjonsovervaking pa Svalbard*. Tromso: Norsk Polarinstitutt Meddelelser 122.

Wüthrich, C. (1989) *Die Bodenfauna in der arktischen Umwelt des Kongsfjords Spitzbergen. Versuch einer integrativen Betrachtung eines Ökosystems*, Materialien zur Physiogeographie 12. Basel: Geographicshes Institut.

Wüthrich, C., Möller, I., and Thannheiser, D. (1998) 'Soil carbon losses due to increased cloudiness in a high Arctic tundra watershed (Spitsbergen)', in A.G. Lewkowicz and M. Allard (eds.), *7th International Conference on Permafrost*. Yellowknife: Collection Nordicana, Université Laval. pp. 1165–72.

Wüthrich, C., Möller, I., and Thannheiser, D. (1999) 'CO_2-fluxes in different plant communities of a high Arctic tundra watershed (West Spitsbergen)', *Journal of Vegetation Science*, 10: 413–20.

Wüthrich, C., and Schaub, D. (1997) 'Reaktionspotential unterschiedlicher Moortypen für Änderungen von Licht und Bodenwasserhaushalt', *Bodenkundl.Ges. Schweiz Bull* 21: 41–48.

14

Forest Ecosystem Structure and Disturbance Dynamics across the Circumboreal Forest

Daniel Kneeshaw, Yves Bergeron, and Timo Kuuluvainen

14.1 INTRODUCTION

The northern circumpolar forest belt covers an area of 12 million km² (Burton et al., 2003), which is about one-quarter of the world's forested area. In countries in the circumboreal zone, the boreal forest accounts for a vast proportion of the total land area, such as in Finland where about 70% of the country is covered by boreal forest. Further east, the Russian boreal forest contains more than half of all the conifers found on the planet, 20% of the forested area and more than 10% of the world's forest biomass (Engelmark, 1999). Altogether the circumpolar boreal forests and peatlands are estimated to contain more than five times the amount of carbon found in the world's temperate forests and almost double the carbon in tropical forests (Goodale et al., 2002; Kasischke, 2000).

Unlike temperate and tropical forests, tree species richness is quite low in the boreal with only nine dominant forest-forming species in North America, five in Fennoscandia, and 12 across Eurasia (Table 14.1). There is, however, a larger number of minor tree species (e.g., 24 indigenous species in Finland). Although species differ, the dominant genera remain the same across this circumpolar region: *Abies, Larix, Picea, Pinus,*

Populus, and *Betula.* Differences in species composition are due to postglacial migration history and variations in climate, topography, soil texture, and disturbance regimes.

Fire is an important part of boreal forest ecology (Zackrisson, 1977; Payette, 1992). However, it is currently understood that considerable variation in fire regimes exists within the boreal forest and that boreal forests with apparently no fire influence since the last glaciation can be found (Zackrisson et al., 1995; Pitkänen et al., 2003). In areas with long fire return intervals, nonfire disturbances (e.g., insect outbreaks, fungi, windthrow, senescence) become important (Kuuluvainen et al., 1998; McCarthy and Weetman, 2006). With short return intervals of stand-replacing fire, the forest is dominated by pyrogeneous tree species such as jack pine (*Pinus banksiana*), black spruce (*Picea mariana*), or trembling aspen (*Populus tremuloides*) in boreal North America. These species all have adaptations to crown fire, such as serotinous cones or the ability to resprout from root suckers. In Eurasia, species such as Scots pine (*Pinus sylvestris*) and larches (*Larix* spp.) have thick bark, which helps them resist low-intensity surface fires (Agee, 1998, Shorohova et al. 2009). Some tree species that are shade-intolerant (e.g., *Populus* and *Betula* spp.) benefit

Table 14.1 Some characteristics of major tree species dominating the canopy in the circumboreal forest

Continent	Species		Regeneration strategy (pyrogeny)	Shade tolerance	Average longevity (years)	Main distribution/ typical site
	Scientific name	Common name				
North America	Pinus banksiana	Jack pine	Serotinous cones	Very shade-intolerant	100–150	Nova Scotia to Alberta (exposed mineral soil)
	Picea glauca	White spruce	Wind dispersal	Mid-tolerant	200–300	Maritime provinces to Alaska (various soil types)
	Picea mariana	Black spruce	Semi-serotinous cones and layering	Shade-tolerant	200	Newfoundland to Alaska (various soil types)
	Picea rubens	Red spruce	Wind dispersal	shade to very shade-tolerant	200–300	Eastern North America (southern boreal) (various soil types)
	Abies balsamea	Balsam fir	Wind dispersal (limited layering)	Very shade-tolerant	60–100	Newfoundland to Alberta (more dominant in east and in southern boreal) (various soil types)
	Larix laricina	Larch or tamarack	Wind dispersal and basal sprouts	Very shade-intolerant	150–180	Newfoundland to Alaska (various soil types)
	Populus tremuloides	Trembling aspen	Root suckers and wind dispersal	Very intolerant	80–100	Newfoundland to Alaska, less abundant in far north (various soil types)
	Populus balsamifera	Balsam poplar	Root suckers and wind dispersal	Intolerant	60–200	Newfoundland to Alaska (various soil types)
	Betula papyrifera	Paper birch	Wind dispersal	Intolerant	70 (rarely past 140 yrs)	Newfoundland to Alaska (well-drained, sandy loams on cool, moist sites)
Eurasia	Picea abies	Norway spruce	Wind dispersal	Shade-tolerant	200–250	Scandinavia and Russia, west of Urals (fertile to medium fertile, well-drained sites)
	Picea abies subsp. Obovata*	Siberian spruce	Wind dispersal, limited layering	Shade-tolerant	Up to 300	Northern Scandinavia to Northern China (various soil types)

Continued

Table 14.1 Cont'd

Continent	Species		Regeneration strategy (pyrogeny)	Shade tolerance	Average longevity (years)	Main distribution/ typical site
	Scientific name	Common name				
	Pinus sibirica	Siberian cedar, Siberian stone pine	Wind and bird dispersal	Moderately shade-tolerant	Up to 400	Siberia to Northern China (800–2500-m altitude) (moist sites)
	Pinus pumila	Dwarf Siberian pine, also known as Japanese stone pine	Animal and bird dispersal	Moderately shade-tolerant	150	Eastern Russia to Japan (1000–2300 m alt) (rocky acidic soils)
	Pinus sylvestris	Scots pine	Wind dispersal/ thick, fire-resistant bark	Shade-intolerant	300–400	Europe, Russia, and Southwest Asia (sandy well-drained sites, pine bogs)
	Larix Sukaczewii	Russian larch	Wind dispersal	Shade-intolerant	250	Central European Russia (various soil types)
	Larix gmelinii	Dahurian larch	Wind dispersal/ thick bark	Shade-intolerant	300	Central and eastern Siberia to Korea (various soil types)
	Larix sibirica, Larix russica	Siberian larch	Wind dispersal/ thick bark	Shade-intolerant	300	Eastern Urals to North China (lowland, great variety of soil)
	Abies sibirica	Siberian fir	Wind dispersal	Shade-tolerant	100	East of Ural to North China (various soil types)
	Betula pubescens	Arctic white birch	Wind dispersal and root suckering	Shade-intolerant	60–80	Scandinavia, European Russia, Western Siberia (especially peatlands and paludified sites)
	Betula pendula	Silver birch; European white birch	Wind dispersal and root suckering	Shade-intolerant	Up to 150	Scandinavia, European Russia, western Siberia (fertile to medium-fertile sites)
	Populus tremula	Eurasian aspen	Wind dispersal and root suckering	Shade-intolerant	Up to 200	All of boreal Eurasia (fertile, well-drained sites)

*In Western literature this is considered to be a subspecies, but in Russia it is considered to be an independent species.

from large areas with no overhead competition and mineral seedbeds on which to establish.

The boreal forest has a relatively low mammal and vascular plant diversity although the diversity of algae, moss, lichen, mushroom, and arthropod species is considered to be high (De Grandpré et al., 2003). Diversity must therefore be considered across taxa and various scales, both temporal and spatial. For example, due to the most recent glaciation (c. 11,000 years B.P.) most of the boreal tree species only arrived 2,000 to 9,000 years ago, making it one of the youngest forest biomes in the world. Glaciation also had different effects on different tree species. For example in boreal North-America white spruce found refugia in northern locations (Anderson et al., 2006), while other species were forced south and thus had long migration routes before recolonizing boreal habitats (Davis, 1981). Barriers created between different populations also led to speciation, as was the case for red spruce, which was probably originally a population of black spruce isolated by glaciers (Perron et al., 2000). Red and black spruce still readily hybridize, suggesting that the glaciation may not have lasted long enough to cause true speciation. In Europe the east-west orientation of the Alps prevented the southward migration of plants during glaciations and this led to the extinction of many plant and animal species (Stehlik, 2003).

At present the boreal zone contains some of the last and largest extents of nonexploited forests in the world outside of tropical forests in the Amazon (Aksenov et al., 2002). However, both regions share the same problem of increasing human pressure through logging and deforestation, and this is changing the structure and dynamics of these forests (Mery et al., 2010).

14.1.1 The boreal climate

The boreal zone is characterized by a cold climate in which during six to eight months of the year there are average minimum temperatures below 0°C and only three to five months have average temperatures greater than 10°C (Woodward, 1995). Daylength (with low sun angles) is quite variable, ranging from 15–24 hours in the summer to almost complete darkness in the winter. The boreal forest can thus be differentiated from other forest types by its climatic setting. In addition to the short growing season, the high seasonality, and the long summer days this region can be characterised by the occurrences of permafrost and low biological productivity (Wein and MacLean, 1983). For example, it has been estimated that net primary productivity in the boreal forest

ranges from 1.2 to 4.3 tons per hectare per year (t ha^{-1} yr^{-1}) whereas in temperate deciduous forests net primary production can be more than double the values (0.8 to 9.8 t ha^{-1} yr^{-1}) found in the boreal (Landsberg and Gower, 1997). Despite the features that are general to the boreal climate, it is important to realize that considerable variability exists, from semi-maritime climates with cool summers and relatively mild winters (e.g., Fennoscandia, Russian Far East, eastern Quebec) to extremely continental ones with cold winters and short, hot summers (e.g., central Siberia and central Canada).

14.1.2 Disturbances

Throughout much of the western North American boreal forest, crown fires have long been considered to be the primary disturbance type (Bergeron, 1991; Heinselman, 1981; Johnson, 1992; Payette, 1992). In continental parts of the boreal forest with dry summer climates, such as in parts of Alberta or Saskatchewan, crown-fire cycles may be as short as 50 years (Heinselman, 1981; Hirsch, 1991) with large crown fires burning more than 100,000 hectares of forest. However, other disturbances such as surface fires, insect outbreaks, windthrow, and gap dynamics may also play important roles (Bonan and Shugart, 1989; Van der Maarel, 1993). In fact where crown-fire cycles are long, patch or gap disturbances drive forest dynamics.

It has been suggested that the role of fire may have been overemphasized in some boreal ecosystems (Engelmark, 1999; Kuuluvainen, 2002). In Scandinavia, China, and Russia there has been recognition of the importance of gap dynamics in boreal forests that are little affected by fire (Ban et al., 1998; Drobyshev, 1999; Kuuluvainen, 1994; Leemans, 1991; Liu and Hytteborn, 1991) as well as in pine forests affected by low-severity surface fires (Rouvinen et al., 2002; Shorohova et al., 2009). In North America this realization has been longer in coming due to the pervasive nature of stand-replacing fire and fire research, and the fact that only a handful of papers have been published in recent years (Coates and Burton, 1997; Cumming et al., 2000; Kneeshaw and Bergeron, 1998, 1999; McCarthy, 2001).

The main objectives of this chapter are thus to provide an understanding of the role and relative importance of different disturbances in boreal forests, to link the primary disturbances to species composition and climate, and to show how these interactions may have changed through time. In doing so, we will compare and contrast boreal forest dynamics amongst the different regions of the circumpolar boreal forest. The implications arising from the research we review are important for those who have advocated clear cutting as a surrogate for fire (which, at least in one way imitates fire by opening large tracts of forest), and even-age forest management as a uniform recipe for forest management of the boreal forest.

14.2 DISTURBANCE REGIMES AND THEIR EFFECTS ON VEGETATION

14.2.1 Fire as a disturbance factor in the boreal forest

Forest fires differ in their severity, size, and return time. Indeed, fires are not all equal in their characteristics and impacts on forests. Most fires burn only a small area and have little impact at a regional scale, although at a local scale their effects may be important. Fire regimes (characterized in terms of fire frequency, severity, and area burned) in the boreal forest differ greatly between continents and across continents.

Climate affects forest fires directly by controlling lightning, fuel moisture levels, and wind regimes. Climatic effects can be observed from more maritime, coastal areas where the influence of fire is reduced compared to interior boreal forests where continental climate effects lead to drier summer conditions and larger and more severe fire events. However, the impact of climate is more complicated, because climate also affects the composition of the vegetation (i.e., the fuel that burns in these fires). Changes in fire regimes also occur through time as climate changes.

North America: The case of crown fire

With the exception of some pine stands for which nonlethal surface fires are locally reported (Smirnova et al., 2008; Bergeron and Brisson, 1990) fire regimes in the North American boreal forest are mainly characterized by stand-replacing fires (Johnson, 1992; Payette, 1992). This contrasts with Eurasian boreal forests where nonlethal surface fires are common (see below). Moreover, the landscape in Canadian boreal forests is mainly characterized by a small proportion of very large lightning-ignited fires. Fires greater than 200 ha represent 97% of the area burned in Canada (Stocks et al., 2002). These large fires control the fire interval (the time interval between fires burning the same point) and, inversely, the fire cycle (the time needed to burn an area equivalent to the total area being studied). This parameter is of major importance as it controls the proportion of the landscape, which is dominated by early successional postfire species (when fire cycles are short) and late successional species more associated with gap disturbances (when fire cycles are long).

However, natural fire frequency is difficult to estimate as it may vary greatly depending on the spatial and temporal extents considered. Table 14.2 presents current (over the last 50 years) and historical (last 300 years) fire frequency for Canada's boreal ecozones. The current burn rate was estimated from the Canadian government's large-fire database (Stocks et al., 2002); this includes all fires 200 ha and larger, which as mentioned earlier represent over 97% of all area burned. Historical burn rates were determined from a literature review (Bergeron et al., 2004a) using available forest fire-history studies in the North American boreal forest (Figure 14.1). To estimate the historic burn rates, the average age of the forest (time since fire) or, if not available, the fire cycle before large clear-cutting activities began were used. The average age of the forest was preferred to the historic fire cycle because it integrates climatically induced changes in fire frequency over a long period and because it is easier to evaluate than a specific fire cycle (Bergeron et al., 2004a). The inverse of the average age (or fire cycle) was used as an estimator of the annual historic burn rate. An average for all studies belonging to a specific ecozone is presented in Table 14.2.

Historical data show a clear distinction between western Canada, which is characterized by short fire cycles, and eastern Canada, where fire cycles are longer. Differences are mainly due to a drier climate in the west, which leads to shorter intervals between fires and thus more aspen and spruce and less fir. Fire frequency also tends to decrease as one moves toward alpine or taiga environments, probably because of a decrease in fuel availability (Payette et al., 1989). Current fire frequencies show a similar trend but values are significantly lower for all ecozones, suggesting a common change affecting the entire North American boreal forest. In Quebec, the decrease in fire frequency has been related to a reduction in the frequency of drought events since the end of the Little Ice Age (Bergeron and Archambault, 1993; Lauzon et al., 2007). It is hypothesized that the warming that started at the end of the Little Ice Age is associated with an important change in the circulation of global air masses that may have affected boreal forest fire regimes (Girardin et al., 2006). This decrease has been exacerbated in the last part of the twentieth century by an increase in effective fire suppression (Lefort et al., 2003).

Table 14.2 Current and historical fire frequencies (percentage of a region burning expressed per year for Canadian boreal ecozones). Its inverse (the fire cycle) is indicated in parentheses

Ecozones	Historical %/year (fire cycle, yrs)	Current %/year, (fire cycle, yrs)	% Changes
Montane cordillera	0.99 (101)	0.058 (1724)	−95
Boreal cordillera	Unknown	0.392 (255)	
Taiga cordillera	Unknown	0.202 (495)	
Taiga plain	Unknown	0.702 (142)	
Boreal plain	1.48 (68)	0.418 (239)	−71
Taiga shield west	0.85 (118)	0.763 (131)	−10
Boreal shield west	1.92 (52)	0.761 (131)	−60
Hudson plains	Unknown	0.123 (813)	
Boreal shield east	0.77 (131)	0.145 (690)	−81
Taiga shield east	0.6 (166)	0.241 (415)	−60

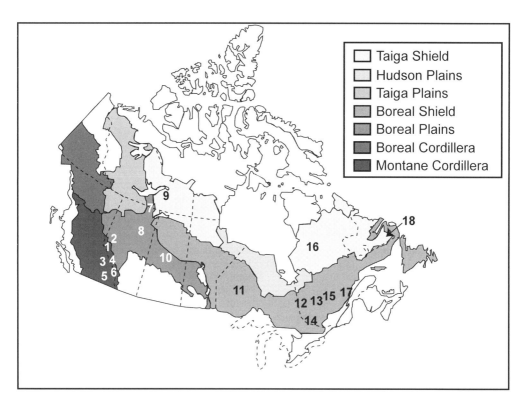

Figure 14.1   Location of the 18 study areas throughout ecozones of the Canadian boreal forests. Numbers refer to the following studies: 1. Tande (1979); 2. van Wagner (1978); 3. Johnson et al. (1990); 4. Johnson and Wowchuk (1993); 5. Masters (1990); 6. Johnson and Larsen (1991); 7. Larsen (1997); 8. Cumming (1997); 9. Johnson (1979); 10. Weir et al. (2000); 11. Suffling et al. (1982); 12. Lefort et al. (2003); 13. Bergeron et al. 2004b; 14. Cwynar (1977); 15. Bergeron et al (2001); 16. Payette et al. (1989); 17. Lesieur et al. (2002); 18. Foster (1983)

Eurasia: Dominance of surface fires

In general, in boreal Eurasia low- to medium-severity fires, such as surface fires, are the most common; although large-scale stand-replacing fires exist they are generally less prevalent (Wein and MacLean, 1983; Gromtsev, 2002; Lampainen et al., 2004, Shorohova et al., 2009). This applies especially not only to boreal Fennoscandia and western Russia but also to a large extent to Siberia and the Russian Far East (Sannikov and Goldammer, 1996; Gromtsev, 2002; Kondrashov, 2004. It has, for example, been reported that in European Russia 76–86% of fires are surface fires while crown fires account for 16–24% of the fires (Melekhov, 1947, cited in Gromtsev, 2002).

The general view that low-intensity surface fires are more common in Eurasia when compared to North America, where crown fires are common, has also been verified using infrared remotely sensed measurements (Moderate Resolution Imaging Spectroradiometer, MODIS) (Wooster and Zhang, 2004). The prevalence of low-intensity nonstand-replacing fires in boreal Eurasia is evidently related to climate, landscape characteristics, and the fire-resistant character of important tree species like pines (e.g., *Pinus sylvestris, P. sibirica*) and larches (e.g., *Larix gmelinii, L. sibirica*) (Helmisaari and Nikolov, 1989; Babintseva and Titova, 1996). Moreover, serotinity does not occur as a life-history strategy in Eurasian conifers, which suggests that nonlethal surface fire regimes have dominated in boreal Eurasia over evolutionary timescales.

Because boreal Eurasia covers a vast area, there are differences in fire regimes both at coarser geographic and finer spatial scales. In Siberia, large stand-replacing crown fires are relatively common and associated with extended summer drought periods in the continental climate. In more humid areas with longer fire intervals, spruce (in Fennoscandia) and fir (in Russia east of the Ural Mountains) may dominate landscapes. However, since they have more fine fuel in the form of foliage they are more prone to stand-replacing fires during extreme drought periods than open pine or larch forests in which trees have less foliage and crowns are located high above the ground level (Ban et al., 1998; Engelmark, 1999).

In Fennoscandia and northwestern Russia (western Eurasia) natural fire regimes are variable also due to the diversity and heterogeneity of landscape conditions, which are characterized by mosaics of forests and peatlands and the semi-maritime climate (Gromtsev, 2002; Wallenius et al., 2005). Even the largest burnt areas are therefore generally relatively small due to climate and landscapes fragmented by natural fire breaks such as peat bogs and water bodies (Wallenius et al., 2004). However, fires larger than 100,000 ha have been documented before settlement in northern Sweden (Niklasson and Granström, 2000). In Siberia large fires are more common than in western Eurasia, where they can cover several hundred thousand hectares.

In Fennoscandia the fire frequency is generally higher in dry upland sites dominated by *Pinus sylvestris* than in moister lowland sites dominated by *Picea abies* (e.g., Zackrisson, 1977; Wallenius et al., 2004). There is also a general south-to-north gradient of decreasing fire frequency and area burned, due to increasing humidity and decreasing duration of the fire season (Granström, 1993; Sannikov and Goldammer, 1996; Larjavaara et al., 2004, 2005).

Recent studies indicate that natural fire return intervals in Fennoscandia are longer than hitherto assumed (Kuuluvainen, 2009; Wallenius, et al., 2010). For example, in eastern middle-boreal Fennoscandia, paleoecological studies show that before 1500 A.D., when human impact was low, upland dry *Pinus sylvestris*-dominated forests had a surface fire interval on the order of 150–250 years (Pitkänen, 1999, Pitkänen et al., 2002). However, on moister Norway spruce-dominated lowland landscapes the natural fire return interval was even longer, up to several hundred years and even over a thousand years (Pitkänen et al., 2003; Wallenius, 2002; Wallenius et al., 2005). Some sites having these environmental characteristics have not burned since the last glaciation over 10,000 years before (Pitkänen et al., 2003).

In Siberia, the climate becomes increasingly continental and natural fire return intervals in general decrease compared to western Russia (Sannikov and Goldammer, 1996). This is due to extended drought periods and the high number of lightning-ignited fires (Sannikov and Goldammer, 1996). In dry pine and larch forests the natural fire cycle (mostly surface fires) is on the order of 60–70 years, but it is longer (100–200 years) in forests characterized by *Abies sibirica* and *Picea obovata*: species that characterize the *dark taiga* as it is known in Russian terminology (Walter and Breckle, 1989). However, it has been suggested that some areas of the northern dark taiga experience fire cycles exceeding 400 years (Schulze et al., 2005). However in the Russian Far East, the climate again becomes more maritime and fire cycles become longer and resemble those found in Fennoscandia (Kondratshov, 2004).

During the past few hundred years fire regimes in many parts of boreal Eurasia have been strongly impacted by human activity that have increased ignitions (Sannikov and Goldammer, 1996; Niklasson and Granström, 2000; Wallenius et al., 2004), and this continues to be the case in European Russia and Siberia. In contrast, the impact of humans during recent decades in countries such as Sweden and Finland has been an exclusion of fires since suppression policies have been efficiently implemented. As a consequence a large number of fire-dependent species have declined or become threatened (Wikars, 2004).

In many parts of Eurasia the long-term human impact on fires makes it difficult to define natural fire regimes. For example, studies based on fire scars have shown that in middle-boreal Fennoscandia, *Pinus sylvestris* forests commonly had a fire interval of 30–60 years during the nineteenth century (Pitkänen and Huttunen, 1999; Niklasson and Granström, 2000), at least three times shorter than the existing estimates for fire intervals (Pitkänen 1999, 2002). However, even during this period moist Norway spruce-dominated landscapes burned less frequently (Wallenius et al., 2005; Wallenius, 2002; Pitkänen et al., 2003).

14.2.2. Nonfire disturbances

In regions where fire cycles are long, nonfire gap disturbances (which are characteristic of old-growth forests) are important in controlling forest dynamics (Kuuluvainen, 1994; Kuuluvainen et al., 1998; Gromtsev, 2002; Kneeshaw and Gauthier, 2003). There is thus a general pattern of an increasing proportion of old forests as one moves from continental areas, such as central North America and central Siberia, to maritime areas, such as Alaska, Eastern Canada, and Fennoscandia. Such a pattern also occurs with latitude as shorter growing and shorter fire seasons in the north also lead to greater proportions of old-growth forest compared to in the south. Although old-growth forest landscapes may be fine-scale mosaics of different forest developmental stages, overall a greater continuous forest cover and fewer large openings are observed (Kuuluvainen et al., 1998; McCarthy and Weetman, 2006). It can be concluded that before significant human impact old-growth forests were a dominant feature of much of the boreal forest, especially in Eurasia where fire cycles are long and tree species longevity is greater than in North America (Table 14.1).

Although forests that experience small-scale disturbances and openings are ubiquitous, nonfire disturbances that lead to even-aged conditions in relatively large openings can exist due either to severe drought, insect outbreaks, or large blow-downs (Syrjänen et al., 1994, Aakala et al., 2011). These events can be generalized to specific forest conditions, with some events such as spruce bud-worm outbreaks being associated with eastern North America and others, such as windthrow,

having important local effects but not being as easy to associate with a geographic region (Figure 14.2).

Nonfire disturbances that open up large areas

The first type of disturbance we consider in this section is insect outbreaks. In the boreal coniferous forests of western Eurasia there are no records of large-scale outbreaks of defoliators, comparable to those caused by the spruce budworm (SBW) in North America. Needle-defoliating insects such as pine sawflies (*Diprion pini, Neodiprion sertifer*) can cause significant damage locally, but no large-scale forest outbreaks have been reported. However, in the Fennoscandian monodominant mountain birch (*Betula pubescens* subsp. *czerepanovii*) forests, the autumnal moth (*Epirrita autumnata*) has occasionally triggered large-scale dieback in forests forming the treeline. Also in Siberia large-scale defoliation and forest dieback can occur, especially in southern boreal *Larix* forests due to the Siberian moth (*Dendrolimus spp.*).

Occasionally, bark beetles (mainly *Ips typographus*) can, in association with large-scale tree mortality, for example, due to storms or extreme drought events, develop population densities high enough to attack and kill healthy trees across large areas. For example, there was an outbreak in Sweden and Norway in the 1970s, following extensive tree blowdowns in winter storms, in which millions of spruce trees died. A similar event associated with extreme drought has been documented in the Archangelsk region, North-Western Russia (Aakala et al., 2011). Other insects are important disturbance agents, but at scales that lead to small openings and single tree mortality or growth losses. Insects may also simply predispose trees to other agents of mortality.

The situation is different in North America, where large-scale outbreaks of species such as the Eastern spruce budworm (*Choristoneura fumiferana*) and the hemlock looper (*Lambdina fiscellaria*) occur in eastern forests and outbreaks of species such as the mountain pine beetle (*Dendroctonus ponderosae*) occur primarily in forests found in the Rockies. An evaluation of the relative importance of different disturbance types in the Canadian boreal forest shows that the average area affected by such outbreaks greatly exceeded the area affected by fire (Kneeshaw, 2001). The impacts of such disturbances are, however, very different. Fires kill most of the standing trees immediately while disturbances such as insect outbreaks may take years to kill trees, and some trees always survive (Kneeshaw and Bergeron, 1999; MacLean, 1980).

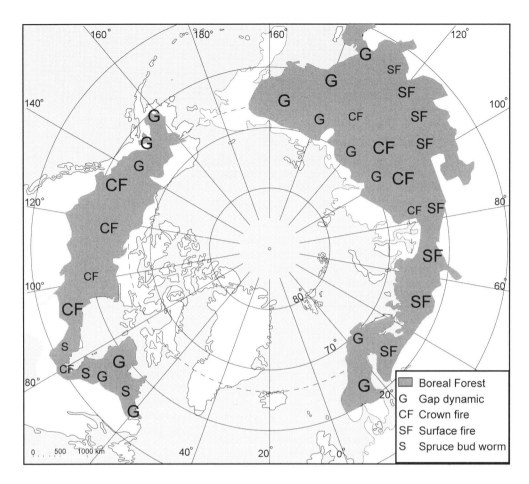

Figure 14.2   Illustration of the gradient of the main disturbances across the circumboreal forest. The main disturbances considered are surface fires, crown fires, and spruce budworm outbreaks

Furthermore, insects are often host-specific, attacking only one species or a small group of species. For example, the Eastern spruce budworm attacks only balsam fir and the spruces; thus outbreaks do not directly kill hardwood or cedar-dominated stands (Bouchard et al., 2005). At a landscape scale, mortality and thus age-class distributions are much patchier than those following fire.

Natural disturbances affect each other's probability of occurrence. For example there is a direct and obvious relationship between fire and insects such as the SBW so that in areas where fire is a frequent and dominant disturbance agent in the landscape, the primary host species (balsam fir) of the SBW is reduced since it is poorly adapted to fire. Temporally, outbreaks have been found to vary, perhaps in direct relationship to the fire cycle

(Blais, 1983). For example, during the Little Ice Age, which finished around 1870 in the eastern North American boreal forest, the climate was generally colder and drier, leading to an increase and a greater occurrence of fires. SBW outbreaks were thus less important in continental areas with shorter fire cycles and more important in maritime regions of eastern Canada where a moister climate maintained an abundant supply of the host species, balsam fir.

As with fire, there is also a north-south gradient such that balsam fir decreases in abundance as one moves from Maine and New Brunswick into northeastern Quebec and Labrador. In these northern forested areas, the climate is less favorable to the development of the SBW, and balsam fir, the primary host of the SBW, is a smaller component of the forest. Although species such as black spruce (a secondary host) are affected, the mortality caused by the SBW is much less than that for fir. Even at a more regional scale, mortality caused by SBW outbreaks is also affected by stand composition in the surrounding landscape. Fir stands in mixed-wood regions undergo a smaller degree of mortality than in landscapes that are primarily dominated by pure fir forests (Su et al., 1996; Bergeron et al., 1995).

Other insects have also had or continue to have an important impact on the boreal forest. In some cases they cause direct mortality as in the case of the hemlock looper but in other cases mortality is only minor and important effects are often reported only as volume losses (e.g., the jack pine budworm and the forest tent caterpillar). Each of these agents thus acts differently on the forest. A contrast can thus be made between insects that primarily cause volume loss (i.e., that slow tree growth) and those that cause mortality. Insects that do not cause mortality can accelerate successional processes if they cause foliage thinning in early successional stands. In such cases, late-successional shade tolerant species that are often slower growing may be able to more quickly attain the canopy. In the case of insects that cause tree mortality, compositional changes such as a shift to early or to late successional species may occur (Kneeshaw and Bergeron, 1998; Bouchard et al., 2006).

The second type of nonfire disturbance we consider is wind disturbance. A number of generalities about the effects of wind on boreal forest ecosystems are true on both continents. The effects of wind are most pronounced in species with long crowns (increasing the drag force of wind) and shallow root systems (e.g., balsam fir and Norway spruce) growing on shallow soils (such as shallow tills covering bedrock or on sites with a high water table). Topographic variables are also important as windthrow seems to occur more frequently on strong slopes or when topography leads to wind-funnelling. Topographic factors, however, will have a smaller influence when winds have a downward component as in 'derechos' (widespread, long-lived windstorms associated with a band of rapidly moving showers or thunderstorms) and downbursts (Frelich, 2002). In western Eurasia most wind damage is caused by autumn and winter storm fronts, originating in the Atlantic Ocean These storms can cause large-scale damage in boreal Fennoscandia, but tree mortality tends to be partial and patchy. Large-scale stand-replacing wind storm damage has also been reported in the Ural Mountains (Syrjänen et al., 1994). In North America, strong wind damage occurs (or at least has been studied) in coastal regions in the east and west (i.e., nonboreal forest in British Columbia and Alaska).

Boreal regions most affected by wind damage are found in coastal areas and where strong thunderstorms are frequent. Strong winds can also be expected where the polar jet stream and the subtropical jet stream meet in summer months (Frelich, 2002).

Based on available studies, the areas most affected by windthrow are around the Great Lakes in North America, in coastal areas of Fennoscandia, and in the central part of Russia. Return intervals in the studied forests are quite long (1,000–2,000 years: Frelich and Lorimer, 1991; or 450 to 10,500 years for heavy windthrows: Schulte and Mladenoff, 2005). Schulze et al. (2005) suggest that return intervals of winds causing mortality can be on the order of about 150 years for parts of Russia's dark taiga and Gromtsev (2002) suggests return intervals of 150–300 years. The size of windthrows is generally small, ranging from small gaps to openings of a few hectares. It has also been suggested that the size and severity of windstorms may be increasing due to climate change, although such an interpretation is tentative as we still do not have a good understanding of geographic wind disturbance patterns (Frelich, 2002).

Finally, we turn our attention to small-scale local level disturbances such as senescence, root rot, partial windthrow, and insect mortality. As well as causing relatively large openings, disturbances such as windthrow and insect outbreaks can also lead to partial mortality in a stand. In such circumstances, canopy gaps are small, corresponding to the growing space of single to multiple canopy trees. Some studies have suggested that insects and wind are responsible for a large proportion of the mortality observed in areas characterized by a continuous canopy interspersed with gaps (de Römer et al., 2007; Kneeshaw and Bergeron, 1998). However, such individual tree mortality often has multiple causes (Rouvinen et al., 2002; Reyes and Kneeshaw, 2008). For example, windthrown trees are often those close to their biological age and weakened by insects and wood-rotting fungi (Kuuluvainen, 1994; McCarthy, 2001).

Relatively little work has been conducted on the extent of mortality caused by fungi in the North American boreal forest. However, we suggest that in areas where neither fire nor insect outbreaks are important (e.g., Labrador, Canada) wind and fungal attacks would be important agents of disturbance (Lewis and Lindgren, 2000). Boreal trees are primarily affected by *Armillaria* spp. (primarily *Armillaria ostoyae*) (Mallett, 1992) and then secondarily by *Inonotus tomentosus* (Whitney, 2000). These two fungi cause both growth loss and tree mortality. Tree mortality is often caused in combination with other agents such as winds, which topple trees that have been structurally weakened by such fungi at the base of the tree or in their root system (Lännenpää et al., 2008). Death is usually greater in older trees (*I. tomentosus* generally kills trees greater than 50 years in age). There are also other decay fungi, primarily root and butt heartwood decayers, but these require damaged or dead roots with exposed heartwood to cause infection. However, since these fungi also cause structural failure they are important in increasing predisposition of trees to windthrow damage (Garry Warren, Canadian Forest Service, personal communication, 2005).

In Eurasia, common surface fires interact in a complex manner with disturbance factors such as bark beetles (e.g., *Ips typographus* and *Tomicus piniperda*) and wood-rotting fungi (e.g., *Phellinus pini, P. chrysoloma, Coniophora* spp.) (Rouvinen et al., 2002; Lännenpää et al., 2008) in contributing to gap formation. These agents kill individual trees or patches of trees, weakened by old age, surface fire, or wood-rotting fungi, so that gaps are formed due to wind or snow loads (Kuuluvainen, 1994). Occasionally after large-scale windthrow, bark beetles reproduce massively and attack and kill healthy trees at large scales.

Gap dynamics are best characterized by relatively small openings (Pham et al., 2004; Bergeron and Kneeshaw, 1998; Kuuluvainen, 1994) but often by a relatively high fraction of the forest being open. For example, many studies have found that the majority of gaps are smaller than 100 m^2 but that 40–50% or more of the forest is open (McCarthy, 2001; de Römer et al., 2007). Causes are often described as being due to senescence, which suggests that such mortality is most important in old stands (although see Hill et al., 2005) that have not been disturbed by major disturbances for periods exceeding the average life span of the dominant trees (Kneeshaw and Gauthier, 2003). These areas are found in northern sites, and in maritime areas which are little affected by fire or insect outbreaks. Nonetheless, large-scale blowdown may be important on some sites though it is usually a local phenomenon. Although Schulze et al. (2005) suggest that forests avoiding any type of disturbance are rare or do not exist they also state that many species in boreal forests are adapted to disturbance-free periods.

It is important to consider that nonfire disturbances play a major role in shaping large parts of the boreal forest. Stand-replacing disturbances, although important in some regions of the boreal forest, especially in North America, are not the only factor driving the dynamics of boreal forests. Many forests are naturally old and punctuated by smaller partial disturbances, creating complex heterogeneity in stand structures (Rouvinen et al., 2002; McCarthy and Weetman, 2006).

14.2.4 *Effects of disturbance regimes on ecological communities*

Where crown-fire cycles are short the effects of other disturbances will be minor and the boreal landscape will be controlled by the recurrence of large fire events. Under such cycles entire stands of trees are killed before trees attain their longevity (Dix and Swan, 1971), and the North American boreal landscape is often thought of as large patches of trees recovering from burns. The long-term cumulative effect of short fire cycles has been to limit not only successional processes but also the presence of trees with late successional characteristics such as shade tolerance and recruitment from seedling banks (Kenkel et al., 1997). Such species in fact dominate in regions of the boreal forest where fire cycles are long. It has, however, been argued that many dominant boreal trees have what can be considered pioneer characteristics, with frequent fires favoring species that are able to regenerate from aerial seed banks (stored in serotinous cones), root or basal sprouting, long-distance dispersal of small seeds, and from underground seed banks (e.g., pin cherry or ephemeral herbs). In areas where surface fires are common, thick fire-resistant bark and a high branching habit (since low branches act as fuel ladders conducting fire to the crown) are important characteristics. Many of the pine species (e.g., *Pinus sylvstris, P. strobus, P. resinosa*) as well as the larches (e.g., *Larix gmelinii*) have such characteristics. On both continents many boreal tree species (e.g., *Picea* spp, *Populus* spp. *Salix* spp. and *Betula* spp.) that regenerate from seeds require a mineral seed bed for successful establishment (Greene et al., 1999). Such a mineral seed bed is often found in severely burnt areas where both surface fires and crown fires consume accumulated organic litter (Johnson, 1992).

Rowe (1983) used Noble and Slatyer's (1977, 1980) vital attributes theory to demonstrate the effectiveness of many regeneration strategies for boreal forest tree species in relation to recurrent disturbances such as fire. In this model, the important elements used for predicting forest composition are determined from the vital attributes of individual species. These tree attributes are longevity, age at reproduction, and method of reproduction. Trees will thus disappear from an area if they are not able to regenerate given the current conditions or when a local population is eliminated before it attains a reproduction age. Thus trees dependent on wind-dispersed seed must have an unburnt seed source in close proximity if they are to regenerate. Galipeau et al. (1997) have also demonstrated that colonization of shade-tolerant balsam fir and white spruce can occur over time as individuals regenerate in an ever-expanding concentric wave from the initial seed source. In other terms once individuals that have established surrounding a fire-preserved zone have attained an age at which they can begin producing seeds, recruitment will occur at an ever-increasing distance from the original source (Galipeau et al., 1997). In large burns, with few remnant islands of surviving trees, colonization by such species can require excessively long time periods and may not occur before the next fire event. This may thus be one of the important factors ensuring the mono-specific nature of many boreal stands (Carleton and Maycock, 1978).

Species that are able to regenerate in situ following canopy fire (i.e., from seeds conserved in serotinous cones or from sprouting) are thus able to maintain themselves following a crown fire. Species such as jack pine would only be predicted to disappear from a site if recurrent fires occurred before the jack pine became sexually mature (i.e., there was no seed source), or if fire intervals were so long that all individuals died out without having the opportunity to release seed onto a favorable substrate (Rowe, 1983). In Eurasia, crown fires may change a landscape from one with trees resistant to surface fires to those adapted to long-distance dispersal, such as birch and aspen. More shade-tolerant species, such as Norway spruce, which may also have established shortly after fire will wait in the understory for the overstory to die off. Thus compositional change may be linked more to species growth rates and longevity than to recruitment-mediated replacement processes (Bergeron, 2000; Schulze et al., 2005). Aspen or birch when regenerated in abundance, often dominate stands for up to 100 years following fire (Aakala and Keto-Tokoi, 2011), but gradually these stands take on a mixed character during the following century before being dominated by conifers after approximately 200 years. Many authors have thus linked compositional dynamics primarily to time since fire (Bergeron, 2000; Kneeshaw and Bergeron, 1998). In monospecific black spruce forests, it is stand structure rather than composition that changes with time since fire (Harper et al., 2005; Boucher et al., 2006).

Following windthrow or insect outbreaks, tree species composition may not shift as it does after fires. Thus instead of a return to dominance by intolerant species, it is often the shade-tolerant species that have formed a seedling bank and that are recruited. *Abies balsamea* and *A. sibirica* have both been shown to follow cyclical patterns, with preestablished seedlings replacing canopy dominants following windthrows or insect outbreaks (MacLean, 1980; Schulze et al., 2005).

A great body of evidence demonstrates that fire is the agent that greatly controls the dynamics of boreal forests when crown-fire cycles are shorter than the life span of the dominant tree species. In the case of surface fires, compositional change has also been noted when fires are excluded for long periods (Frelich, 2002). Communities that are dependent on surface fires often require relatively short intervals between fires. In cases where fire intervals are increasing, later successional species often invade (Lilja and Kuuluvainen, 2006). The maintenance of branches along the stem facilitates fire access to the canopy and the destruction of the seed source of former canopy dominants in forest stands in North America and Siberia. Changes in disturbance regimes can thus have large effects on tree species composition (Romme et al., 1998).

In moister maritime and semi-maritime climates fire intervals are often quite long and shade-tolerant species like balsam fir, Norway spruce, or Siberian fir dominate. These species often have larger seeds and form dense seedling banks that are well adapted to respond to overstory canopy disturbances (Baskerville, 1975; MacLean, 1980). There is also research showing that these species have greater moisture requirements when establishing, conditions that are often found in shaded understory environments, as well as in more humid maritime climates (Kneeshaw and Bergeron, 1999).

14.3 DISCUSSION

The boreal forest is often thought of as a relatively homogeneous biome due to the dominance of a few plant and animal species over large areas. For example, tree species such as black spruce in North America and Scots pine in Eurasia have continent-wide distributions and are almost synonymous with the public image of the boreal forest. Furthermore, many of the tree species form large, intact blocks of forest with similar structure and composition. Animals such as woodland caribou, lynx, and snowshoe hare have continentwide distributions while some lichens and mosses have circumboreal ranges. However, there is substantial regional variation, which, at a biogeographical scale, shows up as patterns in species dominance related to variations in dominant disturbance regimes and climatic patterns. These differences are important for ecosystem processes, levels of biological diversity, and the degree to which a region's ecological integrity is at risk due to often similar and repetitive (across the landscape) types of resource management.

More than 90% of forest management in the boreal forest is based on clear-cutting and even-aged forest management. Justification for this approach has often been linked to the perception that large-scale disturbances such as fire dominate boreal forests and that relatively young even-aged stands are the rule rather than the exception. However, as seen throughout this chapter, disturbance regimes vary between and across continents (Figure 14.2) and this in turn is reflected as variations in forest structure and composition. Variations in the dominant types of disturbance and the intervals between disturbances have important consequences on structure and the species compositions of forests. Short return interval fires are relatively dominant only in the central parts of the continents where drier climates occur. These are thus the regions with fewer late successional shade-tolerant species.

In North America, crown fires have a dominant influence on the forest from Quebec through to eastern British Columbia and southeastern Alaska. However, the intervals between fires become longer as continental climate conditions are replaced by more maritime conditions. However, even in more maritime regions of the boreal zone (i.e., closer to the coasts), dry conditions, which are positively related to fire occurrence, still occur and lead to large burns, although these are infrequent. This in turn leads to large tracts of forest that are of similar composition, age, and structure. In maritime and northern regions with long fire cycles, this is translated as old forest with a complex vertical and horizontal structure. Such a complex structure is not maintained when foresters favor only short-rotation (80–100 years), even-aged management.

In Eurasia, the greater prevalence of surface fires often leads to multi-cohort stands. Such fires lead to a pulse in postfire recruitment and thus to a stand structure that is dominated by large fire-resistant trees with a mix of multiple age cohorts recruited following different fire events. It may thus be expected that in such stands relatively continuous cover was the rule rather than the exception (Pennanen, 2002) and that clear-cut techniques, which favor a single cohort, change the natural forest structure. Much of the natural forest was dominated by old-growth conditions with mixed ages and structures in both continents but especially in Eurasia (Shorohova et al., 2009; Kuuluvainen, 2009).

As previously noted, there is also an important north-south variation in disturbances in the boreal forest. In both North America and Eurasia the natural interval between fires is shorter in the south than in the north, meaning again that there should be a greater proportion of older stands in the north. In North America, spruce budworm outbreaks also create more mortality and larger openings in the southern part of the boreal forest than in the northern part. The reduced influence of these disturbances is in part explained by climate.

The shorter snow-free period limits fire and the shorter growing season limits SBW growth and expansion. Lower productivity in these northern regions also means that fuel or food sources may not always be continuous. Although forest management rotations may thus be somewhat longer to account for the lower productivity and longer period of time taken to grow, they do not approximate the multiple centuries that stands may develop in the absence of large-scale disturbance in natural conditions.

Many boreal tree species are thus generalists with adaptations to fire and nonfire disturbances. Black spruce is perhaps the best example, as it is adapted to fire, nonfire disturbances, and old-growth forest conditions through a combination of cones that are serotinous, cones that are not, and through vegetative layering. *Populus* species are able to maintain themselves in disturbed areas not only through vegetative sprouting and clonal growth but also by dispersing long distances to colonize favorable post-fire mineral seedbeds. Other species such as Dahurian larch have also been shown to be able to regenerate in both old-growth conditions and following fire (Ban et al., 1998). Recent work also shows that shade-intolerant species that were once considered fire obligate are able to recruit second or third cohorts (Cumming et al., 2000; Bergeron, 2000). Even characteristics such as cone serotiny have been shown to decrease under long fire cycles in favor of open cones (Gauthier et al., 1996).

Under natural conditions a large proportion of the boreal forest would have been considered old-growth (Pennanen, 2002; Kneeshaw and Gauthier, 2003). Thus, small-scale disturbances such as gap dynamics are important in much of the boreal forest (Vepakomma et al., 2010). This should be most apparent in maritime and humid northern regions. For example, Labrador in northeastern North America has been reported to have a fire cycle of greater than 500 years (Foster, 1985). However, even in regions where fire cycles are generally short a proportion of old-growth stands can be expected to occur as not all areas burn with equal frequency and because surface fires may only have minor effects on stand structure.

It has been proposed that differences between past and current fire frequencies could be used to determine the proportion of clear-cutting that could be used to emulate the distribution of a proportion of stands originating from stand-replacing fires in the landscape (Bergeron et al., 2006). However, the exclusive use of clear-cut systems (or variants, e.g., variable retention or harvesting that protects advance regeneration) cannot be justified by historical fire frequencies. In most cases, the reported historical fire frequencies are less than the current harvest rotation; or in other terms a large proportion of the preindustrial landscape was composed of forests older than the typical 80- to 100-year commercial forest rotation (Kuuluvainen, 2009). Already, compositional changes due to harvesting are being observed (Carleton and MacLellan, 1994). Alternative silvicultural systems that include extended rotations or a variety of silvicultural treatments have been proposed to maintain a proportion and characteristics of overmature and old-growth forests in the managed landscape (Bergeron et al., 1999; Burton et al., 1999). We may also need to revise the line of thinking that focuses only on optimizing timber supply through forest regulation, since this may, under many scenarios, put us into direct conflict with biodiversity objectives and lead to problems should fire cycles become shorter in the future (Bergeron et al., 2004a).

14.4 SUMMARY

The boreal forest is not as homogeneous as once thought. Considerable variation in disturbance regimes due in part to climatic influences can be found throughout Eurasia and North America. Continental regions in central Eurasia and North America have shorter fire intervals than in more maritime regions. Similarly fires, and spruce budworm in North America, are naturally more important in the southern part of the boreal forest than in the northern part. Such variation has led to a greater dominance of pyrogeneous, shade-intolerant species in the interior of each continent and the dominance of more shade-tolerant species nearer to the coasts. Important differences also exist between continents, due in part not only to species characteristics but also to climate and disturbance regimes. In North America large-scale disturbances include both the spruce budworm in eastern forests and crown fire whereas in Eurasia surface fires are more prevalent than crown fires. Old-growth forests are thus a dominant feature of many natural boreal forests, a condition that has been and is being rapidly modified by human intervention.

REFERENCES

Aakala, T., and Keto-Tokoi, P. (2011) 'The old Norway spruce forests of northern boreal Fennoscandia are alive and well: A review of Sirén (1955)' *Scandinavian Journal of Forest Research*, 26(S10): 25–33.

Aakala, T., Kuuluvainen, T., Wallenius, T., and Kauhanen, H. (2011) 'Tree mortality episodes in the intact *Picea*

abies-dominated taiga in the Arkhangelsk region of northern European Russia', *Journal of Vegetation Science*, (in print). [on line]

Agee, J.K. (1998) 'The landscape ecology of western forest fire regimes', *Northwest Science*, 72(special issue): 24–34.

Anderson, L.L., Feng Sheng, H., Nelson, D.M., Petit, R.J., and Paige, K.N. (2006) 'Ice-age endurance: DNA evidence of a white spruce refugium in Alaska', *Proceedings of the National Academy of Sciences*, 103(33): 12447–50.

Aksenov, D., Dobrynin, D., Dubinin, M., Egorov, A., Isaev, A., Karpachevkiy, M., Laestadius, L., Potapov, P., Atlas of Russia's intact forest landscape. *Global Forest Watch Russia*, Moscow, Russia, see http://forest.ru/eng/publications/intact/. Accessed 1 May 2011.

Babintseva, R.M., and Titova, Y.V. (1998) 'Effects of fire on the regeneration of larch forests in the Lake Baikal Basin', in J.G. Goldammer, and V.V. Furyaev, (eds.), *Fire in Ecosystems of Boreal Eurasia*. Netherlands: Kluwer Academic Publishers. pp. 358–64.

Ban, Y., Xu, H., Bergeron, Y., and Kneeshaw, D.D. (1998) 'Gap regeneration of shade-intolerant *Larix gmelinii* in old-growth boreal forests of northeastern China', *Journal of Vegetation Science*, 9(4): 529–36.

Baskerville, G.L. (1975) 'Spruce budworm: Super silviculturist', *Forestry Chronicle*, 51(4): 138–40.

Bergeron, Y. (1991) 'The influence of island and mainland lakeshore landscapes on boreal forest fire regimes', *Ecology*, 72(6): 1980–92.

Bergeron, Y. (2000) 'Species and stand dynamics in the mixed woods of Quebec's southern boreal forest', *Ecology*, 81(6): 1500–16.

Bergeron, Y., and Archambault, S. (1993) 'Decreasing frequency of forest fires in the southern boreal zone of Québec and its relation to global warming since the end of the "Little ice age"', *The Holocene*, 3(3): 255–59.

Bergeron, Y., and Brisson, J. (1990) 'Fire regime in red pine stands at the northern limit of the species' range', *Ecology*, 71(4): 1352–64.

Bergeron, Y., Flannigan, M., Gauthier, S., Leduc, A., and P. Lefort. (2004a) 'Past, current and future fire frequency in the Canadian boreal forest: Implications for sustainable forest management', *Ambio*, 33(6): 356–60.

Bergeron, Y., Gauthier, S., Flannigan, M. and, Kafka, V. (2004b) 'Fire regimes at the transition between mixedwood and coniferous boreal forest in northwestern Quebec', *Ecology*, 85(7): 1916–32.

Bergeron, Y., Gauthier, S., Kafka, V., Lefort, P., and Lesieur, D. (2001). 'Natural fire frequency for the eastern Canadian boreal forest: consequences for sustainable forestry', *Canadian Journal of Forest Research*, 31(3): 384–91.

Bergeron, Y., Harvey, B., Leduc, A., and Gauthier, S. (1999) 'Forest management guidelines based on natural disturbance dynamics: Stand- and forest-level considerations', *Forestry Chronicle*, 75(1): 49–54.

Bergeron, Y., Morin, H., Leduc, A., and Joyal, C. (1995) 'Balsam fir mortality following the last spruce budworm outbreak in northwestern Quebec', *Canadian Journal of Forest Research*, 25(8): 1375–84.

Bergeron, Y. and 10 authors (2006). 'Past, current, and future fire frequencies in Quebec's commercial forests: Implications for the cumulative effects of harvesting and fire on age-class structure and natural disturbance-based management', *Canadian Journal of Forest Research*, 36(11): 2737–44.

Blais, J.R. (1983) 'Trends in the frequency, extent and severity of spruce budworm outbreaks in eastern Canada', *Canadian Journal of Forest Research*, 13(4): 539–47.

Bonan, G.B., and Shugart, H. (1989) 'Environmental factors and ecological processes in boreal forests', *Annual Review of Ecology and Systematics*, 20(1): 1–28.

Bouchard, M., Kneeshaw, D.D., and Bergeron. Y. (2005) 'Mortality and stand renewal patterns in mixed forest stands following the last spruce budworm outbreak in western Québec', *Forest Ecology and Management*, 204(2–3): 297–313.

Bouchard, M., Kneeshaw, D.D., and Bergeron, Y. (2006) 'Forest landscape composition and structure after successive spruce budworm outbreaks', *Ecology*, 87(9): 2319–29.

Boucher, D., Gauthier, S., and De Grandpré, L. (2006) 'Structural changes in coniferous stands along a chronosequence and productivity gradient in the northeastern boreal forest of Québec', *Ecoscience*, 13(2): 172–80.

Burton, P.J., Messier, C., Smith, D.W., and Adamowicz, W.L. (eds.), (2003) *Towards Sustainable Management of the Boreal Forest*. Ottawa: NRC Research Press.

Burton, P.J., Kneeshaw, D.D. and Coates, K.D. (1999) 'Managing forest harvesting to maintain old growth in boreal and sub-boreal forests', *Forestry Chronicle*, 75(4): 623–631.

Carleton, T.J., and MacLellan, P. (1994) 'Woody vegetation responses to fire versus clear-cutting logging: A comparative survey in the central Canadian boreal forest', *Ecoscience*, 1(2): 141–52.

Carleton, T.J., and Maycock, P.J. (1978) 'Dynamics of the boreal forest south of James Bay', *Canadian Journal of Botany*, 6(9): 1157–73.

Coates, D.L., and Burton, P.J. (1997) 'A gap-based approach for development of silvicultural systems to address ecosystem management objectives', *Forest Ecology and Management*, 99(3): 337–54.

Cumming, S.G. (1997) Landscape dynamics of the boreal mixedwood forest. PhD Thesis, University of British Columbia, Vancouver, Canada.

Cumming, S.G., Schmiegelow, F.K.A., and Burton, P.J. (2000) 'Gap dynamics in boreal aspen stands: Is the forest older than we think?' *Ecological Applications*, 10(3): 744–59.

Cwynar, L.C. (1977) 'The recent fire history of Barron Township, Algonquin Park', *Canadian Journal Botany*, 55(11): 1524–38.

Davis, M.B. (1981) 'Quaternary history and stability of forest communities', in D.C. West, H.H. Shugart, and D.B. Botkin (eds.), *Forest Succession: Concepts and Application*. New York: Springer-Verlag. pp.132–53.

De Grandpré, L., Bergeron, Y., Nguyen-Xuan, T., Boudreault, C., and Grondin, P. (2003) 'Composition and dynamics of understory vegetation in the boreal forests of Quebec', in F.S. Gilliam, and M.R. Robert (eds.) *The Herbaceous Layer of Forests of Eastern North America*. New York: Oxford University Press. pp. 238–61.

de Römer, A., Kneeshaw, D., and Bergeron, Y. (2007) 'Small gap dynamics in the southern boreal forest of eastern Canada: Do canopy gaps influence stand development?' *Journal of Vegetation Science*, 18(6): 815–26.

Dix, R.L., and Swan, J.M.A. (1971) 'The roles of disturbance and succession in upland forest at Candle Lake Saskatchewan', *Canadian Journal of Botany*, 49(5): 657–76.

Drobyshev, I.V. (1999) 'Regeneration of Norway spruce in canopy gaps in Sphagnum-Myrtillus old-growth forests', *Forest Ecology and Management*, 115(1): 71–83.

Engelmark, O. (1999) 'Boreal forest disturbances', in L.R. Walker (ed.), *Ecosystems of Disturbed Ground*. New York: Elsevier. pp. 161–86.

Foster, D.R. (1985) 'Vegetation development following fire in *Picea mariana* (Black spruce)-*Pleurozium* forests of southeastern Labrador, Canada', *Journal of Ecology*, 73(2): 517–34.

Foster, D.R. (1983) 'The history and pattern of fire in the boreal forest of southeastern Labrador', *Canadian Journal of Botany*, 61(9): 2459–71.

Frelich, L.E. (2002) *Forest Dynamics and Disturbance Regimes: Studies from Temperate Evergreen-Deciduous Forests*. New York: Cambridge University Press.

Frelich, L.E., and Lorimer, C.G. (1991) 'Natural disturbance regimes in hemlock–hardwood forests of the Upper Great Lakes region', *Ecological Monographs*, 61(2): 145–64.

Galipeau, C., Kneeshaw, D.D., and Bergeron, Y. (1997) 'Colonisation of white spruce and balsam fir as observed 68 years after fire in the south eastern boreal forest', *Canadian Journal of Forest Research*, 27(2): 139–47.

Gauthier, S., Bergeron, Y., and Simon, J.-P. (1996) 'Effects of fire regime on the serotiny level of Jack pine', *Journal of Ecology*, 84(4): 539–48.

Girardin, M.-P., Tardif, J., Flannigan, M.D., and Bergeron, Y. (2006) 'Synoptic scale atmospheric circulation and boreal Canada summer drought variability of the past three centuries', *Journal of Climate*, 19(10): 1922–47.

Goodale, C.L. and 12 authors (2002) 'Forest carbon sinks in the northern hemisphere', *Ecological Applications*, 12(3): 891–99.

Granström, A. (1993) 'Spatial and temporal variation of lightning ignitions in Sweden', *Journal of Vegetation Science*, 4(6): 737–744.

Greene, D., Zasada, J., Sirois, L., Kneeshaw, D.D., Morin, H., Charron, I., and Simard, M.-J. (1999) 'A review of regeneration dynamics of boreal forest tree species', *Canadian Journal of Forest Research*, 29(6): 824–39.

Gromtsev, A. (2002) 'Natural disturbance dynamics in the boreal forests of European Russia: A review', *Silva Fennica*, 36(1): 41–55.

Harper, K.A, Bergeron, Y., Drapeau, P., Gauthier, S., and De Grandpré, L. (2005), 'Structural development following fire in black spruce boreal forest', *Forest Ecology and Management*, 206(1–3): 293–306.

Heinselman, M.L. (1981) 'Fire and succession in the conifer forests of North America', in D.C. West, H.H. Shugart, and D.B. Botkin (eds.), *Forest Succession: Concepts and Application*. New-York: Springer Verlag. pp. 374–406.

Helmisaari, H., and Nikolov, N. (1989) 'Survey of ecological characteristics of boreal tree species in Fennoscandia and the USSR', Biosphere Dynamics Project, Publication No. 95 IIASA, Vienna, Austria.

Hill, S.B., Mallik, A.U., and Chen, H.Y.H. (2005) 'Canopy gap disturbance and succession in trembling aspen dominated boreal forests in northeastern Ontario', *Canadian Journal of Forest Research*, 35(8): 1942–51.

Hirsch, K.G. (1991) 'A chronological overview of the 1989 fire season in Manitoba', *Forestry Chronicle*, 67(4): 358–65.

Johnson, E.A. (1979) 'Fire recurrence in the subarctic and its implications for vegetation composition', *Canadian Journal of Botany*, 57(12): 1374–79.

Johnson, E.A. (1992) *Fire and Vegetation Dynamics—Studies from the North American Boreal Forest*. Cambridge: Cambridge University Press.

Johnson, E.A., Fryer, G.I., and Heathcott, M.J. (1990) 'The influence of man and climate on frequency of fire in the interior wet belt forest, British Columbia', *Journal of Ecology*, 78(2): 403–412.

Johnson, E.A., and Larsen, C.P.S. (1991) 'Climatically induced change in fire frequency in the southern Canadian Rockies', *Ecology*, 72(1): 194–201.

Johnson, E.A., and Wowchuk, D.R. (1993) 'Wildfires in the southern Canadian Rocky Mountains and their relationship to midtropospheric anomalies', *Canadian Journal of Forest Research*, 23(6): 1213–22.

Kasischke, E.S. (2000). Boreal ecosystems in the global carbon cycle. In: Kasischke, E.S., and Stocks, B.J. (eds.). Fire, Climate Change and Carbon Cycling in the Boreal Forest. Ecological Studies Series, Springer-Verlag, New York. pp. 440–452.

Kenkel, N.C., Walker, D.J., Watson, P.R., Caners, R.T., and Lastra, R.A. (1997) 'Vegetation dynamics in boreal forest ecosystems', *Coenoses*, 12(2–3): 97–108.

Kneeshaw, D.D. (2001) 'Are non-fire gap disturbances important to boreal forest dynamics?' in S.G. Pandalarai (ed.), *Recent Research Developments in Ecology 1*. New York: Transworld Research Press. pp. 43–58.

Kneeshaw, D.D., and Bergeron, Y. (1998) 'Canopy gap characteristics and tree replacement in the southern boreal forest', *Ecology*, 79(3): 783–94.

Kneeshaw, D.D., and Bergeron, Y. (1999) 'Spatial and temporal patterns of seedling recruitment within spruce budworm caused canopy gaps', *Ecoscience*, 6(2): 214–22.

Kneeshaw, D.D., and Gauthier, S. (2003) 'Old-growth in the boreal forest at stand and landscape levels', *Environmental Reviews*, 11(1 suppl.): s99–s114.

Kondrashov, L.G. (2004) 'Russian Far East forest disturbances and socio-economic problems of restoration', *Forest Ecology and Management*, 201(1): 65–74.

Kuuluvainen, T. (1994) 'Gap disturbance, ground microtopography, and the regeneration dynamics of boreal coniferous

forests in Finland: A review', *Annales Zoologici Fennici*, 31(1): 35–51.

Kuuluvainen, T. (2002) 'Natural variability of forests as a reference for restoring and managing biological diversity in boreal Fennoscandia', *Silva Fennica*, 36(1): 97–125.

Kuuluvainen, T. (2009) 'Forest management and biodiversity conservation based on natural ecosystem dynamics in northern Europe: The complexity challenge', *Ambio*, 38(4): 309–15.

Kuuluvainen, T., Syrjänen, K., and Kalliola, R. (1998) 'Structure of a pristine *Picea abies* forest in northeastern Europe', *Journal of Vegetation Science*, 9(4): 563–74.

Lampainen, J., Kuuluvainen, T., Wallenius, T.H., Karjalainen, L., and Vanha-Majamaa, I. (2004) 'Long-term structure and regeneration after wildfire in Russian Karelia', *Journal of Vegetation Science*, 15(2): 245–56.

Landsberg, J.J., and Gower S.T. (1997) *Applications of Physiological Ecology to Forest Management*. Toronto: Academic Press.

Lännenpää, A., Aakala, T., Kauhanen, H., and Kuuluvainen, T. 2008. 'Tree mortality agents in pristine Norway spruce forests in northern Fennoscandia', *Silva Fennica*, 42(2): 151–63.

Larjavaara, M., Kuuluvainen, T., and Rita, H. (2005) 'Spatial distribution of lightning-ignited forest fires in Finland', *Forest Ecology and Management*, 206(1–3): 77–88.

Larjavaara, M., Kuuluvainen, T., Tanskanen, H., and Venäläinen, A. (2004) 'Variation in forest fire ignition probability in Finland', *Silva Fennica*, 38(3): 253–66.

Larsen C.P.S. (1997) Spatial and temporal variations in boreal forest fire frequency in northern Alberta', *Journal of Biogeography*, 24(5): 663–73.

Lauzon, È., Kneeshaw, D., and Bergeron, Y. (2007) 'Reconstruction of fire history (1680–2003) in Gaspesian mixedwood boreal forests of eastern Canada', *Forest, Ecology and Management*, 244(1): 41–49.

Leemans, R. (1991) 'Canopy gaps and establishment patterns of spruce (*Picea abies* [L.] Karst.) in two old-growth coniferous forests in central Sweden', *Vegetatio*, 93(2): 157–65.

Lefort, P., Gauthier, S., and Bergeron, Y. (2003) 'The influence of fire weather and land use on the fire activity of the lake Abitibi area, Eastern Canada', *Forest Science*, 49(4): 509–21.

Lewis, K.J., and Lindgren, B.S. (2000) 'A conceptual model of biotic disturbance ecology in the central interior of BC: How forest management can turn Dr. Jekyll into Mr. Hyde', *Forestry Chronicle*, 76(3): 433–43.

Lilja, S., and Kuuluvainen, T. (2006) 'Stand structural characteristics of old *Pinus sylvestris*-dominated forests along a geographic and human influence gradient in boreal Fennoscandia', *Silva Fennica*, 39(3): 407–28.

Liu, Q.H., and Hytteborn, H. (1991) 'Gap structure, disturbance and regeneration in a primeval *Picea abies* forest', *Journal of Vegetation Science*, 2(3): 391–402.

MacLean, D.A. (1980) 'Vulnerability of fir-spruce during uncontrolled spruce budworm outbreaks: A review and discussion', *Forestry Chronicle*, 56(5): 213–21.

Masters, A.M. (1990) 'Changes in forest fire frequency in Kootenay National Park, Canadian Rockies', *Canadian Journal of Botany*, 68(8): 1763–67.

McCarthy, J. (2001) 'Gap dynamics of forest trees: A review with particular attention to boreal forests', *Environmental Reviews*, 9(1): 1–59.

McCarthy, J., and Weetman, G. (2006) 'Age and size structure of gap dynamic, old-growth boreal forest stands in Newfoundland', *Silva Fennica*, 40(2): 209–30.

Melekhov, I.S. (1947) [Nature of the forest and forest fires] *Archangelsk*, 60 p. (in Russian).

Mery, G., Katila, P., Galloway, G., Alfaro R.I., Kanninen, M., Lobovikov, M., and Varjo, J. (eds.) (2010) *Forests and Society – Responding to Global Drivers of Change*. International Union of Forest Research Organizations (IUFRO), Vienna, Austria.

Niklasson, M., and Granström, A. (2000) 'Numbers and sizes of fires: long-term spatially explicit fire history in a Swedish boreal forest landscape', *Ecology*, 81(6): 1484–99.

Noble, I.R., and Slatyer, R.O. (1977) 'Post-fire succession of plants in mediterranean ecosystems', in H.A. Mooney, and C.E. Conrad (eds.), *Proc. Symp. On the Environmental Consequences of Fire and Fuel Management in Mediterranean Ecosystems*. USDA Forest Service General Technical Report WO-3. pp. 27–63.

Noble, I.R., and Slatyer, R.O. (1980) 'The use of vital attributes to predict successional changes in plant communities subject to recurrent disturbance', *Vegetatio*, 43(1): 5–21.

Payette, S. (1992) 'Fire as a controlling process in the North American boreal forest', in H. H. Shugart, R. Leemans, and G. B. Bonan (eds.), *A Systems Analysis of the Boreal Forest*. Cambridge: Cambridge University Press. pp. 144–69.

Payette, S., Morneau, C., Sirois, L., and Desponts, M. (1989) 'Recent fire history of the northern Quebec biomes', *Ecology*, 70(3): 656–73.

Pennanen, J. (2002) 'Forest age distributions under mixed-severity fire regimes—a simulation-based analysis for middle boreal Fennoscandia', *Silva Fennica*, 36(1): 213–31.

Perron, M., Perry, D.J., Andalo, C., and Bousquet, J. (2000) 'Evidence from sequence-tagged-site markers of a recent progenitor-derivative species pair in conifers', *Proceedings of the National Academy of the Sciences of the United States of America*, 97(21): 11331–36.

Pham, A.T., De Grandpré, L., Gauthier, S., and Bergeron, Y. (2004) 'Gap dynamics and replacement patterns in gaps of northeastern boreal forest of Quebec', *Canadian Journal of Forest Research*, 34(2): 353–64.

Pitkänen, A. (1999) *Palaeoecological Study of the History of Forest Fires in Eastern Finland*.' PhD Thesis, University of Joensuu, Finland.

Pitkänen, A., and Huttunen, P. (1999) 'A 1300-year forest-fire history as a site in eastern Finland and pollen records in laminated lake sediment', *The Holocene*, 9(3): 311–20.

Pitkänen, A., Huttunen, P., Jugner, K., and Tolonen, K. (2002) 'A 10 000 year local forest fire history in a dry heath forest site in eastern Finland, reconstructed from charcoal layer records of a small mire', *Canadian Journal of Forest Research*, 32(10): 1875–80.

Pitkänen, A., Huttunen, P., Tolonen, K., and Jugner, K. (2003) 'Long-term fire frequency in the spruce-dominated forests of Ulvinsalo strict nature reserve, Finland', *Forest Ecology and Management*, 176(1): 305–19.

Reyes, G.P., and Kneeshaw, D.D. (2008) 'Intermediate-scale disturbance dynamics in boreal mixedwoods: The relative importance of disturbance type and environmental factors on vegetation response', *Ecoscience*, 15(2): 241–49.

Romme, W.H., Everham, E.H., Frelich, L.E., Morizt, M.A., and Sparks, R.E. (1998) 'Are large, infrequent disturbances qualitatively different from small, frequent disturbances?' *Ecosystems*, 1(6): 524–34.

Rouvinen, S., Kuuluvainen, T., and Siitonen, J. (2002) 'Tree mortality in a *Pinus sylvestris* dominated boreal forest landscape in Vienansalo wilderness, eastern Fennoscandia', *Silva Fennica*, 36(1): 127–45.

Rowe, J.S. (1983) 'Concepts of fire effects on plant individuals and species', in R.W. Wein, and D.A. MacLean (eds.), *The Role of Fire in Northern Circumpolar Ecosystems*. SCOPE Repost 18. New York: John Wiley. pp. 135–54.

Sannikov, S.N., and Goldammer, J.G. (1996) 'Fire ecology of pine forests of northern Eurasia' in J.G. Goldammer, and V.V Furyaev, (eds.) *Fire in Ecosystems of Boreal Eurasia*, Netherlands: Kluwer Academic Publishers, pp.151–67.

Schulte, L.A., and Mladenoff, D.J. (2005) 'Severe wind and fire regimes in northern forests: Historical variability at the regional scale', *Ecology*, 86(2): 431–45.

Schulze, E.D., Wirth, C., Mollicone, D., and Ziegler, W. (2005) 'Succession after stand replacing disturbances by fire, wind throw and insects in the dark Taiga of central Siberia', *Oecologia*, 146(1): 77–88.

Shorohova, E., Kuuluvainen, T., Kangur, A., and Jõgiste, K. (2009) 'Natural stand structures, disturbance regimes and successional dynamics in the Eurasian boreal forests: a review with special reference to Russian studies', *Annals of Forest Science*, 66(2): 201.

Smirnova, E., Bergeron, Y., and Brais, S. (2008) 'Influence of fire intensity on structure and composition of jack pine stands in the boreal forest of Quebec: Live trees, understory vegetation and dead wood dynamics', *Forest Ecology and Management*, 255(7): 2917–27.

Stehlik, I. (2003) 'Resistance or emigration? Response of alpine plants to the ice ages', *Taxon*, 52(3): 499–510.

Stocks, B.J. and 11 authors (2002) 'Large forest fires in Canada, 1959–1997', *Journal of Geophysical Research*, 10(D1): 8149.

Su, Q., Needham, T.D., and MacLean, D.A. (1996) 'The influence of hardwood content on balsam fir defoliation by spruce budworm'. *Canadian Journal of Forest Research*, 26(9): 1620–28.

Suffling, R., Smith, B., and Molin, J.D. (1982) 'Estimating past forest age distributions and disturbance rates in North-western Ontario: a demographic approach', *Journal of Environmental Management*, 14(1): 45–56.

Syrjänen, K., Kalliola, R., Puolasmaa, A., and Mattson, J. (1994) 'Landscape structure and forest dynamics in subcontinental Russian European taiga', *Annales Zoologici Fennici*, 31(1): 19–36.

Tande, G.F. (1979) 'Fire history and vegetation pattern of coniferous forests in Jasper National Park, Alberta', *Canadian Journal of Botany*, 57(18): 1912–31.

Van der Maarel, E. (1993) 'Some remarks on disturbance and its relations to diversity and stability', *Journal of Vegetation Science*, 4(6): 733–36.

Van Wagner, C.E. (1978) 'Age-class distribution and the forest cycle', *Canadian Journal of Forest Research*, 8(2): 220–27.

Vepakomma, U., D. Kneeshaw, D.D. and St-Onge, B. (2010). 'Interactions of multiple disturbances in shaping boreal forest dynamics – a spatially explicit analysis using multi-temporal lidar data and high resolution imagery', *Journal of Ecology*, 98(3). 536–539.

Wallenius, T.H. (2002) 'Forest age distribution and traces of past fires in a natural boreal landscape dominated by *Picea abies*', *Silva Fennica*, 36(1): 201–211.

Wallenius, T.H., Kuuluvainen, T., and Vanha-Majamaa, I. (2004) 'Fire history in relation to site type variation in Vienansalo wilderness in eastern Fennoscandia, Russia', *Canadian Journal of Forest Research*, 34(7): 1400–9.

Wallenius, T.H., Pitkänen, A., Kuuluvainen, T., Pennanen, J., and Karttunen, H. (2005) 'Fire history and forest age distribution of an unmanaged *Picea abies* dominated landscape', *Canadian Journal of Forest Research*, 35(7): 1540–52.

Wallenius, T., Kauhanen, H., Herva, H., and Pennanen, J. (2010) 'Long fire cycles in northern boreal *Pinus* forests in Finnish Lapland', *Canadian Journal of Forest Research*, 40(10): 2027–35.

Walter, H., and Breckle, S.-W. (1989) *Ecological Systems of the Geobiosphere Vol. 3. Temperate and Polar Sonobiomes of Northern Eurasia*. Toronto: Springer-Verlag.

Wein, R.W., and MacLean, D.A. (1983) 'An overview of fires in northern ecosystems', in R.W. Wein, and D.A. MacLean (eds.), *The Role of Fire in Northern Circumpolar Ecosystems*. New York: John Wiley. pp. 1–18.

Whitney, R.D. (2000) *Forest Management Guide for Tomentosus Root Disease*. Ontarion Ministry of Natural Resources, Ontario, Canada.

Weir, J.M.H., Johnson, E.A., and Miyanishi, K. (2000) 'Fire frequency and the spatial age mosaic of the mixed-wood boreal forest in western Canada', *Ecological Applications*, 10(4): 1162–77.

Wilkars, L.-O. (2004) 'Brand beroende insekter—respons pa tio ars naturvardsbranningar', *Fauna Flora*, 99(2): 28–34. (in Swedish with summary in English).

Woodward, F.I. (1995) 'Ecophysiological controls of conifer distribution', in W.K. Smith and T.M. Hinckley (eds.), *Ecophysiology of Coniferous Forests*. Toronto: Academic Press. pp. 79–94.

Wooster, M.J., and Zhang, Y.H. (2004) 'Boreal forest fires burn less intensely in Russia than in North America', *Geophysical Research Letters*, 31(1): 1–3.

Zackrisson, O. (1977) 'Influence of forest fires on the North Swedish boreal forest', *Oikos*, 29(1): 22–32.

Zackrisson, O., Nilsson, M., Steijlen, I., and Hörnberg, G. (1995) 'Regeneration pulses and climate-vegetation inter-actions in nonpyrogenic boreal Scots pine stands', *Journal of Ecology*, 83(3): 469–83.

The Ecosystem Dynamics of Tropical Savannas

Jayalaxshmi Mistry

15.1 INTRODUCTION

Savannas cover an eighth of the world's land surface (Scholes and Archer, 1997), and support a rich variety of plants, animals, and human populations. They can be loosely defined as tropical and subtropical ecosystems with a continuous and important grass/herbaceous stratum, a discontinuous layer of trees and shrubs of variable height and density, and where growth patterns are closely associated with alternating wet and dry seasons (Bourlière and Hadley, 1983). Productivity of vegetation is governed by the highly seasonal (e.g., dry seasons can vary from two to nine months) and the unpredictable nature of the rainfall distribution. This is exacerbated by the nutrient-poor status of many savanna soils.

Savannas cover large areas of the Americas, Africa, Asia, and Australia (Figure 15.1) and range in form from almost pure grasslands to dense woodlands. This heterogeneity in form led to much confusion over the definition of savannas, which historically took either a climatic or vegetation approach (Table 15.1). However, there were limitations to these definitions; for example, savanna vegetation can extend further into arid zones on soils of high (as opposed to low) moisture-holding capacity. Under a weak seasonal humid climate, savanna vegetation can occur due to strong edaphic influences causing either impeded drainage or low moisture-holding capacity. The differences in definitions brought together a group of savanna ecologists from around the world, in an attempt to identify a framework from which global as well as local comparisons could

be made between savanna types. The outputs from this meeting were a landmark in the history of savanna ecology (Frost et al., 1986; Goldstein et al., 1988). Four key ecological determinants were recognized as controlling the structure and function of tropical savannas, namely Plant Available Moisture (PAM), Plant Available Nutrients (PAN), fire, and herbivory (Stott, 1991). The primary controllers are PAM and PAN, which, as biologically meaningful measures, enable ready comparisons of various savanna sites and plants. For example, PAM could be measured using factors such as the number of ecologically humid days (i.e., the period when growth is not limited by water availability) and/or the number of days in which rainfall exceeds evapotranspiration, and PAN could be based on the sum of exchangeable bases (i.e., calcium, magnesium, potassium, and sodium). Although as yet, there have been few studies that have attempted to characterize PAM and PAN (see Walker and Langridge, 1997; Williams et al., 1996 for some examples), the PAM/PAN concept has been useful in focusing research in savanna ecology internationally. Table 15.2 outlines the main savanna formations found throughout the world and the characteristics of the key determinants.

In addition to establishing the four savanna determinants, the meeting of savanna ecologists also highlighted the importance of scale and the fundamental idea that the four determinants function differently over discrete levels of space and time. This helped to reinforce the view of the tropical savanna landscape not as a homogeneous unit but as being composed of a mosaic of

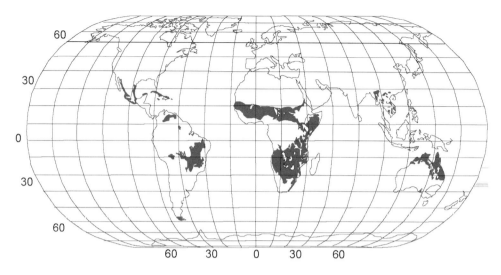

Figure 15.1 The global distribution of tropical savannas (taken from Mistry, 2000)

Table 15.1 Some definitions of savannas (taken from Mistry, 2000)

Definitions of savanna	Authors
Climatic	
Ecosystems that lie in the tropical savanna (*Aw*) and monsoon (*Am*) climatic zones, largely between the latitudes of Cancer and Capricorn, where annual precipitation is between 250–2,000 mm, most of which falls in the wet season.	Köppen (1884, 1900)
Any formation or landscape within the region experiencing a winter dry season and summer rains.	Jäger (1945), Troll (1950), and Lauer (1952)
Ecosystems bound by dry forests at higher rainfall (>1,000 mm), by thorn forests at lower rainfall (<500 mm), and by thorn steppe and temperate savannas at lower temperatures (<18°C).	Holdridge (1947)
Ecosystems with low to moderate rainfall (500–1,300 mm) and with high mean annual temperatures (18–30°C).	Whittaker (1975)
Vegetation	
A mixed physiognomy of grasses and woody plants in any geographical area.	Dansereau (1957)
A mixed tropical formation of grasses and woody plants, excluding pure grasslands.	Walter (1973)
Open formations dominated by grasses, in the lowland tropics, where trees and shrubs, if present, are of little physiognomic significance.	Beard (1953) and Whittaker (1975)

ecosystems and habitat types, such as tropical forest, riparian forest, swamps, shrublands, and grassland. Savannas form the major landscape unit but are set within a rich patchwork of vegetation types, within and between which flows of energy and material occur. While the key ecological determinants, PAM and PAN, create the template for savanna heterogeneity, fire and herbivory play a critical role in modifying the savanna landscape (Stott, 1991). People are yet another dimension of the savanna landscape. With over one-fifth of the world's population living in or around savannas (Frost et al., 1986), and with many communities relying on natural resources for their livelihoods, land use and management have a profound effect on savanna heterogeneity.

Table 15.2 Different tropical savanna formations of the world and their ecological characteristics (see Mistry, 2000 for detailed descriptions of key determinants of different savannas at the regional scale)

Location	Name	Ecological characteristics
Americas		
Brazil	Cerrado	High levels of PAM and low levels of PAN. Fire is a major determinant, but herbivory by native mammals is very low. The dominant physiognomy is grassland with moderately scattered trees and shrubs.
Colombia, Venezuela	Llanos	
Venezuela	Gran sabana	
Guyana, Brazil	Rupununi savannas	
Guianas	Coastal savannas	
Amazonian region	White sand savannas	
Brazil, Bolivia, Paraguay	Gran Pantanal	
Bolivia	Llanos de Mojos	
Southern Mexico, Central America, and the Caribbean (notably Cuba)	Patches of savanna	
Southern Texas and northern Mexico	Rio Grande Plains savanna	
Western Paraguay, eastern Bolivia, northern Argentina, and part of southeastern Brazil	Savannas of Chaco region	
Africa		
Central and southern Africa (Zambezian zone)	Miombo woodlands	Moderate to high levels of PAM and low levels of PAN. Herbivory by native mammals present but low to moderate levels (dominated by elephants). Fire is frequent and an important factor. The dominant physiognomy is savanna woodland.
	Acacia savanna	
	Mopane woodland	
West Africa (Sudanian zone)	Sahelian savannas	PAM ranges from low in northern Sahelian savannas to high in southern Guinean savannas. PAN low to moderate levels. Herbivory by native mammals at moderate levels and fire is an important factor. The Sahelian savannas are dominated by grassland with widely scattered bushes and small trees, and the Sudanian and Guinean savannas are dominated by savanna woodlands.
	Sudanian savannas	
	Guinean savannas	
East Africa (Somalia-Masai zone)	Range of savanna types from *Acacia-Commiphora* woodlands to grass dominated plains	Low levels of PAM and PAN, supporting range of savanna forms. Topography is important for PAM, which in turn determines different grass types. Large presence of native mammalian herbivores. Fire is also an important factor.
Asia		
Mainland Southeast Asia	Dry dipterocarp savanna woodland	High levels of PAM and low levels of PAN. Herbivory by native mammals very low; fire is an important factor. The dominant physiognomy is savanna woodland.
Indian subcontinent	Sal savanna woodland	
Australia and Oceania		
Northern Australia, and a small proportion of southern Papua New Guinea	*Eucalyptus* and *Acacia* savanna woodlands	Moderate to high levels of PAM and low levels of PAN. Herbivory by native mammals very low; fire is an important factor. The dominant physiognomy is savanna woodland.
Timor	*Eucalyptus* savanna woodlands	

The African and Asian savannas in particular have had a long-documented history of human use (Mistry, 2000).

The intimate link in savannas between the abiotic, biotic, and anthropogenic, although seen and practiced in the daily realities of savanna inhabitants, has not been fully recognized in the scientific study and management of the savanna landscape (Mistry and Berardi, 2006). Grounded in the paradigm of people and nature being separate entities, many strands of ecological enquiry and their subsequent management applications have been based on the notion of ecosystem stability (Scoones, 1999). These views see ecosystems as isolated, closed biotic systems, gradually equilibrating to stable, external conditions until a climax state is reached. Stability in the biotic closed system is maintained through intra- and interspecific interactions among species, and any disturbance to this system is perceived as being external to the dynamic functioning of the ecosystem. For example, Clements' (1916) theory of succession, where vegetation assemblages change toward a stable climax, and Lotka (1925) and Volterra's (1926) predator-prey models, where carrying capacities and maximum sustained yields in animal populations were identified, became the guide for managing rangelands in semiarid and arid savannas (Behnke and Scoones, 1993). The Equilibrium Theory of Island Biogeography (MacArthur and Wilson, 1967) and its ideas of a stable relationship between species diversity and area became a basis upon which biodiversity policy could be created and protected areas designed, most notably in the case of national parks in African savannas (Homewood, 2004; Leach and Mearns, 1996).

With the realization from the 1980s that savannas were not 'closed', internally driven ecosystems, there began increasing support for alternative 'open system' paradigms to interpret savanna ecosystem functioning and dynamics. The recognition of the importance of historical effects, spatial heterogeneity, stochastic factors, and environmental perturbations in ecosystem functioning and dynamics led to a paradigm shift to a nonequilibrium ecology (Holling, 1973; Warren, 1995). The concept of variability was particularly true for arid savannas, where biological processes are moisture- or PAM-limited, and rainfall is inherently variable and unpredictable (e.g., Sullivan, 1996). In this model, the biotic component, reflecting factors such as biological productivity, are primarily PAM-limited and driven by aperiodic and idiosyncratic rainfall events.

Savannas are dynamic systems, diverse in their form and functioning. I will begin by reviewing some of the recent work on the age-old question of tree-grass coexistence in savannas. How do the woody and herbaceous life forms coexist in the savanna ecosystem and what factors affect coexistence? I will then go on to explore and illustrate savanna heterogeneity, using key savanna determinants as the building blocks. Finally, I will look at the implications of heterogeneity for savanna management, and I will identify priorities for future research.

15.2 TREE-GRASS COEXISTENCE IN SAVANNAS

What makes savannas unique and distinctive from forests and grasslands is the codominance of trees and grasses. Yet we still know relatively little about what mechanisms permit the woody and herbaceous layers to coexist without displacing one another and what factors determine the proportions of trees and grasses in different savannas around the world (Scholes and Archer, 1997; House et al., 2003; Sankaran et al., 2004). Theories to explain the coexistence of trees and grasses can be grouped into those that emphasize the fundamental role of competitive interactions and niche separation with respect to limiting resources such as water (competition-based models, sensu Sankaran et al., 2004) and those that focus on demographic mechanisms and factors such as fire, herbivory, and PAM variability to promote tree-grass persistence through their dissimilar effects on different life-history stages of trees (demographic-bottleneck models, sensu Sankaran et al., 2004). Whereas competition-based models highlight the roles of both trees and grasses in resource acquisition, demographic-bottleneck models focus only on the role of trees, and how climatic variability and/or fire and grazing limit the germination of trees, their establishment, and/ or their transition to mature-size classes (see Enright, this volume, for a detailed discussion of fire-controlled ecosystem properties). The latter models can also be divided into those that emphasize spatial variation (e.g., Jeltsch et al., 1996, 1998, 2000) and those that emphasize temporal variation (e.g., Higgins et al., 2000; van Wijk and Rodriguez-Iturbe, 2002) for limiting tree germination, establishment, maturity, and mortality.

The debate to resolve the 'savanna question' (Sarmiento, 1984) continues. Sankaran et al. (2004) carried out a useful evaluation of the contrasting theories, the empirical evidence used to support each, and the limitations of the different approaches. The competition-based models, mostly grounded in the equilibrium paradigm (where the competitive balance between trees and grasses is not affected by rainfall variation, fire, or grazing), are exemplified by the root niche separation model. This is the classic equilibrium model, also known as the 'Walter hypothesis' (Walter, 1971). It predicts that PAM is the most important limiting factor in savannas and that there is differential access to PAM by the roots of trees and grasses—grasses are stronger competitors for PAM in the topsoil, whereas trees are able to access moisture in the subsoil, thereby allowing coexistence. Although empirical evidence for root differentiation between adult trees and grasses does exist for some savanna systems in eastern and southern Africa (e.g., Helsa et al., 1985; Knoop and Walker, 1985), data from other savanna formations indicate an overlap in rooting profiles between adult trees and grasses (e.g., Johns, 1984; Belsky, 1990, 1994; Le Roux et al., 1995; Seghieri et al., 1995; Mordelet et al., 1997; Smit and Rethman, 2000; Hipondoka et al., 2003).

It is without doubt that individual events as well as the interactive forces of drought, fire, and herbivory play significant roles in influencing the dynamics of trees and grasses in savannas (e.g., Archer et al., 2001; Bowman et al., 2001; Roques et al., 2001; Burrows et al., 2002; Fensham et al., 2003; Groen et al., 2008). However, there is little empirical evidence of the effects of demographic bottlenecks on trees imposed by these forces, and the relative roles of climate and management for driving changes in woody cover in savannas over the past century are subject to active debate. Data from long-term fire exclusion experiments (25–50 years) have shown an increase in woody cover, particularly for sites receiving less than 650 mm of rainfall per year (e.g., Trapnell, 1959; Bond et al., 2003). In South American savannas, tree densities have been increasing throughout a matrix of land-use histories ranging from heavily disturbed by fire and grazing to ungrazed and unburnt, suggesting that climate rather than disturbance may be driving increases in woody life-forms (Silva et al., 2001). Fensham et al. (2005), looking at woody plant increases in the semiarid savannas of Queensland, Australia, using aerial photography from the 1940s to the 1990s, found that fire and grazing did not explain the increase in woody plants. In general, rainfall was positively related to rates of change in woody plants. The interaction between rainfall and initial woody cover was particularly significant, reflecting the fact that increases in cover coincided with low initial cover when rainfall was higher than average, whereas decreases in cover typically occurred with high initial cover, regardless of rainfall. Their findings highlight the importance of interactions of rainfall fluctuations and density dependence as determinants of large-scale, long-term woody plant-cover dynamics.

Although there is empirical evidence to support the different coexistence models, to date it has been largely small-scale, short-term, and site-specific, and has therefore precluded any global consensus (House et al., 2003). Both competition-based and demographic-bottleneck models have helped to increase our understanding of savanna dynamics, but as Sankaran et al. (2004) point out, "the ecological reality of tree-grass competition is more likely to lie somewhere in between, with the competitive dominance of trees with respect to grasses changing with life-history stage, over time and across environmental gradients." They put forward a framework that integrates demographic and competitive approaches to tree-grass coexistence issues in savannas (Figure 15.2). Demography is expressed in terms of recruitment rates between different life-history stages of trees, and both tree-on-tree and tree-grass competition occurs at each life-history stage. The different factors influencing savanna dynamics, rainfall variation, fire, and herbivory are then considered in terms of their influence on recruitment and competition. For example, rainfall variability and fire seasonality can directly affect tree recruitment through mortality of seedlings and through changes to the competitive balance between trees and grasses.

Scanlon et al. (2005) have recently shown that on sandy soils along the Kalahari Transect in southern Africa, savanna systems are maintained by differential responses to fluctuating water resources, with trees responding to variability in rainfall over longer time periods and grasses responding rapidly over shorter time scales. Sankaran et al. (2004) suggest using their proposed framework as a conceptual tool for developing synthetic models of savanna systems, as an aid to identifying the kinds of empirical data needed for the validation and testing of models, and in the context of applied problems of savanna management. As land use in savannas is expanding and intensifying, and climate is changing, it will be extremely important to understand savanna dynamics in relation to the relative proportions of woody and herbaceous vegetation, and their potential implications for resource conservation and management.

15.3 HETEROGENEITY

Savannas are spatially and temporally dynamic heterogeneous landscapes, determined by a range of factors that are hierarchically linked. The importance attached to scale and hierarchy in savanna studies was developed in the late 1980s when hierarchy theory was used as a conceptual

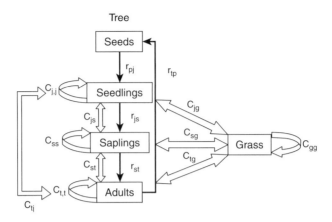

Figure 15.2 A framework for integrating demographic and competitive approaches to tree-grass coexistence issues in savannas (taken from Sankaran et al., 2004). Four recruitment rates r_{xx} express transitions between relevant life-history stages of trees (demography). Both inter- and intralife-form competition occur at each stage and are indicated by competitive indices c_{xx}. All interactions are embedded within a local environmental context (soil moisture, soil texture, nutrient status) that regulates the recruitment and competitive indices shown. Finally, the environment varies in both space and time

tool for positioning ecological determinants of savannas over time and space (Solbrig, 1991). However, it was only after the paradigm shift from equilibrium dynamics to a disequilibrium view of savanna functioning that the idea of hierarchy theory was taken a step further with hierarchical patch dynamics (Wu and Loucks, 1995; Wu and David, 2002). Hierarchical patch dynamics conceptualizes ecological systems as nested hierarchies of patch mosaics while integrating aspects of equilibrium, nonequilibrium, and multiple equilibria systems. A patch, defined as a spatial unit differing from its surroundings in nature or appearance (Kotliar and Wiens, 1990), can be characterized in terms of size, shape, content, duration, structural complexity, boundary characteristics, and dependence on organisms. Different causes and mechanisms operate on different spatial, temporal, and organizational scales, which determine their equilibrium/nonequilibrium properties, thus creating a hierarchical framework of patch determinants. In her study using ^{13}C isotopes from soil profiles and fossil pollen from sedimentary sequences, Gillson (2004a) investigated the changes in savanna tree abundance at micro (10^{-2} m^2), local (10^{-1} km^2), and landscape (10^2 km^2) scales in Tsavo National Park, Kenya, over hundreds of years. She found that the pattern and process of vegetation change were different at the three spatial scales studied (Figure 15.3), consistent with the suggestion that savanna systems

are hierarchically organized. Again using palaeoecological techniques, Gillson (2004b) further showed that over the last 1,400 years in Tsavo National Park, the vegetation surrounding Kanderi Swamp had gone through a series of changes explained by different nonequilibrium models. However, the 'equilibrium' status of the vegetation depended on the scale of observation, supporting the hierarchical patch dynamics notion of scale-dependent dynamics.

Patches are manifestations of heterogeneity. According to Kolasa and Rollo (1991), heterogeneity can be deterministic and random or chaotic in origin, and many kinds of processes and agents (both physical and biological) can produce heterogeneity. Although agents create heterogeneity on specific scales, the effects of heterogeneity can appear either on the scale of action or on different scales (i.e., sources and patterns of heterogeneity can be hierarchically arranged). Heterogeneity may be continuous or discontinuous and be expressed as gradients, patchworks, or graded patchworks. Using this idea of an 'agent' of heterogeneity, Pickett et al. (2003) propose four key features of heterogeneity that interact to create patterns of savanna heterogeneity: (a) 'agents' are drivers of heterogeneity, can be physical or biological, and can act in the short-term or be persistent or slowly changing factors; (b) 'substrates' are the entities upon which the agents act; (c) 'controllers' can modify the behavior of

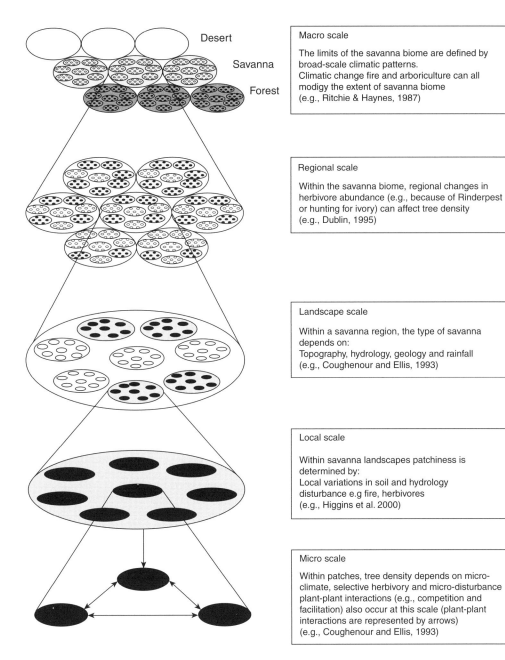

Macro scale

The limits of the savanna biome are defined by broad-scale climatic patterns.
Climatic change fire and arboriculture can all modigy the extent of savanna biome
(e.g., Ritchie & Haynes, 1987)

Regional scale

Within the savanna biome, regional changes in herbivore abundance (e.g., because of Rinderpest or hunting for ivory) can affect tree density
(e.g., Dublin, 1995)

Landscape scale

Within a savanna region, the type of savanna depends on:
Topography, hydrology, geology and rainfall
(e.g., Coughenour and Ellis, 1993)

Local scale

Within savanna landscapes patchiness is determined by:
Local variations in soil and hydrology disturbance e.g fire, herbivores
(e.g., Higgins et al. 2000)

Micro scale

Within patches, tree density depends on micro-climate, selective herbivory and micro-disturbance plant-plant interactions (e.g., competition and facilitation) also occur at this scale (plant-plant interactions are represented by arrows)
(e.g., Coughenour and Ellis, 1993)

Desert

Savanna

Forest

Figure 15.3   A hierarchy of processes that determine tree density in savannas through their effects on tree dispersal, germination, recruitment, and mortality (taken from Gillson, 2004a)

agents or the susceptibility of substrates; (d) 'responders' may be processes or organisms that are sensitive to the state of the substrate, and they can differ compositionally (containing different species or processes), or quantitatively (in magnitude or rate). These four features of

heterogeneity can be used to explain the causes and mechanisms of patch creation within the hierarchical patch dynamics concept (see Pickett et al.'s 2003 example of how patches are created by fire). At any one scale, fire is the agent of heterogeneity. It affects a patch (as defined by

Kotliar and Wiens, 1990) of vegetation or substrate and changes the state of that vegetation by consuming the fuel biomass. The factors controlling fire include the characteristics of the fuel, most notably the biomass (also known as fuel load) and moisture content of the fuel, the topography of the patch, the degree of herbivory taking place on the vegetation, and the local climate. The responders, which are sensitive to the new state of the patch, include vegetation, in terms of the differential response of plant species to the consumed vegetation, PAM, PAN, herbivory, and local climate. Over time, these responders will affect the controllers, which in turn will affect the susceptibility of the new state of the patch to further fire. At a level higher in patch dynamics are PAM and PAN; both PAM and PAN can be 'agents' (maintaining or transforming patches of vegetation), while PAM can also be a controller (modifying the susceptibility of the vegetation to fire).

Factors such as geology and climate produce a heterogeneous savanna physical template upon which biologically generated heterogeneity is layered. Pickett et al. (2003) emphasize the role of ecological engineering and boundary function as ecological processes by which organisms add new layers of heterogeneity to the physical template. Ecological engineering is the direct or indirect modulation via physical, chemical, or transport processes, by one kind of organism, of the availability of resources to other organisms (Jones et al., 1994; Lawton and Jones, 1995). For example, the transformation of a physical structure has ecological consequences because the structure directly controls or indirectly modulates the flow of consumable resources (including energy, materials, and space) used by other species (Pickett et al., 2000). Through consuming plants and causing soil erosion, savanna grazers directly and indirectly accentuate the contrast between sodic sites with their sparse, low-stature woody vegetation and adjacent taller, denser savanna or riparian zones. This change in physical state is essential in the heterogeneity model.

Boundaries exist as discontinuities between contrasting patches. They can be gradual in response to transitions in soil, water, or other variables at various scales (e.g., the continuum of formations in the Brazilian cerrado, Coutinho, 1978), or sharp and distinct, mainly a result of human activity, ecological engineering, or intense physical disturbance (Pickett et al., 2003; Hennenberg et al., 2008). Whether gradual or sharp, boundaries can affect the functioning of populations, landscapes, and ecosystems by modulating fluxes of materials, energy, organisms, or information (Pickett and Cadenasso, 1995). For example, recent studies show that the African grass, *Melinis minutiflora*, invasive in the

Neotropics, forms a boundary zone between cerrado and gallery forest patches (Hoffman et al., 2004). Due to its high flammability and high biomass, one major function of this *M. minutiflora* boundary is as a corridor of propagation through which fire can move extremely fast (Mistry and Berardi, 2005). Boundaries have a functional role in the heterogeneity generated by organisms (Pickett and Cadenasso, 1995) in that they control the expression of engineering over space. However, because organisms as mobile engineers can interact with boundaries, boundaries can also be controllers affecting the agent, substrate, and responder in Pickett et al.'s (2003) heterogeneity model.

The model shown in Figure 15.4 is very simple and in no way illustrates the complexity of patch dynamics. However, if you imagine the process in Figure 15.4 taking place over various spatial scales and over time, and then imagine several other agents going through a similar process of patch creation, and all of these models interacting to create further patches of different spatial and temporal grain, we may be somewhere near being able to understand the heterogeneity of the savanna landscape. It is also important to note that what is an agent in one place or scale may be a substrate, controller, or responder to heterogeneity at other places or scales (Allen and Hoekstra, 1992). For example, vegetation in the form of an individual tree can be a biotic agent of heterogeneity when it is a node of changing soil and climatic conditions around it, it can be a substrate when acted on by fire or herbivory, it can be a controller when it can modify fuel characteristics and hence fire, and it can be a responder when establishing on new soil.

Hierarchical patch dynamics and the model of savanna heterogeneity proposed by Pickett et al. (2003) are particularly illustrative of the dynamics of heterogeneity over space, but less so over time. The temporal nature of patch dynamics is probably best exemplified by Holling's (1986, 2001) 'adaptive renewal cycle' (Figure 15.4). This heuristic model attempts to capture the regular cycles of organization, collapse, and renewal, which are features of both ecological and social systems. The cycle has four phases—exploitation, conservation, release, and reorganization—which repeat themselves over and over again and can be compared to the dynamics of heterogeneity. If we imagine the cycle representing a patch, then in the first phase of exploitation, pioneer savanna plants establish in the patch. As nutrients and biomass are consolidated, the patch goes into the conservation phase of the cycle. It remains in this phase until an agent of heterogeneity occurs, for example, fire, when the accumulated capital in the patch is suddenly released, producing other kinds

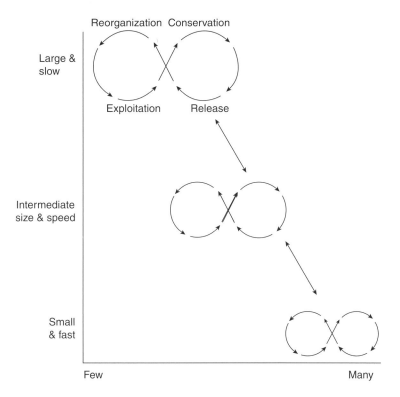

Figure 15.4   The adaptive renewal cycle and panarchy (adapted from Berkes et al., 2003)

of opportunity, which is termed creative destruction (Holling, 2001). This third phase of release, which is very rapid, is followed by reorganization when the new state of the patch is created and responders allow the cycle to repeat itself, albeit in a new trajectory. As shown in Figure 15.4, the adaptive renewal cycle can work at many scales, corresponding to scales of patch dynamics. At the smallest scale, there are many cycles occurring during a short time space. As the patches get larger, the cycles become bigger and slower, but also fewer in number. It is also important to remember that many of the cycles, at all scales, will be overlapping. These hierarchically scaled cycles, nested within one another across space and time, constitute what Gunderson and Holling (2002) term a 'panarchy'. Central to the panarchy idea is resilience—that the magnitude or scale of disturbance that can be absorbed before these nested systems change in structure by changes in variables and processes that control system behavior (Holling, 1973). In this context, resilience is a measure of robustness and buffering capacity of these systems to changing conditions.

The idea that savannas go through regular cycles of organization, collapse, and renewal links to, for example, recent ideas of tree-grass coexistence in savannas (e.g., Higgins et al., 2000; Jeltsch et al., 2000; House et al., 2003; Sankaran et al., 2004). Berkes et al. (2003) have suggested that ecological systems build resilience through disturbance, in that the experiences of disturbances allow the system to 'problem-solve', innovate, and adapt. In savannas, disturbances such as fire may indeed be vital for the long-term resiliency of the system as reflected by various studies of the effects of fire on the coexistence of trees and grasses (e.g., Jeltsch et al., 1998) and on patch mosaic fire management (Parr and Brockett, 1999: see below).

15.3.1 Factors causing savanna heterogeneity

Savanna heterogeneity comes about through various abiotic and biotic factors and processes, working at different spatial scales over varying temporal scales. Trying to unpick this intricate web of connections is a difficult task. However, looking at some examples can help us understand the complexity of the heterogeneous savanna system.

The following section will look at the role of large trees, termitaria, and ground-level vegetation as agents of heterogeneity, causing changes in the soil substrate, with primary responses from PAM and PAN. Fire is an important agent of patch creation and is discussed below, as is herbivory, which can play the role of agent in its direct effects through vegetation consumption, responder through its differential response to, for example, PAM, and controller through its influence on grass biomass and therefore fire.

The coexistence of trees and grasses in savannas has already been discussed above. Savanna trees undoubtedly have a large impact on the herbaceous vegetation underneath, by changing resource availability, namely moisture, nutrients, and light (Belsky, 1989, 1994; Weltzin and Coughenour, 1990; Mordelet and Menaut, 1995; Barnes and Archer, 1999; Dean et al., 1999; Treydte et al., 2008), and acting as agents of heterogeneity. Although the process is not clear (see Ludwig et al., 2004), trees could facilitate increased PAM under canopies through decreased transpiration of understory plants, lower shade temperatures, and/or hydraulic lift (the process of water movement from deep, wet layers to shallow drier layers through plant roots) (e.g., Ludwig et al., 2001). Higher PAN has also been reported under savanna trees (Kellman, 1979; Belsky, 1989; Ludwig et al., 2004; Hagos and Smit, 2005), possibly through trees taking up nutrients from deeper soil layers or outside canopies and depositing them under their canopies through leaching or litter fall (Kellman, 1979; Scholes, 1990; Mlambo et al., 2005). Another hypothesis is that trees attract mammals that may deposit their dung under their canopies (Belsky, 1994; Georgiadis, 1989).

As with large trees, termitaria can also act as patches in savannas, influencing soil characteristics, particularly soil water and nutrient properties, and consequently plant distribution and vegetation dynamics. Konaté et al. (1999) investigated texture, structure, water availability, and water potential of the soil in a large termite mound at different depths and in surrounding areas, and the leaf-shedding patterns of two coexisting deciduous shrubs in the humid West African savanna. They found that soil texture and structure were strongly modified on the termite mound, resulting in a higher soil water content availability for plants in the upper soil layers of the mound, compared to the surrounding area. In the middle of the dry season, *Crossopteryx febrifuga* (which has limited access to soil layers below 0.6 m) had less pronounced leaf shedding on mounds compared to surrounding areas. In contrast, *Cussonia barteri* (which has good access to deep soil layers) was not influenced by the

termite mound. Termite-built structures can also increase PAN and consequently vegetation. Both below-ground, fungus-comb chambers (scale of $1–100$ cm^2; Abbadie and Lepage, 1989) and above-ground lenticular mounds (scale of $1–100$ m^2; Konaté et al., 1999) are often enriched in small particles, organic matter, and mineral nitrogen (Abbadie and Lepage, 1989; Jouquet et al., 2002). Jouquet et al. (2004) has recently shown that specific grass species have preferences for nutrient-rich fungus-comb chamber patches, with particular grasses taking advantage of the nutrients available from the wall of the fungus-comb chambers by developing more roots in these areas.

Vegetation features of the savanna landscape may be very important for capturing and retaining water resources in patches, helping to dampen or amplify temporal variability in PAM, especially in the more arid savannas of the world. Tongway and Ludwig (1997) found that in the dry savannas of northern Australia, rainfall was concentrated into fine-scale vegetation patches (less than 5 m in width) rather than being uniformly dispersed over the landscape. Sampling sites for patchiness along a 1,000-mm rainfall gradient in the Australian savannas, Ludwig et al. (1999a) suggest that in higher rainfall savannas, the size and spacing of ground-layer patches are controlled by the tree layer through competitive interference. However, as rainfall decreases, this control by trees decreases, and runoff-runon processes, whereby rainfall is collected through runoff from large open areas (termed 'sources, fetches') and captured within small runon areas (termed 'sinks, patches'), increasingly structure the landscape.

Looking more closely at patterns of landscape patchiness, Ludwig et al. (1999b) identify three types: stippled (dispersed patches), striped (larger, elongated patches), and stranded (long linear, basal patches). These may influence how efficiently the savanna captures and stores scarce soil resources. Using a landscape simulation model and validation data from a semiarid savanna woodland site in eastern Australia, Ludwig et al. (1999b) found that stripes and strands captured about 8% more rainfall as soil water than a stippled pattern, increasing plant production by about 10%. If landscape patchiness was lost, say through severe land degradation, this reduced the capacity of the landscape to capture rainfall as soil water by about 25%, reducing net primary productivity in these systems by about 40%. The importance of understanding smaller-scale spatial heterogeneity was emphasized by Caylor and Shugart (2004) in a study simulating annual productivity along the Kalahari Transect in southern Africa. With the aim of assessing the effect of patch-scale heterogeneity in savanna vegetation structure on estimates of annual net primary productivity and annual tree/grass water-use efficiency, they found that when aggregation was performed on the spatial mosaic, simulations resulted in drastically reduced distributions of productivity due to reduced structural heterogeneity. The grouping of patches was seen to eliminate information regarding demographic processes such as regeneration and mortality and the dependence of grass productivity on over-story density (Caylor and Shugart, 2004).

Working at a larger patch size, Wu and Archer (2005) compared the variation in topography-based hydrologic features along catenas in Texan savannas using the topographic wetness index (TWI). The TWI represents the potential relative moisture content of a given location in a catchment based on slope inclination and upslope catchment area (Beven and Kirkby, 1979, cited in Wu and Archer, 2005). They found that at the catena scale, there were significantly greater TWI values for woodland areas on the slope foothills than for upland savanna parklands on the higher parts of the catena. They suggest that patterns of runoff and runon are a significant determinant of woody plant cover, stature, and pattern of distribution at the landscape or catena scale. It is hypothesized that over time runoff-runon relationships interact with soil texture and annual rainfall patterns to create unique assemblages of savannas. In the Texan savannas, Wu and Archer (2005) suggest that runoff reduces the effectiveness of rainfall and limits woody plant growth so that parklands establish on the coarse-textured soils of the uplands, whereas the runoff from the uplands is sufficient to support the establishment of woodlands on the fine-textured soils of the lowland landscape locations. This topography-soil texture-rainfall template is overlaid with disturbances that include grazing, fire atmospheric CO_2 enrichment, and nitrogen deposition, to produce further heterogeneity.

Fire has a major impact on savanna structure and functioning and is an important agent of heterogeneity. Using remote sensing analyses, Hudak et al. (2004) have recently shown that, compared to fire exclusion, regular fire occurrence in the savannas of southern Africa promoted landscape heterogeneity. Interestingly, much of the work highlighting the role of fire in savanna heterogeneity has come from understanding the way indigenous and traditional savanna communities burn their landscapes. Australian Aborigines (Haynes, 1985, 1991; Lewis, 1989; Russell-Smith et al., 1997), Brazilian indigenous groups (Posey, 1985; Anderson and Posey, 1985, 1989; Mistry et al., 2005), and traditional communities in West Africa (Mbow et al., 2000; Laris, 2002) all burn

vegetation in what has been termed a 'patch-mosaic fire regime' (Parr and Brockett, 1999). This involves burning from the early dry season through to the late dry season—burning throughout the dry season results in a landscape pattern composed of patches in various stages of fire succession intermingled with unburned patches. Producing a mosaic of patch types of differing fire histories within the landscape (i.e., a heterogeneous landscape) promotes the creation of natural firebreaks within the landscape where burned patches act to 'fireguard' particular vegetation patches (Laris, 2002; Mistry et al., 2005), and maintains and enriches biodiversity through the creation and preservation of a variety of micro-habitats that support different species (Mentis, 1978; Mentis and Bigalke, 1981; Rowe-Rowe and Lowry, 1982; Braithwaite, 1996; Vigilante and Bowman, 2004). In addition, evidence from Kakadu National Park, Australia, shows that early dry season fires excluded from savanna areas for several years had little impact on wet season streamwater quality (Townsend and Douglas, 2004), compared to late dry season fires, which had statistically significant higher storm runoff concentrations of total and volatile suspended sediment, iron, and manganese (Townsend and Douglas, 2004. Creating fire patchiness may therefore also be important for savanna riparian communities.

Animals significantly affect savanna heterogeneity through their actions (see Table 15.3), in particular through the consumption of vegetation and the interaction between herbivory and other savanna determinants. Most work in this area has been carried out in African savannas, where large herbivores are a major modifier of the savanna landscape. The high species richness among large herbivores in many African savannas has been linked to the high spatial heterogeneity at the ungulate habitat scale (du Toit, 2003). It has been suggested that through palaeoclimate change, the savanna biome expanded and contracted over evolutionary time, driving cyclic generation of spatial heterogeneity and concurrent episodes of allopatry and sympatry in ungulates. Large mammals are agents of heterogeneity through their grazing, browsing, and pushing over of vegetation. In addition, they can disturb the soil and change its characteristics through the creation of trails, latrines, and burrows (Belsky, 1985, 1988).

The distribution and density of African savanna herbivores is primarily determined by PAM, namely distance from water, and the composition, quality, and structure of the vegetation (McNaughton and Georgiadis, 1986; Skarpe, 1991; Ogutu et al., 2008). PAM and vegetation are therefore the main controllers of herbivory as a

Table 15.3 The processes by which different animal groups affect savanna heterogeneity in Kruger National Park, South Africa (adapted from Naiman et al., 2003)

Group	Process	Examples
Insects	Pollination	Honeybees, fig wasps
	Decomposition and nutrient cycling	Termites, wood-boring beetles, dung beetles, carrion-frequenting insects such as blowflies
	Herbivory	Termites, grasshoppers
Birds	Seed dispersal	Green pigeon (*Treron calva*) and tinkerbirds (Capitonidea)
	Pollination	Tinkerbirds (Capitonidea)
	Dispersal of nutrients and disease from carrion	Vultures (*Gyps* species)
	Removal of ectoparasites from large herbivores	Oxpeckers (Sturnidae)
	Predation pressure on small animals	Diurnal raptors
Large herbivores	Herbivory	Various, including elephants, giraffes, wildebeest
	Nutrient deposition	Various, including elephants, giraffes, wildebeest
	Soil erosion	Various, including elephants, giraffes, wildebeest
	Seed dispersal	Various, including elephants, giraffes, wildebeest
	Woody plant death	Elephants
Small mammals	Seed dispersal	Springhares, mole-rats
	Nutrient cycling	Aardvarks
	Woody plant death	Bark-gnawing porcupines
Large carnivores	Death of herbivores	Lions, cheetahs, spotted hyenas

heterogeneity agent, and herbivory can be a responder to a modified patch. It has been known for some time that resource separation by African savanna herbivores takes place (e.g., Bell, 1970; Prins and Beekman, 1987; Prins and Iason, 1988; De Bie, 1991), but recent work has emphasized the functional aspect of landscape heterogeneity. Using modeling, Owen-Smith (2004), for example, explored how resource heterogeneity in African savannas might determine herbivore population dynamics. He found that functional heterogeneity requires having high-quality resources to enable herbivore populations to grow and also having reserve or buffer resources to sustain the population after favorable resources have been depleted or are no longer accessible (Owen-Smith, 2002). The lower-quality resources act as a buffer against starvation during critical periods of the seasonal cycle. However, if resources are heterogeneous, but all nutritionally favorable, food quantity becomes the limiting factor and can lead to population crashes.

The spatial heterogeneity of PAM is a major controller of herbivore distributions, especially

during the dry season (Western, 1975; Fryxell and Sinclair, 1988; Bergstrom and Skarpe, 1999; Redfern et al., 2005; Shannon et al., 2008). To investigate the importance of surface water availability on herbivore distributions, Redfern et al. (2005) assessed the contribution of perennial and ephemeral water sources to surface water availability in the dry season in Kruger National Park, South Africa. Perennial water sources included rivers, springs, dams, and boreholes. Ephemeral water sources varied over a range of temporal scales, from small pans created by dry season rainstorms that were only available for a few weeks, to pools in seasonal rivers available throughout the dry season for one or more years, formed by the cumulative effect of the preceding years' rainfall.

Redfern et al. (2005) propose that surface water availability in savannas exists along a continuum of being determined primarily by perennial water sources or primarily by ephemeral water sources. Rather than being placed at a particular point of this continuum, savannas can be placed within a band determined by the amount and variability of rainfall. The left end of the band will occur toward the perennial end of the continuum, but the right end of the band will expand toward ephemeral water sources in savannas experiencing higher annual rainfall, during extremely wet years or when multiple years of above average annual rainfall coincide. Herbivore species can be placed along this continuum according to relative expectations of whether perennial or ephemeral water sources exert a greater influence on their distributions. For example, water-dependent herbivores that are large and mobile, such as buffalo, zebra, and wildebeest, will be less limited by the locations of perennial water sources and more influenced by ephemeral water sources. On the other hand, herbivores that are smaller, have a fixed territory, or are associated with a water habitat, will be more limited by perennial water sources. Nevertheless, whether a species will disperse to an ephemeral water source will depend on many factors that will vary over time, including the number of individuals the water source can support, the duration of water containment, the quality of the water, and the quality and cover of the forage (Redfern et al., 2005), highlighting the fact that there are many attributes of PAM as a controller of herbivory and therefore savanna heterogeneity.

The interaction between fire and grazing, although acknowledged as important for patch dynamics, has been little researched in savannas (see Belsky, 1992). Archibald et al. (2005) recently investigated the interactions of fire and grazing on grass communities in South Africa. Within the savanna landscape, grazing patches are maintained through positive feedback between grazing and palatability—previously grazed areas tend to get regrazed and kept short (Adler et al., 2001), and a patchy mosaic of palatable (short) and less palatable (tall) grasses can develop. Fire can alter the persistence of these patches by attracting herbivores into burned areas where regrowth after fire is palatable and nutritious (e.g., van de Vijver et al., 1999; Gureja and Owen-Smith, 2002; Tomor and Owen-Smith, 2002; Hassan et al., 2008). After fire, the whole area consists of short new regrowth, allowing herbivores to spread more evenly through the environment and have a homogenizing effect (Archibald and Bond, 2004). Over the longer term and with frequent fires, the effect of fire drawing herbivores off grazed patches (termed 'magnet effect', sensu Archibald et al., 2005) could lead to the domination by tall, fast-growing, fire-adapted grass species. However, modeling work (Archibald, 2003, cited in Archibald et al., 2005) suggests that the extent to which fire-grazing interactions influence savanna systems will depend also on abiotic factors, particularly rainfall. The effect of fire-grazing interactions will probably be less pronounced during very wet periods as grazed patches will be less persistent because of high grass productivity. It is during drier years that fire could be used to manipulate grazing areas with fire. While fire is advocated for promoting savanna heterogeneity (e.g., Fuhlendorf and Engle, 2001; also see above), understanding its interaction with grazing animals will be vital, especially for savanna rangelands.

15.4 MANAGING TROPICAL SAVANNAS

Maintaining and responding to savanna heterogeneity will be essential for the long-term sustainability of savannas and the ecosystem services they provide. PAM, PAN, fire, and herbivory are the main ecological agents, controllers, and responders of savanna heterogeneity, but the most important determinant of savanna ecosystems is human activity. Sala et al. (2000) identified the main agents of change in savanna ecosystems as land-use change, elevated CO_2, increased nitrogen deposition, climate change, and alien biota introductions, all anthropogenic in nature. Understanding how humans interact with savanna heterogeneity is vital if we are to make recommendations for effective savanna management. For example, human management interventions for wildlife management can significantly affect savanna heterogeneity. The introduction of artificial water sources, such as boreholes, can seriously compromise savanna heterogeneity by causing zones of bush encroachment (Roques et al., 2001), which over time have a tendency to coalesce, causing relatively extensive homogeneous areas to have reduced quality in terms of potential grazing. Redfern et al. (2005) have suggested that borehole removal from Kruger National Park (which is currently underway) will increase the temporal and spatial heterogeneity of surface water availability, and consequently habitat heterogeneity created by water-related herbivore impacts such as soil trampling, nutrient concentration, and alteration of vegetation composition and structure.

Land-use change is one of the most important factors affecting savannas today, with large areas being converted for agricultural and other small- and large-scale activities. However, as recent studies have emphasized (e.g., Reenberg, 2001; Elliott and Campbell, 2002; Wardell et al., 2003; Brannstrom et al., 2008), land-use change is multidirectional, event-driven, and determined by patch dynamic processes. Understanding how these changes in land use affect the heterogeneity of the key determinants—PAM, PAN, fire and herbivory—is vital for managing these savanna landscapes. For example, elephants can live within agricultural landscapes where agricultural fields divide natural habitats into discontinuous patches of different spatial arrangement, but it is important to understand how elephants respond to this spatial heterogeneity in landscape properties such as vegetation cover. Using remote sensing of vegetation cover and point data sets of elephant populations from the northwest savanna region of Zimbabwe, Murwira and Skidmore (2005) investigated whether and how the spatial distribution of the African elephant responded to spatial heterogeneity of vegetation cover based on data of the early 1980s and early 1990s. Their results indicated that the peak probabilities of elephant presence are associated with high variability in vegetation cover that occurs at intermediate heterogeneity scales, with peaks at 734 m in the early 1980s and 457 m in the early 1990s. Murwira and Skidmore (2005) suggest that the 457–734 m dominant scales of heterogeneity may define the optimal range at which elephant persistence in agricultural landscapes can be ensured, but elephants may be threatened when they are above and below these thresholds. Looking more closely at the changes that occurred between the early 1980s and early 1990s, they deduced that in the early 1980s when agriculture was not so intensive, elephants could roam freely across the agricultural fields or patches. However, with intensification of activities in the 1990s, the patch size associated with peak elephant presence shifted downward to 457 m, suggesting that elephants were being restricted by the relatively smaller dimensions of agricultural patches. Murwira and Skidmore (2005) went on to investigate whether elephants respond to changes in the intensity and dominant scale of spatial heterogeneity over time. Their findings indicate that elephant presence persisted and even increased in land units where there was no change in terms of spatial heterogeneity, and elephant presence increased in situations when there was an increase in both intensity and dominant scale of spatial heterogeneity. This suggests that environments with constant levels of spatial heterogeneity are the most conducive to elephant persistence, a requirement that is probably incompatible with rapidly changing savanna land uses.

Whereas in the past many studies treated humans as outsiders or as an impact, more recent examples show that in fact, humans are integral components of savanna systems (e.g., Behnke and Scoones, 1993; Tiffen et al., 1994; Warren, 1995; Fairhead and Leach, 1996; Sullivan, 1996; Scoones, 1997; Cline-Cole, 1998; Mortimore, 1998; Dougill et al., 1999; Laris, 2002; Bassett and Crummey, 2003; Igoe, 2004). Within the arid, pastoral-dominated savannas, where the debate over equilibrium and nonequilibrium dynamics has been largely focused (e.g., Illius and O'Connor, 1999, 2000; Sullivan and Rohde, 2002), management following the equilibrium model led to government interventions such as destocking schemes, conversion of communal areas into individually managed units, and the settling of nomadic pastoralists into group ranches (Sandford, 1983; Ellis and Swift, 1988). However, as nonequilibrium ideas gained momentum, the focus has shifted to considering spatial heterogeneity and climatic variability, mobility, and variable stocking rates as essential for effective and sustainable savanna management (e.g., Gillson and Lindsay, 2003; Vetter, 2005). This new approach to understanding savannas has also made us rethink degradation, which was thought to be widespread in arid savannas through overgrazing. Different areas and parts of the landscape differ in their susceptibility to transformation, so although degradation does not occur everywhere, it can occur in arid savanna landscapes, particularly if grazing pressure is concentrated on certain parts of the landscapes for prolonged periods of time (see Vetter, 2005 for a review of recent developments in this debate).

Management of savannas must look toward emphasizing the coequal interactions of both natural and human factors in what Berkes and Folke (1998) term 'social-ecological systems'. Social-ecological systems are human-in-nature systems where there are no boundaries between the human and the ecological, and where interactions between the two domains over a range of scales sustain the system over space and time (Berkes and Folke, 1998). The shift from equilibrium to nonequilibrium views in savannas places complexity and unpredictability at the forefront of management approaches. To handle the increasing complexity and unpredictability of social-ecological systems, management must be adaptive and has to deal with the unpredictable interactions between people and ecosystems as they evolve together (Holling, 1978; Walters, 1986). As Rogers (2003) states, "because the heterogeneous nature of savannas is in a continual state of flux and our understanding of ecosystem function is poor, we have to deal with uncertainty from an imperfect knowledge base—this is the basis of adaptive management in savannas."

The complexity involved in tropical savanna management, characterized by multiple and often conflicting objectives, inherent unpredictability, and decentralized control, calls for an approach to management where learning is at its center (Berardi and Mistry, 2006). The notion of adaptive management suggests that more local-level, participatory, community-based forms of management involving various stakeholders are required (Berkes, 2004). Although there is a shift toward local-level, community involvement in various issues surrounding savanna management (e.g., wildlife utilization, protected areas management), the one-way, top-down, command-and-control style of management is still predominant. This is a major inhibition to social learning because feedback given by people who are intimately linked to savanna systems to researchers and institutionalized decision makers (mainly sitting in urban offices) is severely curtailed (Berardi and Mistry, 2006).

15.5 FUTURE RESEARCH IN SAVANNAS

So what of the future for savanna management and research? Throughout this chapter, I have referred to various types of models used to try to understand savanna functioning and dynamics.

There is now a growing and urgent need to validate and test these models using empirical data across diverse savanna systems. In 2000, I listed the major gaps in our knowledge about savanna processes according to savanna formation (Mistry, 2000, see Table 15.4). I would say that most of these still exist. However, what is lacking from this list is a sense of space and time. In a recent review of savanna research, Furley (2004) makes three observations: first, there is a need for more long-term studies of savannas on all aspects of savanna ecology and management; second, even though there is a greater emphasis on spatial aspects, studies incorporating both space and time are required; and third, there is a need for more comparative studies between species, processes, and areas and at a variety of scales. Although I would agree with all of these points, I would also argue that future studies need to focus not only on the topic but also the approach.

Within the social-ecological system and adaptive management framework, a multi-perspective, multidisciplinary approach is necessary for

Table 15.4 The gaps in our knowledge about the major tropical savanna formations of the world (taken from Mistry, 2000)

Savanna	Some major knowledge gaps
Cerrado	Faunal population dynamics, and interactions with savanna functioning. Their responses to fire
	Microorganisms, mychorrhizal associations
	Herbivory, especially invertebrate
	Indigenous peoples' savanna management
	Effect of land conversion to agriculture on savanna species
	The ecology of invasive grasses
Llanos	The ecology of woody plants
	Faunal population dynamics, and interactions with savanna functioning
	Nutrient cycling and adaptations to PAN
	Herbivory
	Factors affecting fire and fire behaviour
	Indigenous peoples' savanna management
	The effect of dikes on savanna ecology
	The ecology of invasive grasses
Miombo	Faunal population dynamics, and interactions with savanna functioning
	The role of termites
	Herbivory by invertebrates and mammals other than elephants
	Fire behaviour
West African savannas	The ecology of the Sahelian, Sudanian, and Guinean savanna vegetation types (e.g., phenology)
	Faunal population dynamics, and interactions with savanna functioning
	Invertebrate herbivory
	Factors affecting fire and fire behaviour
	Indigenous peoples' savanna management
	The effect of hunting on wildlife populations
	Wood consumption

Continued

Table 15.4 Cont'd

Savanna	Some major knowledge gaps
East African savannas	The ecology of savanna vegetation (e.g., grass phenology) Insect herbivory Fire ecology (e.g., factors affecting fire, fire behavior, effects of vegetation and animals) The effect of tourism on savanna ecology
Southern African savannas	Plant phenology The role of soil fauna, such as termites, in nutrient cycling Factors affecting fire such as fuels. Effect of fire on faunal populations The effects of alien plants The ecology of woody encroachment
Dry dipterocarp savannas of Southeast Asia	General ecology of savanna vegetation Faunal population dynamics, and interactions with savanna functioning. Their responses to fire Nutrient cycling Herbivory Rates of deforestation, effects on savanna ecology
Australian savannas	Faunal population dynamics, and interactions with savanna functioning. Their responses to fire The role of the El Niño Southern Oscillation on savanna dynamics Invertebrate herbivory The effect of tourism on savanna ecology The effect of invasive plants and animals

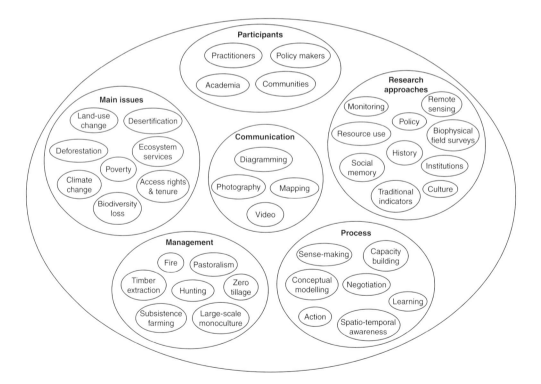

Figure 15.5   A system to link people with nature in savanna research (taken from Berardi and Mistry, 2006)

understanding savanna systems and their management (Berardi and Mistry, 2006). 'Hybrid knowledge' (Forsyth, 1996), where both scientific and local land-user views and values are integrated for understanding management issues, and 'hybrid science' (Batterbury et al., 1997)—a multidisciplinary research approach using plural methodologies—are the way forward. Figure 15.5 presents the range of components considered significant within a multidisciplinary and participatory research approach (Berardi and Mistry, 2006). Tropical savannas are dynamic ecosystems, determined by PAM, PAN, fire, herbivory, and humans, at different spatial and temporal scales. Understanding the heterogeneous nature of these determinants and their interactions is a challenge for future savanna management.

REFERENCES

Abbadie, L., and Lepage, M. (1989) 'The role of subterranean termite fungus-comb chambers (Isoptera, Macrotermitinae) in soil nitrogen cycling in a preforest savanna (Côte d'Ivoire)', *Soil Biology and Biochemistry*, 21: 1067–71.

Adler, P.B., Raff, D.A., and Lauenroth, W.K. (2001) 'The effect of grazing on the spatial heterogeneity of vegetation', *Oecologia*, 128: 465–79.

Allen, T.F.H., and Hoekstra, T.W. (1992) *Towards a Unified Ecology*. New York: Columbia University Press.

Anderson, A.B., and Posey, D.A. (1985) 'Manejo de cerrado pelos indios Kayapó', *Boletim do Museu Paraense Emilio Goeldi Botânica*, 2(1): 77–98.

Anderson, A.B., and Posey, D.A. (1989) 'Management of a tropical scrub savanna by the Gorotire Kayapó of Brazil', *Advances in Economic Botany*, 7: 159–73.

Archer, S., Boutton, T.W., and Hibbard, K.A. (2001) 'Trees in grasslands: Biogeochemical consequences of woody plant expansion', in E.D. Schulze, M. Heimann, S. Harrison, E. Holland, J. Lloyd, I. Prentice and D. Schimel (eds.), *Global Biogeochemical Cycles in the Climate System*. San Diego: Academic Press. pp. 115–38.

Archibald, S., and Bond, W.J. (2004) 'Grazer movements: Spatial and temporal responses to burning in a tall-grass African savanna', *International Journal of Wildland Fire*, 13: 377–85.

Archibald, S., Bond, W.J., Stock, W.D., and Fairbanks, D.H.K. (2005) 'Shaping the landscape: Fire-grazer interactions in an African savanna', *Ecological Applications*, 15: 96–109.

Barnes, P.W., and Archer, S.R. (1999) 'Tree-shrub interactions in a subtropical savanna parkland: Competition or facilitation?' *Journal of Vegetation Science*, 10: 525–36.

Bassett, T.J., and Crummey, D. (eds.). (2003) *African Savannas: Global Narratives and Local Knowledge of Environmental Change*. Oxford: James Currey.

Batterbury, S., Forsyth, T., and Thomson, K. (1997) 'Environmental transformations in developing countries: Hybrid research and democratic policy' *Geographical Journal*, 163(2): 126–32.

Beard, J.S. (1953) 'The savanna vegetation of northern tropical America', *Ecological Monographs*, 23: 149–215.

Behnke, R., and Scoones, I. (1993) 'Rethinking range ecology: Implications for rangeland management in Africa', in R.H. Behnke, I. Scoones, and C. Kerven (eds.), *Range Ecology at Disequilibrium: New Models of Natural Variability and Pastoral Adaptation in African Savannas*. London: Overseas Development Institute and International Institute for Environment and Development. pp. 1–30.

Bell, R.H.V. (1970) 'The use of the herb layer by grazing ungulates in the Serengeti', in A. Watson (ed.), *Animal Populations in Relation to Their Food Resources*. Oxford: Blackwell. pp. 111–24.

Belsky, A.J. (1985) 'Long-term ecological monitoring in the Serengeto National Park, Tanzania', *Journal of Applied Ecology*, 22: 449–60.

Belsky, A.J. (1988) 'Regional influences on small-scale vegetational heterogeneity within grasslands in the Serengeti National Park, Tanzania', *Vegetatio*, 74: 3–10.

Belsky, A.J. (1989) 'Landscape patterns in a semi-arid ecosystem in East Africa', *Journal of the Arid Environment*, 17: 265–70.

Belsky, A.J. (1990) 'Tree/grass ratios in Eat African savannas: A comparison of existing models', *Journal of Biogeography*, 17: 483–89.

Belsky, A.J. (1992) 'Effects of grazing, competition, disturbance and fire on species composition and diversity of grassland communities', *Journal of Vegetation Science*, 3: 187–200.

Belsky, A.J. (1994) 'Influences of trees on savanna productivity: Tests of shade, nutrients and tree-grass competition', *Ecology*, 75: 922–32.

Berardi, A. and Mistry, J. (2006) 'A multidisciplinary and participatory research approach in savannas and dry forests' in J. Mistry and A. Berardi (eds.), *Savannas and Dry Forests: Linking People with Nature*. Aldershot, UK: Ashgate. pp. 265–71.

Bergstrom, R., and Skarpe, C. (1999) 'The abundance of large wild herbivores in a semi-arid savanna in relation to season, pans and livestock', *African Journal of Ecology*, 37: 12–26.

Berkes, F. (2004) 'Rethinking community-based conservation', *Conservation Biology*, 18(3): 621–30.

Berkes, F., Colding, J., and Folke, C. (2003) 'Introduction', in F. Berkes, J. Colding and C. Folke (eds.), *Navigating Social-Ecological Systems. Building Resilience for Complexity and Change*. Cambridge: Cambridge University Press. pp. 1–29.

Berkes, F., and Folke, C. (1998) 'Linking social and ecological systems for resilience and sustainability', in F. Berkes and C. Folke (eds.), *Linking Social and Ecological Systems. Management Practices and Social Mechanisms for Building Resilience*. Cambridge: Cambridge University Press. pp. 1–29.

Bond, W.J., Midgley, G.F., and Woodward, F.I. (2003) 'What controls South African vegetation—climate or fire?' *South African Journal of Botany*, 69: 79–91.

Bourlière, F., and Hadley, M. (1983) 'Present-day savannas: an overview', in F. Bourlière (ed.), *Tropical Savannas. Ecosystems of the World 13*. Amsterdam: Elsevier Scientific. pp. 1–17.

Bowman, D.M.J.S., Walsh, A., and Milne, D.J. (2001) 'Forest expansion and grassland contraction within a Eucalyptus savanna matrix between 1941 and 1994 at Litchfield National Park in the Australian monsoon tropics', *Global Ecology and Biogeography*, 10: 535–48.

Braithwaite, R.W. (1996) 'Biodiversity and fire in the savanna landscape', in O.T. Solbrig, E. Medina, and J. Silva (eds.), *Biodiversity and Savanna Ecosystem Processes*. Berlin: Springer. pp. 121–40.

Brannstrom, C., Jepson, W., Filippi, A.M., Redo, D., Xu, Z.W., and Ganesh, S. (2008) 'Land change in the Brazilian Savanna (Cerrado), 1986–2002: Comparative analysis and implications for land-use policy', *Land Use Policy*, 25(4): 579–95.

Burrows, W.H., Henry, B.K., Back, P.V., Hoffman, M.B., Tait, L.J., Anderson, E.R., et al. (2002) 'Growth and carbon stock change in eucalypt woodlands in northeast Australia: Ecological and greenhouse sink implications', *Global Change Biology*, 8: 769–84.

Caylor, K.K., and Shugart, H.H. (2004) 'Simulated productivity of heterogeneous patches in Southern African savanna landscapes using a canopy productivity model', *Landscape Ecology*, 19: 401–15.

Clements, F.E. (1916) *Plant Succession: An Analysis of the Development of Vegetation*. Publication 242. Washington, DC: Carnegie Institute.

Cline-Cole, R. (1998) 'Knowledge claims and landscape: Alternative view of the fuelwood-degradation nexus in northern Nigeria', *Environment and Planning D Society and Space*, 16: 311–46.

Coutinho, L.M. (1978) 'O conceito de cerrado', *Revista Brasileira de Botânica*, 1: 17–23.

Dansereau, P. (1957) *Biogeography: An Ecological Perspective*. New York: Ronald Press.

De Bie, S. (1991) *Wildlife Resources of the West African Savanna*. Wageningen Agricultural University Papers No. 91–92. The Netherlands: Wageningen Agricultural University.

Dean, W.R.J., Milton, S.J., and Jeltsch, F. (1999) 'Large trees, fertile islands, and birds in arid savanna', *Journal of Arid Environments*, 41: 61–78.

Dougill, A.J., Thomas, D.S.G., and Heathwaite, A.L. (1999) 'Environmental change in the Kalahari: Integrated land degradation studies for nonequilibrium dryland environments', *Annals of the Association of American Geographers*, 89(3): 420–42.

du Toit, J.T. (2003) 'Large herbivores and savanna heterogeneity', in J.T. du Toit, K.H. Rogers and H.C. Biggs (eds.), *The Kruger Experience. Ecology and Management of Savanna Heterogeneity*. Washington, DC: Island Press. pp. 292–309.

Elliott, J.A., and Campbell, M. (2002) 'The environmental imprints and complexes of social dynamics in rural Africa: Cases from Zimbabwe and Ghana', *Geoforum*, 33: 221–37.

Ellis, J.E., and Swift, D.M. (1988) 'Stability of African pastoral ecosystems: Alternate paradigms and implications for development', *Journal of Range Management*, 41: 450–59.

Fairhead, J., and Leach, M. (1996) *Misreading the African Landscape: Society and Ecology in a Forest-Savanna Mosaic*. Cambridge: Cambridge University Press.

Fensham, R.J., Fairfax, R.J., and Archer, S.R. (2005) 'Rainfall, land use and woody vegetation cover change in semi-arid Australian savanna', *Journal of Ecology*, 93: 596–606.

Fensham, R.J., Fairfax, R.J., Bowman, D.M.J.S., and Butler, D.W. (2003) 'Effects of fire and drought in a tropical eucalypt savanna colonised by rain forest', *Journal of Biogeography*, 30: 1405–14.

Forsyth, T. (1996) 'Science, myth and knowledge: Testing Himalayan environmental degradation in Thailand', *Geoforum*, 27: 375–92.

Frost, P.G., Medina, E., Menaut, J.C., Solbrig, O., Swift, M., and Walker, B. (eds.). (1986) *Responses of Savannas to Stress and Disturbance*. Biology International Special Issue 10. Paris: International Union of Biological Sciences.

Fryxell, J.M., and Sinclair, A.R.E. (1988) 'Seasonal migration by white-eared kob in relation to resources', *African Journal of Ecology*, 26: 17–31.

Fuhlendorf, S.D., and Engle, D.M. (2001) 'Restoring heterogeneity on rangelands: Ecosystem management based on evolutionary grazing patterns', *Bioscience*, 51: 625–32.

Furley, P. (2004) 'Tropical savannas', *Progress in Physical Geography*, 28: 581–98.

Georgiadis, N.J. (1989) 'Microhabitat variation in an African savanna: Effects of woody cover and herbivores in Kenya', *Journal of Tropical Ecology*, 5: 93–108.

Gillson, L. (2004a) 'Evidence of hierarchical patch dynamics in an East African savanna?' *Landscape Ecology*, 19: 883–94.

Gillson, L. (2004b) 'Testing non-equilibrium theories in savannas: 1400 years of vegetation change in Tsavo National Park, Kenya', *Ecological Complexity*, 1: 281–98.

Gillson, L., and Lindsay, K. (2003) 'Ivory and ecology—changing perspectives on elephant management and the international trade in ivory', *Environmental Science and Policy*, 6: 411–19.

Goldstein, G., Menaut, J.C., Noble, I., and Walker, B.H. (1988) 'Exploratory research', in B.H. Walker and J.C. Menaut (eds.), *Research Procedure and Experimental Design for Savanna Ecology and Management*. Paris: International Union of Biological Sciences. pp. 13–20.

Groen, T.A., van Langevelde, F., de Vijver, C.A.D.M.V., Govender, N., and Prins, H.H.T. (2008) 'Soil clay content and fire frequency affect clustering in trees in South African savannas', *Journal of Tropical Ecology*, 24(3): 269–79.

Gunderson, L.H., and Holling, C.S. (eds.). (2002) *Panarchy: Understanding Transformations in Systems of Humans and Nature*. Washington, DC: Island Press.

Gureja, N., and Owen-Smith, N. (2002) 'Comparative use of burned grassland by rare antelope species in a lowveld game ranch, South Africa', *South African Journal of Wildlife Research*, 32: 31–38.

Hagos, M.G., and Smit, G.N. (2005) 'Soil enrichment by Acacia mellifera subsp. detinens on nutrient poor sandy soil in a semi-arid southern African savanna', *Journal of Arid Environments*, 61: 47–59.

Hassan, S.N., Rusch, G.M., Hytteborn, H., Skarpe, C., and Kikula, I. (2008) 'Effects of fire on sward structure and grazing in western Serengeti, Tanzania', *African Journal of Ecology*, 46(2): 174–85.

Haynes, C.D. (1985) 'The pattern and ecology of munwag: Traditional Aboriginal fire regimes in north-central Arnhemland', *Proceedings of the Ecological Society of Australia*, 13: 203–14.

Haynes, C.D. (1991) 'Use and impact of fire', in C.D. Haynes, M.G. Ridpath, and M.A.J. Williams (eds.), *Monsoonal Australia: Landscape, Ecology and Man in the Northern Lowlands*. Rotterdam: A.A. Balkema. pp. 61–71.

Helsa, B.I., Tieszen, H.L., and Boutton, T.W. (1985) 'Seasonal water relations of savanna shrubs and grasses in Kenya, East Africa', *Journal of Arid Environments*, 8: 15–31.

Hennenberg, K.J., Goetze, D., Szarzynski, J., Orthmann, B., Reineking, B., Steinke, I., and Porembski, S. (2008) 'Detection of seasonal variability in microclimatic borders and ecotones between forest and savanna', *Basic and Applied Ecology*, 9(3): 275–85.

Higgins, S.I., Bond, W.J., and Trollope, W.S.W. (2000) 'Fire, resprouting and variability: A recipe for grass-tree coexistence in savanna', *Journal of Ecology*, 88(2): 213–29.

Hipondoka, M.H.T., Aranibar, J.N., Chirara, C., Lihavha, M., and Macko, S.A. (2003) 'Vertical distribution of grass and tree roots in arid ecosystems of southern Africa: Niche differentiation or competition?' *Journal of Arid Environments*, 54: 319–25.

Hoffman, W.A., Lucatelli, V.M.P.C., Silva, F.J., Azeuedo, I.N.C., Marinho, M.D., Albuquerque, A.M.S., and Moreira, S.P. (2004) 'Impact of the invasive alien grass Melinis minutiflora at the savanna-forest ecotone in the Brazilian Cerrado', *Diversity and Distributions*, 10: 99–103.

Holdridge, L.R. (1947) 'Determination of world plant formations from simple climatic data', *Science*, 105: 367–68.

Holling, C.S. (1973) 'Resilience and stability of ecological systems', *Annual Review of Ecological Systems*, 4: 1–23.

Holling, C.S. (1978) *Adaptive Environmental Assessment and Management*. Chichester, UK: John Wiley.

Holling, C.S. (1986) 'The resilience of terrestrial ecosystems: Local surprise and global change', in W. Clark and R. Munn (eds.), *Sustainable Development of the Biosphere*. Cambridge: Cambridge University Press. pp. 292–317.

Holling, C.S. (2001) 'Understanding the complexity of economic, ecological and social systems', *Ecosystems*, 4(5): 390–405.

Homewood, K.M. (2004) 'Policy, environment and development in African rangelands', *Environmental Science and Policy*, 7(3): 125–43.

House, J.I., Archer, S., Breshears, D.D., Scholes, R.J., and NCEAS. (2003) 'Tree-grass interactions participants', *Journal of Biogeography*, 30: 1763–77.

Hudak, A.T., Fairbanks, D.H.K., and Brockett, B.H. (2004) 'Trends in fire patterns in a southern African savanna under alternative land use practices' *Agriculture, Ecosystems and Environment*, 101: 307–25.

Igoe, J. (2004) *Conservation and Globalization*. Belmont, CA: Wadsworth/Thompson.

Illius, A.W., and O'Connor, T.G. (1999) 'On the relevance of nonequilibrium concepts to semi-arid grazing systems', *Ecological Applications*, 9: 798–813.

Illius, A.W., and O'Connor, T.G. (2000) 'Resource heterogeneity and ungulate population dynamics', *Oikos*, 89: 283–94.

Jäger, F. (1945) 'Zur gliederung und benenming des tropischen graslandgürtels', *Verhandlunger der Naturforschenden Gesellschaft in Basel*, 56: 509–20.

Jeltsch, F., Milton, S.J., Dean, W.R.J., and van Rooyen, N. (1996) 'Tree spacing and coexistence in semiarid savannas', *Journal of Ecology*, 84: 583–95.

Jeltsch, F., Milton, S.J., Dean, W.R.J., van Rooyen, N., and Moloney, K.A. (1998) 'Modelling the impact of small-scale heterogenities on tree-grass coexistence in semi-arid savannas', *Journal of Ecology*, 86: 780–93.

Jeltsch, F., Weber, G.E., and Grimm, V. (2000) 'Ecological buffering mechanisms in savannas: A unifying theory of long-term tree-grass coexistence', *Plant Ecology*, 161: 161–71.

Johns, G.G. (1984) 'Soil water storage in a semi-arid Eucalyptus populnea woodland invaded by woody shrubs, and the effects of shrub clearing and tree ring-barking', *Australian Rangeland Journal*, 6: 75–85.

Jones, C.G., Lawton, J.H., and Shachak, M. (1994) 'Organisms as ecosystem engineers', *Oikos*, 82: 253–64.

Jouquet, P., Boulain, N., Gignoux, J., and Lepage, M. (2004) 'Association between subterranean termites and grasses in a West African savanna: Spatial pattern analysis shows a significant role for Odontotermes n. pauperans', *Applied Soil Ecology*, 27: 99–107.

Jouquet, P., Lepage, M., and Velde, B. (2002) 'Termite soil preferences and particle selections: Strategies related to ecological requirements', *Insectes Sociaux*, 49: 1–7.

Kellman, M. (1979) 'Soil enrichment by neotropical savanna trees', *Journal of Ecology*, 67: 565–77.

Knoop, W.T., and Walker, B.H. (1985) 'Interactions of woody and herbaceous vegetation in a southern African savanna', *Journal of Ecology*, 73: 235–53.

Kolasa, J., and Rollo, C.D. (1991) 'Introduction: The heterogeneity of heterogeneity: a glossary', in J. Kolasa (ed.), *Ecological Heterogeneity*. New York: Springer Verlag. pp. 1–23.

Köppen, W. (1884) 'Die wärmezonen der erde, nach der dauer der heissen, gemässigten und kalten zeit und nach der wirkung der wärme auf die organische' *Welt betrachtet*. *Meteorologische Zeitschrift*, 1: 215–26.

Köppen, W. (1900) 'Versuch einer klassifikation der klimate, vorzugsweise nach ihren beziehungen zur pflanzenwelt', *Geographische Zeitschrift*, 6: 593–611.

Konaté, S., Le Roux, X., Tessier, D., and Lepage, M. (1999) 'Influence of large termitaria on soil characteristics, soil water regime, and tree leaf shedding pattern in a West African savanna', *Plant and Soil*, 206: 47–60.

Kotliar, N.B., and Wiens, J.A. (1990) 'Multiple scales of patchiness and patch structure: A hierarchical framework for the study of heterogeneity', *Oikos*, 59: 253–60.

Laris, P. (2002) 'Burning the seasonal mosaic: Preventative burning strategies in the wooded savanna of southern Mali', *Human Ecology*, 30(2): 155–86.

Lauer, W. (1952) 'Humide und aride jahreszeiten in afrika und Sudamerika und ihre beziehungen zu der vegetationsgürteln', *Bonner Geographische Abhandlungen*, 9: 15–98.

Lawton, J.H., and Jones, C.G. (1995) 'Linking species and ecosystems: Organisms as ecosystem engineers', in C.G. Jones and J.H. Lawton (eds.), *Linking Species and Ecosystems*. New York: Chapman and Hall. pp. 141–50.

Le Roux, X., Bariac, T., and Mariotti, A. (1995) 'Spatial partitioning of the soil water resource between grass and shrub components in a West African humid savanna', *Oecologia*, 104: 147–55.

Leach, M., and Mearns, R. (eds.). (1996) *The Lie of the Land. Challenging Received Wisdom on the African Environment*. London: James Currey.

Lewis, H.T. (1989) 'Ecological and technological knowledge of fire: Aborigines versus park rangers in Northern Australia', *American Anthropologist*, 91: 940–61.

Lotka, A.J. (1925) *Elements of Physical Biology*. Baltimore: Williams and Wilkins.

Ludwig, F., de Kroon, H., Berendse, F., and Prins, H.H.T. (2004) 'The influence of savanna trees on nutrient, water and light availability and the understorey vegetation', *Plant Ecology*, 170: 93–105.

Ludwig, F., de Kroon, H., Prins, H.H.T., and Berendse, F. (2001) 'The effect of nutrients and shade on tree-grass interactions in an East African savanna', *Journal of Vegetation Science*, 12: 579–88.

Ludwig, J.A., Tongway, D.J., Eager, R.W., Williams, R.J., and Cook, G.D. (1999a) 'Fine-scale vegetation patches decline in size and cover with increasing rainfall in Australian savannas', *Landscape Ecology*, 14: 557–66.

Ludwig, J.A., Tongway, D.J., and Marsden, S.G. (1999b) 'Stripes, strands or stipples: Modelling the influence of three landscape banding patterns on resource capture and productivity in semi-arid woodlands, Australia', *Catena*, 37: 257–73.

MacArthur, R., and Wilson, E. (1967) *The Theory of Island Biogeography*. Princeton, NJ: Princeton University Press.

Mbow, C., Nielson, T.T., and Rasmussen, K. (2000) 'Savanna fires in east-central Senegal: Distribution patterns, resource management and perceptions', *Human Ecology*, 28(4): 561–83.

McNaughton, S.J., and Georgiadis, N.J. (1986) 'Ecology and African grazing and browsing mammals', *Annual Review of Ecology and Systematics*, 17: 39–65.

Mentis, M.T. (1978) 'Population limitation in grey rhebuck and oribi in the Natal Drakensberg', *The Lammergeyer*, 26: 19–28.

Mentis, M.T., and Bigalke, R.C. (1981) 'The effect of scale of burn on the densities of grassland francolins in the Natal Drakensberg', *Biological Conservation*, 21: 247–61.

Mistry, J. (2000) *World Savannas. Ecology and Human Use*. Harlow, UK: Longman (Pearson Education).

Mistry, J., and Berardi, A. (2005) 'Assessing fire potential in a Brazilian savanna nature reserve', *Biotropica*, 37(3): 439–51.

Mistry, J., and Berardi, A. (eds) (2006) *Savannas and Dry Forests: Linking People with Nature*. Aldershot, UK: Ashgate.

Mistry, J., Berardi, A., Andrade, V., Krahô, T., Krahô, P., and Leonardos, O. (2005) 'Indigenous fire management in the cerrado of Brazil: the case of the Krahô of Tocantíns', *Human Ecology*, 33(3): 365–86.

Mlambo, D., Nyathi, P., and Mapaure, I. (2005) 'Influence of *Colophospermum mopane* on surface soil properties and understorey vegetation in a southern African savanna', *Forest Ecology and Management*, 212: 394–404.

Mordelet, P., and Menaut, J.C. (1995) 'Influence of trees on above-ground production dynamics of grasses in a humid savanna', *Journal of Vegetation Science*, 6: 223–28.

Mordelet, P., Menaut, J.C., and Mariotti, A. (1997) 'Tree and grass rooting patterns in an African humid savanna', *Journal of Vegetation Science*, 8: 65–70.

Mortimore, M. (1998) *Roots in the African Dust. Sustaining the Sub-Saharan Drylands*. Cambridge: Cambridge University Press.

Murwira, A., and Skidmore, A.K. (2005) 'The response of elephants to the spatial heterogeneity of vegetation in a Southern African agricultural landscape', *Landscape Ecology*, 20: 217–34.

Naiman, R.J., Braack, L., Grant, R., Kemp, A.C., du Toit, J.T. and Venter, F.J. (2003) 'Interactions between species and ecosystem characteristics', in J.T. du Toit, K.H. Rogers, and H.C. Biggs (eds.), *The Kruger Experience. Ecology and Management of Savanna Heterogeneity*. Washington, DC: Island Press. pp. 221–41.

Ogutu, J.O., Piepho, H.P., Dublin, H.T., Bhola, N., and Reid, R.S. (2008) 'Rainfall influences on ungulate population abundance in the Mara-Serengeti ecosystem', *Journal of Animal Ecology*, 77(4): 814–29.

Owen-Smith, N. (2002) *Adaptive Herbivore Ecology. From Resources to Populations in Variable Environments*. Cambridge: Cambridge University Press.

Owen-Smith, N. (2004) 'Functional heterogeneity in resources within landscapes and herbivore population dynamics', *Landscape Ecology*, 19: 761–71.

Parr, C.L., and Brockett, B.H. (1999) 'Patch-mosaic burning: A new paradigm for savanna fire management in protected areas?', *Koedoe*, 42: 117–30.

Pickett, S.T.A., and Cadenasso, M.L. (1995) 'Landscape ecology: Spatial heterogeneity in ecological systems', *Science*, 269: 331–34.

Pickett, S.T.A., Cadenasso, M.L., and Benning, T.L. (2003) 'Biotic and abiotic variability as key determinants of savanna heterogeneity at multiple spatiotemporal scales', in J.T. du Toit, K.H. Rogers, and H.C. Biggs (eds.), *The Kruger Experience. Ecology and Management of*

Savanna Heterogeneity. Washington, DC: Island Press. pp. 22–40.

Pickett, S.T.A., Cadenasso, M.L., and Jones, C.G. (2000) 'Generation of heterogeneity by organisms: creation, maintenance and transformation' in M. Hutchings (ed.), *Ecological Consequences of Habitat Heterogeneity*. New York: Blackwell. pp. 33–52.

Posey, D.A. (1985) 'Indigenous management of tropical forest ecosystems: The case of the Kayapó Indians of the Brazilian Amazon', *Agroforestry Systems*, 3: 139–58.

Prins, H.H.T., and Beekman, J.H. (1987) 'A balanced diet as a goal of grazing: The food of the Manyara buffalo', in H.H.T. Prins (ed.), *The Buffalo of Manyara*. *The Netherlands*: Meppel: Krips Repro. pp. 69–98.

Prins, H.H.T., and Iason, G.R. (1988) 'Dangerous lions and nonchalant buffalo', *Behaviour*, 108: 262–97.

Redfern, J.V., Grant, C.C., Gaylard, A., and Getz, W.M. (2005) 'Surface water availability and the management of herbivore distributions in an African savanna ecosystem', *Journal of Arid Environments*, 63: 406–24.

Reenberg, A. (2001) 'Agricultural land use pattern dynamics in the Sudan-Sahel—towards an event-driven framework', *Land Use Policy*, 18: 309–19.

Rogers, K.H. (2003) 'Adopting a heterogeneity paradigm: Implications for management of protected savannas', in J.T. du Toit, K.H. Rogers, and H.C. Biggs (eds.), *The Kruger Experience. Ecology and Management of Savanna Heterogeneity*. Washington, DC: Island Press. pp. 41–58.

Roques, K.G., O'Connor, T.G., and Watkinson, A.R. (2001) 'Dynamics of shrub encroachment in an African savanna: Relative influences of fire, herbivory, rainfall and density dependence', *Journal of Applied Ecology*, 38: 268–80.

Rowe-Rowe, D.T., and Lowry, P.B. (1982) 'Influence of fire on small-mammal populations in the Natal Drakensberg', *South African Journal of Wildlife Research*, 16: 32–35.

Russell-Smith, J., Lucas, D., Gapindi, M., Gunbunuka, B., Kapirigi, N., Namingum, G., Lucas, K. and Chaloupka, G. (1997) 'Aboriginal resource utilisation and fire management practice in western Arnhem Land, monsoonal northern Australia: Notes for prehistory, lessons for the future', *Human Ecology*, 25(2): 159–95.

Sala, O.E., and 18 authors. (2000) 'Global biodiversity scenarios for the year 2100', *Science*, 287: 1770–74.

Sandford, S. (1983) *Management of Pastoral Development in the Third World*. New York: John Wiley.

Sankaran, M., Ratnam, J., and Hanan, N.P. (2004) 'Tree-grass coexistence in savannas revisited—insights from an examination of assumptions and mechanisms invoked in existing models', *Ecology Letters*, 7: 480–90.

Sarmiento, G. (1984) *The Ecology of Neotropical Savannas*. Cambridge, MA: Harvard University Press.

Scanlon, T.M., Caylor, K.K., Manfreda, S., Levin, S.A., and Rodriguez-Iturbe, I. (2005) 'Dynamic response of grass cover to rainfall variability: Implications for the function and persistence of savanna ecosystems', *Advances in Water Resources*, 28: 291–302.

Scholes, R.J. (1990) 'The influence of soil fertility on the ecology of Southern African savannas', *Journal of Biogeography*, 17: 415–19.

Scholes, R.J., and Archer, S. (1997) 'Tree-grass interactions in savannas', *Annual Review of Ecology and Systematics*, 28: 517–44.

Scoones, I. (1997) 'Landscapes, fields and soils: Understanding the history of soil fertility management in southern Zimbabwe', *Journal of South African States*, 23: 615–34.

Scoones, I. (1999) 'New ecology and the social sciences: What prospects for a fruitful engagement?' *Annual Review of Anthropology*, 28: 479–507.

Seghieri, J., Floret, C., and Pontanier, R. (1995) 'Plant phenology in relation to water availability: Herbaceous and woody species in the savannas of northern Cameroon', *Journal of Tropical Ecology*, 11: 237–54.

Shannon, G., Druce, D.J., Page, B.R., Eckhardt, H.C., Grant, R., and Slotow, R. (2008) 'The utilization of large savanna trees by elephant in southern Kruger National Park', *Journal of Tropical Ecology*, 24(3): 281–89.

Silva, J.F., Zambrano, A. and Farinas, M.R. (2001) 'Increase in the woody component of seasonal savannas under different fire regimes in Calabozo, Venezuela' *Journal of Biogeography*, 28: 977–83.

Skarpe, C. (1991) 'Impact of grazing in savanna ecosystems', *Ambio*, 20: 351–56.

Smit, G.N., and Rethman, N.F.G. (2000) 'The influence of tree thinning on the soil water in a semi-arid savanna of southern Africa', *Journal of Arid Environments*, 44: 41–59.

Solbrig, O.T. (1991) *Savanna Modelling for Global Change*. Biology International Special Issue No. 24. Paris: International Union of Biological Sciences.

Stott, P. (1991) 'Recent trends in the ecology and management of the world's savanna formations', *Progress in Physical Geography*, 15(1): 18–28.

Sullivan, S. (1996) 'Towards a non-equilibrium ecology: Perspectives from an arid land', *Journal of Biogeography*, 23: 1–5.

Sullivan, S., and Rohde, R. (2002) 'On non-equilibrium in arid and semi-arid grazing systems', *Journal of Biogeography*, 29: 1595–618.

Tiffen, M., Mortimore, M., and Gichuki, F. (1994) *More People, Less Erosion. Environmental Recovery in Kenya*. New York: John Wiley.

Tomor, B.M., and Owen-Smith, N. (2002) 'Comparative use of burnt and unburnt grassland by grazing ungulates in the Nylsvley nature reserve, South Africa', *African Journal of Ecology*, 40: 201–4.

Tongway, D.J., and Ludwig, J.A. (1997) 'The conservation of water and nutrients within landscapes', in J. Ludwig, D.J. Tongway, D. Freudenberger, J. Noble, and Hodgkinson, K. (eds.), *Landscape Ecology, Function and Management: Principles from Australia's Rangelands*. Melbourne: CSIRO. pp. 17–26.

Townsend, S.A., and Douglas, M.M. (2004) 'The effect of a wildfire on stream water quality and catchment water yield in a tropical savanna excluded from fire for 10 years (Kakadu National Park, North Australia)', *Water Research*, 38: 3051–58.

Trapnell, C.G. (1959) 'Ecological results of woodland burning in northern Rhodesia', *Journal of Ecology*, 47: 161–72.

Treydte, A.C., van Beeck, F.A.L., Ludwig, F., and Heitkoenig, I.M.A. (2008) 'Improved quality of beneath-canopy grass in South African savannas: Local and seasonal variation', *Journal of Vegetation Science*, 19(5): 663–70.

Troll, C. (1950) 'Savannentypen und das problem der primar-savannen', *Proceedings of the 7th Botanical Institute Congress*, Stockholm. pp. 670–4.

van de Vijver, C.A.D.M., Poot, P., and Prins, H.H.T. (1999) 'Causes of increased nutrient concentrations in post-fire regrowth in an East African savanna', *Plant and Soil*, 214: 173–85.

van Wijk, M.T., and Rodriguez-Iturbe, I. (2002) 'Tree-grass competition in space and time: Insights from a simple cellular automata model based on ecohydrological dynamics', *Water Resources Research*, 38: 18.11–15.

Vetter, S. (2005) 'Rangelands at equilibrium and non-equilibrium: Recent developments in the debate', *Journal of Arid Environments*, 62: 321–41.

Vigilante, T., and Bowman, D.M.J.S. (2004) 'Effects of individual fire events on the flower production of fruit-bearing tree species, with references to Aboriginal people's management and use, at Kalumburu, North Kimberley, Australia', *Australian Journal of Botany*, 52: 405–15.

Volterra, V. (1926) 'Fluctuations in the abundance of a species considered mathematically', *Nature*, 118: 558–60.

Walker, B.H., and Langridge, J.L. (1997) 'Predicting savanna vegetation structure on the basis of plant available moisture (PAM) and plant available nutrients (PAN): A case study from Australia', *Journal of Biogeography*, 24(6): 813–25.

Walter, H. (1971) *Ecology of Tropical and Subtropical Vegetation*. Edinburgh, UK: Oliver and Boyd.

Walter, H. (1973) *Vegetation of the Earth*. 2nd ed. Berlin: Springer-Verlag.

Walters, C.J. (1986) *Adaptive Management of Renewable Resources*. New York: McGraw-Hill.

Wardell, D.A., Reenberg, A., and Tøttrup, C. (2003) 'Historical footprints in contemporary land use systems: Forest cover changes in savanna woodlands in the Sudano-Sahelian zone', *Global Environmental Change*, 13(4): 234–54.

Warren, A. (1995) 'Changing understandings of African pastoralism and the nature of environmental paradigms', *Transactions of the Institute of British Geographers*, 20: 193–203.

Weltzin, J.F., and Coughenour, M.B. (1990) 'Savanna tree influence on understory vegetation and soil nutrients in northwestern Kenya', *Journal of Vegetation Science*, 1: 325–34.

Western, D. (1975) 'Water availability and its influence on the structure and dynamics of a savannah large mammal community', *East African Wildlife Journal*, 13: 265–86.

Whittaker, R.H. (1975) *Communities and Ecosystems*. 2nd ed. New York: Macmillan.

Williams, R.J., Duff, G.A., Bowman, D.M.J.S., and Cook, G.D. (1996) 'Variation in the composition and structure of tropical savannas as a function of rainfall and soil texture along a large-scale climatic gradient in the Northern Territory, Australia', *Journal of Biogeography*, 23(6): 747–56.

Wu, X.B., and Archer, S.R. (2005) 'Scale-dependent influence of topography-based hydrologic features on patterns of woody plant encroachment in savanna landscapes', *Landscape Ecology*, 20: 733–42.

Wu, J., and David, J.L. (2002) 'A spatially explicit hierarchical approach to modelling complex ecological systems: Theory and applications', *Ecological Modelling*, 153: 7–26.

Wu, J., and Loucks, O.L. (1995) 'From balance of nature to hierarchical patch dynamics: A paradigm shift in ecology', *The Quarterly Review of Biology*, 70(4): 439–66.

Tropical Forests: Biogeography and Biodiversity

Kenneth R. Young

16.1 INTRODUCTION

Exploration of the lands and seas in tropical latitudes provided one of the historical bases for the development of biogeography as a discipline. By the fifteenth century it was clear that the Western world placed a high priority on the identification of plants and animals that could provide useful products for their colonies or that could be brought under production in Europe. In turn, those and other efforts built upon natural history information collected and collated by Greek, Roman, and Arab scholars (e.g., Glacken, 1967). Part of the legacy of biogeography is as a repository of distributional information, both of the original native ranges of the species involved, and of their climatic tolerances, which would indicate their potential ranges before being introduced elsewhere in the world.

Tropical explorers always marveled at the diversity of species and adaptations to be found among the various tropical environments (e.g., Wallace, 1895). Slowly, enough information was gathered so that geographical and other spatial patterns in that biodiversity were revealed (Chazdon and Whitmore, 2002; Hillebrand, 2004; Bermingham et al., 2005). In addition, it became possible to find generalities useful for characterizing the structure and functioning of tropical ecosystems even for distant places that shared similar environmental conditions of rainfall and temperatures but had no species in common (Walter, 1985; Richards, 1996; Whitmore, 1998).

The goal of this chapter is to provide an overview of the major characteristics of humid tropical forest environments located in Central and South America, southern Mexico, Africa, tropical Asia, northeastern Australia, and on numerous small Pacific and Caribbean islands. This is done in relationship to important biophysical factors that are associated with change in ecosystem structure along spatial environmental gradients. In addition, much of the uniqueness of particular tropical locales originates with historical and evolutionary events affecting biological diversity (biodiversity) over millennia and millions of years. Thus, I begin with an evaluation of the consequences of these past conditions and occurrences. Finally, biogeography is often nowadays one of the disciplines often utilized for making decisions for the protection and conservation of biodiversity. This relationship is particularly crucial for the biodiverse tropics where species are numerous but financial resources are limited for spending on protected areas or other conservation efforts. The documentation of biogeographical patterns assists conservation decision making, while the understanding of biogeographical processes helps make possible predictions of conditions in places yet to be studied.

16.2 LEGACIES OF PLATE TECTONICS

In a sense, the entire Earth once was tropical. During the Mesozoic, carbon dioxide levels in the atmosphere were perhaps three or four times what they are today. As a result of the increased greenhouse effect, even high latitudes were warm and species of large-bodied reptiles roamed the world, representing the most massive herbivores and predators ever found on land, the sauropods and tyranosaurids (Rees et al., 2004). Amid the flora, especially by the middle Cretaceous, were to be found species of almost all the modern vascular plant families, with forests formed by conifers and other gymnosperms (Behrensmeyer et al., 1992; Schneider et al., 2004).

In the meanwhile, beginning in the Jurassic, the supercontinent of Pangaea splintered over 100 million years into smaller landmasses in the northern and southern hemispheres, first as Laurasia and Gondwana. Regional differences developed among the biota of newly separated marine environments, and on land in the Laurasian northern hemisphere landmass contrasted with that of Gondwana found south of the equator, and eventually among the pieces of Gondwana that separated into what today are India, Australia, New Zealand, Africa, South America, and Antarctica. The world's greatest extensions of tropical forest, found today in South America and Africa, were at one point in the past to be found in Laurasia and were called the Boreotropics. As an example of evolutionary biogeography, it is possible to use genetic and DNA evaluations of some of the current tropical plant lineages to reveal ancient ties to the northern hemisphere tropics (Davis et al., 2004), which occupy only a relatively small area in today's world.

The cataclysm of 65 million years ago caused by an asteroid impact was followed by severe and abrupt climate change. Global mass extinction resulted, with particularly direct impacts in the northern hemisphere and variable effects elsewhere (Pearson et al., 2001; Vajda et al., 2001). Dinosaur lineages only survived as birds, and the lands of the northern hemisphere were recolonized by plant types adapted to progressively cooler and drier conditions.

Legacies of these past conditions and connections can be found today amid the world's most diverse temperate forests in the mountains of China and of the eastern United States (Qian and Ricklefs, 2000), and through the reshaping of the world's grasslands and associated fauna as herbaceous and silica-rich grass lineages coevolved with their herbivores beginning at least 30 million years ago (MacFadden, 2000). Looking at today's tropical forests, there are (a) the statuesque rain forests of southeast Asia dominated by one plant family, the Dipterocarpaceae; (b) Pacific islands like Papua New Guinea with highly endemic conifers descended from Gondwanan lineages; (c) tropical rain forests of Australia that grade into drier forests and woodlands dominated by *Eucalyptus*; and (d) the tropical rain forests of South America and Africa, sharing many of their plant families, but with quite unrelated birds and mammals that evolutionarily converged later into similar niches and adaptations as the Atlantic Ocean spread and South America and Africa drifted apart (George and Lavocat, 1993; Morley, 2000). The zoogeography of tropical birds also shows spatial patterns originating with ancient land connections (Newton and Dale, 2001).

Historical approaches to biogeography explain patterns in reference to evolutionary processes, including speciation, divergence, convergence, and species extinction. In addition, biogeographers eagerly use information from Earth system science because the timing of climate change and of plate movements and orogenies are critical phenomena to understand. History often acts to create unique places in terms of their assemblages of plants and animals.

Biogeography arose from observations made as explorers moved from continent to continent, island to island, documenting the types and adaptations of the plants and animals, and also typically evaluating their possible uses in commerce and for agriculture. The startling differences that Alfred Wallace found when crossing from the southeast Asian tropics over water to Sulawesi revealed that the Asian mainland tropics sent many plants and animals that far, only to be stopped by water barriers; much later, the theory of plate tectonics further clarified the multiple and sometimes distant sources of species on that island and other landmasses (Morley, 2000; Evans et al., 2003). In the meantime, plants and animals further east had mostly island hopped amid Australia and the Pacific islands or else had evolved in place. Hundreds of examples of place-to-place differences were revealed elsewhere in the tropics, many of which have roots in deep geological time and fit the general outlines of what is now known about the shifts of landmasses over dozens of millions of years. Many biogeographical studies have striven to clarify the relative importance in different places and for different lineages of dispersal versus vicariances (i.e., separations of ancestral populations due to the formation of oceanic or topographic barriers).

New data from the study of DNA combined with other data (phylogeography after Avise, 2000; see Riddle, this volume and also Young, 2003; de Queiroz, 2005) show that long-distance dispersal may be rare in an ecological sense, but it does occur over long time periods and so leaves an evolutionary imprint in the sense of having new lineages appear in places where they did not previously have an evolutionary history. For example, in Venezuela, Givnish et al. (2004) demonstrated that the unique plants adapted to growing on sandy soils perched 1,000 m.a.s.l. high on tabletop sandstone mountains ('tepuis') were assembled from plant lineages that originated in the nearby lowlands and colonized the uplands; in one case, there was also evidence of ancient dispersal across the Atlantic Ocean to similar habitats in the mountains of western Africa. Undoubtedly the floras and faunas of many places in the tropics are of similarly heterogeneous origins: some evolved in place, others colonized from afar, and still others differentiated along environmental gradients or due to barriers to gene flow (e.g., Pennington and Dick, 2004). Biogeographers thus need to study the history both of places and of evolutionary lineages.

16.3 THE RECENT GEOLOGICAL PAST

In the last 900,000 years the Earth has gone through repeated climatic cycles of increased glaciations reaching a maximum of ice coverage and a minimum of atmospheric carbon dioxide, followed by interglacial warming temperatures, receding glaciers, and increasing carbon dioxide (Lambeck et al., 2002). The highest elevations of tropical mountains have been repeatedly affected in the Quaternary by these bouts of cooler temperatures and increased ice fields, altering landforms and soils, and shifting plant and animal distributions vertically up and down the elevation gradients.

Long mountain chains, such as the Andes, which extend more than 50° of latitude, have also permitted species to colonize along dispersal corridors formed by linear belts of similar elevation. In fact, some species with highly specific environmental needs may nevertheless have very long (but quite narrow) distributions (Young, 1995). In combination with climate change, topographic relief provides repeated changes in the extent, connectivity, and accessibility of habitats to the upland tropical biota. As a result, numerous endemic species exist with small ranges, some restricted to specific mountain peaks or intermontane valleys, others to narrow altitudinal zones.

Sea levels drop during glacial periods and rise during interglacials as permanent ice melts. During the Quaternary, the locations of coastlines, the sizes of islands, and the connectivity of topographic rises in shallow seas have all changed repeatedly, altering the locations of barriers to dispersal and the ability of land organisms to move from place to place. Islands are especially interesting to biogeographers because they contain many unique (endemic) species that evolved in situ, in addition to others that were able to disperse across the water barriers. Islands that

have never been physically connected to continental land masses typically have fewer species than those islands that were part of large landmasses in past glacials (MacArthur and Wilson, 1967).

The evidence for what happened to biophysical conditions in the lowland tropics during the Quaternary and to the distribution of the biota is less clear than for the tropical highlands or coasts. One possibility often suggested is that less moisture was available during glacial periods, increasing seasonality and soil moisture stresses, which in turn increased nonforest vegetation (Whitmore and Prance, 1987) or areas of tropical dry forests (Pennington et al., 2000). This would also have the effect of isolating rain forests in smaller areas, creating refugia (see Willis et al., this volume). However, other evidence such as detailed studies of fossil pollen from dated lake sediments suggests this was not so (Colinvaux and De Oliveira, 2000). Given the sizes of the areas of lowland tropics and the multiple climatic influences at play, numerous scenarios are plausible and thus more than one explanation will be required (Bush, 1994).

Older scientific views of the lowland tropics characterized their environmental conditions as stable—warm and humid over hundreds of thousands of years. This equilibrium viewpoint could be labeled a museum hypothesis of tropical diversity wherein a lack of change maintained many species, with total diversity accumulating over time to current maxima. However, evolutionary and climate-change research show in fact that the lowland tropics are and have been constantly changing, across a range of spatial and temporal scales (Flenley, 1998; Moritz et al., 2000). As stated earlier, over global scales and many millions of years, there can be speciation and origin of new lineages due to plate tectonics. In addition, new plant and animal lineages arrive from elsewhere over time due to dispersal and can cause additional alterations in species composition of particular continental landmasses. However, there are also geographic shifts on landscape scales that may be important sources of new biodiversity. For example, selective pressure acting upon the color morphs of *Heliconius* butterflies may vary over distances as short as several hundred meters (Mallet et al., 1990), while gene flow in *Anolis* lizards is altered by habitat changes over a few kilometers (Ogden and Thorpe, 2002). In many lowland sites, it is likely that such constant dynamism is the norm and that current environmental conditions have only been in place for a couple of centuries or millennia in a region and may always be changing at the scale of a particular patch of forest. Many biogeographers are now doing multiscale investigations in order to clarify the relative roles of local and regional processes.

16.4 CURRENT BIOPHYSICAL CONSTRAINTS ON BIOTA AND ECOSYSTEM PROCESSES

Lowland tropical rainforests are warm year round (Walter, 1985; Richards, 1996; Whitmore, 1998) due to their equatorial position. Near the Equator, average temperatures often are 20° to 30°C; only in the subtropics do equatorward-moving cold air masses bring in freezing or near-freezing temperatures for several weeks a year. Some tropical forests are aseasonal in terms of precipitation, with monthly totals always near or above 75 mm. Others are affected by monsoonal air-circulation systems, which produce dry seasons that are about four months long; these are especially common in the tropical forests of India and Southeast Asia. Still other tropical lowland forests are deciduous, with bare branches during dry seasons that can last eight months. These more seasonal forests grade over distances of tens to hundreds of kilometers into open woodlands with a grass understory or into tropical savannas, with only isolated trees found amid extensive grasslands. Africa in particular has large areas of savannas and woodlands with smaller core areas of humid tropical forests in equatorial western Africa and scattered places in eastern Africa. South America has the world's largest extent of rain forest in the Amazon Basin, plus Atlantic rain forest remnants on the coast. The Amazon's rainforests are bracketed to the north and south by savannas and to the west by Andean montane forests. Other tropical rainforests of the world have smaller tracts, often near to upland montane forests or grading into drier or more seasonal vegetation formations.

Many of the tropical rainforests of the lowland tropics have multiple strata and so are structurally complex. This vertical stratification adds to the number of species found in one particular place, particularly when the forest canopy is high, often 20 to 35 meters tall, and when emergent trees can tower to 60 meters or even higher. There are lianas, woody vines, that climb to the top of the trees, able to photosynthesize in full sunlight but without the need to develop strong wood for their stems. There are epiphytes growing on high branches. There are thousands of species of insects at all levels of the forest, consuming leaves and tree sap. And then there are hundreds of bird, lizard, spider, and frog species eating the insects. Much of the larger animal life in fact is up in the productive canopy strata, with monkeys searching for fruits and monkey eagles searching for monkeys.

The study of dynamism predominates in current scientific paradigms used to study ecological processes in tropical environments, especially in response to seasonality, to intradecadal variations in climate, and to disturbances to the vegetation (Leigh et al., 1996). Many biogeographers interested in spatial changes in tropical environments try to explain the similarities by asking these two questions:

1. How do plants adapt to rainforests that have abundant moisture but often poor soils?
2. How does vegetation structure change as seasonality differs from place to place or with unusual climatic extremes?

Other biogeographers are more interested in differences and ask questions such as the following:

1. How do the Old and New World tropics differ in terms of vegetation response to seasonality and disturbance?
2. Is uniqueness caused by slightly different biophysical controls or is it due to evolutionary legacies?

Disturbances frequently alter biophysical conditions in closed forests, causing maybe a quarter of the surface area of old-growth forests to be in some successional state, recovering from the immediate damage caused by treefalls and responding to increased light conditions in the several decades after the falling of a tree. Entire suites of woody plant species are dependent on treefall gaps to complete their life cycles—they are present as seeds in the soil or as small seedlings until a canopy opening allows them to grow quickly up into the middle or upper canopy (Turner, 2001). Additionally, a set of animal species is associated with the environmental conditions found in treefall gaps, and these animals consume the young plants there or their fruits, which often contain seeds that are inadvertently dispersed by the frugivores.

Bats and birds provide much of the seed dispersal for tropical forest plants, while bats, birds, and bees pollinize most of the flowers. In some cases, there are highly specific and dependent relations among plant and animal symbionts; in other cases, the relations are more general, with a range of species involved in pollination and seed dispersal. In general, the biota is characterized by much mutualism and interdependence of species. As just one example, the leaf-cutting ants (*Atta*) of the Neotropics utilize the leaves of a variety of plants growing in gaps on which to cultivate a fungus only found associated with the ants; effectively, the ants are farmers that have coevolved with a fungus they raise to eat that is specifically adapted to life in an ant colony. Some ant species have a mutualism with plant species that involves their providing protection from herbivores that would attack the plant, while receiving a nesting site for their colony within the hollow limbs of the plant.

In addition to treefall gaps, some areas are prone to canopy disturbances due to dramatic tropical storms, to floods, and to fluvial erosion and river meandering. In places that are near ecotonal transition zones with tropical grasslands and savannas, disturbance may be due to fire. In regions with hurricanes or typhoons, for example, on the Caribbean islands and parts of the Central American mainland, or with typhoons in the Asian oceanic tropics, forests are struck by hurricanes with frequencies that range from less than once a decade to up to one in 40–50 years, which has the effect of eliminating particularly tall trees, those with weak wood, or plants with a limited capacity to regrow following defoliation by wind. Scatena and other researchers working in the Luquillio Experimental Forest in Puerto Rico have shown not only the importance of hurricanes in affecting forested landscapes, but they have contrasted those perturbations with alterations due to human activity and seasonal fluxes (e.g., Scatena and Lugo, 1995; Scatena, 2001). The researchers put in large permanent plots, with large trees marked and mapped so they can follow change through time and after storms and hurricanes. They measure response to dynamism in terms of plant species composition, plant heights and growth rates, shifts in land cover and accompanying animals, and alterations to soils and nutrient cycles.

An especially dramatic disturbance type is to be found in volcanically active areas, such as on Hawaii, in the Philippines, or on other Pacific islands where lava flows or explosive eruptions can lead to primary succession. Whittaker (1998) assembled data from such situations to show likely changes in species diversity of both plants and animals following catastrophic disturbances. He was particularly inspired by data from Krakatau Island, which exploded in 1883 and since has been studied by generations of biogeographers who recorded the species of plants and animals that arrived and colonized, and the others that arrived, colonized, but then later disappeared. The continued existence of those species and their relative abundances are affected by ecological succession and by other disturbances that take place through time.

MacArthur and Wilson (1967) proposed the Equilibrium Theory of Island Biogeography, which allowed predictions of numbers of species given data on the size and isolation of the island. Many subsequent researchers have quantified and tested the predictions of that theory. More recently, biogeographers have been examining the consequences when disturbances occur and equilibrium conditions do not pertain (Whittaker, 1995), or if the 'islands' are not literal islands in the ocean but instead are islands of habitat surrounded by a matrix of another land-cover type (Kupfer et al., 2006).

Furley and colleagues (Furley et al., 1992; Furley, 2004) studied the ecotonal transition zone between tropical rain forests and savannas worldwide and concluded that the location, width, and dynamics of that zone were functions of the interaction of herbivory, fire, and edaphic processes. Fire becomes an important ecological disturbance in these environments, as do the trampling, grazing, and browsing actions from herbivorous mammals, such as elephants or wildebeest. Some spatial changes, however, are ultimately due to changes belowground, as soil type or soil moisture vary along a transect. Many biogeographers document these spatial changes, often by mapping or by detailed field studies of plants and soils. Still others study the synergistic effects on animals, or else they work to develop predictive statistical or computer models of ecotones and ecotonal shifts.

The principal environmental gradient associated with regional-scale change in the tropics is the amount and seasonality of precipitation, and in particular how soil moisture and other edaphic characteristics are affected. The places where tropical rain forest becomes tropical savanna are also currently foci for tropical deforestation (Achard et al., 2002), in part because the dry season allows for the ready use of fire to burn, cut, slash, and thus decrease forest cover. Tropical grasslands and shrublands are on the increase worldwide, although sustainable use of pastures or croplands often is not possible due to invasive weeds and poor soil nutrients (Kellman and Tackaberry, 1997). In fact, many tropical savannas and rain forests grow on ultisols and oxisols, which are heavily weathered soils with relatively few nutrients. The native plants are mostly mycorrhizal, using their root-fungi mutualisms to acquire needed chemicals from the soil. Biogeographical explanations of forest cover in Africa are especially complex given that human presence there is truly ancient, meaning that human populations have increased and decreased over time, leading to deforestation and reforestation perhaps multiple times.

Yet another important biophysical gradient found in tropical coastal areas is associated with salinity, which results in fringes of tropical mangrove forests along coastlines consisting of trees that are tolerant of salinity. The mangroves in the Asian tropics are particularly diverse, with 30 or 40 tree species and a series of dependent animals, including the proboscis monkey that is specialized for eating mangrove leaves.

Tropical uplands on tall mountains extend from diverse foothill forests that blend lowland and montane species together, to middle elevation cloud forests immersed in daily fog originating from the cooling of rising air masses, and ending in timberlines with relatively short trees and often understories filled with clambering bamboos (Cavelier et al., 2000). Lower temperatures with rising elevation put limits on photosynthesis, while other factors limit soil nitrogen availability. Steep slopes tend to have shorter trees and increased instability leading to landslides. Above timberline, herbaceous plants or dwarf shrubs predominate, although in some places large rosette plants can also be common. Treelines on large landmasses tend to occur at higher elevations than treelines on islands. In addition, most treelines show evidence of having been depressed by human influences, due to land-use practices such as grassland burning. In the tropical mountains of Africa, Java, South America, and Central America, the highlands are in places densely settled by people, used for terraced agriculture or for the grazing of livestock.

Students of ecological biogeography are considering many research projects that involve the tracking of current and future distributional shifts originating from global climate changes. Species and treelines will often shift up in mountains in a warming world. Interactions with human land use are likely but complex enough that predictions of outcomes are hard to make. Islands will become smaller and in some cases more isolated. Many climate change predictions suggest drier conditions for lowland rain forests, meaning that biogeographers using ecological methods will also need to be cognizant of what happened during past glacial periods.

16.5 TROPICAL FORESTS AND HUMANS

Humans (*Homo sapiens*) are thought to have originated from primate lineages dwelling in the tropical forests and woodlands of sub-Saharan Africa. Erect posture, dexterous hands, speech, and social organization all helped this species to survive subsequent climatic changes while pursuing a hunting and gathering lifestyle.

Evidence of multiple dispersals by humans out of Africa to elsewhere in the Old World can still be discerned in certain traits and gene frequencies of populations today (Hammer et al., 1998). The tropical Asian areas in particular were places where plants and animals were domesticated, eventually leading to complex kinds of agriculture by about 5,000 years B.P. Australia was reached about 70,000 years ago and its tropical landscapes, especially the drier environments, were reshaped through the use of fire, the hunting of animals, and the harvesting of plants. Different past and current land-use types and intensities underlay important geographical differences in the states and usefulness of tropical forests (Primack and Corlett, 2005).

People colonized the New World tropics in the late Pleistocene. Many large mammals became extinct at about this time, perhaps caught amid changing environmental conditions and new pressures brought about by people (Johnson, 2002). Whole new sets of wild plants and animals were shaped genetically by humans, with agriculture developing several thousand years ago in separate places in Central and South America. Five hundred years ago many Old World plants and animals were introduced as these lands were colonized by people from Europe.

It is interesting to note that people living by traditional means in lowland rain forests around the world use agricultural and silvicultural practices that mimic the disturbance regimes of the forests (Gliessman, 1998). Areas cleared for agriculture are often similar in size to compound treefall gaps, meaning that forest recovery after abandonment of the field is fairly quick. In this cyclical process, the first crops planted are often quick growers, such as maize and beans, replaced after several months by longer-lived plants such as cassava, and perhaps by the planting of fruit trees. In addition, it has often been a practice to shift the sites of villages, thereby allowing vegetation and soils a chance to recover after some period of exploitation. The movement of settlements also means that places with the most hunting and fishing pressures shift through time. Along river floodplains, the fishing tends to be best as water levels drop and fish become concentrated, while hunting seasonality is the opposite, with better hunting when water levels are high as the animals are pushed into higher and more restricted dry lands (Goulding et al., 1996).

Land tenure practices that encourage tropical lowland people to stay in one place lead inexorably to larger local populations, shorter fallow and regrowth cycles for cut forest, and eventually to a type of agriculture that exhausts the soil and eliminates the forest, that requires subsidies in the form of chemical pesticides and fertilizers, or that requires the outright abandonment of that area and subsequent colonization and deforestation elsewhere. These tragic outcomes can be seen in virtually all the world's tropics (Young, 2007 and references therein). Even though the native plants have numerous methods to maintain high productivity despite low soil nutrients and numerous insect pests, these adaptations are little recognized or utilized by modern agronomists and agriculturalists. The fast growth possible for some tropical trees following disturbance has been overlooked by modern methods of tropical forestry, which are still mostly premised on one-time timber extraction. Many native forest animals are displaced after these activities, often to be replaced by cattle grazing on poor-quality pasture.

Today the most endangered floras and faunas are found on tropical islands where species are highly endemic and land-cover conversions continue to alter the extent of forests. To clarify aspects of rarity, Rigg and colleagues (Rigg et al., 2002; Rigg, 2005) have striven to elucidate the ecological role and ecophysiological adaptations of the conifers restricted to the uplands of New Caledonia. These species grow very slowly and live for centuries; they also glean moisture from fog interception, resist fires, and use soils low in nutrients and high in metals such as aluminum. These adaptations allow the endemic species to persist on these islands despite their global rarity and the presence of landscape-scale disturbances.

Elsewhere in the tropics, even large continental areas are changing rapidly, at rates without precedence in the biogeographical record (Young, 2007). Some projections suggest that tropical forest degradation can cause biodiversity loss and can alter regional and perhaps global climatic and chemical cycles, including precipitation, carbon dioxide levels, and nitrogen fixation (Miles et al., 2004). The amount and rapidity of change is outside the range of historical variability for most tropical organisms that are so affected. The implication is that extinctions are likely, although species disappearances are notoriously hard to document. Biogeographers can use small distributions and low dispersal or reproductive rates to predict extinctions, but there are often exceptions, with species that can persist despite apparent disadvantages. An open question is if the ongoing extinctions are similar to the mass extinction that ended the Cretaceous in terms of magnitude and consequences. Another is if tropical extinction events will involve additional coupled extinctions due to the existence of many mutualistic and trophically interdependent species. Research to study the geography of extinctions is urgently needed for tropical areas and biota, especially in the context of global environmental change.

16.6 BIODIVERSITY MAXIMA

On land, maximum numbers of flowering plants, ferns, birds, lizards, frogs, spiders, and insects are to be found in humid tropical forests. In addition, tropical waterways have the world's highest numbers of freshwater fish species, while offshore the tropical coral reefs contain the biodiversity equivalents of tropical rainforests with thousands of species and complex physical structures. Historical hypotheses to explain these diversity patterns relate the long time periods that tropical conditions have existed on the planet to the diversification found in some tropical lineages (Hillebrand, 2004; Lomolino et al., 2006). That said, there are also many intriguing exceptions, with some tropical forests in Asia, Africa, and the Amazon dominated by relatively few tree species.

In this context, it becomes important to consider how to define and quantify biodiversity, with possibilities including (a) the number of species in a given area (alpha diversity); (b) that number weighted by the relative abundances of the species (several diversity indices are available for this purpose); (c) the change in species composition from place to place or along a biophysical gradient (beta diversity); and (d) the total number of species in a region (gamma diversity). Theoretical expectations and explanations will differ with the measure of interest.

Other theoretical explanations of biodiversity maxima are ecological in their scope (Chesson, 2000), for example, relating the high amounts of solar energy in the tropics to the possibility that high productivity supports more species than elsewhere. Still other explanations stress the abundance of different environmental conditions due to edaphic or elevation changes as a way to pack many different biotic communities into particular tropical locales. Finally, other explanations emphasize the climatic and disturbance dynamics of the tropics, which over short time periods keep many sites in successional sequences and over long time periods shift the location and characteristics of transition zones that connect the diverse lowland tropics to the dryland tropics, and also to the upland tropics. In all cases, disentangling the role of spatial extent and the degree and kind of habitat or landscape spatial heterogeneity will be important (Rahbek, 2005).

Biogeographers will be especially drawn to these kinds of studies because of the need to evaluate patterns and processes in terms of scale. Many landscape-scale studies give insights into alpha diversity and its relation to forest dynamism and turnover. Studies with regional to continental scales are often designed to document species turnover (beta diversity) or to relate gamma diversity to size or antiquity of an ecoregion or biome (Fine and Ree, 2006). The loss of place-to-place variation is a conservation issue that can be addressed using different solutions than would be used to prevent the loss of a species of concern in one particular location.

16.7 CONCLUSIONS

Biogeography is a core academic discipline for understanding the patterns of biodiversity distribution, most of which point to maxima in tropical latitudes. Given the unique attributes of current human domination of the biosphere, both in terms of altered land covers and of biogeophysical processes (Laurance and Bierregaard, 1997; Goldsmith, 1998; Achard et al., 2002; Miles et al., 2004), biodiversity conservation and planning efforts are needed for tropical landscapes. The places with the highest numbers of species can be mapped. In addition, the uniqueness of those species can be evaluated by documenting the sizes of the species' ranges and by characterizing the biota of particular places in terms of degree of endemism.

Much current research on conservation biogeography in the tropics is concerned with transforming centuries of observations of distribution patterns into predictive models and theories that can also have practical implications for designing systems of nature reserves and of sustainable land-use systems. As a general rule, the humid lowland tropics have the most alpha diversity in particular places, while the upland tropics have the highest species turnovers from place to place and tropical islands have the most endemism. Phylogeography, particularly when used with recent genetic research tools, can help provide a temporal perspective for these efforts by clarifying how long certain traits or organisms have existed. Conservation efforts need to combine a biodiversity perspective with longer term knowledge coming from historical biogeography.

Finally, there are a variety of methods, some using sophisticated computer algorithms (e.g., Peterson, 2001) that can help choose priority areas for conservation and management, based on the relative weights ascribed to diversity, uniqueness, remaining forest, and the spatial layout of proposed protected lands (Millington et al., 2001; Norris and Harper, 2004).

Better understanding of the spatial dimensions of ecological dynamism is also important, in terms of shifts along elevation gradients, in response to natural disturbances, and as climatic change imposes alterations in the abundance and presence of species. As an example of needed research, Benning et al. (2002) used landscape analyses to show the likely future effects of land use and disease spread on endemic birds in remnant highland forests in Hawaii.

Innovative efforts utilizing ecological restoration in combination with succession can help repair altered and damaged tropical forests (Perrow and Davy, 2002). Intact forests have considerable potential economic value in terms of the environmental services they provide in regards to hydrological and carbon cycles, not to mention the numerous chemicals in native plants, animals, and fungi that may have medicinal and industrial uses. Sustainable use of tropical forests is an ideal to strive for, perhaps helped by payments for ecosystem services such as carbon sequestration and storage.

For that matter, human-environment research interests in utilized tropical ecosystems also have a biogeographical component, including the distribution and availability of useful plants and animals. Voeks (2004) found that secondary forests, regrowing after forest clearance, were often more valuable sources of useful plants than undisturbed mature forests. This insight offers innovative ways to promote wise use of tropical species, as do other investigations on how wildlife can be used sustainably (Robinson and Bennett, 2000). Students of biogeography can thus help to improve living conditions of people found in tropical landscapes while being sensitive to potential inequalities (Zerner, 2000).

Tropical forests are changing in dramatic ways. Many of the studies carried out by biogeographers are useful for understanding the historical contexts of those changes (i.e., are current shifts without precedent in the geological record?), for evaluating the consequences of ongoing fragmentation (e.g., how much deforestation results in significant local extinctions?), for identifying species at risk of extinction, and for promoting sustainable uses of landscapes.

REFERENCES

Achard, F., Eva, H.D., Stibig, H.-J., Mayaux, P., Gallego, J., Richards, T. and Malingreau, J.-P. (2002) 'Determination of deforestation rates of the world's humid tropical forests', *Science,* 297: 999–1002.

Avise, J.C. (2000) *Phylogeography: The History and Formation of Species.* Cambridge, MA: Harvard University Press.

Behrensmeyer, A.K., Damuth, J.D., DiMichele, W.A., Potts, R., Sues, H.-D. and Wing, S.L. (eds.). (1992) *Terrestrial Ecosystems through Time: Evolutionary Paleoecology of Terrestrial Plants and Animals.* Chicago: University of Chicago Press.

Benning, T.L., LaPointe, D., Atkinson, C.T. and Vitousek, P.M. (2002) 'Interactions of climate change with biological invasions and land use in the Hawaiian Islands: Modeling the fate of endemic birds using a geographic information system', *Proceedings of the National Academy of Sciences of the United States of America*, 99: 14246–49.

Bermingham, E., Dick, C.W. and Moritz, C. (eds.). (2005) *Tropical Rainforests: Past, Present, and Future*. Chicago: University of Chicago Press.

Bush, M.B. (1994) 'Amazonian speciation: A necessarily complex model', *Journal of Biogeography*, 21: 5–17.

Cavelier, J., Tanner, E. and Santamaría, J. (2000) 'Effect of water, temperature and fertilizers on soil nitrogen net transformations and tree growth in an elfin cloud forest of Colombia', *Journal of Tropical Ecology*, 16: 83–99.

Chazdon, R.L., and Whitmore, T.C. (eds.). (2002) *Foundations of Tropical Forest Biology: Classic Papers with Commentaries*. Chicago: University of Chicago Press.

Chesson, P. (2000) 'Mechanisms of maintenance of species diversity', *Annual Review of Ecology and Systematics*, 31: 343–66.

Colinvaux, P.A., and De Oliveira, P.E. (2000) 'Palaeoecology and climate of the Amazon basin during the last glacial cycle', *Journal of Quaternary Science*, 15: 347–56.

Davis, C.C., Fritsch, F.W., Bell, C.D. and Mathews, S. (2004) 'High-latitude Tertiary migrations of an exclusively tropical clade: Evidence from Malpighiaceae', *International Journal of Plant Science*, 165: S107–S121.

de Queiroz, A. (2005) 'The resurrection of oceanic dispersal in historical biogeography', *Trends in Ecology and Evolution*, 20: 68–73.

Evans, B.J., Supriatna, J., Andayani, N., Setiadi, M.I., Cannatella, D.C. and Melnick, D.J. (2003) 'Monkeys and toads define areas of endemism on Sulawesi', *Evolution*, 57: 1436–43.

Fine, P.V.A., and Ree, R.H. (2006) 'Evidence for a time-integrated species-area effect on the latitudinal gradient in tree diversity', *American Naturalist*, 168: 796–804.

Flenley, J.R. (1998) 'Tropical forests under the climates of the last 30,000 years', *Climatic Change*, 39: 177–97.

Furley, P. (2004) 'Tropical savannas', *Progress in Physical Geography*, 28: 581–98.

Furley, P.A., Proctor, J. and Ratter, J.A. (eds.). (1992) *Nature and Dynamics of Forest-Savanna Boundaries*. London: Chapman and Hall.

George, W.B., and Lavocat, R. (eds.) (1993) *The Africa-South America Connection*. Oxford: Oxford University Press.

Givnish, T.J., Millam, K.C., Evans, T.M., Hall, J.C., Pires, J.C., Berry, P.E. and Sytsma, K.J. (2004) 'Ancient vicariance or recent long-distance dispersal? Inferences about phylogeny and South American-African disjunctions in Rapateaceae and Bromeliaceae based on *ndh*F sequence data', *International Journal of Plant Sciences*, 165: S35–S54.

Glacken, C.J. (1967) *Traces on the Rhodian Shore; Nature and Culture in Western Thought from Ancient Times to the End of the Eighteenth Century*. Berkeley: University of California Press.

Gliessman, S.R. (1998) *Agroecology: Ecological Processes in Sustainable Agriculture*. Ann Arbor, MI: Ann Arbor Press.

Goldsmith, F.B. (ed.). (1998) *Tropical Rain Forest: A Wider Perspective*. London: Chapman and Hall.

Goulding, M., Smith, N.J.H. and Mahar, D. (1996) *Floods of Fortune: Ecology and Economy along the Amazon*. New York: Columbia University Press.

Hammer, M.F., and 8 authors. (1998) 'Out of Africa and back again: Nested cladistic analysis of human Y chromosome variation', *Molecular Biology and Evolution*, 15: 427–41.

Hillebrand, H. (2004) 'On the generality of the latitudinal diversity gradient', *American Naturalist*, 163: 192–211.

Johnson, C.N. (2002) 'Determinants of loss of mammal species during the Late Quaternary "megafauna" extinctions: Life history and ecology, but not body size', *Proceedings of the Royal Society of London Series B*, 269: 2221–27.

Kellman, M., and Tackaberry, R. (1997) *Tropical Environments: The Functioning and Management of Tropical Ecosystems*. London: Routledge.

Kupfer, J.A., Malanson, G.P. and Franklin, S.B. (2006) 'Not seeing the ocean for the islands: The mediating influence of matrix-based processes on forest fragmentation effects', *Global Ecology and Biogeography*, 15: 8–20.

Lambeck, K., Esat, T.M. and Potter, E.-K. (2002) 'Links between climate and sea levels for the past three million years', *Nature*, 419: 199–206.

Laurance, W.F., and Bierregaard, R.O. Jr. (eds.). (1997) *Tropical Forest Remnants: Ecology, Management, and Conservation of Fragmented Communities*. Chicago: University of Chicago Press.

Leigh, J., E.G., Rand, A.S. and Windsor, D.M. (eds.). (1996) *The Ecology of a Tropical Forest: Seasonal Rhythms and Long-Term Changes*. Washington, DC: Smithsonian Institution Press.

Lomolino, M.V., Riddle, B.R. and Brown, J.H. (2006) *Biogeography*. 4th ed. Sunderland, MA: Sinauer.

MacArthur, R.H., and Wilson, E.O. (1967) *The Theory of Island Biogeography*. Princeton, NJ: Princeton University Press.

MacFadden, B.J. (2000) 'Cenozoic mammalian herbivores from the Americas: Reconstructing ancient diets and terrestrial communities', *Annual Review of Ecology and Systematics*, 31: 33–59.

Mallet, J., Barton, N., G. Lamas M., J. Santisteban C., Muedas M. and Eeley, H. (1990) 'Estimates of selection and gene flow from measures of cline width and linkage disequilibrium in Heliconius hybrid zones', *Genetics*, 124: 921–36.

Miles, L., Grainger, A. and Phillips, O. (2004) 'The impact of global climate change on tropical forest biodiversity in Amazonia', *Global Ecology and Biogeography*, 13: 553–65.

Millington, A.C., Walsh, S.J. and Osborne, P.E. (eds.). (2001) *GIS and Remote Sensing Applications in Biogeography and Ecology*. Norwell, MA: Kluwer Academic.

Moritz, C., Patton, J.L., Schneider, C.J. and Smith, T.B. (2000) 'Diversification of rainforest faunas: An integrated molecular approach', *Annual Review of Ecology and Systematics*, 31: 533–63.

Morley, R.J. (2000) *Origin and Evolution of Tropical Rain Forests*. Chichester, UK: John Wiley.

Newton, I., and Dale, L. (2001) 'A comparative analysis of the avifaunas of different zoogeographical regions', *Journal of Zoology*, 254: 207–18.

Norris, K., and Harper, N. (2004) 'Extinction processes in hot spots of avian biodiversity and the targeting of pre-emptive conservation action', *Proceedings of the Royal Society of London Series B*, 271: 123–30.

Ogden, R., and Thorpe, R.S. (2002) 'Molecular evidence for ecological speciation in tropical habitats', *Proceedings of the National Academy of the United States of America*, 99: 13612–15.

Pearson, D.A., Schaefer, T., Johnson, K.R. and Nichols, D.J. (2001) 'Palynologically calibrated vertebrate record from North Dakota consistent with abrupt dinosaur extinction at the Cretaceous-Tertiary boundary', *Geology*, 29: 39–42.

Pennington, R.T., and Dick, C.W. (2004) 'The role of immigrants in the assembly of the South American rainforest tree flora', *Philosophical Transactions of the Royal Society of London Series B*, 359: 1611–22.

Pennington, R.T., Prado, D.E. and Pendry, C.A. (2000) 'Neotropical seasonally dry forests and Quaternary vegetation changes', *Journal of Biogeography*, 27: 261–73.

Perrow, M.R., and Davy, A.J. (eds.). (2002) *Handbook of Ecological Restoration. Volume 2: Restoration in Practice*. Cambridge: Cambridge University Press.

Peterson, A.T. (2001) 'Predicting species' geographic distributions based on ecological niche modeling', *The Condor*, 103: 599–605.

Primack, R.B., and Corlett, R.T. (2005) *Tropical Rain Forests: An Ecological and Biogeographical Comparison*. New York: Blackwell.

Qian, H., and Ricklefs, R.E. (2000) 'Large-scale processes and the Asian bias in species diversity of temperate plants', *Nature*, 407: 180–82.

Rahbek, C. (2005) 'The role of spatial scale and the perception of large-scale species-richness patterns', *Ecology Letters*, 8: 224–39.

Rees, P.M., Noto, C.R., Parrish, J.M. and Parrish, J.T. (2004) 'Late Jurassic climates, vegetation, and dinosaur distributions', *Journal of Geology*, 112: 643–53.

Richards, P.W. (1996) *The Tropical Rain Forest*. 2nd ed. Cambridge: Cambridge University Press.

Rigg, L.S. (2005) 'Disturbance processes and spatial patterns of two emergent conifers in New Caledonia', *Austral Ecology*, 30: 363–73.

Rigg, L.S., Enright, N.J., Perry, G.L.W. and Miller, B.P. (2002) 'The role of cloud combing and shading by isolated trees in the succession from maquis to rain forest in New Caledonia', *Biotropica*, 34: 199–210.

Robinson, J.G., and Bennett, E.L. (eds.). (2000) *Hunting for Sustainability in Tropical Forests*. New York: Columbia University Press.

Scatena, F.N. (2001) 'Ecological rhythms and the management of humid tropical forests: Examples from the Caribbean National Forest, Puerto Rico', *Forest Ecology and Management*, 154: 453–64.

Scatena, F.N., and Lugo, A.E. (1995) 'Geomorphology, disturbance, and the soil and vegetation of two subtropical wet steepland watersheds of Puerto Rico', *Geomorphology*, 13: 199–213.

Schneider, H., Schuettpelz, E., Pryer, K.M., Cranfill, R., Magallón, S. and Lupia, R. (2004) 'Ferns diversified in the shadow of angiosperms', *Nature*, 428: 553–57.

Turner, I.M. (2001) *The Ecology of Trees in the Tropical Rain Forest*. Cambridge: Cambridge University Press.

Vajda, V., Raine, J.I. and Hollis, C.J. (2001) 'Indication of global deforestation at the Cretaceous-Tertiary boundary by New Zealand fern spike', *Science*, 294: 1700–1702.

Voeks, R.A. (2004) 'Disturbance pharmacopaeias: Medicine and myth from the humid tropics', *Annals of the Association of American Geographers*, 94: 868–88.

Wallace, A.R. (1895) *Natural Selection and Tropical Nature*. London: Macmillan.

Walter, H. (1985) *Ecological Systems of the Geobiosphere*. Berlin: Springer-Verlag.

Whitmore, T.C. (1998) *An Introduction to Tropical Rain Forests*. 2nd ed. Oxford: Oxford University Press.

Whitmore, T.C. and Prance, G.T. (1987) *Biogeography and Quaternary History in Tropical America*. Oxford: Clarendon Press.

Whittaker, R.J. (1995) 'Disturbed island ecology', *Trends in Ecology and Evolution*, 10: 421–25.

Whittaker, R.J. (1998) *Island Biogeography: Ecology, Evolution, and Conservation*. Oxford: Oxford University Press.

Young, K.R. (1995) 'Biogeographical paradigms useful for the study of tropical montane forests and their biota', in S.P. Churchill, H. Balslev, E. Forero, and J.L. Luteyn (eds.), *Biodiversity and Conservation of Neotropical Montane Forests*. New York: New York Botanical Gardens. pp. 79–87.

Young, K.R. (2003) 'Genes and biogeographers: Incorporating a genetic perspective into biogeographical research', *Physical Geography*, 24: 447–66.

Young, K.R. (2007) 'Causality of current environmental change in tropical landscapes', *Geography Compass*, 1: 1299–314.

Zerner, C. (ed.). (2000) *People, Plants, and Justice: The Politics of Nature Conservation*. New York: Columbia University Press.

Dynamics of Mountain Ecosystems

Udo Schickhoff

17.1 MOUNTAIN ECOSYSTEMS: GLOBAL SIGNIFICANCE AND THREATS

Mountain ecosystems exist on every continent, from the equator almost to the poles, and from close to sea level to the highest elevation on Earth (Figure 17.1). Mountains differ considerably from one region to another; the connective salient feature, however, is the complexity of their topography. Consequently, they are characterized by sharp environmental gradients, that is, distinct variations of climatic, edaphic, and other environmental factors over very short distances. Given their three-dimensional nature, the world's mountains encompass a very extensive array of climate, soils, and biota, as well as cultural differences. If large plateaus are excluded and a lower boundary of 1,000 meters above sea level (m.a.s.l.) at the equator with a linear reduction to 300 m.a.s.l at high latitudes assumed, mountains cover 20 to 25% of the global land surface (cf. Kapos et al., 2000). With about a quarter of the world's terrestrial surface higher than 1,000 m.a.s.l, and 11% higher than 2,000 m.a.s.l., mountains greatly influence regional and continental atmospheric circulation and water and energy cycles, and they provide goods and services to about half of humanity (Ives et al., 1997). Mountains are important sources of water, energy, forest and agricultural products, minerals, and other natural resources; they are biodiversity hotspots and centers of cultural diversity, provide opportunities for recreation and tourism, and are of spiritual significance. Ecosystem services such as water

purification or climate regulation extend far beyond their geographical boundaries. Körner et al. (2005) argue that water supply is the key function of mountains for humanity because all of the world's major rivers and many smaller ones originate in mountains, and over 40% of the world's population lives in the watersheds of rivers that have their sources in mountains. Thus, they influence the lives of billions of people living either in mountains or adjacent lowlands. Approximately one-tenth of the human population derives their life-support directly from mountains (Price, 1998), including numerous groups of distinct sociocultural identity. Mountains are often called "water towers" since they store immense amounts of fresh water as ice and snow and in lakes and reservoirs, thus playing a crucial role for supplying water to adjoining lowlands. In humid regions, the proportion of water generated in the mountains may already provide more than half of the total fresh water available. In semiarid and arid regions, this percentage is much higher. For instance, more than 95% of the water in the basin of the Aral Sea comes from the Tian Shan and the Pamir Mountains (Messerli, 1999). The drier the lowland, the greater the importance of the wet mountain areas that provide disproportionately high and reliable discharge (Viviroli et al., 2003). The enormous hydropower potential further substantiates the importance of mountain water. The global significance of mountain water resources will increasingly become apparent in the next decades as a growing proportion of the world's population will experience water scarcity.

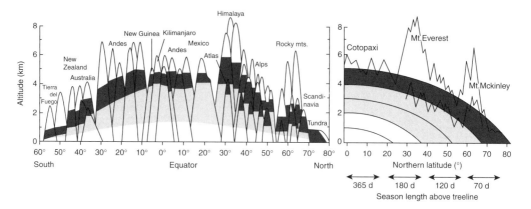

Figure 17.1   Profile of the latitudinal position of alpine belts in mountains across the globe and compression of thermal zones on mountains, altitude for latitude (after Körner, 2003)

Mountains are globally significant as core areas of biodiversity. They are characterized by higher species richness than adjacent lowlands, although species richness generally decreases with elevation. The compression of climatic life zones with altitude (cf. Figure 17.1), the small-scale habitat diversity (geodiversity) caused by the fragmented and topographically diverse terrain, and the differences in substrate, slope, aspect, and topoclimate are the basic causes of this exceptional diversity. The small sizes of organisms, isolation combined with effective reproduction systems, and moderate disturbance further contribute to mountain biota diversity (Körner, 2002). In particular, tropical and subtropical mountains that extend up to glaciated terrain contain highly diverse and rich ecosystems including, for example, the major centers of global species diversity (Mutke and Barthlott, 2005). Mountains are also important as centers of crop diversity and as places where the wild ancestors of many significant food crops, for example, potatoes, maize, wheat, rice or barley, had originated. There is often a high proportion of endemic species since many mountain ranges or peaks represent isolated habitats. Mountain areas of high species richness and endemism are often designated as national parks or nature reserves. Almost a third (c. 260 million hectares [ha]) of all protected areas in the world are in the mountains, an area slightly larger than Sudan, the tenth largest country in the world. However, many mountain ranges are still unrepresented, and the global network of mountain protected areas is in need of additions and redesign in order to achieve the goals of the Convention on Biological Diversity (Hamilton, 2002).

Mountain forests account for more than a quarter of closed forests by area globally (Kapos et al., 2000) and contribute to the provision of ecosystem goods and services from mountainous areas. Within highland-lowland interactions, mountain forests play a critical role because of their multifunctionality, that is, they provide productive, protective, and cultural and amenity functions (Price and Butt, 2000). Mountain forests not only provide timber, fuelwood, and other wood and nonwood products, but they are also vitally important for protection against natural hazards such as landslides, rockfalls, avalanches, and floods. Carbon storage and carbon sequestration in mountain forests is likely globally significant and particularly so in semiarid and arid areas. Moreover, mountain forests greatly contribute to the attractiveness of mountain landscapes that are key tourism destinations in many parts of the world. As tourism is the fastest growing global industry, the significance of mountains as centers for recreation will increase in the coming decades.

However, mountain ecosystems are exceptionally fragile and susceptible to global environmental change. Climate and land-use changes in particular will increasingly threaten the integrity of these ecosystems and alter their capacity to provide goods and services (Bowman, 2006). High relief, complex topography, and condensed vertical ecological gradients make them very vulnerable to changes in environmental conditions. Mountain plants and animals are adapted to relatively narrow ranges of temperature and precipitation (Körner, 2003). High elevation environments with glaciers, snow, permafrost, water, and a complex altitudinal zonation of vegetation and fauna are now widely recognized as being among the most sensitive to climatic changes (e.g., Beniston, 2000; Grabherr et al., 2001). The fragility of mountain ecosystems also represents a considerable challenge to sustainable land use and natural resource management. Unsustainable mining, forestry, agriculture, and tourism often have more rapid and drastic consequences than environmental change, and their impacts are more difficult to correct than in many other parts of the world.

This chapter assesses the available knowledge about the responses of mountain ecosystems to recent climate and land-use changes. Knowing how structures and functions of mountain ecosystems are affected is of fundamental importance since the response to environmental change has significant implications for mountain people as well as millions living elsewhere on the planet. Understanding system responses is also vital for adaptation planning. The growing prominence of global environmental change issues as well as the increased awareness of the essential roles that mountain systems play in the geo-biosphere has stimulated reinforced scientific interest in mountain environments, further supported by international efforts to establish mountains as a research priority (e.g., Chapter 13 of Agenda 21, Rio Earth Summit, 1992; International Year of the Mountains, 2002; Bishkek Global Mountain Summit, 2002). From a scientific point of view, the high sensitivity of mountain regions provides unique opportunities to detect, model, and analyze global change processes and their effects on biophysical and socioeconomic systems (Hofer, 2005). The growing number of national research initiatives and international research programs reflects the topicality and increasing importance of mountain research.

17.2 CLIMATE CHANGE IN MOUNTAIN REGIONS: IMPACTS ON PHYSICAL SYSTEMS

Mountain regions worldwide provide increasing evidence of ongoing impacts of climate change on physical and biological systems. The climate has warmed substantially since the end of the Little Ice Age. Between 1906 and 2005, global average surface temperatures increased by 0.70–1.8°C, and they are projected to increase by 1.8–4.0°C between 2090–2099 relative to 1980 to 1999 (Trenberth et al., 2007). Temperature increases on mountains were higher than the global mean during the twentieth century (Beniston et al., 1997) and will likely continue to be so this century (Nogués-Bravo et al., 2007). The greater than average warming on mountains is largely attributed to increased mean minimum temperatures, which follows a general northern hemisphere trend identified by Diaz and Bradley (1997).

In the European Alps, changes in the annual mean surface temperature from the 1961 to 1990 climatological mean for the twentieth century reveal that the warming rate since the early 1980s, averaged for eight sites in Switzerland, has been much higher (more than 1.5°C) compared to the global mean (Figure 17.2). The marked increases after 1980 are consistent for all central European mountains (Weber et al., 1997). The most intense warming occurred since the 1990s, though there was also significant warming during the 1940s.

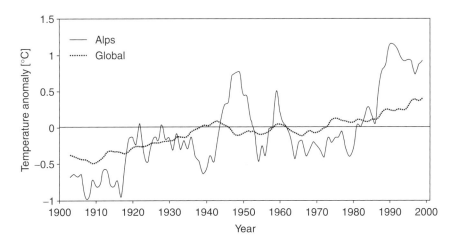

Figure 17.2   Temperature departures from the 1961–90 climatological mean in the Swiss Alps compared to global temperature anomalies for the period 1901–2000 (after Beniston, 2005)

From 1975 to 2004, mean temperatures in Switzerland increased by 0.57°C per decade (Rebetez and Reinhard, 2008). Increases in minimum temperatures of up to 2°C during the twentieth century and a smaller increase in maximum temperatures align with global trends. The Alps have become drier in the last few decades, but there have also been more extreme precipitation events. Climate models predict warmer and more humid winters and much warmer and drier summers for the latter part of the twenty-first century (Christensen et al., 2007).

Warming in the European Alps is consistent with those in mountain ranges on other continents. Observational data show temperatures at high-elevation stations have risen by 1–2°C in the last century (Diaz and Bradley, 1997). North and South American mountains have experienced significant and recently accelerating increases in temperature (Vuille and Bradley, 2000; Bonfils et al., 2008). Significantly warmer conditions since about 1930 have been prevalent on Asian mountains, exemplified by pronounced warming with a particular strong rise in autumn and winter temperatures at higher altitudes in the Himalayas (Shrestha et al., 1999) and over the Tibetan Plateau (Liu and Chen, 2000) and its surrounding areas (Batima et al., 2005; Bolch, 2007).

Global warming on mountains impacts physical systems in a number of ways. Climate-driven changes in the cryosphere and hydrosphere— degrading permafrost, decreasing snow cover, and glacier retreat—will have cascading effects on regional biophysical systems with serious implications for supraregional ecosystem services and socioeconomic systems. At a global scale, snow and ice cover has decreased, especially since 1980 and the decrease continues to accelerate, despite growth in some places and little change in others (Lemke et al., 2007).

The permafrost temperature regime is a sensitive indicator of decadal to centennial climatic variability. Permafrost temperatures have experienced the strongest increase on Arctic mountains, for example, up to 2–3°C on the Alaskan North Slope since the late 1980s (Osterkamp, 2005). In boreal and temperate mountains it has warmed at a slower, but still considerable rate. Data from the Murtèl-Corvatsch borehole in Switzerland indicate a warming rate of 0.6°C at 11.6 m depth from 1987 to 2002 (Harris et al., 2003) and at this depth, permafrost temperatures in 2001 and 2003 were only slightly below −1°C (Vonder Mühll et al., 2004). Permafrost warming and thawing implies a deepening of the active layer as well as decreases in the thickness and/or areal extent of permafrost. In the European Alps, active layer thickness during the hot summer of 2003 was significantly deeper than in previous years, and

degraded ground ice did not recover in the years that followed (Hilbich et al., 2008). Permafrost degradation on steep mountainsides increases the probability of mass movements (see Jentsch and Beierkuhnlein, this volume) because thawing reduces the stability of slopes formed with frozen soils or bedrock.

Fluctuations in the sizes of mountain glaciers are a valuable element in international climate monitoring programs (Haeberli et al., 2007). Glaciers react in a highly sensitive manner to climatic forcing as long as they do not have thick debris covers, calve, or surge. Whereas a change in glacier length is an indirect, delayed, but also enhanced signal of climate change, glacier mass balance is the direct and undelayed response to the annual atmospheric conditions (Haeberli and Hoelzle, 1995). Globally, glacier tongues have retreated from Little Ice Age moraine positions since 1850 to the present day (Barry, 2006), though some glaciers show intermittent readvances. Accelerated melting was observed in the 1980s and 1990s. The 30 World Glacier Monitoring Service reference glaciers showed an average annual mass loss of 0.58 m water equivalent (w.e.) between 1996 and 2005 compared to 0.25 m w.e. between 1986 and 1995 and 0.14 m w.e. in the previous interdecadal period. New, record annual mass losses were measured in 2003, 2004, and 2006, with the 2004 and 2006 losses being almost twice as high as high in 1998. The cumulative average ice loss over the past six decades (Figure 17.3) corresponds to between one-fifth and one-ninth of the global average ice thickness (WGMS, 2008). Under present climate-change conditions, glacier shrinkage may lead to the deglaciation of large parts of many mountain ranges in the coming decades (e.g., Zemp et al., 2006).

Particularly vast ice losses have been recorded between 1996 and 2005 for glaciers in the Andes and in arctic and boreal mountains of North America, Europe, and Central Asia (WGMS, 2008). The glacier cover in the European Alps diminished by about 35% from 1850 to the 1970s and declined a further 22% by 2000 (Zemp et al., 2007) (Figure 17.4). It is likely that many Himalayan glaciers will disappear by 2030–2040 if warming continues at the current rate (Cruz et al., 2007). With the exception of some Karakoram glaciers (Hewitt, 2005), Himalayan glaciers in Nepal, Bhutan, India, and China are retreating at rates ranging from c. 10 to 60 m per year. Rates of 30 m per year are commonplace. The Imja Glacier in the Khumbu Himal, Nepal, retreated at an unprecedented 74 m per year between 2001 and 2006, feeding the growth of a hazardous glacial lake (Ren et al., 2006; Bajracharya et al., 2007).

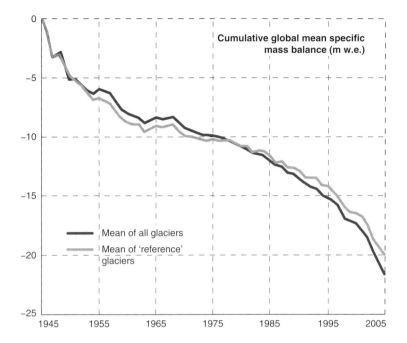

Figure 17.3   The cumulative specific mass balance curves for 1945–2005 showing the mean of all glaciers and 30 'reference' glaciers with (almost) continuous data series since 1976 (after WGMS, 2008)

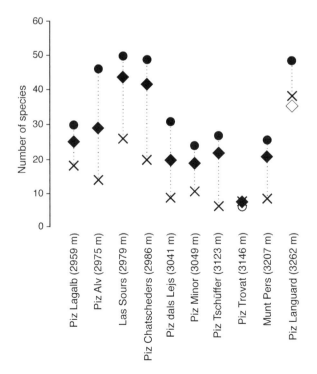

Figure 17.4   Changes in species richness of 10 mountain summits in the Swiss Alps. X = 1900s; ◆ = 1980s; ● = 2003; open symbols (P. Trovat, P. Languard) indicate a (temporary) decrease in species number (after Walther et al., 2005)

Glacier retreat impacts lowlands. For instance, Himalayan glaciers form the largest body of ice outside the polar icecaps and are the sources of rivers flowing across the Indo-Gangetic plains. Mountain discharge, especially dry-season river flows into rivers fed largely by ice melt, is particularly affected by the reduction of glacier volumes. Initially, enhanced glacier melting will result in increased river runoff, discharge peaks, and an increased melt season, but water yield will decrease as the glaciers shrink to the point where glacier area reduction prevails over glacier thinning. Summarizing observed changes in glacier runoff trends of major mountain regions in recent decades, Casassa et al. (2009) assessed significant runoff increases in many mountain ranges (Alberta, northwest British Columbia, Yukon, highly glacierized basins in the Swiss and Austrian Alps, the Tien Shan Mountains, Tibet, and the tropical Andes of Peru), that were closely related to observed temperature increases and enhanced glacier melting. Decreasing runoff trends that are interpreted to be a combined effect of reduced melt from seasonal snow cover and relevant glacier area losses are reported from south-central British Columbia, at low elevations in the Swiss Alps, and in central Chile. Overall, glacier wastage is currently increasing global sea levels by c. 1 mm per year, potentially affecting ocean circulation and ocean ecosystems (Dyurgerov and Meier, 2005).

Changes in mountain glaciers change natural hazard conditions. Glacier hazards include glacial lake outbursts and associated floods and debris flows; ice break-offs and ice avalanches; and destabilization of debris slopes and rock walls (Kääb et al., 2005). The general tendency is toward a shift in hazard zones and a widespread decrease in the stability of high-mountain slopes. A particular threat originates from glacier floods that are triggered by the outburst of glacial water reservoirs and that potentially cause immense damage to ecosystems and human habitation. The formation of large lakes with a high potential for glacial lake outburst floods (GLOFs) associated with glacier retreat is seen in several mountain ranges. More than 200 potentially dangerous glacial lakes have been documented across the Himalayan region, and 21 GLOFs have occurred in Nepal in the recent past (Bajracharya et al., 2007). Overall GLOF frequency in Nepal, Bhutan, and Tibet increased from 0.38 events per year in the 1950s to 0.54 events per year in the 1990s, and it is anticipated to further increase to one significant GLOF each year by 2010 (Richardson and Reynolds, 2000).

Snowfall and snow cover are vulnerable to climate change. Mountain snowpack constitutes a natural reservoir for cold-season precipitation storage and a major portion of the water budget. Snowpack and snowmelt are thus important predictors for summer streamflow that benefits downstream ecosystem functioning. Snowmelt often produces the major discharge event of the year and can be the source of extreme flows, and it floods the distribution of snow, and snow-cover duration is also important as a habitat factor for vegetation and fauna (see Möller and Thannheiser, this volume). The impacts of rising temperatures and changing precipitation on mountain snowpack and snowmelt, however, are complex and strongly influenced by geographic location, latitude, elevation, vegetation, and the general hydroclimatic regime (Beniston et al., 2003; Stewart, 2009). Nevertheless, coherent and consistent spatial and temporal patterns of trends in the volume, extent, and seasonality of snowpack and snowmelt in major mountain regions of the world emerge. Stewart (2009) summarized these patterns and stressed that (i) warmer temperatures at mid-elevations have resulted in decreased snowpack and earlier snowmelt in spite of precipitation increases; (ii) that precipitation increases at high elevations have resulted in increased snowpack; and (iii) that increasingly higher elevations are projected to experience declines in snowpack accumulation and snowmelt under continued warming. The general decrease in snowpack and snow-cover extent in the northern hemisphere, which has accelerated since the mid-1980s, has been consistently observed in low- and mid-elevation regions in western North America and the European Alps, with the greatest response in areas that remain close to freezing throughout the winter season (Laternser and Schneebeli, 2003; Knowles et al., 2006). Changes in snowpack and snow-cover extent have important consequences for the hydrological cycle, particularly in regions where the land surface hydrology is dominated by winter snow accumulation and spring melt (Barnett et al., 2005). There is abundant evidence from North America and northern Eurasia that spring peak river flows have been occurring one to two weeks earlier during the last 65 years, and that winter-base flow has increased in basins with important seasonal snow cover (Rosenzweig et al., 2007). Ongoing substantial shifts in streamflow seasonality of snow-dominated basins will have considerable impacts on water availability. Higher winter and spring runoff rates due to lower snow to total precipitation (S/P) ratios and earlier melt increase the risk of winter and spring floods, while increased rain precipitation may lead to slope instabilities, mass movements, and accelerated erosion. In addition to hydrological, climatological, and geomorphological consequences, changes in snow cover will greatly affect the tourism development of mountain regions.

17.3 RESPONSE OF MOUNTAIN BIOTA TO CLIMATE CHANGE

As ecosystems defined by low temperatures, high mountain regions, along with polar regions, are considered to show particular sensitivity and rapid, visible ecosystem responses to ongoing climate warming and related environmental changes. Alpine plant species have characteristics that have enabled them to pass various environmental filters associated with the harsh environment such as freezing tolerance. However, many of the adaptations of montane and alpine species to their current environments, such as slow and low growth, are likely to limit their responses to climate warming and other environmental changes. Thus, montane and alpine species will have to change their distributions rather than be able to evolve significantly in response to warming. Evidence from the past indicates that geographic ranges of terrestrial species in general are well correlated with bioclimatic variables. Due to steep temperature and precipitation gradients in mountains and associated rapid changes with elevation in altitudinal life zones and their environmental conditions, the often narrow altitudinal species ranges are particularly affected by relatively small climatic changes. The imperative to migrate to a more suitable climate or to adapt, persist, and compete with, or to give way to invading species under current warming potentially leads to biodiversity losses. Species that migrate from lower to higher altitudes exert competitive pressure on preexisting species, which in turn may be restricted from shifting upward due to reduced available habitats, unrealizable niche requirements, or the lack of migration corridors (Grabherr et al., 1995). Shifting upward in elevation involves the risk that equivalent surface areas with similar habitat conditions are no longer available. Thus, high alpine species are extremely vulnerable to global warming. The risk of extinction is lower for species that have a great potential for adaptive responses through genetic diversity, phenotype plasticity, high abundance, or significant dispersal capacities. A particular high risk of extinction has been confirmed for many endemic species in mountain ecosystems as their populations become easily fragmented (Pauli et al., 2003). In addition, shifts in climatic extremes such as drought events as well as other factors of global change such as atmospheric CO_2 concentration and eutrophication, and increasing disturbances such as fire, let alone the impacts of land-use changes, threaten mountain biodiversity.

As a response to anthropogenic climate change, significant changes in biological systems are occurring on all continents and in most oceans, including shifts in spring events, species distributions, and community structure (see Sparks et al., this volume; Walther et al., 2002; Parmesan, 2006). These responses that have potentially severe biological and economic consequences have been assessed to an increasing degree in mountain ecosystems. The evidence that some Tertiary high-elevation relict species have outlived the Quaternary glaciation in situ, that is, in ice-free refugia such as nunataks, may not obscure the fact that species are more likely to migrate in response to climatic changes rather than to persist and perhaps adapt genetically. A more or less consistent pattern of upslope range expansion by plant and animal species to cooler elevations becomes apparent, including a wide range of taxonomic groups and geographical locations (Rosenzweig et al., 2007). However, since old-world mountain regions have been subjected to human land use (grazing, agriculture, forestry) for millennia, climate change impacts, for example, on species distributions, are sometimes hard to separate from other influences, and it is often uncertain whether the present vegetation is in a natural equilibrium between climate and ecological factors (Theurillat and Guisan, 2001). In this respect, paleoecology is an important tool to develop deeper understanding of unaltered climate-vegetation relationships. Assuming that climate change has differential effects on the organizational levels of ecosystems and will be first visible in terms of species migration and species composition, the following review of impact studies is largely restricted to the (plant) species/community level. Actually, a recent analysis of altitudinal range change of the complete native vascular flora of a sub-Antarctic island over 40 years revealed great variations in individual range expansion rates and accordingly considerable changes in community composition (le Roux and McGeoch, 2008). Species-specific physiological tolerances, niche requirements, and biotic interactions determine to a large extent individualistic responses, which are further accentuated by specific responses to concurrent changes in nonclimatic environmental factors. Asynchronous responses to external pressures will lead to the formation of no-analog communities with modified competitive relationships between species and plant-functional types, and thus deviating ecosystem structure and function. Since species along the elevational gradient are differentially affected by climate warming and may respond to a varying degree, altitudinal zones are examined separately with particular attention to ecotones. These transition zones are considered highly sensitive to climate-change impacts.

Vascular plants of higher elevations, that is, of the alpine-nival ecotone and the nival belt, show comparatively pronounced signals of range varia-tions. An increasing number of observational studies, primarily from the European Alps and the Scandes, indicate recent upward shifts of high-elevation plants related to climate warming. Long-term vegetation monitoring series, including detailed surveys dating back to the nineteenth century, are available from high mountain peaks in the Alps. Extensive resurveys of 26 nival summit sites in the 1990s revealed increasing spe-cies richness on most of the summits as well as increased population sizes of long-established nival plant species (Grabherr et al., 1994; Pauli et al., 2001). Both the establishment of new species in the subnival and nival zones and large expansions of populations of already existing species are considered a strong indication of an environmental alteration. Upward shifts of alpine plants, however, obviously depend on appropriate migration corridors. For example, those summits that show stagnating species richness but expand-ing populations of subnival/nival species exhibit conditions (e.g., gravelly, unstable slopes) that largely restrict the arrival of propagules of lower-elevation species (Grabherr, 2003). Repeated resurveys of summits in the Engadine, Swiss Alps, yielded similar results. Figure 17.5 illustrates that nine out of ten reinvestigated summits that were first surveyed during 1903–9, and resurveyed in the 1980s and in 2003, showed a strong trend toward increasing species numbers. Only one summit, Piz Trovat, which is covered with coarse scree unfavorable to species migration and estab-lishment, showed a more or less steady number of vascular plant species. The increase of species richness is accelerating as the rate of change was significantly greater in the later period (3.7 species/decade) compared to the first interval (1.3 species/decade) (Walther et al., 2005).

Another resurvey of 12 summits in the alpine-nival ccotone in the Swiss Alps showed an increase in vascular plant species richness of 11% per decade over the last 120 years, and a general trend of upward migration in the range of several meters per decade (Holzinger et al., 2008). The increase in species richness was higher on calcareous bedrock compared to siliceous bedrock due to the greater variety of microhabitats. Upslope migra-tion of species, mostly anemochorous pioneer species, has been assessed at other summits in the

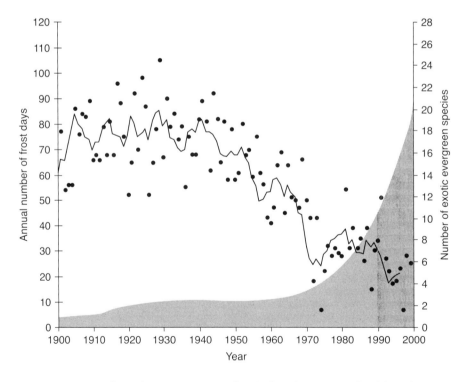

Figure 17.5 Increase of exotic evergreen species during the period of milder winter temperatures at the lower south slope of the Alps. Dots = annual number of frost days; solid line = smoothed 5-year averages (after Walther et al., 2002, modified)

Alps as well (Bahn and Körner, 2003; Burga et al., 2007). Parolo and Rossi (2008) documented a significant upward migration of vascular plant species along a continuous altitudinal transect of 730 m in high-alpine and nival environments of the Italian Alps from 1954–58 to 2003–5. About one-third of the surveyed species were recorded from altitudes 30–430 m higher than their 1950s limit, representing a median migration rate of 23.9 m per decade. Upslope shift was connected to wind dispersal of diaspores. No single species was recorded at a lower altitude in the second than in the first survey. Comparable plant species' shifts on mountain peaks have been reported from the Scandes (Klanderud and Birks, 2003; Kullman, 2006). Upward migration processes to be attributed to climate warming are the most likely explanation for the consistent results across diverse mountain ranges. However, recolonization processes of species with low dispersal potential in the aftermath of the Little Ice Age may have contributed to the observed changes (cf. Kammer et al., 2007).

The first results of the Global Observation Research Initiative in Alpine Environments (GLORIA) monitoring program provide deeper insight into the vegetation dynamics of the alpine-nival ecotone and first evidence of warming-induced species declines in the European Alps (Pauli et al., 2007). Ten years after the establishment of c. 1,100 permanent plots in 1994, average vascular plant species richness increased by 11.8%, but species cover showed differential changes with significant declines of all subnival to nival plants and pronounced increases of alpine pioneer species. The results suggest an ongoing range contraction of subnival to nival species at their rear (lower) edge and a concurrent expansion of alpine pioneer species at their leading edge, which is primarily caused by recent warming as other potential drivers of vegetation changes such as natural primary succession, grazing, or nitrogen deposition can be ruled out or are of minor importance. Thus, an enhancement of local species richness by plant invasions in the upper alpine-subnival belt is obviously accompanied by reduced richness of cold-adapted species from higher elevations that are driven out of their distribution range. This trend could become a major threat to biodiversity in high mountains. In addition, adverse effects on biodiversity and related ecosystem functioning could emanate from the homogenization of alpine summit vegetation (Jurasinski and Kreyling, 2007). High-elevation species are not only threatened to decline at their lower margin but also at the southern margin of their range as has been reported from the Rocky Mountains in Montana (Lesica and McCune, 2004).

In contrast to the alpine-nival ecotone, observed changes in species composition and species abundances in grasslands and dwarf-shrub heaths of the lower and midalpine belt are less pronounced, at least in the European Alps. Late successional vegetation types in this altitudinal zone, characterized by the dominance of long-lived perennials (often clonal perennials), show low internal warming-induced dynamics. Single clones of *Carex curvula* in the Alps are several thousand years old and have persisted on the same spot obviously unaffected by Holocene temperature oscillations (Steinger et al., 1996). Results from the world's oldest permanent plots in alpine grasslands, established in the Swiss National Park in 1914, only indicate minor variations (Grabherr, 2003). The same holds for alpine heaths in the Scandes over the last 70 years (Virtanen et al., 2003). During the same period, a shift to more thermophilic plants was assessed in alpine swards of the Bernese Alps, Switzerland (Keller et al., 2000). Greater climate-related changes in species composition of alpine grasslands have been reported from Mediterranean high mountains (Sanz-Elorza et al., 2003), but the results are difficult to generalize due to varied interrelationships between temperature increase, dispersal, biotic resistance to invasion, and other abiotic and biotic factors in this altitudinal zone across diverse mountain ranges. Nevertheless, new climatic climax communities are likely to develop. Invasion processes inducing greater changes in species composition affect particular azonal habitats and communities such as snowbeds or mountaintops. For example, the effects of declining snowpack and snow cover lower the competitive ability of plant species belonging to snowbed communities, resulting in a high sensitivity to species invasions, and respective shifts in species composition (e.g., Virtanen et al., 2003). Kullman (2007a) assessed a substantial replacement of alpine and subalpine snowbed communities by dense alpine meadows in the Swedish Scandes over the past 30 years. It has been suggested that changing snow-distribution patterns are a major driver of rearrangements in alpine vegetation mosaics (Grabherr et al., 1995). Earlier snow melt has major phenological implications and may also lead to increased frost damages in alpine plants (Inouye, 2008). Even though shifts in species composition of zonal communities in the lower and midalpine belts of many mountain ranges are somewhat less pronounced compared to higher elevations, warming seems to be generally favorable to the growth, development, and reproduction of most plant species. Climate warming is expected to basically cause an amelioration of environmental conditions such as growing season length and nitrogen supply in the soil. The increase in NDVI (normalized difference vegetation index; index of vegetation greenness) of alpine grasslands over the last decades is consistent with results for high-latitude tundra vegetation, and it corresponds to expanding growing seasons, increasing primary productivity, and above-ground biomass observed for northern vegetation in general.

The current retreat of mountain glaciers involves a large increase of recently deglaciated terrain, which represents a highly dynamic alpine habitat. Receding glaciers have induced strongly increasing scientific interest in vegetation succession on glacier forelands. A growing number of studies in the European Alps and other mountain regions (e.g., Erschbamer et al., 2008) substantiate the dynamic response of glacier foreland ecosystems to climate change where successional pathways lead to complex vegetation patterns according to terrain age, but also to meso- and microtopographic variability, seed availability, species-specific life-history traits, local climatic and edaphic conditions, and altitudinal vegetation zonation. Recent evidence suggests currently accelerating ecological responses to climate change on glacier forelands (Cannone et al., 2008).

The subalpine-alpine ecotone, that is, the treeline ecotone in forested mountains, has been attracting particular interest in studies of the impacts of climate change in mountains since treeline advance to higher elevations would result in fundamental physiognomic, structural, and functional changes of mountain landscapes and ecosystems. Although various specific abiotic and biotic parameters influence the treeline, it is generally agreed that the position of natural alpine treelines is caused mainly by heat deficiency (i.e., insufficient air and soil temperatures during the growing season) (Holtmeier, 2009; Körner, 2003). Thus, treelines are assumed to be sensitive indicators of climate changes. However, linear extrapolations of the relationships between tree growth, metabolism, regeneration, and climate conditions according to predicted increases in temperatures are not acceptable. In recent reviews, Holtmeier and Broll (2005, 2007) stressed that treeline ecotones have to be seen as space-time-related phenomena and that treelines do not respond linearly to an altitudinal shift of any isotherm. Instead, climatically driven changes in position and structure are modified to a great extent by regional, local, and temporal variations. Factors such as the effects of varied topography on local site conditions, aftereffects of historical disturbances (extreme events, fire, insect pests, human impact, etc.), inertia of tree populations, species-specific traits, and/or biotic interactions and feedback systems may override or overcompensate the impact of slightly higher average temperatures (e.g., see Stueve et al., 2009,

who analyzed treeline recovery to fire at the subalpine-treeline ecotone in Mount Rainier National Park). It has also to be emphasized that various treeline-forming species will respond differently to a changing climate, in particular with regard to seed-based regeneration processes. Holtmeier and Broll (2007) conclude that treelines will advance in the long term under continued global warming, but not in a closed front parallel to the shift of an isotherm. A synchronous adjustment of environmental conditions and treeline elevation to a changing climate is not to be expected. Results of modeling studies (e.g., Dullinger et al., 2004) provide corroborative evidence that responses of individual treeline systems are not solely climate-driven but depend on a multitude of factors. Nevertheless, observed climate-related changes in treeline ecotones are ubiquitous, and climate warming might be considered a universal forcing mechanism that is amplified or attenuated by local site conditions.

Generally, recent empirical treeline studies show enhanced tree growth (e.g., Rolland et al., 1998), tree establishment and infilling of gaps within the treeline ecotone during the past decades, but often either no treeline advance so far or merely minor elevational shifts lagging behind actual warming (e.g., Klasner and Fagre, 2002; Daniels and Veblen, 2004; Kullman, 2007b). Locally increasing numbers of tree seedlings above the current upper tree limit, occasionally far above, are observed in many mountain ranges (my own observations), but whether the young growth survives the seedling stage and the critical phase when projecting above the winter snow cover, and whether this finally leads to tree establishment far beyond the present tree limit, remains to be seen. Comparatively few studies document a substantial treeline shift during the twentieth century. Meshinev et al. (2000) and Kullman (2001, 2007a) assessed an obviously climatically driven elevational tree-limit rise by more than 100 m (up to 170 m) in the Balkan Mountains and in the southern Swedish Scandes. Repeat photographic studies in the Uinta Mountains, Utah, documented an upward shift of the treeline between 60 and 180 m since 1870 (Munroe, 2003). Peñuelas and Boada (2003) and Shiyatov et al. (2007) reported treeline advances of up to 70–80 m in the Montseny Mts., Northeast Spain, and in the Urals, respectively. Danby and Hik (2007) and Baker and Moseley (2007) detected similar elevational advances in southwest Yukon, Canada, and in the Hengduan Mountains, Yunnan, China. But in the latter case, treeline shift has to be attributed to interactions of a warming climate with the cessation of historical disturbances (grazing, fire). Treeline advance as a result of both land-use change and climate change are presently observed in old-settled mountain regions such as the European Alps (e.g., Gehrig-Fasel et al., 2007).

Contemporary treeline studies give evidence of both advancing alpine treelines and insignificant treeline responses to climate warming, spanning the entire gradient from rapid dynamics to apparently complete inertia (Dullinger et al., 2004). The often conservative behavior of the treeline position, however, may not hide the fact that the warming trend is generally favorable to growth, development, and reproduction of tree species within the treeline ecotone. When assessing the sensitivity to climate warming, one has to distinguish between the different types of treeline. Orographic treelines are not very susceptible to the effects of higher temperatures, whereas anthropogenically depressed treelines may show greatest sensitivity after the cessation of land use. The sensitivity and response of climatic treelines vary according to local and regional topographical conditions and respective interacting abiotic and biotic factor complexes, and thus shows differential extent and intensity of change processes (Holtmeier and Broll, 2005).

The colline, montane, and subalpine altitudinal belts, which are potentially forested in humid mountains, have received comparatively less attention to date regarding the response to climate warming. However, in light of the multiple functions of mountain forests, major reorganizations of forests in terms of species composition and stand structure potentially have severe ecological, economic, and societal consequences. Modeling studies (e.g., Bugmann et al., 2005) suggest major warming-induced effects on tree distributions along altitudinal gradients, leading to considerable range changes of tree species and shifts in species composition and abundances in forest communities. Empirical studies have indeed detected significant recent range shifts of forest communities and forest plant species to higher elevations across a wide range of geographical locations. A comprehensive analysis of large-scale (across temperate and Mediterranean mountain forests in West Europe and along the entire elevation range from 0 to 2,600 m.a.s.l.), long-term (over the twentieth century), and multispecies (through an assemblage of 171 species) climate-related responses in forest plant altitudinal distributions showed that climate warming has resulted in a significant upward shift in species optimum elevation averaging 29 m per decade (Lenoir et al., 2008). The shift was larger for species restricted to mountain habitats. This study provided convincing evidence that climate change affects not only distributional margins but also the spatial core of the distributional range of plant species

along elevational gradients. With respect to biodiversity conservation, a declining potential of mountain protected areas to be able to host migrating species within reserve limits has to be anticipated.

In addition, notable impacts of climate warming are reflected by the upward shift of single forest species, which are known to be temperature-sensitive. An example is the pine mistletoe (*Viscum album* ssp. *austriacum*) in the pine forests of the European Alps. A comparison of current mistletoe occurrences with records from a survey in 1910 showed that the current upper limit is roughly 200 m above the limit found 100 years ago (Dobbertin et al., 2005). Temperate forests at lower elevations on the south slope of the European Alps have experienced substantial increases in the abundance and frequency of indigenous evergreen broadleaved (laurophyllous) species, which become increasingly competitive with decreasing number of frost days and cumulative length of the growing season (Walther, 2001). Moreover, exotic evergreen species including palms (*Trachycarpus fortunei*) have succeeded in colonizing forest areas and establishing stands in the shrub layer (Figure 17.6), driven by changes in winter temperatures and growing season length (Walther et al., 2007). Recent warming has also induced responses of ecotones between forest communities along elevational gradients. Beckage et al. (2008) resurveyed forest plots established across the northern hardwood-boreal forest ecotone in the Green Mountains (Vermont) and assessed climatically induced substantial changes in dominance patterns of tree species and an ecotone upslope shift of c. 100 m from 1962 to 2005. Surprisingly little inertia to climatically induced range shifts has also been documented from medium altitudes in the Montseny Mountains (Northeast Spain), where beech (*Fagus sylvatica*) forests are being replaced by holm oak (*Quercus ilex*) forests in recent decades (Peñuelas and Boada, 2003). Compared to temperate and boreal regions, evidence for range shifts is scarce to date for the tropics where upslope range shifts are far more likely than latitudinal shifts. Recently, Colwell et al. (2008) suggested a substantial attrition in species richness in tropical lowlands since species that are driven upslope by warming cannot be replaced by species adapted to higher temperatures. However, range limits of tropical lowland species are probably more strongly influenced by precipitation (Svenning and Condit, 2008).

In addition to temperature-induced vegetation dynamics, shifts in precipitation may cause extensive changes in species composition and abundances in mountain forests. Allen and Breshears (1998) detected a drought-induced retreat of ponderosa pine (*Pinus ponderosa*) forest

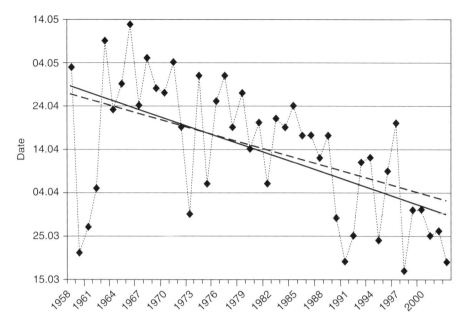

Figure 17.6 Larch (*Larix decidua*) needle appearance in Sargans, Switzerland, 1958–2002, showing earlier occurrence of 28 days over 50 years for the period 1958–98 (dashed line), and of 33 days over 50 years for the period 1958–2002 (solid line). The trend has been reinforced by the mild climate during 1999–2002 (after Defila and Clot, 2005)

leading to the advance of piñon juniper (*Pinus edulis, Juniperus monosperma*) woodland in northern New Mexico. Increased drought stress may also be a key factor in reducing tree resistance against pathogens leading to higher tree mortality and the decline of tree populations (Bigler et al., 2006). At lower limits of altitudinal ranges, higher temperatures, higher evapotranspiration, and drought stress may result in significant growth declines as evidenced for mature beech (*Fagus sylvatica*) trees in the mountains of Northeast Spain (Jump et al., 2006). Even at treeline elevations, growth declines because of drought effects have been detected at some locations (e.g., D'Arrigo et al., 2004). The upward trend in disturbance resulting from forest fires has also been linked to increased spring and summer temperatures, drought episodes, and an earlier spring snowmelt (Westerling et al., 2006).

Across all altitudinal zones in mountains, phenological changes (Figure 17.7) represent an unmistakable and most conspicuous response to climate warming. It has been generally observed that the timing of many life-cycle events such as blooming, migration, and insect emergence had shifted earlier in the spring and often later in the autumn (Rosenzweig et al., 2007). The temperature-driven growing season in mid- and high northern latitudes has been extended by up to two weeks in the second half of the twentieth century. Based on an immense phenological network data set, including 542 plant and 19 animal species in 21 European countries, Menzel et al. (2006) showed that 78% of all leafing, flowering, and fruiting records advanced during 1971 to 2000, and that the species' phenology is responsive to temperature of the preceding months. Assuming that the phenological response follows the spatial variability in climatic trends, a relatively greater shift to an earlier onset of spring phenophases in mountain regions would be expected. A comparison between lowland and alpine regions in Switzerland (Defila and Clot, 2005) revealed a stronger shift toward earlier spring events in the lowland. The spring phenophases occur 20 days in advance in the lowland and 15 days in alpine regions. The proportion of significant trends, however, is clearly higher in alpine regions, suggesting that the development of vegetation at higher altitudes is more influenced by climate warming. Moreover, the photosynthetically active period has been extended to a

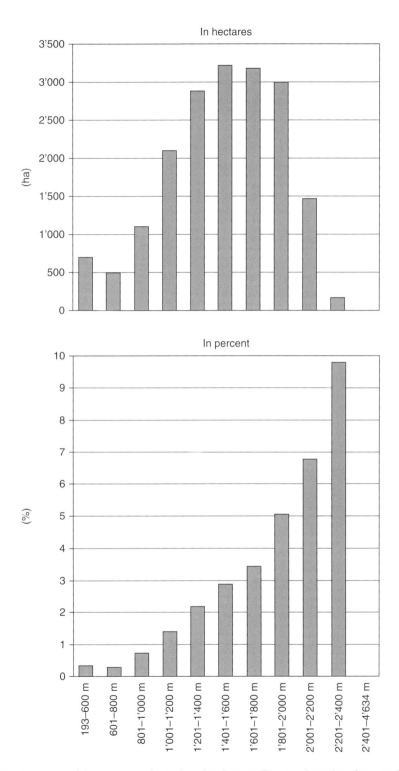

Figure 17.7   Increase of forest cover in Switzerland according to elevational zones during 1979–85 and 1992–97. The percentage increase was highest in the uppermost elevational zones (after Stöcklin et al., 2007)

relatively larger extent in the mountains given the comparatively short growing season. Thus, mountain plants may benefit from a prolonged vegetation period to a comparatively higher degree, and a more distinctive overall response may be expected. At high elevations, however, the photoperiodic control of phenology limits the use that plants are able to make of earlier warmer springs. Keller and Körner (2003) found that half of the tested alpine species are sensitive to photoperiod and may thus not be able to utilize fully periods of earlier snowmelt.

The observations of climate-change responses of mountain plant species and communities could be easily expanded to climate-sensitive animal species. Changes in phenology, species composition, and shifting ranges have been documented for a wide range of taxonomic groups (numerous examples in Parmesan, 2006; Rosenzweig et al., 2007). As both plant and animal species have individualistic traits such as physiological tolerances, life-history strategies, and dispersal abilities, and they show highly variable responses to warming, severe implications for feedback mechanisms and other interactions across trophic levels are to be expected. For example, climate change potentially disrupts the coordination in timing between the life cycles of predators and their prey, herbivorous insects and their host plants, parasitoids and their host insects, or insect pollinators and flowering plants. A recent review (Visser and Both, 2005) confirmed an increasing asynchrony in species interactions that potentially result in population crashes and extinctions. Mountain-restricted harlequin frogs at mid-elevations (1,000–2,400 m.a.s.l.) in Central and South America provide the first examples of global warming—related extinctions of entire species. Obviously, recent trends toward warmer nights and increased daytime cloud cover have shifted these elevations into thermally optimum conditions for chytrid fungi (*Batrachochytrium dendrobatidis*) attacking the frogs (Pounds et al., 2006). Generally, species-specific responses to climate warming imply major changes in population dynamics and species compositions with the consequence of unpredictable impacts on community structure and ecosystem functioning.

17.4 EFFECTS OF CHANGING LAND USE ON HIGHLAND ECOSYSTEMS

Humans have managed and transformed the landscapes of highland regions for millennia. Old-world mountain regions, in particular in Asia and Europe, experienced the foundation of permanent settlements and the development of associated land-use systems since the mid-Holocene or even earlier. Palynological findings suggest that subalpine and alpine pastoralism in the European Alps dates back to at least 6,000–7,000 years B.P. (Bortenschlager, 2000). Recent paleoecological research in southeast Tibet indicates nomadic use of alpine grasslands at least since 8,500 years B.P. (Schlütz et al., 2007). In old-settled mountain regions, humans have developed sophisticated resource utilization strategies including a wide spectrum of farming and pastoral practices. Mountain people have depended through generations on various types of mountain agriculture for their sustenance, usually based on a combination of crops and livestock (Ehlers and Kreutzmann, 2000). Essential natural environmental resources such as forests and pastures are connected with this mixed mountain agriculture through complex linkages in order to incorporate resources of various altitudinal zones and different seasons into the land-use system. The colonization of mountain valleys involved first interferences in forests as well as first encroachments on the upper treeline that can be traced back in the European Alps to c. 5,000 years B.P. (e.g., Tinner et al., 1996). However, overall impacts on alpine environments seem to have been negligible for many centuries and remote mountain ranges remained relatively undisturbed. Owing to the great inaccessibility of mountains and the seclusion from the outside world, subsistence economies had evolved, characterized by limited trade and exchange with the plains or other mountain regions.

Only with the increase in population and the intensification of land use since the Middle Ages, mountain landscapes in central Europe have been subjected to profound alterations. The growing demand for cultivable and pasture land as well as for timber and firewood led to an extensive clearing of mountain forests. Gentle and easily accessible slopes were often completely deforested, in particular to meet the energy demand of ore mining and salt mines. Forest areas recovered and forest conditions were improved only after coal was available as an extensive alternative energy source in the nineteenth century. At higher elevations, downslope extension of alpine pastures and the burning and logging of subalpine forests were concentrated on south-facing slopes. It has been suggested that the average anthropogenic treeline depression in the Alps amounts to 150 to 300 m below the uppermost postglacial level (Holtmeier, 2009). In the course of time, forest clearings and subsequent agricultural use created a patchwork of forests, meadows, grazinglands, crop fields, and other landscape elements, providing habitats for the development of novel animal and plant communities (Stöcklin et al., 2007). A cultural landscape evolved from centuries of human occupation, characterized by sustainable productivity and significantly enhanced biodiversity, and it was considered a cultural heritage of outstanding conservational and historical value. With the creation and careful maintenance of this cultural landscape, being of higher aesthetic attractiveness compared to the natural landscape, the mountain peasants of the Alps established the basis for all economic activities, and in particular for the emerging tourism in the late nineteenth and early twentieth centuries (Bätzing, 2003).

In other old-settled mountain environments, cultural landscapes associated with traditional land management similarly evolved over long time periods. Mountains in the Mediterranean region are well known for strong human influence lasting for millennia, even though the conspicuous degradation of mountain environments is often an outcome of recent human activities since 1750–1800 (McNeill, 1992). Mountain landscapes in central Asia have been partially completely transformed as recently evidenced for the Fergana Range in south Kyrgyzstan, where extensive walnut-fruit forests proved to be a true anthropogenic forest type (Beer et al., 2008). In other less-developed mountain regions, traditional land uses such as shifting cultivation; terracing of farmlands; the use of fodder, fuel, timber, and other forest resources; and transhumant pastoralism have influenced mountain environments for centuries only on a modest scale. Substantial transformations of landscapes have often occurred only since colonial times when natural resources of upper forest belts and alpine grazing lands have been integrated into regional and supraregional economies.

For example, there is no evidence in the Himalayas for a phase in the evolution of the cultural landscape during the Middle Ages that would be comparable, in scale and importance, to the extension of settlements and land-use intensity in the European Alps between the eleventh and fourteenth centuries (Schickhoff, 1995). The forests of the Himalayas were considered to be more or less untouched and inexhaustible until the late eighteenth century when the first logging operations are documented from the Himalayan foothills of Kumaon (Schickhoff, 2007). With the exception of fertile basin landscapes such as the Kathmandu or Kashmir valleys, the Himalayas had been very sparsely populated until the nineteenth century. Accordingly, the resource requirements and the spatial expansion of forest clearings had been unimportant in previous centuries, and human interferences have been insignificant almost throughout the entire mountain system. Rapid landscape transformations

occurred only after the British annexation of Himalayan regions in the first half of the nineteenth century. Subsequently, high population growth, the prospering economic development with the intensification of cultivation and animal husbandry, and the growing timber demand of the lowland resulted in large-scale deforestation and rapid changes in the distribution of forests and agricultural lands. In the following decades, the protective influence of the British-Indian Forest Service led to a retardation of landscape transformation processes. At the beginning of the twentieth century, the cultural landscape of many Himalayan regions corresponded in its basic patterns to the present-day scenery. This holds also for the Nepal Himalaya, where the present distribution of forests, arable lands, and rangelands in the mountainous regions is not much different from the situation 100 years ago (Gilmour and Fisher, 1991). Only some subregions of the mountain system, for example, the Karakorum, experienced considerable recent changes to the forest cover (Schickhoff, 2002, 2005).

Likewise, mountain environments of the Andes have been reshaped to a large extent by colonial impacts (Ellenberg, 1979). However, 12,000 years of continuous human occupation preceded European contact, and increasingly sedentary pastoralist-agriculturalist communities significantly modified the natural landscape since 6,000–7000 years B.P. (Baied and Wheeler, 1993). Advanced civilizations such as the Tiahuanaco and Incan empires developed a highly successful agriculture including the sophisticated management of ridged fields and irrigated terraced slopes, and they thus remodeled the landscape of whole valley systems. Nevertheless, the Spanish conquest in the sixteenth century involved the collapse of the High Andean culture; the elimination of traditional knowledge related to natural resource utilization and agricultural practices; the introduction of European grazing animals, crops, and weeds; new land-use methods; and intensified and diversified land-use impacts. The unprecedented level of human interference has precipitated the Andean environment into a crisis with lasting consequences up to the present (Seibert, 1983). Africa's mountains have long since experienced a much higher population density compared to the lowlands, associated with a constant decrease of natural resources over the last centuries. Since colonial times, land-use systems have been fundamentally transformed, and increased population pressure and deforestation by mountain farmers as well as by commercial forest-mining companies have aggravated problems of land degradation arising from long-lasting land use, in particular on commonly owned land (Hurni et al., 1992).

In contrast to old-settled mountains, human impacts on high mountain environments in North America, Australia, and New Zealand by native people prior to the arrival of the Europeans were mostly insignificant. In the Rocky Mountains, with the exceptions of Spanish settlement in New Mexico and California, most permanent European settlement began in the mid-1800s and has been mainly driven by extractive industries such as logging, mining, and grazing (Riebsame et al., 1996). More recently, human interferences included water diversions and hydroelectricity development as well as tourism and recreation. Even though, for example, the devastation of watersheds by overgrazing and considerable biotic invasions has been reported, agricultural land use has certainly not entailed such fundamental ecological change as in many old-world mountain regions. In New Zealand, pre-European Maori environmental impacts included vertebrate extinctions and deforestation, but the late eighteenth-century arrival of the Europeans, who lacked traditions in highland management, and the subsequent reinforced exploitation of mountain altitudinal zones through grazing, mining, quarrying, and logging resulted in disparately larger environmental transformation, aggravated by the various introduced species partly considered now as severe pests (Pawson and Brooking, 2002).

In the course of rapid global socioeconomic transformation, the cultural landscapes of both developed and less developed mountain regions have undergone significant changes during the twentieth century, in particular during the past 50 years. Developed and developing mountain regions have been passing, however, through converse processes of change. Whereas the former are often characterized by the extensification of traditional land use and land abandonment as well as by the concurrent exploitation of high-mountain environments for tourism, mining, power generation, or industrial-scale farming in favorable areas, the situation in the latter is usually affected by high population growth, poverty, lack of economic opportunities, increased land-use pressure, and increased integration into the economy of the lowlands. For example, mountain farmers in the European Alps have abandoned traditional forms of sustainable agricultural use and are getting increasingly absorbed in the tourist economy representing a substantial shift from the primary to the tertiary sector. Agriculture becomes more and more concentrated on small land areas on valley floors, which are intensively used, whereas the remaining land lies fallow or is used only for nonintensive cultivation or grazing. On the other hand, the increasing degree of commercialization in industrially less advanced countries results in a still growing importance of the primary sector in mountains, and land-use intensification by local mountain farmers and foresters is the inevitable consequence of internally (population growth) and externally (lowland markets) generated pressure. Referring to these contrary developments in mountain regions, Schickhoff (2004) opposed a tropical to an extratropical type of resource flow from highlands to lowlands.

What are the consequences of these contrary patterns of recent land-use changes for mountain ecosystems? To what extent do mountain regions in developed countries differ from those in developing countries with respect to discernible effects on landscape patterns, ecosystem functioning, and dynamics? For mountains in industrially advanced countries, the extensive natural reforestation caused by the widespread decrease of agriculture is the most conspicuous effect at the landscape scale, along with the trend toward more monotonous landscapes and reduced landscape diversity. In the European Alps, an average of approximately 20%, in some areas as much as 70%, of the agricultural land has been abandoned (Tasser et al., 2005). Succession on unused areas inevitably leads to the establishment of new forest areas. Assessments of forest-cover increase in Switzerland range from 50% to 100% in the last 150 years, with a large proportion of this increase having occurred in montane and subalpine altitudinal zones (e.g., Mather and Fairbairn, 2000). During the past two decades, the forest regrowth increased, in percentage terms, continuously through all elevational zones up to the upper treeline (Figure 17.8), and in each case the regrowth was largely confined to marginal sites where cultivation costs are not covered by the yields. Gehrig-Fasel et al. (2007) confirmed the close relationship between land abandonment and forest regrowth that has become apparent in other alpine regions as well (e.g., Tasser et al., 2007).

At the ecosystem level, the general decline of agriculture (or the partial intensification) in developed mountain regions has caused changes in biodiversity, biogeochemical cycles, climatic and hydrological processes, and related feedback effects on, for example, erosion rates, magnitude of floods, snow gliding, and avalanches. Since the 1950s, the formerly positive impact of land use on the diversity of species and landscapes in the Alps has been increasingly reversed into a negative effect (Stöcklin et al., 2007). Traditional land use appears to be a key driver for sustaining high levels of biodiversity, both at the ecosystem and landscape scales. Tasser and Tappeiner (2002) and Tasser et al. (2005) showed for different alpine ecosystems along a south-north transect that both intensification and abandonment reduce plant

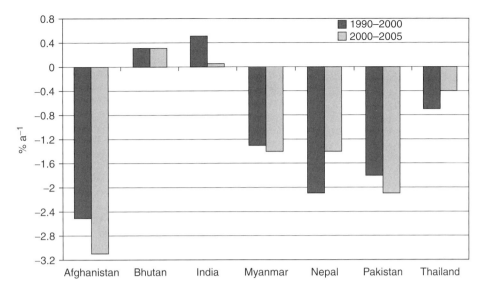

Figure 17.8   Annual change rate (in %) of forest cover in Himalayan countries for 1990–2000 and 2000–2005 (after data in FAO, 2009)

species richness relative to traditional land-use patterns. Likewise, Spiegelberger et al. (2006) assessed that both intensification and abandonment changed species composition and reduced plant species richness in grasslands of the Swiss and French Alps. A significant long-term decline of plant species richness following abandonment at the landscape scale was evident from resampling subalpine and alpine grasslands in the northern calcareous Alps (Dullinger et al., 2003). Investigating grasslands from the montane to the alpine belt in the Swiss Alps, Maurer et al. (2006) corroborated the result that ongoing land-use changes reduce plant species richness within plots and at the landscape level, and they concluded that a high diversity of land-use types had to be maintained in order to preserve plant species diversity. In addition, abandonment of mountain grasslands affects major ecosystem processes, leading in the long term, for example, to decreases in nitrogen mineralization, decomposition rates, nutrient availability, and soil respiration (Tasser et al., 2005). Both the reduction and intensification of agricultural use can increase the vulnerability of Alpine ecosystems to landslides, hillslope erosion, and snow-gliding processes. Land-use effects on vegetation and soil properties (e.g., root density, vegetation cover, canopy characteristics) largely determine the vulnerability to such natural hazards (Tasser et al., 2003).

In contrast to the mountains of industrially advanced countries, less developed mountain regions have been mainly transformed by land-use intensification. Rapid population growth and the increasing demand for agricultural land and forest products such as firewood, fodder, and timber have accelerated the clearing of mountain forests and have intensified farmland production and grazing in many developing countries, often at the expense of degrading environmental conditions. Forest areas in mountainous low- and middle-income countries showed a net loss of more than 4 million hectares between 1990 and 2000, whereas those in developed countries experienced a net gain of more than 2 million hectares (Schickhoff, 2004). Forest statistics at the national level, however, may obscure more differentiated patterns at regional or local scales. For instance, with the exception of Bhutan and India (non-Himalayan areas included), the Himalayan countries witnessed continuing declines in forest areas during 1990–2000 and 2000–2005 (Figure 17.9). The estimated overall rate of forest loss is about 1% per year across the entire Himalaya (Zurick and Pacheco, 2006). On the other hand, there are many examples at the district level of a forest situation having remained stable or even improved. From a total of 120 Himalayan districts, about one-fourth experienced an increase in forest area, while about one-third showed forest loss during recent decades (Figure 17.9).

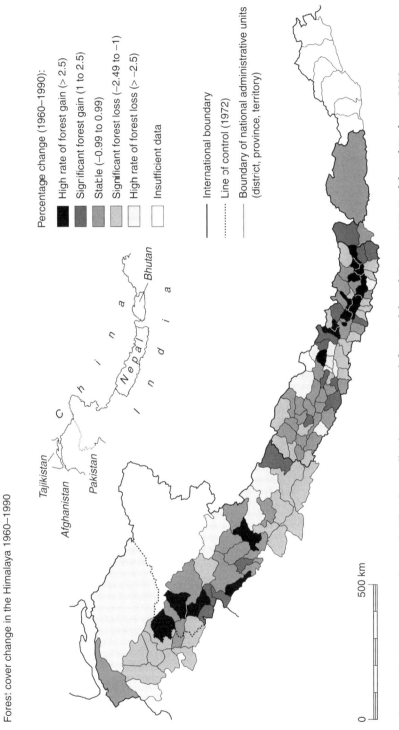

Forest cover change in the Himalaya 1960–1990

Percentage change (1960–1990):

■ High rate of forest gain (> 2.5)
▨ Significant forest gain (1 to 2.5)
▤ Stable (−0.99 to 0.99)
▦ Significant forest loss (−2.49 to −1)
▧ High rate of forest loss (> −2.5)
□ Insufficient data

—— International boundary
········· Line of control (1972)
—— Boundary of national administrative units
    (district, province, territory)

Tajikistan
Afghanistan
Pakistan
China
Nepal
Bhutan
India

500 km
0

Figure 17.9   Forest-cover change in Himalayan districts, 1960–90 (after Zurick and Karan, 1999; Zurick and Pacheco, 2006; Schickhoff, 2007)

Forest resource depletion occurred in some high-elevation districts, but primarily in the Himalayan foothills and in the Terai lowlands where immigration rates are high and forests have been converted to farms and other nonforest uses. Adjoining districts with disparate forest dynamics point to a great complexity of the drivers of change, implying that rapid generalizations on the extent of deforestation or root causes are not appropriate (Ives, 2004; Schickhoff, 2007).

Forest area statistics, however, do not provide any information on the structural integrity of forest stands. Although forest areas have expanded in some Himalayan districts, there is increasing evidence that the all-pervading human impact has been generally deteriorating forest conditions at the stand level during recent decades (e.g., Singh and Singh, 1992; Schickhoff, 2006, 2009). In many forest stands, stem numbers, basal areas, and stand volumes decrease to such an extent that all subsystems of forest ecosystems are adversely affected. It has been demonstrated for the Karakoram Mountains (Schickhoff, 2002) that the advanced loss of structural complexity in degraded stands results in a decrease of species richness and biodiversity, and in a reinforced erosion of humus horizons, followed by increased surface runoff, soil erosion, and nutrient export, and ultimately by lowered site productivity. The maintenance of crucial ecosystem functions and processes is thus jeopardized. This pattern of environmental degradation is discernible in many subtropical and tropical mountain forests. In addition to forest degradation, the increased pressure on grazing lands threatens highland integrity and biodiversity in developing countries. Whereas moderate grazing often exerts no destructive impact on vegetation and site, heavy grazing may lead to dramatic losses of biological richness, soil degradation and erosion, and reduced site productivity (Körner et al., 2006). In High Asia, rising livestock populations, the extension of agriculture into rangelands, the transformation of traditional pastoral production systems, and inappropriate management practices have led to a general downward spiral in the productivity of many rangeland areas, loss of biodiversity, and increased marginalization of pastoral people (Miller, 1997). Moreover, land-use intensification in subtropical and tropical highlands has brought about increased fire frequency and intensity that inevitably leads to major shifts in species composition and ecosystem functioning (Spehn et al., 2006).

In addition to the threats to biodiversity, ecosystem functioning, and ecosystem services arising from the specific pathways of recent land-use changes in developed and less-developed mountain regions, there are overall threats such as atmospheric pollution and biological invasions affecting highlands in general. For example, nitrogen emissions, which have increased dramatically during the past 50 years and are projected to further increase and impact mountain regions worldwide, lead to shifts in species compositions, favoring nitrophilous species at the expense of acidophilous species in nutrient-limited high-elevation communities (e.g., Hülber et al., 2008). Biological invasions represent another major driver of biodiversity decline and related alterations of ecosystem services. To date, mountain environments appeared to be less susceptible to invasions. However, there is increasing evidence that many nonnative plant species have colonized high-elevation habitats across wide climatic and latitudinal ranges and that some invading species have already had a major impact on mountain ecosystems (Pauchard et al., 2009). Prominent examples of exotics that modified ecosystem functioning include *Pinus contorta*, which invaded the timberline ecotone in southern Chile (Pena et al., 2008) and *Pinus radiata*, which invaded the high-elevation *Sophora* shrublands on the Hawaiian island of Maui (Daehler, 2005). It has to be assumed that ongoing climate and land-use changes weaken the resistance of alpine ecosystems to plant invasions that become thus a more serious threat to mountain biodiversity in the near future.

17.5 CONCLUDING REMARKS

The ongoing extensive modification of mountain ecosystems in recent decades, to be attributed to strongly changing climatic, economic, and societal drivers, constitutes a significant threat to the wide range of ecosystem services mountains provide to human communities in highlands and in distant lowland regions. The capacity of mountain ecosystems to regulate water quality, retention, and storage; to maintain forest and soil resources as well as slope stability and erosion control; and to sustain food, wood, and fiber production is being degraded by altered land-use regimes, climate change, atmospheric pollution, and biological invasions. Synergistic interactions and aggravating effects of these factors and disturbances result in a loss of native biodiversity, including species richness and landscape diversity, which is crucial not only for ecosystem functioning but also for the well-being of regional and national economies. At present, changes in land use, climate, and other impacts are going to exceed the historic range of variability, thus challenging the sustainability of mountain ecosystems to provide goods and services. Given the sloping terrain and the typically shallow and immature soils, recovery from disturbances is extremely slow on mountains. As nearly half of the world's population depends on the ecological integrity and resources of mountains, environmental conservation and sustainable land use in mountains and highlands are of global concern. It is imperative to meet the challenge of satisfying immediate human land-use needs while reducing the negative environmental impacts of land use and climate change. A wide array of interdisciplinary research and integrated management skills based on strengthened linkages between highland and lowland stakeholders will be needed to ensure sustainability in mountain environments.

REFERENCES

Allen, C.D., and Breshears, D.D. (1998) 'Drought-induced shift of a forest-woodland ecotone: Rapid landscape response to climate variation', *Proceedings of the National Academy of United States of America*, 95: 14839–42.

Bahn, M., and Körner, C. (2003) 'Recent increases in summit flora caused by warming in the Alps', in Nagy, L., Grabherr, G., Körner, C. and D.B.A. Thompson (eds.), *Alpine Biodiversity in Europe*. Ecological Studies 167. Berlin-Heidelberg: Springer. pp. 437–41.

Baied, C.A., and Wheeler, J.C. (1993) 'Evolution of high Andean Puna ecosystems: Environment, climate, and culture change over the last 12,000 years in the central Andes', *Mountain Research and Development*, 13: 145–56.

Bajracharya, S.R., Mool, P.K., and Shrestha, B.R. (2007) *Impact of Climate Change on Himalayan Glaciers and Glacial Lakes*. Kathmandu: ICIMOD.

Baker, B.B., and Moseley, R.K. (2007) 'Advancing treeline and retreating glaciers: Implications for conservation in Yunnan, P.R. China', *Arctic, Antarctic, and Alpine Research*, 39: 200–209.

Barnett, T.P., Adam, J.C., and Lettenmaier, D.P. (2005) 'Potential impacts of a warming climate on water availability in snow-dominated regions', *Nature*, 438: 303–9.

Barry, R.G. (2006) 'The status of research on glaciers and global glacier recession: A review', *Progress in Physical Geography*, 30: 285–306.

Batima, P., Natsagdorj, L., Gombluudev, P., and Erdenetseg, B. (2005) 'Observed climate change in Mongolia', AIACC Working Paper No. 12. http://www.aiaccproject.org. Accessed April 1, 2009.

Bätzing, W. (2003) *Die Alpen. Geschichte und Zukunft einer europäischen Kulturlandschaft*. Munich: C.H. Beck.

Beckage, B., Osborne, B., Gavin, D.G., Pucko, C., Siccama, T., and Perkins, T. (2008) 'A rapid upward shift of a forest ecotone during 40 years of warming in the Green Mountains of Vermont', *Proceedings of the National Academy of the United States of America*, 105: 4197–202.

Beer, R., Kaiser, F., Schmidt, K., Ammann, B., Carraro, G., Grisa, E., and Tinner, W. (2008) 'Vegetation history of the walnut forests in Kyrgyzstan (Central Asia): natural or anthropogenic origin?' *Quaternary Science Reviews*, 27: 621–32.

Beniston, M. (2000) *Environmental Change in Mountains and Uplands*. London: Edward Arnold.

Beniston, M. (2005) 'Mountain climates and climatic change: An overview of processes focusing on the European Alps', *Pure and Applied Geophysics*, 162: 1587–606.

Beniston, M., Diaz, H.F., and Bradley, R.S. (1997) 'Climatic change at high elevation sites: An overview', *Climatic Change*, 36: 233–51.

Beniston, M., Keller, F., Koffi, B., and Goyette, S. (2003) 'Estimates of snow accumulation and volume in the Swiss Alps under changing climatic conditions', *Theoretical and Applied Climatology*, 76: 125–40.

Bigler, C., Bräker, O.U., Bugmann, H., Dobbertin, M., and Rigling, A. (2006) 'Drought as an inciting mortality factor in Scots pine stands of the Valais, Switzerland' *Ecosystems*, 9: 330–43.

Bolch, T. (2007) 'Climate change and glacier retreat in northern Tien Shan (Kazakhstan/Kyrgyzstan) using remote sensing data', *Global and Planetary Change*, 56: 1–12.

Bonfils, C., and 11 authors. (2008) 'Detection and attribution of temperature changes in the mountainous western United States', *Journal of Climate*, 21: 6404–24.

Bortenschlager, S. (2000) 'The Iceman's environment', in Bortenschlager, S., and K. Oeggl (eds.), *The Iceman and his Natural Environment*. Vienna: Springer. pp. 11–24.

Bowman, W.D. (2006) 'Life on a slope: Biodiversity and ecological functioning in mountains', in Price, M.F. (ed.), *Global Change in Mountain Regions*. Kirkmahoe, Scotland: Sapiens Publishing. pp. 9–10.

Bugmann, H., Zierl, B., and Schumacher, S. (2005) 'Projecting the impacts of climate change on mountain forests and landscapes', in Huber, U.M., Bugmann, H.K.M., and M.A. Reasoner (eds.), *Global Change and Mountain Regions. An Overview of Current Knowledge*. Dordrecht: Springer. pp. 477–87.

Burga, C.A., Frei, E., Reinalter, R., and Walther, G.R. (2007) 'Neue Daten zum Monitoring alpiner Pflanzen im Engadin', *Berichte der Reinhold-Tüxen-Gesellschaft*, 19: 37–43.

Cannone, N., Diolaiuti, G., Guglielmin, M., and Smiraglia, C. (2008) 'Accelerating climate change impacts on alpine glacier forefield ecosystems in the European Alps', *Ecological Applications*, 18: 637–48.

Casassa, G., López, P., Pouyaud, B., and Escobar, F. (2009) 'Detection of changes in glacial run-off in alpine basins: Examples from North America, the Alps, central Asia and the Andes', *Hydrological Processes*, 23: 31–41.

Christensen, J.H., and 16 authors. (2007) 'Regional climate projections', in *Climate Change 2007: The Physical Science Basis. Contribution of Working Group I to the Fourth Assessment Report of the Intergovernmental Panel on Climate Change*. Cambridge: Cambridge University Press. pp. 847–940.

Colwell, R.K., Brehm, G., Cardelus, C.L., Gilman, A.C., and Longino, J.T. (2008) 'Global warming, elevational range shifts, and lowland biotic attrition in the wet tropics', *Science*, 322: 258–61.

Cruz, R.V., and 9 authors. (2007) 'Asia', in *Climate Change 2007: Impacts, Adaptation and Vulnerability. Contribution of Working Group II to the Fourth Assessment Report of the Intergovernmental Panel on Climate Change.* Cambridge: Cambridge University Press. pp. 469–506.

Daehler, C.C. (2005) 'Upper-montane plant invasions in the Hawaiian islands: Patterns and opportunities. *Perspectives in Plant Ecology', Evolution and Systematics*, 7: 203–16.

Danby, R.K., and Hik, D.S. (2007) 'Variability, contingency and rapid change in recent subarctic alpine treeline dynamics', *Journal of Ecology*, 95: 352–63.

Daniels, L.D., and Veblen, T.T. (2004) 'Spatiotemporal influences of climate on altitudinal treeline in northern Patagonia', *Ecology*, 85: 1284–96.

D'Arrigo, R.D., Kaufmann, R.K., Davi, N., Jacoby, G.C., Laskowski, C., Myneni, R.B., and Cherubini, P. (2004) 'Thresholds for warming-induced growth decline at elevational treeline in the Yukon Territory, Canada', *Global Biogeochemical Cycles* 18: GB3021, doi:10.1029/2004GB002249.

Defila, C., and Clot, B. (2005) 'Phytophenological trends in the Swiss Alps, 1951–2002', *Meteorologische Zeitschrift*, 14: 191–96.

Diaz, H.F., and Bradley, R.S. (1997) 'Temperature variations during the last century at high elevation sites', *Climatic Change*, 36: 253–79.

Dobbertin, M., Hilker, N., Rebetez, M., Zimmermann, N.E., Wohlgemuth, T., and Rigling, A. (2005) 'The upward shift in altitude of pine mistletoe (*Viscum album* ssp. *austriacum*) in Switzerland—the result of climate warming?' *International Journal of Biometeorology*, 50: 40–47.

Dullinger, S., Dirnböck, T., and Grabherr, G. (2004) 'Modelling climate-change driven treeline shifts: Relative effects of temperature increase, dispersal and invasibility', *Journal of Ecology*, 92: 241–52.

Dullinger, S., Dirnböck, T., Greimler, J., and Grabherr, G. (2003) 'A resampling approach for evaluating effects of pasture abandonment on subalpine plant species diversity', *Journal of Vegetation Science*, 14: 253–62.

Dyurgerov, M.B., and Meier, M.F. (2005) *Glaciers and the Changing Earth System: A 2004 Snapshot.* Boulder, CO: Institute of Arctic and Alpine Research, University of Colorado, Occasional Paper No. 58.

Ehlers, E., and Kreutzmann, H. (2000) 'High mountain ecology and economy. Potentials and constraints', in Ehlers, E. and H. Kreutzmann (eds.), *High Mountain Pastoralism in Northern Pakistan. Erdkundliches Wissen* 132: 9–36.

Ellenberg, H. (1979) 'Man's influence on tropical mountain ecosystems in South America', *Journal of Ecology*, 67: 401–16.

Erschbamer, B., Niederfriniger Schlag, R., and Winkler, E. (2008) 'Colonization processes on a central Alpine glacier foreland', *Journal of Vegetation Science*, 19: 855–62.

FAO (Food and Agriculture Organization of the United Nations). (2009) *State of the World's Forests 2009.* Rome: FAO.

Gehrig-Fasel, J., Guisan, A., and Zimmermann, N.E. (2007) 'Tree line shifts in the Swiss Alps: Climate change or land abandonment?' *Journal of Vegetation Science*, 18: 571–82.

Gilmour, D.A., and Fisher, R.J. (1991) *Villagers, Forests and Foresters. The Philosophy, Process and Practice of Community Forestry in Nepal.* Kathmandu: ICIMOD.

Grabherr, G. (2003) 'Alpine vegetation dynamics and climate change—a synthesis of long-term studies and observations', in Nagy, L., Grabherr, G., Körner, C., and D.B.A. Thompson (eds.), *Alpine Biodiversity in Europe. Ecological Studies 167.* Berlin-Heidelberg: Springer. pp. 399–409.

Grabherr, G., Gottfried, M., Gruber, A., and Pauli, H. (1995) 'Patterns and current changes in alpine plant diversity', in Chapin, F.S. III and C. Körner (eds.), *Arctic and Alpine Biodiversity: Patterns, Causes and Ecosystem Consequences. Ecological Studies 113.* Berlin-Heidelberg: Springer. pp. 167–81.

Grabherr, G., Gottfried, M., and Pauli, H. (1994) 'Climate effects on mountain plants', *Nature*, 369: 448.

Grabherr, G., Gottfried, M., and Pauli, H. (2001) 'High mountain environment as indicator of global change', in Visconti, G., Beniston, M., Iannorelli, E.D., and D. Barba (eds.), *Global Change and Protected Areas.* Dordrecht: Kluwer. pp. 331–45.

Haeberli, W., and Hoelzle, M. (1995) 'Application of inventory data for estimating characteristics of and regional climate-change effects on mountain glaciers: A pilot study with the European Alps', *Annals of Glaciology*, 21: 206–12.

Haeberli, W., Hoelzle, M., Paul, F., and Zemp, M. (2007) 'Integrated monitoring of mountain glaciers as key indicators of global climate change: The European Alps', *Annals of Glaciology*, 46: 150–60.

Hamilton, L.S. (2002) 'Conserving mountain biodiversity in protected areas', in Körner, C., and E.M. Spehn (eds.), *Mountain Biodiversity: A Global Assessment.* London: Parthenon. pp. 295–306.

Harris, C., Mühll, D.V., Isaksen, K., Haeberli, W., Sollid, J.L., King, L., Holmlund, F., Dramis, F., Guglielmin, M., and Palacios, D. (2003) 'Warming permafrost in European mountains', *Global and Planetary Change*, 39: 215–25.

Hewitt, K. (2005) 'The Karakoram anomaly? Glacier expansion and the "elevation effect," Karakoram Himalaya', *Mountain Research and Development*, 25: 332–40.

Hilbich, C., Hoelzle, M., Scherler, M., Schudel, L., Völksch, I., Mühll, D.V., and Mäusbacher, R. (2008) 'Monitoring mountain permafrost evolution using electrical resistivity tomography: A 7-year study of seasonal, annual, and long-term variations at Schilthorn, Swiss Alps', *Journal of Geophysical Research* 113: F01S90, doi:10.1029/2007JF000799.

Hofer, T. (2005) 'The International Year of Mountains: Challenge and opportunity for mountain research', in Huber, U.M., Bugmann, H.K.M., and M.A. Reasoner (eds.), *Global Change and Mountain Regions. An Overview of Current Knowledge*. Dordrecht: Kluwer. pp. 1–8.

Holtmeier, F.K. (2009). *Mountain Timberlines. Ecology, Patchiness, and Dynamics*. Dordrecht: Springer.

Holtmeier, F.K., and Broll, G. (2005) 'Sensitivity and response of northern hemisphere altitudinal and polar treelines to environmental change at landscape and local scales', *Global Ecology and Biogeography*, 14: 395–410.

Holtmeier, F.K., and Broll, G. (2007) 'Treeline advance— driving processes and adverse factors', *Landscape Online* 1: 1–21.

Holzinger, B., Hülber, K., Camenisch, M., and Grabherr, G. (2008) 'Changes in plant species richness over the last century in the eastern Swiss Alps: Elevational gradient, bedrock effects and migration rates', *Plant Ecology*, 195: 179–96.

Hülber, K., Dirnböck, T., Kleinbauer, I., Willner, W., Dullinger, S., Karrer, G., and Mirtl, M. (2008) 'Long-term impacts of nitrogen and sulphur deposition on forest floor vegetation in the northern limestone Alps, Austria', *Applied Vegetation Science*, 11: 395–404.

Hurni, H., Bagoora, F.D.K., Laker, M.C., Mössmer, M., Ofwono-Orecho, J.K.W., Ojany, F.F., Rakotomanana, J.L., Wachs, T. and Wiesmann, U. (1992) 'African mountain and highland environments: Suitability and susceptibility', in Stone, P.B. (ed.), *The State of the World's Mountains. A Global Report*. London: Zed Books. pp. 11–44.

Inouye, D.W. (2008) 'Effects of climate change on phenology, frost damage, and floral abundance of montane wildflowers', *Ecology*, 89: 353–62.

Ives, J.D. (2004) *Himalayan Perceptions. Environmental Change and the Well-Being of Mountain Peoples*. London: Routledge.

Ives, J.D., Messerli, B., and Spiess, E. (1997) 'Mountains of the world—a global priority', in Messerli, B., and J.D. Ives (eds.), *Mountains of the World—A Global Priority*. London: Parthenon. pp. 1–15.

Jump, A.S., Hunt, J.M., and Peñuelas, J. (2006) 'Rapid climate change-related growth decline at the southern range edge of *Fagus sylvatica*', *Global Change Biology*, 12: 2163–74.

Jurasinski, G., and Kreyling, J. (2007) 'Upward shift of alpine plants increases floristic similarity of mountain summits', *Journal of Vegetation Science*, 18: 711–18.

Kääb, A., Reynolds, J.M., and Haeberli, W. (2005) 'Glacier and permafrost hazards in high mountains', in Huber, U.M., Bugmann, H.K.M., and M.A. Reasoner (eds.), *Global Change and Mountain Regions. An Overview of Current Knowledge*. Dordrecht: Kluwer. pp. 225–34.

Kammer, P.M., Schöb, C., and Choler, P. (2007) 'Increasing species richness on mountain summits: Upward migration due to anthropogenic climate change or re-colonisation?' *Journal of Vegetation Science*, 18: 301–6.

Kapos, V., Rhind, J., Edwards, M., Price, M.F., and Ravilious, C. (2000) 'Developing a map of the world's mountain forests', in Price, M.F., and N. Butt (eds.), *Forests in Sustainable Mountain Development: A State of Knowledge Report for 2000*. Wallingford: CABI Publications. pp. 4–9.

Keller, F., and Körner, C. (2003) 'The role of photoperiodism in alpine plant development', *Arctic, Antarctic, and Alpine Research*, 35: 361–68.

Keller, F., Kienast, F., and Beniston, M. (2000) 'Evidence of response of vegetation to environmental change on high-elevation sites in the Swiss Alps', *Regional Environmental Change*, 1: 70–77.

Klanderud, K., and Birks, H.J.B. (2003) 'Recent increases in species richness and shifts in altitudinal distributions of Norwegian mountain plants', *The Holocene*, 13: 1–6.

Klasner, F.L., and Fagre, D.B. (2002) 'A half century of change in alpine treeline patterns at Glacier National Park, Montana, U.S.A.', *Arctic, Antarctic, and Alpine Research*, 34: 49–56.

Knowles, N., Cayan, D.R., and Dettinger, M.D. (2006) 'Trends in snowfall versus rainfall in the western United States', *Journal of Climate*, 19: 4545–59.

Körner, C. (2002) 'Mountain biodiversity, its causes and function: An overview', In Körner, C., and E.M. Spehn (eds.), *Mountain Biodiversity: A Global Assessment*. London: Parthenon. pp. 3–20.

Körner, C. (2003) *Alpine Plant Life. Functional Plant Ecology of High Mountain Ecosystems*. 2nd ed. Berlin-Heidelberg: Springer.

Körner, C., Nakhutsrishvili, G., and Spehn, E. (2006) 'High-elevation land use, biodiversity, and ecosystem functioning', in Spehn, E.M., Liberman, M., and C. Körner (eds.), *Land Use Change and Mountain Biodiversity*. Boca Raton, FL: CRC Press. pp. 3–21.

Körner, C., and 16 authors. (2005) 'Mountain systems', in Hassan, R., Scholes, R., and N. Ash (eds.), *Ecosystems and Human Well-being: Current State and Trends, Vol. 1*. Washington, DC: Island Press. pp. 681–716.

Kullman, L. (2001) '20th century climate warming and tree-limit rise in the southern Scandes of Sweden', *Ambio*, 30: 72–80.

Kullman, L. (2006) 'Increase in plant species richness on alpine summits in the Swedish Scandes—impacts of recent climate change', in Price, M.F. (ed.), *Global Change in Mountain Regions*. Kirkmahoe, Scotland: Sapiens. pp. 168–69.

Kullman, L. (2007a) 'Modern climate change and shifting ecological states of the subalpine/alpine landscape in the Swedish Scandes', *Geo-Oeko*, 28: 187–221.

Kullman, L. (2007b) 'Treeline population monitoring of *Pinus sylvestris* in the Swedish Scandes 1973–2005: Implications for treeline theory and climate change ecology', *Journal of Ecology*, 95: 41–52.

Laternser, M., and Schneebeli, M. (2003) 'Long-term snow climate trends of the Swiss Alps (1931–99)', *International Journal of Climatology*, 23: 733–50.

Lemke, P., and 10 authors. (2007) 'Observations: Changes in snow, ice and frozen ground', in *Climate Change 2007: The Physical Science Basis. Contribution of Working Group I to the Fourth Assessment Report of the Intergovernmental Panel on Climate Change*. Cambridge: Cambridge University Press. pp. 337–83.

Lenoir, J., Gégout, J.C., Marquet, P.A., de Ruffray, P., and Brisse, H. (2008) 'A significant upward shift in plant species optimum elevation during the 20th century', *Science*, 320: 1768–71.

le Roux, P.C., and McGeoch, M.A. (2008) 'Rapid range expansion and community reorganization in response to warming', *Global Change Biology*, 14: 2950–62.

Lesica, P., and McCune, B. (2004) 'Decline of arctic-alpine plants at the southern margin of their range following a decade of climatic warming', *Journal of Vegetation Science*, 15: 679–90.

Liu, X., and Chen, B. (2000) 'Climatic warming in the Tibetan Plateau during recent decades', *International Journal of Climatology*, 20: 1729–42.

Mather, A.S., and Fairbairn, J. (2000) 'From floods to reforestation: The forest transition in Switzerland', *Environment and History*, 6: 399–421.

Maurer, K., Weyand, A., Fischer, M., and Stöcklin, J. (2006) 'Old cultural traditions, in addition to land use and topography, are shaping plant diversity of grasslands in the Alps', *Biological Conservation*, 130: 438–46.

McNeill, J.R. (1992) *The Mountains of the Mediterranean World. An Environmental History.* Cambridge: Cambridge University Press.

Menzel, A., and 30 authors. (2006) 'European phenological response to climate change matches the warming pattern', *Global Change Biology*, 12: 1969–76.

Meshinev, T., Apostolova, I., and Koleva, E. (2000) 'Influence of warming on timberline rising: A case study on *Pinus peuce* Griseb in Bulgaria', *Phytocoenologia*, 30: 431–38.

Messerli, B. (1999) 'The global mountain problematique', in Price, M.F. (ed.), *Global Change in the Mountains.* London: Parthenon. pp. 1–3.

Miller, D.J. (1997) 'Rangelands and pastoral development: An introduction', in Miller, D.J., and S.R. Craig (eds.), *Rangelands and Pastoral Development in the Hindu Kush-Himalayas.* Kathmandu: ICIMOD. pp. 1–5.

Munroe, J.S. (2003) 'Estimates of Little Ice Age climate inferred through historical rephotography, northern Uinta Mountains, U.S.A.', *Arctic, Antarctic, and Alpine Research*, 35: 489–98.

Mutke, J., and Barthlott, W. (2005) 'Patterns of vascular plant diversity at continental to global scales', *Biologiske Skrifter*, 55: 521–37.

Nogués-Bravo, D., Araujo, M.B., Errea, M.P., and Martinez-Rica, J.P. (2007) 'Exposure of global mountain systems to climate warming during the 21st century', *Global Environmental Change*, 17: 420–28.

Osterkamp, T.E. (2005) 'The recent warming of permafrost in Alaska', *Global Planetary Change*, 49: 187–202.

Parmesan, C. (2006) 'Ecological and evolutionary responses to recent climate change', *Annual Review of Ecology, Evolution, and Systematics*, 37: 637–69.

Parolo, G., and Rossi, G. (2008) 'Upward migration of vascular plants following a climate warming trend in the Alps', *Basic and Applied Ecology*, 9: 100–107.

Pauchard, A., and 16 authors. (2009) 'Ain't no mountain high enough: Plant invasions reaching new elevations', *Frontiers in Ecology and the Environment* 7: doi:10.1890/080072.

Pauli, H., Gottfried, M., Dirnböck, T., Dullinger, S., and Grabherr, G. (2003) 'Assessing the long-term dynamics of endemic plants at summit habitats', in Nagy, L., Grabherr, G., Körner, C., and D.B.A. Thompson (eds.), *Alpine Biodiversity in Europe. Ecological Studies 167.* Berlin-Heidelberg: Springer. pp. 195–207.

Pauli, H., Gottfried, M., and Grabherr, G. (2001) 'High summits of the Alps in a changing climate. The oldest observation series on high mountain plant diversity in Europe', in Walther, G.R., Burga, C.A., and P.J. Edwards (eds.), *"Fingerprints" of Climate Change—Adapted Behaviour and Shifting Species Ranges.* New York: Kluwer Academic/Plenum. pp. 139–49.

Pauli, H., Gottfried, M., Reiter, K., Klettner, C., and Grabherr, G. (2007) 'Signals of range expansions and contractions of vascular plants in the high Alps: Observations (1994–2004) at the GLORIA master site Schrankogel, Tyrol, Austria', *Global Change Biology*, 13: 147–56.

Pawson, E., and Brooking, T. (eds.). (2002) *Environmental Histories of New Zealand.* Melbourne: Oxford University Press.

Pena, E., Hidalgo, M., Langdon, B., and Pauchard, A. (2008) 'Patterns of spread of *Pinus contorta* Dougl ex Loud invasion in a natural reserve in southern South America', *Forest Ecology and Management*, 256: 1049–54.

Peñuelas, J., and Boada, M. (2003) 'A global change-induced biome shift in the Montseny Mountains (NE-Spain)', *Global Change Biology*, 9: 131–40.

Pounds, J.A., and 13 authors. (2006) 'Widespread amphibian extinctions from epidemic disease driven by global warming', *Nature*, 439: 161–67.

Price, M.F. (1998) 'Mountains: Globally important ecosystems', *Unasylva*, 195: 3–12.

Price, M.F., and Butt, N. (eds.). (2000) *Forests in Sustainable Mountain Development: A State of Knowledge Report for 2000.* Wallingford, UK: CABI.

Rebetez, M., and Reinhard, M. (2008) 'Monthly air temperature trends in Switzerland 1901–2000 and 1975–2004', *Theoretical and Applied Climatology*, 91: 27–34.

Ren, J., Jing, Z., Pu, J., and Qin, X. (2006) 'Glacier variations and climate change in the central Himalaya over the past few decades', *Annals of Glaciology*, 43: 218–22.

Richardson, S.D., and Reynolds, J.M. (2000) 'An overview of glacial hazards in the Himalayas', *Quaternary International*, 65/66: 31–47.

Riebsame, W.E., Gosnell, H., and Theobald, D.M. (1996) 'Land use and landscape change in the Colorado Mountains I: Theory, scale, and pattern', *Mountain Research and Development*, 16: 395–405.

Rolland, C., Petitcolas, V., and Michalet, R. (1998) 'Changes in radial tree-growth for *Picea abies*, *Larix decidua*, *Pinus cembra* and *Pinus uncinata* near the alpine timberline since 1750', *Trees*, 13: 40–53.

Rosenzweig, C., and 8 authors. (2007) 'Assessment of observed changes and responses in natural and managed systems', in *Climate Change 2007: Impacts, Adaptation and Vulnerability. Contribution of Working Group II to the Fourth Assessment Report of the Intergovernmental Panel* on *Climate Change*. Cambridge: Cambridge University Press. pp. 79–131.

Sanz-Elorza, M., Dana, E.D., Gonzalez, A., and Sobrino, E. (2003) 'Changes in the high-mountain vegetation of the central Iberian Peninsula as a probable sign of global warming', *Annals of Botany*, 92: 273–80.

Schickhoff, U. (1995) 'Himalayan forest-cover changes in historical perspective. A case study in the Kaghan Valley, northern Pakistan', *Mountain Research and Development*, 15: 3–18.

Schickhoff, U. (2002) *Die Degradierung der Gebirgswälder Nordpakistans. Faktoren, Prozesse und Wirkungs-zusammenhänge in einem regionalen Mensch-Umwelt-System* Erdwissenschaftliche Forschung. 41. Stuttgart: Franz Steiner Verlag.

Schickhoff, U. (2004) 'Highland Lowland Interactions und Gebirgswälder: Dynamik und Risiken von Ressourcen- und Stoffflüssen', in Gamerith, W., Messerli, P., Meusburger, P., and H. Wanner (eds.), *Alpenwelt—Gebirgswelten. Inseln, Brücken, Grenzen*. Tagungsbericht und wissenschaftliche Abhandlungen, 54, Berne: Deutscher Geographentag Bern. pp. 181–90.

Schickhoff, U. (2005) 'The upper timberline in the Himalayas, Hindu Kush and Karakorum: A review of geographical and ecological aspects', in Broll, G. and B. Keplin (eds.), *Mountain Ecosystems. Studies in Treeline Ecology*, Berlin and Heidelberg: Springer pp. 275–354.

Schickhoff, U. (2006) 'The forests of Hunza valley: Scarce resources under threat', in Kreutzmann, H. (ed.), *Karakoram in Transition. Culture, Development, and Ecology in Hunza Valley*, Karachi: Oxford University Press. pp. 123–44.

Schickhoff, U. (2007) 'Die Gebirgswälder des Himalaya und Karakorum: Sinnbild für Ressourcenübernutzung und Umweltdegradierung?' in Glaser, R., and K. Kremb (eds.), *Planet Erde—Asien*, Darmstadt: Wissenschaftlichte Buchgesellschaft. pp. 136–49.

Schickhoff, U. (2009) 'Human impact on high-altitude forests in northern Pakistan: Degradation processes and root causes', in Singh, R.B. (ed.). *Biogeography and Biodiversity*. Jaipur, India: Rawat. pp. 76–90.

Schlütz, F., Miehe, G., and Lehmkuhl, F. (2007) 'Zur Geschichte des größten alpinen Ökosystems der Erde: Palynologische Untersuchungen zu den *Kobresia*-Matten SE-Tibets', *Berichte der Reinhold-Tüxen-Gesellschaft*, 19: 23–36.

Seibert, P. (1983) 'Human impact on landscape and vegetation in the central High Andes', in Holzner, W., Werger, M.J.A., and I. Ikusima (eds.), *Man's Impact on Vegetation*. The Hague: Dr. W. Junk. pp. 261–76.

Shiyatov, S., Terent'ev, M., Fomin, V., and Zimmermann, N. (2007) 'Altitudinal and horizontal shifts of the upper boundaries of open and closed forests in the Polar Urals in the 20th century', *Russian Journal of Ecology*, 38: 223–27.

Shrestha, A.B., Wake, C.P., Mayewski, P.A., and Dibb, J.E. (1999) 'Maximum temperature trends in the Himalaya and its vicinity: An analysis based on temperature records from Nepal for the period 1971–94', *Journal of Climate*, 12: 2775–86.

Singh, J.S., and Singh, S.P. (1992) *Forests of Himalaya. Structure, Functioning, and Impact of Man*. Nainital, India: Gyanodaya Prakashan.

Spehn, E.M., Liberman, M., and Körner, C. (2006) 'Fire and grazing—a synthesis of human impacts on highland biodiversity', in Spehn, E.M., Liberman, M., and C. Körner (eds.), *Land Use Change and Mountain Biodiversity*. Boca Raton, FL: CRC Press. pp. 337–47.

Spiegelberger, T., Matthies, D., Müller-Schärer, H., and Schaffner, U. (2006) 'Scale-dependent effects of land use on plant species richness of mountain grasslands in the European Alps', *Ecography*, 29: 541–48.

Steinger, T., Körner, C., and Schmid, B. (1996) 'Long-term persistence in a changing climate: DNA analysis suggests very old ages of clones of alpine Carex curvula', *Oecologia*, 105: 94–99.

Stewart, I.T. (2009) 'Changes in snowpack and snowmelt runoff for key mountain regions', *Hydrological Processes*, 23: 78–94.

Stöcklin, J., Bosshard, A., Klaus, G., Rudmann-Maurer, K., and Fischer, M. (2007) *Landnutzung und biologische Vielfalt in den Alpen*. Zurich: Swiss National Science Foundation.

Stueve, K.M, Cerney, D.L., Rochefort, R.M. and Kurth, L.L. (2009) Post-fire tree establishment patterns at the alpine treeline ecotone: Mount Rainier National Park, Washington, USA. *Journal of Vegetation Science*, 20: 107–120.

Svenning, J.C., and Condit, R. (2008) 'Biodiversity in a warmer world', *Science*, 322: 206–7.

Tasser, E., Mader, M., and Tappeiner, U. (2003) 'Effects of land use in alpine grasslands on the probability of landslides', *Basic and Applied Ecology*, 4: 271–80.

Tasser, E., and Tappeiner, U. (2002) 'Impact of land use changes on mountain vegetation', *Applied Vegetation Science*, 5: 173–84.

Tasser, E., Tappeiner, U., and Cernusca, A. (2005) 'Ecological effects of land-use changes in the European Alps', in Huber, U.M., Bugmann, H.K.M., and M.A. Reasoner (eds.), *Global Change and Mountain Regions. An Overview of Current Knowledge*. Dordrecht: Kluwer. pp. 409–20.

Tasser, E., Walde, J., Tappeiner, U., Teutsch, A., and Noggler, W. (2007) 'Land-use changes and natural reforestation in the eastern central Alps'. *Agriculture, Ecosystems and Environment*, 118: 115–29.

Theurillat, J.P., and Guisan, A. (2001) 'Potential impact of climate change on vegetation in the European Alps: A review', *Climatic Change*, 50: 77–109.

Tinner, W., Ammann, B., and Germann, P. (1996) 'Treeline fluctuations recorded for 12,500 years by soil profiles, pollen, and plant macrofossils in the central Swiss Alps', *Arctic and Alpine Research*, 28: 131–47.

Trenberth, K.E., and 11 authors. (2007) 'Observations: Surface and atmospheric climate change', in *Climate Change 2007: The Physical Science Basis. Contribution of Working Group I to the Fourth Assessment Report of the Intergovernmental Panel on Climate Change*. Cambridge: Cambridge University Press. pp. 235–336.

Virtanen, R., Eskelinen, A., and Gaare, E. (2003) 'Long-term changes in alpine plant communities in Norway and Finland', in Nagy, L., Grabherr, G., Körner, C., and D.B.A. Thompson (eds.), *Alpine Biodiversity in Europe. Ecological Studies 167*. Berlin-Heidelberg: Springer. pp. 411–22.

Visser, M.E., and Both, C. (2005) 'Shifts in phenology due to global climate change: The need for a yardstick', *Proceedings of the Royal Society of London Series B*, 272: 2561–69.

Viviroli, D., Weingartner, R., and Messerli, B. (2003) 'Assessing the hydrological significance of the world's mountains', *Mountain Research and Development*, 23: 32–40.

Vonder Mühll, D., Nötzli, J., Makowski, K., and Delaloye, R. (2004) *Permafrost in Switzerland 2000/2001 and 2001/2002.* Glaciological Report (Permafrost) No. 2/3, Zurich: Glaciological Commission of the Swiss Academy of Sciences.

Vuille, M., and Bradley, R.S. (2000) 'Mean annual temperature trends and their vertical structure in the tropical Andes', *Geophysical Research Letters*, 27: 3885–88.

Walther, G-R. (2001) 'Laurophyllisation—a sign for a changing climate?' in Burga, C.A., and A. Kratochwil (eds.), *Biomonitoring: General and Applied Aspects on Regional and Global Scales.* Dordrecht: Kluwer. pp. 207–23.

Walther, G.R., Beißner, S., and Burga, C.A. (2005) 'Trends in the upward shift of alpine plants', *Journal of Vegetation Science*, 16: 541–48.

Walther, G.R., Gritti, E.S., Berger, S., Hickler, T., Tang, Z., and Sykes, M.T. (2007) 'Palms tracking climate change', *Global Ecology and Biogeography*, 16: 801–9.

Walther, G.R., Post, E., Convey, P., Menzel, A., Parmesan, C., Beebee, T.J.C., Fromentin, J-M., Hoegh-Guldberg, O., and Bairlein, F. (2002) 'Ecological responses to recent climate change', *Nature*, 416: 389–95.

Weber, R.O., Talkner, P., Auer, I., Böhm, R., Gajic-Capka, M., Zaninovic, K., Brazdil, R., and Fasko, P. (1997) '20th century changes of temperature in the mountain regions of central Europe', *Climatic Change*, 36: 327–44.

Westerling, A.L., Hidalgo, H.G., Cayan, D.R., and Swetnam, T.W. (2006) 'Warming and earlier spring increases western U.S. forest fire activity', *Science*, 313: 940–43.

WGMS (World Glacier Monitoring Service). (2008) *Global Glacier Changes: Facts and Figures.* Zurich: United National Environmental Program/World Glacier Monitoring Service.

Zemp, M., Haeberli, W., Hoelzle, M., and Paul, F. (2006) 'Alpine glaciers to disappear within decades?' *Geophysical Research Letters* 33: L13504, doi:10.1029/2006GL026319.

Zemp, M., Paul, F., Hoelzle, M., and Haeberli, W. (2007) 'Glacier fluctuations in the European Alps 1850–2000: An overview and spatio-temporal analysis of available data', in Orlove, B., Wiegandt, E., and B.H. Luckman (eds.), *The Darkening Peaks: Glacial Retreat in Scientific and Social Context.* Berkeley: University of California Press. pp. 152–67.

Zurick, D., and Karan, P.P. (1999) *Himalaya. Life on the Edge of the World.* Baltimore: John Hopkins University Press.

Zurick, D., and Pacheco, J. (2006): *Illustrated Atlas of the Himalaya.* Lexington KY: University Press of Kentucky.

18

Biogeography of Agricultural Environments

Chris Stoate

18.1 INTRODUCTION

Agrarian culture developed independently in seven regions across the world between 4,000 to 10,000 years ago, changing dramatically the nature of both human society and global ecology. According to Food and Agricultural Organization of the United Nations (FAO), the land devoted to crop and livestock production today occupies nearly 40% of the world's land area and its impact on the planet's ecology is therefore enormous. The change from hunter-gatherer to farmer lifestyles was perhaps the greatest single cultural shift our species has experienced (Boyd and Richerson, 2005), and it has had equally substantial effects on other species, yet it has taken place in our very recent evolutionary history. In this chapter I describe this change in human ecology in the context of the wider environment and consider in some detail the impacts on today's environment and species, the processes involved in present-day agricultural ecosystems, and the implications for the future.

Previous chapters have described a range of environments outside the area that is managed directly by humans. As described by Jentsch and Beierkuhnlein (this volume), natural disturbances play an important role in ecosystem functioning, even in forests and wetlands, and even in the absence of human impacts. Fire, tornadoes, drought, treefall, animal burrowing, flooding, erosion, and sedimentation are all processes of disturbance that influence plant and animal distributions at different scales. Enright (this volume) describes how, in exploiting fire as a management tool, humans are often accelerating a naturally

occurring phenomenon. The preagricultural use of fire by Native Americans to encourage grazing for wild herbivores, and the regeneration of *Camassia quamash*, a culturally, economically, and nutritionally important bulb plant, provides a typical and once widespread example (Stripen and DeWeerdt, 2002). Such management prevented the regeneration of oak forest while stimulating the spread of (now rare) species such as the Willamette daisy (*Erigeron decumbens*) and the Oregon silverspot butterfly (*Speyeriza zerene hippolyta*). Disturbance of 'natural' ecosystems is, and has always been, central to the distribution of plant and animal species, especially when such disturbance is associated with agriculture. Even in early Neolithic Europe, the landscape was not one of continuous forest but was of a heterogeneous nature with open areas of varying scales, maintained by large herbivores such as aurochs (*Bos primigenius*) (Hodder et al., 2005). Vaughan et al. (2005) describe the evolution and dispersal of rice (*Oryza* spp.), aided initially by birds and large herbivores associated with wetland habitat and, in its more recent evolutionary history, with human domestication. Countless species are therefore adapted to disturbed environments and it is these species that have accompanied us through our cultural development as managers of plant and animal species for food and fiber.

Plant species richness under disturbed conditions tends to be relatively high (e.g., Harrison et al., 2001). Small- and dormant-seeded species tend to predominate, but large- and nondormant-seeded species coexist with these. However, Harrison et al. (2001) have shown how types of disturbance and soil type interact in their effects on plant communities, and therefore on the animals associated with them. In their study, the effects of grazing and fire differed between serpentine and nonserpentine soils. The ecological consequences of disturbance therefore vary between regions. Some species of disturbed environments themselves provide food for humans; others compete with plant or animal species managed for food, and others have taken on a cultural significance independent of any perceived practical value. *Papaver rhoeas* has become a symbol of remembrance in Britain since the large-scale appearance of this species on the First World War battlefields, and *Guiera senegalensis*, a shrubby species of farmland across West Africa, is valued as a talisman for safe journeys, including that into the afterlife (Stoate et al., 2001a).

18.2 DEVELOPMENT OF AGRICULTURAL ECOSYSTEMS

The earliest center for the development of agriculture is in the Levant region of the eastern Mediterranean, approximately 10,000 years ago (Smith, 1995; Harris, 1996). Here, by 12,500 years ago, there was already a switch from a largely nomadic hunting and gathering way of life to a relatively sedentary one in which the harvesting of wild annual plants contributed increasingly to the diet. The end of the Pleistocene Ice Age saw the expansion of open forests and grasslands and increased abundance of food species for humans. These included grazing mammals such as gazelles; large-seeded annual grasses such as wild barley (*Hordeum vulgare*), and wild emmer (*Triticum turgidum*) and einkorn (*T. monococcum*) wheats; and perennials such as almond (*Amygdalus orientalis*) and hawthorn (*Crataegus aronia*). The social organization necessary for village life, and for the development of an early agriculture, was therefore already in place. This emergence of agriculture is thought to have coincided with a period of a cold, dry climate (the Younger Dryas), which reduced the abundance of wild food in the Mediterranean region. This forced people to concentrate in the most productive areas where innovative practices such as cultivation of food plants developed. As warmer, wetter conditions returned, soil erosion increased and alluvial fans developed, increasing the area of productive land.

Some details of the processes by which this early form of agriculture developed remain contested (Boyd and Richerson, 2005). However, only in the productive alluvial areas would there have been sufficient food for hunter-gatherers to enable them to adopt a sufficiently sedentary existence to allow them to cultivate crops (Smith, 1995; Harris, 1996). Archaeological evidence from a site on the Euphrates inhabited from 11,000 to 10,000 years ago reveals the harvesting of 157 seed-bearing plants by hunter-gatherers. An ability to exploit a wide range of plant species was important when productivity of any one species could be variable. In times of uncertainty, harvesting a surplus of food for storage, whether from wild or cultivated plants, would have increased food security and the social standing of individuals and families who were in a position to redistribute food during times of shortage. Settlements based on cultivation were two to six times larger than earlier settlements, and buildings were larger, both to accommodate families and to store food. This period also saw the emergence of rituals and belief systems, which contributed to social cohesion and identity with territory.

The three main crop plants of early cultivation—emmer, einkorn, and barley—were widely distributed as wild plants on fine-grained, often basaltic soils on the ecotone between open grassland and scrub oak woodland. Their domestication, however, involved their translocation to alluvial soils in riparian areas in the Levant 9,600 to 10,000 years ago. The domestication of emmer is revealed by the development of larger plumper grains and a tougher rachis so that grain was less easily shed from the ear. Whereas grain that shed quickly would be advantaged in the wild, harvesting by cutting of already ripe crops would favor a greater selection of grain that was held firmly on the plant. Large-seeded species would also be favored as being relatively easy to cultivate.

Animal bone assemblages have been used to inform our understanding of human ecology in the Neolithic period (e.g., Marciniak, 2005). Livestock did not feature in the agriculture of the Levant, but wild sheep were hunted in the area to the east, and wild goats in the Zagros Mountains were still further east. Both were subsequently domesticated in these areas, as evidenced by smaller sizes and an increase in the proportion of immature males (killed for food) and adult females (kept for breeding) at archaeological sites. Transhumance is thought to have been adopted by goat herders, exploiting the natural mountain habitat of the goats in the summer, and the lower grasslands during the winter. Subsequently, sheep and goat herders in the relatively dry eastern part of the Fertile Crescent adopted the cultivation of domesticated cereals from the western alluvial plains so that mixed arable and livestock farming was well established in the area 8,500 years ago. Settlements with livestock, and drinking points established for relatively sedentary livestock in dry areas, may well have influenced the distribution of plants, as revealed in modern Australian rangeland where plant species richness declines significantly with proximity to water because of increased grazing pressure (Landsberg et al., 2003). In contrast, the abundance of exotic plant species increases with proximity to drinking points, and along with arable cultivation, Neolithic grazing may have contributed to an increase in annual plants close to settlements. Weeds such as wild einkorn, ryegrass (*Lolium*), and goatface grass (*Aegilops*) were also spread with crops as their ranges expanded through trade.

Pigs were also domesticated in the northwest part of the Fertile Crescent at around this time, followed by cattle (from wild aurochs) 7,500 to 8,000 B.P. The combination of four domestic livestock species, three major cultivated grass species, plus others developed in the region, such as lentil (*Lens culinaris*), linseed (*Linum usitatissimum*), chickpea (*Cicer arietinum*), and pea (*Pisum sativum*), provided a diverse agricultural system. This diversity made the Fertile Crescent extremely adaptable to a wide range of geomorphologic and climatic circumstances. Agriculture first spread along the southern edge of Europe 7,800 to 8,000 years ago by the gradual diffusion of knowledge and species, and their integration into existing lifestyles, including hunting and gathering. Trade, including the movement of livestock and crops by sea-faring boats, was instrumental and the archaeological record also reveals the colonization of apparently unoccupied low-lying areas by agriculturalists from the east.

Northern Europe was not immediately occupied by farmers as it was largely forested and cold, making it unsuitable for grazing animals and winter-growing crops. The adoption of spring-sown crops and a switch from sheep and goats to cattle 6,700 years ago marked a major ecological transition for agriculture and enabled a northern farming system to be adopted. Cattle made an important contribution to the development of agriculture in northern Europe as they were more tolerant of the colder, wetter conditions than were sheep and goats, and they provided meat and milk, as well as draught for ploughing and eventually for transport.

Farming spread rapidly across northern Europe between 6,000 and 6,500 years ago, concentrating mainly on easily cultivated loess soils. The people are characterized by their timber-framed long houses and by distinctive bands on their pottery, giving them the name 'LBK' from the German 'Linearbandkeramik'. Although open areas within forest were already abundant, maintained by large herbivores such as aurochs, and supporting numerous plant and insect species of grassland and glades (Hodder et al., 2005), the expansion of agriculture favored the spread of these species.

The forest landscape started to open up as agriculture became established and spread into other areas, eventually encompassing Scandinavia and the British Isles where variable climate, topography, and soils demanded more diverse application of domestic species and farming technology. As in the spread of agriculture through southern Europe, indigenous people adopted various parts of the agricultural societies' cultures, according to their own cultural and economic needs and adapted them to their own circumstances. For example, the cereals barley, oats, wheat, and rye would have been adopted differentially across the region, according to soil moisture as they differ in the length of their growing seasons, and therefore in the period through which they require water. Within such physical constraints, some domesticates are likely to have been adopted for largely prestigious or symbolic cultural reasons, as a 'material language' through which people expressed identity and allegiance.

A similar spread of agriculture took place through Africa where other farming methods and species also developed independently. The Sahel region was considerably wetter 4,000 to 7,000 years ago than it is now and provided good grazing and areas of seasonally flooded land associated with shallow lakes. Domestic cattle were kept in the central Sahel 5,000 years ago and in the west, with goats, about 3,500 years ago. Although the goats would have originated in the Fertile Crescent, cattle were Zebu (*Bos indicus*) originating from Asia, rather than the European *B. taurus* (Plate 18.1) However, as in Europe, cattle were to make a major contribution to the farming system and the way land was managed, and Zebu-type cattle continue to predominate across sub-Saharan Africa in numerous forms. As in northern Europe, cattle would have helped to open up the forest landscape, making it more suitable for smaller livestock and for arable cropping. Corralling cattle (and other livestock) at night protected them from predation but also restricted their feeding time, contributing to the smaller size of domestic animals than their contemporary wild co-generics. Work by Augustine (2003) in East Africa also shows how corralling of cattle results in the depletion of soil nutrients through transfer of nitrogen and phosphorus to corrals where nitrogen is lost (through leaching and volatilisation) and phosphorus accumulates, encouraging the long-term development, once corrals are abandoned, of treeless areas dominated by *Cynodon plectostachyus*. This implies that relatively sedentary communities, based largely on livestock, could contribute to a change in the species and structure of vegetation at the landscape scale through changes in nutrient distribution and concentrations alone.

Plate 18.1   Zebu cattle feeding on *Piliostigma* in Nigerian millet field

Three indigenous crops—sorghum (*Sorghum bicolor*), pearl millet (*Pennisetum typhoides*), and African rice (*Oryza glaberrima*)—were developed 3,000 to 4,000 years ago. This coincides with a period of desert expansion and, as has been suggested in the Levant, domestication and cultivation of cereals may have been a response to a more uncertain climate. As is still the case in many parts of West Africa, farmers are thought to have planted cereals on the receding water's edge at the start of the dry season (Plate 18.2).

As in northern Europe, expansion of agriculture in Asia was associated with clearance of forest to create open farmland. Two centers of domestication have been identified in China: one was for millet (8,000 years ago) and the other for Asian rice (6,500 to 8,500 years ago). Like domestic rice (*Oryza sativa*), its wild ancestors were associated with shallow water and the domestication and cultivation took place in the low-lying plains of the Hupie Basin and Yangtze River, sowing seed into receding water, as in West Africa. More sophisticated methods of irrigation and terracing developed, and the area of rice cultivation expanded over the following millennia in response to population pressure, with much of the increase in area and intensification of management occurring in the past two centuries. Rice cultivation expanded into Southeast Asia, India, and Pakistan around 5,000 years ago and into the Mediterranean with Islamic traders

2,500 years later. In contrast to rice, millet (*Panicum miliaceum* and *Setaria italica*) was domesticated from dryland species such as *Setaria viridis* in northern China and became, and remains, an important staple there. Chickens (*Gallus gallus domesticus*) were domesticated in China 7,200 to 7,400 years ago, and water buffaloes (*Bubalus bubalis*) were domesticated about 6,500 years ago.

Other crops were developed independently in South, Central, and North America 4,000 to 4,500 years ago. These included maize (*Zea mays*), beans (Phaseolus spp.), pumpkins (*Cucurbita pepo*), potato (*Solanum tuberosum*), marsh elder (*Iva annua*), quinoa (*Chenopodium quinoa*), sunflower (*Helianthus anuus*), and numerous other tuber and small-seed crops. As with the wild ancestors of linseed and rice in Europe and Africa, those of pumpkin and sunflower were associated with riparian areas, and sunflower and various *Chenopodium* species may well have become associated with human activity as weeds that were subsequently domesticated. Maize became central to the diet and cultural identity of people throughout North, Central, and South America, demonstrating the integral position taken by agricultural crops in the development of human civilization and land use (Staller et al., 2006).

Animals domesticated in South America included the llama (*Lama glama*), alpaca (*Lama pacos*), and guinea pig (*Cavia porcellus*). As has

Plate 18.2   Rice continues to be planted, tended and harvested in West African wetland areas, much as it was three or four thousand years ago

been demonstrated in modern Amazonian agriculture (Naughton-Treves et al., 2002), the slash-and-burn approach can change the composition of large mammal communities through the hunting of large herbivores, carnivores, and primates for food and to reduce losses of crops and livestock, so that 'weedy' species such as brown agouti (*Dasyprocta variegatal*), armadillo (*Dasypus novemcinetus*), and red brocket deer (*Mzama gauazoubira*) predominate. In a modern pastoral setting in the Indian trans-Himalaya, Mishra et al. (2004) demonstrate that direct competition between domestic livestock and wild herbivores can also contribute to lower numbers of the latter through lower reproductive performance, and this may have contributed to the extinction of four wild herbivores during the three millennia of pastoral farming in this area.

As with crop and livestock species from Europe and Asia, many of those from the Americas subsequently became widely adopted within farming systems across the world through a combination of trade and colonization. Today, farm crops and livestock are widespread across much of the world, making them among the most successful species. They also demonstrate considerable genetic diversity within species, as crops and livestock have been adapted to meet the diverse needs of people across the millennia and the varying

climatic and geomorphologic constraints across the regions. Landscapes have also been modified to meet their needs. As a result, wild plant and animal species associated with these modified landscapes have benefited to an almost equal degree.

Weed species that are known to have occurred in Neolithic arable fields in Western Europe included *Papaver rhoeas*, *Fumaria officinalis*, *Sinapis arvensis*, *Raphanus raphanistrum*, *Thlaspis arvense*, *Spergularia marina*, *Polygonum lapathifolium*, and *Vicia hirsuta*. *Agrostemma githago*, *Adonis aesitvalis*, *Bupleurum rotundifolium*, *Ranunculus arvensis*, and *Scandix pectinveneris* were present initially in the Balkans and eastern Europe during the Neolithic period and were transported with grain to western Europe during the Iron Age to Medieval periods (Bogaard and Hynd, 2000). Some weed species (e.g., *Polygonum persicaria*, *Chenopodium album*, and *Avena fatua*) are known to have been used as food, with wild oat being the progenitor of its cultivated equivalent. In more recent times, *Matricaria matricarioides* has come to Europe from China, and *Veronica persica* has colonized Europe from North America. As illustrated for *Echium plantagineum* by Grigulis et al. (2001), plant species are often more successful as aliens than as indigenous species. In the case of *E. plantagineum*, abundance was higher in Australia than in Portugal, in the native Mediterranean range, due to higher seed-bank incorporation and seedling establishment. Pauchard and Alaback (2002) identified roads as a primary pathway for weeds into cultivated areas within a largely forested region of south-central Chile where 85% of the weed species were of Eurasian origin. Clements et al. (2004) note that most North American arable weeds exhibit considerable adaptability to changing management practices. Weeds have therefore shown enormous changes in both global and local distributions in response to agricultural activity.

Weeds play a crucial role in arable ecosystem functioning. Potts (1986; and in Pain and Pienkowski, 1997) demonstrated in Britain that there was a very strong link between the abundance of phytophagous invertebrates associated with arable weeds and the breeding success and subsequent breeding abundance of gray partridges (*Perdix perdix*). Even for this essentially granivorous species, insect food is essential for the survival of young chicks and many of the insects are, in turn, dependent on arable weeds. For a species that is strongly associated with the agricultural environment, this was an important finding. Such a relationship between arable weeds, insects, and birds has subsequently been found for other species. In the case of the gray partridge, Potts showed that there was also a requirement for perennial grassy nesting habitat in the farmland landscape and that the control of nest predators by a gamekeeper influenced nesting success and recruitment to the breeding population in subsequent years. This example clearly illustrates the interactions between ecosystem functioning, landscape structure, and human cultural activities on the abundance and distribution of a wildlife species.

18.3 AGRICULTURAL ECOSYSTEMS TODAY

A very wide range of species is strongly associated with farmland, many of them having evolved with agriculture as farming systems have changed over the millennia. There is therefore a strong relationship between the farming systems adopted and the abundance and distribution of wildlife species. Traditional farming methods support many species, including those with restricted global distribution and high conservation status. In Europe, corn bunting (*Emberiza miliaria*) abundance and distribution is strongly associated with farmland ecosystems, its reproductive performance is influenced by an abundance of phytophagous insects (Brickle et al., 2000), and its distribution is linked to that of extensive (low external input) cropping such as that of the Portuguese Alentejo (Stoate et al., 2000). In this area, bird species richness and abundance are highest in structurally diverse habitats represented by 'montado' (a parklike landscape of *Quercus* species and arable and fallow land (Stoate et al., 2003) (Figure 18.1). However, the abundance of bird species of greatest global conservation value such as great bustard (*Otis tarda*) and lesser kestrel (*Falco naumii*) is associated with simpler, more open, extensively managed (low external input) farming systems. Intensive (high external input) farming systems support lowest bird abundance and species richness. An area of montado can therefore support high bird abundance and species richness at the farm scale, but extensively managed pseudo-steppe, with lower numbers of birds, contributes most to bird species richness at the global scale. Extensively managed landscapes are associated with higher plant species richness and greater landscape diversity. The Spanish 'dehesa', ecologically similar to the Portuguese montado, also supports high species richness. Dehesa is often part of a traditional transhumance system by which herders moved (and in some cases continue to move) livestock between high summer pasture and low-lying winter grazing in dehesa. Such corridors provide routes for the dispersal of plant and animal species between areas of similar habitat.

Beaufoy et al. (1994) describe a number of extensively managed agricultural systems in Europe, all of them contributing to biodiversity at local and global scales. These traditional systems are generally characterized by the minimal use of fertilizer and pesticides and low livestock densities in grazed systems, often with traditional livestock breeds that are characteristic of the region, and by diverse crop rotations, often including a fallow stage in arable systems. In fact, many traditional farming systems combine livestock and crop production, increasing further the diversity of products and habitats for wildlife. Many systems also incorporate perennial crops such as vines, olive, and oak. In Portugal, while *Quercus suber* provides cork as a cash crop, *Q. ilex* provides acorns as food for the traditional Alentejo black pigs, which are culturally important for their gastronomic value. Extensive mixed grazing by sheep and goats is important in much of Greece for the mosaic of evergreen scrub, conifer forest, and rough pasture, which characterizes much of the landscape. As in Spain transhumance still links and contributes to the maintenance of such landscapes in mountain and lowland areas. Raptors such as lanner falcon (*Falco biarmicus*), Egyptian vulture (*Neophron percnopterus*), and golden

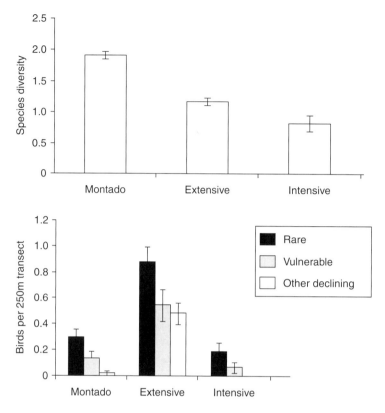

Figure 18.1   Bird abundance data for different Portuguese farming systems—(a) species diversity, and (b) Relative abundance (modified from Stoate et al., 2003)

eagle (*Aquila chrysaetos*) are among the rarer birds to benefit from this management system.

There is therefore a trend, at least in some regions, for species richness to decline with agricultural productivity. Although there is considerable debate about relationships between species richness and productivity (normally measured as evapotranspiration) (e.g., Whittaker and Heegaard, 2003), these are thought to follow either hump-shaped or positive trends, with highest species richness at highest productivity. However, superimposed on this general trend is the exploitation of resources by humans, primarily the appropriation of energy in the form of crops and livestock. This can reverse the trend, so that human appropriation of net primary production is negatively correlated with species richness (Haberl et al., 2003). For example, wildlife abundance and species richness in rangelands are generally inversely related to grazing intensity, which increases with human population pressure and the appropriation of seasonally grazed land by cropping systems (Blench, 2001).

Landscape and habitat structure, and in particular heterogeneity at a range of scales, plays an important role in influencing species abundance and distribution (Benton et al., 2003; Tews et al., 2004). Tews et al. (2004) identified the influence of 'keystone structures' within agricultural landscapes on species richness, illustrating this with the example of increased beetle species richness associated with the presence of temporary ponds in arable fields. Ditches across Europe provide valuable wet vegetated noncropped habitats to both aquatic and terrestrial taxa, supply food resources lacking in cropland, and perform connectivity functions within a wider landscape and harbor a suite of plants and animals not found in other farmland habitats (Herzon and Helenius, 2008). Even small and ephemeral water bodies can contribute to regional biodiversity in farmland (Biggs et al., 2007). The review by Tews et al. found significant positive relationships between animal species richness and habitat heterogeneity in all of the agricultural ecosystems studied. Weibull et al. (cited in Benton et al., 2003) found a positive relationship between butterfly abundance in 5-km squares and habitat heterogeneity. Seed-eating birds occur in higher numbers in pastoral areas that contain small areas of arable land than in a purely pastoral landscape (Robinson et al., cited in Benton et al., 2003) and bird species richness in Ontario, Canada, is higher in heterogeneous landscapes than in landscapes associated with large fields and intensive agriculture (Freemark and Kirk, cited in Benton et al., 2003).

Coffee that is grown under the shade of indigenous trees in South America supports higher numbers of birds and bird species than more intensively managed single-species coffee plantations that lack tree species and structural diversity (Donald, 2004). Shaded coffee yields can be as high as those in monocultural plantations except on the best soils, and the additional trees provide additional products. However, bird communities in shaded coffee plantations are not the same as those in the pristine forest that plantations normally replace, and they vary according to the accompanying tree species present and the distance from surviving undisturbed forest. The more isolated that shaded coffee plantations are from a forest, the less likely they are to support specialist forest bird species.

A major historical impact of the spread of agriculture has been landscape fragmentation, normally in the form of partial replacement of woodland or forest by cropped land or pasture. This can benefit some species while having a negative effect on others (such as forest species), depending on the ratio of habitats. Brown-headed cowbirds (*Molothrus ater*) benefit from landscape fragmentation in North America, exploiting livestock feed sites and other agricultural sources of food and parasitizing nests of passerines along woodland edges. This species is thought to have contributed to the population declines of other passerines through the reduced nesting success of the host species (Smith et al., 2000).

Economies of scale have led to amalgamation of small farms in many regions, reducing landscape heterogeneity and leading to a loss of noncrop features such as hedges, ditches, streams, and small woods. Associated with this, development of large machinery has contributed to scaling up of fields, with consequent loss of field boundary features. Heterogeneity of agricultural habitats influences movement at a number of scales, including movement of individuals, dispersal, and migration, all of which can influence population viability and therefore abundance and distribution. Dover (1996) showed that butterflies in British farmland made considerable use of gaps in linear hedgerows to access sources of food on either side of the hedge, without risking exposure to wind and predation by crossing over the hedge top. Heterogeneity at the landscape scale influences butterfly dispersal, as illustrated for marsh fritillary (*Eurodryas aurinia*) in southern England (Hobson et al., 2001). Devilsbit scabious (*Succisa pratensis*), the larval food plant, is patchily distributed within open grazed grassland landscapes and local populations of marsh fritillaries associated with these patches are susceptible to extinction and recolonization. The viability of the metapopulation depends on the ability of this species to disperse between larval food plant patches, and therefore on the connectivity of patches. High livestock densities can reduce the number of food plant patches, increasing the isolation of butterfly colonies and reducing population viability.

Just as large areas of agricultural land can represent barriers to dispersal, they can also restrict routes of seasonal migration. Wetland areas in sub-Saharan West Africa such as Lake Chad and the River Senegal provide important feeding areas for migratory passerines, prior to their spring migration across the Sahara to breeding grounds in northern Europe, but such areas have been severely affected by abstraction of water and expansion of the agricultural area for rice production (Stoate, 2004). Loss of seasonally productive sources of fruit and insect food could reduce survival on migration, and therefore subsequent breeding abundance. In Kenya, expansion of the land area used for wheat production in the Loita Plains has severely reduced wet-season grazing for wildebeest and constrained their migration routes, contributing to population decline (Serneels and Lambin, 2001). In such cases, agricultural practice in one area can have substantial consequences for wildlife abundance and distribution in another.

On a smaller scale, examples from the United Kingdom illustrate how species associated with managed grassland or arable crops are also influenced by the heterogeneity of those crops in terms of crop structure and their spatial and temporal distribution. Yellowhammers (*Emberiza cintrinella*) provisioning young gather invertebrate food from crops but switch seasonally between the crops used as foraging habitats (Stoate et al., 1998). Yellowhammer territories with several crops within their foraging range (about 300 metres) therefore have a longer effective provisioning period and potentially greater nesting success, than those with just one crop available to them. Seasonal changes in utilization such as this can arise through seasonal changes in insect abundance in different crops, or in the structure of the vegetation that restricts or facilitates access to those insects as food. A combination of spring-sown and autumn-sown crops results in a mosaic of habitats throughout the year. For example, spring-sown crops permit the retention of previous crop stubbles through the winter, providing a source of winter seed food for birds and small mammals. This combination also permits the survival of both spring- and autumn-germinating weed species. Brown hares (*Lepus europaeus*) in arable landscapes become nutritionally stressed when crops are mature, but they will graze grass where this is present as pasture in more diverse, mixed arable, and livestock systems. Abundance of hares is therefore higher in mixed arable and livestock systems (Smith et al., 2004).

The same relationship between wildlife abundance and crop diversity applies to agricultural systems elsewhere, although the mechanisms might vary over quite small spatial scales. Skylarks (*Alauda arvensis*) favor spring-sown cereals over autumn-sown cereals in most of northern Europe because of the more open structure of the former for nesting and foraging, but in Finland Piha et al. (2007) have shown that autumn sowing provides an open structure and supports more skylarks than spring-sown cereals, which are associated with field operations at the start of the nesting season. Spatial and temporal differences in crop management and ecology mean that high crop diversity is associated with high wildlife species diversity. Because there are differences in the way livestock graze pasture or rangelands, diversity of livestock species and breeds can also influence wildlife species richness and abundance through their influence on plant species and vegetation structure.

18.4 INTEGRATED CROP MANAGEMENT

Habitat diversity can also be beneficial to cropping. In the United Kingdom, perennial grassy vegetation in field boundaries has been shown to support high wintering densities of beneficial predatory insects (Carabid and Staphylinid beetles), which control crop pests such as aphids when the crop is growing. The same habitat also reduces competitive annual weeds in field edges and supports many other invertebrates, mammals, and bird species (Stoate et al., 2001b). Low banks of perennial grasses (Beetle Banks) in field centres have been designed to increase numbers of beneficial invertebrates within the crop, increasing control of aphids by natural predators and reducing the need for summer aphicides (Thomas et al., cited in Stoate et al., 2001c) (Plate 18.3). This is an example of integrated crop management, where conventional crop management inputs are combined with natural control measures. Crop rotation, associated with crop diversity,

is also crucial in many cropping systems for facilitating control of crop pests and weeds.

Other wild species provide environmental services to agriculture in a wide range of ways across the world. There are countless examples of invertebrate predators of crop pests, while wild bees provide important pollination for some crops. The role of wild bee species is increasingly important in areas where varoa mites, accidentally imported from Asia, are reducing the numbers of domestic honey bees, which are otherwise needed for crop pollination. Termites, widely distributed and perceived as destructive pests of crops and timber structures, are used in some areas to improve soil structure where soils have become severely impacted. Earthworms play an important role in maintaining soil structure, while soil fungi also bind the soil and facilitate uptake of nutrients through mychorrizal associations with roots of crops and wild plants. Crop management, and particularly soil management methods, can determine the abundance and distribution of soil invertebrates and fungi.

In parts of West Africa (e.g., Hoffman, cited in FAO, 2003) and elsewhere, mixtures of crops (intercropping) are an established means of reducing crop pests, or the impact of them, on overall productivity. With numerous millet cultivars to draw on, farmers in marginal arid areas can adjust their sowing regimes according to current or predicted rainfall. Different cultivars or mixtures of cultivars are sown when rains are early from when rains are late or unpredictable. Rice is perhaps the most genetically diverse crop, with many hundreds of varieties ensuring that combinations of varieties increase disease resistance, reducing the need for pesticides that are detrimental to aquatic ecosystems and the abundance of species within them (Donald, 2004). Intercropping two rice cultivars (a disease-resistant variety and a more valuable but susceptible one) in China reduced overall susceptibility to rice blast fungus by 94% and increased crop yield by 89% and crop value by 40% compared with a monoculture crop (Zhu et al., cited in Mae-Wan and Ching, 2003).

In some countries, such as Cambodia, rice production is highly integrated with management and use of other natural resources such as fish production (Balzer et al., cited in FAO, 2003). One estimate puts the amount of fish produced from rice fields at 15–25% of Cambodia's total fish catch and farmers exploit about 40 aquatic animal species derived from rice fields, including fish, crabs, snakes, frogs, snails, and insects, as well as some plants. In addition, fish and other predators help to control pests of rice. Aquatic animals depend on sympathetic management of the rice fields, including minimal use

Plate 18.3   Field strips in arable cultivation, Poland

of pesticides, and fish are also dependent on nearby mangrove and other marginal vegetation for spawning habitat. There is some threat to this habitat through charcoal production and clearance for more intensive, monocultural farming, which would reduce the diversity of species and products from farmland managed in a more integrated way.

Integration of biodiversity conservation with agriculture and other resource management and use was investigated by the UNEP 'PLEC' program (People, Land-management and Environmental Change, 1992–2002), which linked a series of six case-study clusters across the topics and subtropics. These revealed that economic viability was compatible, and normally greatest, with high agricultural diversity and often with high natural biodiversity as well. In China, the most successful farmers were the ones with the greatest agricultural and natural biodiversity on their farms. Seventy varieties of rice were found in Jinou Township villages. Jinou's agroforestry system in Baka supported 289 plant species, including 50 cultivated and semicultivated native species, 6 endangered species, 3 endemic species,

and 12 species valued for cultural reasons. High species richness was also reported from other sites, with many species being 'cultivated' from the wild and with uses including medicine, vegetable, fruit, ornamental, and herbs. In this case, farming activity was complementary to, and contributed to conservation objectives of nature reserves, whereas proximity of agriculture to nature reserves more often results in conflict. Vegetable diversity, in combination with staple crops, is also important for human nutrition. In Africa, traditional food plants such as green leafy vegetables supply an estimated 80% of vitamin A and more than two-thirds of the vitamin C consumed (FAO, 2001).

In the Brazilian PLEC cluster, wild species were selectively encouraged in long fallows within a cropping rotation, increasing income for those farmers who did so above the level of their neighbours. In Ghana, where large areas of forest have been cleared for monocultural maize and cassava, some farmers are concentrating on native yams, of which there are 23 local varieties, and which can be grown at high density (about 3,000 plants per hectare) under the forest canopy. Farming wild snails provides additional income. In Kenya, gardens contain 46 species of food crop, 23 fodder crops, 22 medicinal species, 29 fuelwood species, and 13 other species for sale. The large tree species *Ficus sycamorous* is valued by farmers for increasing available phosphorus and improving coffee yields under the tree canopy, especially under drought conditions. The fruit of this tree is attractive to cattle and to wildlife species such as primates and many frugivorous birds (personal observation). This principle is adopted on a larger scale in Senegal, where *Faidherbia albida* is widely encouraged for its nitrogen-fixing role in improving millet and groundnut yields, and where the canopy supports large numbers of invertebrates, which in turn provide food for insectivorous birds during the dry season (Stoate et al., 2001a) (Plate 18.4). Stoate and Jarju (2008) demonstrate how indigenous tree species that have cultural, medicinal, and other uses can be reintroduced to farming systems from which they have been eliminated, improving soil organic matter and water retention during drought, and consequently increasing crop yields.

Plate 18.4   Cattle search for seed pods under a *Faidherbia albida* tree, *Senegal*

A number of points are apparent from these case studies:

1. Crop species diversity, and therefore agricultural habitat heterogeneity, can be high in economically and environmentally sustainable farming systems;
2. Heterogeneous diverse agricultural systems tend to be associated with relatively high wild species abundance and diversity; and
3. Crop species diversity also optimizes production and increases resilience to changing market and climate conditions.

18.5 CULTURAL INFLUENCES

Although cultures (especially in industrialised countries) tend to consider crop species as distinct from wild ones, this is far from the case in many regions. Farmers are constantly investigating the potential of wild species for a wide range of consumptive, functional, and cultural uses, thereby increasing species diversity on farmland. Farmers' cultural backgrounds can also contribute to the evolution of crop species without intentions for deliberate plant breeding. Building on the pioneering work in Richards (1985), Longley (1999) compares rice-growing systems in two ethnolinguistic groups (the Limba and the Susu) in Sierra Leone where a semiwild rice type, 'Salli Foreh', grows sympatrically with domestic rice. Limba farmers have an affinity with the land and attach considerable importance to rice growing. They spend time in the fields rogueing 'Salli Foreh' from their rice crops, and sometimes they gather domestic rice panicles by hand prior to the main harvest in order to collect the best seed for subsequent sowing. Susu farmers assume a dichotomy between human culture and nature, attach importance to trade and life within settlements, and associate 'Salli Foreh' with spirits and witches. They harvest the semiwild type with the rest of the crop, only occasionally weeding out individual panicles from harvested sheaves. The two ethnolinguistic groups are therefore likely to contribute differently to the evolution of rice cultivars through differential hybridization between 'Sallie Foreh' and sown cultivars.

'Gurdi' and 'Madishe' are traditional Nepali rice varieties that are culturally important because of their use in the baking of 'selroti' (Pant, 2002), a doughnutlike preparation made from ground rice and nuts. These varieties are preferred by farmers because the resulting softness and puffiness of selroti are better than is possible using modern rice cultivars. Both cultivars are also high-yielding and excellent for boiled rice. They are relatively more tolerant of drought and grow better in less fertile soil (requiring less fertilizer use) than do modern cultivars. Gurdi is also shade-tolerant and grows well in agro-forestry systems that, in turn, support more wildlife. In Costa Rica, as elsewhere, modern bean (*Phaseolus vulgaris*) cultivars respond to high inputs such as fertilizer. However, traditional cultivars that have been developed by generations of local farmers are much more diverse in their nutrient requirements and in food quality and use. In addition to contributing to food security and diversity in diet, one cultivation system, 'Frijol Tapado', involves planting bean seeds in newly cleared ground, the germinating seeds being able to derive sufficient water and nutrients from the breakdown of the cleared vegetation (Felipe Greenheck, University of Costa Rica, personal communication). Such a system permits the continuing regeneration of secondary forest and associated fauna, while also reducing soil erosion from cultivated ground.

In Senegambia, Stoate and Jarju (2008) reported that farmers perceived *Faidherbia albida*, along with other naturally regenerating woody species, as weeds on arable land and removed saplings during crop-weeding operations. Facilitated by a research project, one farmer recalled a story about a prophet making a boat from *F. albida* and who encouraged the regeneration of this species on his farm. Within four years numerous farmers in five adjoining villages were doing the same, recognizing the benefits of this species to soil fertility and crop yields, while incidentally improving dry season habitat for insectivorous birds by providing an insect-rich foraging habitat. In Asia, where vulture numbers have declined very dramatically following the use of domestic animal medication (Swan et al., 2006), the Indian Zoroastrian Parsi community might be expected to be foremost in changing to alternatives to the newly introduced anti-inflammatory livestock drug Diclofenac as vultures play a vital cultural role in the form of disposal of human corpses (Houston, 1990).

Cultural influences also operate in industrialized countries, both on cropping systems, landscapes, and wildlife habitats, influencing the abundance and distribution of species on farmland. For some, wildlife conservation is a moral imperative. For others, landscape management is more important but simultaneously benefits wildlife. Stoate (2002) showed that modification of farmland management, inspired by the cultural activity of wild pheasant (*Phasianus colchicus*) shooting, can result in substantial increases in numbers of nationally declining passerines. In this case study from lowland England, songbird numbers increased alongside those of the species for which management was implemented, while yields of commercial crops were not compromised beyond annual fluctuations (Figure 18.2). In other cases, wildlife conservation is a consequence of the production of traditional niche food products. However, such cultural values are often at odds with more generally recognized economic and socioeconomic motivations, and management for wildlife, especially on productive land, is only possible when income derived from the farm business, or nonfarm income, is relatively high.

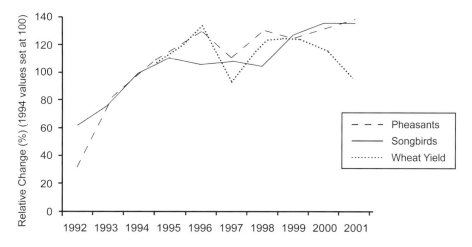

Figure 18.2   Relative changes in autumn pheasant numbers, breeding songbird numbers, and wheat yield. Data are from the GWCT Allerton Project research and demonstration farm, Leicestershire, United Kingdom. Values set to 100 in 1994

Purely economic influences increasingly determine the way agricultural land is managed and the consequent suitability of that land for wildlife. The consequences for species abundance and distribution are generally negative and operate in a wide range of ways.

18.6 AGRICULTURAL INTENSIFICATION

Expansion of the agricultural area has been substantial in primarily agricultural countries, driven largely by human population increase, although food production continues to lag behind population growth (UNEP, 2002). The increase in the agricultural area has resulted mainly in the replacement of forest and wetland ecosystems with cropped and grazed land. For example, the area of arable land increased by 35% in South America, by 21% in Meso-America, and by 32% in the Caribbean between 1972 and 1999 (UNEP, 2002). Unsustainable management of agricultural land has also led to land degradation and abandonment on a large scale. Land degradation takes the form of increased soil erosion by wind and water and of salinization of irrigated land. For example, although the total area of irrigated rice is rising, previously irrigated land has simultaneously been abandoned because of salinization (at a rate of 10 million hectares annually in the 1980s; UNEP, 2002).

In industrialized countries, there has been a slight decline in agricultural area but yields per productive hectare have increased in response to increased inputs (UNEP, 2002). For example, the arable area in Western Europe declined by 6.5% in the last three decades of the previous century, while the use of nitrogen fertilizer quadrupled over a similar period (Potter, in Pain and Pienkowski, 1997). Such changes in land use have been driven by the Green Revolution, the development of crop cultivars and livestock breeds that are able to respond to high inputs, coupled with state-funded economic incentives for agricultural production such as the EU Common Agricultural Policy (CAP). Animal and plant species that are specific to agricultural environments have declined in abundance and distribution as farming has become more intensive in recent decades across the world. This is best documented in industrialized countries, especially in Europe where population declines in birds have been shown to be correlated with levels of agricultural intensification, with cereal yield explaining 30% of the variation in bird population trends (Donald et al., 2001). Population declines were greatest in EU countries and are likely to increase in those eastern European countries that have more recently joined the EU and where agricultural systems are becoming more intensive (Stoate et al., 2009). Figure 18.3 illustrates a separate trend for farmland birds in Central and Eastern Europe to that in the rest of the continent, with a relative increase in abundance since the collapse of the communist era and initial signs of a decline coinciding with accession to the EU (Gregory et al., 2005). Stoate et al. (2001c, 2010) review other ecological

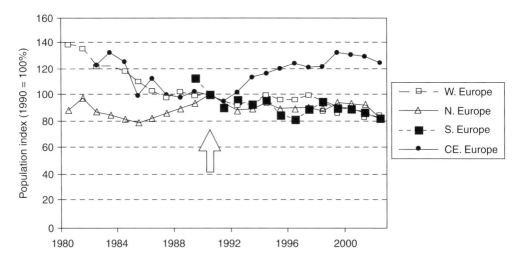

Figure 18.3   Pan-European common bird indicators for farmland in four regions of Europe (from Gregory et al., 2005)

impacts of agricultural intensification in Western Europe.

Artificial fertilizer use has increased in most countries from the 1950s, often supported by government subsidies, but it declined slightly, especially in Europe, during the 1990s (UNEP, 2002). Fertilizers have influenced landscape heterogeneity in some regions by replacing the need for a fallow fertility-building stage in arable rotations. In many agricultural ecosystems throughout the world, fallow land serves an important function by providing a noncropped habitat for many species and the loss of fallow land from the agricultural landscape can result in the depletion of species abundance and richness.

Fertilizer use has affected animal and plant species abundance and distribution through a wide range of mechanisms. Within the area to which fertilizer is applied, increased height and density of arable crops or grasses change radically the ways in which they are used by animals and the species composition of associated plants. Vigorous crops and grasses exclude plant species associated with low soil fertility and fertilizer application has been shown to have a dramatic influence on, for example, the distribution of some now rare arable plants in the United Kingdom such as *Galeopsis angustifolia* and *Centaurea cyanus* (Wilson, 1994, cited in Stoate et al., 2001c). Fertilizer deposited outside the cropped area can also change the composition of plant communities and structure in uncropped habitat such as grass field margins, for example, increasing the prevalence of cleavers (*Galium aparine*) and sterile brome (*Bromus sterilis*) (Boatman et al., 1994, cited in Stoate et al., 2001c). Elevated nitrogen levels in ancient woodlands up to 100 m from arable farmland are associated with an increase in ruderal and nitrophilous plant species and lower cover of ancient woodland indicator species such as *Viola reichenbachiana* (Willi et al., 2005). Hautier et al. (2009), working with grassland, demonstrated that such a loss of botanical diversity associated with eutrophication is due to competition for light.

Considerable impacts occur when fertilizer leaches into watercourses or standing water adjoining cropped land, and nitrogen and other nutrients can be washed into streams via surface runoff, subsurface flow, and field drains. In New South Wales, Australia, the range of the endangered green and golden bell frog (*Litoria aurea*) contracted following an exceptional pulse of fertilizer during heavy rains, while the range of some more abundant species was unaffected (Hamer et al., 2004). Laboratory studies revealed that the survival of *L. aurea* tadpoles was lower than that of other frog species when exposed to ammonium nitrate and calcium phosphate fertilizers. Such an influence of agriculturally derived nutrients in water may explain the differential distribution of other amphibian communities, for example, those of Maine, in the United States, where species richness is considerably lower in agriculturally dominated landscapes (Guerry and Hunter, 2001).

Nitrogen and phosphorus, two major agricultural nutrients, can cause eutrophication: phosphorus mainly in inland waters and nitrogen in coastal waters such as the Baltic and North Sea (Saull, 1990, cited in Stoate et al., 2001c; Baldock et al., 1996, cited in Stoate et al., 2001c). Such eutrophication can result, through excessive growth of phytoplankton, in depletion of oxygen from water bodies (hypoxia) and the subsequent death of aquatic animals. The hypoxic area of the Gulf of Mexico has been recorded as reaching more than 18,000 km^2 in recent summers, with potentially severe negative consequences for marine fish and crustacean communities, including important fisheries (Rabalais and Grimes, 2000). In this case, excessive transport of nitrogen to the sea results from high application rates on agricultural land throughout the Mississippi river basin of the central United States. Nitrogen fertilizer has additional far-reaching effects on both terrestrial and marine plant and animal abundance and distribution through climate change as its manufacture is an energy-intensive process resulting in the emission of greenhouse gasses such as CO_2, and deposition of atmospheric nitrogen (emitted as NO and NO_2) can lead to acidification of seminatural environments. The agricultural sector worldwide accounts for about 5% of anthropogenic CO_2 emissions (ECAF, undated, cited in Stoate et al., 2001c), while Cormack and Metcalf (2000) (cited in Stoate et al., 2001c), estimate that 50% of total energy input to cereal crops is accounted for by the manufacture and transport of fertilizer and pesticides.

As well as depletion of soil invertebrates and micro fauna and fungi, ploughing results in soil erosion, causing the transport of soil and associated phosphorus to water. Subsequent sedimentation of watercourses can severely degrade spawning habitats of salmonid fish, resulting in population declines in arable catchments. Sedimentation of lakes can reduce their capacity for water storage and can alter their ecology. Phosphorus adsorbed to soil particles is a major cause of eutrophication in lakes and watercourses, resulting in excessive algal growth and hypoxic conditions, reducing species richness of aquatic animals and plants across the world. In the Florida Everglades, agriculturally derived phosphorus causes algal blooms and dense growth of macrophytes (e.g., *Typha*), changing radically the ecology of this wetland area and threatening populations of already endangered species such as snail kite (*Rostrhamus sociabilis*), which hunts apple snails (Ampullariidae) in open vegetation.

Large reservoirs constructed for irrigation can cause changes in estuarine ecosystems and adjacent coastal waters, as illustrated by the Alqueva dam in southern Portugal, which causes changes in the estuarine environment about 150 km downstream (Morais et al., 2009). The dam reduced the freshwater flow reaching the estuary, thereby increasing saline penetration upstream (Domingues and Galvão, 2007) and resulting in a penetration of marine and brackish water fish into the upper estuary and a concurrent decline of previously dominant freshwater species, including endangered fish species (Chícharo et al., 2006).

Pesticide use increased globally during the second half of the twentieth century, with a slight decline in some countries in recent years (Stoate et al., 2001c). Impacts on wildlife can be either through direct mortality or indirect effects such as depletion of insect food for insectivorous birds. Direct mortality resulting from pesticide use in the United States inspired Rachel Carson's book *Silent Spring* (1963), which documented the loss of a wide range of wildlife from North America and, in turn, largely inspired the development of the ecology movement. The organochlorines that were largely responsible for these wildlife deaths have subsequently been banned from most countries but continue to be used in some primarily agricultural countries. A contemporary example of direct effects of pesticide use is provided by the veterinary anti-inflammatory drug, Diclofenac, which is widely used in cattle in Pakistan and Nepal. As a result of eating the carcasses of treated cattle, the numbers of three vulture species declined by 97% in 12 years so that their populations have become seriously threatened (Swan et al., 2006).

In terms of indirect effects, Potts (1986) documented the effect of herbicides on gray partridge numbers in the United Kingdom through the removal from farming systems of arable weeds that supported crucial invertebrate food of partridge chicks. In southeast England, the abundance of gray partridge, corn bunting, and skylark (*Alauda arvensis*) was inversely related to herbicide use (Ewald and Aebischer, 1999, cited in Stoate et al., 2001c) and monitoring over more than 30 years revealed declines in broad-leaved weeds associated with herbicide use, and declines in invertebrate groups associated with insecticide use and (for four of the five groups studied) with fungicide use. Some broad-leaved arable weeds such as *Adonis aestivalis* and *Legousia speculum* are now among the rarest plants in the United Kingdom as a result of herbicide use, fertilizer use, grain-cleaning technology, and increased autumn sowing of crops (Wilson, 1994, cited in

Stoate et al., 2001c). Arable weeds that have increased their abundance and distribution include cleavers (*Galium aparine*) and black grass (*Alopecurus myosuroides*). These are highly competitive, and in the case of black grass, they are often herbicide-resistant but have little value to wildlife species. While genetically modified crops could potentially have advantages to wildlife (e.g., through reduced insecticide use on insect-resistant varieties), there is currently considerable concern about negative ecological implications of genetically modified crops, for example, through increased weed control in herbicide-tolerant crops (Firbank et al., 2003), the erosion of regionally adapted crop diversity, and the loss of wildlife, culture, and food security associated with these.

Common themes running through the development of modern farming systems across the world are increased economic efficiency and increased negative environmental externalities such as those described above. Immediate economic gains are effectively subsidized through the long-term ecological and social costs such as climate change induced by greenhouse gas emissions associated with fertilizer manufacture and use, the global transport of agricultural inputs and products, and on-farm operations. Consequences for wildlife abundance and distribution may operate locally (within the same farming system or adjacent habitats) or at some distance from the farming operation responsible.

18.7 WORLD TRADE POLICY

Trade between regions has long been a feature of agriculture. The exchange of agricultural products was responsible for the spread of many arable weed species across Europe about 6,000 years ago and the spread of vertebrates such as rodents. More recently, agricultural trade has been associated with the global spread of animal and plant diseases such as avian influenza and swine flu. The introduction of foot and mouth disease into the United Kingdom in 2001 resulted in the slaughter of thousands of cattle, sheep, and pigs as a preventative measure, altering the ecology, economic stability, and social cohesion within the areas affected.

Accompanying the Green Revolution over the past half century has been a move in primarily agricultural countries toward increased production of cash crops for export, very often as part of structural adjustment plans for debt repayment imposed by the World Bank and International Monetary Fund. Such increased crop production for export has numerous consequences for animal and plant abundance and distribution. It inevitably results in reduced crop diversity and therefore landscape diversity, with the main focus being on crops such as rice, wheat, and maize. For example 'Golden Rice', a genetically modified cultivar, has been promoted as a means of addressing vitamin A deficiency and food security in preference to a diversity of crops including leafy vegetables, which increase food security and provide higher levels of vitamin A as well as other important dietary nutrients. Trade liberalization, in which primarily agricultural countries must open their markets to imports of subsidized crops from industrialized countries, can contribute to the erosion of the diversity of crop cultivars and species, cropping systems, and therefore the wildlife associated with crop and landscape heterogeneity. For example, in Mexico where maize was first developed and where countless varieties have been created and are at the heart of local cultural identity (Staller et al., 2006), hundreds of thousands of tonnes are now imported from American multinational companies each year (Sardo, in Petrini, 2002). As well as undermining local production, imported maize threatens the genetic integrity of indigenous varieties through hybridization. With patenting of genetic material by multinational companies, concerns are increasing for regional food sovereignty and security, and cultural and wildlife diversity in many countries across the world.

The expansion of cash crops also displaces seasonal grazing of nomadic and transhumant livestock. A result of this is that other traditional grazing areas that are often wildlife-species rich come under increasing pressure, and legislation is then introduced to deny herders access to grazing areas. As the cultural background of herders is not that of owner or occupier, their rights are often not recognized and traditional local sources of meat and dairy products, as well as traditional vegetation management and social systems, break down to the detriment of wildlife and society (Blench, 2001).

In Europe and the United States, negative environmental implications of intensive farming systems such as biodiversity loss, soil erosion, and water pollution have been addressed through direct funding mechanisms for environmental improvement. In the European Union, these take the form of the Rural Development Programme under which agri-environment schemes enable farmers to be paid to carry out specific management to meet environmental objectives (European Commission, 2004). Payments for environmental management are currently increasing as subsidies for crop production are being reduced in line with free trade objectives under the Uruguay and Doha Agriculture Agreement Rounds. In the United

States, programmes initiated under the Farm Bill include the Conservation Reserve Program, the Wetland Reserve Program, and the Wildlife Habitat Incentives Program (Heard et al., 2000). Such schemes have been challenged within the World Trade Organisation for indirectly supporting agricultural production. However, where free trade is adopted in the absence of support for environmental management, the impact on the environment and wildlife has been negative (e.g., New Zealand; MacLeod and Moller, 2006). As described above, the abundance and distribution of wildlife species are closely integrated with the management of crops and livestock, and the compatibility of long-term environmental sustainability with short-term economic objectives for free trade is a major challenge for negotiators of world trade policy (Potter and Burney, 2002).

National and international targets for biofuel production and use place increasing demands on land across the world, increasing food prices, displacing food production to agriculturally marginal land, or occupying such land for production of these fuels. As consumption by industrialized countries cannot be met by production within them, there is a reliance on primarily agricultural countries for production of crops such as palm oil and jatropha. In addition to the loss of wildlife habitat, destruction of Malaysian rainforest and associated peat soils contributes to the release of carbon dioxide into the atmosphere and reduces the potential to sequester it in future (Hooijer et al., 2006). The climate-change impact of such agricultural activity has serious ecological implications.

Agriculture is historically a cultural as well as an economic activity, and the objectives of food production and sustainable environmental management are intricately linked with the regional cultures that have evolved alongside these activities (Pretty, 2002). It is becoming increasingly accepted that future policy must embrace cultural diversity and food sovereignty and security if food production is to be sustainable and consistent with the conservation of wildlife species associated with agricultural systems.

REFERENCES

Augustine, D.J. (2003) 'Long-term livestock mediated redistribution of nitrogen and phosphorus in an East African savanna', *Journal of Applied Ecology*, 40: 137–49.

Beaufoy, G., Baldock, D., and Clark, J. (1994) *The Nature of Farming: Low Intensity Farming Systems in Nine European Countries*. London: Institute for European Environmental Policy.

Benton, T.G., Vickery, J.A., and Wilson, J.D. (2003) 'Farmland biodiversity: Is habitat heterogeneity the key?' *Trends in Ecology and Evolution*, 18: 182–88.

Biggs, J., Williams, P., Whitfield, M., Nicolet, P., Brown, C., Hollis, J., Arnold, D., and Pepper, T. (2007) 'The freshwater biota of British agricultural landscapes and their sensitivity to pesticides', *Agriculture, Ecosystems and Environment*, 122: 148–48.

Blench, R. (2001) *You Can't Go Home Again—Pastoralism in the New Millennium*. London: Overseas Development Institute.

Bogaard, A., and Hynd, A. (2000) 'A long-term perspective on arable weed floras: Archaeobotanical weed evidence', in Wilson, P., and King, M. (eds.), *Fields of Vision: A Future for Britain's Arable Plants*. London: Plantlife. pp. 20–29.

Boyd, R., and Richerson, P.J. (2005) *The Origin and Evolution of Cultures*. New York: Oxford University Press.

Brickle, N.W., Harper, D.G.C., Aebischer, N.J., and Cockayne, S.H. (2000) 'Effects of agricultural intensification on the breeding success of corn buntings *Miliaria calandra*', *Journal of Applied Ecology*, 37: 742–55.

Carson, R. (1963) *Silent Spring*. Harmondsworth, UK: Penguin.

Chícharo, M.A., Chícharo, L., and Morais, P. (2006) 'Inter-annual differences of ichthyofauna structure of the Guadiana estuary and adjacent coastal area (SE Portugal/SW Spain): Before and after Alqueva dam construction', *Estuarine, Coastal and Shelf Science*, 70: 39–51.

Clements, D.R., DiTommaso, A., Jordan, N., Booth, B.D., Cardina, J., Doohan, D., Mohler, C.L., Murphy, S.D., and Swanton, C.J. (2004) 'Adaptability of plants invading North America cropland', *Agriculture, Ecosystems and Environment*, 104: 379–98.

Domingues, R.B., and Galvão, H. (2007) 'Phytoplankton and environmental variability in a dam regulated temperate estuary', *Hydrobiologia*, 586: 117–34.

Donald, P.F. (2004) 'Biodiversity impacts of some agricultural commodity production systems', *Conservation Biology*, 18: 17–37.

Donald, P.F., Green, R.E., and Heath, M.F. (2001) 'Agricultural intensification and the collapse of Europe's farmland bird populations', *Proceedings of the Royal Society of London Series B*, 263: 25–29.

Dover, J. (1996) 'Factors affecting the distribution of Satyrid butterflies on arable farmland', *Journal of Applied Ecology*, 33: 723–34.

European Commission. (2004) *Biodiversity Action Plan for Agriculture: Implementation report*. Brussels: European Commission, Agriculture Directorate General.

FAO. (2001) *Improving Nutrition Through Home Gardening—A Training Package for Preparing Field Workers in Africa*. Rome: Food and Agriculture Organisation.

FAO. (2003) *Biodiversity and the Ecosystem Approach in Agriculture, Forestry and Fisheries*. Rome: Food and Agriculture Organisation.

Firbank, L.G., and 18 authors, (2003) *The Implications of Spring-Sown Genetically Modified Herbicide-Tolerant Crops for Farmland Biodiversity: A Commentary on the Farm Scale Evaluations of Spring-Sown Crops*. Merlewood, UK: Centre for Ecology and Hydrology.

Gregory, R.D., van Strien, A., Vorisek, P., Gmelig Meyling, A.W., Noble, D.G., Foppen, R.P.B., and Gibbons, D.W. (2005) 'Developing indicators for European birds', *Philosophical Transactions of the Royal Society Series B*, 360: 269–88.

Grigulis, K., Sheppard, A.W., Ash, J.E., and Groves, R.H. (2001) 'The comparative demography of the pasture weed *Echium plantagineum* between its native and invaded ranges', *Journal of Applied Ecology*, 38: 281–90.

Guerry, A., and Hunter, M.L. (2001) 'Amphibian distributions in a landscape of forests and agriculture: An examination of landscape composition and configuration', *Conservation Biology*, 16: 745–54.

Haberl, H., Plutzar, C., Erb, K-H., Gaube, V., Pollheimer, M., and Schulz, N.B. (2003) 'Human appropriation of net primary production as determinant of avifauna diversity in Austria', *Agriculture, Ecosystems and Environment*, 102: 213–18.

Hamer, A.J., Makings, J.A., Lane, S.J., and Mahony, M.J. (2004) 'Amphibian decline and fertilizers used on agricultural land in south-eastern Australia', *Agriculture, Ecosystems and Environment*, 101: 299–305.

Harris, D.R. (ed.). (1996) *The Origins and Spread of Agriculture and Pastoralism in Eurasia*. London: UCL Press.

Harrison, S., Inouye, B.D., and Safford, H.D. (2001) 'Ecological heterogeneity in the effects of grazing and fire on grassland diversity', *Conservation Biology*, 17: 837–45.

Hautier, Y., Niklaus, P.A., and Hector, A. (2009) 'Competition for light causes plant biodiversity loss after eutrophication', *Science*, 324: 636–38.

Heard, L.P., and 13 authors. (2000) *A Comprehensive Review of Farm Bill Contributions to Wildlife Conservation, 1985–2000*. NRCS Wildlife Habitat Management Institute Technical Report. United States Department of Agriculture, Madison, Mississippi.

Herzon I., and J. Helenius. (2008) 'Agricultural drainage ditches, their biology and functioning: Literature review', *Biological Conservation*, 141: 1171–83.

Hobson, R., Bourn, N.A.D., Warren, M.S., and Bereton, T.M. (2001) *The Marsh Fritillary in England: A Review of Status and Habitat Condition*. Butterfly Conservation Report No. S01–31. Wareham, UK: Butterfly Conservation.

Hodder, K.H., Bullock, J.M., Buckland, P.C., and Kirby, K.J. (2005) *Large Herbivores in the Wildwood and Modern Naturalistic Grazing Systems*. English Nature Research Report 648. Peterborough, UK: English Nature.

Hooijer, A., Silvus, M., Wüsten, H., and Page, S. (2006) *PEAT-CO$_2$, Assessment of CO$_2$ Emissions from Drained Peatlands in SE Asia*. Delft Hydraulics Report, Delft.

Houston, D. (1990) 'The use of vultures to dispose of human corpses in India and Tibet', in Newton, I., and Olsen, P. (eds.), *Birds of Prey*. London: Merehurst Press. p. 80.

Landsberg, J., James, C.D., Morton, S.R., and Müller, W.J. (2003) 'Abundance and composition of plant species along grazing gradients in Australian rangelands', *Journal of Applied Ecology*, 40: 1008–24.

Longley, C. (1999) 'On-farm rice variability and change in Sierra Leone: Farmers' perceptions of semi-weed types. *Agricultural Research and Extension Network Paper* 96. London: Overseas Development Institute.

MacLeod, C., and Moller, H. (2006) 'Intensification and diversification of New Zealand agriculture since 1960: An evaluation of current indicators of sustainable land use', *Agriculture, Ecosystems and Environment*, 115: 201–18.

Mae-Wan, H., and Li Ching, L. (2003) *The Case for a GM-Free Sustainable World*. London: Independent Science Panel.

Marciniak, A. (2005) *Placing Animals in the Neolithic*. Routledge Cavendish: University College London.

Mishra, C., van Wieren, S.E., Ketner, P., Heitkönig, M.A., and Prins, H.T. (2004) 'Competition between domestic livestock and wild bharal *Pseudois nayaur* in the Indian trans-Himalaya', *Journal of Applied Ecology*, 41: 344–54.

Morais, P., Chícharo, M.A., and Chícharo, L. (2009) 'Changes in a temperate estuary during the filling of the biggest European dam', *Science of the Total Environment*, doi:10.1016/j.scitotenv.2008.11.037.

Naughton-Treves, L., Mena, J.L., Treves, A., Alvarez, N., and Radeloff, V.K. (2002). 'Wildlife survival beyond park boundaries: The impact of slash and burn agriculture and hunting on mammals in Tambopata, Peru', *Conservation Biology*, 17: 1106–17.

Pain, D.J., and Pienkowski, M.W. (1997) *Farming and Birds in Europe*. London: Academic Press.

Pant, L.P. (2002) *Linking Crop Diversity with Food Traditions and Food Security in the Hills of Nepal*. MSc Thesis. Norgaric, Centre for International Environment and Development Studies. Agricultural University of Norway.

Pauchard, A., and Alaback, P.B. (2002) 'Influence of elevation, land use, and landscape context on patterns of alien plant invasions along roadsides in protected areas of south-central Chile', *Conservation Biology*, 18: 238–48.

Petrini, C. (ed.). (2002) *Slow Ark* 6. Bra, Italy: Slow Food.

Piha, M., Tiainen, J., Holopainen, J., and Vepsäläinen, V. (2007) 'Effects of land-use and landscape characteristics on avian diversity and abundance in a boreal agricultural landscape with organic and conventional farms', *Biological Conservation*, 140: 50–61.

Potter, C., and Burney, J. (2002) 'Agricultural multifunctionality in the WTO—legitimate non-trade concern or disguised protectionism?' *Journal of Rural Studies*, 18: 35–47.

Potts, G.R. (1986) *The Partridge, Pesticides, Predation and Conservation*. London: Collins.

Pretty, J. (2002) *Agri-Culture—Reconnecting People, Land and Culture*. London: Earthscan.

Rabalais, N., and Grimes, C. (2000) *CoastView Volume 2: Northern Gulf of Mexico Imagery and Information*. Charleston: NOAA Coastal Services Center.

Richards, P.W. (1985) *Indigenous Agricultural Revolution*. London: Hutchinson.

Serneels, S., and Lambin, E.F. (2001) 'Impact of land-use changes on the wildebeest migration in the northern part of the Serengeti-Mara ecosystem', *Journal of Biogeography*, 8: 391–407.

Smith, B.D. (1995) *The Emergence of Agriculture*. New York: Scientific American Library.

Smith, J.N.M., Cook, T.L., Rothstein, S.I., Robinson, S.K., and Sealy, S.G. (2000) *Ecology and Management of Cowbirds and Their Hosts*. Austin: University of Texas Press.

Smith, R.K., Jennings, N.V., Robinson, A., and Harris, S. (2004) 'Conservation of European hares *Lepus europeaus* in Britain: Is increasing habitat heterogeneity in farmland the answer?' *Journal of Applied Ecology*, 41: 1092–102.

Staller, J., Tykot, R., and Benz, B. (2006) *Histories of Maize—Multidisciplinary Appraches to the Prehistory, Linguistics, Biogeography, Domestication, and Evolution of Maize*. London: Academic Press.

Stripen, C., and DeWeerdt, S. (2002) 'Old science, new science', *Conservation in Practice*, 3: 20–27.

Stoate, C. (2002) 'Multifunctional use of a natural resource on farmland: Wild pheasant (*Phasianus colchicus*) management and the conservation of farmland passerines', *Biodiversity and Conservation*, 11: 561–73.

Stoate, C., Moreby, S.J., and Szczur, J. (1998) 'Breeding ecology of farmland Yellowhammers *Emberiza citrinella*', *Bird Study*, 45: 109–21.

Stoate, C. (2004) 'Links with West Africa: People, policy and migratory birds', *British Wildlife*, 15: 258–64.

Stoate, C., Araújo, M., and Borralho, R. (2003) 'Conservation of European farmland birds: abundance and species diversity', *Ornis Hungarica*, 12–13: 33–40.

Stoate, C., Báldi, A., Boatman, N.D., Herzon, I., van Doorn, A., de Snoo, G., Rakosy, L., and Ramwell, C. (2009) 'Ecological impacts of early 21st century agricultural change in Europe', *Journal of Environmental Management*, 91: 22–46.

Stoate, C., Boatman, N.D., Borralho, R.J., Rio Carvalho, C., de Snoo, G.R., and Eden, P. (2001c) 'Ecological impacts of arable intensification in Europe', *Journal of Environmental Management*, 63: 337–65.

Stoate, C., Borralho, R.J., and Araújo, M. (2000) 'Factors affecting corn bunting *Miliaria calandra* abundance in a Portuguese agricultural landscape', *Agriculture, Ecosystems and Environment*, 77: 219–26.

Stoate, C., and Jarju, A.K. (2008) 'A participatory investigation into multifunctional benefits of indigenous trees in West African savanna farmland', *International Journal of Agricultural Sustainability*, 6: 122–32.

Stoate, C., Morris, R.M., and Wilson, J.D. (2001a) 'Cultural ecology of Whitethroat (*Sylvia communis*) habitat management by farmers: Trees and shrubs in Senegambia in winter', *Journal of Environmental Management*, 62: 343–56.

Stoate, C., Morris, R.M., and Wilson, J.D. (2001b) 'Cultural ecology of Whitethroat (*Sylvia communis*) habitat management by farmers: Field boundary vegetation in lowland England', *Journal of Environmental Management*, 62: 329–41.

Swan, G., and 16 authors. (2006) 'Removing the threat of dicofenac to critically endangered Asian vultures', *PLoS*

Biology, 4(3): e66. doi:10.1371/journal.pbio.0040066 www.plosbiology.org.

Tews, J., Brose, V., Grimm, V., Tielborger, K., Wichmann, M.C., Schwager, M., and Jeltsch, F. (2004) 'Animal species diversity driven by habitat heterogeneity/ diversity: The importance of keystone structures', *Journal of Biogeography*, 31: 79–92.

UNEP. (2002) *People, Land Management and Environmental Change (PLEC)—Final Evaluation*. Nairobi: United Nations Environmental Programme.

Vaughan, D.A., Kadowaki, K, Kaga, A., and Tomooka, N. (2005) 'On the phylogeny and biogeography of the Genus Oryza', *Breeding Science*, 55: 113–22.

Whittaker, R.J., and Heegaard, E. (2003) 'What is the observed relationship between species richness and productivity? Comment', *Ecology*, 84: 3384–90.

Willi, J.C., Mountford, J.O., and Sparks, T.H. (2005) 'The modification of ancient woodland ground flora at arable edges', *Biodiversity and Conservation*, 14: 3215–33.

The Biogeography of Built Environments

Claire Freeman

19.1 WHAT IS THE BUILT ENVIRONMENT?

The general perception conjured up by the phrase 'the built environment' is precisely that: an environment of buildings. It is an image supported not only by popular perception but by informed sources. CUBE—The Centre for Understanding the Built Environment—which describes itself as "one of Europe's most exciting architecture and design centres, dedicated to broadcasting the ideas and issues that lie behind the buildings, spaces and environments that make up the built environment" states that the built environment

> refers to all buildings, and the spaces between them such as streets and squares as well as civil engineering works such as roads and railways. Most of us spend about 90% of our lives surrounded by this environment, so its quality can have a significant effect on our lives. (http://www.cube.org.uk/index.asp)

While much of the urban environment may indeed be 'built' and all of it will be influenced by human activities, the biological environment is ever present and in many urban environments may in fact be larger physically than the built area. To see urban environments purely with reference to their 'builtness' is to overlook the reality of naturalness and greenness that is preeminent in urban environments across the world and ignores the presence of the biophysical systems on which the 'urban' environment depends.

This interdependency was recognized by Bridgman et al.'s (1995) study of the urban biophysical environment of Australian cities:

> The physical dimensions of urban areas occupy air space, land, rivers and estuaries, with impacts that greatly modify natural vegetation, atmospheric make up and water quality. Processes of urbanisation have strong impacts on the elements of the atmosphere, the geosphere, the hydrosphere and the biosphere. (Bridgman et al., 1995: 1)

The urban environment also has its own distinctive biogeography and associated biota. This biota reflects the combination of built and natural structures present in urban areas. The built, with the associated human and disturbance factors, have resulted in the creation of a distinct typology of urban habitats and mix of species. In some respects these habitats and the development processes they follow can be seen as analogues of similar habitats present in rural and less disturbed areas. However, in other respects these habitats and their dependent species represent new and emerging habitat mixes accompanied by species characterized by behaviors and interactions that differ from those seen in nonurban environments.

In this chapter then, I explore the biogeography of urban areas; in particular, I focus on interactions between the built and the natural environment. I reveal the enormous levels of biodiversity present in the built environment and the potential the built environment can offer both to alienate and embrace nature. I will also look at the ways that the biodiversity potential of urban areas can be realized and existing biodiversity supported in the often difficult battle against encroaching hard surfaces and buildings. The built environment offers a complex, challenging, changing, highly variable, and accessible environment for biogeographical investigation. The paradox is that to date few biogeographers have focused their attention on this exciting field of study.

19.2 WHY THE BUILT ENVIRONMENT MATTERS

The world is becoming increasingly urban and in 2007 for the first time the majority of the world's population lived in urban areas. The growth has been especially rapid in the latter half of the twentieth century from 29.7% in 1950 to 37.9% (1975) and 47% in 2000, and it is expected to rise to 60.3% in 2030 (World Global Trends, 2009). The urban transformation is most evident in the developing world. It is a process that mirrors, albeit one that occurred perhaps at a slower pace, that which has already occurred in the developed countries. The impact of urbanization is extensive; even countries that appear to be predominantly rural, such as Australia and New Zealand, in fact have 85% of their population living in urban areas. The movement toward urbanization is one that World Global Trends (2009) predicts is likely to continue.

The importance of this urbanization trend for biogeographers is twofold. First, the physical growth of urban areas has resulted in the transformation of huge tracts of land from rural to 'built'. For example, in England between 1990 and 1998 approximately 65,000 hectares of land became classed as urban (DEFRA, 2000). In many developing countries, far more rapid and substantial transformations are occurring. Second, for increasingly large sectors of the population their experience of the 'natural' world will be an urban one. The critical question then for those planners, architects, engineers, housing managers, park managers, and other professionals responsible for managing the urban environment is 'how can the urban environment best be developed and managed to support the natural environment?' There is an important role for biogeographers in providing the research, science, and understanding required to naturalize the urban environment; in many cases there is a substantial ecological basis from which to work. Contrary to the popular view that urban areas are not deficient in terms of their biodiversity, the situation in fact can be quite the reverse.

The physical character of the urban environment is made of a mix of built and biological elements. In the context of this chapter, the terms 'natural area', 'natural space', and 'natural habitat' refer to land that contains predominantly biological elements, namely vegetation and wildlife. While wildlife such as pigeons may use building ledges as nesting sites, a ledge would not be seen as a 'natural' habitat as it is not composed of predominantly biological elements. The makeup of natural spaces in urban areas is highly varied as indicated in Table 19.1. They contain a huge range of greenspaces, all with their own characteristic species, habitats, and interactions with their human populations. If the breakdown of built and nonbuilt land is examined, it becomes clear that much of what is generally perceived as built is in fact host to a range of habitats of varying degrees of naturalness. Pauleit and Duhme (2000) estimated that in Munich, Germany, actual built land covered 16% of the city, 19% was asphalt or pavement, 6% bare soil (including gravel of railway lines), and 59% was vegetation. The breakdown into built and vegetated is provided for Sheffield, United Kingdom, and offers some surprising results (Table 19.2). For large parts of the nineteenth and twentieth century it was one of Britain's most heavily industrial towns containing important steelworks supplied by the neighboring coal pits. In recent years, with the decline of its industrial base, the city has been undergoing something of a green transformation. The data in Table 19.2 show that approximately 55% of the land is greenspace. However, the real proportion of the city that is greenspace or vegetated is substantially higher as land within 'built' categories will incorporate vegetated areas. For example, schools include grassed playing fields; railways, motorways, and roads include vegetated embankments and verges; airports include swathes of grassland beside runways and many hospitals, offices, and industrial sites include grass lawns, flowerbeds, and trees. The most interesting statistic, however, is that related to housing. Again this is usually classed as 'built' but in fact 23% of the total housing land area (32.52%) contains domestic gardens that are predominantly vegetated. From the data in Table 19.2 it can be estimated that only around 30% of the built areas will be predominantly built structures. Sheffield, being a highly vegetated city, is not atypical for a number of cities in Europe, North America, and Australasia.

In the following section I examine how the natural components of city structure have been derived. While the focus is primarily on the development

Table 19.1 A classification of urban greenspaces (from A.R. Beer, 2000) prepared for COSTC11 Research group, European Cooperation in the field of Scientific and Technical Research (http://www.map21ltd.com/COSTC11/gloss.htm#comp)

	Formally designated Open Spaces			Other Actual Greenspaces							
Woods	Paved City Spaces with Plants	Parks, Gardens, and Sports Grounds	Burial Places	Private, Open Spaces	Domestic Gardens	Allotments	Derelict Land and Garbage Dumps	Farmland and Horticulture	Transport corridor verges	Water margins	Water
Ornamental woodlands	Courtyards and patios	Public parks and gardens	Crematorium	Educational institution grounds	House gardens	Allotments	Derellict land	Arable	Road	Riversides	Still water
Timber/biofuel woodland	Roof gardens and balconies	Public sports grounds	Burial grounds	Residental home grounds	Communal semipublic gardens	Allotments with summer hut	Wasteland	Pasture	Cycleway	Lakesides	Running water
Wild-wood	Tree-lined allees	Public recreation areas	Churchyard	Health services grounds	Communal private garden	Disused allotments	Refuse dumps	Orchard	Footpath verges/embankments	Canalsides	Balancing 'lakes'
Semi-natural woodland	Promenades	Public playgrounds		Private sports grounds				Vineyard	Rail embankments		Wetland
	City squares			Private estate grounds				Moorland			
	School yards			Local authority services grounds							
				Industry, warehousing, and commerce grounds							

Table 19.2 Land-use types in Sheffield, United Kingdom (http://www.map21ltd.com/COSTC11/arb-sheff.htm#city)

Land-use type	% City land
Housing 7,407 ha	
Urban Housing	2.59
Terraced	2.64
High-density (local authority)	5.23
Total land-urban housing within the city	
Suburban housing with private gardens	15.41
High density	5.58
Medium density	2.66
Low density	3.64
Very low density	27.29
Total land-suburban housing within city	
Other uses	4.78
Industry and commerce	1.31
Shopping centers and offices	1.31
Hospitals and schools and universities	0.27
Transport interchanges and airports	0.15
Sports centers	0.32
Disused hospital and college sites	2.78
Reservoirs and rivers	7.6
Woods	3.93
Roads, motorways, and railways	45.02
Parks, greenspaces, and farms	

Green structure

55% of Sheffield city is greenspace (excluding private
  gardens).
23% of the built-up area are private domestic gardens
  (84% of the housing land is suburban in style).
  Sheffield has 175,000 private domestic gardens.

of the city in a western context, and then primarily a British perspective, the process is one that is not confined to Britain.

19.3 DEVELOPING INTEREST IN THE NATURAL CITY

There has been an upsurge in interest in recent years in the natural city, largely through the growth of sustainable cities, green cities, and eco-city movements but also through the growth of interest in urban wildlife. While these movements are relatively new, dating initially from the 1980s and taking off in the 1990s, the history of green cities and natural spaces is much longer, with much of its early history being associated with Europe, particularly England, and with the United States.

19.3.1 The early years; eighteenth-and nineteenth-century developments

The great parks movement of the eighteenth and nineteenth centuries saw the creation of planned greenspaces combining formal landscapes and natural, if contrived, landscapes. One of the key exponents of this movement were landscape architects who worked in England. The best known were Capability Brown (1716–1783) and Humphrey Repton (1752–1818), with their focus on the grand country-house landscapes favored by the aristocracy. In the urban areas the focus was on grand master plans inclusive of both grand landscape and building designs. Two of the most celebrated were Baron Haussman's re-creation of Paris under the aegis of Napoleon III and John Nash's master plan for the Regent's-Marylebone area of London. The focus was on greenspace as a supporting backcloth for the strength, grandeur, and power of government and monarchy. However, within this backcloth the beginnings of the natural landscape movement in cities can be seen. The major park elements of the grand urban designs, the Bois de Boulogne in Paris and Regents Park, Hyde Park, Green Park, and St James in London, were impressive in size and construction having lakes, grand avenues, and walkways, even race courses such as the Longchamps in the Bois de Boulogne. There were also places large enough to encompass plantings that mimicked countryside features such as woodlands, meadows, and streams. The idea of planning cities with reference to natural elements was one that had its early realization in the work of one of the most enduringly influential of all urban landscape architects—Frederick Law Olmsted (1822–1903)—in the United States. His physical legacy can be seen in Boston's 'Emerald Necklace'; Riverside, Illinois; the parks networks in Buffalo and Seattle; the Capitol grounds in Washington, DC; and his best-known work, Central Park in New York. Olmsted strongly believed that nature uplifts people and is essential to their well-being; these ideas are echoed in the sustainable cities literature of today.

19.3.2 Take off: The twentieth century

These planned 'natural landscapes' were complemented and indeed dependent on the existence of a range of other natural and less planned spaces. These spaces form a network that dissects the urban fabric and includes streams, rivers, canals, estuaries, 'left over' spaces such as remnant fields, commons (many now themselves parks such as Hampstead Heath in London), cemeteries, cliff faces, clay pits, and other open spaces, many of which are still present in urban areas today. It was and still is these 'unplanned' spaces that form the essential framework in which nature in the urban area has developed and which have for the most part been completely disregarded by biogeographers and ecologists. It was only around the middle of the twentieth century that any attention began to be paid to the natural wealth and well-being of cities through organizations such as the London Natural History Society (Bevan, 1992). In 1942, for example, the society was approached by the then Ministry of Town and Country Planning with a request for a list of potential nature reserves in the London area to be included in Abercrombie's Greater London Plan of 1945. The plan addressed the vexed question of what level of interaction there could be between people and wildlife before wildlife values are themselves compromised and whether such interaction should be controlled.

It is this interaction between people and wildlife and the recognition of the value of urban nature that has been at the forefront of the urban wildlife movement. In 1981 the inauguration of the London Wildlife Trust ushered in a new era for urban nature different to that which had been dominated until then by naturalists concerned about protecting rare and usually remnant habitats. In part the support for a new approach came from two parallel developments. The first was a growing realisation that wildlife in the countryside was itself endangered (see, for example, Shoard, 1980; UK Wildlife and Countryside Act of 1981) and second, there was a growing recognition of 'unofficial nature' present in urban areas (Mabey, 1973; Teagle, 1978). Since the 1980s

the urban wildlife movement has continued to gather strength and become a more legitimate and indeed a preferred focus of concern for many conservationists, ecologists, planners, urban designers, managers, and residents. Though the focus here has been on urban nature as it developed in the United Kingdom, similar patterns of occurrence of natural habitats can be found in cities elsewhere.

This movement toward a more popular engagement with nature continued in the 1990s and this decade also saw the take-off of the natural cities movement, in particular, the sustainable cities idea. Central to both was the notion of cities as urban ecosystems in which the natural, human, and physical are interrelated and interdependent as indicated in the simplified model in Figure 19.1. The urban ecosystem approach is one that has important repercussions for the biogeography of the built environment in that it recognizes the interrelation between natural and human processes. The sustainable cities movement similarly provides support and opportunity for biogeographers using their geographical skills—both human and physical—to contribute to the understanding and creation of better, more sustainable environments.

19.4 ROLES OF PROFESSIONALS AND ACADEMICS

Urban ecology as a natural science is a young discipline. For a long time, it was thought that urban areas were not worth studying with regard to ecology. Cities were seen as antinature.

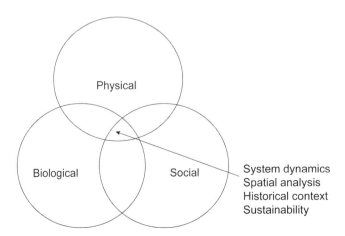

Figure 19.1   A simplified model for understanding urban ecosystems (Nilon et al., 2003)

The importance of figureheads such as Bill Oddie (a British comedian who has become a major TV personality and whose nature programs have helped to inspire a new generation of birdwatchers) is not to be underestimated. The British Broadcasting Corporation's very successful TV program 'The Wild Side of Town', fronted by Chris Baines, brought urban wildlife to a wide British audience in the 1980s. A number of influential books (summarized in Table 19.3) began to be published, starting with McHarg's classic 'Design with Nature' in 1969. Although these books have different emphases, they all celebrated urban nature in a way that had not been done before. McHarg's was the first significant attempt to promote the design of cities and towns with regard to the natural environment, a groundbreaking idea in 1969. Later books by Spirn, Hough, Gilbert, and Sukopp et al. (see Table 19.3) challenged the focus on formal urban landscapes by celebrating the informal, natural diversity of less managed landscapes. Laurie (1979) and Platt et al. (1994) brought together growing numbers of researchers and practitioners interested in urban nature. Douglas's book is especially important because he provides one of the earliest examples of taking a 'biogeographic' approach to an urban environment. He looks not only at land use and vegetation cover but also identifies what he calls 'biogeographic conditions' in urban areas.

The interest in the relationship between the urban environment and its biodiversity and natural character has been largely generated by popular interest, a small group of urban naturalists, and the work of a few academics from the biological sciences. The cause of urban biodiversity owes relatively little in its early years to the work of scientific researchers and academics, with the noted exceptions. Its development appears to be one in which biogeographers and indeed academics and professionals in general have played a quite modest role. They, like those in the wider conservation movement, have preferred to focus attention on the rare, the spectacular, and the purely 'natural' and indigenous characteristics not typically found in urban areas.

An analysis of the mainstream biological and ecological journals is indicative of this lack of interest in urban environments. Collins et al. (1988) calculated that "a mere 0.4 percent—25 of 6157—of the papers published in nine leading ecological journals in the past five years dealt with cities or urban species" (Collins et al., 1988: 416). This relative disinterest has continued. A search of papers published in some of the key ecology/biology journals using the words 'urban', 'city', 'town', and 'built' between 1970 and 2000 is revealing (Table 19.4). *Biological Conservation* had only six such published papers prior to 1990; *Journal of Biogeography* and *Journal of Ecology* even less—three—and the *Journal of Applied Ecology* marginally better at seven papers. *Landscape and Urban Planning* is a different case as in 1986 it incorporated *Urban Ecology* (at that time the key outlet for urban ecology papers) and in total has 111 papers. Even then, most were published in recent years. The search used the words 'nature' and 'ecology'.

Table 19.3 Influential books on urban nature

Author/s	Date of first publication	Title
McHarg, I.L.	1969	Design with Nature
Laurie, I.C. (Ed)	1979	Nature in Cities: The Natural Environment in the Design and Development of Urban Green Space
Douglas, I	1982	The Urban Environment
Hough, M	1984	City Form and Natural Process
Spirn, A.W.	1984	The Granite Garden: Urban Nature and Human Design
Baines, C.	1986	The Wild Side of Town
Gilbert, O.L.	1989	The Ecology of Urban Habitats
Sukopp, H., Hejný, S., and Kowarik I. (eds.)	1990	Urban Ecology: Plants and Plant Communities in Urban Environments
Platt, R.H., Rowntree. R.A, and Pamela, C. (eds.)	1994	The Ecological City: Preserving and Restoring Urban Biodiversity

Table 19.4 Urban-focused biology/ecology papers for selected journals, 1970–2007

Journal	Papers, 1970s	Papers, 1980s	Papers, 1990s	Papers, 2000–2007	Total
Biological Conservation	3	3	22	49	77
Journal of Biogeography	3		10	5	18
Journal of Ecology	3			3	6
Journal of Applied Ecology	1	4	2	1	8
Landscape and Urban Planning		3	36	72	111

*Note that the journals have different search mechanisms so the search results are not directly comparable and are indicative rather than absolute.

**In 1986 *Urban Ecology* was incorporated into *Landscape and Urban Planning*, papers from before 1986 were excluded.

Table 19.5 Urban-focused biology/ecology journals

Journal	Publishers/publishing organisation	Date
Urban Ecology	Elsevier	1976–1986
Urban Ecosystems	Springer	2001
Urban Habitats	Center for Urban Restoration Ecology (CURE), a collaboration between Brooklyn Botanic Garden and Rutgers University	2003
Urban Forestry and Urban Greening	Elsevier	2003

An exactly comparable word search for the other journals was not possible given their different search mechanisms. While these figures are obviously only a snapshot, they clearly indicate the general lack of focus until recently on urban nature by major scientific journals in relevant fields and in particular the very low levels of interest prior to the later 1990s.

Fortunately, there appears to be an upward trend and urban-based papers are becoming far more prevalent. This trend is best seen in *Biological Conservation*, the *Journal of Biogeography*, and *Landscape and Urban Planning*. In particular the interest is demonstrated by the journal publications in Table 19.5, which indicate that the production of urban-based research papers in recent years has been sufficient to encourage the development of a small number of journals dedicated to publishing urban biology/ecology research. As well as showing that there certainly is a growing body of applied research especially since the mid-1990s, the journal search also reveals that there are strong geographical trends relating to where biogeography-related research is being carried out. Europe and North America are the strongest geographical areas producing urban ecological research, in particular Britain, Germany, and Scandinavia. However, there are increasing numbers of articles from Asia, particularly China,

Japan, and Australia but only occasional articles from South America, the Middle East, the Pacific Region, Central Asia, and the Indian subcontinent. It appears that as a general rule in the regions of the world experiencing the most rapid urban growth there is the least focus and in regions of relative urban stability, namely Europe and North America, there is a greater focus.

Thus far, I have traced the developing interest by the research community on urban biogeography. This background is important as it provides a definition of the physical urban environment and the theoretical and research backdrops on which today's urban areas are developing, being interpreted, and are managed. The second part of the chapter explores the concept of the natural city as currently understood and outlines its character. It also identifies some of the key influences on habitats and the species associated with contemporary built environments.

19.5 THE NATURAL CITY TODAY

Cities are invariably green, and few are devoid of natural interest but rarely is the wealth of biodiversity recognised and supported in any substantive way. Urban areas are characterized by very different biotopes and ecological interactions between people, plants, and animals, and they function differently in 'rural' or 'natural' habitats. The successful species in urban habitats have made particular adaptations to survive in the urban environment, while others are unable to adapt to the ecological constraints that human interventions have been placed on them. This section identifies the characteristics that create habitats that maintain or encourage species and that make species successful in cities. Initially, three key indicators of urban biodiversity are examined: diversity, adaptability, and scale.

19.5.1 Diversity

In biological terms the urban environment often has levels of biodiversity in excess of those found in more traditional natural environments. Relatively small areas can combine a wide range of habitats supporting a similarly wide range of species. To illustrate, Camley Street Nature Park, situated just a few minutes' walk from the Kings Cross railway station in London, has been transformed from a derelict site used as an unofficial rubbish ground into a thriving wildlife haven. Within its 0.8 ha it contains ponds and wetlands, grassland/meadows, woodland, scrub, and bare earth habitats. Camley Street demonstrates the high habitat diversity characteristic of many urban sites. The urban environment contains a mosaic of different habitat types, the scale of which can be exceedingly small, as, for example, a wetland habitat in a small garden pond. The diversity of species and number of individuals within a species can be impressive. Scott (2004), states that "in and around Belfast we can expect to find over 200 species of bird, 21 different mammals, 20 species of butterfly, 11 types of dragonfly . . . nearly 600 plant species". More prosaically,

Matthews (2001) asserts that "even in Manhattan you are never more than three feet from a spider". However, for the most part these high levels of species and habitat diversity are often overlooked and undervalued as Brockie states:

> Watching TV or reading the papers these days, we could be excused for thinking the Earth is inhabited by plants and animals on the verge of extinction, and that cities are biological deserts. The media so focus on the endangered, the rare and the remote that the successful everyday plants and animals on our doorsteps have become almost invisible (1997: 6).

As I will demonstrate, the built environment is one that is very much alive with biological diversity but how does it compare with that found in less artificial environments. Given and Meurk (2000) compared the number of plant species found in national parks and in the cities in New Zealand. The results reveal that city environments are important for indigenous species and can show, as in Christchurch and Dunedin (Table 19.6), indigenous species counts per hectare in excess of those found in national parks. There is no simple urban = exotic, national parks = indigenous distribution where plants are concerned. City-dwelling indigenous species may, however, be located outside their 'normal' distributional range, possibly making urban areas important in the conservation of indigenous species: a point that needs further research.

The reality concerning species occurrences and distributions is more complex and exciting, especially for those interested in urban biogeography, than can be indicated by simple species counts. While increased diversity is often a goal there are some habitats that will exhibit low species diversity such as rye grass dominated sports fields. However, there are also some types of forests and grasslands that will be dominated by a single plant

Table 19.6 Minimum estimates for vascular plant (vp) biodiversity in selected New Zealand cities and national parks (adapted from Given and Meurk, 2000)

	Auckland	Christchurch	Dunedin	Arthur's Park National Park	Mt. Cook National Park	New Zealand
Indigenous vp. spp.	559	350	470	660	437	2500
Adventive vp. spp.	615	>500	211	154	137	2500
Area (ha x 10³ha)	265.2	40	37.5	94.5	70	35700
Indigenous vp spp per 1,000 ha	2.11	8.75	12.5	6.98	6.24	0.07
Indigenous vp spp as % of vp spp in entire country	22.4	14.0	18.8	26.4	17.5	100

species where that species and its attendant habitat, though relatively low in diversity, have very high ecological value unlike the sports field example. Particularly, where species have highly specialized requirements the focus will need to be on providing for such needs, for example, clearance of encroaching woody species onto grasslands where, though scrub and woodland would increase diversity, it could be to the detriment of the grasslands. Diversity is therefore but one indicator of interest to those seeking to explain natural distribution and process in the urban environment; another key indicator is that of adaptability.

19.5.2 Adaptability

Some species have the ability to adapt more readily to some urban environments than others. Examples of good adapters include the red fox (*Vulpes vulpes*) found throughout cities in Britain, the brown rat (*Rattus norvegicus*), raccoons (*Procyon lotor*) in North America, seagulls that have adapted to life scavenging off waste dumps, and birds that have become international urban dwellers such as the Indian mynah (*Acridotheres tristis*), the house sparrow (*Passer domesticus*), the starling (*Sturnus vulgaris*), and the feral pigeon (*Columba livia*). The starling has made one of the most spectacular of all adaptations with their habit of coming in to the city in massive flocks as dusk settles, to roost for the night, taking advantage of warmer urban microclimates. Scott (2004) traces the starling's adaptation to city life in Belfast, Northern Ireland. He notes that in 1844 naturalists thought it worthy to comment that a flock of starlings frequented one of the docks and that they were breeding in the city. By the

mid-1960s around 500,000 were roosting in the city, about 14,500 of them on the ledges of City Hall. Since then their numbers have fallen but currently around 80,000 still roost under Queen Elizabeth Bridge and 20,000 under Albert Bridge.

Some adaptations have occurred because the built environment mimics the natural environment, thus feral pigeons whose preferred nesting site is on cliff ledges find in the built environment wonderful artificial cliffs and ledges. The adaptability of species can clearly be seen also in the process of natural colonization that takes place on derelict or undeveloped land. Here it is possible to see an analogy with the process of old-field succession in the countryside. Gilbert (1992) charted the colonization process occurring on bare ground on a site in Sheffield. After eight months, 41 species had established; after 20 months, a further 13 species had arrived and only one had disappeared. Typically, after four to six years such sites are dominated by tall perennial herbs, after 8 to 10 years there are flower meadows with scattered clumps of tall herbs, and by 40 years closed woodland can be found. Bradshaw (1999) identified seven steps in the process of succession in urban areas (Table 19.7).

Where the natural colonisation process occurs, the result is a habitat adjusted to the particular environmental characteristics of that site and one that also needs no external intervention or management. It is a situation far different from that of, say, a flower bed, which needs repeated external inputs to maintain itself and whose survival is dependent on the continuation of human management. In effect such management is designed to retard the natural process of plant succession and development. Costs from Rotterdam for 2002 show significant differences in the costs required

Table 19.7 The essential steps in the process of natural succession in urban areas (after Bradshaw, 1999)

Ecosystem attribute	Processes involved
1. Colonization by species	Immigration of plant species. Establishment of species adapted to local conditions.
2. Growth and accumulation of resources	Surface stabilization and accumulation of fine mineral material and nutrients.
3. Development of the physical environment	Accumulation of organic matter. Immigration of flora and fauna causing changes in soil structure and function.
4. Development of the recycling process	Development of soil microflora and microfauna, possible difficulties in built-up areas.
5. Occurrence of the replacement process	Negative interactions between species through competition for resources and by allelopathy.
6. Full development of the ecosystem	Further growth. New immigration, including aliens or neophytes.
7. Arrested succession	Effect of external factors. Reduction of development.
8. Final diversification	City as a mosaic of environments, high biodiversity.

to maintain different habitats, ranging from 0.5 to 150 hours per 100m² for naturalized plantings and flower beds respectively (Hough, 2004).

However, while some species have indeed made spectacularly successful adaptations, not all have and intervention will be required to support less robust species and their dependent habitats to survive. Habitats such as meadows and their attendant species can be especially vulnerable in the face of encroachment by the built environment, competition from aggressive species, excessive people, animal pressure, and processes such as grassland to woodland succession in the absence of grazing animals. The expansion of the urban environment can act to hasten the decline of species that require large areas of natural space, or have specific needs, require freedom from disturbance, or cannot coexist with urban environments. For example, rabbits (*Cuniculus* spp.) can adapt to living in urban environments whereas their larger relative the hare (*Lepus* spp.), which needs large open spaces, has not adapted to urban environments. Most large mammals would fall in to this category as would more sensitive plants and bird species. Dedicated protection and management will be needed to ensure the survival of more vulnerable habitats and their dependent species in the face of urban growth.

19.5.3 Small-scale, fragmented

Urban habitats are characteristically small in area and have extremely high levels of fragmentation. Examples of small habitats include canals, railways, and motorway corridors which are characterized by long, narrow, and discontinuous habitats, planted traffic islands, and garden ponds. Successful species are those that can adapt to small, fragmented patches and their associated edge effects. A number of factors are associated with small urban habitats, the most significant of which include high levels of human and natural disturbance. Human-induced disturbances include noise, various forms of pollution, changes in hydrology such as channelling runoff into stormwater drains, weeding, and applications of weed killer. Disturbances that verge on 'the natural' include invasion by and competition from exotic plant and animal species, predation, and disturbance by domestic animals, mainly cats and dogs and the introduction of diseases. The small scale of habitats can also be directly related to human activities, notably development.

The most diverse of small-scale habitats are domestic gardens. An average suburban garden can incorporate features that could be labelled a wetland (garden ponds), woodland and shrubland (trees and understory shrubs, and associated ground layers), hedgerow (boundary hedges), cliffs (garden walls), pasture (lawn), as well as flower beds. While many if not most garden species and habitats will not be rare, gardens are nonetheless important for the survival of both indigenous and exotic plants, birds, reptiles, and invertebrates. One long-term study undertaken in a Leicester garden sympathetically managed for wildlife recorded the presence over a 15-year period of 1,757 species of animal and 422 plant species (Owen, 2005). Statistics like these indicate the tremendous potential that urban areas have in supporting biodiversity. Substantive information on the flora and fauna of urban areas and particularly on the biogeographic factors influencing distribution patterns within urban areas, however, is generally poor. Research that has been undertaken tends to be on parks, nature areas, and reserves rather than on smaller habitats and habitat fragments whose role and significance in the larger urban ecosystem remain largely unknown. One exception to this was the UK's National Environmental Research Council—funded URGENT programme 'Urban domestic gardens and creative conservation'. This program explored the nature of urban ecology in gardens in U.K. cities, focusing on Sheffield. Results from the study have been widely published and are an exemplar of the potential for focused ecological studies in urban areas (Gaston et al., 2005).

Any study of the urban environment thus demands a different approach and set of criteria to that used for study of more 'natural' systems and one that recognises the specifically urban character of the biodiversity, its species, and habitats. The built environment has developed its own characteristic ecological morphology, character, and ecological survival mechanisms. Its ecology has a mix of both exotic and indigenous species and exhibits good examples of adaptation and robustness in the face of what can be quite stressful biogeographical conditions. Attitudes to urban biogeography are highly variable and naturalists, ecologists, biologists, zoologists, and others with an interest in the natural world are only now devoting attention to understanding urban biodiversity. There are, however, a number of critical areas of debate that arise in the process of seeking this understanding and on deciding how best to support urban biodiversity.

19.6 SUPPORTING URBAN BIODIVERSITY

Left to themselves, urban areas will develop their own intrinsic biodiversity and provided the degree of hard surface development is controlled, high levels of biodiversity will be achieved without any need for additional human intervention. To take two examples, a garden for a new house and the land in the center of a traffic circle, if both are left unplanted and undisturbed they will in a fairly short space of time become vegetated following the process outlined in Table 19.7. Places where this natural urban succession can be seen are on uncleared bombsites in Europe, old quarries and pits, and old undeveloped industrial sites. However, this process of natural colonization and succession is rare, not because it is deficient but because the process and results are invariably at odds with human perceptions of what urban landscapes should look like and how they should develop. Thus, gardens and traffic circles will be planted with 'desirable' plants and resources invested in maintaining their aesthetically attractive form. Urban biogeographic character is determined by the perceived purpose of any land area, the values placed on that land and its biota, and as a consequence the resources allocated to its survival. Any assessment of the biogeographic character of an urban area is also implicitly a study of the values predominating in the human population. Plants reflect their cultural context as reflected, for example, in the different landscaping and species planting associated with parks and gardens in different parts of the world. For example, in African cities such as Accra (Asomani-Boateng, 2002) the focus in residential areas is on planting vegetable gardens whereas in European cities it is more usual to plant flowers, lawns, and ornamental shrubs and trees.

19.6.1 Conflicting values

The huge range of vegetated spaces found in urban areas, each with their own specific biota and interactions with people, can be seen in Table 19.1. The primary division is between formal and informal natural areas/features or between what Hough (1995) terms the 'pedigree' and 'non pedigree' landscapes. Formal landscapes are generally those that are given high values in human terms (e.g., cemeteries, parks, tree-lined boulevards, and flower beds, all of which require high inputs in terms of resources and maintenance for their survival in their planned form). Informal landscapes are given low values and are consistently overlooked in any urban planning and management programs (e.g., estuaries, railway embankments, canal sides, disused allotments, quarries, and railyards). It is only recently that the spontaneous vegetation found on these sites has been recognized as having existing and potential biodiversity value and that these values should be included in ecological surveys of urban biota. Such recognition though, is not universal even amongst ecologists, whose penchant is still for relict habitats encaptured by urban growth. There is little popular recognition of informal places, which are often seen as messy, neglected, and potential sources of danger and infestation. Authors such as Breuste et al. (1998), Breuste, 2004), Gilbert (1989), and Hough (1995) have devoted much attention to challenging this notion and to revealing the wealth, contribution, and distinctiveness of biodiversity present in these undervalued landscapes.

One of the key debates regarding urban nature is the aliens/exotics-indigenous/natives debate. In some countries, such as Australia and New Zealand, where the impacts of exotics on vulnerable and often endemic flora and fauna have been so manifest this debate is particularly vociferous. The pro-indigenous lobby has been especially influential in urban areas. The strength of support for indigenous biota can be seen in policies to promote native-only plantings in landscaping schemes, and in the proliferation of native regeneration and planting schemes in public parks. Kings Park in Perth, Australia, for example, has undertaken extensive native bush regeneration schemes while Pinaroo Valley Memorial Park Cemetery in Perth uses its native bush setting and the presence of kangaroo grazing to promote itself as an ecologically appropriate burial place (Plate 19.1). There are a number of generally accepted benefits of native priority planting and support policies: (a) natives grow better and are hardier and disease-resistant; (b) natives support more associated species; (c) native genetic diversity needs to be protected; and (d) native plants are regionally appropriate and define landscape character.

However, as Kendle and Rose (2000) argue, none of these arguments are irrefutable and in fact there are equally persuasive arguments for the integration and contribution of nonnatives. The primary concern in urban areas should not be natives or nonnatives but how the various combinations can be used to support the ecological, aesthetic, and natural processes important to both the human and nonhuman populations in the built environment.

19.6.2 Understanding urban habitats and biota

As has already been indicated, understanding of urban habitats and their biogeographical relations is limited compared to that of many more natural systems. In part this is due to the aforementioned lack of interest but it is also due to a lack of

Plate 19.1   The promotion of Pinaroo Valley Memorial Park Cemetery, Perth, as an ecologically appropriate burial place based on its location in native bush and the presence of kangaroo grazing

methodologies for studying urban habitats and species. Urban areas also tend to be more complex, and less easy to classify, lacking the relatively clear typologies and vegetation classifications used to map and evaluate the ecological character of rural areas. Urban areas have suffered from a lack of ecological data and where data have been available it has tended to relate to relict habitats and to designated natural areas. A number of attempts have been made to redress this situation. In England, the cities of Birmingham, Leicester, and London have been at the forefront in creating urban ecological databases, with Leicester Ecology Trust releasing its influential methodology for mapping the ecology of urban habitats. In Germany, through the activities of a small group of highly influential German researchers working in the cities such as Mainz (Frey, 1998), Berlin (Seidling, 1998), and Leipzig (Breuste and Wohlebber, 1998), the diversity and significance of urban wildlife have become clear (Sukopp and Weiler, 1988). In their study of Dusseldorf, for example, Wittig and Schreiber (1983) identified 124 valuable biotopes. Their work has assisted in the process of bringing urban nature to academic notice and giving substance to the study of urban wildlife. In the United States, as part of the National Science Foundation's Long Term Ecological Research Network, comprehensive studies of various aspects of urban ecology have been conducted in Phoenix and Baltimore (the diversity of themes can be seen from these examples from the Baltimore study: riparian 'hotspots' (Groffman et al., 2003); urban-rural gradients (McDonnell and Pickett, 1990); nutrient fluxes (Law et al., 2004); and the development of new research frameworks for ecologists and biogeographers in urban areas (Pickett et al., 2008).

Freeman and Buck (2003), working in Dunedin, New Zealand (Table 19.8), developed an urban land-use and habitat typology for an environment with a complex mix of introduced species and indigenous species. They mapped the entire urban environment, including gardens (spaces usually omitted from standard urban open space mapping studies). In order to understand, protect, and enhance urban habitats, it is imperative that there are databases that identify and quantify the species in habitats and important factors in the urban biogeography such as soil, climate, and hydrology (Clarkson et al., 2007). These data are available for few urban areas. For example, while a number of New Zealand cities do have ecological data relating to designated natural open spaces, no city other than Dunedin has data that fully cover all land in the urban area. Given the importance of gardens, sports pitches, informal natural areas, street trees, and riverside and estuarine

Table 19.8 Habitat types and relative quantities in the city of Dunedin, New Zealand (Freeman and Buck, 2003)

Habitat type	Area (ha)	% Total land area of city
Built	3277.5	54.48
Pasture	840.5	13.97
Bush and forest	690.3	11.48
Exotic lawn	373.0	6.20
Plantation	195.7	3.25
Exotic scrub	180.3	3.00
Tree group	73.0	1.21
Park/woodland	68.3	1.14
Mixed scrub	45.9	0.76
Bare ground	40.8	0.68
Exotic shrubland	40.3	0.67
Rough grassland	37	0.62
Coastal sand and gravel	31.8	0.53
Native scrub	31.0	0.52
Coastal water	30.1	0.50
Dune grassland	14.1	0.23
Native shrubland	11.5	0.19
Mixed shrubland	9.4	0.16
Sand dune	9.3	0.15
Cliff/rock outcrop	5.7	0.10
Standing water	4.3	0.07
Vineland	2.2	0.04
Wetland	2.6	0.04
TOTAL AREA	6015.537	100.00

environments to supporting urban biota and wildlife this means that any understanding of the urban biodiversity will be at best partial and incomplete. A comprehensive ecological database is vital for effective understanding and planning for urban biodiversity. Such databases can be resource-heavy and labor-intensive to produce. The development of new remote sensing, aerial photograph-based methodologies using combinations of object-oriented GIS methodologies, spatial indicators, and advanced classification techniques is already proving to be a substantial step forward in this regard (Mathieu et al., 2007, *Journal for Nature Conservation*, 13: 2005; Myeong et al., 2003).

19.6.3 Managing urban species and habitats

The city can both alienate and support nature. These processes are dependent not only on the presence and type of vegetated space available but also on the management and use regime for that particular space. Thus, for example, a cemetery can contain short, cropped grass and tarmac walkways. Alternatively, if a more flexible management regime is adopted it can support mature trees with regenerating understory, shrubby areas, ponds, unmown grassy meadow areas with wildflowers, and soft surface walkways with mown grass with tarmac walkways confined to high-use areas. The management of vegetated areas is critical to their receptivity to nature. Management is, however, just one of several key factors supporting the presence or absence of urban habitats, others include the following:

1. Urban design; the integration of diverse habitats in the urban environment, creation of functional habitat networks, and maximization of opportunities for creating natural places.
2. Commitment to the conservation, protection, and enhancement for natural areas, formal and informal especially when faced with development proposals.
3. Resources such as practical resources needed for managing natural areas but also support from relevant agencies, community groups, and local government.
4. A policy and legislative base to support the protection and enhancement of natural areas.
5. An information base, which is a combined and coordinated database on habitats, species, and the availability of environmental records and information to relevant agencies, communities, and individuals.

Managing for urban biodiversity also involves managing people. The growth of interest in urban nature over the last 20 years and the practical development of community-based wildlife projects have been enormous (Decker et al., 2005). If the community is on board then a tremendous amount can be achieved. A prime example of this is the example of Belfast's Bog Meadows (Plate 19.2). Not only was the last remnant of the meadows saved from destruction by concerted action by the local community but the meadows have also brought much-needed positive recognition to a community that has experienced decades of strife and tragedy during the 'troubles' in Ulster. Many local nature reserves and natural areas owe their preservation and development to local action by the community. The importance of nature to urban dwellers can be seen in the massive increase in designated natural sites and in the number and vibrancy of urban wildlife groups and agencies.

It can also be seen in the exponential growth in the market for wild bird food, bird feeders, and nesting boxes. Bird feeding has become so widespread in the United Kingdom, for example, that it is a serious factor in supporting urban bird

Plate 19.2   Bog Meadows local nature reserve, Belfast

populations and in encouraging the survival and growth of some bird species that were hitherto struggling. Any explanation of bird numbers and distribution in the United Kingdom would have to incorporate human factors into its analysis. Bird feeding, however, is a matter of controversy, with some naturalists seeing it as undesirable interference because it upsets the natural balance between birds and the natural ecosystem (Sterba, 2002).

Another critical factor to be taken into account in explaining urban biodiversity patterns is the growing support for native plantings by landscapers, horticulturalists, and particularly by individual homeowners. Many councils have released native planting guides, and many gardening books and garden nurseries encourage native plantings as well as the more usual exotics. The availability of native plant species and the widespread use of 'wildflower' seed mixes may lead to interesting changes in the patterns of plant distribution.

While the desirability of artificially encouraging particular species will no doubt continue to be debated, the rising interest in urban wildlife and its support at both the popular level and from environmental agencies and local government is unlikely to abate in the near future. In a number of cities local councils have allied themselves closely to the promotion of green city ideals, including support for urban biodiversity. In the Great Britain an example is Leicester, the first 'Environment City' in the United Kingdom. In New Zealand, Waitakere City in Auckland has vigorously promoted itself as a green city through its green network program. Beijing, China, has in development a network of green wedges, parks, and green corridors and has as its long-term goal the development of an 'eco-city status' (Li et al., 2005). Currently Beijing claims to have around 45% of its urban area covered in trees and grass. In 2002, 51 greenspaces and 10 square kilometers of green belt were developed and over 2,667 ha were planted with trees, such that by the end of 2002 there were 7,907 ha of public greenspace (Beijing International, 2005). While the catalyst for this green growth may well have been the hosting of the 2008 Olympics, it does reveal the increasing significance of greenspace in urban planning. The built environment is home to people, flora, and fauna. They are indivisibly linked, they impact on each other, and in urban areas they adjust to each other. The relationship between people and biological diversity in built environments is at the heart of any urban biogeographic study.

19.7 CONCLUSION

Collins et al. (1998), in their paper on urban ecology in *American Scientist*, stated:

> Whatever the reasons for ecologists' persistent tendency to focus on pristine environments, human dominated ecosystems represent a problem for the field. We lack a method of modeling ecosystems that effectively incorporates human activity and behaviour. And the processes and dynamics within cities largely elude an understanding based on traditional ecological theories. (p. 416)

While their frustration was directed at ecologists, geographers and biogeographers also continue to preferentially focus on natural, primarily rural environments. They still lack methods and models for studying urban biodiversity and the patterns of distribution of urban habitats and species. In this sense scientists and academics are lagging behind popular consensus, which has embraced the urban wildife cause and which is characterized by a quite open attitude to the character of urban biodiversity. The sharp divides and value systems that academics and researchers place on the biological world, the divides into good = indigenous species and bad = exotic species, are not necessarily shared by people living in the urban areas and by planners, educationalists, landscapers, and others whose work shapes the urban environment. Urban environments are composed of built and 'natural' environments. Naturalness will always be a human construct in urban areas; it will never be a copy of the pure natural habitat and neither will the behaviour of an urban species be the same as that of its cousin in the pure natural habitat. And therein lies the challenge for biogeographers: embracing this difference, studying it, seeking to explain and understand it, and assisting in devising ways for the natural to survive and indeed thrive in these ever-increasing and harsh urban environments.

REFERENCES

Asomani-Boateng, R. (2002) 'Urban cultivation in Accra: An examination of the nature, practices, problems, potentials and urban planning implications', *Habitat International*, 26: 591–607.

Baines, C. (1986) *The Wild Side of Town*. London: BBC and Elm Tree Books.

Beijing International. (2005) Beijing Foreign Information Centre, http://www.ebeijing.gov.cn/About%20Beijing/ Cons&Modern/Article/t20030922_1007.htm. Accessed December 9, 2005.

Beer, A.R. (2000) A classification of urban greenspaces prepared for COSTC11 Research group, European Cooperation in the field of Scientific and Technical Research (http://www.map21ltd.com/COSTC11/gloss.htm#comp).

Bevan, D. (1992) 'Conserving nature in London: Presidential address delivered at the annual general meeting of the London Natural History Society', *London Naturalist*, 72: 9–14.

Bradshaw, A.D. (1999) 'Natural ecosystems in cities: A model for cities as ecosystems,' in Berkowitz, A.R. Nilon, C.H. and Hollweg., K.S. (eds.), *Understanding Urban Ecosystems*. New York: Springer. pp. 77–94.

Breuste, J., Feldman, H., and Uhlman, O. (eds.), (1998) *Urban Ecology*. Berlin: Springer.

Breuste, J., and Wohleber, S. (1998) 'Goals and measures of nature conservation and landscape protection in urban cultural landscapes of central Europe—examples from Leipzig', in Breuste, J., Feldman, H., and Uhlman, O. (eds.), *Urban Ecology*. Berlin: Springer.

Bridgman, H., Warner, R., and Dodson, J. (1995) *Urban Biophysical Environments*. Melbourne: Oxford University Press.

Brockie, B. (1997) *City Nature*. Auckland: Viking.

Breuste, J.H. (2004) 'Decision making, planning and design for the conservation of indigenous vegetation within urban development', *Landscape and Urban Planning*, 68: 439–52.

Centre for Understanding of the Built Environment, CUBE, http://www.cbe.org.uk. Accessed 1 May 2011.

Clarkson, B., Wehi, P., and Brabyn, L. (2007) 'A spatial analysis of indigenous land cover patterns and implications for ecological restoration in urban centres, New Zealand', *Urban Ecosystems*, 10: 441–57.

James P. Collins, J.P., Kinzig, A., Grimm, N.B., Fagan, W.F., Hope, D., Wu, J., and Borer, E.T. (1988) 'A New Urban Ecology: Modeling human communities as integral parts of ecosystems poses special problems for the development and testing of ecological theory', *American Scientist*, 88: 416–425.

Decker, D.J., Raik. D.B., Carpenter. L.H., Organ, J.F., and Schusler, T.M. (2005) 'Collaboration for community-based wildlife management', *Urban Ecosystems*, 8: 227–36.

DEFRA (Department for Environment, Food and Rural Areas). (2000) *Accounting for Nature Assessing Habitats in the UK Countryside*. London: DEFRA.

Freeman, C., and Buck, O. (2003) 'Development of an ecological mapping methodology for urban areas in New Zealand', *Landscape and Urban Planning*, 63(3): 161–73.

Frey, J. (1998) 'Comprehensive biotope mapping in the city of Mainz—a tool for integrated nature conservation and sustainable urban planning', in Breuste, J., Feldman, H., and Uhlman, O. (eds.), *Urban Ecology*. Berlin: Springer.

Gaston, K.J., Warren, P.H., Thompson, K., and Smith, R.M. (2005) 'Urban domestic gardens (IV): The extent of the resource and its associated features', *Biodiversity and Conservation*, 14: 3327–49.

Gilbert, O. (1992) *The Flowering of the Cities: The Natural Flora of the Urban Commons*. Peterborough, UK: English Nature.

Gilbert, O.L. (1989) *The Ecology of Urban Habitats*. London: Chapman and Hall.

Given, D., and Meurk, C. (2000) 'Biodiversity of the urban environment: The importance of indigenous species and the role urban environments can play in their preservation', in Stewart, G.H., and Ignatieva, M.E., *Urban Biodiversity and Ecology as a Basis for Holistic Planning and Design, Proceedings of a Workshop held at Lincoln University*. Christchurch: Wickliffe Press. pp. 22–33.

Groffman, P.M., Bain, D.J., Band, L.E., Belt, T.K., Brush, G.S., Grove, J.M., Pouyat, R.V., Yesilonis, I.C., and Zipperer, W.C. (2003) 'Down by the riverside: Urban riparian ecology', *Frontiers in Ecology and Environment*, 1(6): 315–21.

Hough, M. (1984) *City Form and Natural Process*. London: Croom Helm.

Hough, M. (2004) *Cities and Natural Processes: A basis for sustainability*. New York: Routledge

Kendle, A.D., and Rose, J.E. (2000) 'The aliens have landed! What are the justifications for 'native' only policies in landscape plantings', *Landscape and Urban Planning*, 47: 19–31.

Laurie, I.C. (ed.). (1979) *Nature in Cities: The Natural Environment in the Design and Development of Urban Green Space*. New York: John Wiley.

Law, N.L., Band, L.E., and Grove, J.M. (2004) 'Nutrient inputs from residential lawn care practices in suburban watersheds in Baltimore County, MD', *Journal of Environmental Planning and Management*, 47: 737–75.

Li, F., Wang, R., Paulussen, J., and Liu, X. (2005) 'Comprehensive concept planning of urban greening based on ecological principles: A case study in Beijing, China', *Landscape and Urban Planning*, 72: 325–36.

Mabey, R. (1973) *The Unofficial Countryside*. London: Collins.

Mathieu, R., Freeman, C., and Aryal, J. (2007) 'Mapping private gardens in urban areas using object-oriented techniques and very high resolution satellite imagery', *Landscape and Urban Planning*, 81: 179–92.

Matthews, A. (2001) *Wild Nights: Nature Returns to the City*. New York: North Point Press.

McDonnell, S.T., and Pickett, S.T.A. (1990) 'Ecosystem structure and function along rural-urban gradients: A unexploited opportunities for ecology', *Ecology*, 71: 1232–37.

McHarg, I. (1969) *Design with Nature*. New York: Natural History Press.

Myeong, S., Nowak, D.J., and Hopkins, P.F. (2003) 'Urban cover mapping using digital, high spatial resolution aerial imagery', *Urban Ecosystems*, 5: 243–56.

Nilon, C.H., Berkowitz, A.R., and Hollweg. K.S. (2003) 'Introduction: Ecosystem understanding, is a key to understanding cities', in Berkowitz, A.R., Nilon, C.H., and Hollweg., K.S. (eds.), *Understanding Urban Ecosystems*. New York: Springer. pp. 1–13.

Owen, J. (2005) *The Ecology of a Garden: The First Fifteen Years*. Cambridge: Cambridge University Press.

Pauleit, S., and Duhme, F. (2000) 'Assessing the environmental performance of land cover types for urban planning', *Landscape and Urban Planning*, 52: 1–20.

Pickett, S.T.W., and 16 authors. (2008) 'Beyond urban legends: An emerging framework of urban ecology, as illustrated by the Baltimore ecosystem study', *BioScience*, 58(2): 141–52.

Platt, R.H., Rowntree. R.A., and Pamela, C. (eds.). (1994) *The Ecological City: Preserving and Restoring Urban Biodiversity*. Amherst MA: University of Massachusetts Press.

Scott, R. (2004) *Wild Belfast: On Safari in the City*. Belfast: Blackstaff Press.

Seidling, W. (1998) 'Derived vegetation map of Berlin', in Breuste, J. Feldman, H., and Uhlman, O. (eds.), *Urban Ecology*. Berlin: Springer.

Sheffield City Council. (2005) 'Sheffield', http://www.map21ltd.com/COSTC11/arb-sheff.htm#city.

Shoard, M. (1980) *The Theft of the Countryside*. London: Temple Smith.

Spirn, A.W. (1984) *The Granite Garden: Urban Nature and Human Design*. New York: Basic Books.

Sterba, J.P. (2002) 'American backyard feeders may do harm to wild birds', *Wall Street Journal Online*, December 27, 2002.

Sukopp, H., Henjy, S., and Kowarik, I. (eds.). (1990) *Urban Ecology: Plants and Plant Communities in Urban Environments*. The Hague: Academic Publishing.

Sukopp, H., and Weiler, S. (1988) 'Biotope mapping and nature conservation strategies in urban areas of the Federal Republic of Germany', *Landscape and Urban Planning*, 15: 39–58.

Teagle, W.G. (1978) *The Endless Village: The Wildlife of Birmingham, Dudley, Sandwell, Walsall and Wolverhampton*. Shrewsbury. UK: Nature Conservancy Council.

Wittig, R., and Schreiber, K.F. (1983) 'A quick method for assessing the importance of open space in towns for urban nature conservation', *Biological Conservation*, 26: 57–64.

World Global Trends. (2009) *Key Findings Urbanization*, http://t21.ca/urban/index.htm. Accessed April 9, 2009.

Utilizing Mapping and Modelling

In Chapter 1, we noted that David Watts (1978) did not foresee the application of remote sensing in his review of the emerging subdiscipline of biogeography. This is not that surprising, because at that time remote sensing was limited to aerial photography and early Landsat 1 images, and GIScience as we define it today did not exist. Yet four decades later we are able to devote an entire section of this handbook to this topic and can point to weightier treatments in terms of research compilations (e.g., Millington et al., 2003) and textbooks (e.g., Wadsworth and Treweek, 1999).

The growth in remote sensing applications might have been predicted as vegetation and land-use mapping was already using aerial photography by the early 1970s (Millington and Alexander, 2000) and one of the two broad applications of the first earth resources satellite (Landsat 1, then called ERTS 1) was in the field of ecology. What has made remote sensing such an important tool in biogeography and ecology is that it has increasingly been able to detect biogeographical entities and ecological phenomena at a variety of spatial resolutions, from approximately 80 m in 1972 to anything between meter- to kilometer scales at the present time; or if we are using airborne platforms centimeter-scale. Moreover, advances in image processing have enabled us to expand these scales by examining subpixel phenomena (where a pixel equates to the base resolutions quoted above) and aggregated pixels. Second, as data acquired from satellite and airborne sensors have been stored we have created archives of valuable documents, not just images

but maps, photographs, images, and databases that show changes in landscapes, ecosystems, and biomes back to the 1930s if we include air photos, and longer if we include old maps and other types of historical data. Such studies have been pursued in regions ranging from New England (Foster, 1992) to the Belgian Ardennes (Petit and Lambin, 2002). What is even more important is that the methods to extract information from these data (and here we refer not only to advances in image processing, but also spatial analytical tools in GIScience, but related areas like geostatistics, pattern recognition, and landscape ecology) has advanced tremendoulsy.

We can revisit the documents and reanalyze then to fit new or better data than we have been able to do previously. As electromagnetic sensor design has evolved, we have explored different parts of the electromagnetic spectrum beyond the early limitations of visible and near infrared, which has offered new information to biogeographers. An important example is the recent development of LiDAR remote sensing, which has enabled detailed vertical vegetation structures to be elucidated, something that was essentially impossible before the application of LiDAR (Popescu and Wynne, 2004). Remote sensing applications are considered by Giles Foody and Andrew Millington (Chapter 20) and by Joanne Nightingale, Stuart Phinn, and Michael Hill (Chapter 21).

As equally important as the advances in sensor and satellite design, and in archiving capability have been those in the field of software development. We have mentioned some of the important roles of image processing above. Another area has been in landscape ecology, not least the development of metrics to measure the extent and character of landscape fragmentation, most notably the FRAGSTATS suite. Yet issues remain for the application of landscape ecology and one area—the spatial aspects—are considered by Thomas Albright, Monica Turner, and Jeffery Cardille (Chapter 22).

Other important developments have come from spatial analysis and modeling in the domain of GIScience. Modeling in particular has been a major effort in biogeography, and in some respects it has paralleled developments in ecology as George Malanson illustrates in the area of simulation modeling in Chapter 24. Many would argue that the most geographical of contributions has been in spatially explicit modeling: a point illustrated concisely by Niall Burnside and Stephen Waite (Chapter 23). However, perhaps the most important role of geographers in this area has been their contribution in biocomplexity, not because it is often spatially explicit but as Stephen Walsh, George Malanson, Daniel Brown, Joseph Messina, and Carlos Mena illustrate in Chapter 25 because it links biological and socioeconomic systems and speaks to geography's core value of human-environment interaction.

The consideration of remote sensing applications in this book is divided falsely—as Giles Foody and Andrew Millington note in Chapter 20—between static applications in mapping biogeographical entities (Chapter 20) and using remotely sensed data as inputs to models (Chapter 21). In terms of mapping Giles Foody and Andrew Millington show how, with advances in satellite and sensor design, image processing, and data archiving, we can spatially locate and map biogeographical phenomena ranging from the spatial distributions of individual tree species to the changing global distributions of key ecosystems and biomes in relation to climate fluctuations and land-use transformations. Joanne Nightingale and her coauthors provide a comprehensive review of ecological models and the ways that remotely sensed data can be input into these models.

Spatial heterogeneity—the uneven, nonrandom distribution of objects—is inherent in virtually all biogeographical distributions. Thomas Albright and his coauthors, while noting its critical importance, argue that the characterization of spatial heterogeneity has been a challenge relegated to the subjective realm and that techniques and theory designed to understand spatial information has generally been underutilized in biogeography. This is partly because techniques and theories have been drawn from landscape ecology, spatial statistics, geostatistics, signal processing, and pattern recognition. The challenge in characterizing and quantifying spatial heterogeneity in biogeographic data emanate for other reasons: data sources vary in type and quality; it can be difficult to decide which of the multitude of metrics and indices to use; and distribution patterns and organism-environment associations are scale-dependent.

To illustrate these challenges they review the tools available to characterize spatial heterogeneity and examine its potential to generate new insights in biogeography. They tackle the themes of spatial reference and measurement scale of biogeographical data by focusing on spatial point patterns, point samples of a continuously varying field, categorical raster and polygon data, and continuous raster and polygon data.

Niall Burnside and Stephen Waite point out in the first sentence of Chapter 23 that modeling the distribution and dispersal of species is long established in biogeography and ecology. Historically, modeling relied on predictive correlations or rule-based systems that linked species occurrences with environmental and ecological factors and was underpinned by the fundamental biogeographical tenet that species distributions are determined by abiotic and biotic factors. GIS coupled with ecological theory and statistical techniques has provided the ability to develop predictive tools to understand the complex processes that determines distribution and abundance (Franklin, 1995). In fact, she further argues that predictive mapping is largely grounded in ecological niche theory and vegetation gradient analysis.

Fundamentally important is the predictive capacity. Concerns over climate- and land-use change, and biodiversity loss, emphasize the need for this type of modeling. Not surprisingly then it is a thriving area in biogeography and ecology and has been applied at a range of scales from individual (often invasive and rare) species, through plant communities to even larger biogeographic entities such as biomes. The effects of land-use change, agricultural intensification, and conservation have also modeled predictivity at the landscape scale. The predicted futures are employed in counteracting or mitigating adverse and negative effects.

George Malanson discusses simulation as a key modeling paradigm in biogeography in Chapter 24. In doing so he contextualizes the links between more empirically and/or process-based, and spatially explicit modeling, cellular automata and agent-based modeling; he contrasts this with the statistically and mathematically driven predictive and theoretical models in biogeography, the former being exemplified by Niall Burnside and Stephen Waite in the preceding chapter.

Malanson makes a key argument for the importance of all types of modeling in biogeography, which is that models simplify systems—systems that are inherently complex and systems that we do not fully understand. Simulations are computational expressions of this conceptual simplification. An interesting question he poses is why use simulation modeling? He answers his own question as follows. First, simulations can be used for prediction if they represent important dynamics of ecological systems in the same way that weather forecasts do. Second, they can be used to test hypotheses by comparing the simulation to reality. Third, in discovery and understanding, in which context simulations can provide us with a better sense of the way in which a system works by examining the consequences of alternative inputs or relations in the system.

Some scientists argue that simulation has developed simultaneously with computers as a new approach in science. Part of this argument lies in the fact that computation is the tool of investigation in complexity theory, which is the theme of the final chapter written by Steve Walsh and his co-authors.

Biocomplexity considers the complex interactions within and among ecological systems, the physical systems underpinning them, and the human systems they interact with. Biocomplexity arises from the fields of complexity theory and complex systems. A complex system is a system in which the many components interact in ways that link patterns and processes across scales. The large numbers of interactions involving feedbacks, and which occur at one or more lower levels within the system, lead to nonlinearities, nonequilibrium conditions, and interactions that maintain its organization of the system through negative feedbacks or alter subsequent alternatives in state space through positive feedbacks. These lead to emergent multiscale phenomena.

Biocomplexity draws on theories and practices from across the social, natural, and spatial sciences, and it emphasizes complexity sciences in the study of people, places, and environment, especially in the paradigm of coupled natural-human systems. Many geographers, such as the authors who wrote Chapter 25, consider biocomplexity to be a core area of geographical inquiry. Yet that potential has not been realized and many other disciplines occupy this ground. Steve Walsh and his co-authors highlight its application in studying biodiversity, ecological services, and sustainability; fisheries and game hunting; urban systems; climate change; and vegetation and biological modeling—while drawing many of their arguments from studies about tropical deforestation and land change in the Americas.

REFERENCES

Foster, D.F. (1992) 'Land-use history (1730–1990) and vegetation dynamics in Central New England, USA', *Journal of Ecology*, 80(4): 753–71.

Franklin J. (1995) 'Predictive vegetation mapping: Geographic modeling of biospatial patterns in relation to environmental gradients', *Progress in Physical Geography*, 19(4): 474–99.

Millington, A.C., and Alexander, R.W. (2000) 'Vegetation mapping in the last three decades of the twentieth century', in Alexander, R.W. and Millington, A.C. (eds.), *Vegetation Mapping from Patch to Planet*. Chichester, UK: John Wiley.

Millington, A.C., Walsh, S.J., and Osborne, P.E. (2003) *GIS and Remote Sensing Applications in Biogeography and Ecology*. Boston: Kluwer.

Petit, C.C., and Lambin, E.F. (2002) 'Impact of data integration technique on historical land-use/land-cover change: Comparing historical maps with remote sensing data in the Belgian Ardennes', *Landscape Ecology*, 17(2): 117–32.

Popescu, S.C., and Wynne, R.H. (2004) 'Seeing the trees in the forest: Using LiDAR and multispectral data fusion with local filtering and variable window size for estimating tree height', *Photogrammetric Engineering and Remote Sensing*, 70(5): 589–604.

Wadsworth, R., and Treweek, J. (1999) *GIS for Ecology*. Harlow, UK: Addison Wesley Longman.

Watts, D.R. (1978) 'The new biogeography and its niche in physical geography', *Geography*, 63: 324–37.

Remote Sensing for Mapping Biogeographical Distributions: Actualities and Potentials

Giles M. Foody and Andrew C. Millington

20.1 INTRODUCTION

In this handbook, the applications of remote sensing in biogeography are split into two. In this chapter we deal with 'relatively static' concepts, for example, the use of remote sensing for habitat and species mapping. This resonates with other chapters in this book, most notably those on landscape ecology (Kupfer; Albright et al.) and classification (Schwabe and Kratochwil). In the next chapter, Nightingale et al. consider remotely sensed data as inputs to models. This is somewhat of a false division, for instance, five decades ago Board (1969) considered maps to be models. But here we draw a line between remotely sensed data that are being used to produce vegetation and land-cover maps as an end use, and remotely sensed data being used directly as data inputs for modelling.

Initially we provide a general review of how the use of remote sensing in biogeography has evolved since the 1970s. We then consider some of the main ways in which remote sensing has been used within biogeographical research. For brevity, we focus on three major issues: (a) land-cover mapping and monitoring; (b) the derivation of information on specific habitats and species; and (c) contributions to biodiversity assessment and conservation. Emphasis is placed on some of the current problems and possible avenues for future research. In particular, we stress that methods to increase the level of information extracted from remotely sensed imagery are required in order to fulfill its potential in biogeography. This is an important theme in this chapter, because although remote sensing has often been used 'successfully' in biogeography, problems have been encountered. For instance, land-cover maps derived from remote sensing have sometimes been viewed as insufficiently accurate (Wilkinson, 1996) or are considered too generalized for habitat-based work; or relationships for estimating vegetation properties from remotely sensed data may not be applicable much beyond the sites at which they were derived (Foody et al., 2003).

20.2 REMOTE SENSING IN BIOGEOGRAPHY: A TEMPORAL PERSPECTIVE

Since the 1970s, remote sensing has increasingly become a major information source for many environmental disciplines at a range of spatial and temporal scales. These include biogeography and ecology, though the bias appears to be towards the latter. An ISI Web of Science key word search (May 2009) had 427 records for 'remote sensing' and 'ecology', compared to only 33 for 'remote sensing' and 'biogeography'. However, as we will show in this chapter, much of what we consider biogeography in this book is well served by remote sensing. Remote sensing is much more important in biogeography than the above metrics suggest. This observation can be reinforced by the argument that some aspects of spatial ecology (i.e., habitat fragmentation in landscape ecology) and global perspectives on phenological changes in vegetation could not have evolved to their current levels of sophistication without parallel developments in remote sensing. The uses of remotely sensed data in biogeography and ecology are extensive (Muchoney, 2008; Gillespie et al., 2008; Vierling et al., 2008, Newton et al., 2009 and references therein) and are increasing. What is driving this increase? Is it advances in remote sensing technologies, new questions derived from biogeography and ecology, or just an increase in the number of researchers with access to remotely sensed data, processing systems, and the desire to publish?

Newton et al. (2009) in a review of 158 remote sensing-based papers published in 2004–8 in the journal *Landscape Ecology* show that 46 and 42% relied on Landsat TM and MSS data, respectively. Only 1% used IKONOS data, with half that number using Quickbird data. Low-spatial resolution data were poorly represented (e.g., AVHRR only accounted for 4% of studies), and data from space agencies outside Europe and North America were also poorly represented (e.g., IRS-LISS, 1%). Microwave data were only used in 0.5% of studies. This is hardly strong support for new remote sensing technologies driving the increase in applications. They also stress that many of the applications answered already existing questions in landscape ecology. In summary, this suggests that it is the availability of data and processing software, and the need to publish that is leading to increased applications. If this argument holds true in other areas of biogeography and ecology, many opportunities exist for this and future generations of biogeographers to grasp the potential of new sensors and look at applying remotely sensed data in novel ways to answer fundamental biogeographical and ecological questions.

As far back as the early days of satellite remote sensing in the 1970s, imagery acquired by remote sensors was used in biogeographical and ecological studies. Often, it has provided thematic information such as land cover (e.g., Gould, 2000; Kerr and Ostrovsky, 2003; Parmenter et al., 2003) or estimates of the biophysical and biochemical properties of vegetation (e.g., Running et al., 1986; Wessman et al., 1988; Steininger, 2000; Ollinger et al., 2002; Cohen and Goward, 2004; Clark et al., 2004a). Remotely sensed data have also been used to provide information on major abiotic variables, such as fire and wind, which impact biogeographic processes and patterns (e.g., Bowman and Franklin, 2005). The use of remote

sensing is evident in all major biomes. Considerable attention has focused on tropical forests (e.g., Achard et al., 2002; Hansen et al., 2008; Lambin, 1999a, 1999b Sader et al., 1990), savannas (e.g., Jepson, 2005), and grasslands and deserts (e.g., Baldi and Paruelo, 2008; Sobrini and Raissouni, 2001; Tucker et al., 1994; Tucker and Nicholson, 1999; Xu et al., 2008). With tropical forests, for example, there has been great interest in activities such as the monitoring of biome-level change (e.g., Achard et al., 2002), drought impacts (e.g., Asner et al., 2003; Siegert et al., 2001), fragmentation (e.g., Millington et al., 2003; Jha et al., 2005; Skole and Tucker, 1993), and biomass (e.g., Foody et al., 2003; Dong et al., 2003; Drake et al., 2003; Houghton et al., 2001; Santos et al., 2003; Steininger, 2000).

20.3 MAPPING AND MONITORING LAND COVER

Remote sensing has been widely used as a source of information on land cover and land-cover dynamics: in fact, it is the main data source (e.g., Lambin and Geist, 2006; Millington and Jepson, 2008 and references therein). Significant funding and effort have gone into the development of global land-cover data sets over the last two decades. Examples are the global land-cover data sets at spatial resolutions ranging from one to eight kilometers such as those based on AVHRR data developed by IGBP (Loveland et al., 2000), the University of Maryland Global Land Cover Facility (e.g., Hansen et al., 2000; Tucker et al., 2005; DeFries and Townshend, 1994), MODIS-Terra data (Friedl et al., 2002), and a long-term AVHRR data set (Pedlety et al., 2007). There have also been important continental-scale data sets, primarily based on SPOT-4 data, developed under the European Union's Global Land-Cover (GLC) 2000 Project for South America (Eva et al., 2004), Africa (Mayaux et al., 2004), and Southeast Asia (Stibig et al., 2004). There are two major objectives of these continental and global-scale efforts:

1. classify land cover into broad categories to identify contemporary hotspots of land-cover changes (e.g., deforestation hotspots—Achard et al., 2002; Etter et al., 2006; Ingram and Dawson, 2005); and
2. derive biophysical parameters, such as leaf area index (e.g., Asner et al., 2003), fraction of photosynthetically active radiation (FPAR) (e.g., Yang et al., 2006), and biomass (e.g., Hese et al., 2005).

The number of land-cover data sets at these scales pale into insignificance compared to those produced for place-based studies, which by now must rank in the thousands. Place-based land-cover research covers all biomes ranging from the high Arctic tundra (e.g., Stow et al., 2004) to low-latitude tropical forests (e.g., Hansen et al., 2008), though some regions, for example, hyperarid, continental interiors and southern hemisphere dry, seasonal woodlands have been much less frequently studied than many other biomes (e.g., Hof, 2006; Lepers et al., 2005). If just the smallest 'coherent' tropical region—southern Mexico and Central America—is considered, Redo et al. (2009) identified 33 articles on land-cover change focusing solely on forest environments. The volume of research in this area has increased exponentially since the 1990s, stimulated in large part by global conservation and environmental concerns, and the IGBP land-use land-cover change (LUCC) project, which ran from the mid-1990s to the mid-2000s (Lambin and Geist, 2006).

Many reasons may be proffered for the widespread popularity of the use of remote sensing for land-cover and land-cover change mapping, but two in particular are pertinent. First, the response measured by sensors mounted on aircraft and satellites is a function of the surface type (regardless of the wavelength) and, therefore, is a function of the land-cover properties. For example, in parts of the reflected visible and near-infrared wavelengths the measured response relates to leaf cell structure and foliar moisture. At microwave wavelengths response can be related to canopy architecture of trees, grasses, and other vegetation and landscape elements. Consequently, the resulting remotely sensed image is a manifestation of land-cover properties and so it should be possible to accurately extract land-cover information and relevant biophysical parameters from image data. Secondly, remotely sensing offers many practical advantages as a means of data acquisition over field survey in particular (though field measurements are needed to calibrate and validate most remotely sensed data). In particular, remote sensing offers a highly consistent, digital, and maplike view of the land surface at a wide (and increasing) range of spatial and temporal scales. Remote sensing, therefore, is an attractive source of data on land cover at scales ranging from the very local to the global, which may, if required, be updated frequently.

Remote sensing offers biogeographers a potentially powerful source of information on land cover and land-cover dynamics. This is especially useful as many biogeographical concerns relate to spatial patterns of land-cover classes and their temporal variations. These can be direct relationships ranging from seasons (phenological studies) to longer periods (e.g., habitat transformation and succession) or indirect, for example, the influence of changing spatial patterns of land cover on biogeochemical cycling. Knowledge of how land cover is mapped from a particular sensor, the limitations of a sensor for land-cover mapping, and what a land-cover class may include is fundamental to understanding the many biogeographical processes that may be inferred from land-cover maps. Moreover, information on land cover is required for practical applications of biogeography. For example, habitat transformation due to land-cover change is the greatest contemporary threat to biodiversity (e.g, Sala et al., 2000). Knowledge of the location and rates of habitat transformation, and the 'style of fragmentation' that different human actions generate (Imbernon and Branthomme, 2001) through land-cover mapping (as well as its underlying human drivers) can go a long way in substantially enhancing biodiversity conservation and promoting sustainable development. Though to be successful here, biogeographers and remote sensors need to team up with social scientists.

Land-cover mapping is one of the most common applications of remote sensing. Early studies using large-scale aerial photography exploited visually identifiable tonal and textural image characteristics in order to map land-cover classes (Sayn-Wittgenstein, 1961). Although visual interpretation of aerial photographs can yield accurate maps, the inherent subjectivity of the method combined with the difficulty of applying it to large multispectral datasets has resulted in increasing reliance on digital image classification techniques for the derivation of thematic maps (Mather, 2004). Frequently, for example, each case (e.g., an image pixel, a tree, a field, or other selected spatial unit) is allocated to the class to which it has the highest likelihood of membership. In the minds of many biogeographers, land-cover mapping equals land-cover classification, using some kind of unsupervised or supervised classification. Although classification procedures can be an appropriate means for accurate mapping, many significant problems have been encountered with the use of such classifications: we discuss these below.

We illustrate the extremely broad range of studies that have used remote sensing to provide land-cover information for biogeographical research with just three major areas. Two examples consider mapping land cover at each end of the continuum of spatial scale that ranges from the small cartographic (regional-global) scale to the large cartographic (local) scale. The third is monitoring land-cover change.

20.3.1 Small cartographic (regional-global) scale

Remote sensing has unrivalled potential as a source of land-cover maps at regional to global scales mainly because it provides an extremely economical basis for gathering consistent and repeatable observation of continents to the entire globe. Indeed nowadays it is hard to imagine any viable alternative to remote sensing as a source of many kinds of vegetation-focused biogeographical maps over continents, except for plant and animal species distributions (see Mutke, this volume).

The potential of remote sensing for mapping very large regions was noted early in its history. In the 1980s maps showing the land cover of continental areas were derived. These were first produced for Africa (Tucker et al., 1985) and subsequently for other continents. The methods for deriving these maps are based on the daily to monthly variations in syntheses of vegetation responses—vegetation indices—initially over a single year but, as more data became available, over runs of years. These temporal sequences are termed 'remote sensing phenologies', a term first coined by Justice et al. (1985). Tucker et al.'s pioneering work was soon applied to estimate woody biomass (fuelwood) stocks for energy planning for Africa (Millington and Townshend, 1989; Millington et al., 1994)—the first application of small cartographic scale mapping using these data at the biogeography-society interface. Other applications, most notably for rangelands, appeared soon after (e.g., Kennedy, 1989). Maps produced using the remote sensing phenology method typically showed broad classes at a relatively coarse spatial resolution that were hybrids of land cover and land use based on existing maps of biomes and ecosystems (e.g., White, 1983) and a priori knowledge of land use. Research on large-area mapping has developed enormously since the mid-1980s. A decade later, maps of regional-global land cover were commonly being created using remotely sensed data (e.g., DeFries et al., 1998) and major international programmes have focused on providing global land-cover maps.

The production of large-area land-cover maps is far from trivial and numerous challenges remain to be addressed (Rindfuss et al., 2004). Moreover, there are problems in map comparison for land-cover change detection, particularly if the legends differ or differ in meaning. There is now some level of agreement on this, at least in the land-cover community (Millington, 2005). However, land-cover mapping at small cartographic scales has advanced to the position that maps are now available on a frequent basis. For example,

researchers at Boston University use MODIS data to update a global land-cover map with a 1 km spatial resolution every 96 days. However, an oft-heard criticism of such maps from many biogeographers is that they are too generalized. Paradoxically, the global climate-change community has (until recently) found them to be too detailed! This paradox may be on the way to being solved as, in recent years, there have been two parallel trends in global land-cover maps: first, an increase in legend detail, and second, a decrease in the spatial resolution (pixel size). However, a fundamental issue remains. It is that remote sensors have often produced these maps for a wide range of users rather than specifically for biogeographers (or any other group of scientists), and end users (we include many biogeographers here) do not understand the limitations of satellite-sensor systems and image-processing chains.

Another major problem in mapping large areas has been the negative impact of mixed pixels, which can be a major source of error in analyses of remotely sensed data. For example, in early continental-scale land-cover maps of Africa, coastline pixels composed of water, bare mud, sand, and/or coastal mangrove forests were labeled 'mangrove'. Mixed pixels of Lake Chad—which were composed of vegetation, sediment, and water and had similar remote sensing phenologies to mangrove swamps (Millington et al., 1994), causing end-user confusion. To reduce problems such as these, soft classification techniques have been used and continuous field products derived (e.g., fuzzy neural networks—Gopal et al., 1999; spectral unmixing—Cross et al., 1991). Alternatively, as increased computing power and data storage have become availabe and affordable, data acquired by sensors with a finer spatial resolutions than the NOAA-AVHRR and MODIS systems that are widely used for large-area mapping, such as the Landsat TM or ETM+ sensor (spatial resolution ~30m) have been used for small cartographic scale mapping. Nonetheless, to map a very large area this way requires considerable pre-processing and data mosaicking.

20.3.2 Large cartographic (local) scale

Although mapping from remote sensing has its origins in the interpretation of aerial photography and so tended to be undertaken at a relatively local scale, recent developments in satellite remote sensing have seen a proliferation of fine spatial resolution systems that have opened the doors to a vast array of new large-scale mapping applications. The last decade has seen the launch of a series of commercial satellite systems that can provide very fine spatial resolution imagery

(Aplin, 2004). The spatial resolution of these systems is considerably finer than that available from the main civilian systems such as SPOT HRV (with spatial resolution typically in the range of 10–20 m) or Landsat ETM+ (with a spatial resolution typically in the range of ~30 m) that have been used widely in biogeographical applications. For example, the IKONOS system provides imagery with a spatial resolution of 1 m (panchromatic) and 4 m (multispectral) that enables the study of local scale features from satellite orbit (Read et al., 2003). Thus contemporary satellite remote sensing systems allow ecosystems to be characterised over a range of scales, including the very local (Wulder et al., 2004). For example, IKONOS data have been used to map mangrove assemblages (Dahdouh-Guebas et al., 2002), coastal wetland communities (Wei and Chow-Fraser, 2007), and vegetation communities in urban ecology studies (e.g., Grove et al., 2006). Similar vegetation mapping applications have been explored with Quickbird data (e.g., Wang et al., 2004; Wolter et al., 2005; Johansen et al., 2007).

But the provision of fine spatial resolution imagery from space has not just been limited to the production of finer-scale and 'more accurate' maps as listed above. It has allowed researchers to address questions that previously were impractical to study from space or on the ground—one of the questions we asked in the introduction. With IKONOS data it is now possible, for instance, for studies to be undertaken at the scale of individual tree crowns over large areas (Hurtt et al., 2003; Clark et al., 2004a; Palace et al., 2008). For example, fine spatial resolution remote sensing data have been used to quantify tree mortality in a tropical rain forest (Clark et al., 2004b) and so may contribute usefully to contentious debates on the issue (Phillips et al., 2004). Fine spatial resolution imagery acquired by airborne sensors is also widely used in biogeographical research. This may involve the use of inexpensive video systems or large multisensor campaigns. Critically, these approaches allow relatively small areas to be studied in considerable detail. This is beneficial for a wide range of applications, particularly those focused on introduced and invasive species (Everitt et al., 2002; Ramsey et al., 2002; Underwood et al., 2003; Lass et al., 2005; Lawrence et al., 2006; Wu et al., 2006; Pengra et al., 2007; Hestir et al., 2008; Walsh et al., 2008).

20.3.3 Land-cover change

Land-cover change is widely recognised as one of the most important components of environmental change, perhaps having a larger effect on the environment than climate change. Remote sensing is very attractive as a source of information on land-cover changes, especially over large areas. Much research of relevance to biogeography has addressed the monitoring of major biome level of land-cover changes. In this regard, remote sensing has been used to study major changes such as those associated with deforestation, desertification, thicketisation of savannas, and impacts associated with urbanisation and other human activities.

Remote sensing has been used to monitor land-cover change over a range of spatial and temporal scales. For example, in relation to tropical forests, remote sensing has been used to monitor local-scale changes (Read and Lam, 2002) through to regional and global-scale assessments (e.g., Achard et al., 2002). These assessments often differ from those derived by other means (e.g. Achard et al., 2002), highlighting major concerns and often leading to adjustments in estimates of deforestation and other processes (DeFries et al., 2002; Achard et al., 2002).

There are many ways in which remote sensing may be used to monitor land-cover change (Singh, 1989; Mas, 1999; Lambin, 1999). One of the most commonly used approaches is based on post-classification comparison. With this approach, a land-cover map representing one time period is compared directly with that for another time period, enabling both the location and nature of change to be observed. Although this is a commonly used approach to change detection there are numerous concerns. These range from technical issues such as concern over the impacts of spatial misregistration of the two maps, which can lead to substantial misinterpretation of change (Roy, 2000), through the need to distinguish land-cover change from land-cover variability (Bradley and Mustard, 2006) to the constraint imposed by this approach to indicate only land-cover conversion. Considerable recent research has sought to address these concerns. For example, the reduction of misregistration error is important, especially when interested in characterizing edges, which may have considerable impacts on species (Harris and Reed, 2002) and much research has addressed this issue (e.g., Wrigley et al., 2007). The limitation of conventional postclassification comparison analyses to indicate only conversions of land cover is a major problem, leading to misestimation of the extent and severity of change that occurs. Critically, land-cover conversion is only one component of land-cover change. Many land-cover changes involve a change in the quality or condition of the land-cover class and not a complete change in class type. These land-cover modifications are believed to be at least as important as land-cover conversions, yet there is often very little information on them. Modifications such as the degradation of habitats can markedly impact biogeographical variables (Warkentin and Reed, 1999), and so techniques to monitor such change from remotely sensed data would be useful. Possible approaches are to base the analysis on the comparison of soft classifications or to adopt change detection techniques based on image radiometry such as change vector analysis (Lambin, 1999b).

20.4. MAPPING HABITATS AND SPECIES

Although land-cover maps derived from remote sensing are used by a broad range of users with diverse needs, it may sometimes be more appropriate to tailor the mapping process to the specific requirements at hand. This may often be the case when interest is in a small subset of the classes contained within a region, which is often the situation in biogeography.

Traditional land-cover maps are often designed to be multipurpose, thus meeting the needs of a large number of users. However, the classes shown may not be those of interest for a specific application. Within this context, an advantage of remote sensing is that the analyst may specify the classes of interest to be mapped from the imagery. However, conventional approaches to mapping from remotely sensed data (e.g., unsupervised and supervised classification, overreliance on NDVI, remote sensing phenology) may not be the most appropriate means for extracting the information desired from the imagery. Conventional classification approaches to mapping often aim to maximise the overall accuracy of the map (i.e., the accuracy calculated over all classes). However, if interest is on a subset of the classes this may be inappropriate and an alternative approach should be adopted (Lark, 1995; Foody et al., 2005). Many biogeographical studies are often focused on a subset of the classes and directing resources to the other classes is wasteful and inefficient (Boyd et al., 2006).

20.4.1 Mapping habitats

Habitat conservation is embedded in major international and national conservation efforts, including those related to the Convention on Biological Diversity (Griffiths and Vogiatzakis, this volume). Remote sensing has a long history in mapping habitats, dating back to the presatellite, aerial photography era where its niche was an economical means of mapping and monitoring habitats in land-cover mosaics. The European Union's Habitats Directive (1992)—one of the international conservation initiatives—aims to protect biodiversity through conserving natural habitats and wild fauna and flora. A network of protected areas has been developed and a no-net-loss policy adopted. Thus, any loss of a protected habitat must be compensated by restoration or creation of new ones of at least the same surface area and of the same ecological value. Habitat monitoring has become a priority issue and may benefit from remote sensing inputs. Indeed, remote sensing is acknowledged by many to be well suited to monitoring land-cover dynamics for conservation activities at a range of scales (e.g., Amarnath et al., 2003; Gross et al., 2009 and references therein; Kerr and Ostrovsky, 2003; Lengyel et al., 2008 and references therein; Rouget, 2003; Roy and Tomar, 2000).

Assuming that there are relationships between land cover and habitat type, the standard approaches to thematic mapping from remotely sensed imagery are normally used to derive habitat information. There are, however, problems with this type of analysis (Foody, 2002; Wilkinson, 2005), and they have driven research to increase mapping accuracy. Much of this has focused on the potential of new classification approaches. For example, the limitations of conventional approaches such as the maximum likelihood classification have been recognized and alternatives such as artificial neural networks, decision trees, and support vector machines have been promoted (Mather, 2004; Pal and Mather, 2005).

Problems remain with standard classification procedures, specifically the assumption that classes are discrete, mutually exclusive, and exhaustively defined. Soft or fuzzy classifications have been promoted as a means of accommodating problems arising from the presence of continuous classes that result in the failure to satisfy the first two assumptions. This is a particular issue with ecotones, which, by definition, comprise elements from different land-cover types and are often key habitats in terms of conservation and management, hybridization, and evolution (Andrefouet and Roux, 1998; Arnot et al., 2004; Krishnaswamy et al., 2004). Failure to exhaustively define classes can lead to error and misinterpretation (Foody, 2004), making it important that all classes are included in an image classification to ensure that the assumption of an exhaustively defined set is satisfied. If this is the case, there are implications for training the classifier, with the size of the training set required commonly linked to the number of classes and the number of wavebands used. This can place substantial demands on the analysts. For example,

Mather (2004) considered the number of training samples required to be in the order of 10–30 samples per class per waveband. Commonly, however, interest is focused on a subset of the classes— maybe those of high conservation importance— and the resources devoted to the training set requirement should be directed toward this subset.

Additionally, standard probabilistic approaches to classification aim to maximize the overall probability that a case is allocated correctly to a class. This is inappropriate as the analysis is not focused on the class(es) of interest. Boyd et al. (2005) discuss this in relation to the mapping in the English fenlands. To map the fen class with a conventional classifier an analyst has to train all classes, not just the fen class, and the conventional classifier will seek to optimize overall accuracy, to which the fen class is only one contributor. It may be more efficient and appropriate to focus on the single class of interest to provide the targeted information required for conservation actions that fit with the evidence-based approach promoted for much nature conservation (Pullin et al., 2004).

A variety of methods could be used to accurately map a single class. For example, a standard probabilistic classifier could be optimized to minimize error associated with that class (Lark, 1995). This approach, however, typically trades one type of error with another and may require further analyses to produce a final map of the class of interest (Foody et al., 2005). Alternatively, a soft classifier could be used to indicate the spatial distribution of the class. However, a more appropriate approach may be to adopt a nonparametric binary classification analysis that simply seeks to separate the class of interest from all others. Attractive means to achieve this are through the adoption of simple binary classifiers such as a support vector machine (SVM) or decision tree to simply discriminate the single class of interest from all others or adopt one-class classifiers (Sanchez-Hernandez et al., 2007).

20.4.2 Mapping species

Recently, some remote sensers have turned their attention away from the mapping of broad land-cover classes and focused on mapping species. Interest in particular is being directed at mapping of particular species, especially identifying and mapping tree species in forest environments (e.g., Franklin, 2000; Sanchez-Azofelfa et al., 2003; Turner et al., 2003; Goodwin et al., 2005).

In addition, the ability to map tree species accurately would facilitate biogeographical studies ranging from those focused on the testing of hypotheses of biodiversity (for a review of these hypotheses, see Griffin, this volume) through those helping to address problems arising from taxonomic aggregation on environmental modelling (MacKay et al., 2002) to assisting conservation (Landenberger et al., 2003, Wilson et al., 2004) and forest management (Erikson, 2004). This is because the spatial distribution of a tree species within a community influences basic processes such as reproduction, competition, dispersal, and damage or death from pests and pathogens of that species (e.g., Phillips et al., 2004) and the impacts of, and recovery from, disturbances like hurricanes and landslides: it is a very important variable in biogeography. Knowledge of the spatial distribution of tree species over large areas may, therefore, substantially enhance both fundamental biogeographical understanding and aid applications.

However, untangling the causes of spatial distribution patterns of trees is difficult, and various hypotheses have been put forward to explain observed tree distributions. These include the impacts of disturbance events, temporal niche axes, and environmental heterogeneity. Remote sensing may be a useful source of information to test these hypotheses. For example, knowledge of the spatial distribution and dynamics of sycamore (*Acer pseudoplatanus*) will aid understanding of competitive interactions between it and other tree species in the United Kingdom. Since its introduction to the British Isles in the 1500s, it has been displacing the native ash (*Fraxinus excelsior*) in regions where the two species coexist (Waters and Savill, 1992). Since sycamore is a superior competitor to ash, the coexistence of these species over a long period indicates the possible action of a temporal niche axis dictating the competitive dynamic between the species (Kelly and Bowler, 2002, 2005). Testing this hypothesized scenario requires information on the spatial distribution of the tree species in the region of coexistence, which may be derived from an analysis of remotely sensed data tailored to the specific needs of a study (Atkinson et al., 2007; Foody et al., 2005).

Many studies have addressed the mapping of specific species in this decade and they have often highlighted the value of fine spatial resolution imagery (Nagendra, 2001; Brandtberg, 2002; Haara and Haarala, 2002; Brandtberg et al., 2003; Holmgren and Persson, 2003; Carleer and Wolff, 2004; Wang et al., 2004; Clark et al., 2005). In some instances, multitemporal data have proven beneficial in separating species (Key et al., 2001). For this kind of research, the spatial, spectral, and temporal resolutions of the data sets used for classification should be selected with care as they may greatly impact classification accuracy

(Foody et al., 2005; Clark et al., 2005). However, high accuracies have been noted in the literature. For example, Martin et al. (1998) derived a species classification with an overall accuracy of 75% (individual species ranged from 16.6 to 100%) while Clark et al. (2005) achieved an overall accuracy of up to 92% with those of individual species ranging from 70 to 100%. With recent developments in fine spatial resolution sensors, it may be possible to achieve such levels of accuracy for some species from satellite as well as aerial sensor data (Carleer and Wolff, 2004; Wang et al., 2004). As higher accuracies may be obtained from leaf rather than canopy-level classifications (Clark et al., 2005), it seems that there is the potential for further increases in the accuracy of tree species classification from remotely sensed data.

20.5 MAPPING BIODIVERSITY

The shortage of adequate data is one of the most fundamental and widespread problems in biodiversity assessment and conservation (e.g., Griffiths et al., 2000; Wilson et al., 2004). This problem has many components, which include detail, accuracy, and completeness, as well as the spatial and temporal resolution of data sets. The vast majority of information on forest biodiversity, for example, is derived from small < 1 ha plots (Innes and Koch, 1998), yet large plots, of at least 2 hectares in extent, may be required to describe tropical forest biodiversity (Newbery et al., 1992). Moreover, it is difficult to scale ecological knowledge (Loreau et al., 2001) and so use data from small plots to inform landscape-scale activities. Additionally biodiversity is a difficult variable to measure and express. The most widely used index of biodiversity is species richness and, although useful, it reflects just one aspect of biodiversity (Purvis and Hector, 2000). To cite but one example, information on the species composition is required to prioritize sites for conservation (Debinski and Humphrey, 1997).

Remote sensing then may be useful in general biodiversity assessments, especially in providing data at appropriate spatial and temporal scales. For example, the biodiversity intactness index, which is proposed as a general indicator of the overall state of biodiversity to aid monitoring and decision making (Scholes and Biggs, 2005) is based on land-cover data. Remote sensing is an important source of data for its derivation. Generally, remote sensing has considerable potential as a source of information on biodiversity, particularly at landscape- and coarser scales (Nagendra, 2001; Nagendra et al., 2004).

Yet despite its well-established attractions and potential, it has been relatively underutilized historically in studies of biodiversity (Innes and Koch, 1998; Trisurat et al., 2000). Indeed, the common inability to resolve individual organisms has meant that the remote sensing of biodiversity has been seen as a fool's errand by some (Turner et al., 2003). Recently, however, there has been an increase in studies of biodiversity using remote sensing data (e.g., Kerr et al., 2001; Oindo and Skidmore, 2002; Seto et al., 2004), often taking advantage of advances in sensor technology or by focusing on broad patterns in biodiversity (Turner et al., 2003).

The chief attraction of remote sensing as a source of information on biodiversity is that it offers an inexpensive means of deriving complete spatial coverage of environmental information for large areas in a consistent and regularly updatable manner (Muldavin et al., 2001). Information on biodiversity may be extracted from remotely sensed data in a variety of ways (Nagendra, 2001). Typically, studies have focused on assessments of species richness, with limited attention to other aspects such as species abundance and composition, which are difficult to detect (Foody and Cutler, 2003; Schmidtein and Sassin, 2004). Often studies have either related species richness to information about the land-cover mosaic of the test site derived from the imagery (e.g., Imhoff et al., 1997; Nagendra and Gadgil, 1999a, 1999b; Gould, 2000; Griffiths et al., 2000; Oindo and Skidmore, 2002; Kerr et al., 2001) or through a direct relationship with the remotely sensed response, often in the form of a vegetation index such as NDVI (e.g., Oindo and Skidmore, 2002; Seto et al., 2004; Rocchini and Vannini, 2010).

There is scope for further development. Too many studies have focussed on temperate environments, used surrogate variables such as land cover and biomass, and underutilised the spectral information contained in imagery by overreliance on vegetation indices. This is unfortunate for three reasons: (a) only about 10% of species-rich ecoregions are located in temperate biomes (Kier et al., 2005); (b) undertilised spectral regions such as middle and thermal infrared and, to a lesser extent, microwave, may contain important and useful information; and (c) the underlying assumptions of the techniques used may be unsatisfied. New approaches to more fully extract information on biodiversity from remotely sensed imagery are, therefore, necessary if remote sensing is to have a role in this area of biogeography.

Much recent conservation research has focused on the potential of forested lands that are actively used by humans. While forest use is sometimes viewed as being inconsistent with the aim of maintaining biodiversity, it is possible that actively used forests may have a major role to play in biodiversity conservation (Lugo, 1995; Webb and Paralta, 1998; Vasconcelos et al., 2000). Although the present state of knowledge is generally too poor for quantitative prediction of the effects of actions such as logging on biodiversity (Bawa and Seidler, 1998), even relatively severely logged forests may represent a significant resource for biodiversity conservation (Cannon et al., 1998). The active management of forests may be one way of meeting the needs of humans and conservation (Lugo, 1999). Thus, rather than focus attention on just the small proportion of forest that has designated protected status it may be more appropriate to consider the larger area of unprotected forest that may play a useful role in biodiversity conservation as well (Pimentel et al., 1992). In particular, it is becoming increasingly apparent that the landscape mosaic is important in biodiversity conservation. Many have, therefore, suggested that biodiversity conservation activities should be undertaken at the level or scale of the landscape (Nagendra and Gadgil, 1999a; Margules and Pressey, 2000; Potvin et al., 2000; Hannah et al., 2002). This activity may benefit from remote sensing as its synoptic overview provides information on the entire landscape.

The spatial resolution of the imagery used has a major impact on the nature and accuracy of biodiversity assessments by remote sensing. Using fine spatial resolution imagery it has been possible, for example, to accurately identify some tree species (Martin et al., 1998; Haara and Haarala, 2002; Carleer and Wolff, 2004). With a coarsening of the spatial resolution such detailed assessment is untenable, but broad classes of forest may be spectrally separable (Foody and Hill, 1996; Franklin and Wulder, 2002). Moreover, through relationships with land cover it is possible to assess the diversity of species that do not directly affect the remotely sensed response (e.g., birds and butterflies), and assess impacts associated with changes in the habitat mosaic such as fragmentation (e.g., Kerr et al., 2001; Luoto et al., 2002, 2004; Cohen and Goward, 2004). With such indirect approaches to biodiversity assessment, spatial resolution still has an influence as it impacts land-cover classification accuracy. Variation in spatial resolution also influences the estimation of summary indices of biodiversity as well as estimates of composition (Kerr et al., 2001; Oindo et al., 2003). Trends such as these are well known in geography (Openshaw, 1984) but perhaps are not so among conservation biologists and this opens up a role for biogeographers.

Although the scale dependence of the relationships is a concern and indicate that the scale of a study must be selected appropriately, it is apparent that even with relatively coarse spatial resolution imagery it is possible to derive useful information on biodiversity (Kerr et al., 2001; Foody and Cutler, 2003; Foody, 2004; Cohen and Goward, 2004). Some problems of working with large areas, however, have not been addressed. Critically, it is generally assumed that relationships between the biodiversity variable of interest and the remotely sensed response are spatially stationary and hence transferable between sites within the region of study. The spatial resolution and scale dependence of relationships noted in the literature, however, indicate that the relationships assessed may be spatially nonstationary (Fotheringham et al., 2002; Foody, 2004). The commonly made assumption of spatial stationarity of the relationship may impact on the generalizabilty of remote sensing methods and is an issue discussed below.

20.6 FUTURE DIRECTIONS IN MAPPING

Although remote sensing has already played an important role as a data source for biogeographers, much research is required still to realise its full potential. Many issues require attention, and these can be grouped in the following categories:

1. Utilize the full information content of existing datasets—something that is often neglected;
2. Migrate away from analyses based upon simple summary indices such as NDVI;
3. Pay more attention to parts of the electromagnetic spectrum beyond the visible and reflected infrared, and also focus on LiDAR and hyperspectral remote sensing; and
4. Conduct further research on techniques for information extraction.

Building on these four categories we highlight three issues that urgently require the attention of remote sensers working within biogeography.

20.6.1 The assumption of spatially stationary relationships

There is a need to address important underlying assumptions, particularly the widely made assumption that relationships are spatially stationary. For example, it is common for species-area relationships to be used in studies of biodiversity with uniform parameter values when in reality they may vary (Kier et al., 2005). The assumption of spatial stationarity may often be untenable in biogeographical research. Although strong relationships between biogeographical variables have been reported in the literature (e.g., Kerr et al., 2001; Jetz and Rahbek, 2002), there are problems in determining general models to explain observed patterns. Numerous authors have noted that patterns of species richness appear to vary as a function of spatial scale (Gaston, 2000; Godfray and Lawton, 2001; Rahbek and Graves, 2001; Blackburn and Gaston, 2002; Chase and Leibold, 2002; Bond and Chase, 2002; Adler and Lauenroth, 2003; Loreau et al., 2003; Mora et al., 2003). This has important implications for understanding, and ultimately for conserving biodiversity, particularly as a consequence of the desire to upscale or downscale the perceived underlying explanatory relationship in a simple manner (van Gardingen et al., 1997; Willis and Whittaker, 2002). Thus, scale needs to be accounted for in biodiversity studies but it remains a major challenge because of its complex effects (Whittaker et al., 2001). Given the importance of the spatial dimension to biogeographical research (Millington et al., 2003), scale-related issues are likely to be a major component of future study.

Scale dependence may indicate that the explanatory processes and variables in operation differ between spatial scales. Alternatively, the scale effect could be, in whole or in part, a consequence of the relationship between the variables varying in space. It has typically been assumed implicitly that the relationships under study are spatially stationary, and so conventional, aspatial regression analysis that assume spatially stationary have been used (Fotheringham et al., 2002). However, the relationships between the variables may be spatially nonstationary (i.e., they may vary between locations); because of this the parameters of a statistical model describing a relationship may vary greatly in space, thereby limiting the descriptive and predictive utility of global models.

20.6.2 Accuracy assessment

Accuracy assessment remains a major priority (Rindfuss et al., 2004; Strahler et al., 2006). For example, there are many concerns with land-cover mapping and monitoring (Foody, 2002; Rindfuss et al., 2004) such as the validation of the maps derived from remote sensing and the assessment and interpretation of map accuracy (Foody, 2002). Often, the target in thematic mapping from remotely sensed imagery is to attain an overall accuracy of at least 85% in which each class is classified to a similar accuracy. Despite the great potential of remote sensing these targets are often not met (Wilkinson, 1996). Although this has driven considerable research to increase map accuracy, it need not mean that biogeographically useful information cannot be derived at 'lower' accuracies. Researchers differ in their needs and while some may reject a map as inaccurate, the information provided may be sufficient for others. For example, the IGBP's DISCover global land-cover map has an overall accuracy of ~67%, but for some users its accuracy is >90% (DeFries and Los, 1999). Moreover, accuracy may also be expressed and interpreted in various ways, making accuracy assessment dependent on issues such as the sampling design adopted to acquire the data used in the accuracy assessment, the different types of errors, and their weighting subsequent analyses. Commonly, the accuracy with which a specific class has been mapped is expressed either from the perspective of a map user or of a map producer (Story and Congalton, 1986). Through recognition of such different components of accuracy and knowledge of the specific requirements of a study it may be possible to optimize the production of the map for the specific application in hand (Lark, 1995).

20.6.3 Developments in sensor technology

Finally, developments in sensor technology are important. There are important concerns in maintaining continuity in data products. For example, there is a strong desire to extend the archive of Landsat sensor products into the future since such data have played a major role in many biogeographical studies and are already available for a 30-year history (Cohen and Goward, 2004; Boyd and Danson, 2005). Continuity issues, therefore, need consideration in the development of new sensors (Janetos and Justice, 2000). However, developments in sensor technology mean it is now possible to acquire data at enhanced spatial, spectral, and radiometric resolutions. The differences between sensor data need to be addressed. For example, the spatial resolution of the data used impacts on landscape indices (Millington et al., 2003; Saura, 2004). However, new sensors open up new avenues for research. New and proposed sensors could soon provide greater information on spectral responses and from a range of angular geometries. New sensors have much to offer. For example, developments in imaging spectrometers should help provide more detailed characterizations of canopy chemistry (e.g., Ustin et al., 2004; Asner and Vitousek, 2005). Similarly moderate spatial resolution systems and the developments of multiangular sensors together with LiDAR and interferometric synthetic aperture radar will help characterize the vertical dimension that has often been missing in studies (Lefsky et al., 2002; Treuhaft et al., 2004). New sensors are likely to help enhance existing applications as well as allow the development of new areas of research. Application areas likely to grow include epidemiological studies where the links between landscape and disease are important and form part of a growing field of study (Lobitz et al., 2000; Graham et al., 2005) as well as more traditional areas of research that can benefit from advances in remote sensing and data analysis technologies. Additionally, the ability of remote sensing to provide data on any site on the planet may help fill gaps in knowledge on major biomes, notably the flooded grassland and flooded savannas on which knowledge is currently very limited (Kier et al., 2005).

REFERENCES

Achard, F., Eva, H.D., Stibig, H.J., Mayaux, P., Gallego, J., and Malingreau, J-P. (2002) 'Determination of deforestation rates of the world's humid tropical forests', *Science*, 297: 999–1002.

Adler, P.B., and Lauenroth, W.K. (2003) 'The power of time: spatiotemporal scaling of species diversity', *Ecology Letters*, 6: 749–756.

Amarnath, G., Murthy, M.S.R., Britto, S.J., Rajashekar, G. and Dutt, C.B.S. (2003) 'Diagnostic analysis of conservation zones using remote sensing and GIS techniques in wet evergreen forests of the Western Ghats - An ecological hotspot, Tamil Nadu, India', *Biodiversity and Conservation*, 12: 2331–2359.

Andrefouet, S., and Roux, L. (1998) 'Characterisation of ecotones using membership degrees computed with a fuzzy classifier', *International Journal of Remote Sensing*, 19(16): 3205–11.

Aplin, P.A. (2004) 'Remote sensing: Land cover', *Progress in Physical Geography*, 28(2): 283–93.

Arnot, C., Fisher, P.F., Wadsworth, R., and Wellens, J. (2004) 'Landscape metrics with ecotones: Pattern under uncertainty', *Landscape Ecology*, 19(2): 181–95.

Asner, G.P., and Vitousek, P.M. (2005) 'Remote analysis of biological invasion and biogeochemical change', *Proceedings of The National Academy of Sciences of The United States of America*, 102: 4383–4386.

Asner, G.P., and Vitousek, P.M. (2005) 'Remote analysis of biological invasion and biogeochemical change', *Proceedings of the National Academic of the United States of America*, 102(12): 4383–86.

Asner, G.P., Jones, M.O., Martin, R.E., Knapp, D.E. and Hughes, R.F. (2008b) 'Remote sensing of native and invasive species in Hawaiian forests', *Remote Sensing of Environment*, 111(5):1912–26.

Asner, G.P., Scurlock, J.M.O., and Hicke, J.A. (2003) 'Global synthesis of leaf area index observations: Implications for ecological and remote sensing studies', *Global Ecology and Biogeography*, 12: 191–205.

Atkinson, P.M., Foody, G.M., Gething, P.W., Mathur, A., and Kelly, C.K. (2007) 'Investigating spatial structure in specific tree species in ancient semi-natural woodland using remote sensing and marked point pattern analysis', *Ecography*, 30(1): 88–104.

Baldi, G., and Paruelo, J.M. (2008) 'Land-use and land cover dynamics in South American temperate grasslands', *Ecology and Society*, 13(2): 6.

Bawa, K.S., and Seidler, R. (1998) 'Natural forest management and conservation of biodiversity in tropical forests', *Conservation Biology*, 12: 46–55.

Blackburn, T.M., and Gaston, K.J. (2002) 'Scale in macroecology', *Global Ecology and Biogeography*, 11: 185–189.

Board, C. (1969) 'Maps and models', in Chorley, R.J., and Haggett, P., *Physical and Information Models in Geography*. London: Methuen. pp. 671–726.

Bond, E.M., and Chase, J.M. (2002) 'Biodiversity and ecosystem functioning at local and regional spatial scales', *Ecological Letters*, 5: 467–470.

Bowman, D.M.J.S., and Franklin, D.C. (2005) 'Fire ecology', *Progress in Physical Geography*, 29(2): 248–55.

Boyd, D.S., and Danson, F.M. (2005) 'Satellite remote sensing of forest resources: three decades of research development', *Progress in Physical Geography*, 29: 1–26.

Boyd, D.S., Sanchez-Hernandez, C., and Foody, G.M. (2006) 'Mapping a specific class for priority habitats monitoring from satellite sensor data', *International Journal of Remote Sensing*, 27: 2631–2644.

Bradley, B.A., and Mustard, J.F. (2006) 'Characterizing the landscape dynamics of an invasive plant and risk of invasion using remote sensing', *Ecological Applications*, 16(3): 1132–47.

Brandtberg, T., Warner, T.A., Landenberger, R.E. and McGraw, J.B. (2003) 'Detection and analysis of individual leaf-off tree crowns in small footprint, high sampling density LiDAR data from the eastern deciduous forest in North America', *Remote Sensing of Environment*, 85: 290–303.

Brandtberg, T. (2002) 'Individual tree-based species classification in high spatial resolution aerial images of forests using fuzzy sets', *Fuzzy Sets and Systems*, 132: 371–387.

Cannon, C.H., Peart, D.R., and Leighton, M. (1998) 'Tree species diversity in commercially logged Bornean rainforest', *Science*, 281: 1366–1368.

Carleer, A., and Wolff, E. (2004) 'Exploitation of very high resolution satellite data for tree species identification', *Photogrammetric Engineering and Remote Sensing*, 70: 135–140.

Chase, J.M., and Leibold, M.A. (2002) 'Spatial scale dictates the productivity-biodiversity relationship', *Nature*, 416: 427–430.

Clark, D.B., Castro, C.S., Alvarado, L.D.A. and Read, J.M. (2004a) 'Quantifying mortality of tropical rain forest trees using high-spatial-resolution satellite data', *Ecology Letters*, 7: 52–59.

Clark, D.B., Read, J.M., Clark, M.L., Cruz, A.M., Dotti, M.F., and Clark, D.A. (2004b) 'Application of 1-M and 4-M resolution satellite data to ecological studies of tropical rain forests', *Ecological Applications*, 14: 61–74.

Clark, M.L., Roberts, D.A., and Clark, D.B. (2005) 'Hyperspectral discrimination of tropical rain forest tree species at leaf to crown scales', *Remote Sensing of Environment*, 96: 375–398.

Cohen, W.B., and Goward, S.N. (2004) 'Landsat's role in ecological applications of remote sensing', *Bioscience*, 54: 535–545.

Cross, A.A., Settle, J.J., Drake, N.A., and Paivinen, R.T.M. (1991) 'Subpixel measurement of tropical forest cover using AVHRR data', *International Journal of Remote Sensing*, 12(5): 1119–1129.

Dahdouh-Guebas, F., Van Hiel, E., Chan, J. C-W., Loku Pulukkuttige Jayatissa, and Koedam, N. (2002) 'Qualitative distinction of congeneric and introgressive mangrove species in mixed patchy forest assemblages using high spatial resolution remotely sensed imagery (IKONOS)', *Systematics and Biodiversity*, 2: 113–19.

Debinski, D.M., and Humphrey, P.S. (1997) 'An integrated approach to biological diversity assessment', *Natural Areas Journal*, 17: 355–365.

DeFries, R.S., Hansen, M., Townshend, J.R.G. and Sohlberg, R. (1998) 'Global land cover classifications at 8 km spatial resolution: the use of traning data derived from Landsat imagery in decision tree classifiers', *International Journal of Remote Sensing*, 19(6): 3141–3168.

DeFries, R.S., and Los, S.O. (1999) 'Implications of land-cover misclassification for parameter estimates in global land-surface models: An example from the simple biosphere model (SiB2)', *Photogrammetric Engineering and Remote Sensing*, 65: 1083–1088.

DeFries, R.S., and Townshend, J.R.G. (1994) 'NDVI-derived land-cover classifications at a global scale', *International Journal of Remote Sensing*, 15: 3567–3586.

Dong, J., Kaufmann, R.K., Myneni, R.B., Tucker, C.J., Kauppi, P.E., Liski, J., Buermann, W., Alexeyev, V. and Hughes, M.K. (2003) 'Remote sensing estimates of boreal and temperate forest woody biomass: Carbon pools, sources, and sinks', *Remote Sensing of Environment*, 84(3): 393–410.

Drake, J.B., Dubayah, R.O., Knox, R.G., Clark, D.B., and Blair, J.B. (2003) 'Above-ground biomass estimation in closed canopy Neotropical forests using LiDAR remote sensing: Factors affecting the generality of relationships', *Global Ecology and Biogeography*, 12(2): 147–159

Erikson, M. (2004) 'Species classification of individually segmented tree crowns in high-resolution aerial images using radiometric and morphologic image measures', *Remote Sensing of Environment*, 91: 469–477.

Etter, A., McAlpine, C., Wilson, K., Phinn, S., and Possingham, H. (2006) 'Regional patterns of agricultural land use and deforestation in Colombia', *Agriculture, Ecosystems and Environment*, 114(2–4): 369–86.

Eva, H.D., Belward, A.S., de Miranda, E.E., di bella, C.M., Gond, V., Huber, O., Jones, S.D., Sgrenzaroli, M. and Fritz, S.

(2004) 'A land cover map of South America', *Global Change Biology*, 10(5): 731–744.

Everitt, J.H., Yang, C., Helton, R.J., Hartmann, L.H., and Davis, M.R. (2002) 'Remote sensing of giant salvinia in Texas waterways', *Journal of Aquatic Plant Management*, 40: 11–16.

Foody, G.M., and Cutler, M.E.J. (2003) 'Tree biodiversity in protected and logged Bornean tropical rain forests and its measurement by satellite remote sensing', *Journal of Biogeography*, 30: 1053–1066.

Foody, G.M., and Hill R.A. (1996) 'Classification of tropical forest classes from Landsat TM data', *International Journal of Remote Sensing*, 17: 2353–2367.

Foody, G.M. (2002) 'Status of land cover classification accuracy assessment', *Remote Sensing of Environment*, 80: 185–201.

Foody, G.M. (2004) 'Supervised image classification by MLP and RBF neural networks with and without an exhaustively defined set of classes', *International Journal of Remote Sensing*, 25: 3091–3104.

Foody, G.M. (2005) 'Mapping the richness and composition of British breeding birds from coarse spatial resolution satellite sensor imagery', *International Journal of Remote Sensing*, 26: 3943–3956.

Foody, G.M., Atkinson, P.M., Gething, P.W., Ravenhill, N.A., and Kelly, C.K. (2005) 'Identification of specific tree species in ancient semi-natural woodland from digital aerial sensor imagery', *Ecological Applications*, 15: 1233–1244.

Foody, G.M., Boyd, D.S., and Cutler, M.E.J. (2003) 'Predictive assessment of tropical forest biomass from Landsat TM data and their transferability between regions', *Remote Sensing of Environment*, 85(4): 463–74.

Fotheringham, A.S., Brunsdon, C., and Charlton, M. (2002) '*Geographically Weighted Regression. The analysis of spatially varying relationships*, Chichester: John Wiley.

Franklin, S.E., and Wulder, M.A. (2002) 'Remote sensing methods in medium spatial resolution satellite data land cover classification of large areas', *Progress in Physical Geography*, 26:173–205.

Franklin, S.E. (2000) *Remote Sensing for Sustainable Forest Management*, Boa Raton FL: Lewis Publishers.

Friedl, M.A., and 12 authors. (2002) 'Global land cover mapping from MODIS: algorithms and early results', *Remote Sensing of Environment*, 83(1–2): 287–302.

Gaston, K.J. (2000) Global patterns in biodiversity', *Nature*, 405: 220–227.

Gillespie, T.W., Foody, G.M., Ricchini, D., Giorgi, A.P., and Saatchi, S. (2008) 'Measuring and modelling biodiversity', *Progress in Physical Geography*, 32(2): 203–21.

Godfray, H.C.J., and Lawton, J.H. (2001) 'Scale and species numbers', *Trends in Ecology and Evolution*, 16: 400–404.

Goodwin, N., Turner, R., and Merton, R. (2005) 'Classifying Eucalyptus forests with high spatial and spectral resolution imagery: an investigation of individual species and vegetation communities', *Australian Journal of Botany*, 53: 337–345.

Gopal, S., Woodcock C.E., and Strahler, A.H. (1999) 'Fuzzy neural network classification of global land cover from a 1° AVHRR data set', *Remote Sensing of Environment*, 67: 230–43.

Gould, W. (2000) 'Remote sensing of vegetation, plant species richness, and regional biodiversity hotspots', *Ecological Applications*, 10: 1861–1870.

Graham, A.J., Danson, F.M., and Craig, P.S. (2005) 'Ecological epidemiology: the role of landscape structure in the transmission risk of the fox tapeworm Echinococcus multilocularis (Leukart 1863) '(Cestoda: Cyclophyllidea: Taeniidae)', *Progress in Physical Geography*, 29: 77–91.

Griffiths, G.H., Lee, J., and Eversham, B.C. (2000) 'Landscape pattern and species richness; regional scale analysis from remote sensing', *International Journal of Remote Sensing*, 21: 2685–2704.

Gross, J.E., Goetz, S.J., and Cihlar, J. (2009) 'Application of remote sensing to parks and protected area monitoring: Introduction to the special issue', *Remote Sensing of Environment*, 113: 1343–1345

Grove, M.J., Troy, A.R., O'Neil-Dunne, J.P.M., Burch Jr., W.R., Cadenasso, M.L., and Pickett, S.T.A. (2006) 'Characterization of households and its implications for the vegetation of urban ecosystems', *Ecosystems* 9(4): 578–97.

Haara, A., and Haarala, M. (2002) 'Tree species classification using semi-automatic delineation of trees on aerial images', *Scandinavian Journal of Forest Research*, 17: 556–565.

Hannah, L., Midgley, G.F., and Millar, D. (2002) 'Climate change-integrated conservation strategies', *Global Ecology and Biogeography*, 11: 485–495.

Hansen, M.C., Defries, R.S., Townshend, J.R.G., and Sohlberg, R. (2000) 'Global land cover classification at 1km spatial resolution using a classification tree approach', *International Journal of Remote Sensing*, 21: 1331–1364.

Hansen, M.C., Roy, D.P., Lindquist, E., Edusei, B., Justice, C.O., and Altstatt, A. (2008) 'A method for integrating MODIS and Landsat data for systematic monitoring of forest cover and change in the Congo Basin', *Remote Sensing of Environment*, 112(6): 2495–513.

Harris, R. J., and Reed, J. M. (2002) 'Behavioral barriers to non-migratory movements of birds', *Annales Zoologici Fennici*, 39: 275–290.

Hese, S., Lucht, W., Schmullius, C., Barnsley, M., Dubayah, R., Knorr, D., Neumann, K., Riedel, T, and Schröter (2006) 'Global biomass mapping for an improved understanding of the CO_2 balance—the Earth observation mission Carbon-3D', *Remote Sensing of Environment*, 94(1): 94–104.

Hestir, E.L., Khanna, S., Andrew, M.E., Santos, M.J., Viers, J.H., Greenberg, J.A., Rajapakse, S.S. and Ustin, S.L. (2008) 'Identification of invasive vegetation using hyperspectral remote sensing in the California Delta ecosystem', *Remote Sensing of Environment*, 112(11): 4034–47.

Hof, A. (2006) Remote sensing and GIS data sources and data availability for land use/land cover change research in Sudano-Sahelian landscapes', in Kappas, M., Kleinn, C., and Sloboda, B. (eds.), *Global Change Issues in Emerging and Developing Countries*. Göttingen: Universitätsverlag Göttingen. pp. 163–74.

Holmgren, J., and Persson, A. (2003) 'Identifying species of individual trees using airborne laser scanner', *Remote Sensing of Environment*, 90: 415–423.

Houghton, R.A., Lawrence, K.T., Hackler, J.L., and Brown, S. (2001) 'The spatial distribution of forest biomass in the Brazilian Amazon: A comparison of estimates', *Global Change Biology*, 7: 731–46.

Hurtt, G., and 19 authors. (2003) 'IKONOS imagery for the Large Scale Biosphere-Atmosphere Experiment in Amazonia (LBA)', *Remote Sensing of Environment*, 88: 111–127.

Imbernon, J., and Branthomme, A. (2001) 'Characterization of landscape patterns of deforestation in tropical rain forests', *International Journal of Remote Sensing*, 22(9): 1753–65.

Imhoff, M.L., Sisk, T.D., Milne, A., Morgan, G., and Orr, T. (1997) 'Remotely sensed indicators of habitat heterogeneity: Use of synthetic aperture radar in mapping vegetation structure and bird habitat', *Remote Sensing of Environment*, 60: 217–227.

Ingram, J.C., and Dawson, T.P. (2005) 'Inter-annual analysis of deforestation hotspots in Madagascar from high temporal resolution satellite observations', *International Journal of Remote Sensing*, 26(7): 1447–61.

Innes, J.L., and Koch, B. (1998) 'Forest biodiversity and its assessment by remote sensing', *Global Ecology and Biogeography*, 7: 397–419.

Janetos, A.C., and Justice, C.O. (2000) 'Land cover and global productivity: a measurement strategy for the NASA programme', *International Journal of Remote Sensing*, 21: 1491–1512.

Jepson, W. (2005) 'A disappearing biome? Reconsidering land-cover change in the Brazilian savanna', *Geographical Journal*, 171: 99–111.

Jetz, W., and Rahbek, C. (2002) 'Geographic range size and determinants of avian species richness', *Science*, 297: 1548–1551.

Jha, C.S., Goparaju, L., Tripathi, A., Gharai, B., Raghubanshi, A.S., and Singh, J.S. (2005) 'Forest fragmentation and its impact on species diversity: An analysis using remote sensing and GIS', *Biodiversity and Conservation*, 14(7): 1681–98.

Johansen, K., Coops, N.C., Gergel, S.E., and Stange, Y. (2007) 'Application of high spatial resolution satellite imagery for riparian and forest ecosystem classification', *Remote Sensing of Environment*, 110(1): 29–44.

Justice, C.O., Townshend, J.R.G., Holben, B.N., and Tucker, C.J. (1985) 'Analysis of the phenology of global vegetation using meteorological satellite data', *International Journal of Remote Sensing*, 6(8): 1271–318.

Kelly, C.K., and Bowler, M.G. (2002) 'Coexistence and relative abundance in forest trees', *Nature*, 417: 437–440.

Kelly, C.K., and Bowler, M.G. (2005) 'A new application of storage dynamics: differential sensitivity, diffuse competition, and temporal niches', *Ecology*, 86: 1012–22.

Kennedy, P. (1989) 'Monitoring the vegetation of Tunisian grazing lands using the normalized difference vegetation index', *Ambio*, 18: 119–23.

Kerr, J.T., Southwood, T.R.E., and Cihlar, J. (2001) 'Remotely sensed habitat diversity predicts butterfly species richness and community similarity in Canada', *Proceedings of The National Academy of Sciences of the United States of America*, 98: 11365–11370.

Kerr, J.T., and Ostrovsky, M. (2003) 'From space to species: Ecological applications of remote sensing', *Trends in Ecology and Evolution*, 18(6): 299–305.

Key, T., Warner, T.A., McGraw, J.B., and Fajvan, M.A. (2001) 'A comparison of multispectral and multitemporal information in high spatial resolution imagery for classification of individual tree species in a temperate hardwood forest', *Remote Sensing of Environment*, 75: 100–112.

Kier, G., Mutke, J., Dinerstein, E., Ricketts, T.H., Kuper, W., Kreft, H., and Barthlott, W. (2005) 'Global patterns of plant diversity and floristic knowledge', *Journal of Biogeography*, 32: 1107–1116.

Krishnaswamy, J., Kiran, M.C., and Ganeshaiah, K.N. (2004) 'Tree model based eco-climatic vegetation classification and fuzzy mapping in diverse tropical deciduous ecosystems using multi-season NDVI', *International Journal of Remote Sensing*, 25(6): 1185–205.

Lambin, E.F. (1999a) 'Modelling and monitoring land-cover change processes in tropical regions', *Progress in Physical Geography*, 21(3): 375–93.

Lambin, E.F. (1999b) 'Monitoring forest degradation in tropical regions by remote sensing: Some methodological issues', *Global Ecology and Biogeography*, 8(3/4): 191–98.

Lambin, E.F., and Geist, H.F. (eds.). (2006) *Land-Use and Land-Cover Change: Local Processes and Global Impacts*. New York: Springer.

Landenberger, R.E., McGraw, J.B., Warner, T.A., and Brandtberg, T. (2003) 'Potential of digital color imagery for censusing Haleakala silverswords in Hawaii', *Photogrammetric Engineering and Remote Sensing*, 69: 915–923.

Lark, R.M. (1995) 'Components of accuracy of maps with special reference to discriminant-analysis on remote sensor data', *International Journal of Remote Sensing*, 16: 1461–1480.

Lass, L.W., Prather, T.S., Glenn, N.F., Weber, K.T., Mundt, J.T., and Pettingill, J. (2005) 'A review of remote sensing of invasive weeds and example of the early detection of spotted knapweed (*Centaurea maculosa*) and babysbreath (*Gypsophila paniculata*) with a hyperspectral sensor', *Weed Science*, 53(2): 242–51.

Lawrence, R.L., Wood, S.D., and Sheley, R.L. (2006) 'Mapping invasive plants using hyperspectral imagery and Breiman Cutler classifications (randomForest)', *Remote Sensing of Environment*, 100(3): 356–62.

Lefsky, M.A., Cohen, W.B., Parker, G.G., and Harding, D.J. (2002) 'LiDAR remote sensing for ecosystem studies', *BioScience*, 52(1): 19–30.

Lengyel, S., Déri, E., Varga, Z., Horváth, R., Tóthmérész, B., Henry, P-Y., Kobler, A., Kutnar, L., Babij, V., Seliškar, A., Chritia, C., Papastergiadou, E., Gruber, B. and Henle, K. (2008) 'Habitat monitoring in Europe: A description of current practices', *Biodiversity and Conservation*, 17(14): 3327–39.

Lepers, E., Lambin, E.F., Janetos, A.C., DeFries, R., Achard, F., Ramankutty, N., and Scholes, R.J. (2005) 'A synthesis of information on rapid land-cover change for the period 1981–2000', *Bioscience*, 55(2): 115–24.

Lobitz, B., Beck, L., Huq, A., Wood, B., Fuchs, G., Faruque, A.S.G., and Colwell, R. (2000) 'Climate and infectious disease: Use of remote sensing for detection of Vibrio cholerae by indirect measurement', *Proceedings of the National Academy of Sciences of the United States of America*, 97: 1438–1443.

Loreau, M., and 11 authors. (2001) 'Ecology - Biodiversity and ecosystem functioning: Current knowledge and future challenges', *Science*, 294: 804–808.

Loreau, M., Mouquet, N., and Holt, R.D. (2003)' Meta-ecosystems: a theoretical framework for a spatial ecosystem ecology', *Ecology Letters*, 6: 673–679.

Loveland, T., Reed, B.C., Brown, J.F., Ohlen, D.O., Zhu, Z., Yang, L., and Merchant, J.W. (2000). 'Development of a global land cover characteristics database and IGBP DISCover from 1 km AVHRR data', *International Journal of Remote Sensing*, 21 (6&7): 1303–1330.

Lugo, A. E. (1995) 'Management of tropical biodiversity', *Ecological Applications*, 5: 956–961.

Lugo, A. E. (1999) 'Will concern for biodiversity spell doom to tropical forest management'? *Science of the Total Environment*, 240: 123–131.

Luoto, M., Kuussaari, M., and Toivonen, T. (2002) 'Modelling butterfly distribution based on remote sensing data', *Journal of Biogeography*, 29: 1027–1037.

Luoto, M., Virkkala, R., Heikkinen, R.K., and Rainio, K. (2004) 'Predicting bird species richness using remote sensing in boreal agricultural-forest mosaics', *Ecological Applications*, 14: 1946–1962.

MacKay, D.S., Ahl, D.E., Ewers, B.E., Gower, S.T., Burrows, S.N., Samanta, S., and Davis, K.J. (2002) 'Effects of aggregated classifications of forest composition on estimates of evapotranspiration in a northern Wisconsin forest', *Global Change Biology*, 8: 1253–1265.

Margules, C.R., and Pressey, R.L. (2000) 'Systematic conservation planning', *Nature*, 405: 243–253.

Martin, M.E., Newman, S.D., Aber, J.D., and Congalton R.G. (1998) 'Determining forest species composition using high spectral resolution remote sensing data', *Remote Sensing of Environment*, 65: 249–254.

Mas, J-F. (1999) 'Monitoring land-cover changes: A comparison of change detection techniques', *International Journal of Remote Sensing*, 20(1): 139–52.

Mather, P.M. (2004) '*Computer-Processing of Remotely Sensed Images*, third edition, Chichester UK: John Wiley.

Mayaux, P., Bartholome, E., Fritz, S., and Belward, A. (2004) 'A new land-cover map of Africa for the year 2000', *Journal of Biogeography*, 31: 861–877.

Millington, A.C., Velez-Liendo, X.M., and Bradley, A.V. (2003) 'Scale dependence in multitemporal mapping of forest fragmentation in Bolivia: implications for explaining temporal trends in landscape ecology and applications to biodiversity conservation', *ISPRS Journal of Photogrammetry and Remote Sensing*, 57: 289–299.

Millington, A.C. (2005) 'Land-cover change', in Geist, H.J. (ed.), *Our Earth's Changing Land*. Westport, CT: Greenwood Press. pp. xl–xlviii.

Millington, A.C., and Jepson, W.E.J. (2008) '*Land Change Science in the Tropics: Changing Agricultural Landscapes.* New York: Springer.

Millington, A.C., and Townshend, J.RG. (1989) *Biomass Assessment: Woody Biomass in the SADCC Region.* London: Earthscan.

Millington, A.C., Douglas, T.D., Critchley, R.W., and Ryan, P. (1994) *Estimating Woody Biomass in Sub-Saharan Africa.* Washington DC: World Bank.

Mora, C., Chittaro, P.M., Sale, P.F., Kritzer, J.P., and Ludsin, S.A. (2003) 'Patterns and processes in reef fish diversity', *Nature*, 421: 933–936.

Muchoney, D.M. (2008) 'Earth observations for terrestrial biodiversity and ecosystems', *Remote Sensing of Environment*, 112(8): 1909–11.

Muldavin, E.H., Neville, P., and Harper, G. (2001) 'Indices of grassland biodiversity in the Chihuahuan Desert ecoregion derived from remote sensing', *Conservation Biology*, 15: 844–855.

Nagendra, H., and Gadgil, M. (1999) 'Biodiversity assessment at multiple scales: Linking remotely sensed data with field information', *Proceedings of the National Academy of Sciences of the United States of America*, 96: 9154–9158.

Nagendra, H., and Gadgil, M. (1999) 'Satellite imagery as a tool for monitoring species diversity: an assessment', *Journal of Applied Ecology*, 36: 388–397.

Nagendra, H. (2001) 'Using remote sensing to assess biodiversity', *International Journal of Remote Sensing*, 22: 2377–2400.

Nagendra, H., Tucker, C., Carlson, L., Southworth, J., Karmacharya, M., and Karna, B. (2004) 'Monitoring parks through remote sensing: studies in Nepal and Honduras', *Environmental Management*, 34: 748–760.

Newbery, D.M., Campbell, E.J.F, Lee, Y.F, Ridsdale, C.E., and Still, M.J. (1992) 'Primary lowland dipterocarp forest at Danum Valley, Sabah, Malaysia - structure, relative abundance and family composition', *Philosophical Transactions of The Royal Society of London Series B*, 335: 341–356.

Newton, A.C., Hill, R.A., Echeverría, C., Rey Benayas, J.M., Golicher, D., Cayuela, L., and Hinsley, S.A. (2009) 'Remote sensing and the future of landscape ecology', *Progress in Physical Geography*, 33: 528–546.

Oindo, B.O., and Skidmore, A.K. (2002) 'Interannual variability of NDVI and species richness in Kenya', *International Journal of Remote Sensing*, 23: 285–298.

Oindo, B.O., Skidmore, A.K., and De Salvo, P. (2003) 'Mapping habitat and biological diversity in the Maasai Mara ecosystem', *International Journal of Remote Sensing*, 24: 1053–1069.

Ollinger, S.V., Smith, M.L., Martin, M.E., Hallett, R.A., Goodale, C.L., and Aber, J.D. (2002) 'Regional variation in foliar chemistry and N cycling among forests of diverse history and composition', *Ecology*, 83: 339–355.

Openshaw, S. (2004) *The Modifiable Areal Unit Problem*, CATMOG 38, Norwich UK: Geo Abstracts.

Pal, M., and Mather, P.M. (2005) 'Support vector machines for classification in remote sensing', *International Journal of Remote Sensing*, 26(5): 1007–11.

Palace, M., Keller, M., Asner, G.P., Hagen, S., and Braswell, B. (2008) 'Amazon forest structure from IKONOS satellite data and the automated characterization of forest canopy properties', *Biotropica*, 40(2): 141–50.

Parmenter, A.W., Hansen, .A, Kennedy, R.E., Cohen, W., Langner, U., Lawrence, R., Maxwell, B., Gallant, A., and Aspinall, R. (2003) 'Land use and land cover change in the Greater Yellowstone Ecosystem: 1975–1995', *Ecological Applications*, 13: 687–703.

Pedlety, J., and 15 authors. (2007) 'Generating a long-term land data record from AVHRR and MODIS instruments', *Geoscience and Remote Sensing Symposium, IGARRS 2007, IEEE Symposium,* 1021–1025.

Pengra, B.W., Johnston, C.A., and Loveland, T.R. (2007) 'Mapping an invasive plant, *Phragmites australis*, in coastal wetlands using the EO-1 Hyperion hyperspectral sensor', *Remote Sensing of Environment*, 208(1): 74–81.

Phillips, O.L., and 33 authors. (2004) 'Pattern and process in Amazon tree turnover, 1976–2001', *Philosophical Transactions of The Royal Society of London Series B*, 359: 381–407.

Pimentel, D., Stachow, U., Takacs, D.A., Brubaker, H.W., Dumas, A.R., Meaney, J.J., O'Neil, J.A.S., Onsi, D.E. and Corzilius, D.B. (1992) 'Conserving biological diversity in agricultural forestry systems - most biological diversity exists in human-managed ecosystems', *Bioscience*, 42: 354–362.

Potvin, F., Belanger, L., and Lowell, K. (2000) 'Marten habitat selection in a clearcut boreal landscape', *Conservation Biology*, 14: 844–857.

Pullin, A.S., Knight, T.M., Stone, D.A., and Charman, K. (2004) 'Do conservation managers use scientific evidence to support their decision-making'? *Biological Conservation*, 119: 245–252.

Purvis, A., and Hector, A. (2000) 'Getting the measure of biodiversity', *Nature*, 405: 212–219.

Rahbek, C., and Graves, G.R. (2001) 'Multiscale assessment of patterns of avian species richness', *Proceedings of The National Academy of Sciences of The United States of America,* 98: 4534–4539.

Ramsey, E.W., Nelson, G.A., Sapkota, S.K., Seeger, E.B. and Martella, K.D. (2002) 'Mapping Chinese tallow with color-infrared photography', *Photogrammetric Engineering and Remote Sensing*, 68: 251–255.

Read, J.M., and Lam, N.S.N. (2002) 'Spatial methods for characterising land cover and detecting land-cover changes for the tropics', *International Journal of Remote Sensing*, 23: 2457–2474.

Read, J.M., Clark, D.B., Venticinque, E.M., and Moreira, M.P. (2003) 'Application of merged 1-m and 4-m resolution satellite data to research and management in tropical forests', *Journal of Applied Ecology*, 40: 592–600.

Redo, D., Millington, A.C., and Bass, J.O. (2009) 'Forest dynamics and the importance of place in western Honduras', *Applied Geography*, 29 (1): 91–110.

Rindfuss, R.R., Walsh, S.J., Turner, B.L., Fox, J., and Mishra, V. (2004) 'Developing a science of land change: Challenges and methodological issues', *Proceedings of The National Academy of Sciences of the United States of America*, 101: 13976–13981.

Rocchini, D., and Vannini, A. (2010) 'What is up? Testing spectral heterogeneity versus NDVI relationship using quantile regression', *International Journal of Remote Sensing*, 31: 2745–2756.

Rouget, M. (2003) 'Measuring conservation value at fine and broad scales: implications for a diverse and fragmented region, the Agulhas Plain', *Biological Conservation*, 112: 217–232.

Roy, D.P. (2000) 'The impact of misregistration upon composited wide field of view satellite data and implications for change detection', *IEEE Transactions of Geoscience and Remote Sensing*, 38(4): 2017–32.

Roy, P.S., and Tomar, S. (2000) 'Biodiversity characterisization at landscape level using geospatial modelling technique', *Biological Conservation*, 95: 95–109.

Running, S.W., Peterson, D.L., Spanner, M.A., and Teuber, K.B. (1986) 'Remote-sensing of coniferous forest leaf-area', *Ecology*, 67: 273–276.

Sader, S.A., Stone, T.A., and Joyce, A.T. (1990) 'Remote-sensing of tropical forests – an overview of research and applications using nonphotographic sensors', *Photogrammetric Engineering and Remote Sensing*, 56: 1343–1351.

Sader, S.A., Stone, T.A., and Joyce, A.T. (1990) 'zRemote sensing of tropical forests: An overview of research and applications using non-photographic sensors', *Photogrammetric Engineering and Remote Sensing*, 56(10): 1343–51.

Sala, O.E., and 18 authors. (2000) 'Global biodiversity scenarios for the year 2100', *Science*, 287: 1770–74.

Sanchez-Azofelfa, G.A., Castro, K.L. Rivard, B., Kalascka, M.R., and Harriss, R.C. (2003) 'Remote sensing research priorities in tropical dry forest environments', *Biotropica*, 35:134–142.

Sanchez-Hernandez, C., Boyd, D.S., and Foody, G.M. (2007) 'One-class classification for mapping a specific land-cover class: SVDD classification of fenland', *IEEE Transactions on Geoscience and Remote Sensing*, 45(4): 1061–73.

Santos, J.R., Freitas, C.C., Araujo, L.S., Dutra, L.V., Mura, J.C., Gama, F.F., Soler, L.S., and Sant'Anna, S.J.S. (2003) 'Airborne P-band SAR applied to the aboveground biomass studies in the Brazilian tropical rainforest', *Remote Sensing of Environment*, 87(4): 482–93.

Saura, S. (2004) 'Effects of remote sensor spatial resolution and data aggregation on selected fragmentation indices', *Landscape Ecology*, 19: 197–209.

Sayn-Wittgenstein, L. (1961) 'Recognition of tree species on air photographs by crown characteristics', *Photogrammetric Engineering*, 27: 792–809.

Schmidtlein, S., and Sassin, J. (2004) 'Mapping of continuous floristic gradients in grasslands using hyperspectral imagery', *Remote Sensing of Environment*, 92: 126–138.

Scholes, R.J., and Biggs, R. (2005) 'A biodiversity intactness index', *Nature*, 434: 45–49.

Seto, K.C., Fleishman, E., Fay, J.P., and Betrus, C.J. (2004) 'Linking spatial patterns of bird and butterfly species richness with Landsat TM derived NDVI', *International Journal of Remote Sensing*, 25: 4309–4324.

Siegert, F., Ruecker, G., Hinrichs. A., and Hoffmann, A.A. (2001) 'Increased damage from fires in logged forests during droughts caused by El Niño', *Nature*, 414: 437–40.

Singh, A. (1989) 'Digital change detection techniques using remotely-sensed data', *International Journal of Remote Sensing*, 10(6): 989–1003.

Skole, D., and Tucker, C. (1993) 'Tropical deforestation and habitat fragmentation in the Amazon: Satellite data from 1978 to 1988', *Science*, 260: 1905–10.

Sobrini, J.A., and Raissouni, N. (2001) 'Toward remote sensing methods for land cover dynamic monitoring: Application to Morocco', *International Journal of Remote Sensing*, 21(2): 353–66.

Steininger, M. (2000) 'Satellite estimation of tropical secondary forest above-ground biomass: Data from Brazil and Bolivia', *International Journal of Remote Sensing*, 21(6–7): 1139–57.

Stibig, H.J., Achard, F., and Fritz, S. (2004) 'A new forest cover map of continental southeast Asia derived from SPOT-VEGETATION satellite imagery', *Applied Vegetation Science*, 7: 153–162.

Story, M., and Congalton, R.G. (1986) 'Accuracy Assessment – A Users Perspective', *Photogrammetric Engineering and Remote Sensing*, 52: 397–99.

Stow, D., et al. (2004) 'Remote sensing of vegetation and land cover in Arctic tundra ecosystems', *Remote Sensing of Environment*, 89(3): 281–308.

Strahler, A. H., Boschetti, L., Foody, G. M., Friedl, M. A., Hansen, M. C., Herold, M., Mayaux, P., Morisette, J. T., Stehman, S. V., and Woodcock, C. E., 2006. *Global Land Cover Validation: Recommendations for Evaluation and Accuracy Assessment of Global Land Cover Maps*, Euopean Commission, Joint Research Centre, Ispra, Italy, EUR 22156 EN, 48pp.

Treuhaft, R.N., Law, B.E., and Asner, G.P. (2004) 'Forest attributes from radar interferometric structure and its fusion with optical remote sensing', *Bioscience*, 54: 561–571.

Trisurat, Y., Eiumnoh, A., Murai, S., Hussain, M.Z., and Shrestha, R.P. (2000) 'Improvement of tropical vegetation mapping using a remote sensing technique: a case of Khao Yai National Park, Thailand', *International Journal of Remote Sensing*, 21: 2031–2042.

Tucker, C.J., Townshend, J.R.G., and Goff, T.E. (1985) 'African land-cover classification using satellite data', *Science*, 227: 369–375.

Tucker, C.J., Pinzon, J.E., Brown, M.E., Slayback, D., Pak, E.W., Mahoney, R., Vermote, E. and El Saleous, N. (2005) 'An extended AVHRR 8-km NDVI dataset compatible with MODIS and SPOT vegetation NDVI', *International Journal of Remote Sensing*, 26: 4485–4498

Tucker, C.J., and Nicholson, S.A. (1999) 'Variations in the size of the Sahara desert from 1980–1997', *Ambio*, 28(7): 587–91.

Tucker, C.J., Newcomb, W.W., and Dregne, H.E. (1994) 'AVHRR data sets for determination of desert spatial extent', *International Journal of Remote Sensing*, 15(17): 3547.

Turner, W., Spector, S., Gardiner, N., Fladeland, M., Sterling, E., and Steininger, M. (2003) 'Remote sensing for biodiversity science and conservation', *Trends in Ecology and Evolution*, 18, 306–314.

Underwood, E., Ustin, S., and DiPietro, D. (2003) 'Mapping nonnative plants using hyperspectral imagery', *Remote Sensing of Environment*, 86: 150–161.

Ustin, S.L., Roberts, D.A., Gamon, J.A., et al. (2004) 'Using imaging spectroscopy to study ecosystem processes and properties', *Bioscience*, 54: 523–534.

van Gardingen, P.R., Foody, G.M., and Curran P.J. (eds.) (1997) *Scaling-up: From Cell to Landscape*, Cambridge: Cambridge University Press.

Vasconcelos, H.L., Vilhena, J.M.S., and Caliri, G.J.A. (2000) 'Responses of ants to selective logging of a central Amazonian forest', *Journal of Applied Ecology*, 37: 508–514.

Vierling, K.T., Vierling, L.A., Gould, W.A., Martinuzzi, S., and Clawges, R.M. (2008) 'LiDAR—shedding light on habitat characterization and modeling', *Frontiers in Ecology and Environment*, 6(2): 90–98.

Walsh, S.J., McCleary, A.L., Mena, C.F., Shao, Y., Tuttle, J.P, González, A., and Atkinson, R. (2008) '*QuickBird* and *Hyperion* data analysis of an invasive plant species in the Galapagos Islands of Ecuador: Implications for control and land use management', *Remote Sensing of Environment*, 112(5): 1927–41.

Wang, L., Sousa, W.P., Gong, P., and Biging, G.S. (2004) 'Comparison of IKONOS and QuickBird images for mapping mangrove species on the Caribbean coast of Panama', *Remote Sensing of Environment*, 91(3–4): 432–40.

Warkentin, I.G., and Reed, J.M. (1999) 'Effects of habitat type and degradation on avian species richness in Great Basin riparian habitats', *Great Basin Naturalist*, 59: 205–212.

Waters, T.L., and Savill, P.S. (1992) 'Ash and sycamore regeneration and the phenomenon of their alternation', *Forestry*, 65: 417–433.

Webb, E.L., and Peralta, R. (1998) 'Tree community diversity of lowland swamp forest in Northeast Costa Rica, and changes associated with controlled selective logging', *Biodiversity and Conservation*, 7: 565–583.

Wei, A., and Chow-Fraser, P. (2007) 'Use of IKONOS imagery to map coastal wetlands of Georgian Bay', *Fisheries*, 32(4): 167–73.

Wessman, C.A., Aber, J.D., Peterson, D.L., and Melillo, J.M. (1988) 'Remote-sensing of canopy chemistry and nitrogen cycling in temperate forest ecosystems', *Nature*, 335: 154–156.

White, F. (1983) *Vegetation of Africa—A Descriptive Memoir to Accompany the Unesco/AETFAT/UNSO Vegetation Map of Africa*. Natural Resources Research Report XX. Paris: UNESCO.

Whittaker, R.J., Willis, K.J., and Field, R. (2001) 'Scale and species richness: towards a general, hierarchical theory of species diversity', *Journal of Biogeography*, 28: 453–470.

Wilkinson, G.G. (2005) 'Results and implications of a study of fifteen years of satellite image classification experiments', *IEEE Transactions on Geoscience and Remote Sensing*, 43: 433–440.

Wilkinson, G.G. (1996) 'Classification algorithms – where next? Soft Computing in Remote Sensing Data Analysis, (E. Binaghi, P. A. Brivio and A. Rampini, eds.), Singapore: World Scientific, pp. 93–99.

Willis, K.J., and Whittaker, R.J. (2002) 'Ecology – Species diversity – Scale matters', *Science*, 295: 1245–48.

Wilson, R.J., C.D. Thomas, R. Fox, D.B. Roy and W.E.S. Kunin (2004) 'Spatial patterns in species distributions reveal biodiversity change', *Nature*, 432: 393–396

Wolter, P.T., Johnston, C.A., and Niemi, G.J. (2005) 'Mapping submergent aquatic vegetation in the US Great Lakes using Quickbird satellite data', *International Journal of Remote Sensing*, 26(23): 5255–74.

Wrigley, R.C., Card, D.H., Hlvaka, C.A., Hall, J.R., Mertz, F.C., Archwamety, C. and Schwengerdt, R.A. (2007) 'Thematic mapper image quality: Registration, noise, and resolution', *IEEE Transactions on Geoscience and Remote Sensing* GE-22(3): 263–71.

Wu, Y., Rutchey, K., Wang, N., and Godin, J. (2006) 'The spatial pattern and dispersion of Lygodium microphyllum in the Everglades wetland ecosystem', *Biological Invasions*, 8(7): 1483–93.

Wulder, M.A., Hall, R.J., Coops, N.C., Franklin, S.E. (2004) 'High spatial resolution remotely sensed data for ecosystem characterization', *Bioscience*, 54: 511–521.

Xu, B., Yang, X.C., Tao, W.G., Qin, Z.H., Liu, H.Q., Miao, J.M., and Bi, Y.Y. (2008) 'MODIS-based remote sensing monitoring of grass production in China', *International Journal of Remote Sensing*, 29(17–18): 5313–27.

Yang, W., Huang, D., Tan, B., Stroeve, J.C., Shabanov, N.V., Knyazikhin, Y., Nemani, R.R. and Myneni, R.B. (2006) 'Analysis of leaf area index and fraction of PAR absorbed by vegetation products from the terra MODIS sensor: 2000–2005', *IEEE Transactions on Geoscience and Remote Sensing*, 44(7): 1829–42.

21

Remote Sensing for Modeling Biogeographic Features and Processes

Joanne M. Nightingale, Stuart R. Phinn, and Michael J. Hill

21.1 INTRODUCTION

Escalating human population coupled with global changes in climate, impel the need to accurately and efficiently monitor and manage the Earth's resources. As the human footprint upon our planet expands in synchrony with the requirement for renewable and nonrenewable energy and land resources, biogeographical studies become increasingly challenging. Atmosphere–biosphere interactions are very important processes of the Earth system. Climate and ecosystems interact with each other through energy, water, and material transfer and these interactions form the primary mechanism for the development of ecosystems. The study of biogeography involves spatial and temporal analyses of geographic patterns and the processes producing such patterns, at spatial levels from individuals, populations, and communities up to ecosystems. Remotely sensed data and derived products offer spatially comprehensive and temporally repeatable biophysical information that is able to quantify the patterns and dynamics of biogeographic features and processes.

In this chapter we discuss the utility of remotely sensed data and derived products for the modelling of spatial and temporal dynamics of terrestrial and aquatic components of the biosphere. We build on the general overview of remote sensing applications in biogeography presented by Foody and Millington (this volume), and we provide examples of how remote sensing is used to drive and validate the model types outlined in the succeeding chapters on modelling and biocomplexity (see Burnside and Waite, Malanson, and Walsh et al., this volume). Approaches for measuring and monitoring the surface energy-exchanges and vegetation growth are conducted via field assessments and are scaled to larger spatial extents using mathematical models. Because of the range of spatial extents and repetitive data collection requirements associated with regional-global scale biospheric monitoring efforts, remotely sensed data and derived parameters have been employed to initialize, drive, and assess model performance.

The role of remote sensing in models that are used to evaluate changes in biogeographic features and processes has not been considered in detail across a range of environments and scales. Here we will outline the types of biophysical information that can be extracted from remotely sensed data and explain how these spatial-temporal information datasets have been, and may be, used to calibrate and validate models that simulate biogeographic changes or processes. In section 21.2 we discuss the current status of remote sensing for biogeographic change analysis in both terrestrial and aquatic environments. Section 21.5 provides a more detailed overview of the types of remotely sensed data and derived products that are utilized within different classes of ecosystem models, while section 21.4 outlines the future that remotely sensed data and derived parameters and products will play in regional- to global-scale modeling activities in both terrestrial and aquatic environments. We also discuss anticipated developments in satellite sensors and derived products to improve the modeling and monitoring of biogeographic features and the changes in these features through time.

21.1.1 Models for assessing and monitoring biogeographic change

Biological activities occur over a wide range of time, size, and spatial scales. Difficulty in quantifying the role of the terrestrial biosphere in the global carbon cycle arises because of the complex biology underlying carbon storage, the heterogeneity of vegetation and soils, the effects of anthropogenic disturbance (land use, land-use change and management), natural disturbance (climate, edaphic, fire, pests) as well as atmospheric deposition of pollutants and other compounds (Prasad et al., 2000; Schimel, 1995;

Schimel et al., 2000). In order to connect these process levels and produce prediction at high levels of organization, various forms of models continue to be a useful tool for defining, integrating, linking, and scaling biological data (Bazzaz, 1993; Nightingale et al., 2004).

Over the past two decades, interest in issues such as global climate change has increased development and application of ecosystem-scale primary production models (White et al., 2000). A number of models based on biochemical and physiological principles have been developed. Several hundred models of ecosystem productivity and function exist with varying levels of organisation, scope, temporal resolution (seconds to years) and spatial explicitness (single leaf to global), for example, the Register of Ecological Models (Schwalm and Ek, 2001). Tables 21.1 and 21.2 provide an overview of commonly utilized terrestrial (21.1), and aquatic models (21.2). Environment of application and model outputs are provided, as well as the type of remotely sensed data and derived products that are currently utilized within these models. During this time a number of reviews have assessed the role and

Table 21.1 Models that are currently used for assessing the distribution and change in terrestrial and aquatic biogeographical phenomena. This list does not claim to be exhaustive and is limited by access to published literature on these existing models

Model	Current Remote Sensing Inputs	Main Output	Application	References
ECOSYSTEM MODELS—Stand scale (1ha resolution)				
1. GEM 2. HYBRID 3. CENTURY	• IRS-LISS-I derived soil map • MODIS	Carbon and nitrogen cycles in terrestrial ecosystems	General plant functional types	1. (Rastetter et al., 1991) 2. (Friend and Schugart, 1993) 3. (Parton et al., 1992)
ECOSYSTEM MODELS—Biome scale (8km—2° grid resolution)				
1. TEM 2. BIOME-BGC 3. SiB 4. CASA 5. FBM 6. CARAIB 7. DEMETER 8. GLOCO 9. GLO-PEM 10. IBIS 11. LSM 12. BIOME 2 and 3 13. Global Biome Model	• AVHRR NDVI and derived LAI, fPAR, and land cover • MODIS NDVI, EVI, LAI, and fPAR	Carbon, nitrogen, and water fluxes ; biomass and productivity	General plant functional types	1. (Raich et al., 1991) 2. (Running and Hunt, 1993) 3. (Sellers et al., 1986) 4. (Potter et al., 1993) 5. (Ludeke et al., 1994) 6. (Warnant et al., 1994) 7. (Foley, 1994) 8. (Hudson et al., 1994) 9. (Prince and Goward, 1995) 10. (Woodward et al., 1998) 11. (Bonan, 1995) 12. (Prentice et al., 1992)

Continued

Table 21.1 Cont'd

Model	Current Remote Sensing Inputs	Main Output	Application	References
FOREST MODELS—Stand scale (1ha resolution)				
1. FVS	• AVHRR NDVI and derived	Forest growth,	Models designed	1. (Dixon, 2003)
2. BEX	LAI, fPAR and land cover	productivity,	for specific	2. (Bonan, 1991)
3. PIPESTEM	• MODIS NDVI, EVI, LAI,	biomass,	types of	3. (Valentine et al., 1997)
4. DAYTRANS/PSN	and fPAR	energy,	forests	4. (Running et al., 1975)
5. FORGRO		water, and		5. (Mohren and J.R.,
6. FORMIND/FORMIX		carbon		1995)
7. FORECAST		fluxes		6. (Huth et al., 1996)
8. FOREST–BGC				7. (Kimmins et al., 1999)
9. JABOWA				8. (Running and
10. FORET				Coughlan, 1988)
11. LINKAGES				9. (Botkin et al., 1972)
12. 3-PG, 3-PGS				10. (Shugart and West,
13. G'DAY				1977)
14. BEPS				11. (Pastor and Post, 1986)
15. PnET / Day				12. (Landsberg et al.
				1997b)
				13. (Comins et al., 1993)
				14. (Liu et al., 1997)
				15. (Aber and Federer,
				1992)
RANGELAND, GRASSLAND SAVANNA, AGRICULTURAL—Landscape scale (km², ha—0.5°)				
1. SPUR 2.4	• Landsat TM/ETM NDVI	Forage	Rangelands	1. (Teague and Foy, 2004)
2. Dynamic Automata	• AVHRR NDVI	production	Pastures	2. (Witten et al., 2005)
Model	• Landsat	and plant	Grassland	3. (Donnelly et al., 1997)
3. GRASSGRO	* For the most part, these	community	Savanna	4. (Hunt et al., 1991)
4. Hurley Pasture	models do not integrate	dynamics,		5. (Thornley and Cannell,
Model	remotely sensed data.	carbon,		1997)
5. The Sustainable		nitrogen,		6. (Johnson et al., 2003)
Grazing Systems		and water		7. (Nouvellon et al., 2001)
Pasture Model		cycles		8. (Ludwig et al., 2001)
6. Ecosystem Model				9. (Seaquist et al., 2003)
7. LUE model for				
Grassland				
8. SAVANNA				
9. MAPSS				
FIRE AND BIOMASS BURNING—Landscape scale (m², ha—acres)				
1. Seiler and Crutzen	• SPOT–VEGETATION	Biomass	Geographic	1. (Palacios-Orueta et al.,
Model	burned area	burning and	regions	2005)
2. FOFEM	• AVHRR NDVI	emissions	and plant	2. (Reinhardt, 1997)
3. CONSUME			functional	3. (Ottmar et al., 1993)
4. FARSITE			types	4. (Finney, 1998)

All acronyms listed in tables and throughout the text are defined in Appendix 1.

capability of models for scaling biological data at various temporal (hours to decades) and spatial (single leaf to ecosystem) resolutions (e.g., (Bazzaz, 1993; Bonan, 1995; Landsberg et al., 1997a; Lucas and Curran, 1999; Nightingale et al., 2004; Sklar and Costanza, 1990)). We will assess three categories of terrestrial ecosystem models: (a) global water-energy-carbon models; (b) carbon and biochemistry models; and (c) ecosystem structure and function models, as well as generic ocean/aquatic models, and define the types of inputs they require. Model parameters that are currently derived and remain to be derived from remotely sensed data are outlined.

Table 21.2 A description of models currently utilized and with potential use for modeling distribution and change in biogeographical phenomena for aquatic systems

Model	Current Remote Sensing Inputs	Main Output	Application	References
AQUATIC				
Coral Reefs bleaching; hotspots and locations	AVHRR and MODIS SST	Maps of likely bleaching locations	Global coral reefs	http://www.osdpd.noaa.gov/PSB/EPS/SST/ methodology.html (Mumby et al., 2004)
Coral reef zonation/habitat	High spatial resolution images (Ikonos, QuickBird, airborne digital cameras)	Maps of reef habitat type zonation	Individual reefs	(Garza-Perez et al., 2004; Mumby et al., 2004)
Coral reef disturbance impacts	High spatial resolution images (Ikonos, QuickBird, airborne digital cameras)	Maps of reef habitat type zonation	Individual reefs	(Scopélitis et al., 2007; Tanner et al., 1996)
Coral reef fish habitat	High spatial resolution images (Ikonos, QuickBird, airborne digital cameras)	Abundance of fish species	Coastal and reef environments	(Pittman et al., 2004, 2007)
Seagrass growth of seagrass patches	High spatial resolution photographs and field data	Extent of seagrass patches	Seagrass meadows	(Fonseca et al., 2002; Kendrick et al., 2005)
Wetland—saltmarsh	Landsat TM, SPOT, airborne imaging spectrometers, and multispectral cameras	Aboveground biomass and productivity	Saltmarsh	(Hardisky et al., 1986; Phinn et al., 1999)
Wetland—mangrove	Imaging radar	Mangrove structure, composition, and flooding	Mangrove	(Costa, 2005; Hess et al., 2003; Lucas et al., 2007)
Oceanic color and productivity	Seawifs, CZCS, OCTS, and MODIS	Maps of ocean color parameters (chlorophyll; colored, dissolved organic matter; suspended inorganic matter) and productivity	Coastal and oceanic water bodies	http://oceancolor.gsfc.nasa.gov/ (Bukata, 2005; Esaias et al., 1998; Pribble et al., 1994) http://web.science.oregonstate.edu/ocean. productivity/ (Behrenfeld and Falkowski, 1997; Behrenfeld et al., 2005; Campbell et al., 2002)

21.1.2 Utility of remotely sensed data and products for ecosystem modelling

A number of models have been developed at a range of scales, from leaf and tree physiology levels, to general global circulation models that encompass biological feedbacks. While they may vary in their conceptual approach and scale of operation, the majority of models outlined in this review are simplified versions of reality and are typically aspatial (Goetz et al., 1999; Lucas and Curran, 1999; Nightingale et al., 2004). The aspatial nature of many process models has made it difficult to scale up biological and physiological data or processes to understand the functioning of ecosystems and their link to atmospheric transport processes. Many of these models are physiological in nature and represent purely biophysical processes. They do not explicitly represent higher-order processes (i.e., changes in vegetation structure). It has been noted that many of the changes in carbon sinks attributed to CO_2 fertilization, nitrogen deposition, or change in respiration are more likely to be a result of changes in the ecosystem structure (e.g., vegetation thickening and land-use change) (Moorcroft et al., 2001). Remotely sensed data enables the inclusion of changes in land surface vegetative cover into the modeling process. Process models driven by spatial datasets such as interpolated climate variables as well as remotely sensed estimates of key variables may be used to assess the spatial impact of global environment changes on forest processes (Lucas and Curran, 1999). The stability, repeat measurement capability, and global coverage of satellite and airborne instruments have led to the inclusion of their data and products into studies of land surface and atmospheric processes (Prince, 1999).

21.2 REMOTE SENSING IN BIOGEOGRAPHICAL ANALYSIS: CURRENT STATUS

In this section we provide a summary of current biogeographical applications of remote sensing for terrestrial and aquatic systems (Table 21.3 and Table 21.4, respectively). A number of reviews have addressed, in detail, the remote sensing applications for a range of terrestrial features and regions and these are used to mark the broad categories of biogeographical properties listed in Tables 21.3 and 21.4. Enhanced predictions of future patterns and behaviours of biogeographical phenomena may depend on the success of remote sensing in delivering the quantitative and qualitative discrimination required and on the match between the remote sensing products and model input requirements.

21.2.1 Terrestrial systems including wetlands

The scale of biogeographical change in response to forcing by climate and human disturbance necessitates the use of remote sensing data with broad spatial and frequent temporal coverage (Kerr and Ostrovsk, 2003; Table 21.3, I). Multitemporal, multispectral, optical and multifrequency/multipolarization microwave data from satellites are a fundamental resource for detecting changes in the terrestrial surface (Coppin et al., 2004; Lambin and Linderman, 2006; Table 21.3, I). Change detection depends on initial recognition of the target, error, and uncertainty in this, and then the ability to deal with robust and repeatable recognition of the target at a new time, such that the error of recognition does not swamp the detection of change. The main requirements are to detect modifications such as vegetation thinning; rapid and abrupt changes such as flood, fire, drought; separate interannual variation from longer-term trends; correct scale dependencies in statistical change estimates; and match temporal sampling rates to intrinsic temporal scales of processes (Coppin et al., 2004). The two main methods are bitemporal change detection and temporal trajectory analysis. The latter can be enhanced by applying patch panel metrics, which follow spatial properties of individual landscape patches through time (Crews, 2007). For example, Etter et al. (2006) use edge metrics to predict the likelihood of deforestation activity in the Amazon. Stow et al. (2004) review the use of remote sensing of vegetation and landcover change in Arctic tundra systems. This is potentially very informative as the Arctic is subject to the most rapid rates of global warming and subsequent biogeographical changes may be equally rapid. Here, multitemporal remote sensing using AVHRR NDVI, Landsat TM, IKONOS high-resolution imagery, and very high-resolution spectroscopy from a ground-based tramline reveal the rapidly occurring changes at scales from individual rock (moss and lichen patches) between individual seasons, to changes in regional vegetation photosynthetic activity over almost 20 years (Stow et al., 2004).

The role of satellite-derived NDVI in assessing ecological responses to environmental change has been summarized by Pettorelli et al. (2005; Table 21.3, II). They emphasize the value of NDVI for defining temporal and spatial trends and

Table 21.3 Application of remote sensing to measurements and mapping of terrestrial biogeographical phenomena

Imagery	Analysis	Recent Examples
I. Land cover/Land-cover change		
AVHRR MODIS SPOT Product time-series (250m, 500m, 1km, 8km)	• Univariate decision tree and regionally tuned classifications • Image-based cultivated land cover with hindcast modeling to back to 1700 • Linear mixture modeling and decision tree classifications • Deforestation analysis	• Land cover (Bartholome and Belward, 2005; Friedl et al., 2002) • Land-cover change—historical cropland (Goldewijk and Ramankutty, 2004) • Global percentage tree cover and vegetation continuous fields (Defries et al., 2000) • Forest cover change (Iverson et al., 2004)
ASTER Landsat SPOT IKONOS Aerial photographs	• Binary forest—nonforest classifications • Classifications using spectral properties, patch analysis, panel metrics, texture-based segmentation, diversity indices, and generalized linear models	• Forest cover change (Jha et al., 2005; Trigg et al., 2006) • Vegetation diversity, structure, patch structure, and floristics (Crews, 2007; Foody and Cutler, 2003)
II. Regional Dynamics and Phenology		
AVHRR NDVI MODIS NDVI EVI MODIS NBAR	• Simple curve thresholds, trend analysis, and derivatives • Forwards and backwards lagged median filters to define NDVI metrics; piecewise logistic functions fitted to curves	• Growing season length, leaf onset, surface phenology (Rasmussen et al., 2006; Tateishi and Ebata, 2004) • Seasonal phenological metrics (Reed et al., 1994; Zhang et al., 2003)
III. Vegetation Structure, Physiological Function and Productivity		
AVHRR NDVI and spectral reflectance	• NDVI inputs (to derive LAI) to CASA, BGC models • NDVI time trends in South America	• NPP, LUE, transpiration, and photosynthesis (Kimball et al., 2006; Lobell et al., 2002) • Canopy radiation interception (Paruelo et al., 2004)
MODIS spectral reflectance and products MODIS and MISR BRDF products	• Biome-based spectral reflectance pattern • Algorithm derived from BIOME-BGC • Radiative transfer modelling, BRDF angular indexes	• Leaf area index (Myneni et al., 2002) • NPP and photosynthesis (Running et al., 2004) • Canopy and sunlit/shaded fractions and clumping (Chen et al., 2003; Widlowski et al., 2004)
Hyperion AVIRIS hyperspectral imagery	• Differences in growth rates at high VPD from retrieval of canopy pigments	• Detection of physiological basis for invasive species competition (Asner et al., 2006)
Landsat ETM	• Landsat to characterize vegetation distribution and LAI	• Evapotranspiration around flux tower sites (Chen et al., 2005)
Radarsat ERS-1/2 JERS-1 ERS Wind Scatterometer	• Empirical analysis • Simple soil/vegetation scattering model	• Forest structure (Townsend, 2002) • Sahelian pastoral dynamics (Zine et al., 2005)
Airborne LiDAR	• LiDAR-derived height	• Vegetation structural change (Hurtt et al., 2004)

Continued

Table 21.3 Cont'd

Imagery	Analysis	Recent Examples
IV. Inundation, Flooding, and River Flow		
ERS-1 ERS-2 ASAR Radarsat JERS-1	• Inundation patterns from microwave backscatter sensitivity to surface water • Combination of surface water detection and digital elevation data	• Dynamics of wetland inundation and vegetation type (Hess et al., 2003; Kasischke et al., 2003) • Discharge from braided rivers, surface velocity, and width (Smith et al., 1998)
Passive microwave—SMMR and AMSR-E	• Passive microwave emission changes between land and water • Linked passive microwave data to calibrated water-balance model	• Dynamics of inundation—passive microwave (Hamilton et al., 2004; Sippel et al., 1998) • Hydrological and hydraulic modelling (Townsend and Walsh., 19989; Vorosmarty et al., 1998)
SAR Interferometry	• High interferometric sensitivity to changes in surface height	• Dynamics of inundation (Alsdorf et al., 2000)
V. Fire Dynamics and Distribution		
AVHRR AATSR ATSR MODIS SPOT	• Hotspots from thermal channel thresholds; burn-scar mapping from spectral and BRDF changes	• Global active fires and burn scars (Dwyer et al., 2000; Roy et al., 2002)
Landsat TM IRS ALI Quickbird IKONOS ERS2	• Spectral detection and microwave backscatter change	• Regional burn scars and patterns (Gimeno et al., 2004; Miettinen and Liew, 2005; Russell-Smith et al., 2003)
Hyperion AVIRIS	• Quantitative estimation of vegetation water content and spectral discrimination of fractional cover and plant types	• Fuel properties and postfire responses (Riano et al., 2002; Roberts et al., 2003)

Table 21.4 Application of remote sensing to measurement and mapping of aquatic biogeographical phenomena

Imagery	Analysis	Recent Examples
Seagrass/benthic species and corals cover mapping		
Large-format digital aerial photography and airborne multispectral and hyperspectral image data	Manual digitising around patch boundaries Supervised classification from field survey data Inversion of radiative transfer models to extract water depth concentrations of absorbing and scattering particles and reflectance of benthic features Empirical and deterministic predictive models for species distribution and patch growth	(Dekker et al., 2006; Dierssen et al., 2003; Goodman and Ustin, 2003; Kendrick et al., 2002; McKenzie et al., 2001; Mumby et al., 1998)
High spatial resolution multispectral satellite images (Ikonos/Geoeye, QuickBird/Worldview, EROS, ALOS-AVNIR, Landsat ETM, ASTER, Hyperion)	Supervised classification from field survey data Inversion of radiative transfer models to extract water depth concentrations of absorbing and scattering particles and reflectance of benthic features	(Ahmad and Neil, 1994; Dekker et al., 2005, 2006; Dustan et al., 2001; Goodman and Ustin, 2003; Hochberg and Atkinson, 2000; McKenzie et al., 2001; Mumby and Edwards, 2000; Mumby et al., 2004; Purkis et al., 2002)
Coastal topography and Bathymetric mapping		
Landsat TM/ETM ASTER Airborne hyperspectral (AVIRIS, Hymap, CASI, PHILLS, Hydice)	Supervised classification from field survey data Inversion of radiative transfer models to extract water depth concentrations of absorbing and scattering particles and reflectance of benthic features Empirical and deterministic predictive models for depth based on attenuation coefficient estimation	(Bierwirth et al., 1993; Dierssen et al., 2003; Hedley and Mumby, 2003; Louchard et al., 2003; Stumpf et al., 2003)
LiDAR—airborne laser-imaging detection and ranging	Direct processing of signal response to identify water surface and sea floor	(Hardy, 1999; Lyzenga, 1985; Quinn, 2000)
Optical properties of water bodies		
Landsat TM/ETM ASTER Airborne hyperspectral (AVIRIS, Hymap, CASI, PHILLS, Hydice)	Supervised classification from field survey data Inversion of radiative transfer models to extract water-depth concentrations of absorbing and scattering particles and reflectance of benthic features Empirical and deterministic predictive models for depth based on attenuation coefficient estimation	(Brando and Dekker, 2003; Bukata, 2005; Dekker et al., 2001; Dierssen et al., 2003; Lee et al., 1998)
CZCS SeaWiFS MODIS Ocean Products	Inversion of radiative transfer models to extract water-depth concentrations of absorbing and scattering particles and reflectance of benthic features Empirical and deterministic predictive models for depth based on attenuation coefficient estimation	(Carder et al., 1999; Dierssen et al., 2003; Esaias et al., 1998; Hooker and Maritorena, 2000; Smith et al., 1998)
Ocean surface roughness—wave train dimensions		
Quickscat, TOPEX, JASON, Radarsat, ERS½, ADEOS, ALOS PALSAR	Direct processing of signal response	Full datasets from Quickscat, TOPEX/JASON, ADEOS, NSCAT, SMI, Nimbus 7, and Sea-Sat can be found at the NASA Physical Oceanography DAAC
SST and climatology time-series		
AVHRR, MODIS—SST	Direct processing of signal response	Full data sets from AVHRR, GOES, MODIS, and in situ can be found at the NASA Physical Oceanography DAAC

variation in vegetation distribution, productivity, and dynamics to monitor habitat degradation and fragmentation, and the ecological effects of drought and fire. A subset of this is the role of remote sensing in detection of phenology and spring-growth onsets (Badeck et al., 2004; Table 21.3, II). A comparison of 2,853 dates of bud-burst from ground and satellite observations found that satellite green-up was 3.3 days ahead of observed bud-burst, but correlation between satellite and ground data was low (Badeck et al., 2004). Until recently, the coarse to moderate resolution of long-term times-series placed limits on the scale and magnitude of changes and trends detectable in heterogeneous landscapes (Coppin et al., 2004). However, the increased availability of the historical Landsat archive in 2009 (USGS technical announcement, April 21, 2008) should enable more widespread combination with moderate resolution time-series to improve sensitivity at finer scales.

Biodiversity is a key biogeographical measure of ecosystem condition. It is a difficult variable to model (since definitions vary), but remote sensing can make a large contribution to its quantification and modeling (Turner and Urbanski, 2003; Table 21.3, III). They cite species composition, landcover, biochemistry (specifically chlorophyll), ocean color and circulation, rainfall, soil moisture, phenology, topography, and vertical canopy structure as key ecological variables measurable to varying degrees by remote sensing.

The patterns of seasonal inundation in the landscape have a major influence on ecosystem dynamics and species distributions (Chacon-Moreno et al., 2004). Microwave remote sensing can be used to monitor the seasonal dynamics and temporal trends in inundation, river flows, and flooding using active, passive, and interferometry approaches (Hess et al., 2003; Mertes, 2002; Table 21.3, IV).

The application of remote sensing to the assessment of active fires, fire impact, and postfire effects has been summarized by Lentile et al. (2006; Table 21.3, V). There is a relatively long record of active fire hotspot data from several instruments; however, burn scar mapping is less universally available and the MODIS product is only now being produced since 2000. Hyperspectral sensors have already demonstrated powerful capability for the retrieval of quantitative biochemical, physiological, and physical properties (Table 21.3, V).

The examples in Table 21.3 are divided into the following categories: land cover and land-cover change, regional dynamics and phenology, vegetation structure, physiological function and productivity, inundation and river flow, and fire dynamics and distribution. For the most part,

these examples provide biogeographical data of the following forms: vegetation class, binary changes, time trends, diversity and patch structure, manifestation of a time-based physiological process, manifestation of a time-based disturbance, dynamics of water, and snapshots or time courses of a surface biochemical or physical property (Table 21.3). For modeling purposes, by far the most common inputs to analysis or process models are a land-cover classification or a time course of a light interception property such as LAI or fPAR. This suggests that even for analytical studies, let alone modeling, remote sensing is still in its infancy in terms of provision of sophisticated surface and process description into modeling and analysis of biogeographical phenomena and change. However, current advances in satellite sensors and derived data products provide reason to believe that this will rapidly change over the next decade.

21.2.2 Aquatic systems including coasts

Environments considered in this section include coastal margins, which are defined as areas between the high and low tidal levels, as these areas are subject to both aquatic and terrestrial processes. Other environments covered include lacustrine/riverine (lakes and rivers) and oceanic waters, as model development in these areas for both vegetation distributions and processes, have been extensive in the fields of limnology and biological oceanography. Coral reefs and seagrass environments are included as separate application areas here as each field has had extensive development of modelling from the individual plant to whole reef and regional scales, assessing biogeographic composition and process issues. In Table 21.4, the broad categories of aquatic applications have been defined as benthic, seagrass and cover mapping; coastal topographic and seafloor bathymetric mapping; optical properties of water bodies at a range of scales; sea surface roughness; and time-series of sea surface temperatures and climatology. The table provides some recent examples of remote sensing applications and methods.

The coastal margins included in this review extend from the shoreward limit of mean high-tide to depth limit at which benthic vegetation occurs or the limit of the photic zone. The seaward limit is variable and depends on water clarity and several other factors. The biogeographic features mapped, monitored, and modeled in this zone cover the water surface, depth, and contents of the water column, and the benthos and substrate in the inter- and subtidal areas. Water surface features in the coastal zone include algal blooms observable from airborne and satellite multispectral images (Richardson, 1996) and sea-surface elevation and water depth at fine spatial scales from airborne laser systems (Stumpf et al., 2003). Mapping water depth using satellite and airborne images and airborne laser data is confined to shallow (< 10 m for multispectral, < 20 m for airborne laser) and clear water bodies. Due to the variable water clarity and dynamic geomorphology of many coastal environments, there are limited 'standard' image data sets of near-shore bathymetry extending to the mean high-tide line, with the notable exception of the U.S. Army Corps of Engineers SHOALS data for the United States and its territories and the Reefbase global coral reef database.

The color of coastal water bodies is controlled by the scattering and absorption of different wavelengths of sunlight by suspended inorganic material and suspended and dissolved organic material (Kirk, 1994). A range of empirical, semianalytic, and analytic approaches has been developed from hydrologic radiative transfer theory to map the concentrations of specific types of suspended sediment responsible for the scattering and the pigments responsible for the suspended and dissolved organic matter (Dekker et al., 2006). The majority of organic matter mapping focuses on chlorophyll-a pigments associated with various species of phytoplankton. Hyperspectral image data have been used to map concentrations of nonphotosynthetic, accessory pigments, some of which are associated with harmful algal blooms (Richardson, 1996). The semianalytic and analytic approaches for mapping water column attributes require measurement or estimation of inherent optical properties of the water bodies they are applied to, and in addition to estimating concentrations of scattering and absorbing materials, are used to estimate depth, substrate reflectance, and derived properties such as attenuation coefficients (Kd) for light, PAR, and secchi depth (Lee et al., 1998). Standard image map products showing concentrations of suspended sediment load, chlorophyll-a concentration, PAR, and Kd have been developed from a number of long-running global monitoring satellite programs, all of which are accessible as daily, weekly, monthly, and annual data sets, along with climatologies from the Ocean Colour website (Esaias et al., 1998; Hooker and Maritorena, 2000), extending back to 1997. However, the algorithms applied to these data sets only work reliably in open ocean areas where there is no substrate visible. In coastal areas, which may be highly turbid and have substrate visible, these algorithms provide incorrect data and locally developed algorithms should be used (Qin et al., 2007).

Techniques have been developed for mapping and monitoring benthic features, such as the type, density, and biomass of submerged aquatic vegetation (SAV), including corals, seagrass, macro-algae, and micro-algae using various forms of remote sensing. Mumby et al. (2004) provide an overview of what can actually be mapped using airborne and satellite image data sets, and the environmental limits to mapping these features related to depth, water clarity, and level of cover of SAV. In areas that are optically clear, with the sea floor visible, a range of techniques have been applied to aerial photography, and multispectral and hyperspectral images from aircraft and satellites (Green et al., 2000). Multispectral image datasets are able to provide information for mapping seagrass, algae, and coral extents and variations in their density or cover. Mapping down to the species levels is possible in some seagrass environments, from multispectral and hyperspectral images (Dekker et al., 2006). However, mapping corals to genus or species is not possible from these types of data. Mapping of the composition of reef environments has focused more on morphologic and geomorphic differences between corals and zones of the reef, with airborne laser and underwater acoustic images being used to map surface forms and integrate this with image data (Mishra et al., 2005; Purkis et al., 2002). A similar approach is used in optically deep and turbid environments where remotely operated vehicles with acoustic imaging systems are the primary form of data used to assess benthic habitat composition and changes over time.

We have considered oceanic environments last as they have the most developed set of mapping, monitoring, and modeling applications for using remotely sensed data. The size and scale of temporal dynamics in these environments has meant that satellite remotely sensed data were one of the only suitable data sets for this environment. Biophysical variables routinely assessed from remotely sensed data over oceanic areas, and used in short- and long-term studies of physical and biological oceanography, include ocean color (concentrations of suspended inorganic and suspended and dissolved organic material), calcite ($CaCO_3$) concentration, sea-surface elevation, sea-surface roughness, wave height/period/direction, winds, skin temperature, algal blooms, and sea-ice cover. The principal difference of these parameters to those estimated for coastal and riverine/lakes discussed above is that there is no influence of substrate/benthos on the recorded images and that oceanic environments are truly hemispheric to global-scale features with highly dynamic two- and three-dimensional processes, which vary from hourly to decadal periods. The remotely sensed data used are significantly different to the applications discussed above in terms of their spatial and temporal scales, with 1-km pixels, coupled with image extents, which provide daily global coverage and daily revisit frequency. The majority of biophysical parameters, with the exception of ocean-color parameters, are derived from direct physical measurements, as opposed to application of a model to the recorded reflectance value in an individual band. A large community of scientists have developed many of the sensors used to map and monitor the parameters discussed above over the last 40 years. Hence the image products are supported by an extensive literature on the design and validation of the products (e.g., for ocean color, see Gregg and Casey, 2004; Hooker and McClain, 2000; Werdell and Bailey, 2005).

21.3 REMOTE SENSING FOR TERRESTRIAL MODELING

Remote sensing offers a valuable source of spatially comprehensive and temporally repeatable information that may be useful to ecological modellers. It has been recognized since the early 1990s that global ecological understanding is an impossible task to achieve without extensive and intensive use of remotely sensed data (Hall et al., 1995; Plummer, 2000; Sellers et al., 1995). Plummer (2000) indicates four ways in which remotely sensed data can be combined with ecological models: (a) to use remotely sensed data to provide estimates of variables required to drive ecological process models; (b) to use remotely sensed data to test, validate, or verify predictions of ecological process models; (c) to use remotely sensed data to update or adjust ecological process model predictions; and (d) to use ecological process models to understand remotely sensed data. In this section, we will discuss the two most common uses for remotely sensed data and derived products: (1) to provide parameters to drive models and (2) to verify model predictions.

21.3.1 Remotely sensed data and derived parameters for terrestrial ecosystem models

Remotely sensed data sets are most frequently used to provide estimates of variables required to calibrate and drive ecological process models. Satellite and airborne sensors have the advantage that they can provide synoptic data coverage over long periods of time, enabling land surface properties and changes in these to be monitored

temporally. These data may be processed to provide quantitative information on numerous biophysical variables. Sellers et al. (1995) provide a detailed overview of the parameters required for three distinct modeling areas including water-energy-carbon, carbon and biochemistry, and ecosystem structure and function. We have summarized and updated this list, following advances in the derivation of parameters from new and existing satellite datasets, Table 21.3. In addition we furnish our discussion with examples of the most innovative real-time suite of global land and ocean products derived from MODIS. This instrument on the NASA Terra Platform is being used to provide a new generation of land data products in support of the NASA Earth Science Enterprise, global change research, and natural resource management.

21.3.2 Water-energy-carbon models

Broad-scale ecosystem process and general global circulation models (GCMs) that assess water-energy and carbon dynamics are designed to simulate seasonal patterns in live biomass, annual production, and soil carbon and nitrogen levels, as well as the sensitivity of global ecosystems to changes in climate. The model time-step can range from hourly, daily, weekly, and biweekly up to annually and the model may provide simulations for hundreds of years. Ecosystem models generally operate at regional scales ($1-10$ km^2) while GCMs are run on spatial scales on the order of kilometers up to five-degree resolution. Such models include SiB (Sellers et al., 1986); DEMETER (Foley, 1994); G'DAY (Comins and McMurtrie, 1993); CENTURY (Parton et al., 1992); BIOME-BGC (Running et al., 1993); GLOPEM (Prince and Goward, 1995); and GLOCO (Hudson et al., 1994). These models rely on a biogeographical classification of vegetation as the main input parameter driving model calculations (Foley, 1994). They require only a few broad average monthly meteorological variables (such as rainfall, temperature, and evaporation) as well as generalized soil characteristics (type or texture) and may be applied globally for all terrestrial biomes as well as the ocean.

Land use and land cover: At the simplest level, remotely sensed datasets and their derived products are used to delimit the location, extent, and changes over time of vegetation communities and ecosystems (Ustin et al., 1993). For broad-scale GCMs, maps of land use, cover type, and more specifically biome/vegetation types can be derived using supervised or unsupervised classification of coarse resolution image data from sensors such as NOAA's AVHRR and MODIS. There are several

Table 21.5 Parameter requirements for enhanced biogeographic modelling along with potential source sensors and current or future usage

	Input/Validation Parameters	Sensors/Products Optical	Microwave	Usage/Status
Water–Energy–Carbon Models	**Land cover/use** Land cover/use/change	QuickBird, IKONOS, CASI, ADAR, HYMAP, Landsat TM, ETM, MSS, NOAA AVHRR, MODIS, SeaWiFS, SPOT, MERIS, Vegetation, MSG/EVIRI, POLDER, MSG-SEVIRI	JERS, ERS, RADARSAT	Input and validation
	Climate variables Albedo, solar radiation, PAR, long-wave radiation, precipitation, evapotranspiration, cloudiness, humidity, wind speed	MODIS, AVHRR, POLDER, Vegetation, MSG-SEVIRI	METEOSAT, GOES, GMS, INSAT, TRMM TOVS, AMSR-E	Surrogate variables may be used if weather station data are inadequate. Products in development.
	Soils and geophysical properties Soil moisture, texture, type		AMSR, Altimeter, ERS/Scatt, Vegetation L-band	Products in development
	Hydrological Snow/ice, SST, topography/bathymetry	MODIS, MSG-SEVIRI	C/X band radar, RADARSAT, ERS, SAR, LiDAR	
	Vegetation classifications Vegetation class, species types	AVHRR NDVI, MODIS, SeaWiFS		Input and validation
	Fire Fire scar, burnt area, fuel load	SPOT, ATSR-2, SPOT-VEG, MODIS, AVHRR (LAC and GAC)	RADARSAT, ERS JERS, SAR C, X and L bands	Input and validation
Carbon and biochemistry	**Biophysical** Leaf area index, fPAR, SLA, land surface temperature,	SPOT, Landsat MSS, TM, ETM, AVHRR, MISR, POLDER, MSG/SEVIRI, MERIS		Input and validation
	Phenology, radiation use efficiency	MODIS VI's, PRI		Not currently utilised
	Biochemistry Leaf—tree chemistry, water, nutrients Structural attributes and yield	CASI, HYMAP, AVIRIS		Not currently used in models
Structure and function	Gross photosynthesis, gross primary production, net primary production, biomass (aboveground), canopy structure, tree height, tree density, stand volume, basal area, tree diameters, crown width	MODIS SPOT	LiDAR, GLAS, IceSAT	Input and validation

global satellite-derived land and vegetation-cover products. The MODIS product, currently available through 2011 requires a combination of nadir-adjusted, BRDF-adjusted surface reflectances, vegetation index, surface temperature, and surface texture information in conjunction with a global set of land-cover training data to provide global classifications of the land surface at 500 m resolution (Friedl et al., 2010). Classifications range from broad land-cover types consisting of as few as nine classes to more detailed schemes used for the IGBP, which define up to 17 land-cover classes. POSTEL provides a global land-cover map at 300 m resolution that discriminates the land surfaces in 22 classes of the continental ecosystems as part of ESA's GLOBCOVER project. The global land-cover map is produced for the year 2005–2006, using the MERIS ensor onboard ENVISAT (POSTEL, 2006). Land-cover products are also produced by the ALOS and JERS (JAXA, 2007).

The MODIS Vegetation Continuous Fields product contains three layers, percent tree cover, percent herbaceous cover, and percent bare cover. This product is generated from monthly composites of 500-m resolution MODIS surface reflectance data (Hansen et al., 2003). Fractional Vegetation Cover (FVC) defines an important structural property of a plant canopy, which corresponds to the complement to unity of the gap fraction at nadir direction, accounting for the amount of vegetation distributed in a horizontal perspective. The FVC product is currently generated daily at the full spatial resolution of the MSG/SEVIRI instrument and is provided on a 10-day and monthly basis (EUMETSAT, 2008).

Soil Moisture: Soil water-holding capacity is important for understanding the global hydrologic cycle and its effect on weather and climate (Bindlish et al., 2006). On a global scale it is important as a boundary condition for hydrologic and climate models and at regional scales for agricultural assessment and flood control. Soil moisture is a difficult variable to measure on a consistent and spatially comprehensive basis and soils maps delineated at scales of 1 km^2 or larger generally mask significant spatial variation in physical and chemical properties. Global remote sensing of soil moisture has been a goal of research for the past two decades. Remote sensing of soil moisture can be accomplished using L-band (1.4 GHz) AMSR microwave radiometry, as demonstrated by Jackson (1999). There are currently new satellite sensors operating at somewhat higher frequencies that show promise for soil-moisture mapping under some conditions (Njoku et al., 2003). Soil-moisture products produced from AMSR (2003 to 2004) and the ERS Scatterometer (1992 to 2000) are provided at 0.5° spatial, and

10-day temporal resolution are provided through POSTEL (POSTEL, 2006). A surface soil-moisture product is derived from the ASCAT data and given in swath geometry. This product provides an estimation of the water saturation of the 0–2 cm topsoil layer, in relative units between 0 and 100%. The processor has been developed by the Institute of Photogrammetry and Remote Sensing of the Vienna University of Technology (EUMETSAT, 2008). New global soil moisture mapping satellites including ESA's SMOS (Soil Moisture and Ocean Salinity) and NASA's Decadal Survey mission, SMAP (Soil Moisture Active Passive) will provide much needed soil water content data for initializing and driving ecosystem models in the future.

Topography and Bathymetry: Topography is a major factor that determines the amount of solar radiation reaching any particular location on the Earth's surface and has a strong influence over land-surface hydrology. As part of the Shuttle radar topography mission, C-band and X-band radar sensors provided global topographic information with a ground resolution of 30 x 30 m and a vertical (elevation) resolution of 16 m (Rabus et al., 2003). The LiDAR sensor is an airborne terrain-mapping system used principally in the production of digital elevation models. LiDAR is an attractive topographic and bathymetric mapping tool due to its high accuracy, with routine 10–20-cm height (Z) errors in unvegetated low-slope terrain (Haugerud et al., 2003).

Meteorological Parameters: Ecosystem models can also operate with satellite-derived climate parameters such as albedo, solar radiation, photosynthetically active radiation, precipitation, humidity, wind speed, and evapotranspiration, if gridded climate-station data are not available. Daily meteorological data are produced by a global circulation model, GEOS-4, which incorporates both satellite-based and ground observations (Bloom et al., 2005). Data are made available through the GMAO at a 1° by 1.25° resolution every six hours. GEOS-4 utilizes three major types of satellite data: height and moisture profiles obtained from layer mean retrievals using TOVS radiance data; single-level cloud motion vector winds obtained from geostationary satellite images; and column Total Precipitable Water obtained from the SSM/I instrument onboard the DMSP (Defense Military Satellite Program) series of satellites. POSTEL provides a global scale daily precipitation product based on existing multisatellite products (TOVS, NOAA-12 and NOAA-14) and bias-corrected precipitation gauge analyses (POSTEL, 2006). The Japanese Aerospace Exploration Agency produces precipitation data derived from the TRMM (JAXA, 2007).

21.3.3 Carbon and biochemistry models

The most common application of remotely sensed data and derived products are as variables used to drive regional-scale carbon and biochemistry-based ecosystem process models. Biogeochemical models are increasingly being run in a spatially explicit mode, requiring moderate- to high-spatial resolution surfaces of land cover and other biophysical parameters such as LAI derived from satellite imagery (Cohen et al., 2003). The majority of process-based forest growth or ecosystem models simulate the growth of stands or forest plots on a daily to monthly time-interval at spatial scales of several square kilometers to hectares and include models such as DAYTRANS/PSN (Running et al., 1975); FORGRO (Mohren and J.R., 1995); FORMIND (Huth et al., 1996); JABOWA (Botkin et al., 1972); HYBRID (Friend and Schugart, 1993); FOREST-BGC (Running et al., 1988); 3-PG (Landsberg et al. 1997b); 3-PGS (Coops et al., 1998); and FORECAST (Kimmins et al., 1999). These models are designed to examine the effects on stand growth related to climate, soil water, pollutants, and nutrient uptake; calculate stand carbon balance based on photosynthesis and respiration; and simulate community dynamics, growth, biomass, photosynthesis, hydrological balance, and productivity of forest stands.

While the majority of these models do not usually incorporate remotely sensed data or derived parameters, many of the input parameters can be provided from satellite datasets. General input parameters driving model simulations include average monthly meteorological variables such as maximum and minimum temperature, relative humidity, vapor pressure deficit, evapotranspiration, precipitation, and incident photosynthetically active radiation, as derived for the water-energy-carbon models. Site-specific measurements of biophysical properties (LAI, ƒPAR), phenological metrics, canopy radiation-use efficiency and canopy biochemistry, in addition to soil type, texture, water-holding capacity, land use and land-cover change and more specifically, vegetation type, are required and development of these spatially explicit datasets is being refined.

Biophysical Parameters: Biophysical parameters commonly required and often provided by remotely sensed data include: LAI and ƒPAR (fraction of photosynthetically active radiation), Table 21.5. LAI and ƒPAR are biophysical variables, that describe canopy structure and are related to functional process rates of energy and mass exchange. These two parameters are estimated from remote sensing data using empirical relationships between values of LAI/ƒPAR and vegetation indices, which include near-infrared to red-band ratios such as the NDVI. These empirical relationships are site- and sensor-specific and are unsuitable for application to large areas or in different seasons (Stenberg et al., 2003; Tian et al., 2000). Global LAI/ƒPAR products are currently derived from several satellite data sets including MODIS (Myneni et al., 2003); MISR (Hu et al., 2003); SPOT-Vegetation and ATSR (GEOsuccess, 2005); POLDER and ADEOS (POSTEL, 2006); MSG/SEVIRI (EUMETSAT, 2008); and microwave and optical data from ENVISAT (Manninen et al., 2005).

Phenology: Plant vegetative cycles define the onset (start of season) and duration of a photosynthetically active canopy that influences the magnitude of carbon and water fluxes between the atmosphere and biosphere (Jolly et al., 2005). A number of methods have been employed to extract phenological events from a temporal vegetation index (VI) trace over one or multiple years, including greenness threshold values (Boles et al., 2004; Myneni et al., 2001), seasonal-midpoint VI (Schwartz et al., 2002), trend derivative (Gao et al. 2008), and curvature change rate/inflection point methods. The accuracy of these estimates may be questioned given the coarse temporal resolution of VI data. Recently, the TIMESAT program (Jonsson and Eklundh, 2004) has been used to analyze time-series satellite data and is able to derive 11 phenology parameters including the start, end, and length of the growth season (Gao et al., 2008).

Canopy Radiation-Use Efficiency: Radiation-use efficiency (RUE) relates biomass production to the PAR intercepted by a plant. It varies widely with different vegetation types, and accurate determination of the spatial and temporal distribution of photosynthetic CO_2 uptake by terrestrial vegetation is vital for understanding the dynamics of the global carbon cycle. Eddy covariance (EC) flux towers are the most direct means of estimating the daily and seasonal fluctuations in the RUE, however, are limited to single sites in representative ecosystems. The determination of RUE from satellite observations is currently being explored through the measurement of vegetation spectral reflectance changes associated with physiologic stress responses. This approach utilizes the Photochemical Reflectance Index (PRI) derived from MODIS narrow-band measurements of high radiometric accuracy (Drolet et al., 2005).

Fire Detection and Monitoring: Perturbations in biogeochemical cycling processes occur in fire-prone environments having both positive and negative implications for the sustained productivity of natural ecosystems. With the operational availability of satellite-derived fire information relating to the location, timing of fires and area burned, as well as providing information on areas susceptible to wildfire outbreaks, it will be feasible to run improved ecosystem models to accurately determine vegetation succession and productivity. Several satellite systems having different capabilities in terms of spatial resolution, sensitivity, spectral bands, and frequency of overpasses have been utilised in fire detection and monitoring studies (AVHRR, GOES and Landsat). More recently the MODIS fire and thermal anomalies products contain information unique to understanding the timing and spatial distribution of fires and characteristics such as the energy emitted from the fire and is available for both day and night (Giglio et al., 2003). Burn-scar and burned-area products are also produced from MODIS (Roy et al., 2008), SPOT-Vegetation (POSTEL, 2006), SEVIRI (Radiative Fire Power product) (EUMETSAT, 2008), and from a combination of SPOT-Vegetation, ATSR, and AATSR datasets (GEOsuccess, 2005).

Canopy Biochemistry: Information relating to the amount and spatial distribution of canopy biochemicals (chlorophyll content, lignin, nitrogen, cellulose, and starch concentration) is of importance for describing the vegetation process such as photosynthesis and productivity, nutrient cycling, and vegetation stress (Curran et al., 1997). Efforts to map canopy biochemicals have been limited to high spectral resolution hyperspectral sensors such as AVIRIS, HYMAP, and CASI (Coops et al., 2003). This information cannot be generalized or assumed to be constant across sites or biomes, thus limiting current use in broad-scale ecosystem models (Reich et al., 1999).

21.3.4 Ecosystem structure and yield models

Models that assess ecosystem structure are generally similar to those that assess ecosystem function related to carbon and biochemistry outlined above. Several of these models, such as 3-PG incorporate subroutines that determine stand structural attributes (tree density, mortality, basal areas, and tree diameters) based on empirical equations determined from field measurements. Remotely sensed estimates of stand structural attributes are not required as input into these models but rather can be used to validate model output.

21.3.5 Remote sensing to test, validate, and verify model predictions

Remote sensing can also be used to test, validate, or verify predictions of ecological process models as well as update or adjust ecological process model predictions. This is becoming increasingly common as more satellite-derived biophysical products are being derived such as those from the MODIS sensor. Many global products including *f*PAR, LAI, PSN (net photosynthesis), GPP (gross primary production), NPP (net primary production), and LST (land-surface temperature) are being produced regularly and continually validated and improved. Satellite-derived estimates of above-ground vegetation productivity, biomass accumulation, and stand-structural components such as tree height, density, and diameters are vital to understanding carbon sequestration and provide an essential tool for validating empirical forest growth models. LiDAR is an emerging remote sensing technology that directly measures the three-dimensional distribution of plant canopies and can accurately estimate vegetation structural attributes and may provide more accurate estimates than those determined from microwave or optical radiance measurements (Lefsky et al., 2002, 2005).

Satellite-derived measurements of terrestrial productivity such as the MODIS GPP and NPP products provide a relatively high-quality, current, and spatially comprehensive understanding of the global carbon cycle. Ecosystem models tend to define seasonal and regional patterns in biophysical features more accurately, given they can be more highly parameterised. This also means that models are more difficult to calibrate and apply over large areas and temporal resolution is limited. The concern associated with utilizing biophysical information derived from satellite data for model calibration/validation purposes pertains to confirming the accuracy of remotely sensed data products. This is a challenging task because of the difficulty in making direct and temporally repeatable measurements of biophysical phenomena at the appropriate scale (Coops et al., 2003; Nightingale et al., 2007; Turner and Urbanski, 2003). In regard to using satellite-derived vegetation indices or estimates of GPP from highly parameterised ecosystem models as surrogates for potential productivity, a recent study by Nightingale et al. (2008) concluded, the simpler the formulation, the better.

21.4 FUTURE DIRECTIONS

Remote sensing provides well-quantified and calibrated data sets of the Earth's surface, corrected for instrument radiometry, geometric distortions, atmospheric attenuation, and cloud effects. As these data are used they will improve our understanding of global dynamics and processes occurring on the land surface and in the oceans and atmosphere (Cohen and Justice, 1999). Many of the parameters that were desired for biogeographical modeling purposes in the early 1990s could not be derived from remotely sensed datasets. Most of the satellite-based ecosystem research focused on data derived from the NOAA AVHRR series of satellites (1980 to present). In as little as one decade, there has been an evolution of a suite of satellites (both optical and radar), remotely sensed data products, and advanced processing techniques that provide spatially explicit and accurate estimates of ecosystem structure (e.g., LiDAR) and function (e.g., phenology) at multiple temporal and spatial resolutions.

There are a number of modeling contexts within which remote sensing, both calibrated and corrected reflectances, and higher level products can provide new key data inputs:

1. Biogeographical predictive modeling in relation to environmental and species niches (Phillips et al., 2006). In this context, quantitative surface properties derived from remote sensing time-series can add greatly to the sophistication of the environmental and resource layers.
2. Digital global vegetation models (DVGM) such as CASA are being supported with time-series products from the MODIS sensor (Potter et al., 2007).
3. Regional land-use change modeling can utilize the global 30+ year archive of Landsat data to enhance mapping and account for nutrient flows and emissions (Lambin and Linderman, 2006).
4. Hydrological models can utilize archives of C-band SAR data over key global regions, such as Amazonia, enable monitoring and prediction of inundation and water dynamics.
5. Biogeochemical models can make explicit use of canopy chemistry information supplied by a new generation of global hyperspectral imagers (Hill et al., 2006).
6. Multiple constraints model data assimilation approaches (e.g., Barrett et al., 2005; Renzullo et al., 2008) utilize a variety of algorithms from weather prediction and mathematical and geographical analysis to combined diverse spatial and point data sources to optimize the estimation and prediction of dynamics in land surface properties including carbon, nitrogen, water, and energy. These approaches can make use of both raw and processed, and snapshot and time-series products of Earth observation to pragmatically provide best estimates of key dynamical fluxes such as carbon and dust influencing climate, vegetation distribution, and detection, impact, and consequences of human actions.

Considerable progress has been made in both remote sensing of land-surface properties and in understanding ecosystem dynamics. Despite the coupling of remotely sensed data and ecosystem models, the integration of this data within these models has not yet been fully exploited. Problems with satellite data that continue to limit their inclusion into ecosystem models and as standalone tools include continuity of datasets (e.g., AVHRR, MODIS), user overhead and pre-post processing, signal distortions due to cloud and sensor dropouts, data costs (high resolution image acquisitions), and difficulty in collecting reference/validation data sets (Phinn et al., 2002). In response to these known issues, there is a requirement from the user community and satellite data providers to improve data products through coordinated validation efforts and enhanced algorithm improvement (Morisette et al., 2002). Increases in the spatial and temporal resolution of imaging satellites are occurring as new satellites are launched. Future advances in satellite and airborne datasets and derived products will see improvement in the quality of biophysical variables able to be determined from these kinds of data. These sensors will offer new opportunities and challenges for mapping and monitoring the Earth's surface at a range of spatial and temporal resolutions and provide advanced information and products that may be incorporated into ecosystem process models. This will lead to the increased inclusion of these data in global ecosystem modelling activities. It is essential to explore and incorporate potential remotely sensed data sets to calibrate, drive, and validate ecosystem process models with the technology available today, so that full advantage can be taken of future advances in remote sensing (Waring and Running, 1999).

REFERENCES

Aber, J.D., and Federer, C.A. (1992) 'A generalised, lumped-parameter model of photosynthesis, evapotranspiration and net primary production in temperate and boreal forest ecosystems', *Oecologia*, 92: 463–74.

Ahmad, W., and Neil, D.T. (1994) 'An evaluation of Landsat Thematic Mapper (TM) digital data for discriminating coral reef zonation: Heron Reef (GBR)', *International Journal of Remote Sensing*, 15: 2583–97.

Alsdorf, D.E., Melack, J., Dunne, R., Mertes, L., Hess, L., and Smith, L. (2000) 'Interferometric radar measurements of water level changes on the Amazon floodplain', *Nature*, 404: 174–77.

Asner, G.P., Martin, R.E., Carlson, K.M., Rascher, U., and Vitousek, P.M. (2006) 'Vegetation-climate interactions among native and invasive species in Hawaiian rainforest', *Ecosystems*, 9: 1106–17.

Badeck, F.W., Bondeau, A., Bottcher, K., Doktor, D., Lucht, W., Schaber, J., and Sitch, S. (2004) 'Responses of spring phenology to climate change', *New Phytologist*, 162: 295–309.

Barrett, D.J., Hill, M.J., Hutley, L., Beringer, J., Xu, J., Cook, G., Carter, J., and Williams, R. (2005) 'Prospects for improved savanna biophysical models by using multiple-constraints model-data assimilation methods', *Australian Journal of Botany*, 53: 689–714.

Bartholome, E., and Belward, A. (2005) 'GLC2000: A new approach to global land cover mapping from Earth observation data', *International Journal of Remote Sensing*, 26: 1959–77.

Bazzaz, F.A. (1993) 'Scaling in biological systems: Population and community perspectives', in J.R. Ehleringer and B. Field (eds.), *Scaling Physiological Processes: Leaf to Globe*. New York: Academic Press. pp. 233–54.

Behrenfeld, M.J., Boss, E., Siegel, D.A., and Shea, D.M. (2005) 'Carbon-based ocean productivity and phytoplankton physiology from space', *Global Biogeochemical Cycles*, 19: 1–14.

Behrenfeld, M.J., and Falkowski, P.G. (1997) 'A consumer's guide to phytoplankton primary productivity models', *Limnology and Oceanography*, 42: 1479–91.

Bierwirth, P.N., Lee, T.J., and Burne, R.V. (1993) 'Shallow sea-floor reflectance and water depth derived by unmixing multispectral imagery', *Photogrammetric Engineering and Remote Sensing*, 59: 331–38.

Bindlish, R., Jackson, T., Gasiewski, A., Klein, M., and Njoku, E. (2006) 'Soil moisture mapping and AMSR-E validation using the PSR in SMEX02', *Remote Sensing of Environment*, 103: 127–39.

Bloom, S. and 10 authors (2005) 'Documentation and Validation of the Goddard Earth Observing System (GEOS) Data Assimilation System—Version 4', in *Technical Report Series on Global Modeling and Data Assimilation*. Greenbelt, MD: NASA. p. 187.

Boles, S.H., Xiao, X., Liu, J., Zhang, Q., Munkhtuya, S., Chen, S., and Ojima, D. (2004) 'Land cover characterisation of Temperate East Asia using multi-temporal VEGETATION sensor data', *Remote Sensing of Environment*, 90: 477–89.

Bonan, G.B. (1991) 'Atmosphere-biosphere exchange of carbon dioxide in boreal forests', *Journal of Geophysical Research*, 96: 7301–12.

Bonan, G.B. (1995) 'Land-atmosphere CO_2 exchange simulated by a land surface process model coupled to an atmospheric general circulation model', *Journal of Geophysical Research*, 100: 2817–31.

Botkin, D.B., Janak, J.F., and Wallis, J.R. (1972) 'Some ecological consequences of a computer model of forest growth', *Journal of Ecology*, 60: 849–73.

Brando, V.E., and Dekker, A.G. (2003) 'Satellite hyperspectral remote sensing for estimating estuarine and coastal water quality', *IEEE Transactions on Geoscience and Remote Sensing*, 41: 1378–87.

Bukata, R.P. (2005). *Satellite Monitoring of Inland and Coastal Water Quality: Retrospection, Introspection and Future Direction*. Boca Raton, FL: CRC Press/Taylor and Francis.

Campbell, J., and 22 authors. (2002) 'Comparison of algorithms for estimating ocean primary production from surface chlorophyll, temperature, and irradiance', *Global Biogeochemical Cycles*, 16: 1–15.

Carder, K.L., Chen, F.R., Lee, Z.P., Hawes, S.K., and Kamykowski, D. (1999) 'Semianalytic moderate-resolution imaging spectrometer algorithms for chlorophyll a and absorption with bio-optical domains based on nitrate-depletion temperatures', *Journal of Geophysical Research*, 104: 5403–21.

Chacon-Moreno, E., Naranjo, M., and Acevedo, D. (2004) 'Direct and indirect vegetation-environment relationships in the flooding savanna of Venzuela', *Ecotropicos*, 17: 25–37.

Chen, J.M., Chen, X., Jua, W., and Geng, X. (2005) 'Distributed hydrological model for mapping evapotranspiration using remote sensing inputs', *Journal of Hydrology*, 305: 15–39.

Chen, J.M., Liu, J., Leblanc, S.G., Lacaze, B., and Roujean, J. (2003) 'Multi-angular optical remote sensing for assessing vegetation structure and carbon absorption', *Remote Sensing of Environment*, 84: 516–25.

Cohen, W., and Justice, C. (1999) 'Validating MODIS terrestrial ecology products: Linking in situ and satellite measurements', *Remote Sensing of Environment*, 70: 1–3.

Cohen, W.B., Maiersperger, T.K., Yang, Z., Gower, S.T., Turner, D.P., Ritts, W.D., Berterretche, M. and Running, S.W. (2003) 'Comparison of land cover and LAI estimates derived from ETM+ and MODIS for four sites in North America: A quality assessment of 2000/2001 provisional MODIS products', *Remote Sensing of Environment*, 88: 233–55.

Comins, H.N., and McMurtrie, R.E. (1993) 'Long-term response of nutrient-limited forests to CO_2 enrichment; equilibrium behaviour of plant-soil models', *Ecological Applications*, 3: 666–81.

Coops, N.C., Smith, M.L., Martin, M.M., and Ollinger, S. (2003) 'Prediction of Eucalypt foliage nitrogen content from satellite derived hyperspectral data', *IEEE Transactions on Geoscience and Remote Sensing*, 41: 1338–46.

Coops, N.C., Waring, R.H., and Landsberg, J.J. (1998) 'Assessing forest productivity in Australia and New Zealand using a physiologically-based model driven with averaged monthly weather data and satellite-derived estimates of canopy photosynthetic capacity', *Forest Ecology and Management*, 104: 113–27.

Coppin, P., Jonckheere, I., Nackaerta, K., Muys, B., and Lambin, E. (2004) 'Digital change detection methods in ecosystem monitoring: A review', *International Journal of Remote Sensing*, 25: 1565–96.

Costa, M. (2005) 'Estimating net primary productivity of aquatic vegetation of the Amazon Floodplain using Radarsat and JERS-1 imagery', *International Journal of Remote Sensing*, 26: 4527–36.

Crews, K.A. (2007) 'Landscape dynamism: Disentangling thematic versus structural change in northeast Thailand', in R. Aspinall and M.J. Hill (eds.), *Land Use Change: Science, Policy and Management*. New York: Taylor and Francis. pp. 95–118.

Curran, P.J., Kupiec, J.A., and Smith, G.M. (1997) 'Remote sensing the biochemical composition of a slash pine canopy', *IEEE Transactions on Geoscience and Remote Sensing*, 35: 415–20.

Defries, R.S., Hansen, M.C., Townshend, J.R.G., Janetos, A.C., and Lovelands, T.R. (2000) 'A new global 1-km dataset of percentage tree cover derived from remote sensing', *Global Change Biology*, 6: 247–54.

Dekker, A., Brando, V., and Anstee, J. (2005) 'Retrospective seagrass change detection in a shallow coastal tidal Australian lake', *Remote Sensing of Environment*, 97: 415–33.

Dekker, A., Brando, V., Anstee, J., Fyfe, S.K., Malthus, T., and Karpouzli, E. (2006) 'Remote sensing of seagrass systems: Use of spaceborne and airborne systems', in A.W.D. Larkum, R. Orth and C.M. Duarte (eds.), *Seagrasses: Biology, Ecology and Conservation*. Dordrecht: Springer. pp. 347–59.

Dekker, A.G., Brando, V.E., Anstee, J., Pinnel, N., and Held, A. (2001) 'Preliminary assessment of the performance of Hyperion in coastal waters. Cal/Val activities in Moreton Bay, Queensland, Australia', *IGARSS 2001. Scanning the Present and Resolving the Future. Proceedings. IEEE 2001 International Geoscience and Remote Sensing Symposium Cat. No.01CH37217. 2001*. Piscataway, NJ: IEEE. pp. 2665–67.

Dierssen, H.M., Zimmerman, R.C., Leathers, R.A., Downes, T.V., and Davis, C.O. (2003) 'Ocean color remote sensing of seagrass and bathymetry in the Bahamas Banks by high-resolution airborne imagery', *Limnology and Oceanography*, 48: 444–55.

Dixon, G. (2003) 'Essential FVS: A user's guide to the forest vegetation simulator', in U.S. Department of Agriculture, Forest Service, Forest Management Service Center. Fort Collins, CO. p. 193.

Donnelly, J.R., Moore, A.D., and Freer, M. (1997) 'GRAZPLAN: Decision support systems for Australian grazing enterprises. I. Overview of the GRAZPLAN project, and a description of the MetAccess and LambAlive DSS', *Agricultural Systems*, 54: 57–76.

Drolet, G.G., Huemmrich, K.F., Hall, F.G., Middleton, E.M., Black, T.A., Barr, A.G., and Margolis, H.A. (2005) 'A MODIS-derived photochemical reflectance index to detect inter-annual variations in the photosynthetic light-use efficiency of a boreal deciduous forest', *Remote Sensing of Environment*, 98: 212–224.

Dustan, P., Dobson, E., and Nelson, G. (2001) 'Landsat thematic mapper: Detection of shifts in community composition of coral reefs', *Conservation Biology*, 15: 892–902.

Dwyer, E., Pinnock, S., Gregoire, J., and Pereira, J.S. (2000) 'Global spatial and temporal distribution of vegetation fire as determined from satellite observation', *International Journal of Remote Sensing*, 21: 1289–302.

Esaias, W.E., and 12 authors. (1998) 'An overview of MODIS capabilities for ocean science observations', *IEEE Transactions on Geoscience and Remote Sensing*, 36: 1250–65.

Etter, A., McAlpine, C.A., Phinn, S., Pullar, D., and Possingham, H. (2006) 'Unplanned land clearing of Colombian rainforests: Spreading like disease?' *Landscape and Urban Planning*, 77: 240–54.

EUMETSAT (2008). Land Surface Analysis Satellite Applications Facility. http://landsaf.meteo.pt/products/prods.jsp. Accessed 1 April 2009.

Finney, M.A. (1998) 'FARSITE: Fire Area Simulator—model development and evaluation'. USDA Rocky Mountain Research Station. Ogden, UT. p. 47.

Foley, J.A. (1994) 'Net primary productivity in the terrestrial biosphere: The application of a global model', *Journal of Geophysical Research*, 99: 20773–83.

Fonseca, M., Whitfield, P.E., Kelly, N.M., and Bell, S.S. (2002) 'Modeling seagrass landscape pattern and associated ecological attributes', *Ecological Applications*, 12: 218–37.

Foody, G.M., and Cutler, M. (2003) 'Tree biodiversity in protected and logged Bornean tropical rain forests and its measurement by satellite remote sensing', *Journal of Biogeography*, 30: 1053–66.

Friedl, M. A., Sulla-Menashe, D., Tan, B., Schneider, A., Ramankutty, N., Sibley, A., & Huang, X. (2010). 'MODIS Collection 5 global land cover: Algorithm refinements and characterization of new datasets', *Remote Sensing of Environment*, 114: 168–182.

Friend, A.D., and Schugart, H.H. (1993) 'A physiology-based gap model of forest dynamics', *Ecology*, 74: 792–97.

Garza-Perez, J.R., Lehmann, A., and Arias-Gonzalez, J.E. (2004) 'Spatial prediction of coral reef habitats: Integrating ecology with spatial modeling and remote sensing', *Marine Ecology-Progress Series*, 269.

Gao, F., J. Morisette, R. Wolfe, G. Ederer, J. Pedelty, E. Masuoka, R. Myneni, B. Tan, and J. Nightingale. 2008. An algorithm to produce temporally and spatially continuous remote sensing time series data: An example using MODIS LAI. *IEEE Geosciences and Remote Sensing Letters*. 5(1): 60–64.

GEOsuccess (2005). GEOsuccess Products. http://geofront.vgt. vito.be/geosuccess/relay.do?dispatch=products. Accessed 1 April 2009.

Giglio, L., Descloitres, J., Justice, C., and Kaufman, Y. (2003) 'An enhanced contextual fire detection algorithm for MODIS', *Remote Sensing of Environment*, 87: 273–82.

Gimeno, M., San-Miguel-Ayanz, J., and Schmuck, G. (2004) 'Identification of burnt areas in Mediterranean forest environments from ERS-2 SAR time series', *International Journal of Remote Sensing*, 25: 4873–88.

Goetz, S.J., Prince, S.D., Goward, S.N., Thawley, M.M., and Small, J. (1999) 'Satellite remote sensing of primary production: An improved production efficiency modeling approach', *Ecological Modelling*, 122: 239–55.

Goldewijk, K.K., and Ramankutty, N. (2004) 'Land cover change over the last three centuries due to human activities: The availability of new global data sets', *GeoJournal*, 61: 335–44.

Goodman, J., and Ustin, S. (2003) 'Airborne hyperspectral analysis of coral reef ecosystems in the Hawaiian Islands', in *International Symposium on Remote Sensing of Environment*. Honolulu, HI: International Symposium of Remote Sensing of Environment.

Green, E.P., Mumby, P.J., Edwards, A.J., and Clark, C.D. (2000). *Remote Sensing Handbook for Tropical Coastal Management*. Paris: UNESCO.

Gregg, W., and Casey, N. (2004) 'Global and regional evaluation of the SeaWiFS chlorophyll data set', *Remote Sensing of Environment*, 93: 463–79.

Hall, F.G., Townshend, J.R., and Engman, E.T. (1995) 'Status of remote sensing algorithms for estimation of land surface state parameters', *Remote Sensing of Environment*, 51: 138–56.

Hamilton, S., Sippel, S., and Melack, J. (2004) 'Seasonal inundation patterns in two large savanna floodplains of South America: the Llanos de Moxos (Bolivia) and the Llanos del Orinoco (Venezuela and Colombia)', *Hydrological Processes*, 18: 2103–16.

Hansen, M.C., Defries, R.S., Townsend, J., Carroll, M., Dimiceli, C., and Sohlberg, R. (2003) 'Global percent tree cover at a spatial resolution of 500 meters: First results of the MODIS Vegetation Continuous Fields Algorithm', *Earth Interactions*, 7: 1–15.

Hardisky, M.A., Gross, M.F., and Klemas, V.V. (1986). 'Remote sensing of coastal wetlands', *BioScience*, 36: 453–60.

Hardy, J.T. (1999) 'Coral reef monitoring with airborne LIDAR', in *International Workshop on the Use of Remote Sensing Tools for Mapping and Monitoring Coral Reefs*. Honolulu HI, Hawaii.

Haugerud, R., Harding, D., Johnson, S., Harless, J., and Weaver, C. (2003) 'High-resolution Lidar topography of the Puget Lowland, Washington—A bonanza for Earth science', *Geological Society of America Today*, 13: 4–10.

Hedley, J.D., and Mumby, P.J. (2003) 'A remote sensing method for resolving depth and subpixel composition of aquatic benthos', *Limnology and Oceanography*, 48: 480–88.

Hess, L., Melack, J., Novob, E., Barbosac, C., and Gastil, M. (2003) 'Dual-season mapping of wetland inundation and vegetation for the central Amazon basin', *Remote Sensing of Environment*, 87: 404–28.

Hill, M.J., Senarath, U., Lee, A., Zeppel, M., Nightingale, J.M., Williams, R., and McVicar, T. (2006) 'Assessment of the MODIS LAI product for Australian ecosystems', *Remote Sensing of Environment*, 101: 495–518.

Hochberg, E.J., and Atkinson, M.J. (2000) 'Spectral discrimination of coral reef benthic communities', *Coral Reefs*, 19: 164–71.

Hooker, S.B., and Maritorena, S. (2000) 'An evaluation of oceanographic radiometers and deployment methodologies', *Journal of Atmospheric Oceanography Techniques*, 17: 811–30.

Hooker, S.B., and Mc Clain, C.R. (2000) 'The calibration and validation of SeaWiFS data', *Progress in Oceanography*, 45: 427–65.

Hu, J., Shabanov, N., Crean, K., Martonchik, J., Diner, D., Knyazikhin, Y., and Myneni, R.B. (2003) 'Performance of the MISR LAI and FPAR algorithm: a case study in Africa', *Remote Sensing of Environment*, 88: 324–340.

Hudson, R.J.M., Gherini, S.A., and Goldstein, R.A. (1994) 'Modeling the global carbon cycle: Nitrogen fertilisation of the terrestrial biosphere and the "missing" CO_2 sink', *Global Biogeochemical Cycles*, 8: 307–33.

Hunt, J.E.R., Martin, F.C., and Running, S.W. (1991) 'Simulating the effects of climatic variation on stem carbon accumulation of a ponderosa pine stand: Comparison with annual growth increment data', *Tree Physiology*, 9: 161–71.

Hurtt, G., Dubayah, R., Drake, J., Moorcroft, P., Pacala, S., Blair, J., and Fearon, M. (2004) 'Beyond potential vegetation: Combining Lidar data and a height-structured model for carbon studies', *Ecological Applications*, 14: 873–83.

Huth, A., Ditzer, T., and Bossell, H. (1996) 'Simulation of the growth of tropical rain forests, Final Report to GTZ'. Kassel: Centre for Environmental Systems Research, University of Kassel. p. 180.

Iverson, L., Schwartz, M.W., and Prasad, A. (2004) 'How fast and far might tree species migrate in the eastern United States due to climate change?' *Global Ecology and Biogeography*, 13: 209–19.

Jackson, R.B. (1999) 'The importance of root distributions for hydrology, biogeochemistry and ecosystem functioning', in J.D. Tenhunen and P. Kabat (eds.), *Integrating Hydrology, Ecosystem Dynamics, and Biogeochemistry in Complex Landscapes*. London: John Wiley. pp. 217–38.

JAXA (2007). Japanese Aerospace Exploration Agency. http://www.jaxa.jp/projects/index_e.html. Accessed 1 May 2009.

Jha, C., Goparaju, L., Tripathi, A., Gharai, B., Raghubanshi, A., and Singh, J. (2005) 'Forest fragmentation and its impact on species diversity: An analysis using remote sensing and GIS', *Biodiversity and Conservation*, 14: 1681–98.

Johnson, I.R., Lodge, G.M., and White, R.E. (2003) 'The sustainable grazing systems pasture model: Description, philosophy and application to the SGS national experiment', *Australian Journal of Experimental Agriculture*, 43(8): 711–728.

Jolly, W.M., Nemani, R.R., and Running, S.W. (2005) 'A generalised, bioclimatic index to predict foliar phenology in response to climate', *Global Change Biology*, 11: 619–32.

Jonsson, P., and Eklundh, L. (2004) 'TIMESAT—a program for analysing time-series of satellite sensor data', *Computers and Geosciences*, 30: 833–45.

Kasischke, E., Smith, K., Bourgeau-Chavez, L., Romanowicz, E., Brunzella, S., and Richardson, C. (2003) 'Effects of seasonal hydrologic patterns in south Florida wetlands on radar backscatter measured from ERS-2 SAR imagery', *Remote Sensing of Environment*, 88: 423–41.

Kendrick, G., Aylward, M., Hegge, B., Cambridge, M., Hillman, K., Wyllie, A., and Lord, D. (2002) 'Changes in seagrass coverage in Cockburn Sound, Western Australia between 1967 and 1999', *Aquatic Botany*, 73: 75–87.

Kendrick, G.A., Duarte, C.M., and Marba, N. (2005) 'Clonality in seagrasses, emergent properties and seagrass landscapes', *Marine Ecology Progress Series*, 290: 291–96.

Kerr, J., and Ostrovsk, M. (2003) 'From space to species: Ecological applications for remote sensing', *Trends in Ecology and Evolution*, 18: 299–305.

Kimball, J., Zhao, M., McDonald, K., and Running, S. (2006) 'Satellite remote sensing of terrestrial net primary production for the pan-arctic basin and Alaska', *Mitigation and Adaptation Strategies for Global Change*, 11: 783–804.

Kimmins, J.P., Mailly, D., and Seely, B. (1999) 'Modelling forest ecosystem net primary production: The hybrid simulation approach used in FORECAST', *Ecological Modelling*, 122: 195–224.

Kirk, J.T.O. (1994) 'Estimation of the absorption and the scattering coefficients of natural waters by use of underwater irradiance measurements', *Applied Optics*, 33: 3276–78.

Lambin, E., and Linderman, M. (2006) 'Time series of remote sensing data for land change science', *IEEE Transactions on Geoscience and Remote Sensing*, 44: 1926–28.

Landsberg, J.J., and Gower, S.T. (1997). *Applications of Physiological Ecology to Forest Management*. San Diego: Academic Press.

Landsberg, J.J., and Waring, R.H. (1997) 'A generalised model of forest productivity using simplified concepts of radiation-use efficiency, carbon balance and partitioning', *Forest Ecology and Management*, 95: 209–28.

Lee, Z.P., Carder, K.L., Steward, R.G., Peacock, T.G., Davis, C.O., and Patch, J.S. (1998) 'An empirical algorithm for light absorption by ocean water based on color', *Journal Geophysical Research*, 103: 27967–78.

Lefsky, M., Cohen, W., Parker, G., and Harding, D. (2002) 'Lidar remote sensing for ecosystem studies', *BioScience*, 52: 19–30.

Lefsky, M., Harding, D.J., Keller, M., Cohen, W.B., Carabajal, C.C., Del Bom Espirito-Santo, F., Hunter, M.O. and de Oliveira Jr., R. (2005) 'Estimates of forest canopy height and aboveground biomass using ICESat', *Geophysical Research Letters*, 32: L22S02 21–24.

Lentile, L., Holden, Z.A., Smith, A.M.S., Falkowski, M.J., Hudak, A.T., Morgan, P., Lewis, S.A., Gessler, P.E. and Benson, N.C. (2006) 'Remote sensing techniques to assess active fire characteristics and post-fire effects', *International Journal of Wildland Fire*, 15: 319–45.

Liu, J., Chen, J.M., Cihlar, J., and Park, W.M. (1997) 'A process-based boreal ecosystem productivity simulator using remote sensing inputs', *Remote Sensing of Environment*, 62: 158–75.

Lobell, D., Hicke, J., Asner, G.P., Field, C.B., Tucker, C.J., and Loss, S. (2002) 'Satellite estimates of productivity and light use efficiency in United States agriculture', *Global Change Biology*, 8: 722–35.

Louchard, E.M., Reid, R.P., Stephens, F.C., Davis, C.O., Leathers, R.A., and Downes, T.V. (2003) 'Optical remote sensing of benthic habitats and bathymetry in coastal environments at Lee Stocking Island, Bahamas: A comparative spectral classification approach', *Limnology and Oceanography*, 48: 511–21.

Lucas, N.S., and Curran, P.J. (1999) 'Forest ecosystem simulation modelling: The role of remote sensing', *Progress in Physical Geography*, 23: 391–423.

Lucas, R.M., Mitchell, A.M., Rosenqvist, A., Proisy, C., Melius, A., and Ticehurst, C. (2007) 'The potential of L-band SAR for quantifying mangrove characteristics and change: Case studies from the tropics', *Aquatic Conservation: Marine and Freshwater Ecosystems*, 17: 245–64.

Ludeke, M.K.B., and 12 authors. (1994) 'The Frankfurt Biosphere Model: A global process-oriented model of seasonal and long-term CO_2 exchange between terrestrial ecosystems and the atmosphere. 1. Model description and illustrative results for cold deciduous and boreal forests', *Climate Research*, 4: 143–66.

Ludwig, J.A., Coughenour, M.B., Liedloff, A.C., and Dyer, R. (2001) 'Modelling the resilience of Australian savanna systems to grazing impacts', *Environment International*, 27: 167–72.

Lyzenga, D.R. (1985) 'Shallow water bathymetry using combined Lidar and passive multispectral scanner data', *International Journal Remote Sensing*, 6: 115–25.

Manninen, T., Stenberg, P., Rautiainen, M., Smolander, H., Voipio, P., and Ahola, H. (2005) 'Boreal forest retrieval using both optical and microwave data of ENVISAT', *IEEE Transactions on Geoscience and Remote Sensing*, 6: 5033–36.

McKenzie, L., J, Finkbeiner, M.A., and Kirkman, H. (2001) 'Seagrass mapping methods', in F.T. Short and R.G. Coles (eds.), *Global Seagrass Research Methods*. Amsterdam: Elsevier. pp. 101–22.

Mertes, L. (2002) 'Remote sensing of riverine landscapes', *Freshwater Biology*, 47: 799–816.

Miettinen, J., and Liew, S. (2005) 'Connection between fire and land cover change in Southeast Asia: A remote sensing case study in Riau, Sumatra', *International Journal of Remote Sensing*, 26: 1109–26.

Mishra, D.R., Narumalani, S., Rundquist, D., and Lawson, M. (2005) 'High-resolution ocean color remote sensing of Benthic habitats: A case study at the Roatan Island, Honduras' *IEEE Transactions on Geoscience and Remote Sensing*, 43: 1592–604.

Mohren, G.M.J., and J.R., van der Veen (1995) 'Forest growth in relation to site conditions: Application of the model FORGRO to the Solling spruce site', *Ecological Modelling*, 83: 173–83.

Moorcroft, P., Hurtt, G., and Pacala, S. (2001) 'A method for scaling vegetation dynamics: The ecosystem demography model', *Ecological Monographs*, 71: 557–86.

Morisette, J., Privette, J., and Justice, C. (2002) 'A framework for the validation of MODIS land products', *Remote Sensing of Environment*, 83: 77–96.

Mumby, P., Clark, C.D., Green, E.P., and Edwards, A.J. (1998) 'Benefits of water column correction and contextual editing for mapping coral reefs', *International Journal Remote Sensing*, 19: 203–10.

Mumby, P.J., and Edwards, A.J. (2000) 'Remote sensing objectives of coastal managers', in A.J. Edwards (ed.), *Remote Sensing Handbook for Tropical Coastal Management*. Paris: UNESCO. p. 361.

Mumby, P.J., Skirving, W., Strong, A.E., Hrady, J.T., LeDrew, E.F., Hochberg, E.J., Strumpf, R.P. and David, L.T. (2004) 'Remote sensing of coral reefs and their physical environment', *Marine Pollution Bulletin*, 48: 219–28.

Myneni, R.B. and Dong, J., Tucker, C.J., Kaufmann, R.K., Kauppi, P.E., Liski, J., Zhou, L., Alexeyev, V. and Hughes, M.K. (2001) 'A large carbon sink in the woody biomass of Northern forests', *Proceedings of the National Academy of Sciences of the United States of America*, 98: 14784–89.

Myneni, R.B. and 15 authors (2002) 'Global products of vegetation leaf area and fraction absorbed PAR from year one of MODIS data', *Remote Sensing of Environment*, 83: 214–31.

Myneni, R.B. and 15 authors (2003) 'FPAR, LAI 8-day composite NASA MODIS Land Algorithm User's Guide', Boston University. p. 17. http://cliveg.bu.edu/modismisr/index.html. Accessed 1 May 2009.

Nightingale, J.M., Coops, N.C., Waring, R.H., and Hargrove, W.W. (2007) 'Comparison of MODIS gross primary production estimates for forests across the U.S.A. with those generated by a simple process model, 3-PGS', *Remote Sensing of Environment*, 109: 500–509.

Nightingale, J.M., Fan, W., Coops, N.C., and Waring, R.H. (2008) 'Predicting tree diversity across the United States as a function of modeled gross primary production', *Ecological Applications*, 18: 93–103.

Nightingale, J.M., Phinn, S.R., and Held, A.A. (2004) 'Ecosystem process models at multiple scales for mapping tropical forest productivity', *Progress in Physical Geography*, 28: 241–81.

Njoku, E., Jackson, T.J., Lakshmi, V., Chan, T.K., and Nghiem, S.V. (2003) 'Soil moisture retrieval from AMSR-E', *IEEE Transactions on Geoscience and Remote Sensing*, 41: 215–29.

Nouvellon, Y., Moran, M.S., Lo Seen, D., Bryant, R., Ni, W., Bégué, A., Chehbouni, A.G., Emmerich, W.E., Heilmann, P. and Qi, J. (2001) 'Coupling a grassland ecosystem model with Landsat imagery for a 10-year simulation of carbon and water budgets', *Remote Sensing of Environment*, 78: 131–49.

Ottmar, R., Burns, M., Hall, J., and Hanson, A. (1993) *CONSUME Users Guide*. Seattle: USDA.

Palacios-Orueta, A., Chuvieco, E., Parra, A., and Carmona-Moreno, C. (2005) 'Biomass burning emissions: A review of models using remote sensing data', *Environmental Monitoring and Assessment*, 104: 189–209.

Parton, W.J., McKeown, B., Kirchner, V., and Ojima, D.S. (1992) *CENTURY Users Manual*. Fort Collins, CO: Natural Resource Ecology Laboratory, Colorado State University.

Paruelo, J., Garbulsky, M., Guerschman, J., and Jobbagy, E. (2004) 'Two decades of Normalized Difference Vegetation Index changes in South America: Identifying the imprint of global change', *International Journal of Remote Sensing*, 25: 2793–806.

Pastor, J., and Post, W.M. (1986) 'Influence of climate, soil moisture, and succession on forest carbon and nitrogen cycles', *Biogeochemistry*, 2: 3–27.

Pettorelli, N., Olav, V., Atle, M., Gaillard, J.M., Tucker, C.J., and Stenseth, N. (2005) 'Using the satellite-derived NDVI to assess ecological responses to environmental change', *Trends in Ecology and Evolution*, 20: 503–10.

Phillips, S., Anderso, R., and Schapire, R. (2006) 'Maximum entropy modeling of species geographic distributions', *Ecological Modelling*, 190: 231–59.

Phinn, S., Hess, L., and Finlayson, C.M. (1999) 'An assessment of the usefulness of remote sensing for wetland inventory and monitoring in Australia', in C.M. Finlayson and A.G. Speirs (eds.), *Techniques for Enhanced Wetland Inventory and Modelling*, *Supervising Scientist Report 147*. Canberra: CSIRO. pp. 44–83.

Phinn, S., Nightingale, J.M., and Stanford, M. (2002) 'A national survey of remote sensing for environmental monitoring and management applications in Australia', *GIS User*, 51: 26–27.

Pittman, S., Christensen, J.D., Caldow, C., Menza, C., and Monaco, M. (2007) 'Predictive mapping of fish species richness across shallow-water seascapes in the Caribbean', *Ecological Modelling*, 204: 9–21.

Pittman, S.J., McAlpine, C.A., and Pittman, K.M. (2004) 'Linking fish and prawns to their environment: A hierarchical landscape approach', *Marine Ecology Progress Series*, 283: 233–54.

Plummer, S.E. (2000) 'Perspectives on combining ecological process models and remotely sensed data', *Ecological Modelling*, 129: 169–86.

POSTEL. (2006) POSTEL land cover. http://postel.mediasfrance.org/en/BIOGEOPHYSICAL-PRODUCTS/Land-Cover/. Accessed 1 May 2009.

Potter, C., Klooster, S.A., Huete, A., and Genovese, V. (2007) 'Terrestrial carbon sinks for the United States predicted from MODIS satellite data and ecosystem modeling', *Earth Interactions*, 11: 1–21.

Potter, C., Randerson, J.T., Field, B., Matson, P.A., Vitousek, P.M., Mooney, H.A., and Klooster, S.A. (1993) 'Terrestrial ecosystem production: A process model based on global satellite and surface data', *Global Biogeochemical Cycles*, 7: 811–41.

Prasad, V.K., Rajagopal, T., Kant, Y., and Badarinath, K.V.S. (2000) 'Quantification of carbon fluxes in tropical deciduous forests using satellite data', *Advances in Space Research*, 26: 1101–4.

Prentice, C., Cramer, W., Harrison, S., Leemans, R., Monserud, R., and Solomon, A.M. (1992) 'A global biome model based on plant physiology and dominance, soil properties and climate', *Journal of Biogeography*, 19: 117–34.

Pribble, J.R., Walsh, J., Dieterle, D., and Müller-Kargar, F. (1994) 'A numerical analysis of shipboard and coastal zone color scanner time series of new production within Gulf Stream cyclonic eddies in the South Atlantic Bight', *Journal of Geophysical Research*, 99: 7513–38.

Prince, S.D. (1999) 'What practical information about land-surface function can be determined by remote sensing? Where do we stand?' in J.D. Tenhunen and P. Kabat (eds.), *Integrating Hydrology, Ecosystem Dynamics and Biogeochemistry in Complex Landscapes*. London: John Wiley. pp. 39–59.

Prince, S.D., and Goward, S.N. (1995) 'Global primary production: A remote sensing approach', *Journal of Biogeography*, 22: 815–35.

Purkis, S., Kenter, J.A.M., Oikonomou, E.K., and Robinson, I.S. (2002) 'High-resolution ground verification, cluster analysis and optical model of reef substrate coverage on Landsat TM imagery (Red Sea, Egypt)', *International Journal of Remote Sensing*, 23: 1677–98.

Qin, Y., Brando, V., Dekker, A., Oubelkheir, K., and Blondeau-Patissier, D. (2007). 'Validity of SEADAS water constituent retrieval algorithms in Australian tropical coastal waters', *Journal of Geophysical Research*, 34: L21603.

Quinn, R. (2000) 'Bathymetry with airborne Lidar and videography', *Backscatter*, 8–17.

Rabus, B., Eineder, M., Rothe, A., and Bamler, R. (2003) 'The shuttle radar topography mission—a new class of digital elevation models acquired by spaceborne radar', *ISPRS Journal of Photogrammetry and Remote Sensing*, 57: 241–62.

Raich, J.W., Rasteller, E.B., Mellilo, J.M., Kicklighter, D.W., Steudler, P.A., Peterson, B.J., Grace, A.L., Moore, B. and Vorosmarty, C.J. (1991) 'Potential NPP in South America: Application of a global model', *Ecological Applications*, 1: 399–429.

Rasmussen, H., Wittemyer, G., and Douglas-Hamilton, I. (2006) 'Predicting time-specific changes in demographic processes using remote-sensing data', *Journal of Applied Ecology*, 43: 366–76.

Rastetter, E.B., Ryan, M.G., Shaver, G.R., Melillo, J.M., Nadelhoffer, K.J., Hobbie, J.E., and Aber, J.D. (1991) 'A general biogeochemical model describing the responses of the C and N cycles in terrestrial ecosystems to changes in CO_2, climate and N deposition', *Tree Physiology*, 9: 101–26.

Reed, B.C., Brown, J.F., VanderZee, D., Loveland, T.R., Merchant, J.W., and Ohlen, D.O. (1994) 'Measuring phenological variability from satellite imagery', *Journal of Vegetation Science*, 5: 703–14.

Reich, P.B., Turner, D.P., and Bolstad, P. (1999) 'An approach to spatially distributed modeling of net primary production (NPP) at the landscape scale and its application in validation of EOS NPP products', *Remote Sensing of Environment*, 70: 69–81.

Reinhardt, E. (1997) 'Using FOFEM 5.0 to estimate tree mortality, fuel consumption, smoke production and soil heating from wildland fire'. USDA: Missoula Fire Sciences Laboratory. Missoula MT. p. 7.

Renzullo, L., Barrett, D.J., Marks, A.S., Hill, M.J., Guerschmann, J.P., Mu, Q. and Running, S.W. (2008) 'Application of multiple constraints model-data assimilation techniques to coupling satellite passive microwave and thermal imagery for estimation of land surface soil moisture and energy fluxes in Australian tropical savanna', *Remote Sensing of Environment*, 112:1306–19.

Riano, D., Chuvieco, E., Ustin, S., Zomer, R., Dennison, P., Roberts, D. and Salas, J. (2002) 'Assessment of vegetation regeneration after fire through multitemporal analysis of AVIRIS images in the Santa Monica Mountains', *Remote Sensing of Environment*, 79: 60–71.

Richardson, L.L. (1996) 'Remote sensing of algal bloom dynamics', *BioScience*, 46: 492–501.

Roberts, D., Dennison, P., Gardner, M., Hetzel, Y., Ustin, S.L., and Lee, C. (2003) 'Evaluation of the potential of Hyperion for fire danger assessment by comparison to the Airborne Visible/Infrared Imaging Spectrometer', *IEEE Transactions on Geoscience and Remote Sensing*, 41: 1297.

Roy, D., Boschetti, L., Justice, C., and Ju, J. (2008) 'The collection 5 MODIS burned area product—global evaluation by comparison with the MODIS active fires product', *Remote Sensing of Environment*, 112: 3690–707.

Roy, D., Lewis, P., and Justice, C. (2002) 'Burned area mapping using multi-temporal moderate spatial resolution data—a bi-directional reflectance-based expectation approach', *Remote Sensing of Environment*, 83: 263–86.

Running, S., Nemani, R., Heinsch, F., Zhao, M., Reeves, M., and Hashimoto, H. (2004) 'A continuous satellite-derived measure of global terrestrial primary productivity: Future science and applications', *BioScience*, 56: 547–60.

Running, S.W., and Coughlan, J.C. (1988) 'A general model of forest ecosystem processes for regional applications. I. Hydrologic balance, canopy gas exchange and primary production processes', *Ecological Modelling*, 42: 125–54.

Running, S.W., and Hunt, J.E.R. (1993) 'Generalisation of a forest ecosystem process model for other biomes, BIOME-BGC, and an application for global scale models', in J.R. Ehleringer and B. Field (eds.), *Scaling Physiological Processes: Leaf to Globe*. New York: Academic Press. pp. 141–48.

Running, S.W., Waring, R.H., and Rydell, R.A. (1975) 'Physiological control of water flux in conifers: A computer simulation model', *Oecologia*, 18: 1–16.

Russell-Smith, J., Yates, C., Edwards, A., Allan, G., Cook, G., Cooke, P., Craig, R., Heath, B., and Smith, R. (2003) 'Contemporary fire regimes of northern Australia, 1997–2001: Change since Aboriginal occupancy, challenges for sustainable development', *International Journal of Wildland Fire*, 12: 283–97.

Schimel, D.S. (1995) 'Terrestrial ecosystems and the carbon cycle', *Global Change Biology*, 1: 77–91.

Schimel, D.S., and 6 authors. (2000) 'CO_2 and the carbon cycle (extracted from the 1995 Intergovernmental Panel on Climate Change (IPCC) "Second Assessment Report," Climate Change 1995: The Science of Climate Change), in T.M.L. Wigley and D.S. Schimel (eds.), *The Carbon Cycle*. Melbourne: Cambridge University Press. pp. 37–49.

Schwalm, C.R., and Ek, A.R. (2001) 'Climate change and site: Relevant mechanisms and modelling techniques', *Forest Ecology and Management*, 150: 241–57.

Schwartz, M.D., Reed, B.C., and White, M.A. (2002) 'Assessing satellite-derived start-of-season measures in the conterminous USA', *International Journal of Climatology*, 22: 1793–805.

Scopélitis, J., Andréfouët, S., and Largouët, C. (2007) 'Modelling coral reef habitat trajectories: Evaluation of an integrated timed automata and remote sensing approach', *Ecological Modelling*, 205: 59–80.

Seaquist, J.W., Olsson, L., and Ardo, J. (2003) 'A remote sensing-based primary production model for grassland biomes', *Ecological Modelling*, 169: 131–55.

Sellers, P., and 15 authors (1995) 'Remote sensing of the land surface for studies of global change: Models-algorithms-experiments', *Remote Sensing of Environment*, 51: 3–26.

Sellers, P., Mintz, Y., Sud, Y.C., and Dalcher, A. (1986) 'A simple biosphere model (SiB) for use within general circulation models', *Journal of the Atmospheric Sciences*, 43: 505–31.

Shugart, H.H.J., and West, D.C. (1977) 'Development of an Appalachian deciduous forest succession model and its application to assessment of the impact of the chestnut blight', *Journal of Environmental Management*, 5: 161–79.

Sippel, S., Hamilton, S., Melack, J., and Novo, E. (1998) 'Passive microwave observations of inundation area and the area/stage relation in the Amazon River floodplain', *International Journal of Remote Sensing*, 19: 3055–74.

Sklar, F.H., and Costanza, R. (1990) 'The development of dynamic spatial models for landscape ecology: A review and prognosis', in M.G. Turner and R.H. Gardner (eds.), *Quantitative Methods in Landscape Ecology: The Analysis and Interpretation of Landscape Heterogeneity*. New York: Springer-Verlag. pp. 239–88.

Smith, R.C., Baker, K.S., Byers, M.L., and Stammerjohn, S.E. (1998) 'Primary productivity of the Palmer Long Term Ecological Research area and the Southern Ocean', *Journal of Marine Systems*, 17: 245–59.

Stenberg, P., Rautiainen, M., Manninen, T., Voipio, P., and Smolander, H. (2003) 'Reduced simple ratio better than NDVI for estimating LAI in Finnish Pine and Spruce stands', *Silva Fennica*, 38: 3–14.

Stow, D., and 23 authors. (2004) 'Remote sensing of vegetation and land-cover change in Arctic Tundra ecosystems', *Remote Sensing of Environment*, 89: 281–308.

Stumpf, R., Holderied, K., and Sinclair, M. (2003) 'Determination of water depth with high resolution satellite image over variable bottom types', *Limnology and Oceanography*, 48: 547–56.

Tanner, J.E., Hughes, T.P., and Connell, J.H. (1996) 'The role of history in community dynamics: A modelling approach', *Ecology*, 77: 108–17.

Tateishi, R., and Ebata, M. (2004) 'Analysis of phenological change patterns using 1982–2000 Advanced Very High Resolution Radiometer (AVHRR) data', *International Journal of Remote Sensing*, 25: 2287–300.

Teague, W.R., and Foy, J.K. (2004) 'Can the SPUR rangeland simulation model enhance understanding of field experiments?', *Arid Land Research and Management*, 18: 217–28.

Thornley, J.H.M., and Cannell, M.G.R. (1997) 'Temperate grassland responses to climate change: an analysis using the Hurley Pasture Model', *Annals of Botany*, 80: 205–21.

Tian, Y., Zhang, Y., Knyazikhin, Y., Myneni, R.B., Glassy, J.M., Dedieu, G. and Running, S.W. (2000) 'Prototyping of MODIS LAI and FPAR algorithm with LASUR and Landsat data', *IEEE Transactions on Geoscience and Remote Sensing*, 38: 2387–401.

Townsend, P. (2002) 'Estimating forest structure in wetlands using multitemporal SAR', *Remote Sensing of Environment*, 79: 288–304.

Townsend, P., and Walsh, S. (1999) 'Modeling floodplain inundation using an integrated GIS with radar and optical remote sensing', *Geomorphology*, 21: 295–312.

Trigg, S., Curran, L., and McDonald, A. (2006) 'Utility of Landsat 7 satellite data for continued monitoring of forest cover change in protected areas in Southeast Asia', *Singapore Journal of Tropical Geography*, 27: 49–66.

Turner, D.P., and Urbanski, S. (2003) 'A cross-biome comparison of daily light-use efficiency for gross primary production', *Global Change Biology*, 9: 383–95.

Ustin, S.L., Smith, M.O., and Adams, J.B. (1993) 'Remote sensing of ecological processes: A strategy for developing and testing ecological models using spectral mixture analysis', in J.R. Ehleringer and B. Field (eds.), *Scaling Physiological Processes: Leaf to Globe*. New York: Academic Press. pp. 339–56.

Valentine, H.T., Gregoire, T.G., Burkhart, H.E., and Hollinger, D.Y. (1997) 'A stand-level model of carbon allocation and growth, calibrated for loblolly pine', *Canadian Journal of Fish and Aquatic Science*, 27: 817–30.

Vorosmarty, C.J., Federer, C.A., and Schloss, A. (1998) 'Potential evaporation functions compared on U.S. watersheds: Implications for global-scale water balance and terrestrial ecosystem modeling', *Journal of Hydrology*, 207: 147–69.

Waring, R.H., and Running, S.W. (1999) 'Remote sensing requirements to drive ecosystem models at the landscape and regional scale', in J.D. Tenhunen and P. Kabat (eds.), *Integrating Hydrology, Ecosystem Dynamics and Biogeochemistry in Complex Landscapes*. London: John Wiley. pp. 23–38.

Warnant, P., Frangois, L., Strivay, D., and Gerard, J.C. (1994) 'CARAIB: A global model of terrestrial biological productivity', *Global Biogeochemical Cycles*, 8: 255–70.

Werdell, P.J., and Bailey, S.W. (2005) 'An improved in-situ bio-optical data set for ocean color algorithm development and satellite data product validation', *Remote Sensing of Environment*, 98: 122–40.

White, J.D., Coops, N.C., and Scott, N.A. (2000) 'Estimates of New Zealand forest and scrub biomass from the 3-PG model', *Ecological Modelling*, 131: 175–90.

Widlowski, J., Pinty, B., Gobron, N., Verstraete, M., Diner, D., and Davis, A. (2004) 'Canopy structure parameters derived from multi-angular remote sensing data for terrestrial carbon studies', *Climatic Change*, 67: 403–15.

Witten, G.Q., Richardson, F.D., and Shenker, N. (2005) 'A spatial-temporal analysis of pattern formation around water points in a semi-arid rangeland system', *Journal of Biological Systems*, 13: 59–81.

Woodward, F.I., Lomas, M.R., and Betts, R.A. (1998) 'Vegetation-climate feedbacks in a greenhouse world', *Philosophical Transactions of the Royal Society of London Series B* 353: 29–38.

Zhang, X.Y., and 7 authors. (2003) 'Monitoring vegetation phenology using MODIS', *Remote Sensing of Environment* 84: 471–75.

Zine, S., Jarlan, L., Frison, P., Mougin, E., Hiernaux, P., and Rudant, J. (2005) 'Land surface parameter monitoring with ERS scatterometer data over the Sahel: A comparison between agro-pastoral and pastoral areas', *Remote Sensing of Environment* 96: 438–52.

Glossary of Acronyms

Satellites

AATSR	Advanced Along-Track Scanning Radiometer
ADAR	Airborne Data Acquisition and Registration
ADEOS	ADvanced Earth Observing Satellite
ALI	Advanced Land Imager
ALOS	Advanced Land Observing Satellite
AMSR-E	Advanced Microwave Scanning Radiometer
ASAR	Advance Synthetic Aperture Radar
ASCAT	Advanced Scatterometer
ASTER	Advanced Spaceborne Thermal Emission and Refraction Radiometer
ATSR	Along Track Scanning Radiometer
AVHRR	Advanced Very High Resolution Radiometer
AVIRIS	Airborne Visible InfraRed Imaging Spectrometer
AVNIR	Advanced Visible and Near-Infrared Radiometer
CASI	Compact Airborne Spectrographic Imager
CZCS	Coastal Zone Color Scanner
ENVISAT	ENVIronment SATellite
EROS	Earth Resources Observation Systems
ERS	European Remote Sensing satellite number 2
ERS-1	European Remote Sensing satellite number 1
ESA	European Space Agency
EVIRI	Enhanced Visible and Infrared Imager
GEOS	Goddard Earth Observing System
GLAS	Geoscience Laser Altimeter System
GMAO	Goddard Modelling and Assimilation Office
GMS	Geostationary Meteorological Satellite
GOES	Geostationary Operational Environmental Satellite
HYDICE	Hyperspectral Digital Imagery Collection Experiment
HYMAP	Hyperspectral Mapper
IceSAT	Ice, Cloud and Land Elevation Satellite
INSAT	Indian National Satellite
IRS	Indian Remote Sensing Satellite

IRS—LISS-1	IRS Linear Imaging Self Scanning Sensor
JASON	Follow on from TOPEX
JERS-1	Japanese Earth Resources Satellite
Landsat ETM	Enhanced Thematic Mapper sensor
Landsat MSS	Multi Spectral Scanner sensor
Landsat TM	Thematic Mapper sensor
LiDAR	Light Detection and Radar
MERIS	Medium Resolution Imaging Spectometer
METEOSAT	Meteorological Satellites
MISR	Multi-angle Imaging SpectroRadiometer
MODIS	Moderate Resolution Imaging Spectroradiometer
MSG	Meteosat Second Generation / Spinning Enhanced Visible and Infrared Imager
NASA	National Aeronautics and Space Administration
NOAA	National Oceanographic and Atmospheric Administration
OCTS	Ocean Colour and Temperature Scanner
PALSAR	Phased Array type L-band Synthetic Aperture Radar
PHILLS	Portable Hyperspectral Imager for Low-Light Spectroscopy
POLDER	Polarization and Directionality of the Earth Reflectances
POSTEL	Pôle d'Obeservation des Surfaces continentales par Telédétection
Radarsat	Canadian C band radar satellite
SAR	Synthetical Aperture Radar
SeaWiFS	Sea-viewing Wide Field-of-view Sensor
SEVIRI	Spinning Enhanced Visible and Infrared Imager
SHOALS	Scanning Hydrographic Operational Airborne Lidar Survey
SMMR	Scanning Multi-channel Microwave Radiometer
SMAP	Soil Moisture Active Passive
SMOS	Soil Moisture Ocean Salinity
SPOT	Satellite Pour l'Obervation de la Terre
SRTM	Shuttle Radar Topography Mission
SSM/I	Special Sensor Microwave/Imager
SWIR	Short-Wave Infrared
TOPEX	Topography Experiment

TOVS	TIROS (Television and Infrared Observatory Spacecraft) Operational Vertical Sounder		Models	
TRMM	Tropical Rainfall Measuring Mission		3-PG	Physicological Principles Predicting Growth
			3-PGS	3-PGS using Satellite data
Biophysical			BEPS	Boreal Ecosystem Productivity Simulator
BRDF	Bi-directional Reflectance Distribution Function		BIOME-BGC	Biome Bio-Geochemical Cycles
CO_2	Carbon Dioxide		CASA	Carnegie-Ames Simulation Approach
EVI	Enhanced Vegetation Index		FBM	Frankfurt Biosphere Model
fPAR	Fraction of Photosynthetically Active Radiation		FOFEM	First Order Fire Effects Model
			FOREST-BGC	Forest Bio-Geochemical Cycles
GAC	Global Area Coverage		FVS	Forest Vegetation Simulator
GPP	Gross Primary Production		G'DAY	Generic Decomposition and Yield
IGBP	International Geosphere-Biosphere Programme		GEM	General Ecosystem Model
LAC	Local Area Coverage		GLO-PEM	Global Production Efficiency Model
LAI	Leaf Area Index		IBIS	Integrated Biosphere Simulator
LUE	Light Use Efficiency		LSM	Land Surface Model
NBAR	Nadir BRDF Adjusted Reflectance		MAPSS	Mapped Atmosphere Plant Soil System
NDVI	Normalized Difference Vegetation Index		PnET/Day	Photosynthetic / Evapotranspiration Model
NPP	Net Primary Production			
PAR	Photosynthetically Active Radiation		SIB	Simple Biosphere model
PSN	Net Photosynthesis		SPUR	Simulating Production and Utilization of Range Land
SST	Sea-Surface Temperature		TEM	Terrestrial Ecosystem Model
VI	Vegetation Index			

Characterizing Spatial Pattern in Biogeographical Data

Thomas P. Albright, Monica G. Turner, and Jeffrey Cardille

22.1 INTRODUCTION

In their quest to understand the geography of life, biogeographers document, compare, and analyze the distributions of various taxa and the resources relevant to them. Spatial heterogeneity, the uneven, nonrandom distribution of objects across the landscape (Forman, 1995a), is inherent in virtually all distributions that biogeographers encounter. While of critical importance, the characterization of spatial heterogeneity is a challenging task that has too often been restricted to the subjective realm. Even today, techniques and theory designed to impart understanding from spatial information are often underutilized by biogeographers and ecologists. Part of the explanation for this lies in the diversity of disciplines that have contributed to such techniques. In landscape ecology, the development of methods and conceptual frameworks for quantifying spatial patterns, comparing and contrasting different patterns, and relating them to processes has been particularly rapid in recent decades (Turner et al., 2001). The field of spatial statistics has also evolved rapidly, and it offers a variety of new approaches for analyzing spatial data (Fotheringham et al., 2000; Fortin and Dale, 2005). Other disciplines, such as geostatistics, signal analysis, and pattern recognition have also contributed tools, such as variograms, wavelets, and texture analysis, having a great potential to aid biogeographic understanding (Dale and Mah, 1998; Tuceryan and Jain, 1998).

Characterizing and quantifying spatial heterogeneity in biogeographic data is challenging for a variety of reasons. First, data sources may vary in type and quality. Species distributions may be derived from point data, GPS tracking data, or from maps based on continuous variation in abundances. Moreover, spatial data may be converted from one framework or data scale to another and may be analyzed in one, two, or even three spatial dimensions. Second, there is a sometimes-bewildering array of metrics and indices available for quantifying spatial patterns, and it can be difficult to determine which ones to employ in a given study (Li and Wu, 2004). Third, distribution patterns and organism-environment associations are scale-dependent (Wiens, 1989). For example, a species may be densely distributed on a landscape at broad scales, but at fine scales, some locations may have large populations whereas others will lack the species entirely. Furthermore, organisms relate to their landscapes at different spatial scales, and studies must employ appropriate or even multiple scales for a given species or question. Collectively, these challenges (and opportunities) mean that biogeographers must be aware of a variety of techniques and approaches to characterizing spatial pattern and employ the most appropriate ones according to the data and objectives.

Tobler's first law of geography states that "everything is related to everything else, but near things are more related than distant things" (Tobler, 1970). Perhaps the most fundamental way in which spatial pattern may be characterized is to show the degree to and manner in which it conforms to Tobler's law. This is the question of spatial dependence. Biogeographers may ask, for example, whether the presence of an organism in one location makes the occurrence of another individual nearby more likely. If it does, over what distance is this true and does this depend on direction? Other questions may deal with comparing the spatial patterns of two or more ecological variables. For example, do two putatively co-evolved species share a similar distribution? Are the configurations of agricultural/woodland landscapes of England and France similar? Or does the distribution of a predator species affect the habitat selection of a prey species? Of course, once patterns of distribution are characterized, the questions in biogeography often focus on the whys or involve prediction. For example, what factors explain the distribution of a species? Or, if environmental conditions change, where might a particular taxon be distributed in the future? Application of new methods of spatial analysis should facilitate the ability of biogeographers to characterize, explain, and predict distributions.

In this chapter, we provide an overview of tools for characterizing spatial heterogeneity and highlight their potential use in the field of biogeography, focusing on how these methods

might generate new insights. The chapter is organized around the types of spatial reference and measurement scale of data that biogeographers commonly encounter and use: spatial point patterns, point samples of a continuously varying field, categorical raster and polygon data, and continuous raster and polygon data (Figure 22.1). While other schemes are possible, organizing around data types is practical for nonspecialists in spatial analysis. The categorization employed is not exact: In some cases, techniques may apply to several data types. We group techniques according to the data type with which they are most commonly associated and note when techniques can be readily applied to other data types.

22.1.1 Key concepts

Of the several important recurring concepts in this chapter, the most fundamental is that of spatial

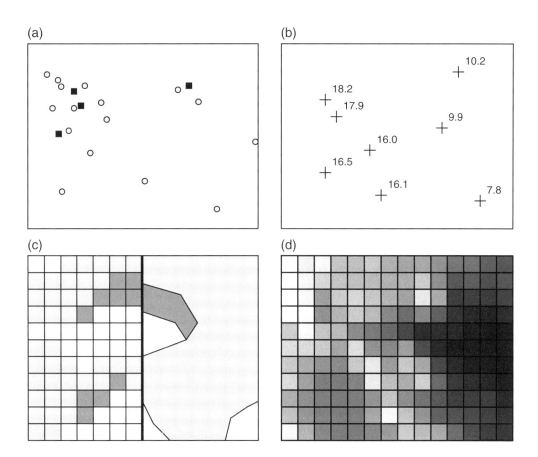

Figure 22.1   The four data sets around which this chapter is organized: (a) spatial point patterns, (b) point samples on a continuous field, (c) categorical raster/polygon data, (d) continuous raster/polygon data

dependence, which may also be referred to as spatial association. We adopt the terminology of Fortin and Dale (2005), employing a general definition of spatial dependence: lacking independence among observations from nearby locations. It is important to note that such spatial dependence may be due to a combination of two types of factors. It may result from an endogenous process of spatial-autocorrelation whereby values of a variable are self-correlated at nearby locations. Or spatial dependence may be induced by a response to an exogenous variable that is itself spatially dependent (Fortin and Dale, 2005). Consider a location with several stands of trees, with one in particular having trees that are exceptionally tall. An observer would note that the height of the trees is spatially dependent because, for instance, tall trees are clustered within the same stand. This may be due to the favorable conditions (which themselves are spatially auto-correlated) at that site—an example of induced spatial dependence. Alternatively, the height might be related to genetic factors and attributed to the endogenous processes of reproduction and limited propagule dispersal, which generated the spatially autocorrelated pattern of genotypic similarity.

Stationarity is another important concept and is a frequent assumption of spatial analysis techniques described in this chapter. Stationarity means that the properties of the processes that generate a spatial pattern are uniform across the area in consideration (Diggle, 2003). This results in a constant mean and variance across the region of study. For instance, if an analysis of the spatial pattern of a tree spans two very different ecoregions, the seed disperser that predominates in one may result in a different type of spatial dependence than in the other ecoregion where another disperser predominates. Finally, we will distinguish between methods that are global, yielding one indicator for the entire area of study, and methods that are local, yielding results calculated for a portion of the data set, and sometimes run repeatedly across the entire study area to reveal spatial variation in results. Global methods are typically more able to detect patterns due to their obtaining a larger sample size and have historically been more common, but they require the assumption of stationarity. In addition to relaxing this assumption, local methods hold the advantage of allowing discrimination of different types of processes in space (Anselin, 1995; Fotheringham and Brunsdon, 1999). Local methods can also be used to assess stationarity across the extent of the study area in order to evaluate the appropriateness of the use of more powerful global methods.

22.2 SPATIAL POINT PATTERNS

A basic form of data available to biogeographers consists of locations where a taxon or other phenomenon is present. Such occurrences can be considered to be discrete 'events' in space, creating a point pattern driven by a particular spatial statistical process (Cressie, 1993). Examples of events include locations of mature trees within a forest stand, wasp nests in a study area, or disease outbreaks in a country. The relationships among different types of events are also of interest to biogeographers. These can include locations of allelopathic plants and their competitors, different age classes of a tree species, or locations of hosts that are either free of or colonized by parasites. Treatment of spatial point patterns is divided according to whether events are of a single type (univariate) or of two or more types (multivariate).

22.2.1 Univariate point pattern analysis

The most common spatial question addressed using event data is whether the events are distributed in a dispersed, random, or clustered manner (Figure 22.2). Dispersed (or regularly spaced) point patterns exhibit negative spatial dependence, wherein the presence of an event at one location depresses the likelihood of events at neighboring locations. Point patterns that lack spatial dependence are said to exhibit complete spatial randomness (CSR) and appear as an even intensity of independent events throughout the area considered. Finally, clustered (or aggregated) point patterns exhibit positive spatial dependence among individual points. As is nearly always the case in spatial analysis, the observed pattern may be scale dependent, and the scale employed should thus always be stated when reporting results. For instance, Wells and Getis (1999) found that Torrey pines (*Pinus torreyana*) in sites in Southern California exhibited clustering at fine spatial scales, which supported theoretical predictions based on adaptation to fire. In a general sense, evaluating spatial dependence in point patterns is usually done by graphically or statistically comparing the point pattern to one of CSR. Complete spatial randomness presents an intuitive 'null model' that is statistically as well as biogeographically useful but is not the only possible frame of reference against which univariate point patterns may be compared (Diggle, 2003).

A common, relatively straightforward approach to the analysis of point patterns counts the number of events in quadrats, which are subregions that either sample from or completely cover a given

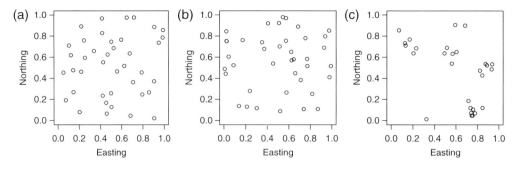

Figure 22.2   Three basic types of spatial point patterns generated by different spatial processes: (a) dispersed, (b) complete spatial randomness, and (c) clustered

study area. Commonly, the quadrats, m, are defined as a square grid. Under CSR, the number of events, X, in each quadrat follows a Poisson distribution, with $E(X) = Var(X) = A$, where $E(X)$ is the intensity, or expected number of events per unit area, and A is the area of a quadrat. Calculating the variance to mean ratio (Clapham, 1936) of the quadrat counts (known as the index of dispersion, ID) indicates whether a spatial point pattern tends toward aggregation (ID >1), complete spatial randomness (ID ~ 1), or dispersion (ID < 1). The appropriate two-sided test statistic is $(n - 1)$ ID and follows a Chi-squared distribution with $n - 1$ degrees of freedom. While this approach is intuitive and simple, it is not robust with small numbers of expected events per quadrat (e.g., $\bar{x} < 5$) and is not very powerful in detecting dispersion (Diggle, 2003). It should also be stressed that this test is very sensitive to the size (and to a lesser degree, shape) of the quadrats (Diggle, 2003). This can be both a peril, if the analysis is conducted or interpreted naïvely, and a useful source of scale-specific information on pattern. In fact, the related Greig-Smith technique (Greig-Smith, 1952) uses a series of nested grids spanning a wide range of sizes in order to identify the scale of vegetation patches.

The use of quadrats to analyze spatial point patterns is not confined to grids covering the entire study area. Quadrat samples have also been used to characterize spatial point patterns when a complete count was impractical (Dacey, 1968). The technique is directly drawn from the index of dispersion, but random effects of sampling must be accounted for. The popularity of quadrat methods is partly rooted in their less-stringent field data collection requirements, but they offer less flexibility and power than methods based on distance, which we will now discuss.

Spatial point patterns may also be characterized by measuring distances between events or between events and randomly placed points. Of the

distance-based methods, some, such as T-square sampling, require only a sample of event locations and are well suited for rapid field assessment, but they come at the expense of robustness and power (Diggle, 2003). Others, such as the point-to-nearest-event based $\hat{F}$, the nearest-neighbor based $\hat{G}$, and the interevent-distance based $\hat{H}$ require a complete mapping of events and consider either distances between events or between events and randomly located points (Diggle, 2003 and references therein). The most widely used distance-based methods are Ripley's K-function, and the related L-function, which both examine the distribution of interevent distances in a spatial point pattern (Ripley, 1977). This is done by calculating curves of the observed number of event pairs within a distance t of each other and of the number expected under CSR. Where the empirical curve is greater than or less than the theoretical curve, the events tend toward aggregation or dispersion, respectively, at that distance. For Ripley's K, the curve is defined as:

$$K(t) = \frac{1}{\lambda} E(t)$$

where λ is the intensity rate of events per unit area and $E(t)$ is the number of events within a distance t of an event. Under CSR, the curve follows the form:

$$K(t) = \pi t^2$$

The empirical function $\hat{K}(t)$ is highly sensitive to edge effects and several methods for accounting for this are available (Goreaud and Pelissier, 1999; Yamada and Rogerson, 2003). One of the more common was proposed by Ripley (1976):

$$\hat{K}(t) = \frac{1}{\hat{\lambda}} \sum_i \sum_{j \neq 1} \left(\frac{1}{w(s_i, s_j)} \right) \frac{I(d_{ij} < t)}{N}$$

where i and j are events, $w(s_i, s_j)$ is a weight proportional to the circumference of a circle centered at s_i and passing through s_j inside of the region A, and $I(d_{ij} < t)$ is an indicator of whether the toroidal distance between i and j is less than the distance t. The toroidal correction removes edges by expanding the area, A, by wrapping both dimensions into the shape of a doughnut (torus).

The related L-function simply transforms the theoretical curve to a horizontal line to facilitate interpretation:

$$L(t) = \sqrt{\frac{K(t)}{\pi}} - t$$

The resulting curves can be examined for departure from CSR. Monte Carlo simulations based on randomly generated CSR point patterns can be used to assess the significance of the point pattern's departure from CSR (Figure 22.3). For each of these patterns generated s times, $\hat{K}$ or $\hat{L}$ can be calculated for a range of distances. The maximum and minimum simulated $\hat{K}$ or $\hat{L}$ values found for each distance evaluated are retained and form the bounds of the resulting simulation envelope. The number of simulations, s, can be selected to meet a desired alpha significance level according to:

$$s = \frac{2}{\alpha}$$

22.2.2 Multivariate point pattern analysis

Biogeographers and ecologists are also interested in whether the distribution of one species (or age class, or resource, etc.) influences the distribution of another (or a number of others). For instance, does a purportedly allelopathic tree repel other tree species? A variety of approaches to bivariate (and multivariate) point pattern analysis offers tools for identifying spatial evidence for such interactions (Smith, 2004 and references therein). While discussion will focus on bivariate analysis, these techniques may be generalized to the multivariate case. The most commonly used is the cross-K function (or its variant, cross-L), which is a generalization of Ripley's K (Ripley, 1981; Cressie, 1993). Here, the theoretical curve becomes

$$K_{12}(t) = \frac{1}{\lambda_2} E(t)$$

where λ_2 is the intensity of pattern 2 and $E(t)$ is the expected number of events of type 2 within distance t of a randomly chosen event of type 1. Thus, for CSR,

$$K_{12}(t) = \pi t^2$$

Incorporating an edge correction analogous to the univariate case, the cross-K function can be calculated empirically as

$$\hat{K}_{ij}(t) = \frac{\sum_{i=1}\sum_{j=1} w(s_i^{(1)}, s_j^{(2)})^{-1} I(d_{i(1),j(2)} < t)}{\hat{\lambda}_i \hat{\lambda}_j A}$$

where $w(s_i^{(1)}, s_j^{(2)})$ is the proportion of the circumference of a circle centered at the event i in pattern one passing through the event j in pattern two that

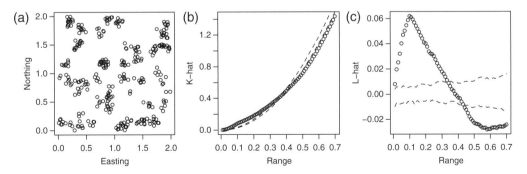

Figure 22.3   (a) Point data representing nest locations in a 2x2 sample square, (b) corresponding to $\hat{K}(t)$, and (c) corresponding to $\hat{L}(t)$. Dotted lines on (b) and (c) represent simulation envelopes and demarcate significance at the 95% level. Note that the clustering at fine scales and dispersion at coarser scales is easily perceived in (c) due to the transformation involved in calculating $\hat{L}(t)$ from $\hat{K}(t)$

is inside the study area A and $I(d_{i(1),j(2)} < t)$ is the indicator of whether the distance between events i and j are less than distance t. Lotwick and Silverman (1982) developed a Monte Carlo procedure for testing the significance of $\hat{K}_{ij}(t)$.

Bivariate point patterns may also arise when there is a single spatial point process that is then 'marked' by some random process resulting in two or more possible types or a quantitative attribute. In many cases dispersal of organisms is difficult to observe and instead must be inferred based on evidence including spatial pattern. Lancaster and Downes (2004) used bivariate point pattern analysis to determine whether a marine lichen, *Pyrencollema halodytes* (Nyl.) R.C. Harris, spread vegetatively or via propagules among the intertidal barnacles, *Semibalanus balanoides* (L.), that it colonized. Both *S. balanoides* and *P. halodytes* were highly aggregated with respect to CSR, but was aggregation in *P. halodytes* due to a cluster-producing process such as vegetative spread or simply an artifact of aggregation in their hosts? By examining the differences between the $L_{(t)}$ functions of *P. halodytes* and *S. balanoides* using Monte Carlo simulations, they found that lichen location was randomly distributed among barnacles. The result suggested that dispersal via propagules was the dominant form of colonization. A further benefit of this approach is that it appears to eliminate the need to make edge corrections, because both patterns being compared are subject to the same edge effects (Lancaster and Downes, 2004). Examples and discussion of quantitatively marked point patterns (e.g., tree diameters) can be found in Cressie (1993).

22.3 POINT SAMPLES OF A CONTINUOUSLY VARYING FIELD

Another type of data frequently encountered by biogeographers consists of spatially referenced samples of a field that varies continuously in space. Examples of this may include chlorophyll concentration in a water body, nitrogen mineralization rates in soils, or allele frequency across a landscape. Note that this data type may include phenomena that are literally continuous in space, such as elevation, as well as those that may be conceptualized as such, as with population data. Often, a suite of tools from the discipline of geostatistics is used to interpolate the data to make a contour or density map of the attribute of interest (Isaaks and Srivastava, 1989). While obtaining such a map may be the goal of the analysis, investigators may also wish to understand how the sampled phenomenon varies in space and at what scale. For example, Miller et al. (2002) found the spatial autocorrelation in density of forest

understory vegetation to be strongly related to dispersal guild, with those suited for long-distance dispersal exhibiting less spatial autocorrelation. Fraterrigo et al. (2005) found that past agricultural or logging land use greatly increased the distance of spatial autocorrelation among several soil resources, indicating homogenization. Statistics on spatial autocorrelation in point sampled data have also been used with increasing frequency to infer evolutionary and genetic process (Heywood, 1991; Parker and Jorgensen, 2003). The most common way to depict and analyze the nature of spatial dependence in point sampled data is through variograms. Moran's I and Geary's c may also be used to detect spatial autocorrelation in point sampled data, but these are covered in the final section.

22.3.1 Variography

Variograms portray the amount of variation among sampled points in the variable of interest over a range of distances (Isaaks and Srivastava, 1989). Similar information can be obtained through correlograms, but we focus on variograms due to their prevalence in biogeography. The theoretical variogram, $\gamma(h)$, is given as:

$$\gamma(h) = \frac{1}{2}\mathrm{var}(Z(s+h) - Z(s))$$

where $Z(s)$ is the value of the variable of interest at point s and $Z(s + h)$ is the value at points of a distance from s indicated in the 'lag' distance h. While this is technically the semivariogram because of the 1/2 coefficient, the term variogram is often used. On the other hand, the terms semivariance and variance are retained based on whether the coefficient is included or not. We use the term variance for describing general aspects that do not depend on whether semivariance or variance were calculated.

Several key terms are used to describe aspects of the various forms that variograms can take (Isaaks and Srivastava, 1989) (Figures 22.4a-c). The nugget indicates the amount of variance inherent in the measurement and sampling of the phenomenon as the lag distance approaches zero. The sill is the level of variance that the variogram approaches as lag distance approaches infinity. It can be thought of as the amount of variance in the data when samples are sufficiently far apart as to be considered unrelated spatially. The range is the distance at which this sill is reached. Thus, samples spaced at distances less than the range are considered to be spatially autocorrelated. Estimates of the range are sometimes used as a dependent variable in studies comparing spatial structure

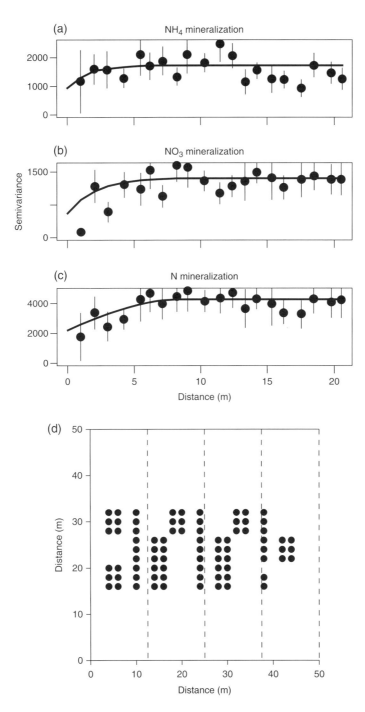

Figure 22.4   Variograms for (a) ammonium (NH₄) mineralization, (b) nitrate (NO₃) mineralization, and (c) total nitrogen mineralization for postfire black spruce (*Picea mariana*) stands in central Alaska. For total N mineralization, the nugget variance is estimated at ~2100, the range is 8.3 m, and the sill is approximately 4200. (d) Locations of soil cores in a cyclic sampling design employed in order to obtain comparable power at different lag distances, maximizing sampling efficiency. Figures used and modified with permission from Smithwick et al. (2005)

among multiple sites, biota, or processes (Smithwick et al., 2005).

In practice, an empirical variogram, $\hat{\gamma}(h)$, must be estimated from a relatively large sample (e.g., $n > 50$) of points as follows:

$$\hat{\gamma}(h) = \frac{1}{2N(h)} \sum (Z(s_i) - Z(s_j))^2$$

where $N(h)$ is the number of pairs (s_i, s_j) that are in the distance class h apart. The distance class h necessarily must encompass a range of distances in order to have multiple samples in each class. Ideally, h will be defined as broad enough to include a sufficient number of pairs to enable sound estimation of variance but be fine enough to capture the desired spatial resolution. The resulting empirical variogram often does not cleanly match one of the theoretical variograms discussed above. Instead, there is usually a great deal of variation in variance among neighboring values of h. Further, larger lags (say, the last half of possible distances) will likely show erroneously small amounts of variance due to the small number of samples that are typically obtained at these distances. It is common practice, therefore, to ignore this portion of the empirical variogram. Preferably, one employs a cyclic sampling design, which ensures adequate and comparable sample sizes in all of the desired distance classes (Burrows et al., 2002; Smithwick et al., 2005) (Figure 22.4d). Once the empirical variogram has been calculated, one typically fits a modeled variogram to it. In this way, the researcher may gather evidence supporting a model that is predicted in theory and may determine the best estimates for the values of the nugget, sill, and range.

Variography is a powerful tool, but several issues confront potential users. First, theoretical variograms are only valid under the assumption of stationarity. Second, if there is reason to suspect that spatial dependence may depend on direction (as in the case of wind-dispersed seeds, for example), directional variograms (Le Corre et al., 1998) should be used. Finally, the empirical variogram is known to be sensitive to outliers. While other versions have been proposed to be more robust (Cressie and Hawkins, 1980), they are not covered in this chapter.

22.4 CATEGORICALLY SCALED RASTER AND POLYGON DATA

A predominant concept in the discipline of landscape ecology and biogeography is that of the patch (see Kupfer, this volume). A patch is a relatively homogeneous contiguous area differing from its surroundings (Forman, 1995b). Most commonly, for geographic data to portray patches, space is partitioned into two or more categories and mapped in either vector or raster format. These data are widely used due both to the technology and methods that make categorical classification of landscape more convenient (e.g., as in GIS and remote sensing) and to the clarity that such a simplification often brings to theory and analysis. As a result, techniques for quantifying and characterizing the nature of heterogeneity in categorical raster and vector data are particularly well developed and abundant in the literature, including numerous reviews (Haines-Young and Chopping, 1996; Gustafson, 1998; Turner et al., 2001; Li and Wu, 2004).

Interest in quantifying heterogeneity is fundamentally driven by the need to better understand the often-reciprocal relationship between landscape pattern and ecological process. For example, Fahrig (2003) reviews research on the link between landscape fragmentation and biodiversity. At the species level, certain patch sizes and configurations are associated with preferred habitat for birds and other animals (Bailey et al., 2002). Patch-based analysis of heterogeneity has also been used as a means of gauging humans' impacts on a landscape (Saura and Carballal, 2004). These approaches have not been confined to the terrestrial landscape, with several applications in marine and riverine systems (Wiens, 2002; Sleeman et al., 2005).

22.4.1 Join count statistics

The simplest question to ask of categorical data is whether there is evidence of spatial dependence. Moran (1948) developed a method to assess spatial dependence in binary-coded (e.g., black and white) polygon or raster data based on whether the number of like-coded 'joins' departed from an expected value. A higher-than-expected number of black-to-black joins, for example, indicates spatial association. This is further generalized to categorical data with more than two classes, basing the statistic on tallies of like-coded joins:

$$J_{rr}(d) = \frac{1}{2}\left[\sum_{\substack{i=1 \\ i \neq j}}^{n} \sum_{\substack{j=1 \\ j \neq i}}^{n} \delta_{ij}(d) x_{ri} x_{rj} \right]$$

where r is a category indicator, $\delta_{ij}(d)$ is an indicator of membership in distance class d (either based on Euclidian distance thresholds or in terms of neighbors) for locations i and j, and x_{ri} and x_{rj}

are binary indicator functions based on whether locations i or j, respectively, belong to category r (Epperson, 2003).

22.4.2 Landscape metrics

Landscape metrics or indices offer an alternative and more flexible manner of quantifying heterogeneity in categorical data. These can be used to answer questions such as, How much diversity of land covers is there in a landscape? How do two landscapes differ? How has a landscape changed over time? and Is a landscape configured in such a way as to meet the needs of a certain organism? It is useful to consider two broad categories of landscape metrics (Dunning et al., 1992), commonly called composition and configuration. Landscape composition refers to the number and abundance of different land-cover categories. Landscape configuration is concerned with the spatial arrangement of these categories.

Composition metrics include basic measures such as the proportion of a landscape occupied by a certain land cover. They also include indicators of richness and diversity based on these proportions, which are inspired by information theory and conservation biology. Richness is simply the number of cover categories in the landscape. Diversity is a function of the number of categories and their relative evenness. A common metric, Shannon's diversity index (Shannon and Weaver, 1949), H, is formulated as:

$$H = -\sum_{i=1}^{s} (p_i) \ln(p_i),$$

where p_i is the proportion of category i in the landscape and s is the number of categories present (McGarigal, 2007). This metric can be normalized by the number of cover types, resulting in an indicator of evenness scaled between 0 and 1 (Turner et al., 2001):

$$E = \frac{-\sum_{i=1}^{s} (p_i) \ln(p_i)}{\ln(s)}.$$

Ricklefs and Lovette (1999) examined faunal diversity on a set of islands in the Lesser Antilles and found that habitat diversity was more important than island area in determining species richness for butterflies and reptiles, which include many habitat specialists.

Configuration is considerably more difficult to quantify and, as a result, the number of metrics developed is far greater than for composition.

Configuration metrics may be calculated at the landscape, cover class, or patch levels and can be further placed into categories such as patch shape, spatial arrangement of patches, texture, and contrast based on the aspect of configuration that they are meant to address. For example, identifying and quantifying habitat areas is very useful for studies of metapopulation dynamics and conservation planning (Hanski, 1998). Core-area metrics provide information on the effective usable area of chosen patch types after a user-specified edge effect is taken into account (Andren and Angelstam, 1988; McGarigal et al., 2002). Contagion, a metric that assesses texture, can be thought of as the degree to which cover types are aggregated. One formulation of contagion is as follows (Li and Reynolds, 1993):

$$C = \left[1 + \frac{\sum_i \sum_k \left((P_i) \left(\dfrac{g_{ik}}{\sum_k g_{ik}} \right) \right) \cdot \left[\ln(P_i) \left(\dfrac{g_{ik}}{\sum_k g_{ik}} \right) \right]}{2 \ln(m)} \right] * 100$$

where i and k are cover types, P_i is the proportion of the landscape covered by cover type i, and g_{ik} is the number of pixel adjacencies between i and k (counting them twice—once for each type). All summation is carried out to the number of cover types present on the landscape, m. Contagion approaches 100% as the number of patches per cover type approaches one. Alternatively, it approaches 0% in landscapes composed of pixels in single-pixel patches. Applications of contagion include pathology and fire ecology, based on the idea that phenomena that spread among similar cover types will be sensitive to the connectedness that contagion and other texture metrics identify (Franklin and Forman, 1987; Maurice et al., 2004).

This discussion and examples of specific landscape metrics have been intentionally brief, due to the abundance of accessible reviews (Haines-Young and Chopping, 1996; Gustafson, 1998; Turner et al., 2001; Li and Wu, 2004), tutorials (Gergel and Turner, 2002; McGarigal, 2007), and available software (Baker, 2001; McGarigal et al., 2002). Readers of this chapter are better served by a more thorough discussion of issues surrounding the use of landscape metrics for characterizing landscape pattern.

The spatial scale of analysis and data collection is a critical issue for the interpretation and use of landscape metrics. Alteration of both grain and extent of categorical map data can have large effects on landscape metric values (Wickham and

Riitters, 1995; Gustafson, 1998). Analysts should also recognize that the appropriate scale of analysis is fundamentally limited by the scale of data collection (Li and Wu, 2004). Another set of cautions relates to the data from which landscape metrics are to be calculated. It is critical that the cover types be appropriately defined for the organisms or process being studied, as an inappropriate classification scheme will result in irrelevant landscape metrics (Turner et al., 2001). Furthermore, the definition of patch boundaries may not always be clear (e.g., ecotones) and values of landscape metrics may be highly dependent on how class membership is defined (Arnot et al., 2004). Finally, landscape metrics may respond to errors in land-cover mapping in unpredictable and nonlinear ways (Shao et al., 2001; Langford et al., 2006). Errors that are considered acceptable for land-cover mapping may produce much larger errors in landscape metrics derived from them.

Perhaps more fundamentally, users of landscape metrics should be mindful of whether there is theoretical justification for quantification of landscape pattern (and specific aspects of landscape pattern) with respect to the organism or phenomenon under consideration (Li and Wu, 2004). The ease with which large numbers of landscape metrics can be calculated makes the hazards of finding spurious correlations especially pertinent (Gustafson, 1998). Furthermore, it has been widely noted that there is a high degree of correlation among the values of multiple landscape metrics (Riitters et al., 1995). In light of these concerns, biogeographers should carefully select appropriate metrics a priori and examine their selections for collinearity.

Due in part to the difficulty in effectively linking landscape metrics to ecological process, enthusiasm for landscape metrics has waned somewhat in recent years (Li and Wu, 2004; Haines-Young, 2005). Nonetheless, fruitful areas of application and research remain. Adjustments to metrics have been proposed for topographically complex areas (Dorner et al., 2002). Techniques to locally quantify landscape metrics, either calculated through moving windows (McGarigal et al., 2002) or by subdividing areas (Cardille et al., 2005), hold the promise of providing a better spatial and statistical context for interpretation and application of landscape metrics (Figure 22.5). Related to this are important efforts to characterize the statistical properties of landscape metrics (Remmel and Csillag, 2003). Finally, for some applications, momentum seems to be building in favor of the analysis of continuously scaled raster data sets in lieu of categorically scaled data (see the next section).

22.5 CONTINUOUS RASTER AND POLYGON DATA

While it has perhaps not received the same level of attention as categorical data among landscape ecologists or biogeographers, spatial analysis of raster data may also be carried out in an ordinal or continuous framework. Two of the most common sources of such data encountered in biogeography

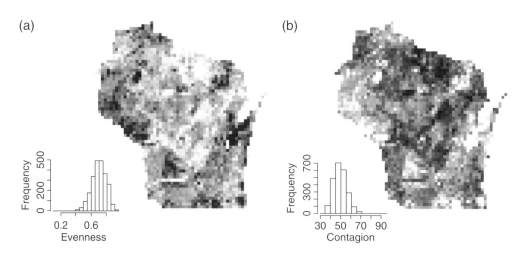

Figure 22.5 The spatial and statistical distributions of two landscape metrics in the state of Wisconsin (United States). Shown are (a) Shannon's evenness and (b) contagion, with higher values appearing in lighter tones

are remotely sensed imagery and digital elevation models. While it is always possible, and in some cases advantageous, to convert these data to categories (as in the case of land use/cover classifications), this is a costly and time-consuming process that necessarily entails a loss of information related to heterogeneity both within and among patches and classes. Thus biogeographers using raster data may be wise to consider analysis in an alternative, continuous framework (Kupfer et al., 2006). Examples of questions that may be posed of data in such a framework include, Does canopy cover exhibit a characteristic spatial texture? Does an organism select habitat in areas with a certain type of vegetation heterogeneity? Can image spatial heterogeneity be used to infer habitat heterogeneity? and Is there spatial dependence in the data and over what scales? In this section, we consider three sets of tools for analyzing continuous raster data: global and local indicators for spatial dependence, wavelet analysis, and texture analysis. The discussion of indicators for spatial dependence also pertains to vector polygon data with continuously scaled attributes.

Indicators of spatial dependence among raster or polygon data are analogous to both Pearson's correlation coefficient and the index of dispersion for spatial point patterns (above). That is, the distribution of values in space is characterized in terms of whether locations closer in space are more likely to have similar values (positive spatial autocorrelation), dissimilar values (negative spatial autocorrelation), or whether there is no spatial effect evident (no spatial autocorrelation). Spatial dependence in continuous raster and polygon data may be evaluated at both the global and local levels, with biogeographic applications including identification of spatial clusters of infectious diseases (Clennon et al., 2004; Chen et al., 2005; de Castro et al., 2006) and invasive species (Whitmire and Tobin, 2006), the analysis of forest cover heterogeneity (Southworth et al., 2004), and the detection of genotypic association in plants (Hardesty et al., 2005).

The global statistic most commonly used by biogeographers for spatial dependence is Moran's I (Moran, 1950). This statistic can be formulated as:

$$I = \left(\frac{n}{W}\right) \frac{\sum_{i=1}^{n}\sum_{j=1}^{n} w_{ij}(x_i - \bar{x})(x_j - \bar{x})}{\sqrt{\sum_{i=1}^{n}(x_i - \bar{x})^2}},$$

where w_{ij} is a weight matrix, the elements of which may be either a binary indicator of neighborhood membership or a value related to

distance, x_i and x_j are the values at sampling locations i and j, respectively, n is the number of sampling locations, and W is the sum of the weights for all possible combinations of sampling locations. Significance of I at each distance class may be tested by either using a normal approximation, with appropriate multiple comparison adjustments, such as the Bonferroni correction, or through Monte Carlo simulations. A similar coefficient, c, was developed by Geary (1954) but is less powerful under most circumstances (Upton and Fingleton, 1985). Note that the calculation of these coefficients is readily performed on polygon data, points (with neighborhoods defined through Theissen polygons or other means), or a raster matrix.

Local versions of such statistics have been developed and reviewed, notably by Anselin (1995), Getis and Ord (1996), and Boots (2002). Key uses of these include the identification of local areas of spatial autocorrelation (some of which may be contrary to the corresponding global index) and assessing whether a larger area is characterized by stationarity. A local version of Moran's I, called I_i, can thus be specified using the same notation

$$I_i = (x_i - \bar{x})\sum_j w_{ij}\left(x_j - \bar{x}\right)$$

(modified from Anselin, 1995).

A powerful advantage in calculating I_i and other local indicators is that it becomes possible to create a map of spatial association, making it valuable for exploratory spatial data analysis and evaluation of stationarity. Complicating the evaluation of their significance, however, is the fact that the normal distribution does not offer an adequate approximation. Once again, Monte Carlo—type simulations can be used, but their reliability has been questioned in the presence of global spatial association and alternatives have been proposed (Boots, 2002; Fortin and Dale, 2005).

Wavelet analysis, a relative newcomer to biogeography and ecology, is a powerful tool that can be used in cross-scale assessment of spatial dependence (Dale and Mah, 1998). In wavelet analysis, data are transformed according to the degree to which values conform to a mathematical wavelet form within a moving window of a given size. The analyst may specify the particular form and scales to be used, resulting in a method that, in many ways, combines the advantages of local and global methods (Csillag and Kabos, 2002). While the mathematics are beyond the scope of this chapter, suitable introductions for biogeographers and ecologists can be found in Dale and Mah (1998) and Csillag and Kabos (2002).

Camarero et al. (2006) used wavelets to identify fine-scale boundaries of plant communities in the Pyrenees. Bradshaw and Spies (1992) used wavelet variance to analyze canopy gap closure data collected along transects and related the resulting diverse characterizations to a rich variety of site conditions and stand-history events. Keitt and Urban (2005) present a particularly effective illustration of the use of wavelets to show correlation among biophysical variables at different scales in mountainous terrain (Figure 22.6). They found that the characteristic scale of variation differed among the variables examined. Underscoring the importance of the locally varying, multiscale capability of wavelet analysis, they also discovered that the results of an analysis conducted at a single scale or portion of the data set would often be highly dependent (even to the point of contrary results) on the particular scale or location chosen.

A third approach to analyzing continuously scaled raster data involves characterizing image texture. The use of image texture analysis techniques offers a promising alternative to traditional classification-based methods for characterizing spatial heterogeneity (Musick and Grover, 1991). While notoriously difficult to define (Tuceryan and Jain, 1998), texture can be considered as the variability of image values in a given area. Most commonly used methods for analyzing

texture consider either first- or second-order statistics (Tuceryan and Jain, 1998). First-order statistics, such as mean and standard deviation, characterize the statistical distribution of pixel values within the area, usually a moving window, but ignore spatial relationships among those pixels. Second-order statistics, on the other hand, are based on the gray level co-occurrence matrix (GLCM), which characterizes the spatial arrangement of pixel values within the area. The GLCM tracks the frequency of occurrence of each possible pairing of values for pixels at specified relative locations (e.g., adjacent East-West pixel pairs, Figure 22.7) (Haralick et al., 1973; Tuceryan and Jain, 1998). In a recent study in an arid area of New Mexico, St.-Louis et al. (2006) found a strong relationship between texture measures derived from digital orthophotos and Landsat imagery and bird species richness, suggesting that image texture measured at both 1-m and 30-m resolution acts as a surrogate for habitat heterogeneity. Kayitakire et al. (2006) used similar texture measures to infer a variety of forest stand parameters from 1-m resolution satellite imagery in Belgium. In addition to texture, McGarigal and Cushman (2005) argue persuasively that approaches based on surface metrology, a field developed from microscopy and molecular physics, offer promising alternatives that may often be seen as continuous analogues to

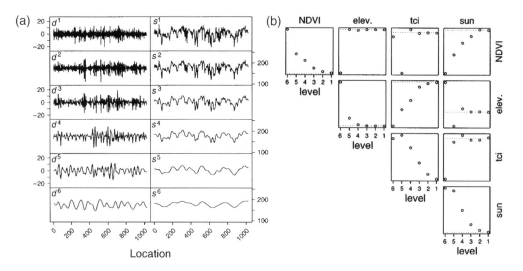

Figure 22.6 **Multiresolutional decomposition of remotely sensed data. The left-hand column shows the decomposition at each scale. The right-hand column shows the sum of the preceding scale components. (a) Wavelet covariances plotted as a function of scale. Each row and column correspond to a different variable. Dotted lines indicate zero covariance. (b) Used and modified with permission from Keitt and Urban (2005)**

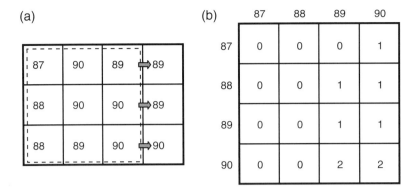

Figure 22.7 **(a) A sample of a continuous raster data set, with dashed box indicating starting position and arrows indicating direction of movement of window. (b) The resulting gray level co-occurrence matrix (GLCM). Columns are starting pixel values and rows are pixel values after movement of window one pixel to the right. Values along the diagonal indicate that the corresponding neighbor has the same value, while cells above and below indicate increases and decreases, respectively**

categorical landscape metrics. Still other approaches, such as those based on fractals (Emerson et al., 2005) or spatial domain filters (Rio and Lozano-Garcia, 2000), are possible.

22.6 CONCLUSIONS AND FRONTIERS

This review serves as a brief introduction to the analysis of spatial pattern in biogeographic data. A great deal more information about the methods we have introduced here is available. There are also important areas we did not address in this short review. We have made little mention of the temporal aspect of pattern variation, yet much of biogeography focuses on dynamism in species distributions and historical factors that have been influential in shaping distributions. Indeed the understanding and characterization of spatiotemporal dynamics is an important area of research in biogeography and landscape ecology (Turner, 2005), with approaches including simply repeating spatial analyses over time, temporal extensions of join counts to assess spread of organisms, and analyses of animal movements (Forester et al., 2007; Fortin and Dale, 2005). Most spatial analysis literature focuses on two-dimensional data, yet many important processes and, increasingly, data sets are in three spatial dimensions. These include important flows of nutrients and biota in marine systems (Cox and Moore, 2000) and forest canopy profiles derived from LIDAR (Dubayah and Drake, 2000). Fortunately, most methods are readily generalized to n-dimensions.

Data models other than the ones we discussed may be important for certain processes and ecosystems. For example, network analysis and graph theoretical approaches are particularly relevant for organisms and processes related to stream networks (Fagan, 2002; Wiens, 2002).

Several unifying lessons emerge from consideration of methods for analysis of spatial patterns in biogeography. First and foremost, we hope that the examples cited make clear that spatial heterogeneity provides a rich source of information on processes of interest in biogeography and need not merely be a nuisance. Second, implicit in this discussion is the value of visually displaying biogeographic data and plots of diagnostic tools such as variograms. Many of the techniques discussed may make some of their greatest contributions in the area of exploratory data analysis. Third, it should be clear that there are many techniques for analyzing and characterizing various types of spatial data. Biogeographers must therefore be able to choose appropriate types of data and analysis to address specific research questions. Exploration of several approaches might be warranted, as some techniques may not correctly identify important patterns or may elucidate complementary information (Fortin and Dale, 2005).

Biogeographers today are enviably positioned amidst a convergence of abundant spatially referenced biological data, a mature body of statistical practice for spatial data, and inexpensive, yet extremely powerful, computation. It is no wonder that interest in biogeography is expanding and theoretical advances are accelerating (Lomolino and Heaney, 2004). Appropriate characterization of spatial patterns will continue to contribute to these advances. While the focus has been on these spatial tools and approaches, we close with a warning issued by Fortin and Dale (2005) that "multiple processes may give rise to the same observed pattern." Indeed the most important lesson may be that, while characterizing, comparing, and analyzing spatial patterns, biogeographers should never lose sight of theoretical underpinnings of their discipline.

REFERENCES

Andren, H., and Angelstam, P. (1988) 'Elevated predation rates as an edge effect in habitat islands—experimental-evidence', *Ecology*, 69(2): 544–47.

Anselin, L. (1995) 'Local indicators of spatial association—LISA', *Geographical Analysis*, 27(2): 93–115.

Arnot, C., Fisher, P.F., Wadsworth, R., and Wellens, J. (2004) 'Landscape metrics with ecotones: Pattern under uncertainty', *Landscape Ecology*, 19(2): 181–95.

Bailey, S.A., Haines-Young, R.H., and Watkins, C. (2002) 'Species presence in fragmented landscapes: Modelling of species requirements at the national level', *Biological Conservation*, 108(3): 307–16.

Baker, W.L. (2001) *r.le: A Set of GRASS Programs for the Quantitative Analysis of Landscape Structure.* Laramie: University of Wyoming.

Boots, B. (2002) 'Local measures of spatial association', *Ecoscience*, 9(2): 168–76.

Bradshaw, G.A., and Spies, T.A. (1992) 'Characterizing canopy gap structure in forests using wavelet analysis', *Journal of Ecology*, 80(2): 205–15.

Burrows, S.N., Gower, S.T., Clayton, M.K., MacKay, D.S., Ahl, D.E., Norman, J.M. and Diak, G. (2002) 'Application of geostatistics to characterize leaf area index (LAI) from flux tower to landscape scales using a cyclic sampling design', *Ecosystems*, 5(7): 667–79.

Camarero, J.J., Gutierrez, E., and Fortin, M.J. (2006) 'Spatial patterns of plant richness across treeline ecotones in the Pyrenees reveal different locations for richness and tree cover boundaries', *Global Ecology and Biogeography*, 15(2): 182–91.

Cardille, J., Turner, M., Clayton, M., Gergel, S., and Price, S. (2005) 'METALAND: Characterizing spatial patterns and statistical context of landscape metrics', *Bioscience*, 55(11): 983–88.

Chen, H.Y., Dennis, J.W., Caraco, T.B., and Stratton, H.H. (2005) 'Epidemic and spatial dynamics of Lyme disease in New York state, 1990–2000', *Journal of Medical Entomology*, 42(5): 899–908.

Clapham, A.R. (1936) 'Over-dispersion in grassland communities and the use of statistical methods in plant ecology', *Journal of Ecology*, 24(1): 232–51.

Clennon, J.A., King, C.H., Muchiri, E.M., Kariuki, H.C., Ouma, J.H., Mungai, P. and Kitron, U. (2004) 'Spatial patterns of urinary schistosomiasis infection in a highly endemic area

of coastal Kenya', *American Journal of Tropical Medicine and Hygiene*, 70(4): 443–48.

Cox, C.B., and Moore, P.D. (2000) *Biogeography: An Ecological and Evolutionary Approach.* Oxford: Blackwell Science.

Cressie, N.A.C. (1993) *Statistics for Spatial Data.* New York: Wiley.

Cressie, N., and Hawkins, D.M. (1980) 'Robust estimation of the variogram: I', *Journal of the International Association for Mathematical Geology*, 12(2): 115–25.

Csillag, F., and Kabos, S. (2002) 'Wavelets, boundaries, and the spatial analysis of landscape pattern', *Ecoscience*, 9(2): 177–90.

Dacey, M.F. (1968) 'An empirical study of the areal distribution of houses in Puerto Rico', *Transactions of the Institute of British Geographers*, (45): 51–69.

Dale, M.R.T., and Mah, M. (1998) 'The use of wavelets for spatial pattern analysis in ecology', *Journal of Vegetation Science*, 9(6): 805–14.

de Castro, M.C., Monte-Mor, R.L., Sawyer, D.O. and Singer, B.H. (2006) 'Malaria risk on the Amazon frontier', *Proceedings of the National Academy of Sciences of the United States of America*, 103(7): 2452–57.

Diggle, P.J. (2003) *Statistical Analysis of Spatial Point Patterns.* London: Arnold.

Dorner, B., Lertzman, K., and Fall, J. (2002) 'Landscape pattern in topographically complex landscapes: Issues and techniques for analysis', *Landscape Ecology*, 17(8): 729–43.

Dubayah, R.O., and Drake, J.B. (2000) 'Lidar remote sensing for forestry', *Journal of Forestry*, 98(6): 44–46.

Dunning, J.B., Danielson, B.J., and Pulliam, H.R. (1992) 'Ecological processes that affect populations in complex landscapes', *Oikos*, 65(1): 169–75.

Emerson, C.W., Lam, N.S.N., and Quattrochi, D.A. (2005) 'A comparison of local variance, fractal dimension, and Moran's I as aids to multispectral image classification', *International Journal of Remote Sensing*, 26(8): 1575–88.

Epperson, B.K. (2003) 'Covariances among join-count spatial autocorrelation measures', *Theoretical Population Biology*, 64(1): 81–87.

Fagan, W.F. (2002) 'Connectivity, fragmentation, and extinction risk in dendritic metapopulations', *Ecology*, 83(12): 3243–49.

Fahrig, L. (2003) 'Effects of habitat fragmentation on biodiversity', *Annual Review of Ecology Evolution and Systematics*: 34487–515.

Forester, J.D., Ives, A.G., Turner, M.G., Anderson, D.P., Fortin, D., Beyer, H.L., Smith, D.W., and Boyce, M.S. (2007) 'Using state-space models to link patterns of elk (*Cervus elaphus*) movement to landscape characteristics in Yellowstone National Park', *Ecological Monographs*, 77(2): 285–299.

Forman, R.T.T. (1995a) *Land Mosaics: The Ecology of Landscapes and Regions.* Cambridge: Cambridge University Press.

Forman, R.T.T. (1995b) 'Some general-principles of landscape and regional ecology', *Landscape Ecology*, 10(3): 133–42.

Fortin, M.J., and Dale, M.R.T. (2005) *Spatial Analysis. A Guide for Ecologists.* Cambridge: Cambridge University Press.

Fotheringham, A.S., and Brunsdon, C. (1999) 'Local forms of spatial analysis', *Geographical Analysis*, 31(4): 340–58.

Fotheringham, A.S., Brundson, C., and Charlton, M. (2000) *Quantitative Geography: Perspectives on Spatial Data Analysis*. London: Sage.

Franklin, J.F., and Forman, R.T.T. (1987) 'Creating landscape patterns by forest cutting: Ecological consequences and principles', *Landscape Ecology*, 1(1): 5–18.

Fraterrigo, J.M., Turner, M.G., Pearson, S.M. and Dixon, P. (2005) 'Effects of past land use on spatial heterogeneity of soil nutrients in southern Appalachian forests', *Ecological Monographs*, 75(2): 215–30.

Geary, R. (1954) 'The contiguity ratio and statistical mapping', *The Incorporated Statistician*: 5115–145.

Gergel, S.E., and Turner, M.G. (eds.). (2002) *Learning Landscape Ecology: A Practical Guide to Concepts and Techniques*. New York: Springer-Verlag.

Getis, A., and Ord, J.K. (1996) 'Local spatial statistics: An overview', in Longley, P. and Batty, M. (eds.), *Spatial Analysis: Modelling in a GIS Environment*, Cambridge: Geoinformation International. pp. 261–78.

Goreaud, F., and Pelissier, R. (1999) 'On explicit formulas of edge effect correction for Ripley's K-function', *Journal of Vegetation Science*, 10(3): 433–38.

Greig-Smith, P. (1952) 'The use of random and contiguous quadrats in the study of the structure of plant communities', *Annals of Botany*, 16: 293–316.

Gustafson, E.J. (1998) 'Quantifying landscape spatial pattern: What is the state of the art?' *Ecosystems*, 1(2): 143–56.

Haines-Young, R. (2005) 'Landscape pattern: Context and process', in Wiens, J.A. and Moss, M.R. (eds.), *Issues and Perspectives in Landscape Ecology*. Cambridge: Cambridge University Press. pp. 103–111.

Haines-Young, R., and Chopping, M. (1996) 'Quantifying landscape structure: A review of landscape indices and their application to forested landscapes', *Progress in Physical Geography*, 20(4): 418–45.

Hanski, I. (1998) 'Metapopulation dynamics', *Nature*, 396(6706): 41–49.

Haralick, R.M., Shanmugam, K., and Dinstein, I. (1973) 'Textural features for image classification', *IEEE Transactions on Systems, Man, and Cybernetics*: SMC-3610–621.

Hardesty, B.D., Dick, C.W., Kremer, A., Hubbell, S., and Bermingham, E. (2005) 'Spatial genetic structure of Simarouba Amara Aubl. (Simaroubaceae), a dioecious, animal-dispersed neotropical tree, on Barro Colorado Island, Panama', *Heredity*, 95(4): 290–97.

Heywood, J.S. (1991) 'Spatial-analysis of genetic-variation in plant-populations', *Annual Review of Ecology and Systematics*, 22: 335–55.

Isaaks, E.H., and Srivastava, R.M. (1989) *Applied Geostatistics*. Oxford: Oxford University Press.

Kayitakire, F., Hamel, C., and Defourny, P. (2006) 'Retrieving forest structure variables based on image texture analysis and IKONOS-2 imagery', *Remote Sensing of Environment*, 102(3–4): 390–401.

Keitt, T.H., and Urban, D.L. (2005) 'Scale-specific inference using wavelets', *Ecology*, 86(9): 2497–504.

Kupfer, J.A., Malanson, G.P. and Franklin, S.B. (2006) 'Not seeing the ocean for the islands: The mediating influence of matrix-based processes on forest fragmentation effects', *Global Ecology and Biogeography*, 15(1): 8–20.

Lancaster, J., and Downes, B.J. (2004) 'Spatial point pattern analysis of available and exploited resources', *Ecography*, 27(1): 94–102.

Langford, W.T., Gergel, S.E., Dietterich, T.G., and Cohen, W. (2006) 'Map misclassification can cause large errors in landscape pattern indices: Examples from habitat fragmentation', *Ecosystems*, 9(3): 474–88.

Le Corre, V., Roussel, G., Zanetto, A., and Kremer, A. (1998) 'Geographical structure of gene diversity in Quercus Petraea (Matt.) Liebl. III. Patterns of variation identified by geostatistical analyses', *Heredity*, 80: 464–73.

Li, H.B., and Reynolds, J.F. (1993) 'A new contagion index to quantify spatial patterns of landscapes', *Landscape Ecology*, 8(3): 155–62.

Li, H.B., and Wu, J.G. (2004) 'Use and misuse of landscape indices', *Landscape Ecology*, 19(4): 389–99.

Lomolino, M.V., and Heaney, L.R. (2004) 'Introduction: Reticulations and reintegration of modern biogeography', in Lomolino, M.V. and Heaney, L.R. (eds.), *Frontiers of Biogeography*. Sunderland, MA: Sinauer. pp. 1–3.

Lotwick, H.W., and Silverman, B.W. (1982) 'Methods for analyzing spatial processes of several types of points', *Journal of the Royal Statistical Society Series B-Methodological*, 44(3): 406–413.

Maurice, K.R., Welch, J.M., Brown, C.P., and Latham, R.E. (2004) 'Pocono mesic till barrens in retreat: Topography, fire and forest contagion effects', *Landscape Ecology*, 19(6): 603–20.

McGarigal, K. (2007) *Fragstats Documentation*. Amherst, MA: University of Massachusetts.

McGarigal, K., and Cushman, S.A. (2005) 'The gradient concept of landscape structure', in McGarigal, K. and Moss, M.R. (eds.), *Issues and Perspectives in Landscape Ecology*. Cambridge: Cambridge University Press. pp. 112–19.

McGarigal, K.S., Cushman, S.A., Neel, M.C., and Ene, E. (2002) *FRAGSTATS: Spatial Pattern Analysis Program for Categorical Maps*. Amherst MA: University of Massachusetts.

Miller, T.F., Mladenoff, D.J., and Clayton, M.K. (2002) 'Old-growth northern hardwood forests: Spatial autocorrelation and patterns of understory vegetation', *Ecological Monographs*, 72(4): 487–503.

Moran, P.A.P. (1948) 'The interpretation of statistical maps', *Journal of the Royal Statistical Society. Series B (Methodological)*, 10(2): 243–51.

Moran, P.A.P. (1950) 'Notes on continuous stochastic phenomena', *Biometrika*, 37(1/2): 17–23.

Musick, H.B., and Grover, H.D. (1991) 'Image textural measures as indices of landscape pattern', in Turner, M.G. and Gardner, R.H., (eds.), *Quantitative Methods in Landscape Ecology: The Analysis and Interpretation of Landscape Heterogeneity*, New York: Springer Verlag. pp. 77–103.

Parker, K.C., and Jorgensen, S.M. (2003) 'Examples of the use of molecular markers in biogeographic research', *Physical Geography*, 24(5): 378–98.

Remmel, T.K., and Csillag, F. (2003) 'When are two landscape pattern indices significantly different?' *Journal of Geographical Systems*, 5: 331–51.

Ricklefs, R.E., and Lovette, I.J. (1999) 'The roles of island area *per se* and habitat diversity in the species–area relationships of four Lesser Antillean faunal groups' *Journal of Animal Ecology*, 68(6): 1142–1160.

Riitters, K.H., and 7 authors. (1995) 'A factor-analysis of landscape pattern and structure metrics', *Landscape Ecology*, 10(1): 23–39.

Rio, J.N.R., and Lozano-Garcia, D.F. (2000) 'Spatial filtering of radar data (Radarsat) for wetlands (brackish marshes) classification', *Remote Sensing of Environment*, 73(2): 143–51.

Ripley, B. (1976) 'The second-order analysis of stationary point processes', *Journal of Applied Probability*, 13: 255–66.

Ripley, B. (1977) 'Modeling spatial patterns', *Journal Royal Statistical Society. Series B-Methodological*, 39(2): 182–212.

Ripley, B.D. (1981) *Spatial Statistics*. New York: Wiley.

Saura, S., and Carballal, P. (2004) 'Discrimination of native and exotic forest patterns through shape irregularity indices: An analysis in the landscapes of Galicia, Spain', *Landscape Ecology*, 19(6): 647–62.

Shannon, C., and Weaver, W. (1949) *The Mathematical Theory of Communication*. Urbana, IL: University of Illinois Press.

Shao, G., Liu, D., and Zhao, G. (2001) 'Relationships of image classification accuracy and variation of landscape statistics', *Canadian Journal of Remote Sensing*, 27(1): 35–45.

Sleeman, J.C., Kendrick, G.A., Boggs, G.S., and Hegge, B.J. (2005) 'Measuring fragmentation of seagrass landscapes: Which indices are most appropriate for detecting change?' *Marine and Freshwater Research*, 56(6): 851–64.

Smith, T.E. (2004) 'A scale-sensitive test of attraction and repulsion between spatial point patterns', *Geographical Analysis*, 36(4): 315–31.

Smithwick, E.A.H., Mack, M.C., Turner, M.G., Chapin, F.S., Zhu, J., and Balser, T.C. (2005) 'Spatial heterogeneity and soil nitrogen dynamics in a burned black spruce forest stand: Distinct controls at different scales', *Biogeochemistry*, 76(3): 517–37.

Southworth, J., Munroe, D., and Nagendra, H. (2004) 'Land cover change and landscape fragmentation—Comparing the utility of continuous and discrete analyses for a Western Honduras region', *Agriculture Ecosystems Environment*, 101(2–3): 185–205.

St-Louis, V., Pidgeon, A.M., Radeloff, V.C., Hawbaker, T.J., and Clayton, M.K. (2006) 'High-resolution image texture as a predictor of bird species richness', *Remote Sensing of Environment*, 105(4): 299–312.

Tobler, W.R. (1970) 'A computer model simulation of urban growth in the Detroit region', *Economic Geography*, 46(2): 234–40.

Tuceryan, M., and Jain, A. (1998) 'Texture analysis', in Chen, C.H., Pau, L.F. and Wang, P.S.P. (eds.), *Handbook of Pattern Recognition and Computer Vision*. Singapore: World Scientific Publishing Company. pp. 207–48.

Turner, M.G. (2005) 'Landscape ecology: What is the state of the science?' *Annual Review of Ecology Evolution and Systematics*, 36: 319–44.

Turner, M.G., Gardner, R.H., and O'Neill, R.V. (2001) *Landscape Ecology in Theory and Practice: Pattern and Process*. New York: Springer-Verlag.

Upton, G.J.G., and Fingleton, B. (1985) *Spatial Data Analysis by Example*. Chichester, UK: John Wiley.

Wells, M.L., and Getis, A. (1999) 'The spatial characteristics of stand structure in Pinus torreyana', *Plant Ecology*, 143(2): 153–70.

Whitmire, S.L., and Tobin, P.C. (2006) 'Persistence of invading gypsy moth populations in the United States', *Oecologia*, 147(2): 230–37.

Wickham, J.D., and Riitters, K.H. (1995) 'Sensitivity of landscape metrics to pixel size', *International Journal of Remote Sensing*, 16(18): 3585–94.

Wiens, J.A. (1989) 'Spatial scaling in ecology', *Functional Ecology*, 3(4): 385–97.

Wiens, J.A. (2002) 'Riverine landscapes: Taking landscape ecology into the water', *Freshwater Biology*, 47(4): 501–15.

Yamada, I., and Rogerson, P.A. (2003) 'An empirical comparison of edge effect correction methods applied to K-function analysis', *Geographical Analysis*, 35(2): 97–109.

23

Predictive Modeling of Biogeographical Phenomena

Niall G. Burnside and Stephen Waite

23.1 INTRODUCTION

Attempts to model the distribution and dispersal of species have a long and established history in biogeography and ecology (Grinnell, 1917; Hutchinson, 1957). That interactions between biotic and abiotic phenomena determine species distributions is a fundamental tenet of biogeography (Pulliam, 2000), and it is one that helps us understand the patterns and limits of distribution at the species, community, and habitat scales. Historically, traditional approaches to modelling have relied on the ability to establish predictive correlations or rule-based systems linking species occurrence and environmental and ecological factors. With the advent and growth of geographical information systems, the potential power of these approaches has been increased, particularly at regional and global scales. When coupled with ecological theory and appropriate statistical techniques, GIS offer the ability to develop and build predictive tools (Franklin, 1995), which are valuable in understanding the complex processes that determine distribution and abundance. In predictive modelling, the identification of these relationships, and their quantification, enable us to predict the effects and results of losses and changes within the environment (Fairbanks et al., 1996) and assist in the development of tools and the identification of mechanisms to counteract or mitigate adverse and negative effects. To understand why a particular organism is found in a specific area requires appropriate knowledge of the organism's ecology and an ability to quantify the complex interaction and relationship between

that organism and its environment (Pulliam, 2000; Cox and Moore, 2005).

Growing concern over issues including climate change, land-use change, and biodiversity loss highlight the need to examine, understand, and quantify the relationships they encompass (Fairbanks et al., 1996; Whittaker and Fernandez-Palacios, 2007). Predictive modelling is a thriving biogeographical and ecological discipline and has been the focus of recent journal special issues and editorials concerned about international biodiversity conservation and conservation planning (see Rodriguez et al., 2007). Many researchers have used predictive modelling techniques to examine the distribution of individual species (be they invasive or rare) (Waite and Burnside, 2001; Zaniewski et al., 2002; Robertson et al., 2004; Hurme et al., 2005), plant communities (Cherrill et al., 1995; Burnside et al., 2002; Palo et al., 2005; Ferrier and Guisan, 2006), or faunal species (Aspinall, 1996; Chefaoui et al., 2005; Mörtberg and Karlström, 2005; Finch et al., 2006), while others have examined biogeographic groups (Ellis et al., 2007) or the effects of land-use change, agricultural intensification, and conservation efforts on the landscape in a predictive sense (Lee et al., 2002; Gkaraveli et al., 2001). The models use the relationship between the occurrence of the target species, communities, or habitats and relate these to a set of predictor variables (Segurado and Araújo, 2004).

Franklin (1995) states that predictive mapping is largely founded on ecological niche theory and vegetation gradient analysis. Those abiotic and biotic factors that control the distribution of a species or community can be used to predict their spatial distribution. Outputs typically vary in relation to the modelling target, the strength of the biotic and abiotic relationships, and the number of variables included (Segurado and Araújo, 2004).

23.2 A TAXONOMY OF MODELS

Although used at differing scales, and for different but related purposes, the techniques or methodologies employed in predictive modelling have many similarities. In fact, inspection shows that the techniques generally fall into two main types, although the definition of these varies between authors.

Robertson et al. (2003) describe the two broad methodological approaches as correlative and mechanistic, while Skidmore (2002) refers to the techniques as inductive and deductive, and Carpenter et al. (1993) suggested pattern-based and process models. Despite using differing terms, the taxonomy is similar, as correlative models are somewhat inductive, and work on the premise of relationships and patterns between the species distribution and environmental predictor variables derived from a population or sample population. The method employs the concept of association and the assumption that a relationship established for n preceding values of the factor(s) and species will hold for the $n + 1$ and subsequent values. Thus, for example, probabilities may be used to predict the distribution of a species in a landscape, based upon the actual or existing distributional data from a similar landscape area (Robertson et al., 2004). Conversely, mechanistic models are by nature deductive and use knowledge about the ecological niche to model the processes and/or mechanisms that result in the observed species distribution within the environment. For example, a strong understanding of the autecology of a plant species could be used to develop a mechanistic or ecophysiological model (sensu Robertson et al., 2003) that predicts the response of a species to long-term climatic change (Woodward and Lomas, 2004; Ellis et al., 2007).

This model taxonomy is largely supported by other authors who, despite using slightly different terms, offer similar classifications (Vaughan and Ormerod, 2005; Ottaviani et al., 2004; Segurado and Araújo, 2004; Woodward and Lomas, 2004; Robertson and Palmer, 2002; Guisan and Zimmermann, 2000; Yates et al., 2000). Although Guisan and Zimmermann (2000), when referring to the early work of Levin (1966), include an additional classification of analytical, suggesting that another branch of models exist that are largely theoretical in design, and seek to predict outcomes in a simplified reality (Figure 23.1) and are typified by modeling approaches outlined by Maynard-Smith (1974).

A useful example of more mathematical and theoretical models that could be described as analytical would include MacArthur and Wilson's (1967) Equilibrium Theory of Island Biogeography. MacArthur and Wilson (1967) hypothesised that a dynamic balance between species immigration and extinction determines the number of species present on oceanic islands. Within their model they showed that the number of species on a given island is usually approximately related to the area of the island by the equation $S = C A^z$, where S represented the number of species, A was the area, and C and z are taxa-specific constants. The model, in a simplified manner, facilitated the analysis and prediction of relative abundances of species and the examination of island biota more generally (Whittaker and Fernandez-Palacios, 2007).

23.3 GEOGRAPHICAL INFORMATION SYSTEMS

Predictive modeling, by whichever method, typically has a spatial dimension, particularly when examining the geographical distribution of species, communities, or habitats. Therefore predictive modeling and GIS are often discussed in association (Franklin, 1995; Goodchild et al., 1996; Skidmore, 2002). Franklin (1995) offers a useful early review and summary of the development of predictive vegetation mapping, while Skidmore (2002) provides a broader comment on the application of GIS and remote sensing in environmental modeling (for details on remote sensing inputs to modelling biogeographical phenomena, see Nightingale et al., this volume).

GIS offer the ability to build and develop predictive tools that examine and describe spatial characteristics and outcomes. The coupling of predictive modelling techniques and GIS offers a means of exploring the processes of distribution and abundance. For example, if the presence of species x and its distribution was described by a particular elevational range and soil type, GIS could be used in either a simple Boolean approach (Franklin et al., 1986 cited in Franklin, 1995) or a more complex fuzzy-envelope classification method (Robertson et al., 2004) to identify where we are likely to encounter the species in a given area of varying topography and pedology. This coupling of predictive modelling and GIS is in itself a major area of research (good summaries

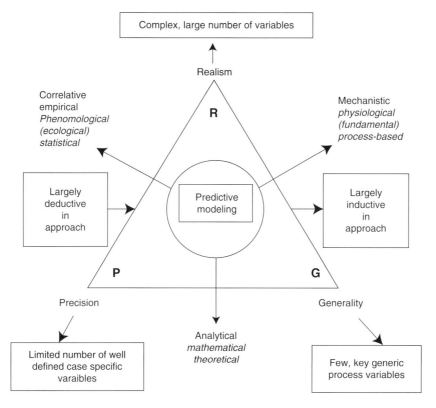

Figure 23.1   A taxonomy of models based on their methodological approaches (adapted from Levin, 1966; Guisan and Zimmermann, 2000)

are provided in Goodchild et al., 1993; Goodchild et al., 1996; and Skidmore, 2002). As an example, Burnside et al. (2002) used a GIS approach and landscape ecological perspective to identify and prioritize landscape areas for selective restoration of internationally important calcareous grassland communities in southeast England. They employed a correlative approach and provided a simple method to assist in targeting programmes to achieve practical and sustainable outcomes. The correlative approach used showed a strong association between the unimproved calcareous grassland communities and particular topographic parameters (e.g., soil, slope, aspect, and elevation) on the South Downs, a range of chalk hills. Model validation was achieved by applying the model to an unsampled portion of the landscape that had not been included as a model training set. The model output was compared with actual grassland sites using the Sørenson's unweighted index (Waite, 2000). The results showed that the model could effectively predict around 80% of sites (for moderately and highly suitable areas) and suggested a sufficient similarity between modelled

habitats and existing patches to proceed with the analysis. This relatively high level of similarity was achieved despite the inaccuracies associated with the coarse resolution of the digital terrain model (DTM). Problems of model accuracy appeared to relate to zones of transition between different community types and false absences. More specifically, the analysis suggested that the main difficulty lay in determining the properties of sites at the transitional boundary of calcareous and mesotrophic communities and how this could best be represented within the model.

23.4 MODELLING METHODS

In the next section we will outline and evaluate the main generic approaches most frequently used to develop predictive models. Where possible we comment on the advantages, limitations, and data requirements of each approach (Figure 23.2). A critical stage during the development of a species distribution model is the selection of an

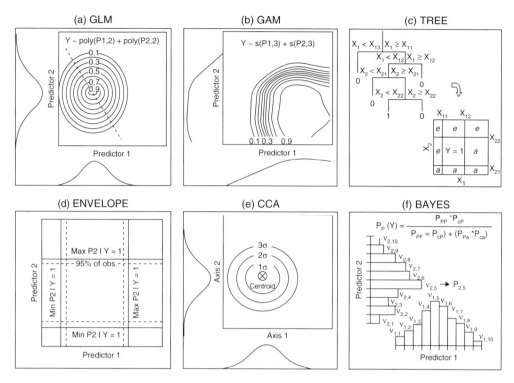

Figure 23.2   Examples of response curves for the six main statistical approaches used in predictive modelling (from Guisan and Zimmermann, 2000). (a) generalised linear models, (b) general additive models, (c) tree models, (d) envelope models, (e) ordinations (canonical correspondence analysis), (f) Bayesian probability models

appropriate statistical model that links environmental factors to either the abundance or the occurrence of the species. The choice of modelling methodology and statistical algorithm should be governed by the nature of the available environmental data, the predictor explanatory valuables, and the response variables (i.e., the abundance or probability of species occurrence at a particular location). The variables may be continuous (i.e., theoretically capable of taking any value within a specified range) or discrete (i.e., categorical variables capable of taking only a limited number of specified values) (Waite, 2000).

In an effort to improve model realism there is a tendency to include too many variables. It is not the case that complicated models always perform better than simple models with a limited number of variables (Hilborn and Mangel, 1997). Linhart and Zucchini (1986) distinguish between two types of model prediction errors. One source of prediction error is due to the necessary approximation or simplification of the complex system by the model; these errors decrease with model complexity. The second type of prediction error is due to parameter estimation; these kinds of errors increase with model complexity. Hilborn and Mangel (1997) consider the problem of balancing these two sources of error and suggest that "optimum model size [i.e., complexity] is much smaller than intuition dictates."

The nature of the available data sets will constrain the range of modelling methods that can be used for effective prediction. The type of the response variable must be matched against the multivariate explanatory data set by the appropriate choice of statistical algorithm. There is a large and growing literature covering these topics, that is, selection of statistical models and testing of model output and assumptions against the characteristics of the dataset. A good review is provided in Manly (2008), while Manly (1994) provides an excellent introduction into the general concepts of multivariate statistical procedures and their application in environmental sciences. Kent and Coker (1992) provide a good introduction to the application of multivariate statistics in plant ecology, and Waite (2000) provides a more general and comprehensive introduction to the use of statistics and multivariate techniques in ecology. Basic reviews of the spatial techniques and modelling procedures that have been used for distributional modelling and predictive mapping within a GIS context are given by Franklin (1995), Goodchild et al. (1993, 1996), and Skidmore (2002).

Guisan and Zimmermann (2000) provide an excellent critical review of the different statistical approaches that have been used to produce predictive models of species and habitat distribution, while Segurado and Araújo (2004) and Vaughan and Ormerod (2005) discuss some of the challenges and issues within particular approaches. The methods used are typically grouped into six major classes (Figure 23.2): (a) generalized linear models (GLM); (b) generalized additive models, (GAM); (c) classification trees (CART); (d) the drawing of environmental envelopes; (e) ordination-based methods, and lastly methods based on (f) Bayesian probability modelling. Because of their frequency of use, and the generality of approach, we will concentrate on GLM, GAM, classification models, environmental envelopes, multivariate analysis and ordination techniques, while providing only an outline of Bayesian models and introducing the concept of dynamic models.

23.4.1 Generalized Linear Models (GLM)

Species are often shown to vary in abundance or presence—absence along an environmental gradient (Pulliam, 2000). This relationship or association, if quantified, can be used to model species distribution in relation to the environmental variable defining the gradient or in relation to a set of variables. Because of the stationary nature of plants we would expect the relationship between vegetation and environmental conditions to be more easily or precisely defined than for more mobile animal species (Austin, 2002). Generalized linear modeling (GLM) presents a means of utilizing these relationships and using environmental variables in a predictive manner. GLM combined with maximum likelihood estimations of model parameters provide a more flexible and robust range of statistical procedures than the traditional approaches, of analysis of variance (ANOVA) and simple and multiple least squares regression (MR) to which it is closely related (Crawley, 1993; Guisan, 2002). The recent popularity of GLM procedures in predictive modelling partly reflects the greater availability of maximum likelihood-based procedures among mainstream general statistical packages (e.g., MINITAB and SPSS). However, they also have

the advantage of being able to deal with response variables that are not normally distributed and do not show homoscedasticity, that is, GLM does not require the variance of the response variable to be constant, unaffected by the size of the independent variable. However, it is assumed that the distribution is exponential in form (e.g., normal, Poisson, binomial, or gamma).

The technique simplifies the relationship between the target group and the environmental variables. The characteristic form of GLM has a response variable y, linked to a linear combination of independent variables $(x_1, x_2 \cdots x_n)$ and an error term ε and a link function f.

$$y = f(\beta_0 + \beta_1 x_1 + \beta_2 x_2 + \cdots + \beta_n x_n) + \varepsilon$$

If the errors are assumed to be normally distributed, and the function f is set to unity, the general form of the model equations becomes

$$y = \beta_0 + \beta_1 x_1 + \beta_2 x_2 + \cdots + \beta_n x_n + \varepsilon$$

which corresponds to the standard model equation for least-squares multiple regression (MR). The model suits a range, and combination, of environmental variables as GLM has by adopting different link functions, the ability to accommodate a variety of different response variable distributions. For example, if y follows a Poisson distribution, a common distribution associated with count data (Manly, 1994; Waite, 2000), for example, the number of individuals of a particular species in a given grid area, the link function f would be set so $f = e^x$ provides a log-linear regression model.

$$y = e^{(\beta_0 + \beta_1 x_1 + \beta_2 x_2 + \cdots + \beta_n x_n)}$$
$$\ln y = \beta_0 + \beta_1 x_1 + \beta_2 x_2 + \cdots + \beta_n x_n$$

An important point to note is that in order to choose the appropriate link function the distribution of the response variable must either be assumed or tested. GAM or simpler fitting/smoothing procedures may be used in an exploratory way to investigate the overall shape of the response variable and help to guide the form of the GLM equation used (i.e., polynomial order) and link function selected (Brown, 1994; Franklin, 1998). What is clear, however, is that the distribution of the response variable, that is, the manner in which species abundance varies along an environmental gradient, is crucial to determining the form of the GLM selected.

When considering GLMs it is important to consider the nature of the relationship between the target and the response variables. Following the empirical work of Whittaker (1956, 1960), it has become commonplace to assume that species follow 'bell-shaped' Gaussian distributions along environmental continua or gradients. Species distribution along hypothetical environment gradients are typically illustrated in ecological texts as a series of smoothly drawn bell-shaped symmetrical or occasionally skewed curves spaced along the length of the gradient. Accepting this dominant model of species distribution means that a linear relationship between species abundance and environmental factors will only hold for restricted sections of the gradient. However, it should be noted that the Gaussian model has been subjected to little empirical testing, and in those studies that have attempted to determine the pattern of species abundance along the entire length of the gradient relatively few species have been shown to have the bell-shaped distribution curve we might expect (Austin, 2002, 2005). In addition, because of the complexity and range of factors that determine the presence and abundance of species, the distribution will not be 'solid'. Along any section of the gradient the species will be absent from some locations or will have abundances below those depicted by the graphed line of the distribution. Austin (2002, 2005) and McCune and Grace (2002) provide detailed reviews of these issues and consider the effects of various possible species-environmental relationships on the performances of the statistical methods used in the analysis of ecological communities. It is worth noting that where the distribution and frequency of ecological communities, groups of species, or habitats are being modelled along an environmental gradient there is no a priori reason to assume that they will follow similar response distribution to individual species.

Logistic regression (LR) is the commonest form of GLM used to model species distribution (Guisan and Zimmermann, 2000; Rushton et al., 2004; Garcia-Ripolles et al., 2005). The probability of species occurrence is linked to a set of environmental variables by the logit link function. The logit function is the appropriate link function when the response variable in binomial, that is, for any trial, locations, where there are only two possible outcomes, in this case species presence or absence. If the probability of species presence is P_A, the probability of the species not being present will be $1 - P_A$ and the odds ratio for these two outcomes is given by $P_A / 1 - P_A$. If the odds were three, this would mean that species A was three times more likely to occur along that part of the gradient than not occur. If the natural logarithm of the odds ratio is plotted against an environmental

variable (x) the relation is typically linearized. To reject the null hypothesis, that the probability of species occurrence is independent of the x, we have only to show that the slope of the plot is not equal to zero. This is the basis of logistic regression, expressing these ideas more formally we have:

$$\text{Odds} = \text{Probability of event/Probability of nonevent} = P_A / 1 - P_A$$

$$\ln(\text{odds}) = \text{logit}(P_A) = \ln (P_A / 1 - P_A)$$

$$\ln\left(\frac{P_A}{1 - P_A}\right) = \beta_0 + \beta_1 x$$

$$\frac{P_A}{1 - P_A} = e^{\beta_0 + \beta_1 x}$$

and

$$P_A = \frac{e^{\beta_0 + \beta_1 x}}{1 + e^{\beta_0 + \beta_1}}$$

These equations describe the situation where the response variable (P_A) is linked by the logit transformation to a single independent variable analogous to simple linear regression. The model may readily be extended to apply to more complex situations involving multiple independent variables.

$$P = \frac{e^z}{1 + e^z} = \frac{1}{1 + e^{-z}}$$

$$z = \beta_0 + \beta_1 x_1 + \beta_2 x_2 + \cdots + \beta_n x_n$$

$$\ln\left(\frac{P}{1 - P}\right) = \beta_0 + \beta_1 x_1 + \beta_2 x_2 + \cdots + \beta_n x_n$$

In comparison to least-squares estimations of parameters, maximum likelihood methods associated with GLM are more sensitive to sample size and need larger sample sizes to be reliable. Peduzii et al. (1996) suggest that as a useful rule of thumb that there should be at least 10 observations for each parameter in the model.

23.4.2 General Additive Models (GAM)

General additive models (GAMs) may be seen as essentially a nonparametric extension of the GLM, and they exploit the idea that particular parameters determine the presence/absence or abundance of a particular species or community. In GAM, however, the response variables are modelled through a link function as the sum of a series of separate functions, each of which relates an independent explanatory variable to the response variable. Therefore the assumption is that the explanatory variables are linked but that each variable has potentially a different function or (strength of) relationship with the target group. If, however, the link function is set at unity the GAM is:

$$y = \beta_0 + f_1(x_1) + f_2(x_2) + \cdots + f_n(x_n) + \varepsilon.$$

The strength of GAMs is in the ability to tailor the model to each specific dataset (Lehmann, 1998). This results in very strong correlations and good reproducibility of the modelled data. It does, however, also mean that each model is somewhat custom-made for the target and response variables, and it lacks a generality often heralded as a quality of predictive modelling (Vaughan and Ormerod, 2005). Therefore each model is comparatively specific in nature and prone to overfitting as the technique applies independent smoothing functions to each predictor variable (Guisan and Zimmermann, 2000). The statistical algorithm used to fit the model has to perform two interactive processes. First, selecting in isolation (i.e., for each individual variable) and, second, in combination for the complete set of variables appropriate polynomial smoothing spline functions that together best describe the relations between the set of explanatory variables and the response variable *y*. It is the smoothing aspect that compounds the overfitting but adds to their versatility and strength of descriptive fit. Examples of the use of GAM to predictively model species distribution and species richness are described in detail by Lehmann (1998), Lehmann et al. (2003), and Moisen et al. (2006). A more comprehensive account of model fitting and statistical basis of the GAM method may be found in Hastie and Tibshirani (1990).

23.4.3 Classification and decision-based methods

Given the nature of the problem, for instance, predicting the presence or absences of a species at a location with known environmental characteristics, an obvious approach would be to produce a key or decision tree that allows the set of environmental characteristics at each location to be mapped to the presence or abundance of the species. This approach links well within applied GIS, and the information embodied in the decision tree is typically captured as a series of

"IF . . . THEN . . ." statements (Urban et al., 2002) and often represented through traditional Boolean-type procedures and cartography (Burrough and McDonnell, 1998). Skidmore (2002) states that such an approach typically represents an example of what he refers to as an inductive model. In practice, however, it may be a synthesis of both inductive and more mechanistic approaches, and a hierarchical decision tree developed from both expert knowledge and the careful quantitative inspection of the relationships between species occurrence and environmental factors (Skidmore, 2002). In such an example, a decision tree may be used to implement the results from more formal and objective attempts to model the data set. For example, where it has been possible to delimit the ecological niche of the species for a range of different environmental factors and the sequential comparison of these layers of information, identifying regions of overlap can allow the development of a simple rule-based decision tree.

Burnside et al. (2003), in a study of nationally important habitat types in the United Kingdom, implemented a rule-based decision-tree approach to initially identify areas of potential habitat opportunity in southeast England (Figure 23.3). Nine target habitat types were chosen and a hierarchical approach was used to identify, select, and quantify areas of opportunity for enhancing and restoring the network of key habitats within an area of conservation interest. While such ad hoc and subjective approaches can successfully produce GIS-based spatial models (Burnside et al., 2003), when used to predict species distributions they have two inherent weaknesses. Because they are subjective they lack generality (and may be landscape specific, as in the example provided), but more importantly because of the hierarchical nature, the explanatory variables can typically only be used to make decisions at one level in the tree. This restriction means that simple rule-based decision trees can fail to capture the complicated patterns of variable interactions frequently encountered in actual biological and physical systems.

These problems can be overcome, in some cases, by the implementation of classification and regression tree models (CART). CART algorithms recursively divide the data into subsets of increasing homogeneity with respect to defined group characteristics (e.g., species presence—absence). The process continues until all end groups are homogeneous or a defined level of within-group similarity is achieved. Because at each partition the data set is divided into two groups on the basis of the single variable, which at that stage in the process produces the most homogeneous subsets, a dichotomous key is produced that allows classification of unknown data points

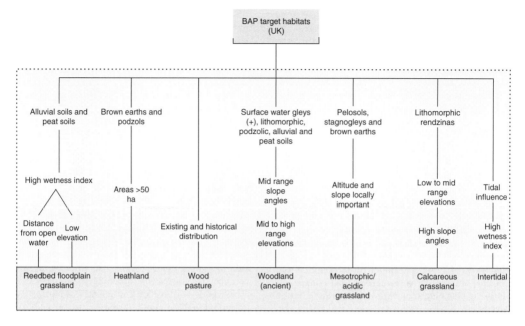

Figure 23.3   Modified and generalized hierarchical decision tree for nine U.K. nationally important BAP habitat types in the southeast of England (adapted from Burnside et al., 2003)

(Brieman et al., 1984; Skidmore, 2002; McCune and Grace, 2002).

The CART approach has a number of important advantages over other classification procedures. In particular the approach is not restricted by particular data types, and the predictor or explanatory variables can be a mixture of continuous and categorical variables. Furthermore, the procedure is nonparametric and does not require the data to follow a specified functional form. Moreover, the same explanatory variable may be used at more than one level in the tree, which allows the context dependency between variables to be identified. Interestingly, Guisan and Zimmermann (2000) in their extensive review were unable to find any examples of the CART procedure being used to model habitat or species distribution. More recent papers and reviews, however, suggest that the use of CART models in predictive modelling has and is likely to increase (e.g., Seoane et al., 2005; Pearson et al., 2006). McCune and Grace (2002) and De'Ath and Fabricius (2000) consider that CART procedures are particularly suitable for, and able to efficiently deal with, complex multivariate ecological data. However, what is apparent is that one principal limitation is the scarcity of readily available software, but this appears to be changing and some routines may be downloaded

free from the web (e.g., QUEST a public domain CART package; www.quest.com).

23.4.4 Environmental envelopes

The 'environmental envelope' method is conceptually the simplest approach to modelling species distribution and has been used extensively to analyse species presence and absence data (Guisan and Zimmermann, 2000; Beaumont et al., 2005). The available training data set is used to establish a multidimensional species 'climate envelope'. In a simple case where the climate envelope is defined by two climate variables (mean annual rainfall and temperature), the distribution data for the species would be used to establish the species range for each variable, which might for example be 150–250 mm and 10–15°C. All areas that satisfy both of these requirements (i.e., have a mean annual rainfall of between 150–250 mm *and* a mean annual temperature of between 10–15°C) are contained in the climate envelope and represent potentially suitable areas for the occurrence of the species. As more dimensions are added to the model, the envelope decreases in size, becoming more precisely defined and the number of locations that meet all of the criteria decreases.

The method, particularly if coupled with GIS techniques, follows a general Boolean logic approach and offers value through its simplicity. This approach mirrors that used by Holdridge (1967) to delimit the biogeographical distribution of ecosystems and is an example of what Skidmore (2002) describes as an inductive empirical model. This approach is increasingly referred to as species 'niche modelling', the concept of an environmental envelope is directly analogous to Hutchinson's (1957) concept of a niche as an n-dimensional space that delimits the ecological hyper-volume in which the species is able to exist.

Several computer programes have been developed to define the environmental envelopes of species. BIOCLIM, developed by Busby (1986, 1991), has been used extensively to model the distribution of the fauna and flora of Australia (cf. Beaumont et al., 2005, and references cited therein) but has also been used successfully in South Africa (Eeley et al., 1999) and South America (Téllez-Valdés and Dávila-Aranda, 2003). BIOCLIM calculates the minimum multi-dimensional rectilinear species envelope using up to 35 climate-related variables. The variables are initially ordered, and then whether the grid cell falls within a user-defined range (normally 5th–95th percentiles) for the two variables being considered is assessed. Locations where all of the values for the selected parameters fall within the 5th–95th percentiles of the species envelope are considered 'core' regions of the species envelope and distribution. Although the BIOCLIM model and related approaches have been successfully used to model species distribution at large spatial scales (e.g., 2.5 x 2.5-km grid squares), the method is prone to 'overfitting' when too many parameters are used and conversely underfitting if inappropriate or insufficient parameters are incorporated. Beaumont et al. (2005) examined the effects of parameter selection when using BIOCLIM to model the distribution of 25 Australian butterfly species. They found that size of the predicted distribution decreases with the number of parameters incorporated into the model, that there was redundancy among the parameters, and that the set of parameters that best modelled distribution varied between species.

Environmental envelope predictions are not only sensitive to the choice of variables, but also to the data points included in the training set. Removing a point can result in a more narrowly defined envelope, while adding data points may overpredict the range of the species. In some cases the predictions can also be based on existing datasets, which may not have been collected for the purposes of modelling, therefore leading to sampling biases. The problem, thus faced by the modeller is how to select the best combination of predictive variables (parameters) and data points. In order to tackle this problem the goodness of fit of the predicted environmental envelope to the original species distribution needs to be assessed. Various measures for doing this are available (Manel et al., 2001). However, while it is feasible to compare the goodness of fit of models produced by incorporating different combinations of variables, it is not computationally practical to produce and test fit all of the possible models parameterized using all possible combinations of data points. McClean et al. (2005) and Termansen et al. (2006) tackled this problem by using a heuristic computational genetic algorithm (GA) that in a process analogous to evolution derives the optimal environmental envelope. The GA begins with an initial population of climate envelopes defined by a set of variable ranges. Each of these possible envelopes can be considered as analogous to a 'chromosome', containing a unique set of environmental trait values. Each trait is represented as a gene that codes for each environmental variable. In this analogy, the initial environmental trait values are generated randomly and are drawn from the data set encompassing the environmental conditions in which the species occurs and the climate range for the study area. The 'fitness' of environmental envelopes coded by each 'chromosome' is assessed using Sørenson's similarity index. This compares the predicted environment generated by the 'chromosome' with the actual environment.

$$S = \frac{2a}{2a + b + c}$$

where a = number of correctly predicted occurrences, b = number of falsely predicted occurrences, and c is the number of unpredicted occurrences. The higher the value of S, the greater the fitness of the chromosome and the more precisely the environmental envelope coded for in the 'chromosome' provides a description of species distribution. In a process analogous to evolution by natural selection, chromosomes with the highest fitness are allowed to reproduce, that is, increase in numbers. The reproductive process includes two means of generating new chromosomes and hence variation. Random pairs of chromosomes undergo crossover, swapping sections of chromosomes and randomly chosen genes on random chromosomes are mutated, the values of gene being randomly increased or decreased. The algorithm continues this iterative process of reproduction and selection, based on the chromosome fitness, until the best possible fit between the evolved environmental envelope and species distribution has been achieved.

Genetic algorithms of the type outlined above are one example of the machine-learning methods available that include artificial neural networks (ANN) and CART models. These approaches provide a potentially useful alternative to traditional statistical modelling (Anderson et al., 2003; Pearson et al., 2002, 2006). ANN have the advantage of being able to cope with nonlinear responses to environmental factors and have been used to model species distribution (Linderman et al., 2004). However, in contrast to GAs and CARTs the biological and ecological basis of solutions generated by ANN can be difficult to identify, it being difficult to assess the relative importance and impact of the environmental factors incorporated in the model.

23.4.5 Multivariate analysis and ordinations

Within the context of biogeography and ecology the terms 'multivariate analysis' and 'ordination' normally refer to a particular group of statistical procedures used to identify and describe patterns in community and environmental datasets (Kent and Coker, 1992). The most frequently used procedures produce either site or species ordinations (e.g., Principal Component Analysis, PCA) or otherwise group samples together using various resemblance of similarity measures to produce a hierarchical classification of data points (Waite, 2000; Kent and Coker, 1992; Fowler et al., 1998). Apart from the initial sorting of very large complex heterogeneous data sets into manageable uniform subsets of related samples for further more detailed analysis and modelling, multivariate clustering-grouping procedures will rarely, if ever, be appropriate for modelling species distributions.

The most commonly used ordination procedures of PCA, Correspondence Analysis (CA, also known as Reciprocal Averaging), Detrended Correspondence Analysis (DCA) and Non Multidimensional Scaling (NMDS) all involve the initial generation of a resemblance matrix that summarizes the relationships between all samples in the data set. This matrix is then subject to eigenanalysis to extract independent (orthogonal) linear combination of variables, eigenvectors, explaining the maximum amount of variation. These combinations of variables are commonly referred to as components or axes and take the form:

$$z = \beta_0 + \beta_1 x_1 + \beta_2 x_2 + \cdots + \beta_n x_n$$

The results of the ordination are typically presented graphically by plotting the calculated sample z scores for the axes that describe the most variation (normally the first two or three axes extracted). The amount of variation explained by each extracted axis is reflected by the size of the associated eigenvalue. The beta values, or variable loadings, provide an indication of the importance of individual variables in determining sample unit scores. The values provide a measure of the correlation between the variable and the axis.

PCA requires variables to be linear and normally distributed, and while this may apply to environmental variables, the distribution of species along environmental gradients rarely meets these requirements (as discussed earlier). Because of this, PCA is only appropriate for community analysis where samples are drawn from short environmental gradients. The more recently developed methods of CA, DCA, and NMDS can be seen as attempts to overcome these limitations. DCA and NMDS are now the methods more frequently used for the analysis of community datasets (Waite 2000; Jongman et al., 2000; McCune and Grace, 2002).

When attempting to model the distribution of species, ordination methods have most often been used as part of a two-stage modelling process. Typically the environmental variable matrix is subjected to PCA, the resulting linear combination of variables summarize the data matrix and allow the original set of environmental variables to be replaced by principal component scores in a GLM or LR model of species distribution (Crawley, 1993). Using PCA in this way as a data or variable reduction tool has two advantages: (a) component scores are statistically well behaved being uncorrelated and normally distributed; and (b) this approach allows large numbers of environmental variables to be summarized and incorporated into the final model.

Although the eigenvalue associated with each vector is a measure of the variation accounted for by that particular linear combination of variables there are no hard and fast rules for deciding whether a component explains sufficient variation to warrant inclusion in any subsequent model. The number of components to be included in this type of analysis is often decided subjectively from inspection of a plot of component eigenvalues against component—a scree plot. The amount of variation accounted for decreases progressively with each component. In a typical scree plot a clear break is obvious, eigenvalues decreasing steeply and then leveling off, components associated with the first steep part of the plot would be included in the model. An alternative approach is to adopt the 'broken stick' stopping rule (Jackson, 1993), where components are only considered to have predictive value if their associated eigenvalues are greater than those obtained from a random set of data with the same characteristics as the experimental dataset (Robertson et al., 2001, 2003).

One indirect strategy to model species and community distribution following ordination of the species data set is to relate the extracted axes scores to a secondary matrix of environmental variable. This can be achieved by correlating or regressing site sample scores against the environmental variables (Waite, 2000). Canonical correspondence analysis (CCA) and the variant detrended canonical correspondence (DCCA) perform this in one step, simultaneously analyzing the species and environmental variable matrices to derive a set of canonical species ordination axes that are 'restricted-constrained' by the environmental variables. The extracted axes allow the species, site, and environmental variable ordinations to be produced. The individual species scores represent the centre of the species distribution with respect to the extracted linear combination of environmental variables. Species ecological tolerance or range values are derived from the standard deviation of score values (Jongman et al., 2000). Because they quantify the relationship between species distribution and abundance to combinations of environmental variables CCA and DCCA have been extensively used in community ecology.

An interesting alternative PCA-based method has been suggested by Robertson et al. (2001) in which a training data set of environmental variables for sites where the species of interest occurs is ordinated using PCA. The center point of this ordination and the scatter of points around it characterizes the distribution of the species in relation to the environmental variables included in the data matrix. The center point defines the species origin in the ordination hyperspace. The experimental data set is then ordinated using the component axes derived from the first training ordination. The distance of ordinated sites to the point of the species origin provides a measure of site habitat suitability. Sites closest to the species origin provide the most suitable environmental habitat for the species.

A procedure closely related to ordination is discriminant analysis (DA), which attempts to identify a linear combination of variables that maximally separates a fixed number of user-defined groups (e.g., two, sites where the species does and does not occur; Figure 23.4). The method is mathematically the same as multivariate analysis of variance (MANOVA) and is also referred to as 'canonical analysis' or 'canonical variance analysis' (McCune and Grace, 2002).

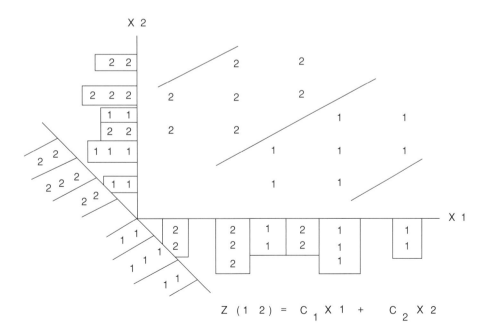

Figure 23.4 An illustration of linear combination of variables by means of discriminant analysis. The DA axis Z offers the best separation of spp 1 and 2 and minimises the overlap in the space of the environmental variables (X1, X2) (from Ludwig and Reynolds, 1988)

DA is supported by most commercially available general statistical software.

23.4.6 Bayesian probability

Where species presence and environmental variables have been recorded at a sufficient number of sites, a simple probability model of species distribution can be produced. The probabilities of species presence at sites with different characteristics can be estimated directly from their frequencies within the training data set and the results summarised in a probability tree. This would be an example of the 'frequentists' or classical approach to probability, where probabilities are estimated directly from empirical data. In contrast, Bayesian statistics involve an element of subjectivity in selecting a starting probability, which is then refined as more evidence becomes available (Lee, 1989; Wade, 2000; Clark, 2005).

Bayesian models involve the calculation of conditional probabilities $(P(P|H)$ where probability of species presence $P(P)$ is conditional on a particular set of environmental or habitat conditions (H). In essence a prior estimate of the overall probability of the species occurrence is modified using information on the nature of the environment. The basic probability calculations are deceptively simple, following Wadsworth and Treweek (1999):

$$P(P|H) = \frac{P(P) \times P(H|P)}{P(P) \times P(H|P) + P(A) \times P(H|A)}$$

where $P(P)$ is the prior probability of the species presence, for example, if the species occurs in 600 out of the 1,200 locations present in the training dataset, $P(P)$ might be set at 0.5. The overall prior probability of the species being absent is $P(A) = 1 - P(P)$.

$P(H|P)$ and $P(H|A)$ are, respectively, the probabilities that habitat condition, H, occurs when the species is present and absent. These probabilities can be estimated from the training set data using basic probability theory.

$$P(H,P) = P(H) \times P((P)$$
$$P(H|P) = \frac{P(H,P)}{P(H)}$$

where $P(H,P)$ is equal to the probability of the common occurrence of the species and habitat (H) assuming independence, and $P(H|P)$ is the

conditional probability of the habitat occurring, given that the species is present. The basis of these calculations is illustrated diagrammatically in Figure 23.5, taken from Tucker et al. (1997), who used a simple Bayesian approach to model the distribution of three bird species in northeast England. In this study the prior probabilities of species occurrence was estimated from national distribution data, which was refined by combining information on habitat preference to produce estimates of habitat suitability.

In practice the modeller will be interested in the probability of the species occurring conditional not on a single habitat attribute (H) but on a set of n attributes. In the simplest case, where each habitat characteristic exists in two states, for instance, the data set is binary, at any location species and $H_i \ldots H_n$ are either present or absent, the ratio of the probabilities of species presence and absence under these circumstances can be estimated from:

$$\frac{P(P|H_1, H_2, \ldots H_n)}{P(A|H_1, H_2, \ldots H_n)} = \frac{P(P)}{P(A)} \times \frac{P(H_1|P)}{P(H_1|A)} \times \frac{P(H_2|P)}{P(H_2|A)}$$
$$\times \ldots \times \frac{P(H_n|P)}{P(H_n|A)}$$

Bayesian analysis involves four stages: (a) selection of a prior parameter probability distribution, reflecting prior belief or previous estimates; (b) data; (c) a model linking collected data to the parameter(s)—this stage involves generating new parameter distributions and estimates that are used in the final stage; and (d) generation of a posterior estimate of the parameter (McCarthy and Masters, 2005). It is important to note that Bayesian analysis should not be seen as an alternative procedure for modelling species distributions. There is no unique modelling procedure associated with stage (c) as the model used to link the collected data to the parameters of interest may take any form, for example, LR. Bayesian statistics provide a way of combining different estimates to improve the final estimates of the parameter.

Advocates of the Bayesian models point to the flexibility of the approach, their ability to deal with complexity and to make efficient use of existing data and prior estimates of key parameters and probabilities (e.g., Wade, 2000; Gelfand et al., 2003; Clark, 2005; Bayliss et al., 2006). It seems likely that the application of Bayesian methods in ecology and biogeography will increase (Clark, 2005; McCarthy and Masters, 2005), although the computational difficulties that arise when species occurrence is modelled as

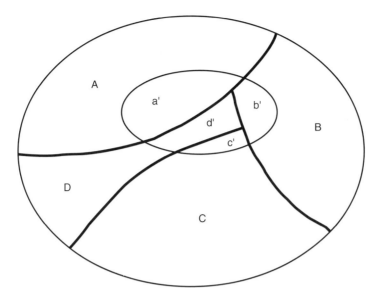

Figure 23.5   Graphical depiction of a Bayesian model of bird distributions in northeast England. The total area of four habitat types are shown as A+B+C+D and bird habitation is shown by the inner oval _a'+b'+c'+d'_ . The probability of a site being of habitat A, and the species being absent (SpA) can be calculated from the associated areas: _P(H$_A$ |SpA) = A − a'/ (A+B+C+D) − (a'+b'+c'+d')_. Other required probabilities can be derived in a similar way (from Tucker et al., 1997)

conditional on many factors may inhibit widespread adoption of this approach.

23.4.6 Dynamic models

The development of dynamic models that accurately mimic or simulate the temporal and spatial distribution of biological entities offers an exciting research field, and requires the adoption of a more realistic correlative or robust mechanistic model structure (Johnston et al., 1996). Two broad approaches are clearly feasible. One approach, the most straightforward, is to start by initially developing a 'static' correlative or statistical model linking environmental variables to the spatial occurrence and/or abundance of the biological entity (Johnston, 1998). An element of dynamic behaviour can then be introduced by developing a secondary model in which key environmental variables are made a function of time. This secondary environmental model can then be used to generate a distinct temporal sets of environmental variable values, which when fed into the original static model can be used to generate a sequence of distribution maps of the biological entity at different time points and under different environmental conditions. This approach has been used extensively to model the impact of climate change on the distribution of vegetation and habitat systems under predicted future climates (Beerling et al., 1997; Bonan et al., 2003).

To date, static GIS-based models have dominated attempts to depict species distribution; however, many researchers are now beginning to suggest that the development of truly dynamic models may be more achievable (Johnston et al., 1996; Johnston, 1998; Bonan et al., 2003). Complex models incorporating large numbers of variables are not necessary and with the advent of cheap computing power-dynamic modelling of species distributions is increasingly practical.

23.5 APPLICATION AND LIMITATIONS

Identifying the ecological niche and available areas for target species, communities, and habitats is fundamental to biogeographers, particularly in light of the current international incentives for prudent and practical nature conservation. Species faced with dwindling distributions or on the brink of extinction are often reliant upon the identification of both former and new potential areas for recolonization, and the development of core populations, metapopulation stepping stones, and wildlife corridors to increase connectivity (Forman, 1997). A suite of methods and techniques are available and are largely based on the underlying concept that species distribution is related to the ecological niche and species requirements (Grinnell, 1917; Hutchinson, 1957; Franklin, 1995; Pulliam, 2000). The methods, as described previously, can be separated into two types: correlative and mechanistic. Both methods have found application in biogeography and related disciplines and both have merits and weaknesses.

In comparative terms the correlative approach has been shown to use existing distributional data to predict alternative distributional patterns. The method develops, via various statistical procedures, an estimate of the probability of the species being present using correlated environmental criteria. The assumption, therefore, is that the existing distribution delineates where the species is found, and an absence of the species from a location reflects an unsuitability or incompatibility. Conversely, the mechanistic approach uses general information on the ecological niche of the species and assumes that the distribution of the species can be predicted through appropriate environmental variables, irrespective of existing distributional patterns. As some workers suggest, existing distributions are likely to result from historical, current and/or forced management practices and therefore do not reflect the full ecological niche. Some, therefore, may be uncomfortable with the assumptions of the correlative technique and suggest that 'false absences' will decrease the reliability of the correlative approach (Chefaoui et al., 2005) or result in spurious predictions (Carpenter et al., 1993). In many predictive scenarios, these statistical approaches suffer due to a lack of absence data and the effect that this has on the robustness of modelled outputs. It is not only important to know where species are found, but it is equally important to identify where species are *not* found. For statistical processes to be valid this consideration of absence data is somewhat fundamental. Current methods to overcome a lack of absence data include the generation of pseudo-absence data (Zarnetski et al., 2007). Pseudo-absence points are generated at random across broadly defined species ranges in an attempt to include additional biological knowledge concerning the species-habitat relationship.

A further assumption associated with correlative models is that species are typically considered to be in equilibrium with their environment.

Austin (2002) and Robertson et al. (2004) state that that this may not true for species whose distributions are currently contracting or still expanding. This is becoming increasingly relevant with current projections for climate change at a global level and, if current predictions are accurate, is likely to include a very large number of species. Carpenter et al. (1993) in a paper implementing a correlative approach, describe the need for field sampling of the predicted distribution data to provide feedback to refine correlation coefficients for target species and increase the accuracy of subsequent predictions. Guisan and Zimmermann (2000), Robertson and Palmer (2002), and Piñeiro et al. (2008) highlight the importance of model evaluation and insist that this evaluation must be through objective means using independent records.

Austin (2007), in a review of recent papers, suggests that ecological theory is rarely explicitly considered. This highlights the need for research into more ecologically relevant models able to incorporate important aspects of ecological theory. Austin (2007) adds that what may actually be required, to progress the field, is a testing of ecological realism for the various statistical methods currently proposed and used by workers. In practice, however, model predictions are often required when we have less than the preferred compliment of data, knowledge, and time. This can be particularly true in conservation and restoration programes, where little is known about the ecological niche and processes surrounding rare or threatened species, therefore correlative approaches can be more applicable. The correlative approach has modest data requirements and can be implemented when little process information is known (Yates et al., 2000).

Mechanistic techniques are often perceived as more appropriate when the existing and historical distributional data are limited, at a poor resolution, or are considered unreliable. The technique does, however, require a thorough understanding of process information and biological and ecological insight. Ottaviani et al. (2004) suggest that this approach is more appropriate when modelling species where the ecological niche is understood, species with low detectability and/or characteristically limited distributions. Therefore, the nature of the target species and the predictor variables is central to the modelling process and can affect the selection of the modelling methodology itself. It is, however, important to recognize that the two techniques are, in fact, related. Mechanistic models often incorporate correlative components. Skidmore (2002: 20), when defining process-driven models, states that "induction is also frequently used to support the development of process driven models particularly to estimate the value of model parameters, or to refine the underlying concepts (or factors) on which the model is constructed."

The generalizability of predictive models is also regarded as a basic requirement (Vaughan and Omrmerod, 2005). The 'overfitting' of a model can, however, easily occur during model development. This is where idiosyncrasies of the training data become incorporated into the model and may mask the true species-environment relationship. Flexible modelling approaches such as GAM are more prone to overfitting than their generalized linear equivalents, for example, LR which allows a greater degree of variable selection (e.g., all subset regression). Vaughan and Omerod (2005), citing Harrell et al. (1996), suggest that the general linear models are also less sensitive to overfitting when only limited training sets are available. Clearly all models are susceptible to overfitting where variables with weak explanatory power are incorporated into the model without question and if the training data sets are either small or not representative.

In modelling the validation process is crucial (Ottaviani et al., 2004; Piñeiro et al., 2008). Verification of the inputs, process, and outcomes is fundamental, and they consider that the validation phase of the modeling process is potentially the most difficult. Appropriate 'goodness of fit' test statistics should be used and are likely to vary between the different procedures. The initial dataset used to formulate the statistical model equation is often referred to as the training dataset. It is important to understand that testing the 'goodness of fit' of the model developed against the training data is different from, and not a substitute for, the evaluation of the final GIS-based species or habitat distribution model. The fact that the statistical model provides a good description of the training data set does not necessarily mean that the final species distribution model will provide a reliable predictor of species distribution in the field. The converse is also possible. The statistical algorithm embedded in the GIS model may provide a poor fit of the training data, but the output of the complete model may provide a good model of species distribution, but clearly in an ideal world we should strive to use robust static algorithms that provide good descriptions of the training datasets.

Johnson and Gillingham (2008) discuss the sensitivity of species-distribution models to error, bias, and model design. Model performance may be affected by species range, prevalence, and/or geographical error, bias in occurrence data, or sample size used to formulate or 'train' the model. However, the sensitivity of models to species range may reflect ecological characteristics of the species and not sampling arte facts. Where species have large and disjointed geographical ranges the species habitat association may be disrupted by the evolution of localized ecotypes or subspecies (Stockwell and Peterson, 2002). If appropriate sampling scales are used, one would expect models based on species with narrow ranges to be the most reliable. McPherson et al. (2004) consider the impact of range size, sample size, and sampling prevalence on the accuracy of distribution models for artificial data and data for 32 endemic South African bird species. The models were built using LR and nonlinear DA. They concluded that 'effects of species rarity or range size on model accuracy appear to be largely artefactual' and did not reflect any meaningful ecological patterns. Both model algorithms were found to be comparable in their sensitivity to sample size and species prevalence, the performance of both decreased with range size where this affected either sample size or species prevalence. Models derived from data sets where species prevalence is atypical will tend to either over- or underestimate species occurrence (Pearce and Ferrier, 2000; Vaughan and Ormerod, 2003). McPherson et al. (2004) advocate the use of training data sets where species prevalence is intermediate, for instance, around 50%, even if subsampling is required to achieve this.

In an early paper, Davis and Goetz (1990) wisely caution that there is a trade-off between model complexity (both variables and operations) and model applicability and reliability. Often models are simplified to aid in their application and avoid overly complex procedures that are difficult to replicate or re-run. A balance is struck between complexity, applicability, and reliability. This does not lessen the value of the model, but the modeller must appreciate that any outputs will in part reflect the series of assumptions made and principles applied in developing the model. It is important to realise that a model is an abstraction of reality and used to provide a simplified reality.

REFERENCES

Anderson, R.P., Lew, D., and Peterson, A.T. (2003) 'Evaluating predictive models of species' distributions: criteria for selecting optimal models', *Ecological Modelling*, 162: 211–32.

Aspinall, R. (1996) 'Use of geographical information systems for interpreting land-use policy and modelling effects of land-use change', in Haines-Young R., Green D.R. and Cousins S.H. (eds.), *Landscape Ecology and GIS*. London: Taylor and Francis.

Austin, M.P. (2002) 'Spatial prediction of species distribution: An interface between ecological theory and statistical modelling', *Ecological Modelling*, 157: 101–18.

Austin, M.P. (2005) 'Vegetation and environment: Discontinuities and continuities', in E. van der Maarel (ed.), *Vegetation Ecology*. Oxford: Blackwell. pp. 52–84.

Austin, M. (2007) 'Species distribution models and ecological theory: A critical assessment and some possible new approaches', *Ecological Modelling*, 200: 1–19.

Bayliss, J., Simonite, V., and Thompson, S. (2006) 'An innovative approach to targeting sites for wading bird assemblages in the UK', *Journal for Nature Conservation*, 14.

Beaumont, L.J., Hughes, L., and Poulsen, M. (2005) 'Predicting species distributions: Use of climate parameters In BIOCLIM and its impact on predictions of species' current and future distributions', *Ecological Modelling*, 186: 250–69.

Beerling, D.J., Woodward, F.I., Lomas, M., and Jenkins, A.J. (1997) 'Testing the responses of a dynamic global vegetation model to environmental change: A comparison of observations and predictions', *Global Ecology and Biogeography*, 6(6): 439–50.

Bonan, G.B, Levis, S., Sitch, S., Vertenstein, M., and Oleson, K.W.A. (2003) 'Dynamic global vegetation model for use with climate models: concepts and description of simulated vegetation dynamics', *Global Change Biology*, 9(11): 1543–66.

Brieman, L., Friedmann, J.H., Olshen, R.A., and Stone, C.J. (1984) *Classification and Regression Trees*. New York: Chapman and Hall.

Brown, D.G. (1994) 'Predicting vegetation types at treeline using topographical and biophysical disturbance variables', *Journal of Vegetation Science*, 4: 499–508.

Burnside, N.G., Smith, R.F., and Waite, S. (2002) 'Habitat suitability modelling for calcareous grassland restoration on the South Downs, United Kingdom', *Journal of Environmental Management*, 65(2): 209–21.

Burnside, N.G., Joyce, C.B., Metcalfe, D., Smith, R.F., and Waite, S. (2003) *South Downs Lifescapes: Habitat Suitability Modelling for BAP Target Habitats in the Proposed South Downs National Park*. Report for English Nature, (18/25/P/02–03). Brighton, UK: University of Brighton.

Burrough, P.A., and McDonnell, R.A. (1998) *Principles of Geographical Information Systems*. Oxford: Oxford University Press.

Busby, J.R. (1986) 'A bioclimate analysis of *Nothofagus cunninghamia* in south eastern Australia', *Australian Journal of Ecology*, 11: 1–7.

Busby, J.R. (1991) 'A bioclimatic analysis and predictive system', in C.R. Margules and M.P. Austin (eds.), *Nature Conservation: Cost Effective Biological Surveys and Data Analysis*. Canberra: CSIRO. pp. 64–68.

Carpenter, G., Gillison, A.N., and Winter, J. (1993) 'DOMAIN: A flexible modelling procedure for mapping potential distributions of plants and animals', *Biodiversity and Conservation*, 2: 667–80.

Chefaoui, R.M., Hortel, J., and Lobo, J.M. (2005) 'Potential distribution modelling, niche characterization and conservation status assessment using GIS tools: A case study of Iberian Copris species', *Biological Conservation*, 122: 327–38.

Cherrill, A.J., McClean, C., Watson, P., Tucker, K., Rushton, S.P., and Sanderson, R. (1995) 'Predicting the distributions of plant species at the regional scale: A hierarchical matrix model', *Landscape Ecology*, 10(4): 197–207.

Clark, J.S. (2005) 'Why environmental scientists are becoming Bayesians', *Ecological Letters*, 8: 2–14.

Cox, C.B., and Moore P.D. (2005) *Biogeography: An Ecological and Evolutionary Approach*. 7th ed. Oxford: Blackwell Scientific.

Crawley, M.J. (1993) *GLIM for Ecologists*. Oxford: Blackwell Scientific.

Davis, F.W., and Goetz, S. (1990) 'Modelling vegetation pattern using digital terrain data', *Landscape Ecology*, 4: 69–80.

De'Ath, G., and Fabricius, K.E. (2000) 'Classification and regression trees: A powerful yet simple technique for ecological data analysis', *Ecology*, 81: 3178–92.

Eeley, H.A., Lawes, M.J., and Piper, S.E. (1999) 'The influences of climate change on the distribution of indigenous forest in Kwa-Zulu-Natal, South Africa', *Journal of Biogeography*, 26: 595–617.

Ellis, C.J., Coppins, B.J., Dawson, T.P., and Seaward, M.R.D. (2007) 'Response of British lichens to climate change scenarios: Trends and uncertainties in the projected impact for contrasting biogeographic groups', *Biological Conservation*, 140: 217–35.

Fairbanks, D., McGwire, K., Cayocca, K., LeNay, J., and Estes, J. (1996) 'Sensitivity to climate change of floristic gradients in vegetation communities', in Goodchild M.F., Steyaert L.T. and Parks B.O. (eds.), *GIS and Environmental Modelling: Progress and Research Issues*. Fort Collins, CO: GIS World Books.

Ferrier, S., and Guisan, A. (2006) 'Spatial modelling of biodiversity at the community level', *Journal of Applied Ecology*, 43(3): 393–404.

Finch, J.M., Samways, M.J., Hill, T.R., Piper, S.E., and Taylor, S. (2006) 'Application of predictive distribution modelling to invertebrates: Odonata in South Africa', *Biodiversity and Conservation*.

Forman, R.T. (1997) *Land Mosaics: The Ecology of Landscapes and Regions*. Cambridge: Cambridge University Press.

Fowler, J., Cohen, L., and Jarvis, P. (1998) *Practical Statistics for Field Biology*. Chichester, UK: John Wiley.

Franklin, J. (1995) 'Predictive vegetation mapping: Geographic modelling of biospatial patterns in relation to environmental gradients', *Progress in Physical Geography*, 19(4): 474–99.

Franklin, J. (1998) 'Predicting the distribution of shrub species in southern California from climate and terrain-derived variables', *Journal of Vegetation Science*, 9: 733–48.

Franklin, J., Logan, T.L., Woodcock, C.E., and Strahler, A.H. (1986) 'Coniferous forest classification and inventory using Landsat and digital terrain data', *IEEE Transactions on Geoscience and Remote Sensing* GE-24: 139–49.

Garcia-Ripolles, C., Lopez-Lopez, P., Garcia-Lopez, F., Aguilar, J.M., and Verdejo, J. (2005) 'Modelling nesting habitat preferences of Eurasian Griffon Vulture Gyps fulvus in eastern Iberian Peninsula', *Areola*, 52(2): 287–304.

Gelfand, A.E., Silander Jr., J.A., Wu, S., LAtimer, A., Lewis, P.O., Rebelo, A.G. and Holder, M. (2003) 'Explaining species distribution patterns through hierarchical modelling', *Bayesian Analysis*, 1(1): 1–35.

Gkaraveli, A., Williams, J.H., and Good, J.E.G. (2001) 'Fragmented native woodlands in Snowdonia (UK): Assessment and amelioration', *Forestry*, 74(2): 89–103.

Goodchild, M.F., Parks, B.O., and Steyaert, L.T. (1993) *Environmental Modelling with GIS*. New York: Oxford University Press.

Goodchild, M.F., Steyaert, L.T., and Parks, B.O. (1996) *GIS and Environmental Modelling: Progress and Research Issues*. Fort Collins, CO: GIS World Books.

Grinnell J. (1917) 'Field tests of theories concerning distributional control', *American Naturalist*, 51: 115–28.

Guisan, A. (2002) 'Semi-quantitative response models for predicting the spatial distribution of plant species', in Scott J.M., Heglund P.J., Morrison M.L., Haufler J.B., Raphael M.G., Wall W.A., and Samson F.B., (eds.) *Predicting Species Occurrences: Issues of Accuracy and Scale*, Washington, DC: Island Press. pp. 315–26.

Guisan, A., and Zimmermann, N.E. (2000) 'Predictive habitat distribution models in ecology', *Ecological Modelling*, 135: 147–86.

Harrell, F.E., Lee, K.L., and Mark, D.B. (1996) 'Multivariate prognostic models: Issues in developing models, evaluating assumptions and adequacy, and measuring and reducing errors', *Statistics in Medicine*, 15: 361–87.

Hastie, T.J., and Tibshirani, R.J. (1990) *Generalized Additive Models*. London: Chapman and Hall.

Hilborn, R., and Mangel, M. (1997). *The Ecological Detectives. Confronting Models with Data*. Monographs in Population Biology, 28. Princeton, NJ: Princeton University Press.

Holdridge, L.R. (1967) *Life Zone Ecology*. San José, Costa Rica: Tropical Sciences Center.

Hurme, E., Mönkkönen, M., Nikula, A., Nivala, V., Reunanen, P., Heikkinen, T., and Ukkola, M. (2005) 'Building and evaluating predictive occupancy models for the Siberian flying squirrel using forest planning data', *Forest Ecology and Management*, 216: 241–56.

Hutchinson, G.E. (1957) 'Concluding remarks', *Cold Spring Harbor Symposium on Quantitative Biology*, 22: 415–57.

Jackson, D.A. (1993) 'Stopping rules in principle components analysis: A comparison of heuristical and statistical approaches', *Ecology*, 74: 2204–14.

Johnson, C.J., and Gillingham, M.P. (2008) 'Sensitivity of species distribution models to error, bias, and model design: An application to resource selection functions for woodland caribou', *Ecological Modelling*, 213: 143–55.

Johnston, C.A. (1998) *Geographic Information Systems in Ecology*. Oxford: Blackwell Science.

Johnston, C.A., Cohen, Y., and Pastor, J. (1996) 'Modelling of spatially static and dynamic ecological processes', in Goodchild M.F., Steyaert L.T., and Parks B.O. (eds.), *GIS and Environmental Modelling: Progress and Research Issues*. Fort Collins, CO: GIS World Books. pp. 149–54.

Jongman, R.H.G., Ter Braak C.J.F. and Van Tongeren O.F.R. (2000) Data analysis in community and landscape ecology. Cambridge: Cambridge University Press.

Kent, M., and Coker, P. (1992) *Vegetation Description and Analysis: A Practical Approach*. Chichester, UK: John Wiley.

Lee, P.M. (1989). *Bayesian Statistics: An introduction*. London: Oxford University Press.

Lee, J.T., Bailey, N., and Thompson, S. (2002) 'Using Geographical Information Systems to identify and target sites for creation and restoration of native woodlands: A case study of the Chilterns', *Journal of Environmental Management*, 64: 25–34.

Lehmann, A. (1998) 'GIS modelling of submerged macrophyte distribution using generalized additive models', *Plant Ecology*, 139: 113–24.

Lehmann, A., Overton, J.McC., and Leathwick, J.R. (2003) 'GRASP: Generalized regression analysis and spatial prediction', *Ecological Modelling*, 160: 165–83.

Levin, R. (1966) 'The strategy of model building in population biology', *American Scientist*, 54(4): 421–31.

Linderman, M., Liu, J., Qi, J., An, L., Ouyang, Z., Yang, J., and Tan, Y. (2004) 'Using artificial neural networks to map the spatial distribution of understorey bamboo from remote sensing data', *International Journal of Remote Sensing*, 25(9): 1685–700.

Linhart, H., and Zucchini, W. (1986) *Model Selection*. New York: John Wiley.

MacArthur, R.H., and Wilson E.O. (1967) *The Theory of Island Biogeography*. Monographs in Population Ecology 1, Princeton, NJ: Princeton University Press.

Manly, B.F.J. (1994) *Multivariate Statistical Methods; A Primer*. 2nd ed. London: Chapman and Hall.

Manly, B.F.J. (2008) *Statistics for Environmental Science and Management*. 2nd Edition London: Chapman and Hall.

Manel, S., Williams, H.C., and Ormerod, S.J. (2001) 'Evaluating presence-absences models in ecology: The need to account for prevalence', *Journal of Applied Ecology*, 38: 921–31.

Maynard-Smith, J. (1974) *Models in Ecology*. Cambridge: Cambridge University Press.

McCarthy, M.A., and Masters, P. (2005) 'Profiting from prior information in Bayesian analyses of ecological data', *Journal of Applied Ecology*, 42(6): 1012–19.

McClean, C.J., Lovett, J.C., Küper, W., Hannah, L., Henning Sommer, J., Barthlott, W., Termansen, M., Smith, G.F., Tokumine, S. and Taplin, J.R.D. (2005) 'African plant diversity and climate change', *Annals of the Missouri Botanical Garden*, 92: 139–52.

McCune, B., and Grace, J.G. (2002) *Analysis of Ecological Communities*. Gleneden Beach, OR: M.j.m Software Design.

McPherson, J.M., Walter, J., and Rodgers, D.J. (2004) 'The effects of species' ranges size on the accuracy of distribution models: Ecological phenomenon or statistical artefact?' *Journal of Applied Ecology*, 41: 811–23.

Moisen, G.G., Freeman, E.A., Blackard, J.A., Frescino, T.S., Zimmermann, N.E., and Edwards, T.C. (2006) 'Predicting tree species presence and basal area in Utah: A comparison of stochastic gradient boosting, generalised additive models, and tree based methods', *Ecological Modelling*, 199(2): 176–87.

Mörtberg, U., and Karlström, A. (2005) 'Predicting forest grouse distribution taking account of spatial auto-correlation', *Journal for Nature Conservation*, 13: 147–59.

Ottaviani, D., Lasinio, G.J., and Boitani, L. (2004) 'Two statistical methods to validate habitat suitability models using presence-only data', *Ecological Modelling*, 179: 417–43.

Palo, A., Aunap, R., and Mander, Ü. (2005) 'Predictive vegetation mapping based on soil and topographical data: A case study from Saare County, Estonia', *Journal for Nature Conservation*, 13: 197–211.

Pearce, J., and Ferrier, S. (2000) 'Evaluating the predictive performance of habitat models developed using logistic regression', *Ecological Modelling*, 133: 225–45.

Pearson, R.G., Dawson, T.P., Berry, P.M., and Harrison, P.A. (2002) 'SPECIES: A spatial evaluation of climate impact on the envelope of species', *Ecological Modelling*, 154(3): 289–300.

Pearson, R.G., Thuiller, W., Araújo, M.B., Martinez-Meyer, E., Brotons, L., McClean, C., Miles, L., Segurado, P., Dawson, T.P. and Less, D.C. (2006) 'Model-based uncertainty in species range predictions', *Journal of Biogeography*, 33: 1704–11.

Peduzzi, P., Concato, J., Kemper, E., Holford T.R., and Feinstein, A.R. (1996) A simulation study of the number of events per variable in logistic regression analysis, *Journal of Clinical Epidemiology*, 49: 1373–1379.

Piñeiro, G., Perelman, S., Guerschman, J.P., and Paruelo, J.M. (2008) 'How to evaluate models: Observed vs. predicted or predicted vs. observed?' *Ecological Modelling*, 216: 316–22.

Pulliam, H.R. (2000) 'On the relationship between niche and distribution', *Ecology Letters*, 3: 349–61.

Robertson, M.P., and Palmer, A.R. (2002) 'Predicting the extent of succulent thicket under current and future climate scenarios', *African Journal of Range and Forage Science*, 19: 21–28.

Robertson, M.P., Caithness, N., and Villet, M.H. (2001) 'A PCA-based modelling technique for predicting environmental suitability for organisms from presence records', *Diversity and Distributions*, 7: 15–27.

Robertson, M.P., Peter, C.I., Villet, M.H., and Ripley, B.S. (2003) 'Comparing models for predicting species' potential distributions: A case study using correlative and mechanistic predictive modelling techniques', *Ecological Modelling*, 164: 153–67.

Robertson, M.P., Villet, M.H., and Palmer, A.R. (2004) 'A fuzzy classification technique for predicting species' distributions: applications using invasive alien plant species and indigenous insects', *Diversity and Distributions*, 10: 461–74.

Rodriguez, J.P., Brotons, L., Bustamante, J., and Seoane, J. (2007) 'The application of predictive modelling of species distributions to biodiversity conservation', *Diversity and Distributions*, 13: 243–51.

Rushton, S.P., Ormerod, S.J., and Kerby, G. (2004) 'New paradigms for modelling species distributions?' *Journal Applied Ecology*, 41: 193–200.

Segurado, P., and Araújo, M.B. (2004) 'An evaluation of methods for modelling species distributions', *Journal of Biogeography*, 31: 1555–68.

Seoane, J., Carrascal, L.M., Alonso, C.L., and Palomino, D. (2005) 'Species-specific traits associated to predictions errors in bird habitat suitability modelling', *Ecological Modelling*, 185: 299–309.

Skidmore, A. (2002) *Environmental Modelling with GIS and Remote Sensing*. London: Taylor and Francis.

Stockwell, D.R.B., and Peterson, A.T. (2002) 'Effects of samples size on accuracy of species distribution models', *Ecological Modelling*, 148: 1–3.

Téllez-Valdés, O., and Dávila-Aranda, P. (2003) 'Protected areas and climate change: A case study of the Cacti in the Tehuacán-Cuiarlán Biosphere Reserve, México', *Conservation Biology*, 17: 646–853.

Termansen, M., McClean, C.J., and Preston, C.D. (2006) 'The use of genetic algorithms and Bayesian classification to model species distributions', *Ecological Modelling*, 193(3–4): 410–24.

Tucker, K., Rushton, S.P., Sanderson, R.A., Martin E.B., and Blaiklock, J. (1997) 'Modelling bird distributions—a combined GIS and Bayesian rule-based approach', *Landscape Ecology*, 12(2): 77–93.

Urban, D., Goslee, S., Pierce, K., and Lookingbill, T. (2002) 'Extending community ecology to landscapes', *Ecoscience*, 9(2): 200–212.

Vaughan, I.P., and Ormerod, S.J. (2003) 'Improving the quality of distribution models for conservation by addressing short-comings in the field collection of training data', *Conservation Biology*, 17: 1601–11.

Vaughan, I.P., and Ormerod, S.J. (2005) 'The continuing challenges of testing species distribution models', *Journal of Applied Ecology*, 42: 720–30.

Wade, P.R. (2000) 'Bayesian methods in conservation biology', *Conservation Biology*, 14(5): 1308–16.

Wadsworth, R., and Treweek, J. (1999) *Geographical Information Systems for Ecology*. Harlow, UK: Longman.

Waite, S. (2000) *Statistical Ecology: A Practical Guide*. London: Prentice Hall.

Waite S., and Burnside, N.G. (2001) 'The radiate capitulum morph of Senecio vulgaris L. within Sussex: The use of GIS in establishing origins', in Millington, A.C., Walsh S.J., and Osborne, P.E. (eds.), *GIS and Remote Sensing Applications in Biogeography and Ecology*. Norwell MA: Kluwer Academic Kluwer Books, Dordrecht, The Netherlands Norwell MA: Kluwer Academic pp. 179–92.

Whittaker, R.H. (1956) 'Vegetation of the Great Smoky Mountains', *Ecological Monographs*, 26: 1–80.

Whittaker, R.H. (1960) 'Vegetation of Siskiyou Mountains, Oregon and California', *Ecological Monographs*, 30: 279–338.

Whittaker, R.J., and J.M. Fernandez-Palacios. (2007) *Island Biogeography, Ecology, Evolution and Conservation*. 2nd ed. Oxford: Oxford University Press.

Woodward, F.I., and Lomas, M.R. (2004) 'Vegetation dynamics—simulating responses to climatic change', *Biological Reviews*, 79: 643–70.

Yates, D.N., Kittel, T.G.F., and Cannon, R.F. (2000) 'Comparing the correlative Holdridge model to mechanistic biogeographical models for assessing vegetation distribution', *Climatic Change*, 44: 59–87.

Zaniewski, A.E., Lehmann, A., and Overton, J.McC. (2002) 'Predicting species spatial distributions using presence-only data: A case study of native New Zealand ferns', *Ecological Modelling*, 157: 261–80.

Zarnetski, P.L., Edwards, T.C., and Moisen, G.G. (2007) 'Habitat classification modelling with incomplete data: Pushing the habitat envelope, *Ecological Applications*, 17(6): 1714–26.

24

Simulation

George P. Malanson

This imaginary of representation, which simultaneously culminates in and is engulfed by the cartographer's mad project of the ideal coextensivity of map and territory [see Borges], disappears in the simulation whose operation is nuclear and genetic, no longer specular or discursive.

Baudrillard, 1994 (italics added; translated from 1981 edition)

24.1 INTRODUCTION

Much of the work discussed in the following chapter on biocomplexity (Walsh et al., this volume) includes simulation. Here, the nature of simulation as a modeling paradigm per se is discussed. The purpose of this chapter is to provide a context for the linkage between more empirically and/or process-based spatially explicit modeling and cellular automata and agent-based modeling and to contrast this context with the statistically and mathematically driven predictive and theoretical models in biogeography.

Simulations are particular kinds of models, which are themselves abstractions of a system, expressed within a computer program. Here, I use "system" loosely but it embodies the proposal that our idea of reality can be represented formally and approximated, in abstracted and simplified form, in a simulation. Models simplify systems, and simulations are computational expressions of a conceptual simplification. It may be that the concept itself is not computable, but for the simulation to be useful a computable approximation is

necessary; conversely, this caveat may be unnecessary. The computer program should embody the states of factors or variables and their effects on each other.

24.1.1 Why use simulations in biogeography?

The first thought that comes to mind is that simulations can be used for prediction. If a simulation faithfully represents important dynamics of our system, then we might be able to use it to anticipate what will happen in the future. Weather forecast simulations are the most widely recognized application of this type. In biogeography, actual predictions with associated probabilities or error estimates are rare. Instead, a common futurecast of biogeographic simulations is to produce alternate scenarios for different inputs or different relations in the system. Various factors that are difficult to include with precision, for example, human activities, make actual prediction too risky.

Simulations are used to test hypotheses by comparing the simulation, which represents the hypothesis, to reality. However, as Parsons and Knight (2005) explain, this approach is subject to alternative interpretations of whether we are testing the reality of our equations or the reality of an environmental process.

To discover and to understand are other reasons for simulation. We can get a better sense of how a system works by examining the consequences of alternative inputs or relations in the system. Sometimes the simulations can be too abstract for prediction or scenario generation but still be informative about the importance of relations or variables. Because building a simulation is usually more precise than a verbal expression, simulation for discovery and understanding is also a way to develop theory. Some have proposed that simulation has developed with computers as a new approach in science. Physicists who argue this line of reasoning do so because they see computation as the tool of investigation in complexity theory (see Malanson, 1999; Walsh et al., in this volume for discussion of complexity theory in biogeography) that is different from actual experiment or from theory, which has been mathematical modeling but not necessarily simulation. By allowing us to explore alternative worlds for which there is no empirical grounding and to combine various approaches in mathematics, simulation opens new ways of thinking in science.

Here, I will examine how simulations are conceived and created and then review important classes of biogeographical simulations in these terms.

24.2 DIMENSIONS OF SIMULATION

Simulations have been characterized as occupying a number of dimensions. These are not orthogonal dimensions, and some of them are describing essentially the same thing. These dimensions can be grouped into three aspects of the resolution of the simulation: its phenomenological, temporal, and spatial resolutions.

24.2.1 Phenomenological resolution

First, the degree of opacity of the simulation in ranging from the unknown to fully apparent detail is described as a gradient from black box to white box simulations. In a black box approach, the operation of the system itself is not known, and the outcomes of the simulation are related directly to the inputs without knowledge of the processes; a simulation that is based directly on a regression equation from observations would be a black box; a simple species-area graph is such a black box. In a white box, the actual processes by which the inputs lead to the outcomes are represented in great detail; the CENTURY model cited below is a white box. Gray box simulations, represented by forest-stand models, occupy most of the gradient. Reynolds et al. (1993) described this type of detail in simulations in terms of a range from empirical

to mechanistic. Empirical models, like black box models and dark models, link inputs and outputs without necessarily knowing how they are related. Mechanistic models represent the relationships of inputs and outputs in the system of interest by characterizing the processes at a lower hierarchical level. For example, the changes in the structure and diversity of a forest may be represented by the interactions of trees. Between empirical and mechanistic simulations they identified phenomenological simulations. Phenomenological simulations include a general understanding of the process at the level of interest; for example, population growth can be simulated using the logistic equation without simulating the births and deaths of individuals.

What appears to be a distinction in understanding may be a distinction in hierarchical level of interest. For example, while one may calculate tree growth using a mechanistic equation that computes photosynthesis and allocation of elements versus finding an empirical relationship between annual diameter increment and temperature, the former calculations are based on equations derived from empirical relationships at more detailed levels of a hierarchy. It makes sense to use the most mechanistic model that can be applied, but in some cases parts of the problem will have more hierarchical levels than others, and it may be necessary to keep the model from spanning too many levels.

Other ways of dividing the amount of understanding in a simulation can be seen among types of simulations, which are often differentiations of the type of mathematics used. For example, systems dynamics simulations can be used to represent the movement of an animal species in a geographic area depending on detailed representation of its behavior and the environment. This type of simulation might use difference equations parameterized from stochastic empirical functions. The same process could be represented in a cellular automaton as an expression of a few rules wherein the mathematics would be simple Boolean algebra. Another dimension of difference is whether the simulation is deterministic or stochastic. A set of computer instructions will be carried out identically for multiple runs of the computer program, that is, the outcome is determined by the code, and so a computer simulation is naturally deterministic. A deterministic approach is satisfactory for systems for which we have reasonably complete knowledge of their initial conditions and behavior; for example, the fall of a cannonball from the Leaning Tower of Pisa can be simulated deterministically. The fall of a winged seed (samara) of a maple tree on a windy day cannot (should not) be simulated deterministically; we do not know enough about wind turbulence and the small differences in the aerodynamics of the many samaras to compute a single certain answer. The alternative is stochastic simulation, in which what we do not know is represented by chance. We may know that maple samaras have a range of dispersal distances from which we can derive a theoretical distribution (e.g., negative exponential); we can then compute the distance for a single samara by randomly drawing from the theoretical distribution and we can repeat this for as many seeds as we believe would fall in a forest. Alternatively, we might calculate the growth of a tree for a given climate input by computing the empirical equation but including some error by randomly drawing from a theoretical distribution of the observed error of this equation. In both cases it is necessary to have a random number at hand.

24.2.2 Temporal resolution

Temporally, textbook authors (e.g., Brooks and Robinson, 2001) have differentiated simulations between continuous versus discrete events, but here continuous means that time proceeds in regular units and can be represented by differential equations or difference equations with finite time-steps. Analytical models using differential equations to capture time usually are solved for one point and so are not simulations. When differential equations are used to represent a system but are solved at regular time intervals, the effect is similar to using difference equations. Thus within the continuous category another level of continuous (differential equation—based) and discrete (difference equation—based) exists, and it is important to recognize the potential for confusion in the existing terminology. At the higher order, continuous simulations are ones in which time proceeds in regular units, either infinitely small in the case of differential equations or in some larger units using difference equations, while discrete event simulations do not have regular time-steps and the passage of time may be in terms of events. Essentially all simulations include time; calculations of relations for a single time-step usually are not addressed as simulations, but I will consider a possible exception.

In biogeography, most simulations are of continuous time and they use difference equations. If one uses different equations, the time step—or actual resolution—must be chosen. The time-step must capture the important processes, given their parameterization. For many biogeographic processes an annual time-step will capture the relevant processes because plants and animals function in many ways (e.g., reproduction) on an annual basis or are commonly parameterized with annual data (e.g., tree-ring widths). For some simulations shorter steps will be needed to capture the effects in intra-annual effects, such as disturbance or climatic extremes. Longer time-steps, such as for century-scale responses of biomes to climate change, may be appropriate, but parameterization with long-term averages may not produce the same results as a shorter-term, more mechanistic resolution.

24.2.3 Spatial resolution

Spatially, simulations can be lumped or distributed. Lumped simulations treat the system as a single spatial unit, that is, they are aspatial. Distributed simulations include space, and in parallel with temporal resolution, spatial representation can vary between continuous space represented by differential equations (usually partial differential equations because continuous space is taken together with continuous time) or difference equations in which spatial units are identified. Most simulations in biogeography treat space as discrete units because partial differential equations are difficult to solve or do not adequately capture discontinuities in the system. Space is usually represented as a regular grid of squares, or rarely hexagons, in simulations in biogeography. These simulations are called cell-based or cellular, but in some cases the spatial units are represented abstractly. Malanson (1996) distinguished spatially explicit from pseudo-spatially explicit simulations, with the latter containing multiple spatial units that are simulated independently, without reference to any other unit. Many simulations are pseudo-spatially explicit in some parameters while fully or partially spatially explicit in others. Starting with the JABOWA (Botkin, 1993) and FORET (Shugart, 1998) forest-stand models described below, for example, Hanson et al. (1990) added dispersal to the JABOWA simulation and thus were spatially explicit in population dynamics but did not include the spatially explicit distributions of light seen in the ZELIG version (Urban et al., 1991) of FORET or of carbon seen in STANDCARB (Harmon and Marks, 2002).

The actual spatial resolution of a spatially explicit simulation is the size of the cells. In most biogeographical simulations the cells are the same area or dimension (for some global models, holding the dimensions constant in terms of degrees of latitude and longitude means difference in area). The cells need to be small enough so that the organisms and the environment being studied can have reciprocal effects, rather than being averaged out, but large enough so that the interactions between the organism and other factors being computed or represented in the cell are realistic and not too constrained.

Given the cellular approach, it is possible to progressively coarsen the resolution, and much can be learned by doing so (e.g., Obeysekera and Rutchey, 1997; Gao et al., 2001).

24.2.4 Courant-Friedrichs-Lewy (CFL) criterion

The temporal and spatial resolutions of a simulation must be linked. The CFL criterion states that for a given spatial resolution, the time-step must be short enough so that the fastest fluxes being simulated—and these are supposed to be affected by the underlying environment—cannot skip over a spatial cell in a single time-step (cf. Martin, 1993). The purpose is so that the process that should be affected by the environment in a cell will be so affected. The reverse of the criterion, that spatial cells be small enough so that the fluxes do proceed from one to the next in any time-step primarily favors efficiency, but if cells are large enough that a flux is not moving, this relationship could also affect computation.

24.3 HOW TO'S OF SIMULATION

A solid approach to simulation is to use the standard scientific method as a basis for simulation (Neelamkavil, 1987; McHaney, 1991). We can break it into five vaguely discrete steps: problem definition, hypothesis statement, experiment, generation of results, and analysis. Between hypothesis statement and experiment come the construction steps described below, which are specific to computer simulations and not to science.

The first step in simulation, as in other areas of biogeography, is to define the problem. Simulation cannot be everything, and so the system must be simplified in order to focus on what it is one wants to know about it. The scope of the problem is its domain, which arises from recognition of the current frontier of knowledge and an assessment of where the frontier can be advanced (but think of a frontier in multiple dimensions with interior holes). It is important to realize that in a model, especially using simulation, it is not possible to simultaneously maximize realism, generality, and precision. It is necessary to sacrifice at least one of these, and sometimes there is a necessary trade-off between realism and generality. This step and/or the next will often determine the phenomenological resolution of the simulation. In simulation, the problem statement needs to be organized with a diagram of the steps of the simulation.

This organization is a boxes-and-arrows picture of how the simulation will work. In addition to outlining the actual sequence of computer code that will be written, the boxes and arrows embody the cause-and-effect relations that are the heart of the problem definition.

The statement of a hypothesis involves refining the problem definition to include scale and detail. The scale of the problem is its range in time and space, and so it determines the spatial and temporal resolution. Detail is also a decision for the investigator in terms of a hierarchy of components from the microscopic to the global that biogeographers address. The hypothesis is a statement of relations, particularly of cause and effect, that represents an idea of how a biogeographic system works. In simulation, the actual structure of the simulation itself is the expression of the hypothesis. If one hypothesizes that dispersal across continents affects the assemblage of species in response to climate change, then the simulations must include continents, dispersing species, and changing climate.

The simulation equivalent of an experiment is the running of the simulation in a computer. To mimic an experiment, simulations need controls, treatments, and replications thereof. It is not necessary to specify which simulation run is the control and which are the treatments, but they have to differ. In the example above, one might alter the spatial structure of the continent, the dispersal characteristics of the species, and/or the rate or type of climate change. As with observational/experimental research, in order to test for differences among treatments, and thus for effects of hypothesized causes, it is necessary to have replicates for each condition (this does not apply to deterministic simulations which should have no difference among replications and are not subject to Maxwell's demon). In stochastic simulations, the effects of chance will cause any two runs of a simulation to differ. We want to know if the differences among the treatments are greater than those caused by stochasticity alone, and so we must compare the differences among treatments to those within treatments—and so we need replications within treatments. Replications are just simulation runs with identical initial conditions and computer code, differing only in the random numbers that introduce stochasticity.

The results are the output of the simulations. The simulation must simply have output of some variables that have been chosen to answer the initial question and test the hypothesis, and thus they must match those in scale and detail. For our problem, the diversity of species in particular spatial units as well as for the continent as a whole might be the output, but additionally some measure of the abundance of each species in the simulation might be output for either spatial resolve.

The analysis of simulation results is usually statistical as it is with other types of experiments. One wishes to know if the treatments differ from one another, that is, if the hypothesized cause and effect actually show up in the simulation. One might statistically compare some output, such as continental species diversity, directly with a statistical test, while other results might require additional processing of the simulation output. For example, the abundances of species in different places computed in the simulation might be ordinated and then the ordination scores of the places would be compared statistically among the treatments.

24.3.1 Construction of a simulation

Between the problem definition and hypothesis statement comes the construction of the simulation. The first step is the model translation from idea to computer code, often with a flow diagram (e.g., Figure 24.1) to guide the process. Within the code there may be equations or relations based on empirical data as well as theory, and so this can include a variety of inputs. Part of the construction also includes verification and validation.

Input: There must be some input to a simulation. It starts with initial conditions. In biogeography, the initial condition is often a spatial pattern of the phenomenon of interest: that of trees in a forest or carbon or biomes across a continent. Input can come from several possible sources. Observations, either direct field measurements or interpreted remote sensing imagery, can serve. Estimation and interpolation are also used to fill in initial conditions. Expert opinion or projections from other models are also sources; for example, the response of biomes to climate change may use the output from climate models as input. The input can itself be hypothetical or assumed; one can simulate the consequences of hypothetical human impacts. Neutral landscapes are one way to create hypothetical initial conditions (Wang and Malanson, 2008). Choices are made in deriving actual inputs. One may know the ages of trees of a species in a forest. To simulate this forest one could start with this actual age distribution, but one would have more generality if one could fit a theoretical distribution to the data and then draw ages from it because the theoretical distribution will include unobserved extremes that one might want to represent in alternative simulation runs in order to produce more generalizable results.

Random numbers: In stochastic simulations, random numbers must be used to introduce variance. If we know that the relations between two variables have variance, it is introduced by using a random number to alter the mean relationship. In simulations, a series of pseudorandom numbers are generated because a computer cannot produce true random numbers, but these should (but may not) pass tests for randomness (e.g., Neelamkavil, 1987).

Verification: Throughout the process of constructing a simulation, steps must be double-checked. This process covers the range from checking computer code to the overall organization and running of the simulation. Verification is thought of as a test that tells the modeler whether or not the simulation is behaving reasonably. If numbers that are expected to stay within a certain range increase toward infinity, then one knows that there is a problem. This stage is preliminary to validation and running the simulation as an experiment and is meant to be sure the model is internally consistent, not that it actually represents the system.

Validation: Before examining the meaning of a simulation, its faithfulness to the system that it represents must be assessed. While the simulation, as a model, is not expected to reproduce every aspect of a system, it should produce results that can be observed in the system. At times, the validation may be the endpoint of a simulation—if the simulation produces results that are faithful to the system, then it has contributed to the understanding of the system. In other cases, the validation precedes experimental analyses.

The primary validation of both inputs and outputs of a simulation is through observation. If the inputs can be derived from observation and the outputs compared to an observed system, then one can have confidence in model validity if they match. There may be cases when the validation is not against observation per se, but for general theory the validation may be sufficient if the simulation is judged useful by its creators or, preferably, by other experts; this is called face validity.

Simulations can also be tested via sensitivity analysis. One can alter inputs or assumptions (relationships built into the model) slightly and then examine the output. If the simulation output is very sensitive to small changes in an input or relationship, it may need to be rethought—or it may need to be treated as important.

24.3.2 Computational questions

One aspect of simulation design is confronting the limits of computation. Some problems, as with the mad cartographer cited at the beginning of this chapter, would require infinite computing ability and time. While faster computers have

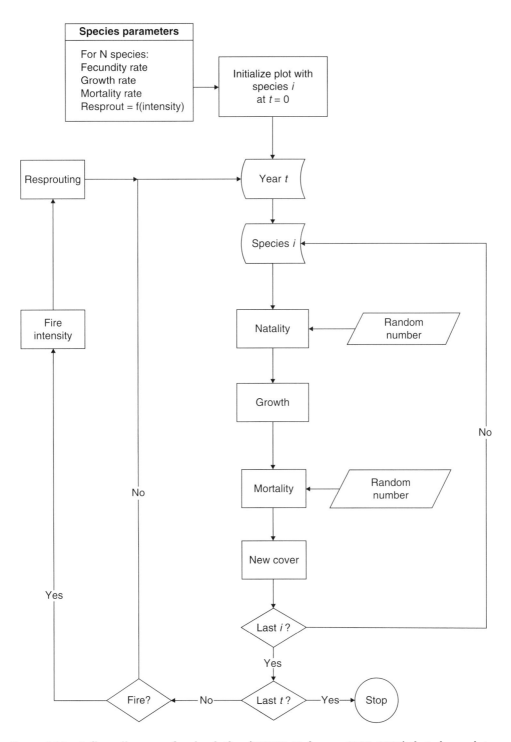

Figure 24.1 A flow diagram of a simulation (FINICS; Malanson 1984, 1985) that shows data, process, and decision steps

progressively lessened the problem, multiple computers, or at least processors, have also reduced it. Geographic problems are particularly well suited to parallel processing when at least some of the computations per cell are independent and can be assigned to different processors (Armstrong and Marciano, 1995; Armstrong, 2000).

Another approach is evolutionary computation (EC). Computer simulations are able to change themselves to better approximate the system they are supposed to represent (e.g., Downing, 1998). Given that the computational changes in EC are specifically modeled on those of biological evolution (e.g., mutation, crossover, selection), it might seem intuitive to adapt this approach for simulation in biogeography, but some aspects of biogeographic simulation are not at the relevant scale. Anderson et al. (2006) have used genetic algorithms to produce predictions of geographic ranges, not simulations. Malanson and Zeng (2004) found that a genetic algorithm was useful in a limited way in attempting a simulation of treeline advance in which the match of simulated to observed spatial pattern was the fitness criterion; their simulation quickly reached a fair but not better match.

24.4 SIMULATION THEMES IN BIOGEOGRAPHY

There are two major themes in biogeographic simulation: carbon and the occupation of space by elements in the hierarchy of biota ranging from species to biomes. There is some crossover between these two themes, and the latter has several levels that are addressed differently in simulations. Here, major examples of a carbon model, a forest-stand-to-landscape model, and a biome model will be used as reference points. More types are possible, but these represent the majority of those used in self-identified biogeography (e.g., published in the *Journal of Biogeography*). Many of the simulations in all three areas were developed in order to address questions related to biogeographic responses and interaction with climate change.

24.4.1 Carbon dynamics

Simulating carbon dynamics is a major part of understanding biogeography's contribution to Earth system science. While the biosphere interacts with the atmosphere in other ways (mostly through hydrology and albedo), the exchange of carbon between the two has attracted attention because of the role of CO_2 in global climate change. Many models have been developed, and broad comparisons of models are a useful introduction to their relative strengths and weaknesses (e.g., Cramer et al., 1999; Adams et al., 2004). These comparisons include those oriented toward net primary productivity (NPP) (e.g., TEM; McGuire et al., 1995) and those oriented toward biome change (e.g., DOLY; Woodward et al., 1995). While the models generally perform similarly, the degree to which nutrients were included influenced the results, different methods of calculating the water balance affected NPP outcomes, and the models tended to differ in seasonality (Cramer et al., 1999; see Adams et al., 2004 for other results). Bonan (2008) reported on linkages of carbon, albedo, and evapotranspirative changes in current models and further concluded that "much of our knowledge of forest influences on climate, and our ability to inform climate change mitigation policy, comes from models. . . . Global models of the biosphere-atmosphere system are still in their infancy, and processes not yet included in the models may initiate unforeseen feedbacks."

Two widely used carbon simulations are the CENTURY series (e.g., Parton et al., 1988, 1993, 2004) and FOREST-BGC (Running and Coughlan, 1988) and its descendants (e.g., Running and Hunt, 1993; Keane et al., 1996; Kang et al., 2006). CENTURY has been used hundreds of times, but its focus, beginning with an emphasis on Great Plains grasslands, has limited its wider application although it can be parameterized and run for forests. FOREST-BGC is also used broadly although it was designed for forests. Its RHESSys version (Band et al., 1993; Tague et al., 2004) is run for a region with a topographically based hydrological model. RHESSys is now being used extensively to examine the link between carbon and water fluxes in forest ecosystems. Studies originally based in the Rocky Mountains (e.g., Band et al., 1993) have expanded Californian shrublands (Tague et al., 2004) and to Europe (Zierl et al., 2007). BIOME-BGC (Running and Hunt, 1993; Thornton, 1998), which allows an application up to global scale, provides further biogeographical interest (e.g., Turner et al., 2006).

Carbon balance models are relatively mechanistic. They compute carbon fluxes based on known relations established from detailed studies at lower hierarchical levels. These models essentially solve difference equations. They are spatially explicit usually only in their hydrological functions. Often they are run for single cells, and so are not truly spatially explicit (Malanson, 1996), but some are (e.g., water moves among cells in RHESSys). Because of its regional application and for applications linked to other models,

I will expand on BGC models (BioGeochemical Cycles models from the Numerical Terradynamics Simulation Group).

The phenomenological focus of the BGC models is usually on carbon, water, and nitrogen dynamics. BGC models compute NPP and nitrogen allocation. These simulations compute photosynthesis and respiration, evapotranspiration, and carbon allocation within vegetation, litterfall, and decomposition. The canopy is treated as a single large leaf—which sounds bizarre but functions well when the canopy depth and LAI (leaf area index) is used to calculate light incident on the leaf. The temporal resolution of most computations in the BGC simulations is daily. Some computations are done annually. The daily computational step for the BGC simulations is appropriate for a mechanistic model that attempts to represent the process of photosynthesis and respiration in some detail. While other simulations may compute the process at much smaller time-steps, on the order of minutes, they often do so for a very short time, such as for a single year. The spatial resolution of the BGC simulations is not set. Given the 'big leaf' approach, this computes well, but one must have confidence that the inputs and assumptions of the model are fairly homogeneous over the area simulated. For application at the alpine timberline, Cairns (2005) added features of multiple vertical dimensions and edge to a BGC simulation and these altered the results at his chosen spatial resolution of 30 m.

The calculations in BGC are deterministic and use difference equations. The actual equations for a given time-step are, then, simply algebraic. For example, from Running and Coughlan (1988): "Canopy stomatal conductance to water vapor is then computed sequentially as a function of first leaf water potential modified by absolute humidity deficit . . . :

$$CC_W = CC_{max} - DCC_W(LWP - LWP_{min})$$

where CC_W is canopy H_2O conductance (m s^{-1}); CC_{max} maximum canopy conductance (m s^{-1}); DCC_W slope of CC vs. LWP (m s^{-1} MPa^{-1}); LWP daily maximum leaf water potential (MPa); LWP_{min} minimum leaf water potential inducing stomatal closure (MPa)."

The primary inputs to the original FOREST-BGC are of two types: driving variables and parameter constants. The latter include indicators for such things as leaf area, soil-water capacity, maximum canopy conductance, and leaf nitrogen concentration. The driving variables are primarily daily meteorological data. The original model was run for seven locations across North America, ranging from Jacksonville, Florida, west to Tucson, Arizona, and north to Fairbanks, Alaska. Key results analyzed were hydrological balances, NPP, and decomposition.

An example of the FOREST-BGC simulation is its use at the alpine treeline in answering the following question: what are the controls of the position of alpine treeline? An earlier application of FOREST-BGC found that it predicted that trees would grow much higher on slopes than they are found in the Sierra Nevada (Scuderi et al., 1993). Cairns (1994) proposed modification of FOREST-BGC to take into account the particular processes that occur at the alpine treeline, in particular winter desiccation and loss of tissue by abrasion. Cairns and Malanson (1998) hypothesized that a model that determined the point of zero carbon balance for trees could predict the location of alpine treeline. The basic FOREST-BGC model was modified so that carbon could be lost via changes in respiration and through damage. The model was run for slopes in Glacier National Park using extrapolated recent climate data. The results showed that the elevation of the ecotone could not be predicted exactly due to a lack of equilibrium between the vegetation and the climate, but the differences in patterns and the sensitivity of the simulation to moisture-related variables indicated that much of the spatial variability in local treeline elevation could be attributed to conditions causing high-elevation sites to be relatively xeric. More recent work modifying ATE-BGC has focused on adding vertical detail by departing from the single big leaf approach. In order to differentiate between prostrate krummholz and dwarf tree forms at the alpine treeline, Cairns (2005) modified the vertical distribution of light and the thermal regimes within the BGC canopy representation.

24.4.2 Forest-stand models

Forest-stand modeling in biogeography developed from the JABOWA simulation produced for the Hubbard Brook, New Hampshire, study site by Botkin et al. (1972); the FORET variant developed by Shugart and West (1977) although similar enough that they class is often referred to as JABOWA-FORET, has become the starting point for more branches of this model (Shugart and Smith, 1996).

The forest-stand models are a mix of degrees of opacity ranging from empirical to mechanistic. In growth terms, most of the past applications of the stand models used phenomenological equations that related growth to size and age relative to the larger and older trees of a species. Relating size to age requires less information, and the inputs to this type of model consist primarily of more recent applications, based on the work developed in the FORSKA models (Leemans, 1991), is more mechanistic. Inputs to FORSKA models require more information than BGC models. FORCLIM is an attempt to simplify these models by using fewer functional types to replace many species (Bugmann, 1996).

The temporal resolution in these models is almost always one year. Much of the work in forest-stand modeling is developed aspatially. A single stand was taken to represent a unit in the middle of a hypothetical forest. The replicate runs could be thought to represent other locations, but no locational variation was introduced. Even then, the area that a cell could represent was an issue. Busing (1991) demonstrated that a cell of about 0.1 hectare (ha) was optimal for forests in the eastern United States because this was the area that could be dominated by a single large tree but not to the exclusion of others while not so large that the effects of the fall of that single large tree would not be apparent. Later, pseudo-spatially explicit versions, in which multiple cells that differed in their environmental characteristics were simulated and mapped, were applied. The ZELIG versions were the first to adopt a degree of spatial process, where the light environment on a cell was modified by the vegetation on surrounding cells. More spatial detail was added when dispersal of species among cells was treated explicitly, as opposed to an assumption of ubiquitous dispersal.

The phenomenological resolution of forest-stand models is a stand of several trees in a forest. The simulation computes their establishment, growth, and death. Thus such a simulation can be used to compute aspects of carbon balance and allocation (e.g., Malanson and Kupfer, 1993), but this type of simulation is best suited to examining the relative abundance of different tree species. Most forest-stand models have a temporal resolution of one year, which is usually one cycle of seed production and growth for a tree. The average growth over a year is incremented as increased diameter and height. For temperate and boreal areas, this temporal resolution makes sense because the trees, when seen at this phenomenological resolution, do function on annual cycles. The spatial resolution for a forest-stand model in the temperate zone forests of the United States, for which the key models were designed, is 0.1 ha. The choice of this spatial resolution illustrates how spatial resolution and phenomenological resolution are related. At much larger resolutions, the dynamics of the interaction of individual trees are lost and the simulations will show little change over the course of centuries, while at smaller resolutions it is possible for a single large tree to so dominate a site that its life and death are the only dynamics simulated. In order to capture the dynamics of a forest, the importance of single trees, especially when they fall and create gaps, must be apparent in the simulation but not exclusive. The 0.1 ha size approximates this compromise. Malanson and Armstrong (1997) investigate to what extent a 0.1 ha—simulated cell could represent a larger area in a spatially explicit simulation in which the cells were arrayed on environmental gradients; they found that the spatial representation made a significant difference in two ways. First, the number of cells of the same environment increase diversity while a larger step-size in an environmental gradient between adjacent cells reduces diversity. Busing and Mailly (2004) reviewed forest-stand models with particular attention to spatial structure, giving new emphasis to three-dimensional structure. They concluded that while models without full spatial complexity do a reasonable job in simulating forest dynamics realistically, by missing disturbance patch size variability and heterogeneous dispersal, they may miss pattern-process relations that are important for some questions.

While some of the components of the forest-stand models are deterministic, there is considerable stochasticity. Much of the stochasticity enters into spatially explicit models if dispersal is included. The primary calculation is the annual diameter growth of each tree on the plot. In the standard JABOWA and FORET versions, an optimal growth, derived from observations of the largest and oldest trees of the species, was calculated and then reduced by less than optimal environmental conditions. For example, the temperature effects were represented as a parabola with its peak at 1.0 and reaching 0 where the species ranges ended on a gradient of growing-degree days (GDD) (problems with this approach were discussed by Malanson and Armstrong, 1996; Malanson et al., 1992); the value from this function is then multiplied by the optimum to get a reduced growth for a particular site, and this multiplication is done for several factors (nutrients, drought, high water table, but not all in any one simulation). Competition is primarily for light, with taller trees lowering the growth of all shorter trees on the plot. In FORSKA, a mechanistic calculation of optimal growth is made based on tree physiology, and the growth is calculated based on the distribution of the tree in different vertical components. A variant of this forest-stand approach considers differences in the important competition for light (Pacala et al., 1996).

Inputs to the JABOWA and FORET models were maximum age, height, and diameter and empirical values for constants relating them, and values to describe the functions by which the environment reduced growth below optimum (e.g., the maximum and minimum GDD for the range of each species. In FORSKA, some of the same or derived parameters are used (e.g., maximum tree height), but more mechanistic parameters are also needed, such as the light compensation point and the sapwood maintenance cost factor. The original models were validated against forest change in the eastern United States during the approximately 100 years preceding their development. They reproduced these forest changes well, but the lack of mechanism, especially in response to climate, brought their applicability to climate-change issues into question.

An example of the application of the JABOWA-FORET models illustrates their elaboration in spatial details. The SEEDSCAPE model for the U.S. Great Plains is an example of an application of this level of stand modeling for realistic geography (Easterling et al., 2001; Guo et al., 2004). Linear riparian forests in the Great Plains might respond differently to climate change because of their spatial patterns. The JABOWA-FORET and FORSKA models have been used extensively in studies of climate-change impacts, but because they simplify or ignore the problem of how seeds get to different places in the landscape, they may not accurately capture the dynamics. Borrowing a dispersal modeling adaptation of JABOWA developed by Hanson et al. (1989, 1990), Easterling et al. (2001) and Guo et al. (2004) simulated detailed landscapes in Nebraska. They hypothesized that the spatial pattern of the forest would result in different dynamics during simulated climate change. The simulation was run for actual landscape patterns including variation in their width. The simulation was not very sensitive to change in width less than 100 m, but it is possible that if wider riparian forests are examined that are not all edge the results will differ significantly.

As indicated by the developments in SEEDSCAPE, a major effort has been put into incorporating dispersal and/or migration processes into forest-stand models. Given the recognition of the importance of this spatial process for vegetation response to climate change (Malanson, 1993; Pitelka et al., 1997), it is not surprising. Efforts range from adding dispersal among cells on a hypothetical grid (e.g., Hanson et al., 1989; Ribbens et al., 1994; Malanson and Armstrong, 1996) to landscape (Dyer, 1994; Collingham et al., 1996; Guertin et al., 1997) and regional approaches (e.g., Mladenoff et al., 1996). The regional approaches have also incorporated spatially explicit disturbance (e.g., He et al., 1999, 2005; Waldron et al., 2007; Lafon et al., 2007; Cairns et al., 2008), a topic of landscape simulation in its own right (e.g., Baker et al., 1991; Mladenoff and Baker, 1999).

22.4.3 Biome change models

Biome change models cover a wide range of approaches, and many are not used as simulations but instead compute a single instantiation of response to climate change. Some of these include simulation components within them, and all have the ability to add temporal time-steps of climate change and thus become dynamic simulations. The aim of these models is to re-map global types of biomes for different future climates.

Descended from BIOME (Prentice, 1992), the focus of BIOME3 is the change of an area from one type of biome to another, but it does so by calculating productivity and then changing biomes according to rules (Haxeltine and Prentice, 1996). In one sense the model is not a simulation in that it is often run for a single condition and an output generated, which may be compared to the output for a different condition, but it is capable of dynamics. The development of the original BIOME focused on plant functional types, in order to find distinct environmental constraints for each by using the concept of an environmental sieve. This would be done by applying an environmental sieve (sensu van der Valk, 1981), that is, an analysis of factors that would eliminate potential functional types from a community at multiple time-steps. An environmental sieve is a trans-temporal and multifactorial expansion of Liebig's Law of the Minimum. The environmental sieve must include the type's requirements, its tolerances, and its ability to avoid what it cannot tolerate. In the original BIOME model, if more than one functional type survives the sieving process, a dominance hierarchy is applied (i.e., trees always supersede nontrees; more productive biomes [tropical evergreen] supersede those that are less so [temperate summer-green]). In BIOME3, a calculated optimal NPP is used to create an index of competitiveness.

Most calculations are at a monthly time-step, although some are computed only annually. The change in biomes, the primary focus, occurs only annually. The output for the model is for a single year, in part because it is possible in this type of model to have cells jump back and forth between biomes in a string of years. As such, the model may not meet our overall temporal criteria for a simulation per se, but it contains simulation. BIOME3 simulates the terrestrial vegetation of the entire planet. The spatial resolution of BIOME3 is 0.5° of latitude and longitude. The climate data were interpolated to the grid from about 1,000 stations.

The inputs for BIOME3 consist of a global meteorological and soils data sets and parameters or constants derived from earlier studies (e.g., the Michaelis constants for CO_2 and O_2). The model was validated by statistically comparing the output maps of biomes to previous maps of biomes and by comparing the model output to observations for NPP and FPAR (fraction of photosynthetically active radiation). The calculations in BIOME are Boolean applications of the environmental sieve. For each cell, environmental constraints are input or calculated from climatological inputs. For example, which functional types can pass the sieve of minimum temperature? Of minimum growing-degree-days? BIOME uses a scalar that represents the drought stress on plants. The sieve thresholds for the climate factors are derived from observed geographical limits of the functional types—but only for those where a mechanism is understood. BIOME3 then computes a maximum LAI and NPP for each possible functional type and then selects among them by using the optimal NPP index. The calculations of LAI and NPP are where BIOME3 crosses from a simple model based on semi-empirical relations between biome location and mapped climate to a process-based simulation. In calculating LAI and NPP, the simulation becomes a combined floristic and carbon model. As with a model such as the BGC series, BIOME3 can create detailed representations of canopy conductance with different parameters for C3 and C4 plants. However, BIOME3 goes back to an empirical base to capture shading of grass by trees and favoring trees over grass with less fire; in moist environments, grass functional types are just excluded.

A series of reports using BIOME3 to study vegetation change in China illustrates this modeling application. Ni et al. (Ni, 2000, 2001) used BIOME3 in two stages to look at change in the vegetation of China. Ni et al. (2000) ran the model for the average climates of 1931–1960 and 2070–2099 as computed with a climate simulation. The hypothesis is that the model can accurately produce the current pattern of vegetation, and so can inform us about the likely changes that would result from the simulated future climatic change. The model for the present climate predicted the locations of 17 unique biomes reasonably well (if one accepts that many biomes). The analysis was the comparison of the biome maps produced by the simulation with existing maps using a statistic that takes into account a number of vegetation characteristics, not just whether or not the biome is the same. Additional work has used BIOME3 to compute carbon stocks (Ni, 2000, 2002).

24.4.4 Examples of combinations

An active area of model combination shows promise in biogeographic simulations. At a fine scale, combinations of an individual tree-growth model, TREGRO, with the ZELIG forest model (which has incorporated the more mechanistic growth

functions from FORSKA in some uses) is producing advances in understanding forest responses to atmospheric change (Laurence et al., 2001). At a medium scale, in order to make the changing species combinations of forest-stand models operate on a more mechanistic base of the growth of the individual trees, attempts have been made to link models such as FORET with ones such as BGC. Friend et al. (1993) attempted a direct linkage, which worked reasonably well but was not computationally efficient for its time. At a larger scale, forest-stand modeling has been combined with a carbon model in FIRE-BGC (Keane et al., 1996). The development of FIRE-BGC is probably one of the best examples of a linkage in that it is solidly based in BGC, it has stand-altering competition, and it is spatially explicit for both dispersal of species among cells and for the spread of fire.

24.4.5 Other simulations

Many other types of simulations in biogeography contribute to the field. Two can be mentioned briefly here as they are covered in other chapters. First, grid-based spatially explicit simulations of the spread of species or disturbance, sometimes linked to one of the above categories as cited (e.g., FIRE-BGC, Keane et al., 1996; Perry and Enright, 2002), bring spatial processes into biogeographic simulations. These begin with continuous field approaches (e.g., Greene and Johnson, 1989) and have evolved to a wider variety of spatial representations (Skarpaas et al., 2005). The specific use of grid cells, probably due to the usefulness of raster-based GIS and remote sensing, now dominates (e.g., Mooij and DeAngelis, 2003). Linking dispersal to landscape patterns is an active area (e.g., Malanson, 2003; Vuilleumier and Metzger, 2006), and it is sometimes linked to studies of invasives (Higgins and Richardson, 1996, Lavorel et al., 1999).

Also not covered here are many types of simulations in theoretical ecology that apply to biogeography. The potential overlap is large because much of the recent simulation in this area has been focused on the spatial (e.g., Durrett and Levin, 1994; Tilman and Kareiva, 1997) and again has often been grid-based. The enormous field of metapopulation dynamics (e.g., Hanski, 1989; Hanski and Ovaskainen, 2000) and its offshoots (e.g., Tilman et al., 1997; Malanson et al., 2007) is the best single example.

24.5 CONCLUSIONS AND MUSINGS

Simulations allow us to create alternate worlds. A fast developing area of simulating alternative worlds is in agent-based modeling (see Walsh et al., this volume), wherein the 'agents' occupy, are affected by, and affect the environment (often spatially explicit in some degree) and each other (often the agents are mobile) (Green and Sadedin, 2005; Brown et al., 2006). Most work related to biogeography has been in land-use change (Brown et al., 2006; see Yadav et al., 2008 for a link to CENTURY), with humans as agents, but other models extend the ideas reviewed above. Savage et al. (2000) simulated alternative forest communities subject to varying lightning frequency, illustrating alternative world outcomes. Simulations can even parameterize themselves using genetic programming in which the model mimics the evolution of the ecology as well as the outcomes (Holland, 1992).

Such alternative worlds present pitfalls and opportunities. The pitfalls are obvious but not always avoided: we can spin yarns that are of interest only within themselves. The opportunities, however, are found in walking a line close to these pitfalls so that we can explore possible worlds as well as the one we think we are in at the moment. In particular, simulation allows us to consider the relations among multiple scales and examine some rather abstract patterns, and here we may find commonalities with other geographers who see real geography as unevenly specified (cf. Massey, 1999). Simulations emphasize, through multiple runs, the myriad alternatives that help us comprehend the spatial and historical contingencies of systems (cf. Phillips, 1999). I have elsewhere alluded to a view of multiple simulation results as 'geographical foam,' borrowing from the view of physical reality as based on quantum foam (Malanson et al., 2006). Thus, simulations can lead us to a view of the world that is more open to complex dynamics (e.g., Zeng and Malanson, 2006).

Two sides of an issue arise in evaluating the foam of alternative worlds: the importance of rare events or individuals that may determine common as well as rare outcomes and the significance, if any, of rare outcomes. We are not really sure whether or not we need to simulate every tree on earth in order to capture global dynamics (cf. Purves and Pacala, 2008). The universe as a single run of a simulation is a recurring theme in science fiction, and our aim must be to maximize our science while minimizing our fiction.

REFERENCES

Adams, B., White, A., and Lenton, T.M. (2004) 'An analysis of some diverse approaches to modelling terrestrial net primary productivity', *Ecological Modelling*, 177: 353–91.

Anderson, R.P., Peterson. A.T., and Egbert, S.L. (2006) 'Vegetation-index models predict areas vulnerable to purple loosestrife (Lythrum salicaria) invasion in Kansas', *Southwestern Naturalist*, 51: 471–80.

Armstrong, M.P. (2000) 'Geography and computational science', *Annals of the Association of American Geographers*, 90: 146–56.

Armstrong, M.P., and Marciano, R. (1995) 'Massively-parallel processing of spatial statistics', *International Journal of Geographical Information Systems*, 9: 169–89.

Baker, W.L., Egbert, S.L., and Frazier, G.F. (1991) 'A spatial model for studying the effects of climatic change on the structure of landscapes subject to large disturbances', *Ecological Modelling*, 56: 109–25.

Band, L.E., Patterson, P., Nemani, R., and Running, S.W. (1993) 'Forest ecosystem processes at the watershed scale—incorporating hillslope hydrology', *Agricultural and Forest Meteorology*, 63: 93–126.

Baudrillard, J. (1994) *Simulacra and Simulation*. Ann Arbor MI: University of Michigan Press.

Bonan, G.B. (2008) 'Forests and climate change: Forcings, feedbacks, and the climate benefits of forests', *Science*, 320: 1444–49.

Botkin, D.B. (1993) *Forest Dynamics*. Oxford: Oxford University Press.

Botkin, D.B., Janak, J.F., and Wallis, J.R. (1972) 'Rationale, limitations, and assumptions of a northeastern forest growth simulator', *IBM Journal of Research and Development*, 16: 101–16.

Brooks, R.J., and Robinson, S. (2001) *Simulation*. Basingstoke, UK: Palgrave.

Brown, D.G., Aspinall, R., and Bennett, D.A. (2006) 'Landscape models and explanation in landscape ecology—A space for generative landscape science?' *Professional Geographer*, 58: 369–82.

Bugmann, H.K.M. (1996) 'A simplified model to study species composition along climate gradients', *Ecology*, 77: 2055–74.

Busing, R.T. (1991) 'A spatial model of forest dynamics', *Vegetatio*, 92: 167–79.

Busing, R.T., and Mailly, D. (2004) 'Advances in spatial, individual-based modelling of forest dynamics', *Journal of Vegetation Science*, 15: 831–42.

Cairns, D.M. (1994) ' Development of a physiologically mechanistic model for use at the alpine treeline ecotone', *Physical Geography*, 15: 104-24.

Cairns, D.M. (2005) 'Simulating carbon balance at treeline for krummholz and dwarf tree growth forms', *Ecological Modelling*, 187: 314–28.

Cairns, D.M., and Malanson, G.P. (1998) 'Environmental variables influencing carbon balance at the alpine treeline ecotone: a modeling approach', *Journal of Vegetation Science*, 9: 679-692.

Cairns, D.M., and 7 authors. (2008) 'Simulating the reciprocal interaction of forest landscape structure and southern pine beetle herbivory using LANDIS', *Landscape Ecology*, 23: 403–15.

Collingham, Y.C., Hill, M.O., and Huntley, B. (1996) 'The migration of sessile organisms: A simulation model with measurable parameters', *Journal of Vegetation Science* 7: 831–46.

Cramer, W., Kicklighter, D.W., Bondeau, A., Moore III, B., Churkina, G., Nemry, B., Ruimy, A., Schloss, A.L. and the participants of the Potsdam NpP Model Intercomparison (1999) 'Comparing global models of terrestrial net primary productivity (NPP): Overview and key results', *Global Change Biology*, 5: 1–15.

Downing, K. (1998) 'Using evolutionary computational techniques in environmental modeling', *Environmental Modelling and Software*, 13: 519–28.

Durrett, R., and Levin, S. (1994) 'The importance of being discrete (and spatial)', *Theoretical Population Biology*, 46: 363–94.

Dyer, James M. (1994) 'Implications of habitat fragmentation on climate change-induced forest migration', *Professional Geographer*, 46: 449–59.

Easterling, W.E., Brandle, J.R., Hays, C.J., Guo, Q.F., and Guertin, D.S. (2001) 'Simulating the impact of human land use change on forest composition in the Great Plains agroecosystems with the Seedscape model', *Ecological Modelling*, 140: 163–76.

Friend, A.D., Schugart[sic], H.H., and Running, S.W. (1993) 'A physiology-based gap model of forest dynamics', *Ecology*, 74: 792–97.

Gao, Q., Yu, M., Yang, X.S., and Wu, J.G. (2001) 'Scaling simulation models for spatially heterogeneous ecosystems with diffusive transportation', *Landscape Ecology*, 16: 289–300.

Green, D.G., and Sadedin, S. (2005) 'Interactions matter—complexity in landscapes and ecosystems', *Ecological Complexity*, 2: 117–30.

Greene, D.F., and Johnson, E.A. (1989) 'A model of wind dispersal of winged and plumed seeds', *Ecology*, 70: 339–47.

Guertin, D.S., Easterling, W.E., and Brandle, J.R. (1997) 'Climate change and forests in the Great Plains—issues in modeling fragmented woodlands in intensively managed landscapes', *BioScience*, 47: 287–95.

Guo, Q.F., Brandle, J., Schoeneberger, M., and Buettner, D. (2004) 'Simulating the dynamics of linear forests in Great Plains agroecosystems under changing climates', *Canadian Journal of Forest Research*, 34: 2546–72.

Hanski, I. (1989) 'Metapopulation dynamics: Does it help to have more of the same?' *Trends in Ecology & Evolution* 4: 113–14.

Hanski, I., and Ovaskainen, O. (2000) 'The metapopulation capacity of a fragmented landscape', *Nature*, 404: 755–58.

Hanson, J.S., Malanson, G.P., and Armstrong, M.P. (1989) 'Spatial constraints on the response of vegetation to climate change', in Malanson, G.P. (ed.), *Natural Areas Facing Climate Change*. The Hague: SPB Academic. pp. 1–21.

Hanson, J.S., Malanson, G.P., and Armstrong, M.P. (1990) 'Landscape fragmentation and dispersal in a model of riparian forest dynamics', *Ecological Modelling*, 49: 277–96.

Harmon, M.E., and Marks, B. (2002) 'Effects of silvicultural practices on carbon stores in Douglas-fir—western hemlock forests in the Pacific Northwest, USA: Results form

a simulation model', *Canadian Journal of Forest Research*, 32: 863–77.

Haxeltine, A., and Prentice, I.C. (1996) 'BIOME3: An equilibrium terrestrial biosphere model based on ecophysiological constraints, resource availability, and competition among plant functional types', *Global Biogeochemical Cycles*, 10: 693–709.

He, H.S., Mladenoff, D.J., and Crow, T.R. (1999) 'Linking an ecosystem model and a landscape model to study forest species response to climate warming', *Ecological Modelling*, 114: 213–33.

He, H.S., Mladenoff, D.J., and Crow, T.R. (2005) 'Simulating forest ecosystem response to climate warming incorporating spatial effects in north-eastern China', *Journal of Biogeography*, 32: 2043–56.

Higgins, S.I., and Richardson, D.M. (1996) 'A review of models of alien plant spread', *Ecological Modelling*, 87: 249–65.

Holland, J.H. (1992) 'Genetic algorithms', *Scientific American*, 267: 66–72.

Kang, S., Lee, D., Lee, J., and Running, S.W. (2006) 'Topographic and climatic controls on soil environments and net primary production in a rugged temperate hardwood forest in Korea', *Ecological Research*, 21: 64–74.

Keane, R.E., Ryan, K.C., and Running, S.W. (1996) 'Simulating effects of fire on northern Rocky Mountain landscapes with the ecological process model FIRE-BGC', *Tree Physiology*, 16: 319–31.

Lafon, C.W., Waldron, J.D., Cairns, D.M., Tchakerian, M.D., Coulson, R.N., and Klepzig, D.D. (2007) 'Modeling the effects of fire on the long-term dynamics and restoration of yellow pine and oak forests in the southern Appalachian Mountains', *Restoration Ecology*, 15: 400–411.

Laurence, J.A., Retzlaff, W.A., Kern, J.S., Lee, E.H., Hogsett, W.E., and Weinstein, D.A. (2001) 'Predicting the regional impact of ozone and precipitation on the growth of loblolly pine and yellow-poplar using linked TREGRO and ZELIG models', *Forest Ecology and Management*, 146: 247–63.

Lavorel, S., Smith, M.S., and Reid, N. (1999) 'Spread of mistletoes (Amyema preissii) in fragmented Australian woodlands: A simulation study', *Landscape Ecology*, 14: 147–60.

Leemans, R. (1991) 'Sensitivity analysis of a forest succession model', *Ecological Modelling*, 53: 247–62.

Malanson, G.P. (1984) 'Linked Leslie matrices for the simulation of succession', *Ecological Modelling*, 21: 13–20.

Malanson, G.P. (1985) 'Simulation of competition between alternative shrub life history strategies through recurrent fires', *Ecological Modelling*, 27: 271–83.

Malanson, G.P. (1993) 'Comment on modeling ecological response to climatic change', *Climatic Change*, 23: 95–109.

Malanson, G.P. (1996) 'Modelling forest response to climatic change: issues of time and space', in S.K. Majumdar, E.W. Miller, and F.J. Brenner (eds.), *Forests—A Global Perspective*. Easton, PA: Pennsylvania Academy of Sciences. pp. 200–211.

Malanson, G.P. (1999) 'Considering complexity', *Annals Association of American Geographers*, 89: 746–53.

Malanson, G.P. (2003) 'Dispersal across continuous and binary representations of landscapes', *Ecological Modelling*, 169: 17–24.

Malanson, G.P., and Armstrong, M.P. (1996) 'Dispersal probability and forest diversity in a fragmented landscape', *Ecological Modelling*, 87: 91-102.

Malanson, G.P., and Armstrong, M.P. (1997) 'Issues in spatial representation: Effects of number of cells and between-cell step size on models of environmental processes', *Geographical and Environmental Modelling*, 1: 47–64.

Malanson, G.P., and Kupfer, J.A. (1993) 'Simulated fate of leaf litter and woody debris at a riparian cutbank', *Canadian Journal of Forest Research*, 23: 582–90.

Malanson, G.P., Wang, Q., and Kupfer, J.A. (2007) 'Ecological processes and spatial patterns before, during and after simulated deforestation', *Ecological Modelling*, 202: 397–409.

Malanson, G.P., Westman, W.E., and Yan, Y-L. (1992) 'Realized versus fundamental niche functions in a model of chaparral response to climatic change', *Ecological Modelling*, 64: 261–77.

Malanson, G.P., and Zeng, Y. (2004) 'Uncovering spatial feedbacks at alpine treeline using spatial metrics in evolutionary simulations', in P.M. Atkinson, G. Foody, S. Darby, and F. Wu (eds.), *GeoDynamics*. Boca Raton, FL: CRC Press. pp. 137–50.

Malanson, G.P., Zeng, Y., and Walsh, S.J. (2006) 'Landscape frontiers, geography frontiers: lessons to be learned', *Professional Geographer*, 58: 383-96.

Martin, P. (1993) 'Vegetation responses and feedbacks to climate—a review of models and processes', *Climate Dynamics*, 8: 201–10.

Massey, D. (1999). 'Space-time, "science" and the relationship between physical geography and human geography', *Transactions of the Institute of British Geographers*, 24: 261–76.

McGuire, A.D., Melillo, J.M., Kicklighter, D.W., and Joyce, L.A. (1995) 'Equilibrium responses of soil carbon to climate change: empirical and process-based estimates', *Journal of Biogeography*, 22: 785-96.

McHaney, R. (1991) *Computer Simulation: A Practical Perspective*. San Diego: Academic Press.

Mladenoff, D.J., and Baker, W.L. (eds.). (1999) *Spatial Modeling of Forest Landscape Change: Approaches and Applications*. Cambridge: Cambridge University Press.

Mladenoff, D.J., Host, G.E., Boeder, J., and Crow, T.R. (1996) 'LANDIS: A spatial model of forest landscape disturbance, succession, and management', in M.F. Goodchild, L.T. Steyaert and B.O. Parks (eds.), *GIS and Environmental Modeling: Progress and Research Issues*. Ft. Collins, CO: GIS Word Books. pp. 175–80.

Mooij, W.M., and DeAngelis, D.L. (2003) 'Uncertainty in spatially explicit animal dispersal models', *Ecological Applications*, 13: 794–805.

Neelamkavil. F. (1987) *Computer Simulation and Modeling*. New York: John Wiley.

Ni, J. (2000) 'Net primary production, carbon storage and climate change in Chinese biomes', *Nordic Journal of Botany*, 20: 415–26.

Ni, J. (2001) 'Carbon storage in terrestrial ecosystems of China: Estimates at different spatial resolutions and their responses to climate change', *Climatic Change*, 49: 339–58.

Ni, J. (2002) 'Effects of climate change on carbon storage in boreal forests of China: A local perspective', *Climatic Change*, 55: 61–75.

Ni, J., Sykes, M.T., Prentice, I.C., and Cramer, W. (2000) 'Modelling the vegetation of China using the process-based equilibrium terrestrial biosphere model BIOME', *Global Ecology and Biogeography*, 9: 463–79.

Obeysekera, J., and Rutchey, K. (1997) 'Selection of scale for Everglades landscape models', *Landscape Ecology*, 12: 7–18.

Pacala, S.W., Canham, C.D., Saponara, J., Silander, J.A., Kobe, R.K., and Ribbens, E. (1996) 'Forest models defined by field measurements: Estimation, error analysis and dynamics', *Ecological Monographs*, 66: 1–43.

Parsons, A.J., and Knight, P. (2005 *How to Do Your Dissertation in Geography and Related Disciplines*. London: Routledge.

Parton, W., Tappan, G., Ojima, D., and Tschakert, P. (2005) 'Ecological impact of historical and future land-use patterns in Senegal', *Journal of Arid Environments*, 59: 605–23.

Parton, W.J., and 11 authors. (1993) 'Observations and modeling of biomass and soil organic matter dynamics for the grassland biome worldwide', *Global Biogeochemical Cycles*, 7: 785–809.

Parton, W.J., Stewart, J.W.B., and Cole, C.V. (1988) 'Dynamics of C, N, P and S in grassland soils—a model', *Biogeochemistry*, 5: 109–31.

Perry, G.L.W., and Enright, N.J. (2002) 'Humans, fire and landscape pattern: Understanding a maquis-forest complex, Mont Do, New Caledonia, using a spatial 'state-and-transition' model', *Journal of Biogeography*, 29: 1143–58.

Phillips, J.D. (1999) 'Spatial analysis in physical geography and the challenge of deterministic uncertainty', *Geographical Analysis*, 31: 359–72.

Pitelka, L.F., and Plant Migration Workshop. (1997) 'Plant migration and climate change', *American Scientist*, 85: 464–73.

Prentice, I.C. (1992) 'A global biome model based on plant physiology and dominance, soil properties and climate', *Journal of Biogeography*, 19: 117–34.

Purves, D., and Pacala, S. (2008) 'Predictive models of forest dynamics', *Science*, 320: 1452–53.

Reynolds, J.F., Hilbert, D.W., and Kemp, P.R. (1993) 'Scaling ecophysiology from the plant to the ecosystem; a conceptual framework', in *Scaling Processes Between Leaf and Landscape Levels*. San Diego, CA: Academic Press. pp. 127–40.

Ribbens, E., Silander, J.A., and Pacala, S.W. (1994) 'Seedling recruitment in forests—calibrating models to predict patterns of tree seedling dispersion', *Ecology*, 75: 1794–806.

Running, S.W., and Coughlan, J.C. (1988) 'A general model of forest ecosystem processes for regional applications. 1. Hydrologic balance, canopy gas exchange and primary production processes', *Ecological Modelling*, 42: 125–54.

Running, S.W., and E.R. Hunt Jr. (1993) 'Generalization of a forest ecosystem process model for other biomes, BIOME-BGC, and an application for global-scale models', in *Scaling Processes Between Leaf and Landscape Levels* San Diego, CA: Academic Press. pp. 141–58.

Savage, M., Sawhill, B., and Askenazi, M. (2000) 'Community dynamics: What happens when we rerun the tape?' *Journal of Theoretical Biology*, 205: 515–26.

Scuderi, L.A., Schaaf, C.B., Orth, K.U., and Band, L.E. (1993) 'Alpine treeline growth variability: Simulation using an ecoysystem process model', *Arctic and Alpine Research*, 25. 175–82.

Shugart, H.H. (1998) *Terrestrial Ecosystems in Changing Environments.* Cambridge: Cambridge University Press.

Shugart, H.H., and Smith, T.M. (1996) 'A review of forest patch models and their application to global change research', *Climatic Change*, 34: 131–53.

Shugart, H.H., and West, D.C. (1977) 'Development of an Appalachian deciduous forest model and its application to assessment of the impact of the chestnut blight', *Journal of Environmental Management*, 5: 161–79.

Skarpaas, O., Shea, K., and Bullock, J.M. (2005) 'Optimizing dispersal study design by Monte Carlo simulation', *Journal of Applied Ecology*, 42: 731–39.

Tague, C., McMichael, C., Hope, A., Choate, J., and Clark, R. (2004) 'Application of the RHESSys model to a California semiarid shrubland watershed', *Journal of the American Water Resources Association*, 40: 575–89.

Thornton, P. E. (1998) 'Description of a numerical simulation model for predicting the dynamics of energy, water, carbon, and nitrogen in a terrestrial ecosystem'. PhD dissertation, University of Montana, Missoula.

Tilman, D., and Kareiva, P. eds. (1997) *Spatial Ecology.* Princeton, NJ: Princeton University Press.

Tilman, D., Lehman, C.L., and Yin, C. (1997) 'Habitat destruction, dispersal, and deterministic extinction in competitive communities', *American Naturalist*, 149: 407–35.

Turner, D.P., and 10 authors. (2006) 'Evaluation of MODIS NPP and GPP products across multiple biomes', *Remote Sensing of Environment*, 102: 282–92.

Urban, D.L., Bonan, G.B., Smith, T.M., and Shugart, H.H. (1991) 'Spatial applications of gap models', *Forest Ecology and Management*, 42: 95–110.

van der Valk, A.G. (1981) 'Succession in wetlands: A Gleasonian approach', *Ecology*, 62: 688–96.

Vuilleumier, S., and Metzger, R. (2006) 'Animal dispersal modelling: Handling landscape features and related animal choices', *Ecological Modelling*, 190: 159–70.

Waldron, J.D., Lafon, C.W., Coulson, R.N., Cairns, D.M., Tchakerian, M.D., Birt, A. and Klepzig, K.D. (2007) 'Simulating the impacts of southern pine beetle and fire on the dynamics of xerophytic pine landscapes in the southern Appalachians', *Applied Vegetation Science*, 10: 53–64.

Wang, Q., and Malanson, G.P. (2008) 'Spatial hyperdynamism in a post-disturbance simulated forest', *Ecological Modelling*, 215: 337–44.

Woodward, F.I., Smith, T.M., and Emanuel, W.R. (1995) 'A global land primary productivity and phyto-geography model', *Global Biogeochemical Cycles*, 9: 471–90.

Yadav, V., Del Grosso, S.J., Parton, W.J., and Malanson, G.P. (2008) 'Adding ecosystem function to agent-based land use models,' *Journal of Land Use Science*, 3: 27-40.

Zeng, Y., and Malanson, G.P. (2006) 'Endogenous fractal dynamics at alpine treeline ecotones', *Geographical Analysis*, 38: 271–87.

Zierl, B., Bugmann, H., and Tague, C.L. (2007) 'Water and carbon fluxes of European ecosystems: An evaluation of the ecohydrological model RHESSys', *Hydrological Processes*, 21: 3328–39.

Biocomplexity

Stephen J. Walsh, George P. Malanson,
Joseph P. Messina, Daniel G. Brown, and
Carlos F. Mena

25.1 INTRODUCTION

A complex system is one in which its multiple components interact in ways that link patterns and processes across scales. Further, complex systems focus on irreducible complexity arising from simplicity. This view sees the complex nature of systems as emerging from nonlinearities due to large numbers of interactions involving feedbacks occurring at one or more lower levels within the system (Cilliers, 1998; Malanson, 1999; Crawford et al., 2005). Complex systems are generally far from equilibrium (Bak, 1998), with a constant set of interactions that maintain the organization of the systems through negative feedbacks or alter subsequent alternatives in state space through positive feedbacks. Thus, complexity theory holds that systems cannot be suitably understood without focusing on the feedbacks and nonlinearities that lead to emergent multiscale phenomena (Matthews et al., 1999; Manson, 2001). A complexity theory analysis aims at understanding feedback mechanisms and changes in state-space through nonlinearities and thresholds, in relation to a dynamic environment with the goal of understanding how simple, fundamental processes combine to produce complex holistic systems (Luhman, 1985; Gell-Mann, 1994). Endogenous and exogenous factors combine in complex ways to alter the vulnerability and resilience of system components (White and Engelen, 1993). Complex systems not only evolve through time, but their past is co-responsible for their present behavior (Cilliers, 1998).

Biocomplexity encompasses the complex interactions within and among ecological systems, the physical systems on which they depend, and the human systems with which they interact (Michener et al., 2001). Biocomplexity is the interdisciplinary and integrated study of coupled human-natural systems, often approached from the perspective of land-use and land-cover (LULC) change, to address the causes and consequences of landscape dynamics. Often considered within a hierarchy theory context, pattern-process relations are explicitly considered at ranges of spatial and/or temporal scales. As such, biocomplexity is scale-sensitive in that the identification of optimum scales involves considering the scale at which pattern-process relations are best represented and understood, as well as the most effective scale for representing and understanding human activity and its impact on the environment (Meentemeyer, 1989; Phillips, 1999; Walsh et al., 1999; Sheppard and McMaster, 2004). Moreover, biocomplexity recognizes the importance of multiscale interactions in a small range of a hierarchy of scales.

Biocomplexity draws on theories and practices from across the social, natural, and spatial sciences, and it emphasizes complexity sciences in the study of people, places, and environments. For instance, it has been applied to the study of biodiversity, ecological services, and sustainability (e.g., Freeman et al., 2001; Desrochers and Anand, 2005); fisheries and game hunting (e.g., Bousquet et al., 2001; Badalamenti et al., 2002); urban systems (e.g., Clarke et al., 1996, 1997;

Batty et al., 1999; White and Engelen, 1993; Webster and Wu, 1999); nutrient cycling (e.g., Jenerette and Wu, 2004); climate change (e.g., Solecki and Oliveri, 2004); vegetation and biological modeling (e.g., Ermentrout and Edelstein-Keshet, 1993; Freeman et al., 2001; Pausas, 2003; Zeng and Malanson, 2006); tropical deforestation (e.g., Silveira et al., 2002; Deadman et al., 2004; Messina and Walsh, 2005; Entwisle et al., 2008); and LULC change in coupled human-natural systems (e.g., Messina and Walsh, 2001; Lambin et al., 2003; Evans and Kelley, 2004; An et al., 2005; Walsh et al., 2006, 2008b). Of particular interest has been the characterization of spatial patterns and the linkages to process, feedback mechanisms and system dynamics, and space-time lags and scale dependence of relationships, processes, actors, and models (Evans and Kelley, 2004; Green et al., 2005; Malanson et al., 2006a, 2006b).

Agent-based models (ABMs) have recently been used to explore complex systems in LULC change (Deadman et al., 2004; Evans and Kelley, 2004; Brown et al., 2004, 2005); rural to urban migration (Jaylson et al., 2006); ecosystem management (Nute, 2004; Bousquet et al., 2001); and agricultural economics (Berger, 2001). Individual-based models (IBMs) have been developed and used in ecology to represent population dynamics (DeAngelis and Gross, 1992; Grimm and Railsback, 2005) and used specifically to model the spread of invasive species (Buckley et al., 2003; Prévosto et al., 2003; Shea et al., 2006). These models use a bottom-up approach to allow the testing of alternative theories to explain specific observable patterns (Grimm et al., 2005).

Among the important places being examined, within the context of biocomplexity, are frontier ecotones—zones of transition in the form, pattern, and density of human settlement and the associated LULC changes (e.g., the Ecuadorian Amazon frontier) (Rindfuss et al., 2007). Global changes exert great exogenous forces on these places, but their systems have their own spatially contingent endogenous dynamics. The application of complexity theory is providing insights on the dynamics occurring in such settings (and others) by looking for universal properties in spatially extended systems (see the special issue of *GeoForum*, edited by Walsh and McGinnis, 2008). Feedbacks between people, places, and the environment constrain or even reverse some of the original changes in LULC through system dynamics (Matthews et al., 1999; Manson, 2001). In this way, properties emerging from local nonlinear feedbacks constrain the evolving patterns of land use (Wolfram, 1984; Blackman, 2000). Critical points in the spatial structure of LULC patterns

and feedbacks can produce a system with identifiable future alternative states in which instabilities can 'flip' a system into another regime of behavior by changing the patterns and processes that control LULC change (Parker et al., 2003). As such, we are motivated by questions that seek understanding of broad areas of biocomplexity:

1. How does a complex approach help explore the internal mechanisms of systems and provide plausible explanations?
2. How do results derived from applying complexity theory help in understanding decision making across levels of social organization, ranging from individual households to national governments?
3. How do fundamental characteristics of complex dynamics of coupled human-natural systems and the limits of predictability pertain to sustainable development?
4. Do positive and negative feedbacks, and feedback switches, produce a system with a critical point subject to small or large effects of exogenous factors functioning through space and time lags?
5. How can nonequilibrium systems, with feedbacks leading to nonlinearity, evolve into systems that exhibit criticality and capture key dynamics?
6. How do space and time affect nonlinearities— do location, spatial properties, and space play important roles in complexity by allowing time lags to be scale-dependent?
7. What are the emergent patterns or trajectories of LULC change, are fractal characteristics evident, and do they organize around fronts of change and development?

This chapter is organized as follows. First, the central concepts, tenets, perspectives, and approaches to biocomplexity will be explored. Second, spatial modeling approaches that are being used to address the integration of people, places, and environment in coupled human-natural systems and the causes and consequences of such changes will be reviewed. Third, the role of genetic algorithms and stylized or simulated environments for considering critical variables and scenarios of landscape dynamics will be examined. Finally, model validation and different perspectives and approaches to biocomplexity will be considered, along with some of the challenges that they present to the biocomplexity community. Our intent is to select only a few key elements of biocomplexity for comment, using case studies that we are actively engaged in to showcase some applications from geography. As a rapidly evolving field of study being integrated into the social, natural, and spatial sciences, we review key elements of biocomplexity and point to areas of challenge and opportunity for biogeographers and geographers more broadly.

25.2 BIOCOMPLEXITY: SOME FUNDAMENTAL CHARACTERISTICS

Complex systems, of the type referred to as aggregate complexity by Manson (2001), are characterized by aggregation, nonlinearity, flows, and diversity (Holland, 1995; see Malanson, 1999, for a description of the relevance of these characteristics within biogeography). Aggregation refers to the creation of system behaviors or patterns through the interaction of multiple, diverse actors. Sometimes referred to as emergent behaviors or patterns, such aggregate characteristics, it can be argued (Parker and Meretsky, 2004), include fragmentation and other spatial-pattern characteristics of common interest within landscape ecology. To understand how these patterns come about, it is important to understand the interactions of multiple diverse organisms, whether of plants, animals, or people. The diversity of the actors, for example, through multiple adaptive strategies can be an important determinant of a particular pattern of system, as well as of system resilience under perturbation. Flows of matter, genetic material, and information provide interaction mechanisms that can be the means through which systemwide characteristics emerge. The interactions can also produce nonlinear dynamics, such that small changes in system properties or modeling dynamics can yield vastly different system outcomes, sometimes from among a limited set of actors (see Figure 1.1 in Holling et al., 2002). Often the system itself can adapt (Holland, 1995).

The study of spatial landscape patterns in biogeography has demonstrated important ecological effects due to habitat connectivity and fragmentation. In addition to information about landscape pattern, landscape management requires information about the formation of properties and dynamics of these patterns. Recent research on landscape dynamics has come to view landscapes as complex systems, consisting of interactions of human and natural processes, in which landscape patterns are important emergent properties of complex dynamics (Walsh et al., 2008a). Complexity theory conceives the world as consisting of self-organized systems, either reproducing their state—a stable state—through negative feedbacks with their environment or by moving along trajectories from one state to another as a result of positive feedbacks. Dynamics emerging from these local nonlinear feedbacks can constrain the evolving patterns or create the emergence of new landscape structures that create

additional feedback mechanisms on subsequent activities. Critical points in the formation of spatial structures and patterns can result in a system that can undergo phase changes, alternate between system states, or develop reinforced pattern trajectories through path dependence. Exogenous forces can influence the overall system behavior by changing the rates and directions of LULC dynamics, whereas endogenous factors may mediate the system by how it responds to exogenous factors (see Rindfuss et al., 2008).

Balanced positive feedbacks, not tightly meshed negative feedbacks or wild, uncontrolled positive feedbacks, produce systems at the edge of chaos. Balanced positive feedbacks produce 'critical points'—where slight changes can alter system behavior sensitively—leading to possible system 'phase changes' or 'self-organized criticality' (Bak, 1998). Critical points can also be created by spatial patterns in percolation theory. Percolation theory addresses a subset of studies of random walks in random environments (Hughes, 1995). The most important result of percolation theory is that there is a critical threshold that will allow connectivity across a lattice. Connection is less likely below the threshold and is nearly certain above it. With smaller lattices, the threshold is at approximately the same point, but the variability is far greater (Varnakovida and Messina, 2005). Other aspects of the geometry of the lattice change nonlinearly in the vicinity of this threshold; notable differences are seen in the number of clusters, the size and fractal dimension of the largest cluster, and the mean edge of all clusters. Much of the literature on percolation theory has developed with the scale of analysis being in the range of electrons to molecules, but most early and review works cite potential applications at geographic scales such as the spread of diseases or wildfires.

Self-organized complexity is a general umbrella for work that addresses power-law distributions (Turcotte and Rundle, 2002). Power laws are basic theoretical explanations for emergent ecological phenomena, such as scaling relationships that are self-similar or show fractal behavior across spatial and temporal scales. Power laws, for example, might explain some underlying process or set of processes that generate biodiversity, inductively, as empirical patterns that suggest how universal principles of ecology arise from the laws of physics, chemistry, and biology (Brown et al., 2002). Self-organization encompasses several explanations of spatial and temporal fractal scaling in spatially extended dynamical systems, such as self-organized criticality, inverse cascade modeling, self-organized instability, and self-organized percolation (Bak et al., 1988; Alencar et al., 1997; Turcotte et al., 1999; Sole et al., 1999). Self-organization implies that a system may converge on an attractor, usually of low dimension, without external tuning (Bak et al., 1988). Highly optimized tolerance—HOT— (Carlson and Doyle, 1999) implies some type of tuning.

HOT is an alternative to self-organization, also using percolation-related models, to explain power-law features in systems with some degree of design or optimization. Optimization occurs where humans engineer a system or where natural selection optimizes processes and/or relations. Carlson and Doyle (1999) described HOT for a forest-fire model. They report that HOT systems have four qualities: power laws, specialized spatial configurations (i.e., compact differentiated structures as opposed to the self-similar structures of self-organization), efficiency of performance (in their case, maintaining trees in the model), and robustness to anticipated perturbation coupled with hypersensitivity to perturbations not anticipated in the design. In HOT systems, subtle details are not overpowered by the generality imposed by self-organization. Although humans optimize their land-use patterns in varying degrees due to imperfect knowledge and uncertainty, even small degrees of optimization can result in HOT states (Carlson and Doyle, 2000).

25.3 BIOCOMPLEXITY AND LANDSCAPE PATTERN

Beyond addressing LULC dynamics and their causes and consequences, biocomplexity is being used to examine the linkages between patterns and processes across diverse landscapes. While much work has been accomplished to identify the effects of landscape pattern on ecological processes, more work is needed to explain the processes that give rise to those patterns (Farina and Belgrano 2004). By developing reasonable explanations of those processes, we can pose management structures or policies that can guide the development of landscape patterns in more ecologically beneficial directions. Because of the complex dynamics associated with landscape pattern formation, for example, feedbacks between the behavior of organisms and the environment, such explanations need to include both the exogenous and endogenous factors that influence pattern formation (Malanson, 2003). Furthermore, models of pattern formation need to be process-based. While a number of modeling approaches have been attempted to describe landscape dynamics, more are based only on descriptions of those pattern

dynamics (e.g., Markov chains) and contain little in the way of process description.

The science of complexity recognizes that systems with spatial and temporal degrees of freedom (e.g., agents on a landscape) lead to nonlinear, nonequilibrium systems that exhibit criticality or phase transitions. Criticality is a condition in a system where any outcome is possible, response to perturbation is of any size, and correlation extends across space (cf. Rescher, 1998; Phillips, 1999). The nonequilibrium functioning of ecosystems is a foundation for the argument that ecosystems exhibit emergence (Holland, 1998) or self-organization (Bak, 1998).

While human socioeconomic systems in general have been characterized as emergent and self-organized, the spatial pattern of land use has received considerable attention. Weidlich (1994), Weidlich and Munz (1990), and Haag (1994) have modeled settlement as a self-organizing process leading to spatially heterogeneous patterns. White and Engelen (1994) characterize urban dynamics as critical phenomena. These works, based in physics, provide a supplement to a body of work in social science that has extended early models based on central place theory (e.g., Allen and Sanglier, 1979) by adopting a complexity viewpoint and computational approaches such as cellular automata (CA) and multiagent models (e.g., Batty et al., 1997; Benenson, 1999; Parker et al., 2003). We might expect fractal, power-law distributions in space and time in human land use. HOT is a reasonable possible explanation.

In addition, issues such as scale-free behavior of complex systems have been associated with power-law distributions; self-organized criticality to fractals and power laws; emergence of patterns to model complexity and to the fundamental building blocks of systems; and complex adaptive systems to pathways of landscape change (see Manson, 2001; Parker et al., 2003, for reviews of complexity and multiagent models; and Malanson, 1999, for a description of complexity in biogeography). These and other nonlinear, dynamic systems' perspectives and approaches are being used to assess the spatial and temporal dynamics of plant populations, the nature and degree of perturbation in plant communities, evolution and state changes of biological systems, and more (Malanson 2003). In short, biocomplexity has established itself as an emerging paradigm for the study of nonlinear and dynamic systems in which pattern-process relations in biogeography, ecology, and coupled human-natural systems are emphasized, and where feedbacks and system dynamics are examined by using a variety of methods including cellular automata, ABMs, genetic algorithms, and stylized or simulated environments. That is not to say that the application of biocomplexity to ecological issues and human-environment interactions is not without its many challenges. Validating model outcomes for historical, contemporary, and future periods; testing 'what if' scenarios of change through spatial simulations; assessing predictive and process accuracy of models; and examining the path dependence and independence of models are among some of the more important considerations in spatial modeling and biocomplexity that are yet to be fully addressed (Brown et al., 2005).

25.4 SPATIAL MODELING: LINKING PATTERN-PROCESS RELATIONS

A model is a set of assumptions or approximations about how the system works. Physical models might be either iconic models (things that look like reality) or analog models (things that act like reality). Symbolic models are either verbal or mathematical. However, all models must be defined by purpose, mode, randomness, and generality. Broadly interpreted, complex systems are modeled via computer simulation addressing: (a) static versus dynamic; (b) continuous change versus discrete change; and (c) deterministic versus stochastic. Unfortunately, computer systems are not precisely capable of modeling complex systems because they cannot handle continuous dynamics, so most operational complex systems models are dynamic, discrete-change, and stochastic. Complex system simulations are based on the global consequences of local interactions of members of a population. These entities or individuals might represent plants and animals in ecosystems, people in crowds, or any autonomous acting or decision-making agents. The models typically consist of an environment or framework in which the interactions occur and some number of individuals defined in terms of their behaviors (procedural rules) and characteristic parameters. In an IBM, the characteristics of each individual are tracked through time. This contrasts to modeling techniques where the characteristics of the population are averaged together and the model attempts to simulate changes in these averaged characteristics for the whole population (Judson, 1994). IBMs include ABMs and are not necessarily different in simulations (e.g., Olson and Sequeira, 1995; Ginot et al., 2002; Park and Sugumaran, 2005; Reynolds, 2005). Some spatially explicit IBMs also exhibit mobility, where the individuals can move around in their environments, for example, animals in an ecological simulation versus plants in the same simulation. Further, some IBMs are not directly spatially explicit, for example, a simulation of a computer network. However, spatially explicit models may use either continuous (object) or discrete (integer-valued, grid-like) space.

An autonomous agent can exist in isolation, or it can be situated in a world shared by other entities. This situated agent can be reactive (responding to conditions) or it can be deliberative (rule-developing). An autonomous agent can deal exclusively with abstract information or it can be embodied in a physical manifestation. Combinations of situated, reactive, and embodied define several distinct classes of autonomous agents (Reynolds, 2000). However, for most biogeographic research, situated, embodied (real agents in a virtual world), reactive, virtual agents are used.

IBMs are a subset of multiagent systems, which are any computational system whose design is fundamentally composed of a collection of interacting parts. Individual-based models are distinguished by the fact that each 'agent' corresponds to an autonomous individual in the simulated environment. Cellular automata (CA) are similar to spatially explicit, grid-based, immobile IBMs. Traditional CA models are homogeneous and dense (all cells are identical), whereas a grid-based IBM might occupy only a few grid cells, and more than one distinct type of individual might live on the same grid. Of course, a CA can have cells in various states and so represent concepts like 'empty' or 'occupied by type x'. One significant difference is whether the simulation's inner loop proceeds cell by cell, or individual by individual, although that distinction is mitigated by parallel-processing hardware. Another significant difference is in reversibility; an ABM is reversible, but CA is not. The philosophical issue is whether the simulation is based on a dense and uniform dissection of the space (as in CA) or based on specific individuals distributed within the space.

Spatially explicit modeling approaches such as cellular automata and agent-based models are highly suited to the exploration and explanation of landscape dynamics and how landscape patterns form and evolve through interactions with heterogeneous places and actors, both human and nonhuman. In CA models, the rules of the cellular automaton replicate transition functions, emergence occurs in generated systems, and patterns may be persistent with changing components. In ABMs, agents consist of autonomous decision-making entities that have unique characteristics, are represented by unique algorithms, an environment through which agents interact with each other, and rules that define the relationships between agents and their environments. The CA and ABM models allow us to develop candidate explanations for specific landscape patterns, spatially simulate landscape patterns over space and time, examine likely future scenarios of change, and examine endogenous factors and exogenous shocks that can alter trajectories of landscape change, resulting in possible shifts in the composition and spatial structure of the landscape.

By coupling these models with various ways of describing spatial patterns, including standard metrics of landscape pattern that describe local, global, and cross-scale characteristics, a number of projects have demonstrated an ability to pose candidate explanations for the dynamics that give rise to various patterns. In addition, the dynamic nature of landscape patterns can be evaluated through models and experiments in ways that are much less difficult and expensive than through direct observation. Model patterns can be compared through controlled experiments with observed patterns mapped through multitemporal remote sensing and other field-based mapping approaches. The challenge in such efforts is to develop methods and data sets that can be used to further refine explanations and validate microlevel processes in the models through field and other empirical investigations. The models, therefore, serve additional important roles in helping to identify critical field data that are needed to represent processes, characterize patterns, and validate findings.

25.5 AGENT-BASED MODELS

ABMs model the activities of individual agents (e.g., individuals or households) as basic building blocks. An ABM may have multiple copies of the same type of agent, for example, multiple copies of one type of plant or human actor, or multiple copies of multiple agents of different types (e.g. households, individuals, and government agencies). Agents differ in important characteristics. Their interactions may be dynamic, in that the characteristics of the agents change over time as the agents adapt to their environments, learn from experiences through feedbacks, or 'die' as they fail to alter behavior relative to new conditions and/or factors. The dynamics that describe how systems change are generally nonlinear. They are sometimes even chaotic but are seldom in any long-term equilibrium. However, for systems that do reach equilibrium, the mechanisms that lead to such a condition are of central interest. Agents may be organized into groups of individuals or into nested hierarchies that may influence how the underlying system evolves over time (Bak, 1998). They are emergent and self-organizing in that macrolevel behaviors emerge from the actions of individual agents as agents learn through

experiences and change and develop feedbacks with finer-scale building blocks.

In the context of understanding landscape change, for instance, LULC change, agents can include landowners, farmers, management agencies, and/or policy-making bodies, all of whom make decisions or take actions that affect land-cover patterns and processes (Parker et al., 2003). By simulating the individual actions of many diverse actors, and measuring the resulting system behavior and outcomes over time (e.g., the changes in patterns of land cover), ABMs can be useful tools for studying the effects on processes that operate at multiple scales, organizational levels, and their effects.

ABMs belong to a category of models known as discrete event simulations, which run with some set of starting conditions over some period of time, allowing the programmed agents to carry out their actions until some specified stopping criterion is satisfied, usually indicated by either a certain amount of time or a specified system state (cf. Brown, 2006). The actions of agents can be scheduled to take place synchronously (i.e., every agent performs actions at each discrete time-step) or asynchronously (i.e., agent actions are scheduled with reference to a clock or to the actions of other agents). The behaviors of agents can vary from being completely reactive, for instance, agents only perform actions when triggered to do so by some external stimulus (e.g., actions of another agent) to being goal-directed (e.g., through seeking a particular goal). For example, a farmer as an 'agent' could be programmed to plant corn every spring (i.e., a relatively reactive agent), choose whether to plant and which crop to maximize return on investment, and plant at the time that is expected to produce the highest yield (i.e., a goal-directed agent).

ABMs have a number of strengths that contrast with traditional methods for modeling landscape change (cf. Brown, 2006). In addition to the richer behavioral representations afforded by ABMs, because an ABM is a dynamical system, it can incorporate positive and negative feedbacks, such that the behavior of an agent has an influence on the subsequent behavior of other agents. These feedbacks can be used to represent the endogeneity of various driving forces of landscape change (Walsh, 2007).

ABMs are philosophically armed for the development of a model-centered science (Henrickson and McKelvey, 2002). Experimentation is the key to understanding similar and different environments, environments shaped by similar and different processes, and environments influenced by the context of place where the synthesis or distillation of pattern-process relations across sites may be difficult to discern. By taking into account their commonalities and differences in structure, function, and evolution across time and space, ABMs and experimentation offer the possibility of considerable insights into system dynamics and behaviors. Agent-based experiments provide flexibility and considerable analytical power to examine pattern and process relations, including policy issues. For instance, current efforts to model LULC change through ABMs involve the use of single and multiple agents, rules, and environments to reproduce stakeholders and interactions in certain study area conditions (Lansing, 2002).

An agent-based approach to biocomplexity often uses data libraries already compiled for study site generated through fieldwork, interpretation of aerial photography, and/or satellite imagery, or by framing the study in established theories to describe system behaviors and key variables and pathways involved. Agents are differentiated by key characteristics and a spatially explicit landscape grid is used where the actions of agents affect the conditions of cells. Feedbacks between the value of a cell and its spatial conditions (e.g., location relative to other changes and features, such as roads and communities) determine the choices made by agents (Silveira et al. 2006). The movement and interaction of agents of differing types and characteristics respond to their 'home' landscape, in turn, altering the probabilities of change.

25.6 CELLULAR AUTOMATA

CA models belong to a family of discrete, connectionist techniques used to investigate fundamental principles of dynamics, evolution, and self-organization (White et al., 1997). CA models are examples of mathematical systems constructed from many simple identical components that together are capable of complex behavior. CA approaches can be used to develop specific models for particular systems and to abstract general principles applicable to a wide variety of complex systems (Wolfram, 1984). CA models do not describe a complex system with complex equations (e.g., differential equations, multilevel statistical modeling) but allow the complexity to emerge from interactions of basic building blocks of systems (e.g., individuals and households represented at the cell level) that follow simple rules. Complexity theory concepts and hierarchical relationships are infused into the CA models for generating simulations to match observed states or for future periods by allowing the model to iterate within the expected bounds of the defined rules.

The essential properties of CA are a regular n-dimensional lattice where each cell of the lattice has a discrete state, and a dynamical behavior described through growth or transition rules. These rules describe the state of a cell for the next time-step, depending on the states of the cells in the defined spatial neighborhood (Wolfram, 1984). The essential components of a CA model are (1) the cell—the basic element of CA that is capable of storing defined states; (2) the lattice, or cells arranged in a spatial matrix; and (3) neighborhoods defined by growth or transition rules that perform changes to the state of the cells depending on neighboring cells and their conditions. Four classes of behavior are recognized in CA models: fixed, periodic, chaotic, and complex (Wolfram, 1984). Lambda (λ) is often used to relate the nature of the rules to the overall behavior of CA models (Langton, 1991).

25.7 APPROACHES TO MODEL VALIDATION

The true measure of spatial complexity as applied in a complex model context is one not yet fully realized in the literature (Messina and Walsh, 2001). One of the challenges is to use spatial simulations in general and complexity-based methods, such as CA and ABMs in particular, in answering the question of 'what is a good fit?' when spatial simulations are developed for antecedent and future time periods. Unanswered questions about the effects of the ecological fallacy and the modifiable areal unit problem can influence model outcomes by effecting the apparent strength and magnitude of relationships between variables (Bian, 1997; Rindfuss et al., 2004). Beyond the composition and pattern of model outcomes is the need to understand complex processes and their characteristics. 'Complexity as property' defines a system as complex if it exhibits certain characteristics of complexity, such as fractal dimension or scale invariance (Manson, 2003a). Often, the outputs of model runs are compared (i.e., expected versus observed) by using satellite observations to represent actual conditions. In essence, we are concerned about our ability to replicate observed spatial and compositional patterns, and hence trend lines and pattern metrics are used to assess the certainty or plausibility of model outcomes. However, other approaches to assess model performance are under development (Manson, 2003b), including the study of pattern-invariant areas in simulations (Brown et al., 2005) and emergent patterns and the creation of development fronts through the actions of individuals or some set of base actors on the landscape (Malanson et al., 2006a, 2006b).

Therefore, a challenge facing the biocomplexity community is the direct analysis of the complexity of system dynamics, and the ability to quantify key indicators of self-organization (e.g., power law slopes) and system trajectories in state space. Lyapunov analysis (Weisstein, 2004a) can be used to compute system dynamics, the divergence or convergence of a system, and Kolmogorov entropy analysis (Weisstein, 2004b) can be used to compute information loss. State space axes (e.g., measures of aggregation, the power-law slope of patch sizes, road network measures), temporal metrics (e.g., population growth rates and LULC rates), economic indicators (e.g., percentage of farm households engaged in off-farm employment), and LULC change trajectories (e.g., forest to nonforest pixel histories) can be used to assess system dynamics. Analyses involving biocomplexity often examine the nature of the state space that models create and the trajectories within it. The most general hypotheses are that the system is sensitive to initial conditions and has bifurcations and/or critical thresholds. Complexity theory has alternative approaches, and it allows us to consider long-term system trajectories that may guide future hypothesis generation. Scaling properties can be assessed through Fourier analysis, wavelet analysis, and renormalization. Sensitivity and uncertainty analyses (e.g., Hammonds et al., 1994), considered computational experiments in ABMs (cf. Kydland and Prescott, 1996), can be used to assess pattern-process relations across space-time scale. Differences among treatments, as well as settings of parameters or inclusion of processes, can be tested for replicate runs of the simulations. Experimental treatments/sensitivity analyses include variation in model assumptions, as well as scenarios varying exogenous factors, and for the variability seen in our preliminary analyses.

25.8 CASE STUDIES: BIOCOMPLEXITY IN COUPLED HUMAN-NATURAL SYSTEMS

25.8.1 Deforestation in a frontier environment of northeastern Ecuador

A complexity theory analysis of LULC change at frontier settings aims at understanding feedbacks and changes in state space through nonlinearities, in relation to a dynamic and coupled human-natural system. As seen in frontier environments,

LULC change patterns are not random but are self-organized around development fronts that are shaped by geographic accessibility into the region and the constraints of resource endowments (Messina and Walsh, 2001, 2005; Malanson et al., 2006a).

The northern Ecuadorian Amazon is a region of drastic change and constant adaptation (Figure 25.1). Complex interactive socioeconomic, demographic, and biotic processes are developing over relatively fine and coarse temporal and spatial scales, which make this region an optimal laboratory to study feedbacks, adaptation, and nonlinear relationships between social and natural systems (Walsh et al., 2002). This region covers approximately 20,000 km^2 and encompasses exceptional biological and cultural diversity. It has been characterized as a region of very high biodiversity (Myers, 1990; Pitman, 2003). Cultural diversity is also high—the northern Ecuadorian Amazon is home to many indigenous communities from different ethnic groups that have adapted to the Amazonian environment over hundreds of years. Even with the rapid and extensive land-cover changes that have occurred, the region conserves approximately 60% of pristine natural formations: montane and cloud forests, lowland *terra firme* forest, forested wetlands, and black- and whitewater riparian environments. The discovery of oil in 1964 triggered infrastructure development and spontaneous agricultural colonization. More oil infrastructure development is expected due to the large petroleum reserves located in this region.

The tropical rain forest of northeastern Ecuador is an area of complex interactions among a number of important and diverse stakeholders— (a) spontaneous colonists who have in-migrated from other regions of the country and settled on household farms; (b) newly emerging communities and market centers that have consolidated services, offer off-farm employment to colonists, and affect LULC through direct and indirect ways; (c) indigenous people who follow traditional practices but are affected by the rise of commercial agriculture, oil production within their historical territories, and a transition to a consumer-based economy; (d) oil exploiters who have built roads and laid pipelines for petroleum extraction in colonist and indigenous areas; and (e) conservation and protected areas established by the government to impede development and retain biodiversity in a rapidly transforming frontier environment (Messina et al., 2006). The greatest changes on the land are those created by agricultural colonists following in the wake of oil exploration (see Plate 25.1), who gained access on roads that made isolated areas accessible for development (Walsh et al., 2003).

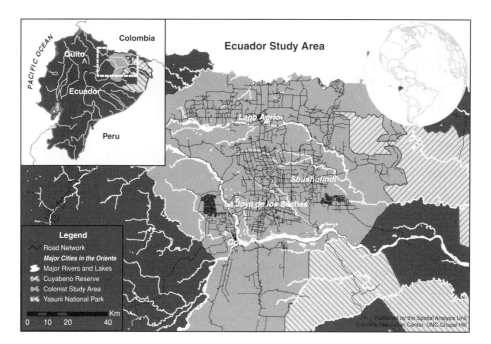

Figure 25.1 The northern Ecuadorian Amazon is a biodiversity 'hotspot' and a region that is being considerably impacted by population in-migration, colonization of a frontier environment, significant land use/land cover change, and the feedbacks between people and environment

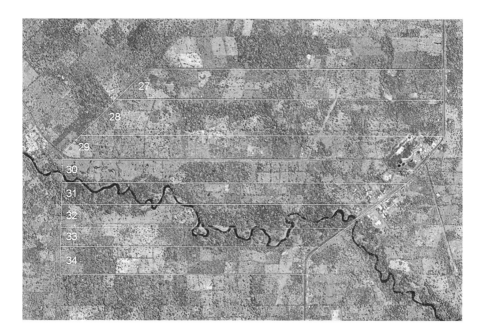

Plate 25.1 An IKONOS, panchromatic satellite image (acquired in 1999) that shows a group of household farms, or *fincas* (measuring approximately 250 x 2,000 meters, or 50 hectares), and land use/land cover patterns that exhibit deforestation and agricultural extensification in the northern Ecuadorian Amazon

However, interactions among the groups and the regions that they've settled are complex, because feedbacks between spatial patterns and rates of change are known to occur at advancing fronts of settlement and land development that have implications for LULC patterns.

The preliminary simulations developed for the northern Ecuadorian Amazon by Messina and Walsh (2001, 2005) using CA have thus far suggested a more homogenous landscape with time, a scenario that fits the theoretical understanding of how in-migration of farmers into existing farms through resale, subdivision of farms to those engaged primarily in the burgeoning service sector, and the establishment of new development sectors alters the natural landscape through deforestation and agricultural extensification (Nepstad et al., 2001). In subsequent models, we are now including additional processes, for instance, that represent social (e.g., labor supply and off-farm employment), demographic (e.g., population density and household income), biophysical (e.g., terrain settings and site suitability for agriculture), and geographical (e.g., spatial linkages between farms and communities and geographic accessibility) factors. Further, the intent is to generate the tools, methods, and protocols for generating CA, as well as ABMs for assessing spatial simulations of LULC change that can be implemented within a planning and policy environment so that 'what if' scenarios can be considered and the social, economic, and ecological implications of landscape change examined. Walsh et al. (2008b) explore the effects of an increase in the income of farm households on land-use patterns, achieved through improvements in the geographic accessibility of farmers to roads and market and service communities and increased off-farm employment as an alternative household livelihood strategy.

A stylized or simulated modelling environment, derived characteristics of agents, and a grid of cells can be used to examine various scenarios of LULC change that include issues of land fragmentation through land sales and household kinship ties, as well as the impact of protected areas and land tenure (Messina et al., 2005) and indigenous communities and other stakeholder groups (Mena et al., 2006) on LULC change patterns and trajectories of change.

In sum, our research in the northern Ecuadorian Amazon seeks to understand complexity in a human settlement frontier by considering the role of positive feedbacks in the spatial pattern of LULC change (Figure 25.2). We integrate exogenous and endogenous drivers to represent a complex and diverse set of forces and factors operating in the region, which together affect LULC change patterns in fundamental ways. Alternative approaches in self-organized complexity, including self-organized percolation, and the inverse cascade model, and an approach to complexity involving optimization, highly optimized tolerance, are considered relative to population-environment interactions at settlement frontiers in the Ecuadorian Amazon rain forest.

25.8.2 Urban-rural land-use change in southeastern Michigan

The complex dynamics giving rise to patterns of development at the urban-rural fringe were the subject of a recent study, called Project SLUCE, set in southeastern Michigan, in the United States, and which was focused on the city of Detroit (Figure 25.3). The landscapes of urban-rural fringe areas are relatively complex in terms of both composition and configuration (Plate 25.2). In this particular environment, rural landscapes are dominated by cropped agricultural land interspersed with patches of forest and wetland. Distributed development grades into densely settled suburban landscapes that are composed of the impervious surfaces of streets, driveways and buildings, urban forests, and intensely managed grasslands.

We can conceive of and model the landscape patterns arising from land development in this environment as a product of two fundamental processes involving the interactions between people and the environment. First, development decisions are made by individuals selecting where to live, a decision that is mediated by professional developers acting within a framework of regulatory environments. The residential location decision is affected by a range of factors (Fernandez et al., 2005), but they include aesthetic characteristics of the landscape that create a feedback between residential development, which affects the environment, and the effect of the environment on the development process. The second major interaction process is in the way people decide to manage land cover on the land they own. This is especially important in determining landscape patterns in exurban areas, where lots tend to be fairly large (e.g., 1 to 10 acres). These decisions are understood less well but also involve interactions between the environment, social networks, and landowner decisions.

A simple ABM was created to represent the residential location decision using the *Swarm* software libraries in Objective-C (Brown et al., 2004, 2005). In the model, residents (i.e., agents) decide where to locate on the basis of such factors as nearness to service centers, intrinsic aesthetic quality of the landscape, and population density.

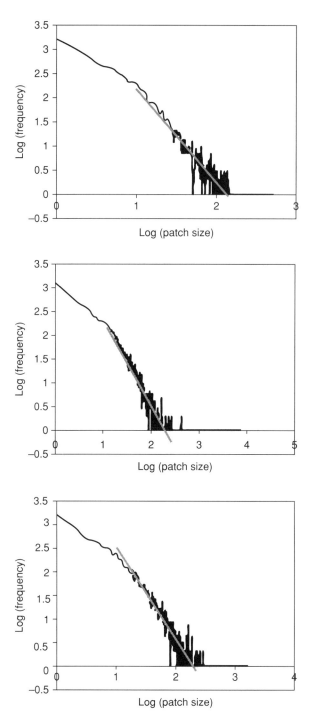

Figure 25.2 Frequency distribution of land use/land cover patch sizes in three 90,000-ha intensive study areas in the northern Ecuadorian Amazon during 1999–2002; the black line is the observed frequency of patch sizes; the gray line is the regression line for observed frequency of patch size (after Malanson et al., 2006a)

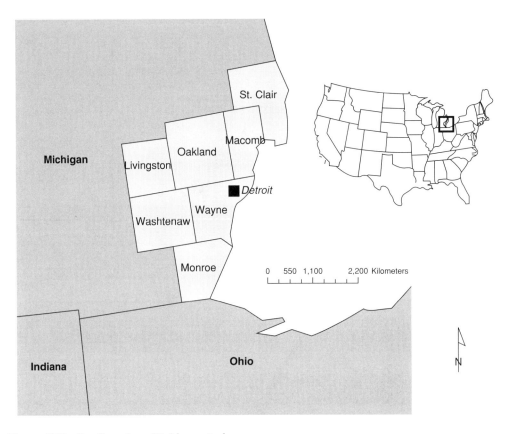

Figure 25.3   Southeastern Michigan study area

Service centers are located near residents. Residents evaluate a limited number of cells within an area (sampled to represent the incomplete information available as residents make decisions) and locate in the cell that maximizes their utility, which is calculated by evaluating the factors with agent-specific factor weights. Because it is agent-based, the model represents heterogeneity among locations and among agents, and interactions and feedbacks between agent decisions and the environment. As a result, the behavior of the model (and presumably the system it represents) exhibits significant nonlinear dynamics, including path dependence that makes location-specific predictions difficult (Brown et al., 2005). In addition, the model demonstrated significant sensitivity to heterogeneity in agent preferences (Rand et al., 2002), largely because of the influence of agents with extremely high or low values of preference for one or more factors. This sensitivity argues for the use of ABMs, because they can represent heterogeneity, and requires that we understand the heterogeneity of resident preferences within the population, which we are

doing through social surveys (Fernandez et al., 2005). The model performed well in comparison with an analytical model of the influence of a greenbelt on land-development patterns but was much more easily extended to two dimensions: a heterogeneous environment and heterogeneous agents (Brown et al., 2004).

An additional model was developed using the *RePast* software libraries in Java, in order to add agent types to represent developers and township policy-making bodies, and resident decisions about land cover (Brown et al., 2008). This model relies on the identification of three different types of subdivisions that differ in terms of both characteristics that residents evaluate (cost and aesthetic quality) and the landscape patterns they produce. Additionally, townships can restrict various kinds of subdivisions, through the use of lot-size restrictions, and/or preserve land from development (e.g., through fee-simple purchase or purchase of development rights). We evaluated three different strategies for locating preserved lands: (a) random—farms were purchased without regard to landscape characteristics; (b) farms with the most

A

B

Plate 25.2   Low-altitude, oblique aerial photographs of two different residential developments in southeastern Michigan, near Ann Arbor

tree cover (i.e., a preservation strategy); and (3) farms with the least tree cover (i.e., a restoration strategy). We assumed that trees would eventually (after 10 time-steps) cover all lands set aside for preservation. Based on data on the effects of subdivision types of tree cover, we assumed that remnant subdivisions increased tree cover by 20% and the other types either resulted in the removal of all trees or no change in tree cover.

Despite the early stages of development and analysis of this particular model, the results from multiple model runs under the three different strategies of locating lands for preservation, and with and without remnant subdivisions available as an option (Figure 25.4), provide useful information about the processes that produce habitat on exurban landscapes. For example, the land that is developed can be as important in determining the resulting landscape patterns as land that is preserved. In particular, remnant subdivisions, that is, those that incorporate natural features into their designs, can produce positive effects on habitat availability, even in the presence of development. In addition, restoration of tree cover to previously nonforested locations may be a more fruitful strategy for improving overall habitat availability.

25.9 CONCLUSIONS

Grand Challenges in Environmental Sciences (National Research Council, 2000) identified an

environmental research agenda for the next decade that has important implications for science, technology, and policy. Central to the recommendations of the council were the interactions of people, places, and environment and the space-time linkages to landscape dynamics. This 'challenge' was built on several international- and U.S.-based initiatives that defined the role of landscape change first within global environmental change science (International Geosphere-Biosphere Program and the International Human Dimensions Program [IGBP-IHDP, 1995, 1999]), NASA's Land Cover/Land Use Change Program, and the Biocomplexity Program, the Human and Social Dynamics Program and now the Coupled Natural Human Systems Program of the U.S. National Science Foundation) and through research programs that subsequently developed in biodiversity and ecosystem initiatives. Through these programs and within the broader scientific community, it is increasingly being recognized that human behavior and agency and the feedbacks between human population and the environment are among the most critical drivers of ecosystem change. As a result, various branches of science and policy are requesting realistic models of ecosystem functions and services at multiple and interacting spatial and temporal scales (Dale et al., 1994; Mertens and Lambin, 1997). Biocomplexity research, using ABMs, CA, and spatial simulation models, figures prominently in the study of coupled human-natural systems and the examinations of the causes and

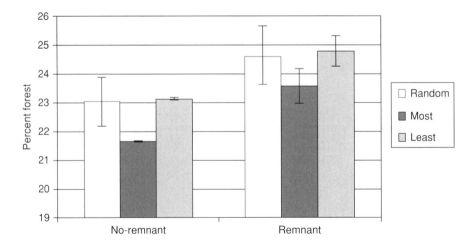

Figure 25.4   Differences in the amount of forest on the landscape: An agent-based model of a township with and without remnant subdivisions was run 100 times, using three different strategies for locating preserved lands on 10% of the landscape. Bars indicate means, and error bars indicate standard deviations across 30 model runs

consequences of ecosystem change. Realistic scenarios of land development, land-change trajectories, and exogenous and endogenous factors affecting ecosystems are under study in an array of environments, often being interpreted within a policy-relevant context and generally involving a number of interacting stakeholder groups. Key feedbacks in these models relate to the pattern and spatial structure of how population and environment interact.

Applying complexity theory to the study of LULC dynamics in coupled human-natural systems suggests that while explaining or predicting pattern-process relations, for instance, the deforestation of a particular parcel of land in the northern Ecuadorian Amazon or landscape changes on the rural-urban fringe of Detroit, Michigan, may be difficult unless the dynamics and the spatial patterns are examined at higher orders of organization, because emergence is a key factor in biocomplexity. It is at these higher orders where our knowledge can build and where policy derived from complexity science should focus. Spatial simulation models that integrate a dynamic landscape, agents of different characteristics who engage the environment and each other in various manners, and who behave in adaptive (or static) ways can add considerable power to the study of ecological systems and biogeography. Models generated through agent-based and CA approaches should be used to define and explain patterns and to relate them to processes. Thresholds and feedback mechanisms are key considerations, as are endogenous factors and exogenous shocks that combine in complex ways to perturb systems and to alter future trajectories. A challenge is to develop and encode rules and pattern-process relations in our biocomplexity models that address fundamental issues of ecosystem functions and services, as well as scenarios of landscape change that are framed within a policy-relevant context. Further, our models must be validated with particular attention given to processes and their relation to observed and modeled patterns. Complexity offers a theoretical framework to the study of the interactions among people, places, and environment. Comparisons made across settings, systems, scales, and processes should be embodied in biocomplexity and how it relates to coupled human-natural systems, landscape dynamics, and biogeography more broadly.

REFERENCES

Alencar, A.M., Andrade, J.S., and Lucena, L.S. (1997) 'Self-organized percolation', *Physical Review E*, 56: R2379–82.

Allen, P.M., and Sanglier, M. (1979) 'Dynamic model of growth in a central place system', *Geographic Analysis*, 11: 256–72.

An, L., Linderman, M., Qi, J., Shortridge, A., and Liu, J. (2005) 'Exploring complexity in a human-environment systems: An agent-based spatial model for multidisciplinary and multiscale integration', *Annals of the Association of American Geographers*, 95(1): 54–79.

Badalamenti, F., D'Anna, G., Pinnegar, J.K., and Polunin, N.V.C. (2002) 'Size-related tropho-dynamic changes in three target fish species recovering from intensive trawling', *Marine Biology*, 141: 561–570.

Bak, P. (1998) *How Nature Works*. New York: Copernicus/Springer-Verlag.

Bak, P., Tang, C., and Wiesenfeld, K. (1988) 'Self-organized criticality', *Physical Review A*, 38: 364–74.

Batty, M., Couclelis, H., and Eichen, M. (1997) 'Urban systems as cellular automata', *Environment & Planning B*, 24: 159–64.

Batty, M., Xie, Y., and Sun, Z. (1999) 'Modeling urban dynamics through GIS-based cellular automata', *Computers, Environment and Urban Systems*, 23(3): 205–33.

Benenson, I. (1999) 'Modeling population dynamics in the city: From a regional to a multi-agent approach', *Discrete Dynamics in Nature and Society*, 3: 149–70.

Berger, T. (2001) 'Agent-based spatial models applied to agriculture: A simulation tool for technology diffusion, resource use changes and policy analysis', *Agricultural Economics*, 25(2–3): 245–60.

Bian, L. (1997) 'Multiscale nature of spatial data in scaling up environmental models', in D.A. Quattrochi and M.F. Goodchild (eds.), *Scale in Remote Sensing and GIS*. New York: Lewis Publishers. pp. 13–26.

Blackman, T. (2000) 'Complexity theory', in G. Browning, A. Halcli, and F. Webster (eds.), *Understanding Contemporary Society: Theories of the Present*. London: SAGE. pp. 139–51.

Bousquet, F., Le Page C., Bakam I., and Takforyan A. (2001) 'Multi-agent simulations of hunting wild meat in a village in eastern Cameroon', *Ecological Modelling*, 138: 331–46.

Brown, D.G. (2006) 'Agent-based models', in H. Geist (ed.), *The Earth's Changing Land: An Encyclopedia of Land-Use and Land-Cover Change*. Westport, CT: Greenwood. pp. 7–13.

Brown, D.G., Page, S.E., Riolo, R.L., and Rand, W. (2004) 'Agent based and analytical modeling to evaluate the effectiveness of greenbelts', *Environmental Modelling and Software*, 19(12): 1097–109.

Brown, D.G., Page, S.E., Riolo, R.L., Zellner, M., and Rand, W. (2005) 'Path dependence and the validation of agent-based spatial models of land-use', *International Journal of Geographical Information Science*, 19(2): 153–74.

Brown, D.G., Robinson, D.T., An, L., Nassauer, J.I., Zellner, M., Rand, W., Riolo, R., Page, S.E., and Low, B. (2008) 'Exurbia from the bottom-up: Confronting empirical challenges to characterizing complex systems', *GeoForum*, 39(2): 805–18.

Brown, J.H., Gupta, V.K., Li, B., Milne, B.T., Restrepo, C., and West, G.B. (2002) 'The fractal nature of nature: Power laws, ecological complexity, and biodiversity', *Philosophical Transactions of the Royal Society of London Series B*, 357: 619–26.

Buckley, Y.M., Briese, D.T., and Rees, M. (2003) 'Demography and management of the invasive plant species Hypericum perforatum II. Construction and use of an individual-based model to predict population dynamics and the effects of management strategies', *Journal of Applied Ecology*, 40(3): 494–507.

Carlson, J.M., and Doyle J. (1999) 'Highly optimized tolerance: A mechanism for power laws in designed systems', *Physical Review*, 60: 1412–27.

Carlson, J.M., and Doyle, J. (2000) 'Highly optimized tolerance: Robustness and design in complex systems', *Physical Review Letters*, 84: 2529–32.

Cilliers, P. (1998) *Complexity and Postmodernism*. New York: Routledge.

Clarke, K.C., Gaydos, L., and Hoppen, S. (1997) 'A self-modifying cellular automaton model of historical urbanization in the San Francisco Bay area', *Environment and Planning B*, 23: 247–61.

Clarke, K.C., Hoppen, S., and Gaydos, L. (1996) 'Methods and techniques for rigorous calibration of a cellular automation model of urban growth', *Third International Conference & Workshop on Integrating GIS and Environmental Modeling*. Santa Barbara: National Center for Geographic Information and Analysis, Santa Fe, New Mexico. pp. 21–26.

Crawford, T.W., Messina, J.P., Manson, S.M., and O'Sullivan, D. (2005) 'Complexity science, complexity systems and land use research', *Environment and Planning B*, 32: 857–75.

Dale, V.H., O'Neill, R.V., Southworth F., and Pedlowski M. (1994) 'Modeling effects of land management in the Brazilian Amazonian settlement of Rondonia', *Conservation Biology*, 8: 196–206.

Deadman, P., Robinson, D., Moran, E., and Brondizio, E. (2004) 'Colonist household decision-making and land use change in the Amazon rainforest: An agent-based simulation', *Environment and Planning B*, 31: 693–709.

DeAngelis D.L., and Gross, L.J. (eds) (1992) *Individual-Based Models and Approaches in Ecology: Populations, Communities and Ecosystems*. New York: Chapman and Hall.

Desrochers, R.E., and Anand, M. (2005) 'Quantifying the components of biocomplexity along ecological perturbation gradients', *Biodiversity and Conservation*, 14(14): 3437–55.

Entwisle, B., Rindfuss, R.R., Walsh, S.J., and Page, P.H. (2008) 'Modeling population as a factor in the deforestation of Nang Rong, Thailand', *GeoForum*, 39(2): 879–97.

Ermentrout, G.B., and Edelstein-Keshet, L. (1993) 'Cellular automata approaches to biological modelling', *Journal of Theoretical Biology*, 160: 97–133.

Evans, T.P., and Kelley, H. (2004) 'Multi-scale analysis of a household level agent-based model of landcover change', *Environmental Management*, 72 (1–2): 57–72.

Farina, A., and Belgrano, A. (2004) 'The eco-field: A new paradigm for landscape ecology', *Ecological Research*, 19(1): 107–10.

Fernandez, L., Brown, D.G., Marans, R., and Nassauer, J. (2005) 'Characterizing location preferences in an exurban population: Implications for agent based modeling', *Environment and Planning B*, 32(6): 799–820.

Freeman, W., Kozma, R., and Werbos, P. (2001) 'Biocomplexity: Adaptive behavior in complex stochastic dynamical systems', *Biosystems*, 59: 109–23.

Gell-Mann, M. (1994) *The Quark and the Jaguar*. New York: Freeman.

Ginot, V., Le Page, C., and Souissi, S. (2002) 'A multi-agents architecture to enhance end-user individual based modeling', *Ecological Modelling*, 157: 23–41.

Green, J.L. and 10 authors (2005) 'Complexity in ecology and conservation: Mathematical, statistical, and computational challenges', *Bioscience*, 55(6): 501–10.

Grimm, V., and Railsback, S.F. (2005) *Individual-Based Modeling and Ecology*. Princeton, NJ: Princeton University Press.

Grimm, V., Revilla, E., Berger, U., Jeltsch, F., Mooij, W.M., Railsback, S.F., Thulke, H.H., Weiner, J., Wiegand, T., and DeAngelis, D.L. (2005) 'Pattern-oriented modeling of agent-based complex systems: Lessons from ecology', *Science*, 310: 987–91.

Haag, G. (1994) 'The rank-size distribution of settlements as a dynamic multifractal phenomenon', *Chaos, Solutions and Fractals*, 4: 519–34.

Henrickson, L., and McKelvey, B. (2002) 'Foundations of 'new' social science: institutional legitimacy from philosophy, complexity science, postmodernism, and agent-based modelling', *Proceedings of the National Academy of the United States of America*, 99(3): 7288–95.

Holland, J.H. (1995) *Hidden Order: How Adaptation Builds Complexity*. Reading, MA: Addison-Wesley.

Holland, J.H. (1998) *Emergence: From Chaos to Order*. Reading, MA: Helix.

Holling, C.S., Gunderson, L.H., and Ludwig, D. (2002) 'In search of a theory of adaptive change', in L.H. Gunderson and C.S. Holling (eds.), *Panarchy: Understanding Transformations in Human and Natural Systems*. Washington, DC: Island Press. pp 3–24.

Hughes, B.D. (1995) *Random Walks and Random Environments, Volume 2: Random Environments*. Oxford: Clarendon.

Jaylson, J., Silveira, A., Espíndola, L., Penna, T.J.P. (2006) 'Agent-based model to rural-urban migration analysis', *Physica A: Statistical Mechanics and its Applications*, 364: 445–456.

Jenerette, G.D., and Wu, J.G. (2001) 'Analysis and simulation of land-use change in the central Arizona-Phoenix region, USA', *Landscape Ecology*, 16: 611–26.

Judson, O.P. (1994) 'The rise of the individual-based model in ecology', *Trends in Ecology and Evolution*, 9: 9–14.

Kydland, F.E., and Prescott E.C. (1996) 'The computational experiment: An econometric tool', *Journal of Economic Perspectives*, 10: 69–85.

Lambin, E.F., Geist H.J., and Lepers E. (2003) 'Dynamics of land-use and land-cover change in tropical regions', *Annual Review of Environment and Resources*, 28: 205–41.

Langton, C. (1991) *Computation at the Edge of Chaos: Phase Transition and Emergent Computation.* Cambridge, MA: MIT Press.

Lansing, J.S. (2002) 'Artificial societies and the social sciences', *Artificial Life*, 8(3): 279–92.

Luhman, N. (1985) *A Sociological Theory of Law.* London: Routledge and Kegan Paul.

Malanson, G.P. (1999). 'Considering complexity', *Annals of the Association of American Geographers*, 89(4): 746–53.

Malanson, G.P. (2003) 'Habitats, hierarchical scales, and nonlinearities: An ecological perspective on linking household and remotely sensed data on land-use/cover change', in J. Fox, R.R. Rindfuss, S.J. Walsh, and V. Mishra (eds.), *People and the Environment: Approaches for Linking Household and Community Surveys to Remote Sensing and GIS.* Dordrecht: Kluwer. pp. 265–83.

Malanson, G.P., Zeng, Y., and Walsh, S.J. (2006a) 'Complexity at advancing ecotones and frontiers', *Environment and Planning A*, 38: 619–32.

Malanson, G.P., Zeng, Y., and Walsh, S.J. (2006b) 'Landscape frontiers, geography frontiers: Lessons to be learned', *Professional Geographer*, 58(4): 383–96.

Manson, S.M. (2001) 'Simplifying complexity: A review of complexity theory', *Geoforum*, 32(3): 405–14.

Manson, S.M. (2003a) 'Epistemological possibilities and imperatives of complexity research: A reply to Reitsma', *GeoForum*, 34(1): 17–20.

Manson, S.M. (2003b) 'Validation and verification of multi-agent models for ecosystem management', in M. Janssen (ed.), *Complexity and Ecosystem Management: The Theory and Practice of Multi-Agent Approaches.* Northampton, MA: Edward Elgar. pp. 63–74.

Matthews, K.B., Subaald, A.R., and Craw, S. (1999) 'Implementation of a spatial decision support system for rural land use planning: Integrating geographic information systems and environmental models with search and optimization algorithms', *Computers and Electronics in Agriculture*, 23: 9–26.

Meentemeyer, V. (1989) 'Geographical perspectives of space, time, and scale', *Landscape Ecology*, 3: 163–73.

Mena, C.F., Barbieri, A., Walsh, S.J., Erlien, C.M., Holt, F.L., and Bilsborrow, R.E. (2006) 'Pressure on the Cuyabeno wildlife reserve: Development and land use/cover change in the northern Ecuadorian Amazon', *World Development*, 34(10): 1831–49.

Mertens, B. and Lambin, E.F. (1997) 'Spatial modelling of deforestation in southern Cameroon: spatial disaggregation of diverse deforestation processes', *Applied Geography*, 17(2): 143–162.

Messina, J.P., and Walsh, S.J. (2001) '2.5D morphogenesis: Modeling landuse and landcover dynamics in the Ecuadorian Amazon', *Plant Ecology*, 156(1): 75–88.

Messina, J.P., and Walsh, S.J. (2005) 'Dynamic spatial simulation modeling of the population-environment matrix in the Ecuadorian Amazon', *Environment and Planning B*, 32(6): 835–56.

Messina, J.P., Walsh, S.J., Mena, C.F. and Delamater, P.L. (2006) 'Land tenure and deforestation patterns in the Ecuadorian Amazon: Conflicts in land conservation in a frontier setting', *Applied Geography*, 26: 113–28.

Michener, W.K., Baerwald, T.J., Firth, P., Palmer, M.A., Rosenberger, J., Sandlin, E.A., and Zimmerman, H. (2003) 'Defining and unraveling biocomplexity', *Bioscience*, 51(12): 1018–23.

Myers, N. (1990) 'The biodiversity challenge: Expanded hot-spots analysis', *The Environmentalist*, 10(4): 243–56.

National Research Council. (2000) *Grand Challenges in Environmental Sciences.* Washington, DC: National Academy Press.

Nepstad, D., Carvalho, G., Barros, A.C., Alencar, A., Capobianco, J.P., Bishop, J., Moutinho, P., Lefebvre, P., Silva, U.L., and Prins, E. (2001) 'Road paving, fire regime feedbacks, and the future of Amazon forests', *Forest Ecology and Management*, 154: 395–407.

Nute, D. (2004) 'NED-2: An agent-based decision support system for forest ecosystem management', *Environmental Modelling and Software*, 19(9): 831–41.

Olson, R.L., and Sequeira, R.A. (1995) 'An emergent computational approach to the study of ecosystem dynamics', *Ecological Modelling*, 79: 95–120.

Park, S., and Sugumaran, V. (2005) 'Designing multi-agent systems: A framework and application', *Expert Systems with Applications*, 28: 259–71.

Parker, D.S., Manson, S.M., Janssen, M., Hoffmann, M., and Deadman, P. (2003) 'Multi-agent systems for the simulation of land use and land cover change: A review', *Annals of the Association of American Geographers*, 93(2): 314–37.

Parker, D., and Meretsky, V. (2004) 'Measuring pattern outcomes in an agent-based model of edge-effect externalities using spatial metrics', *Agriculture, Ecosystems, and Environment*, 101: 233–50.

Pausas, J.G. (2003) 'The effect of landscape pattern on Mediterranean vegetation dynamics: A modelling approach using functional types', *Journal of Vegetation Science*, 14(3): 365–74.

Phillips, J.D. (1999) 'Methodology, scale, and the field of dreams', *Annals of the Association of American Geographers*, 89: 754–60.

Pitman, N.C. (2003) 'A comparison of tree species diversity in two upper Amazonian forests', *Ecology*, 83(11): 3210–24.

Prévosto, B., Hill, D.R.C., and Coquillard, P. (2003) 'Individual-based modelling of Pinus sylvestris invasion after grazing abandonment in the French Massif', *Plant Ecology*, 168(1): 121–38.

Rand, W., Zellner, M., Page, S.E., Riolo, R., Brown, D.G., and Fernandez, L.E. (2002) 'The complex interaction of agents and environments: An example in urban sprawl', in C. Macal and D. Saacj (eds.), *Proceedings, Agent 2002: Social Agents: Ecology, Exchange, and Evolution.* Chicago, Argonne National Laboratory. pp. 149–61.

Rescher, N. (1998) Complexity: A Philosophical Overview. Transaction Publishers (New Brunswick, New Jersey, USA, 219 p.

Reynolds, C.W. (2000) 'Interaction with groups of autonomous characters', *Proceedings of Game Developers Conference 2000*, San Francisco, CA: CMP Game Media Group. pp. 449–60.

Reynolds, C.W. (2005) 'Individual-based models', http://www.red3d.com/cwr/ibm.html. Accessed 1 May 2011

Rindfuss, R.R. and 21 authors (2008) 'Land use change: Complexity and comparisons', *Journal of Land Use Science*, 3(1): 1–10.

Rindfuss, R.R., Entwisle, B., Walsh, S.J., Mena, C.F., Erlien, C.M., and Gray, C.L. (2007) 'Frontier land use change: Synthesis, challenges, and next steps', *Annals of the Association of American Geographers*, 97(4): 739–54.

Rindfuss, R.R., Walsh, S.J., Turner II, B.L., Fox, J., and Mishra, V. (2004) 'Developing a science of land change: Challenges and methodological issues', *Proceedings of the National Academy of the United States of America*, 101(939): 13976–81.

Shea, K., Sheppard, A., and Woodburn, T. (2006) 'Seasonal life-history models for the integrated management of the invasive weed nodding thistle Carduus nutans in Australia', *Journal of Applied Ecology*, 43(3): 517–26.

Sheppard, E., and McMaster, R. (eds.). (2004) *Scale and Geographic Inquiry: Nature, Society, and Method*. Oxford, UK: Blackwell.

Silveira, J.J, Coutinho, G., and Lopes, C. (2002) 'DINAMICA—a stochastic cellular automata model designed to simulate the landscape dynamics in an Amazonian colonization frontier', *Ecological Modelling*, 154(3): 217–35.

Silveira, J.J., Espíndola, A.L. and Penna, T.J.P. (2006) 'Agent-based model to rural-urban migration analysis', *Physica A: Statistical Mechanics and its Applications*, 364: 445–56.

Sole, R.V., Manrubia, S.C., Benton, M., Kauffman, S., and Bak, P. (1999) 'Criticality and scaling in evolutionary ecology', *Trends in Ecology and Evolution*, 14: 156–60.

Solecki, W.D., and Oliveri, C. (2004) 'Downscaling climate change scenarios in an urban land use change model', *Journal of Environmental Management*, 72(1–2): 105–15.

Turcotte, D.L., Malamud, B.D., Morein, G., and Newman, W.I. (1999) 'An inverse-cascade model for self-organized critical behavior', *Physics A*, 268: 629–43.

Turcotte, D.L., and Rundle, J.B. (2002) 'Self-organized complexity in the physical, biological, and social sciences', *Proceedings of the National Academy of the United States of America*, 99: 2463–65.

Varnakovida, P., and Messina, J.P. (2005) 'Critical thresholds and sensitivity dynamics of percolation', *Proceedings of Geocomputation* (CD-ROM).

Walsh, S.J. (2007) 'Feedbacks,' in P. Robbins (ed.), *Encyclopedia of Environment and Society*. Croton-on-Hudson, NY: SAGE. pp. 655–56.

Walsh, S.J., Bilsborrow, R.E., McGregor, S.J., Frizzelle, B.G., Messina, J.P., Pan, W.K.Y., Crews-Meyer, K.A., Taff, G.N., and Baquero, F.D. (2003) 'Integration of longitudinal surveys, remote sensing time-series, and spatial analyses: Approaches for linking people and place', in J. Fox, R.R. Rindfuss, S.J. Walsh, and V. Mishra (eds.), *People and the Environment: Approaches for Linking Household and Community Surveys to Remote Sensing and GIS*. Boston, MA: Kluwer Academic. pp. 91–130.

Walsh, S.J., Entwisle, B., Rindfuss, R.R., and Page, P.H. (2006) 'Spatial simulation modeling of land use/land cover change scenarios in Northeastern Thailand: A cellular automata approach', *Journal of Land Science*, 1(1): 5–28.

Walsh, S.J., Evans, T.P, Welsh, W.F., Entwisle, B., and Rindfuss, R.R. (1999) 'Scale dependent relationships between population and environment in Northeast Thailand', *Photogrammetric Engineering and Remote Sensing*, 65: 97–105.

Walsh, S.J., and McGinnis, D. (2008) 'Biocomplexity in coupled human-natural systems: Study of population & environment interactions', *GeoForum*, 39(2): 773–75.

Walsh, S.J., Messina, J.P., and Brown, D.G. (2008a) 'Mapping and modeling land use/land cover dynamics in frontier settings—special issue foreword', *Photogrammetric Engineering and Remote Sensing*, 74(6): 677– 679.

Walsh, S.J., Messina, J.P., Crews-Meyer, K.A., Bilsborrow, R.E., and Pan, W.K.Y. (2002) 'Characterizing and modeling patterns of deforestation and agricultural extensification in the Ecuadorian Amazon', in S.J. Walsh and K.A. Crews-Meyer (eds.), *Linking People, Place, and Policy: A GIScience Approach*. Boston, MA: Kluwer Academic. pp. 187–214.

Walsh, S.J., Messina, J.P., Mena, C.F., Malanson, G.P., and Page, P.H. (2008b) 'Complexity theory, spatial simulation models, and land use dynamics in the northern Ecuadorian Amazon', *Geoforum*, 39(2): 867–78.

Webster, C.J., and Wu, F. (1999) 'Regulation, land use mix and urban performance. Part 2: Simulation', *Environment and Planning A*, 31: 1529–45.

Weidlich, W. 1994 "Synergetic modeling concepts for sociodynamics with application to collective political opinion formation. *Journal of Mathematical Sociology*, 18(4): 267–291.

Weidlich, W., and Munz, M. (1990) Settlement formation. 1. A dynamic theory', *Annals of Regional Science*, 24: 83–106.

Weisstein, E.W. (2004a) 'Lyapunov characteristic exponent', MathWorld—A Wolfram Web Resource, http://mathworld.wolfram.com/LyapunovCharacteristicExponent.html. Accessed 1 May 2011

Weisstein, E.W. (2004b) 'Kolmogorov entropy', MathWorld—A Wolfram Web Resource, http://mathworld.wolfram.com/KolmogorovEntropy.html. Accessed 1 May 2011

White, R., and Engelen, G. (1993) 'Cellular automata and fractal urban form: A cellular modeling approach to the evolution of urban landuse patterns', *Environment and Planning A*, 25: 1175–99.

White, R., Engelen, G., and Uljee, I. (1997) 'The use of constrained cellular automata for high-resolution modelling of urban land-use dynamics', *Environment and Planning B*, 24(3): 323–43.

Wolfram, S. (1984) 'Cellular automata as models of complexity', *Nature*, 311: 419–24.

Zeng, Y., and Malanson, G.P. (2006) 'Endogenous fractal dynamics at alpine treeline ecotones', *Geographical Analysis*, 38: 271–87.

Linking Biogeography and Society

Geographers actively research human interactions with both the biotic and abiotic components of ecosystems. But on the broader stage of the ecology-society nexus other disciplines, such as in economics and sociology, researchers also plow their respective, but increasingly overlapping, research furrows. As global environmental concerns grow, so does the list of disciplines involved. Increasingly, though human-environment interactions are to many the 'stuff of Geography', our research in this area is becoming indistinguishable from that of these other disciplines. Perhaps not to us, as geographers, but to other disciplines it is.

We now recognize George Perkins Marsh's (1864) *Man and Nature* as the first scholarly book in this area: though it is interesting to note that this book was 'lost' to academia until its rediscovery by the geographer David Lowenthal in the 1950s (Lowenthal, 1958). Other seminal works act as relatively early milestones, for example, Aldo Leopold's *Sand County Almanac* (1949) and the edited collection *Man's Role in Changing the Face of the Earth* (Thomas, 1956). For many outside geography, Rachel Carson's (1962) *Silent Spring* is the first, seminal commentary on the ecological destruction wrought by modern, industrial society. Regardless of the take-off point in the literature, there is no doubt that academic research and scholarship, and popular literature in the ecology-society nexus, has increased exponentially since the 1960s. It is the contemporary interdisciplinary area par excellence encompassing the natural sciences, social sciences, and humanities: the application of biogeography is an

important part of this scholarship. This context raised four questions to us as authors in compiling this section:

- What has been the contribution of biogeographers (particularly those steeped in a geography tradition)?
- Have their contributions changed over time?
- What opportunities have been missed?
- Where can we be most influential as (bio) geographers?

We use four themes familiar to geographers in this section to illustrate biogeography-society relationships as an attempt to answer these questions: (i) 'biogeographical assets' as resources (Chapter 26); (ii) 'biogeography' as a hazard (Chapter 27); (iii) using 'biogeographical' knowledge (Chapter 28); and (iv) preserving and conserving 'biogeography' (Chapters 29–31).

A major theme in this general area is conservation. In the early biogeography textbooks examined in the introductory chapter, there was no doubting that the term in use was 'nature conservation'. It still exists, but it has been largely replaced by biodiversity conservation. It is a moot point whether there is really much difference between them. We do, however, see some differences in approaches to conservation depending on whether the focus is on species or on broader ecosystem goals. This is an uneasy dichotomy but one that is practiced by conservation bodies. Geoffrey Griffiths and Ioannis Vogiatzakis examine habitat approaches to conservation in Chapter 30; and Patrick Osborne and Pedro Leitão offer a parallel review of species-based approaches in Chapter 31. All four authors point out that many conservation efforts are based around static views of the environments that they are trying to conserve. Many conservation units, for example, have fixed boundaries based on past or contemporary ecosystem or habitat boundaries and do not account for future changes or plan for changes in biogeographical ranges or species' behavior consequent upon climate change. This is an important issue that conservation ecologists will have to tackle alongside biogeographers, an issue introduced by Robert Marchant and David Taylor in Chapter 29.

There are, of course, other avenues to the ecology-society nexus where biogeographers are active, but to a lesser degree in terms of research volume (in Geography) than in conservation. Robert Voeks writes on ethnobotany in Chapter 26, and Mark Blumler examines invasive species in Chapter 27. Ethnobotany is, in essence, a subfield of botany but is one that readily lends itself to geographical investigation, especially by those whose interests lie in the perception and use of nature by humankind. It is surprising then that few geographers, relative to the number of anthropologists and botanists, have contributed to this area. Plant and animal invasions are of economic concern in many parts of the world, perhaps most notably in Australia, the United States, and most island states. Mark Blumler uses California as an example of how a geographical approach to invasive species differs from, and yields different insights, than the approaches of botanists, zoologists, and agricultural scientists. We advocate that geographers should apply their particular skills for both ethnobotany and invasive species as they are potentially fertile ground for geographical inquiry.

For Chapter 28, we asked an ecologist—Yordan Uzonov—to write on bioindication. Biological indicators have been used extensively to monitor environmental conditions and, freshwater ecosystems aside, this area seems to have eluded geographers, though again geographers clearly have the skills to make valuable contributions in this area. The use of biological indicators is again expanding, and we have included this chapter as a kind of clarion call for geographers to engage with bioindication more widely.

A review of introductory biogeography books written in the Anglo-American geography tradition (Table 1.3 in Chapter 1), shows that we have attended to the biogeography-society interface by (i) integrating human 'impact' on ecosystems throughout the books; (ii) by creating a false separation between human 'impact' from biogeography and ecology, which perpetuates the idea of natural ecosystems and those strongly influenced by people, by dealing with relevant material in separate chapters; or (iii) by ignoring it. The missed opportunity was to signal the complexity of the human impacts on ecosystems. Some recent developments in understanding the complexities of the human-ecosystem interface have been considered in other chapters in this book, in approaches that range from those grounded in spatial and temporal modeling (see Walsh et al., this volume) to those that are more specific (see Voeks, and Blumler, this volume).

Is there a disconnect between many contemporary biogeographers and (bio)geographers, and those geographers who research human-ecosystem interactions? Will there be a difference between those who will hopefully read this handbook (the former) and those that might glide by it on the library shelves (the later). The quantitative revolution of the 1960s and 1970s led many physical geographers (amongst whom we can identify 'some' biogeographers) to focus on process-based studies. This led to much closer links with ecologists and other physical geographers, notably geomorphologists and hydrologists. Yet biogeography has always interested a much larger proportion of scientists who did not constrain themselves between and in a process-based straitjacket than, say, geomorphology and hydrology. The burgeoning research area driven by concerns about climate change has provided further impetus in the process-driven area, yet simultaneously we are concerned with the biogeographical impacts of the other major global driver— land-use transformation.

In turn, human geography moved away from areas that appeared to interest biogeographers. The cultural turn swung the cultural geography pendulum from something akin to the Berkeley School, which attracted many North American biogeographers, to modern cultural geography, which has more or less been ignored by biogeographers. However, we see signs of change indicating a possible meeting of biogeography as a science, and human-environment relationships espoused by social scientists. Themes that are developing strongly among social scientists that are and will continue to impact biogeography include social science perspectives on nature conservation, people's livelihoods, and development (e.g., Adams, 2008; Brown, 2003; Robbins et al., 2007; Zimmerer, 1997) and people-nature themes in the new cultural geography (where interestingly there is a stronger focus on animals than plants in contrast to the biogeographer's traditional plant preoccupations, and those of Carl Sauer and his converts) (see Millington et al., this volume).

To be effective researchers with clear voices that are heard by social scientists at the ecology-society nexus, biogeographers probably need to develop deeper understandings of the economic, social, cultural, political, and historical elements of human society. Few ecologists have trodden this path. A notable exception is a Harvard group's work on understanding historical contingency in land-use change in New England to understand contemporary plant distributions (e.g., Foster, 1992). Jonathan Philips stresses the role of place and historical contingency in many aspects of physical geography, not only biogeography (Philips, 2004). Place and historical contingency are ignored by many biogeographers, but for those of us who are steeped in geographical traditions, the contributions we can make as (bio) geographers can be both different and valuable in this area.

REFERENCES

Adams, W.M. (2008) *Green Development: Environment and Sustainability in a Developing World*. London: Routledge.

Brown, K. (2003) 'Integrating conservation and development: A case of institutional misfit', *Frontiers in Ecology and Conservation*, 1(9): 479–87.

Carson, R. (1962) *Silent Spring*. New York: Houghton Mifflin.

Foster, D.F. (1992) 'Land-use history (1730–1990) and vegetation dynamics in Central New England, USA', *Journal of Ecology*, 80(4): 753–71.

Leopold, A. (1949) *A Sand County Almanac: With Other Essays on Consrvation from Round River*. New York: Oxford University Press.

Lowenthal, D. (1958) *George Perkins Marsh: Versatile Vermonter*. New York: Columbia University Press.

Marsh, G.P. (1864) *Man and Nature: or Physical Geography as Modified by Human Action*. New York: Charles Scribner.

Philips, J.D. (2004) 'Laws, contingencies, irreversible divergence and physical geography', *The Professional Geographer*, 56(1): 47–53.

Philo, C., and Wilbert, C. (2000) *Animal Spaces, Beastly Places: New Geographies of Human-Animal Relations*. London: Routledge.

Robbins, P., Chhangani, A., Rice, J., Trigosa, E. and Mohnot, S.M. (2007) 'Enforcement Authority and Vegetation Change at Kumbhalgarh Wildlife Reserve, Rajastha, India' *Environmental Management*, 40(3): 365–378.

Thomas, W.L. Jr. (1956) *Man's Role in Changing the Face of the Earth*. Chicago: Chicago University Press.

Zimmerer, K.S. (1997) *Changing Fortunes: Biodiversity and Peasant Livelihood in the Peruvian Andes*. Berkeley: University of California Press.

26

Ethnobotany

Robert Voeks

26.1 INTRODUCTION

Academic ethnobotany has a relatively brief history. Coined in 1895 by the botanist John Harshberger, ethnobotany is defined as the study of the dynamic relationship between plants, people, and the environment. For many, however, the field gravitates toward two marginally related and romanticized cognitive domains. The first, which children often encounter in summer camp and environmental education classes, represents little more than historical anecdotes about how native people employed plants in the distant past. The instructor points to a sprig of horsetail (*Equisetum* sp.) and describes it as 'Indian sandpaper', to stinging nettle (*Urtica dioica*) as 'Indian rope', and to western juniper (*Juniperus occidentalis*) as 'Indian laxative'. This folkloric ethnobotany is descriptive and devoid of cultural relevance. An alternative and more titillating interpretation portrays ethnobotany as something of a scientific extreme sport—rugged, young scientists navigating the forest primeval, plumbing the ancient wisdom of indigenous shamans, persevering personal discomfort and danger, all in the quest for the mysterious healing powers of nature. This latter view is fortified by a steady stream of trade books, documentaries, and Hollywood productions such as *Medicine Man*. However appealing, this line of ethnobotanical inquiry is perceived increasingly by developing world governments and NGOs as biopiracy—the blatant pilfering of plants and genes and intellectual property—and is portrayed as the latest chapter in the seedy saga of neocolonial exploitation.

In spite of these impediments to academic legitimacy, ethnobotany has witnessed a renaissance in recent years. Student interest is on the rise, attendance at professional conferences has grown steadily, the number of professionals calling themselves 'ethnobotanists' has grown, and a few universities now offer advanced degree concentrations in ethnobotany or its equivalent. Most important, practitioners are tackling the challenge of establishing ethnobotany as a rigorous, hypothesis-testing, and ethical scientific endeavor, and at the same time are charting the social and environmental relevance of the field (Bennett, 2005; Martin, 1995; Raven et al., 2007).

While myriad questions are addressed by ethnobotanists, most revolve around two central themes: how do people use plants? and how do people view plants? In this vein, much of traditional ethnobotany has sought to uncover the wealth of material and spiritual resources—foods, fibers, fuels, medicinals, and others—ensconced in the personal and collective memories of traditional societies. When studied for its own sake, without reference to hypothesis testing or theory, this material pursuit is more properly termed 'economic botany'. For better or worse, much of the current ethnobotanical literature continues to consist of lists of useful species employed by one or another group. Simultaneously, ethnobotanical research is oriented increasingly toward understanding how culturally relevant floras are cognitively categorized, ranked, named, and assigned meaning. It seeks to understand the complex strategies employed by folk societies to manage plant taxa and communities, and the degree to which these actions promote or undermine resource conservation. Ethnobotanists seek to translate folk knowledge of nature into policies that enhance the lives of people in rural communities as well as the biological integrity of natural landscapes (Zimmerer, 2002). Although important research is carried out in urban-industrial landscapes, most fieldwork is done in collaboration with people and communities in the rural, developing world. Ethnobotany is decidedly a field-based discipline. Finally, because of its interdisciplinary foundation, academics who call themselves ethnobotanists are drawn from a host of established disciplines—especially anthropology, botany, chemistry, economics, pharmacognosy, and geography (Berlin, 1992; Minnis, 2000; Schultes and von Reis, 1995). Geographers' involvement stems in many respects from the early scholarship of Carl O. Sauer and the Berkeley School of cultural geography. Arguing that much of the so-called natural landscape was the consequence of long-term anthropogenic impacts, whether intentional or inadvertent (cf. Denevan, 1992; Parsons, 1955; Sauer, 1963), geographers set the stage for the more theoretically informed and quantitatively based research to come.

With so many practitioners encompassing such divergent research agendas and paradigms, the conceptual space encompassed by ethnobotany is not surprisingly expansive and growing. Rather than attempt to cover all that could pass as ethnobotanical research, my objective is to review some of the major themes of contemporary research. I focus on 'wild' species—cultivated or uncultivated—with only occasional reference to domesticated plants. After introductory comments on the historical significance of ethnobotanical inquiry, I review plant-naming systems and the breadth of ethnobotanical inventories, particularly as these correspond with modes of subsistence. I examine ongoing questions concerning plant-people relations in wild nature, especially the management of native plant communities with the intent of enhancing the quantity and quality of useful species. I consider the role of successional species and disturbance regimes, anthropogenic and natural, in plant knowledge and utility. The growing literature on gendered ethnobotanical knowledge as well as the value of nontimber forest products is reviewed. I outline recent research in diaspora ethnobotany. And I consider the roles of ethics, intellectual property, and cultural erosion as challenges to the future of ethnobotanical inquiry. Finally, the majority of published ethnobotanical research is forthcoming from rural, tropical landscapes, and this is reflected in the topics and regions presented here. Although not covered in depth in this chapter, there is nevertheless valuable recent research coming out of the North American tundra zone, the Himalayan region, the arid landscapes of North Africa and the Middle East, and the Mediterranean.

26.2 HISTORICAL OVERVIEW

Ethnobotanical knowledge of nature—the ability to perceive, experiment on, genetically modify, name, classify, adopt, and transfer individual plant taxa—was fundamental to the success of the human species. Reports of self-medication by higher primates, and the similarity of simian pharmacopoeias to those used by local human communities, hint at the antiquity of this knowledge domain (Huffman, 2001). During human history and prehistory, whether for food, fiber, fuel, medicine, magic, or the numerous other categories of material and spiritual employ, mastering the properties of the local flora has been a defining trait of our species (Schultes and von Reis, 1995). Written descriptions of useful plants appear in our earliest civilizations, including on Mesopotamian cuneiform tablets, Egyptian papyri, Mayan bas-reliefs, and Han-Dynasty silk scrolls. There have been, it seems safe to speculate, very few successful societies dominated by ineffective ethnobotanists.

Over the millennia, particularly after the dawn of plant domestication, keystone cultivars diffused over space under human agency, from community to community, region to region, and continent to continent in rare cases. Rice moved throughout east and Southeast Asia; maize spread across much of the Americas; yams jumped from island to island in Oceania. However critical to the evolution of ancient cultures, nation-states, and empires, the impacts of these and other pulses of plant migration under human agency were eclipsed almost overnight during the European Age of Exploration. By the early seventeenth century, previously endemic esculents, medicinals, fibers, and weeds were purposely and accidentally introduced to most of the major continents and oceanic islands (Crosby, 1986). By the mid-sixteenth century, Brazil witnessed the successful cultivation of cinnamon from Ceylon, pepper from Malabar, ginger from China, coconuts from Malaysia, and mango from Southeast Asia; later, cacao came from Middle America. In Hispaniola, the East Indian medicinal cannafistula (*Cassia fistola*) was reportedly widely naturalized by the sixteenth century. Asian fruits were so thoroughly acclimated in the West Indies that sixteenth-century missionaries encountered 'whole woods and forests of orange trees'. Within a century of colonization, East Indians were cultivating American cashew, pineapples, soursop, squash, capsicums, and cactus. Jesuit Michael Boym's mid-seventeenth-century illustrations of New World papaya, cashew, pineapple, and avocado being cultivated in Chinese gardens underscored the reciprocal nature of these exchanges (Shaw, 1992). The repercussions of these ethnobotanical movements, in terms of population growth, changing subsistence patterns, integration into global markets, and other features, cannot be overstated, and they continue to be felt today.

This halcyon period of anthropogenic plant diffusion also heralded the emergence of strategies for useful plant identification and exploitation—the ethnobotanical method—which would hold sway through subsequent centuries and to the present day. Colonial authorities were obsessed with mining the memories of the recently conquered for new and useful foods, spices, and medicines, in order to profit by their discoveries. Indeed, the search for novel plant spices and medicines represented a primary catalyst for the early voyages of discovery by Columbus, Vasco de Gama, Magellan, and others (Shaw, 1992). Over time, as intercontinental exchanges of endemic plant taxa proceeded, botanical gardens in Europe and the colonies evolved quickly into globalized gardens of food, medicinal, and other useful species (Grove, 1996). This 'green gold' was sought out especially by colonial authorities in their attempt to fend off the flotilla of tropical microbes that decimated European immigrants. Recognizing the significance of local healers and shamans in identifying the uses and preparations of herbal medicines, colonial 'bioprospectors' appropriated the intellectual property of their subordinates without a thought of compensation (Schiebinger, 2004). The following description of the discovery of guaiacum (*Guaiacum officinale*), the first New World plant species to be reported in Europe after the Columbian landfall, underscores the critical role of local informants:

> There was a Spaniard that suffered great pain of the pox [syphilis], which he had taken by the company of an Indian woman, but his servant, being one of the physicians of that country, gave unto him the water of guaiacan [guaiacum], wherein not only his grievous pains were taken away that he did suffer, but he was healed very well. (Frampton, 1580, fol. 11)

Bioprospecting's sadistic side is illustrated by Burkill (1966: 178) who describes the search for a plant-based antidote to the famous Borneo upas poison tree (*Antiaris toxicaria*), 'the secret of which was tortured out of a native'.

Interest in the search for medicine and food waned in the early twentieth century, as agronomy and pharmacology became increasingly limited to the greenhouse and laboratory. In spite of notable early medicinal successes—quinine derived from the bark of the Peruvian *Cinchona* sp. tree to treat malaria; pilocarpine from the Brazilian herb *Pilocarpus jaborandi* to treat glaucoma; digoxin and digitoxin from European foxglove (*Digitalis purpurea*) to treat heart failure; and diosgenin from winged yams (*Dioscorea alata*) to prevent conception, pharmacology turned away from the forests and fields as sources of novel compounds. This hiatus was followed by a burst of interest in the latter twentieth century, however, with the appearance of valuable chemotherapeutic drug plants such as the Madagascar periwinkle (*Catharanthus roseus*), used to treat childhood leukemia, and the western yew (*Taxus brevifolia*), used to treat ovarian cancer. With the assistance of issue, entrepreneurs, herbalists, and folk healers metamorphosed from purveyors of witchcraft and medical hocus pocus into earthly receptacles of ancient plant wisdom (Balick and Cox, 1996; Voeks, 2004). Likewise, the search for novel crop landraces as grist for Green Revolution high-yielding varieties reestablished the value of peasant and farmer agroecological knowledge. Once again, researchers employed ethnobotanical methods to identify economically useful plants, but they had little or no thoughts of benefit sharing with the local people and communities (Brush and Stabinsky, 1996).

26.3 PLANT NOMENCLATURE AND DEPTH OF ETHNOBOTANICAL KNOWLEDGE

Although people universally name and classify familiar plants, taxa do not receive equal treatment in terms of lexical division and description. A staple crop, such as corn (*Zea mays*) or rice (*Oryza sativa*), will be recognized and specified by dozens of subspecific epiphets—the blue type of corn, the red type of corn, the yellow type of corn that grows well in waterlogged soil, and so on. Other species receive little more than a passing semantic gloss—it is a weed, or it is a vine. Why are some groups lexically important, while others are not? Berlin (1992) and others argue that species salience is the primary feature that leads to plant naming. Manioc is probably more salient to Amazonian subsistence horticulturalists than a rare vine that occasionally invades the swidden. But is salience simply utilitarian, based on its material worth to a people, or is it a product of more cognitive features? Advocates for the former would argue that species are found to have some economic value, and then they are classified. The latter intellectualist view holds that species must be categorized before they can be utilized. Researchers have come down on both sides of the argument (Berlin, 1992; Hunn, 1982; Turner, 1988).

All societies appear to organize their ethnofloras into ranks, often including kingdom, life form, genus, species, and variety (Atran, 1999). Assignment to one or another rank, and inclusion in or exclusion from a particular grouping (e.g. folk genus, folk species), reflects features that are salient to the observer. Ranks tend to be similar across cultures, regardless of cultural significance or meaning. They may be of general purpose, that is, identified by characters such as morphology, aroma, size, color, or biogeography, that are readily apparent to most observers. For example, among Borneo's Dusun people (Voeks and Nyawa, 2006), limpanas purak (*Goniothalamus umbrosus*) is the white (purak) kind of *Goniothalamus* (limpanas), mandap etom (*Polyalthia motleyana*) is the black (etom) kind of *Polyalthis* (mandap), and taguli gantung (*Vanilla griffithii*) is the hanging (gantung) type of vanilla orchid (*taguli*). Plants may have culture-specific utilitarian features, such as for medicine or magic or taboo, which may or may not have meaning to other groups (Brown, 2000; Zent and Zent, 1999).

The genus represents the most recognizable rank of plant perception among humans, and in most classification schemes the genus parallels its Western scientific counterpart (Balée, 1994; Berlin, 1992). Major differences arise between folk and Western taxonomy, however, in the degree to which taxa are recognized below the genus rank, and in the development of higher levels of classification, such as the plant family.

Rural residents, particularly those pursuing subsistence-type strategies, often maintain mammoth plant lexicons. Berlin (1992: 7) considered the sheer quantity of plant names and uses maintained by unlettered forest people to be "awe inspiring". Ethnobotany monographs, such as Alcorn's (1984) work with the Huastec Maya, Balée's (1994) work with Brazil's Ka'apor, Bennett et al.'s (2002) work with Ecuador's Shuar, and Christensen's (2002) work with Borneo's Iban and Kelabit remind us of the profound depth and breadth of this knowledge of nature. The possibility that variation in quantitative knowledge of plants between different culture groups might exist was proposed by Brown (1985 and comments) and later Berlin (1992), who reported sharp differences in the total numbers of labeled plant taxa when subsistence strategy was factored into the calculation. Small-scale agriculturalists, according to Brown (1985), maintained a magnitude of named plant categories that was some five times greater on average than that of foragers. Cultivators averaged 890 named plant classes, whereas hunter-gatherers exhibited a mean of only 179. In addition to less familiarity with domesticated crop plants, which one anticipates

for noncultivating societies, foragers are hypothesized to be less prone to food shortage and disease than their cultivating counterparts. This, Brown suggested, led to a relatively limited breadth of ethnobotanical knowledge concerning famine foods and medicinal species among hunter-gatherers.

Brown's results were criticized on the grounds that many of the small-scale cultivators in his inventory inhabit species-rich, moist tropical regions, where the opportunity to exploit different plant taxa are legion. Most of the hunter-gatherer records he cited, however, were drawn from groups occupying less speciose, temperate zone habitats. Nevertheless, when Berlin (1992) employed folk genera rather than named plant taxa, cultivating groups averaged 520 folk generics, compared to 197 for hunter-gathering societies. Subsequent comparison of sympatric tropical forest foragers and cultivators in Amazonia and interior Borneo lends support to this quantitative lexical disparity. In Brazil, Balée (1999) determined that the Guajá maintained a total of 28 folk-specific names for plants, whereas the Ka'apor recognized 252 folk specifics, or nearly nine times the number of their neighboring foragers. Likewise, in a comparison of ethnobotanical profiles maintained by Borneo foragers (Penan) and cultivators (Dusun), Voeks (2007b) found that three use categories—wood, craft, and medicine—resolved much of the disparity in quantitative plant knowledge maintained by these two groups. The Dusun identified a total of 135 species in these three material categories, compared to 20 species for the Penan. A cultivating lifestyle, at least in these examples, is associated with a profound quantitative knowledge of nature, whereas hunting and gathering as a subsistence choice is not.

A corollary of these findings involves the supposed resource paucity and subsistence limitation in moist tropical habitats. According to this 'green desert' view, advanced by Bailey et al. (1989) and Headland and Bailey (1991 and references therein), hunting and gathering as a subsistence mode was possible only if carbohydrates could be acquired from neighboring sedentary cultivators. Purely foraging societies could only have permanently occupied the forest after the entrance of cultivators. They advanced a clear and testable question—have hunter-gatherers ever lived in tropical rain forest independently of agriculture? Several researchers working on the ethnobotany of contemporary hunter-gatherers corroborate that these groups depend on trading for staples with nearby sedentary farmers. The concept that the vast and diverse tropical forests of Africa, Asia, and the Americas were basically devoid of humans during nearly all of our 100,000-year history has important implications in terms of understanding the history of nature-society relations. Inherent in this hypothesis is that present-day foraging societies are derived from cultivating groups, not vice versa, constituting little more than 'professional primitives', extracting and trading primary forest products with neighboring cultivators for rice, yams, and other starchy staples.

Recent paleo-ethnobotanical findings in the American, Asian, and African wet tropics failed to support this hypothesis. Late Pleistocene/early Holocene foraging of wild plant resources predated horticulture by many millennia in most tropical forest biomes (Mercader, 2003 and references therein). For example, Roosevelt et al. (1996), working in late Pleistocene cave deposits in lowland Amazonia, discovered thousands of plant remains, much consisting of palm and other fruit and nut species that are still harvested by locals today. More recently, Barker et al. (2007) confirm dates of at least 35,000 years B.P. for sophisticated plant use and preparation by early foragers in northern Borneo. Although the climate and vegetation during these periods did not duplicate current conditions, the presence of moist forest species in these deposits indicates that foraging of wild plant resources represented a successful mode of subsistence in tropical landscapes and that it clearly preceded agricultural innovation.

26.4 MANAGEMENT OF ETHNOBOTANICAL RESOURCES

The role of habitat disturbance under human agency, particularly as this reflects on the 'pristine myth' question, has been a fertile area of inquiry. The received wisdom wherein the pre-Columbian Americas are depicted as virgin floristic landscapes has yielded to the view that most natural landscapes constitute in part biocultural reflections of past and present human actions. Indeed, considering the pedigree and pervasiveness of human impacts, the distinction between natural and cultural landscapes appears for many to be rather arbitrary (Denevan, 1992, 2002; Lentz, 2000; Whitmore and Turner, 2001).

Ethnobotanists have contributed to this research agenda, in particular by investigating indigenous and diaspora strategies in the tropical realm for managing the plant resources of wild nature. On the one hand, especially salient and valuable individual plant taxa are husbanded; on the other, entire habitats are manipulated so as to produce more or better-quality harvests. In the case of tropical foragers, high-demand, nondomesticated species are shepherded by weeding, seed scattering, and replacement planting (Harris, 1996). The nomadic Penan of interior Borneo protect the younger stems of their staple sago palm (*Eugeissona utilis*). Older stems are harvested and processed, but the younger stems, 'uvud', are marked and protected for future exploitation, thus increasing seed production as well as abundance and distribution (Puri, 2005; Sercombe and Sellato, 2007 and articles therein). For the Amazonian Huaorani, who regularly scatter the seeds of useful species, "distinguishing between extraction and management becomes almost impossible" (Rival, 2002: 81).

Small-scale cultivators in the tropical realm, who have been the subject of considerable ethnobotanical inquiry, expend considerable effort managing plant resources and, in the process, affecting biogeographic patterns. Transplanting of useful wild species in swiddens and along trails, sparing valuable trees during clearing, and other forms of encouragement dramatically modify species composition and distribution (Balée, 1994; Ellen, 1998; Posey, 1984). Such species often defy simple classification as domesticated, semidomesticated, or wild, and many appear to be caught in the process of domestication (Christensen, 2002; Clement, 1999). Although many of these reports involve processes carried out in the past, in Cameroon and Nigeria, the two fruit tree species *Irvingtonia gabonensis* and *Dacryodes edulis* are currently in the midst of domestication, with some populations being completely wild while others are nearly domesticated (Leakey et al., 2004).

Fire is employed as a management tool for various tropical species, especially for palms. Because many palms employ a cryptogeal germination pattern, they are able to sprout readily after an area is burned. Within a few years after intentional burning, piassava (*Attalea funifera*) and babaçu (*Attalea speciosa*) from Brazil, coyal (*Acrocomia mexicana*) in Mexico, cohune (*Orbignya cohune*) from Belize, and others form dense, nearly monospecific stands of valuable plant resource (Balick, 1988; Voeks, 2002). Palms and other fruit-bearing trees in the Brazilian savanna are burned early in the flowering season to enhance later fruit production (Mistry et al., 2005). Throughout much of Latin America, the presence of fire-tolerant palms in older-growth forest indicates previous agricultural activity (Balée, 1994).

In temperate latitudes, the manipulation of native vegetation, especially by intentional burning, is especially well documented. The degree to which these actions created anthropogenic landscapes is still open to debate (Vale, 2002 and references therein). Although the objective of burning is usually attributed to indigenous people's attempts to enhance forage for game, the incineration of vegetation also produced positive impacts on useful undomesticated plants. Gott (2005) notes that aboriginal peoples of southeastern Australia subsisted predominantly on starch from underground tubers. She argues that widespread burning was carried out to enhance the growth of edible herbaceous plants at the expense of trees. For First Peoples of British Columbia, Turner (1994) reports that burning was carried out specifically to enhance the quality and quantity of 17 edible berry and tuberous species. In California, large expanses of oak woodlands were burned by native peoples so as to reduce insect predation on acorns, which represented the primary carbohydrates for many aboriginal groups, as well as to reduce competition from invading conifers (Mensing, 2006). In the southern Andes, monkey puzzle trees (*Araucaria araucaria*) were managed by fire as well to enhance their growth and productivity (Aagesen, 2004).

Indigenous management of wild species by means other than fire was likely common in temperate latitudes as well, although evidence is more historical and anecdotal. In California, redbud (*Cercis occidentalis*) represented a principal basketry material utilized by at least 20 Indian tribes. In addition to the use of fire, individual plants were pruned and coppiced to increase their quality and accessibility (Anderson, 2000, 2005). Nabhan et al. (2000) compared restored and indigenous-managed parts of the Sonora in northern Mexico. They reported that local burning, planting, and harvesting of wild and domesticated species by the Papago people modify both plant and animal biodiversity. Deur and Turner (2005 and references therein) argue that wild plant cultivation was carried out by many of northwest America's First Peoples. Actions such as "selective harvesting, digging, weeding and pruning, controlled burning, and the rotation of harvesting locales" significantly increased the diversity and productivity of local plant resources (Deur and Turner, 2005: 17).

26.5 WEEDS AND SUCCESSIONAL HABITATS

The ecology and diffusion of weeds and invasive species have intrigued biogeographers at least since the publication of Elton's (1958) classic *Ecology of Invasion by Animals and Plants* and Baker and Stebbins's (1965) *The Genetics of Colonizing Species*. Interest gave way to alarm in the late twentieth century, however, as the problem of invasive organisms emerged as a critical environmental challenge. Transported by humans and favored by environmental conditions similar to their origins, weedy invasives often compete successfully with native species, eventually changing entire community assemblages (Van Driesche and Van Driesche, 2004). In spite of the significant negative environmental consequences of biological invasions, ethnobotanists and others are reexamining these unwanted immigrants through a biocultural lens (Crosby, 1986). Biological invasions affect culturally important flora, including food, medicinal, and ceremonial plants and habitats, in myriad and often unpredictable ways. For indigenous people, they provide novel ethnobotanical opportunities and challenges; for diaspora communities, they facilitate continuity of cultural traditions. For example, weedy plants often represent important nutritional adjuncts to staples and critical famine foods during periods of food scarcity (Minnis, 2000 and references therein). Other weeds, such as *Merremia dissecta*, serve as food, medicine, condiments, and ornamentals throughout the tropical world (Austin, 2007). Thus, in spite of our obsession with virgin habitats and native species, the significance of disturbed habitats and nonnative species to rural communities is a promising area of ethnobotanical research. The (bio)geography of invasives is considered in detail in Blumler's chapter in this volume.

Weeds have a lengthy pedigree of contribution to traditional healing practices and to western pharmacology. Much of Dioscorides' ancient *Materia Medica* is a paean to the glories of healing weeds. Abundant, readily accessible, and easy to harvest, weedy plants would seem to be, from a collector's perspective, ideal medicinal plants. Indeed, according to Albuquerque and Lucena (2005), there exists a significant correlation between species' apparency and utility in Brazil's caatinga biome. Weeds and other successional species are, in addition, often rich in potentially bioactive secondary compounds. Coley and Barone (1996) suggest that weedy and often herbaceous taxa are more likely to employ quantitative strategies against herbivorous predators, such as alkaloids and phenols, whereas large woody species are more likely to develop qualitative defense mechanisms, such as lignin, tannin, and cellulose. Reviewing the number of plant-based compounds that have made the transition to pharmaceutical drugs indicates that a significant portion are derived from weedy herbs, shrubs, climbers, garden cultigens, or gap-dependent trees, and that most were used by local healers to treat similar types of ailments (Lewis, 2003; Stepp, 2004). The statistical association between anthropogenic landscapes and medicinal plant species has been recorded in various locales (Aguilar-Støen and Moe, 2007; Begossi et al., 2002; Chazdon and Coe, 1999; Stepp and Moerman 2001). Contrary to the rhetorical motif provided by environmental issue entrepreneurs, old-growth habitats are unlikely to have served as the primary repository of folk medicinals in the past, and they do not at present (Voeks, 2004).

The relationship between domesticated crop plants and folk medicine is a fertile area of research (Pieroni and Price, 2006), particularly as botanical supplements gain legitimacy in modern society. Logan and Dixon (1996) argue, for example, that because food plants represent the flora with which people are most familiar, their effects can be trusted and locating them is never a challenge. Kitchen gardens in particular are notoriously rich in plants that serve as both food and medicine. Etkin and Ross (1991) note that 49% of plants that serve as gastrointestinal medicine for the African Hausa are also foods. One-third of the food plants censused in Vietnam by Ogle et al. (2003) also have medicinal properties.

Pfeiffer and Voeks (2008) proposed a conceptual framework delineating the differential cultural impacts of weeds and invasive biota. *Culturally enriching* invasive species represent nonnative biota that have been incorporated into local cuisines, pharmacopeias, rituals, and other traditional practices, resulting in cultural expansion as new species are adopted and new traditions generated. For example, invasive plant introductions by Spanish, Russian, and Euro American explorers and settlers in the Western United States, such as wild oats (*Avena fatua*), ripgut brome (*Bromus diandrus*), and Himalayan blackberry (*Rubus discolor*) have been incorporated into the diets and medicinal practices of several dozen Native American tribes (Moerman, 1998; Strike, 1994). *Culturally facilitating* invasive species, which often precede or accompany human migrations, allow continuity and synthesis of traditional ethnobiological practices. For example, purple nutsedge (*Cyperus rotundus*) represents an invasive plant of almost global dimensions, but it also retains a legacy of ethnobotanical utility reaching to nearly all points of the compass. Its colonization of the South American tropics allowed African diaspora to reconstitute a suite of ethnobotanical traditions in a landscape that was otherwise altogether alien (Duke, 2001; Voeks, 1997). *Culturally impoverishing* invasive species, on the other hand, lead to cultural contraction, that is, the displacement of important native species and associated traditions. In California, physiologically armed invasive plants such as starthistle (*Centaurea solstitialis*) and cocklebur (*Xanthium strumatium*) present a physical barrier to accessing and collecting native basketry plants (Pfeiffer and Ortiz, 2007). On Hopi Indian lands, salt cedar (*Tamarix* spp.) invasions are such a serious threat to species of cultural and ceremonial importance that tribal elders are supervising the uprooting and replanting of plants including sand reed (*Calamovila gigantea*), willow (*Salix* spp.), and yucca (*Yucca* spp.) closer to reservation lands where they can be better conserved (Salmon, 2003). Thus, in spite of their consistently deleterious environmental impacts, weeds and invasive plants elicit a range of cultural impacts—from positive to negative.

26.6 GENDERED ETHNOBOTANY

Gendered divisions of labor and space are common features of subsistence-oriented communities. Men are often engaged in hunting, fishing, livestock herding, and timber extraction, activities that take them to relatively undisturbed habitats that are distant from their settlements. Women are more likely to be involved in managing local resources, such as kitchen gardens, swiddens, and other anthropogenic habitats relatively nearby their homes (Momsen, 2004). Unsurprisingly, there are significant variations on this theme. Fadiman (2005) reports, for example, that Mestizo women in Ecuador manage plant resources close to home, whereas women in nearby indigenous Chachi communities are in charge of plant resources farther from home. In Kenya, whereas husbands and wives occupy the same garden space, the men own the land while the women, who carry out most of the labor and management, are more familiar with its vegetative properties (Rocheleau and Edmunds, 1997). Nevertheless, because men and women more often than not travel and toil in different geographical spaces, their familiarity with nature is bound to vary (Howard, 2003 and articles therein).

Men and women learn the properties of and assign names to plants that are highly visible, familiar, and accessible. For males, these characteristics are associated most often with habitats under the influence of ecological as opposed to anthropogenic processes. For females, their knowledge of nature is often derived from more successional landscapes, ecosystems with ecological properties and patterns of cognitive accessibility that are often under the control of human habitat modification (Voeks, 2007a). Although in earlier literature, men were often portrayed as possessing superior knowledge of botanical nature, this was the result largely of past gender-imbalanced fieldwork (Howard, 2003; Pfeiffer and Butz, 2005). The situation has changed significantly in recent years, in part because many more female ethnobotanists have entered the field.

The ethnobotanical consequence of gendered space and habitat divisions is, according to recent research, fairly consistent. On the island of Flores, Pfeiffer (2002) identifies significant spatial partitioning of plant knowledge of native fruit species between women and men. In a rural Brazilian community, Begossi et al. (2002) report that women know a greater number of medicinal species than men, although men show an overall higher diversity and heterogeneity of plant citations than women. Working among rural Mestizo communities in the Peruvian Amazon, Stagegaard et al. (2002) note that men, who often work in old-growth forests, are more knowledgeable about trees than women are. Regarding medicinal plants, they suggest that men know more about tree and liana medicine derived from forest species, whereas women are more proficient in regards to medicinal weeds, herbs, and crops. Coe and Anderson (1996) report that Garifuna women in Nicaragua are much more knowledgeable about medicinal species than men are. They relate this to the life history of the local healing flora, which is mostly successional, and thus within the dominant purview of women. Likewise among indigenous groups in northwestern Amazonia, woman control domesticated crops and kitchen gardens, whereas men have dominion over nonmanaged forest resources (Reichell, 1999). Luoga et al. (2000) also find that women in eastern Tanzania know more about herbaceous plants, whereas men are more knowledgeable about trees. Koizumi and Momose (2007) report that male Penan hunter-gatherers in Kalimantan, Indonesia, have become more knowledgeable about the names and uses of plants than females have. They suggest that this is a consequence of women now being greatly restricted to the vicinity of the longhouse, which was not the situation during their nomadic past. Thus, with few exceptions, studies suggest that men are better acquainted with useful species in old-growth, less disturbed habitats, especially arboreal species, whereas women tend to be more informed regarding disturbance species, that is, those associated with kitchen gardens, swiddens, and other products of human habitat change.

Gendered plant knowledge also becomes deeply ingrained in symbols and rituals. For South Africa's amaXhosa, the 'igoqu' symbolic woodpile is a powerful and highly visual cultural artifact. For women, it functions as a sanctuary, a meeting place, and a sacred site in which to perform ritual sacrifices (Cocks et al., 2006). Among Colombia's Tanimuka and Yukuna, knowledge of complex cosmological rituals is passed from father to son, and from mother to daughter. Girls acquire secret knowledge of plants, seeds, and tubers from their mothers, while men learn the properties of mind-altering cultigens like coca, tobacco, and pineapples for fermenting into alcohol (Reichell, 1999). Because the role of shaman-healer is so often reserved for males, secretive plant-based magical rituals are often under men's domain. Nevertheless, because women so often tend to the everyday health issues of the family, they are often better versed in the collection and preparation procedures of the immediate plant pharmacopoeia (Voeks, 2004).

26.7 THE VALUE OF PLANTS

What is the value of ethnobotanical knowledge and resources? The various stakeholders in rural landscapes hold differing views on the topic. On the one hand, the battle over customary rights to land and resources is waged in part by contrasting the economic value of petty extractive resources, usually managed by locals, with commercial commodities, often controlled by outside elites. The collection of nontimber forest products (NTFPs), such as latex, nuts, fruits, resins, rattans, fibers, medicinals, and others, long portrayed as an impediment to economic progress, is now viewed by ethnobotanists as a strategy that integrates environmental conservation with the rights of local people. On the other hand, indigenous and diaspora peoples have quite different means of valuing local resources. These are based in part on market values, because more and more resources are destined for sale. For example, medicinal plant collection in South Africa is estimated at 20,000 tons per year, with an annual value of perhaps USD $60 million (Mander, 1998). But they are also derived from perceptions of nature's worth that are circumscribed more by everyday subsistence needs and culture-bound relations with place, such as sacred species, or the myriad wild snack foods that never enter into commerce. In these cases, in addition to straightforward material value, plants possess cultural worth that goes far beyond monetization.

The objective of the economic valuation of NTFPs is often to contrast seemingly sustainable habitat management practices with those that are more destructive. An early and persuasive paper was written by Peters et al. (1989), who estimated the extractive values of latex, nuts, and fruits from a single-hectare Peruvian forest plot. They identified 72 species with marketable value in the plot—60 timber, 11 food, and 1 latex (natural rubber). Using net present value (NPV), which assumes the ability to collect resources on a continual basis, they discovered that petty extraction compared favorably to profits generated by more destructive land-use options, such as timber removal and pasture development. For example, the NPV of extraction totaled USD $6,330, whereas timber removal totaled USD $490. Using similar methods in Belize, Balick and Mendelsohn (1992) estimated that the net value of folk medicinal species rivaled the value of conversion to agriculture or to pine plantations.

Whether economic estimates positively influence land-use decisions remains to be seen. Monetary figures are derived from different methods and assumptions, with the result that estimates vary wildly. Thus, Godoy et al. (2000) used barter values and actual market value in nearby villages to estimate the value of NTFPs to the Yapuwas and Krausirpe Indians in Honduras. Their figures ranged from a mere USD $2.50 to $9.05 per hectare, underscoring why local groups are drawn to more immediately lucrative means of resource exploitation. Moreover, monetizing extractive resources creates its own set of environmental dilemmas. For example, Coomes (2004) notes that increasing reliance on chambira palm fiber (*Astrocaryum chambira*) in the Peruvian Amazon has led to rapacious overexploitation of wild populations. Delang (2006) likewise argues that market-driven extraction of NTFPs result often in overexploitation and biological deterioration. He suggests that studies should focus more on those NTFPs that are not destined for markets but that serve significant functions in the daily economic lives of rural people. He reports that people in the Karen Hill tribe of Thailand, for example, earn the equivalent of USD $3.00 per hour by collecting wild food plants that they consume, some 10 times the regional wage. Given the gravity of these questions, the various impacts of emerging markets on NTFPs should be a focus of future ethnobotanical research.

Plants possess significant meaning for local people and communities above and beyond that which can be consumed or sold or in any way quantified monetarily. How valuable is a tree-god to an animist people, or the Bodhi tree under which the Buddha received enlightenment? For their profound place in the history of human affairs, as well as their possible contribution to habitat conservation, sacred species continue to draw the interest of ethnobotanists. For Muslims, sitting under the Christ's thorn jujube (*Ziziphus spina-christa*) is considered providential, since the Prophet Mohammed is believed to have seen the tree in Paradise (Dafni et al., 2005).

Votive offerings are placed between the buttresses of eastern Brazil's sacred Iroko (*Ficus* spp.) by Candomb adherents, propitiating Yoruba deities from distant Africa (Rashford, 2007). Sacred groves and species represent a recent adjunct to conservation policéy because they are less anthropogenic than surrounding habitats and because their sacred status translates to limited opposition for protection in local communities. Thus, cacao (*Theobroma cacao*) has retained deep ritualistic significance to the Maya since at least 600 B.C.E. Because it requires good soils and humid forested habitats, Kufer et al. (2006) suggest that protection of cacao's sacred status translates to habitat conservation. Likewise in Ghana, time-series analysis using GIS and air photos indicate that the level of protection afforded sacred forests with baobab (*Adansonia digita*) is significantly higher than forests without baobab (Campbell, 2005).

Aside from plants with deep spiritual significance, cultures often attach particular importance to one or a few secular but highly salient species. Garibaldi and Turner (2004), drawing parallels with ecology's keystone species concept, developed the 'cultural keystone species' concept. These plants and animals, according to the authors, are "culturally salient species that shape in a major way the cultural identity of people" (Garibaldi and Turner, 2004: 4). The resources could be staples, famine foods, medicine, or of other use. Examples include sago palm (*Eugeissona utilis*) among Borneo's Penan, wild rice (*Zizania aquatica*) among Wisconsin's Menominee, red laver seaweed (*Porphyra abbottiae*) among Canada's Tsimshian, or the akhee fruit (*Blighia sapida*) among Jamaicans (Rashford, 2001). In spite of arguments that the keystone species metaphor is inappropriate for biocultural systems (Davic, 2004), others have found the construct useful. For example, Brosi et al. (2007) employ the concept in their exploration of declining canoe-making knowledge among Pohnpeians.

People also differentially value the myriad plant species that they use during their daily activities. The degree of value, as noted earlier, is often reflected in lexical distinction. Turner (1988) developed a 'cultural significance index' (CSI) to determine the value or importance of particular species to indigenous people. This was based on 'quality of use', 'intensity of use', and 'exclusivity of use'. Revisions of this valuation system, as well as alternatives, have been forthcoming from various authors. Silva et al. (2006), for example, introduced a correction factor to the calculation that incorporates degree of consensus among informants. Using this revision, medicinal plant species received the highest rank among northeastern Brazil's Fulni-ô Indian people. Lykke et al. (2004) used a rank of usefulness index based on perceived usefulness, importance, abundance, and conservation status to rank Sahelian trees by Fulani herders. They discovered that nearly all species—food, firewood, construction, and shade—were important enough to the Fulani to be classified as very important.

26.8 DIASPORA ETHNOBOTANY

Ethnobotanical narratives frequently allude to the antiquity and continuity of knowledge profiles among indigenous people. Particularly where cultures and their connections with nature are threatened by environmental and social disruption, reference is made often to the 'thousands of years' over which this ethnoecological wisdom had accumulated. But this view of the nature-society interface implies a degree of temporal and spatial stasis in human populations that is radically inconsistent with the geographical record. It is based, according to Alexiades (2009 and references therein), "on the untested assumption that environmental know-how is largely a long-term empiricist enterprise." Is ethnobotanical expertise space- and time-contingent? Are mobility and migration by individuals, communities, or even entire culture groups inconsistent with retention and reformulation of ethnobotanical traditions?

Given the preconception that immigrants lack ethnobotanical skills, there have been relatively few studies directed at migrant populations. But this is changing, particularly with the growth of diaspora studies in academe. Pieroni and Vandebroek (2007 and chapters therein) uncover some of the dynamics and subtleties of the ethnobotanical diffusion process. In some cases, immigrants go to considerable lengths to continue using their traditional healing flora, particularly where these represent distinctive cultural markers for oppressed diaspora communities. At the same time, given concerns with invasive species introductions, international customs regulations in many cases represent political barriers to useful plant introduction. Regarding recent migrants, Nesheim et al. (2006) discuss the ethnobotanical impact of Guatemalan emigration and repatriation. Guatemalan exiles lost much of their subsistence plant knowledge during their years in Mexico, but they actively sought to increase their understanding of commercially valuable timber species upon return. Their original knowledge of medicinal species was better preserved, possibly because many are disturbance species with wide distributions. Lacuna-Richman (2006) noted likewise that recent Philippine immigrants to Palawan were quickly acquiring knowledge of commercially valuable NTFPs, often from interaction with the indigenous Tagbanua. As a result of this active acquisition process, the difference between indigenous and diaspora knowledge of locally useful species was statistically insignificant. Pieroni and Quave (2005) examined similarities of medicinal and magical plant use among resident southern Italians and ethnic Albanian immigrants who had arrived from the fifteenth to eighteenth centuries. Whereas the plant pharmacopoeias of the two groups were quite similar, ethnic Italians more often employed these species in naturopathic healing, whereas ethnic Albanians used them to treat illness of magical origin, such as evil eye. In New York City, botánicas dispense the healing traditions and medicinal plants introduced by generations of Hispanic immigrants. In addition to plant sales and healing advice, botánicas are fertile areas of medicinal and magical plant knowledge transmission. The continuity of medicinal plants and healing traditions for Mexican migrants to Atlanta, Georgia (Waldstein, 2006), and Sikhs to London (Sandhu and Heinrich, 2005) has been investigated. Preliminary results suggest that commercially valuable species are especially salient to immigrant peoples and that traditional use of medicinal species perseveres through the process of human migration more than other plant use categories.

Ethnobotanical transfer over space, whether of species or knowledge, is suggested often in the names and descriptors of useful plants. Where plant names are derived from different languages, the likelihood of their acquisition from other groups is high. Thus, throughout the Americas, the frequency of indigenous plant names in Spanish, Portuguese, and French lexicons points to the colonial transfer of plant knowledge. In contemporary communities in which plants retain high use value, lexical features retain particular significance. For example, the Amazonian Ese Eja, who migrated into their present location over the last century, refer to a suite of species by their perceived ancestry. Species long employed for subsistence are referred to as 'of the ancestors' (etiikianaha), whereas those assimilated following migration are known as plants 'of the outsiders' (dejaha) (Alexiades, 2009). Likewise, among Brazil's African diaspora, plant names followed by the phrase 'da costa' (from the coast), refer not to littoral species but rather to plants that are known to originate from the coast of West Africa. Among the forebears of Brazil's forced immigrants, these resistance floras have attained almost sacred status (Voeks, 1997). Finally, Schiebinger

(2004) argues that there is much to be learned from the study of ethnobotanical knowledge that was not transferred. During the early colonial period, enslaved Africans and Indians in Dutch Suriname shared the abortifacient properties of the peacock flower (*Caesalpinia pulcherrima*) in order to prevent the birth of children into servitude. However, while the plant itself was introduced and thrived in Old World gardens, Europe's increasingly pro-natalist policies blunted the transatlantic transference of its abortive properties.

As testimony to the intimate relationship between people and plants, there is a powerful impulse elicited by many immigrants to recreate a semblance of their ancestral flora in their new environment. While some introductions become destructive invasives, many others encourage the process of cultural recovery among diaspora groups. As noted earlier, this feature is documented especially among the descendants of Africa's enslaved peoples in the Americas. Routes of recovery are sometimes straightforward, such as in the case of purple nutsedge (*Cyperus rotundus*). This globally significant invasive traces a formidable ethnobotanical legacy from the present day to pharaonic Egypt, where its aromatic rhizomes were brewed into as special perfume for scenting pig and goose fat, at the time ubiquitous Egyptian medicines (Fahmy, 2005). In West Africa, the aromatic tuber is chewed by people who seek to influence others in court cases. This and other magical applications diffused to Brazil during the slave trade, where the plant maintained one of its African lexemes—'ewe danda'. In addition to its magical powers of influence, Brazil's African diaspora learned to mix nutsedge with jurema (*Mimosa humilis*) in preparing a hallucinogenic wine. The use of jurema was resurrected from native Brazilian Indians, who had abandoned it long ago (Albuquerque, 2002). Other ethnobotanical geographies navigate more circuitous routes. The American peanut (*Arachis hypogaea*), domesticated in southern South America, had been naturalized by the Portuguese in West Africa by the mid-sixteenth century. It soon replaced the native Bambara nut (*Vigna subterranea*), a domesticated but less productive African tuber, and was incorporated into several regional pharmacopoeias (Blench, 1997). As a provision on slave ships, peanuts were (re)introduced over time into African-American kitchen gardens in the New World, where they retained their Bantu lexeme— goober nut. For members of the African diaspora who labored in the fields and forests, peanuts constituted an iconic food and medicinal species, one that from their perspective was native to both coasts of the Atlantic (Grimé, 1979; Carney, 2003).

26.9 FUTURE CHALLENGES: ETHICS AND ETHNOBOTANICAL EROSION

Ethnobotany is a growing interdisciplinary subfield. It has established itself in recent years as a legitimate and relevant academic endeavor. Although much of its literature continues to be of a baseline type, ethnobotanical research increasingly employs rigorous field methods and hypothesis testing. And it is truly an international enterprise, with important contributions coming from researchers from all parts of the globe.

In spite of growing student and researcher enthusiasm, ethnobotany currently confronts two convergent threats. The first is a product of its own entrepreneurial success. In the 1980s, tropical deforestation gathered steam as a pressing environmental claim roughly coeval with a resurgence of interest in the ethnobotanical method for drug plant discovery. Encouraged by seminal works such as Norman Myers's *The Primary Source* (1984) and Schultes and Raffauf's *The Healing Forest* (1990), scientists turned their gaze to exploration and exploitation of tropical healers and habitats. Fantastic claims regarding the potential economic value of preserving these 'pharmaceutical factories' elevated the narrative considerably in the public eye. Principe (1996) placed the total value of undiscovered drug plants at USD $15.5 billion per year, while Mendelsohn and Balick (1995) calculated the value to pharmaceutical companies of each undiscovered tropical taxon at USD $96 million. Although these heady values have not weathered careful scrutiny in the literature (Costello and Ward, 2006; Mendelsohn and Balick, 1997), the received wisdom that the torrid zone is brimming with green gold has persevered.

The notion that miracle drug discovery could help justify habitat protection has made for a compelling environmental narrative and in some ways has put ethnobotany on the map. Its conservation value was enhanced by the belief that indigenous conservators could benefit by sharing their deep understanding of these resources. Scientists would use this knowledge to encourage sustainable development of the 'fragile tropics' and to develop new foods and drugs. Local communities would be compensated for sharing their intellectual property. However, bioprospecting's mixed legacy, combined with the image touted by environmental entrepreneurs of natural habitats endowed with exorbitant monetary values, attracted the attention of developing world governments and human rights advocates. Stories circulated orally and in the press of North Americans who ruthlessly exploited the intellectual property of indigenous folk, procured patent rights, and then sought to enforce their ill-got hegemony over 'their' innovation. The much-cited predatory patents issued for the Amazon's hallucinogenic ayahuasca vine (*Banisteriopsis caapi*) and the Mexican Enola bean (*Phaseolus vulgaris*) did much to fuel the controversy. In the latter case, an American businessman was able to patent a bean cultigen he had purchased in Sonora, Mexico, and then demanded that companies compensate him for importation to the United States from Mexico of 'his beans'. Combined with the nineteenth-century theft of South American *Cinchona* seeds (for quinine) and *Hevea* seedlings (for rubber) by agents of Britain's Kew Gardens, bioprospecting in the eyes of many morphed from a politically benign strategy that encouraged natural area conservation to a 'sophisticated form of biopiracy' (Shiva, 2007).

Efforts to equalize the playing field have been equivocal. The International Society of Ethnobiology's Declaration of Belém (1988) and its later Code of Ethics (2006) specified that indigenous peoples would receive fair economic compensation for sharing their ecological knowledge. This was further elaborated by the 1992 UN Convention on Biological Diversity (CBD), which required benefit sharing between the source nation and the bioprospecting entity. It also stipulated benefit sharing for the local communities from which the knowledge was acquired. Further ethics declarations by relevant professional societies, such as the Society of Economic Botany, International Society for Ethnobiology, and the Society for Ethnopharmacology, outlined ethical guidelines for research, including issues such as prior informed consent, honesty with informants, anonymity, and equitable economic compensation.

Rulings in patent law and international trade agreements at about the same time undermined equitable benefits objectives. After much debate, the U.S. Patent Act, which protects the ownership of novel objects of human invention, was expanded (Chakrabarty decision, upheld by the U.S. Supreme Court in 1980) to allow individuals and companies to patent and monopolize the sales of genetically modified organisms. Provided that it had been sufficiently altered by humans, nature could now be patented. The 1993 Trade Related Aspects of Intellectual Property Agreement enforced this ruling by requiring that developing countries provide patent protection for all technology, including plant or gene patents that were exploited from these same countries. In response, most developing world governments have enacted legislation severely restricting ethnobotanical research, both by foreigners and nationals. Given the justifiable suspicion and bureaucratic red tape, ethnobotanical research is nearly impossible in

several of the world's most biodiverse countries (Berlin and Berlin, 2004; Conforto, 2004; Gómez-Pompa, 2004; Hayden, 2003; Soejarto et al., 2005).

Ironically, increasing restrictions on ethnobotanical fieldwork comes at a time when the future existence of traditional ethnobotanical knowledge is in question. Reports from nearly all rural communities concur that knowledge of botanical nature is eroding at an alarming rate. Habitat destruction in Africa, Asia, and the Americas is limiting access to plant species, as is unsustainable levels of useful plant collection, often to meet the demands of national and even international markets (Brosi et al., 2007; Bussmann et al., 2006; Kala, 2005; Kiringe, 2005). Global climate change, particularly in northern regions, already threatens traditional relations with nature (Dounias, 2008; Ford et al., 2006; Garibaldi and Turner, 2004).

The most pressing challenges to traditional ethnobotanical knowledge, however, appear to be the effects of culture change and modernization. Missionaries and religious zealots school their rural converts to abandon the use of medicinal plants (Steinberg, 2002; Voeks and Sercombe, 2000). Likewise, the immediate effectiveness of commercial drugs, as well as the status associated with pills and injections, translates to waning attraction for herbal remedies (Edwards and Heinrich, 2006; Lewis, 2003). Westman and Yongvanit (1995) report, for example, that increasing wealth is inversely associated with the knowledge of indigenous food plants in Thailand. Benz et al. (2000), working in Mexico's Sierra de Manantlan, note that socioeconomic marginality in some cases correlates with ethnobotanical importance values. Voeks and Leony (2004) suggest, alternatively, that relative prosperity in a rural Brazilian community is not associated with the degree of knowledge of the local pharmacopoeia, but that literacy and increasing access to formal education are negatively correlated with knowledge of medicinal plants. Reyes-Garcia et al. (2005) likewise report a weak association between economic status and ethnobotanical knowledge, but they conversely note a slightly positive association between years of formal education and plant knowledge. Quinlan and Quinlan (2007) also report an inverse relationship between years of formal education and medicinal plant knowledge in Dominica, although other indices of modernization, such as consumerism, exhibited a positive association. There appears to be, in any case, almost universal disinterest among young people in rural areas to assimilate the traditional ethnobotanical knowledge retained by elders (Begossi et al., 2002; Luoga et al., 2000). For many young people, their ethnoecological legacy connects them with a 'primitive' past with which they are keen to disassociate. Zarger and Stepp (2004), however, working with an ethnobotanical trail in a Tzeltal Maya community in Chiapas, Mexico, discovered that children's ability to name culturally significant plants had not diminished significantly over three decades, in spite of major socioeconomic changes in the community. Likewise, Lykke et al. (2004) failed to detect any age-dependent change in ethnobotanical knowledge among Sahelian Fulani. In spite of these studies to the contrary, there can be little doubt that ethnobotanical knowledge is declining at a precipitous rate throughout much of the world. Ironically, as the academic study of ethnobotany gathers momentum, the likelihood that there will be much left to study becomes increasingly remote.

REFERENCES

Aagesen, D. (2004) 'Burning monkey puzzle: Native fire ecology and forest management in northern Patagonia', *Agriculture and Human Values*, 21: 233–42.

Aguilar-Støen, M., and Moe, S. (2007) 'Medicinal plant conservation and management: Distribution of wild and cultivated species in eight countries', *Biodiversity and Conservation*, 16: 1973–81.

Albuquerque, U. (2002) 'A jurema nas practicas dos descendentes culturais do africano no Brasil', In C.N. Mota and U.P. Albuquerque (eds.), *As Muitas Faces da Jurema: De Espécie Botânica a Divindade Afro-Indígena*. Recife, Brazil: Bagaço. pp. 171–92.

Albuquerque, U., and Lucena, R. (2005) 'Can apparency affect the use of plants by local people in tropical forests?' *Interciencia*, 30: 506–11.

Alcorn, J. (1984) *Huastec Mayan Ethnobotany*. Austin: University of Texas Press.

Alexiades, M. (ed.). (2009) *Mobility and Migration in Indigenous Amazonia: Contemporary Ethnoecological Perspectives*. Oxford, UK: Berghahn.

Anderson, M.K. (2000) 'California Indian horticulture: Management and use of redbud by the southern Sierra Miwok', in P. Minnis, (ed.), *Ethnobotany: A Reader*. Norman OK: University of Oklahoma Press. pp. 29–40.

Anderson, M.K. (2005) *Tending the Wild: Native American Knowledge and the Management of California's Natural Resources*. Berkeley: University of California Press.

Atran, S. (1999) 'Itzaj Maya folkbiological taxonomy: Cognitive universals and cultural particulars', in D.L. Medin and S. Atran (eds.), *Folkbiology*, Cambridge, MA: MIT Press. pp. 113–203.

Austin, D. (2007) '*Merremia dissecta* (Convolvulaceae): Condiment, medicine, ornamental, and weed—a review', *Economic Botany*, 61: 109–20.

Bailey, R.C., G. Head, M. Jenike, B. Owen, R. Rechtman, R., and Zechenter, E. (1989) 'Hunting and gathering in tropical rain forest: Is it possible?' *American Anthropologist*, 91: 59–82.

Baker, H.G., and Stebbins, G.L. (eds.). (1965) *The Genetics of Colonizing Species*. New York: Academic Press.

Balée, W. (1994) *Footprints of the Forest: Ka'apor Ethnobotany—The Historical Ecology of Plant Utilization by an Amazonian People*, New York: Columbia University Press.

Balée, W. (1999) 'Mode of production and ethnobotanical vocabulary: A controlled comparison of Guajá and Ka'apor', in T.L. Gragson and B.G. Blount, (eds.), *Ethnoecology: Knowledge, Resources, and Rights*. Athens: University of Georgia Press. pp. 24–40.

Balick, M.J. (ed.). (1988) *The Palm—Tree of Life. Biology, Utilization and Conservation*, New York: New York Botanical Garden.

Balick, M.J., and Cox, P. (1996) *Plants, People, and Culture: The Science of Ethnobotany*. New York: Scientific Publications.

Balick, M., and Mendelsohn, R. (1992) 'Assessing the economic value of traditional medicines from tropical rain forests', *Conservation Biology*, 6: 128–30.

Barker, G.W. and 27 authors. (2007) 'The 'human revolution' in lowland tropical Southeast Asia: The antiquity and behavior of anatomically modern humans at Niah Cave (Sarawak, Borneo)', *Journal of Human Evolution*, 52: 243–61.

Begossi, A., Hanazaki, N., and Tamashiro, J.Y. (2002) 'Medicinal plants in the Atlantic forest (Brazil): Knowledge, use, and conservation,' *Human Ecology*, 30: 281–99.

Bennett, B. (2005) 'Ethnobotany education, opportunities, and needs in the US', *Ethnobotany Research and Applications*, 3: 113–21.

Bennett, B., Baker, M.A., and Gomez Andrade, P. (2002) *Ethnobotany of the Shuar of Eastern Ecuador*. New York: New York Botanical Garden.

Benz, F.F., Cevallos E.J., Santana M.F., Rosales A.J., and Graf M.S. (2000) 'Losing knowledge about plant use in the Sierra de Manantlan Biosphere Reserve, Mexico', *Economic Botany*, 54: 183–91.

Berlin, B. (1992) *Ethnobiological Classification: Principles of Categorization of Plants and Animals in Traditional Societies*, Princeton, NJ: Princeton University Press.

Berlin, B., and Berlin, E.A. (2004) 'Community autonomy and the Maya ICBG Project in Chiapas, Mexico: How a bioprospecting project that should have succeeded failed', *Human Organization*, 63: 472–86.

Blench, R.M. (1997) 'A history of agriculture in northeastern Nigeria', in D. Barreteau, R. Dognin, and C. von Graffenried (eds.), *L'Homme et le Milieu Végétal dans le Bassin du Lac Tchad*. Paris: ORSTROM. pp. 69–112.

Brosi, B.J., Balick, M.J., Wolkow, R., Lee, R., Kostka, M., Raynor, W., Gallen, R., Rayhor, A., Paynor, P. and Ling, D.A. (9 other authors). (2007) 'Cultural erosion and biodiversity: Canoe-making knowledge in Pohnpei, Micronesia', *Conservation Biology*, 21: 875–79.

Brown, C.H. (1985) 'Mode of subsistence and folk biological taxonomy', *Current Anthropology*, 26: 43–53.

Brown, C.H. (2000). 'Folk classification: An introduction'. in P. Minnis (ed). *Ethnobotany: A Reader*, Norman: University of Oklahoma Press, pp. 65–70.

Brush, S., and Stabinsky, D. (eds.). (1996) *Valuing Local Knowledge: Indigenous People and Intellectual Property Rights.* Washington, DC: Island.

Burkill, I. (1966) *A Dictionary of the Economic Products of the Malay Peninsula.* Kuala Lumpur: Government of Malaysia and Singapore.

Bussmann, R., Gilbreath, G., Solio, J., Lutura, M., Lutuluo, R., Kunguru, K., Wood, N., and Mathenge, S. (2006). 'Plant use of the Maasai of Sekenani Valley, Maasai Mara, Kenya'. *Journal of Ethnobiology and Ethnomedicine,* 2: 1–7.

Campbell, M. (2005) 'Sacred groves for forest conservation in Ghana's coastal savannas: Assessing ecological and social dimensions', *Singapore Journal of Tropical Geography,* 26: 151–69.

Carney, J. (2003) 'African traditional plant knowledge in the circum-Caribbean region', *Journal of Ethnobiology,* 23: 167–85.

Chazdon, R., and F.G. Coe. (1999) 'Ethnobotany of woody species in second-growth, old-growth, and selectively logged forests of northeastern Costa Rica', *Conservation Biology,* 13: 1312–22.

Christensen, H. (2002) *Ethnobotany of the Iban and Kelabit.* Kuala Lumpur: Forestry Department of Sarawak.

Clement, C. (1999) '1492 and the loss of Amazonian crop genetic resources: The relation between domestication and human population decline.' *Economic Botany,* 53: 188–202.

Cocks, M.L., Bangay, L., Wiersum, K.F., and Dold, A.P. (2006) 'Seeing the wood for the trees: The role of woody resources for the construction of gender specific household cultural artifacts in non-traditional communities in the Eastern Cape, South Africa', *Environment, Development and Sustainability,* 8: 519–33.

Coe, F.G., and Anderson, G.J. (1996) 'Ethnobotany of the Garífuna of eastern Nicaragua', *Economic Botany,* 50: 71–107.

Coley, P., and Barone, J. (1996) 'Herbivory and plant defenses in tropical plants', *Annual Review of Ecology and Systematics,* 27: 305–35.

Conforto, D. (2004) 'Traditional and modern day biopiracy: Redefining the biopiracy debate', *Journal of Environmental Law and Litigation,* 19: 358–97.

Coomes, O. (2004) 'Rainforest 'conservation through use?' Chambira palm fibre extraction and handicraft production in a land-constrained community, Peruvian Amazon', *Biodiversity and Conservation,* 13: 351–60.

Costello, C., and Ward, M. (2006) 'Search, bioprospecting and biodiversity conservation', *Journal of Environmental Economics and Management,* 52: 615–26.

Crosby, A.W. (1986) *Ecological Imperialism: The Biological Expansion of Europe, 900–1900.* Cambridge, UK: Cambridge University Press.

Dafni, A., Levy, S., and Lev, E. (2005) 'The ethnobotany of Christ's thorn jujube (*Ziziphus spina-christi*) in Israel', *Journal of Ethnobiology and Ethnomedicine,* 1: 1–8.

Davic, R. (2004) 'Epistemology, culture, and keystone species: A response to Garibaldi and Turner. 2004', 'Cultural Keystone Species: Implications for Ecological Conservation and Restoration'. *Ecology and Society* 9: http://www.ecologyandsociety.org/vol9/iss3/

Delang, C. (2006) 'Not just minor forest products: The economic rationale for the consumption of wild food plants by subsistence farmers', *Ecological Economics,* 59: 64–73.

Denevan, W.M. (1992) 'The pristine myth: The landscape of the Americas in 1492', *Annals of the Association of American Geographers,* 82: 369–85.

Denevan, W.M. (2002) *Cultivated Landscapes of Native Amazonia and the Andes.* Oxford: Oxford University Press.

Deur, D., and Turner, N. (2005) *Keeping it Living: Traditions of Plant Use and Cultivation on the Northwest Coast of North America.* Seattle: University of Washington Press.

Dounias, E. (2008). *Unravelling signals: The perceptions of climate fluctuations by forest dwellers.* XI Annual Meeting, International Congress of Ethnobiology, Cuzco, Peru, June 25th–30th.

Duke, J. (2001) *Handbook of Edible Weeds.* Boca Raton, FL: CRC Press.

Edwards, S., and Heinrich, M. (2006) 'Redressing cultural erosion and ecological decline in a far northern Queensland aboriginal community (Australia): The Aurukun ethnobiology database project', *Environment, Development and Sustainability,* 8: 569–83.

Ellen, R. (1998) 'Indigenous knowledge of the rainforest: Perception, extraction, and conservation', in B. Maloney, (ed.) *Human Activities in the Tropical Rainforest: Past, Present, and Possible Future.* Dordrech: Kluwer. pp. 87–99.

Elton, C.S. (1958) *The Ecology of Invasions by Animals and Plants.* London: Methuen.

Etkin, N., and Ross, P. (1991) 'Should we set a place for diet in ethnopharmacology?' *Journal of Ethnopharmacology,* 32: 25–36.

Fadiman, M. (2005) 'Cultivated food plants: Culture and gendered spaces of colonists and the Chachi in Ecuador', *Journal of Latin American Geography,* 4: 43–57.

Fahmy, A.G. (2005) 'Missing plant macro remains as indicators of plant exploitation in Predynastic Egypt', *Vegetation History and Archaeobotany,* 14: 287–94.

Ford, J., Smit, B., and Wandel, J. (2006) 'Vulnerability to climate change in the Arctic: A case study from Arctic Bay, Canada', *Global Environmental Change,* 16: 145–60.

Frampton, J. (1580) *Joyfull Newes Out of the New Found World.* (Translation of N. Monardes). London: William Norton.

Garibaldi, A., and Turner, N. (2004) 'Cultural keystone species: Implications for ecological conservation and restoration', *Ecology and Society,* 9: 1 http://www.ecologyandsociety.org/vol9/iss3/art1.

Godoy, R., Wilkie, D., Overman, H., Cubas, A., Cubas, G., Demmer, J., McSweeney, K. and Brokaw, N. (2000) 'Valuation of consumption and sale of forest goods from a Central American rain forest', *Nature,* 406: 62–63.

Gómez-Pompa, A. (2004) 'The role of biodiversity scientists in a troubled world', *BioScience,* 54: 217–25.

Gott, B. (2005) 'Aboriginal fire management in south-eastern Australia: Aims and frequency, *Journal of Biogeography*, 32: 1203–8.

Grimé, W.E. (1979) *Ethno-botany of the Black Americans*. Algonac, MI: Reference Publications.

Grove, R. (1996) *Green Imperialism: Colonial Expansion, Tropical Island Edens and the Origins of Environmentalism, 1600–1860*. Cambridge, UK: Cambridge University Press.

Harris, D. (1996) 'Domesticatory relationships of people, plants and animals', in R. Ellen and K. Fukui (eds.), *Redefining Nature: Ecology, Culture and Domestication*. Oxford, UK: Berg. pp. 437–63.

Hayden, C. (2003) *When Nature Goes Public: The Making and Unmaking of Bioprospecting in Mexico*. Princeton, NJ: Princeton University Press.

Headland, T., and Bailey, R. (1991) 'Introduction: Have hunter-gatherers ever lived in tropical rain forest independently of agriculture?' *Human Ecology*, 19: 115–22.

Howard, P. (ed.). (2003) *Women and Plants: Gender Relations in Biodiversity Management and Conservation*. London: Zed.

Huffman, M.A. (2001) 'Self-medicative behavior in the African great apes: An evolutionary perspective into the origins of human traditional medicine', *BioScience*, 51: 651–61.

Hunn, E. (1982) 'The utilitarian factor in folk biological classification', *American Anthropologist*, 84: 830–47.

Kala, C. (2005) 'Indigenous uses, population density, and conservation of threatened medicinal plants in protected areas of the Indian Himalayas', *Conservation Biology*, 19: 368–78.

Kiringe, J. (2005) 'Ecological and anthropological threats to ethno-medicinal plant resources and their utilization in Maasai communal ranches in the Amboseli region of Kenya', *Ethnobotany Research and Applications*, 3: 231–41.

Koizumi, M., and Momose, K. (2007) 'The Penan Banalui wild plant use, classification, and nomenclature', *Current Anthropology*, 48: 454–59.

Kufer, J., Grube, N., and Heinrich, M. (2006) 'Cacao in eastern Guatemala: A sacred tree with ecological significance', *Environment, Development and Sustainability*, 8: 597–608.

Lacuna-Richman, C. (2006) 'The use of non-wood forest products by migrants in a new settlement: Experiences of a Visayan community in Palawan, Philippines', *Journal of Ethnobiology and Ethnomedicine,* 2: 36, http://www.ethnobiomed.com/home/.

Leakey, R. and 11 authors. (2004) 'Evidence that subsistence farmers have domesticated indigenous fruits (*Dacryodes edulis* and *Irvingia gabonensis*) in Cameroon and Nigeria', *Agroforestry Systems*, 60: 101–11.

Lentz, D.L. (ed.). (2000) *Imperfect Balance: Landscape Transformations in the Precolumbian Americas*. New York: Columbia University Press.

Lewis, W.H. (2003) 'Pharmaceutical discoveries based on ethnomedicinal plants: 1985 to 2000 and beyond', *Economic Botany*, 57: 126–34.

Logan, M., and Dixon, A. (1994) 'Agriculture and the acquisition of medicinal plant knowledge', in N. Etkin, (ed.), *Eating on the Wild Side: The Pharmacologic, Ecologic and Social Implications of Using Non-Cultigens*. Tucson: University of Arizona Press. pp. 25–45.

Luoga, E., Witkowski, E., and Balkwill, K. (2000) 'Differential utilization and ethnobotany of trees in Kitulanghalo Forest Reserve and surrounding communal lands, Eastern Tanzania', *Economic Botany*, 54: 328–43.

Lykke, A., Kristensen, M., and Ganaba, S. (2004) 'Valuation of local use and dynamics of 56 woody species in the Sahel', *Biodiversity and Conservation*, 13: 1961–90.

Mander, M. (1998) *Marketing of Indigenous Medicinal Plants in South Africa: A Case Study in Kwazulu-Natal*. Rome: FAO.

Martin, G.J. (1995) *Ethnobotany: A Methods Manual*. London: Chapman and Hall.

Mendelsohn, D., and Balick, M. (1995) 'The value of undiscovered pharmaceuticals in tropical forests', *Economic Botany*, 49: 223–28.

Mendelsohn, D., and Balick, M. (1997) 'Valuing undiscovered pharmaceuticals in tropical forests', *Economic Botany*, 51: 328.

Mensing, S. (2006) 'The history of oak woodlands in California: Part II. The native American and historic period', *The California Geographer*, 46: 1–31.

Mercader, J. (ed.). (2003) *Under the Canopy: The Archaeology of Tropical Rain Forests*. New Brunswick, NJ: Rutgers University Press.

Minnis, P. E. (ed.). (2000) *Ethnobotany: A Reader*. Norman: University of Oklahoma Press.

Mistry, J., Berardi, A., Andrade, V., Krahô, T., Krahô, P., and Leonardos, O. (2005) 'Indigenous fire management in the cerrado of Brazil: The case of the Krahô of Tocantins', *Human Ecology*, 33: 365–86.

Moerman, D.E. (1998) *Native American Ethnobotany*. Portland, OR: Timber Press.

Momsen, J. (2004) *Gender and Development*. London: Routledge.

Myers, N. (1984) *The Primary Source: Tropical Forests and our Future*. New York: W. W. Norton.

Nabhan, G., Rea, A., Reichhardt, K., Mellink, E., and Hutchinson, C. (2000) 'Papago (O'odham) influences on habitat and biotic diversity', in P. Minnis (ed.), *Ethnobotany: A Reader*. Norman: University of Oklahoma Press. pp. 41–62.

Nesheim, I., Dhillion, S., and Stølen, K. (2006) 'What happens to traditional knowledge and use of natural resources when people migrate?' *Human Ecology*, 34: 99–131.

Ogle, B.M., Tuyet, H.T., Duyet, H.N., and Dung, N. (2003) 'Food, feed or medicine: The multiple functions of edible wild plants in Vietnam', *Economic Botany*, 57: 103–17.

Parsons, J.J. (1955) 'The Miskito pine savanna of Nicaragua and Honduras', *Annals of the Association of American Geographers*, 45: 36–53.

Peters, C., Gentry, A., and Mendelsohn, R. (1989) 'Valuation of an Amazonian rainforest', *Nature*, 339: 655–56.

Pfeiffer, J.M. (2002) 'Gendered interpretations of bio-cultural diversity in eastern Indonesia: Ethnoecology in the transition zone', in H. Lansdowne, P. Dearden, and W. Neilson (eds.), *Communities in Southeast Asia: Challenges and Responses.* Victoria, Canada: University of Victoria. pp. 43–63.

Pfeiffer, J.M., and Butz, R.J. (2005) 'Assessing cultural and ecological variation in ethnobiological research: The importance of gender', *Journal of Ethnobiology*, 25: 240–78.

Pfeiffer, J.M., and Ortiz, E.H. (2007) 'Invasive plants impact California native plants used in traditional basketry', *Fremontia*, 35: 7–13.

Pfeiffer, J.M., and Voeks, R. (2008) 'Biological invasions and biocultural diversity: Linking ecological and cultural systems', *Environmental Conservation*, 35: 281–93.

Pieroni, A., and Price, L. (eds.). (2006) *Eating and Healing: Traditional Food as Medicine*, New York: Food Products Press.

Pieroni, A., and Quave, C. (2005) 'Traditional pharmacopoeias and medicines among Albanians and Italians in southern Italy: A comparison', *Journal of Ethnopharmacology*, 101: 258–70.

Pieroni, A., and Vandebroek, I. (eds.). (2007) *Traveling Cultures and Plants: The Ethnobiology and Ethnopharmacy of Human Migrations.* New York: Berghahn.

Posey, D. (1984) 'A preliminary report on diversified management of tropical forests by Kayapó Indians of the Brazilian Amazon', *Advances in Economic Botany*, 1: 112–26.

Principe, P. (1996) 'Monetizing the pharmacological benefits of plants', in M. Balick, E. Elisabetsky, and S. Laird (eds.), *Medicinal Resources of the Tropical Forest: Biodiversity and Its Importance to Human Health.* New York: Columbia University Press. pp. 191–218.

Puri, R.K. (2005) 'Post-abandonment ecology of Penan fruit camps: Anthropological and ethnobiological approaches to the history of a rain forested valley in East Kalimantan', in M.R. Dove, P.E. Sajise, and A. Doolittle (eds.), *Conserving Nature in Culture: Case Studies from Southeast Asia.* New Haven, CT: Yale University Press.

Quinlan, M., and Quinlan, R. (2007) 'Modernization and medicinal plant knowledge in a Caribbean horticultural village', *Medical Anthropology Quarterly*, 21: 169–92.

Rashford, J. (2001) 'Those that do not smile will kill me: The ethnobotany of the ackee in Jamaica', *Economic Botany*, 55: 190–211.

Rashford, J. (2007) 'The species of *Ficus* trees that serve as Candomblé's cosmic tree of life', Paper presented at 48th Annual Society of Economic Botany Meeting, Chicago, IL.

Raven, P. and 44 authors. (2007) 'Declaration of Kaua'I', *Economic Botany*, 61: 1–2.

Reichel, E. (1999) 'Cosmology, worldview and gender-based knowledge systems among the Tanimuka and Yukuna (Northwest Amazon)', *Worldviews: Environment, Culture, Religion*, 3: 213–42.

Reyes-Garcia, V., Vadez, V., Byron, E., Apaza, L., Leonard, W., Perez, E., and Wilkie, D. (2005) 'Market economy and the loss of folk knowledge of plant uses: Estimates from the Tsimané of the Bolivian Amazon', *Current Anthropology*, 46: 651–56.

Rival, L. (2002) *Trekking Through History: The Huaorani of Amazonian Ecuador.* New York: Columbia University Press.

Rocheleau, D., and Edmunds, D. (1997) 'Women, men and trees: Gender, power and property in forest and agrarian landscapes', *World Development*, 25: 1351–71.

Roosevelt, A. and 16 authors. (1996) 'Palaeoindian cave dwellers in the Amazon: The peopling of the Americas', *Science*, 272: 373–84.

Salmon, E. (2003) 'Bringing the clouds home: The Hopi plant redistribution project', Twenty-sixth Society of Ethnobiology Conference, March 26–29, 2003, University of Washington, Seattle.

Sandhu, D., and Heinrich, M. (2005) 'The use of health foods, spices, and other botanicals in the Sikh community in London', *Phytotherapy Research*, 19: 633–42.

Sauer, C.O. (1963) 'Man in the ecology of tropical America', in J. Leighly (ed.), *Land and Life: A Selection from the Writings of Carl Ortwin Sauer.* Berkeley: University of California Press. pp. 182–93.

Schiebinger, L. (2004) *Plants and Empire: Colonial Bioprospecting in the Atlantic World.* Cambridge, MA: Harvard University Press.

Schultes, R., and Raffauf, R. (1990) *The Healing Forest: Medicinal and Toxic Plants of the Northwest Amazonia.* Portland, OR: Dioscorides Press.

Schultes, R., and von Reis, S. (eds.). (1995) *Ethnobotany: Evolution of a Discipline.* Portland, OR: Dioscorides Press.

Sercombe, P., and Sellato, B. (eds.). (2007) *Beyond the Green Myth: Borneo's Hunter-Gatherers in the 21st Century.* Copenhagen: Nordic Institute of Asian Studies.

Shaw, E. (1992) *Plants of the New World: The First 150 Years.* Cambridge, MA: Harvard College Library.

Shiva, V. (2007) 'Bioprospecting as sophisticated biopiracy', *Signs: Journal of Women in Culture and Society*, 32: 307–13.

Silva, V., da Andrade, L., and Albuquerque, U. (2006) 'Revising the cultural significance index: The case of the Fulni-ô in Northeastern Brazil', *Field Methods*, 18: 98–108.

Soejarto, D. and 27 authors. (2005) 'Ethnobotany/ethnopharmacology and mass bioprospecting: Issues on intellectual property and benefit-sharing', *Journal of Ethnopharmacology*, 100: 15–22.

Stagegaard, J., Sorensen, M., and Kvist, L. (2002) 'Estimations of the importance of plant resources extracted by inhabitants of the Peruvian Amazon flood plain forests', *Perspectives in Plant Ecology, Evolution, and Systematics*, 5: 103–22.

Steinberg, M. (2002) 'The second conquest: Religious conversion and the erosion of the cultural ecological core among the Mopan Maya', *Journal of Cultural Geography*, 20: 91–105.

Stepp, J.R. (2004) 'The role of weeds as sources of pharmaceuticals', *Journal of Ethnopharmacology*, 92: 163–66.

Stepp, J.R., and D.E. Moerman. (2001) 'The importance of weeds in ethnopharmacology', *Journal of Ethnopharmacology*, 75: 19–23.

Strike, S.S. (1994) *Ethnobotany of the California Indians. Volume 2: Aboriginal Uses of California's Indigenous Plants.* Champaign, IL: Koeltz Scientific Books.

Turner, N. (1988) ' "The importance of a rose": Evaluating the cultural significance of plants in Thompson and Lillooet Interior Salish', *American Anthropologist*, 90: 272–90.

Turner, N. (1994) 'Burning mountain sides for better crops: Aboriginal landscape burning in British Columbia', *International Journal of Ecoforestry*, 10: 116–22.

Vale, T.R. (2002) *Fire, Native Peoples, and the Natural Landscape.* Washington, DC: Island Press.

Van Driesche, J., and Van Driesche, R. (2004) *Nature Out of Place: Biological Invasions in the Global Age.* Washington, DC: Island Press.

Voeks, R.A. (1997) *Sacred Leaves of Candomblé: African Magic, Medicine, and Religion in Brazil,* Austin: University of Texas Press.

Voeks, R.A. (2002) 'Reproductive ecology of the piassava palm (*Attalea funifera* Mart.) of Bahia, Brazil', *Journal of Tropical Ecology*, 18: 121–36.

Voeks, R.A. (2004) 'Disturbance pharmacopoeias: Medicine and myth from the humid tropics', *Annals, Association of American Geographers*, 94: 868–88.

Voeks, R.A. (2007a) 'Are women reservoirs of traditional plant knowledge? Gender, ethnobotany, and globalization in northeast Brazil', *Singapore Journal of Tropical Geography*, 28: 7–20.

Voeks, R.A. (2007b) 'Ethnobotanical knowledge and mode of subsistence: Between foraging and farming in northern Borneo', in P. Sercombe and B. Sellato (eds.), *Beyond the Green Myth: Borneo's Hunter-Gatherers in the 21st Century.* Copenhagen: Nordic Institute of Asian Studies. pp. 333–52.

Voeks, R.A., and Leony, A. (2004) 'Forgetting the forest: Assessing medicinal plant erosion in eastern Brazil', *Economic Botany*, 58 (supplement): 294–306.

Voeks, R.A., and Nyawa, S. (2006) 'Dusun ethnobotany: Forest knowledge and nomenclature in northern Borneo', *Journal of Cultural Geography*, 23: 1–31.

Voeks, R.A., and Sercombe, P. (2000) 'The scope of hunter-gatherer ethnomedicine', *Social Science & Medicine*, 50: 1–12.

Waldstein, A. (2006) 'Mexican migrant ethnopharmacology: Pharmacopoeia, classification of medicines, and explanations of efficacy', *Journal of Ethnopharmacology*, 108: 299–310.

Westman, L., and Yongvanit, S. (1995) 'Biological diversity and community lore in northeastern Thailand', *Journal of Ethnobiology*, 15: 71–87.

Whitmore, T.M., and B.L. Turner II. (2001) *Cultivated Landscapes of Middle America on the Eve of Conquest.* Oxford: Oxford University Press.

Zarger, R., and Stepp, J. (2004) 'Persistence of botanical knowledge among Tzeltal Maya children', *Current Anthropology*, 45: 413–18.

Zent, E.L., and Zent, S.R. (1999) 'Is the *frailejón* a life form or an unaffiliated generic? Examining the rank of an endemic páramo plant', *Journal of Ethnobiology*, 19: 143–76.

Zimmerer, K. (2002) 'Report on geography and the new ethnobiology', *Geographical Review*, 91: 725–34.

27

Invasive Species, in Geographical Perspective

Mark A. Blumler

27.1 INTRODUCTION

Invasive species are taxa that are spreading spatially, into habitats, places, and regions in which they previously were absent. Normally these are alien species, or alien genotypes of native species, introduced from some other country or region. For the Western Hemisphere, species arriving after 1492 are regarded as introduced; most were dispersed in the company of humans, that is, only a few introductions were natural. For some other regions of the Earth the story is more complicated. For instance, a few aliens came to Hawai'i with the Polynesians, while a much larger contingent has followed the arrival of Europeans. In Europe, the introduction of alien species from the Near East, associated with the diffusion of agropastoralism, goes back many millennia. In all probability, many alien ecotypes of native species also have come to Europe from adjacent regions. Consequently, exotics are divided up into more categories in the European literature than in the United States and other former colonies. For example, 'neophyte' is a European term, only occasionally employed in the United States (Pysek et al., 2004).

Of course, many, perhaps most, introduced taxa never become invasive (Williamson and Fitter, 1996). Moreover, there are native invaders: mesquite (*Prosopis juliflora*), invading the desert grasslands of the southwestern United States, is a well-known example (Harris, 1966); another American example is Lyme disease (*Borrelia burgdorferi*), expanding slowly from three, widely spaced refugia (Blumler, 2003b). As Hengeveld (1990) pointed out, all species must invade (to expand their distribution) from time to time or will suffer extinction due to range retraction. The major concern today, however, is with alien invaders.

27.1.1 Why study invasives?

There are many reasons to study invading species. They are generally rated the second leading cause of the current extinction crisis (Diamond, 1989; Blumler, 2002), and they are the primary cause of extinction in some places such as Hawai'i. The other causes of extinction, such as habitat destruction and overexploitation, should be amenable to the solutions typically proposed for environmental problems, to wit, zero population growth and sustainability. But these would be of only limited help in fighting invasion: as long as humans trade and travel, species will be brought from one place to another, the consequence of which will be that some will invade and displace indigenous species (Blumler, 2002). On a practical level, therefore, management of invasives is a major subject of research, discussion, ecological management and, in the United States, government expenditure (www.invasivesinfo.gov). Some conservationists are also concerned about the homogenization of fauna and flora that the current wave of invasion presumably is causing, along with the potential for ecosystem collapse, but Vermeij (2005) and others have shown convincingly that the latter is unlikely. Still another concern is the possible loss of ecosystem services as biodiversity is reduced (Chapin et al., 2000), but as illustrated by the fact that invasives sometimes increase the productivity of the ecosystems they invade (e.g., Vitousek et al., 1987), this is a complicated issue.

Second, invasion is highly geographical, because invading species come from somewhere and invade (spread spatially across) somewhere else. There are manifold ways that this can happen, manifold types of species to which it can happen, and equally manifold resulting changes in ecosystems. All are of inherent interest to biogeographers.

Third, invasion is an example of a human-environment interaction, which since Carl Sauer's time has been a central concern of biogeography as practiced within American geography. Alien invaders almost always establish first under conditions of human disturbance. Many invading species come in association with agropastoralism, so there is a clear relationship to Sauer's (1952) interest in agricultural origins and dispersals. Crops and domestic livestock have been, and continue to be, introduced all over the planet and sometimes become invasive.

This connects to a fourth reason to study invasives: they include many cases of evolution in action. Crop-weed interactions are turning out to be important in many plant invasions, with some cautionary implications concerning GMOs (Ellstrand, 2003; Ellstrand et al., 1999). An increasing number of cases illustrate that invading species frequently evolve through hybridization and introgression (Blumler, 2003a; Ellstrand and Schierenbeck, 2000; Petit, 2004). Crops, weeds, and other agricultural pests and invaders constitute examples of rapid evolution, in a world otherwise dominated by extinction.

A fifth reason is that invading species have figured prominently in several theoretical debates because they were used as test subjects—because of their abundance, accessibility in weedy spots near or on university campuses, and often rapid completion of the reproductive cycle. But we can ask with some validity, just how representative are alien species? In a similar vein they now are being studied intentionally to see if they can shed light on fundamental ecological and evolutionary questions (e.g., Sax et al., 2005).

Invasive species are also of historical interest. The success of alien species in invading New World habitats played an important role in breaking down acceptance of the biblical creation story. If perfectly created by God, then each species should be best suited to its niche, and aliens should not be able to outcompete natives. Darwin's (1845) astonishment at the sight of masses of cardoon (*Cynara cardunculus*), a Mediterranean native, in the Argentine pampas was one of the observations that led him to his theory of evolution (see Blumler et al., this volume). The 'Columbian exchange' of crops,

weeds, livestock, 'varmints', and diseases after 1492 had huge impacts on human affairs (Crosby, 1972, 1986). The collapse of Native American populations after 1492 under the onslaught of Old World diseases was similar to the effects of chestnut blight, Dutch elm disease, and other pathogens on native plant species.

Natural invasions, such as the Great American Interchange (Marshall et al., 1982), have long been of interest to biogeographers. Darwin's recognition that the Galapagos finches must be the result of dispersal followed by adaptive radiation was the key to the development of his theory. Today, there is considerable interest in what natural invasions can tell us about the present, and vice versa (Vermeij, 2005). Natural invasions are often one-sided, with the larger and/or more diverse ecosystem typically contributing more invaders to the smaller or less diverse one than vice versa. Vermeij (2005) asserted that the larger or more diverse ecosystem 'wins' because its species are more competitive, well-defended, and reproductive, but such claims are difficult to prove. Natural invasions feature a homogenization of biota, which also is to be expected with the current invasions. For instance, in Hawai'i total species diversity has gone up due to the large number of introduced species, even though numerous natives have gone extinct. Thus, local diversity is greater, while global diversity is less. Rozenzweig (2001) has termed this mixing and homogenization of biota a 'New Pangaea', but Cox (2004) pointed out that alien invaders remain geographically separated from their source regions, so they should diverge and speciate in time.

The remainder of this essay primarily concerns invasive plants, reflecting my research interests, and frequently makes reference to California, where invasion has been spectacular (see below). California data and examples are used in this chapter to evaluate beliefs and theories about invasive species. Lomolino et al. (2006) term this approach the 'Comparative Method', a classic methodology in biogeography (see Blumler et al., this volume) that has become less widely utilized as scientists have become more specialized and therefore less able to compare across regions. As will be shown repeatedly in this chapter, generalizations about invasive species that are widely accepted in the literature often fail to hold up in California.

27.2 PATTERNS OF INVASION

The introduction of organisms to new lands or seas depends on trade and transportation patterns, and often on societal decisions regarding intentional import of exotic species as well. For instance, in the northeastern United States herbs from Europe dominated the early introductions. But in the last century or so an increasing number of aliens have come from eastern Asia, reflecting the increase in trade and intentional importation from there (Blumler, 2003b). Also, both Europe and east Asia are similar enough climatically to the northeastern United States that exotics have a chance of succeeding in the wild.

Anthropogenic environments, aquatic ecosystems, and oceanic islands are thought to be the most severely invaded regions of the world, though there are exceptions (e.g., Byrne, 1980). The massive impact on freshwater aquatic environments may be due primarily to the importance of boat transportation, as well as the similar nature of such environments around the globe. Australia has suffered enormously from faunal invasion, especially of placental mammals (Rolls, 1969). Mediterranean-climate grasslands, except in the Mediterranean itself, have suffered extreme levels of plant invasion (Blumler, 1995; di Castri et al., 1990; Groves and di Castri, 1991). The Mediterranean region of winter-wet, summer-dry climate extends from the Iberian Peninsula and Morocco to the western Chinese border—a much larger area than the other Köppen Cs climate regions of California, central Chile, southern and southwestern Australia, and the southwestern tip of Africa. Thus, it fits the pattern of one-sided invasion from larger to smaller ecosystems. The Mediterranean is a high-diversity region especially in annual plants, but it is no more diverse than the Cape Province of South Africa, which has been invaded not only by Mediterranean herbs but also by Australian and other woody plants. The Mediterranean is centrally located, while its climatic analogues are geographically isolated on the peripheries of land masses. This may partially explain the one-sided invasion if, as Darwin (1859) proposed, competition is more severe in continental interiors than on their margins; one would expect, then, greater invasion from centrally located to peripheral regions.

The invasion of California's low-elevation, valley grassland (Heady, 1977) was so great that it is not unusual for 80–90% of the cover to be nonnative (Bartolome, 1979). Thus, it constitutes probably the most spectacular recent case of terrestrial ecosystem replacement on a continental landmass (Blumler, 1995). This fact is not well recognized because of the tendency to compare statistics on the proportion of the flora that is introduced, rather than the extent to which aliens have taken over the landscape (e.g., Lonsdale, 1999; Rejmanek et al., 1991). Several northeastern states in the United States have alien flora about one-third of the total, while California's

alien percentage is only about one-sixth (though as great in absolute terms) (Rejmanek and Randall, 1994). But introduced species have been far more successful at penetrating into natural or seminatural habitats in California than in the northeast; in valley grassland, natives come back slowly or not at all when human disturbance ceases (Sauer, 1988; Bartolome, 1989; Blumler, 1992a, 1995; Seabloom et al., 2003). Floristic analysis speaks to the overseas dispersal phase of invasion. Though a precondition for invasion, obviously, this phase is not terribly relevant to invasion success.

Veblen (1975) noted that many of California's aliens are also present in the Guatemalan highlands but are not nearly so invasive there as on the West Coast. Also, many of the same or closely related bunchgrass species that supposedly dominated California before the arrival of the Spanish are present, and often abundant, in the Guatemalan highlands. Veblen offered a climatic explanation, suggesting that the lack of summer drought in Guatemala enabled the bunchgrasses to resist invasion. Blumler (1995) showed that, in general, herbaceous plant invasion of Crosby's (1986) 'neo-Europes' correlates with the degree to which the climate is winter-wet, summer-dry (i.e., Mediterranean). Even in New Zealand, supposedly vulnerable because of its island status, seminatural grassland areas are not nearly as severely invaded as in California (Allan, 1936; Blumler, 1995).

Darwin (1845) thought the spread of cardoon in the pampas must be the most successful case of invasion. But the cheatgrass (*Bromus tectorum*) invasion of the Intermountain West of North America covers a larger area (Mack, 1989), while the early spread of common wild oat (*Avena fatua*) in California was still more spectacular (Blumler, 1995). Among animals, one of the best known examples in the United States is that of the starling (*Sturnus vulgaris*), from New York's Central Park across the entire continent, despite founder effect and reduction in genetic diversity (Cabe, 1998).

Former British colonies, including California under American rule, have more introduced species than former Spanish ones with similar environments—reflecting more extensive trade by the British, and the fact that the British and their colonial descendants emphasized intentional plant imports, while the Spanish did not (Blumler, 2007; cf. Castro et al., 2005). It may be, too, that developed countries have experienced more introductions of exotics than less developed countries, though this remains to be investigated. Less developed countries are mostly located in tropical and subtropical regions, and none seem to have suffered invasion that can compare with Florida, which is sufficiently developed, and hence wealthy enough, to import enormous numbers of ornamentals, tropical fish, and organisms of potential economic benefit. Perhaps the greatest impact of invading species in poor countries has been from the widespread introduction of African grasses in the Neotropics (Parsons, 1972).

Sometimes, invading species radically alter ecosystems (Vitousek et al., 1996). For example, Vitousek et al. (1987) reported that the introduced shrub *Myrica faya*, which colonizes lava flows in Hawai'i, unlike all native colonizers, is a nitrogen-fixer; the soil fertility now increases much more rapidly after volcanic eruptions than it did before, and as succession proceeds, alien species that can take advantage of the increased soil fertility outcompete natives that 'expect' low fertility conditions to persist for a long time. Another example is the succulent annual ice plant *Mesembryanthemum crystallinum*, which is invasive in California. It concentrates salt in the upper levels of the soil and is more tolerant of these elevated salt levels than natives (Vivrette and Muller, 1977). Yet another well-known example is the riparian invader tamarisk (a hybrid of two *Tamarix* spp. [Gaskin and Schaal, 2002]), widely observed to lower the water table, to its own advantage and to the detriment of natives (Sala et al., 1996).

Plant invasion can interact with changes in biogeochemical cycling, as the example of *M. faya* illustrates. In southern California, extensive areas of native, coastal sage shrub vegetation are converting to alien-dominated, annual grassland (Minnich and Dezzani, 1998; Westman, 1979). Nitrogen deposition from automobile exhaust gases is responsible, so areas undergoing conversion are downwind of Los Angeles, while areas remaining as coastal sage are not. The increased nitrogen in the soil allows greater growth of the introduced annuals, which fosters fire that can kill the shrubs, but perhaps more important, it allows the annuals to use up the soil water more completely during the rainy season, leaving the shrubs severely stressed during the summer drought (Wood et al., 2006).

27.3 PERIODIZING THE STUDY OF INVASIVES

Research into invasives has shifted foci over time, being guided by changing concerns and paradigms, which reflect in part the different disciplines of the researchers who have become interested in the topic. Charles Elton's (1958) volume is universally designated as the seminal work that created the field of invasion ecology, but there were others before him. In fact, his book was based on a collection of case studies carried out by applied biologists in agriculture, forestry, entomology, and so on. Early research on invasion was largely pragmatic, focused on agricultural pests and others of economic concern. Research on plants emanated especially from Australia and California, reflecting the severity of their invasions, and the importance of public universities and agricultural research. Biological control, usually of alien species, dates to this pre-Eltonian period (e.g., Fenner and Ratcliffe, 1965; Huffaker and Kennett, 1959). Some taxa, such as pathogens, insect pests, and fish, have continued to see a strong component of applied research. This applied research, though less theoretical, was often excellent and frequently included an attention to the historical details of dispersal that unfortunately are largely ignored today. Applied research continues to the present day in the United States, but with decreasing public funding.

Around the same time that Elton wrote his classic book on invasion biology, an evolutionary perspective on invasion emerged in California with Baker and Stebbins (1965; Baker, 1955, 1965, 1974). Baker (1955) anticipated the current interest in invasion 'rules' with 'Baker's Law', i.e., self-fertilized species are more likely to establish a population after long-distance dispersal. Baker's (1965) hypothesis that aliens would be characterized by 'general purpose genotypes'— high on plasticity, short on genetic diversity, local differentiation, and adaptation—was based in part on Clementsian successional theory, and attempts to test the hypothesis have produced decidedly mixed results. Introduced species are highly variable in terms of genetic diversity and local adaptation. Genetic diversity reflects the number of times an alien is introduced, the number of individuals introduced, the geographical separation of source regions, and the degree to which the alien is able to hybridize with relatives after introduction (Blumler, 2003a; Cox, 2004; Kolbe et al., 2004).

Baker also emphasized the 'founder effect'; and, anticipating current trends, evolution of adaptation to the new environment via introgressive hybridization (Panetsos and Baker, 1967). Outside California, the evolutionary perspective did not inform invasion research until recently (Richardson and Pysek, 2006). Baker's student Spencer Barrett was an important contributor to current understanding (see below, e.g., Section 27.4).

Meanwhile, because so many plant invaders were weeds associated with disturbance, it is not surprising that for a long time they were stereotyped as 'ruderals' (disturbance-adapted) that would be outcompeted by natives in the absence of humans. In part this reflected the influence of succession theory (Clements, 1916). In California, many of the worst invaders are annuals, which can only be early successional in the Clementsian system. Moreover, Clementsian succession idealized climax, with a concomitant denigration of annuals and other pioneer species (Blumler, 1996). Clements (1934) claimed, without any real supporting evidence, that perennial bunchgrasses were the natural dominants of California's valley grassland, further entrenching the belief that invading annuals could not be very competitive. In part, also, the denigration of invaders reflected 'us versus them' thinking, with the natives as 'good' and the alien invaders as 'bad' (Blumler, 1992b). The extensive literature on invasion of Mediterranean climate regions, including influential volumes in the SCOPE series (di Castri et al., 1990; Drake et al., 1989; Grove and di Castri, 1991), reflects this perspective. It had a profound, and mostly counterproductive, impact on management (see Section 27.7). Because succession theory applies primarily to plants, scientists studying animal or microbe invaders were much less sidetracked by theory.

The Clementsian view persists, but two factors have led to its gradual decline: (a) increasing numbers of invasives are neither annuals nor weeds, but escapees after intentional introduction for horticulture or other purposes; and (b) management, predicated on the presumption that the invaders are disturbance-adapted, has repeatedly failed (Blumler, 1992b, 1998, 2002; Griggs, 2000; Seabloom et al., 2003). For unintentional introductions, disturbance (association with humans) is needed for overseas dispersal, and given the mediation of human dispersal initial planting is likely to be in a disturbed environment. Moreover, if the introduced taxon can survive in an anthropogenic habitat, that environment can serve as a staging area from which it may eventually disperse to a natural habitat to which it is suited (what are the odds that overseas dispersal will land a diaspore precisely in its preferred environment?). In competition with natives, the introduced taxon that can persist in some disturbed environment should benefit from mass effect (Shmida and Ellner, 1984): the rain of seeds from the disturbed into nearby undisturbed places. So it is fair to say that introduced species usually must tolerate human disturbance (even intentionally introduced species typically are planted first in human-affected places). But clearly they do not all require anthropogenic disturbance to persist (Sauer, 1988). For instance, Mensing and Byrne (1997) showed that filaree (*Erodium cicutarium*) invaded southern California before the Spanish-initiated settlement. Both filaree and common wild oat spread far out ahead of Spanish settlement, even ahead of the numerous feral cattle and horses (Blumler, 1995).

As the Clementsian view has declined in importance, there has been increasing emphasis on the advantage invaders obtain when they leave behind their pests and predators, which may control their numbers in their native realm (Wolfe, 2002). Studies have tended to show that the benefit to invaders comes both in reduced mortality and increased reproductive output. Some have argued for the evolution of increased vigor as predator defenses are lost, but there appear to be an equal number of counterexamples (Ricklefs, 2005). Unquestionably, leaving pests behind is an important, and geographical, part of the invasion story— it was the fundamental reason for the economic success of the tropical plantation system, for instance (though few plantation crops became invaders except on islands). But escape from predators cannot explain one-sided invasions, such as plants from the Mediterranean to regions of similar climate, or of European herbivorous insects of woody plants to eastern North America, despite the much lower diversity of woody species in Europe (Niemala and Mattson, 1996).

Recently, in conjunction with the ever-increasing concern over biodiversity loss, there has been a major push to develop predictive invasion rules (Enserink, 1999), model invasive spread (Higgins and Richardson, 1996; Peterson, 2003; Shigasada and Kawasaki, 1997; Thuiller et al., 2005), and improve monitoring through remote sensing and GIS (Asner and Vitousek, 2005). Invasion can be modeled as a wave front or can include jump dispersal, which produces outliers. For species that can vector themselves, such as birds and flying insects, this may be sufficient; for many plants, it may be preferable to incorporate likely dispersal routes such as railroad lines. Today, biogeographers increasingly appreciate that invasive species may be of theoretical interest, shedding light on fundamental ecological and evolutionary questions (Cox, 2004; Sax et al., 2005).

27.4 A GEOGRAPHICAL PERSPECTIVE

Invasion is both a geographical and an ecological process. For any given spatial distribution, there frequently can be both an ecological and a geographical explanation. For instance, Medley (1997) mapped the distribution of an invasive Asian shrub honeysuckle, *Lonicera maackii*, in a woodlot in Ohio. The shrub was abundant, especially around the edges, but was rarer and smaller in the interior. Medley offered an ecological explanation: the shrub cannot establish itself under the low light conditions in the forest interior. The geographical alternative would be that the species is invading from the edges and has not yet fully occupied the interior. In this case, as in many others, it would not be surprising if both explanations were valid. Medley's ecological explanation assumes equilibrium conditions; that is, that the invasion is complete; the geographical explanation assumes the opposite. The geographical component of invasion is more important in the early stages, especially in initial dispersal. Once invasion is complete, the ecological component takes over. (Obviously, neither is completely irrelevant at any stage.) For the most part, invasion biologists have sought ecological explanations, though there is increasing acceptance that propagule pressure (i.e., number of introductions and numbers of individuals introduced) is important (Kolar and Lodge, 2001; Lockwood et al., 2005; Thuiller et al., 2005; Williamson and Fitter, 1996). Biologists also tend to seek explanation in terms of biotic factors, while in geographical perspective the abiotic, especially climate, may be more significant (Blackburn and Duncan, 2001; Moyle and Light, 1996; Roura-Pascual et al., 2004; Thuiller et al., 2005).

With all the focus on invasion from ecologists and conservation biologists, supplemented by population genetics research, the missing link has been study of the actual historical geography of invasion. The focus has been on invasion ecology rather than on how invaders get overseas, and then spread (Kolar and Lodge, 2001). Some biologists even argue that all species already must have been transported everywhere (Jackson, 1985; Roy et al., 1991), but this is clearly contradicted by the empirical evidence that aliens continue to be introduced and to invade (Cohen and Carlton, 1998; Randall et al., 1998). The argument reflects a desire to treat invasion as an equilibrium situation, when by definition it is not. Even Elton (1958) predicted that all species eventually would be taken everywhere by humans, reflecting an unsophisticated understanding of trade and transportation.

Barrett (1983; Barrett and Shore, 1989) and Mack (1991; Mack and Erneberg, 2002; Novak and Mack, 2001) are unusually 'geographical' biologists, in that they have paid considerable attention to means of introduction and routes of spread. So are numerous applied biologists (e.g., Cohen and Carlton, 1998; Ruiz and Carlton, 2003). For instance, Forcella and Wood (1984; Forcella et al., 1986) noted that a plant species with a large native range is more likely to be introduced overseas than congeners with restricted ranges. They suggested that a species with a widespread distribution was more likely to intersect a dispersal route than those with smaller ranges. Using the example of the annual bromegrass (*Bromus*) species, Roy et al. (1991) offered an ecological alternative, that more widely spread

species also are superior in overall adaptation. But in a global analysis, Blumler (2006) showed that each Mediterranean climate region has only those annual bromegrass species found along the early trade routes to that region—evidence that strongly supports Forcella and Wood's hypothesis.

From a geographical perspective, the invasion process involves routes, bottlenecks, and staging areas. To determine these routes, as well as what the aliens have replaced, and how they may have changed native ecosystems, one must pore through historical and documentary evidence, herbarium records, and so on (e.g., Delisle et al., 2003; Minnich, 2008). Blumler (2006, 2007) found that there were three main early routes along which Old World species could disperse to California. Figure 27.1 shows those routes. The earliest route was via a staging area in the Mexican highlands (no ships sailed from Spain to the west coast of what is now the United States, except those of a single scientific expedition). The second route was via temperate South America, especially central Chile with its Mediterranean climate and agricultural exports to California. The third route was from Europe to the northeastern United States and then to the Pacific Coast. The staging areas along the first and third routes are temperate but not Mediterranean in climate, so any Mediterranean species passing along those routes would need to tolerate non-Mediterranean conditions. Unsurprisingly, the earliest invaders were species like filaree and common wild oat that grow in the Mediterranean but also in northwestern Europe. Mediterranean endemics, on the other hand, could only come via Chile. Using the evidence Californian weed biologists had gathered when funding for comprehensive studies about time, place, and manner of introduction was more forthcoming (e.g., Hendry, 1931; Robbins, 1940), one can begin to construct a general chronology. One conclusion is that several 'Mediterranean' species are represented in the state by non-Mediterranean ecotypes and consequently are less invasive than they might be. Another is that there remain a huge number of potential invaders in the Mediterranean region, species that are not distributed in contact with a likely dispersal route to California.

Trade routes and travel are far more complicated today, so current and future patterns of introduction are perhaps best correlated with government trade statistics (e.g., Thuiller et al., 2005). An increasing proportion of introductions are intentional, though perhaps not as much as Mack (2003; Mack and Erneberg, 2002) has claimed, given that his approach was floristic and covered only the United States. Intentional introductions can come from almost anywhere, but even they may subsequently spread along highways,

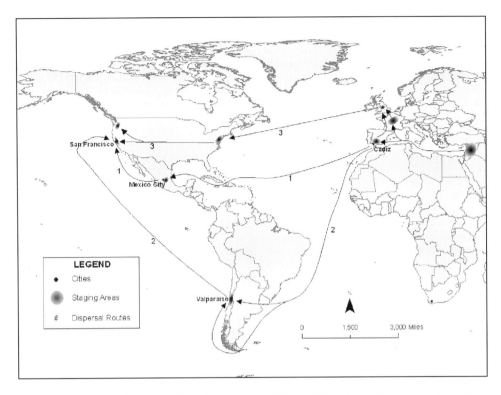

Figure 27.1 Major early routes from Europe to California (Blumler, 2006). Reproduced by permission of the Middle States Geographer.

railroad lines, or other identifiable routes (Gelbard and Belnap, 2003; Sawicki, 1998). It is now possible to use molecular evidence to improve our understanding of origins and dispersals of exotics and of any evolutionary adjustments made along the way (e.g., Novak and Mack, 2001; Tsutsui et al., 2001). While molecular studies are producing fascinating evidence about some taxa, the operative word is 'some'. Such studies are carried out one taxon at a time, so it will be many years before we have a comprehensive record. Thus, the need for traditional information (herbaria, museum specimens, and documentary evidence [Suarez and Tsutsui, 2004]) will continue.

A geographical perspective then does not preclude the ecological; rather, it complements it. While some geographers have traced dispersal routes (e.g., Brothers, 1992), others have engaged in ecological studies of invasives (e.g., Beatty and Licari, 1992). On the whole, though the emphasis on the ecological over the geographical has meant that (bio)geographers have not contributed to the invasion literature as much as potentially they could (and ideally should). Also, painstaking reconstruction from documentary evidence is

time-consuming and seldom fundable, and the results are often difficult to publish given the general denigration of natural history in biology.

27.5 INVASION THEORY AND INVASION RULES

In recent years, some ecologists have attempted to develop invasion rules to predict which species are likely to become invasive if introduced (Enserink, 1999). One could then blacklist the problematic species. Many others doubt that any such rules exist. Still others argue that prediction is not achievable, but statistical relationships can be found (Williamson and Fitter, 1996). The proposed rules are mostly ecological, and many are derived from limited empirical data. Schierenbeck et al. (1998) documented that there seem to be exceptions to all invasion rules.

Williamson and Fitter (1996) proposed a 'tens rule', that about 10% (they suggested a range from 5–20%) of the species dispersed will survive, a similar percentage of the remainder will establish in nature, about a tenth of those become invasive, and so on. Unquestionably, it is true that most alien species do not become invasive; in that sense, surely, the rule is valid. But regions vary: for instance, despite similar numbers of introduced species, plant invasion in the northeastern United States has been serious, while in mediterranean-climate California it was catastrophic. Moreover, the worst invaders into the northeastern United States tend to be Asian in origin, despite the low number of Asian taxa introduced compared to those from Europe. Finally, Williamson and Fitter noted that propagule pressure was a factor: the more individuals introduced, the better the chances of success. But over time, more individuals will be introduced, meaning that the percentage of aliens becoming invasive also should increase. Thus, even though their rule is only statistical, not predictive, it suffers from being time- and space-contingent.

Kolar and Lodge (2001) performed a meta-analysis of the invasion rules literature. Generally speaking, the evidence for ecological rules was weak or conflicting (Blumler, 2007). Their strongest case for an ecological rule was for an inverse relationship between seed or egg weight and invasiveness—based on eight studies, four showing the putative inverse relationship, and four showing no pattern. But Kolar and Lodge did not include Baker's (1972) massive California study, with a greater sample size than all eight Kolar and Lodge studies combined. Baker found a consistent pattern of greater seed size in aliens versus natives, a result that is contrary to the inverse rule proposed by Kolar and Lodge. Blumler (2007) showed that even when a proposed ecological invasion rule seems valid, it is likely to hold only for a specific spatiotemporal context, that is, it will not hold up for other places or for the same place at a later date. Even for the limited context in which it may be valid, there will be exceptions. On the other hand, the few geographical rules that Kolar and Lodge examined were well supported. As Kolar and Lodge (2001: 203) put it, "The most frequent and strong result in these studies was that successful establishment was positively related to propagule pressure." Some biologists are taking note of the better correlation with geographical factors, including climate, than with ecological ones, and they are developing blacklist models for specific regions on that basis (Peterson, 2003; Thuiller et al., 2005).

Darwin's (1859) 'Naturalization Hypothesis' (DNH) postulates that invading species will not be closely related to natives since the struggle for existence between them would be fierce. Several ecologists have tested the DNH recently, with conflicting results (reviewed in Richardson and Pysek, 2006). Rejmanek (1996) provided data on the proportion of grasses introduced from Europe to California that are in native genera and argued that the results supported the DNH. His floristic approach ignored invasion severity. If he had considered only the most invasive grasses, he likely would have reached the opposite conclusion as most of the major invaders of valley grassland are related to paleoinvaders that came to North America from Eurasia (Blumler, 1992b).

Darwin was correct in proposing that competition between congeners is likely to be fierce—that is, that they are likely to occupy the same or closely similar niche—but this suggests the alternative hypothesis that native species might be especially vulnerable to introduced congeners. It is a truism in ecology that competitive exclusion can only occur if two species occupy the same or very similar niche. Examples such as the annual ice plant (Vivrette and Muller, 1977) suggest that an alien invader unrelated to natives may be able to outcompete a wide swath of species if it possesses an adaptation that the natives have never encountered. On the other hand, introduced species closely related to natives may, as discussed above, severely compete with those natives precisely because they are similar in adaptation.

Elton (1958) hypothesized that species-rich communities would be more resistant to invasion than less diverse ones. Recent decades have seen numerous, hopeful attempts to verify this hypothesis (Cox, 2004), with unclear results (Ricklefs, 2005). In contrast, Huston (2004) interprets the literature as supporting the view that species-rich systems are more likely to be invaded, because aliens are not fundamentally different than natives: whatever favors diversity of natives will also favor diversity of introduced species. In the valley grassland, invasion was so complete that there is no way of knowing what the original diversity was. Even when invasion is less severe, it is rare that we have pre-invasion diversity data. But we do know empirically that desert areas have lower diversity than adjacent mediterranean-climate regions (Blumler, 1984), and in California invasion of the deserts has been much less than in valley grassland. This pattern fits with Huston's argument. But on the other hand, it also is well established that serpentine grassland in California is more diverse than adjacent nonserpentine grassland, yet much less invaded (e.g., McNaughton, 1968)—a pattern that supports Elton's hypothesis. These conflicting results are congruent with the notion that dispersal is still the major factor, since few desert species are associated with agriculture or major trade hubs, while the bottlenecks constraining the overseas dispersal of serpentine-adapted species are severe (Blumler, 2007).

Currently, the most popular explanation for invasion is Davis et al.'s (2000) fluctuating resource hypothesis, which states that invasion is favored when resources become available, either due to addition of a resource—nitrogen deposition in southern California, discussed above, is a pertinent example—or when destruction of vegetation frees up existing resources. Mediterranean-climate annual grassland experiences a resource pulse every fall after the first rains, which decompose the previous year's biomass. Thus, under this hypothesis mediterranean annual grassland should be highly invasible, as in fact it has been, everywhere but in the Mediterranean itself. But that single striking exception seems sufficient to raise questions regarding the generality of the fluctuating resource hypothesis.

Thus far, most valley grassland invaders are fertility-adapted, reflecting their dispersal along with agricultural crops. Fertility-adapted species would be expected to benefit from the addition of resources, but would the hypothesis hold up for species adapted to poor soils (serpentine, for instance)? More to the point, all ecosystems must undergo fluctuations in available resources, meaning that all are invasible. Hence, it is difficult to see how the hypothesis could be predictive in any useful sense, despite the authors' claims.

Although many of these rules or theories are fascinating, they amount to putting the ecological cart before the dispersing horse:

1. the data sets upon which the rules are based are too small;
2. information about how invasion occurred is lacking, reflecting the neglect of natural history that pervades biology; and
3. emphasis on the ecological presumes that the invasion process is close to equilibrium, when for most taxa and most places it is not.

No wonder there is no consensus concerning the attributes of a successful invader, as Ricklefs (2005) pointed out. The overgeneralization characteristic of these ecologically based propositions can be severely misleading at a policy level. Since we are still in the early, more dispersal-based stages, the emphasis now needs to be on geographical rules (or better, relationships, as these surely will not be predictive). As for the possibility of ecological invasion rules, we may know better when invasion has run its course (possibly several thousand years from now).

27.5.1 Geographical invasion rules

Blumler (1995) proposed a 'first spreads most' rule: the species that happens to be introduced to a highly invasible region first will essentially be able to invade as a wave front, while subsequent aliens will need to compete with already established invaders. They may be able to outcompete previous invaders in some habitats but probably not all, meaning that there will be bottlenecks or barriers to their dispersal that did not exist for the earlier arrivals. As the number of successful invaders increases, the need to disperse directly to an optimum habitat becomes ever more critical; therefore, invasion becomes more difficult and/or protracted. This may explain the extreme rapidity of some early invasions, such as cheatgrass in the Intermountain West, and common wild oat in California. Blumler (2006) further suggested that ports of entry that formerly were wide open to Mediterranean species invasion have now been largely altered by development to present large areas of irrigated surface (lawns, city parks) or pavement. Thus today, it is less likely that a newly introduced propagule will arrive at an area with a mediterranean water regime.

Blumler (2007) listed additional geographical rules, with the caveat that these are far less predictive than those that invasion biologists seek. Some of these were already recognized: for example, larger ecosystems typically invade smaller ones rather than small ones invading large ones, and climate is an important determinant of invasion success. In addition, he proposed that adaptation to 'normal' conditions, of soil, for example, is likely to reduce bottlenecks, thus enhancing the likelihood of invasion. Also, weediness, or the ability to survive under human impacts, facilitates passage through climatic and other bottlenecks.

27.6 EVOLUTION OF ALIENS

An introduced species is likely to differ genetically from in its source region, if only because of founder effect and genetic drift. Evolution under natural selection is likely, especially for abundant, short-generation time organisms such as insects. The often rapid evolution of insects and weeds within the agricultural context has long been studied. Thus, it should not be surprising that alien species often evolve adaptations to their new environments and even evolve increased invasiveness. More and more examples are being reported (Cox, 2004). Many of these are 'just so stories' that are plausible but poorly discriminated from alternatives such as founder effect or multiple introductions (e.g., Preston, 1993), but some seem clearly due to natural selection (e.g., Huey et al., 2000; Milne and Abbott, 2000). For instance, biological control organisms frequently evolve to attack native (nontarget) species. In all likelihood, invasion forces evolutionary adjustments among the natives, as well (Carroll and Dingle, 1996), though this remains a largely unexplored subject.

Introgressive hybridization is particularly interesting, since it is inherently a spatial process and can cause rapid evolution and dramatic changes in distributions (Blumler, 2003a). Crop geneticists and weed biologists have long appreciated the importance of hybridization and introgression in plant evolution, including speciation, but the continuing influence of the biological species concept until recently has caused many evolutionary biologists to underestimate introgression's importance (Arnold, 1997). In the past two decades introgression has returned to the fore, and the number of cases involving exotic species is ballooning (Abbott et al., 2003; Cox, 2004; Ferdy and Austerlitz, 2002; Reznick and Ghalambur, 2001). Cases of alien species eliminating related natives through introgression and genetic swamping are being reported (Huxel, 1999; Rhymer and Simberloff, 1996; Roush, 1997). Entirely new species sometimes appear through allopolyploidy, for example, in salsify (*Tragopogon*) in the Pacific Northwest (Owenby, 1950), and cordgrass (*Spartina*) in salt marshes in the British Isles. *S. anglica* is the allopolyploid derivative of the hybrid between the native *S. maritima* and the introduced *S. alternifolia*; it is spreading at the expense of the native, and also colonizing areas that formerly were bare of vegetation, extending the marshes seaward (Hacker et al., 2001). This raises the fascinating, but apparently unasked question, why couldn't any native species evolve the ability to colonize those areas?

Unsurprisingly, plant pathogens also may hybridize and introgress as they invade, with increased virulence sometimes being the result (Brasier, 2001). Emerging diseases at times come out of human-dominated environments and then invade nature. Brasier noted that the genus *Phytophthora*, including major pests of several plants of economic importance, flourishes in nursery conditions. Geographically separated races or species are repeatedly brought together in nurseries, then hybridize. This in turn likely leads to new pests. Sudden oak death (*P. ramorum*), which is attacking a remarkable array of woody species in California, may have originated in this fashion.

Barrett and Husband (1990) noted that alien species often pass through a dispersal bottleneck before introduction and are left with reduced genetic variability. Barrett suggested that introgression would be a means of restoring the lost genetic variability, andadapting to a wide range of environments in the new land:

> A more insidious and less appreciated mechanism promoting invasiveness is the potential mixing of genetically differentiated population systems within outcrossing species in their alien ranges. . . . Out of such a diverse 'hybrid soup' inevitably comes genetic combinations with novel phenotypes. While the majority are usually maladapted, some will eventually display high fitness and superior colonizing ability. Further selection aided by abundant genetic variation will refine these phenotypes to local conditions. The expansion and mixing of plant distributions, aided by the globalization of world trade and the burgeoning horticultural industry, seem likely to provide more opportunities for the future genesis of new plant invasions. (Barrett, 2000: 132–33)

Blumler (2003a) commented that there is no reason to think this would apply only to outcrossers. So-called self-fertilizing species, such as most of the invasive Mediterranean annuals in valley grassland, typically exhibit about 2% outcrossing, and given the extremely high number of seedlings in an annual grassland, this should translate to a potential for rapid genetic change. He also pointed out that the new, invaded land is unlikely to be identical to the native range in all respects. The novel phenotypes generated through mixing of genotypes from different source regions, or crossing between related species, should enhance adaptation to those features of the alien environment that are new or different. Finally, Blumler (2003a, 2007) pointed out that opportunities for introgression are especially common between crops and their weed relatives. While some weeds will have reduced genetic variability as a result of dispersal bottlenecks, crops typically are multiply introduced and therefore possess considerable genetic variation. Crop breeders historically imported germplasm from many different regions (and still do), further increasing the likelihood of novel phenotypes appearing after crosses within the crop, or between it and its wild or weedy relatives. Because crops tend to be adapted to mesic, fertile conditions, hybridization with a weed relative tends to increase invasiveness in such environments. Panetsos and Baker's (1967) study of radish is an example. After introgression from wild radish (*Raphanus raphanistrum*), the domestic radish (*R. sativus*) became a major invader of fertile, mesic environments. Ellstrand and Schierenbeck (2000) reported 27 cases of increased invasiveness following hybridization, many involving crops. One of the disquieting implications of the prevalence of introgression is that transgenes (GMOs) are escaping into weed relatives (Ellstrand, 2003; Cox, 2004).

27.7 MANAGEMENT

As pointed out above, invasion is an intractable problem, with no obvious or easy solution; sustainability is irrelevant. Since predictive invasion rules do not seem to hold much promise, there are increasing calls from conservation organizations, such as Nature Conservancy, for a whitelist policy: allowing in only those species likely to be noninvasive. But even a whitelist is unlikely to be problem-free, given the possibilities for evolutionary change once a species is introduced. Biological control is often employed to manage invaders, though usually those that are economic rather than conservation threats. But biological control organisms frequently become invasive themselves. Weevils introduced to control the musk thistle (*Carduus nutans*) in the United States have spread to 23 species of *Cirsium*, also a thistle, including many rare natives (Pemberton, 2000). In Hawai'i, the impact on tropical rain-forest food webs of parasitoids introduced to control Lepidopteran agricultural pests is astonishing (Henneman and Memnott, 2001).

A common approach, often tried in California, is to remove human disturbance under the presupposition that natives will outcompete the invaders, but this has generally backfired (Blumler, 1992b, 1998; Griggs, 2000). Disturbances such as fire and grazing tend to decrease the component of introduced species. Alternatively, native perennial bunchgrasses are planted in a dense sward, which eliminates many exotics but, unfortunately, also eliminates many natives. This approach reflects the 'us versus them' thinking alluded to previously, that all natives are good and all aliens are bad. The management goal in this case is to create native purity, but unwittingly at the expense of native diversity. Native diversity should be the management goal (Blumler, 2002).

Similar impacts of Clementsian successional theory are seen elsewhere. After the eruption of Mt. St. Helens, the introduced legume bird's-foot trefoil (*Lotus corniculatus*) was planted to prevent soil erosion, but it had the unanticipated effect of inhibiting the return of the forest (Lovett, 2000). Abandoned coal mines often are sown with introduced grasses and legumes to prevent soil erosion and acid mine drainage. The assumption, now widely questioned, was that herbs could only facilitate the return of the forest, when in fact, they often inhibit or prevent altogether native tree species regeneration (Brown, 1996).

Undeniably, some introduced species are beneficial. For instance, agricultural productivity would suffer enormously if we were to restrict ourselves to native crops. But even some domesticates, such as feral pigs, can become serious invaders, illustrating that the line between a good and a bad alien is often a thin one. Biological control organisms are another example. Even introduced earthworms, formerly almost universally regarded with favor, are now seen as a major threat to ecosystems that have no native worms, such as the formerly glaciated regions of eastern North America (Hendrix and Bohlen, 2002). Consequently, opinions vary concerning the circumstances that call for invasive species control (Westman, 1990; Chapter 28).

Monitoring and mapping of invaders will remain an important component of management. Remote sensing of aliens is widely attempted. It is most successful when the invader is large, clonal, patch-forming, distinctive, and in low-diversity surroundings; an example is Albright et al.'s (2004) study of water hyacinth (*Eichhornia crassipes*) in Lake Victoria. But remote sensing will probably never be useful for finding small outlier populations of relatively small or indistinct (e.g., many grasses) species; and control of outliers is more tractable and more important than attempting to deal with areas where the alien already is massively spread: 'old-fashioned' fieldwork will continue to be necessary to identify such outliers.

A geographical approach to invasion management would emphasize eradication of outliers and dispersal control. Agricultural pests already are managed in part through dispersal control, by blacklists enforced at state boundaries, and intense control efforts when a known pest is first detected. Dispersal control also could and probably should be applied to invaders that are a conservation threat (Blumler, 2005). One would map known distribution, identify means of spread, and monitor along likely dispersal routes. Control would be attempted where the dispersal routes intersect with nature preserves.

27.8 INVASIVES AND SOME THEORETICAL DEBATES

A surprising number of empirical studies on invasives have been carried out to address theoretical questions or were otherwise influential. This was particularly so in California, with research emanating from prestigious universities such as Berkeley, Stanford, and Davis. Introduced plant species are so dominant in the state that they, rather than natives, frequently became the subject of study. A single example is reviewed here.

27.8.1 The neutralism debate

When Kimura (1968) proposed that many mutations are neutral, hence that much evolutionary change happens without natural selection, a heated debate broke out. Employing the newly developed technique of electrophoresis, selectionists attempted to disprove Kimura by presenting evidence on the distribution of multilocus genotypes of annual grasses. The most dramatic example was that of the slender wild oat (*Avena barbata*), which Robert Allard and his students at Davis investigated. Clegg and Allard (1972) found two genotypes: a xeric genotype occurring almost everywhere, and a mesic genotype concentrated in the San Francisco Bay Area. Recombinants between the two were also concentrated in the Bay Area and were infrequent (Hamrick and Allard, 1972) (Figure 27.2). Allard argued that such a simple pattern must be due to selection,

eliminating the many other genotypes that must have been introduced. Hedrick and Holden (1979) showed mathematically that the pattern could be due to hitchhiking, if only those two genotypes were introduced. Because slender wild oat is self-fertilizing, recombination would be slow, and even after 200 years in California the two introduced genotypes might predominate, even if they were selected against. Allard's explanation was ecological; Hedrick and Holden's was implicitly geographical. Allard presumed an equilibrium distribution of the genotypes—he drew incorrect rainfall maps in his attempts to show that the xeric genotype always grows in drier environments than the mesic—whereas under the hitchhiking hypothesis one would expect an increase in recombinants over time and changes in distribution as some of these succeeded (Blumler, 2000, 2003a).

The crux of the matter was the number of genotypes introduced to California. Allard repeatedly insisted that there had been many introductions, but he never actually investigated the matter. Neither did Hedrick and Holden, who only wished

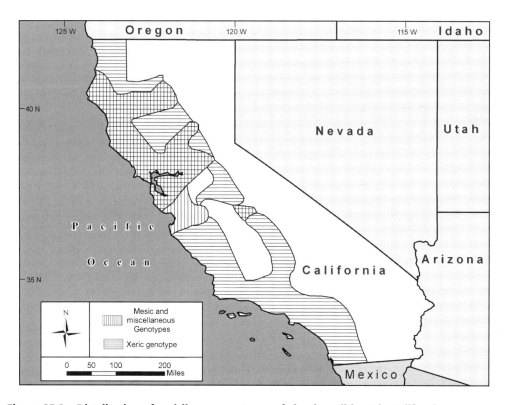

Figure 27.2   Distribution of multilocus genotypes of slender wild oat in California, according to data gathered by Allard and students, circa 1970. Distributions have changed subsequently (Blumler, 2000)

to make a theoretical point concerning the potential significance of hitchhiking. Left unnoticed, though worth pointing out, is that hitchhiking is likely to be pronounced during the early stages of invasion when there are relatively few introduced genotypes, and it should decline in significance as the number of genotypes increases (Blumler, 1995, 2003a).

I examined the massive (over 600 specimens) herbarium record for slender wild oat in the state. Until the 1920s, only the xeric genotype is present. The mesic genotype first appears in the Bay Area, several decades after multiple introductions of a hypermesic, crop mimic ecotype of the closely related sand oat (*A. strigosa*). It turns out that the mesic genotype is a hybrid between the xeric genotype and the sand oat (evidence will be presented elsewhere). The latter could not invade on its own because of its domesticated-type (indehiscent) seeds, but through introgression its mesic traits were passed to slender wild oat. The mesic genotype is spreading rapidly, as are some recombinants between it and the xeric genotype.

This example illustrates Barrett's 'hybrid soup' scenario, quoted above, though in a self-pollinating species. The xeric genotype is essentially monomorphic yet became one of the dominants of the valley grassland, spreading massively from northern Baja California to Oregon. The new genotypes are replacing it in all but the driest areas and are also spreading into habitats and regions that formerly the species was unable to penetrate. This example also illustrates the importance of taking a historical geographical approach to invasion research. Allard's research was carried out before the recent spatial turn in ecology and evolution, and in retrospect, his conceptual model of a gentle rain of numerous slender wild-oat genotypes upon the California landscape was inappropriate (Blumler, 1995). Allard's last paper (Cluster and Allard, 1995) wistfully attempts to convince the reader that the evolution of slender wild oat along the West Coast is a fascinating story. If only he could have grasped the role of introgression, he would indeed have had an astounding story to tell.

27.9 DISCUSSION

The intense pressure on ecologists to develop a body of theory that can stand with that of other sciences has the unfortunate result of an overemphasis on 'theory' that really is hypothesis. For instance, ecological invasion rules are proving to have limited applicability. Geographical rules may be somewhat more useful, since they can direct our attention to the control of flows and the creation of bottlenecks. Although invasion is an intractable problem, dispersal control might at least slow an invader down. The need to control intentional introductions through a whitelist is apparent, though not a magic bullet.

27.9.1 Evolutionary imperialism

A sketch of an explanation for the one-sided Mediterranean invasion may be in order. A possible longer history of evolutionary development under a mediterranean-type climate, and high species diversity, should give the Mediterranean an advantage in invasion, as natural invasions of the past illustrate; however, that is insufficient explanation for the extreme lopsidedness of the present one. For unintentional introductions, the weed niche is the key to dispersal overseas. Once introduced, establishment in a staging area facilitates successful invasion. For weeds, this would be some human-disturbed environment, such as cultivated fields, roadsides, and dumps. For horticultural introductions, these would be the gardens and nurseries in which they are planted. Introduced species that can establish in one or more of these habitats have an advantage in terms of establishment. Repeated seed dispersal may eventually enable the potential invader to find a 'natural' environment that suits it. Even if initially unsuited to any natural habitat, an introduced weed may eventually evolve adaptation to some such habitat through mass effect. Thus, the weed niche is crucial to successful invasion. For return invasion to occur, natives of the analogue regions must evolve weedy species, too. Otherwise, how are they to reach the Mediterranean?

In the Mediterranean, native species almost entirely occupy the weed niche. (Exceptions are summer-irrigated areas.) Ten thousand years of agropastoralism have enabled local species to adapt to various human impacts. Often the same species can be weedy and also exist quite well in undisturbed nature (Blumler, 1994). But in California, there are few native weeds; I believe that even these are decreasing rather than the converse. An example is red maids (*Calandrinia ciliata* var. *menziesii*), a common native annual of fertile, disturbed ground in the nineteenth and early twentieth centuries. Though still widespread and not yet rare, it is nowhere near as common as it once was. As alien weeds come in and compete with the native weeds, it often seems that the alien is already better adapted to the weed niche. By preempting the weed niche, the aliens make it much less likely that natives will adapt and become successful weeds themselves. This in turn drastically reduces the likelihood of return introduction and invasion of the Mediterranean region.

Because California is such an extreme case, it is unlikely that it is entirely generalizable. The remarkable success of many weeds in invading natural habitats in California is seldom repeated elsewhere, except, to a lesser degree, in other neo-Europes and on isolated oceanic islands. Moreover, it is no longer likely that an agricultural system will be introduced wholesale, with its weed associates, to a former hunter-gatherer region. What is generalizable is the importance of attention to patterns of introduction, routes of dispersal, bottlenecks—the historical geography of invasion.

27.10 RESEARCH NEEDS

From a biogeographical standpoint, the greatest research need at the present time is for more attention to invasion histories, to the historical geography of dispersal. Given current funding patterns, this is likely to remain the stepchild of invasion research. Much more also could be done in the area of invading species in the context of global change. Current discussions tend to the apocalyptic and simplistic, and the reality is likely to be more nuanced. The discovery that marine invasions are now being facilitated by floating plastic debris (Barnes, 2002) illustrates the complexity and often geographical nature of the interaction. Transgenes are increasingly significant in American agriculture and even are inserted in wild species such as trees and fish (Cox, 2004); although there is considerable research, it tends to be guided by opposing paradigms, leaving little room for debate. Blumler's (2005) suggestion that dispersal control could be a useful management approach needs to be applied and tested. The need for hybridization to invade vacant niches, in the case of the cordgrass discussed above, raises a host of questions about how evolution occurs.

Finally, this review hopefully has demonstrated the relevance of geographical as well as ecological invasion research. This is not meant to denigrate ecological studies, which also are needed, but to reestablish an appropriate balance of the two.

REFERENCES

Abbott, R.J., James, J.K., Milne, R.I., and Gillies, A.C.M. (2003) 'Plant introductions, hybridization and gene flow', *Philosophical Transactions of the Royal Society of London Series B*, 358: 1123–32.

Albright, T.P., Moorhouse, T.G., and McNabb, T.J. (2004) 'The rise and fall of water hyacinth in Lake Victoria and the Kagera River Basin, 1989–2001', *Journal of Aquatic Plant Management*, 42:73–84.

Allan, H.H. (1936) 'Indigene versus alien in the New Zealand plant world', *Ecology*, 17: 187–93.

Arnold, M.L. (1997) *Natural Hybridization and Evolution*. New York: Oxford University Press.

Asner, G.P., and Vitousek, P.M. (2005) 'Remote analysis of biological invasion and biogeochemical change', *Proceedings of the National Academy of Sciences of the United States of America*, 102: 4383–86.

Baker, H.G. (1955) 'Self-compatibility and establishment after "long-distance" dispersal', *Evolution*, 9: 347–48.

Baker, H.G. (1965) 'Characteristics and modes of origin of weeds', in Baker, H.G., and G.L. Stebbins (eds.), *The Genetics of Colonizing Species*. New York: Academic Press. pp. 148–68.

Baker, H.G. (1972) 'Seed weight in relation to environmental conditions in California', *Ecology*, 53: 997–1010.

Baker, H.G. (1974) 'The evolution of weeds', *Annual Review of Ecology and Systematics*, 5: 1–24.

Baker, H.G., and Stebbins, G.L. (eds.). (1965) *The Genetics of Colonizing Species*. New York: Academic Press.

Barnes, D.K. (2002) 'Invasions by marine life on plastic debris', *Nature*, 416: 808–9.

Barrett, S.C.H. (1983) 'Crop mimicry in weeds', *Economic Botany*, 37: 255–83.

Barrett, S.C.H. (2000) 'Microevolutionary influences of global change on plant invasions', in Mooney, H.A., and R.J. Hobbs (eds.), *Invasive Species in a Changing World*. Washington, DC: Island Press. pp. 115–39.

Barrett, S.C.H., and Husband, B.C. (1990) 'The genetics of plant migration and colonization', in Brown, A.H.D., M.T. Clegg, A.L. Kahler, and B.S. Weir (eds.), *Plant Population Genetics, Breeding, and Genetic Resources*. Sunderland MA: Sinauer. pp. 254–77.

Barrett, S.C.H., and Shore, J.S. (1989) 'Isozyme variation in colonizing plants', in Soltis, D.E., and R.S. Soltis (eds.), *Isozymes in Plant Biology*. Portland, OR: Dioscorides Press. pp. 106–26.

Bartolome, J.W. (1979) 'Germination and seedling establishment in California annual grassland', *Journal of Ecology*, 67: 273–81.

Bartolome, J.W. (1989) 'Local temporal and spatial structure', in Huenneke, L.F., and H.A. Mooney (eds.), *Grassland Structure and Function: California Annual Grassland*. Dordrecht: Kluwer Academic. pp. 73–80.

Beatty, S.W., and Licari, D.L. (1992) 'Invasion of fennel (*Foeniculum vulgare*) into shrub communities on Santa Cruz Island, CA', *Madrono*, 39: 54–66.

Blackburn, T.M., and Duncan, R.P. (2001) Determinants of establishment success in introduced birds. *Nature*, 414: 195–97.

Blumler, M.A. (1984) 'Climate and the annual habit', MA thesis, University of California, Berkeley.

Blumler, M.A. (1992a) 'Seed weight and environment in Mediterranean-type grasslands in California and Israel', PhD dissertation, University of California, Berkeley.

Blumler, M.A. (1992b) 'Some myths about California grasslands and grazers', *Fremontia*, 20(3): 22–27.

Blumler, M.A. (1994) 'Evolutionary trends in the wheat group in relation to environment, Quaternary climate change and human impacts. In Millington, A.C., and K. Pye (eds.), *Environmental Change in Drylands*. Chicester, UK: John Wiley. pp. 253–69.

Blumler, M.A. (1995) 'Invasion and transformation of California's Valley grassland, a Mediterranean analogue ecosystem', in Butlin, R., and N. Roberts (eds.), *Human Impact and Adaptation: Ecological Relations in Historical Times*. Oxford: Blackwell. pp. 308–32.

Blumler, M.A. (1996) 'Ecology, evolutionary theory, and agricultural origins', in Harris, D. R., ed., *The Origins and Spread of Agriculture and Pastoralism in Eurasia*. London: UCL Press. pp. 25–50.

Blumler, M.A. (1998) 'Managing invaders', *Papers and Proceedings of the Applied Geography Conferences*, 21: 450.

Blumler, M.A. (2000) 'Spatial analysis to settle an unresolved question in genetics, with both theoretical and applied implications', *Research in Contemporary and Applied Geography: A Discussion Series*, 24(3): 1–42.

Blumler, M.A. (2002) 'Environmental management and conservation', in Orme, A.R. (ed.), *Physical Geography of North America*. Oxford: Oxford University Press. pp. 516–35.

Blumler, M.A. (2003a) 'Introgression as a spatial phenomenon', *Physical Geography*, 24: 414–32.

Blumler, M.A. (2003b) 'Smokey the bear and Lyme disease', *The Pennsylvania Geographer*, 41(1): 88–105.

Blumler, M.A. (2005) 'Dispersal control of invading organisms', *Papers of the Applied Geography Conference*, 28: 60–69.

Blumler, M.A. (2006) 'Geographical aspects of invasion: The annual bromes', *Middle States Geographer*, 39: 1–7.

Blumler, M.A. (2007) 'Invasion rules: Ecological or geographical?' *Papers of the Applied Geography Conference*, 30: 395–406.

Brasier, C.M. (2001) 'Rapid evolution of introduced plant pathogens via interspecific hybridization', *BioScience*, 51: 123–33.

Brothers, T.S. (1992) 'Postsettlement plant migrations in northeastern North America', *American Midland Naturalist*, 128: 72–82.

Brown, R. (1996) 'Theory versus reality in mine reclamation', MA thesis, Binghamton University.

Byrne, R. (1980) 'Man and the variable vulnerability of island life. A study of recent vegetation change in the Bahamas', *Atoll Research Bulletin*, #240.

Cabe, P.R. (1998) 'The effects of founding bottlenecks on genetic variation in the European starling (*Sturnus vulgaris*) in North America', *Heredity*, 80: 519–25.

Carroll, S.P., and Dingle, H. (1996) 'The biology of post-invasion events', *Biological Conservation*, 78: 207–14.

Castro, S.A., Figueroa, J.A., Munoz-Schick, M. et al. (2005) 'Minimum residence time, biogeographical origin, and life cycle as determinants of the geographical extent of naturalized plants in continental Chile', *Diversity and Distributions*, 11: 183–91.

Chapin III, F.S., Zavaleta, E.S., Eviner, V.T. et al. (2000) 'Consequences of changing biodiversity', *Nature*, 405: 234–42.

Clegg, M.T., and Allard, R.W. (1972) 'Patterns of genetic differentiation in the slender wild oat species *Avena barbata*' *Proceedings of the National Academy of Sciences, of the United States of America*, 69: 1820–24.

Clements, F.E. (1916) 'Plant succession: An analysis of the development of vegetation', *Carnegie Institute of Washington Publications*, 242: 1–512.

Clements, F.E. (1934) 'The relict method in dynamic ecology', *Journal of Ecology*, 22: 39–68.

Cluster, P.D. and Allard, R.W. (1995) 'Evolution of ribosomal DNA (rDNA) genetic structure in colonial California populations of *Avena barbata*', *Genetics*, 139: 941–54.

Cohen, A.N., and Carlton, J.T. (1998) 'Accelerating invasion rate in a highly invaded estuary', *Science*, 279: 555–58.

Cox, G.W. (2004) *Alien Species and Evolution: The Evolutionary Ecology of Exotic Plants, Animals, Microbes, and Interacting Native Species*. Washington, DC: Island Press.

Crosby, A.W. (1972) '*The Columbian Exchange: Biological and Cultural Consequences of 1492*. Westport CT: Greenwood.

Crosby, A.W. (1986) *Ecological Imperialism: The Biological Expansion of Europe, 900–1900*. Cambridge: Cambridge University Press.

Darwin, C. (1845) *Journal of researches into the natural history and geology of the countries visited during the voyage of the H. M. S. Beagle round the world under the command of Capt. Fitz Roy, R. N.* London: John Murray.

Darwin, C. (1859) *On the origin of species by means of natural selection (or the preservation of favoured races in the struggle for life)*. London: John Murray.

Davis, M.A., Grime, J.P., and Thompson, K. (2000) 'Fluctuating resources in plant communities: a general theory of invasibility', *Journal of Ecology*, 88: 528–34.

Delisle, F., Lavoie, C., Jean, M., and Lachance, d. (2003) 'Reconstructing the spread of invasive plants: Taking into account biases associated with herbarium specimens', *Journal of Biogeography*, 30: 10233–42.

Diamond, J. (1989) 'Overview of recent extinctions', in Western, D., and M.C. Pearl (eds), *Conservation for the Twenty-First Century*. New York: Oxford University Press. pp. 71–89.

di Castri, F., Hansen, A.J., and Debussche, M. (eds.). (1990) *Biological Invasions in Europe and the Mediterranean Basin*. Dordrecht: Kluwer.

Drake, J.A., Mooney, H.A., di Castri, F., Groves, R.H., Kruger, F.J., Rejmanek, M., and Williamson, M. (eds.). (1989) *Biological Invasions, a Global Perspective*. New York: John Wiley.

Ellstrand, N.C. (2003) *Dangerous Liaisons? When Cultivated Plants Mate with Their Wild Relatives*. Baltimore: Johns Hopkins University Press.

Ellstrand, N.C., Prentice, H.C., and Hancock, J.F. (1999) 'Gene flow and introgression from domestic plants into their wild relatives', *Annual Review of Ecology and Systematics*, 30: 539–63.

Ellstrand, N.C., and Schierenbeck, K.A. (2000) 'Hybridization as a stimulus for the evolution of invasiveness in plants?' *Proceedings of the National Academy of Sciences of the United States of America*, 97: 7043–50.

Elton, C.S. (1958) *The Ecology of Invasion by Animals and Plants*. London: Methuen.

Enserink, M. (1999) 'Biological invaders sweep in', *Science*, 285: 1834–36.

Fenner, F., and Ratcliffe, F.N. (1965) *Myxomatosis*. Cambridge: Cambridge University Press.

Ferdy, J.B., and Austerlitz, F. (2002) 'Extinction and introgression in a community of partially cross-fertile plant species', *American Naturalist*, 160: 74–86.

Forcella, F., and Wood, J.T. (1984) 'Colonization potentials of alien weeds are related to their native distributions: Implications for plant quarantine', *Journal of the Australian Institute of Agricultural Science*, 50: 35–40.

Forcella, F., Wood, J.T., and Dillon, S.P. (1986) 'Characteristics distinguishing invasive weeds within *Echium* (Bugloss)', *Weed Research*, 26: 351–64.

Gaskin, J.F., and B.A. Schaal. (2002) 'Hybrid *Tamarix* widespread in U.S. invasion and undetected in native Asian range', *Proceedings of the National Academy of Sciences of the United States of America*, 99: 11256–59.

Gelbard, J.L., and J. Belnap. (2003) 'Roads as conduits for exotic plant invasion in a semiarid landscape', *Conservation Biology*, 17: 420–22.

Griggs, F.T. (2000) 'Vina Plains Reserve: Eighteen years of adaptive management', *Fremontia*, 27(4)–28(1): 48–51.

Groves, R.H., and F. di Castri (eds.). (1991) *Biogeography of Mediterranean invasions*. New York: Cambridge University Press.

Hacker, D.D., Heimer, D., Hellquist, C.E., Reader, T.G., Reeves, B., Riordan, T.J., and Dethier, M.N. (2001) 'A marine plant (*Spartina anglica*) invades widely varying habitats: potential mechanisms of invasion and control', *Biological Invasions*, 3: 211–17.

Hamrick, J.L., and Allard, R.W. (1972) 'Microgeographical variation in allozyme frequencies in *Avena barbata*', *Evolution*, 29: 438–42.

Harris, D.R. (1966) 'Recent plant invasions in the arid and semi-arid southwest of the United States', *Annals of the Association of American Geographers*, 56: 408–22.

Heady, H.F. (1977) 'Valley grassland', in Barbour, M.G., and J. Major (eds.), *Terrestrial Vegetation of California*. New York: John Wiley. pp. 491–514.

Hedrick, P.W., and Holden, L.R. (1979) 'Hitch-hiking: An alternative to coadaptation for the barley and slender wild oat examples', *Heredity*, 43: 79–86.

Hendrix, P.F., and Bohlen, P. (2002) 'Ecological assessment of exotic earthworm invasions in North America', *Bioscience*, 52: 801–11.

Hendry, G.W. (1931) 'The adobe brick as an historical source', *Agricultural History*, 8: 64–71.

Hengeveld, R. (1990) *Dynamic Biogeography*. Cambridge: Cambridge University Press.

Henneman, M., and Memnott, J. (2001) 'Infiltration of a Hawaiian community by introduced biological control agents', *Science*, 293: 1314–16.

Higgins, S.I., and Richardson, D.M. (1996) 'A review of models of alien plant spread', *Ecological Modelling*, 87: 249–65.

Huey, R.B., Gilchrist, G.W., Carson, M.L., Berrigan, D., and Serra, L. (2000) 'Rapid evolution of a geographic cline in size in an introduced fly', *Science*, 287: 308–9.

Huffaker, C.B., and Kennett, C.E. (1959) 'A ten-year study of vegetational changes associated with biological control of Klamath weed', *Journal of Range Management*, 12: 69–82.

Huston, M.A. (2004) 'Management strategies for plant invasions: Manipulating productivity, disturbance, and competition', *Diversity and Distributions*, 10: 167–78.

Huxel, G.R. (1999) 'Rapid displacement of native species by invasive species: Effects of hybridization', *Biological Conservation*, 89: 143–52.

Jackson, L.E. (1985) 'Ecological origins of California's Mediterranean grasses', *Journal of Biogeography*, 12: 349–61.

Kimura, M. (1968) 'Evolutionary rate at the molecular level', *Nature*, 217: 624–26.

Kolar, C.S., and Lodge, D.M. (2001) 'Progress in invasion biology: Predicting invaders', *Trends in Ecology and Evolution*, 16: 199–204.

Kolbe, J.J., Klor, R.E., Schettino, L.R.G., Lara, A.C., Larson, A., and Losos, J.B. (2004) 'Genetic variation increases during biological invasion by a Cuban lizard', *Nature*, 431: 177–81.

Lockwood, J.L., Cassey, P., and Blackburn, T.M. (2005) 'The role of propagule pressure in explaining species invasions', *Trends in Ecology and Evolution*, 20: 223–28.

Lomolino, M.V., Riddle, B.H., and Brown, J.H. (2006) *Biogeography* (3rd ed.) Sunderland, MA: Sinauer.

Lonsdale, W.M. (1999) 'Global patterns of plant invasions and the concept of invasibility', *Ecology*, 80: 1522–36.

Lovett, R.A. (2000) 'Mt. St Helens, revisited', *Science*, 288: 1578–79.

Mack, R.N. (1989) 'Temperate grasslands vulnerable to invasions: Characteristics and consequences', in Drake, J.A., H.A. Mooney, F. di Castri, R.H. Groves, F.J. Kruger, M. Rejmanek, and M. Williamson (eds.), *Biological Invasions, a Global Perspective*. New York: John Wiley. pp. 155–79.

Mack, R.N. (1991) 'The commercial seed trade: An early disperser of weeds', *Economic Botany*, 45: 257–73.

Mack, R.N. (2003) 'Plant naturalizations and invasions in the eastern United States: 1634–1860', *Annals of the Missouri Botanical Garden*, 90: 77–90.

Mack, R.N., and Erneberg, M. (2002) 'The United States naturalized flora: Largely the product of deliberate introductions', *Annals of the Missouri Botanical Garden*, 89: 176–89.

Marshall, L.G., Webb, S.D., Sepkoski, J.J. Jr., and Raup, D.M. (1982) 'Mammalian evolution and the Great American interchange', *Science*, 215: 351–57.

McNaughton, S.J. (1968) 'Structure and function of California grasslands', *Ecology*, 49: 962–72.

Medley, K.E. (1997) 'Distribution of the non-native shrub *Lonicera maackii* in Kramer Woods, Ohio', *Physical Geography*, 18: 18–36.

Mensing, S.A., and Byrne, R. (1997) 'Pre-mission invasion of *Erodium cicutarium* in California', *Journal of Biogeography*, 25: 757–82.

Milne, R.I., and Abbott, R.J. (2000) 'Origin and evolution of invasive naturalized material of *Rhododendron ponticum* L. in the British Isles', *Molecular Ecology*, 9: 541–56.

Minnich, R.A. (2008) *California's Fading Wildflowers: Lost Legacy and Biological Invasions.* Berkeley: University of California Press.

Minnich, R.A., and Dezzani, R.J. (1998) 'Historical decline of coastal sage scrub in the Riverside-Perris plain, California', *Western Birds*, 29: 366–91.

Moyle, P.B., and Light, T. (1996) 'Fish invasions in California: Do abiotic factors determine success?', *Ecology*, 77: 1666–70.

Niemala, P., and Mattson, W.J. (1996) 'Invasion of North American forests by European phytophagous insects', *BioScience*, 46: 741–53.

Novak, S.J., and Mack, R.N. (2001) 'Tracing plant introduction and spread: Genetic evidence from *Bromus tectorum* (cheatgrass)', *BioScience*, 51: 114–21.

Owenby, M. (1950) 'Natural hybridization and amphiploidy in the genus *Tragopogon*', *American Journal of Botany*, 37: 487–99.

Panetsos, C.A., and Baker, H.G. (1967) 'The origin of variation in "Wild" *Raphanus sativus* (Cruciferae) in California', *Genetica*, 38: 243–74.

Parsons, J.J. (1972) 'Spread of African pasture grasses to the American tropics', *Journal of Range Management*, 25: 12–17.

Pemberton, R.W. (2000) 'Predictable risk to native plants in weed biological control', *Oecologia*, 125: 489–94.

Peterson, A.T. (2003) 'Predicting the geography of species' invasions via ecological niche modeling', *Quarterly Review of Modeling*, 78: 419–33.

Petit, R.L. (2004) 'Biological invasions at the gene level', *Diversity and Distributions*, 10: 159–65.

Preston, K.P. (1993) 'Selection for sulfur dioxide and ozone tolerance in *Bromus rubens* along the south central coast of California', *Annals of the Association of American Geographers*, 83: 143–55.

Pysek, P.D., Richardson, M., and Williamson, M. (2004) 'Predicting and explaining plant invasions through analysis of source area floras: some critical considerations', *Diversity and Distributions*, 10: 179–87.

Randall, J.M., Rejmanek, M., and Hunter, J.C. (1998) 'Characteristics of the exotic flora of California', *Fremontia*, 26(4): 3–12.

Rejmanek, M. (1996) 'A theory of seed plant invasiveness: The first sketch', *Biological Conservation*, 78: 171–81.

Rejmanek, M., and Randall, J.M. (1994) 'Invasive alien plants in California: 1993 summary and comparison with other areas in North America', *Madrono*, 42: 161–77.

Rejmanek, M., Thomson, C.D., and Peters, I.D. (1991) 'Invasive vascular plants of California', in Groves, R.H., and F. diCastri (eds.), *Biogeography of Mediterranean Invasions*. Cambridge: Cambridge University Press. pp. 81–101.

Reznick, D.N., and Ghalambur, C.K. (2001) 'The population biology of contemporary adaptations: What empirical studies reveal about the conditions that promote adaptive evolution', *Genetica*, 112–13: 183–89.

Rhymer, J.M., and Simberloff, D. (1996) 'Extinction by hybridization and introgression', *Annual Review of Ecology and Systematics*, 27: 83–109.

Richardson, D.M., and Pysek, P. (2006) 'Plant invasions: Merging the concepts of species invasiveness and community invasibility', *Progress in Physical Geography*, 30: 409–31.

Ricklefs, R.E. (2005) 'Taxon cycles: Insights from invasive species', in Sax, D.F., J.J. Stachowicz, and S.D. Gaines (eds.), *Species Invasions: Insights into Ecology, Evolution, and Biogeography*. Sunderland, MA: Sinauer. pp. 165–99.

Robbins, W.W. (1940) 'Alien plants growing without cultivation in California', *California Agricultural Experiment Station Bulletin*, 637: 1–128.

Rolls, E.C. (1969) *They All Ran Wild; the Story of Pests on the Land in Australia*. Sydney: Angus and Robertson.

Roura-Pascual, N., Suarez, A.V., Gomez, C. et al. (2004) 'Geographical potential of Argentine ants (*Linepithema humile* Mayr) in the face of global climate change', *Proceedings of the Royal Society of London Series B*, 271: 2527–34.

Roush, W. (1997) 'Hybrids consummate species invasion', *Science*, 277: 316–17.

Roy, J., Navas, M.L., and Sonie, L. (1991) 'Invasion by annual brome grasses: A case study challenging the homoclime approach to invasions', in Groves, R.H., and F. diCastri (eds.), *Biogeography of Mediterranean Invasions*. New York: Cambridge University Press. pp. 207–24.

Rozenzweig, M.L. (2001) 'The four questions: What does the introduction of exotic species do to diversity?' *Evolutionary Ecology Research*, 3: 361–67.

Ruiz, G.M., and Carlton, J.T. (eds). (2003) *Invasive Species: Vectors and Management Strategies*. Washington, DC: Island Press.

Sala, A., Smith, S.D., and Devitt, D.A. (1996) 'Water use by *Tamarix ramosissima* and associated phreatophytes in a Mojave floodplain', *Ecological Applications*, 6: 888–98.

Sauer, C.O. (1952) *Agricultural Origins and Dispersals*. Washington, DC: American Geographical Society.

Sauer, J.D. (1988) *Plant Migration: The Dynamics of Geographical Patterning in Seed Plant Species*. Berkeley: University of California.

Sawicki, V. (1998) 'Investigation of the distribution of *Lythrum salicaria* and its effect on biodiversity of North American flora: Broome County, New York', MA thesis, Binghamton University.

Sax, D.F., Stachowicz, J.J., and Gaines, S.D. (eds.). (2005) *Species Invasions: Insights into Ecology, Evolution, and Biogeography*. Sunderland, MA: Sinauer.

Schierenbeck, K.A., Gallagher, K.G., and Holt, J.N. (1998) 'The genetics and demography of invasive plant species', *Fremontia*, 26(4): 19–23.

Seabloom, E.W., and 8 authors. (2003) 'Competition, seed limitation, disturbance, and reestablishment of California's native annual forbs', *Ecological Applications*, 13: 575–92.

Shigasada, N., and Kawasaki, K. (1997) *Biological Invasions: Theory and Practice*. New York: Oxford University Press.

Shmida, A., and Ellner, S. (1984) 'Coexistence of plant species with similar niches', *Vegetatio*, 58: 29–55.

Suarez, A.V., and Tsutsui, N.D. (2004) 'The value of museum collections for research and society', *Bioscience*, 54: 66–74.

Thuiller, W., Richardson, D.M., Pysek, P., Midgley, G.F. Hughes, G.O. and Rouget, M. (2005) 'Niche-based modelling as a tool for predicting the risk of alien plant invasions at a global scale', *Global Change Biology*, 11: 2234–50.

Tsutsui, N.D., Suarez, A.V., Holway, D.A., and Case, T.J. (2001) 'Relationships among native and introduced populations of the Argentine ant (*Linepithema humilis*) and the source of introduced populations', *Molecular Ecology*, 10: 2151–61.

Veblen, T.T. (1975) 'Alien weeds in the tropical highlands of Western Guatemala', *Journal of Biogeography*, 2: 19–26.

Vermeij, G.J. (2005) 'Invasion as expectation: A historical fact of life', in Sax, D.F., J.J. Stachowicz, and S.D. Gaines (eds.), *Species Invasions: Insights into Ecology, Evolution, and Biogeography*. Sunderland, MA: Sinauer. pp. 315–39.

Vitousek, P.M., Walker, L.R., Whiteaker, L.D., Mueller-Dombois, D., and Matson, P.A. (1987) 'Biological invasion by *Myrica faya* alters ecosystem development in Hawaii, USA', *Science*, 238: 802–4.

Vitousek, P.M., D'Antonio, C.M., Loope, L.L., and Westbrooks, R. (1996) 'Biological invasions as global environmental change', *American Scientist*, 84: 468–78.

Vivrette, N.H., and Muller, C.H. (1977) 'Mechanism of invasion and dominance of coastal grassland by *Mesembryanthemum crystallinum*', *Ecological Monographs*, 47: 301–18.

Westman, W.E. (1979) 'Oxidant effects on Californian coastal sage scrub', *Science*, 205: 1001–3.

Westman, W.E. (1990) 'Park management of exotic plant species: Problems and issues', *Conservation Biology*, 3: 251–60.

Williamson, M., and Fitter, A. (1996) 'The varying success of invaders', *Ecology*, 77: 1661–66.

Wolfe, L.M. (2002) 'Why alien invaders succeed: Support for the escape-from-enemy hypothesis', *American Naturalist*, 160: 705–11.

Wood, Y.A., Meixner, T., Shouse, P.J., and Allen, E.B. (2006) 'Altered ecohydrologic response drives native shrub loss under conditions of elevated nitrogen deposition', *Journal of Environmental Quality*, 35: 76–92

Bioindicators for Ecological and Environmental Monitoring

Yordan Uzunov

28.1 INTRODUCTION

'Why don't polar bears eat penguins? They inhabit the Arctic and Antarctic respectively and have no chance of meeting each other due to their geographic isolation'.

Though common, this joke bears a biogeographical message that is relevant to this chapter. Both polar bears and penguins are indicative of polar biomes, being indicators of northern and southern polar latitudes, respectively. Living organisms—either single species or assemblages—can be indicative of particular conditions while their spread and distribution depend on environmental factors, for example, conditions and resources, which characterize the environment to which they have adapted in an evolutionary sense. The preferences for a certain environment allow their presence to indicate a set of environmental parameters, which we trivially call "living conditions". Human activities can alter substantially these environmental parameters, which, while influencing living organisms, can result in significant changes of the structure and functioning of individuals, populations, communities, ecosystems.

Thus biological indication can be defined as the use of living organisms to determine and assess the state of environment they inhabit. In general, all the indications could be biotic responses exhibited/expressed at molecular (biochemical, genetic, physiological), morphological (tissue and organ abnormalities), population (changes in phenology or fertility), or ecosystem (changes of dominants, shifts in community structure, loss of biodiversity)

levels if there are established, reliable, and direct links between those responses and the dynamics of the environmental parameters, that is, dose-effect relations.

The ecological background of bioindication stands on Shelford's fundamental principle of ecological tolerance, which states that no organism can live within the full range of an environmental factor, that is, between the freezing (solid phase) and boiling points (gaseous phase) of water. The ecological optimum of a given organism usually lies between some minimum and maximum values beyond which that organism cannot live/exist at all (Figure 28.1). The ecological optimum means conditions at which organisms may reproduce normally and thus maintain a stable population. Species may form a sequence along a factor's gradient depending on their tolerance to the preferable range of that factor, and at each point on this gradient there is a set of coexisting species in a community (Figure 28.2). Thus, if the species tolerance to a factor is known, the value of that factor can be estimated from the presence and level of activity, for example, population density or rate of functioning, of the specific set of represented species. It is noteworthy, however, that some species have a narrow distributional range along a gradient (stenobionts), while others have wider ranges (eurybionts). To this end, stenobionts are considered the best indicators, while eurybionts are less indicative of certain environmental parameters.

So, the biological indicator or bioindicator seems to be an organism (species) known to have particular requirements with regard to some range

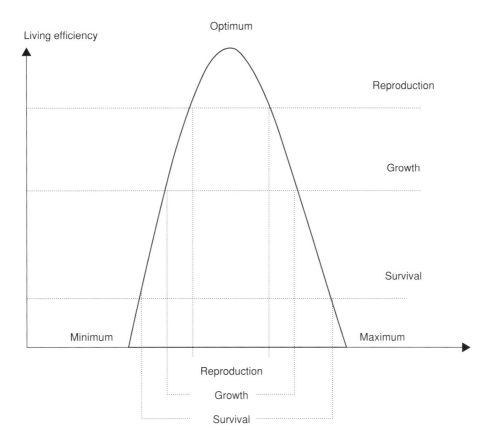

Figure 28.1   Graph of the living efficiency of a species against an external environmental factor (x-axis). The success of the species is contained between the minimum and maximum values of the external environmental factor; beyond these values, the species cannot. The narrower interval in which individuals can reproduce ensures a stable population

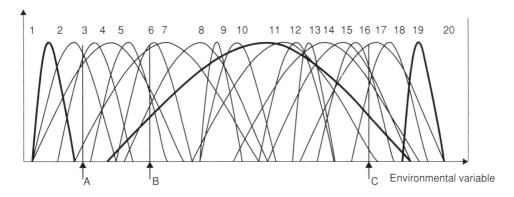

Figure 28.2   Schematic presentation of a sequence of 20 species along an environmental gradient. Note that species #1 and #20 are stenobionts (good indicators), while species #11 is an eurybiont (poor indicator). Each value of the variable (named A, B, or C) could be defined by a specific combination of the species presented and their qualitative parameters (e.g., abundance, rate of process, etc.)

of environmental variables, for example dissolved oxygen, nutrients, temperature, or soil type. Once these are defined, the presence of a particular species in a habitat indicates that the given determinand or parameter is within the ecological tolerance limits of that species. Odum (1971) described three basic requirements of living organisms as bioindicators: (a) to be stenobiotic; (b) to be relatively immobile; and (c) to have a relatively long life cycle. To this end, migratory fish, birds, and mammals are unlikely to serve as good bioindicators except for large-scale ecological evaluations. Hellawell (1986) listed eight key properties for ideal bioindicatorx noting that they are rarely realized, but that there are enough good environmental indicators in all major groups of organisms.

There are several methods and techniques using the physiological or behavioral responses of the organisms for detection and measurement of the state of the environment. Such biotic responses have found practical implementation for bioindication while they send signals for further analysis to be undertaken. In this case such organisms could be defined as biosensors or biomonitors, rather than bioindicators.

Another concept associated with the term 'bioindicator' is that of an organism that accumulates substances in its tissues, which, after chemical analysis, could estimate prevailing environmental concentrations. Such organisms are bioaccumulators of these substances and are useful while they concentrate very low environmental levels of substances, which can be magnified upward through trophic levels, thereby facilitating detection and analysis. The transfer of trace pollutants within the ecological chain is defined also as biomagnification. This is accomplished by concentrations of accumulated substances (e.g., heavy metals or pesticides) that build up (or magnify) as one organism eats another and that substance moves upward along the food chain (Huggett, 2004). Several important criteria need to be met in

choosing an organism as a bioaccumulator (Table 28.1). Both plants and animals have been found to be suitable bioaccumulators in studies of metal, chemical, and radioisotope pollution of the environment (e.g., Ernst et al., 1983; Tyler et al., 1989; Krizek et al., 2003; Kaglyan et al., 2005; Olivares-Rieumont et al., 2007)

It is important to mention that the living efficiency (success) of an organism depends on its level of adaptation to the natural fluctuations of environmental factors and, in general, each organism occupies its own ecological niche differing from others by its ecological tolerances to the set of living conditions in a specific habitat, ecosystem, or biome. To this end, bioindicators do not measure exact values but a range of environmental variables, and contemporary bioindication techniques are not superior to direct measurements if the exact values of a particular variable are required. That is why choosing bioindicators is extremely complicated and requires knowledge of both synecology and autecology. In many cases ecological bioindication focuses on simpler relationships over relatively small temporal and/or spatial scales.

Using indicative organisms for ecological classification of habitats, ecosystems, and biomes has a long history. For example, fish and other riparian animals were used in the development of ecological classification of longitudinal stretches of temperate European rivers (Table 28.2). Later, Vannote et al. (1980) introduced the functional feeding groups of aquatic invertebrates as indicators of stream size as a numerical expression of the main properties of the ecosystem functioning along the river continuum. There are numerous similar examples of using the indicative organisms for biogeographical, topological, and/or ecological classification of smaller or larger units of the biosphere. One of the most well known is the consequent change of forest vegetation along the altitude gradient in mountains (see Mutke, this volume).

Table 28.1 Criteria for selection of bioaccumulators (after Chapman, 1992)

In order to be used as bioaccumulator for direct monitoring of contaminants, an organism must fulfill six criteria:

1. accumulate the contaminants of interest at the levels present in the environment without lethal toxic effects
2. show a simple correlation between the concentration of contaminant in the organism and the average concentration in the ambient environment
3. accumulate the contaminant to a level high enough to enable direct analysis of its body
4. be abundant in and representative of a certain community and/or habitat
5. be easy to sample and survive for long enough in the lab to enable studies of contaminant uptake to be performed
6. be typically immobile or at least unlikely to move far from the area it indicates the contaminant levels of the ambient environment

Table 28.2 Fish and mayflies as indicators for longitudinal zonation of streams and rivers

River stretch	Thienemann (1925)	Carpenter (1928)	Huet (1954)	Zelinka (1960)	Illies and Botosaneanu (1963)
Source				Epeorus	crenon
					hypocrenon
Headwaters/ Upper reach	Trout (Salmo trutta)	Trout (Salmo trutta)	Trout (Salmo trutta)	Ameletus	epirhythron
	Grayling (Thymalus thymalus)	Minnow (Phoxinus phoxinus)	Grayling (Thymalus thymalus)	Rhythrogena	metarhythron
		Gudgeon (Gobio gobio) Chub (Leuciscus cephalus) Bleak (Alburnus alburnus)		Ecdyonurus	hyporhythron
Middle reach	Barbel (Barbus barbus)		Barbel (Barbus barbus)	Oligoneuriella	epipotamon
Lower reach/ Mouth	Bream (Abramis brama)	Bream (Abramis brama) Carp (Cyprinus carpio) Roach (Rutilus rutilus) Tench (Tinca tinca)	Bream (Abramis brama)	Ephemera Polymitarsis Palingenia	metapotamon
	Ruffe (Gymnocephalus cernus) Flounder (Platichthys flesus)				hypopotamon

Modified after Hellawell (1986) and Lampert and Sommer (1997).

The modern field of bioindication, however, is mostly connected with the emerging need of evaluation of environmental health and effects of the human impacts on its components. This approach originates from the middle of the nineteenth century, when some European scientists (e.g., Kolenati, 1848; Cohn, 1853 after Sladeček, 1973) revealed a close relationship between the level of water pollution and specific invertebrate communities. Early in the twentieth century, Kolkwitz and Marson (1902) published their 'System of Saprobic Organisms' (*Saprobiensystem*), which may be considered as the first system of biological indicators to estimate water quality. Later developments in the twentieth century led to the inclusion of aquatic plants and unicellular organisms (algae and protozoa) among biological indicators and to the widespread application of bioindicators in environmental assessment and monitoring.

During the subsequent decades, many scientists developed unique water-quality assessment methods and indices, for instance, Richardson (1928, after Hellawell, 1986) in the United States, Dolgov and Nikitinskii (1927) in Russia, Knöpp (1955) in Germany, Fjerdingstad (1964) in Denmark, and Horaswa (1956) in Japan. A review of existing biological methods for water-quality assessment revealed hundreds of different indices and methods for bioassessment of water quality (see Knoben et al., 1995; Conti, 2008 and references therein). Most of the indices introduced so far explore the total indicative capacity of a community (coexisting species assemblages) rather than that of a single species. Single species are applicable in a few cases, for example, the presence and amounts of the bacteria *Escherichia coli* are widely used for the detection and measurement of contamination of water and soil by fecal masses (*Coli*-titre).

Similarly, the idea of using plants—lichens in particular—to measure the effects of air pollution goes back to the nineteenth century when Nylander (1866, after Hawksworth and Rose, 1976) recognized lichens as potential indicators in Britain and Europe. Possible responses of a plant to air-pollution stress may include chlorophyll degradation, changes in photosynthesis and respiration rates, alterations in nitrogen fixation, membrane leakage, accumulation of toxic elements, and potential changes in spectral reflectance, vegetation cover, morphology, plant community structure, reproduction, and phenology. The methods most widely used to measure these responses are fumigation and gradient studies (Stolte et al., 1993).

Many plants are useful as bioaccumulators and the choice of species depends on the aims of the bioindication application. Mosses and lichens accumulate heavy metals and other compounds very efficiently because of their large specific surface and slow growth. As such they serve mostly as biomonitors to provide an indication of the impact of pollutants at the ecosystem level. On the other hand, field crops and vegetables can serve as an immediate step to detect effects on food and fodder quality and safety. Bioaccumulators are not only used to measure deposits of heavy metals but also radio-nuclides, polycyclic aromatic hydrocarbons, dioxins, and all kinds of aerosols that can also be accumulated efficiently. As far as contaminants of food and fodder crops are concerned, they are a crucial step to evaluate the potential transfer to consumers.

Since then, lichens have played prominent roles in air-pollution studies throughout the world because of their sensitivity to different gaseous pollutants, particularly SO_X and NO_X. They have also been found to act as accumulators of elements, such as trace metals, sulfur, and radioactive elements (Stolte et al., 1993; Ahmadjian, 1993). Between 1973 and 1988, approximately 1,500 papers were published on the effects of air pollution on lichens and many general reviews of lichens and air pollution have been compiled (e.g., Ahmadjian, 1993 and references therein).

28.2 BIOLOGICAL MONITORING AND ENVIRONMENTAL ASSESSMENT

There are many definitions of monitoring. Some people consider each act of observation to be monitoring, while others consider it to be the flow of information obtained during the observations. In general terms, however, monitoring is an important element in the management process that provides back-up information about results achieved and efficiency of managerial decisions that have been made and implemented.

So, environmental monitoring (as a part of environmental management) could be defined as a complex system of activities leading to long-term and regular standardized measurement, observation, evaluation, forecasting, and reporting on the state of the environment and its components whether under anthropogenic pressure or not. Usually, 'environmental monitoring' is also applicable to the relevant information systems for collection, processing, storage, and provision of environmental data. To fit its purposes, each monitoring system should be constructed on several basic principles:

1. large-scale and comprehensive systematic/ regular observations at a set of representative observation sites (a network of monitoring stations);
2. complexity and coherence with similar systems (e.g., meteorological and/or hydrological);
3. unification of measurement methods and parameters/metrics to be monitored; and
4. centralization of obtained data for further processing, storage, and forecasting for management purposes.

There are differences between monitoring (a long-term, regular, and comprehensive measurement program) and other supportive activities such as a survey (an intensive short-term measuring program for a specific purpose) and surveillance (continuous, specific, measurements and/or observations for operational activities).

Biological monitoring represents the systematic registration of the biotic responses of living systems to external impacts. Further discrimination between anthropogenic and natural fluctuations of the environmental variables is possible in biological monitoring. The purpose of the biomonitoring is to provide regular data with the intent to use this information in quality control programs when comparing the results obtained with reference and/or standard values, for instance, biological assessment, of the state of the environment and its components. It seems there is a direct relation between monitoring and assessment using biological indicators, but the opposite is not valid as individual assessments can be performed outside the monitoring programs. A layout of the steps of a monitoring and assessment procedure is shown in Figure 28.3.

These biotic responses can be expressed at all levels of organization of life—from the molecular level to ecosystems and biomes. Even without considering all possible biochemical, cellular, population, and/or community biotic responses, we are aware that an ecosystem has numerous, diverse parameters that characterize its species content and diversity, structural organization and

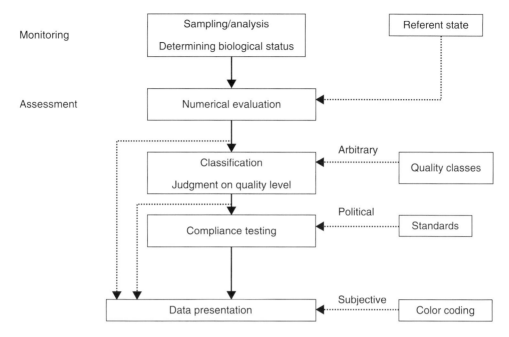

Figure 28.3   Elements of a biological monitoring and assessment procedure (reproduced after Knoben et al., 1995)

functioning, age and maturity, and resilience and integrity. It is evident that the possible biotic responses even at these supra-organismal levels could outnumber the possible impacts, which may differ mostly in their intensity. The knowledge on this overwhelming diversity of possible biotic responses from molecular to ecosystem levels is still insufficient and poorly applied for monitoring observations.

The use of biological and ecological metrics for evaluation, particularly for monitoring of environmental quality and environmental health, has gained substantial advantages over physical and chemical methods. In the best cases, the last may provide the observers with highly accurate and precise data or information about the types, status, and quantities of measured variables. However, such data have little ecological value if the effects of these environmental parameters on living organisms are unknown. Furthermore, physical and chemical variables are so dynamic in time and space (e.g., the diurnal dynamics of the dissolved oxygen within a large lake), and the chemical composition of contaminating substances is so complex (e.g., there are at present more than 10 million artificial, man-made chemical substances in use), that to maintain a comprehensive analysis for purposes of the environmental monitoring and management is extremely expensive. However,

methods of biological indication allow evaluation of the ecological consequences of the environmental contamination and pollution, shifts and alterations of the diversity and abundance of species, rates of community functions, and the degree of ecological modifications within the ecosystems. The main advantages of bioindication over physical and chemical methods are the longer validity and higher integrity of the assessments, which opens the gate for ecological modeling and forecasting and managing environmental health. This is why biological monitoring should be and is a mandatory component of national environmental monitoring systems in most of the developed and less developed countries over the world (e.g., in the European Union, the United States, Japan, and the Russian Federation).

For the purpose of biomonitoring, two groups of evaluations are most applicable:

1. Parameters or indices expressed by an integral in time, for instance, reporting the results of some dynamic processes at the moment of their measurement. These are usually metrics such as species content, the number of species represented, abundance measures such as density and/or biomass, species diversity, dominance, and ratios between taxonomic or trophic groups.

2. Parameters or indices expressed by a derivative in time (differential), for instance, rates of some dynamic processes at the moment of their measurement. These could be metrics such as rates of photosynthesis, productivity, respiration, reproduction, and accumulation, or ratios between production and destruction or between aerobic and anaerobic metabolism.

There is another, third group of evaluations that does not measure quantities but that have general importance in measuring, recording, or evidence of events. These are phenological observations such as the date of first appearance of migrating birds in a lake or the flowering of plants (for further details, see Sparks et al., this volume).

For various historical reasons, most of the widespread and applicable evaluations are from the first group. There are hundreds of biological methods in the first group, which provide criteria and indices for bioassessment and monitoring of the state of environment. These are mostly geared toward water and ambient air quality. The diversity of methods can be summarized into a few basic groups, each of which represents general ways in which these methods and indices are used in bioindication such as species-related methods, community-related methods, or habitat-related methods as discussed below. Attention is paid in this chapter to in situ methods, that is, direct measurements and assessments, in contrast to remote sensing approaches (see Foody and Millington, and Nightingale et al., this volume). In situ methods can be classified as either passive monitoring (when bioindicators are natural elements of an ecosystem) or active monitoring (when organisms are inserted under controlled conditions into the site to be monitored).

28.3 METHODS OF BIOINDICATION (MONITORING AND ASSESSMENT)

Three main groups of methods are commonly used for biomonitoring and bioassessment. They are mainly used for monitoring water and ambient air quality.

28.3.1 Species-related methods

A group of species-related methods explores Shelford's concept of ecological tolerance of the species or their assemblages in relation to an environmental variable or combination of variable. Included in this category are all kinds of responses at lower biological levels (e.g., biochemical,

genetic, morphological, and individual). For example, all enzymes have a temperature optimum; values beyond the optimum may inhibit or even destroy enzyme function and structure. Similarly, the rate of photosynthesis is highest at optimal values of the light spectrum and the amount of nutrients available.

In the scientific literature, one can find numerous similar methods and indices using bioindicators, which represent to some extent modifications of the concept of ecological tolerance of species and species assemblages. There are two basic features of such kinds of indices: (a) numerical presentation of ecological tolerance of species and/or groups; and (b) absolute or relative abundance of a species. It is irrelevant which taxa (all invertebrates or only insects, fish, or plants) are used in these assessments, as each of the community elements involved (e.g., species, genera, families, or trophic groups) is awarded a particular score of its ecological tolerance to the measured variable(s) and/or quality classes. The classic examples are so-called saprobic indices that measure the amount and level of degradation of bio-putrescible organic matter in freshwater. The saprobic index of Pantle and Buck (1955) represents numerically the species preferences and makes it possible to calculate the degree of saprobity as follows:

$$S_{PB} = (\Sigma h_i s_i)/\Sigma h_i, \qquad (1)$$

where h_i is the absolute or relative abundance of i-species, and s_i is the species saprobic value (the maximum of its distribution along the gradient of saprobity). By this way, the valuation of saprobity is represented by a mean value of saprobic values and abundance of species represented in the community. Another saprobic index (Zelinka and Marvan, 1961), is similar to that of Pantle and Buck (1955), but it differs in the way the species saprobic values are expressed. In this case, rather than using the maximum the whole species distribution within a certain saprobic range (represented by saprobic scores from 1 to 10) and they introduced the metric of indicative weight as measure for the range of this distribution. Thus, stenosaprobic organisms received the maximum weight of 5, while eurysaprobic ones had a minimum weight of 1. In this way, the sum of the saprobic scores of the represented species for a certain saprobic degree, X, is expressed in the equation:

$$X = \Sigma(h_i v_i g_i), \qquad (2)$$

where h_i is the species abundance, v_i is the number of valences for the saprobic degree X, and g_i is the species indicative weight. The maximum value of X for each of the saprobic degrees shows the dominant level of saprobity. Later, Rotschein (1962) developed a saprobiological index, $S_{R,}$ which expresses saprobic values in single digits.

Special attention has to be paid to the group of species-related methods and techniques using biosensors. In many countries, for example, Germany and the Netherlands, systems for early warning of pollution, especially of toxic contamination, use these kinds of biomonitors based on biotic signals, for example, through registration of respiration activity (shell contraction) of mussels *Unio spp.*, motion activity of the planktonic crustacean *Daphnia magna*, or fish (e.g., *Leuciscus idus, Lepomis gibosus*), which inhabit automated flowing systems as shown in Figure 28.4. As polluted water enters these systems, changes of the behavior of these species (e.g., acceleration of movement, avoidance of polluted water) are registered and a signal is sent so that the water is automatically sampled for further chemical analyses. Sensitive plants like transgenic tobacco (*Nicotiana tabacum,* variety Bel-W3) or

spiderwort (*Tradescantia sp.* clone # 4430) growing in pots and placed in a contaminated area could act as biosensors for ozone or traffic emissions in the ambient air.

Some species are used as bioaccumulators in a similar manner because of their ability to accumulate, and knowledge of the capacities of particular species to accumulate environmental contaminants enables them to be used for detection and monitoring the levels of substances in the environment. Both animals and plants are used in this kind of monitoring, especially molluscs, fish, and macrophytes for water; and lichens, mosses, and higher vascular plants like lettuce (Nali et al., 2009) or meadow herbs like *Citrus salvifolius* and *Inula viscosa* for ambient air-quality monitoring (Carmo-Freitas, 1995).

28.3.2 Community-related methods

Community-related methods and metrics to quantitatively describe the structural organization of communities have found widespread application in routine practices of environmental

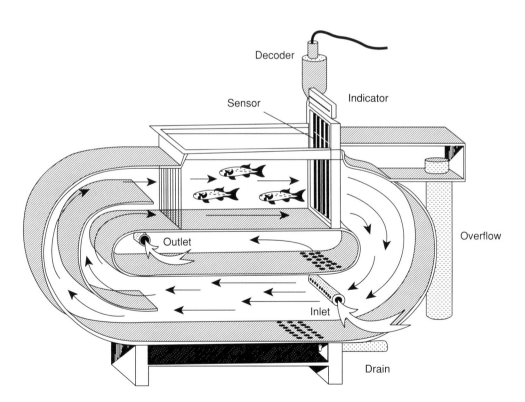

Figure 28.4   An example of using biosensors (biomonitors): a dynamic fish test system using for continuous monitoring of toxicity of water directed from a water body through the test equipment (reproduced from Friedrich et al., 1992: 213)

biomonitoring and assessment. All explore the general, nonspecific reactions of communities to diminished species diversity under external impacts as a reflection of the changing quantitative ratios between species or groups within a stressed community. A common advantage of such methods is a nonmandatory requirement for exact species determination; distinguishing clearly representative species and their relevant abundances is sometimes enough to calculate diversity values. However, the basic weakness of such methods is their low resolution in detecting and demarcating specific impacts on an ecosystem. For example, the discharge of water into a river from a cooler tributary could exhibit values similar or even equal to those of a chronic toxic pollution/contamination. It appears that diversity indices are less applicable for the main purpose of biomonitoring and assessment—discrimination of anthropogenic from natural fluctuations in the environment—than some other types of bioindicators.

There have been errors in the area of community-related methods. Some argue that alpha-diversity parameters of communities are rarely used for direct environmental biomonitoring and assessments. However, this is not so, because all species-related indices of this type use the species diversity of a community. Beta-diversity metrics of communities and habitats are potentially more advantageous, particularly in relation to the ecosystem approach, in biomonitoring and assessment of the ecological status and environmental health as a whole. In contrast, gamma diversity has been widely used in large-scale biogeographical and/or topological assessments for a long time, for instance when delineating the biogeographical regions, provinces, and areas (see Schwabe and Kratchowil, this volume). However, the application of methods for measuring diversity at local, regional, and global scales needs further development in relation to the emerging goals of stopping ongoing biodiversity loss.

28.3.3 Habitat-related methods

There are few habitat-related methods that involve elements of beta- and gamma-diversity. These evaluations combine biological quality (e.g., status or levels of biotic responses as indicators) and habitat quality assessments, and they can be considered as a type of ecological assessment. The main principle behind these methods is the comparison of sample sites with reference site or a set of reference conditions. The results are expressed as percentage resemblance, which can be classified using a range of terms such as poor to excellent. This is the main approach in

ecological classification of water bodies as accepted by the European Water Framework Directive 2000/60/EC (WFD).

The major problem with evaluation of biological effects by assessing biological responses is the often strong nonlinear, dose-effect dependence. Because of this, variables such as species composition, diversity, and community indices are insufficient for describing environmental quality. Structural indices (integrals) cannot provide information about rates of processes because dependencies between community metrics and metabolic rates in an ecosystem are nonlinear (Dechev, 1992). This is why attempts to introduce functional-ecological methods are still limited to single ecosystems and require regular in situ measurements of the rates of basic processes such as biological production, destruction, utilization, transport, and transformation. Consequently, functional approaches seem unlikely to be available for routine biomonitoring purposes in the near future. Attempts to insert functional feeding groups (introduced by Cummins and Klug, 1979) again explore the integral (resulting) metrics like the proportion of presented trophic groups of benthic invertebrates but not the rate of an ecosystem function such as the speed of transition and/or transformation of the organic matter along the food chains/webs of the community.

28.4 BIOINDICATION OF SURFACE WATERS

In the biomonitoring and assessment of water quality, the saprobic indices are typical examples of the species-related approach. They are relatively old but are still widely applied to water-quality assessment in relation to the organic matter loads of natural origin and, particularly, from urban and/or industrial wastewater. Following the definition of Sladeček (1973), saprobity is the ecological situation of the water body with respect to the amount and intensity of decomposition of putrescible organic matter of autochthonous or allochthonous origin. From an ecological perspective, saprobity is considered the total of all metabolic processes, which represent an antithesis of the primary production within aquatic ecosystems. To consider saprobity only as a measure of the organic pollution is incorrect, while the saprobic system recognizes degrees of presence or absence of small loads with biodegradable organic matter (xeno- and oligosaprobity) and heavy loads (poly- and hypersaprobity) usually due to the discharge of polluted effluents into the water bodies. Nowadays, the concept of saprobity is closest to that of ecological status as introduced by the European Water Framework Directive.

The basic measure of the saprobity levels is formed by the communities and direction of processes they indicate. The indices in use explore the indicative potential (tolerance range) of a single species and, in summary, of the total community to the certain level of saprobity. Depending on species preferences, each bioindicator has an indicative value that represents numerically its position along the saprobic gradient. The overall valuation of saprobity is represented by the mean of the saprobic values and abundance of species in a community.

During the last three decades of the twentieth century saprobic indices have been built around bottom-dwelling invertebrates in rivers. Further developments have allowed the definition of basic metrics (saprobic index, saprobic indices, valences, and indicative weight) and have included other aquatic communities like phytobenthos, phytoplankton, and zooplankton in bioassessment. Thus the list of the saprobic bioindicators exceeded 3,000 species (Sladeček, 1973). Of course, the number of species in a single sampling is much lower than 3,000, but nevertheless sufficiently high levels of taxonomic expertise are needed for accurate assessment of saprobity. Lack of taxonomic expertise is one of the commonly recognized deficiencies in the application of saprobic indices. In fact, however, the basic disadvantage is a weak capacity to detect toxic and/or inert impacts on aquatic communities. Due to the biocidal or depressing effects of pollution and/or contamination on the aquatic environment, stenobiotic or simply sensitive species could be removed from the system, and the evaluation of the saprobity level might be not fully adequate for the actual saprobic conditions to be established in a water body.

There are numerous reviews of the different methods developed in recent decades for monitoring and assessment of the quality of aquatic environment (see Knoben et al., 1995; Conti, 2008 and references therein). The common feature of all of them is that the species or higher taxa are used to develop a numerical expression of their tolerance to the state of the aquatic environment. They include the Family Biotic Index (FBI; Hilsenhoff, 1988), Index of Ephemeroptera-Plecoptera-Trichoptera (EPT-Index; Plafkin et al., 1989), Index Biologique Diatomée (IBD), or Trophic Diatom Index (TDI; Prygel et al., 1999), Dutch Quality-Index (Peeters et al., 1994, after Knoben et al., 1995), and many others.

Among community-related methods, species diversity indices have wide application. There are many equations used to describe various elements of alpha and beta diversities that are commonplace in general ecology handbooks (e.g., the indices of McIntosh, Menhinick, Brillouin, Whittaker, and Cody). However, as previously noted in this chapter, the disadvantage of these kinds of indices in bioindication is the lack of specific responses to the environmental factors causing stresses within a surface water system. Nevertheless such indices are able to assess quantitatively the changes in community structure in respect to the biocenoctic principles of Thienemann (1920, after Sladeček, 1973). However, routine application is operational at a smaller scale (local and regional) due to the specific geographical distribution of species and communities' structure in larger geographical units. It is noteworthy that most of the indices in use represent the state of the community monitored as a reflection of the state of the environment they used to inhabit, which is different from measurement of the environmental variables themselves.

The biotic indices approach was developed by Woodiwiss (1964) and takes some features from both species-related (indicative capacity) and community-related (diversity) approaches for water quality biomonitoring and assessment by means of the bottom-dwelling invertebrates. On one hand, there is a set of key indicative species (these can be species, genera, or higher taxonomic units such as families, orders, classes) for each level of pollution or state of the aquatic environment. On the other hand, the diversity of the community is represented by a number of key taxa (these could also be genera, families, orders, or classes). The presence and relative abundance of indicator species and the number of key taxa results in a score ranging from 1 for most polluted to 10 for the cleanest waters. Most of the biotic indices that have been developed since 1964 could be considered derivatives of the original Woodiwiss index: for example, Biotic Scores (Chandler, 1970), Belgian Biotic Index (De Pauw and Vanhooven, 1983), BMWP-Score (Armitage et al., 1983), and its later modifications (Alba-Tercedor and Sanchez-Ortega, 1988; quoted after Knoben et al., 1995, McGarrigle et al., 1992), and the Bulgarian Biotic Index (BGBI; Uzunov et al., 1998). All differ from the original Woodiwss index only by the set of indicative species and number of key groups used, which are justified according to local and/or regional biogeographical features of the invertebrate fauna involved in the assessment as well as by the number of scores awarded to each of the water quality classes (see for example, the Belgian Biotic Index). The reason to develop so many, and generally similar, indices is mainly biogeographical—in their evolution, species and genera occupy different biogeographical regions and could not be used for all the territory of a larger continent. This is why the attempts to use directly the saprobic bioindicators (developed mostly in and for Central Europe) in North America or the Soviet Union failed (CMEA, 1983). This is also why the EU Water Framework Directive introduced as a first and obligatory criterion for setting the river or lake typology the affiliation of a water body to one of the European eco-regions, which represent the biogeographical areas within the continent (Illies, 1978).

The main advantage of these kinds of indices is in their relatively easy performance and the need to identify only a limited number of trivial species and several higher taxonomic categories (classes, orders, and families). This makes biotic indices suitable for rapid evaluation of the state of the environment. However, it is not easy to define the measured phenomena in these indices, for example, biological water quality, which usually reflects various types of environmental stressors like organic pollution, acidification, and hydromorphological disturbances. To this end, scoring systems such as those used in these kinds of indices are adapted to existing systems for water-quality classification in a given country, or even an area within a country, and as a consequence have limited geographical coverage.

There are also a wide range of methods that are based on various ratios (e.g., occurrence, abundance, and frequency) between single taxa (species, genera, or families) and other elements of the community that is being monitored. These methods could also be attached to the community-related methods, for example, the ratio of density of the sludge worms (*Tubifex tubifex*) to other genera, families, and/or total number of oligochete worms (Oligochaeta) or other bottom invertebrates represented (Milbrink, 1983). It is noteworthy that the Ecological Quality Ratio (EQR) introduced in the European Water Framework Directive is a way of expressing data obtained by some biological metrics when comparing monitoring results with a reference value.

There is a very large and diverse group of other indices that explore various relationships between species and groups within monitored communities. The simplest are comparative indices, which use information about the state of environment, usually the level of pollution, to compare community metrics upstream and downstream of a point source of pollution. This group includes all possible indices of similarity, such as the Sorensen, Simpson, Jaccard, and Czekanovsky indices or species deficit (Kothe, 1962), which may be found in applied ecology handbooks. They can use species presence/absence data when comparing sites or involve abundance metrics based on numbers, densities, and/or biomass of each of the

faunal or floral elements being compared. The main disadvantage of these indices is their local applicability and need for parallel information about the levels of impacting factors. Consequently they are used to measure the biological effects of environmental variables, including pollution, rather than the state of environment itself.

There are only a few habitat-related methods that involve elements of beta and gamma diversity. Rapid Bioassessment Protocols (RBP) have been developed and are used in the United States. This approach was first developed by Karr (1981, 1997) as an Index of Biotic Integrity (IBI) for fish communities and later refined to involve invertebrates and algae as well. These combined evaluations imply the simultaneous assessment of water biological quality and of habitat quality, so this method could be considered an ecological assessment method. The latest versions of RBPs consist of a number of comparative indices, which assess the biological condition of benthic communities that are divided in three categories: structure, community balance, and functional feeding groups (Barbour et al., 1992). The main principle of this method is the comparison with a reference site or a set of reference conditions. The results are expressed as a percentage of resemblance, which can be classified from poor to excellent/high. This concept was later accepted by European Water Framework Directive 2000/60/EC. The habitat quality is evaluated in combination with the biological condition of the communities, resulting in an integrated assessment.

RBPs and IBI are in use mostly in the United States, but similar approaches have been developed in some European countries, for example, RIVPACS (River InVertebrate Prediction and Classification System; Wright et al., 1993) in the United Kingdom, and STOWA and AMOEBA (General Method for Ecosystem Description and Evaluation) in the Netherlands (after Knoben et al., 1995). Later developments in the implementation of the European Water Framework Directive also accepted a multihabitat approach leading to an integral assessment of the ecological status of water bodies (see, for example, Mountain Lakes. org, n.d.; STAR, 2006).

The methods outlined above do not exhaust the many opportunities for bioindication applied in monitoring and assessing aquatic environments. Integrated water management, including that of water quality, nowadays requires the involvement of an ecosystem approach and relevant methods for biomonitoring and bioassessment. The European Water Framework Directive introduced the concept of 'ecological status' as an expression of the quality of structure and functioning of the aquatic ecosystems associated with surface waters and requires classifying this ecological status by parameters/metrics of biological quality elements such as phytoplankton, macrophytes or phytobenthos, bottom invertebrates, and fish.

28.5 BIOINDICATION OF AIR QUALITY

Contrary to the aquatic environment, where animals are the basis for biomonitoring and assessment, plants are the main tools of bioindication in terrestrial ecosystems and are used as bioindicators of air pollution because they show clearly the effects of phytotoxic compounds present in the atmosphere.

The significance of biomonitoring the air-pollution burden by plants offers some important results from different abilities. Plants show an integrated response to pollution. Thus they provide information on the potency of complex pollutant mixtures, which can either occur simultaneously or in a stochastic pattern, as they react only to the effective part of a given pollution situation. Plants react to an ambient air-pollution burden (which often has a strongly fluctuating pattern) with an assessable and verifiable reaction, while modeling dose effects renders information with a much lesser degree of confidence due to, for example, the random distribution of pollutants in time and space.

Different levels of organization of the plants can be used for biomonitoring, ranging from a single plant (or even a leaf or plant cell) through a plant association and to the level of the ecosystem. The response that is obtained at the community level, for example, shifts in species composition, results from the integration of different factors over a relatively long time period that have been experienced by competitive plant species and, as such, cannot be detected on the basis of physical and chemical measurements. Some air pollutants have very low ambient concentrations and are difficult to measure accurately with physical and chemical methods. Plants though can accumulate those pollutants to a level that is easier to analyze. Effects are expressed in sensitive plant species as visible injury (e.g., leaf injury or changes in habit), and in less sensitive species, even pollution-tolerant species, in the accumulation of pollutants. Both provide important tools in recognizing air-pollution effects.

The background of using plants as bioindicators lies in molecular mechanisms of their responses resulting in cell degradation and visible injuries to leaves. This is a common, nonspecific reaction of the photosynthetic and transpiration organs of leaves to the external impacts, which can be of natural or anthropogenic origin. Visible damage is usually preceded by deep transformations within fine cell metabolism and structures that are expressed by enzyme activity, membrane leakage, and the degradation of chloroplast structure, pigment content, and ratios. This is the so-called physiological damage or invisible injury (Stoklasa, 1923). At this stage the injuries are reparable but if exposure is prolonged the next steps of visible injuries—chloroses (yellow) and necroses (brown speckles/spots) on the upper surfaces of leaves are not (Plate 28.1). On the one hand some plant species may be problematic as indicators of short-term changes because of their slow growth and because compounds from past pollution exposure may persist in cells over time. On the other hand, some plants have quite differing strategies to survive under conditions of air pollution. Perennial plants, for example, grasses, and deciduous shrubs and trees, may shorten the time of their active vegetation and thus avoid long exposure to damaging environmental factors. This is why lichens, mosses, and coniferous plants are the best indicators of air pollution due to their long life and relatively slow growth.

The bioindicative potential of plants used in biomonitoring depends on their sensitivity to individual or groups of air pollutants. These are mostly gaseous (sulfur and nitrogen oxides, fluorine, ozone, volatile organic compounds), but also include dust particles, heavy metal aerosols, PAHs, dioxins, and organochlorines. Many problems exist in using laboratory observations to predict sensitivity of plants under field conditions that may include long-term exposure to multiple pollutants and dynamic environmental conditions (Ahmadjian, 1993). Fumigation field studies allow environmentally realistic observations of how specific physiological or morphological changes correlate with specific pollutants (Stolte et al., 1993).

Among plants, lichens are known mostly as receptor-based bioindicators and biomonitors in air-quality studies. Lichen characteristics measured in air-pollution studies include morphological, physiological, and population characteristics. Historically, lichens have been used in a qualitative way, with observations of population changes and morphological effects serving as indicators of pollutants. In the last few decades quantitative measurements of the chemical content of lichens and sensitive physiological processes have increasingly been used to indicate pollutants. Lichens, and also many vascular plants, are used to detect bioaccumulation of more than 40 heavy metals and trace elements, for instance, aluminum, arsenic, barium, boron, cadmium, caesium (including the radioactive isotope ^{137}Cs), chlorine, copper, fluorine, iron, lead, nickel, magnesium,

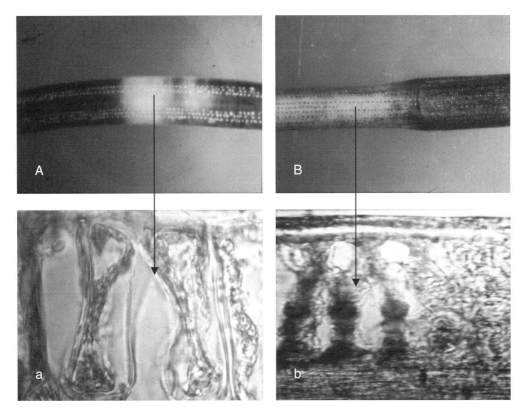

Plate 28.1   Typical visible injuries of coniferous needles affected by industrial air pollution and effects on parenchyma cells. In the early stage of band chlorosis (A) cells are still alive (a) while in the final stage of necrosis (B) cells are dead and empty (b) (reproduced with the kind permission of Professor Dr. Alexandra Uzunova, Plant Physiology Dept, University of Sofia)

manganese, mercury, and zinc. One of the problems in using lichens in routine biomonitoring is that lichen taxonomy depends heavily on determination of lichen chemistry because it produces an abundance of unusual and distinctive secondary metabolites. Some of these metabolites may be determined by relatively simple spot tests or microcrustal techniques, but increasingly, lichen chemistry is determined by thin-layer chromatography (TLC). TLC methods to identify specific lichen metabolite products should be considered necessary to obtain a correct species determination.

However, the effects of air pollution cannot always be clearly differentiated from natural stress factors. Lack of practical experience with certain bioindicators sometimes makes clear interpretation of findings more difficult, especially if no comparable pollutant measurements are available. Nevertheless, there are numerous species-related methods using lichens as bioindicators.

Some of them explore the presence and morphology of lichens represented as a quantitative measure of the level of air pollution, for example, with SO_2 (Table 28.3), while other methods simply classify the ambient air quality in classes. Due to the great variety of species and different ways of expressing their biotic responses, application of such methods is limited to single countries and smaller regions. This is why there is no commonly accepted system for air-pollution biomonitoring in spite of efforts to develop sampling and measurement standards. On the other hand, there are simple biogeographical reasons not to use widely some lichens as bioindicators due to their living requirements. For instance, many common species in northern Europe are found only at high altitudes in southern Europe where usually there is not much industrial air pollution. These boreal and alpine plants and animals, especially insects, are therefore habitat-related indicators rather than of ambient air quality.

Table 28.3 Lichens as bioindicators. The method of Hawksworth and Rose (1976) was developed using the bioindication capacity of 80 lichen species living on trees, which presence detects the concentrations of sulfur dioxide in ambient air as defined in 11 quality classes/zones from 0 (no lichens able to survive) till 10 (clean air)

Zones	Lichen species	SO_2 ($\mu g.m^3$)
0	Epiphytes absent at all.	
1	*Pleurococcus viridis* limited in the base of the stems.	about 170
2	*Pleurococcus s.1.* on the whole stems.	about 150
	Lecanora conizaeoides limited in the base.	
3	*Lecanora conizaeoides* on the whole stem;	about 125
	Lepracia incana frequently on the base.	
4	*Hypogymnia physodes* and/or *Parmelia sulcata* appeared on the base.	about 70
	Lecidea scalaris, Lecanora expallens and *Chaenotheca ferruginea* often presented.	
5	*Hypogymnia physodes* and *P. saxatilis* presented at 2,5 m and more; *P. glabratula, P. subrudecta,*	about 60
	Parmeliopsis ambigua and *Lecanora chlarotera* get appeared; *Calicium viride, Lepraria*	
	candelaris, Pertusaria amara could be presented.	
	Ramalina farinacea and *Evernia prunastri* presented but limited on the base; *Plastismatia glauca*	
	can be presented on horizontal branches.	
6	*Parmelia caperata* represented at least on the base, often presented *Pertusaria (P. albescens,*	about 50
	P. hymenea) and *Parmelia (P. tiliacea, P. exasperatula), Graphis elegans, Pseudevernia*	
	furfuradea and *Alectoria fuscescens* presented in mountain regions.	
7	*Parmelia caperata, P. revoluta, P. tiliacea, P. exasperatula* presented on the whole stem; *P.*	about 40
	hemisphaerica, Usnea subfloridana, Rinodina roboris and *Arthonia impolita* get appeared.	
8	*Usnea ceratina, Parmelia perlata* or *P. reticulata* get appeared.	about 35
	Increasing of *Rinodina roboris; Normandina pulchella* and *U. rubigena* generally presented.	
9	*Lobaria pulmonaria, L. amplissima, Pachyphiale cornea, Dimerella lutea* or *Usnea florida* well	about 30
	represented together with other 25 species developing on well-illuminated trees.	
10	*Lobaria amplissima, L. scrobitulata, Sticia limbata, Pannaria sp., Usnea articulata, Usnea*	clean
	filipendula or *Teleschisyes flavicans*.	

In contrast to the great amount of scientific projects and papers on bioindication of air pollution there are only a few countries such as Germany, Austria and the Netherlands, where some of these methods are also being applied by environmental authorities and private enterprise for routine monitoring of industrial installations and urban centers. Therefore, within Europe, the use of bioindicator plants to assess air-pollution effects is not very well established. The insufficient standardization of techniques is one of the major reasons for the poor acceptance of this effect-related methodology of air-quality monitoring by policy makers, public administrators, and the private sector.

REFERENCES

Ahmadjian, V. (1993) *The Lichen Symbiosis*. New York: John Wiley.

Alba-Tercedor, J., and A. Sanchez-Ortega. (1988): 'Un méthodo rápido y simple para evaluar la calidad biológica de las aguas corrientes basado en el de HELLAWELL' (1978), *Limnética* 4: 51–56. (In Spanish)

Armitage, P.D., Moss, D., Wright, J.F., and M.T. Furse. (1983) 'The performance of a new biological water quality score system based on macroinvertebrates over a wide range of unpolluted running water sites', *Water Resources*, 17(3): 333–47.

Barbour, M.T., Plafkin, J.L., Bradley, B.P., Graves, C.G., and R.W. Wisseman. (1992) 'Evaluation of the EPA's Rapid Bioassessment Benthic Metrics: Metric redundancy and variability among reference stream sites', *Environmental Toxicology and Chemistry*, 11: 437–49.

Carmo-Freitas, M. (1995) 'Elemental bioaccumulators in air pollution studies', *Journal of Radioanalytical and Nuclear Chemistry*, 192(2): 171–81.

Carpenter, K.E. (1928) *Life in Inland Waters*. London: Sedgwick and Jackson.

Chandler, J.R. (1970) 'A biological approach to water quality management', *Water Pollution Control* 69: 415–422.

Chapman. D. (ed.). (1992) *Water Quality Assessment. A Guide to the Use of Biota, Sediments and Water in Environmental Monitoring*. London: Chapman and Hall.

CMEA. (1983) *Unified Methods for Studying the Water Quality. Part III. Methods of Biological Analyses of Water*. Moscow: CMEA (in Russian).

Cohn, F. (1853) Über lebende Organismen im Trinkwasser, *Zeitung der klinischen Medizin* 4: 229–237.

Conti, M.E. (ed.). (2008) *Biological Monitoring: Theory & Application—Bioindicators and Biomarkers for Environmental Quality and Human Exposure Assessment.* Boston: WIT Press.

Cummins, K.W., and M.J. Klug. (1979) 'Feeding ecology of stream invertebrates', *Annual Review Ecology and Systematics*, 10: 147–72.

Dechev, G. (1992) 'The "data-rich but information-poor" insufficiency syndrome in water quality monitoring', in Newman et al. (eds.), *River Water Quality, Ecological Assessment and Control.* Proceedings International Conference on River Water Quality (Brussels, December 1991), p. 751.

De Pauw, N., and G. Vanhooren. (1983) 'Method for biological quality assessment of watercourses in Belgium', *Hydrobiologia*, 100: 153–68.

Dolgov, G., and Y. Nikitinksii. (1927) 'Hydrobiological methods of study', in *Standard Methods for Assessment of Waters and Wastewaters.* Leningrad. pp. 252–89 (in Russian).

Ernst, W.H.O., Verkleij, J.A.C., and R. Vooijs. (1983) 'Bioindicators of a surplus of heavy metals in terrestrial ecosystems', *Environmental Monitoring and Assessment.* 3(3–4): 297–305.

Fjerdingstad, E. (1964) 'Pollution of streams estimated by benthal phyto-micro-organisms I.E. saprobic system based on communities of organisms and ecological factors', *International Review Gesamten Hydrobiologie* 49: 63–131.

Friedrich, G., Chapman, D., and A. Beim. (1992) 'The use of biological material', in D. Chapman (ed.), *Water Quality Assessment.* Cambridge, UK: Chapman & Hall. pp. 171–238.

Hawksworth, D.L., and F. Rose. (1976) *Lichens as Pollution Monitors.* London: Edward Arnold.

Hellawell, J.M. (1986) 'Biological indicators of freshwater pollution and environmental management', in K. Mellanby (ed.), *Pollution Management Series.* London: Elsevier Applied Science.

Hilsenhoff, W.L. (1988) 'Rapid field assessment of organic pollution with a family level biotic index', *Journal of the North American Benthological Society*, 7: 65–68.

Horasawa, I. (1956) 'A preliminary report on the biological index of water pollution', *Zoological Magazine*, (Tokyo) 54(1).

Huet, M. (1954) 'Biologie, profiles en long et en travers des eaux courantes', *Bull. Fr. Pisciculture*, 175: 41–53.

Huggett, R.J. (2004) *Fundamentals of Biogeography.* 2nd ed. London: Routledge.

Illies, J. (ed.). (1978) *Limnofauna Europaea. A Check-list of the Animals Inhabiting European Inland Waters with Account of Their Distribution and Ecology.* 2nd ed. Stuttgart: Gustav Fisher Verlag.

Illies, J., and L. Botosaneanu. (1963) 'Problèmes et methodes de la classification et de la zonation ecologique des eaux curante, considerees surtout du point de vue faunistique', *Mitteilungen Internationale Vereinigung Limnologie* 12: 1–57.

Kaglyan, A., Klenus, V., Kuz'menko, M., Belyaev, V., Nabyvanets, Yu., and D. Gudkov. (2005) 'Macrophytes as bioindicators of radionuclide contamination in ecosystems of different aquatic bodies in Chernobyl exclusion zone', in Brechignac, F. and G. Desmet (eds.), *Equidosimetry— Ecological Standardization and Equidosimetry for Radioecology and Environmental Ecology.* NATO Security through Science Series C. Dordrecht: Springer. pp. 79–86.

Karr, J.R. (1981) 'Assessment of biotic integrity using fish community', *Fisheries* 6: 21–27.

Karr, J.R. (1997) 'Measuring biological integrity', in G.K. Meffe, C.R. Carroll, and contributors (eds.), *Principles of Conservation Biology.* 2nd ed. Sunderland, MA: Sinauer. pp. 483–85.

Knoben, R., Roos, C., and M. van Oirschot. (1995) *Biological Assessment Methods for Watercourses.* UN/ECE Task Force on Monitoring and Assessment, vol. 3. RIZA Report No 95.066. Lelystad.

Knöpp, H. (1955) 'Grundsatzliches zur Frage biologischer Vorfuntersuchungen, erlauert an einem Guttelagssschnitt des Mains', *Arch. Hydrobiol., Suppl. Bd* 2: 363–68.

Kolenati, F.A. (1848) Über Nitzen und Schaden der Trichopteren, *Settiner entomologisches Zeitung* 9: 251–259.

Kolkwitz, R., and M. Marson. (1902) 'Grundsätze fur die biologische Beurteilung des Wassers nach seiner Flora und Fauna', *Kleine Mitteilungen Koniglich Prufungsanstalt Wasserversorgung und Abwässerreinig* 1: 33–72.

Kothe, P. (1962) 'Der "Artenfehlbetrag", ein einfaches Gutekriterium und seine Anwedung bei biologischen Vorfluteruntersuchungen', *Die Gewasserkundlich Mitteilungen* 6: 60–65.

Krizek, B.A., Prost, V., Joshi, R.M., Stoming, T., and T.C. Glenn. (2003) 'Developing transgenic *Arabidopsis*-plants to be metal-specific bioindicators', *Environmental Toxicology and Chemistry*, 22(1): 175–81.

Lampert, W., and U. Sommer. (1997) *Limnoecology.* New York: Oxford University Press.

MacGarrigle M.L.; Lucey, J. and K.C. Clabby (1992): Biological assessment of river water quality in Ireland. In: Newman, P.J.; Piavaux, M.A. and R.A. Sweeting (eds.): *River Water Quality. Ecological Assessment and Control.* Brussels: Commission of the European Community: 371–385.

Milbrink, G. (1983) 'An improved environmental index based on the relative abundance of Oligochaeta species', *Hydrobiologia*, 102: 89–97.

Mountain Lakes. (n.d.) Mountain Lakes.org. http://www.mountain-lakes.org/ Accessed July 22, 2009.

Nali, C. Balducci, E., Frati, L., Paoli, L., Loppi, S., and G. Lorenzini. (2009) 'Lettuce plants as bioaccumulators of trace elements in a community of Central Italy', *Environmental Monitoring and Assessment*, 149: 143–49.

Odum, E. (1971) *Fundamentals of Ecology.* 3rd ed. Philadelphia: W.B. Saunders.

Olivares-Rieumont, S., Sella, S.M., Silva-Filho, E.V., Guimarães Pereira, R. and Miekeley, N. (2007) 'Water hyacinths (*Eichhornia crassipes*) as indicators of heavy metal impact of a large landfill on the Almendares River near Havana, Cuba', *Bull. Environ. Contam. and Toxicol*, 79(6): 583–87.

Pantle, R., and H. Buck. (1955) 'Die biologische Überwachung der Gewässer und die Dastellung der Ergebnisse', *Gas- und Wasserfach* 96: 604.

Peeters, E.T.H.M., Gardeniers J.J.P. and H.T. Tolkamp (1994) 'New methods to assess the ecological status of surface waters in the Netherlands. Part 1: Running waters', *Verhandlungen des Internationalen Verein Limnologie*: 1914–1916.

Plafkin, J.L., Barbour, M.T., Porter, K.D, Sharon, K.G., and R.M. Hughes. (1989) *Rapid Bioassessment Protocols for Use in Streams and Rivers: Benthic Macroinvertebrates and Fish*. United States Environmental Protection Agency, EPA 440/4–89/001.

Prygel, J.; Coste, M. and J. Bukowska (1999): Review of the major diatom-based techniques for the quality assessment of rivers – State of the art in Europe. – In: (Prygel, J.; Whitton, B.A. and J. Bukowska (eds.) *Use of algae for monitoring rivers* III: 224–238.

Rotschein, J. (1962) *Graphical Expression of Biological Data Dealing with Evaluation of the Water Quality*. Bratislava: VUVH (in Slovakian).

Sladeček, V. (1973) 'System of water quality from biological point of view', *Arch. Hydrobiol./Beih. Ergebn. Limnol*, 7(I–IV): 218.

STAR. (2006) 'Standardisation of river classifications', http://www.eu-star.at/frameset.htm. Accessed July 22, 2009.

Stoklasa, D. (1923) *Die Bedschädigung der Vegetation durch Rauchgasexhalationen*. Berlin, P. Parry.

Stolte, K., et al. (1993) *Lichens as Bioindicators of Air Quality*. General Technical Report, RM-224. Fort Collins, CO: Rocky Mountain Forest and Range Experiment Station.

Thienemann, A. (1920) *Die Grundlagen der Biozönotik and Monard's faunistische Prinzipen*. Festschrift für Zschokke (Basel), 4.

Thienemann, A. (1925) 'Die Binnengewässer Mitteleurops', *Die Binnengewässer, Schweizerbart, Stuttgart*, 1: 54–83.

Tyler, G., Balsberg-Påhlsson, A.-M., Bengtsson, G., Bååth, E., and L. Tranvik. (1989) 'Heavy-metal ecology of terrestrial plants, microorganisms and invertebrates', *Water, Air and Soil Pollution,* 47(3–4): 189–215.

Wright, J.F., Furse, M.T., and P.D. Armitage. (1993) 'RIVPACS—a technique for evaluating the biological quality of rivers in the UK', *European Water Pollution and Control* 3(4): 15–25.

Uzunov, Y., Penev, L., Kovachev, S., and P. Baev. (1998) 'Bulgarian biotic index (BGBI)—an express method for bioassessment of the quality of running waters', *Comptes Rendus de l'Academie bulgare des Scences* 5 (No 11–12): 117–20.

Vannote, R.L., Minshall, G.W., Cummins, K.W., Sedel, J.R., and C.E. Cushing. (1980) 'The river continuum concept', *Canadian Journal Fisheries and Aquatic Science,* 37: 130–37.

Woodiwiss, F.S. (1964) 'The biological system of stream classification used in Trent River Board', *Chemistry and Industry* 11: 443–47.

Zelinka, M. (1960) 'A contribution to a more precise classification of clean waters', *Sci. Pap. Inst. Chem. Technol., Prague Technology of Water* 4(1); 419–27 (in Czech).

Zelinka, M., and P. Marvan. (1961) 'Zur Präzizierung der bioilogischen Klassifikation der Reinheit fliessender Gewässer', *Arch. Hydrobiol.* 57: 389–407.

29

Historical Biogeography as a Basis for the Conservation of Dynamic Ecosystems

Rob Marchant and David M. Taylor

29.1 CHAPTER OVERVIEW AND ORGANIZATION

The chapter is divided into three sections. The first section presents concepts of climate space and discusses how the rate and character of climate change impacts ecosystems. The second section presents the forms of evidence upon which the composition and distribution of past ecosystems can be reconstructed and how such evidence can be integrated through a modelling approach to investigate potential future changes. The third section discusses the implications for conservation of dynamic ecosystems under future climate changes, and highlights shortcomings in our present understanding. Areas for future research activity are suggested. Throughout the chapter, a series of case studies from the Neotropics and equatorial Africa are used to illustrate the impacts of environmental shifts on ecosystems during recent geological time. The time period covered is from the Last Glacial Maximum (LGM), 21,000 years before present (B.P.) (18,000 radiocarbon years B.P.), through the present interglacial period, the Holocene, and to the future. This focus on low latitudes throughout this chapter is particularly relevant as ecosystems in the tropics have not received the same attention as their counterparts in more temperate regions, particularly in terms of their responses to global climate changes. Moreover, low latitudes are likely to be amongst the most heavily impacted if predicted levels and

rates of future climate change (McSweeny et al., 2008) prove accurate.

29.2 DYNAMIC ECOSYSTEMS— PRESSURES, EFFECTS, AND FEEDBACKS

Within the global change community there is growing awareness of processes of environmental change and how these are likely to impact our planet and its ecosystems. The current geological period (the Quaternary) spans approximately the last two million years and is characterised by a series of relatively long-lived largely cool, arid (glacial) phases and relatively short-lived, mainly warm, humid (interglacial) phases. There have been at least 20 major glacial phases over the course of the Quaternary during which the global extent of ice was greater than during intervening interglacials (Peltier, 1994). Glacial-interglacial cycles are associated with climate changes but are also characterised by lower sea levels, differences in albedo and changed atmospheric composition, and are largely orbitally driven. The tendency of climates to change relatively suddenly during the Holocene, even prior to the anthropocene (the current period of significant human impact), has been one of the most surprising outcomes of the study of Earth history (Marchant and Hooghiemstra,

2004). Superimposed upon the major glacial-interglacial variations are numerous lower-magnitude, higher-frequency events operating at subdecadel (e.g., the El Niño Southern Oscillation), decadal (e.g., sunspots), and centennial/millennial (e.g., precessional) time scales. Superimposed on these natural cyclic variations are more unpredictable events that also impact on climate (e.g., volcanic eruptions). Factors that drive climate change are also involved in a series of complex feedback mechanisms, such as the interplay between ice sheets, albedo, and ocean circulation (Broeker, 2000).

Perhaps therefore the only constant regarding climate is that it has constantly changed; such changes being unevenly felt over the Earth's surface, with some areas experiencing greater variations in temperature, precipitation, and seasonality than others. Expansion of the massive polar ice sheets during past glacials had a major impact on climate conditions, with sea levels and monsoon-associated precipitation probably greatly lowered relative to the present (Farrera et al., 1999). The maximum extent of ice for the last glacial period in other parts of the world may not have coincided with the polar LGM. For example, the extent of ice on several mountains in eastern Africa reached its maximum in the late glacial, owing to a combination of relatively cool and humid climate conditions (Osmaston, 1975). Indeed, as more data on environmental change and its ecosystem impacts are produced, a different perspective on the spatial and temporal character of abrupt climate shifts and how these impact ecosystem composition emerges (Stocker and Marchal, 2000; Willis et al., 2004). The tropics, rather than following climate changes recorded at temperate latitudes, are thought to record some changes first (Stager et al., 2002) and indeed may act as a pace setter for change (Dunbar, 2003; Kerr, 2003; Turney et al., 2004; Partin et al., 2007). Tropical ecosystems may provide an early warning system to climate change, particularly within the present interglacial period when climatic ties to high latitudes have weakened with the demise of the polar ice sheets (Johnson et al., 2002), a situation that one would expect to continue in the future as ice sheets undergo accelerated contraction.

Climatic parameters, most importantly temperature, moisture, and seasonality (Table 29.1), largely control the composition and distribution of ecosystems: as climates change, the distributional range of component individuals, particularly those in ecotonal locations, respond to the change by migrating (Figure 29.1). Climate conditions exert a strong influence on ecosystems; as climates change over time and space the distribution, composition, structure, and functioning of ecosystems

Table 29.1 Climatic parameters commonly incorporated into a bioclimatic/envelope model

Variable name	Variable description
MTW	Mean temperature of the warmest month
MTC	Mean temperature of the coldest month
ATR	Annual range of mean monthly temperature
MINTC	Mean minimum temperature of coldest month
MAXTC	Mean maximum temperature of warmest month
AMIN	Absolute minimum temperature
MAP	Mean annual precipitation
MAPD	Mean annual precipitation of the driest month
AMI	Annual moisture index

may follow suit. Changes in climate conditions throughout time have resulted in associated ecosystems having a very different past composition relative to today. For example, during glacial periods significant areas that are today occupied by rain forests would have been replaced by more open forms of vegetation, such as grasslands or sparse woodland due to low temperatures, reduced precipitation, low levels of CO_2, increased seasonality, and an altered fire regime, or a combination of these factors (Bush et al., 1990; Hooghiemstra and van der Hammen, 1998; Colinvaux et al., 2000, 2001; Elenga et al., 2000; Bond et al., 2005). Climatic factors, especially moisture in the tropics and temperature in more temperate latitudes or at higher altitudes, can be important explanatory variables for species distribution, ecosystem composition, and hence some diversity patterns (Linder, 1991; O'Brien, 1993). However, vegetation cover is also a function of ecological and local environmental factors. Ecosystems may have remained relatively stable in composition where the impact of global climate variations has been mitigated by local—notably edaphic—environmental conditions. In addition, in some parts of the world climate conditions did not fall below critical thresholds, and as a result there were few, if any, biological effects. For example, in some exceptionally wet areas, such as the Chocó lowland plain that lies between the Pacific and the northern Andes, the levels of rainfall reduction associated with events such as the LGM are unlikely to have forced major changes in vegetation. Even with the highest proposed reduction in precipitation (50% at the LGM) due to high precipitation (>15,000 mm yr^{-1}) for this area

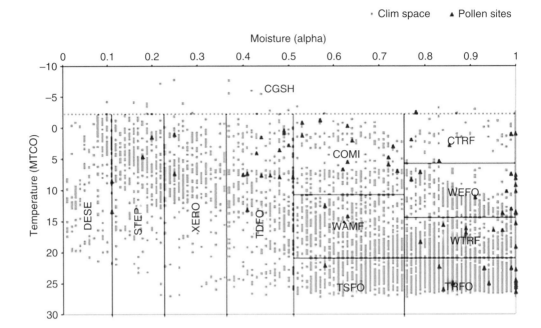

Figure 29.1   Simplified concept of climate space and ecosystem distribution. Two variables of temperature (mean temperature of the coldest month) and plant available moisture (Priestly-Taylor alpha) form two climate gradients that change ecosystem type as these are moved along. For example, under high moisture and temperature, tropical rainforest (TRFO) forms, whereas under drier conditions desert (DESE) is present. The actual distribution is driven by numerous environmental gradients in addition to ecological constraints such as dispersal and history. Such changes are easily seen in montane areas where the altitude gradient is a surrogate for temperature (e.g., Figure 29.3)

rainfall would not fall below the 2000-mm yr^{-1} threshold required for tropical rainforest growth (Marchant et al., 2007).

Areas where forests are believed to have persisted under maximum climate change (at the LGM) are known as forest refuges. The theory of forest refuges was developed largely from results of investigations in South America (Haffer, 1969), which have since been added to following further studies in the Neotropics and in other parts of the world (Haffer, 1997) (see Willis et al., this volume for a detailed discussion of the concept of a refugium). Although there is evidence in support of the existence of forest refuges in some areas (Huntley and Birks, 1983; Overpeck et al., 1992; Colinvaux, 1996; Maley, 1996; Willis, 1996), the evidence is largely circumstantial and based mainly on present-day distributions of plants and animals (Section 29.3.2), genetic studies on populations, the output of coarsely resolved biome response models (Section 29.3.3), or isolated fossil-based studies (Section 29.3.1). Although potentially useful, direct supporting evidence in the form of well-dated fossil remains is

largely lacking. Hence, the controversy about which areas were ecologically more stable, how they were manifested, and what were the climate mechanisms to allow this to happen continues. However, recently produced direct evidence, in the form of pollen sequences from cores of sediment from one of the areas originally proposed as a forest refuge, the Eastern Arc Mountains of Tanzania (Hamilton, 1982), support the existence of relatively stable vegetation communities through the LGM (Mumbi et al., 2008; Finch et al., 2009). The driving mechanism behind such ecosystem stability, compared with other mountains in eastern Africa, is likely a relatively constant supply of moisture from the Indian Ocean (Hamilton, 1982; Marchant et al., 2007).

In contrast to areas that are relatively immune from climate change, the presence of steep climate gradients associated with mountainous areas, where there is not some stabilising influence, provide a range of habitats and these may provide accommodation space for the spatial adjustments of taxa at the local scale (Fjeldså and Lovett, 1997; Haffer, 1997). In such environments, dynamic moisture regimes, local topography, and edaphic conditions are important controls on the form and function of ecosystems across a range of altitudes. The diverse climates characteristic of montane areas would have been equally important in the past and could be crucial in sustaining forest ecosystems through future climate change. As more long-term ecological data and studies into predicting impacts of climate change on species distribution become available (Lewis et al., 2004), it is clear that future ecosystem composition, structure, and functioning will be different from the past (Davis, 1986; Davis et al., 2005; Thomas et al., 2004). In particular, the Earth is entering into a period of high CO_2 levels and temperature. In order to survive, ecosystems and their component species are likely to have to adapt to conditions that have not been experienced for at least two million years, as the majority of the Quaternary period (some 80%) has been spent under an environmental regime of low CO_2 and low temperature. Moreover, the conversion of large tracts of land to agriculture and introductions of alien species have made dispersal of climate change—stressed taxa increasingly difficult (Ellis and Ramankutty, 2008).

29.3 EVIDENCE FOR CHANGED CLIMATE SPACE AND ECOSYSTEM CHARACTER

Ecosystem dynamics can be determined from both direct and indirect sources of evidence. In this section we first discuss the methods used to obtain both forms of evidence. Second, we examine the benefits of an interdisciplinary approach to the conservation of ecosystems under varying climate conditions (Section 29.4).

29.3.1 Palaeoecological evidence

Palaeoecologists generally utilise material accumulating in sedimentary basins such as those containing swamps or lakes as a basis for reconstructing the past composition of ecosystems. Material from both biotic and abiotic sources can be used. Examples of the former are the remains of chironomids, charcoal, diatoms, foraminifera, pollen, and testate amoebae. Sediment chemistry is also potentially an important source of information.

Sediments bearing material of value to palaeoecologists can potentially accumulate under a range of environmental conditions, although for many forms of biological remains preservation is optimised in the absence of oxygen (e.g., as is

found in sedimentary settings associated with wetland environments). For example, the sediment-based remains of plants, notably pollen and spores that can be identified to ecologically meaningful taxonomic levels, can be used to reconstruct vegetation history over a range of geographic scales. Assuming that the remains can be dated accurately, this type of evidence can be used to gain an insight into the vegetation composition in an area at a particular point in time. When these snapshot reconstructions are carried out on samples from different depths in a sedimentary sequence and placed within a time frame provided by an independent dating method such as radiocarbon analysis, information can be obtained on the nature and rate of past vegetation changes, depending on the availability of additional information and the potential causes behind these changes.

A good example of the responsive nature of tropical vegetation to register impacts of climate and human-induced forcing has been produced from Lake Euramoo, Queensland, Australia (Harbele, 2005). Lake Euramoo is a small crater lake at 730 m.a.s.l., located near the boundary of submontane rain forest and schlerophyllous woodland; schlerophyllous tree species occurring as emergent and codominant species in the canopy. Pollen and charcoal data from sediments accumulating in the lake show how the surrounding vegetation has changed over the past 22,000 years or so (Figure 29.2). The palaeoecological record shows that between c. 22,000 and c. 15,000 calibrated (cal) years B.P. the vegetation was a mix of cold and dry schlerophyllous woodland with low species turnover indicative of relatively stable climate and associated ecosystem conditions. In the first past of this period fires were relatively common in the catchment, possible due to the relatively dry climate at this time (Harbele, 2005). Between c. 15,000 and c. 8,300 cal years B.P. a mosaic of schlerophyllous woodland and submontane rain forest established within the catchment characterised by high turnover episodes. During the early to mid-Holocene (c. 8,300 to c. 4,800 cal. years B.P) warm submontane rain forest established with very low turnover and very low incidence of fires. From c. 4800 to c. 120 cal. years BP, cool, dry submontane rain forest established with increasing turnover and the loss of long-lived trees (e.g. *Agathis*). From approximately 1,500 cal. years BP there is a steady increase in fires and diversity of the vegetation. During the last 120 years or so, the cool, dry submontane rain forest continues with high turnover in species and a massive increase in fires. This most-recent period is also characterised by an invasion of exotic plants.

Pollen analysis is a remote sensing tool that enables investigation of long-term ecosystem

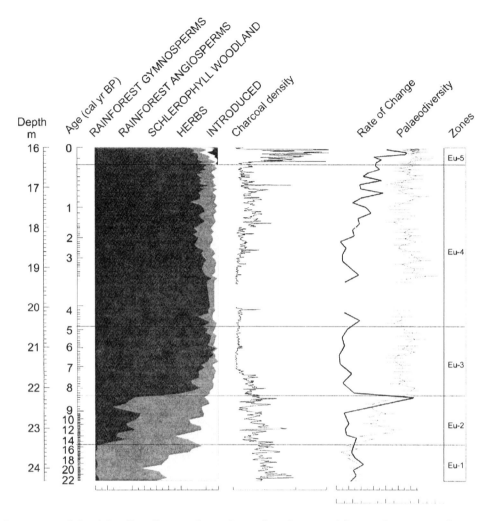

Figure 29.2 'Classic' pollen diagram from the topics: the record from Lake Emaroo shows vegetation and fire history over the past 25,000 years. A series of recognizable vegetation zones document the spread of rainforest taxa as climate became wetter and warmer coming out of the glacial period. The vegetation has become more open in the recent past, particularly with the arrival of European settlers in the area (Harbele, 2005)

dynamics (Prentice, 1985); like all remote sensing tools there is a need to understand constraints on the spatial resolution attainable. One of the perennial problems for interpreting palaeoecological records is the provenance of the pollen taxa (Marchant and Taylor, 2000); how reflective of the surrounding vegetation reconstruction is the pollen accumulating within sediments? This problem of provenance is particularly acute in the tropics where the discipline is relatively new compared with the more intensively studied temperate latitudes (Hicks et al., 2001). Information on the nature of the modern pollen to surrounding vegetation

relationship produces a pollen-vegetation calibration that can feed directly into a modeling tool, such as that developed as part of the European 'PollandCal' network to explore pollen deposition in a landscape scenario (Bunting and Middleton, 2005). Within the model, floristic elements, landscape characteristics, and factors influencing pollen emission, fall speed, and climatic factors influencing the pollen deposition can be changed to investigate 'what if?' scenarios (Bunting et al., 2005).

The harmonizing techniques of the international BIOME-6000 project (Prentice and Webb, 1998), where pollen data from multiple sites are combined in a unifying methodology (Figure 29.3), have fueled collaboration with research groups outside of palaeoecology, particularly the climate and vegetation modeling community. One of the main benefits of this has been to reduce the complexity of palaeoecological datasets by assigning individual pollen taxa to plant functional types and biomes (Figure 29.4), thus providing a means of linking the results of palaeoecological research with bioclimatic/vegetation modelling (Section 29.3.1). As a consequence, evidence of past variations can be used to constrain the modelling of future conditions.

Another example of an international effort to integrate records of the drivers and effects of environmental changes over large geographic areas, in this case incorporating interhemispheric linkages, is the IGBP-PAGES pole-equator-pole (PEP) transects (PEP I, PEP II, and PEP III). The PEP transects incorporate great biological, climatic, edaphic, and topographic diversity (Markgraf et al., 2000; Battarbee et al., 2004; Dodson et al., 2004). One outcome of international collaboration through PEP has been that spatial disparities in both the quantity and quality of information available to the wider scientific community have been made obvious. Major gaps exist on all three transects, but particularly at low latitudes in both hemispheres (Jackson et al., 2000; Davis et al., 2005). Unfortunately, a high density of palaeo sites throughout the tropics is required to underpin studies that can capture, for example, the detail of geographic variations in the past rate and likely source of propagules for recolonisation by forests following major climatic shifts. There is, however, a sufficient amount of palaeoecological evidence to indicate that ecosystems do not respond to climate change as discrete and fixed units; it is clear from the range of palaeoecological archives that biota in certain locations were more responsive to the climatic changes of the Late Quaternary than in others. Evidence from central Africa (Jolly et al., 1997) and Latin America (Urrego et al., 2005) demonstrates the Gleasonian view that ecosystems respond to climate change as a combined series of individual species responses resulting in the formation of novel assemblages of taxa. Thus, as in other parts of the world, the response of ecosystems at low latitudes to climate change in the past appears to have been a product of the reactions of individual taxa, each with its own range of ecological tolerances and therefore sensitivity to change (Figure 29.5).

29.3.2 Inferential evidence

Past ecological changes may also be determined from indirect sources of evidence. Current distributions of species may carry an imprint of past environments, because the distribution patterns of extant species reflect both past and present-day conditions. Two main patterns of species distributions are commonly used as sources of information on past conditions; levels of diversity, or differences in the number of organisms between areas; and levels of endemism, or differences in the degree of biological uniqueness between areas. Loci of high species diversity and endemism have been used as surrogates for forest refuges (Nelson et al., 1990) under the assumption that high diversity and endemism are facilitated by relative environmental stability (Fjeldså et al., 1997; Fjeldså and Lovett, 1997; Küper et al., 2004; Lovett et al., 2000, 2005; Taplin and Lovett, 2003): the corollary being that intervening areas of relatively low species diversity and endemism have been impacted much more severely by past environmental changes (Mutke et al., 2002).

Information on the genetic structure and diversity of populations is an increasingly useful form of evidence, given that evolutionary history may be considered a more inclusive measure of biodiversity than species numbers alone (Purvis and Hector, 2000; Sechrest et al., 2002; Leira et al., 2009). Genetic diversity is the raw material for evolution, and all species have arisen not by a random evolutionary walk, but through a journey where each step depends largely on the variation present at the last (Darwin, 1872). When subject to climate change, genetic diversity therefore has a value that is proportional to its amount: greater diversity being able to offer greater option for species to adapt (Ledig, 1986). However, as we cannot predict the future beyond the extrapolation of current trends, we are unable to determine the evolutionary direction that any species will take. Consequently, we cannot identify which genetic variants will be valuable for species persistence in the natural environment but must assume that all genetic variation is valuable per se (Humphries et al., 1995). Strong parallels exist between processes influencing levels of diversity within species and the ecosystems in which they reside and maintaining diversity during periods of climate change (Jump et al., 2009). Following dispersal, species range changes will have genetic consequences, with new assemblages having new genetic signatures. The advent of DNA technology provides suitable markers to examine these variations (Hewitt, 2000; Jump et al., 2009). Although the research area is in its relative infancy, several good datasets are available to test expectations of and provide insight into species dispersals. One of the best examples on the use of genetic information has been the tracing of human origins to the African continent: phylogenetic analyses of individuals from Africa, Asia, Europe,

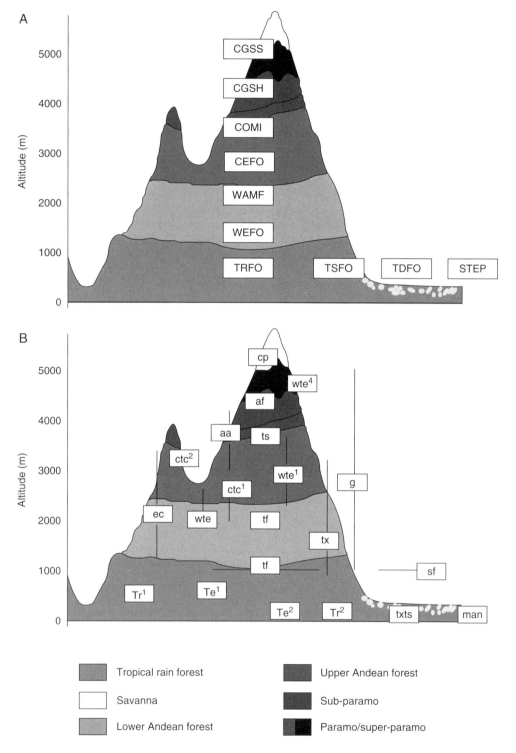

Figure 29.3   Conceptual distribution of plant functional types and relationship to biomes—such different levels of ecological organisation provide opportunities for palaeoecologists to link with other disciplines such as climate modeling groups and vegetation ecologists

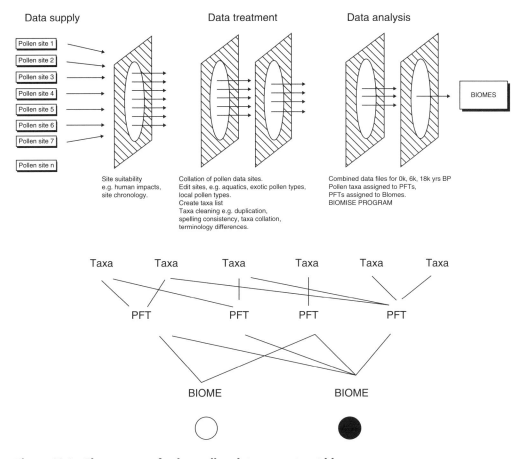

Figure 29.4   The process of using pollen data reconstruct biomes

Oceania, and the New World resulted in a tree for 10 distinct Y haplotypes with a coalescence time of approximately 150,000 years. The 10 haplotypes were unevenly distributed among human populations, with the ancestral haplotype limited to African populations (Hammer et al., 1998). The combination of phylogeographic data for multiple species with data on the geographic distributions of species and plant communities will allow the identification of fundamental, regional biogeographic patterns, providing key information for conservation practitioners (Moritz and Faith, 1998). In addition, such information has the potential to explain correlations between geographic patterns of genetic and species diversity (Moritz and Faith, 1998; Vellend, 2003).

There are numerous problems with inferential evidence, one being the dating of those past conditions that are still assumed to be reflected in current distributions and/or genetic composition of some plants and animals. Furthermore, there are multiple interconnecting variables, with

patterns of diversity likely to have their route in numerous different causes (Williams et al., 1999; Gaston, 2000). Without confirmatory evidence, for example, from the palaeoecological record or information on rates of spread and genetic transformations among the taxa involved, it is impossible to state the time period over which distribution patterns have formed. Furthermore, the impacts of various climate and environmental variables are likely to be different in the future. For example, dramatic increased levels of atmospheric CO_2 and its interrelationships with fire, water-use efficiency, and growth rate are considered to be major determinants of global (and hence local) vegetation patterns, particularly those that are currently relatively water-stressed (Bond et al., 2005). A fourth problem concerns the distribution patterns themselves and whether they actually exist. For example, in the case of Amazonia, Nelson et al. (1990) present a convincing argument that proposed centers of diversity and endemism may in fact be artefacts of

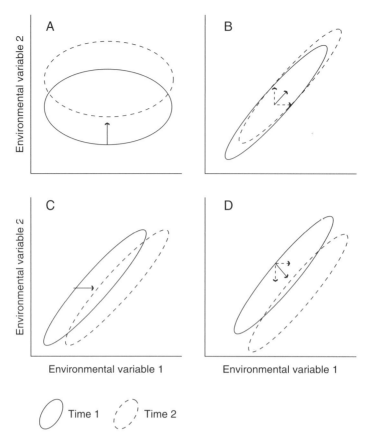

Figure 29.5   As climates change, individual species will respond differently to the new environments creating novel ecosystems

sampling intensity. In short, there is a circularity of reasoning to the use of high levels of species richness and uniqueness as criteria for defining the possible location of diverse ecosystems. Such 'honey pot' locations tend to have disproportionally more research effort, become more species-rich as botanists and zoologists focus their attention there, and therefore attract more research effort in a continual feedback loop. Such areas can then afford improved levels of protection and attract conservation resources on the grounds that they provide a means of conserving large amounts of species/genetic diversity.

29.3.3 A modeled approach to detecting ecosystem impacts of climate change

An insight into how ecosystems respond to environmental variability can be obtained from models (Doherty et al., 2000; Roberts et al., 2002).

Biome response models, such as BIOME-4, simulate vegetation cover by integrating modern climate and environmental data (Prentice et al., 1992). Once verified through comparison with a map of potential vegetation, the models can be used to simulate vegetation patterns under different climate regimes, such as at the LGM (Figure 29.6). This simulation can take two forms. First, possible vegetation cover can be explored by direct manipulation of the input variables, such as CO_2, precipitation, and temperature. Second, vegetation cover can be reconstructed by feeding output from a global climate model (GCM) into a vegetation model (Figure 29.6). Modelled vegetation cover under different climate scenarios can be compared with reconstructions based on empirical (pollen) data obtained from palaeoecological investigations. These comparisons provide a means of verifying output of vegetation response models (Figure 29.6). In addition to being an independent research strand, model outputs used in conjunction with data-based evidence combine to make a

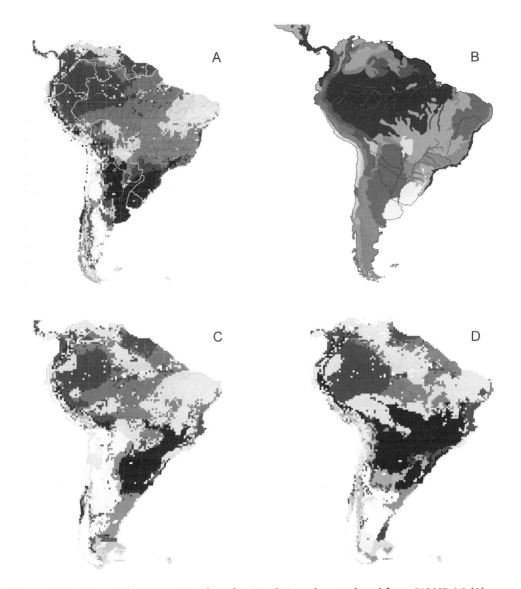

Figure 29.6   Vegetation reconstructions for South America produced from BIOME 4.2 (A) compared with a map of the potential vegetation (B). The vegetation reconstruction is then driven by output from two different climate models (C, D) for the Last Glacial Maximum showing quite different ecosystem response, for example, marked different levels of reduction in the arboreal cover of Amazonia

powerful tool that can be used to examine mechanisms of climate change and ecosystem response under series of 'what if' scenarios, and to investigate the relationship between changed moisture or atmospheric CO_2 concentration (Marchant et al., 2002a). This process of climate modelling generally assumes equilibrium conditions, that is, that there is a unilinear relationship between the dependent variable (vegetation) and the

independent variable (climate). However, models are being developed that are able to incorporate some feedbacks between vegetation and climate and include more dynamic elements, such as changes in grazing or fire regimes (http://www.pik-potsdam.de/ateam/). However, vegetation impacts of edaphic and topographic factors are poorly resolved with the scale such models operate on (over) smoothing local environmental detail.

Distribution modelling approaches—variously described as niche and bioclimatic envelope modelling (MacArthur, 1972; Huntley et al., 1995; Pearson et al., 2002; Peterson et al., 2002; Pearson and Dawson, 2003)—are another promising area for development. For this potential to be realised, model development will need to focus on main limitations. First, it is assumed plant species distribution is solely determined by the macroclimate (such as temperature, precipitation, and seasonal regime), whereas in reality, the distribution of many plant species is also determined by competition, soil type, history, fire, and microclimates. Second, there is a need to include ecological constraints on plant distribution such as dispersal: differential dispersal ability of plants can be parameterised and used to constrain the response of individual taxa, and hence new communities, to climate change scenarios. These multiple factors controlling vegetation distribution are further complicated by the factor of resolution—present work is at best conducted at a grid scale of one degree. In many areas with high habitat heterogeneity a cell of one degree latitude by one degree of latitude longitude would not capture such variation; for example, montane forest is recorded adjacent to hot springs up to 4000 m.a.s.1. on Mount Kenya—well above the treeline (Coe, 1967). The area represented by one degree of latitude also varies; this being much larger at the equator than elsewhere. To overcome these problems we need to move to a finer resolution grid size. Equally important is the need to improve the resolution of the input data; these are frequently derived from relatively few meteorological stations and standard soil maps that do little justice to reality.

Socioeconomic impacts of ecosystems responding to climate change also need to be understood (Hulme et al., 1999). An area that needs development is to place anthropogenic/disturbance factors central within ecocentric understanding of ecosystems, realizing the strong impact that humans have had on the majority of ecosystems (Ellis and Ramankutty, 2008). Such a development will need to combine existing information on the extent and type of human-induced control on the vegetation with a historical perspective, such as that currently available within the HYDE database (http://www.mnp.nl/hyde/). Understanding long-term human-environment interrelationships is highly relevant to determining the full range of factors that have resulted in the present-day composition of ecosystems past ecosystem changes, such as that experienced about 4,000 cal. years B.P., may have been so severe as to result in significant ecosystem impacts with ramifications on the civilisation of the time leading to mass migration, adoption of new technology, and new modes of subsistence (Marchant and Hooghiemstra, 2004). Such severe

societal implications provide salutary lessons on the role that climate change may play in influencing human culture. Past human impacts on ecosystems can be complicated signals to reconstruct: human and climate-induced change sometimes being difficult to separate. For example, one indicator of climate aridity is an increase in grasses as the vegetation becomes more open; such a change could alternatively be attributed to growth in pastoralism and/or agricultural cereals. Relatively simple cause-and-effect relationships levied between climate change and human response (Verschuren et al., 2000) must be treated with caution; thoughtful consideration must be given to how societies might have perceived, been impacted by, and adapted to changing conditions (Robertshaw et al., 2004).

29.4 CONTRIBUTION OF HISTORICAL BIOGEOGRAPHY TO CONSERVATION

This chapter has demonstrated the sensitivity of ecosystems to past climate change and human impacts; here, we discuss the potential of applying this understanding to ecosystem conservation. As more information has become available regarding the nature of past climatic changes it is apparent that climatic change is not a constant process, temporally or spatially, with certain areas experiencing greater changes in temperature and precipitation than others (Midgley et al., 2003). By weaving together and presenting in a suitable format different strands of information on past, present, and future ecosystem responses to climate change, the information can be used to enhance our ability to manage ecosystems as they respond to future climate change. In addition todeveloping new results from crucial areas, there is a pressing need to develop more complete methodologies that integrate past, present, and future perspectives on ecosystem dynamics, particularly those that can move from the site to local to regional scales, and can inform socioeconomic predictions, and hence, inform policy formulation. Hitherto, comparisons between climate reconstructions from data- and model-based research have been carried out in an ad hoc fashion, with little consideration for the methodological problems of the independent research strands and poor resolution of models that make them inappropriate for socioeconomic application and hence policy development. These methodological developments will need to integrate fully the skills of archaeologists, ecologists, and modelers to understand ecosystem responses to climate change; the presence, and character, of stable states; signals of transitions; and likely future scenarios. This is not a particularly new call with an excellent overview of the potential contribution from the palaeoecological and modeling community to conservation practitioners provided by Willis et al. (2007). As Willis et al. (2007) note, the scientific community provide data that are interesting, descriptive, but of little applied use to specific conservation targets and goals, and the former group believes that all long-term ecological data are far too crude in spatial and temporal scale to be of any relevance. It is time to talk. This challenge is being taken up by Antje Ahrends within the International Biogeography Society Newsletter (http://www.biogeography.org/html/newsletter.html), where she contrasts perspectives from different communities (conservation scientists, practitioners, and academics) in a series of interviews. Although still in the early stages it is clear that there is little dialogue and interaction between the various communities and something that is in need of development as the ecosystem effects of climate change start to impact conservation efforts.

The process of ecosystem conservation, through the formation of protected areas, is based on the principle of preserving habitats for future generations. Ecosystem conservation is typically implemented by delimiting a piece of land, providing some legislative protection and demarcation such as national park boundaries (Plate 29.1) within which a set of management rules can be applied to a clearly defined area with known boundaries. There is little dispute that global climates are changing, and the nature of this change is projected to continue even if the most extreme abatement scenarios are implemented (IPCC, 2007). Although the specifics of past climate change are very different to those suggested for the future (IPCC, 2007), the threat to biodiversity posed by global climate change is widely recognised by the global change and conservation communities (Markham, 1996; Thomas et al., 2004). Future climatic change is likely to be different from the past for three reasons. First, the rate of predicted climatic change is rapid and equivalent to those rapid changes of the Younger Dryas transition; second, many natural habitats have become fragmented by human populations, producing isolated habitat islands from which species are unable to disperse;and finally the present situation of high global temperature and high atmospheric CO_2 is an unprecedented situation in the Quaternary history of the Earth. How ecosystems will respond to such an environmental shift is unknown although it is likely that

Plate 29.1   The forest edge of Bwindi-Impenetrable Forest National Park, Uganda, showing the restricted nature of an important ecosystem within the current strict boundaries drawn around national parks

drought-stressed ecosystems, such as savannas, will benefit from high atmospheric CO2 due to increased growth rates fuelled by greater water-use efficiency and concomitant reduced impact of fire (Bond et al., 2005) as trees are able to grow to a height above the size threshold where a fire will result in mortality (Plate 29.2).

Notwithstanding the above, areas that have remained relatively stable during phases of past climate change and human interaction, or are located towards the center of bioclimatic space, and are concomitant with present-day biological hotspots are probably best placed to be buffered against future changes in climate regime, irrespective of

Plate 29.2   Repeat ground photographs from lowland areas in the Lolldaiga Hills, Laikipia Plateau, Kenya, looking north showing long-term presence of open glades between 1935 (upper photograph) and 2002 (lower photograph). The same large tree in the foreground can be recognised as can a general increase in tree densities, thought particularly to be a response to implementation of restrictions on burning of vegetation, with effects possibly enhanced by' historical increases in CO$_2$ levels. Photo credits: (A) C Robert Wells, reproduced with permission. (B) Laragh Larsen, British Institute of East Africa, reproduced with permission

the fact that future climate changes are likely to be different than experienced in the past.

Knowledge of potential impacts will enable policy makers to prepare appropriate strategies in advance of climate change events, and so minimise and manage effects—be they adverse or beneficial. It is for this reason that a comprehensive understanding of ecosystem response to climate change is crucial: it is imperative to understand the full range of natural variability and how this may project to the future. Studies on past ecosystem dynamics in response to climate change and human impacts can be considered by conservation biologists to impart effective long-term management strategies (Burgess et al., 2006; Willis et al., 2004). Future climate shifts (as we have seen in the past) will result in distributional shifts of species, as they respond to new conditions in an individualistic manner and create novel community associations. Such a situation makes existing networks of protected areas at worst redundant and at best inadequate for coping with dynamic populations (Figure 29.7). Strengthening the degree of protection afforded to areas that may be viewed as climatically relatively stable is not sufficient to realise the full conservation value (Cincotta et al., 2000). Successful conservation strategies are more likely to result from a detailed knowledge of the biota and their ecological

requirements than from the simple location of forest reserves coincident with relatively stable areas based on indirect and relatively low-resolution evidence. Such ecosystem responses to climate dynamics have resulted in a proposal for a climate change—integrated conservation strategy (Hannah et al., 2002). Although climate change is an important factor to consider for long-term successful conservation, practical implementation is hindered as conservation resources are usually only adequate for daily management and often assigned to immediate issues such as poaching, encroachment, resolving animal wildlife conflict. However, at an institutional and international level climate change is being incorporated in International Conservations on Biodiversity and is now a recognised threat feeding into the IUCN Red listing process. Natural resource managers need to think in terms of dynamic processes and underlying causes of species richness; different centres of species richness and endemism have very different historical backgrounds and so require similarly diverse management. Future climate and environmental change is predicted to cause major changes to biodiversity for which new conservation paradigms must be established that need predictions of potential future change on which to base conservation strategy (Hannah et al., 2002; Roberts et al., 2002).

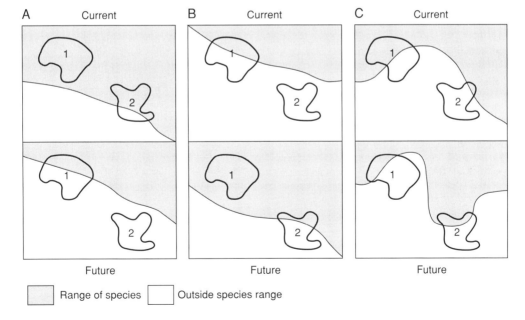

Figure 29.7   Different scenarios on the impact of climate change on statically protected areas. Under scenario A, area 2 loses the presence of a species; under scenario B, area 2 gains the presence of an 'invasive' species; under scenario C, both areas maintain a status quo but with a different location of the species within the park

A switch from static protected area management to a dynamic management system that accounts for climate-change impacts is needed (Figure 29.7c). The fortress conservation and momentum behind static protected area management must be replaced by a system that takes into account ecosystem history and ultimate causes of species richness possibly include low-intensity human impacts.

As we have seen, climate change and subsequent ecosystem response are of major importance to policy makers, but are areas surrounded by uncertainty and controversy. With increasing realisation of the impacts of climate change on ecological, social, and economic levels there is a need to develop a science-led policy framework. To link findings of pure research, policy, and economics, the void in understanding natural and historical processes behind present-day landscapes needs to be filled. When there is sufficient information it will be possible to move away from reactionary response towards informed management of many urgent environmental and development issues. Increased scientific understanding regarding land use, soil and water conservation, climate change, capacity building, and the wider socioeconomic consequences of climate change through likely changes in ecosystem form and function is essential for long-term sustainable development. In tandem to developing such a laudable approach there is a need to develop research capacity for communities living close to important habitats to contribute to and benefit from the services that ecosystems provide. By reconstructing past impacts of climate change in relation to potential future events, it is possible to make an assessment of future risks, thereby helping to guide current policy on the impacts of climate change locally, with manifestations regionally and indeed globally. At the heart of this is a realization that numerous complementary research strands need to be woven together to form a complete understanding of ecosystem dynamics and responses to environmental change; only then will findings on ecosystem dynamics and associated societal impacts be able to move from the scientific to policy arena.

REFERENCES

Ahrends A. International Biogeography Society newsletter (http://www.biogeography.org/html/newsletter.html). Accessed 1 May 2011.

Battarbee, R.W., Gasse, F., and Stickley C.E. (2004) *Past Climate Variability through Europe and Africa*. Dordrecht: Kluwer Academic.

Bond, W.J., Woodward, F.I., and Midgley, G.F. (2005) 'The global distribution of ecosystems in a world without fire', *New Phytologist*, 165: 525–38.

Bunting, M.J., Armitage, R., Binney, H.A., and Waller, M.P. (2005) 'Estimates of "relative pollen productivity" and "relevant source area of pollen" for major tree taxa in two Norfolk (UK) woodlands', *Holocene*, 15: 459–65.

Bunting, M.J., and Middleton, D. (2005) 'Modelling pollen dispersal and deposition using the HUMPOL software: Simulating wind roses and irregular lakes', *Review of Palaeobotany and Palynology*, 134: 185–96.

Burgess, N., Küper, W., Mutke, J., Brown, J., Westaway, S., Turpie, S., Meshack, C., Taplin, J., McClean, C., and Lovett, J.C. (2006) 'Major gaps in the distribution of protected areas for threatened and narrow range Afrotropical plants', *Biodiversity and Conservation*, 14(8): 1877–94.

Bush, M., Colinvaux, P., Wiemann, M.C., Piperno, D.R., and Liu, K. (1990) 'Late Pleistocene temperature depression and vegetation change in Ecuadorian Amazonia', *Quaternary Research*, 34: 330–45.

Cincotta, R.P., Wisnewski, J., and Engleman, R. (2000) 'Human population in the biodiversity hotspots', *Nature*, 404: 990–92.

Coe, M.J. (1967) *The Ecology of Alpine Zone Mount Kenya*. The Hague: Dr W. Junk. pp. 836–39.

Colinvaux, P.A. (1996) 'Quaternary environmental history and forest diversity in the Neotropics', in B.C. Jackson, A.F. Budd and A.G. Coates (eds.), *Evolution and Environment in Tropical America*. Chicago: University of Chicago Press.

Colinvaux, P.A., de Oliveira, P.E., and Bush, M.B. (2000) 'Amazonian and neotropical communities on glacial timescales: The failure of the aridity and refuge hypothesis', *Quaternary Science Reviews*, 19: 141–69.

Colinvaux, P.A., Irion, G., Räsänen, M.E., Bush, M.B., and Nunes de Mello, J.A.S. (2001) 'A paradigm to be discarded: Geological and paleoecological data falsify the Haffer and Prance refuge hypothesis of Amazonian speciation', *Amazoniana*, 16: 609–46.

Darwin, C. (1872) *The origin of species by means of natural selection, or the preservation of favoured races in the struggle for life*. London: John Murray.

Davis, M.B. (1986) 'Climatic instability, time lags, and community disequilibrium', in Diamond, J. and Case T.J. (eds.), *Community Ecology*. New York: Harper and Row. pp. 269–304.

Davis, M.B., Shaw, R.G., and Etterson, J.R. (2005) 'Evolutionary responses to changing climate', *Ecology*, 86: 1704–14.

Dodson, J., Ono, Y, Wang, P., and Taylor, D.M. (2004) 'Climates, human, and natural system of the PEPII transect', *Quaternary International*, 118–119: 3–12.

Doherty, R., Kutzbach, J., Foley, J., and Pollard, D. (2000) 'Fully coupled climate/dynamical vegetation model simulations over northern Africa during the mid-Holocene', *Climate Dynamics*, 16: 561–73.

Dunbar, R.B. (2003) 'Lead, lags and the tropics', *Nature*, 421: 121–22.

Elenga, H., and 23 authors. (2000) 'Pollen-based reconstruction for southern Europe and Africa 18,000 years ago', *Journal of Biogeography*, 27: 621–34.

Ellis, E.C., and Ramankutty, N. (2008) 'Putting people in the map: Anthropogenic biomes of the world', *Frontiers in Ecology and the Environment*, 6: 18–25.

Farrera, I. and 18 authors. (1999) 'Tropical climates at the last glacial maximum: A new synthesis of terrestrial palaeoclimate data. I. Vegetation, lake-levels and geochemistry'. *Climate Dynamics*, 11: 823–56.

Finch, J.M., Leng, M., and Marchant, R. (2009) 'Late Quaternary vegetation dynamics in a biodiversity hotspot: the Uluguru Mountains of Tanzania', *Quaternary Research*, 72(1): 111–22.

Fjeldså, J.A., Ehrlich, D., Lambin, E., and Prins, E. (1997) 'Are biodiversity "hot spots" correlated with current ecoclimatic stability? A pilot study using the NOAA-AVHRR remote sensing data', *Biodiversity and Conservation*, 6: 399–420.

Fjeldså, J., and Lovett, J.C. (1997) 'Geographical patterns of old and young species in African forest biota: The significance of specific montane areas as evolutionary centres', *Biodiversity and Conservation*, 6: 325–47.

Gaston, K.J. (2000) 'Global patterns in biodiversity', *Nature*, 405: 220–27.

Haffer, J. (1969) 'Speciation in Amazonian forest birds', *Science*, 165:131–37.

Haffer, J. (1997) 'Alternative models of vertebrate speciation in Amazonia: An overview', *Biodiversity and Conservation*, 6: 451–76.

Hamilton, A.C. (1982) *The Environmental History of East Africa*. London: Academic Press.

Hammer, M.F., and 8 authors. (1998) 'Out of Africa and back again: Nested cladistic analysis of human Y chromosome variation', *Molecular Biology and Evolution*, 15: 427–41.

Hannah, L., Midgley G.F., and Miller, D. (2002) 'Climate change-integrated conservation strategies', *Global Ecology and Biogeography*, 11: 485–95.

Harbele, S. (2005) 'A 23,000-yr pollen record from Lake Euramoo, wet tropics of NE Queensland, Australia', *Quaternary Research*, 64: 343–56.

Hedberg, O. (1959) 'An "open-air hothouse" on Mount Elgon, Tropical east Africa', *Svensk-Botanisk Tidskerift*, 53: 160–66.

Hewitt, G.M. (2000) 'The genetic legacy of the Quaternary ice ages', *Nature*, 405: 907–13.

Hicks, S., Tinsley, H., Huusko, A., Jensen, C., Hättestrand, M., Gerasimides, A., and Kvavadze, E. (2001) 'Some comments on spatial and temporal variation in arboreal pollen deposition: First records from the PMP (Pollen Monitoring Programme)', *Review of Palaeobotany and Palynology*, 117: 186–204.

Hooghiemstra, H., and van der Hammen, T. (1998) 'Neogene and Quaternary development of the neotropical rain forest: The forest refugia hypothesis, and a literature overview', *Earth Science Reviews*, 44: 147–203.

Hulme, M., Barrow, E.M., Arnell, N.W., Harrison, P.A., Johns, T.C., and Downing, T.E. (1999) 'Relative impacts of human induced climate change and natural variability', *Nature*, 397: 688–91.

Humphries, C.J. Williams, P.H., and Wright R.I.V. (1995) 'Measuring biodiversity value for conservation', *Annual Reviews in Ecology and Systematics*, 26: 93–111.

Huntley, B., Berry P.M., Cramer W., and McDonald A.P. (1995) 'Modelling present and potential future ranges of some European higher plants using climate response surfaces', *Journal of Biogeography*, 22: 967–1001.

Huntley, B., and Birks, H.J.B. (1983) *An Atlas of Past and Present Pollen Naps for Europe: 0–13000 yr BP*. Cambridge: Cambridge University Press.

IPCC. (2007) *Climate Change 2007. Synthesis report, IPCC reports*. Cambridge: Cambridge University Press.

Jackson, S.T., Webb, R.S., Anderson, K.H., Overpeck, J.T., Webb III, T., Williams, J.W., and Hansen, B.C.S. (2000) 'Vegetation and environment in eastern North America during the Last Glacial Maximum', *Quaternary Science Reviews*, 19: 489–508.

Johnson, T.C., Brown, E.T., McManus, J., Barry, S., Barker, P., and Gasse, F. (2002) 'A high-resolution palaeoclimate record spanning the past 25,000 years in southern East Africa', *Science*, 306: 113–17.

Jolly, D., Taylor, D.M., Marchant, R.A., Hamilton, A.C., Bonnefille, R., and Riolett, G. (1997) 'Vegetation dynamics in central Africa since 18,000 yr BP: Pollen records from the interlacustrine highlands of Burundi, Rwanda and western Uganda', *Journal of Biogeography*, 24: 495–512.

Jump, A.S., Marchant, R., and Peñuelas, J. (2009) 'Environmental change and the option value of genetic diversity', *Trends in Plant Science*, 14(1), 51–58.

Kerr, R.A. (2003) 'Tropical Pacific a key to deglaciation', *Science*, 309: 183–84.

Küper, W., Henning Sommer, J., Lovett, J.C., Mutke, J., Linder, H.P., Beentje, H.J., Van Rompaey, R.S.A.R., Chatelian, C., Sosef, M. And Barthlott, W. (2004) 'Africa's hotspots of biodiversity redefined', *Annals of the Missouri Botanical Garden*, 91: 525–35.

Ledig, F.T. (1986) 'Heterozygosity, heterosis, and fitness in outbreeding plants', in Soulé, M.E., (ed.), *Conservation Biology: The Science of Scarcity and Diversity*. Sunderland, MA: Sinauer. pp. 77–104.

Leira, M., Dalton,C., Chen, G., Irvine, K., and Taylor, D. (2009) 'Patterns in freshwater diatom taxonomic distinctness along an eutrophication gradient', *Freshwater Biology*, 54: 1–14.

Lewis, S.L., Malhi, Y., and Phillips, O. (2004) 'Fingerprinting the impacts of global change on tropical forest', *Philosophical Transactions of the Royal Society London Series B*, 359: 437–62.

Linder, H.P. (1991) 'Environmental correlates of patterns of species richness in the south-western Cape Province of South Africa', *Journal of Biogeography*, 18: 509–18.

Lovett, J.C., Marchant, R., Taplin, J., and Küper, W. (2005) 'The oldest rainforests in Africa: Stability or resilience for survival and diversity?' in A. Purvis and J.L. Gittleman (eds.), *Phylogeny and Conservation*. Cambridge: Cambridge University Press. pp. 197–230.

Lovett, J., Rudd, C.S., Taplin, J.R.D., and Frimodt-Møller, C. (2000) 'Patterns of plant diversity in Africa south of the Sahara and their implications for conservation management', *Biodiversity and Conservation*, 9: 33–42.

MacArthur, R. (1972) *Geographical Ecology*. New York: Harper and Row.

McSweeny, C., New, M., and Lizcano. (2008) *UNDP Climate Change Country Profiles*. http://country-profiles.geog.ox. ac.uk. Accessed 1 January 2010.

Maley, J. (1996) 'The African rain forest: Main characteristics of changes in vegetation and climate from the upper Cretaceous to the Quaternary', *Proceedings of Royal Society of Edinburgh*, 104B: 31–73.

Marchant, R.A., and 11 authors. (2002a) 'Colombian vegetation derived from pollen data at 0, 3000, 6000, 9000, 12,000, 15,000 and 18,000 radiocarbon years before present', *Journal of Quaternary Science*, 17: 113–30.

Marchant, R.A., Boom, A., and Hooghiemstra, H. (2002b) 'Pollen-based biome reconstructions for the past 450,000 years from the Funza-2, core, Colombia: Comparisons with model-based vegetation reconstructions', *Palaeogeography, Palaeoclimatology Palaeoecology*, 177: 30–45.

Marchant, R.A., and Hooghiemstra, H. (2004) 'Rapid environmental change in tropical Africa and Latin America about 4000 years before present: A review', *Earth Science Reviews*, 66: 217–60.

Marchant, R., Mumbi, C., Behera, S., and Yamagata. (2007) 'The Indian Ocean dipole—the unsung driver of climatic variability in East Africa', *African Journal of Ecology*, 45: 14–16.

Marchant, R.A., and Taylor, D.M. (2000) 'Numerical analysis of modern pollen spectra and in situ montane rainforest—implications for the interpretation of fossil pollen sequences from tropical Africa', *New Phytologist*, 146: 505–15.

Markgraf, V., Baumgartner, T.R., Bradbury, J.P., Diaz, H.F., Dunbar, R.B., Luckman, B.H., Seltzer, G.O., Swetman, T.W., and Villalba, R. (2000) 'Paleoclimate reconstruction along the Pole-Equator-Pole transect of the Americas (PEP 1)', *Quaternary Science Reviews*, 19: 125–40.

Markham, A. (1996) 'Potential impacts of climate change on ecosystems: A review of implications for policymakers and conservation biologists', *Climate Research*, 6: 179–91.

Midgley, G.F., Hannah, L., Miller, D., Thuiller W., and Booth, A. (2003) 'Developing regional and species-level assessments of climate change impacts on biodiversity: The Cape Floristic Region', *Biological Conservation*, 112: 87–97.

Moritz, C., and Faith, D.P. (1998) 'Comparative phylogeography and the identification of genetically divergent areas for conservation', *Molecular Ecology*, 7: 419–30.

Mumbi, C.T., Marchant, R., Hooghiemstra, H., and Wooller, M.J. (2008) 'Late Quaternary vegetation reconstruction from the Eastern Arc Mountains, Tanzania', *Quaternary Research*, 69: 326–41.

Mutke, J., Kier, G., Braun, G., Schultz, C., and Barthlott, W. (2002) 'Patterns of African vascular plant diversity—a GIS based analysis', *Systematics and Geography of Plants*, 71: 1125–36.

Nelson, B.W., Ferreria, C.A.C., de Silva, M.F., and Kawasaki, M.L. (1990) 'Endemism centres, refuge and botanical collection density in Brazilian Amazonia', *Nature*, 345: 714–16.

O'Brien, E.M. (1993) 'Climatic gradients in woody plant species richness: Towards an explanation based on an analysis of southern Africa's woody flora', *Journal of Biogeography*, 20: 181–98.

Osmaston, H.A. (1975) 'Models for the estimation of firmlines of present and Pleistocene glaciers', in: R.F. Peel, M. Chisholm, and P. Haggett (eds.), *Processes in Physical and Human Geography, Bristol Essays*. London: Heinemann. pp. 218–45.

Overpeck, J. Webb, R.S., and Webb, T. (1992) 'Vegetation maps for eastern North America reconstructed from pollen evidence', *Geology*, 20: 1071–74.

Partin, J.W., Cobb, K.M., Adkins, J.F., Clark, B., and Fernandez, D.P. (2007) 'Millennial-scale trends in west Pacific warm pool hydrology since the Last Glacial Maximum', *Nature*, 449: 452.

Pearson, R.G., and Dawson, T.P. (2003) 'Predicting the impacts of climate change on the distribution of species: Are bioclimate envelope models useful?' *Global Ecology and Biogeography*, 12: 361–71.

Pearson, R.G., Dawson, T.P., Berry P.M., and Harrison P.A. (2002) 'SPECIES: A Spatial evaluation of climate impact on the envelope of species', *Ecological Modelling*, 154: 309–300.

Peltier, W.R. (1994) 'Ice age paleotopography', *Science*, 265: 195–201.

Peterson A.T., et al. (2002) 'Future projections for Mexican faunas under global climate change scenarios', *Nature*, 416: 626–30.

Prentice, I.C. (1985) 'Pollen representation, source area, and basin size: Toward a unified theory of pollen analysis', *Quaternary Research*, 23: 76–86.

Prentice, I.C., Cramer, W., Harrison, S.P., Leemans, R., Monserud. R.A., and Solomon, A.M. (1992) 'A global biome model based on plant physiology and dominance, soil properties and climate', *Journal of Biogeography*, 19: 117–34.

Prentice, I.C., and Webb III, T. (1998) 'BIOME 6000: Reconstructing global mid-Holocene vegetation patterns from palaeoecological records', *Journal of Biogeography*, 25: 997–1005.

Purvis, A., and Hector, A. (2000) 'Getting the measure of biodiversity', *Nature*, 405: 212–19.

Roberts, C., and 11 authors. (2002) 'Biodiversity hotspots and conservation priorities in the sea', *Science*, 305: 1300–1304.

Robertshaw, P., Taylor D., Doyle S., and Marchant, R. (2004) 'Famine, climate and crisis in Western Uganda', in Battarbee, R.W., Gasse, F., and Stickley, C.E. (eds.), *Past Climate Variability through Europe and Africa*. Dordrecht: Kluwer Academic.

Sechrest, W., Brooks, T.M., da Fonseca, G.A.B., Konstant, W.R., Mittermeier, R.A., Purvis, A., Rylands, A.B., and Gittleman, J.L. (2002) 'Hotspots and the conservation of evolutionary history', *Proceedings of the National*

Academy of Sciences of the United States of America, 99: 2067–71.

Stager, J.C., Mayewski, P.A., and Meeker, D.L. (2002) 'Cooling cycles, Heinrich event 1, and the desiccation of Lake Victoria', *Palaeogeography, Palaeoclimatology, Palaeoecology,* 183: 169–78.

Stocker, T.F., and Marchal, O. (2000) 'Abrupt climate change in the computer: Is it real?' *Proceedings of the National Academy of Sciences of the United States of America*, 97: 1362–65.

Taplin, J.R.D., and Lovett J.C. (2003) 'Can we predict centres of plant species richness and rarity from environmental variables in sub-Saharan Africa?' *Botanical Journal of the Linnean Society*, 141: 187–97.

Thomas, C.D., and 18 authors. (2004) 'Extinction risk from climate change', *Nature*, 427: 145–48.

Turney, C.S.M., Kershaw, A.P., Clemens, S.C., Branch, N., Moss, P.T., and Fifield, L.K. (2004) 'Millennial and orbital variations of El Niño/Southern Oscillation and high-latitude climate in the last glacial period', *Nature*, 430: 306–10.

Urrego, D.H., Silman, M.R., and Bush, M.B. (2005) 'The Last Glacial Maximum: Stability and change in a western Amazonian cloud forest', *Journal of Quaternary Science*, 20: 693–701.

Vellend, M. (2003) 'Island biogeography of genes and species', *American Naturalist*, 132: 358–65.

Vellend, M., and Geber, M.A. (2005) 'Connections between species diversity and genetic diversity', *Ecology Letters*, 8: 767–81.

Verschuren D., Laird K.R., and Cumming B.R. (2000) 'Rainfall and drought in equatorial East Africa during the past 1,100 years', *Nature*, 403: 410–14.

Williams, P.H., de Klerk, H.M., and Crowe, T.M. (1999) 'Interpreting biogeographical boundaries among Afrotropical birds: Spatial patterns in richness gradients and species replacement', *Journal of Biogeography*, 26: 459–74.

Willis, K.J., Araujo, M.B., Bennett, K.D., Figueroa-Rangel, B., Froyd, C.A., and Myers, N. (2007) 'How can a knowledge of the past help to conserve the future? Biodiversity conservation and the relevance of long-term ecological studies', *Proceedings of the Royal Society of London Series B*, 362: 175–86.

Willis, K.J., Gillson, L., and Brencic, T.M. (2004) 'How 'virgin' is virgin rainforest', *Science*, 304: 402–3.

30

Habitat Approaches to Nature Conservation

Geoffrey Hugh Griffiths and
Ioannis Vogiatzakis

30.1 INTRODUCTION

The precise meaning and use of the term 'habitat' remains elusive. A working definition is "the resources and conditions present in an area that produces occupancy for a particular species, population or organism" (McDermid et al., 2005; Table 30.1). A prevalent usage equates habitat with land cover—coniferous forest, steppe, evergreen shrubland—a convenient definition that enables land managers to map habitat over large areas often from remotely sensed satellite data (Alexander and Millington, 2000). The difference in definition is significant and potentially confusing. The first is species-specific, and the second is a more general definition that refers to the extent of natural, or seminatural, vegetation cover. However, there can be no doubt that attributes of a habitat, such as its area, shape, productivity, composition, and stability, are important determinants of the richness and abundance of species that live there. Unsurprisingly, therefore, the 'habitat', both as a spatial entity and a concept, has become fundamental to conservation efforts (Table 30.1).

Although the topic of nature conservation is split between two chapters in this book (see also Osborne and Leitão, this volume), the distinction between species-based and habitat approaches is tenuous: habitats are protected to ensure the survival of species and species' protection frequently safeguards the associated habitat. Nevertheless, it is argued in this chapter that the need to maintain and, in some cases, restore ecosystem functioning, develop habitat networks,

and maintain a minimum size of habitat for species' survival, critically depends upon knowledge of the distribution and pattern of habitats at a range of spatial scales.

30.2 THE SPECIES-BASED APPROACH: PROBLEMS AND CHALLENGES

In theory it is possible to identify and protect, at a global scale, the minimum area of land to ensure that global biodiversity is safeguarded sustainably, the 'maximal area coverage problem' (Kiester et al., 1996). In reality, estimates of total species vary between three to ten million (Groombridge and Jenkins, 2002). We have knowledge of the habitats and life cycles for only a minority of species and the designation of protected sites is more frequently driven by political expediency than by scientific principles. Despite the fact that approximately 114,000 sites totaling c. 13% of the global land area (IUCN, 2006) is now protected in some form or other, it is unclear whether this will ensure the survival of even the majority of species at a time of rapidly expanding human populations and climate change.

Most biodiversity conservation effort focuses on species rather than on genetic diversity, communities, or ecosystems. Species are the most recognizable expressions of taxonomic diversity. A species-based approach generally emphasizes the value of individual species that are rare or endangered, or habitats that are characterized by a high degree of richness and/or endemism.

Table 30.1 Habitat terminology (after McDermid et al., 2005)

Term	Definition
Habitat	The sum and location of the specific resources needed by an organism for survival and reproduction
Habitat type	A mappable unit of land considered homogeneous with respect to vegetation and environmental factors
Habitat use	The way an organism uses the physical and biological resources in a habitat
Habitat selection	A series of innate and learned behavioral decisions made by an animal about what habitat it would use
Habitat preference	The consequences of habitat selection
Habitat availability	The accessibility and procurability of the physical and biological components of a habitat
Habitat suitability	The ability of the habitat to sustain life and support population growth
Habitat quality	The ability of the environment to provide conditions that are appropriate for individual and population persistence
Critical habitat	A legal term describing the physical or biological features essential to the conservation of a species, which may require special management considerations or protection

Whilst this is an effective strategy for some species at some spatial and temporal scales, the approach has limits and other complementary approaches need to be developed and applied.

The first limitation is simply that we have insufficient knowledge of the number of species globally, their life cycles, habitat preferences, and distribution with the attendant dangers of 'missing' species that perform important functions within an ecosystem, for example, keystone species (Simberloff, 1998). A critical prerequisite for global protection of biodiversity is the establishment of a well-connected network of protected areas: nature reserves, national parks, and other designations. Whilst the techniques available for the selection and design of nature reserves are relatively advanced (Bedward et al., 1992; Pressey et al., 1996), a number of methodological problems make their application problematic (see Osborne and Leitão, this volume).

One objective of such techniques is to identify biodiversity hotspots (Myers et al., 2000) to protect the optimal number of species in the minimum area (see Osborne and Leitão, this volume). Biodiversity hotspots are regions at a range of scales where diversity of species is high relative to other similar areas. Linked to this concept is complementarity, the identification of sites that complement the protection of different species rather than replicating the role of existing sites. The identification of hotspots is notoriously difficult and there is little agreement about definitions at different scales and for different taxonomic groups. For example, Myers et al. (2000) define their global hotspots on three criteria: (a) high species richness; (b) high levels of endemism; and (c) high levels of threat to the integrity of the ecosystems in the hotspots, rather than, say, simply high species richness. Data at a national scale for Great Britain show that the centers of maximum diversity differ between taxonomic groups—there is no single 'hotspot' that, were it to be protected, protects the majority of species (Prendergast et al., 1993; Williams et al., 1996). The index is also scale-dependent: high levels of diversity and/or rarity at one scale may not be replicated at larger or smaller scales (see Mutke, this volume).

There is no generally accepted scheme for establishing conservation priorities, partly as a result of the complexity and variety of life itself but also because of limited resources of time and money and the need to prioritize. The criteria established by Ratcliffe in the Nature Conservation Review in Britain in the 1970s (Ratcliffe, 1977) retain their currency and have been developed and adopted by the International Union for Conservation of Nature (IUCN). However, the application of the criteria is piecemeal and, in many parts of the world the pressure for development is so intense that it is difficult to apply strict conservation criteria that do not also take into account human welfare. Furthermore, the information required to implement the criteria established by Ratcliffe (1977) often do not exist, and time-consuming and expensive fieldwork may be required to make good the information gaps if the system is to be applied rigorously and effectively.

In Europe, the Birds Directive (1979) and Habitats Directive (1992) provided the legislation to designate Special Protected Areas (SPA) for birds and Special Areas for Conservation (SAC) for habitats (Council of Europe, 1992).

These areas are the backbone of a network of protected sites within the 27 states of the European Union (EU), the so-called Natura 2000 network. This represents an imperfect attempt to achieve an (almost) continental scale of protected sites for nature conservation. There is plentiful evidence that sites have been selected as much for political reasons as for scientific ones. The EU Natura 2000 network has met with criticisms (e.g., Papageorgiou and Vogiatzakis, 2006) since present and predicted future threats appear to be happening at a faster rate and over a larger area than had previously been thought and there is, as yet, no effective method for monitoring the condition of the reserves within the Natura 2000 network.

Global attempts to establish goals for habitat protection have resulted in worldwide recommendations for the percentage of each nation's total area that should be officially designated as protected. However, there is plentiful evidence to indicate that ecologically based goals would be much higher. For example, Brooks et al. (2004) showed that even though the global area of protected land is close to 13%, biodiversity protection is still far from complete.

It may also be impossible to protect sufficient sites given the dynamic nature of land use under the twin pressures of an expanding human population and climate change (see Marchant and Taylor, this volume, for a discussion of climate change, ecosystem response, and conservation planning). Even when an area is designated for protection, the successful conservation of species is not guaranteed. Protected areas are subject to threats of both natural (e.g., drought and flooding) and human (e.g., forest clearance) agency, requiring the replication of similar sites to avoid extinction from such stochastic events. A recent analysis on the state of protected species and habitats in Europe revealed that less than half of those are considered to be in 'favourable conservation status'. For most of the remaining species and habitats, the conservation status is considered either inadequate or bad, while for a significant number of cases, the available data are insufficient to reach any assessment (European Environment Agency, 2008).

Furthermore, networks of protected areas might work today but could become redundant as, for example, global changes in the economics of food production and climate change render the location of protected sites ineffective. Finally, despite the potential of advanced computerized techniques for reserve selection (Kiester et al., 1996; Margules and Pressey, 2000), there is little evidence that such methods are widely used by land managers (Prendergast et al., 1999).

A species-based approach has the advantage that a focus on more endangered or valued (by people) species may have the effect of conserving others. It is also often easier to mobilize public opinion in support of a single species (e.g., the Osprey, *Pandion haliaetus,* in Scotland) than for a habitat, a generally poorly understood concept in the public mind. But knowledge of variation is often poor at the species level and below and, in many cases, by the time it is determined that a species is in danger it can be too late to guarantee its protection.

Alternative approaches are needed to buffer against the uncertainty of global change and, increasingly, to complement an essentially species-based approach with a more holistic one that integrates human use of resources with nature conservation.

30.3 WHAT IS A HABITAT?

The concept of the habitat can be applied at a range of scales from the habitat of a small plant exploiting a microtopographical solution hollow in limestone on the island of Crete, Greece (e.g., *Silene variegata*), contrasting with the range of environments exploited by the European wolf (*Canis lupus*) across the circumboreal forests of Russia and Scandinavia. In practice, in nature conservation the term tends to be used interchangeably with biotope: habitats are generally defined as distinctive vegetation communities characterized by representative sets of species. The term, therefore, defies precise definition because in this strict definition of the term it is species-focused: at which taxonomic level, spatial and temporal scale is the definition to be applied (Hall et al., 1997). This is the basis on which a number of habitat classifications have been developed, including CORINE Biotopes (Coordinated Information on the Environment), a hierarchical system of habitat classification for Europe (European Communities, 1991). It is worth noting, however, that habitat classification systems are more notable for their differences and inconsistencies than for their similarities and transferability. Thus, those habitats that we wish to protect are characterized by a relatively high degree of species richness and may also include rare and/or endemic species.

30.4 WHY ARE HABITATS IMPORTANT?

In many cases habitats represent the relic of a former more continuous vegetation cover, becoming the only refuge for plants and animals associated with that vegetation community. Such relic habitats therefore become 'reservoirs' from which plants and animals can disperse to colonize other suitable habitat 'patches'. This critically depends upon surviving habitat patches being 'connected' to enable organisms to disperse to new sites. In many cases the loss of the original habitat is so extensive and surviving habitats are so small and isolated, that it has become impossible to restore the habitat and associated species without artificial translocation, an expensive and not always successful exercise.

Habitats are also critically important within an ecosystems approach by providing ecological goods and services. Thus, there is evidence that crop pollination, and ultimately yield, is higher in areas where there is sufficient foraging habitat for bees (Potts et al., 2006). The potential for delivering an ecosystem service may vary between habitats and in different places. For example, deciduous broadleaf woodland may have potential for carbon sequestration in one location but provides a buffer against flooding in another. The difficult trade-offs between ecosystem goods and services are likely to become an increasingly important component of nature conservation policy as more attention is devoted to the total range of goods and services provided by ecosystems (Luyssaert et al., 2008).

30.5 MAPPING AND MODELING HABITATS

In many cases defining and delineating a habitat is far from straightforward. In the Lefka Ori of Crete where aridity and topography are limiting factors above 2,000 m.a.s.l. the microscale patches of spiny-cushion plants such as *Acantholimon creticum* and *Berberis cretica* are obvious and well delineated (Vogiatzakis et al., 2003). Not so, for example, in the highlands of Scotland where the small-scale variations in soil type, wetness, exposure, and altitude result in a complex mosaic of upland plants that is not obviously patch forming. This brings us to a fundamental point—we tend to see patches as spatially discrete entities from a human perspective—but the fact that habitats are recognizable entities is a critically important advantage in a habitat-based approach to conservation. Thus, habitats can be identified, mapped, and managed. They are susceptible to intervention in a form that is recognizable to land managers, conservation officers, etc. This is critically important, especially in countries where resources, especially for expensive species-level survey and monitoring, are limited.

The field of vegetation mapping has developed as "a fruit from the union of botany and geography" (Küchler and Zonneveld, 1988). The relative ease with which we can identify, map, and monitor habitats (as opposed to species) favors a habitat approach in some circumstances. Perhaps unsurprisingly there is no single international system that can be applied globally for habitat mapping. In Europe, the closest is the CORINE system (European Communities, 1991), which recognizes the relative roles of field survey, aerial photography, and satellite imagery for mapping and monitoring.

Field survey provides identification to the species level and this can be of critical importance when a system is responding over the short to medium term to small changes in the environment, such as increases in the application of nitrogen-based fertilizer with consequent loss of species. The increasing sophistication of aerial photography with the application of multichannel and Lidar systems is enabling information on both species composition and structure to be derived (Millington et al., 2002). The large area coverage and repeat viewing of satellite imagery and the advent of high-resolution, multispectral, and hyperspectral sensors are making the application of satellite imagery increasingly attractive for habitat mapping and monitoring (e.g., McDermid et al., 2005).

In many situations habitats are too large, too variable, or simply too inaccessible to be mapped effectively. Also, data on the distribution, biology, and habitat requirements of species in danger is often lacking. Where those data are available, GIS provides a means of rapidly reviewing the distribution and conservation status of several components of biodiversity. This information can then be used for detailed resource inventory and for decision-making purposes in nature conservation and management (Scott et al., 2002). Where information is poor or nonexistent, GIS techniques can be used to predict species distribution patterns based on the limited field data available (Austin, 2002).

Thus, with the advent of GIS and remote sensing techniques and the increasing availability of digital maps of environmental variables such as topography, geology, and soils, the development and implementation of predictive models has found a wide range of applications in vegetation studies (see, for example, Millington et al., 2002; Scott et al., 2002). Predictive Vegetation Mapping (PVM) provides the tools to predict the distribution of individual species (Wu and Smeins, 2000), species richness (Luoto et al., 2002), and vegetation communities (Miller and Franklin, 2002). A good example is from Crete (Vogiatzakis and Griffiths, 2006), where the mapped variables of landform, slope, and altitude were used to predict five separate plant communities in the mountains of western Crete, an area known for its high level of rare and endemic plants.

Franklin (1995) defines PVM as predicting the vegetation composition across the landscape from mapped environmental variables. The habitat features assumed to influence species distribution patterns are mapped and subsequently analyzed within a GIS. Literature or empirical data can supply information on species-habitat relationships that can then facilitate the identification of habitat suitability and consequently the creation of potential distribution maps (Wadsworth and Treweek, 1999). A major difficulty with this type of predictive approach is that measurement of accuracy is problematic, requiring intensive sampling, often in remote areas, of an independent dataset (Scott et al., 2002). Predictive modeling is discussed in detail in Burnside and Waite, this volume.

Since vegetation provides food and shelter to wildlife, it can be used as a surrogate for other habitat factors in the modeling process. However, several factors complicate the use of vegetation, imposing limitations to a habitat model. For example, the structure of vegetation rather than its floristic composition can be important to some taxonomic groups such as birds. Species have different habitat requirements in different parts of their range and at different times in their life cycle (Scott et al., 2002). Therefore, the presence of a vegetation type within a species' range does not necessarily mean presence of the species under investigation.

However, mapping and monitoring is only the first part of developing effective policy and practice for nature conservation. Techniques are required to use these data to protect species and habitats at a range of spatial scales and across the range of environments that characterize global biomes.

30.5.1 Habitats at the patch scale

One approach is to treat the habitat as a single entity, the patch. This forces us to return to the dichotomy between the definition of the habitat as a resource for a species or community and the habitat as a patch within the classical patch-matrix-corridor model (Forman and Godron, 1986). This dichotomy is reflected in the continuing debate about patch- versus landscape-scale approaches to nature conservation. At the patch scale, the nature conservation effort is species-focused, analyzing the specific habitat preferences of different species to identify optimal resource needs to ensure persistence. Frustration with this

species-by-species approach is alluded to by Osborne and Leitão (this volume) and has led to the selection of indicator, keystone, and umbrella species as surrogates for other taxonomic groups. The patch-scale approach is also strongly based in community ecology. Fahrig (2007) argues that a modeling approach is needed to estimate species persistence and the models should include parameters relating to individual area requirements, reproductive rate, emigration and dispersal rates, population, and environmental stochasticity. Interestingly, Fahrig (2007) relates these parameters to changes in patch size, recognizing that habitat area is a critical determinant of species persistence. But how much habitat is enough? (Fahrig, 2001). A patch-focused approach suffers from the same limitations as a species-by species approach—every patch is different and generalization for the benefit of nature conservation policy is difficult.

Nevertheless, there is plentiful empirical evidence to demonstrate that the quality of a habitat patch, measured by its age, composition, productivity, etc., is a critical determinant of species richness and abundance (Garcia et al., 2007; Dennis and Eales, 1997). A number of studies have sought to develop indices of habitat quality. These indices are usually developed on the basis of a species' perception of its environment (Garcia et al., 2007), combining a range of habitat attributes relevant to the persistence of the species. But generic habitat quality indices that include species richness as one of the parameters are also common. Maintaining habitat quality, therefore, is a critical component of nature conservation policy.

30.5.2 Habitats at the landscape scale

The loss and fragmentation of seminatural habitats in the wider landscape is a major factor in the decline of biodiversity. Such decline is attributed to reductions in habitat-patch size and increased isolation of habitat patches. Empirical evidence suggests that habitat loss has large, consistently negative effects on species abundance and distribution while habitat fragmentation per se has much weaker effects (Fahrig, 2003). However, until recently most schemes for habitat creation and restoration have been site-based at the scale of individual parcels of land or, at best, the farm level. This ignores the importance of understanding the spatial pattern of habitat patches and the character of the intervening matrix in targeting potential sites for the creation of new habitats and the restoration of former habitats. This is illustrated by the minimum amount of suitable habitat (MASH) concept: the number of well-connected patches (generally estimated to be in the range of 15–20) needed for the long-term survival of a metapopulation (Hanski et al., 1996; Hanski, 1999). There are many good examples of models that incorporate spatially explicit decision rules to identify suitable sites for habitat restoration (Lee and Thompson, 2005; Nikolakaki, 2004; Bailey et al., 2006).

Landscape ecological theory (Forman and Godron, 1986; Kupfer, this volume) has tended to conceptualize habitats as patches: islands of wildlife interest (i.e., with a greater diversity and abundance of species than the surrounding area) embedded in a matrix of intensively managed land that is frequently hostile to characteristic patch species. Forman (1995) states that every point in a landscape is either within a patch, corridor, or the background matrix. This is the typical situation of many intensively farmed landscapes and is a direct consequence of the fragmentation of the original vegetation cover. Nevertheless, habitat patches are created by a range of factors from stochastic events such as landslides, wind-throw, and snow and rock avalanches to long-term processes such as succession.

The debate has tended to become polarized between traditional ecology, with its focus on the site, and landscape ecology, with its spatial focus on the wider landscape. However, the debate is critical for nature conservation, determining where effort and limited resources should be focused at a time of rapid environmental change.

The critical question is whether the classical patch-matrix concept of habitat creation and survival is (a) realistic, and (b) useful in nature conservation. The model has emerged, via much iteration, from the Equilibrium Theory of Island Biogeography (MacArthur and Wilson, 1967). Debate about the validity of the theory for the design and selection of nature reserves goes back a long way (Simberloff and Abele, 1976). In particular, doubt surrounds the validity of oceanic islands as analogues for islands of terrestrial habitats. Terrestrial islands are surrounded by land across which many plants and animals are able to disperse such that species frequently use both the habitat island and the matrix.

The patch-matrix concept is undoubtedly useful for those species that are relatively mobile and exploit a range of different habitats for their survival. For other more sedentary species this is not the case, and habitat quality may be a much more important factor determining long-term breeding success and survival. In particular, the patch-matrix model seems to be most valid for species that operate as a metapopulation, a set of subpopulations that occupy connected patches across a range of spatial and temporal scales (Hanski, 1999).

The metapopulation concept depends on a functioning network of interconnected patches. The role of linear features as habitats in the landscape has been the subject of considerable research (Jongman and Pungetti, 2004) at a range of scales from hedgerows in the British landscape (Hinsley and Bellamy, 2000), corridors in the French landscape (Burel and Baudrey, 1990), to roads in Australia (Bennett, 1998; Bennett and van der Rees, 2001). A recent good example of core areas linked by habitat corridors comes from Wales (Watts et al., 2005). GIS techniques were used to identify areas containing a high concentration of seminatural, broadleaf woodland defined as core woodland areas. Hypothetical but realistic distance values for low-, medium-, and high-dispersing species were used to evaluate the degree of isolation of core woodland areas and to identify potential linking routes based on the distribution of woodland patches between the core areas.

However, considerable debate surrounds the validity of the metapopulation model, with many arguing that it is relevant only for mobile species occupying patchy landscapes (Baguette and Mennechez, 2004). In particular, the model has been applied to animals in recognition that the relatively sedentary nature of most plants means they are less likely to follow a metapopulation model.

30.5.3 Habitats and species

Modeling is important in other respects in that it provides the link between a species of interest and its habitat. Statistical analyses of wildlife–environment relationships have been often employed to predict bird species distribution patterns, particularly in cases where a species is elusive, for example, the Jerdon's courser in India (Jeganathan et al., 2004). A well-known example is the northern spotted owl (*Strix occidentalis caurina*), which inhabits the old-growth forests of the Pacific Northwest of the United States. Lamberson et al. (2002) used a combination of information on forest fragmentation and species data to determine the minimum patch area and number of patches that must be left to ensure the survival of the northern spotted owl. Another good example comes from northeast Scotland where Avery and Haines-Young (1990) demonstrated that a simple index of 'wetness' derived from a single channel of Landsat TM data was highly correlated with the abundance of a small wading bird (the dunlin, *Calidris alpina*) at a time when the extensive peat bog ecosystem was under threat from forest planting of exotic species.

There is an urgent need to develop systems for monitoring change to habitats over large areas. The European Union, for example, is working on a directive to stop biodiversity loss by 2010 (European Environment Agency, 2008). However, we do not currently have the means either to catalogue the type and distribution of habitats, let alone species, or to monitor change subsequent upon damage sustained from development, forestry, and agricultural intensification. At the European level there have been attempts to provide a common framework that standardizes, focuses, and coordinates existing monitoring programs by comparing and integrating existing methods and monitoring schemes of species and habitats of community interests (Lengyel et al., 2008; Bunce et al., 2008).

30.6 THE HABITAT APPROACH TO NATURE CONSERVATION

The habitat approach to nature conservation uses a range of techniques, including restoration, rehabilitation, enhancement, reclamation, and mitigation (Table 30.2). The approach is enshrined in international policy and legislation, as exemplified by the Natura 2000 network of protected areas in Europe (Council of Europe, 1992).

We have the scientific understanding, albeit imperfect, if not the operational tools, to map and monitor habitats and to link habitats to species. But if whole landscapes are to retain their ecological functions, even where they are intensively used, a more holistic approach is required that extends the scope of conservation beyond the simple designation of isolated sites for protection. There is increasing recognition that the benefits from global ecosystem goods and services derive from whole ecosystems and that, for these benefits to be realized, the ecosystem must function effectively (Millennium Ecosystem Assessment, 2005). In this sense, an ecosystem relates both to human-modified ecosystems (e.g., the intensively cultivated lowland agricultural landscapes of northwest Europe) and to the natural ecosystems of primary forest. Both need to be protected, managed, and increasingly rehabilitated if ecological goods and services are to be realized (Kremen et al., 2007). The habitat approach, therefore, differs from a species approach in that the emphasis is holistic, seeking to protect habitats of an ecosystem at a landscape scale.

Increasingly, matrix management is complementary to biodiversity conservation and, critically, it is a recommended adaptation measure for climate change for a number of ecological and practical reasons. This includes measures to increase species movement through the landscape by improving connectivity between suitable habitat patches. In practical terms this might translate into creating buffer zones around existing habitat patches or to retain and restore linear features that may serve as corridors.

Gap analysis (Kiester et al., 1996) is a tool used in nature conservation to identify gaps in protected areas or in the wider landscape where significant plant and animal species and their habitats or important ecological features occur. This tool can be employed by scientists and practitioners to provide recommendations to improve the representativeness of nature reserves or the effectiveness of protected areas in conserving biodiversity. Gap assessments can be done using a GIS platform where available information on topography, biological and geological features, landownership, and land use are compared with the distribution of wildlife (Scott and Schipper, 2006; Pressey et al., 1993).

Table 30.2 Common terms used in habitat restoration

Term	Definition
Restoration	To return a habitat to its original, undisturbed state
Rehabilitation	To restore or improve some aspects of an ecosystem but not necessarily to fully restore all components
Habitat enhancement or improvement	To improve the quality of a habitat through direct manipulation
Reclamation	To return an area to its previous habitat type but not necessarily to restore all functions fully
Mitigation	Actions taken other than habitat rehabilitation to alleviate or compensate for potentially adverse effects on habitats that have been modified or lost through human activity
Habitat creation	Establishment of habitat where it was not known to have existed previously
Habitat re-creation	Creation of habitat in areas where it had been lost

Ever-increasing habitat destruction has attracted growing attention for the restoration of damaged habitats and for the re-creation of lost habitats. In rare cases, such destruction may also include the creation of new habitats in areas affected by harmful activities in order to mitigate the damage caused, either by artificial translocation or by re-seeding and re-planting (Holl et al., 2003). A recent example from the South Australian wheat belt uses state and transition models to develop restoration strategies for salmon gum (*Eucalyptus salmonophloia*) woodlands. The model has the potential to provide a tool for land managers with which they can assess the action and effort needed to undertake woodland restoration in agricultural landscapes (Yates and Hobbs, 1997).

The habitat approach to nature conservation is not a straightforward application of a set of well-defined and recognized procedures but is rather a guiding principle based on habitat attributes: type, extent, pattern, and quality. This is partly because there is no internationally agreed-upon definition of a habitat and no system of habitat classification that can be applied across all environments and at an appropriate range of spatial scales. Spatially, it is difficult to delimit the boundaries of a habitat that are critically scale- and organism-dependent. In many environments, habitats do not conform to the classical patch-matrix model, especially in complex vegetation communities where composition is controlled by environmental gradients that operate at a range of spatial scales.

Nevertheless, despite these drawbacks, the habitat approach is finding widespread application for nature conservation, especially for site selection and protection. We have attempted to show how it can be applied and, in particular, its advantages and disadvantages in comparison with a species-based approach. First, it is important to recognize the significance of scale dependency: the scale at which the approach is applied varies in accordance with the objectives of the nature conservation project in question. Thus, the identification of potential sites for habitat restoration, often as part of the development of a set of targets within a biodiversity strategy, is a landscape-scale problem that requires knowledge of the spatial pattern of habitats and the response of species to habitat isolation and connectivity. The models for this type of analysis, often within a GIS environment, are increasingly sophisticated combining a predictive species component with the tools to analyze and summarize the spatial patterns of habitats. By contrast, the implementation of any biodiversity strategy is a site-level problem, involving the evaluation of biotic and abiotic conditions of a site. This is essentially the realm of restoration ecology.

REFERENCES

Alexander, R.W., and Millington, A.C. (eds.). (2000) *Vegetation Mapping: From Patch to Planet*. Chichester, UK: John Wiley.

Austin, M.P. (2002) 'Spatial prediction of species distribution: An interface between ecological theory and statistical modeling', *Ecological Modelling*, 157: 101–18.

Avery, M.I., and Haines-Young, R.H. (1990) 'Population estimates for the dunlin *Calidris alpina* derived from remotely sensed satellite imagery of the Flow Country of northern Scotland', *Nature*, 344: 860–62.

Baguette, M., and Mennechez, G. (2004) 'Resource and habitat patches, landscape ecology and metapopulation biology: A consensual viewpoint', *Oikos*, 106: 399–403.

Bailey, N., Lee, J.T., and Thompson, S. (2006) 'Maximising the natural capital benefits of habitat creation: Spatially targeting native woodland using GIS', *Landscape and Urban Planning*, 75(3–4): 227–43.

Bedward, M., Pressey, R.L., and Keith, D.A. (1992) 'A new approach for selecting fully representative reserve networks: Addressing efficiency, reserve design and land suitability with an iterative analysis', *Biological Conservation*, 62: 115–25.

Bennett, A.F. (1998) *Linkages in the Landscape: The Role of Corridors and Connectivity in Wildlife Conservation*. Gland, Switzerland: IUCN.

Bennett, A.F., and van der Rees, R. (2001) 'Roadside vegetation in Australia: Conservation values and function of a linear habitat network in rural environments', in *Proceedings of the Tenth Annual IALE (UK) Conference—Hedgerows of the World: Their Ecological Functions in Different Landscapes*. International Association for Landscape Ecology, United Kingdom. pp. 231–40.

Brooks, T.M., and 14 authors. (2004) 'Coverage provided by the global protected-area system: Is it enough?' *BioScience*, 54(12): 1081–91.

Bunce, R.G.H., and 19 authors. (2008) 'A standardized procedure for surveillance and monitoring European habitats and provision of spatial data', *Landscape Ecology*, 23: 11–25.

Burel, F., and Baudrey, J. (1990) 'Structural dynamic of a hedgerow network landscape in Brittany France', *Landscape Ecology*, 4: 197–210.

Council of Europe. (1992) *Council Directive 92/43/EEC of 21 May 1992 on the conservation of natural habitats and of wild fauna and flora*. O. J. Eur. Comm. No. L206/7.

Dennis, R.L.H., and Eales, H.T. (1997) 'Patch occupancy in *Coenonympha. tullia* (Muller, 1764) (*Lepidoptera: Satyrinae*): Habitat quality matters as much as patch size and isolation', *Journal of Insect Conservation*, 1: 167–76.

European Communities. (1991) *CORINE Biotopes: The Design, Compilation and Use of an Inventory of Sites of Major Importance for Nature Conservation in the European Community*. Luxembourg: European Communities.

European Environment Agency. (2008) 'Progress towards halting the loss of biodiversity', http://www.eea.europa.eu/highlights/europe-is-losing-biodiversity-2013-even-in-protected-areas. Accessed April 1, 2009.

Fahrig, L. (2001) 'How much habitat is enough?' *Biological Conservation*, 100: 65–74.

Fahrig, L. (2003) 'Effects of habitat fragmentation on biodiversity', *Annual Review Ecology Evolution Systematics*, 34: 487–515.

Fahrig, L. (2007) 'Estimating minimum habitat for population persistence in managing and designing landscapes for conservation', in Lindenmayer D.B and Hobbs, R.J. (eds.), *Managing and Designing Landscapes for Conservation*. Oxford: Blackwell. pp. 64–80.

Forman, R.T.T. (1995) *Land Mosaics: The Ecology of Landscapes and Regions*. Cambridge: Cambridge University Press.

Forman, R.T.T., and Godron, M. (1986) *Landscape Ecology*. New York: John Wiley.

Franklin, J. (1995) 'Predictive vegetation mapping: Geographic modelling of biospatial patterns in relation to environmental gradients', *Progress in Physical Geography*, 19(4): 474–99.

Garcia, J., Suárez-Seoaneb, S., Miguélez, D., Osborne, P.E., and Zumalacárregui, C. (2007) 'Spatial analysis of habitat quality in a fragmented population of little bustard (*Tetrax tetrax*): Implications for conservation', *Biological Conservation*, 137: 45–56.

Groombridge, B., and Jenkins, M.D. (2002) *World Atlas of Biodiversity*. Berkeley: University of California Press.

Hall, L.S., Krausman, P.R., and Morrison, M.L. (1997) 'The habitat concept and the plea for standard terminology', *Wildlife Society Bulletin*, 25: 173–82.

Hanski, I. (1999) *Metapopulation Ecology*. Oxford: Oxford University Press.

Hanski, I., A. Moilanen, and M. Gyllenberg. (1996) 'Minimum viable metapopulation size', *American Naturalist*, 147: 527–41.

Hinsley, S.A. and Bellamy, P.E. (2000) 'The influence of hedge structure, management and landscape context on the value of hedgerows to birds: A review', *Journal of Environmental Management*, 60: 33–49.

Holl, K.D., Krone, E.E., and Schultz, C.B. (2003) 'Landscape restoration, moving from generalities to methodologies', *BioScience*, 53: 491–502.

IUCN World Commission on Protected Areas. (2006). *WCPA Strategic Plan 2005–12*, Gland, Switzerland: IUCN.

Jenganathan, P., Green, R.E., Norris, K., Vogiatzakis, I.N., Bartsch, A., Wotton, S.R., Bowden, C.G.R., Griffiths, G.H., Pain, D. and Rahmani, A.R. (2004) 'Modelling habitat selection and distribution of the critically endangered Jerdon's courser *Rhinoptilus bitorquatus* in scrub jungle: An application of a new tracking method', *Journal of Applied Ecology*, 41: 224–37.

Jongman, R.H., and Pungetti, G. (2004) *Ecological Networks and Greenways: Concept, Design, Implementation*. Cambridge: Cambridge University Press.

Kiester, A.R., Scott, J.M., Csuti, B., Noss, R.F., Butterfield, B., Shar, K. and White, D. (1996) 'Conservation prioritization using GAP data', *Conservation Biology*, 10(5): 1332–42.

Kremen, C., and 18 authors. (2007) 'Pollination and other ecosystem services produced by mobile organisms: A conceptual framework for the effects of land use change', *Ecology Letters*, 10: 299–314.

Küchler, A.W., and Zonneveld, I.S. (eds.). (1988) *Vegetation Mapping*. Dordrecht: Kluwer.

Lamberson, R.H., McKelvey, R., Noon, B.R., and Voss, C. (1992) 'A dynamic analysis of northern spotted owl viability in a fragmented forest landscape', *Conservation Biology*, 6(4): 505–12.

Lee, J.T., and Thompson, S. (2005) 'Targeting sites for habitat creation: An investigation into alternative scenarios', *Landscape and Urban Planning*, 71(1): 17–28.

Lengyel, S., Kobler, A., Framstad, E., Henry, P-Y., Babij, V., Gruber, B., Schmeller, D. and Henkle, K. (2008) 'A review and a framework for the integration of biodiversity monitoring at the habitat level', *Biodiversity and Conservation*, 17(14): 3341–56.

Luoto, M., Toivonen, T., and Heikkinen, R. (2002) 'Prediction of total and rare plant species richness in agricultural landscapes from satellite images and topographic data', *Landscape Ecology*, 17: 195–217.

Luyssaert, S.E., and 7 authors. (2008) 'Old-growth forests as global carbon sinks', *Nature*, 455: 213–15.

MacArthur, R.H., and Wilson, E.O. (1967) *The Theory of Island Biogeography*. Princeton, NJ: Princeton University Press.

Margules, C.R., and Pressey, R.L. (2000) 'Systematic conservation planning', *Nature*, 405: 243–53.

McDermid, G., Franklin, S.E., and LeDrew, E.F. (2005) 'Remote sensing for large-area habitat mapping', *Progress in Physical Geography*, 29: 449–74.

Millennium Ecosystem Assessment. (2005) *Ecosystems and Human Well-Being: Current State and Trends*. Millennium Ecosystem Assessment Series. Washington, DC: Island Press.

Miller, J., and Franklin, J. (2002) 'Modeling the distribution of four vegetation alliances using generalized linear models and classification trees with spatial dependence', *Ecological Modelling*, 157: 227–47.

Millington, A.C. and Walsh, S.D. (2001) *GIS and Remote Sensing Applications in Biogeography and Ecology*. Norwell MA: Kluwer Academic.

Myers, N., Mittermeier, R.A, Mittermeier, C.G., da Fonseca, G.A.B., and Kent, J. (2000) 'Biodiversity hotspots for conservation priorities', *Nature*, 403: 853–58.

Nikolakaki, P. (2004) 'A GIS site-selection process for habitat creation: Estimating connectivity of habitat patches', *Landscape and Urban Planning*, 68: 77–94.

Papageorgiou, K., and Vogiatzakis, I.N. (2006) 'Nature protection in Greece: An appraisal of the factors shaping integrative conservation and policy effectiveness', *Environmental Science and Policy*, 9: 476–86.

Potts, S.G., Petanidou, T., Roberts, S., O'Toole, C., Hulbert, A., and Willmer, P.G. (2006) 'Plant-pollinator biodiversity and pollination services in a complex Mediterranean landscape', *Biological Conservation*, 129: 519–29.

Prendergast J., Quinn R., and Lawton J.H. (1999) 'The gaps between theory and practice in selecting nature reserves', *Conservation Biology*, 13: 484–92.

Prendergast, J.R., Quinn, R.M., Lawton, J.H., Eversham, B.C., and Gibbons, D.W. (1993) 'Rare species, the coincidence of diversity hotspots and conservation strategies', *Nature*, 365: 335–37.

Pressey, R.L., Humphries, C.J., Margules, C.R., Vane-Wright, R.I., and Williams, P.H. (1993) 'Beyond opportunism: Key principles for systematic reserve selection', *Trends in Ecology and Evolution*, 8(4): 124–28.

Pressey, R. L., Possingham, H.P., and Margules, C.R. (1996) 'Optimality in reserve selection algorithms: When does it matter and how much?' *Biological Conservation*, 76: 259–67.

Ratcliffe, D.A. (1977) *A Nature Conservation Review, Volume I*. Cambridge: Cambridge University Press.

Scott, J.M., Heglund, P.J., Morrison, M.L., Haufler, J.B., Raphael, M.G., Wall, W.A. and Samson, F.B. (eds.). (2002) *Predicting Species Occurrences: Issues of Accuracy and Scale*. Washington, DC: Island Press.

Scott, M.J., and Schipper, J. (2006) 'Gap analysis: A spatial tool for conservation planning', in Groom, M.J. Meffe, G.K., and Carroll R.C. (eds.), *Principles of Conservation Biology* (3rd ed.). Sunderland, MA: Sinauer.

Simberloff, D. (1998) 'Flagships, umbrellas, and keystones: Is single-species management passé in the landscape era?' *Biological Conservation*, 83: 247–57.

Simberloff, D.S., and Abele, L.G. (1976) 'Island biogeography theory and conservation practice', *Science*, 191: 285–86.

Vogiatzakis, I.N., and Griffiths, G.H. (2006) 'A GIS-based empirical model for vegetation prediction in Lefka Ori, Crete', *Plant Ecology*, 184: 311–23.

Vogiatzakis, I.N., Griffiths, G.H., and Mannion, A.M. (2003) 'Environmental factors and vegetation composition Lefka Ori massif, Crete, S. Aegean', *Global Ecology and Biogeography*, 12: 131–46.

Wadsworth, R., and Treweek, J. (1999) *Geographical Information Systems for Ecology: An Introduction*. Harlow, UK: Longman.

Watts, K., Griffiths, M., Quine, C., Ray, D., and Humphrey, J.W. (2005). *Towards a Woodland Habitat Network for Wales*. Contract Science Report No.686, Countryside Council for Wales, Bangor, Wales.

Williams, P., Gibbons, D., Margules, C., Rebelo, A., Humphries, C., and Pressey, R. (1996) 'A comparison of richness hotspots, rarity hotspots and complementary areas for conserving diversity, using British birds', *Conservation Biology*, 10: 155–74.

Wu, X.B., and Smeins, F.E. (2000) 'Multiple-scale habitat modeling for rare plant conservation', *Landscape and Urban Planning*, 51: 11–28.

Yates, C.J., and Hobbs, R.J. (1997) 'Woodland restoration in the Western Australian wheatbelt: A conceptual framework using a state and transition model', *Restoration Ecology*, 5(1): 28–35.

31

Species Approaches to Conservation in Biogeography

Patrick E. Osborne and Pedro J. Leitão

31.1 INTRODUCTION

It is a simple truth that no species occurs everywhere and most have rather restricted ranges (Gaston, 2003). Inevitably, then, the practice of species conservation and knowledge of biogeography are intertwined. Indeed, biodiversity is a spatial phenomenon and it makes little sense to discuss the occurrence and abundance of species divorced from geography. By species conservation we mean actions that are carried out for the benefit of the species level of biodiversity. Although there may be sound theoretical reasons for conservation to be focussed on the genetic level of biodiversity (Mallet, 1996), there are severe practical difficulties in mapping genetic variation, especially at large spatial scales. On the other hand, habitats (land use or ecosystems) are relatively easy to map but when used alone lack the detail of biotic variation needed for effective biodiversity conservation (see Griffiths and Vogiatzakis, this volume, for an analysis of habitat-based approaches to biodiversity conservation). This lack of precision at the habitat level and practical difficulties at the genetic level mean that biodiversity conservation is often centred on the species (but see below for arguments against this).

Species-level actions for conservation take many forms from the very local scale for individual plants or animals to the global scale. The cross-fostering of red kite (*Milvus milvus*) chicks from one nest with two chicks to another that has just lost its clutch is an example of hands-on

species conservation at the local scale, but it does not concern biogeography. Species approaches to conservation within the context of biogeography focus on actions planned from a knowledge of species distributions, usually at larger spatial scales than the local. Actions are usually (but not exclusively) based on the coincidence of species distributions, which identify priority areas for conservation. In this chapter we provide an overview of the main approaches employing knowledge of the distributions of species for conservation planning. There is a vast and growing literature on this topic and it is impossible here to use more than a few examples. We have selected papers from across taxonomic groups and from different parts of the world to illustrate the generality of the approaches, but our choices are inevitably biased by our own interests and experience. Readers wishing to expand the literature should consult the reference list.

31.2 USING SPECIES AS A BASIS FOR CONSERVATION PLANNING

Much of conservation planning has come to depend on counts of species in some form that may be assembled and compared across locations and taxa (Myers et al., 2000). The popular appeal of the species as a unit, coupled with limitations in working with genetics or ecosystems, means that species conservation is widely practiced. The well-known Red Lists published by the

International Union for the Conservation of Nature, local and national conservation action plans, and much legislation refer to species rather than other levels of biodiversity. Yet what is a species? Before examining the biogeographic input to species conservation, it is worth stepping back to ask whether it is really that simple or appropriate to focus on the species as a unit.

Perhaps the first widely accepted definition of the species was expressed in the 'biological species concept', namely that species represent groups of interbreeding natural populations that are reproductively isolated from other such groups (Mayr, 1942, 1970). This definition has been heavily criticised in recent years as being counter-productive to understanding biodiversity, and more than 20 alternative definitions have arisen in its place (Mayden, 1997). Opinions have become divided and the debates have been intense (Eldredge, 1995), casting doubt on the value of the species concept in conservation (Hey et al., 2003). Mayden (1997) has concluded that the 'evolutionary species concept' is the most theoretically sound, that is, that species are lineages that evolved while separated from one another and with their own evolutionary role and tendencies (Simpson, 1961). Others disagree and argue for the 'phylogenetic species concept', which defines a species as the smallest cluster of organisms that possess at least one diagnostic character (Baum, 1992). These are not just academic debates: favoring the phylogenetic approach increases the number of species by an average of 48.7% and could increase the bill for conserving species under the U.S. Endangered Species Act from $4.6 billion to $7.6 billion (Agapow et al., 2004). The radical alternative is to abandon the notion of a species altogether and focus on units based on phylogenetic diversity, genetic or phenotypic diversity, or evolutionary time (Faith, 1994; Crozier, 1997; Avise and Johns, 1999; Owens and Bennet, 2000). However sound this may be from a theoretical standpoint, it is probably true that conservation would not attract the support it currently receives were such a change made. Conservation without flagship species such as the giant panda (*Ailuropoda melanoleuca*) would not be conservation as we know it. Imperfect as it may be, it is difficult to see the species concept disappearing from conservation, and it would be churlish to deny that it has served us well to date. Indeed, the range of uses that species information has been put to in conservation (see details below) is testament to the value of the species unit. We should not forget, however, that alternative definitions or classifications could alter conservation priorities, especially given the limited resources that are available for the protection of biodiversity.

31.3 CONSERVATION PLANNING BASED ON SURROGATE SPECIES

Ideally, conservation planning would be based on detailed survey and ecological data, but this is rarely possible because the data are lacking and are too costly or time-consuming to collect (Favreau et al., 2006). Conservation biologists have therefore sought to use better-known species as surrogates for species assemblages on the assumption that areas protected for them will support many other species that share similar requirements (Caro and O'Doherty, 1999). Unfortunately, the terms used to describe these surrogates, such as indicator, flagship, or keystone species, have not been applied consistently and some confusion exists in the literature. Following Favreau et al. (2006), we recognise five categories of surrogate species used in conservation planning (although we group focal species with umbrella species). Good discussions of the distinctions between these categories are also provided by Caro and O'Doherty (1999).

31.2.1 Indicator species

If species can act as predictors for other species or for environmental conditions, then good knowledge of a few species could help us to conserve many others. Indicator species are those used to indicate characteristics of species or environmental conditions that are too difficult, inconvenient, or expensive to measure (Landres et al., 1988). Despite indicator species or taxa being widely used for conservation planning, there is inconsistent evidence on their effectiveness at indicating overall species richness (Hess et al., 2006).

Suspecting that spatial resolution and geographic extent might be influencing the outcome of studies on the value of indicator species, Hess et al. (2006) performed a detailed analysis to separate their effects. They correlated the species richness between each of seven different taxonomic groups in the United States (amphibians, birds, butterflies, fish, mammals, reptiles, and freshwater mussels) with the total richness of the remaining six groups, for various resolutions and extents. Worryingly, the correlation between the indicator taxon and total richness depended on the resolution (i.e., grain or pixel size used as the unit of analysis), extent, region, and taxon studied (Hess et al., 2006). This suggests that indicator relationships cannot be transferred across scales or geographic locations. If indicator species have to be identified anew for almost every situation, doubt must remain whether they offer an efficient alternative to a fuller enumeration of the species present.

If the occurrence of one species can be used to predict the occurrence of another, it is a small step to ask whether the composition of whole communities can be predicted from indicator species and environmental characteristics. 'Nestedness' is the observed tendency for the biotas of poor habitat patches to be nonrandom subsets of the species present in richer (often larger) patches. This pattern is generally thought to arise from the predictable sequence in which species go extinct as habitats are fragmented. The mechanism may simply be the differing minimum area requirements of the various species: those requiring larger areas will disappear first. The key point here is whether nestedness can be used predictively so that biodiversity may be estimated from partial knowledge of the communities present. Maron et al. (2004) tested this idea by looking at whether changes in the bird communities of forest remnants could be predicted from their nestedness. To measure nestedness, they used Atmar and Patterson's (1993) 'temperature' index, which measures the order in which species extinctions (or colonizations) would occur in a system. The 'colder' a system is, the more fixed the order of extinctions. In a 'warmer' system, extinctions would tend towards a random order. Maron et al. (2004) found that the predictions derived from the temperature index were no more accurate than a second set of predictions generated by use of a simple nonnested model. It therefore appears that nestedness is not sufficiently reliable to generate predictions for biodiversity management (see also Hansson, 1998; Fischer and Lindenmayer, 2005; Azeria et al., 2006) although Maron et al. (2004) recommended that further work needs to be done.

31.3.2 Keystone species

Keystone species are those whose presence is especially crucial to maintaining the organisation and diversity of their ecological communities (Mills et al., 1993). The idea of keystone species is appealing because knowledge of how these species act could provide understanding of the functioning of whole ecosystems that may be useful for management (Simberloff, 1998). Despite this, few studies have attempted to identify quantitatively which species actually act as keystones, because most have been selected by intuition (Fauth, 1999). Furthermore, the identity of keystone species varies geographically: the eastern newt (*Notophthalmus viridescens*) and the lesser siren (*Siren intermedia*) are keystone predators in temporary ponds in North Carolina, but their roles

as keystone species are taken over by the mole salamander (*Ambystoma talpoideum*) in South Carolina (Fauth, 1999). This geographic variability could arise due to differences in population density, community structure, abiotic conditions, and even genetic differences between subpopulations. Whatever the cause, variations in the roles that species play suggest that caution is needed in identifying and using keystone species for conservation planning.

31.3.3 Umbrella (and focal) species

The umbrella species concept appears to date from Frankel and Soule (1981) (cited by Roberge and Angelstam 2004). Although several definitions have been proposed, we prefer Roberge and Angelstam's (2004): an umbrella species is one "whose conservation confers protection to a large number of naturally co-occurring species". In contrast to the keystone species concept, no functional relationship is implied between the surrogate and target taxa (Simberloff, 1998): protection is conferred simply by co-occurrence. In fact, any species whose protection provides for the needs of a larger suite of co-occurring species could be termed an umbrella species (Launer and Murphy, 1994), although it is usual to think of habitat as the key. That is, if a species requires a large area of habitat, its protection will provide resources for a larger suite of species occupying the same habitat (Noss, 1990). For example, it has been argued that the proposed North American Sage-grouse (*Centrocercus* spp.) Management Plan would benefit a suite of species occupying sagebrush-dominated habitats (Braun, 2005). In practice, most taxa cited as umbrella species are large mammals or birds (Roberge and Angelstam, 2004), and there is clearly an overlap between the flagship and umbrella species concepts. This is an issue because objective methods should be used for deciding which species can best act as umbrella species in a particular situation, but in fact different methods select different species (Fleishman et al., 2000). Unless umbrella species are selected objectively, it is often a matter of faith whether many other species really fall under the umbrella of protection (Simberloff, 1998). A good example of this failure to define umbrella species objectively comes from a Habitat Conservation Plan implemented at sites across Montana and Washington, in the United States, designed around one species, the bull trout (*Salvelinus confluentus*). An analysis by Hitt and Frissell (2004) showed that by focusing on this one species, around 75% of all priority sites for a second trout species, *Oncorhynchus clarki lewisi*, were not captured. Thus a valuable opportunity for multispecies conservation was wasted.

Despite examples of failure, research on butterflies in the Great Basin of the western United States suggests that a species can serve as an effective umbrella for regional faunal assemblages, but that a suite rather than a single species is to be preferred (Fleishman et al., 2000). This study is typical of many others in urging the use of more than one species as an umbrella because it is clear that some species under the umbrella could be affected by ecological factors, which would not limit a single umbrella species (Roberge and Angelstam, 2004). This point is made explicit by Lambeck (1997), who argues for conservation based on focal species, that is, a suite of umbrella species that complement each other in their spatial needs, resources used, and management required. He suggests grouping species at risk according to the threats facing them and choosing the most sensitive species to define the minimum acceptable level of threat. Indeed, it should be recognised that umbrella species have limited use for monitoring impacts on the integrity of protected areas because they are not necessarily the species most sensitive to change (Thomas, 1995; Fleishman et al., 2000). In such situations, indicator species are the preferred option, either in the traditional sense (e.g., Landres et al., 1988) or in the umbrella-indicator hybrids that characterise Lambeck's (1997) focal species concept.

31.3.4 Flagship species

Flagship species are those that attract the attention of the public and generate popular support for conservation (Caro and O'Doherty, 1999). According to the Worldwide Fund for Nature, flagship species are those "selected to act as an ambassador, icon or symbol for a defined habitat, issue, campaign or environmental cause". In contrast to indicator, umbrella, and keystone species, they are not identified on ecological grounds but are species that appeal to humans. As a species ourselves, we are incredibly biased about what species we find appealing, usually focusing on the 'charismatic megafauna' such as panda, elephants, and whales while forgetting altogether fungi and microbes. Flagship species are, however, influential because they become umbrella species in the sense that they channel funds towards particular habitats and any co-occurring species. The lack of objectivity in their selection may be an issue although it is not yet clear whether the use of flagship species is generally effective or not for biodiversity conservation. In Portugal, for example, protected-area selection based on charismatic vertebrates with large home ranges (i.e., flagship

and umbrella species) did not capture the diversity in lower plants (Araújo, 1999). This suggests that flagship species may not be appropriate surrogates when representativeness is a key criterion for conservation. As Simberloff (1998) put it, a flagship species is not necessarily a good indicator or umbrella and its conservation is often very expensive. On the other hand, a study by Sergio et al. (2006) showed that top predators may indeed identify good sites for biodiversity conservation. They compared the diversity of birds, butterflies, and trees at the breeding sites of six raptor species with that observed at three types of control sites. Analysis showed that biodiversity levels were higher at the sites used by raptors and that reserve networks based on inclusion of top predators held more biodiversity at fewer sites than selection based on lower trophic levels (Sergio et al., 2006). It should also be remembered that flagship species can channel funds into conservation, which would otherwise not be spent on biodiversity, and they therefore serve a purpose in increasing the extent of protected areas.

31.4 MULTISPECIES APPROACHES TO CONSERVATION PLANNING

While the use of surrogates has an established place in conservation planning, attempts have also been made to use existing knowledge on all species distributions to identify priorities for action. These multispecies approaches often assume that we know species distributions well enough to make judgments on priorities. We question whether this is often true, especially for analyses at finer spatial resolutions, but we defer discussion on the adequacy of our knowledge of species distributions until the next section.

31.4.1 Hotspot maps

Biodiversity loss is proceeding at such a pace that resources are inadequate to tackle the problem at the appropriate global scale (Myers et al., 2000). One solution is to target action on areas which protect the 'most species per dollar' (Myers et al., 2000; Possingham and Wilson, 2005), for example, by focusing on 'hotspots'. The term 'hotspot' (coined by Myers, 1988) is applied both to high concentrations of all species (biodiversity hotspots) and to geographically restricted species (hotspots of endemism). Myers et al. (2000) have shown how conservation of just over 1% (1.4%) of the Earth's surface could protect 44% of all vascular plants and 35% of four vertebrate groups in

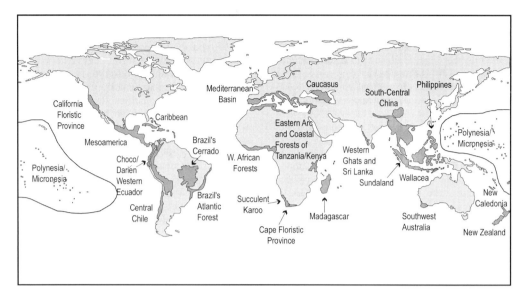

Figure 31.1   One way to define priorities for conservation is to delineate hotspots of endemism. Twenty-five such hotspots lie within the red areas shown, which represent distinct biogeographic units. These are home to 44% of all vascular plant species and 35% of all species in four vertebrate groups. Reproduced from Myers et al. (2000)

25 hotspot areas (Figure 31.1). In a practical application of the hotspot approach, BirdLife International has identified 218 regions of the world where the distributions of two or more bird species with restricted ranges overlap. These regions of overlap, which are relatively rich in endemic bird species compared to other parts of the world, are termed Endemic Bird Areas. An alternative approach is not to work with hotspots but to focus efforts on the countries hosting the highest concentrations of the world's biodiversity, the so-called 17 Megadiversity Countries proposed by Conservation International in the late 1990s, but that have now seemingly dropped from their suite of conservation strategies. While hugely successful as an academic concept, there is a danger that too much attention to hotspot or 'priority country' conservation could overshadow the need for biodiversity protection elsewhere (Kareiva and Marvier, 2003). Furthermore, the lack of data on species distributions means that hotspot maps are always based on an incomplete assessment of biodiversity and are highly selective in what they show. Unfortunately, there is little congruence between maps showing species richness for different biological groups (Prendergast et al., 1993; Lombard, 1995; Reid, 1998; Brummit and Lughadha, 2003), and maps for richness and endemism or threat rarely coincide (Williams et al., 1996; Bonn et al., 2002; Jetz et al., 2004a; Orme et al., 2005; Possingham and Wilson, 2005).

31.4.2 Complementarity

One argument against selecting protected areas on the number of species present, whether total richness or endemics, is that all areas could theoretically protect the same limited range of species. Reserve selection using complementarity seeks to avoid this, maximising the representativeness of the biodiversity that is protected while minimising costs (Margules and Pressey, 2000). Most commonly, the approach is to use computer algorithms to find a set of sites in which all species of conservation concern are represented in the minimum total area (Gaston et al., 2001). To cite a marine example, Awad et al. (2002) sought to identify a minimum set of key sites for benthic invertebrates around the South African coast. They undertook complementarity analyses based on rarity, species richness, and endemicity, each of which showed individual bias. Overall they found that 16 sites in addition to the existing reserve network were necessary to represent the species analysed. Interestingly, they found congruence with the work of others for fish and seaweeds, and they recommended the establishment of a reserve in the Durban area to provide protection for the greatest number of species inadequately covered at present (Awad et al., 2002).

While intuitively attractive, complementarity would fail to serve its purpose if the sites selected were not adequate for the long-term survival of the species concerned. This might happen if the selection algorithms focussed on the margins of species' ranges (Branch et al., 1995; Araújo and Williams, 2001). Using data from the Southern African Bird Atlas Project (Harrison et al., 1997), Gaston et al. (2001) found that complementarity approaches selected areas in ecological transition (i.e., margins) more often than expected by chance. Similarly, Araújo and Williams (2001) confirmed that complementarity had biased selection towards marginal populations for three vertebrate groups in Europe. There is concern over the inclusion of the margins of species' ranges because peripheral populations have traditionally been regarded as having less resilience to stochastic events and hence a higher extinction risk (e.g., Goodman, 1987; Griffith et al., 1989). Araújo and Williams (2001) question, however, whether this really poses a problem for conservation. Channell and Lomolino (2000) found that, contrary to expectation, species undergoing range contraction mostly persist in marginal areas of their historical ranges. Furthermore, genetic variation within populations might be maximised in areas of transition between ecosystems where environmental gradients are sharp (e.g., Kark et al., 1999). If so, marginal populations might contain the genetic variability needed to survive stochastic events. Rather than being poor areas, then, range margins might actually offer valuable opportunities for conservation. The ideal solution would be to protect both core and marginal areas, and complementarity reserve selection algorithms can be constrained to do this. The downside is the significantly increased cost of the reserve network but this is to be preferred if the alternative is a network that fails to meet its conservation purpose (Gaston et al., 2001).

31.4.3 Gap analysis

Gap analysis is a tool for systematic conservation planning, based on assessment of the comprehensiveness of existing protected-area networks and identification of gaps in coverage (Scott et al., 1993). Gap analysis arose in the 1980s in response to a recognised need for a rapid method for assessing the conservation status of biodiversity over large spatial extents. In its original form, gap analysis used data on actual vegetation to produce maps of potential species richness, which could then be compared with landownership and management status to identify 'gaps' in the protection network (Scott et al., 1987, 1993; Caicco et al., 1995). Actual vegetation rather than potential vegetation was used because a main purpose was to infer maps of vertebrate species richness, and species normally respond to what is present rather than to what could potentially be present. In practice, gap analysis nowadays uses a variety of species data sources. For example, Oldfield et al. (2004) used 'natural area' types (which are broadly equivalent to biogeographic zones) and elevation data to assess the representativeness of the protected-area network in England. Maiorano et al. (2006) generated species distribution models from environmental data layers for their gap analysis in Italy. Rodrigues et al. (2004) relied on global assessments of species distributions by organisations such as the International Union for the Conservation of Nature and BirdLife International for their gap analysis of the global protected-area network. In other words, gap analysis suffers from the same limitations as any other conservation-planning approach that requires species data.

Despite this limitation, gap analysis has been widely adopted as an approach to identifying the representativeness and adequacy of the existing protected-area network and has been developed into a formal method applied by the U.S. Geological Survey National Gap Analysis Program (Jennings, 2000). A good example of the ways in which gap analysis may be applied is a global assessment by Rodrigues et al. (2004). They overlaid the ranges of 11,633 vertebrate species on the world distribution of protected areas from the 2003 World Database on Protected Areas. Species were classed as 'covered' if its range overlapped any extent of the protected-area network; otherwise they were classed as 'gap species'. Overall, 12% of species analyzed were found to be gap species and a further 12% of covered species were not represented in any protected area larger than 1,000 hectares and were therefore viewed as vulnerable (Figure 31.2). Rodrigues et al. (2004) went on to show that the current practice of setting percentage targets for identifying countries or regions that are adequately protected is inappropriate. In fact, the regions with the greatest need for an increase in the protected-area network are not those with a lower percentage area protected but those with higher levels of endemism. Thus the concept of identifying gaps in the coverage of the protected network is a powerful tool in conservation and helps to target limited resources into priority areas.

31.5 IMPROVING SPECIES LOCATION DATA FOR CONSERVATION PLANNING

If we are to use species locations in conservation planning (e.g., in hotspot mapping or complementarity approaches for identifying priority areas to

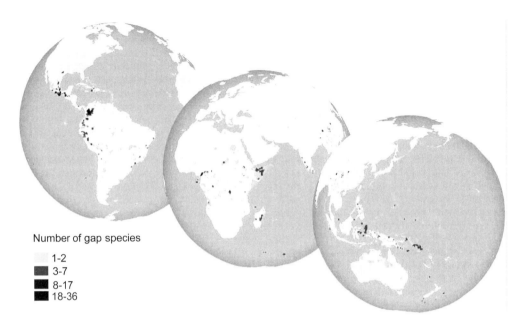

Number of gap species

▨ 1-2
■ 3-7
■ 8-17
■ 18-36

Figure 31.2   Density of 'gap species' per half-degree cell, created by overlaying the ranges of all species not covered by any protected area. Reproduced from Rodrigues et al. (2004)

protect, or in gap analysis), we must ensure that the data used are as reliable and as complete as possible. Yet it is surprisingly difficult to map species accurately and to know the level of accuracy in the mapping. In this section we examine sources of species data and delve into the growing armoury of statistical tools that have been developed to extrapolate partial data sets to full distribution maps to improve their value to conservation.

31.5.1 Data sources

Humans have been recording observations of species for centuries and so the sources of locational data that are available are many and varied. Among historical sources are the labels on museum specimens (e.g., bird skins) or herbarium specimens (so-called passport data); the accounts of gentlemen naturalists such as Gilbert White, explorers, and distinguished scientists (e.g., Charles Darwin); and even bone fragments and fossil finds. Collections worldwide are thought to hold some three billion specimens, which have the potential to provide biogeographic data (Soberon, 1999). More contemporary sources range from casual observations to purpose-designed survey data, the latter often in the form of biotic atlases. One of the first attempts to map species systematically was the *Atlas of the British Flora*,

published in 1962 (Perring and Walters, 1962), which laid the foundations for many successors worldwide. Atlas projects have grown in scale and ambition from national projects at higher spatial resolutions to multinational ventures crossing international boundaries and highly varied ecosystems (e.g., the Southern African Bird Atlas Project—Harrison et al., 1997). It is an impractical ideal to want accurately mapped species data for all taxonomic groups for all regions of the world because both time and resources are in short supply. Indeed, at the current rate of species extinctions, conservation would fail miserably if we waited for perfect data before taking action.

While the existing mapped sources of species data are a treasure trove for biodiversity studies, some caution is needed in their use. Museum specimens are not always correctly identified and, because collectors worked in their favourite locations rather than attempting systematic recording, the locations alone do not reflect true distributions (Graham et al., 2004). Furthermore, distribution mapping is complicated by issues of spatial resolution, extent, and positional accuracy. Historical collectors did not have access to global positioning systems and the locations recorded might refer to the nearest village rather than the point where the species was found. This bias does not matter if the resolution of the mapping is coarse, but it does matter if local-scale maps are needed for conservation. Even in modern-day atlas projects, it is virtually impossible to ensure even geographic coverage when the labour force is predominantly voluntary. Concern over this geographic bias led Osborne and Tigar (1992) to consider how to fill the gaps in atlas data caused by uneven effort. Their solution, inspired by the work of Nicholls (1989), was to extrapolate the known distribution by establishing links with background predictor variables such as terrain and land use. Elsewhere in botanical science, similar approaches had been proposed (see review by Franklin, 1995). These beginnings led the way to what have become standard but involved procedures for modelling the distributions of species from local to global scales based on sample data. It is our view that these approaches have much to offer conservation planning, and the example of Maiorano et al. (2006) in using modeled distributions in gap analysis is illustrative of the potential. Like all models, however, distribution models are not the truth but only representations of the truth based on assumptions and conditioned by bias (and error). Understanding how distribution models are built and their limitations are essential for their effective use in conservation.

31.5.2 Establishing a modelling framework for individual species

Models of species distributions may be dynamic or static, and correlative or mechanistic (process-based). Dynamic models are responsive to differences over time whereas static models capture only a snapshot view at a particular point. Mechanistic models try to mirror the mechanisms that lead to distribution patterns whereas correlative models use observed relationships between occurrence and environmental features. The ideal would be to build dynamic, mechanistic models for all species because they incorporate explanation and should therefore be robust enough to transfer to new situations (time or space). This ideal is, however, a long way off because mechanistic models require extensive details on how the environment limits a species distribution, for example, through the physiological limits of temperature tolerance. Conservation cannot wait for such knowledge to be gathered and, indeed, it may be faster to map species directly than to research the mechanisms of range limitation. In practice, then, most distribution models are static and correlative. It is generally agreed that a useful theoretical framework for species distribution models is provided by niche theory, particularly Hutchinsonian niche theory (but see Guisan and Thuiller, 2005, for arguments on why this is an oversimplification). Hutchinson (1957) envisaged

that for a species using n resources, its niche would comprise n resource axes, which, when plotted in hyperspace, would form an n-dimensional hypervolume. Hutchinson went further by distinguishing between the 'fundamental niche', the full range of conditions under which a species could occur, and the 'realized niche', the reduced hypervolume under which a species is compelled to exist due to biotic interactions. As each species occurs over only a part of each resource axis, a simple plot of abundance, frequency of occurrence, or some measure of fitness is expected to show an optimum for each axis. There has been much debate over the shape of this response curve as it is known but the general consensus is that it is not necessarily unimodal or symmetrical (Austin, 1999, 2002). This is especially true when considering the realised niche: the presence of a competing species could theoretically split the response curve into two (or more) peaks by displacing the subordinate species into less favoured parts of the resource axis. What distribution models essentially do is to characterize the shape of each response curve so that in combination they yield an approximation to the Hutchinsonian niche.

The data used to characterize the resource axes are locations. In other words, positions in geographic space (e.g., latitude, longitude) are translated into positions in resource space by identifying the values for critical resources (e.g., light, nutrient levels) at the geographic location. We rarely know what these critical resources are nor are we always able to measure them directly, and so niche modelling often deals with incomplete characterization of Hutchinson's n dimensions. A well-distributed set of locations in geographic space may characterize the response curve quite well, but biased collecting can mean that only part of the curve is specified (Thuiller et al., 2004; Guisan and Thuiller, 2005). A further difficulty is that observed distributions are the result of many factors and a species may be absent from geographic space due to lack of dispersal, historical conditions such as hunting, or behavioural reasons such as conspecific attraction and site fidelity, as well as biotic interactions (see Pulliam, 2000). At best then, the resource space identified by translation from geographic space represents the realized and not the fundamental niche but is usually constrained further. Phillips et al. (2006) argue that it is more accurate to be conservative and describe the outcome as an approximation to the realized niche within the geographic area being studied and within the resource axis considered.

There are many static, correlative distribution modeling algorithms (e.g., Elith et al., 2006), but they essentially fall into two camps. If both presence and absence data are available (e.g., through structured surveys), it is best to employ group discrimination approaches. These contrast the characteristics of sites with known presence against those with known absence and they fit any new sites to the closest match (often giving a probability of membership). A good example is Generalised Additive Modelling (GAM; Hastie and Tibshirani, 1990). In cases where the locational data come from museum specimens, it is rare to have verified records of absence and alternative approaches are needed that use presence locations alone (Guisan and Zimmermann, 2000). (An alternative is to generate a set of pseudo-absences for use with the presence data, but this introduces other issues. For example, how many pseudo-absence points should be chosen? Where geographically should they be placed? Should we only sample close to where survey data have been collected to minimize the chances of defining missed presence points as pseudo-absences?) Algorithms for presence-only modelling usually start by identifying the range of conditions under which the species occurs (e.g., maximum and minimum temperature) as the limits to an 'envelope'. Once the envelopes are known for each relevant resource axis, the algorithm then predicts whether the species should occur at each geographic location by examining the resources available there and the way the species respond to them. The process effectively re-projects resource space onto geographic space; the identified locations are those that possess the characteristics of locations where the species is known to occur. The result is the species potential distribution (Phillips et al., 2006) but it may be incomplete if nonmodelled factors such as hunting led to the systematic exclusion of parts of the resource space. One difficulty with presence-only modelling is defining a stopping rule during model fitting: a model could fit any dataset simply by assigning presence to every location, yet such a model would be useless. Good examples of presence-only modeling algorithms are GARP (Stockwell and Peters, 1999), Ecological Niche Factor Analysis (Hirzel et al., 2002), and Maximum Entropy modelling (Phillips et al., 2006).

While the niche theoretical basis for species modelling is attractive, it may play only a limited role when species models are actually built. Hutchinson's (1957) niche model employed resource axes, yet few species distribution models are able to make use of true resource axes. Botanists have more success by building models based on direct predictors such as radiation balance and evapotranspiration alone, but animal modelers are forced to rely on proxy variables, which are thought to correlate with the critical resource axes. The problem is that correlations often apply to limited datasets or geographic areas, so proxy variables fail when applied elsewhere (Guisan and Theurillat, 2000; Austin, 2002). It is also false to assume that the statistical approaches used in distribution modelling need the Hutchinsonian niche: they do not. Almost any classifier used in remote sensing, for example, can achieve discrimination between areas used by a species and those that are unoccupied. Indeed, the raster GIS package Idrisi (Andes Edition, 2006: Clark Labs, Clark University, Worcester, Massachusetts) advertises its Mahalanobis Distance classifier as a presence-only modeling tool for species distributions.

There are many other issues to consider in distribution modelling and it is not possible here to give more than a brief overview. For an extensive treatment of distribution modelling, see Guisan and Zimmermann (2000), Scott et al. (2002) and the papers in collections by Guisan et al. (2002), Lehmann et al. (2002), Guisan et al. (2006), and Moisen et al. (2006). Spatial dependence is a particular issue that is gaining attention because most modelling techniques assume data points to be independent whereas they clearly are not. Spatial autocorrelation is common in both the species data and the predictor variables, and relationships may not be stationary over geographic space. See Brunsdon et al. (1998), Osborne and Suárez-Seoane (2002), Segurado and Araújo (2004), and Osborne et al. (2007) for treatments of these issues.

31.5.3 Individual species models in action

A study by Suárez-Seoane et al. (2002) illustrates the application of individual species models in conservation. The authors gathered around 7,500 records of three steppe bird species in Spain (the great bustard, *Otis tarda*; little bustard, *Tetrax tetrax*; and calandra lark, *Melanocorhypha calandra*), including confirmed absence records from field surveys. They modeled these data using Generalised Additive Models against a suite of predictor variables derived from satellite imagery and digital cartography. Distribution maps for the whole of Spain were produced at one-kilometre resolution for each species (Figure 31.3). Importantly, the best fitting models for each species used different sets of predictors even though they occupy similar habitats. This quantification and knowledge of species-habitat associations is often the by-product of distribution models that have a direct benefit to conservation. By overlaying the modeled distribution maps and comparing them to the locations of special protection areas in two key regions, Suárez-Seoane et al. (2002) were able to judge whether an adequate network of

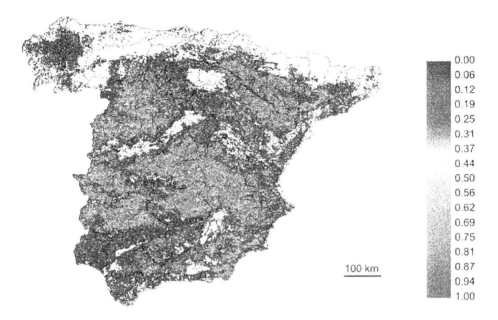

Figure 31.3   Habitat suitability map for the little bustard (*Tetrax tetrax*) in Spain, generated by generalised additive modeling of presence-absence data against a suite of predictors derived from remote sensing and digital cartography. Reproduced from Suárez-Seoane et al. (2002)

sites existed for the conservation of steppe birds.

Once relationships have been established between a species and its environment through modelling, it is possible to look for potential habitats outside the normal range and for shifts in distributions over time. For example, Thuiller et al. (2005) used distribution modelling to identify parts of the world with climatically similar biomes to those in South Africa, and to predict the risk of invasion of 96 South African plant taxa (Figure 31.4). Harrison et al. (2006) and Huntley et al. (2006) have used models to forecast the impacts of climate change on biodiversity at the European scale. Species distribution models have even been useful in identifying where species unknown to science may occur. When Raxworthy et al. (2003) used distribution models to predict the occurrence of chameleons on Madagascar, they found consistent overprediction (i.e., predicted occurrences in areas where the species were known to be absent) that seemed like modelling errors. Subsequent survey work, however, revealed that the overpredicted areas were home to a staggering seven new species of chameleon, compared with just two new species found through intensive surveys elsewhere in the country. These are just a few examples, and the literature is rich in the products of a growing industry charting the present, future, and past distributions of species.

31.5.4 Multispecies and biodiversity modeling

Community-level modelling may confer significant advantages and is especially useful when dealing with large numbers of species, particularly where few records exist for several of them (Ferrier and Guisan, 2006). In their comprehensive review of the subject, Ferrier and Guisan (2006) distinguish three broad modeling strategies for community-level modeling. In the 'assemble first, predict later' strategy, survey data are summarised into community data by using techniques such as ordination or simply by expressing the species present as a count of species richness. The community measure may then be modelled by using techniques similar to those employed for individual species. In an early example, Fjeldså et al. (1997) related the occurrence of biodiversity hotspots in Africa to the Normalised Difference Vegetation Index (NDVI) and brightness surface temperature derived from the Advanced Very High Resolution Radiometer onboard NOAA satellites. This theme was pursued by Foody (2004) with criticism and response over the appropriate

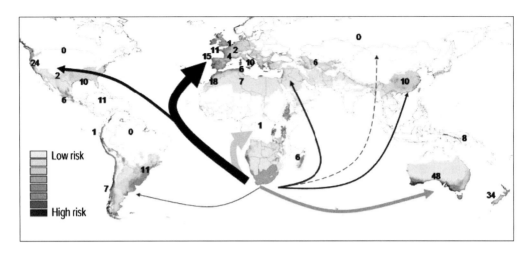

Figure 31.4   An assessment of the risk of invasion of 96 South African plant taxa across the world based on species distribution modeling. Reproduced from Thuiller et al. (2005)

modelling approaches by Jetz et al. (2004b) and Foody (2005). In the second strategy, 'predict first, assemble later', species distributions are modeled individually at first and the resultant maps of potential range are used to calculate community metrics (Ferrier and Guisan, 2006). This approach is very similar to the example given above for individual species models by Suárez-Seoane et al. (2002), where species maps were combined after modeling for conservation purposes, although no community metrics were calculated. The third strategy, 'assemble and predict together', performs the two tasks together and may be achieved using generalized dissimilarity modeling (Ferrier et al., 2002) and multivariate adaptive regression splines (MARS) (Leathwick, 2006). Among its advantages are processing speed (versus the multiple modeling of individual species), the way in which the responses of common species may be used to support predictions for rarer ones, and the fact that the identities of the species present at each locality are predicted by the model as opposed to simply richness (cf. the 'assemble first, predict later' strategy). Community models also perform well in comparison to single species approaches (Elith et al., 2006). Much work remains to be done, however, on the suitability of the different community-level modelling approaches to particular situations.

31.6 CONCLUSION

Looking toward the future, there seems little doubt that conservation actions will continue to be

focussed at the species level despite the theoretical need to consider genetics and the advantages of working with habitats for biodiversity conservation. This means that species locational data will continue to play an integral role in conservation planning despite the inherent errors, omissions, and limitations. In our review we have shown how conservation biologists and biogeographers have sought to make up for the deficiencies in individual species data through the use of surrogates, while modelers have extrapolated full distributions and community metrics from partial datasets. We see an increasing potential for the various approaches to using species distribution data in conservation to come together, for example, in the explicit use of modeled data within gap analysis (Maiorano et al., 2006). As conservation managers grapple to understand the effects of a changing climate and other pressures on wild species, the need for accurate data on species distributions and numbers will grow. For example, while models of climate-change impacts on birds in Europe generally predict a shift of ranges northwards (Huntley et al., 2006), actual data will be needed to judge whether this is happening and how fast it is occurring. There is clearly a research need to improve our understanding of how much survey data are needed for particular tasks and how often these data need to be updated. We must have in place quantitative monitoring (i.e., of population sizes rather than simply presence or absence) that tracks the response of species over vast geographic areas. Whether this can be done using selected species as indicators, or whether community compositions will change in a way that makes the use of indicator species unreliable, remains to be seen. There must also be some doubt whether currently identified hotspots will continue to be hotspots in the future if species distributions shift. Protected-area networks designed around known or modeled species alone may poorly represent ecosystem and environmental diversity for the present day (Bonn and Gaston, 2005) but they will be even worse in the future if species move in response to climate change (see Marchant and Taylor, this volume for a discussion of the potential effects of climate forcing on conservation planning). In short, mapping and understanding the distributions of species is not a one-off task for conservationists and biogeographers but an ongoing necessity if we are to address the conservation challenges of a changing world.

REFERENCES

Agapow, P-M., Bininda-Emonds, O.R.P., Crandall, K.A., Gittleman, J.L., Mace, G.M., Marshall, J.C. and Purvis, A. (2004) 'The impact of species concept on biodiversity studies', *Quarterly Review of Biology*, 79: 161–79.

Araújo, M.B. (1999) 'Distribution patterns of biodiversity and the design of a representative reserve network in Portugal', *Diversity and Distributions*, 5: 151–63.

Araújo, M.B., and Williams, P.H. (2001) 'The bias of complementarity hotspots toward marginal populations', *Conservation Biology*, 15: 1710–20.

Atmar, W., and Patterson, A.D. (1993) 'The measure of order and disorder in the distribution of species in fragmented habitat', *Oecologia, 96: 373–82.*

Austin, M.P. (1999) 'The potential contribution of vegetation ecology to biodiversity research', *Ecography*, 22: 465–84.

Austin, M.P. (2002) 'Spatial prediction of species distribution: An interface between ecological theory and statistical modelling', *Ecological Modelling*, 157: 101–18.

Avise, J.C., and Johns, G.C. (1999) 'Proposal for a standardised temporal scheme of biological classification for extant species', *Proceedings of the National Academy of Sciences of the United States of America*, 96: 7358–63.

Awad, A.A., Griffiths, C.L., and Turpie, J.K. (2002) 'Distribution of South African marine benthic invertebrates applied to the selection of priority conservation areas', *Diversity and Distributions*, 8: 129–45.

Azeria, E.T., Carlson, A., Pärt, T., and Wiklund, C.G. (2006) 'Temporal dynamics and nestedness of an oceanic island bird fauna', *Global Ecology and Biogeography*, 15: 328–38.

Baum, D. (1992) 'Phylogenetic species concepts', *Trends in Ecology and Evolution*, 7: 1–3.

Bonn, A., and Gaston, K.J. (2005) 'Capturing biodiversity: Selecting priority areas for conservation using different criteria', *Biodiversity and Conservation*, 14: 1083–100.

Bonn, A., Rodrigues, A.S.L., and Gaston, K.J. (2002) 'Threatened and endemic species: Are they good indicators of patterns of biodiversity on a national scale?' *Ecology Letters*, 5: 733–41.

Branch, W.R., Benn, G.A., and Lombard, A.T. (1995) 'The tortoises (Testunidae) and terrapins (Pelomedusidae) of southern Africa: Their diversity, distribution and conservation', *South Africa Journal of Zoology*, 30: 91–102.

Braun, C.E. (2005) 'Multi-species benefits of the proposed North American Sage-Grouse Management Plan', USDA Forest Service Gen. Tech. Rep. PSW-GTR-191. pp. 1162–64.

Brummit, N., and Lughadha, E.N. (2003) 'Biodiversity: Where's hot and where's not', *Conservation Biology*, 17: 1442–48.

Brunsdon, C., Fotheringham, S., and Charlton, M. (1998) 'Geographically weighted regression—modelling spatial non-stationarity', *The Statistician*, 47: 431–43.

Caicco, S.L., Scott, J.M., Butterfield, B., and Csuti, B. (1995) 'A gap analysis of the management status of the vegetation of Idaho (U.S.A). *Conservation Biology*, 9: 498–511.

Caro, T.M., and O'Doherty, G. (1999) 'On the use of surrogate species in conservation biology', *Conservation Biology*, 13: 805–14.

Channell, R., and Lomolino, M.V. (2000) 'Dynamic biogeography and conservation of endangered species', *Nature*, 403: 84–86.

Crozier, R.H. (1997) 'Preserving the information content of species: Genetic diversity, phylogeny and conservation worth', *Annual Review of Ecology, Evolution and Systematics*, 28: 243–68.

Eldredge, N. (1995) 'Species, selection and Paterson's concept of the specific mate recognition system', in Lambert, D.M. and Spencer, H.G (eds.), *Speciation and the Recognition Concept*. Baltimore: Johns Hopkins University Press.

Elith, J., and 25 authors. (2006) 'Novel methods improve prediction of species' distributions from occurrence data', *Ecography*, 29: 129–51.

Faith, P.D. (1994) 'Phylogenetic diversity: A general framework for the prediction of feature diversity', in Forey, P.L., Humphries, C.J., and Vane-Wright, R.I. (eds.), *Systematics and Conservation Evaluation*. Oxford: Clarendon Press.

Fauth, J.E. (1999) 'Identifying potential keystone species from field data—an example from temporary ponds', *Ecology Letters*, 2: 36–43.

Favreau, J.M., Drew, C.A., Hess, G.R., Rubino, M.J., Koch, F.H., and Eschelbach, K.A. (2006) 'Recommendations for assessing the effectiveness of surrogate species approaches', *Biodiversity and Conservation*, 15: 3949–69.

Ferrier, S., Drielsma, M., Manion, G., and Watson, G. (2002) 'Extended statistical approaches to modelling spatial pattern in biodiversity in north-east New South Wales. II. Community-level modelling', *Biodiversity and Conservation*, 11: 2309–38.

Ferrier, S., and Guisan, A. (2006) 'Spatial modelling of biodiversity at the community level', *Journal of Applied Ecology*, 43: 393–404.

Fischer, J., and Lindenmayer, D.B. (2005) 'Nestedness in fragmented landscapes: A case study on birds, arboreal marsupials and lizards', *Journal of Biogeography*, 32: 1737–50.

Fjeldså, J., Ehrlich, D., Lambin, E., and Prins, E. (1997) 'Are biodiversity hotspots correlated with ecoclimatic stability? A pilot study using the NOAA-AVHRR remote sensing data', *Biodiversity and Conservation*, 6: 401–22.

Fleishman, E., Murphy, D.D., and Brussard, P.F. (2000) 'A new method for selection of umbrella species for conservation planning', *Ecological Applications*, 10: 569–79.

Foody, G.M. (2004) 'Spatial nonstationarity and scale-dependency in the relationship between species richness and environmental determinants for the sub-Saharan endemic avifauna', *Global Ecology and Biogeography*, 13: 315–20.

Foody, G.M. (2005) 'Clarifications on local and global data analysis', *Global Ecology and Biogeography*, 14: 99–100.

Franklin, J. (1995) 'Predictive vegetation mapping: Geographic modelling of biospatial patterns in relation to environmental gradients', *Progress in Physical Geography*, 19: 474–99.

Gaston, K.J. (2003). *The Structure and Dynamics of Geographic Ranges*. Oxford: Oxford University Press.

Gaston, K.J., Rodrigues, A.S.L., Van Rensburg, B.J., Koleff, P., and Chown, S.L. (2001) 'Complementary representation and zones of ecological transition', *Ecology Letters*, 4: 4–9.

Goodman, D. (1987) 'The demography of chance extinction', in Soule, M.E. (ed.), *Viable Populations for Conservation*. Cambridge: Cambridge University Press.

Graham, C.H., Ferrier, S., Huettmann, F., Moritz, C., and Peterson, A.T. (2004) 'New developments in museum based informatics and applications of biodiversity analysis', *Trends in Ecology and Evolution*, 19: 497–503.

Graham, C.H., Smith, T.B., and Languy, M. (2005) 'Current and historical factors influencing patterns of species richness and turnover of birds in the Gulf of Guinea highlands', *Journal of Biogeography*, 32: 1371–84.

Griffith, B., Scott, J.M., Carpenter, J.W., and Reed, C. (1989) 'Translocation as a species conservation tool: Status and strategy', *Science*, 245: 477–80.

Guisan, A., Edwards, J., Thomas, C., and Hastie, T. (2002) 'Generalised linear and generalised additive models in studies of species distributions: Setting the scene', *Ecological Modelling*, 157: 89–100.

Guisan, A., and Theurillat, J-P. (2000) 'Equilibrium modelling of alpine plant distribution: How far can we go?' *Phytocoenologia*, 30: 353–84.

Guisan, A., and Thuiller, W. (2005) 'Predicting species distributions: Offering more than simple habitat models', *Ecology Letters*, 8: 993–1009.

Guisan, A., and 6 authors. (2006) 'Guest editorial: Making better biogeographical predictions of species' distributions', *Journal of Applied Ecology*, 43: 386–92.

Guisan, A., and Zimmermann, N.E. (2000) 'Predictive habitat distribution models in ecology', *Ecological Modelling*, 135: 147–86.

Hansson, L. (1998) 'Nestedness as a conservation tool: Plants and birds of oak-hazel woodland in Sweden', *Ecology Letters*, 1: 142–45.

Harrison, P.A., Berry, P.M., Butt, N., and New, M. (2006) 'Modelling climate change impacts on species' distributions at the European scale: Implications for conservation policy', *Environmental Science and Policy*, 9: 116–28.

Harrison, J.A., Allan, D.G., Underhill, L.G., Herremans, M., Tree, A.J., Parker, V. and Brow, C.J. (eds.). (1997). *The Atlas of Southern African Birds*. Johannesburg: BirdLife South Africa.

Hastie, T.J., and Tibshirani, R. (1990) *Generalised Additive Models*. London: Chapman and Hall.

Hess, G.R., Bartel. R.A., Leidner, A.K., Rosenfeld, K.M., Rubino, M.J., Snider, S.B. and Ricketts, T.H. (2006) 'Effectiveness of biodiversity indicators varies with extent, grain, and region', *Biological Conservation*, 132: 448–57.

Hey, J., Waples, R.S., Arnold, M.L., Butlin, R.K., and Harrison, R.G. (2003) 'Understanding and confronting species uncertainty in biology and conservation', *Trends in Ecology and Evolution*, 18: 597–603.

Hirzel, A.H., Hausser, J., Chessel, D., and Perrin, N. (2002) 'Ecological-niche factor analysis: How to compute habitat-suitability maps without absence data?' *Ecology*, 83: 2027–36.

Hitt, N.P., and Frissell, C.A. (2004) 'A case study of surrogate species in aquatic conservation planning', *Aquatic Conservation of Marine and Freshwater Ecosystems*, 14: 625–33.

Huntley, B., Collingham, Y.C., Green, R.E., Hilton, G.M., Rahbek, C., and Willis, S.G. (2006) 'Potential impacts of climate change upon geographical distributions of birds', *Ibis*, 148: 8–28.

Hutchinson, G.E. (1957) 'Concluding remarks', *Cold Spring Harbor Symposium on Quantitative Biology*, 22: 415–27.

Jennings, M.D. (2000) 'Gap analysis: Concepts, methods and recent results', *Landscape Ecology*, 15: 5–20.

Jetz, W., Rahbek, C., and Colwell, R.K. (2004a) 'The coincidence of rarity and richness and the potential signature of history in centres of endemism', *Ecology Letters*, 7: 1180–91.

Jetz, W., Rahbek, C., and Lichstein, J.W. (2004b) 'Local and global approaches to spatial data analysis in ecology', *Global Ecology and Biogeography*, 14: 97–98.

Kareiva, P., and Marvier, M. (2003) 'Conserving biodiversity coldspots', *American Scientist*, 91: 344–51.

Kark, S., Alkon, P.U., Safriel, U.N., and Randi, E. (1999) 'Conservation priorities for Chukar partridge in Israel based on genetic diversity across an ecological gradient', *Conservation Biology*, 13: 542–52.

Lambeck, R.J. (1997) 'Focal species: A multi-species umbrella for nature conservation', *Conservation Biology*, 11: 849–56.

Landres, P.B., Verner, J., and Thomas, J.W. (1988) 'Ecological uses of vertebrate indicator species: A critique', *Conservation Biology*, 2: 316–28.

Launer, A., and Murphy, D. (1994) 'Umbrella species and the conservation of habitat fragments: A case of a threatened butterfly and a vanishing grassland ecosystem', *Biological Conservation*, 69: 145–15.

Lehmann, A., Overton, J.M., and Austin, M.P. (2002) 'Regression models for spatial prediction: Their role for biodiversity and conservation', *Biodiversity and Conservation*, 11: 2085–92.

Lombard, A.T. (1995) 'The problems with multispecies conservation: Do hotspots, ideal reserves and existing reserves coincide?' *South African Journal of Zoology*, 30: 145–63.

Maiorano, L., Falcucci, A., and Boitani, L. (2006) 'Gap analysis of terrestrial vertebrates in Italy: Priorities for conservation planning in a human dominated landscape', *Biological Conservation*, 133: 455–73.

Mallet, J. (1996) 'The genetics of biological diversity: From varieties to species', in Gaston, K.J. (ed.), *Biodiversity*. Oxford: Blackwell.

Margules, C.R., and Pressey, R.L. (2000) 'Systematic conservation planning', *Nature*, 405: 243–53.

Maron, M., MacNally, R., Watson, D.M., and Lill, A. (2004) 'Can the biotic nestedness matrix be used predictively?' *Oikos*, 106: 433–44.

Mayden, R.L. (1997) 'A hierarchy of species concepts: The denouement in the saga of the species problem', in Claridge, M.F., Dawah, H.A., and Wilson, M.R. (eds.), *Species: The Units of Biodiversity*. London: Chapman and Hall.

Mayr, E. (1942) *Systematics and the Origin of Species*. New York: Columbia University Press.

Mayr, E. (1970) *Populations, Species and Evolution*. Cambridge, MA: Harvard University Press.

Mills, L.S., Soule, M.E., and Doak, D.F. (1993) 'The keystone species concept in ecology and conservation', *BioScience*, 43: 219–24.

Moisen, G.G., Edwards Jr, T.C., and Osborne, P.E. (2006) 'Further advances in predicting species distributions', *Ecological Modelling*, 199: 129–31.

Myers, N. (1988) 'Threatened biotas: "Hotspots" in tropical forests', *The Environmentalist*, 8: 187–208.

Myers, N., Mittermeier, R.A., Mittermeier, C.G., da Fonseca, G.A.B., and Kent, J. (2000) 'Biodiversity hotspots for conservation priorities', *Nature*, 403: 853–58.

Nicholls, A.O. (1989) 'How to make biological surveys go further with generalised linear models', *Biological Conservation*, 50: 51–75.

Noss, R.F. (1990) 'Indicators for monitoring biodiversity: A hierarchical approach', *Conservation Biology*, 4: 355–64.

Oldfield, T.E.E., Smith, R.J., Harrop, S.R., and Leader-Williams, N. (2004) 'A gap analysis of terrestrial protected areas in England and its implications for conservation policy', *Biological Conservation*, 120: 303–9.

Orme, C.D.L., and 14 authors. (2005) 'Global hotspots of species richness are not congruent with endemism or threat', *Nature*, 463: 1016–19.

Osborne, P.E., Foody, G.M., and Suárez-Seoane, S. (2007) 'Non-stationarity and local approaches to modelling the distributions of wildlife', *Diversity and Distributions*, 13: 313–23.

Osborne, P.E., and Suárez-Seoane, S. (2002) 'Should data be partitioned spatially before building large scale distribution models?' *Ecological Modelling*, 157: 249–59.

Osborne, P.E., and Tigar, B.J. (1992) 'Interpreting bird atlas data using logistic models: An example from Lesotho, Southern Africa', *Journal of Applied Ecology*, 29: 55–62.

Owens, I.P.F., and Bennet, P.M. (2000) 'Quantifying biodiversity: A phenotypic perspective', *Conservation Biology*, 14: 1014–22.

Perring, F.H., and Walters, S.M. (eds.). (1962) *Atlas of the British Flora*. London and Edinburgh: Nelson.

Phillips, S.J., Anderson, R.P., and Schapire, R.E. (2006) 'Maximum entropy modelling of species geographic distributions', *Ecological Modelling*, 190: 231–59.

Possingham, H.P., and Wilson, K.A. (2005) 'Turning up the heat on hotspots', *Nature*, 436: 919–20.

Prendergast, J.R., Quinn, R.M., Lawton, J.H., Eversham, B.C., and Gibbons, D.W. (1993) 'Rare species, the coincidence of diversity hotspots and conservation strategies', *Nature*, 365: 335–37.

Pulliam, H.R. (2000) 'On the relationship between niche and distribution', *Ecology Letters*, 3: 49–361.

Raxworthy, C.J., Martinez-Meyer, E., Horning, N., Nussbaum, R.A., Schneider, G.E., Ortega-Huerta, M.A., Townsend Peterson, A. (2003) 'Predicting distributions of known and unknown reptile species in Madagascar', *Nature*, 426: 837–41.

Reid, W.V. (1998) 'Biodiversity hotspots', *Trends in Ecology and Evolution*, 13: 275–80.

Roberge, J.-M., and Angelstam, P. (2004) 'Usefulness of the umbrella species concept as a conservation tool', *Conservation Biology*, 18: 76–85.

Rodrigues, A.S.L., and 20 authors. (2004) 'Effectiveness of the global protected area network in representing species diversity', *Nature*, 428: 640–43.

Scott, J.M., and 6 authors (eds.). (2002) *Predicting Species Occurrences: Issues of Accuracy and Scale*. Covelo, CA: Island Press.

Scott, M.J., Csuti, B., Jacobi, J.D., and Estes, J.E. (1987) 'Species richness: A geographic approach to protecting future biological diversity', *Bioscience*, 37: 782–88.

Scott, M.J., and 11 authors. (1993) 'Gap analysis: A geographic approach to protection of biological diversity', *Wildlife Monographs*, 123: 1–41.

Segurado, P., and Araújo, M.B. (2004) 'An evaluation of methods for modelling species distributions', *Journal of Biogeography*, 31: 1555–68.

Sergio, F., Newton, I., Marchesi, L., and Pedrini, P. (2006) 'Ecologically justified charisma: Preservation of top predators delivers biodiversity conservation', *Journal of Applied Ecology*, 43: 1049–55.

Simberloff, D. (1998) 'Flagships, umbrellas, and keystones: Is single-species management passé in the landscape era?' *Biological Conservation*, 83: 247–57.

Simpson, G.G. (1961) *Principles of Animal Taxonomy*. New York: Columbia University Press.

Soberon, J. (1999) 'Linking biodiversity information sources', *Trends in Ecology and Evolution*, 14: 291.

Stockwell, D.R.B., and Peters, D. (1999) 'The GARP Modeling System: Problems and solutions to automated spatial prediction', *International Journal of Geographical Information Science*, 13: 143–58.

Suárez-Seoane, S., Osborne, P.E., and Alonso, J.C. (2002) 'Large-scale habitat selection by agricultural steppe birds in Spain: Identifying species-habitat responses using Generalised Additive Models', *Journal of Applied Ecology*, 39: 755–71.

Thomas, J.A. (1995) 'The conservation of declining butterfly populations in Britain and Europe: Priorities, problems and successes', *Biological Journal of the Linnean Society*, 56 (supplement): 55–72.

Thuiller, W., Brotons, L., Araújo, M.B., and Lavorel, S. (2004) 'Effects of restricting environmental range of data to project current and future species distributions', *Ecography*, 27: 165–72.

Thuiller, W., Richardson, M.D., Pysek, P., Midgley, G.F., Hughes, G.O., and Rouget, M. (2005) 'Niche-based modelling as a tool for predicting the risk of alien plant invasions at a global scale', *Global Change Biology*,11: 2234–50.

Williams, P., Gibbons, D., Margules, C., Rebelo, A., Humphries, C., and Pressey, R. (1996) 'A comparison of richness hotspots, rarity hotspots and complementary areas for conserving diversity using British birds', *Conservation Biology*, 10: 155–74.

Index